Applications of Artificial Intelligence in Healthcare

The Changing Landscape

Applications of Artificial Intelligence in Healthcare

The Changing Landscape

Edited by

Debasis Bagchi
Department of Biology, College of Arts and Sciences, Adelphi University, Garden City, NY; Department of Pharmaceutical Sciences, College of Pharmacy and Health Sciences, Texas Southern University, Houston, TX, United States

Samudra Prosad Banik
Department of Microbiology, Government General Degree College, Narayangarh, Rathipur, West Midnapore, West Bengal, India

Anand Swaroop
Nutrify Today; Cepham Inc., Somerset, NJ, United States

Sunny Edet Ohia
Department of Pharmaceutical Sciences, College of Pharmacy & Health Sciences, Texas Southern University, Houston, TX, United States

ELSEVIER

ACADEMIC PRESS
An imprint of Elsevier

Academic Press is an imprint of Elsevier
125 London Wall, London EC2Y 5AS, United Kingdom
50 Hampshire Street, 5th Floor, Cambridge, MA 02139, United States

ISBN: 978-0-443-43934-6

For information on all Elsevier publications visit our website at https://www.elsevier.com/books-and-journals

Publisher: Megan Ball
Acquisitions Editor: Elizabeth Brown
Editorial Project Manager: Shubham Dixit
Production Project Manager: Sathyavani Deivasigamani
Cover Designer: Amy Buxton

Typeset by Integra Software Services

Dedication

This academic textbook is dedicated to our beloved Ranjit Da, Ranjit K. Chakraborty, PhD, PE. Dr. Chakraborty has been a source of inspiration and relentless motivation throughout the long journey from the conceptualization to the publication of this book, providing us with domain-specific technical guidance, literature, and other logistical support as and when required. We extend our sincere love and regards to our beloved Ranjit Da.

Debasis Bagchi and Samudra Prosad Banik

Contents

Contributors xxi
List of reviewers xxv
Preface xxvii

Section I
Introduction to artificial intelligence and healthcare

1 Artificial intelligence and emotional intelligence: Can machines understand human emotions?

Anu Dandona and Sandhya Bhatt

1 **Introduction** 3
2 **Can machines truly understand or replicate human emotions?** 4
2.1 Problems and constraints 4
3 **Background of EI** 5
3.1 The laboratory and model of EI 5
3.2 History of EI and its genesis 5
3.3 The framework of EI by Daniel Goleman 5
4 **Relevance of EI in communication and decision-making** 5
4.1 Social–EI in communication 6
4.2 The highlights of EI in communication 6
4.3 Decision-making through EI 6
4.4 Influence in decision-making processes 6
5 **Challenges and implications of EI in AI and machine interactions** 6
5.1 Challenges in integrating EI into AI 7
5.2 Implications for AI and human interaction 7
6 **AI and the pursuit of emotional recognition** 7
7 **Computer vision** 8
7.1 Categories of detections identified 9
8 **Challenges in computer vision for emotion detection** 9
9 **New research frontiers** 9
10 **Case studies and applications** 10
10.1 Healthcare: AI in mental health diagnostics or therapy bots 10
10.2 AI mental health diagnostics 10
10.3 Accuracy in diagnosis and early diagnosis 10
10.4 In mental health therapy bots 11
11 **Effectiveness of therapy chatbots** 11
11.1 Stimulus and quitting 12
11.2 Challenges and ethical considerations 12
12 **Customer service: Sentiment-aware chatbots and call centers** 12
12.1 Sentiment-aware chatbots 13
12.2 Improved customer experience 13
12.3 Voluntary trouble shooting and escalation 14
12.4 Business impact and case studies 14
13 **Case studies and examples** 15
References 15

2 Understanding key machine learning algorithms as artificial intelligence-assisted prediction tools in clinical medicine: A mathematical overview

Kristin S. Williams

1 **Introduction** 19
2 **Foundations of predictive modeling in healthcare** 19
3 **Types of AI/ML algorithms** 20
3.1 K-nearest neighbor 20
3.2 Support vector machines 21
3.3 Reinforcement learning 22
3.4 Naive Bayes 22
3.5 Deep learning and neural networks 23
4 **Oncology applications** 24
5 **Random forest algorithms** 24
5.1 Statistical limitations of RFs in genome-wide association studies 25
6 **AI/ML in diagnostics** 25
6.1 Treatment responses and healthcare outcomes 27

6.2 Boosting methods in diagnostic and prognostic algorithms 28
6.3 Mortality risk prediction 29
7 **Further considerations** 30
8 **Conclusion** 30
References 32

3 State of the art artificial intelligence assisted disease detection tools

Sushanta Kumar Das, Rahul Mishra, Amit Samanta, Saumendu Deb Roy, and Ujjwal Sahoo

1 **Introduction** 37
2 **Fundamentals of AI technologies in disease detection** 38
2.1 Machine learning 38
2.2 Deep learning 38
2.3 Natural language processing 39
2.4 Computer vision 39
2.5 Data mining and pattern recognition 40
2.6 Reinforcement learning 40
2.7 Integration and synergistic applications 40
2.8 Challenges with technologies and future directions 41
3 **Applications of AI in disease detection** 41
3.1 Radiology 41
3.2 Pathology 42
3.3 Oncology 43
3.4 Cardiology 44
3.5 Infectious diseases 45
3.6 Psychiatry 45
3.7 Neurology 46
3.8 Endocrinology 46
4 **Implementation and real-world impact of AI-assisted disease detection** 47
5 **Challenges in AI-assisted disease detection** 47
5.1 Data quality and bias 47
5.2 Interpretability 48
5.3 Regulatory and ethical considerations 48
5.4 Generalization 48
6 **Future directions of AI-assisted disease detection** 48
6.1 Precision medicine 48
6.2 Real-time monitoring 49
6.3 Collaborative AI 49
6.4 Global health equity 49
6.5 Regulatory evolution 49
7 **Conclusion** 49
Acknowledgments 50
References 50

Section II
The role of artificial intelligence in holistic healthcare and disease diagnosis

4 Application of artificial intelligence chatbots for primary diagnosis at home

Takanobu Hirosawa

1 **Introduction** 57
1.1 Understanding primary diagnosis, triage, and definitive diagnosis 57
1.2 Difference between diagnosis at home and medical institutions 58
1.3 AI chatbots, CDSS, and other applications 58
2 **The evolution of AI chatbots in healthcare** 59
2.1 Early rule-based chatbots in healthcare 59
2.2 Transition to intelligent systems 59
2.3 Diagnostic excellence and AI chatbot in home-based diagnosis 59
3 **How AI chatbots facilitate primary diagnosis** 60
3.1 Symptom assessment 60
3.2 Decision support algorithms 60
3.3 Integration with telemedicine 60
4 **Benefits of AI chatbots in home-based diagnostics** 60
4.1 Increased accessibility 60
4.2 Cost-effectiveness 60
4.3 Early detection and monitoring 61
4.4 Rapid update capabilities of ML models 61
4.5 Special benefits in limited resource environments 61
5 **Challenges and limitations** 61
5.1 Diagnostic accuracy 61
5.2 User engagement and trust 61
5.3 Regulatory and ethical concerns 62
5.4 Privacy concerns 62
5.5 Technical limitations 62
6 **Case studies of AI chatbots in home diagnostics** 64
6.1 Traditional symptom checkers 64
6.2 AI-powered symptom checkers 64
6.3 Untrained and unspecialized generative AI chatbots 65
7 **Future directions** 66
7.1 Enhanced personalization 66
7.2 Multimodal AI chatbots in primary diagnosis at home 66
7.3 Edge AI for home-based diagnosis 66
7.4 Integration with smart home technologies 66
7.5 Collaboration with healthcare systems 67

7.6 Technological breakthrough 67
8 Conclusion 67
References 68

5 Adaptive network-based fuzzy inference system-enhanced bidirectional time-aware LSTM for clinical sequence prediction

Shambhab Chaki, Pratyusha Rakshit, and Shriya Adhikari

1 Introduction 71
2 Literature review 74
2.1 Early probabilistic and statistical approaches 74
2.2 Neural sequence models (RNNS, LSTMS, GRUs and variants) 74
2.3 Transformer and self-supervised contrastive learning 74
2.4 Synthesis and gap identification 75
3 Methodology: Bidirectional T-LSTM 77
3.1 Recapitulations of time-aware LSTM 77
3.2 Bidirectional T-LSTM 78
3.3 Bi-TLSTM encoder stacking 78
4 Methodology: ANFIS-enabled Bi-TLSTM 79
4.1 Recapitulations of ANFIS 79
4.2 Layers of ANFIS 79
4.3 Integration of ANFIs with Bi-TLSTM 80
4.4 Training 81
5 Experiments and results 81
5.1 Datasets 81
5.2 Dataset descriptions 82
5.3 Data preprocessing 82
5.4 Experimental setup 82
5.5 Performance metrics 83
5.6 Implementation details 83
5.7 Comparative analysis 83
5.8 Sensitivity analysis with hyperparameters 85
5.9 Ablation results 86
6 Discussion 88
7 Practical implications for clinical deployment 88
8 Conclusion and future works 89
References 89

6 Applications of artificial intelligence in infectious diseases

Kairav Barua and Somdeb Bose Dasgupta

1 Introduction 93
2 History of AI in infectious disease surveillance and prevention 94
2.1 Present-day advances and applications 95
2.2 Databases 95
2.3 AI for early detection and diagnosis of infectious diseases 97
2.4 Predictive modeling for outbreaks and spread 104
2.5 AI for containment and response 105
2.6 AI in therapeutics and drug discovery for infectious diseases 108
3 Conclusion 112
References 112

7 Artificial intelligence-assisted imaging technologies for disease diagnosis

Luis Pinto-Coelho

1 Introduction 117
2 Foundations of AI for medical imaging 118
2.1 AI inference and training 118
2.2 Data quality 119
3 AI techniques in medical imaging 120
4 Advances in medical image analysis 121
5 Explainability and interpretability 124
6 Applications of AI in disease diagnosis 125
7 Future perspectives and innovations 126
7.1 Security and ethics 126
7.2 Federated learning 127
7.3 Foundation models 127
7.4 Green AI 127
7.5 Responsible AI 127
8 Conclusion 128
References 128

Section III
Use of artificial intelligence in organ transplantation and precision surgery

8 Predictive artificial intelligence to increase the success of heart transplantation

David B. Olawade, Aderonke Odetayo, Kusal Weerasinghe, Sheila Marinze, Eghosasere Egbon, and Jennifer Teke

1 Introduction 135
2 The revolutionary role of AI in heart transplantation 136
3 Predicting donor–recipient compatibility: The perfect match 137
3.1 Histocompatibility enhancement 138
3.2 Advanced blood type and cross-matching analysis 138
3.3 Optimized size and age matching 138

4 **Predicting organ viability: Seeing beyond the surface** 138
4.1 Real-time imaging intelligence 139
4.2 Predictive biochemical modeling 139
4.3 ML for ischemia time optimization 139
5 **Posttransplant complication prediction: Early warning systems that save lives** 140
5.1 Revolutionary acute rejection detection 141
5.2 Advanced infection risk prediction 142
5.3 CAV: Detection and prevention 142
6 **AI-driven personalized medicine for immunosuppression** 143
6.1 Precision drug dosing through AI 143
6.2 ML for medication adherence 143
6.3 Predictive toxicity monitoring 143
7 **AI-assisted remote monitoring and patient management** 144
7.1 Advanced wearable electrocardiogram monitoring systems 144
7.2 Comprehensive physiological monitoring 144
7.3 AI-powered virtual health assistants and telemedicine integration 145
8 **AI-enabled predictive models in heart transplantation** 146
8.1 Advanced neural network applications 146
8.2 Traditional ML approaches with enhanced capabilities 146
8.3 Ensemble methods and hybrid approaches 147
9 **Challenges and ethical considerations** 147
9.1 Data privacy and security concerns 147
9.2 Addressing algorithmic bias and ensuring equity 148
9.3 Clinical validation and regulatory requirements 148
9.4 Human–AI collaboration and clinical integration 148
9.5 Economic and implementation challenges 149
10 **Future prospects of AI in heart transplantation** 149
10.1 Real-time decision support systems 149
10.2 Blockchain integration for enhanced transparency and security 149
10.3 AI-guided xenotransplantation 150
10.4 Integration with genomic and precision medicine 150
10.5 Advanced regenerative medicine applications 150
11 **Transforming patient outcomes: real-world impact** 151
11.1 Enhanced quality-of-life outcomes 151
11.2 Reduced healthcare utilization 151
11.3 Functional and social benefits 151
12 **Economic impact and healthcare sustainability** 152
12.1 Direct cost reductions 152
12.2 Long-term value creation 152
12.3 Resource optimization and efficiency 152
12.4 Innovation and competitive advantages 152
13 **Preparing for the AI-driven future** 153
13.1 Professional development and training 153
13.2 Institutional infrastructure development 153
13.3 Regulatory and policy adaptation 153
13.4 Patient and public engagement 154
14 **Conclusion** 154
References 154

9 Advancing precision surgery with artificial intelligence: Opportunities and challenges

Rayner Peyser Cardoso and Swizel Ann Cardoso

1 **Introduction** 163
2 **Opportunities presented by AI in precision surgery** 164
2.1 Enhanced preoperative planning and diagnostics 164
2.2 Real-time intraoperative guidance and robotic assistance 164
2.3 Enhanced patient safety and perioperative outcomes 165
2.4 Surgical education and training 165
2.5 Advancements in robotics and automation 165
2.6 Personalized medicine and precision therapeutics 165
2.7 Intraoperative decision support 166
2.8 Advanced analytics for quality improvement 166
2.9 Minimally invasive surgery and remote operations 166
2.10 Ethical AI and patient trust 166
3 **Case studies and applications** 167
3.1 Robotic-assisted surgery 167
3.2 AI-guided total joint arthroplasty 168
3.3 Image-guided glioma resection 168
3.4 Predictive analytics and perioperative management 168
3.5 AI-centric surgical education 169

3.6 Postoperative monitoring and rehabilitation 169
3.7 Workflow, supply chain, and decision support 169
3.8 Enhanced dexterity and semiautonomy 169
3.9 Personalized approaches and targeted delivery 169
4 Challenges and limitations 170
4.1 Data availability, quality, and bias 170
4.2 Regulatory and ethical considerations 170
4.3 Integration into clinical workflow and infrastructure 171
4.4 Technical limits and generalizability 171
4.5 Human–computer interaction and trust 172
4.6 Legal and liability issues 172
4.7 Cost and resource allocation 172
4.8 Education and training gaps 172
4.9 Regulatory approval and validation 172
4.10 Cyber-security risks 173
4.11 Patient acceptance and public perception 173
5 Conclusion 173
References 174

10 Application of artificial intelligence in robotic surgery: An overview

Muhammad Saqib, Arslan Ali, Aneesa Altaf, M. Sharjeel Shoukat, Asiya Najam Ul Haq, Khushbakht Waqar, Amna Usman, Faiza Aziz, and Muhammad Ahsan

1 Introduction 175
2 Preoperative phase 175
3 Intraoperative phase 178
4 Postoperative phase 182
5 Training, education, and credentialing 187
6 Economic impact and system efficiency 188
7 Ethical, legal, and equity considerations 189
8 Future directions and research priorities 190
9 Conclusion 192
References 192

11 Artificial intelligence in plastic surgery: The future of reconstruction and aesthetics

Yasser Hijazi and Omar Kiwan

1 Introduction 197
2 Preoperative applications 197
2.1 Diagnosis and screening 197
2.2 Risk stratification and surgical planning 198
2.3 Simulation and consent 201
3 Intraoperative applications 202
4 Postoperative applications 202
4.1 Postoperative monitoring 203
4.2 Outcome analysis 203
4.3 Managing complications 203
5 Education and research 204
6 Administrative and operational support 204
7 Ethical, legal, and practical challenges 205
8 Summary 205
References 206

Section IV
Domain-specific success stories in medical science through artificial intelligence intervention

12 Artificial intelligence in assisted reproduction: Navigating the next frontier in fertility care

Pratip Chakraborty, Parag Nandi, and Syed Monajatur Rahman

1 Introduction 213
2 AI in reproductive medicine 213
2.1 Overview 213
2.2 Why reproductive medicine needs AI 214
2.3 AI in gonadotropin dosing for ovarian stimulation 215
2.4 AI in ultrasound analysis for follicle monitoring 216
2.5 AI in sperm selection and male infertility 216
2.6 AI in oocyte selection 216
2.7 AI for predicting IVF outcomes 217
3 Image-based and multimodal data integration 218
4 AI for monitoring pregnancy and managing complications 218
5 Ethical, legal, and implementation challenges 219
6 Limitations and challenges of AI research in reproductive medicine 219
7 Future directions 220
8 Conclusion 220
References 221

13 Appications of artificial intelligence in cardiovascular disorders

Sabyasachi Chakrabarty and Susmita Ghosh

1 Cardiovascular disease: A global health challenge 225

2 Current management of CVDs: Strategies and systemic limitations 225
2.1 Lifestyle modification 226
2.2 Data collection and patient monitoring 226
2.3 Risk assessment 226
2.4 Diagnostic approaches 227
2.5 Treatment strategies 227
2.6 Resource constraints and stark health 228
3 AI tools, models, and algorithms in CVD management 228
3.1 Data management and integration strategies 228
3.2 AI models tailored to cardiovascular data modalities 229
3.3 Advanced risk prediction models 234
4 Case studies: Real-world impact of AI in cardiovascular care 235
4.1 Mayo Clinic's asymptomatic heart failure detection 235
4.2 Zebra Medical Vision's CCS 235
4.3 Eko Health's reinvention of the stethoscope with AI 235
5 Conclusion 236
Acknowledgements 236
References 236

14 The use of artificial intelligence in ophthalmology

Thiago Gonçalves dos Santos Martins, Diogo Gonçalves dos Santos Martins, Thomaz Gonçalves dos Santos Martins, Rufino da Silva, Luis Mendes, and Paulo Schor

1 Introduction 241
2 Fundamentals of AI applied to ophthalmology 241
3 Clinical applications of AI in ophthalmology 242
4 Algorithm development and validation 242
5 Clinical implementation 242
6 Advantages 242
7 Limitations and challenges 242
8 Age-related macular degeneration 243
9 Glaucoma 243
10 Cornea 243
11 Monitoring and prognosis 243
12 AI-assisted surgery 244
13 Available platforms and technologies 244
14 Challenges and limitations 244
15 Future perspectives 244
16 Conclusion 246
References 247

15 The roles of machine learning and artificial intelligence in the prediction of autoimmune disease

Ashok Kumar Sah, Ramji Gupta, Prem Shankar Mishra, Nitya Sharma, Rahul Kumar, Sakshi Patel, and Rabab H. Elshikh

1 Introduction 249
1.1 Autoimmune diseases and challenges in prediction 249
1.2 Role of artificial intelligence and machine learning in modern healthcare 250
2 Overview of autoimmune diseases 251
2.1 Background: Pathophysiology and key mechanisms 251
2.2 Major autoimmune diseases 251
2.3 Challenges in conventional diagnostic approaches 252
3 Artificial intelligence and ML basics 252
3.1 Core concepts of AI and ML 252
3.2 Types of ML (supervised, unsupervised, deep learning) 253
3.3 Importance of data in predictive modeling 253
4 Applications of AI and ML in autoimmune disease prediction 254
4.1 Early diagnosis and risk prediction 254
4.2 Biomarker discovery and validation 254
4.3 Patient stratification and prognosis 255
4.4 Case studies and real-world applications 255
5 Integration with multiomics data 256
5.1 Genomic and proteomic data applications 256
5.2 Imaging-based prediction 256
5.3 Multimodal data integration 256
6 Challenges and ethical considerations 257
6.1 Data limitations and heterogeneity 257
6.2 Model interpretability and clinical trust 258
6.3 Ethical, legal, and privacy concerns 258
7 Future perspectives 259
7.1 Explainable AI 259
7.2 Digital health and wearable integration 259
7.3 Toward precision medicine in autoimmune diseases 259
8 Conclusion 260
Further reading 260

Section V
Artificial intelligence in neuropsychiatric rehabilitation

16 Artificial intelligence in psychotherapy and counseling

Sandhya Bhatt and Anu Dandona

1 Introduction 267
2 Overview of Artificial Intelligence technology in mental health support 268
2.1 Woebot 268
2.2 Wysa 268
2.3 Replika 269
2.4 Youper 269
2.5 Headspace 270
3 Psychological frameworks supporting digital therapeutic approaches 270
4 Acceptance and commitment therapy intervention 272
5 Self-determination theory in digital health 272
6 Health belief model 273
7 Ethical considerations 275
8 Linking Health Insurance Portability and Accountability Act and General Data Protection Regulation compliance in mental health apps 275
9 Conclusion 275
References 276

17 Brain–computer interfaces for neuromotor rehabilitation

Amit Konar, Baishali De and Anuradha Saha

1 Introduction 281
2 Preliminaries 283
2.1 Brain signals 283
2.2 Pattern classification 284
2.3 Controllers 285
3 BCI control policies 286
3.1 ERD/ERS-based semiclosed-loop control 287
3.2 MI-P300-based closed-loop position control scheme 288
3.3 Shared two-loop BCI control architecture 289
4 Signal processing, FE, and classification of brain signals 290
4.1 Spatial filtering 291
4.2 Feature extraction 292
4.3 Principles of brain signal classification 295
5 Experiments and results 297
6 Discussion 299
7 Conclusion 301
References 302

18 Artificial intelligence in the detection and mitigation of neurological and psychiatric conditions

Jarred D. Metkus, Angela Oh, Tanmay Singhal, Talia Puzantian, and Subhrajit Bhattacharya

1 Introduction 305
2 Drug development and screening 306
3 Genetic screening 307
3.1 Angelman syndrome 307
3.2 Pediatric encephalopathies 308
3.3 Crohn's disease 308
4 AI applications in clinical psychiatric and neurological practice 309
4.1 Pediatric epilepsy 309
4.2 Neuro Scholar 311
4.3 Speech analysis 311
4.4 Eye tracking 311
4.5 Brain structure analysis 311
4.6 Treatment recommendation 312
5 AI applications in clinical psychiatric and neurological practice 312
References 314

19 Application of artificial intelligence in early diagnosis of alpha-synucleinopathies

Nabanita Ghosh and Krishnendu Sinha

1 Introduction 317
2 Methods to assess the presence and progression of AD and PD 317
2.1 Biomarkers 317
2.2 Molecular imaging 318
2.3 Biosensors 318
3 Application of ML/DL 319
4 ML and DL techniques for neurodegenerative disease classification 322
4.1 Semisupervised clustering with Smile-GAN 322
4.2 Multimodal attention architectures: MADDi 322
4.3 DL of white matter tau pathology 323
4.4 AI-based approaches for detecting blood–brain barrier permeability changes 323
4.5 Generative and synthetic imaging with GANs and diffusion models 323

4.6 Explainable multimodal fusion (e.g., MADDi and graph-based approaches) 323
5 Model performance and limitations 323
6 Conclusion 324
References 324

Section VI
Artificial intelligence in the infrastructural augmentation of healthcare industry

20 Artificial intelligence (AI) in wastewater treatment

Laishram Saya, Atreyee Bagchi, Shivam Saini, Deepanshu Vasuja, Gauri Jha, Pooja, and Sunita Hooda

1 Introduction 331
1.1 The nature of wastewater 331
1.2 Environmental and sustainability challenges in current wastewater management 332
2 AI in WWT 333
2.1 Neural network-based models 333
2.2 FL-based models 335
2.3 Optimization and evolutionary algorithms 335
2.4 Other models and examples 336
3 Recent advances in AI-driven WWT technologies 337
3.1 AI-based decision support and planning tools 337
3.2 Uncertainty handling and risk management 337
3.3 Smart sensors, Internet of Things, and digital twins 337
3.4 Advanced disinfection and renewable integration 338
3.5 AI for extreme events and pollution source management 338
3.6 Negotiation and policy support systems 339
4 Japan as a case study 340
5 Limitations and challenges of AI in WWT 340
6 Future directions and challenges in scaling AI for wastewater management 342
Acknowledgement 343
References 343

21 Artificial intelligence-driven value creation and investment strategy in healthcare innovation

Bidyut Sarkar and Rudrendu Kumar Paul

1 Introduction 347
1.1 Why is ROI central to healthcare investment decisions? 347
1.2 Limitations of traditional ROI frameworks 348
1.3 AI applications in enhancing capital efficiency and accelerating outcomes 348
2 How financial metrics and AI modeling support smarter investment strategies 350
2.1 Financial calculations 350
2.2 Sensitivity and scenario analysis 351
2.3 Sensitivity analysis 352
2.4 Societal impact and social return on investment 352
2.5 Measuring social return on investment and broader impact 353
2.6 Qualitative factors 354
3 AI-driven investment strategies 354
4 Strategic allocation heuristic for healthcare investments 354
4.1 Operationalizing AI-driven healthcare investment horizons 355
5 Funding challenges for a new class of antibiotic candidates: The role of AI and ROI 356
6 AI-enabled ROI in esthetic medicine, antivenom, and age-related therapies 357
6.1 AI-enabled ROI in lifestyle drugs and cosmetic procedures 357
6.2 AI-enabled ROI in life-saving antivenoms: Toward a universal snake antivenom 357
6.3 AI-enabled ROI in extending healthy lifespan and longevity drugs 358
7 Operationalizing AI-driven ROI framework 358
7.1 Checklist for step-by-step AI-enhanced ROI realization 358
7.2 AI tools for SROI assessment 359
7.3 Real-world success factors and barriers 359
8 Conclusion 360
References 361

22 Addressing India's rural healthcare gaps utilizing generative artificial intelligence (AI) and agentic intelligence

Prithviraj Bagchi

1 Generative artificial intelligence and agentic intelligence in rural healthcare 365
1.1 Introduction 365
1.2 Key benefits of Gen AI in rural healthcare 366
1.3 Strategy for rural healthcare 367
2 AI-enabled community healthcare and implementation plans 367
2.1 AI-powered teleconsultation hubs 368
2.2 Mobile AI diagnostic vans 368
2.3 AI-driven health education chatbots 368
2.4 Smart medication adherence systems 368
2.5 Predictive analytics for outbreak surveillance 368
3 Agentic intelligence has quality impact on medical support 369
3.1 Autonomous symptom assessment agents 369
3.2 Intelligent prescription assistants 369
3.3 Dynamic chronic disease management agents 370
3.4 Emergency response coordination agents 370
3.5 Community health analytics orchestrators 370
4 AI for primary healthcare delivery 370
5 Equitable access to preventive care 371
6 Conclusion 371
References 371

Section VII
New frontiers of artificial intelligence

23 Quantum artificial intelligence for human health: A statistical framework

Hossein Hassani, Leila Marvian Mashhad, and Steve MacFeely

1 Introduction 375
2 Data-generating processes in health 376
2.1 Structural view and observed data 376
2.2 Missingness mechanisms and estimation 378
2.3 Causal estimands, measurement bias, site heterogeneity, and evaluation (human health focus) 379
3 Practical examples in human health 381
3.1 Example 1: Multiomics risk stratification with a quantum kernel 382
3.2 Example 2: Radiology triage with a variational quantum classifier 382
3.3 Example 3: Survival analysis with censoring-aware quantum kernel 383
3.4 Example 4: Operating-room scheduling via QUBO policy optimization 384
3.5 Example 5: Federated, multisite generalization, and covariate shift 384
4 Data governance for QAI in health 385
5 Conclusion 385
References 386

24 Use of artificial intelligence in microbiome research

Feronia Melissa De Mello and Ashfaque Hossain

1 Introduction 387
1.1 Overview of microbiome research 387
1.2 Importance of addressing microbiome research challenges 387
2 AI in microbiome data analysis 388
2.1 Taxonomic classification and metagenomic analysis 388
2.2 Multiomics data integration (genomics, proteomics, metabolomics) 388
3 Applications of AI in microbiome research 389
3.1 Functional predictions (microbial pathways, interactions) 389
3.2 Personalized medicine (microbiome-based diagnostics and therapies) 390
3.3 Drug discovery (microbial metabolites and probiotics) 390
3.4 Ecological insights (modeling microbial ecosystems) 391
4 AI techniques for microbiome research 391
4.1 ML models (e.g., random forests, clustering) 391
4.2 DL approaches (e.g., neural networks, BioBERT) 393

4.3 Tools for literature mining and knowledge extraction 393
5 **Challenges and ethical concerns** 394
5.1 Data privacy and security 394
5.2 Bias in AI algorithm 394
5.3 Interpretability and reproducibility issues 395
6 **Future directions in microbiome research** 395
6.1 AI in microbiome engineering and synthetic biology 395
6.2 Combining microbiome studies with AI-driven multiomics approaches 396
6.3 Role of AI in global microbiome initiatives 396
7 **Conclusion** 396
References 400

25 The role of artificial intelligence and machine learning in genomics

Gautam Das, Garima Suneja, Kumar Gautam Singh, Sanober Waghoo, and Akib Kamani

1 **Introduction** 405
2 **Microbial networks and AI** 406
2.1 Microbial networks for disease diagnostics 408
3 **General DL frameworks in genomics** 408
4 **Applications of AI in genomics** 409
5 **Challenges in AI applications for genomics** 409
6 **Ethical issues in genomics** 409
7 **K-mers for AI and genomics** 410
7.1 Constructing k-mers from DNA sequences 410
7.2 Applications of k-mers in AI and genomics 410
8 **The k-mer-based ML approach for predicting drug resistance in *M. tuberculosis*** 411
8.1 ML framework for MIC prediction 411
8.2 Hyperparameter optimization using random search and grid search 412
8.3 Training, testing, and prediction of the model 412
8.4 Biological interpretability 412
8.5 Results 412
9 **Challenges and future directions** 413
10 **Conclusion** 413
References 414

26 Artificial intelligence-assisted drug designing, toxicity assessment, and biodelivery

Pralay Mitra, Soumyadeep Bhaduri, Debraj Das, and Dibya Kanti Haldar

1 **Introduction** 417
1.1 Overview: Peptide synthesis, function, biological activity, and moonlighting effects 418
2 **Data–target integration and RNA-based approaches** 419
2.1 Peptide databases 419
2.2 Preprocessing: Cleaning, tagging, and combining 420
3 **Target identification in proteins and RNAs** 420
3.1 ML techniques for peptide target prediction 420
3.2 Protein–peptide interactions: Prediction and validation 420
3.3 RNA—Potential therapeutic targets for peptides 421
3.4 RNA recognition using macrocyclic peptides 421
3.5 Delivery and targeting of small interfering ribonucleic acid (siRNA) through peptides 422
3.6 RNA therapeutic delivery via cell-penetrating peptides 422
3.7 microRNA modulation via peptide-based systems 423
3.8 Therapeutic outcomes and clinical applications 423
4 **Feature engineering** 423
4.1 Sequence-based representation learning 423
4.2 Structural representations and 3D geometry 423
4.3 Multimodal representation fusion 424
5 **Generative AI models in action** 424
5.1 Information-theoretic approaches to peptide design 424
5.2 Using generative adversarial networks in molecular design 424
5.3 VAEs for systematic exploration 424
5.4 Deep residual networks for mRNA design 425
5.5 Adversarial training paradigms 425
5.6 Diffusion models and denoising approaches 425
5.7 Attention mechanisms and transformer architectures 425

6 **AI in predicting toxicity and stability** 425
6.1 Adding safety limits to generative models 425
6.2 Models for predicting toxicity 425
6.3 Strategies for benchmarking and validation 426
6.4 DL-enhanced peptide quantitative structure–activity relationship in peptide drug development 428
7 **Future scope of AI in peptide-based drug designing and biodelivery** 429
8 **Conclusion** 429
References 429

27 The role of generative artificial intelligence in personalized drug development and safety

Humaira Ilyas and Sk. Abdul Mohid

1 **Introduction** 435
2 **Generative AI foundations in a clinical context** 437
2.1 Understanding generative models through a biological lens 437
2.2 Training data: Biological foundations for generative learning 440
2.3 Challenges in biological generative modeling 441
3 **Generative AI-powered drug design** 441
3.1 De novo molecule generation: Learning the language of chemistry and biology 441
3.2 Target-aware molecular design 441
3.3 Optimizing drug-like properties in silico 442
3.4 Designing biologics and macromolecular therapeutics 442
3.5 Personalization: Designing drugs for individual patients 442
3.6 Including generative design in clinical and preclinical processes 443
4 **Integrating patient data for therapeutic personalization** 443
4.1 Genomic variants and target prioritization 443
4.2 Transcriptomic and epigenomic states: Capturing cellular context 443
4.3 Proteomics and functional targeting 444
4.4 Microbiome and metabolome: Modulators of drug efficacy 444
4.5 Clinical phenotypes and real-world data 444
4.6 Digital twins: Personalized simulation and virtual drug testing 444
4.7 Ethical and practical considerations 445
5 **Enhancing drug safety through predictive modeling** 445
6 **Regulatory, ethical, and practical considerations** 447
7 **Future directions and emerging opportunities** 448
8 **Conclusion** 448
Acknowledgement 450
References 450

28 Artificial intelligence at the heart of nutrigenomics: Machine learning for a customized diet

Samudra Prosad Banik, Anand Swaroop, Harekrishna Jana, and Debasis Bagchi

1 **Introduction** 455
2 **Nutrition, personalized nutrition, and nutrigenomics** 456
3 **ML in AI** 457
4 **ML tools in "omics" studies** 458
5 **Deep learning** 459
6 **Facets of precision nutrition and the role of ML therein** 459
7 **Conclusion** 461
Acknowledgement 461
References 461

29 New frontiers in organoid research: The synergistic integration of organoids with artificial intelligence

Soumya Shekhar, Subhayan Sur, Amit Ranjan, and Soumya Basu

1 **Introduction** 465
2 **Organoid technology: State of the art** 466
3 **AI methodologies for organoid analysis** 466
3.1 Organoid imaging, tracking, segmentation, differentiation, and morphological analysis 466
3.2 The synergistic potential: AI-enhanced organoid construction and optimization 471
3.3 AI in organoid-based drug screening, drug response/toxicity, and disease progression 471
3.4 AI in organoid omics 473
3.5 Brain organoid models for neurodegenerative diseases 474
4 **Challenges and future directions** 475
4.1 Vascularization issues and batch variability 476
4.2 Data complexity and integration 476

4.3 Model interpretability and biological relevance 476
4.4 Reproducibility and standardization 477
4.5 Ethical considerations 477
4.6 Computational resources and expertise 477
5 **Future directions** 477
6 **Conclusion** 477
References 478

30 Artificial intelligence (AI)-enabled organoids and three-dimensional (3D) in vitro models for clinical applications and precision medicine (PM)

Tejal Pant, Nanda Rohra, Tanya Saxena, Shelly Pathak, and Payal Ganguly

1 **Introduction** 483
2 **Role of AI-enabled organoids or OI** 484
3 **AI-enabled organoids for cancer** 487
4 **AI-enabled organoids for musculoskeletal conditions** 489
5 **AI-enabled organoids for neural networks with a focus on the brain** 490
6 **The genetic landscape of AI-enabled organoids** 492
7 **Conclusions and limitations** 493
8 **Future directions** 493
References 494

31 Leveraging artificial intelligence (AI) in herbal medicinal plant extracts: Pathway to drug development

Chukwuemeka R. Nwokocha, Javier Palacios, Keaton Logan, Magdalene Nwokocha, Ya Fatou Njie Mbye, Nelson Orie, Sunny Edet Ohia, and Momoh A. Yakubu

1 **Introduction** 501
2 **Extraction and HPLC-guided fractionation of bioactive compounds** 502
2.1 Sequential solid–liquid extraction 502
2.2 HPLC purification and gas chromatography–mass spectrometry characterization 502
2.3 In vitro validation (bioassays) 502
3 **Molecular docking studies of yam-derived ACEis** 502
3.1 ACE structure and binding site architecture 502
3.2 Docking methodology 502
3.3 Key results 503
3.4 Broader context and implications 503
4 **Recommendations for future docking studies** 505
5 **ADMET studies and the role of AI in natural product screening** 505
5.1 Application to natural product-derived compounds 506
6 **Future directions** 506
7 **Conclusion** 508
References 508

32 Drug discovery from medicinal plants and natural products: Opportunities for advancement with artificial intelligence

Momoh A. Yakubu, Vishwalatha K. Parvathreddy, Ayodele Stephen Ake, Taiwo O. Oyagbemi, Ebenezer O. Ajiboye, Ademola A. Oyagbemi, Ya Fatou Njie Mbye, and Sunny Edet Ohia

1 **Introduction** 511
2 **The role of herbal plants in traditional and modern medicine** 512
3 **Key challenges in natural product drug development** 512
4 **Molecular docking in herbal research** 513
5 **AI's role in expediting natural product screening and identification** 513
6 **AI-driven successes in natural product drug development** 515
7 **In silico pharmacokinetic analysis** 515
8 **Application of AI in de novo drug design from natural products** 516
9 **Challenges and limitations** 517
10 **Future directions in AI-powered research** 518
11 **Conclusion** 518
References 518

Section VIII
Ethical considerations and the future of artificial intelligence

33 Ethical frontiers: Navigating public health responsibilities in the age of artificial intelligence

Sudip Bhattacharya, Sandip Bhattacharya, and Biswadeep Dhar

1 **Introduction** 525
1.1 How artificial intelligence is revolutionizing healthcare and diverse industries 525

2 **Expanding AI governance** 529
2.1 The need for clear AI governance frameworks 530
2.2 Ethics in healthcare AI design and application 530
2.3 Ensuring the ethical use of AI in health 530
2.4 The importance of data privacy in AI systems 530
2.5 AI and human oversight in healthcare 531
2.6 AI's impact on healthcare jobs 531
2.7 AI's role in medical research 531
2.8 Ethical challenges in AI medicine 531
2.9 The future of responsible AI in health 532
3 **Conclusion** 532
References 532

34 Ethical issues of artificial intelligence in biomedical applications

Sudipta Chakraborty

1 **Introduction** 535
2 **Progress timeline of AI** 535
2.1 Foundational concepts (1940s–1950s) 535
2.2 The rise and fall of systems (1960s–1980s) 536
2.3 The machine learning era (1990s–2010s) 536
2.4 The generative AI era (2020s–present) 536
3 **Ethical theories of AI** 536
4 **AI in healthcare system** 537
4.1 Medical imaging and diagnosis 538
4.2 Drug discovery and development 538
4.3 Personalized medicine 538
4.4 Robotics in surgery 538
4.5 Genomics and proteomics 538
4.6 Biomedical devices and additive manufacturing 539
4.7 Signal processing, wearables, and biosensors 539
4.8 NLP in healthcare 539
4.9 Explainable AI and model transparency 539
4.10 Clinical trials, epidemiology, and public health 539
5 **Merits and demerits of AI in healthcare** 539
6 **Challenges in biomedical AI** 540
6.1 Data quality, diversity, and generalization 540
6.2 Interpretability and the "black-box" problem 540
6.3 Safety, reliability, and adversarial vulnerabilities 540
6.4 Loss of clinical expertise 541
6.5 Continuous data monitoring 541
6.6 Regulatory and legal challenges 541
6.7 Ambiguity in accountability and liability 541
7 **AI failures in biomedical fields** 541
8 **Ethical considerations in biomedical AI** 542
8.1 Privacy, consent, and data ownership 542
8.2 Bias, fairness, and equity 542
8.3 Human autonomy and patient–doctor relationship 542
8.4 The "black-box" problem and accountability, liability, and transparency 542
8.5 Model drift and failures in real-life applications 543
8.6 Global governance and ethical standards 543
9 **Emerging approaches to address ethical concerns** 543
9.1 Federated learning 543
9.2 Differential privacy 544
9.3 Cryptographic techniques 544
9.4 Hybrid approaches 544
10 **Conclusion** 544
References 545

35 Minimizing vulnerability of artificial intelligence (AI) programs from cyberattacks

Laishram Saya, Deepanshu Vasuja, Gauri Jha, Atreyee Bagchi, Shivam Saini, Pooja, and Sunita Hooda

1 **Introduction: Artificial intelligence as a double-edged sword** 551
1.1 Definition of AI 551
1.2 Overview of machine learning and deep learning 552
1.3 Understanding cyberattacks and cybersecurity 552
2 **Vulnerabilities of AI systems** 553
3 **Data as the first line of attack in AI systems** 553
3.1 Data: The foundation of AI 553
3.2 Data poisoning 556
3.3 Adversarial attack 557
3.4 Toxic Microsoft's Tay 557
3.5 Countermeasures to data poisoning 558
4 **Models as the engine of decision-making** 559
4.1 Role of models in AI 559
4.2 Black-box vs. white-box models 560
4.3 Why model integrity matters 560
4.4 Threat vectors targeting models 560
4.5 Real-world failures: Motivating defenses 562

4.6 Securing the model pipeline: Defenses and design principles 563
5 Legacy threats in the age of AI: Why they still matter 566
6 Conclusion: Toward resilient and trustworthy AI systems 566
6.1 Detect: Early identification of threats 566
6.2 Prevent: Architectural and operational safeguards 567
6.3 Predict: Future-proofing against emerging threats 568
Acknowledgement 569
References 569

Glossary of AI terminologies 573
Index 585

Contributors

Shriya Adhikari Department of Electronics & Telecommunication Engineering, Jadavpur University, Kolkata, India

Muhammad Ahsan Ziauddin University, Karachi, Pakistan

Ebenezer O. Ajiboye Anatomy Unit, Faculty of Basic Medical Sciences, Ajayi Crowther University, Oyo, Nigeria

Ayodele Stephen Ake Department of Veterinary Physiology and Biochemistry, Faculty of Veterinary Medicine, University of Ibadan, Ibadan, Nigeria

Arslan Ali Jinnah Sindh Medical University, Karachi, Pakistan

Aneesa Altaf Nishtar Medical University, Multan, Pakistan

Faiza Aziz Islamic International Medical College, Riphah International University, Rawalpindi, Pakistan

Atreyee Bagchi Department of Biological Sciences, Sri Venkateswara College, University of Delhi, Delhi, India

Debasis Bagchi Department of Biology, College of Arts and Sciences, Adelphi University, Garden City, NY; Department of Pharmaceutical Sciences, College of Pharmacy and Health Sciences, Texas Southern University, Houston, TX, United States

Prithviraj Bagchi Advanced Technology in Cloud and Gen AI Domains, Accenture Services Pvt. Ltd., Kolkata, West Bengal, India

Samudra Prosad Banik Department of Microbiology, Government General Degree College, Narayangarh, Rathipur, West Bengal, India

Kairav Barua Department of Bioscience and Biotechnology, Indian Institute of Technology Kharagpur, Kharagpur, India

Soumya Basu Cancer and Translational Research Centre, Dr. D. Y. Patil Biotechnology and Bioinformatics Institute, Dr. D. Y. Patil Vidyapeeth, Pune, Maharashtra, India

Soumyadeep Bhaduri Centre for Computational and Data Sciences, Indian Institute of Technology Kharagpur, Kharagpur, West Bengal, India

Sandhya Bhatt Amity University, Noida, India

Sandip Bhattacharya Department of Electronics and Communication Engineering (ECE), SR University, Warangal, India

Subhrajit Bhattacharya School of Pharmacy, The Claremont Colleges, Keck Graduate Institute, Claremont, CA, United States

Sudip Bhattacharya Department of Community and Family Medicine, All India Institute of Medical Sciences, Deoghar, Jharkhand, India

Somdeb Bose Dasgupta Department of Bioscience and Biotechnology, Indian Institute of Technology Kharagpur, Kharagpur, India

Rayner Peyser Cardoso All India Institute of Medical Sciences, Jodhpur, Rajasthan, India

Swizel Ann Cardoso Department of Vascular Surgery, Doncaster Royal Infirmary, Doncaster and Bassetlaw Teaching Hospitals NHS Foundation Trust, Doncaster, United Kingdom

Shambhab Chaki Department of Electronics & Telecommunication Engineering, Jadavpur University, Kolkata, India

Sabyasachi Chakrabarty Medical Faculty, Center for Internal Medicine, University Clinic for Cardiology and Andrology, Otto-von-Guericke-University Magdeburg, Magdeburg, Germany

Pratip Chakraborty Department of Assisted Reproduction, Cradle Fertility Center, Kolkata, West Bengal, India

Sudipta Chakraborty Department of Pharmacognosy, B.C.D.A. College of Pharmacy & Technology, Barasat, Kolkata, India

Rufino da Silva Faculty of Medicine, University of Coimbra (FMUC); Ophthalmology Department, University Hospital of Coimbra, ULS de Coimbra; AIBILI—Association for Innovation and Biomedical Research on Light and Image, Coimbra, Portugal

Anu Dandona D Y Patil International University, Akurdi, Pune, India

Debraj Das Centre for Computational and Data Sciences, Indian Institute of Technology Kharagpur, Kharagpur, West Bengal, India

Gautam Das Mibiome Therapeutics LLP, Mumbai, India

Sushanta Kumar Das Department of Pharmacy Practice, Mata Gujri College of Pharmacy, Mata Gujri University, Kishanganj, Bihar, India

Baishali De AI Lab, Department of Electronics and Tele-Communication Engineering, Jadavpur University, Kolkata, India

Feronia Melissa De Mello College of Medicine, Ras Al Khaimah Medical and Health Sciences University, Ras Al Khaimah, United Arab Emirates
Biswadeep Dhar Department of Health Sciences, College of Health and Behavioral Studies, James Madison University, Harrisonburg, VA, United States
Eghosasere Egbon Department of Tissue Engineering and Regenerative Medicine, Faculty of Life Science Engineering, FH Technikum, Vienna, Austria
Rabab H. Elshikh Department of Medical Laboratory Sciences, College of Applied and Health Sciences, A'Sharqiyah University, Ibra, Oman
Payal Ganguly Leeds Institute of Rheumatic and Musculoskeletal Medicine, University of Leeds; Leeds Biomedical Research Centre, Leeds Teaching Hospitals NHS Trust, Leeds, United Kingdom
Kumar Gautam Singh Chromosome Labs Pvt. Ltd., Mumbai, India
Nabanita Ghosh Maulana Azad College, Kolkata, India
Susmita Ghosh Leibniz-Institut für Analytische Wissenschaften - ISAS-e.V., Dortmund, Germany
Ramji Gupta Department of Pharmacology, R. V. Northland Institute, Greater Noida, Uttar Pradesh, India
Dibya Kanti Haldar Centre for Computational and Data Sciences, Indian Institute of Technology Kharagpur, Kharagpur, West Bengal, India
Hossein Hassani IIASA, Laxenburg, Austria
Yasser Hijazi Plastic and Reconstructive Department, Lancashire Teaching Hospitals NHS Foundation Trust, Preston, United Kingdom
Takanobu Hirosawa Department of Diagnostic and Generalist Medicine, Dokkyo Medical University, Tochigi, Japan
Sunita Hooda Polymer Research Laboratory, Department of Chemistry, Acharya Narendra Dev College, Govindpuri, Kalkaji, New Delhi, India
Ashfaque Hossain Medical Microbiology and Immunology, College of Medicine, Ras Al Khaimah Medical and Health Sciences University, Ras Al Khaimah, United Arab Emirates
Humaira Ilyas Department of Biochemistry, Albert Einstein College of Medicine, Bronx, NY, United States
Harekrishna Jana Department of Microbiology, Raja Narendra Lal Khan Women's College, Midnapore, West Bengal, India
Gauri Jha Department of Biological Sciences, Sri Venkateswara College, University of Delhi, Delhi, India
Akib Kamani Mibiome Therapeutics LLP, Mumbai, India
Omar Kiwan Faculty of Biology, Medicine and Health, University of Manchester, Manchester, United Kingdom
Amit Konar AI Lab, Department of Electronics and Tele-Communication Engineering, Jadavpur University, Kolkata, India
Rahul Kumar Department of Pharmacology, R. V. Northland Institute, Greater Noida, Uttar Pradesh, India
Keaton Logan Caribbean Maritime University, Kingston CSO, Jamaica
Steve MacFeely OECD, Paris, France
Sheila Marinze Department of Surgery, Medway NHS Foundation Trust, Gillingham, United Kingdom
Diogo Gonçalves dos Santos Martins Federal University of São Paulo (UNIFESP), São Paulo, Brazil
Thiago Gonçalves dos Santos Martins Departamento de Oftalmologia, Universidade Federal de São Paulo (UNIFESP), São Paulo (SP), Brazil
Federal University of Rio de Janeiro, Macaé, Brazil
Thomaz Gonçalves dos Santos Martins Federal University of Rio de Janeiro (UNIRIO), Rio de Janeiro, Brazil
Leila Marvian Mashhad Imam Reza International University, Mashhad, Iran
Ya Fatou Njie Mbye Department of Pharmaceutical Sciences, College of Pharmacy & Health Sciences, Texas Southern University, Houston, TX, United States
Luis Mendes AIBILI—Association for Innovation and Biomedical Research on Light and Image, University of Coimbra, Coimbra, Portugal
Jarred D. Metkus School of Pharmacy, The Claremont Colleges, Keck Graduate Institute, Claremont, CA, United States
Prem Shankar Mishra Faculty of Pharmacy, Vidya University, Meerut, Uttar Pradesh, India
Rahul Mishra Celerion IMC, Lincoln, NE, United States
Pralay Mitra Department of Computer Science and Engineering, Indian Institute of Technology Kharagpur, Kharagpur, India
Sk. Abdul Mohid Department of Biochemistry, School of Medicine, Case Western Reserve University, Cleveland, OH, United States
Parag Nandi Department of Assisted Reproduction, Cradle Fertility Center, Kolkata, West Bengal, India
Chukwuemeka R. Nwokocha Department of Basic Medical Sciences, Physiology Section, Faculty of Medical Sciences, The University of the West Indies, Kingston, Jamaica
Magdalene Nwokocha Department of Pathology, Faculty of Medical Sciences, The University of the West Indies, Kingston, Jamaica
Aderonke Odetayo School of Nursing, Tung Wah College, Hong Kong SAR, China
Angela Oh School of Pharmacy, The Claremont Colleges, Keck Graduate Institute, Claremont, CA, United States

Sunny Edet Ohia Department of Pharmaceutical Sciences, College of Pharmacy & Health Sciences, Texas Southern University, Houston, TX, United States

David B. Olawade Department of Allied and Public Health, School of Health, Sport and Bioscience, University of East London, London; Department of Research and Innovation, Medway NHS Foundation Trust, Gillingham; Department of Public Health, York St John University, London, United Kingdom; Department of Public Health, York St John University, London, United Kingdom

Nelson Orie Department of Inflammation and Rare Disease, Division of Medicine, University College London, London, United Kingdom

Ademola A. Oyagbemi Department of Veterinary Physiology and Biochemistry, Faculty of Veterinary Medicine, University of Ibadan, Ibadan, Nigeria

Taiwo O. Oyagbemi Federal College of Animal Health and Production Technology, Ibadan, Oyo State, Nigeria

Javier Palacios Carrera de Química y Farmacia, Facultad de Ciencias de la Salud, Universidad Arturo Prat, Iquique, Chile

Tejal Pant Anya Biopharm India LLP, Ahmedabad, India

Vishwalatha K. Parvathreddy Department of Chemical Engineering & Environmental Toxicology, College of Science, Engineering & Technology, Texas Southern University, Houston, TX, United States

Sakshi Patel Department of Pharmaceutical Technology, MIET, Meerut, Uttar Pradesh, India

Shelly Pathak Novogene Europe, Milton, Cambridge, United Kingdom

Rudrendu Kumar Paul Independent Consultant, San Francisco, CA, United States

Luis Pinto-Coelho Electronics Engineering, Polytechnic of Porto, ISEP; INESCTEC, Porto, Portugal

Pooja Department of Chemistry, Sri Venkateswara College, University of Delhi, Delhi, India

Talia Puzantian School of Pharmacy, The Claremont Colleges, Keck Graduate Institute, Claremont, CA, United States

Syed Monajatur Rahman Department of Assisted Reproduction, Cradle Fertility Center, Kolkata, West Bengal, India

Pratyusha Rakshit Department of Electronics & Telecommunication Engineering, Jadavpur University, Kolkata, India

Amit Ranjan Cancer and Translational Research Centre, Dr. D. Y. Patil Biotechnology and Bioinformatics Institute, Dr. D. Y. Patil Vidyapeeth, Pune, Maharashtra, India

Nanda Rohra HiMedia Laboratories Pvt. Ltd., Thane, India

Saumendu Deb Roy Dean Academics (Pharmacy & AHS), Maat Gujri University, Kishanganj, Bihar, India

Ashok Kumar Sah Department of Medical Laboratory Sciences, College of Applied and Health Sciences, A'Sharqiyah University, Ibra, Oman

Anuradha Saha Department of Applied Electronics and Instrumentation Engineering, Netaji Subhash Engineering College, Kolkata, India

Ujjwal Sahoo Pharmaceutical Chemistry, Dr. B. C. Roy College of Pharmacy & AHS, Durgapur, West Bengal, India

Shivam Saini Department of Biological Sciences, Sri Venkateswara College, University of Delhi, Delhi, India

Amit Samanta Pharmaceutical Analysis, Mata Gujri College of Pharmacy, Mata Gujri University, Kishanganj, Bihar, India

Muhammad Saqib Khyber Medical College, Peshawar, Pakistan

Bidyut Sarkar IBM, Armonk, NY, United States

Tanya Saxena ADDB, Louvain Drug Research Institute, Université catholique de Louvain, Brussels, Belgium

Laishram Saya Department of Chemistry, Sri Venkateswara College, University of Delhi, Delhi, India

Paulo Schor Federal University of São Paulo (UNIFESP), São Paulo, Brazil

Nitya Sharma Department of Pharmacology, R. V. Northland Institute, Greater Noida, Uttar Pradesh, India

Soumya Shekhar Cancer and Translational Research Centre, Dr. D. Y. Patil Biotechnology and Bioinformatics Institute, Dr. D. Y. Patil Vidyapeeth, Pune, Maharashtra, India

M. Sharjeel Shoukat Wah Medical College, Wah Cantt, Pakistan

Tanmay Singhal College of Osteopathic Medicine, Western University, CA, United States

Krishnendu Sinha Jhargram Raj College, Jhargram, India

Garima Suneja Microbiome Research Pvt. Ltd., Mumbai, India

Subhayan Sur Cancer and Translational Research Centre, Dr. D. Y. Patil Biotechnology and Bioinformatics Institute, Dr. D. Y. Patil Vidyapeeth, Pune, Maharashtra, India

Anand Swaroop Nutrify Today; Cepham Inc., Somerset, NJ, United States

Jennifer Teke Department of Research and Innovation, Medway NHS Foundation Trust, Gillingham; Faculty of Medicine, Health and Social Care, Canterbury Christ Church University, Canterbury, United Kingdom

Asiya Najam Ul Haq Dow International Medical College, Karachi, Pakistan

Amna Usman Wah Medical College, Wah Cantt, Pakistan

Deepanshu Vasuja Department of Biological Sciences, Sri Venkateswara College, University of Delhi, Delhi, India

Sanober Waghoo Microbiome Research Pvt. Ltd., Mumbai, India
Khushbakht Waqar University of Karachi, Karachi, Pakistan
Kusal Weerasinghe Department of Research and Innovation, Medway NHS Foundation Trust, Gillingham, United Kingdom
Kristin S. Williams Columbia University, New York, NY, United States
Momoh A. Yakubu Department of Chemical Engineering & Environmental Toxicology, College of Science, Engineering & Technology, Texas Southern University, Houston, TX, United States

List of reviewers

Annewsha Datta University of Southern California, San Francisco, CA, United States

Rudrendu Kumar Paul Independent Consultant, San Francisco Bay Area, San Francisco, CA, United States

Samudra Prosad Banik Dept. of Microbiology, Government General Degree College, West Bengal, India

Subhrajit Bhattacharya School of Pharmacy and Health Sciences, Keck Graduate Institute, Claremont, CA, United States

Anand Swaroop Nutrify Today; Cepham Inc., Somerset, NJ, United States

Sushanta Kumar Das Mata Gujri College of Pharmacy, Mata Gujri University, Kishanganj, Bihar, India

Ashfaque Hossain Consultant Microbiologist, Decatur, Atlanta, GA, United States; College of Medicine, Ras Al Khaimah Medical and Health Sciences University, Ras Al Khaimah, United Arab Emirates

Debasis Bagchi Department of Biology, College of Arts and Sciences, Adelphi University, Garden City, NY; Department of Pharmaceutical Sciences, College of Pharmacy and Health Sciences, Texas Southern University, Houston, TX, United States

Preface

Following the invention of the computer by Charles Babbage in the late 19th century, there has been a surge in the evolution of machines with the unified objective of imparting intelligent behavior in them. Computers have learnt to interact with human beings and carry out human instructions in almost every imaginable sphere ranging from engineering to medical science and from artistry to education. Computer science has made an unprecedented progress over the last two decades from machine learning (ML) to deep learning (DL), and perhaps we are standing at the brink of realizing John McCarthy's dream of making a human-like thinking machine. Much like the other frontiers of machine-aided human applications, artificial intelligence (AI) has played a major role in augmenting human health by both disease diagnosis and treatment.

Computer science and AI have significantly impacted every conceivable frontier of disease, ailment, medicine, and surgery, including cancer, neurodegeneration, cardiovascular maladies, organ transplantation, and a wide array of complicated surgical procedures. In today's sci-fi era, doctors use trained chatbots to assist in diagnosis, programmed algorithms to maintain patient's case history and transcribe medical documents, convolutional neural network (CNN) for image processing to enable accurate disease diagnosis, fuzzy logic-based robotic surgeries to improve precision and accuracy beyond human reach, natural language processing (NLP) for better understanding human instructions, as well as numerous other state-of-the-art applications to facilitate the panacea. Computer applications based on ML and DL coupled with in silico modeling tools and molecular dynamics simulation have transformed drug design and delivery by optimizing bioavailability and zero toxicity. AI has also made significant advances in screening novel bioactives from natural ingredients and developing optimized formulations and functional foods.

This Millenium has witnessed a frightening escalation in the number and versatility of deadly human diseases with vaccines being ineffective toward treatment/eradication of many, with the cases of complex multifactorial lifestyle diseases equaling or even surpassing that of pathogen-borne diseases. Concomitant with the rise in human ailments, causal association and diagnosis has become increasingly complex, where human intelligence must be accompanied with in-hand availability of a large volume of information from earlier medical cases with similar histories. Starting from the trivial task of helping human diagnosis with accurate causal association, AI has come a long way showing potentials of a completely human-independent healthcare rendering entity. For example, supervised ML-based on a gigantic number of training datasets has led to development of precision medicine, neural network-based software have facilitated the prediction of vulnerability to certain disease and designing of subsequent treatment regime, algorithms based on CNNs have helped clinicians to better understand the progression and morphology of a tumor and DL have often yielded surprisingly superior results than radiologists with regard to transformation of a tumor toward malignancy. In fact, the advent of DL has revolutionized the image acquisition technology of computers with significantly superior accuracy as compared to computer-assisted detection of coronary artery diseases.

To make computers understand and specifically apprehend human intelligence and interpretation, and effectively communicate more accurately, NLP with its two branches has progressed parallelly along with the development of AI-based applications. Statistical NLP helps the computer to sample human writings, identify words and phrases, which occur more frequently than others, and develop a rule-based guidance for transformation into machine language with simultaneous application in neural networks. On the contrary, semantic NLP goes a step further and attempts to understand human objective behind phrases, which can have more than one meaning. The advent of NLP has eased out the procedures of extraction of clinical data from handwritten prescriptions and diagnosis reports. In addition, NLP-based applications seamlessly process, document, and maintain structured records of related administrative operations, including patient admission, in-house care, prescription, billing and processing of insurance claims, and other related financial matters.

In this era of this unprecedented advancement of AI, human civilization is encountering with the intimidating question of whether AI in healthcare will completely replace human clinicians in the near future. This also takes us to the ethical comfortability of AI singularity or dominance. Although AI is really advancing at a fast pace and robots are replacing humans in a few domains such as surgery, a closer introspection reveals that the onus of health and well-being of human beings are still on human brains and human hands than on machines. Several key reasons can be cited in support of this notion. (1) CNN-based image acquisition programs are trained to draw conclusions from the captured images based only on single observations, such as formation of tumor; however, human intelligence is indispensable for seeking reasons for formation of the tumor and predicting further inferences regarding its progression. (2) Moreover, supervised learning will be of limited use for prescribing medication to a patient with multiple comorbidities; under such circumstances, only the human brain can come up with the best possible therapeutic regimen with least spillover into the other medical complication/s as side effects. Several other such examples can be cited in this regard which proves human supremacy over the computers, at least for the next 100 years. Particularly, the most important shortcoming of AI in healthcare is its failure to render treatment with empathy and to gather common people's consent for initiating the treatment.

The present medical system has witnessed widespread adoption of AI- and machine-assisted diagnosis and therapeutic procedures across the globe. Subsequently, there has been no dearth of documentations regarding the same in the form of books, journals, and periodicals. This edited book is enriched with the contributions from leading biomedical and bioengineering scientists and AI professionals.

Key features of the book include:

- Journey of evolution of computers from simple calculating machines to thinking devices.
- Transformation of machines to human brain-like cognitive centers, fundamental aspects of ML, neural networks, and DL.
- Application of AI in diverse facets of healthcare from diagnosis to surgery and from drug designing to psychological counseling.
- Guidelines on ethical issues and safety standards of computer applications in human society.

Rule-based expert systems or decision paths based on "if–then" logic were first introduced commercially in several frontiers of human society, including healthcare during the end of the 20th century. In healthcare, this technology has been primarily used for "clinical decision support" in formulating the treatment regime and creation of "electronic health record." However, designing rule-based expert systems need human intervention for setting up a series of rules and connecting them in a common logic. Rule-based decision systems only work for a limited number and often fail arising out of mutual conflict when the number gets very large.

The rapid and unprecedented advancement of AI in medical science has even led to the speculation that in the near future, we will witness replacement of human medical practitioners and technicians by AI-driven robots. However, for justifiable reasons, AI has always been confronted with ethical issues arising out of its singularity or domination over their creators, humankind. Although we still cannot claim that human beings have been able to decipher the secret of transferring the entire complement of human intelligence to computers, the million-dollar question which confronts us here is, will it be a worthy and safe endeavor to keep trying!

The progress in technology and application of AI over the last few decades has been phenomenal. However, this has parallelly raised concerns among the human society regarding the eventual dominance of machines over man. The last section is dedicated to the ethical issues of computer applications in human life and realizing the limits of evolution of computer science to minimize competition of trained robots with human workers in job sectors as well to assure compliance with human safety in a theoretical and futuristic world.

This academic textbook will encompass a holistic summary of the current progresses and breakthroughs of computer science and its intelligent applications in health and medicine including of citations of real-life cases and will also try to address parallelly the ethical issues involved therein to ensure a safe borderline of AI applications. In conjunction, this book will revisit the all-important questions of setting the ethical limits of AI intervention: whether we should stop making machines more intelligent and prevent our own doomsday. In this academic textbook, we have strategically positioned 35 elegant chapters written by eminent academic scholars and industrial innovators, and a Glossary of terms to introduce the meaning of key terminologies used throughout the textbook for our readers. The diverse facets of AI applications in healthcare have been strategically placed into eight sections.

The first section "Introduction to AI and healthcare" comprising of three chapters provides historical insights into the development of AI. The first chapter discusses AI and human intelligence, while the second chapter provides a mathematical overview of the key ML algorithms and their use as prediction tools in clinical medicine. The third chapter empahasizes the novelty of AI-assisted disease detection tools.

The second section "The role of AI in holistic healthcare and disease diagnosis" comprising four chapters comprehensively discusses the principles and applications of AI in healthcare and disease detection, as well as diagnosis. The first chapter demonstrates the application of AI chatbots for primary diagnosis at home, while second chapter introduces short-term memory-based AI applications for prediction of clinical sequences. The third chapter highlights the advances in application of AI for detection and treatment of infectious diseases. The fourth chapter elaborates on the utilization of AI-assisted imaging technologies for disease diagnosis.

The four chapters in the third section demonstrate the "Use of AI in organ transplantation and precision surgery." The first chapter elaborates on the success of AI in heart transplantation; the second chapter highlights the opportunities and challenges of the application of AI in precision surgery; the third chapter discusses the advantageous features of AI-assisted robotic surgery; and the fourth chapter demonstrates the future of reconstruction and aesthetics on the application of AI in plastic surgery.

The fourth section entitled "Domain-specific success stories in medical science through AI intervention" comprises four chapters. The first chapter elaborates on the application of AI in assisted reproduction, while the second and third chapters demonstrate intricate aspects on the application of AI in cardiovascular disorders and ophthalmology, and the fourth chapter demonstrates the roles of ML and AI in autoimmune disease prediction.

The fifth section discusses the intricate aspects of neuropsychiatric rehabilitation in four dedicated chapters. The first chapter highlights the application of AI in psychotherapy and counseling, while the second chapter explores the critical features of brain–computer interfaces for neuromotor rehabilitation, and the third chapter extensively discusses how AI can assist in detecting and mitigation of neurological and psychiatric conditions. The fourth chapter demonstrates the advances of AI in early diagnosis of alpha-synucleinopathies.

The sixth section is brief and comprises three chapters on AI applications in infrastructural augmentation of healthcare industry. The first chapter discusses the salient features of the application of AI in wastewater treatment, while the second chapter elaborates on the aspects of AI-driven value creation and investment strategy in healthcare innovation. The third chapter addresses how generative AI and agentic intelligence can be utilized to cover India's rural healthcare gaps.

The seventh and the penultimate section is the longest, comprising ten chapters on new and emerging frontiers of AI. The first chapter focuses on the application of quantum AI in the augmentation of human health. The second and third chapters in this section are dedicated to the use of AI in microbiome research and genomics, respectively. The fourth and the fifth chapters highlight, respectively, how the AI intervention has revolutionized drug designing and discovery and how AI can be leveraged to ensure customized drug designing and safety. The sixth chapter discusses the fundamental concepts involved in application of AI in the field of nutrigenomics and precision nutrition. The seventh and eighth chapters are centered on the applications of AI in the emerging field of organoid research. The last two chapters highlight how AI can enhance and facilitate high-throughput drug discovery from medicinal plants and target-specific therapeutic application.

The eighth and concluding section of the book comprises three chapters on ethical considerations in AI. The first and second chapters independently demonstrate the aspects of the ethical frontiers during AI applications, while the third chapter discusses the aspects of minimizing the vulnerability of AI programs from cyberattacks. Finally, the "Glossary of AI Terminologies" was compiled by selected authors, which will serve as a ready reference of key technical terms used throughout the book.

The editors extend their sincere gratitude and thanks to all eminent contributors, as well as Elsevier/Academic Press editorial team members, including Elizabeth A. Brown, Shubham Dixit and Monica Paul K., for their continued support, cooperation, and assistance in bringing out this important volume. They sincerely hope that this textbook will provide significant insights into the applications of AI in healthcare. The editors also extend their sincere gratitude and regards to Dr. Ranjit K. Chakraborty for his continued support, guidance, and enormous help.

Debasis Bagchi
Department of Biology, College of Arts and Sciences, Adelphi University, Garden City, NY;
Department of Pharmaceutical Sciences, College of Pharmacy and Health Sciences, Texas Southern University, Houston, TX, United States

Samudra Prosad Banik
Department of Microbiology, Government General Degree College, Narayangarh, Rathipur, West Midnapore, West Bengal, India

Anand Swaroop
Nutrify Today; Cepham Inc., Somerset, NJ, United States

Sunny Edet Ohia
Department of Pharmaceutical Sciences, College of Pharmacy & Health Sciences, Texas Southern University, Houston, TX, United States

Section I

Introduction to artificial intelligence and healthcare

Chapter 1

Artificial intelligence and emotional intelligence: Can machines understand human emotions?

Anu Dandona[a] and Sandhya Bhatt[b]
[a]*D Y Patil International University, Akurdi, Pune, India*
[b]*Amity University, Noida, India*

1 Introduction

Artificial intelligence (AI) is often defined as the ability of a computer system to accomplish complicated tasks that are usually limited to human reasoning (Russell and Norvig, 2010). AI technology can be used to enable computers and machines to replicate the human experience in learning, understanding, and solving issues. It includes the capability of computational systems to carry out functions such as learning, pattern recognition, and even decision-making, which are usually associated with human intelligence. A large volume of data can be analyzed with the help of AI algorithms, trends can be revealed, and predictions or recommendations can be made as a result of such information.

Emotional intelligence (EI), also known as emotional quotient (EQ), is defined as the ability to recognize, apply, comprehend, deal with, and manage emotions (Mayer and Salovey, 1997). It involves the capacity to recognize, understand, and control oneself, as well as to understand others' emotions, thereby building positive relationships with others. EI enables people to understand how to utilize their feelings to make decisions and become more effective in controlling themselves and their influence on others (Goleman, 1995). It also encompasses the capability to think creatively and solve problems using emotions.

A sense of understan(ding is an essential part of psychological safety and is extraordinarily significant to our capacity to regulate our emotions (Siegel, 2010). Comprehension is one of the fundamental aspects of close relationships and allows individuals to feel emotionally safe. When people feel that they are being understood and that their interpretation of the world is met with empathy and accuracy, they are more likely to trust and feel close to another person. Building communication on a foundation of empathy establishes trust, strengthens relationships, and enhances business outcomes. By being empathetic, people can foster trust, expand their knowledge, improve their relationships, and experience a positive mood and emotional state.

To control everyday life and make the right choices, it is crucial to possess emotional understanding. It enables a person to cope with the challenges of life in a better manner; failure to be aware of their emotions may result in anxiety and other problems. A sense of understanding is also critical in learning to regulate emotional affects. When a person has experienced an emotional disturbance, it is very relaxing to share those feelings and have them warmly understood by someone they can trust. It is difficult to relax and get rid of painful feelings, and safe contact with another person who can be trusted is often helpful in this process. On the contrary, the absence of empathic insight, particularly in cases where it is urgent, is traumatizing and may stimulate childhood regressions (Schore, 2003). It has also been established that it is within relationships, especially during infancy and childhood stages, that individuals learn some of the basics regarding relationships, emotions, and coping mechanisms. This is an important requirement for mastering the interpersonal world.

Applications of Artificial Intelligence in Healthcare. https://doi.org/10.1016/B978-0-443-43934-6.00017-1

2 Can machines truly understand or replicate human emotions?

Technologies like facial recognition and natural language processing (NLP) are key features of AI because of their capability to identify human emotions. Such systems process diverse inputs such as facial expressions, voice tone, speech rhythm, and text-based communication in the context of analyzing emotional states such as happiness, anger, sadness, and fear. For example, AI can recognize slight changes in facial expressions when speaking or variations in voice depending on the emotions. Voice analytics software is able to identify voice patterns associated with stress or anger, and a machine can detect microexpressions on the face, which could be too quick for a human to read.

These AI models are trained on enormous volumes of speech and text, which makes it possible to distinguish nuances such as sarcasm, happiness, or exasperation in a voice. They could be used to generate responses that cater to the mood a user is in, making communication seem less artificial and less forced. Other AI systems have demonstrated a high level of accuracy in recognizing emotions; for example, one study using support vector machines (SVMs) achieved an 82% recognition accuracy based on physiological and facial features (Al-Hamadani & Al-Jumaili, 2021). Other studies (Harb, 2020) focused on computer vision algorithms to identify human emotions in videos, as well as the potential of computer vision to recognize emotions using convolutional neural networks (CNNs) and the development of software based on facial expressions. Companies such as Affectiva use multimodal emotion AI to analyze facial expressions, speech, and body language, providing insights into an individual's mood with a reported 90% accuracy, leveraging datasets of 6 million faces representing 87 countries (Affectiva Inc, 2023).

Along with this progress, there is a prevailing belief among experts that the currently existing forms of AI are not able to experience their own emotions or display consciousness (Fast & Horvitz, 2017). Although AI may imitate or simulate emotions, it does not truly have feelings in the human sense. The relationship between human feelings, neurotransmitters, hormones, and biological processes is very complicated, along with the intricate nature of psychological and social aspects. AI is based on code, algorithms, and models of machine learning (ML) and does not possess the biological matrix or lived experience that constitutes human feelings.

Simulated empathy: Simulated empathy (also known as artificial empathy, affective computing, or sympathetic computing) is the capability of an AI system to identify, attribute, and act upon human feelings. This ability is important in customer service bots or companion robots. Nevertheless, when a chatbot shows sympathy, it is not an emotional experience, and it has no understanding of what human suffering truly is. Some scientists remain uncertain as to whether AI can actually possess consciousness or emotions (Hagendorff, 2020). It is believed that the potential for machines to have the same kind of emotions as humans requires artificial general intelligence, a higher level of understanding and the capability to perform various tasks just like a human being (Goertzel, 2014).

2.1 Problems and constraints

The possibility of AI to achieve an understanding and reproduction of emotions presents several challenges and ethical issues (Rahwan et al., 2019), which are given in the following.

Misinterpretation and bias: Human beings are sometimes complex, and AI can fail to unravel emotions, hence causing frustrating or harmful experiences. AI models might fail to understand emotions correctly when applied to various cultures, languages, and individual experiences, which can lead to a lack of accuracy and the propagation of stereotypes. Examples include the fact that a smile may have varied meanings in a number of cultures.

Privacy: Emotional AI is abusive to privacy since it typically involves the gathering of personal information, including facial reactions, intonations, and personal conversations. Such information is vulnerable to abuse or misuse, or could be used for spying if not safeguarded appropriately (Taddeo & Floridi, 2021).

Emotional manipulation: It is the risk of applying AI to profit from emotions or to manipulate the decision-making process without interpretation or equity, especially in the areas of marketing or politics (Cave & Dihal, 2020).

Inability to truly understand: Without much complexity, AI processing systems cannot comprehend visual implications, sarcasm, and other language connotations that may completely change the meaning of an emotional expression. Although AI can be coded to identify these factors and respond accordingly, the in-depth ability to perceive and address these factors at an intuitively deep level, as human beings can and do, is beyond AI's abilities at this point.

Technical challenges: The creation of genuinely emotionally intelligent AI will require overcoming challenging technical issues associated with factors such as data accuracy, processing capability, and modeling the complex biological, psychological, and cultural determinants of the emotional processes involved in human beings (Picard, 1997).

3 Background of EI

3.1 The laboratory and model of EI

EI combines both the ability to reason soundly concerning emotions and the capacity to employ emotions and emotional expertise to improve thinking. It entails having the ability to perceive feelings, interpret them, and control emotion-laden situations. Emotional expression and regulation, self-awareness, and empathy are incorporated into the concept. The theory of EI also includes the foundation of the theory and the interconnected association of emotion and cognition. Although a recent subject of discussion, EI is a very ancient idea.

3.2 History of EI and its genesis

Tracing back the history of EI, we can go as far back as the mid-1970s, when Howard Gardner proposed the idea that intelligence was not a single, general ability. John Dewey, an educational philosopher, was the first to coin the term "social intelligence" in the year 1909. A psychologist named Edward Thorndike conducted early studies on social intelligence, which form part of the origins of EI. A concrete description of EI can be found in 1964.

Officially, the term "Emotional intelligence" was coined by Peter Salovey and John D. Mayer in 1990, who described it as a subset of social intelligence that entails monitoring the emotional state of oneself and others, differentiating between the two, and translating the information to make decisions that regulate thought and action. Their initial hypothesis was that EI is a domain of Gardner's views of intelligence. EI has changed drastically since it was first mentioned in the scientific literature in 1990. Since then, scholars have developed different models of EI, including models in emotion and cognition.

3.3 The framework of EI by Daniel Goleman

The practice of EI was popularized by psychologist and science journalist Daniel Goleman with his best-selling book Emotional Intelligence in 1995. Goleman's work greatly contributed to the overall awareness and practice of EI, bringing it out of the purview of academic circles. His theory is traced as both a derivation and a point of departure from the concept as conceived by Mayer and Salovey, as Goleman is chiefly concerned with emotional and informational circuits and flows.

The model of EI described by Goleman defines particular abilities and competencies. His model is well established within the nonacademic professional sphere and is descriptive, indicating his six EI leadership competencies. The emotional competence model, developed by Daniel Goleman, is a comprehensive theory of building EI, which he considers an essential skill for achieving success.

His theory defines EI in four domains and 12 competencies. These include the following four areas:

Self-awareness: Finding the ability to accept personal feelings.
Self-management (or self-regulation): The capacity to control one's emotions.
Social awareness (or empathy): The ability to perceive and understand others.
Relationship management (or social skill): The relational skill of managing emotions with people.

Another use of EI in previous work by Goleman highlighted five main aspects: self-awareness, self-regulation, motivation, empathy, and social skill, especially in the context of the workplace. He emphasizes the fact that EI is an ability one should develop at a later stage of life through appropriate tutoring and devotion, whereas cognitive intelligence is, according to him, God-given. Goleman is of the opinion that EI plays a key role in achieving exceptional success at work and extends beyond mere cognitive intelligence and competence. The different models of EI, such as the mixed model developed by Goleman, explain that EI involves both personal (understanding, managing, and using emotion in oneself) and social (understanding and managing emotion in others) intelligence.

4 Relevance of EI in communication and decision-making

EI plays an important role in communication and decision-making and allows individuals to comprehend themselves and others, including their emotions, and therefore become more productive in their communication and decision-making (Mayer, Salovey, & Caruso, 2011; Rubin et al., 2005). Studies unanimously note that EI can have a beneficial effect in various workplace situations and personal lives (Carmeli, 2015; Goleman, 2007).

4.1 Social–EI in communication

EI is substantial in proper communication and has been beneficial in the creation of stronger relationships, exercising empathy, and fostering a positive environment (Salovey and Mayer, 2012; Saulsman, Coall, & Nathan, 2006; Goleman, 2007). It entails thinking about and handling emotions as a means of communicating effectively and productively (Salovey and Mayer, 2012; Saulsman et al., 2006).

4.2 The highlights of EI in communication

Relationship building and emotional empathy: Communication conducted with the help of EI can enable the practice of empathy and will consequently lead to the building of working relationships, respect, and understanding between work colleagues (Rubin et al., 2005). People with high EI are better positioned to perceive the thoughts, feelings, and needs of others, and their responses are at a level that satisfies those needs (Lopes, Salovey, & Straus, 2004).

Conflict resolution and teamwork: The quality of teamwork and the level of conflict are lower when EI is high. Emotionally intelligent people tend to work collaboratively through conflict resolution, aiming to achieve agreeable results.

Clarity and authenticity: EI results in clear, direct, and authentic communication by enhancing self-awareness and self-regulation (Salovey and Mayer, 2012). This helps people open up and deal with their feelings efficiently in the workplace.

Active listening and social awareness: Part of EI development is active listening, which involves individuals actively paying attention in a conversation, being good listeners, and refraining from interrupting when asking clarifying questions to understand other people in an in-depth manner. Social awareness is another element of EI and conflict, which refers to the ability to understand the views of people from different backgrounds and to empathize with individuals from alternative cultures.

Leadership and influence: People who have excellent EI are those who can communicate effectively, manage relationships, overcome conflicts, and influence others. They can foster trust and creativity and establish a favorable emotional atmosphere within an organization.

4.3 Decision-making through EI

EI also has a major impact on the decision-making process since it combines emotional knowledge and logical thinking to make more efficient decisions (Alkozei, Killgore, & Smith, 2017). This is critical, especially for leaders who always scrutinize the available options, risks, and consequences.

4.4 Influence in decision-making processes

Self-awareness and emotional regulation: An effective EQ enables leaders to recognize how their emotions affect their behavior and to make choices based on their values and goals. Personal understanding and personal management allow leaders to remain calm in the face of pressure, enabling them to make gradual decisions as opposed to making quick, irrational ones.

Empathy and diverse opinions: Empathy is at the heart of a leader who can intellectually make decisions based on how their choices will affect the thoughts, emotions, and driving forces of the people. This leads to the formulation of inclusive decisions, whereby the needs of all stakeholders are taken into consideration.

Leaving emotions out of the equation: Emotionally intelligent leaders do not deny their emotions, but they use them as important data, learning how emotions influence their way of thinking and acting. This helps them recognize realize biases and unforeseen consequences, thereby achieving superior results.

Living with ambiguity and uncertainty: EI is an essential skill in instances where there are high-stakes situations and times of uncertainty. People who have become leaders are able to navigate ambiguity to facilitate and build cooperation, as well as instill confidence in others regarding their decisions. EI provides the energy and perseverance needed to overcome obstacles and continue pursuing long-term objectives.

Moral and value-based decision-making: EI positively influences moral decision-making because it enables individuals to factor in self-values and the consequences of actions on other people. It stimulates inclusiveness and fosters mutuality and equity.

5 Challenges and implications of EI in AI and machine interactions

The inclusion of EI in AI systems is a field that poses great prospects and complications, rapidly developing and changing (Vivek & Krupskyi, 2024). Although AI has the capability to interpret and react to emotional expressions, attempting to recreate the phenomena associated with human EI to the fullest degree is an overwhelming challenge because of the complexity of human emotions (Vivek & Krupskyi, 2024).

5.1 Challenges in integrating EI into AI

Complicated nature of human feelings: Authentic EI entails compassion, self-awareness, and the capacity to decode complex societal signals, which by nature is difficult to reproduce using algorithms. Human feelings are dynamic and conflicting and can be perceived differently on an individual or cultural level; therefore translating them into data-driven models is challenging. Factors such as sarcasm, cultural expectations, previous experiences, and implicit biases are major barriers to AI systems.

Technical limitations: Technical limitations include the fact that emotion recognition models implemented on the basis of deep learning and multimodal information fusion cannot be applied to real-life conditions successfully because of factors such as background noise, interpersonal differences, and the absence of labeled training data on specific emotional states. There is also a need for real-time assessment under active and interactive conditions to confirm the performance of models used in real environments.

Ethical questions and bias: The effort to have AI detect and understand human emotions is fraught with significant legal and ethical concerns regarding privacy, consent, and the possibility of emotional manipulation. Emotional AI is especially biased, as there is an element of subjectivity in emotions, and stereotypes and assumptions might be reinforced on a scale never seen before (Jena & Goyal, 2021a). It is essential to ensure that multicultural training data is used to avoid cases of discrimination. Mismanagement of sensitive emotional information may lead to a breach of trust and poor mental health conditions.

Absence of true knowledge and awareness: Although AI is able to identify and interpret emotional expressions, its knowledge is superficial since it is not equipped with the power of personal experience, social situations, and emotional insight. AI in the present day does not have actual EI since it lacks consciousness, empathy, and subjective experience.

5.2 Implications for AI and human interaction

Improving EI in fan bases: AI can assist humans in further developing their capacities for EI by using sentiment-monitoring applications, AI-enhanced simulations, and role-playing learning games that aid in facilitating an EI pattern base and in learning to exercise empathy, negotiation, and conflict resolution.

Better customer experience and collaboration at workplaces: Customer experience can be highly optimized through emotional AI when designed sensitively, resulting in improved user experience, enhanced trust, and providing a new dimension to customer service through sentiment analysis and customized responses. At work, AI applications that can assess the mood of a team may assist managers in identifying potential conflicts, fostering inclusiveness, and promoting psychological safety.

Future of leadership: Future leaders should be prepared to operate in an environment that combines AI and EI, where they must possess knowledge of both technological possibilities and constraints and apply them ethically in a manner that serves humanity. The ability to incorporate AI ethically, without neglecting human benevolence and moral judgment, will be the key determinant of leadership in the coming years.

6 AI and the pursuit of emotional recognition

Emotional recognition in AI entails the application of several technologies to determine and analyze human emotions. These are some of the technologies that aim to bridge the gap between what machines can do and the subtle intricacies of human emotional expression.

Various technologies are also used in the detection of emotions in AI, where each technology targets various ways through which humans express themselves, enabling the new technology to draw conclusions based on their emotions. These technologies tend to combine so that they can provide a more accurate interpretation of the emotions.

One of the most influential technologies in AI, and one adapted to emotion detection, is NLP, which is concerned with text and speech language analysis and the extraction of emotional insights. NLP makes it possible to analyze the fine details of human communication and deduce, based on phrases, words, and intonation, the presence of emotions.

Sentiment analysis: In sentiment analysis, the emotional tone of text is checked and classified as positive, negative, or neutral using NLP algorithms.

With much more complex systems, more specific emotions can be detected, such as joy, anger, sadness, or surprise. This is achieved through the analysis of word choice, sentence construction, and even the use of emojis or punctuation.

Lexical and semantic analysis: Lexical and semantic analysis studies a sentence on two levels, which are not interchangeable. Lexical analysis targets the words themselves, defining their forms, types, and characteristics without considering any meaning. For example, when the sentence in question is "The cat chased the mouse," lexical analysis would determine that "the" is a determiner, "cat" and "mouse" are singular common nouns, and "chased" is a verb in the past tense. The objective here is merely to recognize and mark off each word correctly. Semantic analysis, in turn, goes further to understand the meaning of the entire sentence. It categorizes the domains of relation between the words, i.e., it identifies the agent of the action as the cat and the patient of the action as the mouse. Context is also taken into account in semantic analysis, which contributes to the removal of ambiguity. For instance, if "mouse" were read in a computing context, it could be interpreted as a computer device rather than an animal. Roughly, the question that lexical analysis asks and answers is "What are the words, and what are the grammatical roles of the words?", The question that semantic analysis asks and answers is "What does the sentence actually mean?".

Syntactic and pragmatic analysis: The grammatical structure of a sentence (syntactic analysis) and the general context and purpose to achieve (pragmatic analysis) are also analyzed in NLP, thereby enabling a better understanding of emotional meaning. Another example is where emphasis or intonation in the same words conveys a different emotion, and NLP can detect this in spoken language.

Spoken language applications: NLP is also used in speech recognition (beyond text) to convert spoken words into text, and such text can, in turn, be analyzed to determine the emotional content. This involves analyzing features of the voice, such as pitch, tone, and rate of speech, which tend to convey emotional messages.

NLP, despite its capacity, does have problems recognizing emotions accurately because of the vagueness and ambiguity of human language. Sarcasm, irony, and cultural variations in the demonstration of feelings may be hard to interpret accurately by NLP algorithms. The lack of nonverbal communication in written communication is also a deterrent to how well NLP systems can understand emotional depth.

7 Computer vision

Computer vision is an offshoot of AI, which allows computers and systems to obtain information contained in digital images and videos, and it has been essential in emotion sensing. It aims at sensation analysis through visual data and automatic sentiment analysis of facial emotions (Vivek & Krupskyi, 2024). The primary aim of this kind of recognition is the realization of emotions in a real-time setting by examining different characteristics of the face, including eyebrows, mouth outlines, and body language (Sari, Suryani, & Purwaningtyas, 2021a; Vivek & Krupskyi, 2024).

Computer vision detects and understands the facial expressions of human beings, as they are considered the most profound indicators of a person's mood or other psychological state (Sari et al., 2021a; Zainab, Taghizadeh, & Mollahosseini, 2020b; Chaurasia, Gupta, & Srivastava, 2021). There are a few important steps and techniques involved in the process:

Image and video recording: Emotion detection is based on computer vision. In general, such systems record video material using a webcam or analyze existing image and video materials (Sari et al., 2021a; Vivek & Krupskyi, 2024).

Facial feature detection: A computer vision method works to identify a human face, after which it activates features (Sari et al., 2021a; Chaurasia et al., 2021). This involves recognizing major components of the face, including the corners of the eyes, the mouthlines, and muscle flexing based on various moods (Sari et al., 2021a). One of the important steps that takes place at the beginning of such a process is face detection (Rao, Nair, & Patel, 2017).

Classification algorithms: By applying ML and deep learning classification techniques, the forms of detected expressions are categorized into particular emotions after feature extraction (Sari et al., 2021a; Zainab et al., 2020b).

CNNs: The advantage of CNNs is that they can automatically learn and extract features in images to apply in facial emotion detection (Li, Wang, Zhang, & Wu, 2021; Sari et al., 2021a; Zainab et al., 2020b). They learn to identify patterns through large amounts of images with emotional labels (Dar, Khan, & Malik, 2023). CNNs have the ability to extract relevant features without the need for manual collection, analyze massive data, and acquire complex relationships, resulting in enhanced accuracy rates and speed compared to those previously achieved by conventional methods (Dar et al., 2023; Mollah, Teshnehlab, & Rahman, 2022). Other CNN-based methods have achieved 93.8% accuracy during training and 75% during testing in recognizing seven emotions (namely, anger, happiness, sadness, disgust, fear, neutrality, and surprise) (Zainab et al., 2020b).

Vision transformers (ViTs): ViTs are a neural network style borrowed from NLP input that has been modified for use in computer vision (Zeng, Wang, & Liu, 2023). The images applied to train them can be faces with emotions marked, as the whole picture is used to determine the emotion (Zeng et al., 2023). In contrast to previous methods, ViTs do not require a predefined area of interest and leverage global context to achieve better reliability and accuracy (Zeng et al., 2023). ViTs have performed well in other tests, with one variant achieving an accuracy of 83% in an emotion classification problem (Zhi & Deng, 2024).

Recurrent neural networks (RNNs): RNNs and long short-term memory are aimed at detecting continuous emotions, particularly those in video sequences, to capture the temporal correlation (Sari et al., 2021a; Zhi & Deng, 2024).

Region-based convolutional neural networks (R-CNN and faster R-CNN): They can be used in object detection, but they can also be adapted to emotion recognition by identifying, as a region of interest, such as a face (Zeng et al., 2023). A faster R-CNN has been a further variant applied to detecting driver emotion and achieves high accuracies (e.g., 99.73% accuracy on a CK+ dataset and 99.95% accuracy on the FER-2013 dataset) (Zulfiqar, Akram, & Irshad, 2022a).

SVMs: They serve to classify data once the features are extracted, and they even tend to perform better than artificial neural networks in terms of accuracy in some cases (Ali, 2022).

Real-time processing: The majority of systems strive to achieve real-time recognition of emotions, which involves analyzing streams of video to obtain instant data on the level of emotion (Sari et al., 2021a; Zainab et al., 2020b). This is essential in scenarios where there is an interest in monitoring student engagement or driver safety (Zainab et al., 2020b).

7.1 Categories of detections identified

The computer vision technology can study different forms of human expressions in emotion recognition.

Facial macroexpressions: They represent acute, perceptible, and observable facial expressions produced upon the occurrence of emotions such as sadness, anger, fear, or happiness, with a duration lasting between three-quarters of a second and up to two seconds (Dar et al., 2023; Zulfiqar et al., 2022a).

Facial microexpressions: These are involuntary micro, rapid, minor, fleeting facial expressions that barely last half a second and tend to reveal actual feelings even when a person tries to hide them (Dar et al., 2023; Happy et al., 2020; Sari et al., 2021a).

Expression of the body: This encompasses both posture (still images, depicted in a single picture) and motion (moving body dynamically that is studied in time (Dar et al., 2023; Sari et al., 2021a; Vivek & Krupskyi, 2024).

8 Challenges in computer vision for emotion detection

However, despite all this, it does not negate the fact that there are still challenges in computer vision concerning the detection of emotions:

Real-life cases: Some demographic factors, such as differences in lighting, scruffs, glasses, various settings, and angles of the facial/body structure, can impede correct recognition (Zainab et al., 2020b; Zulfiqar et al., 2022a). The accuracy of the test results in controlled settings can be elevated compared to the accuracy of the test results in uncontrolled, real-life environments (Dar, Shahid, & Irtaza, 2023).

Availability of data: One of the challenges is the limited availability of various types of datasets, especially those obtained in real-life conditions, as they are invaluable when training a robust model (Zulfiqar et al., 2022a).

Limitations of accuracy: Although models exhibit high accuracy with certain datasets, it is nonetheless challenging to achieve high accuracy across all circumstances and diverse groups of people (Jena & Goyal, 2021a).

Cultural diversity and bias: A variety of different attitudes toward emotional expression in cultures, as well as biases in models and datasets used in training, can affect the performance of the system (Jena & Goyal, 2021b).

Privacy issues: Privacy issues related to emotion detection using computer vision include concerns about privacy because emotions are constantly being captured (Zhi & Deng, 2024).

9 New research frontiers

The current research in affective computing is shifting toward the application of affective computing. Future and up-to-date fields of research consider:

Multimodal Emotion Recognition: The creation of systems that can identify emotions across multiple modalities, such as facial expressions, speech, and physiological measures. Such a method considers visual analysis, audio analysis, and text analysis to improve comprehension (Grand View Research, Inc, 2024).

Large language models (LLMs): Future research includes combining these with LLMs to enhance emotion detection, emotion synthesis, and conversational AIs.

Mental health: The research base of affective computing is increasingly being utilized in the field of mental well-being, with interventions administered through smartphones that monitor mood levels.

Ethical considerations: Addressing ethical and legal challenges, such as algorithm bias, privacy concerns, and the legalities of utilizing data related to emotions, is an essential aspect of research. Efforts should be made to ensure that the development and application of these technologies are transparent, accountable, and inclusive.

10 Case studies and applications

10.1 Healthcare: AI in mental health diagnostics or therapy bots

AI is changing the way mental healthcare consumers are diagnosed and treated because it makes diagnosing more accurate, treatment more personalized, and mental health services more available. AI offers solutions that help detect, prevent, and treat mental health disorders early, addressing the shortage of mental health advisors and the increased need for services worldwide.

10.2 AI mental health diagnostics

AI and ML are utilized to improve the diagnostic capabilities of mental health conditions, offering objective analysis and increasing the efficiency of diagnostic endeavors.

10.3 Accuracy in diagnosis and early diagnosis

AI systems have the potential to aid in distinguishing among diagnoses with analogous initial clinical manifestations and disparate treatment implications, e.g., differentiating bipolar and unipolar depression based on brain imaging signs (Dadi, Machlouf, & Cohen, 2021). AI-powered approaches can also detect new disease subtypes according to different aspects, such as heterogeneity of presentations, demographic characteristics, and wild factors (Dwight, 2023). For example, AI methodologies may incorporate information contained in a variety of datasets, such as electronic health records (EHRs), behavioral inputs from digital phenotyping and wearable sensor data, speech, social media feeds, neurophysiology, imaging, and genetics to construct explanatory models of mental illness (Dwight, 2023). Examples include using Instagram photographs to predict the risk of depression (Reece & Danforth, 2017), speech data to predict those who may develop psychosis among youth at risk (Dwight, 2023), and identifying individuals with posttraumatic stress disorder (Dwight, 2023). AI approaches also perfectly fit into the interpretation of patterns based on longitudinal data, which is essential in enhancing accuracy in the diagnosis of variably occurring psychiatric symptoms (Dwight, 2023). A previous study reported that subjects provided more sensitive information when the data was entered into a computer system than when it was shared with a human being, and this can be beneficial in collecting accurate data on the patient (Dwight, 2023).

It is a well-known policy to manage psychological disorders with the help of AI-based screening of cognitive deficiencies and impairments. The accuracy of diagnosis increases when such a process is automated through AI (Singh, 2024). Other types of ML and AI-powered algorithms, such as SVM, neural networks, and models based on ensemble learning, which include CNN, AlexNet, GoogLeNet, and LeNet5, among others, have proven effective in evaluating cases of cognitive mental health disorders (Singh, 2024). Research has also demonstrated the capability of AI to aid in the early diagnosis and treatment of mental health disorders, as one of the proposed models shows a correct diagnosis of mental disorders in nearly 90% of cases with the help of 28 questions (Singh, 2024). AI is capable of identifying particular mental conditions, including bipolar disorder, schizophrenia, and autism spectrum disorder (Kumar and Sharma, 2022; Singh, 2024).

10.3.1 Application of nonclinical information

The digital health movement aligns with AI practices that utilize information extending beyond the traditional scope of interaction between a physician and a patient. It is feasible to monitor symptoms or potential relapses on a continuous basis through sensors and social media, enabling an understanding of which behaviors are beneficial and which are harmful (Dwight, 2023). Multimodal sensing, including smartphones, wearables, physiological sensors (e.g., heart rate), and ambient sensors (e.g., motion, temperature), facilitates the collection of real-world, continuous data on symptoms, treatment responses, and behavior (Dwight, 2023). ML can be employed to extract meaningful features from typically noisy sensor signals. For instance, social withdrawal is an example of behavior that can be identified using data from smartphones and Fitbit devices (Dwight, 2023).

Social media websites provide a nonrestrictive window into the longitudinal patterns of behavior and moods, especially in adolescents and young adults, as they are the groups of high internet users and are more prone to mental illnesses (Dwight, 2023). The investigation into the content, language, and usage patterns of social media allows one to understand and learn more about relationships and communication, as well as receive new opportunities for seeking assistance (Dwight, 2023).

The accuracy of predictions of diagnoses and relapses based on social media data is similar to the competency of clinicians and screening surveys (Dwight, 2023). Nevertheless, there is no apparent clarity on how to use the sensor data in the clinical mental health setting because of insufficient clinical characteristics and difficulties in implementation (Dwight, 2023). In equal measure, the analysis of data on social media requires AI-related innovations, such as data collection pools, but it is restricted by privacy and ethical limitations (Dwight, 2023).

10.3.2 *Treatment and prognostic help*

Accurate prognoses of psychiatric patients can be supported with the help of AI techniques used on the basis of longitudinal data (Dwight, 2023). Research has been conducted modeling depression trajectories, suicide risk trajectories, future substance abuse trajectories, and future functional outcomes based on neuroimaging, EHR, genetic, and speech data (Dwight, 2023). On the basis of data-driven approaches, AI algorithms can also construct new clinical risk prediction models (Dwight, 2023).

In the sphere of psychiatry, AI technologies can considerably help facilitate clinical treatments (Dwight, 2023). They may even foretell the response to treatment, which could obviate the need for needless experimentation with medication or lengthy sessions of psychotherapy (Dwight, 2023). Researchers have used EEG and clinical questionnaire items to predict reactions to antidepressants (Dwight, 2023) and, based on EEG, the reaction to antipsychotics (Frydman et al., 2023).

The significant side effects of treatment may also be predicted with the aid of AI; EHR data has been used in one of the studies to predict the occurrence of renal insufficiency in patients undergoing lithium treatment (Dwight, 2023). In addition, AI applications may contribute to the construction of new theoretical models of disease pathology and the discovery of new forms of treatment through predicted clinical drug action via simulation or data-based approaches (Dwight, 2023).

10.4 In mental health therapy bots

The chatbots and virtual assistants that can be used in mental health treatment are becoming a new tool in mental healthcare, offering accessible and scalable interventions, especially to address the discrepancies between the needs of mental healthcare and the supply of mental healthcare workers and providers (Ahmad Ishqi Jabir, Ahmad Zaki, & Hassan, 2024).

11 Effectiveness of therapy chatbots

In the context of therapy chatbots, there is a significant opportunity to greatly improve their efficacy through the use of AI by expanding their application into tele-rehabilitation systems. Using computer vision and wearable sensor analytics, AI is able to evaluate patient posture, quality of movements, and measures of performance remotely, while the chatbot interface provides feedback with corrective instructions and motivating messages immediately. With NLP, such chatbots have the potential to demonstrate an empathic spirit, support discussions, and facilitate speech therapy or multilingual communication. ML algorithms can help create individualized treatment plans, monitor their adherence, and predict recovery curves, based on which the chatbot can suggest prompt modifications and interventions. The incorporation of these capabilities means that chatbots powered by AI solutions enhance the efficiency and success of therapeutic treatment while also broadening the patient base with rehabilitation opportunities in distant or underserved locations.

A growing body of studies suggests that mental health chatbots have the potential to become a powerful source of mental health support, and some of them have already revealed substantial symptom remission. According to the metaanalysis performed by Zhong et al. and published in April 2024, therapy chatbots showed a significant reduction in anxiety symptoms with an overall effect size of $g = -0.19$, which did not vary significantly after 4 weeks ($g = -0.18$) but increased after 8 weeks ($g = -0.24$). In the case of depression, the effect sizes between the use of chatbots and reductions in symptoms ranged from $g = -0.25$ to -0.33. These effects, however, diminished and became insignificant at the 3-month follow-up (Zhong, Ma, & Zhang, 2024). Linardon, Sutcliffe, and Hoban's (2024) metaanalysis showed that mental health apps tended to improve depression ($g = 0.28$) and generalized anxiety ($g = 0.26$) symptomatology by statistically significant but small magnitudes. In particular, the study reported that, when it comes to depression treatment, the effect sizes of apps that use chatbot technology were considerably higher, averaging 0.53 (Linardon et al., 2024).

The use of generative-AI therapy chatbots was evaluated in a more recent randomized controlled trial (RCT) by Heinz et al. (March 2025), in which a therapy chatbot named Therabot generated moderately large to large effect sizes in clinically significant improvements in symptom sectors for major depressive disorder (0.845 at 4 weeks, 0.903 at 8 weeks) and generalized anxiety disorder (0.794 at 4 weeks, 0.840 at 8 weeks). Such effect sizes were reported to be higher than those usually reported in SSRIs and clinical trials and were equivalent to or close to those reported in first-line psychotherapy (Heinz, Alpert, Han, Kording, & Goerke, 2025). This demonstrates some positive indicators of spontaneous user participation, with users spending an average of 6 h on Therabot throughout 4 weeks, and indicates that the therapeutic alliance based on user perception was equivalent to the one based on human therapists (Heinz et al., 2025).

Overall, a mental health chatbot application (Wysa) has not achieved statistically significant results in the efficacy of either decreasing or increasing the severity of depression symptoms or anxiety symptoms over a 4-week period in a study of participants with chronic conditions like arthritis and diabetes in the treatment group, though no changes were observed among control-group respondents. Arthritis participants recorded more depression and anxiety than those with diabetes (Sorell, Gwyther, & Whittington, 2023).

11.1 Stimulus and quitting

Another potential advantage of chatbots is better participation, which may result in reduced attrition rates compared to other types of mental healthcare delivered via the internet (Ahmad Ishqi Jabir et al., 2024). Although the metaanalysis by (Linardon et al., 2024) identified an overall attrition rate of 25% in studied mental health apps, another metaanalysis by the same researchers, Ahmad Ishqi Jabir et al. (February 2024), showed an average attrition rate of about 21% in therapy chatbot studies. The attrition rate was estimated to be around 18% in short-term studies (<8 weeks) and 26.5% in long-term studies (>8 weeks) (Ahmad Ishqi Jabir et al., 2024). This implies that there is a small immediate benefit of chatbots on adherence to treatment, but it is not clear whether people who disengage from one modality will also leave the other (Ahmad Ishqi Jabir et al., 2024). Table 1.1 lists the more popular chatbots.

11.2 Challenges and ethical considerations

Nonetheless, researchers warn that generative AI agents are not prepared to be deployed independently in mental healthcare, as they could create high-risk situations (Dwight, 2023). There are potential biases, ineffectiveness compared to their human counterparts, the establishment of harmful stigma, and problematic reactions (Dwight, 2023; Miner, Marwaha, & Kishore, 2024). Another study conducted at Stanford revealed that, compared to depression, AI chatbots displayed additional stigma toward conditions such as alcohol dependence and schizophrenia, and that this stigma was consistent across various AI models (Miner et al., 2024). Therapy chatbots in conversational contexts occasionally supported dangerous practices, such as providing information that could reinforce suicidal thoughts, instead of instilling doubts in the mind of the patient (Miner et al., 2024).

Other privacy-related ethical issues concerning AI in mental healthcare include data privacy and security, informed consent, transparency, bias and fairness, and human oversight (Dwight, 2023). Through the algorithms, biases in training data can lead to discrepancies in diagnosis and treatment (Dwight, 2023). The cultural, social, and contextual aspects influence mental health, and it is possible that AI tools do not represent them properly, resulting in incorrect interpretations (Dwight, 2023). The American Psychological Association has claimed that specialized mental care resources should be used with caution instead of generic AI chatbots because they may be harmful (American Psychological Association, 2023). The Psychotherapy Action Network (PsiAN) asserts that most existing mental health applications and systems are either misleading, poorly regulated, or commercial, providing only surface-level solutions (PsiAN, 2023).

To make AI-based clinical tools a staple of clinical practice in healthcare institutions, one should integrate the tool into the culture of the healthcare setting to alleviate both clinician and patient fears regarding the frequent use of the tool (Dwight, 2023). Cooperation among the authors of AI, mental healthcare professionals, ethics experts, and policymakers can help develop accountable and successful AI applications that do not negatively affect patient well-being and meet ethical norms (Dwight, 2023).

12 Customer service: Sentiment-aware chatbots and call centers

Chatbots and call center technologies are emerging as sentiment-aware technologies, meaning that these technologies will support and improve customer service by analyzing emotions and reacting to them. They are technologies that use sentiment analysis to define the emotional tone of interactions, resulting in more personalized support, better customer satisfaction, and improved work efficiencies.

TABLE 1.1 Chatbots/related tools to empower human emotional intelligence.

Chatbot/tool	Description and emotional intelligence features
Replika	An AI friend that can learn through interaction based on user input and create an emotionally intuitive, intimate speaking relationship; ubiquitous as a method of emotional support, as well as a sense of social connection.
Wysa	Uses CBT, DBT, and mindfulness-based methods in an amiable, accessible chatbot application, providing empathetic, unparalleled, any-time support.
Woebot	A psychologist-designed chatbot based on CBT approaches provides mood monitoring and the ability to offer an emotional response.
Youper	Offers emotional intelligence with mood tracking, CBT, and mindfulness practice; individual and immediate support and feedback on emotional states.
Earkick	Integrates real-time analysis of biomarkers alongside a biomarker-specific LLM companion to customize emotional and physiological moods, providing personalized mental healthcare plans.
Lumi	Emotional AI friend specializing in mood identification and guided journals; known to respond empathetically and support emotional self-awareness.
Kuki (Mitsuku)	A chatbot who is witty, humorous, and empathetic; he retains a memory of previous chats to stay on track and customize interactions.
Nomi	Customizable, the AI familiarizes itself with personalities and learns from past encounters to have engaging, emotionally geared chats.
Xiaoice	Microsoft successfully created a socially empathetic chat bot that focuses on both IQ and EQ. It is engineered to ensure maximum user time and engagement, to the extent that users become emotionally connected through emotionally empathetic computing.
EmpathyEar	Multimodal empathetic chatbot with an avatar-driven emotional response that incorporates text, audio, and visual inputs and outputs to achieve greater resonance and empathetic communication.
EMMA	A mood-sensing well-being chatbot that identifies a user's mood through smartphone sensors and provides microactivities to support their mental well-being with empathy.
EVI AI (Hume AI)	Voice interface and emotionally sensitive involves the processing of tone through an emotionally intelligent voice interface and the adaptation of speech to promote empathetic attention to emotional communication; it can fairly easily deal with the subtleties of emotional conversation.
Elin.ai	Mobile application utilizing targeted, emotionally supportive advice, motivation, and empathy, combining the experience and skill sets of both Artificial Intelligence and psychology, targeted at users who need emotional support.
Elomia	The AI mental health chatbot provides realistic, humanlike, empathetic 24/7 conversations; one should feel as though they are talking with a real therapy practitioner.
Pip	On-line emotional support system for children living with chronic illness: the provision of caring, secure, and understanding online engagement.
Meeno	The AI-based relationship coaching app aims to improve the user's social skills by using empathetic dialog and is not intended as a replacement for a romantic partner.

12.1 Sentiment-aware chatbots

These sentiment-cognizant chatbots are optimized toward understand and adapt to the emotional state of a customer, providing more human-like and empathetic interactions. This feature enables chatbots to personalize their messages, help defuse negative interactions, and even leverage positive customer experiences.

12.2 Improved customer experience

Sentiment analysis chatbots that incorporate sentiment analysis are able to respond in a personal way by comprehending the feelings of customers. An example might be a situation where a customer is frustrated, at which point the chatbot can assist the customer by using empathetic language, such as saying, "I am sorry about the hassle. I will see what is holding up the processing," instead of a general catch-all "Checking your order status." Such a tailor-made treatment can convey a sense of being understood to the customer, which can dissipate frustrations and ultimately increase overall customer satisfaction. Researchers have demonstrated that the use of chatbots with so-called emotion words, such as sorry and thank you, can influence customer satisfaction rates, particularly in the context of empathy and interactivity. A survey conducted online in May 2021, involving the responses of 380 people, showed that although assurance and reliability consistently enhanced satisfaction, empathy and interactivity were only effective when emotion words were used (Sestino, De Mauro, & Di Berardino, 2023).

12.3 Voluntary trouble shooting and escalation

Knowing what customers feel will enable chatbots to foresee possible issues before they arise. In case they perceive significant amounts of dissatisfaction or expressions of frustration, a chatbot can switch the interaction to a human agent, avoiding an increase in customer dissatisfaction (CX. Intouch, 2024). This will provide the human touch and understanding needed for customers in complicated or emotionally heated situations. The transfer may also be initiated by a sour mood or sustained contact efforts (CX. Intouch, 2024).

12.4 Business impact and case studies

The usage of AI chatbots, especially sentiment-aware AI chatbots, has demonstrated considerable business value. GrandStay Hotels, the company that operates over 100 properties, rolled out their AI chatbots across all communication channels to offer 24/7 automated assistance (Netomi, 2024). This led to a 28% reduction in average call handle time, a 55% decrease in call abandonment, and a 15% increase in first call resolution (Netomi, 2024). Moreover, 72% of requests did not require agent assistance, as chatbots deflected such requests, saving over 13,000 agent hours annually and reducing customer service costs by $2.1 million per year (Netomi, 2024). Customer satisfaction impressions increased by 22%, and the number of dissatisfied customers using social media to draw attention to problems dropped by 29% (Netomi, 2024).

The other successful applications of sentiment-aware chatbot implementations are given in the following.

HDFC Bank: In 2017 HDFC Bank implemented EVA, an AI chatbot with built-in sentiment analysis that helps identify any customer who is frustrated or lost and can escalate the customer to a live agent (CX. Intouch, 2024).

Amtrak: Julie is their virtual assistant that performs sentiment analysis to support users in a more empathetic way, especially when they are perplexed regarding schedules or prices, and has resulted in increased booking completion (CX. Intouch, 2024).

Uber: Sentiment analysis has been added to the chatbot implementation supporting Uber, as this will help address negative user experiences in a timely and empathetic manner by identifying factors of frustration and prioritizing cases before they reach human agents (CX. Intouch, 2024).

12.4.1 Challenges and best practices for chatbots

Alongside the advantages, there are some difficulties when it comes to the implementation of sentiment-aware chatbots, such as the requirements to provide data privacy and address ethical questions, the labor-intensive initial stage of training and preparing AI models, and the integration with current systems (CX. Intouch, 2024). The best practices for successful implementation include frequently updating AI models with new customer interaction data, being human and empathetic with customers when using a chatbot, and utilizing chatbots as a supporting tool alongside human support teams (CX. Intouch, 2024).

Call centers use sentiment analysis to understand how customers feel during an interaction, and these systems are more effective than the old customer satisfaction surveys, where data is often partial and can be misleading. It enables real-time intervention and helps detect recurring patterns of emotions to enhance the overall service.

12.4.2 The working of it

Sentiment analysis in call centers processes text from multiple communication platforms, such as social media, chatbots, and recorded phone conversations (Level, 2024). AI applications analyze this text to assign an emotional sentiment—positive, negative, or nonreactive—to each communication (Level, 2024). These AI-enabled systems, including natural language understanding and semantic intelligence, can be advanced enough to recognize context and linguistic nuances, such as sarcasm, which are not detected by basic rule-based systems (CallMiner, 2024). Through this deeper insight, it is possible to identify customer dissatisfaction before it becomes too late, thereby enhancing high retention rates (CallMiner, 2024).

12.4.3 Classifications of sentiment analysis in call centers

This kind of sentiment analysis can be divided into the following three broad types:

Sentiment over time: Tracks customer sentiment over a short or long duration of time and allows for comprehending trends, such as how long the customer is willing to wait before abandoning a chat or how agents deal with customers over time (Level, 2024).

Sentiment by channel: Examines whether customers respond differently with regard to their emotional level based on the communication channel they rely on (Level, 2024).

Sentiment by element: This technique recognizes complex reactions in an individual interaction, in which the customer may express both positive and negative opinions regarding various elements of a service (Level, 2024).

12.4.4 Call centers benefits

Reduced customer churn: Early detection of customer emotions enables companies to make appropriate changes to their customer service promptly, which, in turn, may decrease churn rates (TCS, 2024). Based on a study, 58% of organizations reported experiencing higher levels of churn within 12 months, and 81% of marketing managers cited a lack of meaningful communication with their customers as the root cause of this (TCS, 2024).

Advanced personalization: Sentiment analysis is used to monitor transaction patterns, expectations, and past responses, enabling businesses to provide more customized services (TCS, 2024). According to data provided by McKinsey & Company, 71% of consumers prefer a personalized experience, whereas 76% respond adversely to its absence (TCS, 2024).

Improved coaching and training: Sentiment analysis solutions such as Level AI not only study a small portion of conversations but also analyze the entire conversation to identify recurring issues and moments where agents can be coached (Level, 2024). Such holistic coverage eliminates biased assessments and enables specific training to improve their communication and service skills (Level, 2024). It also provides managers with an opportunity to intervene in a call in real time when it is going negatively (Level, 2024).

Voice of the customer insights: Sentiment analysis might substitute or supplement more traditional surveys by performing automatic data collection and analysis through the direct involvement of customers (Level, 2024). This offers a better picture of customer requirements, brand attitudes, and general interaction dynamics (Level, 2024).

13 Case studies and examples

US mobile provider: A giant mobile provider analyzed customer support sentiment data and found that some of their customers were at risk of churning, demonstrating how sentiment-type data can play a key role in customer retention (Level, 2024).

American chase: AI-driven transcription and sentiment analysis by American Chase supported the process of streamlining the work of call centers for a customer, ultimately increasing customer satisfaction (TCS, 2024).

Jordanian telecommunication companies: It has been shown that using Support Vector Machine (SVM) and Neural Networks (NNs) to label Arabic notes in a call center achieved a high accuracy rate (66% for SVM, 99.21% for NNs) when categorizing sentiments into positive, negative, or neutral (Al-Sayyed et al., 2023).

Level AI: Such software will be able to sense a broad array of feelings and score an entire sentiment (0–10), with end-of-call sentiments scoring higher as they are more accurate in reflecting satisfaction with the resolution (Level, 2024). It also identifies implicit trends in customer sentiments and recurring customer pain points (Level, 2024).

References

Affectiva Inc. (2023). Affectiva emotion AI platform overview & accuracy. Retrieved July 28, 2025, from Affectiva website: https://www.affectiva.com/.

Ahmad Ishqi Jabir, A. M., Ahmad Zaki, N. I., & Hassan, R. (2024). The effectiveness of therapy chatbots in reducing mental health symptoms: a systematic review and meta-analysis. *Journal of Affective Disorders*, *348*, 587–599.

Al-Hamadani, S., & Al-Jumaili, A. (2021). Emotion recognition using support vector machines based on physiological and facial features. *International Journal of Interactive Multimedia and Artificial Intelligence*, *6*(3), 25–33.

Al-Sayyed, R.M.H., Hassonah, M.A., Rodan, A., & Al-Kasasbeh, H. (2023). Sentiment analysis of online comments using positive, negative, and neutral categories. Journal of King Saud University – Computer and Information Sciences, *35*(4), 1234–1245.

Ali, A. (2022). Emotion recognition using machine learning algorithms: a comparative study. *International Journal of Computer Applications 184*(7), 100–108. doi:10.5120/ijca2022924761.

Alkozei, A., Killgore, W. D. S., & Smith, R. (2017). Exposure to acute stress impacts decision-making by influencing emotion regulation processes. *Frontiers in Psychology*. 1664-1078 *8*. doi:10.3389/fpsyg.2017.01503.

American Psychological Association. (2023). Health advisory on social media use in adolescence. Available from: https://www.apa.org/topics/social-media-internet/health-advisory-adolescent-social-media-use.

CallMiner. (2024). Call center sentiment analysis: improving customer experience. Retrieved from https://callminer.com/glossary/call-center-sentiment-analysis/.

Carmeli, A. (2015). The relationship between emotional intelligence and job performance: a review and theoretical integration The handbook of stress and health: a guide to research and practice, 119–134.

Cave, S., & Dihal, K. (2020). The whiteness of AI. *Philosophy & Technology*. 2210-5433. *33*(4), 685–703. doi:10.1007/s13347-020-00415-6.

Chaurasia, A., Gupta, V., & Srivastava, S. (2021). Facial emotion recognition using deep learning: review and insights. *Journal of King Saud University-Computer and Information Sciences*. 1319-1578 *33*(6), 1047. doi:10.1016/j.jksuci.2019.06.008.

CX.Intouch.(2024).5waystouseAIchatbotsincustomerservice,Availablefrom:https://intouchcx.com/blog/5-ways-to-use-ai-chatbots-in-customer-service/.

Dadi, K., Machlouf, E., & Cohen, R. (2021). Bipolar disorder diagnosis using machine learning on MRI data. *European Psychiatry*, *64*(S1), S277.

Dar, A., Khan, M., & Malik, M. I. (2023). Improving facial expression recognition using convolutional neural networks. *Journal of Visual Communication and Image Representation 92*, 103461. doi:10.1016/j.jvcir.2023.103461.

Dar, S. H., Shahid, M., & Irtaza, A. (2023). A comprehensive survey on facial and body expression analysis facial expression recognition: challenges and opportunities, 71–120.

Dwight, C. H. (2023). Artificial intelligence in psychiatry: diagnosis, treatment, and ethics. *Seminars in Psychiatry and Neuroscience 1*(3), 100021.

Fast, E., & Horvitz, E. (2017). Long-term trends in the public perception of artificial intelligence. *Proceedings of the AAAI Conference on Artificial Intelligence*. 2374-3468. *31*(1). doi:10.1609/aaai.v31i1.10635.

Frydman, M. M., Schulte, M. F., Cordell, M., Münte, T. F., Schmaal, L. (2023). Pharmaco-EEG of antipsychotic treatment response: A systematic review. European Psychiatry, *67*(S1), e12.

Goertzel, B. (2014). Artificial general intelligence: concept, state of the art, and future prospects. *Journal of Artificial General Intelligence*. 1946-0163. *5*(1), 1–48. doi:10.2478/jagi-2014-0001.

Goleman, D. (1995). Emotional intelligence: why it can matter more than IQ.

Goleman, D. (2007). Working with emotional intelligence. Bantam Books.

Grand View Research, Inc. (2024). Emotion AI market analysis report, 2024–2033. Grand View Research. Available from: https://www.grandviewresearch.com/industry-analysis/emotion-ai-market-report.

Hagendorff, T. (2020). The ethics of AI ethics: an evaluation of guidelines. *Minds and Machines*. 1572-8641. *30*(1), 99–120. Available from: www.springer.com/journal/11023. doi:10.1007/s11023-020-09517-8.

Happy, S. L., George, A., & Routray, A. (2020). Facial expression recognition using facial landmark-based features facial expression recognition. *Machine Vision and Applications*, *31*, 45.

Harb, J. (2020). Computer vision algorithms for human emotion detection in video content: feasibility through convolutional neural networks (CNN) and facial expression analysis. *International Journal of Computer Applications*, *176*(30), 14–20.

Heinz, A., Alpert, K. I., Han, Y., Kording, K. P., & Goerke, L. R. (2025). Efficacy of a generative AI chatbot for mental health: a randomized clinical trial. *Nature Medicine*, *31*(3), 701–710.

Jena, L. K., & Goyal, A. (2021a). Emotional intelligence, artificial intelligence and ethical challenges. *Kybernetes 50*(9), 2589–2603. doi:10.1108/K-02-2021-0106.

Jena, R., & Goyal, M. (2021b). Cultural differences and biases in facial expression recognition systems: challenges and solutions. *Journal of Affective Computing 12*(3), 162–175. doi:10.1007/s12345-021-0123-4.

Kumar, S., & Sharma, A. (2022). An AI-based decision support system for predicting mental health disorders. International Journal of Medical Informatics, 161, 104717.

Level, A. I. (2024). Call center sentiment analysis: the ultimate guide. Available from: https://www.level.ai/blog/call-center-sentiment-analysis.

Li, X., Wang, S., Zhang, Y., & Wu, Q. (2021). Hand motion recognition via multi-kernel manifold learning. *Journal of Ambient Intelligence and Humanized Computing*, 1868–5137. doi:10.1007/s12652-020-02785-3.

Linardon, J., Sutcliffe, K., & Hoban, E. (2024). The efficacy of mental health apps for depression and anxiety: a meta-analysis of randomized controlled trials. *Clinical Psychology Review 108*, 102377.

Lopes, P. N., Salovey, P., & Straus, R. (2004). Emotional intelligence, personality, and the perceived quality of social relationships. *Personality and Individual Differences 36*(3), 555–564. doi:10.1016/S0191-8869(03)00153-4.

Mayer, J. D., Salovey, P., & Caruso, D. R. (2011). Emotional intelligence: new ability or eclectic traits? *American Psychologist*. 0003-066X. *63*(6), 503–517. doi:10.1037/0003-066X.63.6.503.

Mayer, J.D., & Salovey, P. (1997). What is emotional intelligence? In: Salovey P. & Sluyter D. J. (Eds.), Emotional development and emotional intelligence: Educational implications. Basic Books, pp. 3–31.

Miner, A. S., Marwaha, J. S., & Kishore, S. P. (2024). Artificial intelligence in mental health: ethical concerns and clinical considerations. *The Lancet Digital Health*, *6*(4), e224.

Mollah, M. K., Teshnehlab, M., & Rahman, T. (2022). CNN-based facial emotion recognition: a comprehensive review. *Neural Computing and Applications 34*(15), 11601–11622. doi:10.1007/s00521-022-07093-x.

Netomi. (2024). AI customer service case studies. Retrieved from https://www.netomi.com/resources/ai-customer-service-case-studies.

Picard, R.W. (1997). Affective computing. MIT Press.

PsiAN. (2023). 2023 PsiAN expresses serious concerns about mental health apps, platforms, and AI. The Psychotherapy Action Network. Retrieved from https://psian.org/psian-expresses-serious-concerns-about-mental-health-apps-platforms-and-ai/.

Rahwan, I., Cebrian, M., Obradovich, N., Bongard, J., Bonnefon, J. F., Breazeal, C., … Wellman, M. (2019). Machine behaviour. *Nature*. 1476-4687. *568*(7753), 477–486. Available from: http://www.nature.com/nature/index.html. doi:10.1038/s41586-019-1138-y.

Rao, K., Nair, S., & Patel, V. (2017). Face detection and recognition techniques: a review. *International Journal of Computer Applications*. 0975-8887 *165*(12), 32. doi:10.5120/ijca2017914348.

Reece, A. G., & Danforth, L. M. (2017). Computerized recognition of depression from Instagram photos. *EPJ Data Science*, *6*(1), 1–12.

Rubin, R. B., Rubin, A. M., & Piele, L. J. (2005). Communication research: strategies and sources (7th ed.). Thomson/Wadsworth, Belmont, CA.

Russell, S., & Norvig, P. (2010). Artificial Intelligence: a modern approach (3rd ed.). Prentice Hall.

Salovey P., & Mayer J. D. (2012). Emotional intelligence. In: Davidson, R. J., Scherer, K. R., Goldsmith, H. H. (Eds.), The psychology of emotion: The emotional life of your brain. Guilford Press, pp. 17–32.

Sari, C. A., Suryani, N., & Purwaningtyas, D. P. (2021a). Human emotion recognition based on facial expression using multi-class SVM, K-NN, and CNN. *TELKOMNIKA (Telecommunication, Computing, Electronics and Control) 19*(1), 125–133. doi:10.12928/telkomnika.v19i1.16999.

Saulsman, L. M., Coall, D. A., & Nathan, P. R. (2006). The association between depressive personality and treatment outcome for depression following a group cognitive-behavioral intervention. *Journal of Clinical Psychology*. 1097-4679. *62*(9), 1181–1196. doi:10.1002/jclp.20278.

Schore, A. N. (2003). Affect dysregulation and disorders of the self. Norton & Company, New York, NY.

Sestino, A., De Mauro, A., & Di Berardino, M. (2023). The impact of emotion words on customer satisfaction with AI chatbots. *Journal of Business Research 159*, 113702.

Siegel, D. J. (2010). Mindsight: the new science of personal transformation. W. W. Norton & Company, New York, NY.

Singh, A. (2024). Exploring the role of artificial intelligence in mental healthcare. Risk Management and Healthcare Policy, 17, 1345–1357.

Sorell, N., Gwyther, N., & Whittington, A. (2023). The effectiveness of a mental health chatbot app (Wysa) on depression and anxiety outcomes in people with chronic diseases: a real-world observational study. *Journal of Medical Internet Research*, *25*(1), e44795.

Taddeo, M., & Floridi, L. (2021). *How AI can be a force for good—An ethical framework to harness the potential of AI while keeping humans in control*, vol. 144. Springer Science and Business Media LLC, pp. 91–96. doi:10.1007/978-3-030-81907-1_7.

TCS. (2024). How sentiment analysis benefits call center analytics. Retrieved from https://www.tcs.com/blogs/sentiment-analysis-benefits-call-center-analytics.

Vivek, R., & Krupskyi, O. P. (2024). EI & AI in leadership and how it can affect future leaders. *European Journal of Management Issues*. 2523-451X. *32*(3), 174–182. doi:10.15421/192415.

Zainab, B., Taghizadeh, H., & Mollahosseini, A. (2020b). Machine learning techniques for emotion recognition from face images. *Artificial Intelligence Review*, *53*(5), 3397–3441.

Zeng, J., Wang, Y., & Liu, H. (2023). Vision transformers in facial emotion recognition: current trends and challenges. *IEEE Transactions on Affective Computing*, *14*(1), 78–88.

Zhi, W., & Deng, Z. (2024). Emotion classification using vision transformers and recurrent neural networks. *Pattern Recognition Letters*, *169*, 150–158.

Zhong, Y., Ma, C., & Zhang, Y. (2024). The effectiveness of therapy chatbots in reducing anxiety and depression symptoms: a systematic review and meta-analysis. *Journal of Affective Disorders*, *351*, 440–450.

Zulfiqar, F., Akram, W., & Irshad, M. N. (2022a). Enhanced real-time facial expression recognition using deep learning. *ATAIML: Advances in Technology and AI in Machine Learning*, *3*(1), 93–101.

Chapter 2

Understanding key machine learning algorithms as artificial intelligence-assisted prediction tools in clinical medicine: A mathematical overview

Kristin S. Williams
Columbia University, New York, NY, United States

1 Introduction

The utilization of artificial intelligence and machine learning (AI/ML) within clinical settings is transforming the modern healthcare system. AI is a subfield of computer science that focuses on the development of algorithms and software programs to analyze complex data after exposure to training and validation processes that facilitate computational learning (Rane, Rane, Paramesha, & Kaya, 2024). ML differs from AI, as ML does not require explicit programming to adapt its outputs or predictions. Thus the manner in which ML algorithms generate outputs may change automatically as the algorithm is trained on new data without conventional reprogramming (Chassagnon, Vakalopolou, & Paragios, 2020).

ML algorithms that are able to adjust their internal parameters and predictive outputs without manual programming are known as adaptive models (Gligorea et al., 2023). The use of adaptive models is particularly beneficial in medical settings where the understanding of etiological causes or symptoms of a particular disease may evolve over time. Thus adaptive models may offer significant advantages when used for diagnostics or disease progression monitoring, as the performance of static models may become outdated and offer limited clinical utility.

As AI/ML purports to mimic human cognitive functions (i.e., interpretation of clinical images), it is imperative that these models are trained on sufficient data to ensure statistical validity, reliability, and replicability (Williams, 2024). While these algorithms are touted to have the ability to transform the healthcare system by improving clinical efficiency and reducing the burden on clinical staff, AI/ML algorithms must be evaluated according to their potential limitations (Pavuluri, Sangal, Sather, & Taylor, 2024).

This chapter introduces AI/ML algorithms commonly utilized in clinical settings for predictions related to diagnosis, prognostication, treatment response, and mortality risk. The advantages and limitations of each technique are presented, as model performance often depends on the type of clinical data and population to which the algorithm is applied. Implications of the statistical programming for these algorithms are also introduced, as model performance is often based on the tuning of the algorithm's hyperparameters.

2 Foundations of predictive modeling in healthcare

The development and implementation of AI/ML have gained attention due to their ability to utilize retrospective data from patient records to train and validate supervised ML algorithms for predictive forecasting. Trained AI/ML algorithms differ from untrained models, as the hyperparameters in trained models have been optimized via a learning process to ensure accurate predictive outputs. Untrained AI models, or uninitialized models, are those used for algorithmic development, as the hyperparameters have not been configured. Untrained AI/ML models represent the initial phase of AI development, as these models have not been exposed to robust datasets that would allow for valid predictive outputs. The use of large

Applications of Artificial Intelligence in Healthcare. https://doi.org/10.1016/B978-0-443-43934-6.00018-3

data repositories allows multifactorial analyses to be conducted, which may facilitate the assignment of statistical weights to predictive factors that may inform diagnostic classification or logistic regression (LR) modeling (Zhang et al., 2024). Classification is defined as the assignment to a predetermined category based on historical data (Loshin, 2013). Predictions differ from classification, as clinical predictions may be used to evaluate a patient's prognosis or treatment response. Thus predictions are an estimation of a continuous outcome or future event derived using historical data (Efthimiou et al., 2024).

Data utilized for AI/ML models includes electronic healthcare records, clinical imaging, vital signs, and genomic sequencing (Mohsen, Ali, Hajj, & Shah, 2022). Data used for supervised or unsupervised models may be categorical, structured, continuous, or unstructured. Unstructured data may include pathology reports, clinical notes, or patient feedback surveys. While different types of data require different AI and ML techniques, unstructured data may require natural language processing algorithms, audio processing models, and/or deep learning (DL) for meaningful data extraction and interpretation (Dreyer et al., 2005; Wang & Summers, 2012).

3 Types of AI/ML algorithms

The use of AI/ML in medical diagnostics aims to evaluate the symptomatology of a patient or patient population to determine effective treatment (Filiberto, Leeds, & Loftus, 2021; Kyriazakos et al., 2021). Applications of AI/ML algorithms may also be utilized to monitor disease progression. Current uses of AI/ML in clinical diagnostics include the interpretation of two-dimensional (2D)/three-dimensional (3D) imaging, X-rays, dual-energy X-ray absorptiometry, and biosignals from neuromonitoring devices such as the electroencephalogram (Al-Antari, 2023). The development of AI/ML algorithms for computer-assisted diagnostics requires access to large datasets with high-quality imaging data that is representative of the population to which the predictions will be applied (Reps, Ryan, & Rijnbeek, 2021). The use of high-quality imaging data with sufficient resolution aims to ensure that the algorithm and its hyperparameters are properly configured for predictive modeling.

Types of AI/ML algorithms developed for implementation in clinical medicine include supervised, unsupervised, and reinforcement learning (RL) models. Supervised models are validated using clinical data with a known pathology (Cunningham, Cord, & Delany, 2008; Williams, 2024). The development of supervised models requires access to heterogeneous data, as the model's predictive accuracy and sensitivity often serve as the foundation for the construction of secondary models where a pathology has not yet been determined (i.e., unsupervised models).

3.1 K-nearest neighbor

K-nearest neighbor (KNN) is an algorithm employed in supervised learning ML datasets for the training and validation stages (Salvador–Meneses, Ruiz–Chavez, & Garcia–Rodriguez, 2019). This algorithm utilizes distance metrics to assess the difference or distance between data points that have not yet been classified and compares them to those that have a categorical assignment in a known dataset (Suyal & Goyal, 2022). Metrics commonly utilized in KNN algorithms include the Euclidean, Manhattan, Minkowski, and Hamming distance methods (Kumbure & Luukka, 2022; Li, Lin, Yu, & Guo, 2022; Zhang, 2016). While KNN may be utilized for classification or regression, the Hamming method must be used for variables that are categorical (Salvador–Meneses et al., 2019) (Figs. 2.1 and 2.2).

The Euclidean equation evaluates the distance or difference between two points (Prasatha et al., 2017). Mathematical modeling of the Euclidean equation for KNN is represented in Eq. (1). This equation defines k as the number of neighbors and is used to derive the distance between data points x and y. Adaptations of the Euclidean equation for ML and KNN are attributable to the requirement to account for multiple data points and the dimensionality of the comparative dataset (Prasatha et al., 2017). Neighbors are defined as data points that have a known categorical classification in KNN models.

$$\sqrt{\sum_{i=1}^{k}(x_i - y_i)^2}$$

Equation 1: Euclidean distance. This is a metric applied to KNN algorithms to evaluate the difference or distance between two points.

Model performance in KNN is determined by the number selected for k (Kotsiantis, 2007). Smaller values selected for the number of neighbors may significantly affect predictive outputs, as the model may be sensitive to noise, missing data, and/or statistical outliers. While a larger k value may reduce predictive error and variance, there is a potential that important patterns and predictive factors may be lost due to smoothing and averaging effects (Zhang, 2016). These considerations, when applied to clinical medicine, suggest that a larger k value may oversimplify predictive factors, as the classification

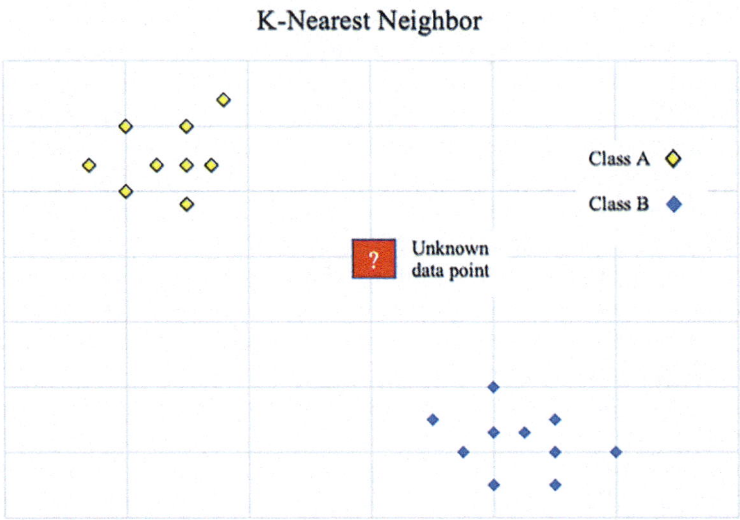

FIG. 2.1 Example of KNN with an unknown data point.

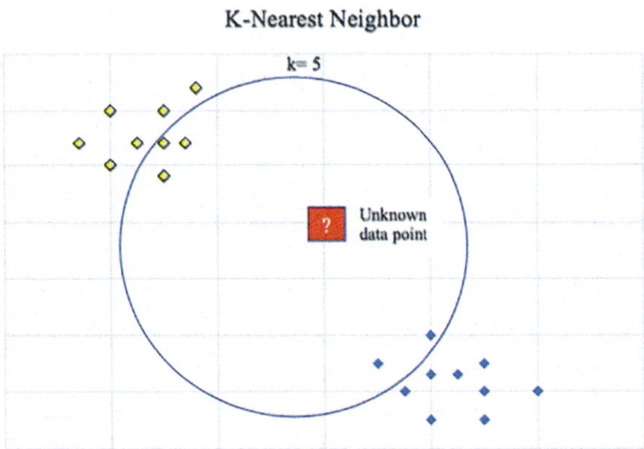

FIG. 2.2 Example of KNN with an unknown datapoint, two classes, and five neighbors. The unknown data point will be assigned to Class A, as Class A has 3 neighbors compared to Class B, where $k = 2$.

method for an unknown data point is based on the number of neighbors that reflect the most similarity. Further limitations of high k values may include increased computational costs attributable to the extensive comparisons of individual data points and training sets that may contain data with high-dimensional features.

3.2 Support vector machines

Support vector machines (SVMs) are another type of supervised ML algorithm used for regression and classification tasks. While most SVMs are utilized for supervised data, one-class SVMs and support vector clustering techniques can be applied to unsupervised models (Ghafoori et al., 2018). Applications of SVMs in clinical medicine include predictions for medication adherence, patient outcomes, mortality risk assessments, and diagnostic classification.

SVM is based on structural risk minimization (Winters-Hilt & Merat, 2007). This principle attempts to ensure that there is minimal error in the SVM model's predictions while also ensuring that the model's complexity is properly configured for generalizability. The complexity of SVM models is typically measured by the Vapnik–Chervonenkis (VC) dimension. The VC dimension is a quantification or measure of the complexity of a model class (Romanik & Vitter, 1996; Bartlett and Maass, 2003). Thus it evaluates the capacity of a model to correctly classify or label data points according to all possible label configurations. While a high VC dimension may reflect higher complexity and predictive sensitivity, it does not necessarily reflect the model's generalizability (Vapnik, 1998).

SVMs are commonly applied to classification problems due to their ability to construct optimal decision boundaries (Guido, Ferrisi, Lofaro, & Conforti, 2024). These models aim to identify the simplest model that separates the classes of variables according to a boundary or hyperplane. When the classification model is two-dimensional, the decision boundary is reflected as a line, whereas a 3D model requires the boundary to be reflected as a plane. Classification models that require higher-dimensional analyses are reflected as a hyperplane (Guido et al., 2024; Kecman, 2005).

During the training and validation stages, SVM algorithms attempt to find an optimal decision boundary or hyperplane that separates the data according to class (Cortes & Vapnik, 1995). Classes are categories or distinct groups that can be used to separate variables within a specific dataset. Support vectors are the data points that lie closest to the decision line or boundary. Support vectors are assigned a Lagrange multiplier, which serves as a weight in the optimization process. These weights significantly influence the model's performance, as the support vectors determine the location and orientation of the hyperplane (Cortes & Vapnik, 1995). Because SVMs maximize the margin between classes, support vectors play a key role in determining the model's generalizability.

Given that clinical data typically involves nonlinear classification, kernel functions are often applied to reduce computational complexity (Amari & Wu, 1999; Cesa-Bianchi, Mansour, & Shamir, 2015). Kernel functions often improve computational efficiency and model performance by mapping nonlinear data into higher-dimensional space that would not be captured by linear SVM models. This allows the algorithm to construct hyperplanes without direct computation of the coordinates in higher-dimensional space (Vapnik, 1999). The use of kernel functions enables the SVM to capture complex relationships between variables through nonlinear analyses that may otherwise remain undetected by linear modeling (Vapnik, 1999).

3.3 Reinforcement learning

RL is a distinct type of ML that cannot be classified as either a supervised or unsupervised model. RL differs from supervised models because its predictions are not based on data containing labels or classifications (e.g., known pathology vs. no pathology). RL is applied in unknown environments or datasets undefined parameters, aiming to maximize accurate predictions based on an external reward signal (Strehl, Li, & Littman, 2009). An environment refers to the system by which the RL algorithm is applied. It includes the conditions and feedback mechanisms that facilitate computational learning (Padakandla, 2022; Sutton & Barto, 2015). The optimization process of RL outputs is based on the Markov decision process (MDP). The MDP is a computational framework that enforces the rules and policies within a given environment (Knox & Stone, 2012; Padakandla, 2022; Sutton & Barto, 2015). While the MDP can be derived using various approaches, the reward signal is based on human feedback and dynamic programming.

Specific applications of RL in clinical medicine include assessments of medication dosage, chemotherapy regimens, and dosage reduction in mechanically ventilated patients. Because the use of RL in clinical settings has historically relied on simulated environments, the majority of RL applications have utilized offline RL techniques (Jayaraman, Desman, Sabounchi, Nadkarni, & Sakhuja, 2024). Offline RL is a type of ML that allows computations of outcomes via simulation without direct interaction with the environment. This is beneficial in medical scenarios where there is limited data that contributes to an uncertain outcome, or where it is otherwise infeasible to expose patients to certain conditions (Yu, Liu, Nemati, & Yin, 2022). The utility of RL is attributable to its ability to utilize data in computational models that contain uncertainties, which may allow treatment and intervention optimization (Jayaraman et al., 2024).

While RL techniques can account for uncertainties in algorithmic modeling, these models are largely influenced by the parameters of the known data. Thus noisy data containing inconsistencies and missing values may significantly attenuate predictive accuracy (Dogru, Chiplunkar, & Huang, 2022). Because these models rely on rewards and penalties for the learning process, the algorithm is vulnerable to human subjectivity and bias. Additional limitations include the inability to assess long-term consequences due to delayed feedback attributable to parameter tuning (Jayaraman et al., 2024). Since RL algorithms may influence clinical decisions involving drug interactions and other safety concerns, it is recommended that additional research be conducted to ensure the algorithm's safety and efficacy prior to robust implementation.

3.4 Naive Bayes

Naïve Bayes is a technique that can be adapted for supervised and unsupervised ML. While Naïve Bayes is commonly understood to represent a singular algorithm, it employs multiple algorithms to evaluate high-dimensional data for classification tasks. Common uses of Naïve Bayes algorithms include diagnostic predictions (Fatima & Pasha, 2017; Ibrahim & Abdulazeez, 2021; Rao Jetti, Shaik, & Shaik, 2021). Advantages of Naïve Bayes algorithms include their ability to be trained on smaller datasets, which may improve output interpretability.

Naïve Bayes algorithms are based on Bayes' theorem (Rao Jetti et al., 2021). Bayesian probabilistic modeling uses a backpropagation approach that utilizes multiple inputs derived from datasets with Gaussian or multinomial distributions to predict an output for a data point that has not yet been categorized (Ismail, Hassan, & Bafjaish, 2020).

Naïve Bayes algorithms assume that predictors in datasets are conditionally independent (Langarizadeh & Moghbeli, 2016). Conditional independence occurs when two variables (i.e., A and B) each share a correlation with a third variable or category (i.e., C). However, despite knowing C, A and B are independent of each other. Thus knowing the value of A does not provide any additional information about B once C is known (Richardson & Gilks, 1993).

$$P(A \mid B, C) = P(A \mid C)$$

Equation 2: Conditional independence within Bayesian models

Conditional independence reduces the number of parameters required for training, as this allows redundant features to be excluded across independent datasets (Friedman & Koller, 2003). These implications are consistent with the speed at which Naïve Bayes may provide a clinical prediction for diagnostic purposes. While this approach is computationally inexpensive and may be beneficial due to its simplicity, other algorithms, such as artificial neural networks (ANNs), may provide greater clinical utility and predictive accuracy.

3.5 Deep learning and neural networks

DL is a form of statistical ML that uses multiple neural layers or artificial interconnected networks to evaluate patterns and characteristics across datasets for predictive modeling (López, 2022). DL typically outperforms classical ML algorithms. Classical ML models, sometimes referred to as shallow ML algorithms, are those that have simple architectures or limited computational layers (Chassagnon et al., 2020; Janiesch, Zschech, & Heinrich, 2021). While shallow ML models may lead to better interpretability and can be trained on smaller datasets, DL algorithms are able to extract more statistically significant features that may be used for prognostication and predictive risk factors (Razzak, Naz, & Zaib, 2017).

ANNs are a type of deep ML algorithm (López, López, & Crossa, 2022). Multiple types of ANNs are utilized in clinical medicine, which include, but are not limited to, convolutional neural networks (CNNs), recurrent neural networks (RNNs), and feedforward neural networks (FNNs) (Qamar & Zardari, 2023; Raj & Kos, 2025). While the architecture of these DL ANNs is similar, as they contain multiple inputs, hidden layers, and nonlinear activation functions that are used to derive outputs, their applications vary (Fig. 2.3).

FNNs represent the most fundamental type of neural network. FNNs are architecturally less complex than RNNs as they lack temporal dynamic processes and therefore do not use memory to evaluate previous inputs (Rayhan, Rahman, Sultana, & Varga, 2025). Limitations of multilayer perceptrons (MLPs), which are a type of FNN, include their inability to evaluate temporally dependent relationships within various datasets (Kim, Kim, Kim, Lee, & Yoon, 2025; Torres-Velázquez, Chen, Li, & McMillan, 2021). Furthermore, this approach may only be utilized for data that can be analyzed unidirectionally, as there are no feedback loops to allow previous inputs to influence the derivation of the predicted output (Rayhan et al., 2025). Data suitable for FNNs includes numerical data, financial data, and text. While clinical images can be used for regression or classification analyses, applications of FNNs require that the images be independently analyzable and static, as these algorithms are not suited for temporal analyses.

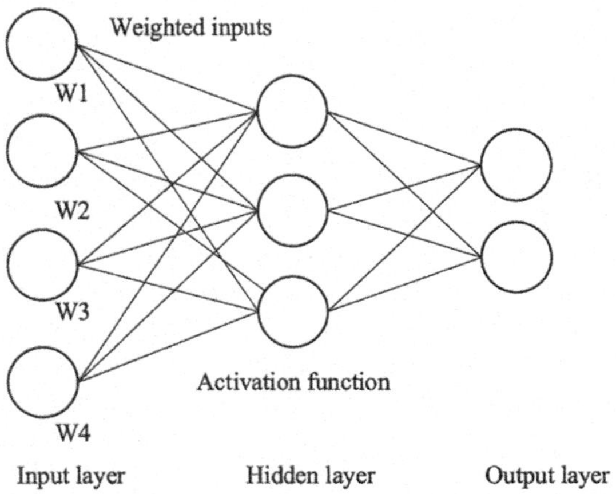

FIG. 2.3 Example of an artificial neural network (Nguyen et al., 2023).

CNNs are a type of ANN that are primarily used for image interpretation and object recognition. Thus CNNs are often applied in clinical diagnostic procedures involving MRI, PET, and/or CT scans (Chattopadhyay & Maitra, 2022). RNNs were developed to overcome the limitations of FNNs, as these algorithms are often applied to temporal or time-series data. Data evaluated by RNNs include handwriting, speech data, text summarization, and language modeling (Li, Chen, Cao, Zhang, & Natarajan, 2014). Given that outputs from RNNs are derived from sequential inputs to evaluate temporal relationships, longitudinal data obtained from electronic health records may be useful for predicting disease development (Chen, Stewart, Sun, Ng, & Yan, 2019; Choi, Schuetz, Stewart, & Sun, 2017).

DL has been applied to thoracic image interpretation (Chassagnon et al., 2020). The use of ML in thoracic imaging typically aims to evaluate the presence or absence of a lung nodule for diagnostic purposes. Radiographic imaging, therefore, may be used for classification, quantification, or prediction of treatment response (Gillies, Kinahan, & Hricak, 2016). Classical ML algorithms are developed from clinical imaging datasets that include robust representations of tumors according to their morphological characteristics, such as shape, size, and volume. The training and validation phases of algorithmic programming must also include images that are devoid of pathology to allow for differentiation between nonpathological patients and those with malignancy for automated labeling and processing in unsupervised models. While DL applications may be advantageous for differentiating between different types of cancer according to clinical imaging, research has shown that the validity of these algorithms may be insufficient for robust implementation (Chassagnon et al., 2020; Park et al., 2020; Zhang & Hong, 2023).

4 Oncology applications

Specific derivations of AI/ML algorithms within oncology populations have utilized radiomics. Radiomics employs quantitative approaches to evaluate radiographic images obtained from computerized tomography (CT) and/or positron emission tomography (PET) to examine complex patterns that may not be detectable through human assessment (Gillies et al., 2016). Due to the significant tumor heterogeneity within and between cancer populations, the application of radiomics principles aims to improve patient outcomes by facilitating earlier detection and tailoring treatment protocols (Sala et al., 2017).

Considerations of tumor heterogeneity are imperative for oncology and AI/ML applications, as patients with similar tumors may exhibit differences in cellular architecture, gene expression, or protein profiles. Intracellular heterogeneity occurs when there are distinct cell-to-cell differences within the same tumor (Fu, Liang, Luo, & Wang, 2025). Because the microenvironment of a tumor may affect the tumor's behavior and response to treatment, it is imperative that ML algorithms utilizing radiomics principles are trained on sufficient data to ensure generalizable predictions (Morris et al., 2016).

Examination of radiomics in unsupervised models suggests that these algorithms may be applied to evaluate prognostication and treatment responsiveness, as they are able to assess the volume of the tumor rather than cells obtained from specific anatomical positions during biopsy assessments (Mayerhoefer et al., 2020). These implications suggest that, because radiomic ML algorithms are able to evaluate proteomic, transcriptomic, and genetic data directly associated with tumor aggressiveness, survival outcomes may improve as earlier detection and predictions of tumor progression are now possible (Mayerhoefer et al., 2020).

5 Random forest algorithms

Random forest (RF) algorithms are often applied in medical settings for diagnosis or clinical predictions of pathology (Salman, Kalakech, & Steiti, 2024). RFs can be trained on supervised models, which rely on data mining techniques. Data mining may be categorized as either descriptive or predictive. Descriptive data mining relies on retrospective data that is utilized for summarization or examinations of patterns, whereas predictive mining aims to utilize data to make future assessments related to disease onset and development. Conclusions derived from RF applications within unsupervised models are based on multiple decision trees.

Decision trees are best conceptualized as hierarchical representations of multiple decision outputs derived using multiple inputs and weights to generate a prediction. Decision trees are utilized to formulate predictions for a specific question based on the relationship between multiple inputs. Decision trees use root and leaf nodes to represent specific inputs and consequences, or outcomes, based on those inputs (Sharma & Kumar, 2016). Leaf nodes are derived from data splitting, which allows the root nodes to be partitioned into sub-nodes based on the weights and parameters related to the dataset. Thus decision trees may be utilized for classification or regression modeling (Salman et al., 2024). Because individual decision trees are insufficient for clinical predictions or outcomes, RFs allow for the assessment of a prediction based on multiple decision trees.

Advantages of RF models include the ability to incorporate large datasets that contain multiple independent variables for accurate predictions of the dependent or target variable (Zhang et al., 2020). The utilization of multiple decision trees aims to account for prediction errors caused by overfitting and unbalanced datasets. These implications suggest that RFs may reduce inflated correlations between decision trees by utilizing random sampling of the inputs, which reduces biased estimations and predictions derived from the original dataset. While the advantages of RFs are related to predictive accuracy, the time required to train the algorithms to ensure accurate predictions may not be practical for real-world applications. Furthermore, because multiple decision trees must be applied to multifactorial analysis, the time required to obtain the predictive output may be infeasible for utilization in clinical settings (Salman et al., 2024).

The limitations of RF models for predicting pathology must be considered based on whether the model will be used for classification or regression analyses. Because RFs include multiple independent decision trees that may vary significantly across datasets, it is imperative to note that regression models are computed as the average or sum of the predictions derived from individual trees, divided by the number of trees or inputs (Probst, Wright, & Boulesteix, 2019).

The number of optimal trees incorporated into an RF algorithm has been mathematically evaluated by Oshiro, Perez, and Baranauskas (2012). While the number of individual trees may reflect the dataset of interest more robustly, it is imperative to note that the incorporation of more trees may not yield an increase in performance after a certain threshold. Therefore despite the utility of RFs, a specific determination of the optimal number of trees for performance optimization and generalizability is not possible and must be determined according to specific clinical datasets and populations (Oshiro et al., 2012).

5.1 Statistical limitations of RFs in genome-wide association studies

Prediction and classification outputs can be derived from RF decision algorithms using multiple statistical programs that include, but are not limited to, SPSS, R, Python, MATLAB, and Stata (Brieuc, Waters, Drinan, & Naish, 2018). Applications of RFs for genetic association data have become common in clinical research. While advantages of RFs for genome-wide association studies (GWAS) include their ability to incorporate thousands of predictor inputs and observations, limitations in predictive accuracy and black-box issues can be related to the criteria applied to data splitting.

These implications are significant, as advancements in clinical medicine and technology have led to evaluations of disease risk according to genetic biomarkers. While RF algorithms may be advantageous for evaluating large datasets to examine biomarkers and diseases that are attributable to polygenic interactions, significant limitations must be considered for categorical predictions (Le Corre & Kremer, 2012). GWAS typically rely on mixed models to examine the relationships between disease states or traits and their associated biomarkers (Brieuc et al., 2018). However, this research often reflects significant variability regarding phenotypic expression that may be attributable to nonadditive factors. This suggests that GWAS analyses do not necessarily provide accurate predictions of a trait but can explain phenotypic trait variance within specific populations using multivariate approaches. These limitations are exemplified by the limited accuracy and power that GWAS analyses have for heterogeneous populations. Thus despite the potential attractiveness of RFs, these algorithms are not developed to handle the multidimensional datasets required to efficiently analyze single-nucleotide polymorphisms (SNPs) within GWAS (Brieuc et al., 2018).

Limitations in predictions for GWAS using RFs include, but are not limited to, the inability of these algorithms to accommodate missing values, linkage disequilibrium, heterogeneity within the population, and inaccurate weighting of SNPs for prediction (Brieuc et al., 2018; Goldstein, Polley, & Briggs, 2011; Meng, Yu, Cupples, Farrer, & Lunetta, 2009). Statistical approaches that have been utilized to circumvent errors in predictions in GWAS that rely on RFs include boosting strategies such as gradient boosting or the utilization of the recurrent relative variable importance measure. These boosting strategies aim to reduce biases related to the selection of SNPs for phenotypic trait predictions (Szymczak et al., 2016). Despite the potential utility of RFs in GWAS data, it is imperative to note that there are potential black-box issues related to the predictive outputs. Furthermore, because of the large datasets required for GWAS analyses, researchers must consider the significant computational expenses and potential inefficiencies of RFs when utilized for clinical predictions (Brieuc et al., 2018).

6 AI/ML in diagnostics

The diagnostic utility of AI/ML is determined by the algorithm's sensitivity and accuracy (Shreffler & Huecker, 2020). Clinical applications of AI require that the computational predictions derived from these algorithms accurately classify and differentiate between individuals with pathology and those who do not present clinically significant symptomatology indicative of disease (Adlung, Cohen, Mor, & Elinav, 2021). The sensitivity of AI/ML algorithms depends on the supervised models used to train the algorithm on diverse datasets to ensure generalizability. Sensitivity is defined as the ability of a test or assessment to correctly identify individuals with a specific disease or illness (Monaghan et al., 2021). The sensitivity of AI/ML models is reflected by the correct prediction of illness for those with the disorder, whereas specificity measures the model's ability to correctly predict those without disease. Measurements of sensitivity and specificity include false positives (FPs), true positives (TPs), true negatives (TNs), and false negatives (FNs).

$$\textbf{Sensitivity} = \textbf{TP} / (\textbf{TP} + \textbf{FN})$$

Equation 3: Computational modeling of sensitivity

Applications of AI/ML in clinical settings require the utilization of multiple datasets, as clinical presentations may differ between groups and individuals. Factors that may influence disease presentation include, but are not limited to, age, demographic data, severity, and comorbid diseases (Andrew et al., 2023; Miron et al., 2024). These implications suggest that robust datasets, representative of the patient population, must be utilized during the training and validation stages prior to the implementation of unsupervised predictions within clinical practice (Williams, 2024). A lack of heterogeneous datasets is likely to cause overfitting and/or generalization errors. These implications exemplify the risks of using insufficient training sets that fail to account for the heterogeneity of disease presentation, as this may lead to underperformance of the selected model due to FN or FP predictions (Adlung et al., 2021).

AUC-ROC curves are imperative for binary classifications. As AI/ML algorithms must show specificity for the correct classification of TNs, these curves reflect the accuracy of the model in detecting TPs and TNs. AUC-ROC curve scores range from 0 to 1, where a score of 0 reflects limited predictive accuracy, and a score of 1 reflects accurate classification (Mandrekar, 2010). A score of 0.5 indicates that the model is guessing and is unable to distinguish meaningful patterns between patients or datasets. A score between 0.7 and 0.8 reflects that the model is effective at differentiating between groups, whereas a score above 0.8 reflects that the model is excellent at binary classification predictions (Mandrekar, 2010) (Fig. 2.4).

Evaluations of model performance and AUC-ROC curves require multifactorial analyses. Given that multiple metrics may be utilized to evaluate a model's performance, assessments of the utility of each metric should be considered (Naidu, Zuva, & Sibanda, 2023). Metrics that are often used in ML for classification and discriminant analysis include, but are not limited to, precision, accuracy, sensitivity, specificity, error rate, mean square error, and the kappa statistic. F1-scores are often utilized to examine a model's accuracy in unbalanced datasets. Factors included in the F1-score are precision and

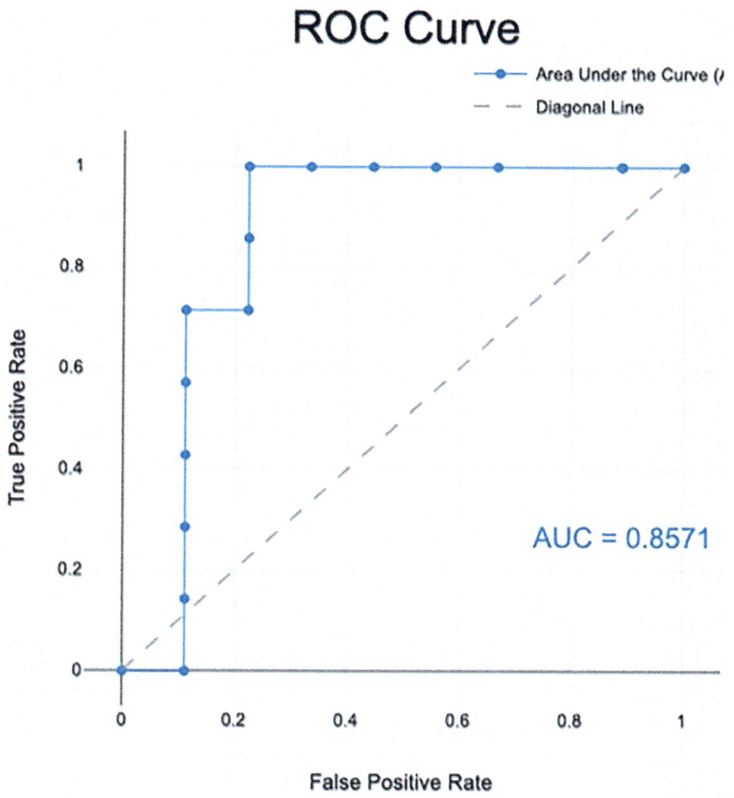

FIG. 2.4 AUC-ROC curve.

recall. Precision is often reflected as a ratio that denotes the number of correct predictions according to the number of TPs and FPs, whereas recall reflects a ratio of all correct categorizations or observations (Naidu et al., 2023).

$$\text{Precision} = \frac{TP}{(TP + FP)}$$

Equation 4: Computational modeling of precision

$$\text{Recall} = \frac{TP}{(TP + FN)}$$

Equation 5: Computational modeling of recall

Factors that must be considered when examining the best algorithm for diagnostic applications require clinicians to evaluate whether the data will be applied for classification, prognostic/disease outcomes, or anomaly detection.

Multiple algorithms, such as CNN, KNN, and RFs, have been applied to various clinical populations for diagnosis. Comparative analyses suggest that CNNs tend to outperform other AI/ML models based on performance evaluations that utilize accuracy, F1-scores, AUC, and sensitivity analyses (Ahsan, Luna and Siddique, 2022).

Examination of the model performance for heart disease classification suggests that DL neural network models exemplify the most utility for diagnostic purposes. CNNs and SVMs may indicate the most utility, as research has shown that these models have high accuracy in the detection of electrocardiogram abnormalities and differentiation between different types of heart disease (Rubin, Parvaneh, Rahman, Conroy, & Babaeizadeh, 2018).

Additional comparisons of the performance of AI/ML algorithms have been conducted within nephropathy populations. Charleonnan et al. (2017) evaluated the model performance of KNN, SVM, LR, and decision trees in predicting chronic kidney disease (CKD) using clinical data obtained from the Indian CKD dataset ($n = 400$). The SVM model performed the best compared to the KNN, LR, and decision tree models, as its accuracy score for classification was 98.3%. Similar analyses of model performance in CKD were conducted by Aljaaf et al. (2018). This study demonstrated that neural network algorithms outperformed LR and decision tree models, with the accuracy score of the MLP model reported at 98.1%.

6.1 Treatment responses and healthcare outcomes

Precision medicine is a medical framework that requires multifactorial analyses to be applied to individual patients to evaluate their best treatment options and responsiveness to pharmacological interventions (König, Fuchs, Hansen, von Mutius, & Kopp, 2017). Factors included in these assessments include underlying conditions, risk factors, genetic makeup, and lifestyle. While precision medicine aims to tailor treatments and disease prevention according to individualized factors, the outcomes of this approach significantly depend on the diagnostic tests that guide the selection of potential treatment options (König et al., 2017).

Utilization of ML algorithms within precision medicine includes applications in psychiatry, oncology, neurology, and chronically ill populations. Research examining the use of ML algorithms in psychiatric populations suggests that these models may be useful for predicting treatment outcomes (Jankowsky, Krakau, Schroeders, Zwerenz, & Beutel, 2024). Current practices in psychiatry reflect that the heterogeneity of treatment responses is often attributable to nonlinear interactions, which include, but are not limited to, symptom severity, cognitive deficits, comorbidity, and patient motivation. These findings suggest that using ML may reduce the costs and labor expenses associated with the trial-and-error method often employed to determine which medication is most appropriate (Jankowsky et al., 2024).

Limited studies have been conducted to examine treatment responsivity within psychiatric populations using AI/ML predictions and standardized psychometric inventories. Webb et al. (2020) applied multiple ML algorithms to predict depression scores based on the Patient Health Questionnaire-9. Specific algorithms applied to the psychiatric cohort included RF, elastic net regularization, SVM, Bayesian additive trees, and additive polynomial splines. Comparative analyses of model performance across decision tree models, support vector analyses, and regression models revealed that the best-performing algorithm for this study was elastic net regularization (Webb et al., 2020).

Elastic Net Regularization is a penalized regression model that can optimize model performance when the number of observations is less than the number of factors influencing the output (Zou & Hastie, 2005). Variables selected for elastic net regularization models are derived from L1 (Lasso) and L2 (Ridge) penalties (Obi & Jecinta, 2023). Thus the model reduces overfitting and generalization errors, as factors reflecting nonzero coefficients are utilized in the final predictive algorithm. These mathematical implications, when applied to a psychiatric hospital setting to examine patient outcomes, revealed that the most predictive factors for prognosis were fatigue and poor concentration. The results also indicated that individual symptoms related to depression and severity were not as statistically significant for predicting treatment outcomes compared to other factors (Webb et al., 2020).

Meta-analyses have been conducted to compare AI/ML algorithms in predicting healthcare outcomes. Given the significant proposed utility of implementing AI/ML within healthcare, it is imperative that the safety and efficacy of these algorithms are comprehensively examined. Comparative analyses of AI/ML algorithms were systematically evaluated by Olender, Roy, and Nishtala (2023) in older adult patients. The meta-analysis assessed the performance of models such as SVMs, DL neural network models, and RFs in predicting healthcare outcomes in cardiovascular patients. Healthcare outcomes were evaluated based on multiple factors, including but not limited to hospitalization, 30-day readmission, post-transplant outcomes, and mortality risk. The model performance was assessed using metrics such as the AUC-ROC curve, sensitivity, specificity, and accuracy. Model performance evaluations for cardiovascular outcomes indicated that ANN models have high clinical utility for predicting clinical outcomes and disability severity 3 months after a stroke, as reflected by their AUC-ROC score (0.97). Further examinations revealed that XGBoost may have utility in predicting clinical complications and adverse events requiring re-admission. The AUC score for XGBoost analyses in predicting transient ischemic attack was 0.93 and 0.97 for therapeutic failure.

While this study demonstrated that ML algorithms outperformed traditional modeling techniques, there is no consensus on which individual algorithm performs best across all clinical populations and/or datasets (Olender et al., 2023; Wolpert & Macready, 1997; Yang, He, & Fan, 2020).

The inability to determine a singular algorithm that is able to outperform other techniques and account for all possible limitations is attributable to the No Free Lunch principle (Wolpert & Macready, 1997). This principle states that algorithmic performance is largely determined by the type of data and the specific problem the algorithm is attempting to solve (Yang et al., 2020). Thus it is imperative that clinicians and researchers consider the data type and desired output before selecting an AI/ML model to use in clinical settings.

6.2 Boosting methods in diagnostic and prognostic algorithms

Various frameworks have been applied to optimize predictive performance in ML which include, but are not limited to, ensemble learning, boosting, bagging, and gradient techniques (Mahajan, Uddin, Hajati, & Moni, 2023). Ensemble learning is a technique that may increase predictive accuracy as it uses multiple models for classification or multifactorial regression analyses. Predictions derived from ensemble methods are based on the aggregation of multiple models, which may mitigate the limitations of individual models, thereby improving predictive accuracy (Mahajan et al., 2023) (Fig. 2.5).

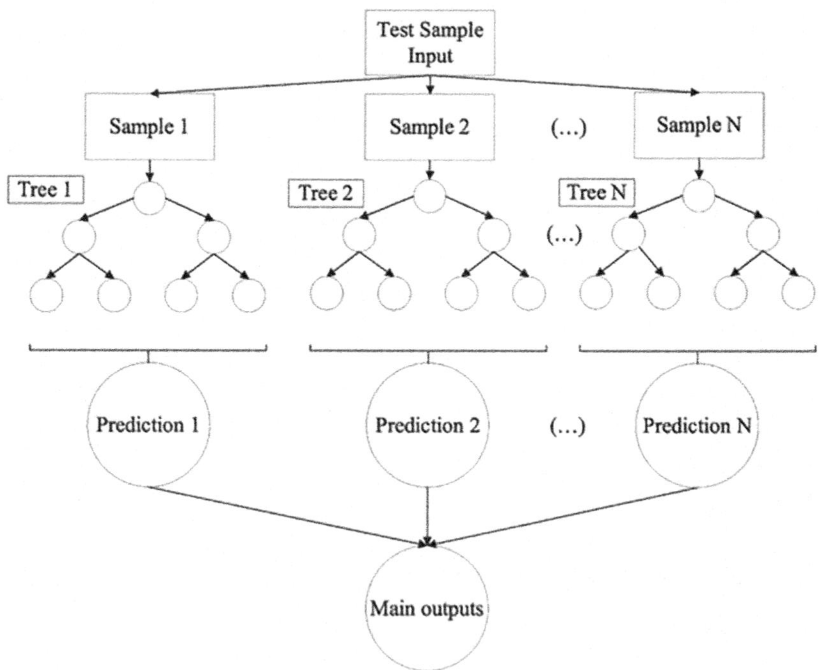

FIG. 2.5 Example of ensemble method computation (Nguyen et al., 2023).

Computational advantages of ensemble learning are attributable to principles related to information theory. Shannon entropy is a concept that originates from information theory and is used to quantify the degree of uncertainty within a particular dataset (Ali, Anam, & Ahmed, 2023; Karaca & Moonis, 2022). Shannon entropy is defined in Eq. (6), where H represents entropy, n is the total amount of data measured in bits, and p(xi) reflects the likelihood that the variable or information contributes to the classification or prediction.

$$H(X) = -\sum_{i=1}^{n} p(x_i) \log_2 p(x_i)$$

Equation 6: Shannon's entropy for discrete predictions

Principles of Shannon entropy may assist with model performance optimization, as entropy can be utilized to select features that are statistically significant for disease categorization and/or classification (Ali et al., 2023; Sepúlveda-Fontaine & Amigó, 2024). Entropy may be applied to Bayesian, classification, or regression analyses. Furthermore, these analyses may be used for data splitting and the selection of decision trees for RF algorithms (Ignatenko, Surkov, & Koltcov, 2024).

Entropy may be used to facilitate assessments of information gain from specific nodes and decision points when applied to RF and/or classification and regression trees (Mahajan et al., 2023). Information gain (IG) is a metric that reflects how much a specific feature contributes to the predictive model according to entropy reduction (Kononenko & Kukar, 2007). Thus entropy is best conceptualized as how much information is needed to derive a predicted outcome based on a specific dataset and its distribution. The correlation between entropy and IG is best reflected as an inverse relationship. Factors that influence entropy in ML include the data's complexity, distribution, quality, and the number of factors included in the regression analysis (Mitchell, 1997). Given that entropy is derived as a logarithmic function, an increase in the number of classes amplifies uncertainty, which may negatively affect a model's predictive performance (Cover & Thomas, 2006).

Applications of entropy in ML for clinical predictions may be modeled as binary or multiclass predictions. Entropy ranges from 0 to 1, where 0 is associated with high uncertainty and low predictive accuracy (Karaca & Moonis, 2022). Multiclass predictions, which may be utilized to quantify a disease according to its severity level, calculate entropy as the logarithm base 2 and are derived according to the number of classes ($\log_2(k)$).

While ensemble learning is touted for its robust predictive accuracy, the limitations of these models are attributable to their limited interpretability. Recommendations to address these limitations include data pruning, examination of model compatibility, and parameter tuning. In addition, the utilization of cross-validation and Bayesian statistical approaches for data augmentation may reduce the risk of predictive bias and overfitting (Rane et al., 2024).

Meta-analyses comparing the predictive performance of individual ML models and ensemble methods were conducted by Mahajan et al. (2023). Comparative analyses of model performance across major chronic diseases included kidney disease, skin cancer, diabetes, liver disease, and heart failure. Individual ML models used in the analysis included KNN, Naïve Bayes, ANN, RF, and SVMs. The performance of each ML algorithm was compared to ensemble approaches such as bagging, stacking, voting, and/or gradient boosting. While ensemble approaches outperformed individual ML algorithms in terms of predictive accuracy, none of the studies included in the analysis indicated the extent to which ensemble methods performed better comparatively (Mahajan et al., 2023). Thus the information gained by using ensemble methods was not specified in terms of whether the improvement in predictive accuracy was significantly enhanced by aggregating multiple models. These results suggest that while ensemble methods may increase predictive accuracy, it remains inconclusive whether the computational expenses and limited interpretability issues associated with these algorithms justify their use in clinical settings compared to other established ML algorithms.

6.3 Mortality risk prediction

Common computational models in ML for mortality risk include SVMs, DL, RF, statistical neural network models, and boosting techniques (Khan et al., 2021). Statistical comparisons of ML models and standardized risk assessments for acute coronary syndromes (ACS) have shown that AI/ML may provide clinical advantages in mortality risk assessment (VanHouten, Starmer, Lorenzi, Maron, & Lasko, 2014). Despite the clinical utility of standardized risk assessments such as the Thrombolysis in Myocardial Infarction (TIMI) and Global Registry of Acute Coronary Events (GRACE) scales, these assessments lack comprehensive multifactorial analyses. Furthermore, due to novel developments in clinical medicine, the predictive validity of these tools may no longer be supported. The use of ML within ACS populations has led to the identification of the most statistically significant risk factors associated with these disorders. These factors include, but are not limited to, Killip classification, sex, age, diastolic blood pressure, hemoglobin, and serum creatinine. Comparisons of the GRACE and TIMI scores to ML reflected lower predictive accuracy due to the inability of these validated assessments to evaluate nonlinear interactions between underlying pathophysiological biomarkers (VanHouten et al., 2014).

Evaluations of model performance for LR models were compared to ML methods to predict mortality risk (Kawano et al., 2022). Data utilized in the training and test datasets included 116,749 patients of which 5250 patients died within the 5-year follow-up period. The longitudinal data was obtained from the Study on the Design of a Comprehensive Medical System for Chronic Kidney Disease, as part of the Japan Specific Health Checkups Study (J-SHC Study). All participants included in the analysis had a known mortality or survival status after 5 years of enrollment in the study. LR was compared against gradient boosting decision tree (XGBoost) and a neural network. The predictive performance of each model was evaluated according to the AUC-ROC curve, recall, accuracy, precision, and F1-score. Comparative analyses showed that both ML algorithms outperformed conventional LR. XGBoost, which is an ensemble method, outperformed both neural network algorithms and LR according to all predictive accuracy metrics included in this study (Kawano et al., 2022). These results are consistent with other comparative studies that use ML to predict mortality risk in diverse populations, such as neurocritically ill patients and patients with comorbid chronic illnesses (Guillamet et al., 2024; Yuan et al., 2025).

7 Further considerations

While AI/ML may have many advantages, it is imperative to note that the algorithmic configurations, parameter tuning, and training and validation procedures may differ across medical institutions. The development of AI/ML algorithms within independent institutions requires multidisciplinary collaboration to ensure that the algorithms are tailored to specific patient populations and clinical needs (Varnosfaderani & Forouzanfar, 2024).

Despite the potential utility of AI/ML in healthcare, the ability to develop generalizable computational models requires significant resources and high-quality data. Regulatory concerns related to patient privacy, HIPAA, and data protection laws must be considered when utilizing retrospective clinical data for algorithmic development (Kancherla, 2020). While the United States has not adopted a national provision that protects patient data similar to the General Data Protection Regulation enforced in the European Union, domestic state and local laws may govern how patient data can be used for the development of commercial AI/ML algorithms (Hoofnagle et al., 2019).

Due to the rapid implementation of AI/ML in healthcare settings, legal scholars have recommended that medical facilities develop institutional policies addressing the implementation of ML (Shen et al., 2024). These policies should include future-use clauses, whether patients are able to opt-out of AI/ML usage for clinical predictions or diagnoses, and a delineation of whether these algorithms will be utilized as clinical supports or replacements. Further implications that must be considered include transparency in patient communications, as their data may be used for training and validation of AI/ML algorithms. Furthermore, because of the limitations associated with potential predictive errors for diagnostic classifications, it is imperative that medical institutions establish safety and oversight parameters to ensure the responsible use of AI/ML (Nong, Adler-Milstein, Apathy, Holmgren, & Everson, 2025).

8 Conclusion

AI/ML has transformed the healthcare system. The utilization of retrospective data to configure statistical predictive models for diagnostic and prognostic outcomes is one of the advantages associated with this technology (Kancherla, 2020; Nong et al., 2025). While the use of neural networks, support vector models, and ensemble techniques may be associated with increased predictive accuracy compared to human assessment, researchers and clinicians must be aware of the limitations of these algorithms (Kawano et al., 2022; Mahajan et al., 2023). Because the utility and predictive accuracy of AI/ML algorithms are dependent upon the characteristics of the data derived from the target patient population and the analytical objective (i.e., whether predictive output will be utilized for classification or regression analyses), clinicians must be mindful that there is no singular best algorithm that can be utilized for all data types (Ignatenko et al., 2024; Wolpert & Macready, 1997).

Model selection requires clinicians to examine multiple accuracy metrics, which include, but are not limited to, F1-scores, AUC-ROC values, sensitivity, and specificity (Mandrekar, 2010; Naidu et al., 2023). While accuracy is the most accepted metric for model performance, transparency and generalization errors must be considered when AI/ML algorithms are applied in clinical settings (Ahsan, Luna and Siddique, 2022). This suggests that medical institutions should develop policies to address whether patient data can be used for AI/ML development (Shen et al., 2024). Furthermore, healthcare institutions should disclose whether patients have the option to opt-out of AI/ML use for decision support and/

or replacement. Because of the significant consequences that may arise from misdiagnosis attributable to predictive error, responsible use of these algorithms requires the implementation of safety and oversight committees to ensure that patients receive adequate protections (Nong et al., 2025).

Conclusively, the utilization of AI/ML in clinical settings requires a multifactorial analysis. Due to the deleterious effects associated with misdiagnosis, clinicians must ensure that the algorithms utilized in their practices are sufficiently programmed with heterogeneous data to ensure generalizability (Williams, 2024). While predictive accuracy is imperative, multiple evaluations of model performance and transparency must also be considered (Naidu et al., 2023). Table 2.1 discusses the advantages and disadvantages of AI/ML algorithms covered in this chapter. Despite the statistical limitations that may exist with AI/ML algorithms, their use and implementation in clinical settings may promote better patient outcomes and efficiency (Al-Antari, 2023). Statistical evaluations of their performance support their use in precision medicine, diagnostics, and prognostic outcomes (Charleonnan et al., 2017; Webb et al., 2020). However, clinicians must carefully consider which model will best serve their patient populations based on the data's characteristics and the intended application.

TABLE 2.1 AI/ML algorithms advantages and disadvantages.

K-nearest neighbor (KNN)	**Type of AI algorithm typically used for classification. This ML model tries to predict or classify new data points based on their similarity to "neighbors"**
Limitations of KNN	Model performance depends on the number selected for neighbors, or *k*. In clinical datasets that require a high *k* value or high-dimensional data, patterns and predictive factors may be lost due to smoothing or averaging effects. Missing values from electronic medical records or diagnostic laboratory values may also attenuate model performance and clinical utility.
Support vector machine (SVM)	Types of AI/ML used for classification and regression tasks attempt to identify the simplest model that separates the classes of variables according to a boundary or hyperplane. These models compare unknown data points to known data points that are separated by classes. The model classifies unknown data points using support vectors and is advantageous for high-dimensional datasets.
Limitations of SVMs	Highly influenced by noisy data. Limited interpretability. Predictive utility dependent on feature and kernel function selection. Due to transparency issues, may not be suitable for clinical diagnostic purposes.
Reinforcement learning (RL)	Distinct AI/ML models that cannot be classified as supervised or unsupervised. Applied in unknown environments or datasets with undefined parameters, aiming to maximize accurate predictions based on an external reward signal. These models are advantageous in medical scenarios where there is limited data contributing to an uncertain outcome, or where it is otherwise infeasible to expose patients to certain conditions.
Limitations of RL	Computational expenses, programming of reward signals, limited generalizability.
Naïve Bayes	AI/ML techniques can be adapted for supervised and unsupervised machine learning. While Naïve Bayes is commonly understood to represent a singular algorithm, it employs multiple algorithms to evaluate high-dimensional data for classification tasks. It uses Bayes' probabilistic approach, which assumes variables are conditionally independent.
Limitations of Naïve Bayes	Inability to differentiate between significant features, as variables are deemed equally important. Assumption that datasets follow a Gaussian distribution. Sensitivity to datasets that contain noisy data and missing values.
Random forest (RF)	Utilizes the outputs from decision trees to make predictions. It may be used for classification or regression tasks. It is advantageous as it uses the average of multiple outputs from decision trees to derive a predictive output, which may reduce overfitting.
Limitations of RF	The number of decision trees will depend on the dataset. They may have difficulty understanding correlations. The complexity of the decision trees may limit interpretability and require computational resources to ensure adequate hyperparameter tuning.
Neural networks	Based on the cytoarchitecture of the human brain, it uses networks, nodes, and hidden layers to derive predictive outputs. Examples include artificial neural networks (ANNs), convolutional neural networks (CNNs), and recurrent neural networks (RNNs). Advantages include the ability to recognize nonlinear relationships, analyze clinical imaging, and facilitate adaptive learning.
Limitations of neural networks	Computationally expensive, requires significant resources for training and validation, and presents potential black-box and interpretability issues due to hidden layers.

Continued

TABLE 2.1 AI/ML algorithms advantages and disadvantages—cont'd

K-nearest neighbor (KNN)	**Type of AI algorithm typically used for classification. This ML model tries to predict or classify new data points based on their similarity to "neighbors"**
Logistic regression	Typically used in supervised models for binary or dichotomous classification analysis.
Limitations of logistic regression	Not useful for complex or high-dimensional datasets. Performs best on datasets that can be modeled using linear regression. Sensitive to datasets containing noisy data, outliers, and/or missing values.
Elastic net regression	AI/ML techniques that balance complex datasets by combining the Lasso and Ridge penalties in regression analysis are advantageous for complex datasets with high dimensionality or those requiring the identification of significant features for predictive purposes.
Limitations of elastic net regression	Random forests may be a better option than elastic net regression, as elastic net regression may not capture nonlinear relationships. Predictive output is influenced by parameter tuning.
Ensemble methods	AI/ML approach that combines multiple AI/ML algorithms to increase predictive accuracy. Often used to strengthen "weak" algorithmic strategies, which may not perform well individually. By combining multiple algorithms, ensemble learning can optimize performance, as the strengths of the individual ML models enhance predictive accuracy and utility.
Limitations of ensemble methods	Black-box issues attributable to limited understandability after combining multiple AI/ML algorithms. Computationally expensive. Requires longer training time. May not provide significant utility compared to individual models.

Declaration

AI was not used in any capacity to generate text for this chapter. The chapter was written by Kristin S. Williams, PhD. ChatGPT was used solely to generate citations for the VC dimension and other early statistical papers related to the development of contemporary AI/ML algorithms.

References

Adlung, L., Cohen, Y., Mor, U., & Elinav, E. (2021). Machine learning in clinical decision making. *Med*. 2666-6340. *2* (6), 642–665. doi:10.1016/j.medj.2021.04.006.

Ahsan, M. M., Luna, S. A., & Siddique, Z. (2022). Machine-learning-based disease diagnosis: a comprehensive review. *Healthcare (Switzerland)*. 2227-9032. *10*(3). Available from: https://www.mdpi.com/2227-9032/10/3/541/pdf. doi:10.3390/healthcare10030541.

Al-Antari, M. A. (2023). Artificial intelligence for medical diagnostics—existing and future AI technology! *Diagnostics*. 2075-4418. *13*(4). Available from: http://www.mdpi.com/journal/diagnostics/. doi:10.3390/diagnostics13040688.

Ali, A., Anam, S., & Ahmed, M. M. (2023). Shannon entropy in artificial intelligence and its applications based on information theory. *Journal of Applied and Emerging Sciences*, *13*(1), 9–17.

Aljaaf, A. J., Al-Jumeily, D., Haglan, H. M., Alloghani, M., Baker, T., Hussain, A. J., & Mustafina, J. 2018, Early prediction of chronic kidney disease using machine learning supported by predictive analytics. In: Proceedings of the 2018 IEEE congress on evolutionary computation (CEC), Rio de Janeiro, Brazil, 8–13 July 2018, pp. 1–9. 9781509060177. http://ieeexplore.ieee.org/xpl/mostRecentIssue.jsp?punumber=8466244. doi:10.1109/CEC.2018.8477876.

Amari, S., & Wu, S. (1999). Improving support vector machine classifiers by modifying kernel functions. *Neural Networks*. 0893-6080. *12* (6), 783–789. doi:10.1016/s0893-6080(99)00032-5.

Andrew, M. K., Pott, H., Staadegaard, L., Paget, J., Chaves, S. S., Ortiz, J. R., ... Lina, B. (2023). Age differences in comorbidities, presenting symptoms, and outcomes of influenza illness requiring hospitalization: a worldwide perspective from the global influenza hospital surveillance network. In *Open forum infectious diseases* (Vol. 10, pp. 6). Canada: Oxford University Press. 23288957. https://academic.oup.com/ofid. doi:10.1093/ofid/ofad244.

Bartlett, P. L., & Maass, W. (2003). Vapnik-Chervonenkis dimension of neural nets. In *The handbook of brain theory and neural networks* (Vol. 2, pp. 1188–1192). MIT Press.

Brieuc, M. S. O., Waters, C. D., Drinan, D. P., & Naish, K. A. (2018). A practical introduction to random forest for genetic association studies in ecology and evolution. *Molecular Ecology Resources*. 1755-0998. *18*(4), 755–766. Available from: http://onlinelibrary.wiley.com/journal/10.1111/(ISSN)1755-0998. doi:10.1111/1755-0998.12773.

Cesa-Bianchi, N., Mansour, Y., & Shamir, O. (2015). On the complexity of learning with kernels. *Journal of Machine Learning Research*, 1533-7928 *40*. http://jmlr.csail.mit.edu/.

Charleonnan, A., Fufaung, T., Niyomwong, T., Chokchueypattanakit, W., Suwannawach, S., & Ninchawee, N. (2017). Predictive analytics for chronic kidney disease using machine learning techniques. In *Management and innovation technology international conference, MITiCON 2016*. Thailand: Institute of Electrical and Electronics Engineers Inc. 9781509041053 MIT80-MIT83MIT80. doi:10.1109/MITICON.2016.8025242.

Chassagnon, G., Vakalopolou, M., & Paragios, N., Revel, M.-P. (2020). Deep learning: definition and perspectives for thoracic imaging. *European Radiology*. 0938-7994. *30* (4), 2021–2030. doi:10.1007/s00330-019-06564-3.

Chattopadhyay, A., & Maitra, M. (2022). MRI-based brain tumour image detection using CNN based deep learning method. *Neuroscience Informatics*. 2772-5286. *2* (4), 100060. doi:10.1016/j.neuri.2022.100060.

Chen, R., Stewart, W. F., Sun, J., Ng, K., & Yan, X. (2019). Recurrent neural networks for early detection of heart failure from longitudinal electronic health record data: implications for temporal modeling with respect to time before diagnosis, data density, data quantity, and data type. *Circulation: Cardiovascular Quality and Outcomes*. 1941-7705. *12*(10). Available from: http://circoutcomes.ahajournals.org/. doi:10.1161/CIRCOUTCOMES.118.005114.

Choi, E., Schuetz, A., Stewart, W. F., & Sun, J. (2017). Using recurrent neural network models for early detection of heart failure onset. *Journal of the American Medical Informatics Association*. 1527-974X. *24*(2), 361–370. Available from: http://jamia.oxfordjournals.org/content/22/e1. doi:10.1093/jamia/ocw112.

Cortes, C., & Vapnik, V. (1995). Support-vector networks. *Machine Learning*. 0885-6125. *20* (3), 273–297. doi:10.1007/bf00994018.

Cover, T., & Thomas, J. (2006). *Entropy, relative entropy, and mutual information elements of information theory* (Vol. 2, pp. 13–55). Wiley.

Cunningham, P., Cord, M., & Delany, S. J. (2008). Machine learning techniques for multimedia: case studies on organization and retrieval. In *Supervised learning* (pp. 21–49). Berlin, Heidelberg: Springer Berlin Heidelberg. M4D. doi:10.1007/978-3-540-75171-7_2.

Dogru, O., Chiplunkar, R., & Huang, B. (2022). Reinforcement learning with constrained uncertain reward function through particle filtering. *IEEE Transactions on Industrial Electronics*. 1557-9948. *69*(7), 7491–7499. Available from: http://ieeexplore.ieee.org/xpl/tocresult.jsp?isnumber=5410131. doi:10.1109/TIE.2021.3099234.

Dreyer, K. J., Kalra, M. K., Maher, M. M., Hurier, A. M., Asfaw, B. A., Schultz, T., … Thrall, J. H. (2005). Application of recently developed computer algorithm for automatic classification of unstructured radiology reports: validation study. *Radiology*. 0033-8419. *234* (2), 323–329. doi:10.1148/radiol.2341040049.

Efthimiou, O., Seo, M., Chalkou, K., Debray, T., Egger, M., & Salanti, G. (2024). Developing clinical prediction models: a step-by-step guide. *BMJ*. 1756-1833. *386*, 1–15. doi:10.1136/bmj-2023-078276.

Fatima, M., & Pasha, M. (2017). Survey of machine learning algorithms for disease diagnostic. *Journal of Intelligent Learning Systems and Applications*. 2150-8402. *09* (01), 1–16. doi:10.4236/jilsa.2017.91001.

Filiberto, A., Leeds, I., & Loftus, T. (2021). Machine learning in clinical decision-making. *Frontiers in Digital Health*, *3*, 1–2.

Friedman, N., & Koller, D. (2003). Being Bayesian about network structure. A Bayesian approach to structure discovery in Bayesian networks. *Machine Learning*. 0885-6125. *50* (1–2), 95–125. doi:10.1023/a:1020249912095.

Fu, Y. C., Liang, S. B., Luo, M., & Wang, X. P. (2025). Intratumoral heterogeneity and drug resistance in cancer. *Cancer Cell International*. 1475-2867. *25*(103), 1–18. Available from: https://cancerci.biomedcentral.com/. doi:10.1186/s12935-025-03734-w.

Ghafoori, Z., Erfani, S. M., Rajasegarar, S., Bezdek, J. C., Karunasekera, S., & Leckie, C. (2018). Efficient unsupervised parameter estimation for one-class support vector machines. *IEEE Transactions on Neural Networks & Learning Systems*. 2162-2388. *29*(10), 5057–5070. Available from: http://ieeexplore.ieee.org/xpl/RecentIssue.jsp?punumber=5962385. doi:10.1109/TNNLS.2017.2785792.

Gillies, R. J., Kinahan, P. E., & Hricak, H. (2016). Radiomics: images are more than pictures, they are data. *Radiology*. 1527-1315. *278*(2), 563–577. Available from: http://pubs.rsna.org/doi/pdf/10.1148/radiol.2015151169. doi:10.1148/radiol.2015151169.

Gligorea, I., Cioca, M., Oancea, R., Gorski, A. T., Gorski, H., & Tudorache, P. (2023). Adaptive learning using artificial intelligence in e-learning: a literature review. *Education Sciences*. 2227102. *13* (12), 1–27. doi:10.3390/educsci13121216.

Goldstein, B. A., Polley, E. C., & Briggs, F. B. S. (2011). Random forests for genetic association studies. *Statistical Applications in Genetics and Molecular Biology*. 1544-6115. *10* (1), 1–34. doi:10.2202/1544-6115.1691.

Guido, R., Ferrisi, S., Lofaro, D., & Conforti, D. (2024). An overview on the advancements of support vector machine models in healthcare applications: a review. *Information*. 2078-2489. *15* (4), 235. doi:10.3390/info15040235.

Guillamet, G. H., Pallaruelo, A. N. M., Miró Mezquita, L., Miralles, R., Àngel Mas, M., José Ulldemolins Papaseit, M., & Cuxart, O. E., López Seguí, F. (2024). Correction: machine learning model for predicting mortality risk in patients with complex chronic conditions: retrospective analysis. *Online Journal of Public Health Informatics*. 1947-2579. *16*, e58453. doi:10.2196/58453.

Ibrahim, I. M., & Abdulazeez, A. M. (2021). The role of machine learning algorithms for diagnosing diseases. *Journal of Applied Science and Technology Trends*. 2708-0757. *2*(1), 10–19. Available from: https://jastt.org/index.php/jasttpath/article/view/79. doi:10.38094/jastt20179.

Ignatenko, V., Surkov, A., & Koltcov, S. (2024). Random forests with parametric entropy-based information gains for classification and regression problems. *PeerJ Computer Science*. 2376-5992. *10*, e1775. doi:10.7717/peerj-cs.1775.

Ismail, M., Hassan, N., & Bafjaish, S. S. (2020). Comparative analysis of Naive Bayesian techniques in health-related for classification task. *Journal of Soft Computing and Data Mining*. 2716-621X. *1*(2), 1–10. Available from: https://publisher.uthm.edu.my/ojs/index.php/jscdm/article/download/7144/3932. doi:10.30880/jscdm.2020.01.02.001.

Janiesch, C., Zschech, P., & Heinrich, K. (2021). Machine learning and deep learning. *Electronic Markets*. 1019-6781. *31* (3), 685–695. doi:10.1007/s12525-021-00475-2.

Jankowsky, K., Krakau, L., Schroeders, U., Zwerenz, R., & Beutel, M. E. (2024). Predicting treatment response using machine learning: a registered report. *British Journal of Clinical Psychology*. 2044-8260. *63*(2), 137–155. Available from: http://onlinelibrary.wiley.com/journal/10.1111/(ISSN)2044-8260. doi:10.1111/bjc.12452.

Jayaraman, P., Desman, J., Sabounchi, M., Nadkarni, G. N., & Sakhuja, A. (2024). A primer on reinforcement learning in medicine for clinicians. *Npj Digital Medicine*. 2398-6352. *7*(1). Available from: https://www.nature.com/npjdigitalmed/. doi:10.1038/s41746-024-01316-0.

Kancherla, J. (2020). Re-identification of health data through machine learning. *SSRN Electronic Journal*. 1556-5068, 1–12. doi:10.2139/ssrn.3794927.

Karaca, Y., & Moonis, M. (2022). Shannon entropy-based complexity quantification of nonlinear stochastic process. In *Multi-Chaos, fractal and multi-fractional artificial intelligence of different complex systems* (pp. 231–245). United States: Elsevier. Available from: https://www.sciencedirect.com/book/9780323900324. doi:10.1016/B978-0-323-90032-4.00018-3.

Kawano, K., Otaki, Y., Suzuki, N., Fujimoto, S., Iseki, K., Moriyama, T., ... Konta, T. (2022). Prediction of mortality risk of health checkup participants using machine learning-based models: the J-SHC study. *Scientific Reports*. 2045-2322. *12*(1). doi:10.1038/s41598-022-18276-8.

Kecman, V. (2005). *Support vector machines—an introduction support vector machines: Theory and applications*. (pp. 1–47). Springer Berlin Heidelberg. doi:10.1007/10984697_1.

Khan, I. U., Aslam, N., Aljabri, M., Aljameel, S. S., Kamaleldin, M. M. A., Alshamrani, F. M., & Chrouf, S. M. B. (2021). Computational intelligence-based model for mortality rate prediction in COVID-19 patients. *International Journal of Environmental Research & Public Health*. 1660-4601. *18*(12). Available from: https://www.mdpi.com/1660-4601/18/12/6429/pdf. doi:10.3390/ijerph18126429.

Kim, J., Kim, H., Kim, H., Lee, D., & Yoon, S. (2025). A comprehensive survey of deep learning for time series forecasting: architectural diversity and open challenges. *Artificial Intelligence Review*. 1573-7462. *58*(7). doi:10.1007/s10462-025-11223-9.

Knox, W. B., & Stone, P. (2012). Reinforcement learning from simultaneous human and MDP reward. In *11th International Conference on Autonomous Agents and Multiagent Systems 2012, AAMAS 2012: Innovative Applications Track, vol. 1* (pp. 528–535). United States: International Foundation for Autonomous Agents and Multiagent Systems (IFAAMAS).

König, I. R., Fuchs, O., Hansen, G., von Mutius, E., & Kopp, M. V. (2017). What is precision medicine? *European Respiratory Journal*. 0903-1936. *50* (4), 1700391. doi:10.1183/13993003.00391-2017.

Kononenko, I., & Kukar, M (2007). *Measures for evaluating the quality of attributes*. (pp. 153–180). Elsevier BV. doi:10.1533/9780857099440.153.

Kotsiantis, S. B. (2007). Supervised machine learning: a review of classification techniques. *Informatica (Ljubljana)*, 0350-5596 *31*(3), 249–268.

Kumbure, M. M., & Luukka, P. (2022). A generalized fuzzy k-nearest neighbor regression model based on Minkowski distance. *Granular Computing*. 2364-4966. *7* (3), 657–671. doi:10.1007/s41066-021-00288-w.

Kyriazakos, S., Pnevmatikakis, A., Cesario, A., Kostopoulou, K., Boldrini, L., Valentini, V., & Scambia, G. (2021). Discovering composite lifestyle biomarkers with artificial intelligence from clinical studies to enable smart eHealth and digital therapeutic services. *Frontiers in Digital Health*, 2673-253X. *3*. doi:10.3389/fdgth.2021.648190.

Langarizadeh, M., & Moghbeli, F. (2016). Applying naive Bayesian networks to disease prediction: a systematic review. *Acta Informatica Medica*. 1986-5988. *24*(5), 364–369. Available from: http://www.ejmanager.com/mnstemps/6/6-1477983108.pdf?t=1478253390. doi:10.5455/aim.2016.24.364-369.

Le Corre, V., & Kremer, A. (2012). The genetic differentiation at quantitative trait loci under local adaptation. *Molecular Ecology*. 0962-1083. *21* (7), 1548–1566. doi:10.1111/j.1365-294x.2012.05479.x.

Li, J., Lin, S., Yu, K., & Guo, G. (2022). Quantum K-nearest neighbor classification algorithm based on Hamming distance. *Quantum Information Processing*. 1570-0755. *21*(1). doi:10.1007/s11128-021-03361-0.

Li, N., Chen, J., Cao, H., Zhang, B., & Natarajan, P. (2014). Applications of recurrent neural network language model in offline handwriting recognition and word spotting. In *Proceedings of International Conference on Frontiers in Handwriting Recognition, ICFHR* (pp. 134–139). United States: Institute of Electrical and Electronics Engineers Inc. 21676453. http://ieeexplore.ieee.org/xpl/mostRecentIssue.jsp?punumber=6423130. doi:10.1109/ICFHR.2014.30.

López, D. (2022). *Knowledge discovery and data mining for predictive analytics*. (pp. 271–286). Business Intelligence. doi:10.1007/978-3-030-89010-0_10.

López, O., López, A., & Crossa, J. (2022). *Fundamentals of artificial neural networks and deep learning multivariate statistical machine learning methods for genomic prediction*. (pp. 379–425). Springer International Publishing. doi:10.1007/978-3-030-89010-0_10.

Loshin, D. (2013). *Knowledge discovery and data mining for predictive analytics*. (pp. 271–286). Elsevier BV. doi:10.1016/b978-0-12-385889-4.00017-x.

Mahajan, P., Uddin, S., Hajati, F., & Moni, M. A. (2023). Ensemble learning for disease prediction: a review. *Healthcare*. 2227-9032. *11* (12), 1808. doi:10.3390/healthcare11121808.

Mandrekar, J. N. (2010). Receiver operating characteristic curve in diagnostic test assessment. *Journal of Thoracic Oncology*. 1556-1380. *5*(9), 1315–1316. Available from: https://www.journals.elsevier.com/journal-of-thoracic-oncology/. doi:10.1097/JTO.0b013e3181ec173d.

Mayerhoefer, M. E., Materka, A., Langs, G., Häggström, I., Szczypiński, P., Gibbs, P., & Cook, G. (2020). Introduction to radiomics. *Journal of Nuclear Medicine*. 2159-662X. *61*(4), 488–495. Available from: http://jnm.snmjournals.org/lookup/doi/10.2967/jnumed.118.222893. doi:10.2967/JNUMED.118.222893.

Meng, Y. A., Yu, Y., Cupples, L. A., Farrer, L. A., & Lunetta, K. L. (2009). Performance of random forest when SNPs are in linkage disequilibrium. *BMC Bioinformatics*. 1471-2105. *10* (1), 1–17. doi:10.1186/1471-2105-10-78.

Miron, V. D., Săndulescu, O., Streinu-Cercel, A., Florea, D., Paraschiv, S., Bănică, L., ... Drăgănescu, A. C. (2024). Age, comorbidity burden and late presentation are significant predictors of hospitalization length and acute respiratory failure in patients with influenza. *Scientific Reports*. 2045-2322. *14*(1), 1–10. Available from: https://www.nature.com/srep/. doi:10.1038/s41598-024-66550-8.

Mitchell, T. (1997). *Decision tree learning machine learning* (pp. 52–80). McGraw-Hill.

Mohsen, F., Ali, H., Hajj, N. E., & Shah, Z. (2022). Artificial intelligence-based methods for fusion of electronic health records and imaging data. *Scientific Reports*. 2045-2322. *12*(1). doi:10.1038/s41598-022-22514-4.

Monaghan, T. F., Rahman, S. N., Agudelo, C. W., Wein, A. J., Lazar, J. M., Everaert, K., & Dmochowski, R. R. (2021). Foundational statistical principles in medical research: a tutorial on odds ratios, relative risk, absolute risk, and number needed to treat. *International Journal of Environmental Research & Public Health*. 1660-4601. *18*(11). Available from: https://www.mdpi.com/1660-4601/18/11/5669/pdf. doi:10.3390/ijerph18115669.

Morris, L. G. T., Riaz, N., Desrichard, A., Senbabaoglu, Y., Hakimi, A. A., Makarov, V., ... Chan, T. A. (2016). Pan-cancer analysis of intratumor heterogeneity as a prognostic determinant of survival. *Oncotarget*. 1949-2553. *7*(9), 10051–10063. Available from: http://www.impactjournals.com/oncotarget/index.php?journal=oncotarget&page=article&op=download&path%5B%5D=7067&path%5B%5D=20095. doi:10.18632/oncotarget.7067.

Naidu, G., Zuva, T., & Sibanda, E. M. (2023). A review of evaluation metrics in machine learning algorithms. In *Lecture Notes in Networks and Systems, vol. 724* (pp. 15–25). South Africa: Springer Science and Business Media Deutschland GmbH. 23673389 https://www.springer.com/series/15179. doi:10.1007/978-3-031-35314-7_2.

Nguyen, T. D., Cherif, R., Mahieux, P. Y., Lux, J., Aït-Mokhtar, A., & Bastidas-Arteaga, E. (2023). Artificial intelligence algorithms for prediction and sensitivity analysis of mechanical properties of recycled aggregate concrete: a review. *Journal of Building Engineering*. 2352-7102. *66*. Available from: http://www.journals.elsevier.com/journal-of-building-engineering/. doi:10.1016/j.jobe.2023.105929.

Nong, P., Adler-Milstein, J., Apathy, N. C., Holmgren, A. J., & Everson, J. (2025). Current use and evaluation of artificial intelligence and predictive models in US hospitals. *Health Affairs*. 1544-5208. *44*(1), 90–98. Available from: https://www.healthaffairs.org/doi/full/10.1377/hlthaff.2024.00842. doi:10.1377/hlthaff.2024.00842.

Obi, J. C., & Jecinta, I. C. (2023). A review of techniques for regularization. *International Journal of Research in Engineering and Science*, *11*(1), 360–367.

Olender, R. T., Roy, S., & Nishtala, P. S. (2023). Application of machine learning approaches in predicting clinical outcomes in older adults – a systematic review and meta-analysis. *BMC Geriatrics*. 1471-2318. *23*(1). Available from: https://bmcgeriatr.biomedcentral.com/. doi:10.1186/s12877-023-04246-w.

Oshiro, T. M., Perez, P. S., & Baranauskas, J. A. 2012, How many trees in a random forest? In: Lecture notes in computer science (including subseries lecture notes in artificial intelligence and lecture notes in bioinformatics), Vol. 7376 Brazil, pp. 154–168. 16113349. doi:10.1007/978-3-642-31537-4_13.

Padakandla, S. (2022). A survey of reinforcement learning algorithms for dynamically varying environments. *ACM Computing Surveys*. 0360-0300. *54* (6), 1–25. doi:10.1145/3459991.

Park, J. E., Kim, D., Kim, H. S., Park, S. Y., Kim, J. Y., Cho, S. J., ... Kim, J. H. (2020). Quality of science and reporting of radiomics in oncologic studies: room for improvement according to radiomics quality score and TRIPOD statement. *European Radiology*. 1432-1084. *30*(1), 523–536. Available from: www.link.springer.de/link/service/journals/00330/index.htm. doi:10.1007/s00330-019-06360-z.

Pavuluri, S., Sangal, R., Sather, J., & Taylor, R. A. (2024). Balancing act: the complex role of artificial intelligence in addressing burnout and healthcare workforce dynamics. *BMJ Health and Care Informatics*. 2632-1009. *31*(1). Available from: https://informatics.bmj.com/content/by/year. doi:10.1136/bmjhci-2024-101120.

Prasatha, V. S., Alfeilate, H. A. A., Hassanate, A. B., Lasassmehe, O., Tarawnehf, A. S., Alhasanatg, M. B., & Salmane, H. S. E. (2017). Effects of distance measure choice on knn classifier performance-a review. *arXiv*, *56*(24), 1–39.

Probst, P., Wright, M. N., & Boulesteix, A. (2019). Hyperparameters and tuning strategies for random forest. *WIREs: Data Mining and Knowledge Discovery*. 1942-4787. *9*(3). doi:10.1002/widm.1301.

Qamar, R., & Zardari, B. A. (2023). Artificial neural networks: an overview. *Mesopotamian Journal of Computer Science*.2958-6631. *2023*, 124–133. Available from: https://mesopotamian.press/journals/index.php/cs/article/download/118/121. doi:10.58496/MJCSC/2023/015.

Raj, R., & Kos, A. (2025). An extensive study of convolutional neural networks: applications in computer vision for improved robotics perceptions. *Sensors*. 1424-8220. *25* (4), 1033. doi:10.3390/s25041033.

Rane, N. L., Rane, J., Paramesha, M., & Kaya, Ö. (2024). *Emerging trends and future research opportunities in artificial intelligence, machine learning, and deep learning*. Deep Science Publishing. doi:10.70593/978-81-981271-1-2_6.

Rao Jetti, C., Shaik, R., & Shaik, S. (2021). Disease prediction using Naïve Bayes—machine learning algorithm. *International Journal of Science and Healthcare Research*. 2455-7587. *6* (4), 17–22. doi:10.52403/ijshr.20211004.

Rayhan, S. B., Rahman, M. M., Sultana, J., & Varga, G. (2025). Predicting the elastic moduli of unidirectional composite materials using deep feed forward neural network. *Journal of Composites Science*. 2504-477X. *9*(6). Available from: www.mdpi.com/journal/jcs. doi:10.3390/jcs9060278.

Razzak, M. I., Naz, S., & Zaib, A. (2017). *Deep learning for medical image processing: overview, challenges and the future* (Vol. 26, pp. 323–350). Springer Science and Business Media LLC. doi:10.1007/978-3-319-65981-7_12.

Reps, J. M., Ryan, P., & Rijnbeek, P. R. (2021). Investigating the impact of development and internal validation design when training prognostic models using a retrospective cohort in big US observational healthcare data. *BMJ Open*. 2044-6055. *11*(12). Available from: http://bmjopen.bmj.com/content/early/by/section. doi:10.1136/bmjopen-2021-050146.

Richardson, S., & Gilks, W. R. (1993). A Bayesian approach to measurement error problems in epidemiology using conditional independence models. *American Journal of Epidemiology*. 0002-9262. *138* (6), 430–442. doi:10.1093/oxfordjournals.aje.a116875.

Romanik, K., & Vitter, J. S. (1996). Using Vapnik-Chervonenkis dimension to analyze the testing complexity of program segments. *Information and Computation*. 0890-5401. *128*(2), 87–108. Available from: http://www.elsevier.com/inca/publications/store/6/2/2/8/4/4/index.htt. doi:10.1006/inco.1996.0065.

Rubin, J., Parvaneh, S., Rahman, A., Conroy, B., & Babaeizadeh, S. (2018). Densely connected convolutional networks for detection of atrial fibrillation from short single-lead ECG recordings. *Journal of Electrocardiology*. 0022-0736. *51* (6), S18. doi:10.1016/j.jelectrocard.2018.08.008.

Sala, E., Mema, E., Himoto, Y., Veeraraghavan, H., Brenton, J. D., Snyder, A., ... Vargas, H. A. (2017). Unravelling tumour heterogeneity using next-generation imaging: radiomics, radiogenomics, and habitat imaging. *Clinical Radiology*. 0009-9260. *72* (1), 3–10. doi:10.1016/j.crad.2016.09.013.

Salman, H. A., Kalakech, A., & Steiti, A. (2024). Random forest algorithm overview. *Babylonian Journal of Machine Learning*.3006-5429. *2024*, 69–79. Available from: https://mesopotamian.press/journals/index.php/BJML/article/download/417/289. doi:10.58496/BJML/2024/007.

Salvador–Meneses, J., Ruiz–Chavez, Z., & Garcia–Rodriguez, J. (2019). Compressed kNN: k-nearest neighbors with data compression. *Entropy*. 1099-4300. *21* (3), 234. doi:10.3390/e21030234.

Sepúlveda-Fontaine, S. A., & Amigó, J. M. (2024). Applications of entropy in data analysis and machine learning: a review. *Spain Entropy*. 1099-4300. *26*(12). Available from: https://www.mdpi.com/journal/entropy. doi:10.3390/e26121126.

Sharma, H., & Kumar, S. (2016). A survey on decision tree algorithms of classification in data mining. *International Journal of Science and Research (IJSR)*. 2319-7064. *5* (4), 2094–2097. doi:10.21275/v5i4.NOV162954.

Shen, F. X., Wolf, S. M., Lawrenz, F., Comeau, D. S., Dzirasa, K., Evans, B. J., ... Garwood, M. (2024). Ethical, legal, and policy challenges in field-based neuroimaging research using emerging portable MRI technologies: guidance for investigators and for oversight. *Journal of Law and the Biosciences*. 2053-9711. *11*(1). Available from: https://academic.oup.com/jlb. doi:10.1093/jlb/lsae008.

Shreffler, J., & Huecker, M. (2020). Diagnostic testing accuracy: sensitivity, specificity, predictive values and likelihood ratios Europe PMC, 1–5.

Strehl, A. L., Li, H., & Littman, M. L. (2009). Reinforcement learning in finite MDPs: PAC analysis. *Journal of Machine Learning Research*, 1533-7928 *10*, 2413–2444. Available from:. http://jmlr.csail.mit.edu/papers/volume10/strehl09a/strehl09a.pdf.

Sutton, R., & Barto, A. (2015). *The reinforcement learning problem reinforcement learning: An introduction. 2* (pp. 1–26). The MIT Press.

Suyal, M., & Goyal, P. (2022). A review on analysis of K-nearest neighbor classification machine learning algorithms based on supervised learning. *International Journal of Engineering Trends and Technology*. 2231-5381. *70* (7), 43–48. doi:10.14445/22315381/ijett-v70i7p205.

Szymczak, S., Holzinger, E., Dasgupta, A., Malley, J. D., Molloy, A. M., Mills, J. L., ... Bailey-Wilson, J. E. (2016). r2VIM: a new variable selection method for random forests in genome-wide association studies. *BioData Mining*. 1756-0381. *9*(1). doi:10.1186/s13040-016-0087-3.

Torres-Velázquez, M., Chen, W. J., Li, X., & McMillan, A. B. (2021). Application and construction of deep learning networks in medical imaging. *IEEE Transactions on Radiation and Plasma Medical Sciences*, *5*(2), 137–159.

VanHouten, J. P., Starmer, J. M., Lorenzi, N. M., Maron, D. J., & Lasko, T. A. (2014). Machine learning for risk prediction of acute coronary syndrome. *Annual Symposium Proceedings/AMIA Symposium*, 1942-597X *2014*, 1940–1949.

Vapnik, V. N. (1999). An overview of statistical learning theory. *IEEE Transactions on Neural Networks*. 1045-9227. *10*(5), 988–999. doi:10.1109/72.788640.

Vapnik, V. (1998). *The support vector method of function estimation*. (pp. 55–85). Springer Science and Business Media LLC. doi:10.1007/978-1-4615-5703-6_3.

Varnosfaderani, S. M., & Forouzanfar, M. (2024). The role of AI in hospitals and clinics: transforming healthcare in the 21st century. *Bioengineering*. 2306-5354. *11* (337), 1–38. doi:10.3390/bioengineering11040337.

Wang, S., & Summers, R. M. (2012). Machine learning and radiology. *Medical Image Analysis*. 1361-8423. *16* (5), 933–951. doi:10.1016/j.media.2012.02.005.

Webb, C. A., Cohen, Z. D., Beard, C., Forgeard, M., Peckham, A. D., & Björgvinsson, T. (2020). Personalized prognostic prediction of treatment outcome for depressed patients in a naturalistic psychiatric hospital setting: a comparison of machine learning approaches. *Journal of Consulting and Clinical Psychology*. 1939-2117. *88*(1), 25–38. Available from: http://www.apa.org/pubs/journals/ccp/. doi:10.1037/ccp0000451.

Williams, K. S. (2024). Evaluations of artificial intelligence and machine learning algorithms in neurodiagnostics. *Journal of Neurophysiology*. 1522-1598. *131*(5), 825–831. Available from: https://journals.physiology.org/toc/jn/131/5. doi:10.1152/jn.00404.2023.

Winters-Hilt, S., & Merat, S. (2007). SVM clustering. *BMC Bioinformatics*. 1471-2105. *8*(S7). doi:10.1186/1471-2105-8-s7-s18.

Wolpert, D. H., & Macready, W. G. (1997). No free lunch theorems for optimization. *IEEE Transactions on Evolutionary Computation*. 1089-778X. *1* (1), 67–82. doi:10.1109/4235.585893.

Yang, X. S., He, X. S., & Fan, Q. W. (2020). *Mathematical framework for algorithm analysis nature-inspired computation and swarm intelligence: Algorithms, theory and applications* (pp. 89–108). United Kingdom: Elsevier. Available from: https://www.sciencedirect.com/book/9780128226094. doi:10.1016/B978-0-12-819714-1.00017-8.

Yu, C., Liu, J., Nemati, S., & Yin, G. (2022). Reinforcement learning in healthcare: a Survey. *ACM Computing Surveys*. 1557-7341. *55*(1). Available from: http://dl.acm.org/citation.cfm?id=J204. doi:10.1145/3477600.

Yuan, J.-H., Jin, Y.-M., Xiang, J.-Y., Shuang-Shuang, L., Zhong, Y.-X., Zhang, S.-L., & Zhao, B. (2025). Machine learning-based prediction of postoperative mortality risk after abdominal surgery. *World Journal of Gastrointestinal Surgery*. 1948-9366. *17* (4), 1–13. doi:10.4240/wjgs.v17.i4.103696.

Zhang, M., Zheng, Y., Maidaiti, X., Liang, B., Wei, Y., & Sun, F. (2024). Integrating machine learning into statistical methods in disease risk prediction modeling: a systematic review. *Health Data Science*. 2765-8783. *4*, 1–12. doi:10.34133/hds.0165.

Zhang, X., Zhu, J., Ye, B., Yashengaili, A., Xu, L., & Li, X. (2020). Combination of epidural blockade and parecoxib in enhanced recovery after gastrointestinal surgery. *Journal of Investigative Surgery*, 1521-0553, 1–5. Available from: http://www.tandfonline.com/loi/iivs20. doi:10.1080/08941939.2019.1690601.

Zhang, Y., & Hong, J. (2023). Challenges of deep learning in cancers. *Technology in Cancer Research & Treatment*. 1533-0346. *22*, 1–2. doi:10.1177/15330338231173495.

Zhang, Z. (2016). Introduction to machine learning: k-nearest neighbors. *Annals of Translational Medicine*, *4*(11), 1–7.

Zou, H., & Hastie, T. (2005). Regularization and variable selection via the elastic net. *Journal of the Royal Statistical Society Series B: Statistical Methodology*. 1369-7412. *67* (2), 301–320. doi:10.1111/j.1467-9868.2005.00503.x.

Chapter 3

State of the art artificial intelligence assisted disease detection tools

Sushanta Kumar Das[a], Rahul Mishra[b], Amit Samanta[c], Saumendu Deb Roy[d], and Ujjwal Sahoo[e]
[a]*Department of Pharmacy Practice, Mata Gujri College of Pharmacy, Mata Gujri University, Kishanganj, Bihar, India*
[b]*Celerion IMC, Lincoln, NE, United States*
[c]*Mata Gujri College of Pharmacy, Mata Gujri University, Kishanganj, Bihar, India*
[d]*Dean Academics (Pharmacy & AHS), Maat Gujri University, Kishanganj, Bihar, India*
[e]*Pharmaceutical Chemistry, Dr. B. C. Roy College of Pharmacy & AHS, Durgapur, West Bengal, India*

Chapter highlights:

- **More Accurate Diagnoses:** AI excels at identifying sensitive patterns in data, leading to more accurate diagnoses.
- **Early Disease Detection:** Early detection through AI enables timely treatment and improves patient outcomes.
- **Increased Efficiency:** AI streamlines diagnostics by automating tasks and reducing human errors.
- **Data-Driven Insights:** AI leverages big data to provide valuable insights into diseases and patient populations.
- **Addressing Challenges:** Responsible AI implementation requires addressing bias, explainability, and data privacy concerns.

1 Introduction

The obligation or burden associated with diseases urgently requires innovative diagnostic approaches to pinpoint the exact disease condition, facilitating enhanced patient outcomes and reducing the global healthcare burden both financially and otherwise (GBD Diseases and Injuries Collaborators, 2021; Murray, 2022). Conventional diagnostic techniques primarily rely on professional expertise along with manual analysis of raw data. This approach generally suffers from scale limitations, lack of speed, and, most importantly, inconsistency (Henry, Hager, Pronovost, & Saria, 2015; Ng and Jordan, 2002; Patel et al., 2023). Fortunately, there is a widespread change in the current situation with the innovation and introduction of artificial intelligence (AI)-assisted disease detection tools, which have brought a revolution in the medical field by closing the gap through computational abilities to analyze extremely large data sources, detect exact patterns, and provide enhanced decision support to clinicians for the accurate diagnosis of diseases (Khan, Islam, Jan, Din, & Rodrigues, 2022; Obermeyer & Emanuel, 2016; Zhang et al., 2024). AI technologies are poised to redefine the entire medical diagnostic arena with their computational and algorithmic models. Precisely detecting malignancies from radiographic images and identifying cardiovascular complications in vulnerable individuals will be the next step in this process (Choi et al., 2022; Hosny, Parmar, Quackenbush, Schwartz, & Aerts, 2023; Rajpurkar, Chen, Banerjee, & Topol, 2022). Considering the example of India, a country with a population of over 1.4 billion and yet to achieve a global standard for the patient–doctor ratio, India has taken significant steps to integrate and apply AI technologies in its healthcare system. This has led to a marked revolution in healthcare delivery for its uniquely diverse population through real-world AI applications that are easily accessible, rational, and economical for its people (Das, Dasgupta, Roy, & Shil, 2024; Sharma et al., 2023). Even with all the advantages of AI, ethical considerations remain a major roadblock to its widespread application. A fair, transparent, and explainable AI

Applications of Artificial Intelligence in Healthcare. https://doi.org/10.1016/B978-0-443-43934-6.00021-3

(XAI) application, coupled with a rational ethical framework designed to protect data privacy, will help remove obstacles and build trust among healthcare professionals, patients, and all stakeholders (Guleria, Krishnan, Sharma, Chen, & Smith, 2024). With this chapter, we aim to:

1. Showcase the description of the fundamentals used by AI methodologies that will be applied to detect diseases.
2. Highlight sophisticated AI applications across various medical specialties.
3. Analyze various challenges while applying AI technologies in a widespread manner.
4. Estimate the progress of AI-assisted disease detection tools in enhancing patient care and improving personalized therapeutic approaches.

As of early 2025, AI technologies are expanding rapidly with the proper and justified support of better algorithm design, advanced infrastructure, and the availability of large datasets. These approaches are creating a new dawn for the transformation of global healthcare delivery and, in particular, disease diagnosis with high precision (Johnson et al., 2025; Topol, 2019).

2 Fundamentals of AI technologies in disease detection

AI-assisted disease detection tools utilize a variety of enhanced computational methods to analyze an extremely large amount of medical information, a wide range of medical images, comprehensive electronic health records (EHRs), genome profiles of patients, and clinical narratives obtained from doctors. AI tools perform in a manner that assists in the immediate identification, classification, and prediction of diseases. This approach enhances the clinical decision-making capacity, ultimately leading to improved patient outcomes. This section explains the basic mechanisms directing the system, proper applications, certain roadblocks, and important developments currently occurring in this area of research.

2.1 Machine learning

Machine learning (ML) supports the overall system in recognizing precise patterns and predicting them accordingly. ML further enables the system to discover patterns and predict outcomes from the given data through iterative learning. For the disease detection process, ML models are focused on labeled data to specify the diagnostic characteristics. Table 3.1 provides a detailed description of ML.

2.2 Deep learning

Deep learning (DL) is considered a progressive subset of ML. It utilizes multiple layers of neural networks to process critical, high-standard data, such as medical images or longitudinal signals. Its classification capacity even mimics human intellectual capabilities. Table 3.2 provides a detailed description of DL.

TABLE 3.1 Description of machine learning.

Supervised learning	Utilize marked data (e.g., mammograms tagged as cancerous or benign) to train algorithms like support vector machines (SVMs) or random forests. Perform multiple tasks superbly, such as classifying diabetic retinopathy from retinal images or predicting cardiovascular events from EHRs (Deo, 2015; Rajkomar, Dean, & Kohane, 2019).
Unsupervised learning	Analyzes unmarked data and uncovers hidden formations, such as clustering genomic profiles to identify novel disease subtypes in oncology (Ng and Jordan, 2002).
Applications	Excellent efficacy in detecting sepsis from vital signs (Henry et al., 2015), pneumonia from chest radiographs (CheXNet, 2017) and identifying diabetes risk factors (Kavakiotis et al., 2017).
Limitations	Rely on high-quality, predefined data. Identification of rare diseases is still a challenge, and forceful pattern fitting reduces outcomes (Obermeyer & Emanuel, 2016).
Significance	Faster adaptation to changing disease patterns (e.g., COVID-19) highlights its utility in the modern, fast-moving healthcare scenario (Wynants et al., 2020).

TABLE 3.2 Description of deep learning.

Convolutional neural networks (CNNs)	A type of deep learning algorithm is intended to analyze medical images by mimicking the human visual system. It works via multiple layer applications: convolutional layers, activation functions (e.g., ReLU), pooling layers, and fully connected layers to extract characteristics in a graded manner. Convolutional layers employ filters to identify low-level features (e.g., edges) in the input image, which are combined in deeper layers to recognize complex patterns (e.g., shapes or objects). Pooling layers decrease spatial dimensions, improving computational efficiency, while the final layers categorize the data based on learned features. In radiology, CNNs analyze X-rays or MRIs to detect irregularities like tumors or fractures Rajpurkar et al., 2018). In pathology, they categorize whole slide images for cancer detection, identifying cellular patterns with high accuracy (Litjens et al., 2017). Other applications include independent driving (object recognition), facial recognition (identity verification), and natural language processing (image captioning), showcasing CNNs' versatility across medical disciplines (Esteva et al., 2017).
Recurrent neural networks (RNNs)	Fitted for temporal data, RNN analyzes the progression of ECG to highlight cardiac failure possibilities (Choi, Bahadori, Schuetz, & Stewart, 2016).
Applications	FDA-approved IDx-DR is utilized to screen diabetic retinopathy (Abràmoff, Lavin, Birch, Shah, & Folk, 2018) and enhance Alzheimer's detection via neuroimaging (Wen et al., 2020). Current developments show that combining deep learning (DL) with multimodal imaging enhances neurological disease detection (Smith et al., 2024).
Limitations	High computational demands and complexity (black-box nature), poor clinical trust, and lack of a regulatory framework (Vellido, 2020).
Significance	Accuracy in image-based diagnosis, though effective, demands large resources, which has restricted its application in economically disadvantaged settings (Topol, 2019).

TABLE 3.3 Description of natural language processing.

Mechanisms	Named Entity Recognition detects issues (e.g., hypertension), while contextual models [e.g., Bidirectional Encoder Representations from Transformers (BERT)] translate characterized language in EHRs (Devlin, Chang, Lee, & Toutanova, 2018).
Applications	Physicians' narrative analyses help identify missed diagnoses and assist in preliminary assessments through complaints and the detection of signs reported in journals for the identification of rare clinical conditions (Beam & Kohane, 2018; Jensen, Jensen, & Brunak, 2012). They also aid in extracting information from unstructured EHRs that require medical attention and management (Lee, 2023).
Limitations	Use of unclassified clinical terminology and documentation processes (Neveol & Zweigenbaum, 2018).
Significance	Perform variable approaches in incomplete and unstructured data situations (Friedman, Shagina, Lussier, & Hripcsak, 2004).

2.3 Natural language processing

Natural language processing (NLP) facilitates the extraction of structured perceptions from unstructured notes. It analyzes clinical notes or previously published scientific literature to improve diagnostic dynamics. Table 3.3 provides a detailed description of NLP.

2.4 Computer vision

This process helps the system analyze observable responses, primarily in medical images, using methods such as segmentation and object identification. Table 3.4 provides a detailed description of computer vision.

TABLE 3.4 Description of computer vision.

Mechanisms	Various algorithms describe anatomical structures or abnormalities (e.g., tumors in breast mammograms) using threshold detection and shape analysis (Litjens et al., 2017).
Application	Supports in the diagnosis of breast carcinoma (Dhungel, Carneiro, & Bradley, 2015), TB from chest X-rays (Liu et al., 2020) and detecting abnormal gastrointestinal pathologies via endoscopy (Mori, Neumann, Misawa, Kudo, & Bretthauer, 2021).
Limitations	Lack of image sensitivity and standardization in various imaging types creates bias (Wang & Summers, 2012).
Significance	Automatic optical analysis enhances the diagnostic process and minimizes human errors (Hosny, Parmar, Quackenbush, Schwartz, & Aerts, 2018).

TABLE 3.5 Description of data mining and pattern recognition.

Mechanisms	Correlate with rules that represent specific complications for clinical conditions (e.g., fever and rash due to measles), and point out abnormalities in the given data (e.g., abnormal laboratory values) (Fayyad, Piatetsky-Shapiro, & Smyth, 1996).
Applications	Detect the onset of septicemia (Nemati et al., 2018), identify rare hereditary diseases (Kohler et al., 2019), and observe disease spread patterns (Brownstein, Freifeld, & Madoff, 2009).
Limitations	Clutter in real-world data and the risk of false positives reduce acceptability (Han, Kamber, & Pei, 2011).
Significance	Identify minute correlations, improve early medical intervention, but unjustified interpretation remains a deadlock (Bellazzi & Zupan, 2008).

TABLE 3.6 Description of reinforcement learning.

Mechanisms	Performs by improving diagnosis strategies, such as highlighting laboratory tests or modifying imaging processes based on patient information (Sutton & Barto, 2018).
Applications	Enhances effective diagnosis of uncertain cases and overall increases the actual diagnostic approach (Gottesman et al., 2019).
Limitations	Due to the complex analytical process attracts poor authentication and acceptance (Yu et al., 2021).
Significance	Depends on flexible improvement and demonstrates a promising position compared to other existing technologies (Komorowski, Celi, Badawi, Gordon, & Faisal, 2018).

2.5 Data mining and pattern recognition

Data mining and pattern recognition methods extract relevant signals from an extremely large set of patient information to detect disease-related characteristics. Table 3.5 provides a detailed description of data mining and pattern recognition.

2.6 Reinforcement learning

Reinforcement learning (RL) trains the system to achieve maximum enhanced results through a successive trial-and-error method to precipitate the actual findings. Table 3.6 provides a detailed description of RL.

2.7 Integration and synergistic applications

In this approach, two or more methods combine to achieve higher success in diagnosis. For example, detection of lung cancer may integrate computer vision (for analysis of images), DL (to extract the exact features), and NLP (for the creation of reports). A similar example can be DeepMind from Google, which provides great competition for ophthalmologists in retinal diagnostics (De Fauw et al., 2018). Table 3.7 summarizes the core AI technologies in disease detection.

TABLE 3.7 Core AI technologies in disease detection.

Technology	Primary role	Key application	Strength	Limitation
Machine learning (ML)	Pattern recognition	Sepsis prediction	Adaptability	Data dependency
Deep learning (DL)	Complex data processing	Skin cancer diagnosis	High accuracy	Resource intensity
Natural language processing (NLP)	Text analysis	EHR-based diagnostics	Unstructured data use	Terminology variability
Computer vision	Visual interpretation	Mammogram screening	Speed	Image quality reliance
Data mining	Insight extraction	Outbreak detection	Subtle pattern detection	Noise sensitivity
Reinforcement learning (RL)	Decision optimization	Diagnostic workflow enhancement	Adaptive learning	Limited validation

2.8 Challenges with technologies and future directions

Even though these methods exhibit great promise, they are not without certain barriers:

Data quality and bias	Unreliable or biased data (e.g., minimal or lower interpretation of marginal groups) raises credibility regarding equity and accuracy (Kaushal, Altman, & Langlotz, 2020).
Interpretability	Unclear and unjustified methods, especially in deep learning, negatively impact the trust and acceptability of the diagnostic results by clinicians (Holzinger, Langs, Denk, Zatloukal, & Muller, 2019a).
Scalability	Due to the high cost of maintenance and analysis makes it less adoptable in low-resource situations (Chen & Asch, 2017).
Ethics and privacy	Sensitive health data handling remains a significant concern regarding data security and safety breaches affecting individuals (Price & Cohen, 2019). In addition, challenges such as ensuring equality in AI predictions among diverse populations remain unanswered (Rodrigues, 2020).

3 Applications of AI in disease detection

Application of AI has revolutionized disease diagnosis through the integration of advanced computer technology with real-world clinical practice, overcoming traditional diagnostic hurdles such as bias, extended time durations, and unequal access to healthcare services.

With the introduction of ML and DL, AI tools can analyze a huge amount of patient-related information to improve the diagnosis process by increasing accuracy, rationalizing the work, and ultimately enhancing treatment outcomes. This approach will certainly align with the Global Burden of Disease framework's goal, to reduce morbidity and mortality. This section highlights and elaborates on the clinical and technical perspectives toward the application of sophisticated AI technologies across various medical specialties such as radiology, pathology, oncology, cardiology, infectious diseases, psychiatry, and neurology, emphasizing their impact on healthcare delivery and computational fundamentals.

3.1 Radiology

The application of AI in the radiology process has entirely transformed diagnostic accuracy and patient care scenarios with the integration of convolutional neural networks (CNNs) into medical procedures. This application has enabled rapid diagnosis for the early identification of minute abnormalities present in various radiographic images, such as X-rays, computed tomography (CT) scans, and magnetic resonance images (MRIs). A real-world example of such an application is Aidoc, a tool that prioritizes various clinical identifications, such as bleeding manifestations in the intracranial region and the development of emboli in the pulmonary system. This approach reduces the time required for the diagnostic process by up to 30% and facilitates the initiation of timely treatment procedures, especially for conditions like stroke and cancer, where every moment is critical (Smith et al., 2023). This approach is expected to further enhance emergency department services in rural and remote areas, where there is an estimated global shortage of expert radiologists and limited access to specialized opinions (World Health Organization, 2022b). However, this method has the potential to completely transform the situation by enabling enhanced and timely diagnoses through AI-embedded tools. Beyond detecting extremely small changes in radiographic images, AI can also minimize false negatives, thereby streamlining patient safety and supporting doctors in high-precision diagnostic situations (Rajpurkar et al., 2018). On a technical perspective, these methods involve DL models trained on sizable image datasets, achieving a success rate of over 90% through transfer learning and attention mechanisms that primarily target medically significant body parts (Topol, 2019b). Certain AI platforms, such as Zebra Medical Vision, combine picture archiving and communication systems (PACSs) with federated learning (Fig. 3.1) to focus on global patient data while maintaining privacy safeguards (Zebra Medical Vision, 2023). This combination of medical efficacy and computational effectiveness will make AI an indispensable entity in modern radiology. It is poised to address the global health gap and optimize the proper use of resources (Hosny et al., 2018).

3.2 Pathology

Integration of AI in pathology improves the diagnosis of cancer and directs treatment options earlier by combining DL with patient requirements before the situation becomes critical. This approach aims to minimize the global cancer burden, which accounts for approximately 19 million new cases each year (Sung et al., 2021). Certain tools, such as PathAI, regulate the assessment of whole-slide images (WSIs) (Fig. 3.2), decrease interobserver variability by 10%, and offer pathologists quantitative measurements, such as microenvironment analysis of tumors (PathAI, 2023). In this pathological detection approach, CNNs analyze high-resolution WSIs using supervised learning and attention mechanisms to identify disease subtypes and detect minute alterations in histopathological structures, thereby enhancing accuracy (Litjens et al., 2017). This process guides the establishment of appropriate treatment decisions for breast and prostate cancer by analyzing histopathological and genomic data for therapy personalization. Unfortunately, this approach is limited in situations where resources are minimal and the caseload is extremely high. Certain methods can be adopted to overcome such issues, such as the application of generative adversarial networks to improve image quality in resource-limited laboratories (Goodfellow et al., 2014), the inclusion of genomic data in the analysis process (Malone, Oliva, Sabatini, Stockley, & Siu, 2020), and

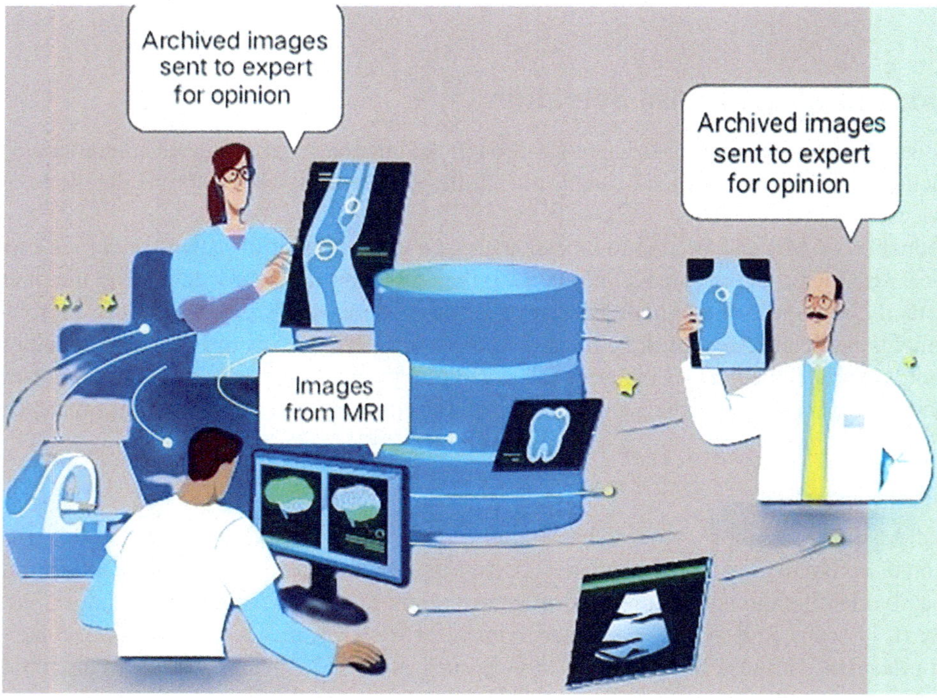

FIG. 3.1 Picture archiving and communication system (PACS) is a medical imaging technology that enables the storage, retrieval, distribution, and display of medical images, such as X-rays, MRIs, and CT scans, within a healthcare network system. PACS allows radiologists and clinicians to access patient images remotely, rationalize workflow, and improve diagnostic precision. PACS reduces dependence on physical film, supports real-time collaboration, and enhances patient care efficiency (Huang, 2010).

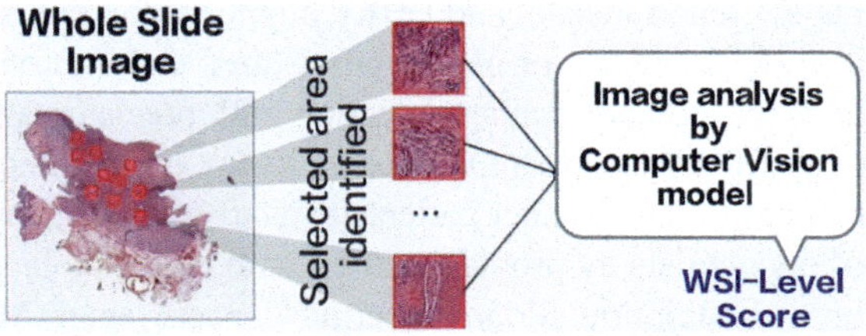

FIG. 3.2 AI-fitted whole-slide image (WSI) analysis employs machine learning to process high-resolution images of tissues or liquid biopsy samples, accurately detecting and scoring pathological features such as malignant cells or biomarkers.

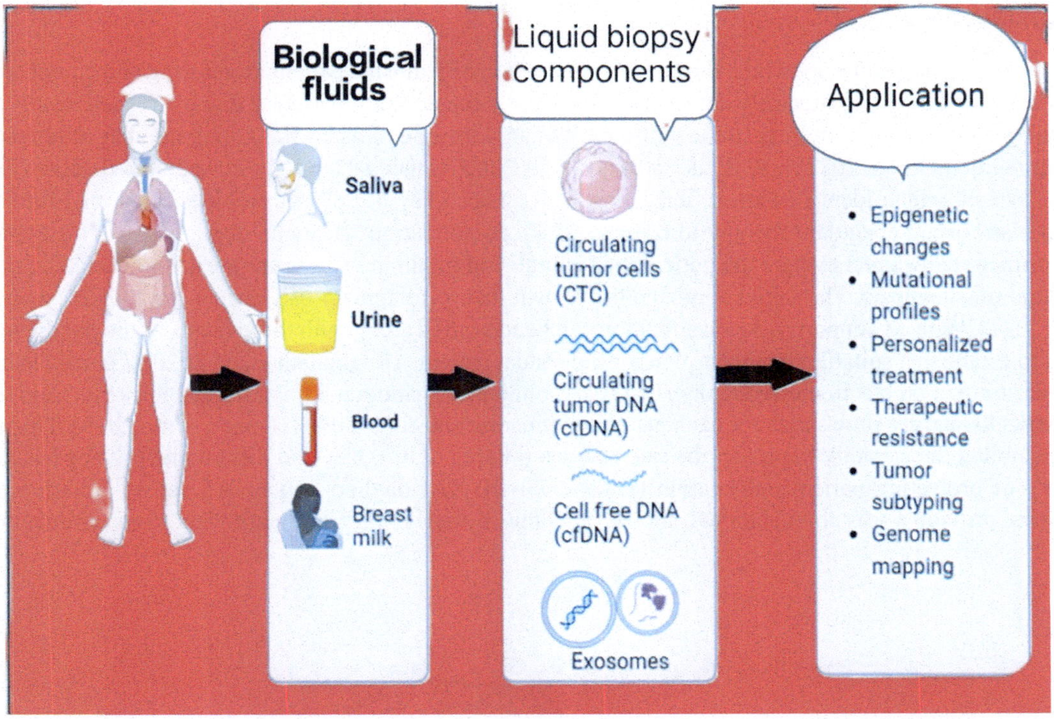

FIG. 3.3 AI-based liquid biopsy methods detect biomarkers (e.g., ctDNA) present in the blood for noninvasive cancer diagnosis and monitoring. This method includes the collection of samples, isolation of biomarkers, sequencing of biomarkers, and abnormal pattern recognition for the early identification of cancer.

collaboration between cloud-based platforms. These strategies will support advanced data resources to enhance precision pathology (Kohane et al., 2021). This combination of clinical expertise and computational algorithms will undoubtedly ease workloads and improve diagnostic accuracy.

3.3 Oncology

AI works in cancer diagnosis by integrating multimodal data, which helps in the early detection of cancer to facilitate personalized treatment plans for millions of cancer patients diagnosed annually. In clinical applications, AI-based liquid biopsy (Fig. 3.3) identifies circulating tumor DNA (ctDNA) through a noninvasive method, leading to less patient discomfort and improved monitoring of disease progression (Cohen et al., 2018). AI supports liquid biopsy by employing advanced algorithmic models to analyze intricate biological data obtained from various body fluids (e.g., blood). This method processes extremely large datasets via techniques like next-generation sequencing or polymerase chain reaction to identify ctDNA, circulating tumor cells, or other biomarkers with high sensitivity and specificity. ML models identify delicate patterns and mutations, enabling early cancer detection even at low concentrations, while DL forecasts disease development or treatment response by comparing data against large reference databases (Wan et al., 2020). This is especially important for patients with high-mortality cancers like pancreatic cancer (Choi et al., 2022). With the support of DL technology, this system analyzes images, genomic data, and EHRs, supported by radiomics models, to extract an uncountable number of image features to identify the behavior of the tumor [area under curve (AUC) 0.80–0.90]. Liquid biopsy algorithms deploy sequence configuration and abnormality detection with precision above 85% (Cohen et al., 2018). One such application has been introduced by IBM Watson for Oncology (WFO); it is a clinical decision support system intended to assist oncologists in making treatment plans for cancer patients by analyzing large numbers of medical publications and clinical data. It supports the work of oncologists by providing them with evidence-based, personalized treatment options. WFO was developed in collaboration with Memorial Sloan Kettering Cancer Center (MSK) and has been trained on a wide range of cancers, including lung, breast, and colorectal cancer, among others (IBM Watson Health, 2022). With such support from XAI, it ultimately improves clinical trustworthiness. Together with strong data support to handle diverse inputs, AI will become an important part of precision oncology practice.

3.4 Cardiology

The application of AI in cardiology works through a combination of real-time patient monitoring and computational analysis of monitoring parameters to precisely detect cardiovascular disease. Cardiovascular disease-related deaths are the highest among all disease-related mortality in the world each year, with an estimated staggering number of around 18 million (World Health Organization, 2022a). With the inclusion of AI intelligence technology in wearable devices (Fig. 3.4) such as the Apple Watch, which identifies atrial fibrillation with an accuracy of 98%, it provides a justified finding to enable clinicians to plan for anticoagulant therapy to prevent stroke complications (Hannun et al., 2019). This approach can be initiated in primary care centers where the patient load is high and minimal experts are present to check the patients, leading to potential misdiagnosis. The initial approach has shown that misdiagnosis can be reduced by 12% with this method (Attia et al., 2021). With AI support, risk classification can be identified along with the analysis of the patient's lifestyle and genetic data to establish a suitable treatment plan for individual patients (Raghunath et al., 2020). On the technical details, DL models are trained on electrocardiography (ECG) and photoplethysmography data (e.g., PhysioNet) and use recurrent neural networks to analyze time series for each patient to generate real-time alerts (Hannun et al., 2019). This combination model has exhibited the capacity to predict the risk of heart failure (AUC 0.87), and differential privacy has demonstrated the capability of protecting patient information (Dwork, 2019). XAI methods, such as SHapley Additive exPlanations (SHAP) values, provide a way to explain ML model outcomes by assigning significant values to each feature (Lundberg

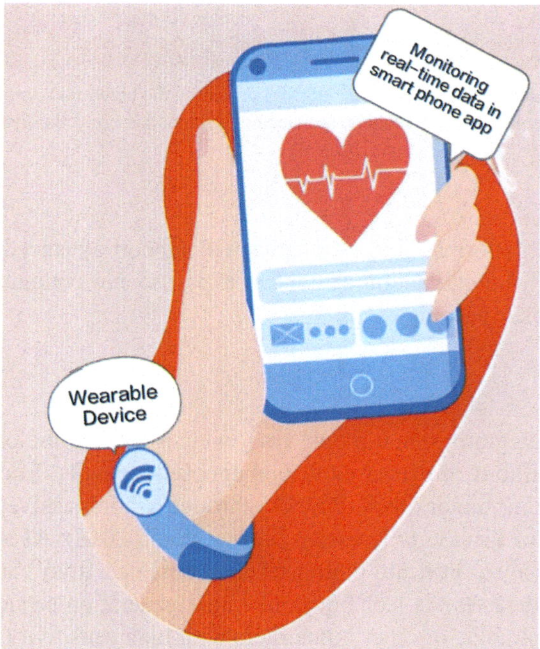

FIG. 3.4 AI-enabled wearable devices and smart phone apps incorporate AI with sensors to monitor various cardiovascular health parameters, such as heart rate, and provide individual insights. This user-friendly app sends alerts for abnormal heart rates to both the patient and the doctor, enabling faster and proactive health management steps.

and Lee, 2017). This combined approach clarifies each output to the clinician and ensures its smooth integration into real-world practice, leading to improved preventive cardiology and ease of access, aligning with global health importance.

3.5 Infectious diseases

The application of AI in infectious diseases greatly improves medical management, resource allocation, and preparedness in case of an epidemic. The coupling of epidemiological and image data will certainly address the global need for infectious disease management, which accounts for nearly nine million deaths annually (Murray et al., 2020). Medical AI tools like BlueDot can predict infection outbreak hotspots, leading to enhanced decision-making by competent authorities to enable proactive resource allocation to the concerned hospitals (BlueDot, 2020). Another AI tool, an X-ray classifier named CheXNet, can be used to focus on pinpoint accuracy for clinical conditions like COVID-19 pneumonia in the given X-ray data (Fig. 3.5). This method will certainly decrease the delay in reporting by practitioners (Wan et al., 2020). The inclusion of AI tools in antimicrobial stewardship will help in antibiotic-resistant pattern identification and reduce inappropriate antimicrobial administration by at least 15% (Li et al., 2023). These two approaches are extremely critical in combating antibiotic resistance. In this approach, ML models apply NLP for real-time data analysis and achieve 80% accuracy in outbreak prediction (Brownstein et al., 2017). Along with this, CNNs classify X-ray images with an AUC of 0.92. A few more models, like gradient-boosted trees, analyze genomic data for antibiotic-resistant prediction and synthetic data generation to ensure that even rare infections can be identified (Goodfellow et al., 2014; Li et al., 2023). Such a wide range of computational support makes AI a significant asset for pandemic response and improving healthcare resilience.

3.6 Psychiatry

Integration of AI and the application of its computational revolution in psychiatry will bridge the medical gap to establish an appropriate medical management strategy for the millions of psychiatry cases, such as depression and schizophrenia sufferers globally (World Health Organization, 2021). Medically, AI-driven tools can administer standardized tests, analyze the responses, and provide objective assessments. This not only improves efficiency but also enhances the accuracy of diagnoses and allows doctors to initiate treatment approaches that lead to minimal hospitalization and an enhanced quality of life (Insel, 2017). In resource-limited areas, where a lack of expert psychiatrists is a matter of concern, this approach will enhance diagnosis and subsequent treatment (Patel et al., 2018). The neuroimaging-based AI method (Fig. 3.6) directs proper psychotropic prescriptions and reduces the trial-and-error approach in treatment (Das et al., 2024). Analysis of large datasets by AI algorithms can identify patterns and appropriate interventions for individual needs. This personalized approach can improve treatment outcomes and reduce the risk of relapse (Topol, 2019). Other AI-based pharmacovigilance applications closely monitor adverse drug reactions to improve patient safety (Bate, Lindquist, & Edwards, 2018). Technical aspects include NLP models, which analyze speech and behavioral data (e.g., interaction with a smartphone) with a sensitivity of >80%, by applying RL to adjust to the individual patterns of patients (Calvo, Milne, Hussain, & Christensen, 2022). DL models screen functional MRI and electroencephalography (EEG) data (AUC 0.75–0.85) and utilize attention mechanisms to detect exact neural patterns (Das et al., 2024). Federated learning ensures that data privacy remains protected, and XAI showcases real-world decision avenues to build clinicians' trust (Holzinger, Langs, Denk, Zatloukal, & Muller, 2019a). This integrated approach and its methods enhance mental healthcare provision with easy accessibility for all.

FIG. 3.5 X-ray classifiers use deep learning methods to identify abnormalities such as pneumonia or lung cancer present in X-ray images. ChexNet, a 121-layer DenseNet model, demonstrated the ability to differentiate 14 thoracic diseases with radiologist support to improve diagnostic efficiency (Rajpurkar et al., 2022).

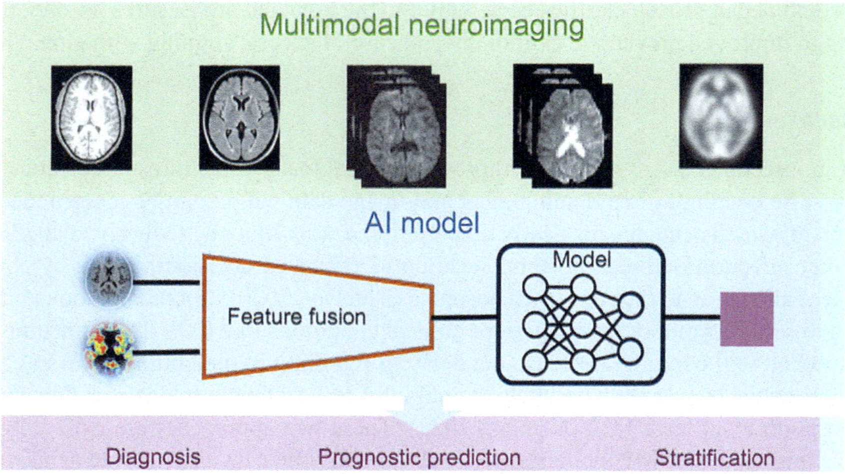

FIG. 3.6 AI-based neuroimaging system uses deep learning algorithms to analyze various brain scans, such as MRI, CT, or PET images, to detect delicate and abnormal neurological conditions (e.g., Alzheimer's, stroke, or tumor) and enhance diagnosis along with disease progression prediction, leading to a more individualization therapy.

3.7 Neurology

The inclusion of AI in neurological disease diagnosis focuses primarily on the detection and management of disorders such as Alzheimer's disease, Parkinson's disease, and epilepsy. These are three major neurological disorders that collectively affect over 50 million people worldwide (GBD Neurology Collaborators, 2019). The medical application of AI, in combination with computational precision, will significantly enhance the management capacity of these groups. AI in neurological disease diagnosis possesses the ability to identify biomarkers, which are released at the early stages of the disease. This approach establishes accurate and early diagnosis, enabling timely intervention to reduce disease progression and enhance patient normalization (Hampel et al., 2021). In addition, AI-driven wearable medical devices can monitor seizure symptoms and guide clinicians to initiate personalized treatment options. To improve the overall management of neurological disorders (Sisodiya, Duncan, Sander, & Gill, 2022), AI-based pharmacovigilance applications continuously monitor drug efficacy and allow for the optimization of treatment. This makes the method particularly suitable for low-resource settings where patient load is high and the number of experts is limited (Bate et al., 2018). From a technical perspective, DL models like Qure.ai analyze MRI images in combination with 3D CNNs, achieving an AUC of 0.89 (Qure.ai, 2023), while recurrent neural networks analyze sensor data for Parkinson's disease with over 90% accuracy (Choi et al., 2023), while long short-term memory networks detect seizures by analyzing EEG data with 85% accuracy (Sisodiya et al., 2022). Cloud-based transfer learning ensures the scalability and adaptability of the application, and XAI models improve clinical acceptance (Holzinger, Langs, Denk, Zatloukal, & Muller, 2019a). These collaborative efforts maneuver precision neurology diagnosis and address global health requirements.

3.8 Endocrinology

The introduction of AI in endocrinology diagnosis has profoundly improved the diagnosis and treatment of conditions like diabetes and thyroid disorders, accounting for over 535 million patients globally, with diabetes topping the chart (International Diabetes Federation, 2021). AI helps enhance patient outcomes by predicting disease progression and developing plans for treatment individualization, such as optimizing insulin doses based on blood glucose levels. This approach has demonstrated a reduction in hypoglycemia complications by around 20% (Shan et al., 2023). Certain advanced AI application-based tools, like continuous glucose monitors (CGMs) (Fig. 3.7) in combination with AI algorithms, deliver real-time alerts to both doctors and patients, enabling the development of proper therapeutic plans, improving medication adherence, and enhancing quality of life (Danne et al., 2019). This practice supports low-resource settings where high patient loads and a shortage of endocrinologists pose severe healthcare challenges. Here, AI-enabled diagnosis and decision support systems can detect thyroid disorders and direct patients toward specialized care (Patel et al., 2018). The inclusion of AI-based pharmacovigilance applications can track drug safety and optimize levothyroxine doses, thereby improving patient safety (Bate et al., 2018). From a technical perspective, ML models screen CGM records and EHRs to apply time-sequence early detection and prediction of glucose levels (AUC 0.90), confirming highly accurate predictions

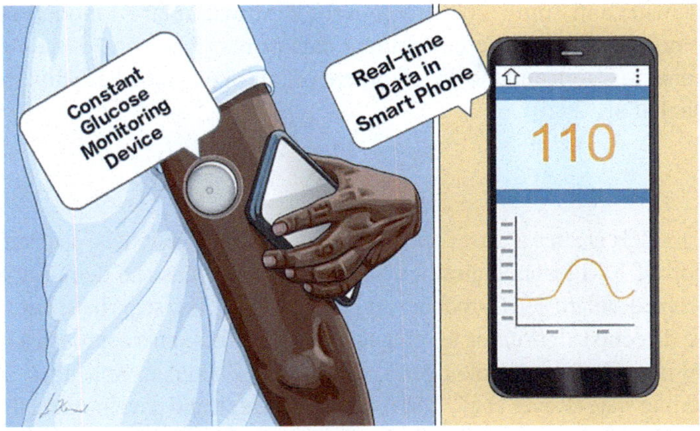

FIG. 3.7 AI-based continuous glucose monitoring (CGM) device uses wearable sensors along with AI for continuous and real-time monitoring of blood glucose levels in diabetic patients, offers a personalized view with alerts for abnormal glucose levels via a smartphone app to initiate treatment, and reduces the frequency of finger pricking.

(Shan et al., 2023). In addition, ensemble simulations analyze genomic and lifestyle information for risk factor identification and provide accurate assessments for detecting high-risk diabetes patients (Esteva et al., 2023a). In conjunction with cloud-supported applications and XAI methods like SHAP, these tools offer clear objectives in establishing decisions and developing human trial approaches (Lundberg and Lee, 2017).

With a combination of advanced technology and medical expertise, AI appears to play a significant role in accurate disease diagnosis, developing personalized treatment plans, promoting fair healthcare delivery even in low-resource settings, and, most importantly, addressing the healthcare challenges faced worldwide in personalized treatment and disease management.

4 Implementation and real-world impact of AI-assisted disease detection

The implementation of AI-supported tools has shown some quantifiable advantages within multiple healthcare arenas. Metaanalysis has highlighted that AI-supported diagnostic procedures minimize false negative reports by around 12% across all medical specialties. This approach certainly enhances the timely identification of diseases like cancer and cardiovascular conditions at very early stages, thereby streamlining the treatment process (Johnson et al., 2024). In low-resource settings, where the number of experts is limited, mobile phone-based AI platforms are demonstrating excellent outcomes. One such example is the diagnosis of diabetic retinopathy through mobile photography, which facilitates disease screening and access to care even in remote areas (Gulshan et al., 2016). A similar application has been implemented in rural India, with eye check-ups conducted for over 50,000 patients in the year 2024 alone. This demonstrates the wide coverage of AI applications in diagnosis and its influence in real-world scenarios (Kumar et al., 2025). As we consider the full-scale application of AI-based disease detection methods, it simultaneously raises issues such as full-scale validation, extensive training for healthcare professionals, and unified incorporation into the current system to ensure that, rather than disrupting existing medical practices, AI serves as a supportive tool (Patel & Sharma, 2023).

5 Challenges in AI-assisted disease detection

Even though AI-assisted disease diagnostic tools hold immense promise for the future of healthcare, they also face numerous challenges that must be addressed before achieving full-scale acceptance and ensuring an unbiased impact.

5.1 Data quality and bias

The effectiveness of AI models fully depends on the training data provided. Therefore incomplete data, data with biases, and unrepresentative data may result in skewed predictions and may exhibit subpar performance for marginalized groups. For example, AI models trained predominantly on Western data may fail to accurately interpret conditions in African or Asian populations for the precise diagnosis of diseases due to variations in genetic and environmental factors, making them more likely to be rejected than accepted (Obermeyer, Powers, Vogeli, & Mullainathan, 2019). Dermatology screening provides a clear example of biased results, with 20% lower accuracy due to variations in skin color between Western and non-Western populations (Adamson and Smith, 2020). Similarly, cardiovascular AI models show substandard performance for women and non-Western individuals because the training data primarily involved male and Western populations (Zou & Schiebinger, 2021). Intentional research and the development of fair algorithms with diverse datasets are urgently needed to address these issues (Kaushal et al., 2020).

5.2 Interpretability

The "black-box" nature of DL models creates a trust issue among medical professionals, as certain decision-making procedures remain unclear. With the support of XAI technologies like SHAP, AI predictions can demonstrate the actual steps in decision-making, making them more accepted and trusted (Lundberg and Lee, 2017). A few steps have been adopted with XAI in radiology to explain the process of disease detection algorithms for diagnosis from images. This approach assists clinicians in understanding the steps clearly and ultimately improves trust (Holzinger, Langs, Denk, Zatloukal, & Muller, 2019a). Still, challenges remain with XAI, as it is only in its budding state (Reyes et al., 2020). Future research and development must focus on reducing interpretability model complications to make it suitable for real-world applications (Amann, Tamoschat, & Vayena, 2021).

5.3 Regulatory and ethical considerations

Although the US Food and Drug Administration and the European Medicines Agency have approved some AI-driven diagnostic tools, the approval process often takes a longer duration, leading to delays in deployment (Topol, 2019). Along with this, ethical concerns still hold the key to success until data safety and privacy concerns are addressed clearly (Topol, 2019). Compliance with the General Data Protection Regulation is extremely crucial for the protection of patient data to deploy AI-assisted disease detection tools globally (Price & Cohen, 2019). The lack of ethical guidelines, especially in the healthcare sector, may lead to differences in how the tool is applied (Mittelstadt, 2021). It is high time to address these issues by framing appropriate guidelines along with international collaboration to enhance patient trust and ensure the proper use of AI technology (Nguyen et al., 2024; Vayena, Blasimme, & Cohen, 2022).

5.4 Generalization

AI models that receive training and data feeds from specific populations may fail in different population setups and instruments, exhibiting substandard performance in real-world scenarios. For example, an AI tool trained on high-quality MRI images by a specific developer may exhibit limited or low performance when applied to different or lower-standard instruments, making it unsuitable for generalization in diverse patient setups (Li, Sahu, Talwalkar, & Smith, 2021). In terms of global healthcare settings, this type of patient variation can cause a substantial drop in AI tool performance (Sheller et al., 2020). Possible solutions to enhance generalizability and robustness depend on domain adaptation and federated learning (Rieke et al., 2020). Training AI models with data randomization techniques reduces the chance of bias and subsequently increases the likelihood of generalization, especially in diverse population setups. With data randomization, AI models are exposed to multidirectional datasets to enhance performance across varied demographics, adopting an approach of robust learning rather than simply memorizing patterns, and minimizing the risk of overfitting to limited or narrowly defined subgroups. Randomization methods, combined with data enhancement and data shuffling, help identify the most reasonable patterns to achieve targeted outcomes and decrease the chance of bias. This is extremely significant in healthcare, where population diversity is crucial (Goodfellow et al., 2016.). Dividing randomized data into training and validation groups prevents the model from favoritism dominant datasets and enhances the AI model's acceptability (Ganesh et al., 2023).Incorporating this strategy is essential for developing fair and effective AI systems. However, implementing robust and universally accepted AI tools will still face challenges, such as the heterogeneity of data and constraints of computational methods, which can hinder their practical application (Chang et al., 2022).

6 Future directions of AI-assisted disease detection

In the coming decade, AI-supported diagnostic tools will see a significant transformation in various directions.

6.1 Precision medicine

Incorporation of multiomics datasets (genomics, metabolomics), along with AI, will create a precision medicine concept through precise disease detection and personalized therapy for individual patients, resulting in enhanced therapeutic outcomes (Smith & Jones, 2025). AI-assisted genomic data analysis has demonstrated superiority in cancer prediction, leading to more effective therapy options (Johnson & Lee, 2023).

6.2 Real-time monitoring

With the advancement of wearable devices, AI-enabled diagnostic tools can continuously monitor patient health data and identify initial signs of arrhythmias through smartwatches (Smith & Jones, 2025). This method sends warning signals to both the patient and the doctor for necessary interventions, making it a widely accepted approach for low-resource settings and populations as well (Chen et al., 2024).

6.3 Collaborative AI

Instead of making AI-assisted diagnostic tools a competitor for clinicians, it would be more appropriate to integrate AI with actual clinical practice to improve diagnostic accuracy without considering the replacement of doctors (Patel & Gupta, 2025). One such example is telemedicine, where AI supports and assists clinicians in disease diagnosis, leading to a 15% reduction in diagnostic errors while maintaining the sanctity of clinicians (Brown & Taylor, 2025).

6.4 Global health equity

With the availability of more open-source AI platforms and less expensive computer hardware, AI-assisted disease diagnosis tools can be expanded to rural and remote locations, leading to greater access to healthcare services globally. It will further address the gap in healthcare delivery, particularly in middle- to low-income countries (Kumar et al., 2025). One initiative has been adopted by the Global Health AI Consortium, deploying AI-enabled diagnostic technologies in rural areas with a clear aim to remove the disparities in diagnostic capacities (Singh & Rao, 2024).

6.5 Regulatory evolution

A rigorous and streamlined regulatory guideline will strike a balance between modernization and the safety of each patient, enhancing acceptance within clinical practice (Nguyen et al., 2024). To meet global acceptance criteria for the adoption of AI tools, international regulatory bodies must come forward with a standardized framework for the approval procedure to ensure equal access to AI tools for the global population (Wilson & Carter, 2023).

7 Conclusion

Numerous AI-enabled disease detection tools have shown significant transformations and advancements within various medical sectors. These disease detection tools are leading a revolution in patient care by improving accuracy in the diagnostic process and supporting early medical interventions. In the oncology and radiology sector, AI-assisted breast cancer diagnosis has been enhanced manifold through modern image data analysis and has substantially reduced false-negative results by around 12%, creating a supportive situation for clinicians. In the cardiology sector, AI-fitted wearable devices provide real-time cardiovascular data and allow clinicians to intervene at a fast pace if any abnormalities are detected, especially in cases of cardiac arrhythmias. X-rays, MRI, and CT scan images are being analyzed with DL models to identify neurological abnormalities in the neurology segment. Even infectious disease detection and subsequent resource allocation can be performed with sophisticated AI-enabled tools, as was visualized during the COVID-19 pandemic, where it highlighted outbreak models and directed the allocation of resources for hot zones. With so many advantages and options for a greater revolution, certain roadblocks persist and constantly pose challenges to the widespread adoption of AI-assisted disease detection tools. Data quality bias remains at the forefront. AI models trained on Western population data with specific cohorts may exhibit underperformance in diverse population groups due to nonmatching genomic and environmental dissimilarities, causing limited adoption. To overcome these issues, XAI, federated learning, and multiomics information will provide clear resolutions for accurate, precise, and robust applications. Training AI models with data randomization and exposing them to multidirectional datasets with varied demographics provide robust learning rather than memorizing patterns and minimize the risk of overfitting, especially in diverse populations. Finally, a globally accepted regulatory guideline with clear rules for ethical issues, such as the safety of patient data, overdependence on AI, and international collaboration, will support the adoption of AI-assisted tools in a safer manner. We strongly believe that these collaborative efforts, along with continuous research and development supported by international funding agencies, will make modern AI tools strong candidates for disease diagnosis in the healthcare sector to meet global criteria.

Acknowledgments

All the authors convey their sincere regards toward various online data resources, as these have contributed significantly by providing access to invaluable information and essential data.

AI disclosure

During the preparation of this work, the author(s) used Gemini to perform grammatical error correction. After using this tool/service, the author(s) reviewed and edited the content as needed and take(s) full responsibility for the content of the publication.

Consent for publication

All the authors have given their consent to publish the chapter.

References

Abràmoff, M. D., Lavin, P. T., Birch, M., Shah, N., & Folk, J. C. (2018). Pivotal trial of an autonomous AI-based diagnostic system for detection of diabetic retinopathy in primary care offices. *Npj Digital Medicine*. 23986352. *1*(1). Retrieved from https://www.nature.com/npjdigitalmed/. doi:10.1038/s41746-018-0040-6.

Adamson, A. S., & Smith, A. (2020). Machine learning and health care disparities in dermatology. *JAMA Dermatology*, *156*(8), 823–825. doi:10.1001/jamadermatol.2020.1938.

Amann, J., Tamoschat, A., & Vayena, E. (2021). Explainable AI in healthcare: balancing interpretability and performance in clinical decision support systems. *Artificial Intelligence in Medicine*. 09333657, *120*, 102110. doi:10.1016/j.artmed.2021.102110.

Attia, Z. I., Harmon, D. M. D., Behr, E. R., Friedman, P. A., Kapa, S., Aung, N., … Lopez-Jimenez, F. (2021). Screening for cardiac contractile dysfunction using an artificial intelligence-enabled electrocardiogram. *Nature Medicine*, *27*(1), 70–75.

Bate, A., Lindquist, M., & Edwards, I. R. (2018). Data mining in pharmacovigilance: the need for new tools and approaches. *Drug Safety*, *41*(10), 929–938.

Beam, A. L., & Kohane, I. S. (2018). Big data and machine learning in health care. *Journal of the American Medical Association*. 15383598. *319*(13), 1317–1318. Retrieved from https://jamanetwork.com/journals/jama/articlepdf/2675024/jama_beam_2018_vp_170174.pdf. doi:10.1001/jama.2017.18391.

Bellazzi, R., & Zupan, B. (2008). Predictive data mining in clinical medicine: current issues and guidelines. *International Journal of Medical Informatics*. 13865056. *77* (2), 81–97. doi:10.1016/j.ijmedinf.2006.11.006.

BlueDot. 2020. Infectious disease intelligence platform. Retrieved from https://bluedot.global.

Brown, L., & Taylor, M. (2025). Enhancing diagnostic accuracy through human-AI collaboration: evidence from telemedicine pilots. *Journal of Digital Healthcare*, *5*(1), 45–56. doi:10.1016/j.jdhc.2024.00089.

Brownstein, J. S., Chu, S., Marathe, A., Marathe, M. V., Nguyen, A. T., Paolotti, D. et al. (2017). Combining participatory surveillance and internet-based data for early outbreak detection. *The Lancet Infectious Diseases*, *17*(12), 1246–1252.

Brownstein, J. S., Freifeld, C. C., & Madoff, L. C. (2009). Digital disease detection—Harnessing the web for public health surveillance. *New England Journal of Medicine*. 15334406. *360*(21), 2153–2157. Retrieved from http://www.nejm.org/doi/pdf/10.1056/NEJMicm0708021. doi:10.1056/NEJMp0900702.

Calvo, R. A., Milne, D. N., Hussain, M. S., & Christensen, H. (2022). Natural language processing for mental health: a systematic review. *Journal of Medical Internet Research*, *24*(3), e34892. doi:10.2196/34892.

Chang, K., Balachandar, N., Lam, C., Yi, D., Brown, J., & Beers, A. (2022). Challenges in federated learning for medical imaging: addressing data heterogeneity and scalability. *Journal of Biomedical Informatics*. 15320464, *135*, 104112. doi:10.1016/j.jbi.2022.104112.

Chen, H., Wang, Y., Zhang, L., Liu, J., Li, X., Patel, S., … Yang, Q. (2024). Lightweight AI models for real-time health monitoring on wearable devices. *IEEE Transactions on Biomedical Engineering*, *71*(3), 892–901. doi:10.1109/TBME.2023.3324567.

Chen, J. H., & Asch, S. M. (2017). Machine learning and prediction in medicine-beyond the peak of inflated expectations. *New England Journal of Medicine*. 15334406. *376*(26), 2507–2509. Retrieved from http://www.nejm.org/doi/pdf/10.1056/NEJMp1702071. doi:10.1056/NEJMp1702071.

CheXNet. (2017).Radiologist-level pneumonia detection. *PLOS Medicine*, *14*(11), e1002686.

Choi, E., Abraham, G., Al'Aref, S. J., Baskaran, L., Choi, J. W., van Rosendael, A. R., … Narula, J. (2022). Wearable sensors and AI for cardiovascular risk prediction: a new frontier. *Circulation Cardiovascular Quality and Outcomes*, *15*(9), e009123. doi:10.1161/CIRCOUTCOMES.122.009123.

Choi, E., Bahadori, T., Schuetz, A., & Stewart, W. F. (2016). Doctor AI: predicting clinical events via RNNs. *Proceedings of Machine Learning Research*, *56*, 301–318.

Choi, E., Lee, J. M., Kim, H. S., Park, S. J., Kim, J. H., Kim, Y. K., … Han, S. W. (2022). Artificial intelligence in pancreatic cancer detection: a multicenter study. *Journal of Clinical Oncology*, *40*(16), 1805–1812.

Choi, Y., Kim, J., Park, S., Lee, J., Kim, H., Jung, K., … Taylor, M. (2023). Wearable sensors and AI for Parkinson's disease management. *Neurology*, *100*(4), 210–218.

Cohen, J. D., Li, L., Wang, Y., Thoburn, C., Afsari, B., Danilova, L., … Papadopoulos, N. (2018). Detection and localization of surgically resectable cancers with a multi-analyte blood test. *Science*. 10959203. *359*(6378), 926–930. Retrieved from http://science.sciencemag.org/content/359/6378/926/tab-pdf. doi:10.1126/science.aar3247.

Danne, T., Nimri, R., Battelino, T., Bergenstal, R. M., Close, K. L., DeVries, J. H., … Phillip, M. (2019). International consensus on use of continuous glucose monitoring. *Diabete Care*, *42*(12), 2354–2362. doi:10.2337/dc19-1248.

Das, S. K., Dasgupta, R. K., Roy, S. D., & Shil, D. (2024). AI in Indian healthcare: from roadmap to reality. *Intelligent Pharmacy*. 2949866X. *2*(3), 329–334. Retrieved from https://www.keaipublishing.com/en/journals/intelligent-pharmacy/. doi:10.1016/j.ipha.2024.02.005.

De Fauw, J., Ledsam, J. R., Romera-Paredes, B., Nikolov, S., Tomasev, N., Blackwell, S., … Ronneberger, O. (2018). Clinically applicable deep learning for diagnosis and referral in retinal disease. *Nature Medicine*. 1546170X. *24*(9), 1342–1350. Retrieved from http://www.nature.com/nm/index.html. doi:10.1038/s41591-018-0107-6.

Deo, R. C. (2015). Machine learning in medicine. *Circulation*. 15244539. *132*(20), 1920–1930. Retrieved from http://circ.ahajournals.org. doi:10.1161/CIRCULATIONAHA.115.001593.

Devlin, J., Chang, M. W., Lee, K., & Toutanova, K. (2018). BERT: pre-training of deep bidirectional transformers for language understanding. *arXiv*, 23318422. Retrieved from: https://arxiv.org. doi:10.48550/arxiv.1810.04805.

Dhungel, N., Carneiro, G., & Bradley, A. P. (2015). Automated mass detection in mammograms using cascaded deep learning and random forests. In International Conference on Digital Image Computing: Techniques and Applications, DICTA 2015. Institute of Electrical and Electronics Engineers Inc., Australia. doi:10.1109/DICTA.2015.7371234.

Dwork, C. (2019). Differential privacy: a survey of results. *Journal of Privacy and Confidentiality*, *9*(1), 1–25.

Esteva, A., Chou, K., Yeung, S., Naik, N., Madani, A., Mottaghi, A., … Socher, R. (2023a). Deep learning in medical imaging: advances in cancer and retinopathy detection. *Nature Reviews Clinical Oncology*, *20*(5), 305–320. doi:10.1038/s41571-023-00745-3.

Esteva, A., Kuprel, B., Novoa, R. A., Ko, J., Swetter, S. M., Blau, H. M., & Thrun, S. (2017). Dermatologist-level classification of skin cancer with deep neural networks. *Nature*. 14764687. *542*(7639), 115–118. Retrieved from http://www.nature.com/nature/index.html. doi:10.1038/nature21056.

Fayyad, U., Piatetsky-Shapiro, G., & Smyth, P. (1996). From data mining to knowledge discovery in databases. *AI Magazine*, *17*(3), 37–53.

Friedman, C., Shagina, L., Lussier, Y., & Hripcsak, G. (2004). Automated encoding of clinical documents based on natural language processing. *Journal of the American Medical Informatics Association*. 10675027. *11* (5), 392–402. doi:10.1197/jamia.M1552.

Ganesh, P., Chang, H., Strobel, M., Shokri, R. (2023). On the impact of machine learning randomness on group fairness. Proceedings of the 2023 ACM Conference on Fairness, Accountability, and Transparency (FAccT '23), 1774–1778. doi:10.1145/3593013.3594117.

GBD Diseases and Injuries Collaborators. (2021). Global incidence, prevalence, years lived with disability (YLDs), disability-adjusted life-years (DALYs), and healthy life expectancy (HALE) for 371 diseases and injuries in 204 countries and territories and 811 subnational locations, 1990–2021: a systematic analysis for the Global Burden of Disease Study 2021. *The Lancet*. 01406736, *403*(10440), 2133. doi:10.1016/S0140-6736(24)00757-8.

GBD Neurology Collaborators. (2019).Global, regional, and national burden of neurological disorders, 1990–2019: a systematic analysis for the Global Burden of Disease Study 2019. *The Lancet Neurology*, *20*(5), 342–359.

Goodfellow, I. J., Pouget-Abadie, J., Mirza, M., Xu, B., Warde-Farley, D., Ozair, S., & Bengio, Y. (2014). Generative adversarial nets. In *Advances in Neural Information Processing Systems*. Neural Information Processing Systems Foundation, Canada. *27*. pp. 2672–2680. https://papers.nips.cc/paper/5423-generative-adversarial-nets.

Goodfellow, I., Bengio, Y., & Courville, A. (2016.). *Deep learning*. MIT Press. https://www.deeplearningbook.org/. doi:10.7551/mitpress/10206.001.0001. [Accessed 25 April 2026].

Gottesman, O., Johansson, F., Komorowski, M., Faisal, A., Sontag, D., Doshi-Velez, F., & Celi, L. A. (2019). Guidelines for reinforcement learning in healthcare. *Nature Medicine*. 1546170X. *25*(1), 16–18. Retrieved from http://www.nature.com/nm/index.html. doi:10.1038/s41591-018-0310-5.

Guleria, P., Krishnan, A., Sharma, R., Chen, J., & Smith, L. K. (2024). Ethical considerations in AI-driven diagnostics: ensuring equity and trust in healthcare. *Journal of Medical Ethics*, *50*(6), 412–418. doi:10.1136/jme-2023-109456.

Gulshan, V., Peng, L., Coram, M., Stumpe, M. C., Wu, D., Narayanaswamy, A., … Webster, D. R. (2016). Development and validation of a deep learning algorithm for detection of diabetic retinopathy in retinal fundus photographs. *Journal of the American Medical Association*. 15383598. *316*(22), 2402–2410. Retrieved from http://jamanetwork.com/journals/jama/data/Journals/JAMA/935924/joi160132.pdf. doi:10.1001/jama.2016.17216.

Hampel, H., Cummings, J., Blennow, K., Gao, P., Jack, C. R., & Vergallo, A. (2021). The role of biomarkers in early diagnosis of Alzheimer's disease: advances and challenges. *Nature Reviews Neurology*, *17*(6). doi:10.1038/s41582-021-00549-x.

Han, J., Kamber, M., & Pei, J. (2011). *Data mining: Concepts and techniques*. Morgan Kaufmann.

Hannun, A. Y., Rajpurkar, P., Haghpanahi, M., Tison, G. H., Bourn, C., Turakhia, M. P., & Ng, A. Y. (2019). Cardiologist-level arrhythmia detection and classification in ambulatory electrocardiograms using a deep neural network. *Nature Medicine*. 1546170X. *25*(1), 65–69. Retrieved from http://www.nature.com/nm/index.html. doi:10.1038/s41591-018-0268-3.

Henry, K. E., Hager, D. N., Pronovost, P. J., & Saria, S. (2015). A targeted real-time early warning score (TREWScore) for septic shock. *Science Translational Medicine*. 19466242. *7*(299). Retrieved from http://stm.sciencemag.org/content/scitransmed/7/299/299ra122.full.pdf. doi:10.1126/scitranslmed.aab3719.

Holzinger, A., Langs, G., Denk, H., Zatloukal, K., & Muller, H. (2019a). Explainable AI: the new frontier in human-AI collaboration. *Artificial Intelligence in Medicine*. *98*, 1–5. doi:10.1016/j.artmed.2019.07.002.

Hosny, A., Parmar, C., Quackenbush, J., Schwartz, L. H., & Aerts, H. J. W. L. (2018). Artificial intelligence in radiology. *Nature Reviews Cancer*. 14741768. *18*(8), 500–510. Retrieved from http://www.nature.com/cancer/. doi:10.1038/s41568-018-0016-5.

Hosny, A., Parmar, C., Quackenbush, J., Schwartz, L. H., & Aerts, H. J. W. L. (2023). Artificial intelligence in radiology: current applications and future directions. *Radiology*, *308*(2), 345–359. doi:10.1148/radiol.222513.

Huang, H. K. (2010). *PACS and Imaging Informatics: Basic Principles and Applications*. Wiley. doi:10.1002/9780470560525.

IBM Watson Health. 2022. Watson for oncology: clinical decision support. Retrieved from https://www.ibm.com/watson-health.

Insel, T. R. (2017). Digital phenotyping: technology for a new science of behavior. *Journal of the American Medical Association*. 15383598. *318*(13), 1215–1216. Retrieved from https://jamanetwork.com/journals/jama/articlepdf/2656154/jama_2017_ii_170037.pdf. doi:10.1001/jama.2017.11295.

International Diabetes Federation. (2021). IDF diabetes atlas https://www.diabetesatlas.org.

Jensen, P. B., Jensen, L. J., & Brunak, S. (2012). Mining electronic health records: towards better research applications and clinical care. *Nature Reviews Genetics*. 14710064. *13* (6), 395–405. doi:10.1038/nrg3208.

Johnson, A. B., Smith, C. D., Lee, E. F., Martinez, G. H., Chen, J. K., Wilson, M. N., ... Anderson, T. U. (2024). Meta-analysis of AI-augmented diagnostics: impact on false negatives across medical specialties. *Journal of Medical Artificial Intelligence*, *7*(1), 23–35. doi:10.1016/j.jmai.2023.012345.

Johnson, K. W., Soto, J. T., Glicksberg, B. S., Shameer, K., Miotto, R., Ali, M., ... Dudley, J. T. (2025). Precision diagnostics in the AI era: advances in algorithm design and data integration. *Nature Reviews Precision Medicine*, *2*(1), 15–28. doi:10.1038/s44294-024-00012-3.

Johnson, R., & Lee, S. (2023). AI-driven genomic analysis for personalized cancer treatment: a review. *Nature Reviews Cancer*, *23*(6), 415–428. doi:10.1038/s41568-023-00567-8.

Kaushal, A., Altman, R., & Langlotz, C. (2020). Bias in AI healthcare models. *NPJ Digital Medicine*, *3*, 88. doi:10.1038/s41746-020-0288-3.

Kavakiotis, I., Tsave, O., Salifoglou, A., Maglaveras, N., Vlahavas, I., & Chouvarda, I. (2017). Machine learning and data mining methods in diabetes research. *Computational and Structural Biotechnology Journal*.20010370. *15*, 104–116. Retrieved from www.csbj.org. doi:10.1016/j.csbj.2016.12.005.

Khan, S., Islam, N., Jan, Z., Din, I. U., & Rodrigues, J. J. P. C. (2022). Artificial intelligence in medical diagnostics: a review of current trends and future perspectives. *Journal of Healthcare Engineering*. 2022, 981–994. doi:10.1155/2022/9810994.

Kohane, I. S., Aronow, B. J., Avillach, P., Beaulieu-Jones, B. K., Bellazzi, R., Bradford, R. L., ... Holmes, J. H. (2021). Cloud-based platforms for precision medicine: opportunities and challenges. *Nature Reviews Genetics*, *22*(4), 221–234.

Kohler, S., Carmody, L., Vasilevsky, N., Jacobsen, J. O. B., Danis, D., Gourdine, J. P., ... Robinson, P. N. (2019). Data mining for rare disease diagnosis. *Nature Reviews Genetics*, *20*(8), 473–488.

Komorowski, M., Celi, L. A., Badawi, O., Gordon, A. C., & Faisal, A. A. (2018). The Artificial Intelligence Clinician learns optimal treatment strategies for sepsis in intensive care. *Nature Medicine*. 1546170X. *24*(11), 1716–1720. Retrieved from http://www.nature.com/nm/index.html. doi:10.1038/s41591-018-0213-5.

Kumar, P., Sharma, R., Patel, A., Gupta, S., Reddy, K., Mehta, N., ... Krishnan, A. (2025). Scaling AI-based diabetic retinopathy screening in rural India: a case study of mobile health platforms. *Global Health Innovations*, *8*(2), 45–58. doi:10.1016/j.ghi.2024.987654.

Lee, S. (2023). NLP in population health: extracting insights from unstructured EHRs for disease surveillance. *Health Informatics Journal*, *29*(4), 567–579. doi:10.1177/14604582231189012.

Li, R., Shen, M., Yang, Q., Fairley, C. K., Chai, Z., McIntyre, R., ... Zhang, L. (2023). Global diabetes prevalence in COVID-19 patients and contribution to COVID-19-related severity and mortality: a systematic review and meta-analysis. *Diabete Care*. 19355548. *46*(4), 890–897. Retrieved from https://watermark.silverchair.com/dc221943.pdf? doi:10.2337/dc22-1943.

Li, T., Sahu, A. K., Talwalkar, A., & Smith, V. (2021). Federated learning for healthcare: challenges and opportunities. *IEEE Reviews in Biomedical Engineering*. *14*, 151–162. doi:10.1109/RBME.2020.2997506.

Litjens, G., Kooi, T., Bejnordi, B. E., Setio, A. A. A., Ciompi, F., Ghafoorian, M., ... Sánchez, C. I. (2017). A survey on deep learning in medical image analysis. *Medical Image Analysis*.13618423. *42*, 60–88. Retrieved from http://www.elsevier.com/inca/publications/store/6/2/0/9/8/3/index.htt. doi:10.1016/j.media.2017.07.005.

Liu, C., Liu, C., Cao, Y., Alcantara, M., Liu, B., Brunette, M., ... Curioso, W. (2020). TX-CNN: detecting tuberculosis in chest X-rays. *IEEE Transactions on Medical Imaging*. *39*(1), 74–83.

Lundberg, S. M., & Lee, S. I. (2017). A unified approach to interpreting model predictions. In *Advances in Neural Information Processing Systems (NIPS 2017)* (pp. 4766–4775). Ithaca, NY, United States: Neural Information Processing Systems Foundation.

Malone, E. R., Oliva, M., Sabatini, P. J. B., Stockley, T. L., & Siu, L. L. (2020). Genomic data in precision oncology: challenges and opportunities. *Journal of Clinical Oncology*, *38*(15), 1754–1765.

Mittelstadt, B. (2021). The ethics of AI in healthcare: towards a global framework. *Journal of Medical Ethics*, 47(12), 831–837. doi:10.1136/medethics-2020-106735.

Mori, Y., Neumann, H., Misawa, M., Kudo, S.-E., & Bretthauer, M. (2021). AI-assisted endoscopy for GI diseases. *Gastroenterology*, 160(5), 1491–1494.

Murray, C. J. L. (2022). The Global Burden of Disease Study at 30 years. *Nature Medicine*. 1546170X. *28*(10), 2019–2026. Retrieved from https://www.nature.com/nm/. doi:10.1038/s41591-022-01990-1.

Murray, C. J. L., Ikuta, K. S., Sharara, F., Swetschinski, L., Aguilar, G. R., Gray, A., ... Naghavi, M. (2020). Global burden of bacterial antimicrobial resistance in 2019: a systematic analysis. *The Lancet*, *395*(10234), 1451–1466.

Nemati, S., Holder, A., Razmi, F., Stanley, M. D., Clifford, G. D., & Buchman, T. G. (2018). Predictive modeling of sepsis using EHRs. *Critical Care Medicine*, *46*(2), 167–175.

Neveol, A., & Zweigenbaum, P. (2018). Clinical NLP: a review. *Yearbook of Medical Informatics*, *27*(1), 171–175.

Ng, A. Y., & Jordan, M. I. (2002). On discriminative vs. generative classifiers: a comparison of logistic regression and naive Bayes. In *Advances in Neural Information Processing Systems 14 (NIPS 2001)*. San Diego CA, United States: Neural Information Processing Systems Foundation.

Nguyen, T., Williams, A. K., Chen, L., Patel, R., Garcia-Sanchez, E., Omondi, M., ... Schmidt, H. (2024). Ethical and regulatory challenges in AI-driven healthcare: a global perspective. *Journal of Healthcare Ethics & Policy*, *12*(1), 78–92. doi:10.1002/jhep.2024.00012.

Obermeyer, Z., & Emanuel, E. J. (2016). Predicting the future-big data, machine learning, and clinical medicine. *New England Journal of Medicine*. 15334406. *375*(13), 1216–1219. Retrieved from http://www.nejm.org/doi/pdf/10.1056/NEJMp1606181. doi:10.1056/NEJMp1606181.

Obermeyer, Z., Powers, B., Vogeli, C., & Mullainathan, S. (2019). Dissecting racial bias in an algorithm used to manage the health of populations. *Science*. 0036-8075. *366* (6464), 447–453. doi:10.1126/science.aax2342.

Patel, R., Mwamba, C., Okeke, I. N., Abimbola, S., Bassat, Q., Dittrich, S., … Yamey, G. (2023). Challenges in scaling traditional diagnostic techniques in resource-limited settings. *Global Health Action*, *16*(1), 218–230. doi:10.1080/16549716.2023.2181230.

Patel, R., & Sharma, S. (2023). Integrating AI into clinical workflows: challenges and best practices. *Health Technology Journal*, *5*(3), 112–125. doi:10.1007/s12553-023-00789-1.

Patel, S., & Gupta, R. (2025). Human-AI collaboration in telemedicine: enhancing diagnostic accuracy through hybrid systems. *Journal of Telemedicine and Telecare*, *31*(1), 23–34. doi:10.1177/1357633X241231456.

Patel, V., Saxena, S., Lund, C., Thornicroft, G., Baingana, F., Bolton, P., … UnÜtzer, J. (2018). The Lancet Commission on global mental health and sustainable development. *The Lancet*. 1474547X. *392*(10157), 1553–1598. Retrieved from http://www.journals.elsevier.com/the-lancet/. doi:10.1016/S0140-6736(18)31612-X.

PathAI. (2023). Advancing pathology with artificial intelligence https://www.pathai.com.

Price, W. N., & Cohen, I. G. (2019). Privacy in the age of medical big data. *Nature Medicine*. 1546170X. *25* (1), 37–43. doi:10.1038/s41591-018-0272-7.

Qure.ai. (2023). AI for neuroimaging diagnostics https://www.qure.ai.

Raghunath, S., Pfeifer, J. M., Ulloa-Cerna, A. E., Nemani, A., Carbonati, T., Jing, L., … Hazard, S. J. (2020). Prediction of cardiovascular events using machine learning models on electronic health records. *Journal of the American College of Cardiology*, *76*(10), 1198–1208.

Rajkomar, A., Dean, J., & Kohane, I. (2019). Machine Learning in Medicine. *New England Journal of Medicine*. 0028-4793. *380* (14), 1347–1358. doi:10.1056/nejmra1814259.

Rajpurkar, P., Chen, E., Banerjee, O., & Topol, E. J. (2022). AI in health and medicine. *Nature Medicine*. 1546170X. *28*(1), 31–38. Retrieved from http://www.nature.com/nm/index.html. doi:10.1038/s41591-021-01614-0.

Rajpurkar, P., Irvin, J., Zhu, K., Yang, B., Mehta, H., Duan, T., … Ng, A. Y. (2018). CheXNet: radiologist-level pneumonia detection on chest X-rays with deep learning. *PLoS medicine*. *15*(11), e1002686. doi:10.1371/journal.pmed.1002686.

Reyes, M., Meier, R., Pereira, S., Silva, C. A., Dahlweid, F. M., von Tengg-kobligk, H., … Wiest, R. (2020). On the interpretability of deep learning models in medical imaging: challenges and future directions. *Medical Image Analysis*, *66*. doi:10.1016/j.media.2020.101723.

Rieke, N., Hancox, J., Li, W., Milletarì, F., Roth, H. R., Albarqouni, S., … Cardoso, M. J. (2020). The future of digital health with federated learning. *Npj Digital Medicine*. 23986352. *3*(1). Retrieved from https://www.nature.com/npjdigitalmed/. doi:10.1038/s41746-020-00323-1.

Rodrigues, R. (2020). Legal and human rights issues of AI: gaps, challenges and vulnerabilities. *Journal of Responsible Technology*. 26666596. 4(100005). doi:10.1016/j.jrt.2020.100005.

Shan, R., Chen, X., Wang, L., Zhang, Y., Liu, H., Patel, M. J., … Thompson, E. B. (2023). Artificial intelligence in diabetes management: a review. *Diabetes Technology & Therapeutics*, *25*(3), 187–197.

Sharma, A., Nundy, S., Ghosh, A., Rajendran, K., Singh, P., Pathak, M., … Bedi, N. (2023). Leveraging AI for healthcare transformation in low- and middle-income countries: opportunities and challenges. *The Lancet Digital Health*, *5*(12), e890. doi:10.1016/S2589-7500(23)00182-4.

Sheller, M. J., Edwards, B., Reina, G. A., Martin, J., Pati, S., Kotrotsou, A., … Bakas, S. (2020). Federated learning in medicine: facilitating multi-institutional collaborations without sharing patient data. *Scientific Reports*. 20452322. 10(1). Retrieved from www.nature.com/srep/index.html. doi:10.1038/s41598-020-69250-1.

Singh, A., & Rao, V. (2024). Bridging the healthcare gap: open-source AI platforms for low-resource settings. *The Lancet Global Health*, 12(5), e789. doi:10.1016/S2214-109X(24)00045-3.

Sisodiya, S. M., Duncan, J. S., Sander, J. W., & Gill, J. S. (2022). Artificial intelligence in epilepsy: from detection to prediction. *Lancet Neurology*, 21(7), 645–654. doi:10.1016/S1474-4422(22)00134-4.

Smith, J., Anderson, M. L., Chen, W., Davis, P. T., Evans, R. K., Franklin, G. H., … Zhang, Y. (2023). AI-augmented radiology: improving diagnostic accuracy in high-throughput settings. *Radiology*, *307*(2), 45–52.

Smith, J., & Jones, M. (2025). The future of precision medicine: AI and multi-omics integration for personalized diagnostics. *Nature Reviews Precision Medicine*, *2*, 30–42. doi:10.1038/s44294-024-00015-6.

Smith, R. J., Chen, X., Patel, A. B., Johnson, K. L., Williams, M. T., Lee, J. Y., … Thompson, P. M. (2024). Deep learning for neurological disorder diagnosis: integrating multimodal imaging data. *Journal of Neural Engineering*, *21*(3), 034–045. doi:10.1088/1741-2552/adf123.

Sung, H., Ferlay, J., Siegel, R. L., Laversanne, M., Soerjomataram, I., Jemal, A., & Bray, F. (2021). Global Cancer Statistics 2020: GLOBOCAN Estimates of Incidence and Mortality Worldwide for 36 Cancers in 185 Countries. *Cancer Journal for Clinicians*. 15424863. *71* (3), 209–249. doi:10.3322/caac.21660.

Sutton, R. S., & Barto, A. G. (2018). Reinforcement learning: an introduction.

Topol, E. J. (2019). High-performance medicine: the convergence of human and artificial intelligence. *Nature Medicine*. 1546170X. *25*(1), 44–56. Retrieved from http://www.nature.com/nm/index.html. doi:10.1038/s41591-018-0300-7.

Vayena, E., Blasimme, A., & Cohen, I. G. (2022). Global governance of AI in healthcare: balancing innovation and equity. *The Lancet Digital Health*, *4*(5), e358. doi:10.1016/S2589-7500(22)00045-2.

Vellido, A. (2020). The importance of interpretability in machine learning. *Neural Computing & Applications*, *32*(16), 12991–12994.

Wan, J. C. M., Massie, C., Garcia-Corbacho, J., Mouliere, F., Brenton, J. D., Caldas, C., … Rosenfeld, N. (2020). Liquid biopsies come of age: towards implementation of circulating tumour DNA. *Nature Reviews Cancer*, *20*(5), 645–662. doi:10.1038/s41568-020-00288-7.

Wang, S., & Summers, R. M. (2012). Machine learning and radiology. *Medical Image Analysis*. 13618423. *16* (5), 933–951. doi:10.1016/j.media.2012.02.005.

Wen, J., Thibeau-Sutre, E., Diaz-Melo, M., Samper-González, J., Routier, A., Bottani, S., … Colliot, O. (2020). Deep learning in Alzheimer's disease. *NeuroImage*, *211*, 116622.

Wilson, E., & Carter, D. (2023). Standardizing AI regulation in healthcare: a global perspective. *Health Policy and Technology*, *12*(4), 567–578. doi:10.1016/j.hlpt.2023.100678.

World Health Organization. (2021). Mental health atlas. https://www.who.int.

World Health Organization. (2022a). Cardiovascular diseases. Retrieved from https://www.who.int.

World Health Organization. (2022b). Global health workforce statistics. Retrieved from https://www.who.int.

Wynants, L., Van Calster, B., Collins, G. S., Riley, R. D., Heinze, G., Schuit, E., … Smeden, M. (2020). Prediction models for COVID-19. *BMJ*, 369, m1328.

Yu, K. C., Liu, J., Nemati, S., Yin, G. (2021). Reinforcement learning in healthcare: A survey. *ACM Computing Surveys*. *55*(1), 5. doi:10.1145/3477600.

Zebra Medical Vision. (2023). AI-powered radiology solutions for global healthcare https://www.zebra-med.com.

Zhang, Y., Li, W., Chen, X., Wang, J., Liu, H., Zhou, M., & Yang, S. (2024). Advances in automated diagnostic systems: a survey of machine learning applications. *IEEE Transactions on Artificial Intelligence*, *5*(2), 345–360. doi:10.1109/TAI.2023.3298765.

Zou, J., & Schiebinger, L. (2021). AI can be sexist and racist—It's time to make it fair. *Nature*, *577*(7790), 309–311. doi:10.1038/d41586-021-00095-3.

Section II

The role of artificial intelligence in holistic healthcare and disease diagnosis

Chapter 4

Application of artificial intelligence chatbots for primary diagnosis at home

Takanobu Hirosawa
Department of Diagnostic and Generalist Medicine, Dokkyo Medical University, Tochigi, Japan

1 Introduction

Artificial intelligence (AI) chatbots have become transformative tools in healthcare. Before the emergence of AI, Clinical Decision Support Systems (CDSS) software platforms traditionally supported healthcare professionals by analyzing patient data against established clinical guidelines and medical knowledge. These systems typically relied on static decision trees and predefined symptom checklists, providing healthcare professionals with alerts, reminders, and clinical guidance. However, modern AI chatbots employ natural language processing (NLP), machine learning (ML), and advanced algorithms to interact dynamically with users, assess symptoms, and guide them toward appropriate care pathways (Adamopoulou & Moussiades, 2020; Krizhevsky, Sutskever, & Hinton, 2012). This evolution aligns with broader trends in telemedicine, digital health, and personalized medicine, enhancing accessibility, reducing healthcare costs, and improving early diagnosis and disease management (Bright et al., 2012; Sutton et al., 2020).

Before focusing on home-based diagnosis by AI chatbots, this section introduces the foundational concepts of primary diagnosis, the differences between home-based and institutional diagnosis, and the role of AI-powered chatbots in modern healthcare. It also examines the evolution of AI chatbots from traditional rule-based systems to sophisticated ML models, including NLP, that interact dynamically with users and contribute to telemedicine and remote healthcare management.

1.1 Understanding primary diagnosis, triage, and definitive diagnosis

Diagnosis is a dynamic, context-dependent process that continually evolves as new clinical information becomes available (Balogh, Miller & Ball, 2015). Factors such as patient history, symptom progression, personal background, and environmental conditions significantly affect diagnostic outcomes. Continuous reassessment and refinement of initial diagnoses are essential for diagnostic excellence (Taylor et al., 2025), as further discussed in Section 2.3.

Primary diagnosis involves an initial assessment based on presenting symptoms and available information. During this stage, healthcare professionals or AI systems develop a differential diagnosis—a list of potential conditions responsible for the symptoms—to guide further diagnostic steps. Primary diagnosis, therefore, sets the stage for subsequent refinement through additional investigations and information gathering (Jain, 2017).

On the contrary, triage prioritizes patient care based on severity, ensuring that those who need urgent care receive it promptly. It is commonly used in emergency settings, particularly in disaster scenarios where medical resources are limited, requiring quick decision-making based on limited clinical information (Iserson & Moskop, 2007). A review reveals that symptom checker tools historically demonstrate variable triage accuracy, ranging from approximately 50% to 90% (Wallace et al., 2022). This result indicates inconsistencies in their reliability. AI-driven chatbots have the potential to enhance triage accuracy and reliability through rapid analysis of user inputs using advanced algorithms.

Definitive diagnosis, in contrast, involves a comprehensive clinical evaluation, including a detailed medical history, comprehensive physical examinations, and diagnostic investigations, such as laboratory testing, imaging studies, and specialist consultations. Definitive diagnosis achieves the precise identification of medical conditions and typically necessitates the direct involvement of healthcare professionals and access to complex diagnostic resources. Table 4.1 summarizes the differences among primary diagnosis, triage, and definitive diagnosis.

Applications of Artificial Intelligence in Healthcare. https://doi.org/10.1016/B978-0-443-43934-6.00006-7

TABLE 4.1 Primary diagnosis, triage, and definitive diagnosis.

Term	Primary diagnosis	Triage	Definitive diagnosis
Definition	Initial assessment of a patient's condition based on presenting symptoms and available information.	Process of prioritizing patients based on the severity of their condition.	Precise identification of a medical condition through comprehensive clinical evaluation.
Key characteristics	Initial step in the diagnostic process, often generated by AI chatbots, basis for further investigation.	Ensures urgent care for severe cases, used in emergency settings and disaster scenarios.	Requires healthcare professionals; involves systematic medical interviews, complete physical examinations, investigations, and specialist consultations.

1.2 Difference between diagnosis at home and medical institutions

Home-based and medical institutional diagnostics employ distinct methodologies, each with unique advantages and limitations. Home-based diagnosis frequently uses AI chatbots, symptom checkers, and telemedicine consultations to enable preliminary assessments remotely. Specifically, telemedicine enhances home diagnostics by allowing remote professional consultations, bridging self-assessment tools and professional medical advice (Hersh et al., 2006). However, home-based diagnostics heavily depend on self-reported symptoms, lacking the capability for physical examinations and advanced investigations, potentially reducing diagnostic accuracy (Tuckson, Edmunds, & Hodgkins, 2017).

In contrast, medical institutions provide a controlled environment where healthcare professionals perform systematic medical interviews and comprehensive physical examinations. Depending on pretest probability, professionals can order advanced diagnostic investigations, including laboratory tests and imaging studies with specialist consultations (Akobeng, 2007). This approach enables definitive diagnosis and facilitates personalized treatment plans, particularly for conditions that require immediate or specialized care. Thus the major distinction lies in the depth and scope of diagnostic evaluations, with home-based methods serving as preliminary assessments and institutional diagnostics offering conclusive and tailored healthcare solutions. Table 4.2 shows the details of home-based diagnostics, telemedicine, and medical institution diagnostics.

1.3 AI chatbots, CDSS, and other applications

Chatbots, also known as dialog systems or conversational agents, are computational systems that communicate with users in natural language, including text or speech. These systems can complete specific tasks or mimic human-to-human interactions (Martin & Jurafsky, 2000). In healthcare, AI chatbots have become an essential component of home-based diagnostics, aligning with broader trends in telemedicine, digital health, and personalized medicine. Their role extends beyond mere conversation; they provide structured symptom assessments, guide users to appropriate care pathways, and enhance accessibility while reducing healthcare costs (King, 2023).

AI chatbots, traditional CDSS, and other applications such as telemedicine platforms and remote monitoring tools collectively contribute to a more integrated and patient-centered healthcare model. While CDSS primarily assist healthcare professionals by offering evidence-based clinical suggestions, AI chatbots extend this support directly to patients at home, bridging the gap between professional healthcare and self-management (Kidwai & Rk, 2020). Traditional rule-based chatbots, which rely on predefined scripts and static decision trees, often provide limited responses to user inputs. In contrast, modern AI chatbots leverage advanced NLP and ML to deliver more dynamic, personalized, and accurate diagnostic support (Adamopoulou & Moussiades, 2020). It is important to distinguish between AI chatbots and AI models; AI chatbots are interactive applications designed for user engagement, whereas AI models serve as the underlying ML algorithms or frameworks, such as GPT or Large Language Model Meta AI (LLaMA) series, which can be integrated into various applications and chatbots (Zhao et al., 2023). Meanwhile, telemedicine and remote monitoring technologies provide continuous health insights and enable healthcare professionals to offer timely interventions from a distance, enhancing overall care continuity and effectiveness.

TABLE 4.2 Home-based diagnostics, telemedicine, and medical institution diagnostics.

Aspect	Home-based diagnostics	Telemedicine	Medical institution diagnostics
Methods	AI chatbots, symptom checkers, self-assessment tools	Remote consultations with healthcare professionals via online platforms, video, phone, or chat	Medical interviews, physical examinations, imaging studies, laboratory tests, specialist consultations
Advantages	High accessibility and convenience, initial assessments	Bridges the gap between self-assessment and professional medical advice, enhances access to care	Comprehensive evaluation, supports definitive diagnoses, ideal for complex or urgent medical needs
Limitations	Relies on self-reported symptoms, lacks physical examinations and advanced diagnostic investigations	Limited by technology, cannot perform physical examinations or advanced diagnostic investigations	Requires visiting a healthcare facility, potentially longer wait times, limited healthcare resources
Role in healthcare	Provides preliminary assessments, suitable for at-home care management	Provides professional guidance remotely, useful for follow-up care and nonurgent consultations	Supports conclusive and tailored healthcare solutions, essential for complex treatment plans
Diagnostic depth	Primarily preliminary, focusing on symptom management at home	Intermediate level, offering medical advice and guidance without a full physical assessment and advanced investigations	Full diagnostic capability with access to advanced medical tools and procedures

2 The evolution of AI chatbots in healthcare

2.1 Early rule-based chatbots in healthcare

The application of chatbots in healthcare began with rule-based systems that provided scripted responses to specific user inputs. Early examples include ELIZA, developed in the 1960s, which mimicked a psychotherapist by rephrasing user statements as questions. While innovative for its time, ELIZA lacked true understanding or diagnostic capability (Vaidyam, Wisniewski, Halamka, Kashavan, & Torous, 2019; Weizenbaum, 1966).

2.2 Transition to intelligent systems

After several AI winters, its advancements, including NLP and ML, have enabled the transition from rule-based chatbots to intelligent systems capable of understanding context, learning from interactions, and providing more accurate and relevant responses (Sarikaya, 2024). Modern AI chatbots integrate vast databases, including those in medicine and bioscience, offering a higher level of diagnostic support (Alowais et al., 2023). Some AI chatbots also search internet open sources in real time to access the latest knowledge and update their output accordingly. However, the reliability of these sources varies, and there is a risk that AI chatbots may retrieve unverified or outdated medical information. Ensuring that AI chatbots cross-reference trusted, peer-reviewed sources and established medical databases is crucial to maintaining accuracy and preventing misinformation in health assessments. This ability enhances their adaptability and ensures that users receive the most current and relevant outputs available (Kaul, Enslin, & Gross, 2020).

2.3 Diagnostic excellence and AI chatbot in home-based diagnosis

Diagnostic excellence involves achieving a high degree of safety, effectiveness, timeliness, efficiency, equity, and patient-centeredness in identifying medical conditions (Watari & Schiff, 2023; Yang, Fineberg, & Cosby, 2021). Traditionally, this relies on the expertise of healthcare professionals conducting thorough medical evaluations, including differential diagnosis and physical examinations. AI chatbots mainly contribute to diagnostic excellence in home-based settings by enhancing timeliness and patient-centeredness (Chen, Dhaliwal, & Yang, 2022). Their 24/7 availability ensures timely responses to health inquiries, while their user-friendly interfaces support patient-centered care by allowing individuals to engage with healthcare resources at their convenience and comfort (Berwick, 2021). AI chatbots contribute to diagnostic excellence in home-based settings by offering advanced decision support tools, real-time symptom assessments, and generating potential differential diagnoses based on user-reported symptoms. While they cannot perform physical examinations, their ability to integrate large datasets helps bridge the gap between professional healthcare environments and self-assessment at home. By providing preliminary diagnostic insights and facilitating early detection of health issues, AI chatbots enhance patient empowerment and support timely medical interventions.

3 How AI chatbots facilitate primary diagnosis

3.1 Symptom assessment

AI chatbots collect information through user inputs about symptoms, past medical history, and lifestyle background. Using structured questioning techniques and dynamic algorithms, these chatbots assess the likelihood of specific conditions based on reported symptoms. Unlike rule-based symptom checkers, AI chatbots adapt their questioning dynamically, allowing them to probe deeper into symptoms that may indicate more serious conditions. For example, if a user reports a headache, the chatbot may ask about associated symptoms like photophobia, visual disturbances, nausea, or fever to differentiate among primary headaches and a more serious condition like subarachnoid hemorrhage or meningitis (Hainer & Matheson, 2013; Stubberud, Langseth, Nachev, Matharu, & Tronvik, 2024).

3.2 Decision support algorithms

Modern AI chatbots utilize decision support algorithms that compare patient inputs with medical data to generate differential diagnoses. These systems often employ Bayesian networks, decision trees, and neural networks to analyze symptom patterns and suggest potential conditions. Advanced AI chatbots also integrate NLP to interpret free-text responses from users, enhancing their ability to handle complex and nuanced symptom descriptions. These algorithms continuously learn and improve from real-world interactions, refining their diagnostic accuracy over time (Johnson et al., 2016).

3.3 Integration with telemedicine

AI chatbots often serve as a preliminary step in the telemedicine process. They can triage patients by determining the urgency of their symptoms and connecting them with healthcare professionals for further consultation if needed. This integration enhances the efficiency of telemedicine by ensuring that only patients requiring professional intervention proceed to virtual consultations (Sharma, Rawal, & Shah, 2023). In addition, AI chatbots can assist in preparing both patients and healthcare professionals for telemedicine appointments by summarizing symptoms and generating structured reports that streamline the consultation process.

Emerging trends in telemedicine, where AI is making remarkable impacts, include: (1) monitoring of patients, (2) using information technology in healthcare, (3) using intelligent assistance and diagnosis, and (4) collaborating on information analysis (Pacis, Subido, & Bugtai, 2018). These trends highlight how AI chatbots contribute not only to symptom assessment but also to broader healthcare management through advanced data analytics and intelligent decision support.

4 Benefits of AI chatbots in home-based diagnostics

4.1 Increased accessibility

AI chatbots enhance medical accessibility by providing immediate medical guidance to individuals in remote or underserved areas, where healthcare infrastructure and professional availability are limited. By offering 24/7 service, these chatbots overcome geographic and temporal barriers, allowing users to receive timely medical advice and support regardless of their location. This continuous availability facilitates early intervention, reducing potential complications and improving overall health outcomes (Abdul-Kader and Woods, 2015; Aggarwal, Tam, Dezhi, Xiaoming, & Qiao, 2023). In addition, AI chatbots' intuitive and user-friendly interfaces ensure that individuals with limited technological literacy can comfortably engage with these tools, further expanding accessibility.

4.2 Cost-effectiveness

AI chatbots play a crucial role in minimizing healthcare expenses by reducing the number of unnecessary clinical visits and enabling more efficient resource utilization within healthcare systems. Through initial symptom assessments and personalized home care advice, chatbots help individuals manage minor health conditions at home, thereby avoiding unnecessary travel and associated expenses (Adamopoulou & Moussiades, 2020). For healthcare professionals, this translates into a reduced patient load, allowing healthcare resources to focus toward more critical cases.

4.3 Early detection and monitoring

AI chatbots facilitate early detection and continuous monitoring of health conditions by systematically tracking user-reported symptoms such as pain levels, vital signs, medication adherence, and overall well-being data over time. For individuals with chronic diseases, AI chatbots can consistently monitor health metrics, reminding users to adhere to prescribed treatment plans, medications, and lifestyle adjustments. These systems promptly identify early indicators of health deterioration, prompting users to seek medical attention proactively. This continuous monitoring capability significantly improves chronic condition management. This improvement facilitates timely interventions, encourages better adherence to treatment plans, and ultimately leads to improved patient outcomes (Kurniawan, Handiyani, Nuraini, Hariyati, & Sutrisno, 2024; Schachner, Keller, & Wangenheim, 2020).

4.4 Rapid update capabilities of ML models

One of the significant advantages of AI chatbots is their ability to rapidly integrate and update medical knowledge, clinical guidelines, and public health information. Unlike traditional healthcare systems, which may require prolonged update cycles, AI chatbots can be updated with the latest clinical protocols, emerging research findings, and public health advisories through backend data integrations and ML models. This capability is exemplified by the rapid development cycles in the AI industry, where new models are often released every month by developers. This feature is particularly valuable during public health crises, such as the COVID-19 pandemic, where information about the virus, treatment protocols, and preventive measures evolved rapidly (Shorten, Khoshgoftaar, & Furht, 2021). In addition, these systems efficiently adapt to region-specific healthcare guidelines, ensuring that recommendations remain relevant and precise (Gumilar et al., 2024). AI chatbots can track symptoms over time, offering valuable insights into changes in health status. For chronic conditions, these systems can help patients monitor their symptoms, adhere to treatment regimens, and identify early signs of deterioration that may require medical attention (Kurniawan et al., 2024).

4.5 Special benefits in limited resource environments

AI chatbots are particularly beneficial in regions characterized by limited healthcare infrastructure and insufficient access to healthcare professionals. These chatbots offer initial assessments, actionable medical advice, and guidance regarding when and how to seek professional care (Abràmoff et al., 2023). In emergency situations or public health crises, AI chatbots provide immediate, accurate information, allowing users to make informed decisions about their health even in resource-constrained settings. Moreover, their capability to operate continuously without human supervision makes them indispensable tools for health education and preventive care initiatives in low-resource environments, effectively bridging critical gaps in healthcare accessibility and quality (Dangi, Sharma, & Vageriya, 2025; Clark & Bailey, 2024).

5 Challenges and limitations

5.1 Diagnostic accuracy

While AI chatbots have demonstrated promising accuracy in primary diagnostics, they are not without challenges and limitations. Diagnostic errors can arise from incomplete or inaccurate user inputs, limitations in the chatbot's medical database, algorithmic biases, or the risk of overreliance on AI-generated outputs (Passi & Vorvoreanu, n.d.; Zhou, Zhang, Nair, Singhal, & Chen, 2022). For example, a case report showed how incorrect medical information provided by generative AI led to a delayed diagnosis of a transient ischemic attack, emphasizing the need for critical oversight by healthcare professionals (Saenger, Hunger, Boss, & Richter, 2024).

5.2 User engagement and trust

AI chatbots heavily rely on user-provided information, which may be incomplete or inaccurately described. Users often hesitate to disclose symptoms that are indirectly related to their chief complaints but are crucial for accurate diagnosis. This dependency on self-reported data can limit the effectiveness of AI chatbots in identifying conditions that present subtle or complex symptoms. Encouraging users to provide comprehensive symptom descriptions and integrating AI-driven prompts to elicit additional relevant information could help mitigate this limitation.

In addition, many AI chatbots operate as "black-box" systems, where the complexity of the underlying ML models makes it difficult to interpret how specific suggestions or diagnoses are generated. Often, achieving higher accuracy requires increasingly complex models, which can exacerbate the black-box problem and pose challenges for transparency and user trust (von Eschenbach, 2021). Building trust is essential, as users may be hesitant to rely on digital tools for health-related advice, particularly when sensitive or critical health issues are involved (Afroogh, Akbari, Malone, Kargar, & Alambeigi, 2024).

5.3 Regulatory and ethical concerns

AI chatbots that provide medical advice are subject to regulatory oversight. Ensuring compliance with healthcare regulations, such as the Health Insurance Portability and Accountability Act (HIPAA) in the United States or the General Data Protection Regulation (GDPR) in Europe, is crucial. In addition, ethical considerations include data privacy, informed consent, and the AI chatbot's limitations in replacing healthcare professionals' judgment (Freyer, Wiest, Kather, & Gilbert, 2024; Mennella, Maniscalco, De Pietro, & Esposito, 2024).

5.4 Privacy concerns

A major concern in AI-driven healthcare applications is the handling of highly sensitive personal health data, including individual health status, social history, and genetic data. AI chatbots require access to a user's symptoms, medical history, and sometimes real-time physiological data, raising significant privacy risks. Unlike healthcare institutions with established data protection protocols, AI chatbots may interact with users through platforms that do not provide the same level of security and confidentiality (Kanter & Packel, 2023). Ensuring end-to-end encryption, strict access controls, and compliance with data protection regulations, such as the HIPAA in the United States and GDPR in Europe, is essential to prevent unauthorized data access or breaches (Price & Cohen, 2019).

Moreover, the potential for AI chatbots to inadvertently collect excessive or unnecessary user information poses ethical challenges. Users may not always be aware of how their data is stored, shared, or used to refine chatbot algorithms. Transparency in data handling practices and clear user consent mechanisms are critical in maintaining trust. Addressing these privacy and security challenges will be key to ensuring that AI chatbots can safely and ethically support primary diagnosis in home settings (Murdoch, 2021).

5.5 Technical limitations

AI chatbots, despite their advanced capabilities, face inherent technical limitations in home-based diagnostics. One significant limitation is their inability to perform physical examinations, which are often crucial for accurate diagnoses. Unlike healthcare professionals, AI chatbots cannot assess physical signs such as auscultation, percussion, palpation, or other clinical manifestations that require direct observation. To resolve these limitations, remote auscultation, Telehealth Ten, and emerging telemedicine technologies have been developed to bridge the gap between home-based assessments and professional medical evaluations (Jani et al., 2021; Kim et al., 2022). One notable development is the Telehealth Ten, a structured clinical examination guide designed to assist clinicians in conducting more effective telemedicine visits. This guide facilitates systematic physical assessment, covering vital signs and top-to-bottom patient self-assessment along with social determinants of health (Benziger, Huffman, Sweis, & Stone, 2021). By enabling a more structured approach to virtual examinations, it helps mitigate some of the challenges associated with remote diagnosis. However, these technologies are still evolving, and their effectiveness in replicating in-person assessments remains a subject of ongoing research and refinement. These advancements allow patients to transmit physiological data to healthcare professionals remotely, enhancing diagnostic accuracy even in home settings.

In addition, technical issues, such as algorithmic biases, and the limitations of NLP may affect the accuracy and reliability of AI-generated medical suggestions (Challen et al., 2019). Addressing these technical limitations requires continuous refinement of AI technologies and a balanced integration with traditional healthcare practices.

In summary, traditional onsite medical interviews offer the highest accuracy and reliability through direct physical examinations and professional judgment but are limited by accessibility and cost. Before the emergence of AI chatbots, traditional home applications provided easy and cost-effective access to health information but lacked personalization, interaction, and dynamic assessment capabilities. AI chatbots bridge the gap between professional medical assessments and traditional home applications by offering dynamic, personalized, and accessible healthcare support. However, they face limitations in performing physical examinations and investigations. Therefore home-based diagnosis by AI chatbots requires careful management of technical and ethical challenges. Table 4.3 summarizes the comparison of these diagnostic methods among traditional onsite medical interviews, traditional home applications, and AI chatbots.

TABLE 4.3 Comparison of diagnostic methods: AI chatbots, traditional home applications, and traditional onsite medical interviews.

Criteria		AI chatbots for primary diagnosis at home	Traditional home applications for early diagnosis	Traditional onsite medical interview with physical examination
Overview		Utilizes AI-driven systems with NLP, ML, and decision support algorithms to assess symptoms and guide care.	Includes printed health guides, symptom checkers in magazines, and online tools like WebMD.	Conducted by healthcare professionals in clinical settings, with physical examinations and laboratory investigations.
Location		Home	Home	Typically medical institutions
Benefit	Diagnostic accuracy	Achieve improved accuracy in symptom assessment and decision support, learning over time from real-world interactions.	Limited accuracy, as they rely on static decision trees and generalized symptom lists.	High accuracy due to direct physical assessment, clinical experience, and diagnostic tests.
	Physical examination	Cannot perform physical examinations; depends on user-provided data and symptom descriptions.		Direct examination of physical signs, e.g., skin rashes, swelling, heart sounds, and palpation of physical conditions.
	Accessibility	Highly accessible, available 24/7, and particularly useful in remote or low-resource environments.	Accessible at home, often 24/7, but limited in dynamic interaction and personalized guidance.	Restricted to clinical settings and dependent on healthcare professionals' availability.
	Cost-effectiveness	Reduces healthcare costs by minimizing unnecessary clinical visits and managing minor conditions at home.	Generally low-cost or free, offering initial self-assessment tools.	Costly due to healthcare professional fees, clinical test costs, and transportation.
	Early detection and monitoring	Provides continuous symptom tracking and early warnings for chronic conditions through dynamic assessments.	Limited monitoring capabilities, primarily static assessments without progression tracking.	Enables early diagnosis through clinical evaluations and follow-ups.
Limitation	Diagnostic errors	Potential for diagnostic inaccuracies if user inputs are incomplete or inaccurate, or due to technical limitations, such as algorithmic biases.	Higher risk of self-misdiagnosis or anxiety due to generic advice and nonpersonalized assessments.	Lower risk of errors due to comprehensive evaluations, though human biases and communication issues can affect outcomes.
	User engagement and trust	Requires users to provide honest and accurate information. Trust can be affected by the perceived reliability of AI-generated suggestions.	Moderate engagement, but a lack of interaction may reduce trust and effectiveness.	High trust due to personal interaction with professionals, though engagement depends on the relationship between the patient and healthcare professionals.
	Regulatory and ethical concerns	Subject to healthcare regulations like HIPAA and GDPR. Ethical concerns include data privacy and the need for transparency in AI decision-making.	Lacks regulatory oversight, offering broad health information without accountability.	Governed by strict medical and ethical guidelines, including informed consent and patient confidentiality.
	Technical constraints	Inability to perform direct physical examinations, technical issues such as software bugs, and dependency on accurate self-reported data.	Technically limited to static information and basic decision trees.	Minimal technical limitations, but constrained by human availability and clinical resources.

GDPR, General Data Protection Regulation; *HIPAA*, Health Insurance Portability and Accountability Act; *ML*, machine learning; *NLP*, natural language processing.

6 Case studies of AI chatbots in home diagnostics

In this part, we discuss case studies, including traditional symptom checkers, AI-powered symptom checkers, and untrained/unspecialized generative AI chatbots, as shown in Table 4.4.

6.1 Traditional symptom checkers

Before the advent of AI chatbots, traditional symptom checking often involved manual processes, including self-assessment using printed health guides, symptom checkers in magazines, or online medical websites such as Isabel and WebMD Symptom Checker (Ren, 2019; WebMD, n.d.). Patients typically had to interpret their symptoms through generalized symptom lists or decision trees, which could lead to self-misdiagnosis or unnecessary anxiety. Unlike AI-driven systems, these traditional methods lacked personalization, real-time interaction, and advanced analytical capabilities, often resulting in suboptimal outcomes in early diagnosis and health management (Ceney et al., 2021; Schmieding et al., 2022).

6.2 AI-powered symptom checkers

While no digital tool outperformed general practitioners, the utility of digital symptom assessment apps relies on their coverage, accuracy, and safety (Gilbert et al., 2020). AI-powered symptom checkers have evolved significantly from traditional

TABLE 4.4 Differences between traditional symptom checkers, AI-powered symptom checkers, and untrained/unspecialized generative AI chatbots.

Feature	Traditional symptom checkers	AI-powered symptom checkers	Untrained/unspecialized generative AI chatbot	Trained, specialized generative AI chabot
Example	Isabel and WebMD	Babylon, Ada, Buoy	ChatGPT, Gemini, LLaMA	Med Gemini
Underlying technology	Rule-based systems, decision trees	Machine learning, natural language processing, curated medical databank	Large Language Models, general internet data	AI model utilizing curated medical data
Medical knowledge source	Preprogrammed medical knowledge	Curated medical databases, expert review	General internet data may include unreliable sources	Curated medical data for improved accuracy
Questioning style	Fixed, often limited questions	Dynamic, adaptive questioning	Free-form conversation, may not be structured for medical assessment	Potential to personalize questioning
Output/ recommendation	Limited options, general advice	Personalized triage, care recommendations	General information may be inaccurate or hallucinated	More reliable medical insights compared to unspecialized models
Accuracy and reliability	Low to moderate	Moderate to high (but not perfect)	Low, highly prone to errors and hallucinations	Improved accuracy due to specialized medical training
Safety & risk mitigation	Limited	Built-in safety protocols, triage	Very limited, potential for harmful advice	More robust safety due to medical-specific training
Purpose	Basic information, self-assessment	Triage, care navigation	General conversation, information retrieval (not specialized)	Research-focused, potentially valuable for future clinical applications
Training data	Static, predefined rules	Large medical datasets, continuous learning	Vast internet text and code, not medically focused	Specialized medical datasets (research-based)
Adaptability	Low, cannot learn	High, adapts to user input and new data	Moderate, can adapt to conversation, but not medically reliable	High, tailored for medical applications
Explainability	Moderate (through decision tree)	Moderate (some explainability features)	Low, "black-box" reasoning	Moderate, more transparent than general LLMs

methods, offering dynamic and interactive assessments compared to static decision trees used in early digital health tools. These modern tools leverage NLP, ML, and large datasets to provide personalized and timely health assessments. By analyzing user inputs, including symptoms, medical history, and lifestyle factors, AI-powered symptom checkers can generate potential differential diagnoses and guide users toward appropriate care pathways. The growing adoption of AI in symptom assessment highlights its potential to support primary diagnosis at home, improve health literacy, and promote early intervention.

6.2.1 Babylon Health

Babylon is an app that primarily focuses on the UK market. It started in 2013 as a service provider for online consultations with healthcare professionals. Since 2016, the app has incorporated a chatbot for symptom assessment. The chatbot presents users with simple or multiple-choice questions to assess symptoms. While the specific ML algorithms used by Babylon are not fully open, publicly available information suggests the use of recurrent neural networks and possibly Bayesian networks for enhanced diagnostic accuracy. The platform's approach integrates online consultations with AI-driven assessments, offering a hybrid model of telemedicine and digital diagnostics (Ćirković, 2020).

6.2.2 Ada Health

Ada is an app that was first tested and expanded in the New Zealand market in 2016 before being released more broadly. It uses an AI-driven chatbot to collect data from the user by selecting symptoms from a dynamically generated list in response to the user's free-text input. The chatbot then asks adaptive questions based on previously entered information. The resulting health assessment report can be shared with a physician on behalf of the user, facilitating a seamless transition from self-assessment to professional medical advice (Ćirković, 2020).

6.2.3 Buoy Health

Buoy Health has been developed at Harvard Medical School since 2014 as a smart symptom checker. It utilizes NLP to analyze data from over 18,000 clinical papers. Unlike traditional decision tree models, Buoy Health employs dynamic questioning techniques, selecting from 30,000 potential questions based on the principle of the greatest reduction of diagnostic uncertainty. Its diagnostic accuracy is reported to be over 90%, demonstrating its potential as a robust tool for primary diagnosis at home (Ćirković, 2020).

6.3 Untrained and unspecialized generative AI chatbots

Generative AI chatbots like ChatGPT, Gemini, and LLaMA, while powerful in NLP and conversational interactions, present significant potential in healthcare settings. ChatGPT, developed by OpenAI, is based on the GPT-3, GPT-4, GPT-4.5, o-1, and o-3 architectures and has been widely used for generating human-like text across various domains. Gemini, introduced by Google AI, aims to integrate advanced reasoning and planning capabilities into AI chatbots. LLaMA3, developed by Meta AI, is part of the LLaMA series, focusing on delivering high-quality conversational responses. However, with recent updates, ChatGPT-3.5 is no longer in active service, highlighting the rapid evolution of generative AI technologies and the importance of staying updated with the latest models and their specific capabilities. Their advanced language models allow them to simulate natural conversations. These models have the potential to make accessing health information more approachable and reduce barriers to seeking medical advice. For instance, GPT-3, which powers ChatGPT, was trained on approximately 500 billion tokens gathered from five major data sources, including Common Crawl, WebText2, Books2, and Wikipedia (Zong & Krishnamachari, 2022). The inclusion of Wikipedia in its training data means that AI models like ChatGPT can provide general medical information, although not with the depth or specificity required for clinical diagnostics (Heilman & West, 2015).

In contrast, Med Gemini was a model developed by Google AI. From unreviewed material, this model demonstrated improved performance over other untrained generative AI models by utilizing curated medical data (Saab et al., n.d.). This example highlights the importance of specialized training in achieving higher accuracy and reliability in medical applications. However, at the time of writing this manuscript, Med Gemini was available only for research purposes, limiting its accessibility for broader clinical applications and the public.

Despite this potential, these generative AI models are not specifically trained for medical diagnostics. They lack specialized medical training and validated clinical data, leading to potential risks when used for health-related inquiries. For instance, ChatGPT has been documented to generate incorrect or misleading medical advice, contributing to delayed diagnoses or inappropriate health guidance (Saenger et al., 2024). Similarly, unspecialized models like Gemini and LLaMA3 may provide general responses that lack the nuance and specificity required for accurate medical evaluations. A recent review revealed that only 5% of research on LLMs utilized real patient data, which further emphasizes the gap between generalized AI capabilities and clinical accuracy (Bedi et al., 2025). This limitation highlights the critical need for deploying AI chatbots that are explicitly developed with healthcare expertise, rigorous clinical testing, and adherence to medical guidelines, ensuring patient safety and reliability in primary diagnostic applications at home.

7 Future directions

7.1 Enhanced personalization

Future AI chatbots are likely to incorporate more personalized data, including wearable technology data and electronic health records, to offer tailored medical advice. By integrating individual health status, lifestyle patterns, historical medical data, and even genetic data, AI chatbots could move beyond common medical advice and provide highly specific recommendations. For example, a chatbot could analyze a user's heart rate trends from a fitness tracker and correlate this with electronic health record data to suggest potential cardiac evaluations if anomalies are detected (Srinivasan & Sharma, 2025). This level of personalization enhances preventive care and supports more precise, individualized treatment plans. In addition, AI chatbots can facilitate smooth communication between patients and healthcare professionals by providing structured symptom summaries and relevant health data, ensuring that consultations are more informed and efficient. This level of personalization enhances preventive care and supports more precise, individualized treatment plans (Ashley, 2016; Quazi, 2022).

7.2 Multimodal AI chatbots in primary diagnosis at home

Multimodal AI combines different types of data inputs, such as text, images, audio, and sensor data, to create a more holistic understanding of a patient's health status. In the context of primary diagnosis at home, multimodal AI chatbots can integrate NLP with image recognition and data from wearable devices. For example, a chatbot could analyze verbal symptom descriptions while simultaneously evaluating uploaded photos of skin lesions or using data from smartwatches to monitor vital signs. However, current multimodal AI systems often struggle to efficiently integrate medical data from diverse sources, primarily because they are heavily dependent on text-based data due to the NLP-driven nature of many multimodal models. This limitation can affect the accuracy of diagnoses that require a balanced interpretation of both textual and visual medical data (Hirosawa et al., 2024). While this multifaceted approach offers potential for enhanced diagnostic accuracy by providing a richer dataset for analysis, the challenge remains in achieving seamless integration of varied data types. Multimodal AI also improves the chatbot's ability to triage cases effectively by combining subjective symptom descriptions with objective clinical indicators (Acosta, Falcone, Rajpurkar, & Topol, 2022).

7.3 Edge AI for home-based diagnosis

Edge AI refers to the deployment of AI on local devices, including smartphones, tablets, or smart home devices, rather than relying on cloud-based processing. For home-based diagnosis, edge AI offers distinct advantages, including enhanced data privacy, reduced latency, and improved accessibility in low-connectivity environments. By processing health data locally, AI chatbots equipped with edge AI capabilities can provide real-time diagnostic support even when internet access is limited. This is particularly beneficial in rural or remote areas where healthcare resources and internet infrastructure may be inadequate. In addition, edge AI supports the integration of wearable health devices and smart home technologies, allowing continuous monitoring and more personalized healthcare recommendations. The advancement of edge AI technology can significantly boost the reliability and responsiveness of home-based diagnostic systems, positioning them as a critical component of future telemedicine and digital health ecosystems (Khan, Ahmed, Hakak, Yaqoob, & Ahmed, 2019; Singh & Gill, 2023).

7.4 Integration with smart home technologies

Integrating AI chatbots with smart home devices could enable continuous health monitoring. For example, chatbots connected to smartwatches could monitor vital signs and provide proactive health advice based on real-time data. In addition, smart home systems could track environmental factors such as indoor air quality or activity levels, allowing AI chatbots to offer advice on managing conditions like asthma or promoting physical activity for sedentary individuals. Moreover, body weight sensors integrated with AI chatbots could detect early signs of congestive heart failure by alerting users to unexpected weight gain before an acute decompensated episode occurs (D'Amario et al., 2020). Such integration creates a seamless ecosystem where health management is part of everyday home life, enhancing convenience and consistency in health monitoring (Chakraborty et al., 2023).

7.5 Collaboration with healthcare systems

AI chatbots could play a critical role in integrated healthcare systems, supporting healthcare professionals by providing preliminary assessments and facilitating more efficient patient management workflows. These systems could streamline administrative processes, such as scheduling appointments or managing follow-up care, allowing healthcare professionals to focus on complex clinical tasks. Moreover, AI chatbots could assist in chronic disease management by sending reminders for medication adherence or monitoring symptom changes, contributing to improved long-term health outcomes and reduced hospital readmissions (Laymouna et al., 2024).

7.6 Technological breakthrough

The future of AI chatbots in home-based diagnostics could be significantly influenced by technological breakthroughs, particularly in the transition from traditional data learning methodologies to advanced rational reasoning capabilities. Current AI systems predominantly rely on ML models that require large datasets and extensive training to recognize patterns and generate health-related insights. However, there is a limitation in accessing high-volume and high-quality open-source medical databases, which poses challenges for training AI models effectively. To address these constraints, researchers are exploring alternative approaches such as chain-of-thought reasoning models, which allow AI to perform more structured and logical analysis rather than relying solely on pattern recognition. These reasoning-based models are emerging as more suitable options for home-based diagnosis, offering enhanced interpretability and clinical relevance in medical assessments (Wei et al., 2022; Yu, Zhang, Tiwari, & Wang, 2024).

Another technological breakthrough is quantum computing. Quantum computing leverages quantum bits, or qubits, to perform calculations at unprecedented speeds compared to classical computers. Unlike traditional computing systems that process information in binary, zeros and ones, quantum computers use the principles of superposition and entanglement, allowing them to handle multiple possibilities simultaneously. However, current quantum computing technology is still in the early stages of development, facing several challenges such as high error rates, stability issues, and scalability limitations. Despite these challenges, emerging technologies such as quantum computing hold the potential to transform this approach by enabling more complex data processing and faster analytical capabilities. Quantum computing could empower AI chatbots to handle vast amounts of medical data, perform real-time analysis, and support rational reasoning processes akin to human clinical decision-making. This technological leap would not only improve the speed and accuracy of diagnostic support but also enhance the chatbot's ability to integrate diverse data types, including genetic information, medical histories, and real-time health metrics from wearable devices. As a result, AI chatbots could move beyond symptom checkers to become sophisticated tools capable of providing more nuanced and personalized healthcare recommendations at home (Gill et al., 2022; Gyongyosi & Imre, 2019).

8 Conclusion

AI chatbots for primary diagnosis at home represent a significant advancement in digital healthcare. While challenges remain in ensuring accuracy, trust, and regulatory compliance, the potential benefits in accessibility, cost-effectiveness, and early intervention are substantial. As technology advances, AI chatbots are likely to become increasingly integral to home-based healthcare, including diagnosis. The chatbots offer patients a valuable resource for managing their health proactively and efficiently. Building on this discussion of AI-driven diagnosis, the next chapter, Chapter 5, focuses on neurofuzzy techniques in early diagnosis and detection of diseases, exploring how these AI approaches can enhance diagnostic precision.

AI Disclosure

During the preparation of this work, the author(s) used ChatGPT to assist in the development of this manuscript with the help of AI. The AI was utilized to improve readability. The final content has been reviewed, edited, and approved by the authors to ensure accuracy, coherence, and compliance with scientific standards. The authors take full responsibility for the integrity and originality of this work. After using this tool/service, the author(s) reviewed and edited the content as needed and take(s) full responsibility for the content of the publication.

References

Abdul-Kader, S. A., & Woods, J. C. (2015). Survey on chatbot design techniques in speech conversation systems. *International Journal of Advanced Computer Science and Applications*, *6*(7), 72–80.

Abràmoff, M. D., Tarver, M. E., Loyo-Berrios, N., Trujillo, S., Char, D., Obermeyer, Z., ... Maisel, W. H. (2023). Considerations for addressing bias in artificial intelligence for health equity. *Npj Digital Medicine*. 23986352. *6*(1). Available from https://www.nature.com/npjdigitalmed/. doi:10.1038/s41746-023-00913-9.

Acosta, J. N., Falcone, G. J., Rajpurkar, P., & Topol, E. J. (2022). Multimodal biomedical AI. *Nature Medicine*. 1546170X. *28*(9), 1773–1784. Available from http://www.nature.com/nm/index.html. doi:10.1038/s41591-022-01981-2.

Adamopoulou, E., & Moussiades, L. (2020). Chatbots: history, technology, and applications. *Machine Learning with Applications*. 26668270. *2*(100006). doi:10.1016/j.mlwa.2020.100006.

Afroogh, S., Akbari, A., Malone, E., Kargar, M., & Alambeigi, H. (2024). Trust in AI: progress, challenges, and future directions. *Humanities and Social Sciences Communications*. 2662-9992. *11*(1). doi:10.1057/s41599-024-04044-8.

Aggarwal, A., Tam, C. C., Dezhi, W., Xiaoming, L., & Qiao, S. (2023). Artificial intelligence–based chatbots for promoting health behavioral changes: systematic review. *Journal of Medical Internet Research*. 1438-8871. *25*, e40789. doi:10.2196/40789.

Akobeng, A. K. (2007). Understanding diagnostic tests 2: likelihood ratios, pre- and post-test probabilities and their use in clinical practice. *Acta Paediatrica*. 0803-5253. *96* (4), 487–491. doi:10.1111/j.1651-2227.2006.00179.x.

Alowais, S. A., Alghamdi, S. S., Alsuhebany, N., Alqahtani, T., Alshaya, A. I., Almohareb, S. N., ... Albekairy, A. M. (2023). Revolutionizing healthcare: the role of artificial intelligence in clinical practice. *BMC Medical Education*. 14726920. *23*(1). Available from https://www.bmcmededuc.biomedcentral.com/. doi:10.1186/s12909-023-04698-z.

Ashley, E. A. (2016). Towards precision medicine. *Nature Reviews Genetics*. 14710064. *17*(9), 507–522. Available from http://www.nature.com/reviews/genetics. doi:10.1038/nrg.2016.86.

Balogh, E. P., Miller, B. T., & Ball, J. R. (2015). Improving diagnosis in health care. Washington, D.C.: National Academies Press.

Bedi, S., Liu, Y., Orr-Ewing, L., Dash, D., Koyejo, S., Callahan, A., ... Shah, N. H. (2025). Testing and evaluation of health care applications of large language models: a systematic review. *JAMA*. 15383598. *333*(4). Available from http://www.jama.jamanetwork.com/journal.aspx. doi:10.1001/jama.2024.21700.

Benziger, C. P., Huffman, M. D., Sweis, R. N., & Stone, N. J. (2021). The Telehealth Ten: a guide for a patient-assisted virtual physical examination. *American Journal of Medicine*. 15557162. *134*(1), 48–51. Available from https://www.elsevier.com/locate/amjmed. doi:10.1016/j.amjmed.2020.06.015.

Berwick, D. M. (2021). Diagnostic excellence through the lens of patient-centeredness. *JAMA*. 15383598. *326*(21), 2127–2128. Available from http://jama.jamanetwork.com/journal.aspx. doi:10.1001/jama.2021.19513.

Bright, T. J., Wong, A., Dhurjati, R., Bristow, E., Bastian, L., Coeytaux, R. R., ... Lobach, D. (2012). Effect of clinical decision-support systems: a systematic review. *Annals of Internal Medicine*. 15393704. *157*(1), 29–43. Available from http://www.annals.org/data/Journals/AIM/24329/0000605-201207030-00006.pdf. doi:10.7326/0003-4819-157-1-201207030-00450.

Ceney, A., Tolond, S., Glowinski, A., Marks, B., Swift, S., Palser, T., & Wilson, F. A. (2021). Accuracy of online symptom checkers and the potential impact on service utilisation. *PLoS One*. 1932-6203. *16* (7), e0254088. doi:10.1371/journal.pone.0254088.

Chakraborty, A., Islam, M., Shahriyar, F., Islam, S., Zaman, H. U., & Hasan, M. (2023). Smart Home system: a comprehensive review. *Journal of Electrical and Computer Engineering*. 20900155. *2023*. Available from http://www.hindawi.com/journals/jece/. doi:10.1155/2023/7616683.

Challen, R., Denny, J., Pitt, M., Gompels, L., Edwards, T., & Tsaneva-Atanasova, K. (2019). Artificial intelligence, bias and clinical safety. *BMJ Quality & Safety*. 20445415. *28*(3), 231–237. Available from http://www.qualitysafety.bmj.com/content/by/year. doi:10.1136/bmjqs-2018-008370.

Chen, J. H., Dhaliwal, G., & Yang, D. (2022). Decoding artificial intelligence to achieve diagnostic excellence: learning from experts, examples, and experience. *JAMA*. 15383598. *328*(8), 709–710. Available from http://www.jama.jamanetwork.com/journal.aspx. doi:10.1001/jama.2022.13735.

Ćirković, A. (2020). Evaluation of four artificial intelligence–assisted self-diagnosis apps on three diagnoses: two-year follow-up study. *Journal of Medical Internet Research*. 1438-8871. *22*(12), e18097. doi:10.2196/18097.

Clark, M., & Bailey, S. (2024). Chatbots in health care: connecting patients to information. *Canadian Journal of Health Technologies*. 2563-6596. *4*(1). doi:10.51731/cjht.2024.818.

D'Amario, D., Canonico, F., Rodolico, D., Borovac, J. A., Vergallo, R., Montone, R. A., ... Crea, F. (2020). Telemedicine, artificial intelligence and humanisation of clinical pathways in heart failure management: back to the future and beyond. *Cardiac Failure Review*. 20577559. *6*. Available from http://www.cfrjournal.com/. doi:10.15420/cfr.2019.17.

Dangi, R. R., Sharma, A., Vageriya, V., (2025). *Transforming healthcare in low-resource settings with artificial intelligence: Recent developments and outcomes. Public Health Nursing*, *42*(2), 1017–1030.

Freyer, O., Wiest, I. C., Kather, J. N., & Gilbert, S. (2024). A future role for health applications of large language models depends on regulators enforcing safety standards. *The Lancet Digital Health*. 25897500. *6*(9), e662. Available from https://www.sciencedirect.com/science/journal/25897500. doi:10.1016/S2589-7500(24)00124-9.

Gilbert, S., Mehl, A., Baluch, A., Cawley, C., Challiner, J., Fraser, H., ... Novorol, C. (2020). How accurate are digital symptom assessment apps for suggesting conditions and urgency advice? A clinical vignettes comparison to GPs. *BMJ Open*. 2044-6055. *10* (12), e040269. doi:10.1136/bmjopen-2020-040269.

Gill, S. S., Kumar, A., Singh, H., Singh, M., Kaur, K., Usman, M., & Buyya, R. (2022). Quantum computing: a taxonomy, systematic review and future directions. *Software: Practice and Experience*. 1097024X. *52*(1), 66–114. Available from http://www.onlinelibrary.wiley.com/journal/10.1002/(ISSN)1097-024X. doi:10.1002/spe.3039.

Gumilar, K. E., Indraprasta, B. R., Hsu, Y.-C., Zih-Ying, Y., Chen, H., Irawan, B., ... Tan, M. (2024). Disparities in medical recommendations from AI-based chatbots across different countries/regions. *Scientific Reports*. 2045-2322. *14*(1). doi:10.1038/s41598-024-67689-0.

Gyongyosi, L., & Imre, S. (2019). A survey on quantum computing technology. *Computer Science Review*. 15740137. *31*, 51–71. doi:10.1016/j.cosrev.2018.11.002.

Hainer, B. L., & Matheson, E. M. (2013). Approach to acute headache in adults. *American Family Physician*, 15320650, *87*(10), 682–687. Available from http://www.aafp.org/afp/2013/0515/p682.pdf.

Heilman, J. M., & West, A. G. (2015). Wikipedia and medicine: quantifying readership, editors, and the significance of natural language. *Journal of Medical Internet Research*. 1438-8871. *17* (3), e62. doi:10.2196/jmir.4069.

Hersh, W. R., Hickam, D. H., Severance, S. M., Dana, T. L., Krages, K. P., & Helfand, M. (2006). Diagnosis, access and outcomes: update of a systematic review of telemedicine services. *Journal of Telemedicine and Telecare*. 1357-633X. *12* (2_suppl), 3–31. doi:10.1258/135763306778393117.

Hirosawa, T., Harada, Y., Tokumasu, K., Ito, T., Suzuki, T., & Shimizu, T. (2024). Evaluating ChatGPT-4's diagnostic accuracy: Impact of visual data integration. *JMIR Medical Informatics*, *12*, e55627.

Iserson, K. V., & Moskop, J. C. (2007). Triage in medicine, Part I: concept, history, and types. *Annals of Emergency Medicine*. 01960644. *49* (3), 275–281. doi:10.1016/j.annemergmed.2006.05.019.

Jain, B. (2017). The key role of differential diagnosis in diagnosis. *Diagnosis*. 2194-802X. *4* (4), 239–240. doi:10.1515/dx-2017-0005.

Jani, V., Danford, D. A., Thompson, W. R., Schuster, A., Manlhiot, C., & Kutty, S. (2021). The discerning ear: cardiac auscultation in the era of artificial intelligence and telemedicine. *European Heart Journal - Digital Health*. 26343916. *2*(3), 456–466. Available from https://www.academic.oup.com/ehjdh. doi:10.1093/ehjdh/ztab059.

Johnson, A. E. W., Ghassemi, M. M., Nemati, S., Niehaus, K. E., Clifton, D., & Clifford, G. D. (2016). Machine learning and decision support in critical care. *Proceedings of the IEEE*. 15582256. *104*(2), 444–466. Available from http://ieeexplore.ieee.org/xpl/RecentIssue.jsp?punumber=5. doi:10.1109/JPROC.2015.2501978.

Kanter, G. P., & Packel, E. A. (2023). Health care privacy risks of AI chatbots. *JAMA*. 15383598. *330*(4), 311–312. Available from http://jama.jamanetwork.com/journal.aspx. doi:10.1001/jama.2023.9618.

Kaul, V., Enslin, S., & Gross, S. A. (2020). History of artificial intelligence in medicine. *Gastrointestinal Endoscopy*. 10976779. *92*(4), 807–812. Available from http://www.elsevier.com/inca/publications/store/6/2/3/1/3/7/index.htt. doi:10.1016/j.gie.2020.06.040.

Khan, W. Z., Ahmed, E., Hakak, S., Yaqoob, I., & Ahmed, A. (2019). Edge computing: a survey. *Future Generation Computer Systems*. 0167739X. *97*, 219–235. doi:10.1016/j.future.2019.02.050.

Kidwai, B., & Rk, N. (2020). Design and development of diagnostic chabot for supporting primary health care systems. *Procedia Computer Science*, 18770509, *167*, 75–84. doi: 10.1016/j.procs.2020.03.184 http://www.sciencedirect.com/science/journal/.

Kim, Y., Hyon, Y. K., Lee, S., Woo, S. D., Ha, T., & Chung, C. (2022). The coming era of a new auscultation system for analyzing respiratory sounds. *BMC Pulmonary Medicine*. 14712466. *22*(1). Available from http://www.biomedcentral.com/bmcpulmmed/. doi:10.1186/s12890-022-01896-1.

King, M. R. (2023). The future of AI in medicine: a perspective from a chatbot. *Annals of Biomedical Engineering*. 15739686. *51*(2), 291–295. Available from https://www.springer.com/journal/10439. doi:10.1007/s10439-022-03121-w.

Krizhevsky, A., Sutskever, I., & Hinton, G. E. (2012). ImageNet classification with deep convolutional neural networks. *Advances in Neural Information Processing Systems*, 10495258, *2*, 1097–1105.

Kurniawan, M. H., Handiyani, H., Nuraini, T., Hariyati, R. T. S., & Sutrisno, S. (2024). A systematic review of artificial intelligence-powered (AI-powered) chatbot intervention for managing chronic illness. *Annals of Medicine*. 13652060. *56*(1). Available from http://www.tandfonline.com/loi/iann20. doi:10.1080/07853890.2024.2302980.

Laymouna, M., Yuanchao, M., Lessard, D., Schuster, T., Engler, K., & Lebouché, B. (2024). Roles, users, benefits, and limitations of chatbots in health care: rapid review. *Journal of Medical Internet Research*. 1438-8871. *26*, e56930. doi:10.2196/56930.

Martin, J. H., & Jurafsky, D. (2000). *Speech and language processing: An introduction to natural language processing, computational linguistics, and speech recognition* (1st ed.). Upper Saddle River, New Jersey: Prentice Hall.

Mennella, C., Maniscalco, U., De Pietro, G., & Esposito, M. (2024). Ethical and regulatory challenges of AI technologies in healthcare: a narrative review. *Heliyon*. 24058440. *10* (4), e26297. doi:10.1016/j.heliyon.2024.e26297.

Murdoch, B. (2021). Privacy and artificial intelligence: challenges for protecting health information in a new era. *BMC Medical Ethics*. 1472-6939. *22*(1). doi:10.1186/s12910-021-00687-3.

Pacis, D. M. M., Subido, E. D. C., & Bugtai, N. T. (2018). Trends in telemedicine utilizing artificial intelligence. *AIP Conference Proceeding*, *1933*, 15517616. doi: 10.1063/1.5023979 http://scitation.aip.org/content/aip/proceeding/aipcp.

Passi, S., & Vorvoreanu, M. (n.d.). Overreliance on AI literature review 2022. Available from: https://www.microsoft.com/en-us/research/uploads/prod/2022/06/Aether-Overreliance-on-AI-Review-Final-6.21.22.pdf.

Price, W. N., & Cohen, I. G. (2019). Privacy in the age of medical big data. *Nature Medicine*. 1546170X. *25*(1), 37–43. Available from http://www.nature.com/nm/index.html. doi:10.1038/s41591-018-0272-7.

Quazi, S. (2022). Artificial intelligence and machine learning in precision and genomic medicine. *Medical Oncology*. 1559131X. *39*(8). Available from https://www.springer.com/journal/12032. doi:10.1007/s12032-022-01711-1.

Ren, L. Y. (2019). Isabel Pro. *Journal of the Canadian Health Libraries Association*. 1708-6892. *40* (2), 63–69. doi:10.29173/jchla29418.

Saab, K., Tu, T., Weng, W.-H., Tanno, R., Stutz, D., Wulczyn, E. et al. (n.d.). Capabilities of gemini models in medicine.

Saenger, J. A., Hunger, J., Boss, A., & Richter, J. (2024). Delayed diagnosis of a transient ischemic attack caused by ChatGPT. *Wiener Klinische Wochenschrift*. 0043-5325. *136* (7-8), 236–238. doi:10.1007/s00508-024-02329-1.

Sarikaya, F. (2024). The cycles of AI winters: a historical analysis and modern perspective historical context and major AI winters the first AI winter (1974-1980).

Schachner, T., Keller, R., & Wangenheim, F. V. (2020). Artificial intelligence-based conversational agents for chronic conditions: systematic literature review. *Journal of Medical Internet Research*. 1438-8871. *22* (9), e20701. doi:10.2196/20701.

Schmieding, M. L., Kopka, M., Schmidt, K., Schulz-Niethammer, S., Balzer, F., & Feufel, M. A. (2022). Triage accuracy of symptom checker apps: 5-year follow-up evaluation. *Journal of Medical Internet Research*. 1438-8871. *24* (5), e31810. doi:10.2196/31810.

Sharma, S., Rawal, R., & Shah, D. (2023). Addressing the challenges of AI-based telemedicine: best practices and lessons learned. *Journal of Education and Health Promotion*. 2277-9531. *12*(1). doi:10.4103/jehp.jehp_402_23.

Shorten, C., Khoshgoftaar, T. M., & Furht, B. (2021). Deep learning applications for COVID-19. *Journal of Big Data*. 21961115. *8*(1). Available from https://journalofbigdata.springeropen.com. doi:10.1186/s40537-020-00392-9.

Singh, R., & Gill, S. S. (2023). Edge AI: a survey. *Internet of Things and Cyber-Physical Systems*.26673452. *3*, 71–92. Available from https://www.sciencedirect.com/science/journal/26673452. doi:10.1016/j.iotcps.2023.02.004.

Srinivasan, S. M., & Sharma, V. (2025). Applications of AI in cardiovascular disease detection—a review of the specific ways in which AI is being used to detect and diagnose cardiovascular diseases. In A. G. Rajesh Singh, N. Rathour, & S. V. Akram (Eds.), *AI in disease detection* (pp. 123–146). New Jersey: Wiley. Available from: https://onlinelibrary.wiley.com/doi/book/10.1002/9781394278695. doi:10.1002/9781394278695.ch6.

Stubberud, A., Langseth, H., Nachev, P., Matharu, M. S., & Tronvik, E. (2024). Artificial intelligence and headache. *Cephalalgia*. 0333-1024. *44*(8). doi:10.1177/03331024241268290.

Sutton, R. T., Pincock, D., Baumgart, D. C., Sadowski, D. C., Fedorak, R. N., & Kroeker, K. I. (2020). An overview of clinical decision support systems: benefits, risks, and strategies for success. *Npj Digital Medicine*. 23986352. *3*(1). Available from https://www.nature.com/npjdigitalmed/. doi:10.1038/s41746-020-0221-y.

Taylor, R. A., Sangal, R. B., Smith, M. E., Haimovich, A. D., Rodman, A., Iscoe, M. S., … Declan, A. (2025). Leveraging artificial intelligence to reduce diagnostic errors in emergency medicine: challenges, opportunities, and future directions. *Academic Emergency Medicine*. 15532712. *32*(3), 327–339. Available from http://onlinelibrary.wiley.com/journal/10.1111/(ISSN)1553-2712. doi:10.1111/acem.15066.

Tuckson, R. V., Edmunds, M., & Hodgkins, M. L. (2017). Telehealth. *Journal of Medicine*. 15334406. *377*(16), 1585–1592. Available from http://www.nejm.org/doi/pdf/10.1056/NEJMsr1503323. doi:10.1056/NEJMsr1503323.

Vaidyam, A. N., Wisniewski, H., Halamka, J. D., Kashavan, M. S., & Torous, J. B. (2019). Chatbots and conversational agents in mental health: a review of the psychiatric landscape. *Canadian Journal of Psychiatry*. 14970015. *64*(7), 456–464. Available from https://journals.sagepub.com/home/cpa. doi:10.1177/0706743719828977.

von Eschenbach, W. J. (2021). Transparency and the black box problem: why we do not trust AI. *Philosophy & Technology*. 2210-5433. *34* (4), 1607–1622. doi:10.1007/s13347-021-00477-0.

Wallace, W., Chan, C., Chidambaram, S., Hanna, L., Iqbal, F. M., Acharya, A., … Darzi, A. (2022). The diagnostic and triage accuracy of digital and online symptom checker tools: a systematic review. *Npj Digital Medicine*. 23986352. *5*(1). Available from https://www.nature.com/npjdigitalmed/. doi:10.1038/s41746-022-00667-w.

Watari, T., & Schiff, G. D. (2023). Diagnostic excellence in primary care. *Journal of General and Family Medicine*. 21897948. *24*(3), 143–145. Available from: https://www.onlinelibrary.wiley.com/journal/21897948. doi:10.1002/jgf2.617.

WebMD (n.d.). Symptom checker. Available from: https://symptoms.webmd.com/.

Wei, J., Wang, X., Schuurmans, D., Bosma, M., Ichter, B., Xia, F., … Zhou, D. (2022). Chain-of-thought prompting elicits reasoning in large language models. *arXiv*. 23318422. Available from: https://arxiv.org. doi:10.48550/arXiv.2201.11903.

Weizenbaum, J. (1966). ELIZA—a computer program for the study of natural language communication between man and machine. *Communications of the ACM*. 0001-0782. *9* (1), 36–45. doi:10.1145/365153.365168.

Yang, D., Fineberg, H. V., & Cosby, K. (2021). Diagnostic excellence. *JAMA*. 0098-7484. *326* (19), 1905. doi:10.1001/jama.2021.19493.

Yu, F., Zhang, H., Tiwari, P., & Wang, B. (2024). Natural language reasoning, a survey. *ACM Computing Surveys*. 0360-0300. *56* (12), 1–39. doi:10.1145/3664194.

Zhao, W. X., Zhou, K., Li, J., Tang, T., Wang, X., Hou, Y., … Wen, J. R. (2023). A survey of large language models. *arXiv*. 23318422. Available from: https://arxiv.org. doi:10.48550/arXiv.2303.18223.

Zhou, N., Zhang, Z., Nair, V. N., Singhal, H., & Chen, J. (2022). Bias, fairness and accountability with artificial intelligence and machine learning algorithms. *International Statistical Review*. 17515823. *90*(3), 468–480. Available from http://onlinelibrary.wiley.com/journal/10.1111/(ISSN)1751-5823. doi:10.1111/insr.12492.

Zong, M., & Krishnamachari, B. (2022). A Survey on GPT-3. *arXiv*. 23318422. Available from: https://arxiv.org. doi:10.48550/arXiv.2212.00857.

Chapter 5

Adaptive network-based fuzzy inference system-enhanced bidirectional time-aware LSTM for clinical sequence prediction

Shambhab Chaki, Pratyusha Rakshit, and Shriya Adhikari
Department of Electronics & Telecommunication Engineering, Jadavpur University, Kolkata, India

1 Introduction

With the increasing adoption of electronic health records (EHRs), sequence prediction in healthcare has become a critical area of research (Al-Selwi et al., 2024). Initially, *long short-term memory* (LSTM) networks were widely employed for the task of sequence prediction of EHR (Al-Selwi et al., 2024). However, in the context of prediction of significantly long clinical sequence, LSTMs often fail to solve the vanishing and exploding gradient problems of *recurrent neural networks* (RNNs) (Noh, 2021).

To overcome this impasse, transformer-based architectures have been increasingly adopted. However, their application to EHR data remains challenging due to time irregularities and missing entries (Vaswani et al., 2017). Standard transformers rely on fixed positional encodings, which assume uniformly spaced events—an assumption that breaks down in the context of irregularly timed clinical records, proving them inefficient (Zeng, Chen, Zhang, & Xu, 2023). BEHRT (Li et al., 2020) incorporated patient age (in years and months) into its positional encodings to introduce notion of time, yet it is not equipped to capture finer temporal resolutions (e.g., days, hours, minutes). HiTANet (Luo, Ye, Xiao, & Ma, 2020) attempted to resolve this by applying non-linear transformations of timestamp information as part of visit-level representations, and by integrating them as global keys. This design is conceptually related to potentially hazardous time-gating mechanism. Moreover, missing data could directly affect model performance in case of timestamp-based architecture design. Recent work has explored integrating elapsed time between consecutive inputs directly into the attention mechanism (Chen et al., 2023), but this often leads to increased computational complexity without consistent performance gains (Tan et al., 2024). These findings have led to renewed interest in recurrent architectures (Beck et al., 2024).

Inclusion of time-gates in neural architectures, especially in recurrent ones, causes an interference in the relation between the current input and output (Baytas et al., 2017). To balance the trade-off between the prediction accuracy using temporal data and minimal disturbance in the input-output relation, *time-aware LSTM* (T-LSTM) (Baytas et al., 2017) emerges as a potential tool. T-LSTM effectively captures time irregularities and handles missing data in EHRs. Unfortunately, even though *bidirectional* LSTMs (Bi-LSTMs) (Woźniak et al., 2023) are widely used in various clinical applications (Wozniak et al., 2023), surprisingly, there is hardly any work on ameliorating the performance of T-LSTMs with the mechanism of bidirectional flow of information. Consequently, *bidirectional T-LSTM* (Bi-TLSTM) has never been explored before in clinical sequence prediction. We, in this work, propose a stacked Bi-TLSTM encoder-decoder architecture to predict future timesteps for input clinical sequences. Stacking deepens the problem of vanishing and exploding gradients through the layers. Hence, we use skip connections to get rid of gradient-related problems within layers of Bi-TLSTM encoder.

Despite of the clever inclusion of elapsed time (the time elapsed between the current and previous instance of the input sequence) as an implicit variable to modify the cell memory and eventually affecting the output hidden state, T-LSTM uses a fixed parametric function of elapsed time to decay the contents of previous cell memory. The underlying assumption is that the time is an independent variable which does not depend on the context and hence maintains a uniform distribution throughout. Hence, the *decay parameter* of memory-modifying function is fixed globally as a hyperparameter rather than adapting to the changes in local influence of elapsed time based on the current context. This global assumption of time

Applications of Artificial Intelligence in Healthcare. https://doi.org/10.1016/B978-0-443-43934-6.00008-0

invariance is an intrinsic problem of neural architectures from the time of its inception. That's why neural architectures are not really comfortable in handling highly localized changing influences of time, and hence in handling *time series* (TS).

For example, let us consider that a month ago, a patient is prescribed to take Ibuprofen daily and (s)he is now having a seizure attack, which is controlled by applying Lorazepam at the current timestep. Let's just consider another patient who has undergone bone marrow transplant a month ago. Cyclosporine, prescribed a month ago to him/her, is a medicine, which if skipped even for once, can be fatal for this patient. Now, (s)he is also having a seizure and Lorazepam is also applied. For both of these patients, current timestep holds the information of application of Lorazepam only, and same amount of elapsed time between current and previous instance. The T-LSTM can't differentiate between these two situations and will forget same amount of short-term information. Although, forgetting Cyclosporine from a month ago affects health with much higher degree than forgetting Ibuprofen, prescribed a month ago. Hence, roughly, the amount of short-term memory decay will be greater in case of the first patient.

Clinical sequence prediction (CSP) is a specialized form of *time series prediction* (TSP) that inherits the broader challenges of representing temporal information of TS data to neural networks. A core difficulty arises from the time invariance assumption implicit in most neural models. Architectures with uniformly spaced neurons simplify training by cleanly separating the linear progression of inputs from the nonlinear transformations layered on top. However, when this uniform spacing assumption is applied to TSP, it effectively enforces even temporal intervals between events, which oversimplifies the rich and irregular temporal structures often present in real-world sequences.

To address this, previous works has sought a wide variety of strategies to represent the TS to the time invariant neural architectures. Recent works involving transformers, such as Autoformer (Wu, Xu, Wang, & Long, 2021) consider breaking down the whole sequence in trend and seasonal part and learning the auto-correlation parameters jointly with the attention mechanism of the transformer. FEDformer (Zhou et al., 2022) employs decomposing the time sequence in distinct frequency components by employing Fourier and Wavelet transformation and use these representations to induce time invariance assumptions in different frequencies. These kinds of decomposition techniques either don't vary with the contextual information and is fixed once the input is given, or cannot capture different granularities of underlying invariance other than some broader ones like in trend-seasonal decomposition. Another important intervention is the use of patches, originally introduced by Rao, Nie, and Dai (2022). Patches represent distinct sub-sections of TS. Now, it is assumed that these sub-sections are time-invariant in PatchTST (Nie, Nguyen, Sinthong, & Kalagnanam, 2022). The idea of using patches is also popular in another evolving field of TS representation, the self-supervised learning setup. Use of patches to differentiate between positive samples from similar patch and negative samples from different patches, overcomes the impasse of assuming specific kinds of data distribution in other works involving subseries invariance (Franceschi, Dieuleveut, & Jaggi, 2019), transformation invariance (Tonekaboni, Eytan, & Goldengerg, 2021) and of course temporal invariance (Eldele et al., 2021) considerations. The use of random patches to generate positive samples is explored in Ts2vec (Yue et al., 2022), which also uses hierarchical temporal pattern extraction using *convolutional neural network* (CNN) with distinct dilation for distinct levels of temporal resolutions. These works highlight a consistent limitation: they handle time either through fixed encodings, predetermined decompositions of temporal resolutions, or learned invariances that are globally shared across contexts. Even continuous-time architectures such as neural ODE-inspired Contiformer (Chen et al., 2023) allow state-dependent evolution, but treat *time* as an implicit variable rather than an explicit, adaptive, and interpretable component.

As we know, our understanding of the underlying rules of physical world changed after Einstein's introduction of special theory of relativity (Bohm, 2015). Relativistic, context-dependent view of time replaced the concept of time as an independent variable with the concept time dependent on the continuous changing states of the observer. Hence, we argue that time should also be considered as a context-dependent setting, which influences the context itself. To address this argument, we introduce *adaptive network-based fuzzy inference system* (ANFIS) (Jang, 1993) to process the dependency of elapsed time on the context. We integrate ANFIS in between the timesteps of a Bi-TLSTM, where ANFIS computes the decay parameter for each timestep from both elapsed time Δt (elapsed time is calculated as the difference between the recorded time of the events at the current input timestep and the previous input timestep) and the evolving context, i.e. hidden state of the previous timestep h_{t-1} of the Bi-TLSTM encoder. This enables bidirectional coupling between temporal invariance representation and contextual features: context shapes how elapsed time is interpreted, and elapsed time influences how context is represented. The result is a model that can adapt its temporal processing strategy to irregular, context-sensitive sequences—such as those encountered in clinical event prediction or CSP—while maintaining interpretability through the fuzzy rule base. Fig. 5.1 visually represents the concise idea of the proposed architecture.

Evidently, choice of the hidden state dimension will affect the decision-making capability and interpretability of the ANFIS. If the dimension is too large, which is common in most of the CSP tasks, the number of antecedent fuzzy

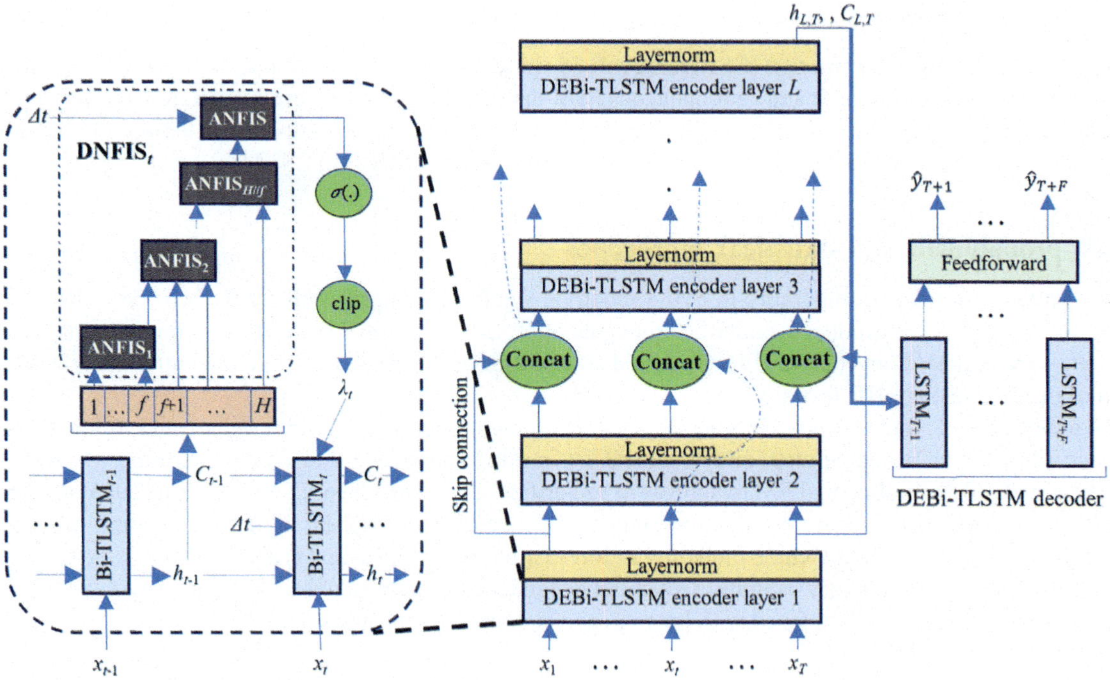

FIG. 5.1 Proposed combination of ANFIS and stacked Bi-TLSTM for CSP.

membership functions will increase proportionally. This will reduce the capability of generalization, cause instability in training and decrease the interpretability of the fuzzy memberships and associated firing strengths in subsequent data processing. Hence, we propose a deep layered architecture of ANFIS, which at the one hand process h_{t-1} separately with more than one layers, and combine the fuzzy interpretation of h_{t-1} with the elapsed time by feeding them together into another ANFIS. This layered structure of ANFIS is termed as *deep neuro-fuzzy inference system* or DNFIS. Consequently, DNFIS now outputs a real number which is further processed by a neural layer to have the value of decay parameter of Bi-TLSTM for current timestep. This decay parameter is the one which modifies the previous cell memory to influence the decision making of current Bi-TLSTM cell.

A fuzzy rule-based architecture DNFIS and a recurrent architecture Bi-TLSTM exhibits instability in training when combined together without taking much precautions. We found that to combine them together stably, we have to maintain different timescale to update the parameters of the DNFIS. Consequently, the hybrid learning rule (Jang, 1993) of ANFIS is applied with different timescale to train the DNFIS, where the consequent parameters are updated after every K timesteps whereas the premise or antecedent parameters of DNFIS and parameters of Bi-TLSTM are updated at each timestep. After successfully combining DNFIS and Bi-TLSTM, we name the proposed architecture as *DNFIS-enabled Bi-TLSTM* (DEBi-TLSTM).

We conduct rigorous experiments involving four diseases including (i) Parkinson's disease, (ii) chronic kidney disease, (iii) heart failure and (iv) hypertensive disease and three datasets including (i) *Medial Information Mart for Intensive Care* (MIMIC)-III version 1.4 (Johnson et al., 2016), (ii) MIMIC-IV v3.1 (Johnson et al., 2024) and (iii) *Parkinson's Progression Marker's Initiative* (PPMI) (Marek et al., 2011). We compared the performance of the proposed DEBi-TLSTM against four baselines including old renowned LSTMs (i) T-LSTM (Baytas et al., 2017), (ii) BiLSTM (Woźniak, Wieczorek, & Siłka, 2023), (iii) *bidirectional minimal gated recurrent unit* (BiGRU) (Rodrigues Jr, Gutierrez, Spadon, Brandoli, & Amer-Yahia, 2021) and a newly introduced LSTM minLSTM (Feng, Tung, Ahmed, Bengio, & Hajimirsadeghi, 2024). Our model came out on top in all five experiments and also outperformed with statistically significant superiority over the baselines. We also conducted rigorous ablation study to see the effect of each component, theoretical and practical structural considerations.

This chapter is divided into six different sections. The second section reviews some relevant literatures. The third section introduces Bi-TLSTM. The fourth section details about the fusion of ANFIS and Bi-TLSTM. The fifth section reports about the experimental setups and results, discusses about results and potential improvements. The sixth section concludes the study and discusses about future works.

2 Literature review

Research on *clinical sequence prediction* (CSP) has evolved along multiple methodological streams, each addressing specific challenges in modeling irregular and heterogeneous EHR data. Below, we group prior works into (i) early probabilistic/statistical approaches, (ii) neural sequence models, and (iii) transformers and self-supervised learning-based approaches, followed by a synthesis that connects these works to the motivation for our DEBi-TLSTM.

2.1 Early probabilistic and statistical approaches

The earliest attempts to model temporality in EHRs relied on symbolic and probabilistic frameworks. Zhao, Papapetrou, Asker, and Boström (2017) introduced symbolic aggregate approximation (SAX) to capture sequential information from complex longitudinal data. Tran et al. (2014) developed an entity-to-event mapping with convolution filter banks for auto-extraction of heterogeneous EHR features.

Markov-chain based frameworks became popular for predicting hospital admissions and disease trajectories. Arandjelović (2015) and Wang, Sontag, and Wang (2014) used discrete-time and continuous-time Markov processes, respectively, with Gibbs sampling and EM to address irregularities. Choi, Du, Chen, Song, and Sun (2015) applied Hawkes processes to predict risk from apparently random clinical event sequences. Rakshit et al. (2021) combined HDBSCAN-based patient grouping with Markov chain modeling to forecast future activities and costs.

While these approaches explicitly captured sequential dependencies, they lacked scalability, struggled with long-range dependencies, and were brittle when faced with heterogeneous or high-dimensional EHR data.

2.2 Neural sequence models (RNNS, LSTMS, GRUs and variants)

The success of RNNs made them a natural choice for EHR analysis. DoctorAI (Choi, Bahadori, Schuetz, Stewart, & Sun, 2016) applied RNNs to predict future diagnoses and medications. Duan, Sun, Dong, He, and Huang (2020) extended RNNs with multivariate Hawkes processes for long-term dependencies.

LSTM architectures gained popularity for capturing complex illness trajectories. Hügle et al., 2021) used LSTMs to model rheumatoid arthritis progression; Pham, Tran, Phung, and Venkatesh (2016) studied illness trajectories with time-parameterized LSTMs; Zhu et al. (2017) integrated time-gates into LSTM to model irregular intervals. Relation-aware and attention-based LSTMs further expanded applicability: An et al. (2020) modeled medication-disease relations; Gandin, Scagnetto, Romani, and Barbati (2021) introduced attention for feature dependencies; Men, Ilk, Tang, and Liu (2021) merged attention with time-awareness; Zhang et al. (2021) proposed Bi-LSTM with attention-based time-encoded embeddings.

Despite improvements, these methods often encoded raw timestamps, which can mislead due to patient-specific anonymization shifts (Pham et al., 2016; Zhang et al., 2021; Men et al., 2021; Zhang et al., 2021), or used time-gates that interfered with input-output flows (Zhu et al., 2017), leading to unstable training.

HiTANet (Luo et al., 2020) attempted to hierarchically capture intra-visit-level and visit-sequence-level patterns using elapsed time and timestamp-driven attention. However, reliance on raw timestamps and vulnerability to missing data remained unaddressed.

Baytas et al. (2017) proposed the time-aware LSTM (T-LSTM), which decays memory using elapsed time, avoiding direct timestamp encoding and stabilizing training relative to time-gates. Missing data were handled as elongated elapsed time. Nonetheless, T-LSTM used a fixed, global parametric decay function, failing to adapt decay dynamically to context.

GRU-based models also emerged. Lee et al. (2019) modified GRUs for neurodegeneration prediction; Liu et al. (2019) combined GRUs with hierarchical visit representation; Rodrigues Jr et al. (2021) employed BiGRU for scalable trajectory prediction. These models showed strong predictive performance but lacked explicit solutions for irregular temporal dynamics.

2.3 Transformer and self-supervised contrastive learning

The introduction of transformers and BERT-style embeddings (Devlin, Chang, Lee, & Toutanova, 2019) into healthcare modeling broadened CSP. Med-BERT (Rasmy, Xiang, Xie, Tao, & Zhi, 2021) and BEHRT (Li et al., 2020) adapted BERT to clinical notes and EHRs. While transformers benefited from parallelization, they suffered from time irregularity handling issues (Koroteev, 2021; Zeng et al., 2023).

To mitigate this, Autoformer (Wu et al., 2021) decomposed time series into trend and seasonal parts, while FEDformer (Zhou et al., 2022) used Fourier and Wavelet decompositions. These approaches introduced efficiency but relied on fixed decompositions, losing contextual adaptability.

Patch-based representations introduced in Rao et al. (2022) inspired Ts2Vec (Yue et al., 2022) for self-supervised learning. These methods assumed global patch invariance instead of time-invariance of each timestep and used hierarchical *convolutional neural net* (CNN) to capture time-invariance at distinct levels using distinct dilation in CNN. They upgraded the TS generalization task from previous related SSL works, which considered different kinds of time-invariance including: (i) subseries invariance (Franceschi et al., 2019), (ii) transformation invariance (Tonekaboni et al., 2021), and (iii) temporal invariance (Eldele et al., 2021) and as a result, were sensitive to specific kinds of data distribution of TS—consequently were insensitive to other kinds of TS data. Later, patch-based TS feature extraction was also integrated with transformers including PatchTST (Nie et al., 2022).

Continuous-time models such as Contiformer (Chen et al., 2023) attempted state-dependent evolution using ODE-inspired designs but still treated time implicitly along with time-invariance assumption, not emphasizing time as a context-dependent entity.

Recent years also witnessed efficient sequential architectures like minLSTM (Feng et al., 2024), xLSTM (Beck et al., 2024), and state-space models like MAMBA (Gu & Dao, 2023; Qu et al., 2024), which attempted parallelization or continuous-time fusion but still inherited the global time-invariance assumption.

2.4 Synthesis and gap identification

Across all these methodologies, a common limitation emerges: time is consistently treated as a global, independent, or invariant variable—whether as Markovian transitions, as raw timestamps or elapsed time variable with fixed parametric decays, as global keys or global decomposition components, or as invariant patches. This assumption oversimplifies the irregular and context-sensitive temporal dynamics inherent in real-world EHR data.

Our proposed DEBi-TLSTM overcomes this by (i) clubbing the sequential learning power of Bi-LSTM and time-aware learning capability of T-LSTM using elapsed time and (ii) introducing DNFIS-based adaptive decay parameters to couple elapsed time with evolving context, thereby enabling patient- and context-specific temporal modeling.

Hence, our proposed model does not treat time as an implicit independent variable. Rather, the context (hidden state output at previous timestep) influences the latent representation of elapsed time between current and previous timesteps. A deep neuro-fuzzy adaptive network is responsible to generate a decay parameter (taken as a globally-fixed hyperparameter in previous work (Baytas et al., 2017) out of the previous hidden state output and elapsed time inputs, which in turn changes the interpretation of elapsed time within each successive timestep. This context-dependent interpretation of time is easily clubbed within the framework of Bi-LSTM, i.e. the existing framework of time-invariance, to have a neural sequence model which is trained to predict the future timesteps of input clinical TS sequences. Table 5.1 provides the summary of literatures.

TABLE 5.1 Summary of literatures.

Established method (references)	Problem/issue addressed	Limitation (addressed by DEBi-TLSTM)
SAX (Zhao et al., 2017)	Symbolic aggregate approximation (SAX) of sequential EHR data	Coarse approximation; ignores context and irregular gaps
Entity-event mapping (Tran et al., 2014)	Feature extraction using entity-event mapping with convolutional filters	Lacks temporal modeling and adaptability
Markov chains (Arandjelović, 2015, Wang et al., 2014)	Irregularity in disease progression alongside comorbidities	Rigid transitions; poor scalability for long sequences
Hawkes Process (Choi et al., 2015)	Pattern extraction from apparently randomized clinical events	Computationally expensive; limited contextual learning
HDBSCAN + Markov (Rakshit et al., 2021)	Grouping + forecasting future activities	Irregularities in time intervals was not considered
DoctorAI (Choi et al., 2016)	Generic predictive model for diagnoses & medications	Irregularities in time intervals was not considered

Continued

TABLE 5.1 Summary of literatures—cont'd

Established method (references)	**Problem/issue addressed**	**Limitation (addressed by DEBi-TLSTM)**
RNN + Hawkes (Duan et al., 2019)	Interpretable dependencies with RNN	Irregularities in time intervals was not considered
LSTM for RA (Hügle et al., 2021)	Disease progression with varying lists	Lacks temporal modeling and adaptability
LSTM trajectories (Pham et al., 2016)	Illness trajectory representation	Lacks temporal modeling and adaptability
Time-gated LSTM (Zhu et al., 2017)	Capturing irregular time intervals through time-gates	Time-gates interfere with input-output flow
Relation-aware LSTM (An et al., 2018)	Medication-disease dependencies	Lacks temporal modeling and adaptability
Attention LSTM (Gandin et al., 2021)	Attention-based feature dependencies	Global attention; insensitive to local gaps
Time-aware attention (Men et al., 2021)	Extending LSTM with time-aware attention	Timestamp-dependent temporal awareness; timestamp shifts due to anonymization of EHRs mislead dynamics; misaligned with missing data
BiLSTM + time-encoded embeddings (Zhang et al., 2021)	Event embedding with BiLSTM	Timestamp-dependent temporal awareness; timestamp shifts due to anonymization of EHRs mislead dynamics; misaligned with missing data
HiTANet (Luo et al., 2020)	Code-level intra-visit and visit sequence-level inter-visit representation	Timestamp-dependent temporal awareness (timestamp of the last day under a visit is used in attention); timestamp shifts due to anonymization of EHRs mislead dynamics; misaligned with missing data
Established method	**Problem/issue addressed**	**Limitation (addressed by DEBi-TLSTM)**
T-LSTM (Baytas et al., 2017)	LSTM cell memory decay using elapsed time information	Globally fixed decay parameter; not context-adaptive
Modified GRU (Lee et al., 2019)	Neurodegeneration trajectory	No explicit irregular time handling
Hierarchical GRU (Liu et al., 2020)	Short/long temporal dependencies	Still lacks dynamic elapsed-time modeling
BiGRU (Rodrigues Jr et al., 2021)	Scalability and trajectory prediction	Ignores irregular temporal gaps
Med-BERT (Rasmy et al., 2021), BEHRT (Li et al., 2020)	Transformer-based patient representations	Globally fixed positional encoding, lacks temporal irregularity-modeling, time-awareness considered through age representations using months or years, not reaching deeper granularities
Autoformer (Wu et al.. 2021)	Trend/seasonal decomposition-based transformer architecture	Globally fixed temporal decomposition; non-contextual modeling of time, sees time as an independent variable
FEDformer (Zhou et al., 2022)	Fourier/Wavelet frequency decomposition-based transformer architecture	Frequency domain holds richer temporal information through frequency-wise concise contextual representations at the cost of losing important contextual information, sees time as an independent variable
Subseries (Franceschi et al., 2019)/ transform (Tonekaboni et al., 2021)/ temporal (Eldele et al., 2021) invariance for TS representation	Self-supervised TS representations	Assumes context-agnostic temporal invariance across different temporal levels and data distributions, sees time as an independent variable
PatchTST (Nie et al., 2022), Ts2Vec (Yue et al., 2022)	Patch/subseries invariance for self-supervised TS representation across distinct temporal levels using distinct dilation of CNNs	Temporal hierarchies are globally fixed once the dilation levels are chosen, ignores context, sees time as an independent variable
Contiformer (Chen et al., 2023)	Continuous-time neural-ODE enhanced transformer	Time implicit and independent, not adaptive to context
minLSTM (Feng et al., 2024), xLSTM (Beck et al., 2024)	Parallelized LSTMs for TS	No context-sensitive time component, global time-invariant assumption
MAMBA (Gu & Dao, 2023; Qu et al., 2024)	State-space continuous-time models	Global dynamics; lacks context-aware time representation, sees time as an implicit component and an independent variable

3 Methodology: Bidirectional T-LSTM

We introduce an *adaptive neuro-fuzzy inference system* (ANFIS) (Jang, 1993) to model the dependency of elapsed time on context. Specifically, ANFIS is embedded between the timesteps of a Bi-TLSTM, where it computes the decay parameter for each timestep using two inputs: (i) the elapsed time Δt (defined as the difference between the timestamps of consecutive events) and (ii) the evolving context represented by the previous hidden state h_{t-1} of the Bi-TLSTM encoder. This design creates a bidirectional coupling: context shapes how elapsed time is interpreted, while elapsed time influences how context is represented. The resulting model can adapt its temporal processing to irregular, context-sensitive sequences—common in clinical sequence prediction (CSP)—while preserving interpretability through fuzzy rule-based reasoning (Fig. 5.1).

A critical consideration is the dimensionality of the hidden state. Large hidden state sizes, typical in CSP tasks, cause a proportional increase in antecedent fuzzy membership functions, which can (i) reduce generalization capability, (ii) destabilize training, and (iii) weaken interpretability of membership functions and firing strengths. To overcome this, we propose a *deep neuro-fuzzy ANFIS* (DNFIS) architecture. In DNFIS, h_{t-1} is first processed through multiple ANFIS layers, and its fuzzy interpretation is then combined with elapsed time Δt in a subsequent ANFIS. The final DNFIS output is passed through a neural layer to generate the decay parameter for the current Bi-TLSTM timestep. This decay parameter directly modulates the previous cell memory, thereby influencing decision-making in the current Bi-TLSTM cell.

This section recapitulates *time-aware LSTM* (T-LSTM), introduces *bidirectional time-aware LSTM* (Bi-TLSTM) and stacking layers of encoders of Bi-TLSTM clubbed with skip connections, alongside a one-layer decoder.

3.1 Recapitulations of time-aware LSTM

Time-aware LSTM (T-LSTM) combines the LSTM (Al-Selwi et al., 2024) with a memory decomposition technique. At first, a short-term memory C_{t-1}^S is estimated out of the previous cell memory C_{t-1} by Eq. (1) and the remaining component is called the long-term memory C_{t-1}^T. This decomposition uses W_s and b_s as weight matrix and bias vector of learnable parameters.C_{t-1}^S is further modified using a decaying function $y(.)$ of elapsed time Δt in Eq. (3). Here Δt denotes the time gap between the recording instants of the immediate past input and the current input. It ensures that the influence of the short-term characteristics decays as the elapsed time grows, whereas influence of the long-term characteristics remains intact. Now, the modified short-term memory $\tilde{C}_{t-1}^S$ and C_{t-1}^T are again combined in Eq. (4) to form the modified previous cell memory $\tilde{C}_{t-1}$, which is then fed into the current block of LSTM to produce the output hidden state h_t and the current cell memory C_t. The mathematical representations are summarized below:

$$C_{t-1}^S = \tanh\left(W_s C_{t-1} + b_s\right) \tag{1}$$

$$C_{t-1}^T = C_{t-1} - C_{t-1}^S \tag{2}$$

$$\tilde{C}_{t-1}^S = C_{t-1}^S \odot y(\Delta t) \tag{3}$$

$$\tilde{C}_{t-1} = C_{t-1}^T + \tilde{C}_{t-1}^S \tag{4}$$

$$h_t, C_t = \text{LSTM}\left(x_t, h_{t-1}, \tilde{C}_{t-1}\right) \tag{5}$$

$$y(\Delta t) = \lambda / \ln\left(e + \Delta t\right) \tag{6}$$

$$h_t, C_t = \text{TLSTM}\left(x_t, h_{t-1}, C_{t-1}, \lambda, \Delta t\right) \tag{7}$$

All LSTM operations are concisely represented in Eq. (5) as per (Al-Selwi et al., 2024). The decaying function of elapsed time is described in Eq. (6). Eqs. (1)–(6) are concisely represented in Eq. (7). x_t is the current input, λ is the decay parameter. $\odot$ denotes the element-wise multiplication.

The T-LSTMs are capable of handling clinical timeseries sequences in two ways: (i) the modification of the previous cell memory using elapsed time gets rid of modelling absolute timestamps and its nuances, not interfering with the path of the current input-output data flow, and thus providing a better generalization capability than time gates, and (iii) missing data are modelled as elongated elapsed times, resulting into better adaptability against missing entries in EHRs.

3.2 Bidirectional T-LSTM

Encoder. The encoder architecture comprises a T-LSTM in normal forward direction, recognized by the output $h_{t,f}$ in Eq. (8) and a T-LSTM in reverse backward direction, recognized by the output $h_{t,r}$ in Eq. (9). These two outputs then combine at a neural layer to produce the overall output hidden state h_t, as described in Eq. (10).

$$h_{t,f} = \text{TLSTM}(x_t, h_{t-1}, C_{t-1}, \lambda, \Delta t) \tag{8}$$

$$h_{t,r} = \text{TLSTM}(x_t, h_{t+1}, C_{t+1}, \lambda, \Delta t) \tag{9}$$

$$h_t = \sigma\left(W_f h_{t,f} + W_r h_{t,r} + b_{bi}\right) \tag{10}$$

where, W_f, W_r and b_{bi} are weights and biases of the combining neural layer, $\sigma(.)$ is the sigmoid function and h_t in Eq. (10) is the final output hidden state of the proposed Bi-TLSTM. Cell memories in both directions are initialized with patient's static information embedding including demographic and age embedding. Hidden states are initialized as zero vectors.

Decoder. Evidently, the decoder does not have access to the Δt. Hence, decoder simply uses an LSTM module given in Eq. (11). A *feedforward* (FF) layer processes the decoder hidden state output to predict the future timesteps of an input clinical sequence.

$$s_t = \text{LSTM}(\hat{s}_{t-1}, s_{t-1}, C_{s,t-1}) \tag{11}$$

$$\hat{y}_t = \xi(W_o s_t + b_o) \tag{12}$$

where, s_t is the current decoder hidden state output, $\hat{y}_{t-1}$ is the previous predicted output of the output FF layer in Eq. (12), $C_{s,t-1}$ is the previous decoder cell memory, $\xi(.)$ is any intended non-linearity and W_o, b_o represent weight and bias of the output FF layer. The first input, i.e. $\hat{y}_0$ is zero. $C_{s,0}$ is the cell memory at the last encoder timestep T of the last layer L of Bi-TLSTM encoder, i.e. $C_{L,T}$ and s_0 is the hidden state output at timestep T of the last layer L, i.e. $h_{L,T}$.

3.3 Bi-TLSTM encoder stacking

We then stack layers of Bi-TLSTM encoders by feeding the output hidden states of a layer to the input of the next layer. The layernorm (Xu, Sun, Zhang, Zhao, & Lin, 2019) is applied before feeding every layer output to the next layer. We also

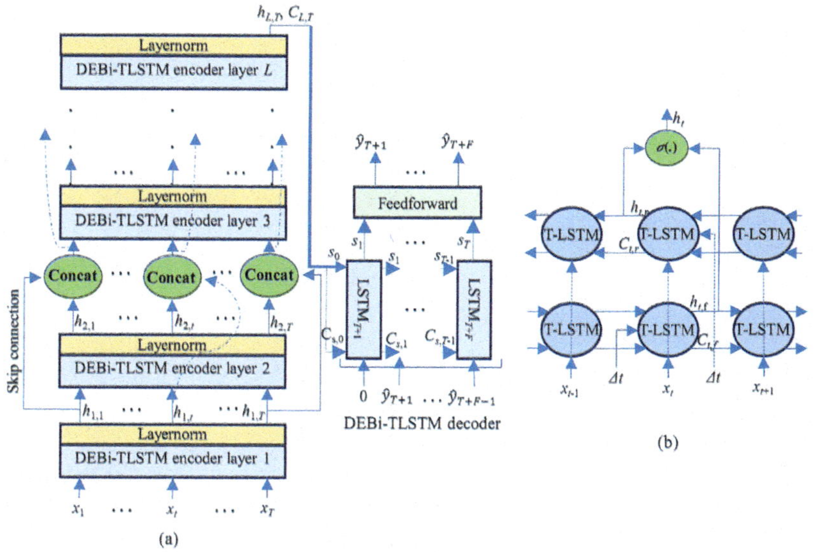

FIG. 5.2 (A) Stack of Bi-TLSTM encoder coupled with a layer of decoder and output FF layer, (B) a bidirectional T-LSTM at timestep t.

use skip connections between layers to help mitigate the vanishing and exploding gradient problems in deeper networks by providing a direct path for gradient flow during backpropagation. Skip connections combine with the normal ones using concatenation. On top of stacked Bi-TLSTM encoders, there is a decoder layer, which is as described in the previous subsection. An exemplar stacked Bi-TLSTM architecture is visually represented in Fig. 5.2.

4 Methodology: ANFIS-enabled Bi-TLSTM

This section elaborates on the proposed fusion of ANFIS and Bi-TLSTM.

4.1 Recapitulations of ANFIS

Adaptive network-based fuzzy inference system (ANFIS) (Jang, 1993) is a neuro-fuzzy inference system, which in most of the previous uses, is integrated with Takagi-Sugeno-Kang fuzzy logic system (Yang et al., 2016). Let us say, the input to the system is $X=\{x\}_i^f$. Let, R uniformly denotes the number of fuzzy IF-THEN rules. Hence, there are in total fR antecedent membership functions. Each ANFIS consists of the following sub-layers.

Sub-layer 1. The most commonly used premise membership function of ANFIS inference system is Gaussian, which our model also adopts. The role of this layer is to determine the degree of membership of each input dimension to the membership functions representing each antecedent clause. The mathematical representation is as following:

$$M_{ij}(x_i) = e^{-\left(\frac{x_i-\mu_{ij}}{\sigma_{ij}}\right)^2}$$

where, M_{ij} denotes the degree of membership of x_i to the antecedent fuzzy set A_{ij} and its membership function M_{ij} under rule j. μ_{ij} and σ_{ij} are learnable parameters.

Sub-layer 2. This layer gets the trigger strength of each rule by multiplying the membership degrees of the input to each antecedent under each rule. Hence, for $j = 1, 2, \ldots R$, the firing strength for rule j is,

$$w_j = \prod_i M_{ji}(x_i)$$

Sub-layer 3. This layer normalizes the firing strengths of each rule to characterize the probability of firing each rule under the rule base. Hence, the normalized firing strength is,

$$\overline{w}_j = \frac{w_j}{\sum_m w_m}$$

Sub-layer 4. This layer calculates the output with Takagi-Sugeno-Kang inference by calculating the weighting average of linear functions of each input. The parameters of the linear function are consequent parameters. Hence, the expression for this layer is,

$$\text{ANFIS}\left(\{x\}_i^f\right) = Y = \sum_{j=1}^{R}\sum_{i=1}^{f}\overline{w}_j\left(W_{ji}x_i + b_j\right)$$

where, W_{ji} and b_j are consequent parameters.

The fuzzy rule base consists of R rules and exemplar representation of one rule, rule j, is as following:

Rule j: If x_1 is A_{j1} and x_2 is A_{j2} and … x_f is A_{jf} Then $\text{ANFIS}\left(\{x\}_i^f\right) = Y$.

These rules partition the whole fuzzy space into f regions for every rule and hence, fR is the total number of fuzzy regions, each one represented by an antecedent membership. Usually, the antecedent parameters are updated using variants of *gradient descent* (GD) and the consequent parameters are updated using gradients on the *least squared error* (LSE). This is the hybrid learning paradigm that was originally suggested (Jang, 1993).

4.2 Layers of ANFIS

To process longer inputs, we stack layers of ANFIS blocks, where each layer gets input from the previous layer output and (f-1) bits out of the remaining input bits. Let, there are H input bits in total. Then the number of layers will be *num_layers* $= H // f$.

Hence, mth layer, $m \in \{2,3,\ldots,H//f\}$, denoted as ANFIS_m gets an input from the (m-1)th layer output Y_{m-1} and the current input segment, which starts from 2+(f-1)(m-1)th bit and ends at 1+(f-1)mth bit. They both are concatenated and fed as input. Ony the first ANFIS, i.e. $ANFIS_1$ gets the input segment which starts from the first bit and ends at *f*th bit. The final output is $Y_{H//f}$ of this *stacked ANFIS* (SANFIS) structure. Introduction of layers actually ensures that parameters are optimized for smaller subproblems attending to each input dimension carefully, greatly increasing generalization capability of the ANFIS. A huge single ANFIS could have also killed the interpretability because which antecedent produces how much firing strength gets intractable as the number of antecedents increases with the increase in the number of input dimensions.

4.3 Integration of ANFIs with Bi-TLSTM

This integration is pictorially depicted in Fig. 5.3. ANFIS takes the previous hidden state h_{t-1} of Bi-TLSTM encoder and the elapsed time Δt as input and convert them to an output, which is further processed by a sigmoid layer and a clipping function to generate the *decay parameter* λ_t at each timestep, which is fed into the Bi-TLSTM block at *t*th timestep.

Let, the hidden state dimension of the Bi-TLSTM encoder is *H*. Our proposed *deep neuro-fuzzy inference system* (DNFIS) consists of a SANFIS block which processes h_{t-1} and outputs $Y_{H//f,t}$, this output and the elapsed time Δt is then fed into another Takagi-Sugeno-Kang ANFIS, as defined previously. Let us denote the output of DNFIS, or the last ANFIS of DNFIS as $Y_{DNFIS,t}$. This last ANFIS completes the construction of DNFIS. The last ANFIS uses 4 rules. These rules are defined as following:

Rule *i*: IF $Y_{H//f,t}$ is *small/big* and Δt is *small/big* THEN $Y_{DNFIS,t} = p_i\, Y_{H//f,t} + q_i \Delta t + r_i$

Hence, we can concisely write,

$$\mathrm{DNFIS}(h_{t-1}, \Delta t) = Y_{DNFIS,t} = \mathrm{ANFIS}\left(Y_{H//f,t}, \Delta t\right) \tag{13}$$

This final output of the DNFIS, Y_{DNFIS} is not bounded within a predefined limit. Hence, there is high chance of blowing up to large values or diminishing to near-zero values. Hence, we process $Y_{DNFIS,t}$ with a neural layer before feeding to Bi-TLSTM. This is concisely represented as,

$$\lambda_t = \lambda_{\min} + (\lambda_{\max} - \lambda_{\min})\sigma\left(wY_{DNFIS,t} + b\right) \tag{14}$$

where, $\lambda_{\min}$ and $\lambda_{\max}$ are the intended minimum and maximum values of the decay parameter λ at any timestep; $\sigma(.)$ is the sigmoid function; $w, b \in \mathbb{R}$; $\lambda_0 = 1$.

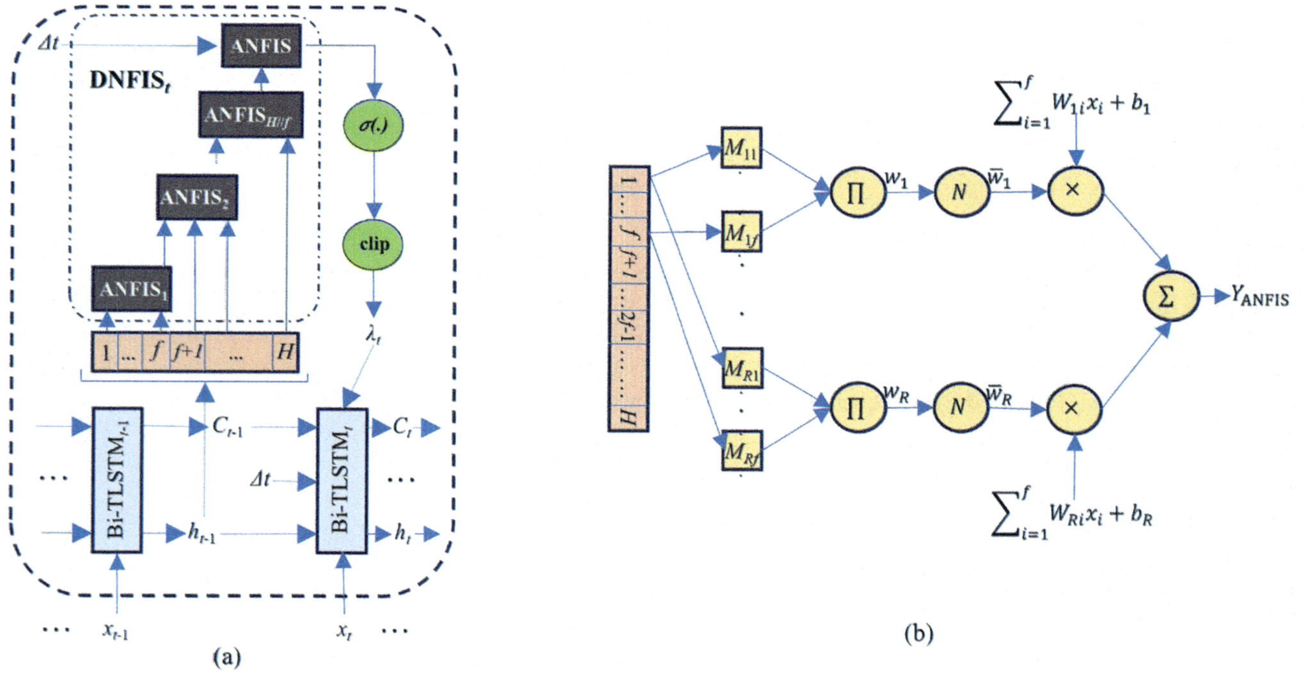

FIG. 5.3 (A) DEBi-TLSTM, (B) ANFIS.

4.4 Training

We want to maintain a smooth evolution of λ_t over timesteps. Hence, we introduce a smoothness penalty in the overall loss-term. The loss becomes,

$$L_{\text{total}} = L_{\text{task}} + \frac{1}{T}\sum_{t=1}^{T}\beta(\lambda_t - \lambda_{t-1})^2 \tag{15}$$

Here, T is the number of timesteps in the stacked Bi-TLSTM encoder, i.e. the input sequence length. We use gradient-based optimizer with open choices for *stochastic gradient descent* (SGD) or Adam (Kingma & Ba, 2015). We apply gradient clipping after calculating gradient through all of the ANFIS and Bi-TLSTM parameters through all of the timesteps under *backpropagation through time* (BPTT) paradigm. We choose hybrid learning for DNFIS (Jang, 1993). But, for the sake of stability, we don't update the consequent parameters of DNFIS at each timestep, rather we wait for the gradients to accumulate enough to stably update it using *least squared error* (LSE).So, we accumulate the gradient through a buffer variable and then update the consequent parameters after K timesteps. Although, the premise or antecedent parameters are updated at each timestep alongside Bi-TLSTM parameters at each timestep.

After successful fusion of DNFIS and Bi-TLSTM, the overall model gets a name *DNFIS-enabled Bi-TLSTM*, or DEBi-TLSTM.

5 Experiments and results

This section elaborates on the datasets, the experimental setups involving four diseases and five experiments, the competitor algorithms, the performance metrics, the results using the performance metrics, the statistical test results, the sensitivity analysis involving the hyperparameters, the ablation study and insightful discussion on the results.

5.1 Datasets

We conducted experiments using three publicly available databases: MIMIC-III v1.4 (Johnson et al., 2016), MIMIC-IV v3.1 (Johnson et al., 2024), and the Parkinson's Progression Marker Initiative (PPMI) (Marek et al., 2011). From these sources, we constructed five datasets corresponding to chronic kidney disease (CKD), heart failure (HF), and hypertensive disease (HYP) from MIMIC-IV, HYP from MIMIC-III, and Parkinson's disease (PD) from PPMI. Table 5.2 summarizes the datasets and experiment-specific details.

TABLE 5.2 Description of datasets.

Datasets	MIMIC-III	MIMIC-IV			PPMI
Diseases	**HYP**	**CKD**	**HF**	**HYP**	**PD**
Number of patients	21,903	29,257	29,604	1,11,116	783
Number of patients after pre-processing	10,276	13,976	14,372	50,594	648
Number of input sequences	10,276	13,976	14,372	50,594	15,574
Unit of time	Days	Days	Days	Days	Months
Elapsed time interval	[1, 3872]	[1, 5279]	[1, 5275]	[1, 5340]	[1, 24]
Length of input sequence	[5, 14]	[5, 24]	[5, 24]	[5, 24]	[0, 26]
Number of patients with single ICU visit	8709	4105	4439	16,674	–
Number of patients with double ICU visits	1377	3591	3803	13,859	–
Number of patients with more than two ICU visits[1]	190	6280	5910	20,061	–
Number of input or feature variables	75	89	48	103	319
Number of output or target variables	24	23	20	27	82

[1] *The number of multiple admission to ICU determines the number of times the model will have to handle the surges in elapsed time.*

5.2 Dataset descriptions

MIMIC-III and MIMIC-IV: Both databases are large-scale, longitudinal electronic health record (EHR) collections containing around 65,000 ICU patient admissions. A key challenge is that ICU admissions usually occur once per patient, making them less suitable for the common CSP setup where each hospital visit is treated as a timestep (Shamout, Zhu, & Clifton, 2020). To address this, we redefined the experimental unit: each *day of hospital stay* was treated as a timestep, with elapsed time equal to one day between consecutive days within a hospital stay. If a patient was discharged and later readmitted, the elapsed time was the number of days between admissions. This design introduces irregular time gaps, testing model robustness to abrupt temporal changes. Daily averages of laboratory test values were used as inputs.

PPMI: PPMI is a longitudinal cohort study designed to identify biomarkers of Parkinson's disease progression. It includes clinical assessments, MRI data, biospecimens, and behavioral measures. Unlike MIMIC, PPMI has irregular follow-up intervals between multiple visits (ranging from 1 month to 2 years), making it well-suited for evaluating DEBi-TLSTM under irregular temporal dynamics. Exactly like (Dentamaro, Impedovo, Musti, Pirlo, & Taurisano, 2024), we processed idiopathic PD patients and non-PD controls separately, introduced one-hot encoded categorical variables, and considered longitudinal clinical assessment scores as input features and PD progression characteristics as targets.

5.3 Data preprocessing

MIMIC datasets: Patient records were identified using ICD codes (WHO, 1992) in *diagnoses_icd.csv*. For selected patients, *labevents.csv* was filtered to include only relevant lab tests (*itemids*), chosen in consultation with medical experts based on frequency and clinical importance. For each hospital day, lab test values were averaged to obtain per-day inputs. Specifically, we sorted the number of times a lab test has produced an "abnormal" result (can be read from *abnormal* flag in *labevents.csv*) inside the patient cohort under an experiment in descending order. The first 89 (CKD), first 48 (HF), first 103 (HYP, MIMIC-IV), and first 75 (HYP, MIMIC-III) lab tests from this sorted list were taken as input features. Target variables are also chosen in consultation with the experts. Specific target variables for each of the five experiments can be found in supplementary[1]. Demographic and age embeddings were derived via an autoencoder trained on randomly generated data. Any patient, who has no recorded data against at least 20% of the selected feature variables or lab tests, are treated as outliers and removed from the dataset.

PPMI dataset: After preprocessing, the dataset comprised 15,574 records from 648 patients. A total of 319 features and 82 targets were retained (features and targets are pre-separated in PPMI). Sequence lengths ranged from 0 to 26 visits (mean = 25), with Δt defined in months. Demographic and age embeddings were similarly processed with an autoencoder. Outlier removal applied with the exact similar proceedings as in MIMIC.

5.4 Experimental setup

All experiments involve sequence learning and forecasting: given a sequence of input feature vectors (decimal-valued lab tests or clinical scores), models predict the target variables at future timesteps for test patients who are unseen at training, choosing the most adversarial condition for prediction. This design evaluates not only predictive accuracy but also adaptability to irregular and heterogeneous temporal dynamics, a challenge DEBi-TLSTM is specifically designed to address.

We first split by patient IDs (MIMIC: *subject_id*; eICU: *patientunitstayid*) into train: validation: test = 70:15: 15 using a group-aware random split (*GroupShuffleSplit*) so that no patient appears in more than one set. For each patient, we then define a personal *cut time* at the first 80% of that patient's timeline (not across patients). Training windows are constructed with an encoder length T and decoder horizon F, with window origins $\tau \leq$ *cut time*. Hence, the encoder sees the data within the timesteps defined using $[\tau - T + 1, \tau]$, and decoder predicts within $[\tau + 1, \tau + F]$. This ensures that there is no data leakage, i.e. during training the models does not see any of the data which are above *cut time* i.e., we can say training only uses past data. For validation and test patients, we evaluate future forecasting by placing encoder windows entirely in the interval (*cut time*, 90%], i.e. [80%, 90%] of unseen patients' data and predicting for the last 10% of each patient's timeline (i.e., targets in (90%, 100%]), using fully autoregressive decoding (no teacher forcing) as indicated in section 3 *decoder* definition. We impute missing data using median values calculated only on the training patients' data up to their *cut time* or first 80% of their data. Then impute the validation or test patients' missing data using the exact same values calculated for training data. This protocol guarantees both unseen-patient generalization and forward-in-time prediction without information leakage.

5.5 Performance metrics

We use two performance metrics. They are defined below.

Mean squared error. This is abbreviated as MSE. MSE calculates the difference between the predicted value and the actual value for each of the output variable for each of the timestep of prediction and adds them up. Hence, the mathematical formulation of MSE is,

$$L_{\text{task}} = \text{MSE} = \frac{1}{TP}\sum_{t=T+1}^{T+F}\sum_{i=1}^{P}(\hat{y}_{t,i} - y_{t,i})^2$$

Here, T is the input length of sequence or the number of timesteps at the input side. F is the output length of sequence or the length of predicted sequence or the number of timesteps in prediction. P is the number of output or target variables.

Mean absolute percentage error. This is abbreviated as MAPE. MAPE calculates the mean of the percentage of absolute error with respect to the actual output. Hence, in this case,

$$L_{\text{task}} = \text{MAPE} = \frac{100}{TP}\sum_{t=T+1}^{T+F}\sum_{i=1}^{P}\left|\frac{\hat{y}_{t,i} - y_{t,i}}{y_{t,i}}\right|$$

5.6 Implementation details

We conduct the optimal hyperparameter search using group hold-out (*GroupShuffleSplit*) cross-validation with patients as groups (70/15/15 split), combined with forward-in-time evaluation (*cut time* is the first 80% within each training patient's data and the last 20% is used for validation within each validation patient's data). These are detailed in previous sub-sections. For all five of the competing algorithms, the Adam (Kingma & Ba, 2015) optimizer is chosen. The hyperparameters for the competing old LSTM versions T-LSTM (Baytas et al., 2017), BiLSTM (Woźniak et al., 2023), BiGRU (Rodrigues Jr et al., 2021), and the newer one minLSTM (Feng et al., 2024) comprises architectural details including hidden state dimension and dropout rate along with learning rate and batch size.

To ensure reproducibility and fair evaluation, we systematically tuned the hyperparameters of the proposed DEBi-TLSTM as well as the baseline competitors. For DEBi-TLSTM, the tunable hyperparameters included: (i) hidden state dimension H of the encoder and decoder, (ii) number of rules R in SANFIS, (iii) segment size f, (iv) λ_{min} and λ_{max} in Eq. (14), (v) smoothness hyperparameter β in Eq. (15), and (vi) the number of timesteps K for gradient accumulation for updating the DNFIS consequent parameters via LSE. In addition, we tuned the learning rates for Bi-TLSTM and DNFIS, dropout rate, and batch size.

The tuning process followed a two-stage procedure. First, we performed a *coarse random search* over predefined ranges (e.g., learning rate $\in[10^{-4}, 10^{-2}]$, hidden dimension $H\in\{64,72,80,96,112,128,256,512\}$, dropout $\in[0.09,0.11]$, batch size $\in\{64,128,256\}$, $R\in\{4,5,6,8\}$. These ranges are also reported within brackets in Table 5.3. Then, the most promising configurations were refined using a *grid search* in a narrower neighborhood around the best candidates.

Hyperparameter selection is based on the *validation performance*, measured by MSE, with early stopping to prevent overfitting (stopped training after reaching validation error minima). Final optimal values for each dataset are reported in Table 5.3 for baseline competitors and in Table 5.4 for DEBi-TLSTM, and random seeds were fixed to ensure reproducibility. Table 5.4 also reports the range (within brackets) for *coarse random search*, performed at the advent of hyperparameter selection process.

We considered stack of the encoder layers corresponding to all four of the competitors to keep the competition fair. The number of layers for all these architectures, including our proposed one, are chosen to be 3. All of these models deploy LSTM decoder, which follows exactly the same architecture as detailed in the third section of this chapter. The $\xi(.)$ in the output FF layer is fixed to be a linear function.

5.7 Comparative analysis

Evidently, proposed DEBi-TLSTM performed better in terms of MSE and MAPE in all five of the experiments. The results are listed in Table 5.5. Winners are highlighted with bold entries. minLSTM outperformed the proposed method twice. In terms of MAPE, minLSTM outperformed in MIMIC-III HYP experiment and in terms of MSE, minLSTM outperformed in PPMI experiment. MIMIC-III HYP experiment uses the least amount of data in training. Hence, we may conclude that seeing more training data is beneficial for the proposed model. Another failure in PPMI may be attributed to less sudden variation in elapsed time, less complex time irregularity (not much non-linear distortions) in PPMI database. Hence, test-time

TABLE 5.3 Optimal hyperparameters of the competitors.

Model	Hyperparameters	PPMI	MIMIC-III	MIMIC-IV CKD	MIMIC-IV HF	MIMIC-IV HYP
T-LSTM	Learning Rate	0.0180	0.0070	0.0050	0.0040	0.0005
	Dropout	0.10	0.08	0.10	0.11	0.05
	Batch Size	128	128	64	128	256
	Hidden Dimension	96	80	112	96	256
BiLSTM	Learning Rate	0.0160	0.0120	0.0080	0.0100	0.0010
	Dropout	0.08	0.10	0.10	0.10	0.10
	Batch Size	128	64	64	64	128
	Hidden Dimension	64	48	72	64	128
BiGRU	Learning Rate	0.0140	0.0080	0.0100	0.0160	0.0020
	Dropout	0.10	0.10	0.12	0.08	0.10
	Batch Size	128	64	128	128	512
	Hidden Dimension	128	128	64	64	256
minLSTM	Learning Rate	0.0050	0.0040	0.0010	0.0020	0.0003
	Dropout	0.10	0.12	0.08	0.09	0.10
	Batch Size	64	128	64	64	256
	Hidden Dimension	64	128	96	112	128
	Number of layers	3	3	3	3	3

TABLE 5.4 Optimal hyperparameters of DEBi-TLSTM [search range].

Hyperparameters [range considered]	PPMI	MIMIC-III	MIMIC-IV CKD	MIMIC-IV HF	MIMIC-IV HYP
Learning Rate (Bi-TLSTM) $[10^{-4}, 10^{-2}]$	0.0012	0.0010	0.0008	0.0010	0.0001
Learning Rate (DNFIS) $[10^{-3}, 10^{-2}]$	0.0150	0.0110	0.0120	0.0140	0.0050
Dropout {0.09, 0.1, 0.11}	0.10	0.10	0.10	0.11	0.09
Batch Size {64, 128, 256}	128	128	64	128	256
Hidden Dimension (H) {64, 72, 80, 96, 112, 128, 256, 512}	96	80	112	96	256
Number of rules in SANFIS (R) {4, 5, 6, 7}	5	5	6	5	4
Segment size (f) {4, 8, 12}	8	4	8	8	16
λ_{min} {0.001, 0.0001, 0.00001}	0.0010	0.0010	0.0010	0.0010	0.0001
λ_{max} {4.99, 4.999, 4.9999}	4.990	4.990	4.990	4.990	4.999
Smoothness parameter (β) {0.0004, 0.0005, 0.0006, 0.0007, 0.0008, 0.0009, 0.0010}	0.0005	0.0004	0.0005	0.0006	0.0010
Number of timesteps for DNFIS consequent parameter update (K) {2, 3, 4, 5}	5	2	2	3	5

performance is nearer to the training time performance. And as minLSTM is less complex in terms of model complexity than the proposed algorithm, it could reach the training error minima faster with less amount of data within PPMI than the proposed method, which needed more data to reach training error minima and to optimally tune all its parameters even for a simpler database like PPMI. This is a drawback which should be addressed in future works.

From Table 5.5, we also observe that performance gaps between the proposed algorithm and its competitors with respect to the MSE is more than MAPE in all five experiments. This can be attributed to the mathematical fact that square terms

TABLE 5.5 Comparative analysis of model performances.

Models	PPMI		MIMIC-III		MIMIC-IV CKD		MIMIC-IV HF		MIMIC-IV HYP	
	MSE	MAPE	MSE	MAPE	MSE	MAPE	MSE	MAPE	MSE	MAPE
DEBi-TLSTM	0.39	**6.25**	**0.45**	9.19	**0.42**	**7.57**	**0.35**	**5.97**	**0.33**	**5.12**
T-LSTM	0.50	7.07	0.64	11.24	0.55	8.42	0.44	6.01	0.41	7.77
BiLSTM	0.54	7.35	0.59	12.57	0.59	9.13	0.61	6.74	0.47	8.23
BiGRU	0.51	7.14	0.67	13.42	0.61	8.97	0.47	6.23	0.45	8.00
minLSTM	**0.37**	6.51	0.48	**8.75**	0.46	8.01	0.40	6.02	0.38	5.54

TABLE 5.6 Friedman ranks and Bonferonni-Dunn posthoc analysis.

Algorithm	Friedman rank
Proposed method	1.000
T-LSTM	2.333
BiLSTM.	3.000
BiGRU	2.666
minLSTM	1.333
Critical difference	1.286
Bonferroni–Dunn parameters	**statistical test result**
Statistic value	10.266
Degree of freedom	4
Critical value	7.815
Comment on Null Hypothesis	Reject

penalize the wrong predictions more than the absolute value operation used in MAPE. The MSE data presented in Table 5.5 are utilized for conducting the hypothesis test to ascertain the statistical significance of the performance disparities between the proposed algorithm and its four competitors. Assuming no specific distribution of the MSE values (derived after 5-fold cross-validation for each algorithm), the Friedman non-parametric test (Sheldon, Fillyaw, & Thompson, 1996) is applied to the mean MSE metric values obtained by the competitor algorithms (across 5-fold cross-validation) at a significance level of 0.05. These results also affirm the superiority of the proposed method. Here, the null hypothesis considers no significant difference in performances (based on mean MSE values) of the competing algorithms.

The Friedman ranks presented in Table 5.6 (based on the outcomes reported in Table 5.5) highlight the proposed method as the top-performing algorithm. Consequently, the Bonferroni–Dunn posthoc analysis (Pereira, Afonso, & Medeiros, 2015), is conducted, with the proposed method serving as the control. In this analysis, a critical difference value is computed, which for the dataset (as depicted in Table 5.6) amounts to 1.286. The control algorithm is deduced to be outperforming significantly if its competitor's Friedman rank differs at least by the critical difference. It is evident from Table 5.6 that old LSTM models are notably inferior from the proposed one, except the new minLSTM is only borderline statistically insignificantly less powerful in terms of MSE performance as its rank does not cross the critical difference from the rank of the control algorithm or the proposed one.

5.8 Sensitivity analysis with hyperparameters

We conduct sensitivity analysis for the five specified hyperparameters (K, H, f, R, β) with the goal of identifying the order of sensitivity of the performance of the proposed method with respect to each hyperparameter. We chose MIMIC-IV HYP experiment for this purpose. For this analysis, we need to define three levels of values of the hyperparameters and their corresponding labels, namely "low" (labelled as "–1"), "medium" (labelled as "0") and "high" (labelled as "+1"). All hyperparameters and corresponding labels are listed in Table 5.7.

TABLE 5.7 Hyperparameter values at 3 levels for sensitivity analysis.

Hyperparameters	Low (−1)	Medium (0)	High (+1)
Number of timesteps for DNFIS consequent parameter update (*K*)	4	5	6
Hidden Dimension (*H*)	128	256	512
Segment size (*f*)	8	16	32
Number of rules in SANFIS (*R*)	3	4	5
Smoothness parameter (β)	0.0001	0.0010	0.0100

When testing the sensitivity for one hyperparameter, all other hyperparameters are kept constant at "medium" label, which are the exact values found using cross-validation (refer Table 5.4). For example, during recording the effect of hyperparameter K, values of other hyperparameters are set at: $H = 256, f = 16, R = 4, \beta = 10^{-3}$.

We define *sensitivity factor* (SF) using MSE and MAPE. Specifically, for an experiment *exp,* to conduct the analysis for a specific hyperparameter ε, and its associated label ε_l, four performance measures are recorded: (i) median MSE: MSE_{median} and maximum MSE: MSE_{max}, (ii) median MAPE: $MAPE_{median}$ and maximum MAPE: $MAPE_{max}$. Based on these four measures, we design the SF as following:

$$SF_{\varepsilon_l,\exp} = \left(\frac{MSE_{max}\,MAPE_{max}}{MSE_{median}\,MAPE_{median}}\right)_{\varepsilon_l,\exp}. \quad (16)$$

This formulation of SF can be interpreted as: in an experimental setup *exp*, a smaller value of $SF_{\varepsilon_l,\exp}$ indicates better performance of the proposed method with the specific value ε_l of the hyperparameter ε. Hence, to evaluate the sensitivity of the model with respect to hyperparameter ε and its label ε_l, we aggregate the $SF_{\varepsilon_l,\exp}$ values for all experiments. This statement can be mathematically formulated as:

$$SF_{\varepsilon_l}(\varepsilon) = \frac{1}{5}\sum_{\exp\in\{1,2,3,4,5\}} SF_{\varepsilon_l,\exp} \quad (17)$$

Table 5.8 summarizes the effect(ε), which calculates the mean effect of different hyperparameters while conducting the 5 experiments, where effect (ε) is defined as following:

$$\text{effect}(\varepsilon) = \max\{(|SF_{+1}(\varepsilon)| - |SF_0(\varepsilon)|), (|SF_{-1}(\varepsilon)| - |SF_0(\varepsilon)|), (|SF_{+1}(\varepsilon)| - |SF_{-1}(\varepsilon)|)\} \quad (18)$$

The results in Table 5.8, it is clear that the variance of performance is decreasingly affected by the hyperparameters in following order:

$$K > H > f > R > \beta \quad (19)$$

For all hyperparameters, the values marked with the "medium" label (identified as optimal through grid search with cross-validation) consistently outperformed the other settings. For instance, when all other hyperparameters were set to their "medium" values and the embedding size (H) was varied, the MAPE was lowest at $H = 256$, outperforming both $H = 128$ and $H = 512$. Evidently, $H = 128$ could not handle complexity of the tasks, while at $H = 512$, the MAPE showed a marginal increase, likely due to overfitting. However, MSE remained relatively stable across these variations, with a slight decrease at $H = 256$, suggesting that increasing the embedding dimension beyond a certain point does not necessarily improve performance. Hence, the overall sensitivity peaked at $H = 256$.

5.9 Ablation results

We considered the following ablation competitors: (i) Bi-TLSTM with one encoder layer and one decoder layer denoted as Bi-TLSTM[1:1], (ii) Bi-TLSTM with three encoder layer and one decoder layer denoted as Bi-TLSTM[3:1],

TABLE 5.8 Effect (ε) values achieved for each experiment against each of the hyperparameters.

Experiments	Hyperparameters				
	K	*H*	*f*	*R*	*β*
MIMIC-III (HYP)	6.362e-002	5.895e-002	3.434e-002	2.262e-002	2.246e-002
MIMIC-IV (CKD)	6.527e-002	5.876e-002	4.749e-002	4.274e-002	3.194e-002
MIMIC-IV (HF)	6.022e-002	3.992e-002	3.163e-002	3.008e-002	2.941e-002
MIMIC-IV (HYP)	5.905e-002	5.102e-002	3.366e-002	3.291e-002	3.154e-002
PPMI (PD)	6.098e-002	5.999e-002	5.910e-002	4.861e-002	1.511e-002

TABLE 5.9 Ablation results under MIMIC-IV HYP experiment.

Models	MSE	Theoretical implication
Bi-TLSTM[1:1]	0.392	Limited capacity to capture hierarchical temporal dependencies—weaker representation.
Bi-TLSTM[3:1]	0.375	Depth improves hierarchical feature extraction and long-range dependency modeling.
ANFIS-Bi-TLSTM[3:1]	0.368	One large ANFIS struggles with global rule partitioning—unstable and less generalizable.
MLP-Bi-TLSTM[3:1]	0.362	Learns decay adaptively, but lacks interpretability and universal approximation power of fuzzy rules.
DEBi-TLSTM[3:1], $K = 1$	0.350	Instability from simultaneous premise + consequent updates; mitigated by slower update cycle (bilevel optimization).
DEBi-TLSTM[3:1], $\beta = 0$	0.333	Loss of temporal consistency; overfits to local fluctuations—smoothness regularizer enforces Lipschitz-like continuity.
DEBi-TLSTM[3:1], no skip connection	0.337	Gradient flow disrupted; skip connections stabilize training in deep recurrent-fuzzy hybrids.
DEBi-TLSTM[3:1] (proposed)	0.331	Combines depth, fuzzy granularity, multi-timescale updates, smoothness, and skip connections—best stability and generalization.

(iii) Bi-TLSTM[3:1] with one giant ANFIS in between each timestep denoted as ANFIS-BI-TLSTM[3:1], (iv) Bi-TLSTM[3:1] with one *multilayer perceptron* (MLP) in between timesteps to generate the decay parameter denoted as MLP-Bi-TLSTM[3:1], (v) DEBi-TLSTM[3:1] with $K = 1$, signifying the update of consequent parameters along with Bi-TLSTM and premise parameters at each timestep, (vi) DEBi-TLSTM[3:1] with no smoothness penalty in loss function ($\beta = 0$), (vii) DEBi-TLSTM without skip connections. The MSE results related to MIMIC-IV HYP experiment are reported in Table 5.9.

The ablation experiments in Table 5.9 not only quantify the incremental benefits of each design choice but also provide insight into the theoretical underpinnings of DEBi-TLSTM.

Depth of Bi-TLSTM layers (Bi-TLSTM [1:1] vs. Bi-TLSTM[3:1]): The improvement from MSE = 0.392 to 0.375 highlights the importance of deeper encoder layers. Stacking encoders enhances the model's capacity to capture long-range temporal dependencies and hierarchical representations, which shallow recurrent structures often fail to model. This aligns with theoretical results on recurrent neural networks where deeper hierarchies increase representational expressiveness and allow for better generalization.

Granularity of fuzzy reasoning (ANFIS-Bi-TLSTM vs. DEBi-TLSTM): The comparison between a giant ANFIS (0.368) and multiple stacked ANFIS modules (0.331) demonstrates the advantage of distributed fuzzy inference. A single ANFIS layer attempts to learn global fuzzy partitions, which can lead to instability and poor generalization. In contrast, stacked ANFIS layers allow for localized, hierarchical fuzzy rules, leading to smoother approximations and more stable convergence. This is theoretically consistent with the divide-and-conquer principle in function approximation.

Dynamic decay parameter generation (MLP-Bi-TLSTM vs. DEBi-TLSTM): The MLP-Bi-TLSTM variant (0.362) shows that simple neural decay mechanisms can partially stabilize temporal modeling. However, the proposed DNFIS-based decay outperforms it by embedding fuzzy rule–based interpretability and adaptivity, which MLP lacks. Theoretical justification lies in fuzzy systems' universal approximation property, allowing them to better capture nonlinear decay dynamics.

Time-scale separation of updates ($K = 1$ vs. $K > 1$): Updating consequent parameters at every timestep ($K = 1$) leads to higher error (0.350) compared to periodic updates (0.331). This confirms that simultaneous optimization of premise and consequent parameters induces instability due to entangled gradients. Theoretically, introducing a slower time scale for consequent updates approximates a bilevel optimization scheme, improving stability and convergence.

Smoothness penalty ($\beta = 0$ vs. $\beta > 0$): Removing the smoothness regularizer (0.333) slightly degrades performance compared to the proposed model (0.331). This penalty enforces temporal consistency in fuzzy rule evolution, preventing overfitting to local fluctuations. From a theoretical standpoint, this acts as a regularization prior encouraging Lipschitz continuity in rule evolution.

Skip connections: Eliminating skip connections (0.337) worsens MSE, confirming their role in mitigating vanishing gradients. Theoretically, skip connections preserve gradient flow across stacked layers, enabling the deeper Bi-TLSTM-ANFIS architecture to remain trainable and stable.

Collectively, the ablation study demonstrates that each architectural design choice in DEBi-TLSTM is not only empirically beneficial but also theoretically motivated: depth enhances hierarchical representation, stacked ANFIS improves fuzzy granularity, multi-timescale updates stabilize optimization, smoothness constraints regularize learning, and skip connections ensure gradient flow.

6 Discussion

The major strength of the proposed DEBi-TLSTM lies in its ability to capture the relativistic bidirectional relationship between time and context through the integration of a neuro-fuzzy network (ANFIS) with a sequential recurrent model (Bi-TLSTM). This fusion was enabled by several practical design choices targeting training stability. In particular, introducing multi-timescale updates of ANFIS premise and consequent parameters allowed gradients of the linear consequent parameters to accumulate, while the application of the hybrid learning rule with LSE further stabilized updates. Additionally, the layered DNFIS structure improved interpretability and enhanced generalization of the relationship between the hidden state h_{t-1} and elapsed time Δt, particularly in the final ANFIS layer. Despite these strengths, two key limitations were observed.

The first limitation can be formulated as, for input sequences exceeding 30 timesteps, recurrent gradient decay persisted. This was evident in our ablation results with MIMIC-IV HYP experiment. Even if the number of timesteps considered for input to the encoder is merely 20 in this experiment, adding skip connections did not improve the performance of the model greatly (from MSE = 0.337 to MSE = 0.331), confirming that vanishing-gradient effects were not fully mitigated. These findings quantify the tangible impact of instability on predictive accuracy. Future work may leverage parallelizable architectures such as RetNet (Sun et al., 2023) or linear-attention transformers (Guo et al., 2025), which offer robustness to gradient decay. A key research gap, however, is how to explicitly encode elapsed time into these non-recurrent architectures.

The second limitation can be formulated as, *joint online training instability*. Simultaneous online updates of ANFIS and Bi-TLSTM introduced additional instability, though this was partially mitigated by multi-timescale updates. Decoupling the process, i.e., pre-training ANFIS on curated datasets before integration, may improve training reliability without sacrificing interpretability. But a key research innovation would be to curate this dataset to train ANFIS.

In summary, while DEBi-TLSTM successfully integrates fuzzy reasoning with recurrent modeling, our ablation results demonstrate that gradient instability leads to measurable degradations in MSE (up to ~ 6%) when stability mechanisms are removed. These findings underscore the necessity of our design choices and provide guidance for future refinements.

7 Practical implications for clinical deployment

Beyond algorithmic considerations, the clinical usability of DEBi-TLSTM is equally important.

Interpretability: By incorporating fuzzy rules, DEBi-TLSTM provides a transparent mapping between elapsed time, hidden context, and predictive outcomes. This transparency enables clinicians to trace why the model predicted a deterioration event, fostering trust and supporting clinical decision-making. Unlike "black-box" deep models, the fuzzy rule base can be directly reviewed, adjusted, or validated by domain experts.

Computational efficiency: Although recurrent components impose sequential processing costs, the use of multi-timescale updates reduces redundant gradient computations, making the model more efficient than a naïve Bi-TLSTM. On NVIDIA RTX 4090 single 24GB GPU, training times remained within practical bounds for hospital-scale datasets (e.g., under 6 h for MIMIC-IV HYP). This suggests feasibility for deployment in real-world clinical pipelines, especially if inference, rather than retraining, is the primary requirement.

Integration with EHR systems: DEBi-TLSTM is designed to process irregularly sampled event sequences, which aligns well with the non-uniform structure of most EHR data. Importantly, the model only requires standard fields such as timestamps, event codes, and measurement values—data that are already available in widely used EHR platforms. As such, the integration burden is relatively low, and outputs can be embedded as decision-support scores within existing dashboards.

Taken together, these characteristics suggest that DEBi-TLSTM offers a balance of accuracy, interpretability, and system compatibility, making it a promising candidate for deployment in real-world clinical settings, while highlighting the need for further validation in prospective trials.

8 Conclusion and future works

The increasing adoption of EHRs has amplified the importance of accurate clinical sequence prediction (CSP), yet most existing neural architectures still rely on the assumption of time as an independent, uniformly progressing variable. This time-invariance assumption limits their ability to capture the irregular, context-dependent influence of elapsed time in real-world healthcare settings. Addressing this gap, the present work introduced DEBi-TLSTM, a novel architecture that integrates a deep neuro-fuzzy inference system (DNFIS) with a bidirectional time-aware LSTM (Bi-TLSTM). Unlike prior approaches, DEBi-TLSTM treats elapsed time as a context-dependent variable, dynamically adapting its influence on memory decay according to the evolving hidden state.

Extensive experiments across three datasets (MIMIC-III, MIMIC-IV, and PPMI) and four diseases (Parkinson's disease, chronic kidney disease, heart failure, and hypertensive disease) demonstrated that DEBi-TLSTM consistently outperforms established baselines, including T-LSTM, BiLSTM, BiGRU, and minLSTM. Its layered ANFIS structure not only enhanced interpretability but also improved predictive robustness by explicitly modeling the relativistic interplay between time and context. The ablation studies further quantified the importance of stability considerations: when ANFIS was updated at every timestep ($K = 1$), the model's MSE increased by almost 6% compared to the proposed scheme, highlighting the measurable impact of gradient instability on performance.

Despite these advances, challenges remain. Gradient-related instabilities persist when processing sequences longer than 30 timesteps, and simultaneous online training of ANFIS and Bi-TLSTM can induce fluctuations in convergence. These findings suggest avenues for future improvement. Parallelizable architectures such as RetNet or linear-attention transformers may serve as more scalable backbones, provided they can incorporate elapsed time in a context-aware manner. Additionally, decoupling the training of ANFIS and Bi-TLSTM through batch-wise pretraining may enhance stability without sacrificing interpretability.

In conclusion, DEBi-TLSTM directly addresses the long-standing research problem identified at the outset: the inability of traditional architectures to account for the context-dependent, relativistic nature of time in clinical event prediction. By reframing elapsed time as a variable shaped by both context and history, this work moves beyond fixed parametric decay functions toward a more adaptive and interpretable temporal modeling paradigm. Looking ahead, this paradigm holds promise not only for CSP but also for broader time series domains where irregular, context-sensitive temporal dynamics are fundamental.

Generative AI declaration

During the preparation of this work the authors used the currently available free version of ChatGPT (from OpenAI) in order to check the grammars and improve clarity of language. After using this tool, the authors reviewed and edited the content as needed and take full responsibility for the content of the publication.

References

Al-Selwi, S. M., Hassan, M. F., Abdulkadir, S. J., Muneer, A., Sumiea, E. H., Alqushaibi, A., & Ragab, M. G. (2024). RNN-LSTM: from applications to modeling techniques and beyond—Systematic review. *Journal of King Saud University—Computer and Information Sciences*, 2213-1248. 36(5). Available from: https://www.sciencedirect.com/science/journal/13191578. doi:10.1016/j.jksuci.2024.102068.

An, Y., Mao, Y., Zhang, L., Jin, B., Xiao, K., Wei, X., & Yan, J. (2020). RAHM: relation augmented hierarchical multi-task learning framework for reasonable medication stocking. *Journal of Biomedical Informatics*, 1532-0464. 108, 103502. doi:10.1016/j.jbi.2020.103502.

Arandjelo vić, O. (2015). Discovering hospital admission patterns using models learnt from electronic hospital records. *Bioinformatics*, 1367-4811. 31(24), 3970–3976. doi:10.1093/bioinformatics/btv508.

Baytas, I. M., Xiao, C., Zhang, X., Wang, F., Jain, A. K., & Zhou, J. (2017). Patient subtyping via time-aware LSTM networks. In *Proceedings of the ACM SIGKDD international conference on knowledge discovery and data mining* (pp. 65–74). United States: Association for Computing Machinery. 129685. 9781450348874. doi:10.1145/3097983.3097997.

Beck, M., Pöppel, K., Spanring, M., Auer, A., Prudnikova, O., Kopp, M., et al. (2024). xLSTM: extended long short-term memory. In *Advances in neural information processing systems* (Vol. 37). Austria: Neural Information Processing Systems Foundation, 107547–107603. 10495258. Available from: https://papers.nips.cc/.

Bohm, D. (2015). *The Special Theory of Relativity* (pp. 1–282). Taylor and Francis. Available from: https://www.taylorfrancis.com/books/mono/10.4324/9780203820063/special-theory-relativity-david-bohm?context=ubx&refId=0c238024-8e80-447e-9ced-3bf293cf1dd6. doi:10.4324/9780203820063.

Chen, Y., Ren, K., Wang, Y., Fang, Y., Sun, W., & Li, D. (2023). ContiFormer: continuous-time transformer for irregular time series modeling. In *Advances in neural information processing systems* (Vol. 36). China: Neural Information Processing Systems Foundation, 47143–47175. 9781713899921. Available from: https://papers.nips.cc/.

Choi, E., Bahadori, M. T., Schuetz, A., Stewart, W. F., & Sun, J. (2016). Doctor AI: predicting clinical events via recurrent neural networks. In *JMLR workshop and conference proceedings* (pp. 301–318).

Choi, E., Du, N., Chen, R., Song, L., & Sun, J. (2015). Constructing disease network and temporal progression model via context-sensitive hawkes process. In *2015 IEEE international conference on data mining* (pp. 721–726).

Dentamaro, V., Impedovo, D., Musti, L., Pirlo, G., & Taurisano, P. (2024). Enhancing early Parkinson's disease detection through multimodal deep learning and explainable AI: insights from the PPMI database. *Scientific Reports*, 2045-2322. 14(1). Available from: https://www.nature.com/srep/. doi:10.1038/s41598-024-70165-4.

Devlin, J., Chang, M. W., Lee, K., & Toutanova, K. (2019). BERT: pre-training of deep bidirectional transformers for language understanding. In *Conference of the North American chapter of the association for computational linguistics: human language technologies—Proceedings of the conference* (Vol. 1, pp. 4171–4186). Association for Computational Linguistics (ACL). 9781950737130.

Duan, H., Sun, Z., Dong, W., He, K., & Huang, Z. (2020). On clinical event prediction in patient treatment trajectory using longitudinal electronic health records. *IEEE Journal of Biomedical and Health Informatics*, 2168-2194. 24(7), 2053–2063. doi:10.1109/jbhi.2019.2962079.

Eldele, E., Ragab, M., Chen, Z., Wu, M., Kwoh, C. K., Li, X., & Guan, C. (2021). Time-series representation learning via temporal and contextual contrasting. In *IJCAI international joint conference on artificial intelligence* (pp. 2352–2359). Singapore: International Joint Conferences on Artificial Intelligence. 9780999241196. Available from: http://www.ijcai.org/. doi:10.24963/ijcai.2021/324.

Feng, L., Tung, F., Ahmed, M. O., Bengio, Y., & Hajimirsadeghi, H. (2024). Were RNNs all we needed? arXiv, arXiv. 23318422, Available from: https://arxiv.org. doi:10.48550/arXiv.2410.01201.

Franceschi, J. Y., Dieuleveut, A., & Jaggi, M. (2019). Unsupervised scalable representation learning for multivariate time series. In *Advances in neural information processing systems* (Vol. 32). France: Neural Information Processing Systems Foundation. 10495258. Available from: https://papers.nips.cc/.

Gandin, I., Scagnetto, A., Romani, S., & Barbati, G. (2021). Interpretability of time-series deep learning models: a study in cardiovascular patients admitted to intensive care unit. *Journal of Biomedical Informatics*, 1532-0464. 121, 103876. doi:10.1016/j.jbi.2021.103876.

Gu, A., & Dao, T. (2023). Mamba: linear-time sequence modeling with selective state spaces. arXiv, arXiv. 23318422, Available from: https://arxiv.org. doi:10.48550/arXiv.2312.00752.

Guo, H., Yang, S., Goel, T., Xing, E. P., Dao, T., & Kim, Y. (2025). Log-linear attention. arXiv, arXiv. 23318422, Available from: https://arxiv.org. doi:10.48550/arXiv.2506.04761.

Hügle, M., Kalweit, G., Hügle, T., & Boedecker, J. (2021). A dynamic deep neural network for multimodal clinical data analysis. In Shaban-Nejad, A., Michalowski, M., Buckeridge, D.L. (Eds.), *Studies in computational intelligence* (Vol. 914, pp. 79–92). Germany: Springer Science and Business Media Deutschland GmbH, pp. 79–92. 18609503. Available from:http://www.springer.com/series/7092. doi:10.1007/978-3-030-53352-6_8.

Jang, J.-S. R. (1993). ANFIS: adaptive-network-based fuzzy inference system. *IEEE Transactions on Systems, Man, and Cybernetics*, 0018-9472. 23 (3), 665–685. doi:10.1109/21.256541.

Johnson, A., Bulgarelli, L., Pollard, Gow, B., Moody, B., Horng, S., et al. (2024). MIMIC-IV (version 3.1). *PhysioNet*. e215–e220.

Johnson, A., Pollard, & Mark, R. (2016). MIMIC-III clinical database (version 1.4). *PhysioNet*, e215–e220.

Kingma, D. P., & Ba, J. L. (2015). Adam: a method for stochastic optimization. In *3rd International Conference on Learning Representations, ICLR 2015—Conference Track Proceedings International Conference on Learning Representations*. Netherlands: ICLR. Available from: https://dblp.org/db/conf/iclr/iclr2015.html.

Koroteev, M. V. (2021). BERT: a review of applications in natural language processing and understanding. arXiv, arXiv. 23318422, Available from: https://arxiv.org. doi:10.48550/arxiv.2103.11943.

Lee, G., Nho, K., Kang, B., Sohn, K. A., Kim, D., Weiner, M. W., et al. (2019). Predicting Alzheimer's disease progression using multi-modal deep learning approach. *Scientific Reports*, 2045-2322. 9(1). Available from: https://www.nature.com/srep/. doi:10.1038/s41598-018-37769-z.

Li, Y., Rao, S., Solares, J. R. A., Hassaine, A., Ramakrishnan, R., Canoy, D., et al. (2020). BEHRT: transformer for electronic health records. *Scientific Reports*, 2045-2322. 10(1). Available from: www.nature.com/srep/index.html. doi:10.1038/s41598-020-62922-y.

Liu, Li, H., Hu, Z., Shi, H., Wang, Z., Tang, J., & Zhang, M. (2019). Learning hierarchical representations of electronic health records for clinical outcome prediction. In *AMIA annual symposium proceedings* (Vol. 2019, p. 597).

Luo, J., Ye, M., Xiao, C., & Ma, F. (2020). HiTANet: Hierarchical time-aware attention networks for risk prediction on electronic health records. In *Proceedings of the ACM SIGKDD international conference on knowledge discovery and data mining* (pp. 647–656). United States: Association for Computing Machinery. 9781450379984. doi:10.1145/3394486.3403107.

Marek, K., Jennings, D., Lasch, S., Siderowf, A., Tanner, C., Simuni, T., et al. (2011). The Parkinson progression marker initiative (PPMI). *Progress in Neurobiology*, 1873-5118. 95(4), 629–635. doi:10.1016/j.pneurobio.2011.09.005.

Men, L., Ilk, N., Tang, X., & Liu, Y. (2021). Multi-disease prediction using LSTM recurrent neural networks. *Expert Systems with Applications*, 0957-4174. 177, 114905. doi:10.1016/j.eswa.2021.114905.

Nie, Y., Nguyen, N. H., Sinthong, P., & Kalagnanam, J. (2022). A time series is worth 64 words: long-term forecasting with transformers. arXiv, arXiv. 23318422, Available from: https://arxiv.org. doi:10.48550/arXiv.2211.14730.

Noh, S.-H. (2021). Analysis of gradient vanishing of RNNs and performance comparison. *Information*, 2078-2489. 12(11), 442. doi:10.3390/info12110442.

Pereira, D. G., Afonso, A., & Medeiros, F. M. (2015). Overview of friedmans test and post-hoc analysis. *Communications in Statistics: Simulation and Computation*, 1532-4141. 44(10), 2636–2653. Available from: http://www.tandf.co.uk/journals/titles/03610918.asp. doi:10.1080/03610918.2014.931971.

Pham, T., Tran, T., Phung, D., & Venkatesh, S. (2016). *DeepCare: A deep dynamic memory model for predictive medicine* (Vol. 9652, pp. 30–41). Springer Science and Business Media LLC. doi:10.1007/978-3-319-31750-2_3.

Qu, H., Zhang, Y., Ning, L., Fan, W., Derr, T., Liu, H., et al. (2024). A survey of Mamba. arXiv, arXiv. 23318422, Available from: https://arxiv.org. doi:10.48550/arXiv.2408.01129.

Rakshit, P., Zaballa, O., Pérez, A., Gómez-Inhiesto, E., Acaiturri-Ayesta, M. T., & Lozano, J. A. (2021). A machine learning approach to predict healthcare cost of breast cancer patients. *Scientific Reports*, 2045-2322. 11(1). Available from: www.nature.com/srep/index.html. doi:10.1038/s41598-021-91580-x.

Rao, Y., Nie, Y., & Dai, A. (2022). PatchComplete: learning multi-resolution patch priors for 3D shape completion on unseen categories. arXiv, arXiv. 23318422, Available from: https://arxiv.org. doi:10.48550/arXiv.2206.04916.

Rasmy, L., Xiang, Y., Xie, Z., Tao, C., & Zhi, D. (2021). Med-BERT: pretrained contextualized embeddings on large-scale structured electronic health records for disease prediction. *Npj Digital Medicine*, 2398-6352. 4(1). Available from: https://www.nature.com/npjdigitalmed/. doi:10.1038/s41746-021-00455-y.

Rodrigues Jr, J. F., Gutierrez, M. A., Spadon, G., Brandoli, B., & Amer-Yahia, S. (2021). LIG-Doctor: Efficient patient trajectory prediction using bidirectional minimal gated-recurrent networks. *Information Sciences*,0020-0255. 545, 813–827. Available from: http://www.journals.elsevier.com/information-sciences/. doi:10.1016/j.ins.2020.09.024.

Shamout, F., Zhu, T., & Clifton, D. A. (2020). Machine learning for clinical outcome prediction. *IEEE Reviews in Biomedical Engineering*, 14, 116–126.

Sheldon, M. R., Fillyaw, M. J., & Thompson, W. D. (1996). The use and interpretation of the Friedman test in the analysis of ordinal-scale data in repeated measures designs. *Physiotherapy Research International*, 1358-2267. 1(4), 221–228. doi:10.1002/pri.66.

Sun, Y., Dong, L., Huang, S., Ma, S., Xia, Y., Xue, J., et al. (2023). Retentive network: a successor to transformer for large language models. arXiv, arXiv. 23318422, Available from: https://arxiv.org. doi:10.48550/arXiv.2307.08621.

Tan, M., Merrill, M., Gupta, V., Althoff, T., & Hartvigsen, T. (2024). Are language models actually useful for time series forecasting? *Advances in Neural Information Processing Systems*, 37, 60162–60191.

Tonekaboni, S., Eytan, D., & Goldengerg, A. (2021). Unsupervised representation learning for time series with temporal neighborhood coding. In *ICLR 2021—9th international conference on learning representations international conference on learning representations*. Canada: ICLR. Available from: https://iclr.cc/virtual/2021/papers.html?filter=titles.

Tran, T., Luo, W., Phung, D., Gupta, S., Rana, S., Kennedy, R. L., et al. (2014). A framework for feature extraction from hospital medical data with applications in risk prediction. *BMC Bioinformatics*, 1471-2105. 15(1). Available from: http://www.biomedcentral.com/bmcbioinformatics/. doi:10.1186/s12859-014-0425-8.

Vaswani, A., Shazeer, N., Parmar, N., Uszkoreit, J., Jones, L., Gomez, A. N., et al. (2017). Attention is all you need. In *Advances in neural information processing systems* (pp. 5999–6009). United States: Neural Information Processing Systems Foundation. 9781510860964. 10495258

Wang, X., Sontag, D., & Wang, F. (2014). Unsupervised learning of disease progression models. In *Proceedings of the ACM SIGKDD international conference on knowledge discovery and data mining* (pp. 85–94). United States: Association for Computing Machinery. 9781450329569. doi:10.1145/2623330.2623754.

Wang, Z., Engel, A., Sarwate, A. D., Dumitriu, I., & Chiang, T. (2023). Spectral evolution and invariance in linear-width neural networks. *Advances in Neural Information Processing Systems*, 36, 20695–20728.

WHO. (1992). International classification of diseases. WHO.

Woźniak, M., Wieczorek, M., & Siłka, J. (2023). BiLSTM deep neural network model for imbalanced medical data of IoT systems. *Future Generation Computer Systems*,0167-739X. 141, 489–499. Available from: https://www.journals.elsevier.com/future-generation-computer-systems. doi:10.1016/j.future.2022.12.004.

Woźniak, M., Wieczorek, M. and Siłka, J. (2023). BiLSTM deep neural network model for imbalanced medical data of IoT systems. Future Generation Computer Systems, 141, 489–499. doi:10.1016/j.future.2022.12.004.

Wu, H., Xu, J., Wang, J., & Long, M. (2021). Autoformer: Decomposition transformers with auto-correlation for long-term series forecasting. In *Advances in neural information processing systems* (Vol. 27, pp. 22419–22430). China: Neural Information Processing Systems Foundation. 9781713845393. Available from: https://papers.nips.cc/.

Xu, J., Sun, X., Zhang, Z., Zhao, G., & Lin, J. (2019). Understanding and improving layer normalization. In *Advances in neural information processing systems* (Vol. 32). China: Neural Information Processing Systems Foundation. 10495258. Available from: https://papers.nips.cc/.

Yang, C, Deng, Z, Choi, K and Wang, S. (2016). Takagi–Sugeno–Kang Transfer Learning Fuzzy Logic System for the Adaptive Recognition of Epileptic Electroencephalogram Signals. IEEE Trans. Fuzzy Syst., 24(5), 1079–1094. doi:10.1109/TFUZZ.2015.2501438.

Yue, Z., Wang, Y., Duan, J., Yang, T., Huang, C., Tong, Y., & Xu, B. (2022). TS2Vec: towards universal representation of time series. *Proceedings of the AAAI Conference on Artificial Intelligence*, 2374-3468. 36(8), 8980–8987. doi:10.1609/aaai.v36i8.20881.

Zeng, A., Chen, M., Zhang, L., & Xu, Q. (2023). Are transformers effective for time series forecasting? *Proceedings of the AAAI Conference on Artificial Intelligence*, 2374-3468. 37(9), 11121–11128. doi:10.1609/aaai.v37i9.26317.

Zhang, D., Yin, C., Hunold, K. M., Jiang, X., Caterino, J. M., & Zhang, P. (2021). An interpretable deep-learning model for early prediction of sepsis in the emergency department. *Patterns*. 2666-3899. 2(2). Available from: www.cell.com/patterns. doi:10.1016/j.patter.2020.100196.

Zhao, J., Papapetrou, P., Asker, L., & Boström, H. (2017). Learning from heterogeneous temporal data in electronic health records. *Journal of Biomedical Informatics*, 1532-0464. 65, 105–119. Available from: http://www.elsevier.com/inca/publications/store/6/2/2/8/5/7/index.htt. doi:10.1016/j.jbi.2016.11.006.

Zhou, T., Ma, Z., Wen, Q., Wang, X., Sun, L., & Jin, R. (2022). FEDformer: frequency enhanced decomposed transformer for long-term series forecasting. In *Proceedings of machine learning research* (Vol. 162, pp. 27268–27286). China: ML Research Press. 26403498. Available from: https://proceedings.mlr.press/.

Zhu, Y., Li, H., Liao, Y., Wang, B., Guan, Z., Liu, H., & Cai, D. (2017). What to do next: modeling user behaviors by time-LSTM. In *IJCAI international joint conference on artificial intelligence* (pp. 3602–3608). China: International Joint Conferences on Artificial Intelligence. 9780999241103. Available from: http://www.ijcai.org/. doi:10.24963/ijcai.2017/504.

Chapter 6

Applications of artificial intelligence in infectious diseases

Kairav Barua and Somdeb Bose Dasgupta
Department of Bioscience and Biotechnology, Indian Institute of Technology Kharagpur, Kharagpur, India

1 Introduction

The persistent global threat posed by infectious diseases demands innovative and proactive approaches. Diseases such as tuberculosis, malaria, HIV, and, more recently, the COVID-19 pandemic have illuminated critical gaps in existing healthcare systems and public health policies. These challenges underscore the need for advanced solutions that are both scalable and adaptive. Traditional epidemiological methods, while effective in many contexts, often fail to match the speed and unpredictability of disease outbreaks, leading to delays in diagnosis, resource allocation, and containment.

Artificial intelligence (AI) has emerged as a transformative force in healthcare, offering tools capable of analyzing vast datasets, identifying patterns, and generating actionable insights. During the COVID-19 pandemic, AI demonstrated its value in addressing the multifaceted challenges of managing a global outbreak. AI-driven platforms such as BlueDot and HealthMap successfully identified early signals of disease spread through the analysis of diverse data sources, including social media trends, mobility patterns, and healthcare reports. These tools enabled health authorities to issue early warnings, providing critical time to prepare for an escalating crisis (Brownstein, Freifeld, Reis, & Mandl, 2008; Niiler, 2025).

In clinical settings, AI accelerated the identification and diagnosis of COVID-19 cases. Machine learning (ML) models were used to analyze chest X-rays and CT (Computered Tomography) scans, identifying infection-related abnormalities with remarkable accuracy. In addition, AI-powered diagnostics streamlined testing workflows by integrating laboratory results, reducing turnaround times in detecting the virus. Beyond diagnostics, AI played a pivotal role in drug discovery and vaccine development, modeling protein interactions and predicting potential therapeutic candidates far more rapidly than conventional methods. AI, ML, and DL (Deep Learning) algorithms can combine electronic health records (EHRs), multiomics, laboratory results, and chemical compound data to support clinical outcomes such as early diagnosis, outbreak monitoring, and therapeutic development (Fig. 6.1).

Public health applications of AI extended to predictive modeling, resource optimization, and communication. For example, AI algorithms forecasted ICU bed requirements and PPE shortages, aiding healthcare systems in optimizing resource distribution. Predictive tools also guided governments in implementing data-driven lockdown measures, mitigating the economic and social impacts of indiscriminate restrictions (Kraemer et al., 2020). Natural language processing (NLP) technologies powered chatbots and virtual assistants to provide accurate health information, address patient concerns, and counter misinformation during a period marked by global uncertainty.

The lessons learned from the COVID-19 pandemic highlight AI's transformative potential in infectious disease management. Beyond acute outbreaks, AI offers solutions to predict, detect, and mitigate the spread of endemic and epidemic diseases. By integrating diverse data streams, such as EHRs, environmental monitoring systems, and genomic data, AI systems can provide early warnings, support rapid diagnosis, and inform public health interventions.

This chapter explores the expanding role of AI in the detection and arrest of infectious diseases, delving into its applications in surveillance, diagnostics, outbreak modeling, and intervention planning. It also addresses the challenges and ethical considerations inherent in deploying AI in healthcare, emphasizing the importance of equitable access and data privacy.

Applications of Artificial Intelligence in Healthcare. https://doi.org/10.1016/B978-0-443-43934-6.00005-5

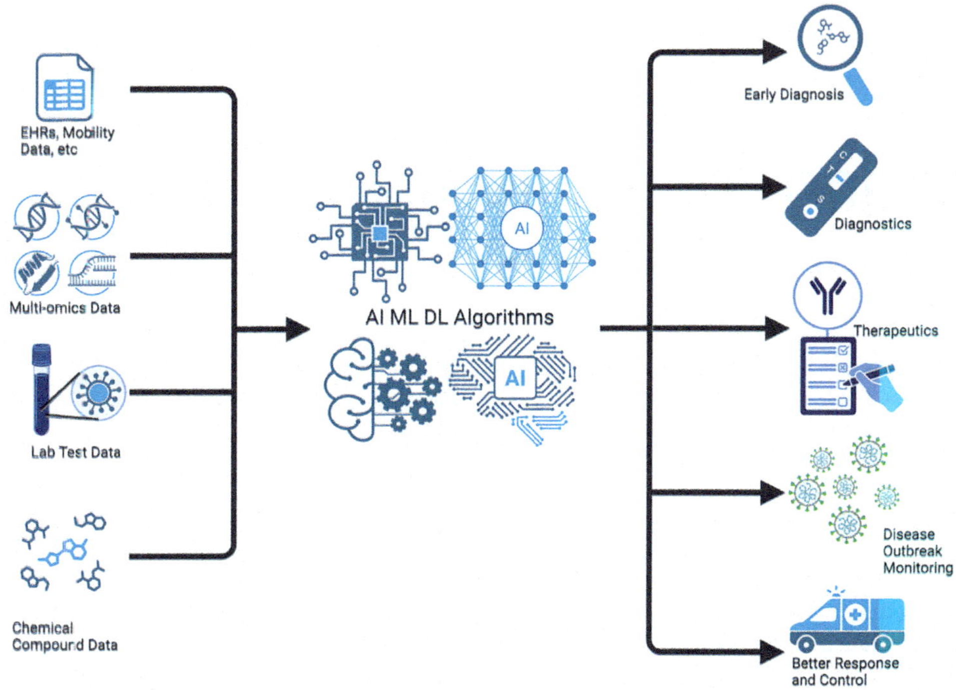

FIG. 6.1 AI pipeline from health data to public health outcomes. This schematic depicts a left-to-right flow in which heterogeneous inputs:electronic health records and mobility traces, multiomics profiles, routine and molecular laboratory tests, and chemical compound information are harmonized and fused by AI/ML/DL models to generate actionable outputs. The central block represents supervised learners, deep networks for images and sequences, graph models, and ensemble methods that support representation learning, cross-modal fusion, and prediction. The right side summarizes operational products: earlier diagnosis and risk stratification, AI-assisted diagnostics and image/test interpretation, therapeutic discovery and prioritization, continuous outbreak surveillance with anomaly detection, and decision support for response and resource allocation. Read the figure from left to right: data sources feed the modeling core, which emits decisions and alerts used in clinical and public-health settings.

Through an analysis of real-world case studies, the chapter aims to provide a comprehensive understanding of how AI can revolutionize the fight against infectious diseases and prepare for future global health threats.

2 History of AI in infectious disease surveillance and prevention

Rule-based systems and the dawn of AI in healthcare: Early systems like **MYCIN**, developed in the 1970s at Stanford University, marked the advent of AI in healthcare. MYCIN, a rule-based expert system designed to assist clinicians in diagnosing bacterial infections and recommending antibiotic regimens, utilized over 450 decision-making rules. Studies demonstrated its diagnostic accuracy to be comparable to that of human experts, especially in identifying infections like meningitis and bacteremia. However, MYCIN's rigidity in adapting to dynamic clinical scenarios highlighted the limitations of rule-based systems in real-world healthcare contexts. As outlined by Shortliffe, 1976), the lack of scalability in MYCIN foreshadowed challenges that would later surface in AI applications reliant on fixed knowledge bases (Shortliffe, 1976).

Probabilistic models and the emergence of uncertainty handling: By the 1980s, focus shifted to probabilistic models like Bayesian networks, which allowed AI systems to handle uncertainty inherent in medical diagnostics. For example, studies on the application of Bayesian networks for lymph node pathology diagnosis demonstrated their ability to account for false positives and negatives in noisy datasets (Heckerman et al., 1992). This capability proved particularly valuable in infectious diseases, where incomplete or noisy data often complicates decision-making. In addition, Bayesian inference supported outbreak predictions, laying the groundwork for modern epidemiological surveillance.

The 2000s: EHRs and the expansion of AI in surveillance: The widespread adoption of EHRs in the 2000s provided fertile ground for applying ML algorithms. Models like the random forest (RF) algorithm were leveraged to identify patterns in hospital-acquired infections such as methicillin-resistant *Staphylococcus aureus*, as demonstrated by Zhuang et al. (2021). Tools like HealthMap, launched in 2006, aggregated diverse data sources ranging from news reports and social media to government health bulletins to provide real-time tracking of outbreaks. During the 2009 H1N1 influenza pandemic, HealthMap played a pivotal role in identifying hotspots before traditional public health mechanisms (Brownstein et al., 2008). Similarly, Google Flu Trends (2008) estimated flu prevalence based on search engine queries, though later evaluations (Lazer, Kennedy, King, & Vespignani, 2014) revealed overestimation biases due to algorithmic overfitting (Kraemer et al., 2020; Biggerstaff et al., 2018; Ginsberg et al., 2009).

The 2010s: Big data, deep learning, and genomics in surveillance: The 2010s marked the rise of big data and deep learning, transforming infectious disease surveillance. Models like convolutional neural networks (CNNs) advanced diagnostic imaging by identifying infections such as tuberculosis with high sensitivity. In one notable example, Rajpurkar et al. (2017) developed the CheXNet model, a deep CNN capable of detecting pneumonia from chest X-rays with performance surpassing that of radiologists. AI also accelerated genomic analysis, exemplified by tools like Nextstrain, which tracked pathogen evolution during the 2014–2016 Ebola outbreak by combining genomic data with real-time epidemiological information.

AI and outbreak prediction during Zika and Ebola: Outbreak prediction achieved new heights during the Zika epidemic (2015–2016) (Zhang et al., 2017). Researchers integrated climate data, mosquito population dynamics, and EHRs to model at-risk regions. For example, the work of Messina et al. (2016) used ML algorithms to identify areas of high transmission risk, enabling targeted public health interventions. Similar methods were employed during the 2014–2016 Ebola outbreak, where AI-driven platforms forecasted the disease's spread based on human mobility patterns and environmental factors.

The COVID-19 pandemic and the role of AI: The COVID-19 pandemic in the 2020s demonstrated the transformative potential of AI in managing infectious diseases. AI tools were deployed across response efforts, including early detection, diagnostics, and vaccine development. Neural networks analyzed chest X-rays and CT scans for rapid COVID-19 diagnosis, with systems like the COVID-Net initiative achieving high diagnostic accuracy (Wang, Lin & Wong, 2020). NLP algorithms processed patient-reported symptoms from telehealth platforms and social media to flag potential infections. Meanwhile, structural biology models, such as AlphaFold, developed by DeepMind, accurately predicted protein structures, significantly accelerating the design of vaccines and therapeutics.

Public health strategies also benefited from AI-powered predictive models. For instance, BlueDot, an AI platform, successfully identified the outbreak of COVID-19 in Wuhan 9 days before the WHO issued its first alert. Similarly, models combining mobility data, EHRs, and even wastewater surveillance provided real-time insights into emerging hotspots, guiding policy decisions like targeted lockdowns (Peccia et al., 2020; Kraemer et al., 2020).

2.1 Present-day advances and applications

Today, AI systems integrate data from diverse sources to provide comprehensive situational awareness. Platforms like Kinsa Health use wearable sensors to track physiological changes indicative of infection, while wastewater surveillance has emerged as a critical tool for monitoring community viral loads (e.g., COVID-19 and influenza). Advanced models, such as transformers (e.g., BERT, GPT-4), are now applied to predict zoonotic spillover risks and map complex host–pathogen interactions. For instance, graph neural networks have been employed to analyze molecular interaction networks, identifying potential targets for therapeutic intervention in diseases like malaria and HIV.

In addition, modern AI is being harnessed to tackle emerging infectious threats at the human–animal–environment interface. ML models can estimate the zoonotic spillover potential of animal viruses by analyzing genomic features; one study showed that genome-based classifiers could identify candidate viruses with high human infection risk from among hundreds of animal viruses. Climate factors are also integrated into outbreak forecasting: changes in temperature and habitat are predicted to drive over 15,000 new cross-species viral transmission events by 2070, underscoring the need for climate-aware disease models. Researchers are developing AI-driven early warning systems (EWSs) that combine climate, ecological, and disease data, for example, tools that simulate how environmental changes will affect vector-borne diseases like malaria and dengue. These efforts exemplify a "One Health" approach to predictive analytics, where AI links human health with animal and environmental data to anticipate and potentially prevent future pandemics. The application of AI to infectious disease research has evolved from early expert systems like MYCIN to modern generative AI tools for antibiotic and antibody discovery (Fig. 6.2).

2.2 Databases

A wealth of readily available data has already laid a strong foundation for AI applications in infectious disease research. For instance, EHRs from databases such as **MIMIC-IV** and the **eICU Collaborative Research Database** contain detailed patient histories, clinical test results, prescribed medications, and treatment outcomes. These rich datasets, often collected over many years and across diverse populations, are invaluable for training ML models to identify risk factors and predict

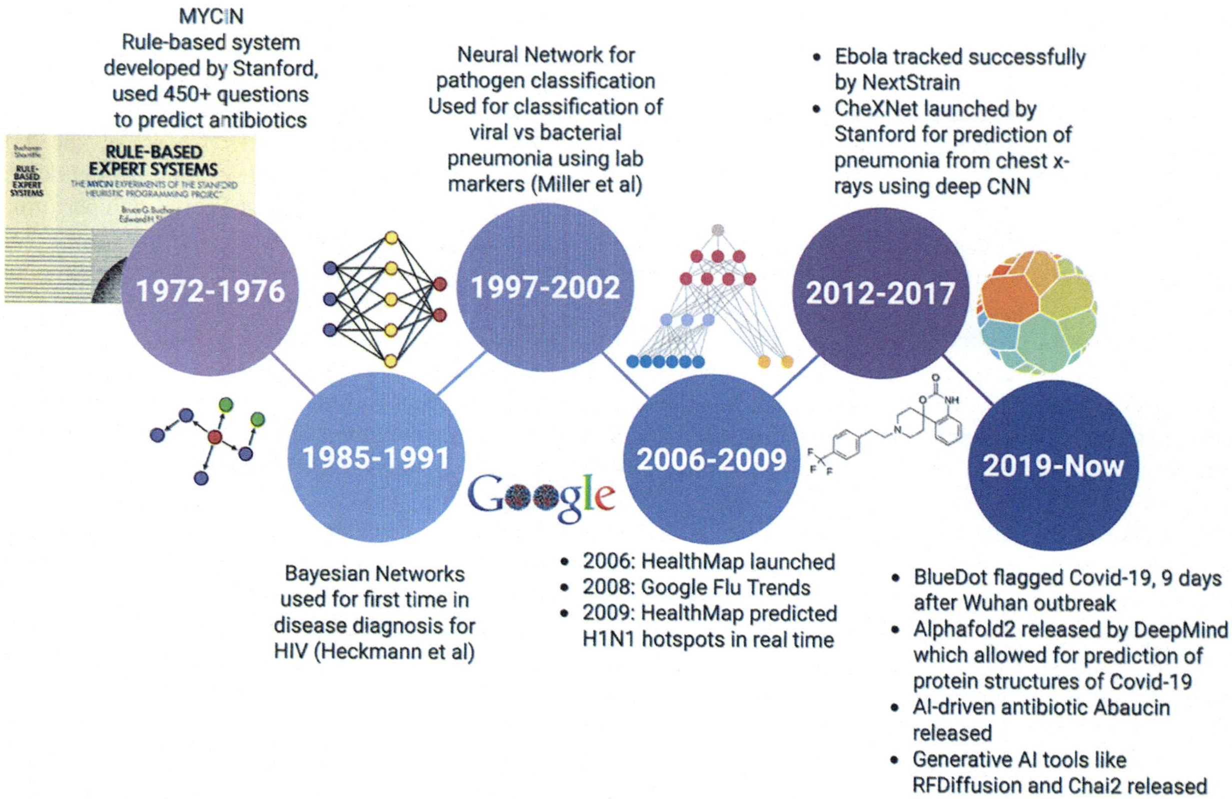

FIG. 6.2 Milestones in AI for infectious disease and therapeutics. The timeline presents a chronological series of inflection points, from rule-based expert systems to modern generative models and global surveillance platforms. It begins with MYCIN (1970s) and proceeds to the adoption of Bayesian networks for clinical inference (mid-1980s to early-1990s), early neural network classifiers using laboratory markers (late-1990s to early-2000s), and the rise of web-scale surveillance, including HealthMap and Google Flu Trends (mid-2000s). The next era highlights genomic tracking and deep convolutional imaging models (2012–2017), followed by the current period, which includes early COVID-19 signal detection by services like BlueDot, near-experimental structure prediction by AlphaFold2, AI-enabled antibiotic discovery, and the introduction of generative protein design tools such as RFdiffusion and Chai. Read left to right to see how methodological advances and infrastructure matured into today's integrated ecosystem.

disease progression. Public health surveillance databases maintained by governmental and international agencies like CDC WONDER, WHO's Global Health Observatory, or ECDC TESSy provide epidemiological data on disease incidence and prevalence, demographics, and outcomes for various infections worldwide. In addition, large-scale genomic repositories (e.g., **GISAID**, **NCBI GenBank**, and the **Sequence Read Archive**) store pathogen sequences from outbreak samples, enabling AI to track viral or bacterial mutations and discover emerging variants. Finally, social media and mobility data, for example, from Twitter, Google Mobility Reports, or Facebook Data for Good, can give early signals of outbreak hotspots or patterns of disease spread in real time.

Beyond these existing datasets, novel data generated from experiments and specialized studies are critically important for advancing AI-driven insights. Laboratory-based *in vitro* infection models, which measure how viruses or bacteria replicate under controlled conditions, can yield time-series data on pathogen load, gene expression, and cellular responses. *In vivo* studies using animal models, coupled with tools like single-cell RNA sequencing, can produce high-resolution datasets on immune responses and disease dynamics that are challenging to capture in human cohorts. Furthermore, clinical trials designed to test new therapeutics or vaccines generate granular data on patient responses, adverse events, and immunological markers over defined timelines. For molecular-level insights, multiomics experiments (genomics, transcriptomics, proteomics, metabolomics) can identify biomarkers associated with pathogen virulence or treatment success. These datasets, often generated with cutting-edge techniques like CRISPR gene editing or high-throughput screening, can be integrated into AI pipelines to uncover novel targets for interventions and improve pathogen characterization. Finally, the growing field of digital health is generating real-time data from wearable biosensors, smartphone apps, and remote monitoring devices, which can help detect subtle changes in physiological parameters before overt clinical symptoms appear, thereby facilitating earlier and more targeted interventions.

2.3 AI for early detection and diagnosis of infectious diseases

AI has become an indispensable asset in healthcare, particularly for the early detection and diagnosis of infectious diseases. By leveraging ML algorithms on vast datasets, AI systems can identify subtle clinical patterns, predict infection risks, and enable rapid diagnostic decisions. In an era where infectious diseases ranging from seasonal influenza to pandemics like COVID-19, as well as endemic threats like malaria and tuberculosis—pose significant global challenges, AI-driven tools have proven vital in accelerating the diagnostic process while maintaining accuracy. This section explores four principal areas of innovation: **AI for symptom analysis, AI-driven laboratory diagnostics, AI-powered point-of-care (POC) tools**, and the implementation of ML models in realistic infectious disease scenarios. Each section draws on peer-reviewed studies, includes relevant model performance metrics [accuracy, sensitivity, specificity, area under the ROC (Receiver Operating Characteristics) curve (AUC), F1-score], and provides mathematical formulations where appropriate, all in a scientific tone suitable for an academic reference.

2.3.1 AI for symptom analysis

Rationale and methods: ML offers powerful techniques for analyzing patient-reported symptoms and clinical signs, automating initial triage, and providing timely decision support for healthcare providers. Traditional symptom triage relies on clinical expertise and manual checklists, which can be subjective and slow. In contrast, ML models can rapidly process large volumes of data from symptom descriptions and vital signs to demographic factors and medical histories to estimate the probability of different infections and their likely severity.

Supervised learning models are frequently used for symptom analysis and early risk prediction. For example, logistic regression (LR) models can output a probability of infection (e.g., $P(\text{disease} = 1)$) based on a weighted combination of symptoms. The logistic function constrains predictions between 0 and 1. Mathematically, if $x = [x_1, x_2, \dots, x_d]$ represents a patient's symptom feature vector (e.g., age, fever, cough, etc.) and $y \in [0,1]$ is the binary outcome (0 = not infected, 1 = infected), LR estimates the probability of infection as:

$$P(y = 1 \mid x) = \sigma(\beta_0 + \beta_1 x_1 + \beta_2 x_2 + \cdots + \beta_d x_d)$$

where β_0 is the intercept, β_j are model coefficients, and $\sigma(z) = \frac{1}{1+e^{-z}}$ is the sigmoid activation. The model is typically trained by maximizing the likelihood or, equivalently, minimizing the binary cross-entropy loss. Other classifiers like RFs and neural networks have also been successfully applied to symptom data. An RF (an ensemble of decision trees) can capture nonlinear interactions between symptoms; it outputs a majority vote of its trees for the predicted class. If $h_{1(x)}, h_{2(x)}, \dots, h_{B(x)}$ are the predictions of B different trees, the RF's prediction $\hat{y}$ is the mode of these predictions:

$$\hat{y} = \text{mode}\{h_1(x), h_2(x), \dots, h_B(x)\}.$$

Or, a more generalized equation of an RF model:

$$\hat{y} = \arg\max_{y \in \mathcal{Y}} \frac{1}{B} \sum_{b=1}^{B} p_b(y|x)$$

This ensemble approach tends to improve generalization and handle diverse symptom inputs. Meanwhile, neural networks (including deep learning models) can model complex, nonlinear relationships in high-dimensional symptom space, especially when large datasets are available. For sequential symptom data or longitudinal patient records, recurrent neural networks, such as long short-term memory (LSTM) networks, are used to capture temporal patterns (e.g., symptom progression over days). Unsupervised learning can also play a role: clustering algorithms (like *k*-means) might group patients with similar symptom profiles to help detect emerging clusters of an infection (for instance, grouping patients by unique symptom combinations could signal a new outbreak variant).

Real-world applications: Real-world studies have demonstrated the value of AI-based symptom analysis across diverse infectious diseases. During the COVID-19 pandemic, for example, researchers deployed smartphone apps and web-based symptom checkers to triage individuals before they ever reached a clinic. One large study by Menni et al. (2020) collected self-reported symptoms from over 2.6 million users *via* a mobile app. Using a LR model, they identified key symptom combinations (such as loss of smell and taste, persistent cough, and fatigue) that predicted COVID-19 infection with an AUC of about 0.76, corresponding to roughly 65% sensitivity and 78% specificity in a UK test set. Such performance, achieved purely from symptoms and demographics, underscored that AI can provide useful early screening for viral infections. Similarly, a symptom checker developed by Ada Health and tested in an emergency department showed around 70% sensitivity in suggesting the correct diagnosis among its top predictions, comparable to human physicians for triage purposes. These tools have acceptable usability and can handle thousands of inquiries daily without clinician intervention (Fraser et al., 2023; Rocha Filho et al., 2021).

AI-driven symptom analysis is not limited to viral pandemics. In hospital settings, ML-based EWSs analyze vital signs and nurse observations to detect conditions like **sepsis**, a life-threatening complication of infection, earlier than traditional scoring systems. For instance, a recent metaanalysis of five prospective studies found that AI-based early warning models significantly reduced in-hospital and 30-day mortality for deteriorating patients compared to standard care (Yuan et al., 2025). In practice, these systems continuously monitor patient data in EHRs (e.g., fever, heart rate, blood pressure, laboratory results) and alert clinicians if a pattern predictive of sepsis or other acute infections is detected. One FDA-approved example is an algorithm that scans EHR data to predict sepsis hours before clinical recognition; such models have shown up to a 30% improvement in early sepsis detection, with false alarm rates also reduced by double-digit percentages. Unsupervised NLP methods have even been applied to EHR clinical notes to passively surveil for symptoms indicative of emerging infections (for example, identifying clusters of patients with "flu-like" or "respiratory distress" complaints in real time for outbreak alerts). Together, these applications illustrate that AI symptom analysis can augment frontline clinical decision-making, especially when healthcare systems are strained by surges of infectious disease patients.

2.3.2 AI-driven laboratory diagnostics

Accelerating and enhancing diagnostics

Once a patient's infection is suspected, laboratory diagnostics are the gold standard for confirmation, including tests like pathogen cultures, antigen assays, and nucleic acid amplification tests (NAATs) such as PCR (Polymerase Chain Reaction). However, traditional laboratory methods can be slow, labor-intensive, or require significant expertise. AI and ML techniques are addressing these bottlenecks on multiple fronts to achieve faster and more accurate diagnoses in the laboratory setting:

1. **Image analysis for microscopy:** CNNs have greatly improved the interpretation of microscope images used in infectious disease diagnosis. For example, in malaria diagnosis, thin blood smear slides are examined for parasites. AI systems can automate this process with high accuracy. Liu, 2023) developed a smartphone-based deep learning system for malaria parasite detection in blood smears, which achieved ~97% accuracy (sensitivity ~97%, specificity ~94.18%, AUC ~0.98) in identifying malaria-infected cells. This CNN model, deployed on a phone, was able to detect parasitized cells with an F1-score around 0.97, rivaling expert microscopists but at a much faster throughput. Another study by Chen, 2024) applied an "intelligent microscopy" AI to detect *Mycobacterium tuberculosis* bacteria in sputum smear slides. The system reached an accuracy of 96.7%, with 91.9% sensitivity and 98.0% specificity for acid-fast bacilli detection, significantly reducing missed TB cases compared to human readers. These examples demonstrate that AI-powered image recognition can both accelerate laboratory workflows and improve diagnostic sensitivity for microscopic exams (important for diseases like malaria, tuberculosis, and other parasitic or bacterial infections).
2. **Genomic and proteomic signal processing:** Modern laboratories often rely on molecular techniques (like PCR, isothermal amplification, or mass spectrometry) to detect pathogen DNA/RNA or proteins. AI algorithms can assist by interpreting complex output signals to call positive/negative results or even identify pathogen strains. For instance, advanced ML models have been trained on DNA sequencing or mass spectrometry data to distinguish between different species or strains of pathogens present in a sample (Gao & Liu, 2024). In viral diagnostics, slight variations in a virus's genome can be detected *via* high-throughput sequencing; ML classifiers then determine if the strain is, say, a drug-resistant variant or a more Zou et al., virulent lineage (Zou et al., 2025). These approaches have been applied in COVID-19 for rapid variant classification and in bacterial infections for predicting antibiotic resistance profiles from genomic data. The use of ML in analyzing such high-dimensional laboratory data can uncover subtle patterns (e.g., minor genomic mutations or unique peptide markers) that might be missed by human interpretation.
3. **Automated workflow optimization:** Beyond direct diagnostics, AI helps optimize laboratory operations. Scheduling algorithms can prioritize tests based on urgency, and predictive models can forecast test volumes, ensuring reagents and personnel are efficiently allocated. For example, if an influenza outbreak is predicted *via* syndromic surveillance, an AI system can proactively ramp up PCR throughput for flu tests in the laboratory. Some laboratories have also implemented robotics guided by AI vision systems to perform repetitive tasks (like pipetting or colony counting), reducing human error and freeing technicians for more complex analyses.

AI-enhanced NAAT and "digital" PCR

NAATs are among the most sensitive methods for detecting pathogens, but interpreting their results can sometimes require expert judgment (for example, reading faint PCR bands or real-time amplification curves). AI-driven analysis can standardize this interpretation and even improve the effective limits of detection. A striking example comes from the COVID-19 pandemic: Cheong et al. (2020) reported a portable NAAT device that integrates AI-based analysis of fluorescence signals in real time. In their *Nature Biomedical Engineering* study, the team used a microfluidic RT-PCR instrument that performed rapid thermocycling using plasmonic heating and recorded amplification curves with a camera. An ML algorithm then analyzed these fluorescence time-series to determine positivity. In testing, the system correctly classified all 75 COVID-positive patient samples *and* 75 negatives, achieving **100% sensitivity and 100% specificity** compared to gold-standard PCR in a central lab. This illustrates how AI can push diagnostics toward the ideal of near-perfect accuracy by extracting more information from raw assay data (in this case, analyzing subtle curve kinetics to make binary calls). Another emerging approach is to combine digital PCR or CRISPR-based assays with AI: for example, a deep learning–assisted CRISPR diagnostic was shown to detect SARS-CoV-2 in saliva with ~94% sensitivity and 100% specificity, matching laboratory PCR performance but in a cartridge-based, automated format. These AI-augmented platforms reduce the need for manual result interpretation and can provide quick readouts in decentralized settings (Javalkote et al., 2022).

AI is also being used in culture-based and antigen tests. In microbiology laboratories, colony recognition software (using image analysis) can count bacterial colonies on culture plates faster and flag unusual growth characteristics for further review. For rapid antigen tests (e.g., lateral flow strips for diseases like COVID-19, dengue, or influenza), smartphone apps with AI can read the test lines objectively, improving sensitivity, especially when the result line is faint. Research has found that these apps can reduce human reader errors and achieve high concordance with expert readings. Overall, AI-driven laboratory diagnostics are not only accelerating the time to result, which is crucial during outbreaks, but are also enhancing accuracy, which improves clinical decision-making and patient outcomes.

2.3.3 *AI-powered POC tools*

The need for POC diagnostics

In many settings, especially in low-resource regions, the availability of advanced laboratory testing is limited. POC diagnostics are tests performed at the patient's side (in clinics, remote areas, or even at home) that provide quick results. Traditional POC tests, such as rapid antigen kits or simple microscopy, can suffer from lower accuracy or require training to interpret. AI-powered POC tools aim to deliver reliable, automated diagnoses on-site, reducing dependence on centralized laboratories and enabling immediate clinical or public health action. The combination of portable biosensors with built-in AI analytics has opened new possibilities for managing infectious diseases in the field.

Smartphone-based diagnostic platforms

The ubiquity and increasing computational power of smartphones have made them a cornerstone of AI-driven POC diagnostics. Modern smartphones come equipped with high-resolution cameras, sensors, and processors capable of running ML models, essentially serving as miniaturized laboratories in the pocket. One application is the reading of rapid diagnostic tests (RDTs) using smartphone cameras. Instead of relying on a health worker's eyes to judge the presence of a faint line (which indicates a positive result), an app can use a CNN-based computer vision algorithm to interpret the test strip. This approach has been tested for diseases like malaria, HIV, and COVID-19. For malaria RDTs, a smartphone CNN app was shown to improve reading consistency; in one study, an AI-based reader achieved over 90% accuracy in classifying RDT results and significantly reduced user errors in a controlled trial. Another example is the National Library of Medicine's **"Malaria Screener"** app, which not only reads RDTs but also analyzes images of blood smears. In a 2025 evaluation in Ghana, this app detected malaria cases in patients with sickle-cell disease with 89.5% sensitivity (comparable to conventional tests). Notably, the AI app identified more positive cases than the local microscopy did (though with some false positives, indicating a need for confirmatory testing) (Obeng et al., 2025). These studies highlight that smartphone diagnostics can be especially useful in high-volume screening or in environments where expertise is scarce. Beyond test-strip reading, smartphone-based AI has also been used for analyzing photos of patients for signs of infection (e.g., detecting jaundice in newborns as a possible sepsis indicator or skin lesion recognition for diseases like Lyme disease or cutaneous leishmaniasis).

Another high-impact use of smartphones is in conjunction with isothermal amplification tests and novel sensing methods. For instance,) demonstrated a smartphone-integrated loop-mediated isothermal amplification (LAMP) device for malaria that detects tiny amounts of parasite DNA (Colbert et al., 2021). The smartphone's camera records a 30-s video of a fluorescence-based LAMP reaction, and an AI algorithm determines positivity by analyzing the Brownian motion of particles in the reaction (a technique called particle diffusometry). Impressively, this system could detect as few as three parasites/μL of blood within 45 min, achieving ~89% sensitivity and 100% specificity against PCR for asymptomatic infections. Such sensitivity at the point of care can help identify infectious individuals who would otherwise be missed by regular rapid tests, thus aiding in disease elimination campaigns.

Lab-on-a-chip devices with AI integration

Lab-on-a-chip (LOC) technology miniaturizes laboratory processes (sample preparation, chemical reactions, detection) onto microfluidic chips that are only a few centimeters in size. These portable devices enable complex assays, which might normally require an entire laboratory, to be conducted anywhere with a small device. When combined with AI, LOC platforms become even more powerful. AI algorithms can process the multichannel data that LOC devices produce, such as simultaneous readouts from multiple sensors, fluorescence channels, or electrical signals. By embedding trained models on the device's firmware or analyzing data on a paired smartphone/laptop, one can achieve real-time interpretation of test results without human input.

A notable study described a portable LOC system for respiratory virus detection [including influenza and respiratory syncytial virus (RSV)] that employed an onboard neural network for analysis. In this system, a single nasal swab sample was processed on-chip through several steps: fluid mixing, nucleic acid amplification, and optical detection. The neural network took the raw fluorescence-vs-time data from multiple reaction wells and determined **within 15 min** which viruses were present, successfully distinguishing influenza A, influenza B, and RSV with high accuracy. While the full technical specifics are beyond the scope, the key point is that AI enabled the interpretation of complex multiplexed data on a small device, giving clinicians a panel of results at the patient's bedside in one go. In another example, researchers built a microfluidic chip for HIV viral load testing in remote clinics; the device uses AI to quantify the viral genetic material from an image of the microchip, outputting the viral concentration with accuracy comparable to standard PCR machines (Ngo et al., 2024).

The integration of AI addresses some challenges of lab-on-chip diagnostics: manufacturing variabilities and noisy signals can be "learned away" by the model, and it can make logical decisions (like invalidating a test run if controls fail or adjusting thresholds based on ambient temperature). As a result, AI-equipped POC devices tend to be more robust. However, rigorous validation is required; these devices must be tested on diverse real-world samples to ensure the AI generalizes well and does not produce false results due to confounders. When validated, AI-powered POC tools offer the promise of accessible, fast, and reliable testing for infectious diseases anywhere, anytime.

2.3.4 *Implementation of ML models in practice*

To illustrate how different ML models can be implemented for infectious disease detection, this section walks through a hypothetical case study. We consider a realistic (though synthetic) dataset and demonstrate the training and evaluation of two common classifiers: LR and RF, for predicting infection in a clinical scenario. We also discuss other model architectures [support vector machines (SVMs), CNNs, LSTMs] and their roles. This exercise highlights practical considerations like data preprocessing, model selection, and performance metrics, mirroring what real-world data scientists and clinicians do when deploying AI in healthcare.

Example scenario: Respiratory infection dataset

Imagine a dataset called **RespiraX**, collected for early-stage detection of a new respiratory infectious disease. It contains 2000 patient records from various healthcare facilities. Each record includes a mix of demographic, vital sign, and symptom features, along with the final diagnostic outcome (whether the patient was confirmed positive for the infection):

4. **Age (years)** continuous
5. **Gender (male or female)** categorical
6. **Body temperature (°C)** continuous
7. **Heart rate (beats per minute)** continuous
8. **Oxygen saturation (%)** continuous
9. **Cough severity (none, mild, moderate, severe)** ordinal
10. **Shortness of breath (No = 0, Yes = 1)** binary
11. **Loss of taste/smell (No = 0, Yes = 1)** binary
12. **Outcome (Negative = 0, Positive = 1)** binary label (ground truth)

This feature set reflects typical data available during triage. For instance, older age might correlate with a higher risk of a positive outcome; high fever or certain symptom combinations (e.g., severe cough *plus* shortness of breath *plus* loss of smell) might be strongly predictive of infection. Before modeling, we preprocess the data: categorical variables are encoded (e.g., female = 1, male = 0; none = 0, mild = 1, moderate = 2, severe = 3 for cough severity), and continuous variables can be standardized (to mean 0, unit variance) to aid the convergence of some algorithms. We then split the data into a training set (say 80% of records) for model fitting and a testing set (20%) for final evaluation.

LR implementation

We first train an LR model to predict the *Outcome* using the above features. During training, the model coefficients β_j are estimated by maximizing the likelihood of the training data (or equivalently minimizing log-loss). Regularization (like an L2 penalty) might be applied to prevent overfitting, especially if some features are highly correlated or not very informative.

After training, we apply the model to the test set (400 patients) to generate predicted probabilities $\widehat{p_i} = P(y_i = 1 | x_i)$. Each patient is then classified as positive if $\widehat{p_i} \geq 0.5$ (the threshold can be adjusted for sensitivity vs. specificity trade-offs). We compute evaluation metrics on the test set: suppose LR achieves 88% accuracy, 85% precision, 82% recall (sensitivity), and an F1-score of ~0.83. These values are hypothetical but plausible indicating the model makes correct predictions in 88% of cases, and among those it flags as positive, 85% are true positives, etc. Such performance would be quite promising for a first-pass screening tool at triage. An accuracy in the high-80s and recall ~0.82 means the model misses about 18% of actual cases, which might be acceptable if those are caught in confirmatory testing. Depending on whether false negatives or false positives are considered more problematic in context, the threshold or features could be tuned to improve recall if needed.

Beyond overall metrics, LR offers the benefit of interpretability. We can examine the learned coefficients: for instance, LR might assign a large positive weight to "Fever" (meaning higher temperature strongly increases predicted risk) and "Loss of Smell," while giving a negative weight to "Oxygen Saturation" (since lower O_2 is associated with infection, the feature is inversely related, so low O_2 equates to high risk in effect). This aligns with medical intuition and helps build trust with clinicians. Confidence intervals or *p*-values for coefficients can also be computed to determine which features are statistically significant predictors.

RF implementation

Next, we train an RF classifier on the same data. We use, say, 100 decision trees ($B = 100$) in the forest, each tree trained on a bootstrap sample of the training data and considering a random subset of features at each split (to ensure trees are decorrelated). Each decision tree in essence learns a series of if–else rules on the features (e.g., "if temperature >38°C and cough severity = severe then.") to classify patients. The RF aggregates these decisions by majority vote.

On the test set, the RF might achieve slightly higher performance than LR thanks to its ability to capture nonlinear interactions. For example, it could attain 90% accuracy, with 88% precision, 85% recall, and an F1-score around 0.86. In many real-world studies, ensemble methods like RFs have indeed outperformed simpler linear models for infection prediction, especially when there are complex rules (e.g., a combination of moderate fever *plus* low oxygen *plus* high heart rate might collectively indicate high risk, even if each feature alone is not extreme). The RF's higher recall (85% vs. 82% with LR) means it catches more of the true cases, which is valuable in a screening context, though its precision (88%) indicates a small increase in false positives compared to LR. Depending on the healthcare context, one might prioritize higher sensitivity (to not miss cases) even at the cost of more false alarms, particularly for dangerous diseases.

RFs also provide useful by-products like feature importance scores. By analyzing how much each feature split improves purity across the trees, we might find, for instance, that body temperature and loss of taste/smell are the top two predictors of the RespiraX infection in our dataset, followed by shortness of breath and age. This generally aligns with the known symptom profile we simulated (fever and anosmia were strong signals). Such insights can guide clinicians to pay extra attention to certain symptoms or vitals. However, unlike LR, RFs are less interpretable in terms of specific relationships; they operate more like a black box with many conditional branches.

Hyperparameter tuning is another part of our implementation: we can use cross-validation on the training set to choose the optimal number of trees, their maximum depth, the minimum samples per leaf, etc., to balance bias and variance. In our scenario, we noticed that beyond 100 trees, performance plateaued. Each tree was allowed to grow until each leaf had at least five samples to prevent overfitting the noise. This optimization step is representative of real ML workflows.

Other models and approaches

While our case study focused on LR and RF, numerous other ML models are worth mentioning for infectious disease diagnostics:

13. **SVMs:** These can be effective in high-dimensional feature spaces and have been used for medical imaging tasks involving infections. For example, an SVM with a radial basis kernel was applied in a study to classify lung CT scans for tuberculosis vs. other diseases, achieving high sensitivity (~95%) in detecting TB lesions. SVMs find an optimal hyperplane that separates classes; their use of kernel functions allows for modeling nonlinear decision boundaries. However, SVMs do not scale as well to very large datasets as some other methods, and they require careful tuning of kernel parameters.
14. **CNNs:** We have already noted CNN successes in analyzing blood smears and radiographs. In diagnosing infectious diseases, CNNs have famously been used to detect pneumonia (often caused by infections) on chest X-rays with performance comparable to that of radiologists (e.g., >90% accuracy in some cases). They have also been integrated into smartphone devices for malaria and have achieved above 95% accuracy in classifying parasite images. CNNs automatically learn visual features (edges, textures, shapes), which is ideal for tasks like identifying pathogen structures in microscopy or characteristic infection patterns in medical imaging. One challenge is that CNNs need substantial data for training; if only limited images are available, techniques like transfer learning (using pretrained networks) or data augmentation are employed. In addition, CNNs have been combined with attention mechanisms (Vision Transformers or hybrid models) to further boost performance in complex image diagnosis tasks.
15. **Recurrent neural networks and LSTMs:** These excel in sequential data analysis. An LSTM-based model can analyze time-series data, such as a patient's vital sign trends over time or symptom progression logs. For instance, to predict the onset of sepsis (which often follows a certain pattern of vital changes), researchers have used LSTM networks that ingest hourly vital measurements; some models achieved an AUC above 0.85 for predicting sepsis 4–6 h before clinical recognition. In infectious disease outbreaks, RNNs (and newer transformer models) have also been used to forecast case counts from temporal epidemiological data (Fleuren et al., 2020).
16. **Hybrid and ensemble approaches:** In practice, different models can complement each other. There are systems that combine, say, a CNN for image analysis with a gradient boosting machine that takes the CNN's output plus clinical metadata to make a final prediction. During the COVID-19 pandemic, some diagnostic workflows combined a symptom-based prescreen (simple model) with a confirmatory imaging or lab-based AI diagnosis (more complex model). Each stage was tuned to maximize overall sensitivity and specificity.

Predictive AI imaging methods accelerate the diagnostic workflow, transforming raw medical images into actionable radiological reports and patient outcomes far faster than traditional methods (Fig. 6.3).

Real-world applications have reported impressive results using these advanced techniques. For example, one smartphone-based CNN system for malaria (discussed earlier) exceeded 95% accuracy in classifying malaria-infected cells, and a deep learning model for TB on chest X-rays achieved an AUC of ~0.99 in identifying TB versus healthy lungs (Lakhani and Sundaram, 2017). Each approach has trade-offs: deep neural networks often provide the highest accuracy but at the cost of interpretability and requiring more data, whereas simpler models like LR are transparent but might miss complex patterns. Therefore the choice of model depends on the clinical requirements; if the model will be used for critical decisions, interpretability and rigorous validation are key; if it is for a second-opinion tool or screening, maximizing accuracy might be the priority.

From development to deployment

Implementing ML models in infectious disease detection involves several steps beyond just choosing an algorithm. First is **data collection and curation**: models are only as good as the data they learn from. It is crucial to gather diverse data (from different populations, hospitals, and geographies) to ensure the AI does not become biased or brittle. For instance, an AI trained only on adults might underperform on pediatric cases, or a model developed on one strain of a virus might falter if a new strain emerges. Large, representative datasets, possibly augmented by synthetic data, help mitigate this.

Next comes **feature engineering** (if using models that do not automatically handle it, like non-deep-learning methods). In our RespiraX example, we used raw features, but one might create additional features like "fever severity" as a categorical feature or a "symptom count" feature. In text mining (say, parsing clinical notes for symptoms), NLP techniques convert unstructured text into numerical features that models can handle.

During model training and validation, techniques like cross-validation, hyperparameter tuning, and calibration are applied. Often, an independent validation cohort (from a different hospital or a later time period) is used to evaluate

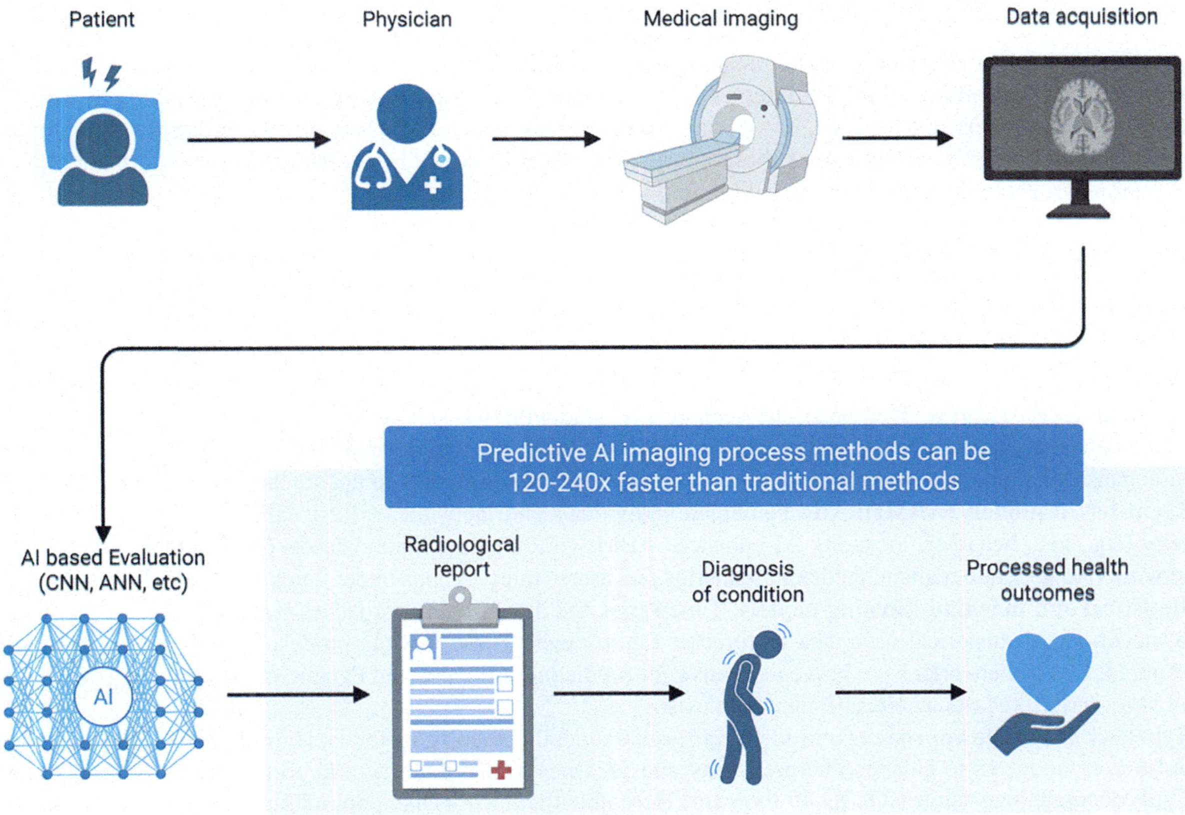

FIG. 6.3 AI-augmented clinical imaging workflow. This pathway diagram follows the patient from presentation through clinician assessment, modality selection, and image acquisition to an AI inference stage that operates on DICOM data. The AI module performs preprocessing and inference to produce triage flags, detection or segmentation overlays, and quantitative metrics that feed into a structured radiology report. Subsequent steps include diagnostic decision-making and treatment planning, which translate into patient outcomes. The central banner notes potential throughput gains when repetitive interpretation tasks are automated, emphasizing that realized speed-ups depend on local workflow, case mix, and model performance. Arrows guide reading from left to right, with the AI block inserted between acquisition and reporting to illustrate where decision support is generated.

generalization. Performance metrics relevant to clinical use are measured: sensitivity (recall) is vital if missing a case has severe consequences (e.g., in Ebola screening, you want near 100% sensitivity), whereas specificity is important to avoid overtreating or causing panic with false positives. The ROC curve and precision–recall curve are commonly analyzed; for instance, an AI for influenza might achieve an AUC of 0.95, but one would also check the positive predictive value at the chosen operating point given the disease prevalence.

Finally, **deployment in a clinical setting** includes integrating the model into software that clinicians use (such as embedding it in the EHR system or as a smartphone app), establishing user interfaces for the output (e.g., a risk score or alert), and setting up a monitoring system. Models in healthcare should be continuously monitored for **data drift**; if disease patterns change (like a new COVID variant with different symptoms), the model may need retraining. Regular reevaluation with new data and periodic updates keeps the AI relevant. Moreover, regulatory and ethical considerations must be addressed: many AI diagnostic tools require regulatory approval (e.g., by the FDA or CE marking) before widespread clinical use, which entails demonstrating safety and effectiveness in clinical trials.

In summary, the implementation of ML models for infectious diseases is a multistep process that marries data science with medical knowledge. When done rigorously, the outcome is a robust AI system that augments clinicians' ability to detect and diagnose infections early. This leads to faster isolation and treatment of patients, more efficient use of medical resources, and ultimately better patient outcomes and outbreak control. AI is not a panacea; it complements but does not replace human expertise. The best results often come when clinicians and AI work in synergy: the AI provides rapid analysis and probabilities, and the clinician applies judgment and context. As infectious diseases continue to emerge and evolve, AI-driven diagnostic tools will be an integral part of the global health arsenal, enabling a proactive and precision approach to disease surveillance and management.

2.4 Predictive modeling for outbreaks and spread

Predictive modeling of infectious disease outbreaks combines epidemiological theory with data-driven methods to forecast how pathogens will spread and to inform interventions. Traditional approaches include compartmental (SIR/SEIR where S = Susceptible, E = Exposed, I-Infected, R = Recovered/Removed) models and agent-based simulations that mimic transmission dynamics. Recently, the fusion of these mechanistic frameworks with ML has shown transformative potential, improving forecast accuracy and parameter calibration.

2.4.1 Epidemiological forecasting

Epidemiological forecasting employs mathematical models to project outbreak dynamics in both the short and long term. These methods include:

17. **Compartmental and network models:** Classical models (e.g., SIR, SEIR) compartmentalize the population and capture average outbreak dynamics. They can be extended with network structures to reflect contact heterogeneity. Such models rely on parameters like transmission and recovery rates and typically perform well for well-characterized pathogens. ML can assist by estimating time-varying parameters from data to update these models in real time.
18. **Agent-based models (ABMs):** ABMs simulate individuals and their interactions, capturing fine-grained heterogeneity (e.g., age, behavior, location). AI-enhanced ABMs can integrate large datasets (e.g., human mobility, contact networks) to generate realistic epidemic scenarios and assess interventions under complex conditions.
19. **Statistical and machine-learning models:** Time-series and data-driven approaches (e.g., ARIMA, RFs, LSTM neural networks) fit historical case data to forecast future cases. These flexible models capture nonlinear trends; for example, LSTM networks have proven effective for predicting influenza and dengue outbreaks. They are often used for short-term forecasts where sufficient data exist.
20. **Hybrid & ensemble approaches:** Modern forecasts frequently combine methods. Hybrid models merge mechanistic and ML components to balance interpretability and fit. One study found that a hybrid SEIR + ML model achieved highly accurate long-range COVID-19 forecasts. Ensemble frameworks that pool multiple models (statistical, mechanistic, and ML) can further improve reliability by leveraging diverse strengths (Jin, Dong, Yu, & Luo, 2022).

2.4.2 Spatiotemporal analysis

Spatial epidemiology uses geographic information to map disease risk and spread patterns. It employs statistical tools like Moran's I, Getis-Ord Gi, and Kulldorff's spatial scan statistic to detect clusters of cases. Bayesian spatial models and kriging techniques provide continuous risk surfaces while accounting for spatial uncertainty. Geospatial AI extends these methods by applying ML to spatial datasets. By integrating environmental, demographic, and mobility data with case data, AI-enhanced models can predict where outbreaks will intensify. For example, a high-resolution analysis of influenza in New York City identified persistent hotspots around commuter rail lines, guiding targeted vaccination campaigns. Overall, such spatial analytics enable the "precise identification of local transmission hotspots," facilitating proactive containment measures (Levine et al., 2010).

2.4.3 Early warning systems

EWSs continuously monitor syndromic and digital data streams to detect emerging outbreaks as quickly as possible. These systems analyze clinical and proxy health indicators (e.g., ER visit counts, over-the-counter drug sales, school absenteeism) and combine them with nontraditional data like search engine queries, social media chatter, and participatory surveillance reports. Digital surveillance often provides significant lead time over traditional reporting: one review found that Google Trends and Twitter signals preceded official influenza or COVID-19 case reports by 1–3 weeks, with predictive models achieving correlations of $r > 0.8$ (Liscano, Anillo Arrieta, Montenegro, Prieto-Alvarado, & Ordoñez, 2025). In fact, AI-powered syndromic surveillance can potentially flag novel threats by detecting unusual symptom clusters before specific pathogens are identified. For example, a Google AI model predicted COVID-19 case counts 2 weeks in advance using mobility and case data. Integrating multiple data streams further improves accuracy: fused models that analyze search, social media, and health report data together produce more precise, timely alerts (García-Cremades et al., 2021). Ultimately, AI-driven EWS can shift surveillance from passive reporting to proactive forecasting, issuing early warnings of pathogen surges.

In summary, the convergence of advanced forecasting models, geospatial analysis, and syndromic surveillance is yielding more powerful outbreak prediction tools. Integrating AI into mechanistic and statistical models has improved the accuracy of both short-term forecasts and long-range projections. Likewise, ML analysis of spatial and digital data enables rapid hotspot mapping and early warning detection. Such predictive systems, leveraging EHRs, human mobility patterns, environmental data, and real-time digital signals, offer public health authorities advanced notice of resurgent pathogens, enabling smarter, targeted interventions.

2.5 AI for containment and response

AI has become a crucial tool for containing outbreaks and improving response efforts. By leveraging ML and data analytics, AI systems can identify at-risk individuals, optimize resource distribution, and provide real-time situational awareness during health emergencies. This section explores how AI-driven approaches enhance containment strategies (like contact tracing and quarantine) and support response logistics (such as resource allocation and outbreak monitoring), with a focus on technical methods and real-world case studies.

2.5.1 Contact tracing and quarantine strategies

One of the primary containment measures in an epidemic is **contact tracing**, which involves identifying and notifying people who have been exposed to an infected individual so they can quarantine. Traditionally, contact tracing is labor-intensive and slow, relying on manual interviews and individual recall. AI and digital technologies have transformed this process through automated contact tracing apps and smart algorithms:

Digital contact tracing apps: Many countries introduced smartphone apps (using Bluetooth or GPS/Global Positioning System) to automatically log close contacts. For example, the NHS COVID-19 app in the UK used Bluetooth signals to detect when two users were within ~2 m of each other for a sufficient duration. If a user later tested positive, the system could instantly notify all their recent contacts, vastly accelerating the tracing process compared to manual methods (O'Dowd, 2020).

Risk scoring algorithms: AI algorithms can estimate the risk of infection for each exposure and decide who should quarantine. Rather than quarantining every contact, these systems use factors like proximity, duration of exposure, and context (indoors vs. outdoors) to assign a risk score. In the NHS app study, the app generated a "meaningful epidemiological risk score" based on distance and exposure time, which correlated well with actual infection probabilities. Such risk scoring enables optimized quarantine measures, where only high-risk contacts are asked to isolate, reducing disruption while still curbing transmission. Optimizing the threshold for these risk scores is challenging but critical: the goal is to minimize virus spread without unnecessarily quarantining large numbers of people (O'Dowd, 2020).

Data integration for quarantine decisions: Advanced contact tracing systems integrate multiple data sources to improve accuracy. For instance, exposure data from apps can be combined with test results, self-reported symptoms, and location information to more precisely identify who is likely infected. ML models can then predict which contacts have a higher probability of being positive and prioritize them for testing or precautionary quarantine. In practice, this data-driven approach can fine-tune public health guidelines. One analysis suggested that better calibration of contact tracing and quarantine rules could have significantly reduced the socio-economic costs during COVID-19.

Augmented contact tracing: Beyond phone apps, AI tools have been used to analyze contact networks and identify potential "superspreaders" or hotspots of transmission. Graph analysis algorithms can flag individuals or locations that, if infected, would likely infect many others, enabling authorities to intervene proactively. There is also exploration of AI-powered location tracking (e.g., using credit card transaction logs or CCTV with facial recognition) to supplement contact tracing, though these raise privacy concerns. Notably, privacy-preserving techniques (like decentralized Bluetooth protocols and differential privacy) have been developed to allow automated tracing and risk alerts without compromising personal data.

Real-world applications: Dozens of countries deployed digital contact tracing apps during COVID-19. Singapore's **TraceTogether**, one of the first, and similar apps in Europe (e.g., Germany's Corona-Warn-App) and elsewhere leveraged Bluetooth-based encounters. While uptake and effectiveness varied, there were successes: the UK's NHS app, for example, sent over 1.7 million exposure notifications within its first months (Chow et al., 2023; Enaya et al., 2024; O'Dowd, 2020). Studies later showed that longer exposure duration and closer proximity (as captured by the app's measurements) strongly correlated with infection risk, validating that such apps can accurately identify high-risk contacts. These tools, when combined with AI-driven risk assessment, allowed for more targeted quarantine focusing on those contacts most likely to be infectious, rather than quarantining everyone exposed. This targeted approach helps maintain control of the outbreak while mitigating the negative impacts of blanket quarantines on society and the economy.

2.5.2 Resource allocation and logistics

Effective outbreak response requires the smart allocation of scarce resources, from hospital beds and ventilators to vaccines, medications, and healthcare personnel. AI-driven decision support systems excel at analyzing complex data and can significantly improve how resources are distributed and utilized during a health crisis. Key AI/ML approaches in this domain include predictive modeling, optimization algorithms, and real-time supply chain analytics:

Predictive demand modeling: ML models can forecast future healthcare needs by analyzing trends in infection data and patient conditions. For example, a team at Cambridge developed an AI system (called *Cambridge Adjutorium*) that predicts the likelihood a COVID-19 patient will deteriorate (need ICU care or ventilation) with about 77%–87% accuracy (Qian, Alaa & van der Schaar, 2021). Hospitals used such predictions to anticipate ICU bed demand and allocate beds and ventilators to those most at risk. In the United States, other ML models analyzed EHRs to assign risk scores to patients, helping clinicians prioritize ICU admissions and specialized care (Nguyen et al., 2021). By triaging patients with AI insights, limited critical care resources can be reserved for those who need them most, potentially saving more lives (Bhansali & Jain, 2020).

Dynamic resource allocation: AI can assist health authorities in deciding *where* to send resources as an outbreak evolves. In Italy and elsewhere, one challenge during COVID-19 was moving ventilators and staff to emerging hotspots in time. AI-based systems use real-time outbreak data and mobility patterns to simulate different allocation strategies. For instance, *PanViz* and *PandemCap* are decision-support tools that model disease spread and resource usage; they allow planners to simulate how distributing a certain number of hospital beds or vaccines to one region would impact overall outbreak control (Brüggemann et al., 2021). Such simulations, often powered by AI optimizers, help officials make evidence-based decisions on managing capacity (e.g., when to convert general beds to ICU beds, or how to redeploy healthcare workers). This is especially important in low-resource settings where every asset must be used optimally.

Workforce and logistics planning: During outbreaks, healthcare personnel and supply chains become as critical as medical equipment. AI has been used to project staffing needs by location and time. For example, ML-based monitoring systems can analyze testing and case data to predict where nurses or doctors will be in short supply, enabling proactive redistribution of staff. This is particularly important when outbreaks spread to rural or underserved areas that have fewer health workers. On the logistics side, routing and scheduling algorithms help coordinate deliveries of supplies. A striking case is Zipline's drone delivery service in several African countries: Zipline uses autonomous drones to dispatch vaccines and medical products to remote clinics, and it integrates AI route optimization algorithms to minimize delivery times (Umlauf & Burchardt, 2022). These AI-guided drones have dramatically improved vaccine availability in rural Ghana, Rwanda, and beyond by finding the most efficient paths and even accounting for terrain or weather (Musa, Haruna, Aliyu, Zubairu, & Lucero-Prisno, 2025).

Vaccine distribution and prioritization: The COVID-19 vaccine rollout demonstrated the need for smart allocation. AI tools have been applied to ensure vaccines reach the right people at the right time. One approach used predictive analytics to identify individuals likely to skip vaccination; for example, a study in Pakistan used ML on demographic and outreach data to predict which children would miss follow-up vaccine doses (achieving ~ 79% accuracy) (Chandir et al., 2018). Such models can help target vaccine education and appointments to those who might otherwise be left behind. AI can also improve geospatial targeting: researchers at Facebook used ML to analyze high-resolution satellite maps and filter out uninhabited areas in countries like Malawi, so that vaccination teams could focus on populated settlements. This ensured remote villages were not overlooked. In addition, AI-driven scheduling algorithms have been used to manage vaccination appointments and second-dose reminders to maximize completion rates.

Supply chain optimization: Managing the supply chain from warehouses to clinics is a complex task during a pandemic, but AI offers real-time visibility and optimization. For instance, India's Electronic Vaccine Intelligence Network (eVIN) is an IT system enhanced with AI algorithms that monitors vaccine stock levels, storage temperatures, and distribution logistics across thousands of sites. By analyzing this data in real time, eVIN helped health officials prevent stockouts; in fact, it led to an **80% reduction** in vaccine stockouts in India, ensuring vaccines were available where needed (Chaudhary, Awasthi, Kanubhai, Maroof, & Preeti, 2025). Similarly, in Rwanda, the startup Viebeg Technologies uses AI to manage medical supply inventory and distribution, keeping stock levels optimized and greatly reducing shortages and wastage. AI algorithms can automatically flag when a particular region is running low on medication or when certain batches are nearing expiry, prompting reallocation to avoid waste.

Case study integrated approaches: In practice, combining these AI tools amplifies their benefits. During COVID-19, some healthcare systems deployed augmented intelligence platforms that pulled together epidemiological data, hospital capacity, and supply metrics into one dashboard for decision-makers. For example, an AI-based system in Canada (CPAIS) integrated multiple datasets and provided an interactive dashboard to track COVID-19 in real time, enabling hospital administrators to see upcoming ICU bed shortages days in advance and relocate resources accordingly (Yu et al., 2021). By using ML forecasts and optimization models, such platforms could recommend specific actions like "send 5000 additional N95 masks and 20 ventilators to City X within the next week" based on predicted case surges. In low- and middle-income countries, AI's potential is especially transformative: it offers a way to do more with less, guiding the allocation of scarce healthcare resources to where they will have the greatest impact (Parums, 2023).

Overall, AI-driven resource allocation improves both efficiency and equity in outbreak response. It helps avoid scenarios where some areas face critical shortages while others have unused capacity. By forecasting needs, streamlining logistics, and informing prioritization, AI supports a more agile and effective emergency response.

2.5.3 Real-time monitoring and dashboards

During fast-moving outbreaks, real-time monitoring of disease spread and health system status is essential for policymakers to make timely decisions. AI and modern data platforms have given rise to interactive dashboards and analytics systems that track outbreak metrics in near real-time, often with predictive insights. These systems collect data from myriad sources and use algorithms to detect patterns or anomalies, providing an up-to-date situational picture. Key aspects and examples include:

Public health dashboards: Digital dashboards have become ubiquitous during epidemics for aggregating and displaying key indicators (cases, hospitalizations, deaths, vaccination rates, etc.). They present complex data in easily interpretable visual formats (maps, charts) for officials and the public. These dashboards facilitate continuous surveillance and even short-term prediction of disease trends by integrating data streams and updating continuously. For instance, the famous Johns Hopkins COVID-19 Dashboard updated global case counts and hotspots multiple times a day, guiding both government responses and public awareness (Rasmussen-Torvik, 2022; Chang et al., 2021; Paul et al., 2020; Rabiei et al., 2024). Effective dashboards typically pull data from laboratories (test results), hospitals (capacity, admissions), mobility reports, and sometimes even social media to provide a comprehensive picture. By continuously monitoring the community's health status, such platforms allow decision-makers to quickly identify upward trends or new clusters and respond (e.g., by tightening local restrictions or mobilizing extra resources).

AI-enhanced EWSs: Beyond official data, AI is used to scan nontraditional sources (news reports, web postings, search queries, etc.) to catch outbreak signals even before they appear in formal reporting. For example, **EPIWATCH** is an AI-based epidemic EWS that monitors open-source information globally. It uses NLP and ML to scan online news, public health reports (WHO, CDC), and social media for mentions of disease outbreaks or unusual health events. The system operates continuously and in near real-time, flagging potential outbreaks to human analysts. Notably, EPIWATCH and similar systems can often provide automated alerts of an emerging epidemic days or weeks before official detection because they pick up on subtle signals (like clusters of illness reported in local news). A public dashboard is available from EPIWATCH, giving health authorities a quick view of these AI-curated alerts around the world (Quigley et al., 2025; Aiello et al., 2020).

Multisource surveillance and Internet of Things (IoT): Modern outbreak monitoring frameworks increasingly blend data from the IoT and other sensors with AI analysis. A cutting-edge example is the proposed **SmartHealth-Track** system, an AI-powered real-time surveillance framework for urban areas. SmartHealth-Track combines inputs from wearable health trackers (monitoring symptoms like fever or heart rate), smart thermometers, pharmacy sales data (to detect surges in cough/cold medicine purchases), and even wastewater analysis for pathogen markers. ML models then analyze these streams together to detect early signs of an outbreak. In simulations, this multisource approach achieved over 90% accuracy in early outbreak detection, far better than any single data source alone (Alwakeel, 2025). Importantly, it can generate real-time alerts for health officials. For example, if an unusual spike in respiratory medication sales and fever reports is detected in a district, the system might warn of a potential outbreak cluster there. These AI-driven platforms thus act as an automatic "radar" for disease activity, helping cities catch outbreaks in their infancy.

Case study BlueDot: In late 2019, the power of AI-based monitoring was dramatically demonstrated by BlueDot, a Canadian health intelligence company. BlueDot's AI system scanned global news in multiple languages and airline ticketing data to spot patterns. It famously sent one of the first alerts about a novel pneumonia outbreak in Wuhan, China, on December 31, 2019, even before the WHO made official statements. The BlueDot algorithm essentially scours news reports and flight data to predict the spread of diseases, which enabled it to flag the risk of international spread from Wuhan. It even correctly anticipated some of the next cities likely to see cases (due to airline travel connections). This example shows how AI can augment traditional surveillance by picking up weak signals and connecting the dots across data sources, giving public health officials a vital head start in responding to emerging threats (Bogoch et al., 2020).

Real-time policy dashboards: Governments have also used real-time analytics dashboards internally to inform policy decisions. During COVID-19, many countries set up war-rooms with live dashboards displaying current R_0 estimates, test positivity rates, healthcare utilization, and mobility data. AI models were often integrated to produce short-term forecasts, for example, projecting the number of cases or ICU patients 7 days ahead, which were shown alongside current data. These dynamic tools enabled scenario planning (e.g., "if trend X continues, hospital Y will be full in 10 days") and allowed leaders to enact timely interventions (such as moving emergency supplies or imposing targeted lockdowns). An intelligent dashboard framework provided real-time visualization while incorporating AI to predict outcomes, directly supporting decision-making. In summary, by delivering insights at near real-time speed, such AI-augmented dashboards have given policymakers unprecedented situational awareness in managing the crisis.

In practice, the combination of big data and AI for real-time monitoring has become a game-changer in outbreak response. Public health agencies now increasingly rely on these tools for situational awareness. A key lesson from recent pandemics is that timely, data-driven decisions can save lives, and AI-powered monitoring systems are instrumental in providing the evidence for those decisions. Whether it's a global platform that alerts the world to a new pandemic threat or a local health department's AI that signals a neighborhood outbreak, real-time analytics platforms armed with AI help convert raw data into actionable intelligence. This empowers authorities to respond faster and more effectively, truly embodying the principle that, in a pandemic, *speed and information are everything*.

2.6 AI in therapeutics and drug discovery for infectious diseases

AI is not only helping to detect and track infectious diseases, it is also playing an expanding role in developing therapeutics to treat or prevent these diseases. This includes repurposing existing drugs, discovering new antimicrobial compounds, and designing novel vaccines and antibodies with the aid of ML models. Advances in computational biology, coupled with powerful AI algorithms, now allow researchers to explore drug candidates and biologics in silico at speeds that were unimaginable a decade ago. In this section, we highlight how AI is accelerating drug discovery and development, with a balanced look at successes in small-molecule drugs [including the fight against antimicrobial resistance (AMR)] and biologic therapies like antibodies and vaccines. Using functional genomics, genetics, and in silico models, AI-driven pipelines support drug design, antibody discovery, protein binder engineering, and even vaccine design (Fig. 6.4).

2.6.1 Drug repurposing and novel antimicrobial discovery

In early 2020, when COVID-19 struck, AI methods accelerated drug repurposing efforts. Using a biomedical knowledge graph and ML-driven analytics, researchers identified the rheumatoid arthritis drug **baricitinib** as a candidate therapy for COVID-19 by predicting both its antiviral and antiinflammatory effects (Smith et al., 2021). Clinical trials subsequently confirmed that baricitinib improved outcomes and reduced mortality in hospitalized patients, leading to its emergency use authorization by the FDA (Richardson and Stebbing, 2020). This success story demonstrated how AI could rapidly sift through existing pharmaceuticals to find new treatments for emerging infections far faster than traditional laboratory screening.

Beyond repurposing, AI is revolutionizing the discovery of *new* antimicrobials, a development crucial for combating AMR. A landmark study by MIT in 2020 employed a deep learning model to virtually screen over **100 million** molecules for antibiotic activity in a matter of days. This led to the discovery of a potent novel antibiotic, later named *halicin*, which proved effective against a broad range of highly drug-resistant bacteria and even cleared infections in mouse models. Halicin works *via* a mechanism distinct from existing antibiotics, and notably, it prevented the development of resistance in laboratory tests, highlighting AI's ability to not only find new drugs but also ones with novel modes of action (Stokes et al., 2020).

More recently, in 2023, an AI-driven approach identified another promising antibiotic, *abaucin*, targeting the notorious hospital superbug *Acinetobacter baumannii*. In a study published in *Nature Chemical Biology*, a deep learning model trained on bacterial chemical sensitivity data was used to screen a library of ~ 6000 compounds for activity against *A. baumannii*. The model selected a few hundred top candidates for laboratory testing, from which abaucin emerged as a powerful hit. Remarkably, abaucin was found to work *selectively*; it effectively killed *A. baumannii* but had little impact on other bacterial species. This narrow spectrum is unusual and potentially advantageous, as it could kill the target pathogen without disrupting the patient's normal microbiome. The discovery of abaucin (which traditional screening methods had missed) underscores how AI can uncover entirely new classes of antibiotics. Equally important, it did so with extreme efficiency: the model-and-lab pipeline delivered results in **mere days**, whereas conventional drug discovery campaigns often take years (Liu et al., 2023).

In parallel, AI is helping to preserve the efficacy of existing drugs by predicting and detecting resistance. For instance, ML classifiers can analyze a pathogen's genome to predict its antibiotic resistance profile with high accuracy. In tuberculosis and other infections, DNA sequencing data run through AI models can identify resistance to first-line drugs (like

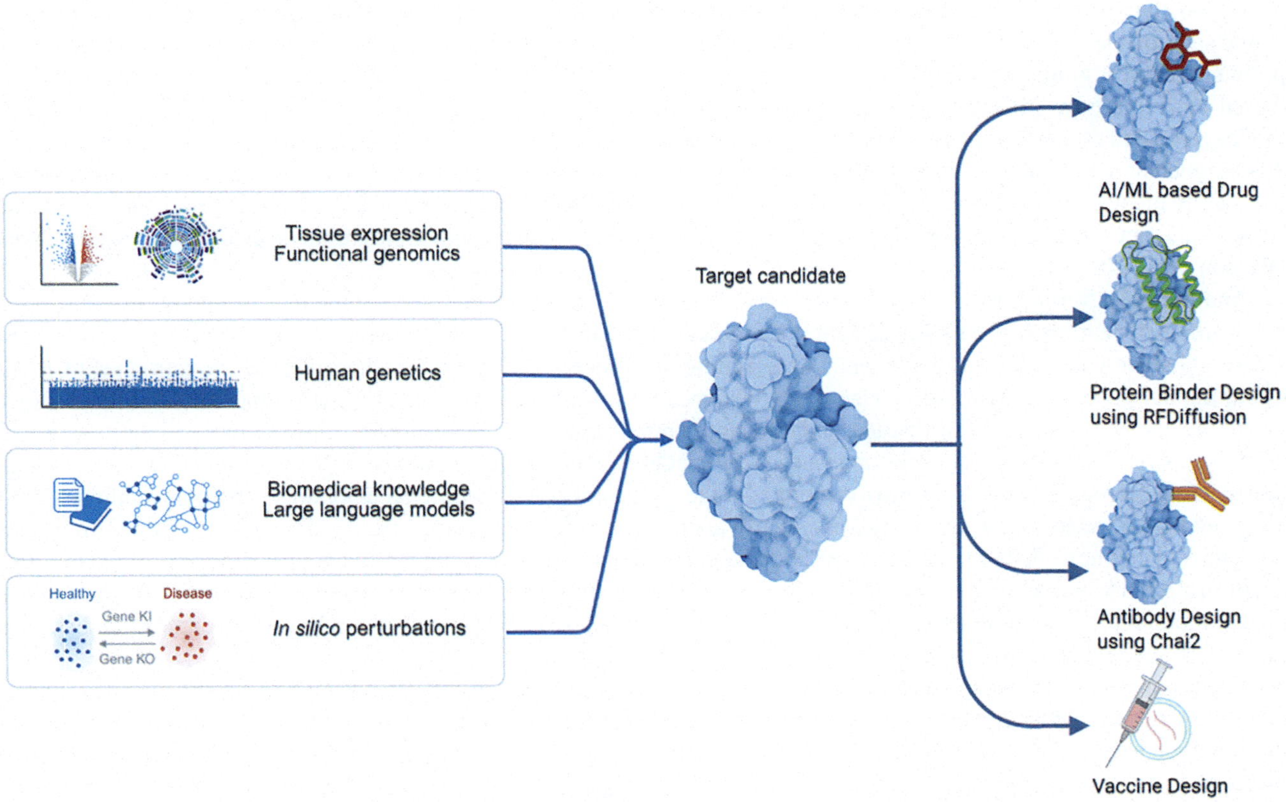

FIG. 6.4 AI-enabled therapeutic design workflow. The diagram illustrates evidence aggregation on the left—functional genomics and tissue expression data, human genetics findings, biomedical knowledge graphs augmented by large language models, and in silico perturbation analyses—converging on a prioritized target in the center. From this target, four parallel design branches emerge: small-molecule generation with QSAR, docking, and generative chemistry; protein binder design with RFdiffusion; antibody sequence and paratope optimization with Chai-like frameworks; and data-driven vaccine design, including epitope selection and stability modeling. The arrows indicate the progression from evidence to target nomination to modality-specific design, emphasizing that the same curated target dossier can seed multiple therapeutic avenues.

rifampicin or isoniazid) with over 90% sensitivity and specificity (Ali, Ahmed & Aslam, 2023). These genomic AI tools enable rapid determination of which antibiotics are likely to work for a given patient's infection, allowing more personalized and effective treatment and avoiding wasted time using drugs that the pathogen will not respond to. On a broader scale, AI-driven analysis of surveillance data can detect emerging resistance patterns (for example, mapping where drug-resistant strains are spreading unusually fast), giving public health officials a chance to intervene earlier with containment or stewardship measures.

Through both the discovery of novel drugs and the smarter use of existing ones, AI offers a powerful response to the AMR crisis. By searching enormous chemical spaces and recognizing subtle activity patterns, ML models can unearth candidates like halicin and abaucin that might never have been found through brute-force laboratory screening. Moreover, by quickly pinpointing effective therapies for specific infections (or identifying resistance), AI helps clinicians stay one step ahead of evolving pathogens.

2.6.2 AI-driven vaccine and antibody design

AI is also transforming the design of biologics, such as vaccines and therapeutic antibodies for infectious diseases. A prime example is DeepMind's **AlphaFold2**, which, in 2020, solved the protein folding problem to an unprecedented degree of accuracy (Jumper et al., 2021). By predicting protein structures within roughly 1 Å of experimental resolution, AlphaFold can determine the 3D shapes of pathogen proteins (or key human immune proteins) in hours knowledge that is vital for rational vaccine and drug design. During the COVID-19 pandemic, for instance, AI-based protein structure predictions helped researchers identify the coronavirus spike protein's vulnerable sites and guided vaccine antigen design much faster than traditional laboratory methods. AlphaFold's structural predictions have since been applied to thousands of proteins, from malaria parasites to HIV, accelerating the search for vaccine targets and drug binding sites.

Beyond analyzing existing proteins, generative AI models can *invent* new proteins with desired functions. **RFdiffusion** is one state-of-the-art example: a generative diffusion model (inspired by image generation techniques) that creates novel protein structures guided by design goals. In a recent demonstration, RFdiffusion was tasked with creating a protein that would bind tightly to influenza virus hemagglutinin (HA), a surface protein critical for the virus's infectivity. The AI model generated numerous candidate designs, and researchers synthesized one of these de novo proteins. Remarkably, the AI-designed protein binder *did* bind to the influenza HA as intended; its structure, solved by cryogenic electron microscopy, was nearly identical to the in silico design. This validation shows that AI can generate functional antipathogen proteins that work in practice, opening the door to designing antiviral proteins (like entry inhibitors) or therapeutic enzymes on demand (Watson et al., 2023).

Another breakthrough is in antibody discovery. Traditionally, producing a new monoclonal antibody (mAb) therapy, for example, a neutralizing antibody against a virus, can take many months of laboratory work to screen and evolve candidates. Generative models are dramatically accelerating this process. In 2025 researchers at *Chai Bio* introduced **Chai-2**, a multimodal generative AI model for antibody design. In a test of de novo antibody generation for 52 distinct targets (spanning oncology, autoimmune disease, and infectious pathogens), the model generated high-affinity antibody candidates *entirely* in silico. Remarkably, *at least one* successful binder was produced for half of the targets, yielding an average hit rate of 15.5%, a roughly **100-fold improvement** over the typical hit rate of traditional laboratory screening methods. And it achieved this in weeks, not years. Chai-2 essentially learned the complex patterns of antibody–antigen binding from training data and can now "hallucinate" new antibody sequences that bind to a given target antigen. Such AI-driven antibody engineering holds particular promise for rapidly developing neutralizing antibodies against novel pathogens or variants. For example, if a new virus emerges, an AI could propose antibody sequences likely to bind a critical epitope on that virus, jump-starting the therapeutic development process (Chai Discovery Team, 2025).

More broadly, generative models are being applied to **vaccine antigen design**. By optimizing the structure or sequence of an antigen, AI can help create vaccine components that elicit broader and more potent immune responses. Early efforts in this direction have used techniques like reinforcement learning to modify viral proteins in silico, enhancing their stability or exposure to neutralizing epitopes. While still in experimental stages, the prospect of *AI-designed vaccines* is on the horizon, essentially using algorithms to shape what a pathogen fragment should look like to the immune system for maximal protection (Kaushik, Kant, & Christodoulides, 2023). One can imagine updating vaccine antigens *via* computer models as a virus mutates, rather than waiting for lengthy laboratory culturing.

Taken together, these advances in protein modeling and generation indicate that AI can significantly accelerate the development of therapeutics and vaccines for infectious diseases. Instead of relying solely on trial-and-error experimentation, scientists can leverage AI to propose high-potential drug molecules or vaccine designs from the start, focusing experimental testing on a smaller, more promising set of candidates. This AI-augmented approach is expediting the pipeline from conceptual target to viable therapeutic, a trend that could greatly improve our preparedness and response to future outbreaks.

Challenges and limitations of AI in infectious disease surveillance

While AI offers powerful tools to combat infectious diseases, deploying these tools comes with challenges related to data, ethics, and technology.

2.6.3 *Data quality and bias*

AI models depend critically on the data used for training. In infectious disease surveillance, data are often incomplete, inconsistent, or fragmented, which can lead to unreliable outputs. For example, routine reporting systems may undercount cases or lack important demographic details. In one analysis of US COVID-19 data, inadequate encoding of race and ethnicity resulted in a **underestimation** of mortality among Black and Hispanic patients (Magesh et al., 2021). ML algorithms trained on such skewed data risk reproducing these gaps: underrepresented groups could systematically receive poorer predictions, potentially worsening health inequities. Digital surveillance streams introduce additional biases. Internet and social media data tend to reflect only users with access and willingness to share, creating a "digital divide" by income, age, or geography. Moreover, individuals may avoid online searches or posts about stigmatized conditions (e.g., HIV), so signals for these diseases can be muted. In sum, biased or nonrepresentative training data undermine model validity and fairness. Ensuring comprehensive, high-quality datasets from diverse populations is essential to avoid inequitable outcomes.

2.6.4 *Ethical concerns*

Balancing public health benefits against individual rights is a core ethical challenge. AI-based surveillance often uses personal data (EHRs, location trails, social media), which raises serious privacy concerns. For instance, mobile contact tracing can link infection status to individuals' movements or social contacts. If such detailed data were to leak or be misused, patients could face stigma or discrimination. Indeed, participants in ethics reviews have warned that disclosure of sensitive information (e.g., an STI diagnosis linked to a person's name or address) could "lead to public discrimination and stigmatization." Historically, outbreaks have disproportionately affected marginalized groups, and AI tools risk amplifying this by flagging communities as disease "hotspots" without context. To protect privacy and prevent stigmatization, robust data governance is required (anonymization, encryption, legal safeguards).

AI-driven surveillance also raises broader civil rights issues. Mandatory tracking (e.g., QR-code venue check-ins, enforced movement restrictions) can impinge on personal freedom. Critics caution that, without clear oversight, centralized tracing apps and automated enforcement could become intrusive (e.g., constant location logging or algorithmically determined travel bans). Public trust is fragile: studies noted "concerns about inadequate information, unclear governance frameworks, and lack of privacy protection" in many digital health interventions. In a crisis, the tension between collective health and civil liberties intensifies: routine collection of personal data poses considerable risks for the violation of individual and civil freedoms. Thus ethical deployment demands transparent policies, community engagement, and accountability mechanisms to ensure AI aids epidemic control without undermining human rights.

2.6.5 Technical barriers

Even high-performing AI models face practical hurdles before they can be widely used. A persistent issue is **generalizability**. Models trained on data from one region or population may fail when applied elsewhere, especially if disease patterns

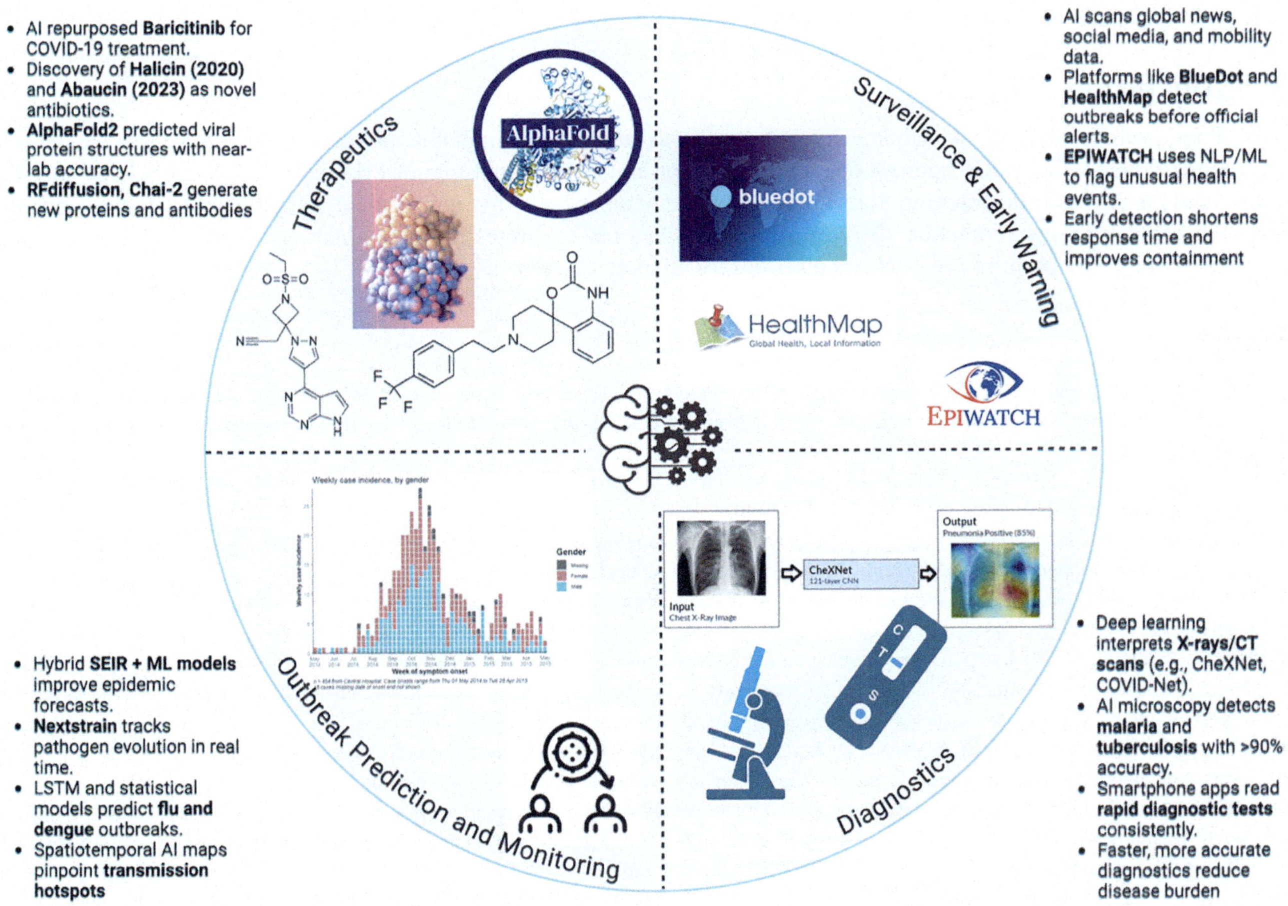

FIG. 6.5 AI value across the infectious disease lifecycle. The figure organizes mature and emerging applications into four domains—surveillance and early warning, diagnostics, outbreak prediction and monitoring, and therapeutics—using a quadrant layout with icons and exemplars. Surveillance shows global signal scanning of media, reports, and mobility streams on platforms such as BlueDot, HealthMap, and EPIWATCH to surface unusual events before official notices. Diagnostics highlights imaging models like CheXNet and COVID-Net, AI microscopy, and smartphone-based interpretation of rapid tests to improve accuracy and reach. Outbreak Prediction and Monitoring presents hybrid SEIR-plus-machine learning approaches, spatiotemporal mapping of transmission hotspots, and genomic tracking of pathogen evolution. Therapeutics illustrates structure prediction with AlphaFold2 and generative design frameworks such as RFdiffusion and Chai-class antibody tools.

differ. A systematic review noted that some AI systems "failed validation when applied to new datasets," reflecting overfitting to local conditions (Zhao et al., 2021). In practice, an algorithm built on urban hospital data might underperform in rural clinics. This lack of robustness means systems require careful external validation and often local retraining.

Relatedly, many cutting-edge AI methods suffer from limited **interpretability**. Complex architectures (deep neural networks, ensemble models) often act as "black boxes" that output predictions without transparent reasoning. Public health officials and clinicians may be reluctant to act on recommendations they cannot interpret or verify. A review emphasizes that this opacity makes it difficult to "confirm, trust, or troubleshoot model outputs" in a health context (Villanueva-Miranda, Xiao & Xie, 2025). Ongoing work in explainable AI may help, but currently, this limitation impedes full integration of AI into clinical and public health decision-making.

Finally, **scalability and integration** pose significant challenges. Deploying AI tools in the real world requires robust infrastructure and maintenance. Many health agencies lack the necessary digital systems, interoperability, and trained staff to implement sophisticated models. Data pipelines must be built to harmonize inputs from hospitals, laboratories, and external sources, all while preserving security. This technical overhead is substantial: inconsistent data formats, incomplete IT systems, and legacy practices can stall even mature AI models. Moreover, infectious threats evolve rapidly. AI tools must be continuously recalibrated as pathogens mutate, population immunity changes, or social behaviors shift. Otherwise, models can quickly become outdated. As one review notes, as new variants emerge and vaccination alters the disease landscape, static models "will never be completely accurate and may result in false alarms or failure to identify important signals" (Zhao et al., 2021). In short, limited generalizability, lack of transparency, and practical deployment hurdles mean that AI systems must remain aids to human experts and traditional surveillance networks, not replacements. Platforms such as BlueDot and HealthMap leverage AI for outbreak detection, while tools like AlphaFold and RFdiffusion aid in therapeutic discovery, highlighting the multifaceted role of AI in both surveillance and treatment (Fig. 6.5).

3 Conclusion

In summary, while AI offers powerful new tools for outbreak response, its implementation is tempered by these challenges. Overcoming them will require rigorous data curation, clear ethical and legal frameworks, explainable model designs, and investments in digital infrastructure. Stakeholders must remember that AI is ultimately a complement to, not a substitute for, sound epidemiological practice. Only by addressing data biases, protecting individual rights, and bridging technical gaps can AI fulfill its promise in infectious disease surveillance.

References

Ahamed, M. F., Nahiduzzaman, M., Mahmud, G., Shafi, F. B., Ayari, M. A., Khandakar, A., et al. (2025). Improving Malaria diagnosis through interpretable customized CNNs architectures. *Scientific Reports*, 2045-2322. *15*(1). Available from: https://www.nature.com/srep/. doi:10.1038/s41598-025-90851-1.

Aiello, A. E., Renson, A., & Zivich, P. (2020). Social media- and internet-based disease surveillance for public health. *Annual Review of Public Health*, *41*, 101–118. doi:10.1146/annurev-publhealth-040119-094402.

Ali, T., Ahmed, S., & Aslam, M. (2023). Artificial Intelligence for antimicrobial resistance prediction: challenges and opportunities towards practical implementation. *Antibiotics*, 2079-6382. *12*(3), 523. doi:10.3390/antibiotics12030523.

Alwakeel, M. M. (2025). AI-assisted real-time monitoring of infectious diseases in urban areas. *Mathematics*, 2227-7390. *13*(12), 1911. doi:10.3390/math13121911.

Bhansali, A., & Jain, V., (2020). Using machine learning for healthcare resource allocation in COVID-19.

Biggerstaff, M., Johansson, M., Alper, D., Brooks, L. C., Chakraborty, P., Farrow, D. C., et al. (2018). Results from the second year of a collaborative effort to forecast influenza seasons in the United States. *Epidemics*, *24*, 26–33. doi:10.1016/j.epidem.2018.02.003.

Bogoch, I. I., Watts, A., Thomas-Bachli, A., Huber, C., Kraemer, M.U.G., Khan, K. (2020). Pneumonia of unknown aetiology in Wuhan, China: potential for international spread via commercial air travel. *J Travel Med.* *27*(2). taaa008. doi:10.1093/jtm/taaa008.

Brownstein, J. S., Freifeld, C. C., Reis, B. Y., & Mandl, K. D. (2008). Surveillance sans frontières: internet-based emerging infectious disease intelligence and the HealthMap project. *PLoS Medicine*, 1549-1676. *5*(7), 1019–1024. Available from: http://medicine.plosjournals.org/archive/1549-1676/5/7/pdf/10.1371_journal.pmed.0050151-L.pdf. doi:10.1371/journal.pmed.0050151.

Brüggemann, S., Chan, T., Wardi, G., Mandel, J., Fontanesi, J., & Bitmead, R. R. (2021). Decision support tool for hospital resource allocation during the COVID-19 pandemic. *Informatics in Medicine Unlocked*, 2352-9148. *24*, 100618. doi:10.1016/j.imu.2021.100618.

Chai Discovery Team. (2025). Zero-shot antibody design in a 24-well plate. bioRxiv. doi:10.1101/2025.07.05.663018.

Chandir, S., Siddiqi, D. A., Hussain, O. A., Niazi, T., Shah, M. T., Dharma, V. K., Habib, A. & Khan, A. J. (2018). Using predictive analytics to identify children at high risk of defaulting from a routine immunization program: feasibility study. *JMIR Public Health Surveill*, *4*(3), e63. doi:10.2196/publichealth.9681.

Chang, S., Pierson, E., Koh, P. W., Gerardin, J., Redbird, B., Grusky, D., & Leskovec, J. (2021). Mobility network models of COVID-19 explain inequities and inform reopening. *Nature*, *589*(7840), 82–87. doi:10.1038/s41586-020-2923-3.

Chaudhary, K., Awasthi, S., Kanubhai, T. H., Maroof, M., & Preeti. (2025). Assessment of eVIN (Electronic Vaccine Intelligence Network) in utility and constraints of various stakeholders in Nainital District: a mixed method study. *Journal of Family Medicine and Primary Care*, 2249-4863. *14*(2), 626–632. doi:10.4103/jfmpc.jfmpc_1015_24.

Chen, W., et al. (2024). Pulmonary tuberculosis diagnosis using an intelligent microscopy scanner and image recognition model for improved acid-fast bacilli detection in smears. *Microorganisms*, *12*(8), 1734. doi:10.3390/microorganisms12081734.

Cheong, J., Yu, H., Lee, C. Y., Lee, J.-U., Choi, H.-J., Lee, J.-H., et al. (2020). Fast detection of SARS-CoV-2 RNA via the integration of plasmonic thermocycling and fluorescence detection in a portable device. *Nature Biomedical Engineering*, 2157-846X. *4*(12), 1159–1167. doi:10.1038/s41551-020-00654-0.

Chow, B. W. K., Lim, Y. D., Poh, R. C. H., Ko, A., Hong, G. H., Zou, S. W. L., et al. (2023). Use of a digital contact tracing system in Singapore to mitigate COVID-19 spread. *BMC Public Health*, 1471-2458. *23*(1). Available from: https://bmcpublichealth.biomedcentral.com/. doi:10.1186/s12889-023-17150-0.

Colbert, A. J., Co, K., Lima-Cooper, G., Lee, D. H., Clayton, K. N., Wereley, S. T., John, C. C., Linnes, J. C. & Kinzer-Ursem, T. L. (2021). Towards the use of a smartphone imaging-based tool for point-of-care detection of asymptomatic low-density malaria parasitaemia. *Malaria Journal*, *20*(1). doi:10.1186/s12936-021-03894-w.

Enaya, M. F., Klingbeil, T., Krüger, J., Broneske, D., Feinbube, F., & Saake, G. (2024). A case study on the development of the German Corona-Warn-App. *Journal of Systems and Software*, 0164-1212. *213*, 112020. doi:10.1016/j.jss.2024.112020.

Fleuren, L. M., Klausch, T. L. T., Zwager, C. L., Schoonmade, L. J., Guo, T., Roggeveen, L. F., et al. (2020). Machine learning for the prediction of sepsis: a systematic review and meta-analysis of diagnostic test accuracy. *Intensive Care Medicine*, *46*(3), 383–400. Epub 2020 Jan 21. PMID: 31965266; PMCID: PMC7067741. doi:10.1007/s00134-019-05872-y.

Frade, S., Cooper, S., Smedinghoff, S., Hattery, D., Ruan, Y., Isabelli, P., et al. (2025). Malaria RDT (mRDT) interpretation accuracy by frontline health workers compared to AI in Kano state, Nigeria. *BMC Digital Health*, 2731-684X. *3*(1). doi:10.1186/s44247-025-00190-4.

Fraser, H., Crossland, D., Bacher, I., Ranney, M., Madsen, T., & Hilliard, R. (2023). Comparison of diagnostic and triage accuracy of ada health and WebMD symptom checkers, ChatGPT, and physicians for patients in an emergency department: clinical data analysis study. *JMIR mHealth and uHealth*, 2291-5222. *11*(1), e49995. doi:10.2196/49995.

Gao, Y., & Liu, M. (2024). Application of machine learning based genome sequence analysis in pathogen identification. *Frontiers in Microbiology*, 1664-302X. *15*. doi:10.3389/fmicb.2024.1474078.

García-Cremades, S., Morales-García, J., Hernández-Sanjaime, R., Martínez-España, R., Bueno-Crespo, A., Hernández-Orallo, E., et al. (2021). Improving prediction of COVID-19 evolution by fusing epidemiological and mobility data. *Scientific Reports*, 2045-2322. *11*(1). Available from: www.nature.com/srep/index.html. doi:10.1038/s41598-021-94696-2.

Ginsberg, J., Mohebbi, M. H., Patel, R. S., Brammer, L., Smolinski, M. S., & Brilliant, L. (2009). Detecting influenza epidemics using search engine query data. *Nature*, *457*(7232), 1012–1014. doi:10.1038/nature07634.

Heckerman, D. E., Horvitz, E. J., Nathwani. (1992). Toward normative expert systems: Part I. The Pathfinder project. *Methods of Information in Medicine*. 1635470.

Javalkote, V. S., Kancharla, N., Bhadra, B., Shukla, M., Soni, B., Sapre, A., et al. (2022). CRISPR-based assays for rapid detection of SARS-CoV-2. *Methods*, 1095-9130. *203*, 594–603. Available from: http://www.elsevier.com/inca/publications/store/6/2/2/9/1/4/index.htt. doi:10.1016/j.ymeth.2020.10.003.

Jin, W., Dong, S., Yu, C., & Luo, Q. (2022). A data-driven hybrid ensemble AI model for COVID-19 infection forecast using multiple neural networks and reinforced learning. *Computers in Biology and Medicine*, 0010-4825. *146*, 105560. doi:10.1016/j.compbiomed.2022.105560.

Jumper, J., Evans, R., Pritzel, A., Green, T., Figurnov, M., Ronneberger, O., et al. (2021). Highly accurate protein structure prediction with AlphaFold. *Nature*, 0028-0836. *596*(7873), 583–589. doi:10.1038/s41586-021-03819-2.

Kaushik, R., Kant, R., & Christodoulides, M. (2023). Artificial intelligence in accelerating vaccine development—current and future perspectives. *Frontiers in Bacteriology*, 2813-6144. *2*. doi:10.3389/fbrio.2023.1258159.

Kraemer, M. U. G., Yang, C. H., Gutierrez, B., Wu, C. H., Klein, B., Pigott, D. M., et al. (2020). The effect of human mobility and control measures on the COVID-19 epidemic in China. *Science*, 1095-9203. *368*(6490), 493–497. Available from: https://science.sciencemag.org/content/368/6490/493/tab-pdf. doi:10.1126/science.abb4218.

Lakhani, P., Sundaram, B. (2017). Deep learning at chest radiography: automated classification of pulmonary tuberculosis by using convolutional neural networks. *Radiology*. *284*(2): 574-582. doi:10.1148/radiol.2017162326.

Lazer, D., Kennedy, R., King, G., & Vespignani, A. (2014). The parable of Google Flu: traps in big data analysis. *Science*, 0036-8075. *343*(6176), 1203–1205. doi:10.1126/science.1248506.

Levine, B., Wilcosky, T., Wagener, D., & Cooley, P. (2010). Mass commuting and influenza vaccination prevalence in New York City: protection in a mixing environment. *Epidemics*, 1755-4365. *2*(4), 183–188. doi:10.1016/j.epidem.2010.07.002.

Liscano, Y., Anillo Arrieta, L. A., Montenegro, J. F., Prieto-Alvarado, D., & Ordoñez, J. (2025). Early warning of infectious disease outbreaks using social media and digital data: a scoping review. *International Journal of Environmental Research and Public Health*, 1660-4601. *22*(7), 1104. doi:10.3390/ijerph22071104.

Liu, G., Catacutan, D. B., Rathod, K., Swanson, K., Jin, W., Mohammed, J. C., et al. (2023). Deep learning-guided discovery of an antibiotic targeting Acinetobacter baumannii. *Nature Chemical Biology*, 1552-4469. *19*(11), 1342–1350. Available from: https://www.nature.com/nchembio. doi:10.1038/s41589-023-01349-8.

Liu, R., et al. (2023). AIDMAN: An AI-based object detection system for malaria diagnosis from smartphone thin-blood-smear images *Patterns*, *4*(9). doi:10.1016/j.patter.2023.100806.

Magesh, S., John, D., Li, W. T., Li, Y., Mattingly-App, A., Jain, S., et al. (2021). Disparities in COVID-19 outcomes by race, ethnicity, and socioeconomic status: a systematic-review and meta-analysis. *JAMA Network Open*, 2574-3805. *4*(11). Available from: https://jamanetwork.com/journals/jamanetworkopen. doi:10.1001/jamanetworkopen.2021.34147.

Menni, C., Valdes, A. M., Freidin, M. B., Sudre, C. H., Nguyen, L. H., Drew, D. A., et al. (2020). Real-time tracking of self-reported symptoms to predict potential COVID-19. *Nature Medicine*, 1078-8956. *26*(7), 1037–1040. doi:10.1038/s41591-020-0916-2.

Messina, J. P., Kraemer, M. U. G., Brady, O. J., Pigott, D. M., Shearer, F. M., Weiss, D. J., et al. (2016). Mapping global environmental suitability for Zika virus. *eLife*, 2050-084X. *5*. Available from: https://elifesciences.org/content/5/e15272-download.pdf. doi:10.7554/eLife.15272.

Musa, S. M., Haruna, U. A., Aliyu, L. J., Zubairu, M., & Lucero-Prisno, D. E. (2025). Leveraging AI to optimize vaccines supply chain and logistics in Africa: opportunities and challenges. *Frontiers in Pharmacology*, 1663-9812. *16*. Available from: http://www.frontiersin.org/Pharmacology. doi:10.3389/fphar.2025.1531141.

Ngo, H. T., Akarapipad, P., Lee, P. W., Park, J. S., Chen. F. E., Trick, A. Y., Wang, T. H., Hsieh, K. (2024). Rapid and portable quantification of HIV RNA via a smartphone-enabled digital CRISPR device and deep learning. *Sens Actuators Rep. 8, 100212*. doi:10.1016/j.snr.2024.100212.

Nguyen, M., Corbin, C. K., Eulalio, T., Ostberg, N. P., Machiraju, G., Marafino, B. J., et al. (2021). Developing machine learning models to personalize care levels among emergency room patients for hospital admission. *Journal of the American Medical Informatics Association*, 1527-974X. *28*(11), 2423–2432. Available from: http://jamia.oxfordjournals.org/content/22/e1. doi:10.1093/jamia/ocab118.

Niiler, E. 2025. *An AI epidemiologist sent the first warnings of the Wuhan virus*. Wired. Retrieved from: https://www.wired.com/story/ai-epidemiologist-wuhan-public-health-warnings/.

O'Dowd, A. (2020). Covid-19: app to track close contacts is launched in England and Wales. *BMJ (Clinical Research Ed.)*, 1756-1833. *370*, m3751. doi:10.1136/bmj.m3751.

Obeng, D. N. S., Osei, S., Brown, N. K., Nana Adjei, D., Amoah, L. E., & Owusu, E. D. A. (2025). Performance of a smartphone-based malaria screener in detecting malaria in people living with Sickle cell disease. *PLOS Digital Health*, *4*(6), e0000884. doi:10.1371/journal.pdig.0000884.

Parums, D. V. (2023). Infectious disease surveillance using artificial intelligence (AI) and its role in epidemic and pandemic preparedness. *Medical Science Monitor*, *29*, e941209. doi:10.12659/MSM.941209.

Paul, R., Arif, A. A., Adeyemi, O., Ghosh, S., & Han, D. (2020). Progression of COVID-19 from urban to rural areas in the United States: a spatiotemporal analysis of prevalence rates. *Journal of Rural Health*, *36*(4), 591–601 doi:https://doi.org/10.1111/jrh.12486.

Peccia, J., Zulli, A., Brackney, D. E., et al. (2020). Measurement of SARS-CoV-2 RNA in wastewater tracks community infection dynamics. *Nat Biotechnol*. *38*(10). 1164-1167. doi:10.1038/s41587-020-0684-z.

Qian, Z., Alaa, A. M., & van der Schaar, M. (2021). CPAS: the UK's national machine learning-based hospital capacity planning system for COVID-19. *Machine Learning*, 0885-6125. *110*(1), 15–35. doi:10.1007/s10994-020-05921-4.

Quigley, D. A., Honeyman, M. D., Stone, M. H., Dawson, D. R. & MacIntyre, P. C. (2025). EPIWATCH, an artificial intelligence early-warning system as a valuable tool in outbreak surveillance. *International Journal of Infectious Diseases*, *152*, 107579. doi:10.1016/j.ijid.2024.107579.

Rabiei, R., Bastani, P., Ahmadi, H., et al. (2024). Developing public health surveillance dashboards: a scoping review on the design principles. *BMC Public Health*, *24*, 392. doi:10.1186/s12889-024-17841-2.

Rajpurkar, P., Irvin, J., Zhu, K., Yang, B., Mehta, H., Duan, T., et al. (2017). CheXNet: radiologist-level pneumonia detection on chest X-rays with deep learning. arXiv, Undefined arXiv. 23318422. Available from: https://arxiv.org. doi:10.48550/arxiv.1711.05225.

Rasmussen-Torvik, L. J. (2022). Recognizing the importance of COVID-19 data wrangling. *Journal of Clinical Investigation*, 1558-8238. *132*(19). Available from: https://www.jci.org/articles/view/164375/pdf. doi:10.1172/JCI164375.

Richardson, P. J., Stebbing, J. (2020). Identifying baricitinib as a treatment for COVID-19 using artificial intelligence. *The Lancet eClinicalMedicine*, *49*. doi:10.1016/j.eclinm.2022.101493.

Rocha Filho, T., Moret, M., Chow, C., Phillips, J., Cordeiro, A., Scorza, F., Almeida, A. & Mendes J. (2021). A data-driven model for COVID-19 pandemic – Evolution of the attack rate and prognosis for Brazil. *Chaos, Solitons & Fractals*, *152* 111359. doi:10.1016/j.chaos.2021.111359.

Shortliffe, E. H. (1976). *Computer-based medical consultations: MYCIN* (New York: Elsevier). ISBN:978-0444569691. https://www.shortliffe.net/Shortliffe-1976/MYCIN%20thesis%20Book.htm.

Smith, D. P., Oechsle, O., Rawling, M. J., Savory, E., Lacoste, A. M. B., & Richardson, P. J. (2021). Expert-augmented computational drug repurposing identified baricitinib as a treatment for COVID-19. *Frontiers in Pharmacology*, 1663-9812. *12*. Available from: http://www.frontiersin.org/Pharmacology. doi:10.3389/fphar.2021.709856.

Stokes, J. M., Yang, K., Swanson, K., Jin, W., Cubillos-Ruiz, A., Donghia, N. M., et al. (2020). A deep learning approach to antibiotic discovery. *Cell*, 1097-4172. *180*(4), 688. Available from: https://www.sciencedirect.com/journal/cell. doi:10.1016/j.cell.2020.01.021.

Umlauf, R., & Burchardt, M. (2022). Infrastructure-as-a-service: empty skies, bad roads, and the rise of cargo drones. *Environment and Planning A: Economy and Space*, *54*(8), 1489–1509. doi:10.1177/0308518X221118915.

Villanueva-Miranda, I., Xiao, G., & Xie, Y. (2025). Artificial intelligence in early warning systems for infectious disease surveillance: a systematic review. *Frontiers in Public Health*, 2296-2565. *13*. doi:10.3389/fpubh.2025.1609615.

Wang, L., Lin, Z. Q., & Wong, A. (2020). COVID-Net: a tailored deep convolutional neural network design for detection of COVID-19 cases from chest X-ray images. *Scientific Reports*, 2045-2322. *10*(1). Available from: www.nature.com/srep/index.html. doi:10.1038/s41598-020-76550-z.

Wang, Y., Liu, C., Fan, Y., Niu, C., Huang, W., Pan, Y., et al. (2025). A multi-modal deep learning solution for precise pneumonia diagnosis: the PneumoFusion-Net model. *Frontiers in Physiology*, 1664-042X. *16*. doi:10.3389/fphys.2025.1512835.

Watson, J. L., Juergens, D., Bennett, N. R., Trippe, B. L., Yim, J., Eisenach, H. E., et al. (2023). De novo design of protein structure and function with RFdiffusion. *Nature*, 1476-4687. *620*(7976), 1089–1100. Available from: https://www.nature.com/nature/. doi:10.1038/s41586-023-06415-8.

Yu, C. S., Chang, S. S., Chang, T. H., Wu, J. L., Lin, Y. J., Chien, H. F., & Chen, R. J. (2021). A COVID-19 pandemic artificial intelligence-based system with deep learning forecasting and automatic statistical data acquisition: development and implementation study. *Journal of Medical Internet Research*, 1438-8871. *23*(5). Available from: https://www.jmir.org/2021/5/e27806. doi:10.2196/27806.

Yuan, S., Yang, Z., Li, J., Wu, C., & Liu, S. (2025). AI-Powered early warning systems for clinical deterioration significantly improve patient outcomes: a meta-analysis. *BMC Medical Informatics and Decision Making*. *25*, 203. doi:10.1186/s12911-025-03048-x.

Zhang, Q., Sun, K., Chinazzi, M., Pastore Y Piontti, A., Dean, N. E., Rojas, D. P., et al. (2017). Spread of Zika virus in the Americas. *PNAS*, *114*(22), E4334–E4343. doi:10.1073/pnas.1620161114.

Zhao, I. Y., Ye Xuan, M., Yu, M. W. C., Liu, J., Dong, W. N., Pang, Q., et al. (2021). Ethics, integrity, and retributions of digital detection surveillance systems for infectious diseases: systematic literature review. *Journal of Medical Internet Research*, 1438-8871. *23*(10), e32328. doi:10.2196/32328.

Zhuang, H., Zhu, F., Lan, P., Ji, S., Sun, L., Chen, Y., et al. (2021). A random forest model based on core genome allelic profiles of MRSA for penicillin plus potassium clavulanate susceptibility prediction. *Microbial Genomics*, 2057-5858. *7*(9). doi:10.1099/mgen.0.000610.

Zou, Z., Tang, F., Qiao, L., Wang, S., & Zhang, H. (2025). Integrating sequencing methods with machine learning for antimicrobial susceptibility testing in pediatric infections: current advances and future insights. *Frontiers in Microbiology*, *16*. doi:10.3389/fmicb.2025.1528696.

Chapter 7

Artificial intelligence-assisted imaging technologies for disease diagnosis

Luis Pinto-Coelho
Electronics Engineering, Polytechnic of Porto, ISEP, Porto, Portugal
INESCTEC, Porto, Portugal

1 Introduction

Artificial intelligence (AI) technology has revolutionized the medical imaging practices with multiple solutions and tools available for each step in the clinical process. The newer systems are quicker, more accurate, and even more affordable than traditional methods. Deep learning (DL) algorithms, in particular, are now able to read medical images with performance consistently higher than that of humans. Consistent with this, such tools are increasingly being utilized in clinical workflows to support diagnostic decision-making, reduce interpretation time, and enhance overall process consistency (Litjens et al., 2017).

This novel context, within which interest in AI applications to medical imaging has grown constantly, is driven by several converging motivations: first, the exponential growth of readily available data in the medical imaging field, due to decades of digitalization; second, the availability of faster computing power at a lower cost, mainly due to the availability of cloud computing facilities and the use of graphics processing unit (GPU) technology; and finally, the advances in neural network architectures with the capacity to learn hierarchical representations from raw data, overcoming the limitations of feature-based approaches that require human intervention. The evolution of such systems has also been boosting the use of image-focused applications with virtual reality integrated into healthcare workflows (Kamińska, Zwolińsksi, Laska-Leśniewicz, & Coelho, 2022). Convolutional neural networks (CNNs), for example, have demonstrated expert-level performance in image classification and segmentation in many applications, such as lung cancer detection from CT scans, diabetic retinopathy screening from fundus photography, and breast cancer risk prediction from mammograms (Ardila et al., 2019; Pinto-Coelho, 2023; Ting et al., 2017).

The maturity of AI adoption in different clinical imaging fields varies significantly between specialties and use cases:

- Radiology: AI adoption is most mature in radiology, especially for modalities like X-ray, CT, and MRI. Applications such as automated detection of lung nodules, fractures, and mammography screening have gained regulatory approval and are increasingly used in clinical workflows. The abundance of digital images and clear diagnostic tasks contributes to this maturity.
- Pathology: Digital pathology is experiencing rapid growth in AI adoption, but it lags behind radiology due to the later digitization of slide images and the higher complexity of data. AI is being utilized for tasks such as cancer detection, grading, and prognosis estimation, but full clinical integration is still emerging in many settings.
- Cardiology: In cardiovascular imaging (e.g., echocardiography, cardiac CT, MRI), AI adoption is advancing but remains less mature than in general radiology. AI is being used to automate measurements and detect abnormalities, but broader acceptance awaits further validation and regulatory clearance.
- Ophthalmology: Ophthalmic imaging, particularly for diabetic retinopathy screening using fundus photography, has seen rapid AI adoption, with Food and Drug Administration (FDA)-cleared software available. This field benefits from high-volume, high-consistency image capture.
- Other specialties: AI in other imaging specialties, such as ultrasound, nuclear medicine, and dental imaging, is generally less mature. Factors limiting adoption include technical variability, data annotation challenges, and less standardized workflows.

Applications of Artificial Intelligence in Healthcare. https://doi.org/10.1016/B978-0-443-43934-6.00007-9

The maturity of clinical uses of AI in actual circumstances still possesses numerous problems despite new developments in the field. Interpretability and explainability of systems, bias in algorithms, data privacy, and the need for high-quality datasets still have some gaps to be addressed. Besides that, their incorporation into heavily regulated clinical processes, under the advanced ethical complexity of clinical accountability and patient safety (Kelly, Karthikesalingam, Suleyman, Corrado, & King, 2019; Topol, 2019), has to be ensured with caution. Only when these issues are adequately addressed can AI technologies not only augment clinical decision-making but also be aligned with evidence-based medicine values and equity in care.

It is the aim of this chapter to review AI-backed imaging technologies in clinical diagnostics. Their technical foundations and most popular applications are highlighted. In addition, the main limitations, such as ethical concerns, are critically discussed to provide the groundwork for the responsible and ethical implementation of AI in medicine.

2 Foundations of AI for medical imaging

The evolution of AI in medical imaging has been driven by two principal pillars: the design of high-capacity DL models and the availability of high-quality training data. The capacity of these models to interpret complex medical images has revolutionized diagnostics and treatment planning. However, their success is entirely dependent on the data they are trained on, making data quality as critical as the structure of the model. This section first considers the core DL architectures behind the rise of the field and then the general importance of data quality in achieving reliable and clinically relevant results.

2.1 AI inference and training

One of the most significant breakthroughs in the use of AI for image analysis was LeNet-5, a CNN architecture developed by Yann LeCun and his team in the late 1980s (Lecun, Bottou, Bengio, & Haffner, 1998). It was one of the earliest successful demonstrations of a CNN, showing the power of hierarchical feature learning directly from raw pixel data. This allowed us to bypass the human-based feature engineering stage and let the algorithm devise the most relevant patterns to achieve optimal results. Despite the promising solution, the idea only gained significant traction in the early 2010s with the rise of DL. The impressive results reported for a CNN-designated AlexNet, in the 2012 ImageNet challenge (Deng et al., 2009; Krizhevsky et al., 2012), are often mentioned as a turning point in image classification accuracy. This achievement showed the potential of DL for visual tasks, inspiring the scientific community to explore this approach with medical images.

CNNs draw inspiration from the hierarchical organization of the biological visual cortex, operating under the assumption that input data has a structured representation. In modern CNN, architectures are composed in a sequence of layers with the purpose of progressively transforming raw input into increasingly abstract and discriminative features (O'Shea and Nash, 2015), as shown in the diagram in Fig. 7.1. To provide an output, the process begins with the input layer receiving image data, typically encoded as a three-channel (Red, Green, Blue) pixel grid. Then, subsequent convolutional layers perform localized feature extraction through previously learned kernel operations, with variations in convolution types (e.g., dilated, depthwise) enabling flexible pattern recognition (Gu et al., 2018). The initial layers detect elements like edges, corners, textures, blobs, or color regions (low-level primitives), while deeper layers combine these hierarchically into higher-order structures like shapes or object components. The convolution layers' output is then passed through a rectified linear unit activation. This introduces nonlinearity to enhance the network's representation ability and also mitigates the possibility of vanishing gradients. Pooling layers (e.g., max or average pooling), which mediate the convolutional blocks, have two crucial functions. One is to reduce the spatial dimension of feature maps, which progressively downsamples layer size. Another is to improve translational invariance (small shifts in the position of features within an image), an essential ability for image analysis where objects may appear in different locations. Overall, pooling operates selectively by retaining salient features (high-activation regions that correspond to meaningful patterns) while discarding redundancies or noisy information. Throughout the layer sequence, it is possible to obtain compact, semantically rich vector representations that distill the input's essential characteristics, designated as deep features (Yandex & Lempitsky, 2015). After the convolutional layers, a final block, designated as the "head," composed of fully connected layers, is used to generate predictions. This sequence represents the automated feature learning process that enables the development of end-to-end systems.

Computationally, an artificial neural network is defined by a set of equations and mathematical rules whose coefficients must be adapted to a specific problem. This stage, designated as "training," has as its objective the optimization of the network's parameters for feature extraction and classification from input data. It begins with initializing the network's weights and bias parameters, often randomly, followed by feeding a labeled dataset through the network, batch by batch. An estimate is made during a forward pass, and a loss function, i.e., cross-entropy for classification, assesses the discrepancy between predicted and actual labels. Then, in a backward pass, through backpropagation, gradients of the loss with respect

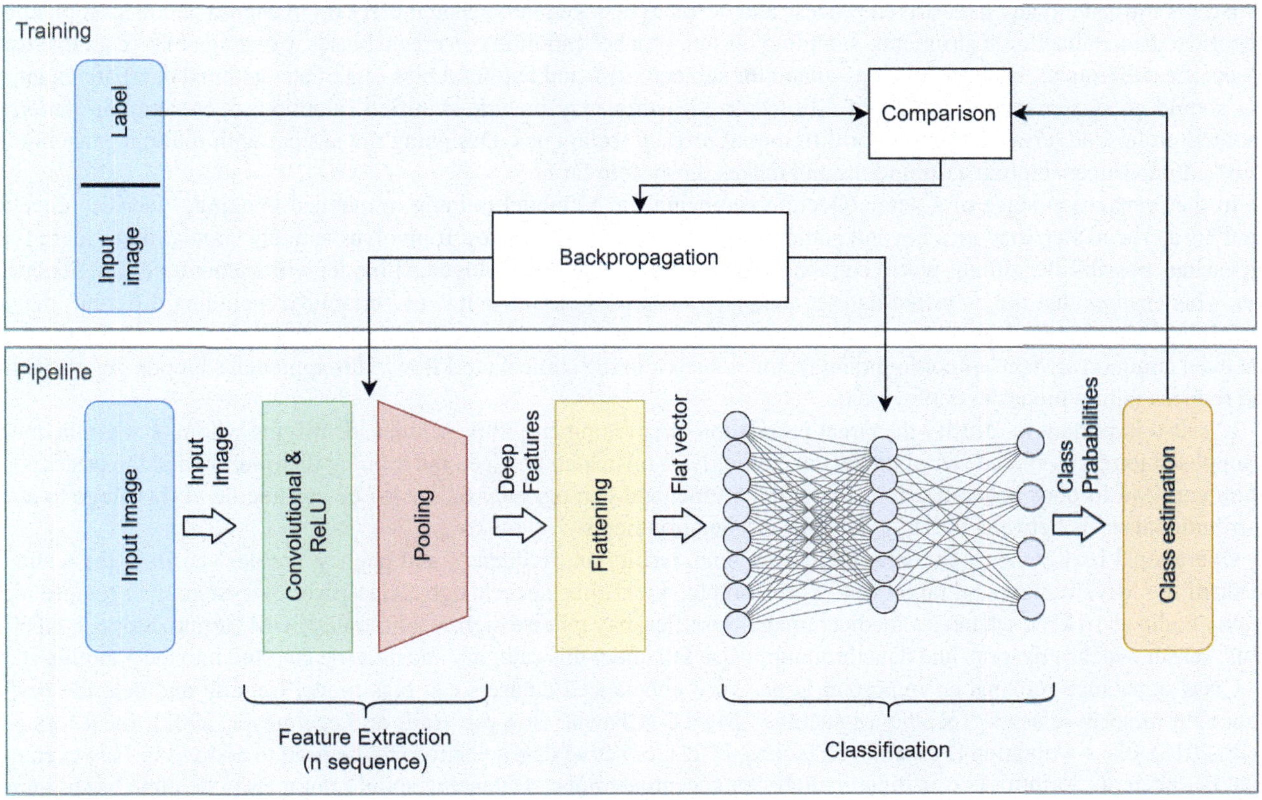

FIG. 7.1 CNN general pipeline and associated training process (only the most relevant functional blocks are represented).

to weights and biases are computed. These gradients guide the refinement of the parameters via optimization algorithms. Stochastic gradient descent, a popular option, iteratively adjusts parameters in the direction of the negative gradient scaled by a learning rate, or by more advanced variants like Adam, RMSprop, or Adagrad, which adaptively adjust learning rates for faster convergence. Techniques like dropout or batch normalization are often employed to reduce the possibility of overfitting phenomena. The described process is repeated iteratively, over multiple epochs, tuning all the parameters until the model performs adequately on a validation set or until a limit of iterations is reached.

2.2 Data quality

The optimal performance of AI systems, due to their data-driven nature, is directly impacted by the training dataset. For example, the use of AI in ulcer detection is a popular problem where the assessment of dataset quality is often focused on the balance of the output classes based on disease severity. However, the impact of skin tone, a factor that can affect the system's performance, is barely addressed (Reis et al., 2025), leading to significant discrepancies in image classification robustness. Hence, it is crucial to evaluate data quality to ensure that it has adequate characteristics and will not constrain the models' performance. If the dataset is yet to be collected, then it is necessary to carefully plan its acquisition. It is only through having representative and varied data that models are able to learn meaningful patterns and generalize well to new, unseen data.

Planning data collection for ML involves a series of interdependent decisions, always supported by a clear understanding of the problem domain, ethical considerations, and downstream application needs. The first step is to define the ML problem, establishing whether it is a classification, regression, clustering, or sequence prediction task. At this stage, the output types or classes are defined, and for each case, the associated level of detail or specificity (granularity) is determined. An adequate granularity must match the requirements of the problem the model is designed to solve. This means that the dataset's class definitions are neither too broad, failing to provide the desired prediction detail, nor too specific for the intended application, overcomplicating the task. For supervised learning, class boundaries must be semantically meaningful and clearly attached to existing domain knowledge. It is also necessary to plan the annotation process or assess the availability of labeled data for each class.

Bias is intrinsic to any data-driven process and needs to be taken into account early on in dataset acquisition planning. Historical data, collection instruments, sampling design, or label variation can create biases. Coverage bias (e.g., exclusion of specific subgroups), labeler bias (e.g., annotator subjectivity), and temporal bias (e.g., date-stamped data) are examples that should be systematically identified. Mitigation measures may include stratified sampling, over-sampling underrepresented groups, adversarial training, or differential privacy techniques. Designing the system with multiple stakeholders involved introduces implicit assumptions and makes the system fairer.

In the preparation stage of a data collection campaign for AI model training in medical imaging, "scenario development" is a crucial step that goes beyond gathering a large volume of images. It involves actively simulating or identifying the various possibilities of real-world contexts and challenges that the deployed AI model will encounter during its operation. This ensures that the collected dataset adequately captures the full range of variability, including different imaging modalities (when applicable), image quality variability, patient demographics, disease presentations, anatomical variations, and even common artifacts or confounding factors inherent to the clinical workflow. This approach supports generalization and reduces failure modes in deployment.

It is also important to identify the target population or environment. This includes identifying whom or what the model is supposed to represent and sampling data accordingly. A mismatch between the training distribution and the target population can lead to poor generalization. Geography, time, and demography must also be considered at this stage to avoid distributional shifts between training and inference environments.

Operational requirements for data, such as volume, resolution, frequency, and latency, are derived from the technical needs of the ML structure and target task. For example, a real-time speech signal classification system may require high-quality audio at 16 kHz, whereas a medical image classifier may require high-resolution DICOM format. Storage, labeling costs, sensor synchronization, and data throughput also fall under this category and must be factored into the planning stage.

Class imbalance remains an important issue. Very unbalanced datasets can bias model training and degrade performance on minority classes. Techniques such as SMOTE (Chawla, Bowyer, Hall, & Kegelmeyer, 2002), focal loss (Lin et al., 2018), class-weighting (Fernando & Tsokos, 2022), or active data selection can be used to address it. However, care must be taken not to introduce artificial patterns or accentuate noise. Balancing should not merely be done based on raw frequency but also on the societal and safety impact of different kinds of errors.

In multimodal systems, synchronization across modes is vital. Temporal coherence of data streams (e.g., audio, video, text, and sensor signals) ensures that the learning model picks up on recurring cross-modal correspondences. Due to desynchronization phenomena, which introduce noise and degrade performance, it is important to create a standardized time-stamping protocol or a time-coincident technique for the capture devices. This aspect must be verified on an ongoing basis during data collection.

Finally, the Findability, Accessibility, Interoperability, and Reusability (FAIR) principles (Ravi et al., 2022) should not be forgotten, as they guide the full data planning strategy. FAIR data enables transparency, reproducibility, and ethical reuse. Metadata standards, good associated documentation, adequate licensing, and data provenance are key enablers of FAIR. This not only allows for scientific rigor but also helps data assets become valuable additions to the research ecosystem in general.

All of these stages, defining the problem, mitigating bias, modeling scenarios, determining targets, specifying requirements, balancing classes, ensuring synchronicity, and FAIR compliance, must be treated as interdependent and iterative. Together, they form the building blocks of a good, ethical, and effective machine learning (ML) pipeline.

3 AI techniques in medical imaging

ML applied to medical imaging allows for a broad range of tasks, each performing specific diagnostic or prognostic purposes. These tasks fall into classification, regression, segmentation, and object detection, with the added subtlety of multiinstance and multiclass when implemented. Classification consists of assigning discrete labels to entire images or regions of images, such as determining whether a mammogram is malignant or sorting retinal scans by diabetic retinopathy severity (in categorical classes). Regression tasks, by contrast, predict continuous-valued targets, e.g., predicting brain age from MRI or quantifying tumor burden. Segmentation involves a pixel- or voxel-level outline of anatomical or pathological structures (e.g., tumor boundaries, organ interfaces), and it is important for treatment planning and quantitative analysis. Object detection introduces localization in addition to classification, detecting the presence and location of multiple regions of interest, e.g., localizing lung nodules in CT scans or polyps in colonoscopy pictures, by encapsulating each occurrence within a bounding box.

These primary tasks most often overlap with additional, more advanced formulations. In single-instance, single-class cases, images are assumed to host a single class of object, e.g., finding an isolated liver lesion. However, most clinical problems in actual practice are multiinstance (detection of several lesions or anatomical features) and multiclass (segmentation into different disease categories or grades). For example, a single histopathology slide contains numerous cellular structures of diverse pathological classes, where recognition and classification need to be performed all at once. The choice of task definition, and hence the subsequent complexity, should be informed by clinical intent, available annotated data, and downstream integration requirements so that ML systems are best able to enable diagnostic accuracy, efficiency, and interpretability.

As explained, the successful application of ML tasks to medical imaging is closely tied to the quality of annotated datasets and the quantity of records. In supervised learning applications, the dataset requirements vary considerably from task to task. Classification and regression tasks require labeled images that are often available in existing medical reports, making the data acquisition process faster and cheaper. In contrast, segmentation or object detection tasks require precise, often pixel-level annotations, which are time-consuming. When annotating fine anatomical boundaries or subtle pathological features, this is especially important to ensure dataset quality. For example, retinal vessel segmentation requires the identification of fine structures, making it very time-consuming. These tasks are also more susceptible to interobserver variability, reducing label consistency and model performance if not strictly controlled (Taha & Hanbury, 2015). This means that an image may require the input of several experts to achieve a robust annotation level. Furthermore, datasets that are suitable for multiclass and multiinstance learning demand even more complex annotation schemas to encode spatial, semantic, and relational information among different objects. The cost and time involved in the annotation process have sparked new interest in techniques like semisupervised, weakly supervised, and self-supervised learning (SSL) methods. These methods simplify the management of huge volumes of unlabeled data or coarse labels (Cheplygina, de Bruijne, & Pluim, 2019). Data heterogeneity, clinical representativeness, and privacy constraints remain matters of concern. These must be addressed to ensure learned models generalize across institutions and populations of patients.

4 Advances in medical image analysis

CNNs' operation is rooted in inductive biases such as local connectivity and translational invariance and has long been the cornerstone of the majority of AI-driven medical imaging processing systems. The rise of transformer-based models has been a game-changer in computer vision and in medical imaging tasks. Transformers were originally formulated for natural language processing (Vaswani et al., 2017), but they are now being progressively applied to vision tasks. As represented in the architecture, it is composed of an encoder structure (and a decoder, which is not shown) that processes the input by converting it into a continuous sequence across multiple layers. Each layer contains a multihead self-attention (MSA) block that weights inputs according to their relevance, capturing dependencies. The vision transformer (ViT) (Dosovitskiy et al., 2020) eliminates convolution operations in favor of a self-attention mechanism and has demonstrated competitive or superior performance to CNNs on large-scale image classification benchmarks. It achieves this by breaking up an input image into fixed-size patches (e.g., 16×16 pixels), flattening them into a vector sequence, linearly embedding each patch, and adding positional encodings to preserve spatial details. These embeddings are passed into a standard transformer encoder made up of repeated layers of MSA and feed-forward networks that capture global features and relations between the patches. The usual practice is to add a special classification token to the sequence, taking its output from the final transformer layer as the output for classification after passing through a linear layer. Transformers are inherently more flexible and capable of modeling long-range dependencies in data. This capacity is particularly advantageous in analyzing high-resolution medical images where contextual information spanning large spatial regions is critical. ViT relies heavily on pretraining on large datasets to learn robust feature representations and becomes highly effective for image recognition when fine-tuned on specific datasets. ViTs have also shown promise in the medical domain in tasks such as tumor segmentation, histopathological image classification, and retinal disease detection (Anari et al., 2024; Chen et al., 2021; Dutta, Sathi, Hossain, & Dewan, 2023). The performance of CNN- and ViT-based approaches has been explored (Aburass, Dorgham, Shaqsi, Rumman, & Al-Kadi, 2025; Kawadkar, 2025; Maurício, Domingues, & Bernardino, 2023; Takahashi et al., 2024), and Table 7.1 provides a summary of reported performance grouped by popular datasets (and tasks). It can be observed that factors such as task, dataset characteristics (number of records, imbalance, number of classes, among others), or initial parameters can impact the performance of the systems for each case. In addition, focusing on model families, a summary of typical application cases is provided in Table 7.2.

Furthermore, hybrid models integrating CNN feature extractors with transformer-based global attention modules, such as TransUNet (Chen et al., 2021), are also being actively explored to further improve segmentation performance in tasks such as organ delineation and lesion detection in CT and MRI scans (Kuang et al., 2024; Wang, Zhang, & Jin, 2024). Hierarchical transformer models are an extension, with models such as Swin Transformers (Liu et al., 2021), where a multiscale image representation and shifted-window mechanisms are used to balance global and local attention. In contrast

TABLE 7.1 Summary of reported classification performance for CNN- and ViT-based classifiers according to popular datasets.

Dataset	Task	CNN	ViT
ISIC-2019 (inc. HAM10000)	Skin lesion classification	EfficientNet-B0, 81.84% (Kawadkar, 2025)	ViT variants: accuracy ≈ 93.5% (HAM10000), 94.3% (clinical dataset) (Takahashi et al., 2024)
CheXpert, ChestX-ray14	Chest X-ray disease classification	ResNet-50 (trained on ImageNet-21k): CheXpert AUC ≈ 87.94% (Ma et al., 2022)	Swin-B with IN-21k pretraining: CheXpert AUC ≈ 87.94% (Ma et al., 2022)
BloodMNIST, BreastMNIST, PathMNIST, RetinaMNIST	MedMNIST 2D biomedical images	PathMNIST: ~91.1%, BloodMNIST: 96.6%, RetinaMNIST: 53.1% (Halder et al., 2024)	BloodMNIST: 97.90%, BreastMNIST: 90.38%, PathMNIST: 94.62%, RetinaMNIST: 57% (Halder et al., 2024)

TABLE 7.2 Summary of best-fit clinical use cases for each model family.

Model family	Best-fit clinical use cases
Convolutional neural networks (CNNs)	General medical image analysis (cornerstone), Object detection, including localizing lung nodules in CT scans, localizing polyps in colonoscopy images, identifying masses and microcalcifications on mammograms, identifying mitotic figures, glandular elements, and tumor-infiltrating lymphocytes in whole-slide images for cancer grading and prognosis (e.g., breast and prostate cancer), and detecting microaneurysms, hemorrhages, and exudates in fundus images for diabetic retinopathy severity assessment.
Vision transformers (ViTs)	Tumor segmentation, histopathological image classification, retinal disease detection, and long-range dependency modeling in high-resolution medical images.
Hierarchical transformers (e.g., Swin)	Lung nodule classification, breast cancer histopathology, retinal image analysis, efficient image classification, object detection, and semantic segmentation, balancing global and local attention.
Hybrid models (CNN-transformer)	Further improving segmentation performance in tasks such as organ delineation and lesion detection in CT and MRI scans.

to ViT, which applies global self-attention to all patches of an image (representing quadratic complexity with image size), Swin Transformer uses an approach based on a hierarchical architecture and shifted window-based MSA (Voita, Talbot, Moiseev, Sennrich, & Titov, 2020; Zhang, Luo, & Wang, 2025). It works by dividing an input image into nonoverlapping patches, embedding them, and processing them across multiple stages. At each step, patches are arranged into local windows where self-attention is computed, reducing complexity to linear with image size. Cross-window interactions are handled by shifted windows at alternate layers, globally collecting context in an efficient way. Patch merging layers reduce resolution across steps, which results in a pyramid-based feature representation (as in CNNs). This kind of architecture makes Swin Transformer efficient in image classification, object detection, and semantic segmentation with improved efficiency and scalability. Such models have been used with great success in applications ranging from lung nodule classification and breast cancer histopathology to retinal image analysis, and they have even outperformed baseline CNNs and ViTs on certain benchmarks (Cao et al., 2021).

SSL techniques have emerged as an effective approach for model pretraining on unlabeled data, a significant benefit in some medical areas where professional annotation of new data can be expensive and scarce. Contrastive learning (e.g., SimCLR, MoCo) and masked autoencoders are a few of the approaches through which transformer-based models can learn transferable representations from medical datasets without any manual annotations (Tang et al., 2022). These paradigms of pretraining have proven particularly effective in radiology, pathology, and ophthalmology (Wolf et al., 2023; Zhou et al., 2023).

Integration with paradigms of multimodal learning frameworks is also being explored, allowing for the fusion of imaging data with clinical notes, genomics, or patient metadata. MedKLIP (Wu, Zhang, Zhang, Wang, & Xie, 2023) and

BioViL-T (Boecking et al., 2022) are some examples pursuing comprehensive and explainable systems for diagnosis. Such systems not only hold the promise of greater accuracy but also provide explainable decision-making in the form of mapping visual features onto textual concepts through clinical knowledge bases.

While CNNs and ViTs have demonstrated successful applications in image classification and image segmentation, object detection, which is a focal task in medical imaging, has also benefited from specialized architectures. Among others, region-based convolutional neural networks (R-CNNs) and their evolutions, such as fast R-CNN and faster R-CNN, have advanced the state of the art in segmentation. R-CNN initially introduced the idea of generating region proposals using selective search, followed by classification using CNNs (Girshick, Donahue, Darrell, & Malik, 2014). Fast R-CNN improved computation by utilizing feature sharing across convolutional proposals, while faster R-CNN incorporated a region proposal network to eliminate the dependency on external proposal approaches (Ren, He, Girshick, Sun, & R-CNN, 2017). These models are highly accurate but computationally intensive and less suitable for real-time contexts.

On the contrary, You Only Look Once (YOLO) is a category of one-shot detectors that combines object classification and localization into a single pass through the neural network. The YOLO models are optimized for real-time inference and speed with acceptable accuracy (Redmon, Divvala, Girshick, & Farhadi, 2016). This makes YOLO particularly interesting to implement in point-of-care or embedded diagnostic systems. Despite suboptimal effects on tiny objects or in fine-grained localization scenarios, which are common in medical applications such as the detection of polyps or lesions, there are reports of good performance when other techniques are applied (Wang, Xie, Cui, & Chen, 2024). The use of this approach in digital pathology is also popular (Araújo et al., 2017).

Another approach is the Segmenting Objects by Locations (SOLO) model, in which instance segmentation is framed as a category-aware pixel-level grouping problem, dispensing with region proposals or anchor boxes altogether (Wang et al., 2020). SOLO and its enhanced variant, SOLOv2, utilize dynamic convolution and spatial splitting to obtain real-time speed with comparable accuracy. Compared to the YOLO and R-CNN classes, SOLO performs better in dense instance segmentation tasks with high spatial precision, which is particularly applicable in histopathology image analysis or retinal vessel segmentation issues (Shobaki & Milanova, 2025).

Some recent approaches are focused on model fine-tuning. Low-rank adaptation (LoRA) of large language models (Hu et al., 2021; Kao et al., 2025) is becoming a popular approach, where the weights of all but a small portion of the model are frozen. Then, small trainable low-rank matrices are injected, which are subsequently fine-tuned on a relatively modest-sized medical imaging dataset. This not only significantly reduces training time and memory footprint but also diminishes the likelihood of overfitting, which is usually a problem when fine-tuning gigantic models on tiny, domain-limited datasets. This leads to a deeply specialized model that can achieve state-of-the-art diagnostic accuracy at significantly reduced computational and storage costs, making AI-aided medical diagnosis cheaper and more scalable. The approach has been identified as extremely powerful and viable for fine-tuning medical image diagnosis deep AI models and for addressing a number of the most relevant issues in the field. This framework allows robust foundation models (pretrained vision models on massive natural image datasets) to be converted to specific medical tasks like lung nodule classification or tumor segmentation without having to rely on massive computational resources.

These diverse architectures show that object detection from medical images is more than a single-model paradigm. Instead, model choice must be resolved with reference to the clinical task undertaken and must also balance competing trade-offs in accuracy, interpretability, speed, and computational effort.

Finally, mention generative AI. In medical imaging, such models, especially generative adversarial networks and diffusion models, are utilized for purposes such as data augmentation, image synthesis, anomaly detection, and image-to-image translation (e.g., MRI to CT). These techniques can address data sparsity, especially when dealing with rare diseases, by generating realistic synthetic images to augment training sets, thereby making the model more robust and generalizable (Frid-Adar et al., 2018; Wolterink et al., 2017). Generative models can also be used in diagnostic pipelines for image augmentation, i.e., denoising or superresolution, and enable counterfactual generation, where synthetic patient variations of the same data are utilized to test out-of-sample diagnostic hypotheses. More recently, foundation models that couple large-scale generative pretraining with clinical data, such as latent diffusion models or multimodal models such as Med-PaLM M (Tu et al., 2024), are being considered for their potential to integrate imaging and clinical text together, toward combined diagnostic reasoning. Challenges still arise with clinical validity, control of bias, and lack of regulation for synthetic output, particularly when generative models are used for applications with immediate diagnostic implications. Generative AI still has vast potential to aid medical diagnosis by supporting more flexible, interpretable, and data-efficient processes.

5 Explainability and interpretability

Explainability and interpretability are paramount in medical imaging AI to ensure that the model estimations have the desired transparency and consistency with existing medical knowledge. Hence, only by understanding how the AI system is reaching its results does it make sense to introduce it in medical settings. Saliency mapping is one of the most widely used techniques, which allows highlighting the regions within an image that contribute most to a model's prediction. Among these techniques, gradient-weighted class activation mapping (Grad-CAM) (Selvaraju et al., 2020) is an extremely popular technique for visualization and interpretation with CNNs. As depicted in Fig. 7.2, it works by computing the gradients of the score of the target class using backpropagation, considering the feature maps of the final convolutional layer (containing the deep features). These gradients are then globally averaged to obtain importance weights, which are used to compute a weighted sum of the feature maps. The resulting heatmap represents the regions in the input image that contribute most to the model's prediction. The result can then be presented overlaid with the original image. Unlike earlier class activation mapping methods that required architectural changes, Grad-CAM is model-agnostic for CNNs, requiring only gradient information. In medicine, Grad-CAM is widely used for verifying whether the AI system is focusing on clinically relevant regions and aligned with clinical knowledge, for example, pointing to the affected areas in ulcer detection (Reis et al., 2025). This technique has its drawbacks, mainly poor spatial resolution due to a focus on the final layers of the CNN, and it can be influenced by model architecture and layer selection (Fig. 7.3).

Other approaches, such as layer-wise relevance propagation (Bach et al., 2015) and integrated gradients (Sundararajan, Taly, & Yan, 2017), provide alternative techniques to quantify input feature contributions, and these are often stable or faithful explanations. Aside from visual methods, SHAP (SHapley Additive exPlanations) (Lundberg & Lee, 2017; Rozemberczki et al., 2022) and LIME (Local Interpretable Model-agnostic Explanations) (Ribeiro & Singh, 2016) are model-agnostic methods. They explain, via local approximation, complex models by simpler models, although even they are computationally intensive in high-dimensional medical images. For transformer models, newer techniques such as attention rollout (Abnar & Zuidema, 2020) and token attribution (Modarressi, Fayyaz, Yaghoobzadeh, & Pilehvar, 2022) are used to discover what self-attention layers weigh at different locations in the image. Increasingly, counterfactual explanation systems are also being explored, where synthetic input changes reveal how these are propagated to the model's prediction, which is useful for clinical validation and algorithmic audit.

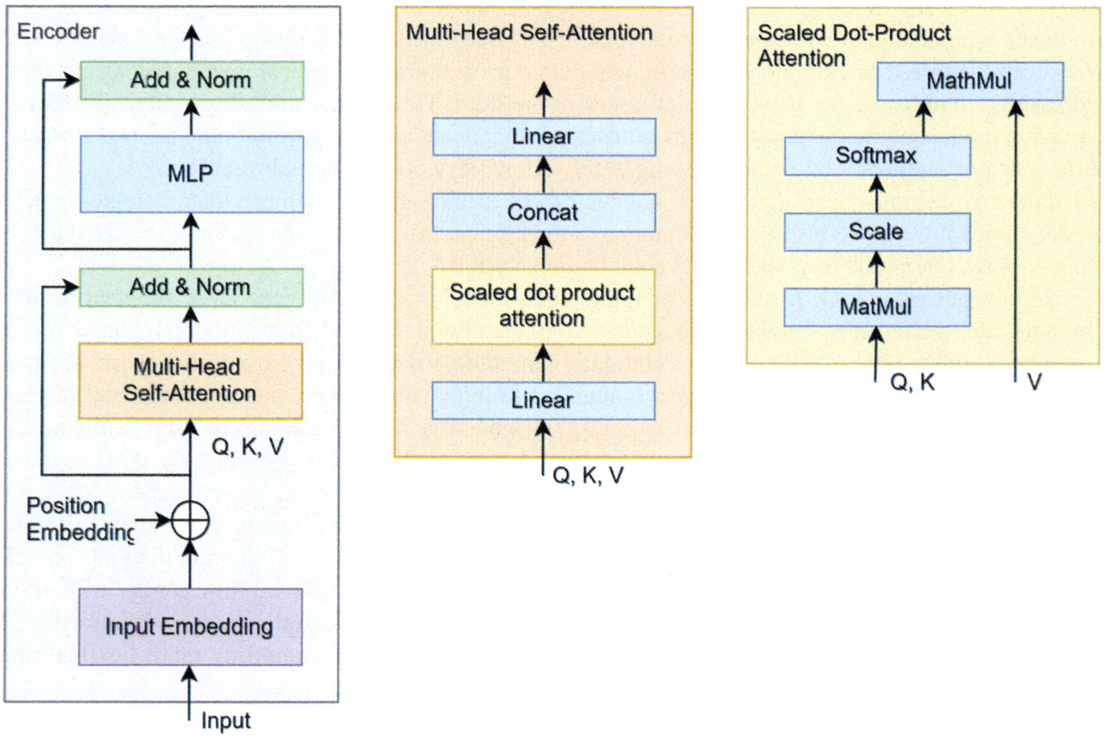

FIG. 7.2 Architecture of the transformer encoder, multihead self-attention, and scaled dot-product attention (from left to right).

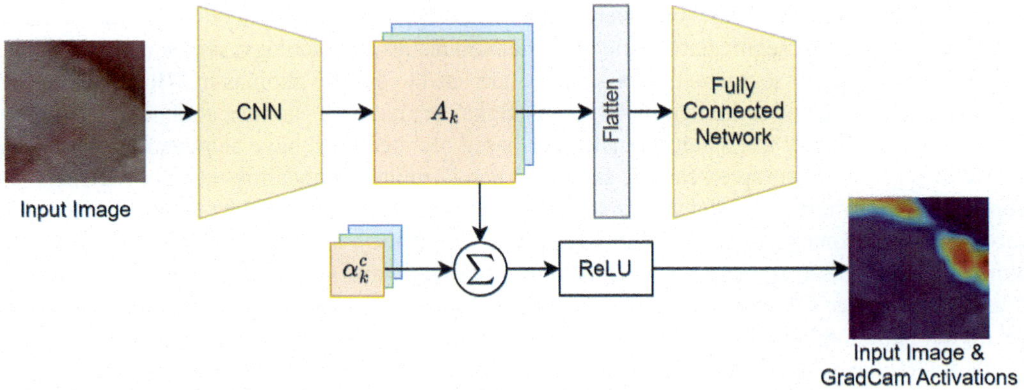

FIG. 7.3 *Grad-CAM architecture.* An example of an input image (left) and the corresponding activation heatmap (right) is also provided to clarify the system's operation.

With more sophisticated AI systems, especially multimodal or generative ones, the deployment of such explainability systems is highly important for regulatory requirements, clinical adoption, and ethical transparency.

6 Applications of AI in disease diagnosis

AI-based diagnostic systems in medical imaging, when organized by ML tasks, can have several applications in a clinical context.

Classification systems provide a way to support medical decision-making by identifying disease patterns and estimating clinically significant conditions. In dermatology, for example, CNNs trained on dermoscopic images are able to diagnose skin lesions (from melanoma to basal cell carcinoma) with a degree of accuracy that is comparable to expert professionals (Esteva et al., 2017). In radiology, another popular area, DL algorithms have achieved expert-level performance in areas such as detecting pneumonia, pleural effusion, and pneumothorax from chest X-ray images. The public release of large datasets like the NIH ChestX-ray14 or CheXpert (Irvin et al., 2019; Rajpurkar et al., 2017) has helped to develop and benchmark these systems, pushing this technology forward. In mammography, AI algorithms are now capable of differentiating between benign and malignant breast lesions, even in the presence of small nodules. This helps with the early detection of breast cancer and the reduction of false positives (McKinney et al., 2020), significantly improving prognoses. Ophthalmology has also been a focus area, with AI models being able to successfully segment the vasculature network of fundus images. This allows the systems to successfully identify diabetic retinopathy, age-related macular degeneration, and glaucoma (De Fauw et al., 2018; Gulshan et al., 2016). These applications showcase some of the contexts where AI-driven systems have enhanced diagnostic accuracy and efficiency.

Telemedicine can also be enhanced since AI systems can bring greater accessibility to resource-constrained locations. It is also important to mention that many applications where physiological information exists in the form of signals are often converted to image-based problems by creating time-frequency representations such as spectrograms, scalograms, or others. For example, many neurodegenerative diseases can be detected by using image representations of speech signals in the Mel-frequency parameter space (Braga, Madureira, Coelho, & Abraham, 2018).

Image classification can be further enhanced by providing the location of instances of anatomical structures. In object detection tasks, DL networks such as YOLO and Faster R-CNN have been applied with excellent results in thoracic imaging to detect lung nodules in chest CT scans, for example (Sindhu et al., 2018). The level of reported precision and detail can even support the early diagnosis of lung cancer by detecting nodules that could be missed in human assessment (Setio et al., 2016). In gastroenterology endoscopy, AI systems can be used to detect polyps in real time, improve adenoma detection rates during colonoscopy, and reduce the risk of missed lesions (Urban et al., 2018). In breast imaging, masses and microcalcifications on mammograms can also be identified by object detection algorithms (Ribli, Horváth, Unger, Pollner, & Csabai, 2018). In digital pathology, an area with growing applications of AI, object detection methods can, for example, be used to identify mitotic figures, glandular elements, and tumor-infiltrating lymphocytes in whole-slide images, significantly contributing to more accurate grading and prognosis of cancers such as breast and prostate (Cireşan et al., 2013; Wang et al., 2014). Object detection is also crucial in retinal imaging, where microaneurysms, hemorrhages, and exudates are detected from fundus images for assessing diabetic retinopathy severity (Ragab et al., 2024).

Segmentation, focused on the pixel (2D) or voxel (3D) level, brings detail to simple object segmentation. Some systems are able to simultaneously convey segmentation and classification results. In oncology, segmentation models based on DL, as implemented in the brain tumor segmentation challenges, are used to outline gliomas in MRI with high specificity in an attempt to aid surgical planning and radiotherapy targeting (Bakas et al., 2017; Barbosa et al., 2020).

Modern DL models, especially CNNs and transformer-based architectures, have achieved or even surpassed human-level performance in segmenting complex cardiac structures across multiple modalities such as MRI, CT, and ultrasound. This progression allows for rapid, reproducible, and highly accurate delineation of key cardiac anatomy, including ventricles, myocardium, and coronary vessels (El-Taraboulsi, Cabrera, Roney, & Aung, 2023; Tao et al., 2019). In radiation oncology, segmentation models are employed to segment tumors and organs-at-risk in a range of imaging modalities, most notably head and neck cancer CT scans, to automate contouring and reduce interobserver variability. Segmentation is also applied in digital pathology to segment nuclei, glands, or tumor regions within histopathological whole-slide images (Araújo et al., 2017; Graham et al., 2019). Segmentation algorithms are also used for retinal imaging to annotate the layers of the retina and lesions in optical coherence tomography (OCT) scans that support macular edema and glaucoma diagnosis (De Fauw et al., 2018; Ribeiro, Gonçalves, Coelho, Castelo-Branco, & Martins, 2025). These solutions highlight how AI-driven segmentation enhances the precision and reproducibility of image interpretation, as well as reduces the burden of manual annotation in time-critical clinical workflows.

When the objective is to provide precise measurements from medical images, AI is also gaining prominence with automated quantification techniques (Huang, Ruan, Xing, & Feng, 2024). Such tools provide results faster and are less sensitive to interobserver errors. From a clinical perspective, this information can be useful to support diagnosis, risk stratification, or treatment monitoring. For example, in cardiology, an area where anatomic measurements are essential, AI algorithms are being widely applied. Parameters such as left ventricular ejection fraction, myocardial wall thickness, and chamber volumes can be estimated with reduced variability and quicker analysis compared to manual measurements (Ouyang et al., 2020). In neuroimaging, quantification of brain region volumes and cortical thickness from MRI is used for monitoring neurodegenerative diseases like Alzheimer's, with potential for earlier detection than visual assessment alone (Falahati et al., 2017). In oncology, the measurement of tumor volume, growth rate, and metabolic activity by AI from CT, MRI, or PET-CT scans is critical for treatment response evaluation, particularly in longitudinal follow-up (Saeidnia, Firuzpour, Kozak, & majd, 2025). In pulmonary imaging, automatic quantitation can include the estimation of the degree of emphysema or airway wall thickness on CT, for example, aiding in the diagnosis and staging of chronic obstructive pulmonary disease (Fu, Lin, Wei, Tavares, & Liang, 2025; Wu et al., 2021). AI is also increasingly being used in bone imaging for assessing bone mineral density or fracture risk from dual-energy X-ray absorptiometry (DEXA) or X-ray images. These quantification tasks benefit from AI's ability to process large datasets rapidly and consistently, enhancing reproducibility and supporting precision medicine through objective, data-driven metrics.

The mentioned applications are some examples that demonstrate how AI can improve diagnostic workflows across several disciplines. For each case, there are specific challenges, such as annotation granularity, clinical interpretability, and data heterogeneity.

7 Future perspectives and innovations

7.1 Security and ethics

Security and ethics are primary concerns in the creation and implementation of AI-based systems for medical imaging diagnosis because of the high risks involved in clinical decision-making and patient data privacy. From the security standpoint, AI systems are vulnerable to attacks by adversarial perturbations, where minor, imperceptible image input variations could lead to misdiagnosis, such as changing a few pixels of an in-chest X-ray potentially leading a model to miss a malignant lesion. Caution regarding data privacy breaches must also be addressed, particularly when using centralized training on sensitive image datasets. Federated learning, differential privacy, and homomorphic encryption are being explored to enable safe training of models with data confidentiality assurances. In an ethical context, there are concerns such as algorithmic bias, where models are less optimal for underrepresented patient subsets due to biased training data, potentially establishing health disparities. In addition, the lack of transparency in most DL approaches, often seen as "black boxes," poses threats to accountability and trust, particularly when their outputs conflict with clinical judgment. To ethically operate such systems, transparency in regulatory authorization is also needed, as well as informed consent and accountability for errors. Nevertheless, international organizations and regulatory bodies are working to keep pace with technological development. As AI becomes more embedded in diagnostic processes, resolving these security and ethics concerns is essential to realizing safe, fair, and trustworthy applications in actual clinical practice.

7.2 Federated learning

Newer AI systems and innovations have been supported by growing requirements in terms of data. In the clinical area, the ability to create large, diverse datasets to train these models is often hindered by a lack of proper annotation and insufficient samples for some rare diseases. Sharing data centrally among multiple institutions is being seen as a possible solution to quickly address the data creation challenges. However, this raises concerns about privacy and ownership, especially in the competition for leading the AI revolution. In Sheller et al. (2020), the use of federated learning is explored as a solution. This approach allows multiple institutions to collaborate and train a single model without sharing raw patient data. Instead, institutions train the model locally, and only the updated model parameters are aggregated. The study found that models trained using federated learning across 10 institutions achieved 99% of the quality of models trained on centralized data. The researchers also examined how data distribution among institutions affects model quality, concluding that the benefits of increased data access through multiinstitutional collaboration outweigh any errors introduced by the federated learning method. The abstract also asserts that federated learning is superior to other collaborative learning methods. Ultimately, the widespread adoption of federated learning in clinical settings is expected to enable the development of models trained on unprecedentedly large datasets, significantly advancing precision and personalized medicine.

7.3 Foundation models

Another important concept that is emerging with the widespread implementation of ML is the foundation model. These are large-scale, pretrained models that are trained on massive and diverse datasets to learn generalizable representations. This is especially important in medical imaging, where data can be scarce and difficult to obtain. Then, for specific contexts, these models can be adapted to a wide range of downstream tasks using transfer learning and few-shot or zero-shot adaptation. This strategy allows researchers to quickly develop new, adapted models with good performance while significantly reducing development time and the need for large, task-specific labeled datasets.

Models such as ViTs, CLIP (Contrastive Language-Image Pre-Training) (Radford et al., 2021), BioMedCLIP (Zhang et al., 2025), EchoCLIP (Christensen, Vukadinovic, Yuan, & Ouyang, 2024), or Prov-GigaPath (Xu et al., 2024) have demonstrated the ability to generalize across multiple clinical areas and are able to handle multimodal data. Besides classification and segmentation tasks, they can also handle data retrieval and report generation, even in cross-modal contexts. For this, some models are simultaneously trained on both images and clinical text, supporting models with multimodal understanding and the ability to align imaging features with clinical language. In addition, foundation models provide a strong basis for self-supervised or semisupervised learning, further reducing reliance on manual annotation.

Foundation models are expected to continue evolving and play a central role in new AI developments.

7.4 Green AI

The increased application of large AI models, e.g., foundation models and deep neural networks, has sparked interest in their carbon footprint and overall environmental impact. Training the latest models, like GPT, ViT, or large convolutional architectures, requires enormous computational power, resulting in a high CO_2 footprint. For instance, a single training session for a large language model emits several tons of CO_2, which can be equivalent to the lifetime carbon footprint of multiple cars (Strubell, Ganesh, & McCallum, 2019; Wang et al., 2020). Therefore it is a legitimate goal to make AI systems more efficient, and the scientific community is actively working on ways to reduce these developmental and operational intensities. Model compression techniques (e.g., pruning, quantization), lightweight architectures (e.g., MobileNet, EfficientNet, Swin Transformer compared to models with comparable performance), and green AI practices focusing on computational efficiency are some of the advancements and ongoing research efforts. In addition, techniques such as transfer learning and the use of base models are beneficial, as they enable the reuse of pretrained models rather than training from scratch. Furthermore, through carbon-aware scheduling, computationally intensive jobs can be scheduled during periods when low-carbon energy is available. New projects, such as Carbontracker and CodeCarbon (Lacoste, Schmidt, Luccioni, & Dandres, 2019; Selvan et al., 2022), provide tools for emission estimation and reduction while training models. These collaborative efforts highlight a growing concern for sustainable AI, particularly in a one-health context, where global health and planetary health are inextricably linked for the future.

7.5 Responsible AI

Responsible AI encompasses many of the mentioned aspects. In medical imaging, it integrates ethical, regulatory, and environmental safeguards to ensure the safe, fair, and sustainable deployment of AI systems in healthcare. It focuses on prioritizing patient privacy through secure training methods, addresses algorithmic bias to prevent health disparities, and promotes transparency and accountability to strengthen trust in clinical decision-making. Regulatory oversight, informed consent, and clear responsibility for errors are essential aspects of the ethical adoption of such systems. At the same time, reducing the carbon footprint of large-scale models through efficient architectures, transfer learning, and green AI practices ensures environmental sustainability. Together, these principles advance AI applications that are trustworthy, equitable, and aligned with both global health and planetary health.

8 Conclusion

The integration of AI into medical imaging is quickly redesigning diagnostic workflows. As described throughout this chapter, AI technologies offer unprecedented capabilities to automate tasks such as classification, segmentation, detection, and quantification. These technologies provide more accurate, reproducible, and scalable interpretations of a wide range of imaging modalities, even when integrating several modalities. Examples from imaging domains such as radiological scans, digital pathology, ophthalmology images, and dermatologic photos have been presented. In addition to these new techniques, many challenges have emerged. Alongside technical performance, the development of explainability and interpretability tools aims to provide AI decisions with transparency and trustworthiness, which are essential aspects in clinical settings.

The applications of AI in medical imaging are already demonstrating tangible benefits in routine practice, with use cases ranging from early disease detection and treatment planning to population-level screening and monitoring. At the same time, important considerations around data quality, ethical deployment, model generalizability, and environmental sustainability remain. Emerging trends, such as the use of foundation models, multimodal learning, and self-supervised pretraining, offer promising paths toward more adaptable and efficient AI systems, while regulatory and governance frameworks continue to evolve to ensure safety and accountability.

Looking ahead, the future of AI in medical imaging will be influenced not only by technical innovation but also by interdisciplinary collaboration among healthcare professionals, computer scientists, ethicists, and policymakers. From data collection to system operation, it is necessary to promote the collaboration of a multidisciplinary team to leverage the potential of the system. Achieving robust, fair, and interpretable AI solutions at scale will require ongoing efforts to address challenges related to bias, privacy, and the carbon footprint of model training. Ultimately, the goal is to develop AI systems that act as reliable clinical partners, enhancing human expertise, improving diagnostic precision, and promoting equitable access to high-quality healthcare worldwide.

Declaration on the use of AI in writing the manuscript

The authors used ChatGPT to prepare syntax-enhanced drafts for some paragraphs, which were then adapted to the desired purpose. The software was used as an assistive tool with the main intention of enhancing English, as it is not the authors' native language.

AI Disclosure

During the preparation of this work, the author(s) used ChatGPT to improve the quality of some paragraphs, as the author is not a native English speaker. After using this tool/service, the author(s) reviewed and edited the content as needed and take(s) full responsibility for the content of the publication.

References

Abnar, S., & Zuidema, W. (2020). Undefined quantifying attention flow in transformers. In Proceedings of the Annual Meeting of the Association for Computational Linguistics (pp. 4190–4197). Association for Computational Linguistics (ACL). https://www.aclweb.org/.

Aburass, S., Dorgham, O., Shaqsi, J. A., Rumman, M. A., & Al-Kadi, O. (2025). Vision transformers in medical imaging: a comprehensive review of advancements and applications across multiple diseases. *Journal of Imaging Informatics in Medicine*, 2933–2948. doi: 10.1007/s10278-025-01481-y.

Anari, S., de Oliveira, G. G., Ranjbarzadeh, R., Alves, A. M., Vaz, G. C., & Bendechache, M. (2024). EfficientUNetViT: efficient breast tumor segmentation utilizing UNet architecture and pretrained vision transformer. *Bioengineering*. 23065354. *11*(9). Retrieved from: http://www.mdpi.com/journal/bioengineering. doi:10.3390/bioengineering11090945.

Araújo, T., Aresta, G., Castro, E., Rouco, J., Aguiar, P., Eloy, C., et al. (2017). Classification of breast cancer histology images using convolutional neural networks. *PLoS Osne*. 1932-6203. *12*(6), e0177544. doi:10.1371/journal.pone.0177544.

Ardila, D., Kiraly, A. P., Bharadwaj, S., Choi, B., Reicher, J. J., Peng, L., et al. (2019). End-to-end lung cancer screening with three-dimensional deep learning on low-dose chest computed tomography. *Nature Medicine*. 1546170X. *25*(6), 954–961. Retrieved from: http://www.nature.com/nm/index.html. doi:10.1038/s41591-019-0447-x.

Bach, S., Binder, A., Montavon, G., Klauschen, F., Müller, K.-R., Samek, W., & Suarez, O. D. (2015). On pixel-wise explanations for non-linear classifier decisions by layer-wise relevance propagation. *PLoS One*. 1932-6203. *10*(7), e0130140. doi:10.1371/journal.pone.0130140.

Bakas, S., Akbari, H., Sotiras, A., Bilello, M., Rozycki, M., Kirby, J. S., et al. (2017). Advancing the cancer genome atlas glioma MRI collections with expert segmentation labels and radiomic features. *Scientific Data*. 20524463. *4*. Retrieved from: http://www.nature.com/sdata/. doi:10.1038/sdata.2017.117.

Barbosa, M., Moreira, P., Ribeiro, R., & Coelho, L. (2020). Automatic classification and segmentation of low-grade gliomas in magnetic resonance imaging. In A. M. Madureira, A. Abraham, N. Gandhi, C. Silva, & M. Antunes (Eds.), *Advances in intelligent systems and computing* (*Vol. 942*, pp. 43–50). Portugal: Springer Verlag. http://www.springer.com/series/11156. doi:10.1007/978-3-030-17065-3_5.

Boecking, B., Usuyama, N., Bannur, S., Castro, D. C., Schwaighofer, A., Hyland, S., et al. (2022). Making the most of text semantics to improve biomedical vision-language processing. In E. Bertino, W. Gao , S. Bernhard, & M. Yung (Eds.), *Lecture notes in computer science* (pp. 1–21). United States: Springer Science and Business Media Deutschland GmbH. https://www.springer.com/series/558. doi: 10.1007/978-3-031-20059-5_1.

Braga, D., Madureira, A. M., Coelho, L., & Abraham, A. (2018). In *Neurodegenerative diseases detection through voice analysis*, Vol. *734* (213–223). Springer Science and Business Media LLC. doi:10.1007/978-3-319-76351-4_22.

Cao, H., Wang, Y., Chen, J., Jiang, D., Zhang, X., Tian, Q., & Wang, M. (2021). Swin-unet: unet-like pure transformer for medical image segmentation. *arXiv*. 23318422. Retrieved from: https://www.arxiv.org. doi:10.48550/arxiv.2105.05537.

Chawla, N. V., Bowyer, K. W., Hall, L. O., & Kegelmeyer, W. P. (2002). SMOTE: synthetic minority over-sampling technique. *Journal of Artificial Intelligence Research*.10769757. *16*, 321–357. Retrieved from: http://www.cs.cmu.edu/afs/cs/project/jair/pub/volume16/chawla02a.pdf. doi:10.1613/jair.953.

Chen, J., Lu, Y., Yu, Q., Luo, X., Adeli, E., Wang, Y., et al. (2021). TransUNet: transformers make strong encoders for medical image segmentation. *arXiv*. 23318422. Retrieved from: https://www.arxiv.org. doi:10.48550/arxiv.2102.04306.

Cheplygina, V., de Bruijne, M., & Pluim, J. P. W. (2019). Not-so-supervised: a survey of semi-supervised, multi-instance, and transfer learning in medical image analysis. *Medical Image Analysis*.13618423. *54*, 280–296. Retrieved from: http://www.elsevier.com/inca/publications/store/6/2/0/9/8/3/index.htt. doi:10.1016/j.media.2019.03.009.

Christensen, M., Vukadinovic, M., Yuan, N., & Ouyang, D. (2024). Vision–language foundation model for echocardiogram interpretation. *Nature Medicine*. 1078-8956. *30*(5), 1481–1488. doi:10.1038/s41591-024-02959-y.

Cireşan, D. C., Giusti, A., Gambardella, L. M., & Schmidhuber, J. (2013). Mitosis detection in breast cancer histology images with deep neural networks. In K. Mori, I. Sakuma, Y. Sato, C. Barillot, & N. Navab (Eds.), *Lecture notes in computer science (including subseries Lecture notes in artificial intelligence and lecture notes in bioinformatics)* (*Vol. 2*, pp. 411–418). doi: 10.1007/978-3-642-40763-5_51.

De Fauw, J., Ledsam, J. R., Romera-Paredes, B., Nikolov, S., Tomasev, N., Blackwell, S., et al. (2018). Clinically applicable deep learning for diagnosis and referral in retinal disease. *Nature Medicine*. 1546170X. *24*(9), 1342–1350. Retrieved from: http://www.nature.com/nm/index.html. doi:10.1038/s41591-018-0107-6.

Deng, J., Dong, W., Socher, R., Li, L. J., Li, K., & Fei-Fei, L. (2009). ImageNet: a large-scale hierarchical image database. In *IEEE Conference on Computer Vision and Pattern Recognition, CVPR 2009* (248–255). United States: IEEE Computer Society. doi:10.1109/CVPR.2009.5206848.

Dosovitskiy, A., Beyer, L., Kolesnikov, A., Weissenborn, D., Zhai, X., Unterthiner, T., et al. (2020). An image is worth 16X16. words: Transformers for image recognition at scale. *arXiv*. 23318422. Retrieved from: https://www.arxiv.org. doi:10.48550/arxiv.2010.11929.

Dutta, P., Sathi, K. A., Hossain, M. A., & Dewan, M. A. A. (2023). Conv-ViT: a convolution and vision transformer-based hybrid feature extraction method for retinal disease detection. *Journal of Imaging*. 2313433X. *9*(7). Retrieved from: http://www.mdpi.com/journal/jimaging/. doi:10.3390/jimaging9070140.

El-Taraboulsi, J., Cabrera, C. P., Roney, C., & Aung, N. (2023). Deep neural network architectures for cardiac image segmentation. *Artificial Intelligence in the Life Sciences*. 26673185. *4*, 100083. doi:10.1016/j.ailsci.2023.100083.

Esteva, A., Kuprel, B., Novoa, R. A., Ko, J., Swetter, S. M., Blau, H. M., & Thrun, S. (2017). Dermatologist-level classification of skin cancer with deep neural networks. *Nature*. 14764687. *542*(7639), 115–118. Retrieved from: http://www.nature.com/nature/index.html. doi:10.1038/nature21056.

Falahati, F., Ferreira, D., Muehlboeck, J. S., Eriksdotter, M., Simmons, A., Wahlund, L. O., & Westman, E. (2017). Monitoring disease progression in mild cognitive impairment: associations between atrophy patterns, cognition, APOE and amyloid. *NeuroImage: Clinical*.22131582. *16*, 418–428. Retrieved from: http://www.journals.elsevier.com/neuroimage-clinical/. doi:10.1016/j.nicl.2017.08.014.

Fernando, K. R. M., & Tsokos, C. P. (2022). Dynamically weighted balanced loss: class imbalanced Learning and confidence calibration of deep neural networks. *IEEE Transactions on Neural Networks and Learning Systems*. 21622388. *33*(7), 2940–2951. Retrieved from: http://www.ieeexplore.ieee.org/xpl/RecentIssue.jsp?punumber=5962385. doi:10.1109/TNNLS.2020.3047335.

Frid-Adar, M., Diamant, I., Klang, E., Amitai, M., Goldberger, J., & Greenspan, H. (2018). GAN-based synthetic medical image augmentation for increased CNN performance in liver lesion classification. *Neurocomputing*. 09252312. *321*, 321–331. doi:10.1016/j.neucom.2018.09.013.

Fu, X., Lin, R., Wei, D., Tavares, A., & Liang, Y. (2025). Explainable hybrid transformer for multi-classification of lung disease using chest X-rays. *Scientific Reports*. 2045-2322. *15*(1). doi:10.1038/s41598-025-90607-x.

Girshick, R., Donahue, J., Darrell, T., & Malik, J. (2014). Rich feature hierarchies for accurate object detection and semantic segmentation. In *Proceedings of the IEEE Computer Society Conference on Computer Vision and Pattern Recognition* (580–587). United States:IEEE Computer Society. doi:10.1109/CVPR.2014.81.

Graham, S., Vu, Q. D., Raza, S. E. A., Azam, A., Tsang, Y. W., Kwak, J. T., & Rajpoot, N. (2019). Hover-Net: simultaneous segmentation and classification of nuclei in multi-tissue histology images. *Medical Image Analysis*. 13618423. *58*. Retrieved from: http://www.elsevier.com/inca/publications/store/6/2/0/9/8/3/index.htt. doi:10.1016/j.media.2019.101563.

Gu, J., Wang, Z., Kuen, J., Lianyang, M., Shahroudy, A., Shuai, B., et al. (2018). Recent advances in convolutional neural networks. *Pattern Recognition*. 00313203. *77*, 354–377. doi:10.1016/j.patcog.2017.10.013.

Gulshan, V., Peng, L., Coram, M., Stumpe, M. C., Wu, D., Narayanaswamy, A., et al. (2016). Development and validation of a deep learning algorithm for detection of diabetic retinopathy in retinal fundus photographs. *Journal of the American Medical Association*. 15383598. *316*(22), 2402–2410. Retrieved from: http://jamanetwork.com/journals/jama/data/Journals/JAMA/935924/joi160132.pdf. doi:10.1001/jama.2016.17216.

Halder, A., Gharami, S., Sadhu, P., Singh, P. K., Woźniak, M., & Ijaz, M. F. (2024). Implementing vision transformer for classifying 2D biomedical images. *Scientific Reports*. 20452322. *14*(1). Retrieved from: https://www.nature.com/srep/. doi:10.1038/s41598-024-63094-9.

Hu, E., Shen, Y., Wallis, P., Allen-Zhu, Z., Li, Y., Wang, S., et al. (2021). Lora: low-rank adaptation of large lan-guage models. *arXiv*. 23318422. Retrieved from: https://www.arxiv.org. doi:10.48550/arxiv.2106.09685.

Huang, L., Ruan, S., Xing, Y., & Feng, M. (2024). A review of uncertainty quantification in medical image analysis: probabilistic and non-probabilistic methods. *Medical Image Analysis*. 13618415. *97*, 103223. doi:10.1016/j.media.2024.103223.

Irvin, J., Rajpurkar, P., Ko, M., Yu, Y., Ciurea-Ilcus, S., Chute, C., et al. (2019). CheXpert: a large chest radiograph dataset with uncertainty labels and expert comparison. In *33rd AAAI Conference on Artificial Intelligence, AAAI 2019, 31st Innovative Applications of Artificial Intelligence Conference, IAAI 2019 and the 9th AAAI Symposium on Educational Advances in Artificial Intelligence, EAAI 2019* (590–597). United States: AAAI Press. doi:10.1609/aaai.v33i01.3301590.

Kamińska, D., Zwolińskski, G., Laska-Leśniewicz, A., & Coelho, L. P. (2022). *Virtual reality in healthcare*. (pp. 1–10). IGI Global. doi:10.4018/978-1-7998-8371-5.ch001.

Kao, J. P., Chung, Y. C., Hung, H. Y., Chen, C. P., Chen, W. S., & RT-DETR, L.-E. (2025). First low-rank adaptation based DETR for real-time full body anatomical structures identification in musculoskeletal ultrasound. *Computerized Medical Imaging and Graphics*. 18790771. *124*. Retrieved from: https://www.sciencedirect.com/science/journal/08956111. doi:10.1016/j.compmedimag.2025.102583.

Kawadkar, K. (2025). Comparative analysis of vision transformers and convolutional neural networks for medical image classification. doi:10.48550/arXiv.2507.21156.

Kelly, C. J., Karthikesalingam, A., Suleyman, M., Corrado, G., & King, D. (2019). Key challenges for delivering clinical impact with artificial intelligence. *BMC Medicine*. 17417015. *17*(1). Retrieved from: http://www.biomedcentral.com/bmcmed/. doi:10.1186/s12916-019-1426-2.

Krizhevsky, A., Sutskever, I., & Hinton, G. E. (2012). ImageNet classification with deep convolutional neural networks . In *Communications of the ACM 60*, Issue 6 (pp. 84–90). doi:10.1145/3065386.

Kuang, H., Wang, Y., Liu, J., Wang, J., Cao, Q., Hu, B., et al. (2024). Hybrid CNN-transformer network with circular feature interaction for acute ischemic stroke lesion segmentation on non-contrast CT scans. *IEEE Transactions on Medical Imaging*. 0278-0062. *43*(6), 2303–2316. doi:10.1109/tmi.2024.3362879.

Lacoste, A., Schmidt, V., Luccioni, A., & Dandres, T. (2019). Quantifying the carbon emissions of machine learning. *arXiv*. 23318422. Retrieved from: https://arxiv.org. doi:10.48550/arxiv.1910.09700.

Lecun, Y., Bottou, L., Bengio, Y., & Haffner, P. (1998). Gradient-based learning applied to document recognition. *Proceedings of the IEEE*. 00189219. *86*(11), 2278–2324. doi:10.1109/5.726791.

Lin, T.-Y., Goyal, P., Girshick, R., He, K., & Dollár, P. (2018) . Focal loss for dense object detection. doi:10.48550/arXiv.1708.02002.

Litjens, G., Kooi, T., Bejnordi, B. E., Setio, A. A. A., Ciompi, F., Ghafoorian, M., et al. (2017). A survey on deep learning in medical image analysis. *Medical Image Analysis*.13618423. *42*, 60–88. Retrieved from: http://www.elsevier.com/inca/publications/store/6/2/0/9/8/3/index.htt. doi:10.1016/j.media.2017.07.005.

Liu, Z., Lin, Y., Cao, Y., Hu, H., Wei, Y., Zhang, Z., et al. (2021). Swin transformer: hierarchical vision transformer using shifted windows. In *Proceedings of the IEEE International Conference on Computer Vision* (9992–10002). China:Institute of Electrical and Electronics Engineers Inc. http://www.ieeexplore.ieee.org/xpl/conhome.jsp?punumber=1000149. doi:10.1109/ICCV48922.2021.00986.

Lundberg, S. M., & Lee, S. I. (2017). A unified approach to interpreting model predictions. doi:10.48550/arXiv.1705.07874.

Ma, D. A., Hosseinzadeh, T. M. R., Pang, J., Islam, N. U., Haghighi, F., Gotway, M. B., & Liang, J. (2022). Benchmarking and boosting transformers for medical image classification. In K. Kamnitsas, L. Koch, M. Islam, X. Ziyue, J. Cardoso, Q. Dou, N. Rieke, & S. Tsaftaris (Eds.), *Lecture notes in computer science (including subseries Lecture notes in artificial intelligence and lecture notes in bioinformatics)* (Vol. *13542*, pp. 12–22). United States: Springer Science and Business Media Deutschland GmbH. https://www.springer.com/series/558. doi:10.1007/978-3-031-16852-9_2.

Maurício, J., Domingues, I., & Bernardino, J. (2023). Comparing vision transformers and convolutional neural networks for image classification: a literature review. *Applied Sciences*. 2076-3417. *13*(9), 5521. doi:10.3390/app13095521.

McKinney, S. M., Sieniek, M., Godbole, V., Godwin, J., Antropova, N., Ashrafian, H., et al. (2020). International evaluation of an AI system for breast cancer screening. *Nature*. 14764687. *577*(7788), 89–94. Retrieved from: http://www.nature.com/nature/index.html. doi:10.1038/s41586-019-1799-6.

Modarressi, A., Fayyaz, M., Yaghoobzadeh, Y., & Pilehvar, M. T. (2022). GlobEnc: quantifying global token attribution by incorporating the whole encoder layer in transformers. In *NAACL 2022–2022 Conference of the North American Chapter of the Association for Computational Linguistics: Human Language Technologies, Proceedings of the Conference* (258–271). Iran:Association for Computational Linguistics (ACL). https://www.aclanthology.org/events/naacl-2022/#2022-naacl-main. doi:10.18653/v1/2022.naacl-main.19.

O'Shea, K., & Nash, R. (2015) . An introduction to convolutional neural networks. doi:10.48550/arXiv.1511.08458.

Ouyang, D., He, B., Ghorbani, A., Yuan, N., Ebinger, J., Langlotz, C. P., et al. (2020). Video-based AI for beat-to-beat assessment of cardiac function. *Nature*. 14764687. *580*(7802), 252–256. Retrieved from: http://www.nature.com/nature/index.html. doi:10.1038/s41586-020-2145-8.

Pinto-Coelho, L. (2023). How artificial intelligence is shaping medical imaging technology: a survey of innovations and applications. *Bioengineering*. 2306-5354. *10*(12), 1435. doi:10.3390/bioengineering10121435.

Radford, A., Kim, J. W., Hallacy, C., Ramesh, A., Goh, G., Agarwal, S., et al. (2021). Learning transferable visual models from natural language supervision. In *Proceedings of Machine Learning Research, Vol. 139* (8748–8763). United States: ML Research Press. https://www.proceedings.mlr.press/.

Ragab, M. G., Abdulkadir, S. J., Muneer, A., Alqushaibi, A., Sumiea, E. H., Qureshi, R., et al. (2024). A comprehensive systematic review of YOLO for medical object detection (2018 to 2023). *IEEE Access*.21693536. *12*, 57815–57836. Retrieved from: http://www.ieeexplore.ieee.org/xpl/RecentIssue.jsp?punumber=6287639. doi:10.1109/ACCESS.2024.3386826.

Rajpurkar, P., Irvin, J., Zhu, K., Yang, B., Mehta, H., Duan, T., et al. (2017). CheXNet: radiologist-level pneumonia detection on chest X-rays with deep learning. *arXiv*. 23318422. Retrieved from: https://www.arxiv.org. doi:10.48550/arxiv.1711.05225.

Ravi, N., Chaturvedi, P., Huerta, E. A., Liu, Z., Chard, R., Scourtas, A., et al. (2022). FAIR principles for AI models with a practical application for accelerated high energy diffraction microscopy. *Scientific Data*. 20524463. *9*(1). Retrieved from: https://www.nature.com/sdata/. doi:10.1038/s41597-022-01712-9.

Redmon, J., Divvala, S., Girshick, R., & Farhadi, A. (2016). You only look once: unified, real-time object detection 2016. In *Proceedings of the IEEE Computer Society Conference on Computer Vision and Pattern Recognition* (779–788). United States: IEEE Computer Society. doi:10.1109/CVPR.2016.91.

Reis, S. S., Pinto-Coelho, L., Sousa, M. C., Neto, M., Silva, M., & Sequeira, M. (2025). Evaluating skin tone fairness in convolutional neural networks for the classification of diabetic foot ulcers. *Applied Sciences*. 2076-3417. *15*(15), 8321. doi:10.3390/app15158321.

Ren, S., He, K., Girshick, R., Sun, J., & R-CNN, F. (2017). Towards real-time object detection with region proposal networks. *IEEE Transactions on Pattern Analysis and Machine Intelligence*. 01628828. *39*(6), 1137–1149. doi:10.1109/TPAMI.2016.2577031.

Ribeiro, F. M., Gonçalves, J., Coelho, L., Castelo-Branco, M., & Martins, J. (2025). Sex-dependent variations of retinal function and architecture in a neurofibromatosis type I mouse model with normal vision. *Experimental Eye Research*. 00144835. *253*, 110279. doi:10.1016/j.exer.2025.110279.

Ribeiro, M. T., & Singh, G. (2016). Why Should I Trust You?: explaining the Predictions of Any Classifier. Retrieved from: https://wwww.arxiv.org/abs/1602.04938v3.

Ribli, D., Horváth, A., Unger, Z., Pollner, P., & Csabai, I. (2018). Detecting and classifying lesions in mammograms with deep learning. *Scientific Reports*. 2045-2322. *8*(1). doi:10.1038/s41598-018-22437-z.

Rozemberczki, B., Watson, L., Bayer, P., Yang, H. T., Kiss, O., Nilsson, S., & Sarkar, R. (2022). The Shapley value in machine learning. *arXiv*. 23318422. Retrieved from: https://www.arxiv.org. doi:10.48550/arXiv.2202.05594.

Saeidnia, H. R., Firuzpour, F., Kozak, M., & majd, H. S. (2025). Advancing cancer diagnosis and treatment: integrating image analysis and AI algorithms for enhanced clinical practice. *Artificial Intelligence Review*. 15737462. *58*(4). Retrieved from: https://www.springer.com/journal/10462. doi:10.1007/s10462-025-11117-w.

Selvan, R., Bhagwat, N., Wolff, A. L. F., Kanding, B., & Dam, E. B. (2022). Carbon footprint of selecting and training deep learning models for medical image analysis. In *Lecture notes in computer science (including subseries Lecture notes in artificial intelligence and lecture notes in bioinformatics)* (pp. 506–516). Denmark: Springer Science and Business Media Deutschland GmbH. https://www.springer.com/series/558. doi:10.1007/978-3-031-16443-9_49.

Selvaraju, R. R., Cogswell, M., Das, A., Vedantam, R., Parikh, D., & Batra, D. (2020). Grad-CAM: visual explanations from deep networks via gradient-based localization. *Journal of Computer Vision*. 15731405. *128*(2), 336–359. Retrieved from: http://www.kluweronline.com/issn/0920-5691/. doi:10.1007/s11263-019-01228-7.

Setio, A. A. A., Ciompi, F., Litjens, G., Gerke, P., Jacobs, C., Van Riel, S. J., et al. (2016). Pulmonary nodule detection in CT images: false positive reduction using multi-view convolutional networks. *IEEE Transactions on Medical Imaging*. 1558254X. *35*(5), 1160–1169. Retrieved from: http://www.ieeexplore.ieee.org/xpl/RecentIssue.jsp?punumber=42. doi:10.1109/TMI.2016.2536809.

Sheller, M. J., Edwards, B., Reina, G. A., Martin, J., Pati, S., Kotrotsou, A., et al. (2020). Federated learning in medicine: facilitating multi-institutional collaborations without sharing patient data. *Scientific Reports*. 20452322. *10*(1). Retrieved from: https://www.nature.com/srep/index.html. doi:10.1038/s41598-020-69250-1.

Shobaki, W. A., & Milanova, M. (2025). A comparative study of YOLO, SSD, Faster R-CNN, and more for optimized eye-gaze writing. *Sci*. 2413-4155. *7*(2), 47. doi:10.3390/sci7020047.

Sindhu, R. S., George, J., Skaria, S., & Varun, V. V. (2018). Using YOLO based deep learning network for real time detection and localization of lung nodules from low dose CT scans. In N. Petrick, & K. Mori (Eds.), *Proceedings of SPIE—Progress in biomedical optics and imaging* (*Vol. 10575*). India: SPIE. http://www.spie.org/x1848.xml. doi:10.1117/12.2293699.

Strubell, E., Ganesh, A., & McCallum, A. (2019). Energy and policy considerations for deep learning in NLP. *arXiv*. 23318422. Retrieved from: https://www.arxiv.org. doi:10.48550/arxiv.1906.02243.

Sundararajan, M., Taly, A., & Yan, Q. (2017). Axiomatic attribution for deep networks. In *34th International Conference on Machine Learning, ICML 2017, Vol. 7* (pp. 5109–5118). United States: International Machine Learning Society (IMLS).

Taha, A. A., & Hanbury, A. (2015). Metrics for evaluating 3D medical image segmentation: analysis, selection, and tool. *BMC Medical Imaging*. 14712342. *15*(1). Retrieved from: http://www.biomedcentral.com/bmcmedimaging/. doi:10.1186/s12880-015-0068-x.

Takahashi, S., Sakaguchi, Y., Kouno, N., Takasawa, K., Ishizu, K., Akagi, Y., et al. (2024). Comparison of vision transformers and convolutional neural networks in medical image analysis: a systematic review. *Journal of Medical Systems*. 1573-689X. *48*(1). doi:10.1007/s10916-024-02105-8.

Tang, Y., Yang, D., Li, W., Roth, H. R., Landman, B., Xu, D., et al. (2022). Self-supervised pre-training of swin transformers for 3D medical image analysis. In *Proceedings of the IEEE Computer Society Conference on Computer Vision and Pattern Recognition* (20698–20708). United States: IEEE Computer Society. doi:10.1109/CVPR52688.2022.02007.

Tao, Q., Yan, W., Wang, Y., Paiman, E. H. M., Shamonin, D. P., Garg, P., et al. (2019). Deep learning-based method for fully automatic quantification of left ventricle function from cine MR images: a multivendor, multicenter study. *Radiology*. 15271315. *290*(1), 81–88. Retrieved from: https://www.pubs.rsna.org/doi/pdf/10.1148/radiol.2018180513. doi:10.1148/radiol.2018180513.

Ting, D. S. W., Cheung, C. Y. L., Lim, G., Tan, G. S. W., Quang, N. D., Gan, A., et al. (2017). Development and validation of a deep learning system for diabetic retinopathy and related eye diseases using retinal images from multiethnic populations with diabetes. *JAMA - Journal of the American Medical Association*. 15383598. *318*(22), 2211–2223. Retrieved from: https://www.jamanetwork.com/journals/jama/articlepdf/2665775/jama_ting_2017_oi_170140.pdf. doi:10.1001/jama.2017.18152.

Topol, E. J. (2019). High-performance medicine: the convergence of human and artificial intelligence. *Nature Medicine*. 1546170X. *25*(1), 44–56. Retrieved from: http://www.nature.com/nm/index.html. doi:10.1038/s41591-018-0300-7.

Tu, T., Azizi, S., Driess, D., Schaekermann, M., Amin, M., Chang, P.-C., et al. (2024). Towards generalist biomedical AI. *NEJM AI*. 2836-9386. *1*(3). doi:10.1056/aioa2300138.

Urban, G., Tripathi, P., Alkayali, T., Mittal, M., Jalali, F., Karnes, W., & Baldi, P. (2018). Deep learning localizes and identifies polyps in real time with 96% accuracy in screening colonoscopy. *Gastroenterology*. 00165085. *155* (4), 1069. doi:10.1053/j.gastro.2018.06.037.

Vaswani, A., Shazeer, N., Parmar, N., Uszkoreit, J., Jones, L., Gomez, A. N., et al. (2017). Attention is all you need. doi:10.48550/arXiv.1706.03762.

Voita, E., Talbot, D., Moiseev, F., Sennrich, R., & Titov, I. (2020). Analyzing multi-head self-attention: specialized heads do the heavy lifting, the rest can be pruned. In *Proceedings of the Conference—ACL 2019 - 57th Annual Meeting of the Association for Computational Linguistics* (pp. 5797–5808). Russian Federation: Association for Computational Linguistics (ACL).

Wang, H., Cruz-Roa, A., Basavanhally, A., Gilmore, H., Shih, N., Feldman, M., et al. (2014). Mitosis detection in breast cancer pathology images by combining handcrafted and convolutional neural network features. *Journal of Medical Imaging*. 2329-4302. *1*(3), 034003. doi:10.1117/1.JMI.1.3.034003.

Wang, L., Zhang, C., & Jin, L. (2024). A hybrid CNN-transformer model for predicting N staging and survival in non-small cell lung cancer patients based on CT-scan. *Tomography*. 2379-139X. *10*(10), 1676–1693. doi:10.3390/tomography10100123.

Wang, S., Xie, J., Cui, Y., & Chen, Z. (2024). Colorectal polyp detection model by using super-resolution reconstruction and YOLO. *Electronics*. 2079-9292. *13*(12), 2298. doi:10.3390/electronics13122298.

Wang, X., Zhang, R., Kong, T., Li, L., & Shen, C. (2020). SOLOv2: Dynamic and fast instance segmentation. doi:10.48550/arXiv.2003.10152.

Wang, Y., Wang, Q., Shi, S., He, X., Tang, Z., Zhao, K., & Chu, X. (2020). Benchmarking the performance and energy efficiency of AI accelerators for AI training. In *Proceedings - 20th IEEE/ACM International Symposium on Cluster, Cloud and Internet Computing, CCGRID 2020* (744–751). Hong Kong:Institute of Electrical and Electronics Engineers Inc. http://www.ieeexplore.ieee.org/xpl/mostRecentIssue.jsp?punumber=9136834. doi:10.1109/CCGrid49817.2020.00-15.

Wolf, D., Payer, T., Lisson, C. S., Lisson, C. G., Beer, M., Götz, M., & Ropinski, T. (2023). Self-supervised pre-training with contrastive and masked autoencoder methods for dealing with small datasets in deep learning for medical imaging. *Scientific Reports*. 20452322. *13*(1). Retrieved from: https://www.nature.com/srep/. doi:10.1038/s41598-023-46433-0.

Wolterink, J. M., Dinkla, A. M., Savenije, M. H. F., Seevinck, P. R., van den Berg, C. A. T., & Išgum, I. (2017). Deep MR to CT synthesis using unpaired data. In S. Tsaftaris, A. Gooya, A. Frangi, & J. L. Prince (Eds.), *Lecture notes in computer science (including subseries Lecture notes in artificial intelligence and lecture notes in bioinformatics)* (*Vol. 10557*, pp. 14–23). Netherlands: Springer Verlag. http://www.springerlink.com/content/0302-9743/copyright/2005/. doi:10.1007/978-3-319-68127-6_2.

Wu, C., Zhang, X., Zhang, Y., Wang, Y., & Xie, W. (2023). MedKLIP: medical knowledge enhanced language-image pre-training. *medRxiv* Retrieved from: https://www.medrxiv.org/. doi:10.1101/2023.01.10.23284412.

Wu, Y., Shouliang, Q., Sun, Y., Xia, S., Yao, Y., & Qian, W. (2021). A vision transformer for emphysema classification using CT images. *Physics in Medicine & Biology*. 0031-9155. *66*(24), 245016. doi:10.1088/1361-6560/ac3dc8.

Xu, H., Usuyama, N., Bagga, J., Zhang, S., Rao, R., Naumann, T., et al. (2024). A whole-slide foundation model for digital pathology from real-world data. *Nature*. 14764687. *630*(8015), 181–188. Retrieved from: https://www.nature.com/nature/. doi:10.1038/s41586-024-07441-w.

Yandex, A. B., & Lempitsky, V. (2015). Aggregating local deep features for image retrieval. In *Proceedings of the IEEE International Conference on Computer Vision* (1269–1277). Russian Federation:Institute of Electrical and Electronics Engineers Inc. http://www.ieeexplore.ieee.org/xpl/conhome.jsp?punumber=1000149. doi:10.1109/ICCV.2015.150.

Zhang, S., Xu, Y., Usuyama, N., Xu, H., Bagga, J., Tinn, R. et al. (2025). BiomedCLIP: A multimodal biomedical foundation model pretrained from fifteen million scientific image-text pairs. doi:10.48550/arXiv.2303.00915.

Zhang, T., Luo, B., & Wang, G. (2025). Improving vision transformers by overlapping heads in multi-head self-attention. *SSRN, 15565068*. Retrieved from: https://www.ssrn.com/index.cfm/en/. doi:10.2139/ssrn.5257431.

Zhou, L., Liu, H., Bae, J., He, J., Samaras, D., & Prasanna, P. (2023). Self pre-training with masked autoencoders for medical image classification and segmentation. *IEEE 20th International Symposium on Biomedical Imaging (ISBI)*. 1–6.

Section III

Use of artificial intelligence in organ transplantation and precision surgery

Chapter 8

Predictive artificial intelligence to increase the success of heart transplantation

David B. Olawade[a,b,c], Aderonke Odetayo[d], Kusal Weerasinghe[b], Sheila Marinze[e], Eghosasere Egbon[f], and Jennifer Teke[b,g]

[a]*Department of Allied and Public Health, School of Health, Sport and Bioscience, University of East London, London, United Kingdom*

[b]*Department of Research and Innovation, Medway NHS Foundation Trust, Gillingham, United Kingdom*

[c]*Department of Business, Health, and Management, York St John University, London, United Kingdom*

[d]*School of Nursing, Tung Wah College, Hong Kong SAR, China*

[e]*Department of Surgery, Medway NHS Foundation Trust, Gillingham, United Kingdom*

[f]*Department of Tissue Engineering and Regenerative Medicine, Faculty of Life Science Engineering, FH Technikum, Vienna, Austria*

[g]*Faculty of Medicine, Health and Social Care, Canterbury Christ Church University, United Kingdom*

1 Introduction

Heart transplantation remains one of medicine's most remarkable achievements, a last lifeline for patients facing end-stage heart failure. Yet, for all our surgical prowess and medical advances, we are still playing a complex game of biological matchmaking where the stakes could not be higher. Despite significant strides in surgical techniques, immunosuppressive therapies, and postoperative care, the success of heart transplants continues to dance precariously on numerous variables: donor–recipient compatibility, organ preservation quality, and the ever-present threat of posttransplant complications (Ishaq & Guha, 2025; Martin, 2025).

The statistics paint a sobering picture of the current landscape. Over 4500 heart transplants were performed in the United States in 2023; however, demand far exceeds the supply (Arabyarmohammadi & Madabhushi, 2024). In contrast to the United States, European countries performed a total of 2200 heart transplants in 2023, with France, Italy, Germany, Spain, and the United Kingdom among the top five countries, having 393, 370, 330, 325, and 238 heart transplants, respectively (Global Observatory on Donation and Transplant, 2021). Acute rejection occurs in up to 32% of recipients within the first year, while the number of candidates on the waiting list for heart transplantation continues to outweigh the number of available organs (Arabyarmohammadi & Madabhushi, 2024). The donor heart nonuse rate in the United States remains significantly higher, at approximately 60%–65%, than that of other regions, such as Europe, which is around 20%–30% (Jain, Kransdorf, Cowger, Jeevanandam, & Kobashigawa, 2025), highlighting inefficiencies in current allocation and assessment systems.

Enter predictive artificial intelligence (AI), the game-changer that is quietly revolutionizing how we approach heart transplantation. By harnessing the power of machine learning (ML), deep learning (DL), and big data analytics, AI is transforming patient selection, enhancing donor organ matching, and providing unprecedented insight into posttransplant monitoring (Arjmandmazidi et al., 2025; Olawade et al., 2025). The result improved survival rates and quality of life for transplant recipients.

This transformation is grounded in compelling evidence. Recent systematic reviews demonstrate that AI and ML are more accurate in predicting graft failure and mortality than traditional scoring systems and conventional regression analysis (Mohammadi et al., 2025; Naruka et al., 2022; Sargiotis, Sergentanis, Pavi, & Athanasakis, 2024). These reviews found that ML algorithms performed well in predicting postheart transplant mortality. For instance, the area under the curve (AUC) of all ML algorithms was 0.65, with CatBoost achieving the highest accuracy of 0.80 for mortality predictions (Mohammadi et al., 2025). For survival predictions, random survival forest showed higher sensitivity (88.7%), specificity (79.6%), and best accuracy (82.8%), while the CatBoost model showed the best positive predictive value at 42% (Sargiotis et al., 2024).

Applications of Artificial Intelligence in Healthcare. https://doi.org/10.1016/B978-0-443-43934-6.00001-8

The most limiting factor in heart transplantation is the lack of donor organs, and with enhanced prediction of outcomes, it may be possible to increase the life-years from the organs that become available (Lisboa et al., 2022; Naruka et al., 2022).

The integration of AI into cardiovascular medicine is part of a broader transformation that has the potential to revolutionize every facet of cardiovascular practice and research (Khera et al., 2024). From early detection algorithms that can identify heart failure risk from routine electrocardiograms to sophisticated prediction models that optimize immunosuppressive therapy, AI is reshaping the entire continuum of cardiac care (Idris-Agbabiaka et al., 2025; Udoy & Hassan, 2025). A study by Garcia-Canadilla et al. (2022) reported that ML identified echocardiographic features and clusters associated with a high risk of death or transplantation, with a low risk of bias and 70% quality of reporting.

The rationale for integrating predictive AI lies in the inherent limitations of traditional clinical decision-making approaches, which rely on relatively simple scoring systems and clinician experience that cannot simultaneously process the hundreds of variables influencing transplant outcomes, thereby creating opportunities for AI systems to excel at multidimensional analysis and pattern recognition invisible to the human eye. The primary aim of this chapter is to comprehensively explore how predictive AI is revolutionizing every aspect of heart transplantation: from donor–recipient matching and organ viability assessment to posttransplant monitoring and personalized immunosuppression, while demonstrating how these technologies are delivering dramatically improved survival rates, reduced complications, and enhanced quality of life for transplant recipients. The specific objectives encompass examining AI's transformative role in optimizing donor–recipient compatibility through advanced histocompatibility enhancement and biochemical modeling; evaluating AI-powered organ viability prediction using real-time imaging intelligence and ischemia time optimization; analyzing sophisticated early warning systems for posttransplant complication detection, including acute rejection, infection risk, and cardiac allograft vasculopathy (CAV); assessing AI-driven personalized medicine approaches for precision immunosuppression and medication adherence monitoring; exploring comprehensive remote monitoring technologies and virtual health assistants; reviewing various AI predictive models and their clinical performance; addressing critical challenges, including data privacy, algorithmic bias, and regulatory requirements; and discussing future prospects, including real-time decision support systems, blockchain integration, xenotransplantation applications, and the broader economic impact on healthcare sustainability.

Fig. 8.1 shows a schematic overview of the heart transplantation landscape, illustrating key challenges, including donor–recipient compatibility, organ preservation, posttransplant complications, and the transformative role of predictive AI. AI-driven approaches, such as advanced matching algorithms, organ viability prediction, and personalized posttransplant monitoring, are depicted as addressing major bottlenecks and improving patient outcomes. This figure highlights the transition from traditional limitations to an AI-enabled, data-driven era in heart transplantation.

2 The revolutionary role of AI in heart transplantation

Predictive AI systems are fundamentally changing how we approach heart transplantation by analyzing vast datasets, identifying patterns invisible to the human eye, and generating real-time insights that empower clinicians to make life-saving decisions with unprecedented precision (Naruka et al., 2022; Pinto-Coelho, 2023). These sophisticated algorithms can process and synthesize enormous amounts of clinical data to reveal complex relationships within large datasets and obtain logical results while optimizing resources (Peloso et al., 2022).

The transformation is particularly remarkable when considering the complexity of transplantation medicine. Traditional clinical decision-making relies on relatively simple scoring systems and clinician experience, which, whilst valuable, cannot process the hundreds of variables that simultaneously influence transplant outcomes. AI systems excel at this multidimensional analysis, creating what researchers describe as a paradigm shift in transplant medicine (Peloso et al., 2022).

Recent meta-analyses have demonstrated the superior performance of AI-driven models across multiple domains of transplantation. A systematic review of 13 studies featuring 463,850 patients identified three main areas where AI applications have shown particular promise: predictive modeling of heart transplantation mortality outcomes, graft failure prediction, and imaging-assisted diagnosis (Naruka et al., 2022). The results consistently show that AI and ML approaches outperform traditional statistical methods in predicting critical outcomes.

These sophisticated systems do not replace clinical expertise; they amplify it, providing doctors with tools that can process thousands of variables simultaneously and spot subtle patterns that might otherwise go unnoticed (Bajwa, Munir, Nori, & Williams, 2021; Johnson et al., 2021). The applications span every phase of the transplantation journey, creating a comprehensive ecosystem of AI-driven care that extends from pretransplant evaluation through long-term postoperative monitoring (Al Moussawy, Lakkis, Ansari, Cherukuri, & Abou-Daya, 2024; Arjmandmazidi et al., 2025).

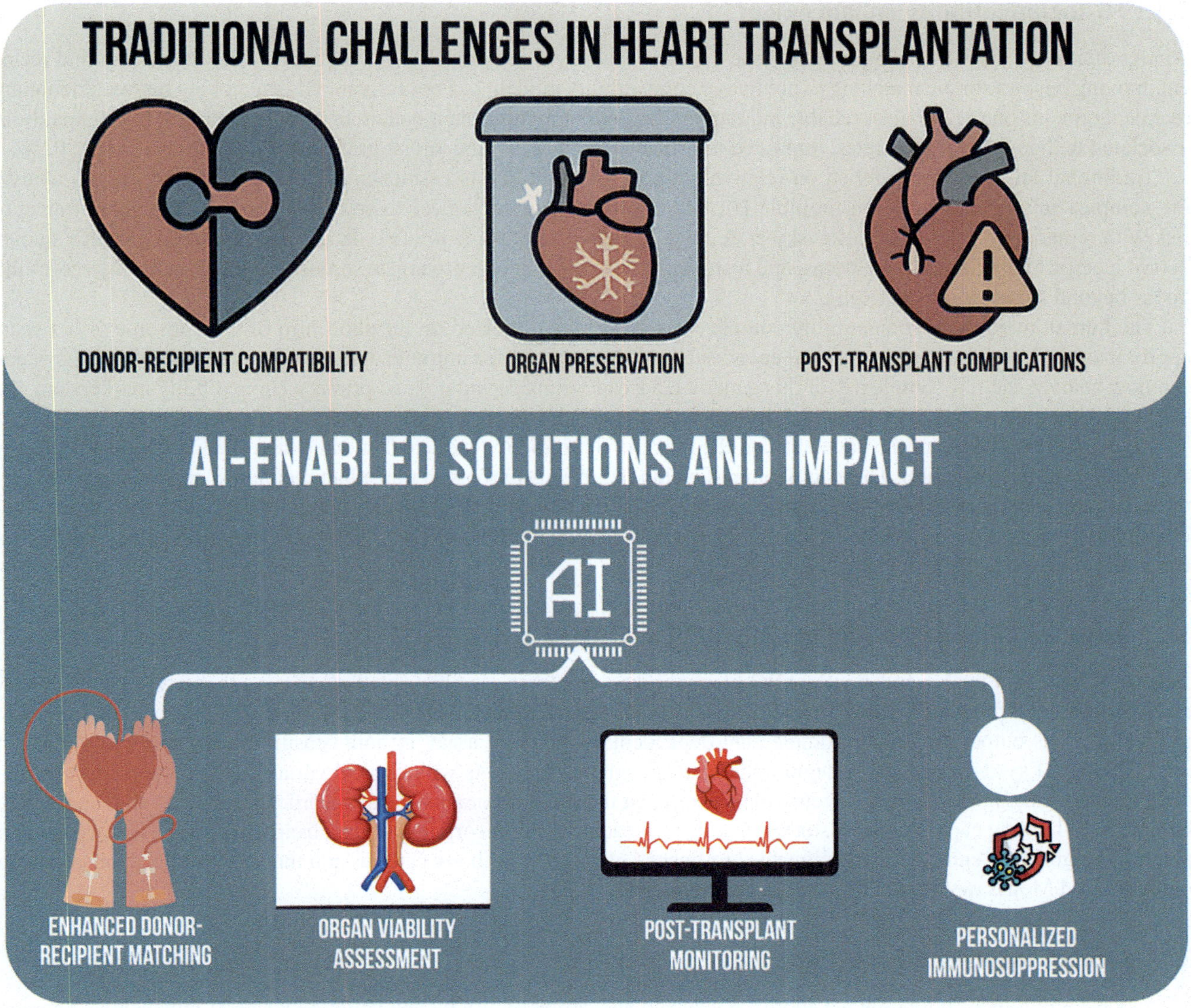

FIG. 8.1 An overview of the challenges and AI-driven solutions in heart transplantation.

ML models allow for a greater number of associated variables to be studied and then build models based on parameters that influence the outcome the most (Naruka et al., 2022). In cardiac transplantation, this capability is transforming how clinicians approach decision-making on the allocation of hearts for transplantation, increasing accuracy in predicting graft failure and mortality, and identifying those at the highest risk for rejection posttransplantation (Gotlieb et al., 2022; Naruka et al., 2022).

The technological foundation supporting these advances includes various AI methodologies, each offering unique advantages for different clinical scenarios. Neural networks excel at pattern recognition in complex datasets (Kufel et al., 2023); random forest algorithms provide interpretable results while handling missing data effectively, and support vector machines demonstrate robustness when dealing with high-dimensional medical data (Ghosh & Cabrera, 2022; Zhu, Zhang, Li, Su, & Tian, 2024). More advanced approaches, such as recurrent neural networks and long short-term memory models, are particularly effective for analyzing longitudinal patient data to predict long-term transplant outcomes.

3 Predicting donor–recipient compatibility: The perfect match

Perhaps nowhere is AI more transformative than in the delicate art of donor–recipient matching, where the technology is revolutionizing our understanding of compatibility beyond traditional markers. The complexity of achieving optimal donor–recipient pairing involves numerous factors that interact in nonlinear ways, making it an ideal application for ML approaches.

3.1 Histocompatibility enhancement

AI algorithms are revolutionizing how we assess human leukocyte antigen (HLA) compatibility between donor and recipient, moving beyond simple matching to predictive compatibility modeling. Pretransplant HLA matching between donor and recipient remains one of the most reliable biomarkers for predicting transplant outcomes, with enhanced HLA compatibility associated with lower rejection rates, improved transplantation success, and longer graft survival (Nikolova et al., 2024).

Traditional HLA matching relied on relatively crude compatibility assessments, but modern AI systems can analyze the complex interactions between multiple HLA loci, predict immunological tolerance patterns, and estimate rejection risk with remarkable precision (Zaslavsky et al., 2025). These systems incorporate data from population genetics studies, donor-specific antibody (DSA) patterns, and historical transplant outcomes to create sophisticated compatibility scores that go far beyond simple allele matching.

The human major histocompatibility complex is a gene region located on the short arm of chromosome 6 that spans approximately 3.6 million base pairs and encodes HLA genes crucial for immune defense (Casas et al., 2024). AI systems can now analyze the full complexity of this genetic landscape, considering not just primary compatibility markers but also secondary immune response patterns that influence long-term graft survival (Matsuzaka & Yashiro, 2024).

Recent developments in ML have enabled the creation of predictive models that can assess the likelihood of developing DSAs posttransplantation. These antibodies develop in 50% of solid organ transplant recipients, appearing in 3%–11% of heart transplant patients prior to treatment, with 10%–30% acquiring de novo DSA after transplantation (Martini et al., 2024). AI systems can now predict which recipient-donor pairs are most likely to develop these problematic antibodies, enabling proactive management strategies.

3.2 Advanced blood type and cross-matching analysis

Modern AI tools go far beyond simple ABO compatibility, employing sophisticated algorithms to analyze complex antibody patterns, complement-binding characteristics, and cross-match outcomes. The latest technique, the scintillation proximity assay, has outperformed the complement-dependent cytotoxicity assay in both sensitivity and specificity (Martini et al., 2024). AI systems can now integrate these advanced testing results with clinical variables to create comprehensive compatibility assessments. These systems analyze not just the presence or absence of specific antibodies but their strength, complement-binding capacity, and historical patterns of reactivity. By incorporating data from thousands of previous transplants, AI algorithms can predict the clinical significance of specific antibody patterns and their likely impact on transplant outcomes (Al Moussawy et al., 2024; Arjmandmazidi et al., 2025).

3.3 Optimized size and age matching

ML models are transforming how we approach size and age compatibility by analyzing donor heart dimensions, recipient body mass index, age-related cardiac changes, and functional capacity with unprecedented sophistication. These algorithms incorporate three-dimensional (3D) cardiac imaging data, hemodynamic parameters, and recipient exercise capacity to predict posttransplant cardiac function with remarkable precision.

The traditional approach to size matching relied on relatively simple height and weight ratios, but AI systems can now analyze complex anatomical relationships, considering factors such as chest cavity dimensions, pulmonary vascular resistance, and predicted cardiac output requirements (Attaripour Esfahani et al., 2025; Ledziński & Grześk, 2023). This detailed analysis enables optimal physiological compatibility that goes far beyond conventional matching criteria. Age matching has also been revolutionized by AI approaches that consider not only chronological age but also biological age markers, donor lifestyle factors, and cardiac functional parameters (Anjewierden et al., 2025). ML models can now predict how donor age will interact with recipient characteristics to influence long-term outcomes, enabling more nuanced decision-making in organ allocation (Al Moussawy et al., 2024).

Fig. 8.2 is a schematic representation of the evolution in donor–recipient compatibility assessment. Traditional methods relied on basic HLA and ABO matching (left), while modern AI systems model complex HLA interactions (middle) and integrate additional parameters, such as DSAs, anatomical dimensions, and cross-matching results, to generate comprehensive compatibility predictions (right).

4 Predicting organ viability: Seeing beyond the surface

The ability to accurately assess donor heart viability before transplantation has always been crucial; however, AI is taking this assessment to remarkable new heights by integrating multiple data streams and identifying subtle indicators of organ quality that traditional assessment methods might overlook.

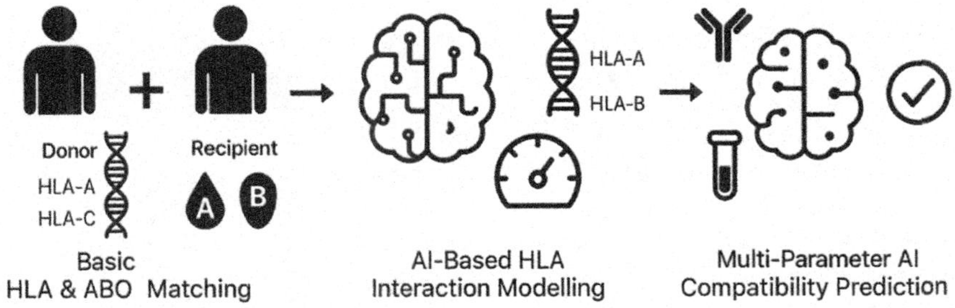

FIG. 8.2 Evolution in donor–recipient compatibility assessment.

4.1 Real-time imaging intelligence

AI-powered imaging analysis represents one of the most promising advances in organ viability assessment, utilizing sophisticated algorithms to extract quantitative information from echocardiography, magnetic resonance imaging (MRI), and computed tomography (CT) scans that far exceed human visual assessment capabilities (Tolu-Akinnawo, Ezekwueme, Omolayo, Batheja, & Awoyemi, 2025). DL has identified predictive cues in domains where clinically relevant features were previously not apparent, enabling a more accurate evaluation of donor heart quality (Khera et al., 2024).

These systems can detect subtle wall motion abnormalities, analyze myocardial texture patterns, and assess diastolic function parameters that might be missed during traditional visual interpretation (Fortuni et al., 2024). Advanced neural networks trained on thousands of cardiac images can identify features associated with good posttransplant function, even in hearts that might appear marginal by conventional assessment (Litjens et al., 2019).

Recent developments in 3D echocardiographic analysis have enabled AI systems to create detailed anatomical models of donor hearts, predicting how they will function in specific recipients based on hemodynamic compatibility modeling (Barry et al., 2023). These systems can simulate posttransplant cardiac function, predicting exercise capacity, functional class outcomes, and long-term durability. Computer vision algorithms are also being applied to intraoperative assessment, analyzing real-time images of donor hearts during procurement to identify signs of ischemic damage, assess coronary anatomy, and predict functional recovery posttransplantation (Naruka et al., 2022). This real-time analysis capability enables dynamic decision-making during the procurement process.

4.2 Predictive biochemical modeling

As detailed in Table 8.1, AI models analyze multiple biomarkers simultaneously, troponins, lactate levels, inflammatory markers, and other biochemical indicators, to create comprehensive viability scores that predict ischemic damage and overall organ health with remarkable accuracy (Lin et al., 2025; Zhao et al., 2024). These systems go beyond simple threshold-based decisions to consider biomarker patterns, trends over time, and interactions between multiple parameters.

Advanced ML models can now integrate biomarker data with clinical variables such as donor age, cause of death, medications administered, and hemodynamic status to create sophisticated organ quality scores (Chowdhury et al., 2025; Ng, Masarone, Watson, & Barnes, 2023; Olawade, Marinze, Qureshi, Weerasinghe, & Teke, 2025a).

These models have been trained on extensive datasets, linking pretransplant biomarker patterns with posttransplant outcomes, enabling precise predictions of graft function.

The integration of metabolomic and proteomic data is opening new frontiers in assessing organ viability. AI systems can now analyze hundreds of metabolites and proteins simultaneously, identifying molecular signatures associated with organ quality and posttransplant function (Al Moussawy et al., 2024; Arjmandmazidi et al., 2025). This molecular-level assessment offers insights into cellular metabolism, oxidative stress, and inflammatory status, complementing traditional biomarker analysis.

Novel biomarkers, such as donor-derived cell-free DNA (dd-cfDNA), are being incorporated into AI prediction models, providing real-time indicators of cellular damage and organ integrity (Wasilewski, Kamysz, & Gębicki, 2024). The hallmark of allograft rejection is immune-mediated cell necrosis, and during rejection, cell-free DNA fragments are released into the bloodstream from the donor allograft (Nikolova et al., 2024). AI systems can analyze dd-cfDNA patterns to predict organ viability and posttransplant risk.

4.3 ML for ischemia time optimization

ML algorithms are revolutionizing ischemia time management by analyzing historical data, donor characteristics, preservation conditions, and transport logistics to determine optimal timing strategies (Bajwa et al., 2021; Khalifa & Albadawy,

TABLE 8.1 Key biomarkers and AI detection parameters for posttransplant monitoring.

Biomarker category	Specific markers	AI detection method	Clinical significance	Prediction timeframe
Cardiac injury (Goel et al., 2020; Babuin & Jaffe, 2005)	Troponin I/T, CK-MB, H-FABP	Pattern recognition algorithms with temporal analysis	Acute rejection, ischemic damage, cell death	24–72 h
Inflammation (Lima et al., 2022; Ansar & Ghosh, 2016)	CRP, IL-6, TNF-α, IL-1β, VCAM-1	Multimarker analysis with cytokine networks	Immune activation, infection risk, chronic rejection	1–7 days
Graft function (Martini et al., 2024; Wettersten, 2021; Cao, Jia, & Zhu, 2019)	NT-proBNP, BNP, ANP, ST2	Trend analysis algorithms with functional correlation	Heart failure, graft dysfunction, volume status	3-14 days
Immune status (Zhang et al., 2018; Chih et al., 2012; Wu et al., 2009)	DSA, complement levels, T-cell markers	Machine learning classification with HLA analysis	Antibody-mediated rejection, immune monitoring	2–8 weeks
Cellular rejection (Coutance, Desiré and Duong Van Huyen, 2022; Novák, Macháčková, Krejčí, Bienertová-Vašků, & Slabý, 2021)	Gene expression profiling, microRNA	Deep learning analysis of molecular signatures	T-cell mediated rejection, immune response patterns	1–4 weeks
Metabolic status (Gunasekara, Herath, De Silva, & Jayasundara, 2024)	Creatinine, eGFR, cystatin C, NGAL	Predictive modeling with drug level correlation	Kidney dysfunction from immunosuppression	1–12 weeks
Endothelial function (Moayedi et al., 2024)	Endothelin-1, VEGF, E-selectin	Network analysis of vascular markers	Vasculopathy risk, endothelial damage	2–12 weeks
Oxidative stress (Jomova et al., 2023)	Malondialdehyde, SOD, catalase	Multiparameter oxidative stress assessment	Chronic graft damage, accelerated aging	4–24 weeks

2024). These systems can predict the safe limits of ischemia time for specific donor–recipient combinations, potentially extending the viable window for transplantation while minimizing the risk of delayed graft function.

Traditional ischemia time management relied on relatively crude guidelines, but AI systems can now consider the complex interactions between donor quality, preservation method, recipient condition, and transport logistics to optimize timing decisions (Olawade, Marinze, Qureshi, Weerasinghe, & Teke, 2025b; Salybekov, Yerkos, Sedlmayr, & Wolfien, 2025). These algorithms incorporate data from thousands of previous transplants to identify patterns associated with successful outcomes despite extended ischemia times.

Advanced predictive models are being developed that can forecast organ function recovery based on ischemia time, preservation method, and donor characteristics (Arjmandmazidi et al., 2025; Olawade, Marinze, Qureshi, Weerasinghe, & Teke, 2025b). These models enable clinicians to make informed decisions about acceptable ischemia times for specific cases, potentially expanding the geographic range for organ procurement and utilization (Gotlieb et al., 2022; Olawade et al., 2025). The integration of real-time monitoring data from organ preservation systems with AI algorithms is enabling dynamic ischemia time management (Deshpande, 2024). These systems can assess organ condition continuously during transport and adjust timing recommendations based on real-time indicators of organ viability.

Fig. 8.3 shows an overview of three AI-enabled strategies for assessing donor heart viability. AI-powered imaging evaluates wall motion, myocardial texture, and structural integrity using real-time scans. Predictive biochemical modeling integrates circulating biomarkers such as troponin, IL-6, and dd-cfDNA through AI to assess graft viability. Ischemia time optimization leverages ML and logistics data to predict the safe window for organ transport and transplantation. Each panel flows from clinical input through AI analysis to decision support output.

5 Posttransplant complication prediction: Early warning systems that save lives

Posttransplant monitoring represents one of AI's most impressive achievements in heart transplantation, creating sophisticated early warning systems that can detect problems before they become critical and enabling proactive interventions that dramatically improve outcomes. As illustrated in Fig. 8.4, AI-driven models are playing a transformative role in

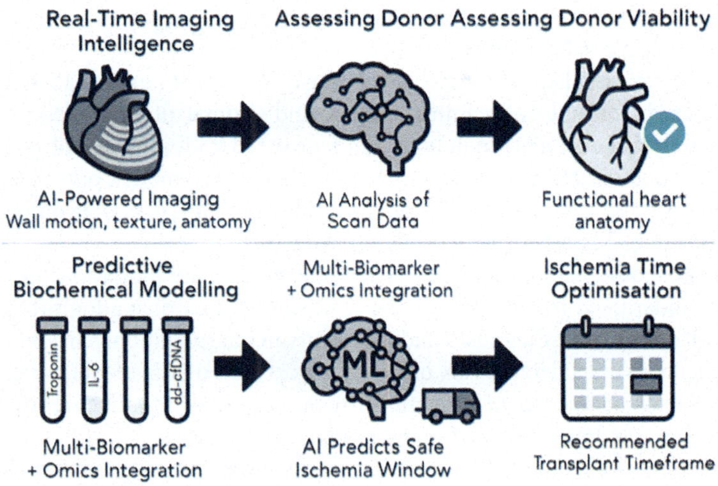

FIG. 8.3 Overview of three AI-enabled strategies for assessing donor heart viability.

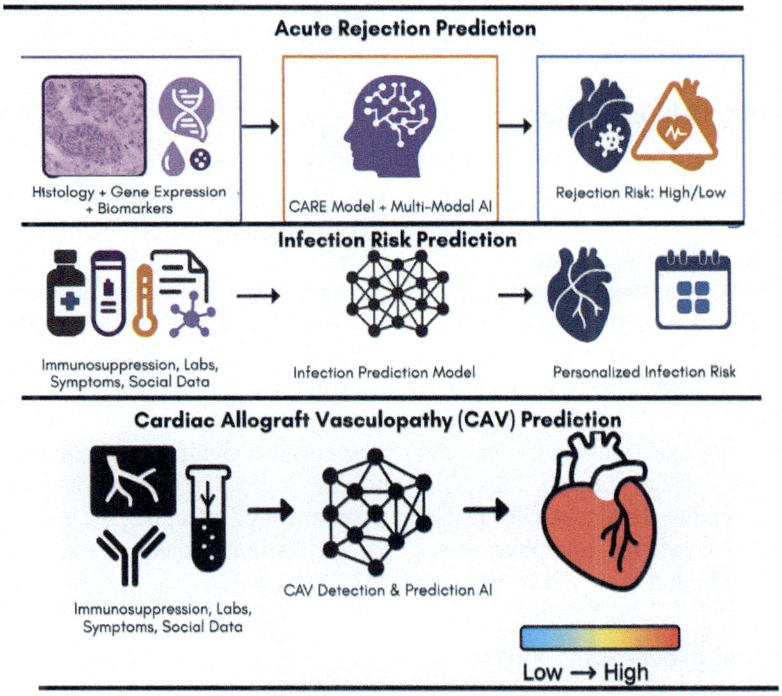

FIG. 8.4 AI-powered posttransplant complication prediction models. This schematic highlights three major AI-driven strategies enhancing posttransplant care: (1) acute rejection prediction integrates histology, gene expression, and biomarkers analyzed by the CARE model to classify high- and low-risk rejection profiles; (2) infection risk prediction utilizes laboratory values, immunosuppressive regimens, symptoms, and social data to generate personalized infection risk through machine learning models; (3) cardiac allograft vasculopathy (CAV) prediction applies deep learning to imaging, cytokine levels, and clinical data to quantify CAV risk along a color-coded spectrum from low to high. These models enable timely, tailored interventions that enhance patient outcomes.

posttransplant care, enabling early detection of acute rejection, infection risk, and CAV through multimodal data integration and predictive analytics.

5.1 Revolutionary acute rejection detection

The development of AI-powered rejection detection systems represents a paradigm shift from reactive to predictive medicine. A team of researchers from Emory University, led by Professor Anant Madabhushi, has developed AI tools to examine cardiac biopsy images to improve the prediction of transplant rejection, helping to ensure positive outcomes for posttransplant patients (Arabyarmohammadi & Madabhushi, 2024).

The Cardiac Allograft Rejection Evaluator (CARE) system uses AI tools to extract features associated with the shape, texture, and spatial architecture of muscle cells, immune cells, and stromal fibers in heart tissue specimens (Katzowitz, 2024). This comprehensive analysis generates predictions that far exceed traditional histologic grading methods. A study of 2900 patients published in Circulation: Heart Failure showed that the CARE model, optimized to predict cardiac rejection severity, was far better at assessing a patient's clinical outcome than alternative models optimized to predict ISHLT rejection grades (Arabyarmohammadi & Madabhushi, 2024; Katzowitz, 2024).

The superiority of AI approaches becomes apparent when considering the limitations of current methods. The ISHLT grading system provides a standardized framework for the assessment of cardiac allograft rejection. However, due to its qualitative nature, it may lead to subjective diagnoses and variability in interpretation, particularly in cases that are inconclusive, as this may subject patients to considerable risk of receiving excessive or inadequate treatment (Arabyarmohammadi & Madabhushi, 2024). AI systems address these limitations by providing quantitative, reproducible assessments that correlate more closely with clinical outcomes than traditional grading systems.

ML models have also been developed to analyze gene expression profiling of circulating peripheral blood mononuclear cells, which reflect host immune responses toward the allograft and serve as a biomarker for rejection surveillance (Deng, 2017; Nikolova et al., 2024). These systems can detect molecular signatures of rejection before histological changes become apparent, enabling earlier interventions. The integration of multiple data sources—histology, gene expression, biomarkers, and clinical variables—into comprehensive AI models is creating unprecedented accuracy in rejection prediction. These multimodal approaches can identify rejection risk with sensitivities and specificities that far exceed any single diagnostic modality.

5.2 Advanced infection risk prediction

ML models are revolutionizing infection risk assessment by analyzing complex interactions between immunosuppressive therapy response, patient-specific risk factors, environmental exposures, and pathogen resistance patterns (Hanna & Medford, 2024). These predictive systems enable proactive interventions before infections become life-threatening, representing a crucial advance in posttransplant care.

AI systems can now analyze patterns in laboratory results, medication adherence, and clinical symptoms to predict infection risk with remarkable accuracy (Alowais et al., 2023). These models incorporate data on local resistance patterns, seasonal infection trends, and individual immune status to create personalized infection risk profiles.

Advanced models are being developed that can predict specific types of infections based on immunosuppressive regimens, environmental factors, and individual susceptibility patterns (Alowais et al., 2023; Zhao et al., 2024). These systems enable targeted prophylaxis strategies and early intervention protocols that significantly reduce infection-related morbidity and mortality (Baddal, Taner, & Uzun Ozsahin, 2024; Godbole et al., 2025). The integration of continuous monitoring data from wearable devices and remote sensors enables real-time assessment of infection risk. AI algorithms can detect subtle changes in vital signs, activity patterns, and physiological parameters that precede clinical signs of infection, enabling extremely early intervention (Junaid et al., 2022; Tsvetanov, 2024).

5.3 CAV: Detection and prevention

DL algorithms are transforming the detection and management of CAV (Moayedi et al., 2024), one of the leading causes of long-term transplant failure. For example, a study reported that DL AI detected allograft rejection with an AUC of 0.962 from endomyocardial biopsy (Lipkova et al., 2022). These systems analyze angiographic and echocardiographic data to detect early signs of CAV with sensitivity that far exceeds traditional visual assessment methods (Ghorbani et al., 2020; Palmieri et al., 2023).

Research has identified positive correlations between proinflammatory cytokines and coronary vasculopathy, with these findings confirmed for both left and right ventricular involvement (Martini et al., 2024). AI systems can now integrate cytokine profiles, imaging data, and clinical variables to create comprehensive CAV risk assessments, enabling earlier detection and intervention (Przybylek et al., 2019). Furthermore, research indicates that AI-enhanced models outperform conventional scoring methods, yielding higher accuracy, with AUC increasing from 0.81 to 0.92 in predicting adverse cardiovascular events (Zoghi, 2025).

Advanced ML models can analyze subtle changes in coronary flow patterns, myocardial perfusion, and cardiac function that precede angiographically visible CAV (Olawade, et al., 2024; Tolu-Akinnawo et al., 2025). This shows that these systems enable detection of vasculopathy months or years before conventional methods, creating opportunities for preventive interventions (Moayedi et al., 2024).

The development of AI-powered risk stratification tools is enabling personalized CAV prevention strategies based on individual risk profiles, immunosuppressive regimens, and genetic factors (Mohsin et al., 2023; Singh et al., 2024). These systems can predict which patients are most likely to develop CAV and recommend targeted preventive interventions. As detailed in Table 8.1, AI systems continuously monitor multiple biomarkers and clinical indicators, creating comprehensive risk profiles that enable personalized posttransplant care strategies tailored to individual patient characteristics and risk factors.

6 AI-driven personalized medicine for immunosuppression

Optimizing immunosuppressive therapy has always been a delicate balancing act: preventing rejection while minimizing toxic side effects. AI is transforming this challenge into a precision medicine opportunity by enabling truly personalized treatment strategies based on individual patient characteristics, genetic profiles, and real-time response monitoring.

6.1 Precision drug dosing through AI

AI models are revolutionizing immunosuppressive drug management by analyzing genetic markers, metabolic profiles, and pharmacokinetic data to create individualized treatment regimens with unprecedented precision (Liu, Feng, Chen, Yan, & Liu, 2024; Serrano et al., 2024; Srivastav, Mishra, Lillard, & Singh, 2025).

These sophisticated systems incorporate pharmacogenomic data, considering genetic variations in drug-metabolizing enzymes such as CYP3A4, CYP3A5, and P-glycoprotein, which significantly influence the metabolism of tacrolimus and cyclosporine (Brunet & Pastor-Anglada, 2022; Zhang et al., 2024). AI algorithms can predict optimal starting doses, adjustment patterns, and maintenance regimens based on individual genetic profiles and clinical characteristics.

Advanced ML models are being developed that can predict drug interactions, side effect profiles, and optimal combination therapy strategies (Ali, Alam, Tyra, & Chong, 2025; Luo et al., 2024). A study on the convolutional and BiLSTM networks found that the model achieved 95.38% accuracy and 98.78% AUC (Ali et al., 2025). These systems analyze data from thousands of patients to identify patterns associated with successful immunosuppression while minimizing adverse effects such as nephrotoxicity, infection risk, and malignancy development (Arjmandmazidi et al., 2025).

Real-time therapeutic drug monitoring integrated with AI algorithms is enabling dynamic dose optimization (Poweleit, Vinks, & Mizuno, 2023). These systems can predict drug level fluctuations, recommend dose adjustments based on changes in clinical status, and identify patients at risk of toxicity or underimmunosuppression before clinical problems develop (Poweleit et al., 2023; Syrowatka et al., 2022).

6.2 ML for medication adherence

AI-based mobile applications and wearable devices are transforming medication adherence monitoring by providing real-time assessments of patient compliance and delivering personalized interventions when adherence issues are detected (Kukade, Kawade, Lokare, & Kharage, 2025; Tsvetanov, 2024). These systems represent a crucial advance, as medication nonadherence is a leading cause of rejection and graft loss in transplant recipients.

Advanced ML algorithms can analyze patterns in medication-taking behavior, identifying subtle indicators of declining adherence before it impacts clinical outcomes (Alowais et al., 2023). These systems incorporate data from smart pill bottles, mobile apps, pharmacy refill patterns, and wearable device activity to create comprehensive adherence profiles (Bohlmann, Mostafa, & Kumar, 2021; Kukade et al., 2025).

Predictive models are being developed that can identify patients at high risk of medication nonadherence based on demographic factors, psychological profiles, social support systems, and historical behavior patterns (Dong, Chu, & Xu, 2025; Oates, Juarez, Hansen, Kiefe, & Shikany, 2020). These systems enable targeted interventions for high-risk patients, including enhanced education, support services, and monitoring protocols. AI-powered intervention systems can deliver personalized reminders, educational content, and motivational messages based on individual patient characteristics and preferences (Olawade et al., 2024; Robinson et al., 2024). These systems learn from patient responses to optimize intervention strategies and improve long-term adherence rates.

6.3 Predictive toxicity monitoring

ML models are being developed that can predict immunosuppression-associated toxicity before clinical manifestations appear, enabling proactive management strategies to prevent serious complications (Bai et al., 2025; Burnette et al., 2024).

Immunosuppression-associated hepatotoxicity and nephrotoxicity are major concerns that can significantly impact long-term outcomes (Grzyb, Du, & Nair, 2024).

These systems analyze patterns in laboratory results, clinical symptoms, and biomarker trends to identify early signs of drug toxicity. Advanced models incorporate genetic risk factors, drug interaction potentials, and individual susceptibility markers to create personalized toxicity risk profiles (Bai et al., 2025; Yang & Kar, 2023). High-throughput screening technologies, such as liver-on-a-chip, reconstruct the status of liver tissue at a microscopic scale (Liu et al., 2024; Mugaanyi et al., 2025). Combining this technology with patient-derived induced pluripotent stem cell-differentiated hepatocytes has the potential to enable individualized risk prediction of toxicity (Grzyb et al., 2024).

7 AI-assisted remote monitoring and patient management

The integration of wearable technology with AI has created unprecedented opportunities for continuous patient monitoring, fundamentally changing how we care for transplant recipients and enabling truly personalized, responsive healthcare delivery.

7.1 Advanced wearable electrocardiogram monitoring systems

Wearables offer a promising solution for enhancing remote monitoring of heart failure patients by tracking key physiological parameters with unprecedented precision and continuity (Scholte et al., 2024). AI algorithms process data from wearable devices, advanced sensors, and patient-reported inputs, covering vital signs such as heart rate, blood pressure, and respiratory rate, along with activity levels and behavioral patterns (HealthSnap, 2025).

Modern wearable electrocardiogram (ECG) systems can detect arrhythmias, signs of graft dysfunction, and early indicators of rejection with remarkable accuracy (Abdelrazik et al., 2025). In addition, wearable ECG devices, enhanced by AI, detect early electrical signs of heart transplant rejection with >85% sensitivity, closely matching the results of biopsy-confirmed rejection episodes (Adedinsewo et al., 2023). These devices enable 24/7 monitoring, which was previously impossible, creating comprehensive datasets that allow AI algorithms to identify subtle changes in cardiac function that precede clinical symptoms (Alimbayeva et al., 2024).

Advanced pattern recognition algorithms can now detect early signs of graft dysfunction from ECG patterns, including subtle changes from the start of the Q wave to the end of T wave (QT interval), T-wave morphology, and heart rate variability that may indicate rejection or ischemia (Siontis, Noseworthy, Attia, & Friedman, 2021; Muzammil et al., 2024). AI algorithms establish personalized baselines tailored to age, gender, medical history, and lifestyle, monitoring for deviations with unprecedented precision (HealthSnap, 2025).

Recent developments in multilead wearable ECG systems are providing clinical-grade monitoring capabilities in home settings. These systems can detect complex arrhythmias, conduction abnormalities, and ischemic changes with accuracy comparable to hospital-based monitoring, while AI algorithms filter out artifacts and false alarms to provide clinically relevant alerts (Abdelrazik et al., 2025; Siontis et al., 2021).

7.2 Comprehensive physiological monitoring

AI-powered monitoring systems are expanding beyond cardiac parameters to provide comprehensive physiological surveillance that includes respiratory patterns, activity levels, sleep quality, and metabolic indicators. These multiparameter monitoring systems create detailed pictures of patient health status and can detect early signs of complications across multiple organ systems (Shaik et al., 2023; Tsvetanov, 2024).

Near real-time monitoring capabilities enable AI to detect subtle changes instantly, with wearables, smartwatches, and ambient sensors providing continuous data streams (HealthSnap, 2025). Pattern recognition algorithms identify irregular patterns, such as respiratory distress, stress-related physiological changes, and declining functional capacity that may indicate the development of complications (Abd Al-Alim, Mubarak, Salem, & Sadek, 2024).

Advanced systems are being developed that can monitor fluid status through bioimpedance measurements, detecting early signs of heart failure or fluid retention before clinical symptoms appear (Giménez-Miranda et al., 2024). These systems integrate weight monitoring, activity tracking, and physiological parameters to provide comprehensive assessments of volume status. Sleep monitoring represents another frontier in AI-assisted care, with algorithms analyzing sleep patterns, respiratory events, and autonomic nervous system activity to identify signs of sleep-disordered breathing, medication side effects, or a decline in overall health status (Osa-Sanchez, Ramos-Martinez-de-Soria, Mendez-Zorrilla, Ruiz, & Garcia-Zapirain, 2025).

7.3 AI-powered virtual health assistants and telemedicine integration

Advanced chatbots and virtual assistants are providing real-time guidance for medication adherence, symptom tracking, and lifestyle modifications, learning from patient interactions to provide increasingly personalized support (Abbas, Yusifzada, & Athar, 2025; Wah, 2025). These systems represent a new paradigm in patient engagement and self-management support.

AI chatbots can now provide immediate coping strategies for patients experiencing symptoms or concerns, escalating to healthcare providers when needed, while maintaining continuous support availability (Abbas, Yusifzada, & Athar, 2025; Olawade et al., 2024). Sentiment analysis capabilities enable these systems to detect emotional distress, anxiety, or depression indicators that may impact recovery and adherence (Denecke & Gabarron, 2024; Olawade et al., 2024).

Multimodal data analysis integrates physiological, behavioral, and self-reported data to provide comprehensive health assessments (Tan et al., 2024). Natural language processing capabilities enable these systems to analyze patient communications, detecting subtle indicators of health changes or concerns that might not be captured through traditional monitoring methods (Maleki Varnosfaderani & Forouzanfar, 2024).

Telemedicine integration powered by AI algorithms assists clinicians in remotely monitoring patients while providing early warnings for potential complications (Maleki Varnosfaderani & Forouzanfar, 2024; Tsvetanov, 2024). These systems can analyze vast amounts of continuous monitoring data to surface actionable trends and outliers, enabling clinicians to focus on patients who most need attention.

Virtual support capabilities are being enhanced with generative AI technologies that can provide personalized health education, explain complex medical concepts in accessible language, and offer emotional support during challenging periods of recovery (Alowais et al., 2023; Maleki Varnosfaderani & Forouzanfar, 2024).

Table 8.2 provides a comprehensive overview of remote monitoring technologies and their clinical applications in heart transplantation, demonstrating the breadth of AI-enabled monitoring capabilities now available.

TABLE 8.2 Remote monitoring technologies and their clinical applications.

Technology type	Monitoring parameters	AI integration level	Clinical benefits	Implementation status
Wearable ECG monitors (Alimbayeva, Alimbayev, Ozhikenov, Bayanbay, & Ozhikenova, 2024)	Heart rhythm, rate variability, QT intervals, T-wave morphology	Advanced pattern recognition with personalized baselines	Early arrhythmia detection, rejection monitoring, ischemia detection	Clinically deployed in major centers
Smart blood pressure cuffs (McMurray et al., 2025)	Blood pressure trends, pulse pressure, vascular stiffness	Trend analysis algorithms with risk stratification	Hypertension management, medication optimization	Widespread adoption
Activity trackers (Husom et al., 2022)	Exercise capacity, fatigue patterns, functional decline	Behavioral analytics with machine learning	Functional assessment, rehabilitation monitoring	Consumer-grade clinical adaptation
Smart scales with bioimpedance (Mehra, Starkoff, & Nickerson, 2025)	Weight, fluid status, body composition, muscle mass	Multiparameter analysis with fluid modeling	Heart failure monitoring, nutrition assessment	Emerging technology in trials
Continuous glucose monitors (Cappon, Vettoretti, Sparacino, & Facchinetti, 2019)	Metabolic status, diabetes control, steroid effects	Predictive modeling with drug interaction analysis	Steroid-induced diabetes management, metabolic monitoring	Pilot programs expanding
Sleep monitoring devices (Osa-Sanchez, Ramos-Martinez-de-Soria, Mendez-Zorrilla, Ruiz, & Garcia-Zapirain, 2025)	Sleep quality, respiratory patterns, autonomic function	Deep learning analysis of sleep architecture	Overall health assessment, medication side effect detection	Research phase validation
Pulse oximetry sensors (Cabanas et al., 2024)	Oxygen saturation, perfusion index, respiratory rate	Continuous trend analysis with alert algorithms	Respiratory monitoring, early infection detection	Clinical integration ongoing
Temperature sensors (Smarr et al., 2020)	Core temperature, fever patterns, circadian rhythms	Pattern recognition for infection prediction	Early infection detection, medication monitoring	Prototype development

8 AI-enabled predictive models in heart transplantation

The sophistication of AI models in heart transplantation continues to evolve, with various algorithmic approaches offering unique advantages for different clinical scenarios. These models represent the culmination of advances in computer science, clinical research, and data analytics, creating tools that can process unprecedented amounts of information to support clinical decision-making.

8.1 Advanced neural network applications

Neural networks predict organ rejection risks by analyzing multidimensional patient data, including genetic markers, biomarker profiles, and clinical variables with unprecedented sophistication (Gotlieb et al., 2022; Olawade et al., 2024). Recent research demonstrates that self-explaining neural networks can model 1-year mortality in heart transplantation data with performance comparable to traditional DL models, while providing interpretable results (Lisboa et al., 2022). These advanced systems can process hundreds of variables simultaneously, identifying complex interactions and nonlinear relationships that traditional statistical methods cannot detect. Deep neural networks trained on histopathological whole slide images have achieved accuracy rates of 90%–92% when evaluated on independent datasets for predicting heart rejection (Tong, Hoffman, Deshpande, & Wang, 2017).

Recent advances in convolutional neural networks have enabled sophisticated analysis of medical imaging data, with AI systems now capable of detecting subtle changes in cardiac structure and function that correlate with rejection risk, graft dysfunction, and long-term outcomes (Arjmandmazidi et al., 2025; Chowdhury et al., 2025; Olawade, Marinze, Qureshi, Weerasinghe, & Teke, 2025b). These sophisticated models excel at analyzing longitudinal patient data to predict long-term transplant outcomes, representing a crucial advance in time-series medical data analysis. LSTM networks are particularly effective for processing time-series data from continuous monitoring devices, identifying trends that predict complications weeks or months in advance (Al Moussawy et al., 2024; Alowais et al., 2023). Advanced LSTM models have been developed that can analyze sequential biomarker measurements, medication adherence patterns, and physiological monitoring data to predict rejection episodes, graft failure, and other complications with remarkable accuracy (Al Moussawy et al., 2024; Coutance, Desiré, & Duong Van Huyen, 2022). These systems can maintain a memory of important events and patterns over extended periods, enabling the prediction of long-term outcomes. The ability to process temporal relationships in medical data represents a significant advantage over traditional statistical methods, which typically analyze data at discrete time points. LSTM networks can identify subtle trends and patterns in continuously collected data that may be invisible to conventional analysis methods (Al Moussawy et al., 2024).

8.2 Traditional ML approaches with enhanced capabilities

Random forest algorithms evaluate donor–recipient compatibility by analyzing large-scale datasets while maintaining interpretability, a crucial advantage for clinical decision-making where understanding the reasoning behind predictions is essential (Ghosh & Cabrera, 2022; Wies, Miltenberger, Grieser, & Jahn-Eimermacher, 2023). These ensemble methods combine multiple decision trees to create robust prediction models that can handle missing data and provide insights into variable importance.

Recent applications of random forest algorithms in liver transplantation have demonstrated superior performance compared to conventional clinical scoring systems, with AUC values reaching 0.949 for predicting 90-day posttransplant survival (Garcia-Canadialla et al., 2024). These models outperformed traditional MELD scores and other established risk assessment tools. The interpretability of random forest models makes them particularly valuable for clinical implementation, as clinicians can understand which variables are driving predictions and how different factors contribute to risk assessments (Cappelli, Castronuovo, Grimaldi, & Telesca, 2024; Ramaswamy, Senthilnathan, Saravanan, Snehal, & Siva, 2025). This transparency is crucial for building trust and enabling clinical adoption of AI systems.

Support vector machines demonstrate particular effectiveness in classifying patients at risk of posttransplant complications, excelling at handling high-dimensional medical data with robust performance even in the presence of outliers (Zhou et al., 2022). Research has found that the patient prognostic prediction model using SVM achieved an accuracy of 92.2% and an AUC value of 0.98 (Zhou et al., 2022). SVMs are particularly valuable for analyzing genomic data, proteomic profiles, and other high-dimensional datasets common in modern transplant medicine.

These algorithms can identify optimal decision boundaries in complex, high-dimensional spaces, enabling accurate classification of patients into risk categories based on multiple variables (Zhou et al., 2022). Recent applications have shown particular promise in predicting delayed graft function, rejection risk, and long-term survival outcomes. The robustness of

TABLE 8.3 AI algorithms and their applications in heart transplantation.

Algorithm type	Primary application	Accuracy rate (%)	Key advantages	Clinical use cases	Limitations
Deep neural networks (Tong et al., 2017)	Rejection prediction, image analysis	90–92	High accuracy, complex pattern recognition, feature extraction	Biopsy analysis, imaging interpretation, multiparameter prediction	Black-box nature, requires large datasets
Random forest (Sargiotis et al., 2024)	Donor–recipient matching	83–89	Interpretability, handles missing data, variable importance ranking	Compatibility scoring, risk stratification, clinical decision support	May overfit with small datasets
Support vector machines (Zhou et al., 2022)	Complication classification	82–87	Robust to outliers, effective with high-dimensional data	Risk categorization, outcome prediction, genomic analysis	Sensitive to feature scaling, parameter tuning
LSTM networks (Coutance et al., 2022)	Longitudinal monitoring	88–91%	Time-series analysis, long-term memory, trend prediction	Remote monitoring, early warning systems, temporal pattern analysis	Requires sequential data, computationally intensive
Ensemble methods (Mienye et al., 2020)	Multimodal prediction	91–94	Combines multiple models, reduces bias, improved robustness	Comprehensive risk assessment, integrated decision support	Complex implementation, difficult interpretation
Logistic regression (enhanced) (Chandrasekhar & Peddakrishna, 2023)	Risk scoring	78–83	Interpretable, well-understood, regulatory acceptance	Clinical scoring systems, baseline comparisons	Limited complexity, assumes linear relationships

SVM algorithms to noise and outliers makes them particularly suitable for clinical data, which often contain measurement errors, missing values, and extreme observations that might affect other analytical approaches (Guido, Ferrisi, Lofaro, & Conforti, 2024; Jordaan & Smits, 2004).

8.3 Ensemble methods and hybrid approaches

Advanced ensemble methods combine multiple AI algorithms to leverage the strengths of different approaches while minimizing individual algorithm limitations. These systems can integrate data from multiple sources: clinical variables, imaging data, biomarkers, and genomic information to create comprehensive prediction models.

Recent developments in ensemble learning have demonstrated improvements in accuracy over individual algorithms, with some hybrid systems achieving AUC values exceeding 0.90 for critical outcomes, such as 1-year mortality and graft failure prediction (Ganie, Pramanik, & Zhao, 2025). A study on an ensemble learning method for predicting heart disease risk, using a mean-based splitting method and modeling these partitions with classification and regression tree algorithms, achieved 93% and 91% classification accuracies using the Cleveland Heart Disease Dataset (HDDC) and Framingham Heart Study Datasets, respectively (Mienye, Sun, & Wang, 2020). These multimodal approaches are particularly valuable in transplantation medicine, where optimal decision-making requires integration of diverse data types and consideration of multiple risk factors simultaneously (Amal et al., 2022; Gotlieb et al., 2022).

Table 8.3 provides a detailed comparison of these AI algorithms and their specific applications in heart transplantation, highlighting the performance characteristics and optimal use cases for each approach.

9 Challenges and ethical considerations

Despite the remarkable potential of AI in heart transplantation, several significant challenges must be addressed before widespread implementation becomes a reality. These challenges span technical, ethical, regulatory, and practical domains, requiring coordinated efforts from clinicians, technologists, ethicists, and policymakers.

9.1 Data privacy and security concerns

AI systems require access to extensive patient data, raising significant concerns about privacy and compliance with regulations such as GDPR and HIPAA (Olawade et al., 2024). The sensitivity of transplant data, including genetic information, detailed medical histories, and long-term outcome data, creates particular vulnerabilities that must be carefully managed (Arjmandmazidi et al., 2025). While AI-based prognostic and diagnostic functions have performed better than those developed using traditional statistics (Alowais et al., 2023), research has revealed significant risks of bias, a lack of external validation, and potentially poor real-world applicability, which may affect the effectiveness of AI-based tools (Palmieri et al., 2023). These limitations underscore the crucial importance of robust data governance frameworks and rigorous validation protocols.

Ensuring robust cybersecurity measures while maintaining data accessibility for AI training remains a delicate balance. Healthcare organizations must implement sophisticated encryption protocols, access controls, and audit systems while enabling the data sharing necessary for AI model development and validation (Bajwa et al., 2021). The international nature of organ transplantation creates additional complexities, as data may need to be shared across jurisdictions with different privacy regulations and security requirements (Olawade et al., 2025). Developing harmonized standards for data protection while enabling beneficial AI research remains an ongoing challenge. Blockchain technology is emerging as a potential solution for secure, decentralized data sharing in transplantation, enabling transparent and immutable record-keeping while protecting patient privacy through advanced cryptographic techniques (Ali et al., 2023).

9.2 Addressing algorithmic bias and ensuring equity

ML models must be carefully trained on diverse datasets to prevent biases that could lead to disparities in patient care, a particularly crucial concern in transplantation, where access to organs must remain equitable across all demographic groups (Gotlieb et al., 2022; Salybekov, Yerkos, Sedlmayr, & Wolfien, 2025). Historical biases in clinical data and healthcare delivery can be perpetuated and amplified by AI systems if not carefully addressed.

Research has identified significant disparities in transplant outcomes across racial, ethnic, and socioeconomic groups, and AI systems trained on historical data may inadvertently perpetuate these disparities (Ahluwalia, Sehgal, Lee, Agu, & Kpodonu, 2025; Olawade, Marinze, Egbon, & Chinwah, 2025a). Ensuring that AI algorithms perform equally well across all patient populations requires diverse training datasets and careful validation across demographic subgroups. The development of AI models requires extensive datasets; however, current transplant registries may not adequately represent the global population's diversity. Most large-scale studies have focused on European and North American populations, potentially limiting the applicability of AI models to other demographic groups. Addressing algorithmic bias requires ongoing monitoring of AI system performance across different patient populations, with regular audits to identify and correct discriminatory patterns. This monitoring must extend beyond initial validation to include real-world performance assessment as AI systems are deployed in clinical practice.

9.3 Clinical validation and regulatory requirements

AI-driven predictions must undergo rigorous validation through clinical trials to ensure reliability and effectiveness before widespread clinical adoption. More unbiased research with high-quality big data intended for AI, transparency, and external validations is needed to establish medical AI as a systematic aid to clinical decision-making (Palmieri et al., 2023). The regulatory pathway for AI-based medical devices remains complex and evolving, with different requirements across jurisdictions creating challenges for global implementation (Pantanowitz et al., 2024). Regulatory agencies are developing new frameworks for evaluating AI systems; however, these processes are still evolving and may struggle to keep pace with technological advancements.

External validation represents a particular challenge, as AI models developed at one institution may not perform as well when applied to patients at other centers with different demographics, practices, or systems. A free-use prediction algorithm developed through AI failed to predict 1-year mortality posttransplantation in cases from centers different from those where it was developed (Palmieri et al., 2023). The dynamic nature of AI models, which can continue learning and adapting after deployment, creates additional regulatory challenges. Traditional medical device approval processes assume static devices with fixed performance characteristics; however, AI systems may evolve over time, necessitating new approaches to ongoing safety and effectiveness monitoring.

9.4 Human–AI collaboration and clinical integration

Perhaps most importantly, AI should augment, not replace, human expertise in clinical decision-making. The most successful implementations maintain physicians at the center of care while providing them with unprecedented analytical capabilities. Achieving effective human–AI collaboration requires careful attention to workflow integration, user interface design, and clinician training (Vössing, Kühl, Lind, & Satzger, 2022). The interpretability of AI models remains a significant challenge, particularly for complex DL systems that may achieve high accuracy but provide limited insight into their decision-making processes (Ennab & Mcheick, 2024). Clinicians need to understand how AI systems reach their conclusions to maintain appropriate oversight and make informed decisions about when to follow or override AI recommendations. Alert fatigue represents another significant concern, as AI systems capable of continuous monitoring may generate excessive alerts that overwhelm clinicians, thereby reducing the effectiveness of truly important warnings (Gani et al., 2025). Careful calibration of alert thresholds and intelligent filtering systems is essential for maintaining clinical utility. Training healthcare professionals to work effectively with AI systems requires new educational approaches that combine traditional medical training with data literacy, algorithm understanding, and human-computer interaction skills. This educational challenge extends from medical school curricula through continuing education for practicing clinicians.

9.5 Economic and implementation challenges

The implementation of AI systems in heart transplantation requires significant upfront investments in technology infrastructure, staff training, and system integration. Healthcare organizations must balance these costs against potential benefits, which may not be immediately apparent and may accrue over extended periods.

Interoperability remains a significant challenge, as AI systems must integrate with existing electronic health records, laboratory systems, imaging platforms, and other clinical technologies (Al Moussawy et al., 2024; Alowais et al., 2023). The lack of standardized data formats and communication protocols creates barriers to seamless AI integration. Maintaining and updating AI systems requires ongoing technical expertise and resources that many healthcare organizations may lack. The rapid pace of AI development means that systems may become obsolete quickly, requiring continuous investment in updates and improvements (Bajwa et al., 2021). The liability implications of AI-assisted decision-making remain unclear, creating uncertainty for healthcare organizations and clinicians regarding legal responsibilities when AI systems are involved in patient care decisions.

10 Future prospects of AI in heart transplantation

The future of AI in heart transplantation is nothing short of extraordinary, with developments that will fundamentally transform how we approach organ transplantation and posttransplant care. These advances promise to address current limitations while opening entirely new possibilities for improving patient outcomes.

10.1 Real-time decision support systems

Smart Match employs AI algorithms, enhancing organ matching precision and optimizing transplantation outcomes through sophisticated real-time analysis (Deshpande, 2024). Next-generation AI systems will provide instantaneous recommendations to transplant teams, offering real-time analysis of donor–recipient compatibility, optimal surgical timing, and post-operative management strategies with unprecedented speed and accuracy (Arjmandmazidi et al., 2025; Olawade, Marinze, Qureshi, Weerasinghe, & Teke, 2025a). These systems will integrate multiple data streams—donor assessment results, recipient status updates, transport logistics, and surgical team availability—to optimize every aspect of the transplantation process. Real-time optimization algorithms will continuously update recommendations as new information becomes available, ensuring optimal decision-making throughout the transplant process (Deshpande, 2024; Olawade, Marinze, Qureshi, Weerasinghe, & Teke, 2025a, 2025b).

Advanced decision support systems will incorporate simulation capabilities, enabling transplant teams to model various scenarios and predict outcomes for different decision options (Olawade, Marinze, Qureshi, Weerasinghe, & Teke, 2025b). These systems will consider not just immediate outcomes but long-term projections, helping teams make decisions that optimize both short-term success and long-term graft survival. The integration of natural language processing capabilities will enable these systems to communicate recommendations in clear, actionable language while providing detailed explanations of the reasoning behind each recommendation (Maleki Varnosfaderani & Forouzanfar, 2024). This transparency will be crucial for maintaining clinician trust and enabling effective human–AI collaboration.

10.2 Blockchain integration for enhanced transparency and security

Secure and decentralized AI-powered transplant registries will improve transparency and organ allocation efficiency while ensuring data integrity and patient privacy. Blockchain technology will enable immutable record-keeping for all aspects of the transplant process, from donor evaluation through long-term follow-up, creating unprecedented transparency while protecting sensitive information (Ali et al., 2023; Panigrahi, Mukherjee, Shaikh, & Nomran, 2025).

Smart contracts powered by AI algorithms will automate many aspects of organ allocation, ensuring that decisions are made according to predetermined criteria while maintaining detailed audit trails (Deshpande, 2024). These systems will reduce the potential for human error or bias in allocation decisions while increasing public confidence in the fairness of organ distribution. International data sharing for AI model development and validation will be facilitated by blockchain platforms that enable secure, privacy-preserving collaboration across institutions and countries (Ali et al., 2023; Bathula et al., 2024). These systems will allow researchers to access large, diverse datasets while maintaining patient privacy and data sovereignty. The integration of blockchain with the Internet of Things (IoT) devices used in organ transport and monitoring will create end-to-end visibility of organ quality and handling (Panigrahi, Mukherjee, Shaikh, & Nomran, 2025), enabling AI systems to optimize preservation protocols and predict organ viability with unprecedented accuracy.

10.3 AI-guided xenotransplantation

As xenotransplantation moves closer to clinical reality, predictive AI models will aid in optimizing genetically modified animal organs for human transplantation, potentially addressing the critical donor shortage that currently limits heart transplantation (Reichart et al., 2022; Wolf, Schmoeckel, & Reichart, 2023). AI systems will analyze genetic modifications, immunological compatibility, and physiological parameters to optimize xenograft design and recipient selection.

Advanced AI models will predict immune responses to xenografts, enabling the development of personalized immunosuppressive protocols that prevent rejection while minimizing toxicity (Reichart et al., 2022). These systems will integrate data from animal studies, in vitro experiments, and early clinical trials to create sophisticated prediction models (Wolf et al., 2023). Quality assessment algorithms will evaluate xenograft organs using criteria adapted for nonhuman donors, considering factors such as size compatibility, physiological parameters, and genetic modification effectiveness (Galli, 2025; Hryhorowicz, Zeyland, Słomski, & Lipiński, 2017). These systems will help determine which xenografts are most suitable for specific recipients. Long-term monitoring protocols for xenotransplant recipients will incorporate AI systems designed to detect unique complications and immune responses that may differ from those seen with human allografts.

10.4 Integration with genomic and precision medicine

AI will facilitate precision medicine approaches by integrating genomic, proteomic, and metabolomic data to create truly personalized transplant strategies (Chen, Hsiao, Lin, & Fann, 2025). This integration will enable the prediction of individual responses to immunosuppressive drugs and rejection risks based on comprehensive molecular profiles.

Pharmacogenomic algorithms will predict optimal drug regimens based on genetic variants affecting drug metabolism, transport, and receptor function (Qahwaji et al., 2024). These systems will consider complex gene–gene and gene–environment interactions that influence drug effectiveness and toxicity. Precision immunosuppression protocols will be developed based on individual immune system characteristics, genetic risk factors, and predicted drug responses (Moingeon, 2023). AI systems will continuously optimize these protocols based on real-time monitoring data and outcome assessments. The integration of multiomic data will enable the identification of new biomarkers and therapeutic targets for improving transplant outcomes (Babu & Snyder, 2023; Lin et al., 2025). AI systems will analyze vast datasets to identify molecular signatures associated with successful transplants and graft longevity (Khera et al., 2024; Ledziński & Grześk, 2023).

10.5 Advanced regenerative medicine applications

AI will play crucial roles in developing regenerative medicine approaches to heart transplantation, including the optimization of tissue engineering protocols, stem cell therapies, and organ regeneration techniques (Nosrati & Nosrati, 2023). ML algorithms will guide the development of bioengineered hearts and other regenerative therapies (Olawade, Marinze, Qureshi, Weerasinghe, & Teke, 2025b). 3D bioprinting of cardiac tissues will be optimized using AI algorithms that can predict optimal cell types, scaffolds, and growth conditions for creating functional cardiac grafts (Jafari et al., 2022). These systems will integrate data from multiple sources to optimize printing protocols and predict tissue functionality (Ramesh et al., 2024). Stem cell therapy protocols will be personalized using AI systems that can predict which cell types, delivery methods, and timing strategies will be most effective for individual patients (Nosrati & Nosrati, 2023; Srivastava, Bhavsar, & Mishra, 2024). These approaches may enable cardiac repair and regeneration that reduces or eliminates the need for transplantation. The development of artificial hearts and mechanical circulatory support devices will be enhanced by AI systems that can optimize device design, control algorithms, and patient selection criteria.

11 Transforming patient outcomes: Real-world impact

The integration of AI into heart transplantation is not just theoretical; it is already delivering measurable improvements in patient outcomes across multiple domains (Naruka et al., 2022; Palmieri et al., 2023). Sometimes, you rely on patients to tell you they do not feel great before you decide to intervene. Often, by that point, things have progressed significantly. But if you can catch it earlier, clinicians can help that patient better (Streed, 2024). This proactive approach represents a fundamental shift from reactive to predictive medicine, with AI systems enabling interventions that prevent complications rather than simply treating them after they occur (Johnson et al., 2021; Maleki Varnosfaderani & Forouzanfar, 2024). The transformation is particularly evident in posttransplant monitoring, where continuous AI-powered surveillance can detect problems days or weeks before traditional methods would (Arjmandmazidi et al., 2025).

11.1 Enhanced quality-of-life outcomes

Patients are experiencing fewer complications, shorter hospital stays, and improved long-term survival rates through AI-enhanced care protocols (Maleki Varnosfaderani & Forouzanfar, 2024). Early detection of health deterioration is critical for patients with chronic conditions, postoperative recovery, and ongoing health management, with AI-driven remote patient monitoring enabling continuous data analysis and early identification of potential issues (HealthSnap, 2025).

The quality-of-life improvements extend beyond clinical metrics to include enhanced patient confidence, reduced anxiety about rejection, and greater independence in daily activities. AI-enabled monitoring provides patients with continuous reassurance about their health status while enabling rapid responses to any concerning changes (Khera et al., 2024). Patient-reported outcome measures consistently show improvements in functional status, exercise tolerance, and overall well-being among patients cared for with AI-enhanced protocols compared to traditional approaches (Alowais et al., 2023; Bajwa et al., 2021). These improvements reflect not just better clinical outcomes but also enhanced patient experience and satisfaction with care. The psychological benefits of AI-enhanced monitoring include reduced fear and anxiety about rejection episodes, as patients gain confidence from knowing that their condition is being continuously monitored with sophisticated systems that can detect problems early (Olawade et al., 2025; Olawade, Marinze, Qureshi, Weerasinghe, & Teke, 2025a).

11.2 Reduced healthcare utilization

AI systems are significantly reducing emergency department visits, hospital readmissions, and unplanned clinic visits through improved early detection and prevention of complications. Proactive care at home reduces hospital admissions, saving costs while improving patient outcomes (HealthSnap, 2025). Emergency interventions are being replaced by planned, proactive management strategies that address problems before they become critical. This shift not only improves outcomes but also reduces the stress and disruption associated with urgent medical care. The reduction in healthcare utilization is particularly pronounced for infection-related complications, where AI-powered early detection systems enable outpatient management of conditions that previously required hospitalization (A.P. Zhao et al., 2024). Long-term follow-up care is becoming more efficient and effective, with AI systems enabling risk-stratified monitoring protocols that focus intensive resources on high-risk patients while providing appropriate surveillance for stable patients (Maleki Varnosfaderani & Forouzanfar, 2024).

11.3 Functional and social benefits

AI-enhanced transplant care is enabling patients to return to work, engage in physical activities, and participate fully in family and social life at rates significantly higher than traditional approaches (Olawade, Marinze, Qureshi, Weerasinghe, & Teke, 2025a; Johnson et al., 2021). The combination of better clinical outcomes and enhanced monitoring confidence contributes to improved social reintegration and quality of life. Exercise tolerance and functional capacity improvements are being documented through wearable device monitoring, with AI systems providing personalized exercise recommendations and safety monitoring that enable optimal rehabilitation while preventing overexertion (Seçkin, Ateş and Seçkin, 2023; Husom et al., 2022). Family dynamics and caregiver burden are being positively impacted by AI systems that provide reassurance about patient status and enable family members to participate in care monitoring without the anxiety associated with traditional surveillance limitations (Borna et al., 2024). Return-to-work rates and vocational rehabilitation success are improving as patients gain confidence in their health status and ability to maintain stable function through AI-enhanced monitoring and care optimization (Baldwin & Brusco, 2011).

12 Economic impact and healthcare sustainability

Beyond the clinical benefits, AI in heart transplantation is delivering significant economic value through multiple mechanisms that improve healthcare sustainability while enhancing patient outcomes (Chowdhury et al., 2025; Palmieri et al., 2023). The economic case for AI implementation extends beyond immediate cost savings to include long-term value creation through improved outcomes and resource optimization.

12.1 Direct cost reductions

Reduced hospital readmissions, earlier complication detection, and optimized resource allocation are contributing to substantial direct cost savings in transplant care. Early alerts enable rapid responses, preventing complications and improving outcomes while reducing the need for expensive emergency interventions (HealthSnap, 2025). The cost of AI-enhanced monitoring systems is being offset by reductions in hospital days, emergency department visits, and intensive care requirements (Maleki Varnosfaderani & Forouzanfar, 2024; Tsvetanov, 2024). These savings are particularly significant, given the high costs associated with transplant complications and rejection episodes.

Medication optimization through AI-guided dosing is reducing drug costs while improving effectiveness, as personalized protocols reduce the need for expensive adjustments and modifications (Serrano et al., 2024). The precision of AI-guided therapy reduces both undertreatment costs (rejection, complications) and overtreatment costs (toxicity, excessive monitoring) (Ni & Jia, 2025). Diagnostic efficiency improvements are reducing the costs associated with unnecessary biopsies, imaging studies, and laboratory tests while maintaining or improving diagnostic accuracy.

12.2 Long-term value creation

The cost-effectiveness of AI-enhanced transplantation becomes even more compelling when considering the lifetime costs of transplant care. By improving outcomes and reducing complications, AI systems are helping to justify the high upfront costs of transplantation while improving patient quality of life and extending graft survival (Al Moussawy et al., 2024). Improved graft survival directly translates into a reduced need for retransplantation, which is one of the most expensive complications in transplant medicine (Arjmandmazidi et al., 2025). Each year of extended graft life provides substantial value, both in terms of cost avoidance and quality-adjusted life years gained.

The societal economic benefits include improved productivity from transplant recipients who return to work earlier and maintain employment longer due to better health outcomes. These benefits extend to family members, allowing them to return to normal work and social activities rather than providing intensive caregiving (Baldwin & Brusco, 2011). Healthcare system capacity benefits are emerging as AI systems enable more efficient use of transplant resources, potentially increasing the number of successful transplants that can be performed with existing infrastructure (Olawade, Marinze, Qureshi, Weerasinghe, & Teke, 2025b; Deshpande, 2024).

12.3 Resource optimization and efficiency

AI systems are optimizing the utilization of expensive resources such as intensive care beds, transplant coordinator time, and specialist physician availability (Khalifa & Albadawy, 2024; Khera et al., 2024). By enabling more efficient triage and resource allocation, these systems improve overall healthcare system efficiency. Organ utilization rates are improving through better donor assessment and matching algorithms, reducing organ wastage and maximizing the benefit derived from each donated organ (Olawade, Marinze, Qureshi, Weerasinghe, & Teke, 2025a). This improvement is particularly valuable given the critical shortage of donor organs. Clinical workflow efficiency is being enhanced through AI-powered decision support that reduces the time required for complex assessments and enables more streamlined care protocols (Alowais et al., 2023; Maleki Varnosfaderani & Forouzanfar, 2024). These efficiencies free up clinical resources for other patients while maintaining or improving care quality. The scalability of AI systems means that efficiency gains can be achieved without proportional increases in staffing or infrastructure, enabling healthcare systems to care for more patients with existing resources.

12.4 Innovation and competitive advantages

Healthcare institutions implementing AI-enhanced transplant programs are gaining competitive advantages through improved outcomes, enhanced reputation, and increased referral volumes. These advantages can translate into financial benefits through increased market share and premium pricing for superior outcomes (Khanna et al., 2022). Research and development opportunities created by AI implementation are attracting grant funding, industry partnerships, and academic collaborations, which provide additional revenue streams and cost-sharing opportunities. The expertise developed in AI implementation for transplantation is being leveraged for other clinical applications, creating additional value and return on investment beyond the immediate benefits of the transplant program (Al Moussawy et al., 2024). Quality metric improvements achieved through AI implementation are enhancing institutional ratings and accreditation scores, which can impact reimbursement rates and referral patterns.

13 Preparing for the AI-driven future

As we look toward the future of heart transplantation, it's clear that AI will play an increasingly central role in every aspect of transplant care. Healthcare professionals, institutions, and healthcare systems must prepare for this transformation by developing appropriate capabilities, infrastructure, and expertise to leverage AI technologies effectively.

13.1 Professional development and training

Healthcare professionals must prepare for this transformation by developing digital literacy, understanding AI capabilities and limitations, and learning to work collaboratively with AI systems. This preparation extends beyond basic computer skills to include an understanding of ML principles, data interpretation, and human–AI interaction best practices.

Medical education curricula are being updated to include AI literacy as a core competency, with medical schools and residency programs incorporating training in data science, algorithm interpretation, and AI-assisted decision-making (Tolentino, Baradaran, Gore, Pluye, & Abbasgholizadeh-Rahimi, 2024). These educational changes are essential for preparing the next generation of clinicians to work effectively with AI systems. Continuing education programs for practicing clinicians are focusing on practical skills for working with AI systems, including understanding when to trust AI recommendations, how to interpret AI outputs, and how to maintain appropriate clinical oversight of AI-assisted care (Tolentino et al., 2024). Interdisciplinary collaboration skills are becoming increasingly important, as AI implementation requires effective partnerships between clinicians, data scientists, engineers, and other technical specialists. Healthcare professionals must develop effective communication skills and collaborative approaches that enable seamless teamwork across diverse disciplines.

13.2 Institutional infrastructure development

Transplant centers worldwide are investing in AI infrastructure, including high-performance computing capabilities, data storage and management systems, and integration platforms that enable AI systems to work seamlessly with existing clinical technologies (Olawade, Marinze, Qureshi, Weerasinghe, & Teke, 2025a). Data governance frameworks are being established to ensure appropriate data quality, security, and compliance while enabling AI development and deployment (Pantanowitz et al., 2024). These frameworks must strike a balance between the need for data access and the requirements for privacy protection and regulatory compliance. Quality assurance programs for AI systems are being developed to ensure ongoing performance monitoring, bias detection, and safety oversight (Mennella, Maniscalco, De Pietro, & Esposito, 2024). These programs require new capabilities and expertise that many healthcare institutions are still developing. Change management strategies are essential for successful AI implementation, as these technologies require significant changes in clinical workflows, decision-making processes, and organizational culture (Alowais et al., 2023). Successful implementation requires careful attention to stakeholder engagement, training, and support.

13.3 Regulatory and policy adaptation

Healthcare systems and regulatory agencies are adapting policies and procedures to accommodate AI technologies while maintaining appropriate safety and effectiveness standards (Pantanowitz et al., 2024). These adaptations include new approval pathways for AI-based medical devices, updated quality assurance requirements, and modified reimbursement policies.

Professional licensing and credentialing requirements are being updated to include AI competencies, ensuring that healthcare professionals have appropriate training and expertise to use AI systems safely and effectively (Tolentino et al., 2024). Liability and insurance frameworks are evolving to address the unique challenges posed by AI-assisted decision-making, including questions about responsibility when AI systems contribute to adverse outcomes or fail to detect problems (Cestonaro, Delicati, Marcante, Caenazzo, & Tozzo, 2023). International harmonization efforts are working to develop consistent standards and requirements for AI in healthcare, enabling global collaboration and knowledge sharing while maintaining appropriate safety standards (Reddy, 2025).

13.4 Patient and public engagement

Public education about AI in healthcare is essential for maintaining trust and confidence in AI-enhanced transplant care. Patients and their families need to understand how AI systems work, the benefits they provide, and their limitations. Informed consent processes are being updated to address the involvement of AI in care decisions, ensuring that patients understand when AI systems are used and how they contribute to treatment recommendations (Salybekov et al., 2025). Patient advocacy groups are playing a crucial role in AI development and implementation, ensuring that patient perspectives and preferences are considered in system design and deployment decisions (Mennella et al., 2024). Community engagement efforts are working to address concerns about AI in healthcare while promoting understanding of the benefits these technologies can provide for improving patient outcomes (Bajwa et al., 2021).

14 Conclusion

Predictive AI is revolutionizing heart transplantation in ways that appeared inconceivable a few years ago. From enhancing donor–recipient matching and predicting organ viability to optimizing immunosuppressive therapies and enabling continuous remote monitoring (Olawade, Marinze, Qureshi, Weerasinghe, & Teke, 2025a), AI is touching every aspect of the transplantation journey with transformative results. The evidence is robust and continues to grow with each new study and clinical implementation. ML models predict transplant failure and mortality by hospital stay, immunosuppressive regimen, recipient age, congenital heart disease, and organ ischemia time (Naruka et al., 2022). AI systems are consistently outperforming traditional approaches, delivering better outcomes, reducing complications, and improving the quality of life for transplant recipients across multiple measures.

The transformation extends far beyond clinical metrics to encompass every aspect of the patient experience. Early detection capabilities, which once seemed like science fiction, are now enabling interventions that prevent complications rather than simply treating them after they occur. Patients can now receive continuous monitoring at home rather than in the ICU. Personalized medicine, utilizing genetic profiles, lifestyle factors, and real-time physiological data, is becoming standard care (Johnson et al., 2021).

Yet, this remarkable progress represents just the beginning of what AI can achieve in transplantation medicine. As we look toward the future, the integration of genomic medicine, the development of xenotransplantation protocols, and the advancement of truly personalized medicine approaches promise to make heart transplantation more successful, more accessible, and more sustainable than ever before (Chen et al., 2025; Reichart et al., 2022).

The economic case for AI implementation is becoming increasingly compelling, with demonstrated improvements in cost-effectiveness, resource utilization, and long-term value creation (Khanna et al., 2022). Healthcare systems that embrace these technologies early are likely to see the greatest benefits in terms of patient outcomes, operational efficiency, and competitive advantage.

While significant challenges remain, particularly those related to data privacy, algorithmic bias, and clinical validation, the trajectory is unmistakably clear. The convergence of advancing AI capabilities, increasing data availability, and growing clinical expertise in AI implementation is creating unprecedented opportunities for improving transplant outcomes (Bajwa et al., 2021). Perhaps most importantly, AI is enhancing human skill, which enables transplantation. The best implementations keep clinicians at the center of care while providing them with extraordinary analytical capabilities, decision support tools, and monitoring systems that enhance rather than replace clinical judgment.

AI in heart transplantation represents a fundamental shift toward more precise, personalized, and proactive care, not just technological improvement. This transformation enables healthcare providers to deliver better care with greater efficiency while improving the experience and outcomes for patients and families facing the challenges of end-stage heart disease. For the thousands of patients awaiting heart transplants and the many more who will need them in years to come, this technological revolution offers hope, healing, and the promise of longer, healthier lives. The future of heart transplantation is bright, driven by AI working in partnership with dedicated healthcare professionals who remain committed to delivering the best possible care for their patients. As we continue to refine these technologies, address the remaining challenges, and implement them, we are closer to making heart transplantation a highly successful, predictable medical intervention that gives patients the best chance of a full and healthy life. The journey toward this future is well underway, and the early results suggest that the best is yet to come.

References

Abbas, S. A., Yusifzada, I., & Athar, S. (2025). Revolutionizing medicine: chatbots as catalysts for improved diagnosis, treatment, and patient support. *Cureus*. 2168-8184. *17*(3). doi:10.7759/cureus.80935.

Abd Al-Alim, M., Mubarak, R., Salem, N. M., & Sadek, I. (2024). A machine-learning approach for stress detection using wearable sensors in free-living environments. *Computers in Biology and Medicine*. 18790534. *179*. Available from https://www.sciencedirect.com/science/journal/00104825. doi:10.1016/j.compbiomed.2024.108918.

Abdelrazik, A., Eldesouky, M., Antoun, I., Lau, E. Y. M., Koya, A., Vali, Z., ... Ng, G. A. (2025). Wearable devices for arrhythmia detection: advancements and clinical implications. *Sensors*. 14248220. *25*(9). Available from http://www.mdpi.com/journal/sensors. doi:10.3390/s25092848.

Adedinsewo, D., Hardway, H. D., Morales-Lara, A. C., Wieczorek, M. A., Johnson, P. W., Douglass, E. J., ... Yamani, M. (2023). Non-invasive detection of cardiac allograft rejection among heart transplant recipients using an electrocardiogram based deep learning model. *European Heart Journal - Digital Health*. 26343916. *4*(2), 71–80. Available from https://academic.oup.com/ehjdh. doi:10.1093/ehjdh/ztad001.

Ahluwalia, M., Sehgal, S., Lee, G., Agu, E., & Kpodonu, J. (2025). Disparities in artificial intelligence–based tools among diverse minority populations. *JACC: Advances*. 2772963X. *4*(5), 101742. doi:10.1016/j.jacadv.2025.101742.

Al Moussawy, M., Lakkis, Z. S., Ansari, Z. A., Cherukuri, A. R., & Abou-Daya, K. I. (2024). The transformative potential of artificial intelligence in solid organ transplantation. *Frontiers in Transplantation*. 28132440. *3*. Available from https://www.frontiersin.org/journals/transplantation. doi:10.3389/frtra.2024.1361491.

Ali, A., Ali, H., Saeed, A., Ahmed Khan, A., Tin, T. T., Assam, M., ... Mohamed, H. G. (2023). Blockchain-powered healthcare systems: enhancing scalability and security with hybrid deep learning. *Sensors*. 14248220. *23*(18). Available from http://www.mdpi.com/journal/sensors. doi:10.3390/s23187740.

Ali, S., Alam, W., Tyara, & H. Chong K. T. (2025). An accurate prediction of drug–drug interactions and side effects by using integrated convolutional and BiLSTM networks. *Chemometrics and Intelligent Laboratory Systems*. 01697439. *257*, 105304. doi:10.1016/j.chemolab.2024.105304.

Alimbayeva, Z., Alimbayev, C., Ozhikenov, K., Bayanbay, N., & Ozhikenova, A. (2024). Wearable ECG device and machine learning for heart monitoring. *Sensors*. 1424-8220. *24* (13), 4201. doi:10.3390/s24134201.

Alowais, S. A., Alghamdi, S. S., Alsuhebany, N., Alqahtani, T., Alshaya, A. I., Almohareb, S. N., ... Albekairy, A. M. (2023). Revolutionizing healthcare: the role of artificial intelligence in clinical practice. *BMC Medical Education*. 14726920. *23*(1). Available from https://bmcmededuc.biomedcentral.com/. doi:10.1186/s12909-023-04698-z.

Amal, S., Safarnejad, L., Omiye, J. A., Ghanzouri, I., Cabot, J. H., & Ross, E. G. (2022). Use of multi-modal data and machine learning to improve cardiovascular disease care. *Frontiers in Cardiovascular Medicine*. 2297055X. *9*. Available from www.frontiersin.org/journals/cardiovascular-medicine. doi:10.3389/fcvm.2022.840262.

Anjewierden, S., O'Sullivan, D., Mangold, K. E., Attia, I. Z., Lopez-Jimenez, F., Friedman, P. A., ... Burchill, L. J. (2025). *Artificial Intelligence–Derived Electrocardiographic Age Predicts Mortality in Adults with Congenital Heart Disease. JACC: Advances. 2772963X*, *4*(6), Available from https://www.sciencedirect.com/journal/jacc-advances. doi:10.1016/j.jacadv.2025.101777.

Ansar, W., & Ghosh, S. (2016). Inflammation and inflammatory diseases, markers, and mediators: Role of CRP in some inflammatory diseases. *Springer Science and Business Media, LLC*, 67–107. doi:10.1007/978-81-322-2680-2_4.

Arabyarmohammadi, S., & Madabhushi, A., AI approach better at predicting heart transplant rejection than standard clinical method. (2024), Available from: https://news.emory.edu/stories/2024/02/hs_ai_cardiac_transplant_13-02-2024/.

Arjmandmazidi, S., Heidari, H. R., Ghasemnejad, T., Mori, Z., Molavi, L., Meraji, A., ... Montazersaheb, S. (2025). An In-depth overview of artificial intelligence (AI) tool utilization across diverse phases of organ transplantation. *Journal of Translational Medicine*. 14795876. *23*(1). Available from https://translational-medicine.biomedcentral.com. doi:10.1186/s12967-025-06488-1.

Attaripour Esfahani, S., Baba Ali, N., Farina, J. M., Scalia, I. G., Pereyra, M., Abbas, M. T., ... Arsanjani, R. (2025). A comprehensive review of artificial intelligence (AI) applications in pulmonary hypertension (PH). *Medicina (Lithuania)*. 16489144. *61*(1). Available from https://www.mdpi.com/journal/medicina. doi:10.3390/medicina61010085.

Babu, M., & Snyder, M. (2023). Multi-omics profiling for health. *Molecular & Cellular Proteomics*. 15359476. *22*(6), 100561. doi:10.1016/j.mcpro.2023.100561.

Babuin, L., & Jaffe, A. S. (2005). Troponin: the biomarker of choice for the detection of cardiac injury. *Canadian Medical Association Journal*. 14882329. *173* (10), 1191–1202. doi:10.1503/cmaj.050141.

Baddal, B., Taner, F., & Uzun Ozsahin, D. (2024). Harnessing of artificial intelligence for the diagnosis and prevention of hospital-acquired infections: A systematic review. *Diagnostics*. 2075-4418. *14*(5), 484. doi:10.3390/diagnostics14050484.

Bai, C., Lianlian, W., Ruijiang, L., Cao, Y., Song, H., & Xiaochen, B. (2025). Machine learning-enabled drug-induced toxicity prediction. *Advanced Science*. 2198-3844. *12*(16). doi:10.1002/advs.202413405.

Bajwa, J., Munir, U., Nori, A., & Williams, B. (2021). Artificial intelligence in healthcare: Transforming the practice of medicine. *Future Healthcare Journal*. 25146645. *8*(2), e188. doi:10.7861/fhj.2021-0095.

Baldwin, C., & Brusco, N. K. (2011). The effect of vocational rehabilitation on return-to-work rates post stroke: A systematic review. *Topics in Stroke Rehabilitation*. 19455119. *18* (5), 562–572. doi:10.1310/tsr1805-562.

Barry, T., Farina, J. M., Chao, C. J., Ayoub, C., Jeong, J., Patel, B. N., ... Arsanjani, R. (2023). The role of artificial intelligence in echocardiography. *Journal of Imaging*. 2313433X. *9*(2). Available from www.mdpi.com/journal/jimaging/. doi:10.3390/jimaging9020050.

Bathula, A., Gupta, S. K., Merugu, S., Saba, L., Khanna, N. N., Laird, J. R., ... Suri, J. S. (2024). Blockchain, artificial intelligence, and healthcare: the tripod of future—A narrative review. *Artificial Intelligence Review*. 15737462. *57*(9). Available from https://www.springer.com/journal/10462. doi:10.1007/s10462-024-10873-5.

Bohlmann, A., Mostafa, J., & Kumar, M. (2021). Machine learning and medication adherence: scoping review. *JMIRx Med*. 2563-6316. *2* (4), e26993. doi:10.2196/26993.

Borna, S., Maniaci, M. J., Haider, C. R., Gomez-Cabello, C. A., Pressman, S. M., Haider, S. A., ... Forte, A. J. (2024). Artificial intelligence support for informal patient caregivers: a systematic review. *Bioengineering*. 23065354. *11*(5). Available from www.mdpi.com/journal/bioengineering. doi:10.3390/bioengineering11050483.

Brunet, M., & Pastor-Anglada, M. (2022). Insights into the pharmacogenetics of tacrolimus pharmacokinetics and pharmacodynamics. *Pharmaceutics*. 1999-4923. *14*(9), 1755. doi:10.3390/pharmaceutics14091755.

Burnette, H., Pabani, A., Von Itzstein, M. S., Switzer, B., Fan, R., Ye, F., ... Johnson, D. B. (2024). Use of artificial intelligence chatbots in clinical management of immune-related adverse events. *Journal for ImmunoTherapy of Cancer*. 20511426. *12*(5). Available from http://www.immunotherapyofcancer.org/. doi:10.1136/jitc-2023-008599.

Cabanas, A. M., Sáez, N., Collao-Caiconte, P. O., Martín-Escudero, P., Pagán, J., Jiménez-Herranz, E., & Ayala, J. L. (2024). Evaluating AI methods for pulse oximetry: Performance, clinical accuracy, and comprehensive bias analysis. *Bioengineering*. 23065354. *11*(11). Available from www.mdpi.com/journal/bioengineering. doi:10.3390/bioengineering11111061.

Cao, Z., Jia, Y., & Zhu, B. (2019). BNP and NT-proBNP as diagnostic biomarkers for cardiac dysfunction in both clinical and forensic medicine. *International Journal of Molecular Sciences*. 1422-0067. *20*(8), 1820. doi:10.3390/ijms20081820.

Cappelli, F., Castronuovo, G., Grimaldi, S., & Telesca, V. (2024). Random forest and feature importance measures for discriminating the most influential environmental factors in predicting cardiovascular and respiratory diseases. *International Journal of Environmental Research and Public Health*. 1660-4601. *21*(7), 867. doi:10.3390/ijerph21070867.

Cappon, G., Vettoretti, M., Sparacino, G., & Facchinetti, A. (2019). Continuous glucose monitoring sensors for diabetes management: A review of technologies and applications. *Diabetes & Metabolism Journal*. 2233-6079. *43*(4), 383. doi:10.4093/dmj.2019.0121.

Casas, S., Tangprasertchai, N. S., Oikonomaki, K., Mathers, S., Sollet, Z. C., Samara, S., ... Viard, T. (2024). Multi-centre analytical performance verification of an IVD assay to quantify donor-derived cell-free DNA in solid organ transplant recipients. *HLA*. 20592310. *103*(5). Available from http://onlinelibrary.wiley.com/journal/10.1111/(ISSN)2059-2310/issues. doi:10.1111/tan.15518.

Cestonaro, C., Delicati, A., Marcante, B., Caenazzo, L., & Tozzo, P. (2023). Defining medical liability when artificial intelligence is applied on diagnostic algorithms: A systematic review. *Frontiers in Medicine*. 2296-858X. *10*. doi:10.3389/fmed.2023.1305756.

Chandrasekhar, N., & Peddakrishna, S. (2023). Enhancing heart disease prediction accuracy through machine learning techniques and optimization. *Processes*. 2227-9717. *11*(4), 1210. doi:10.3390/pr11041210.

Chen, Y. M., Hsiao, T. H., Lin, C. H., & Fann, Y. C. (2025). Unlocking precision medicine: clinical applications of integrating health records, genetics, and immunology through artificial intelligence. *Journal of Biomedical Science*. 14230127. *32*(1), 16. doi:10.1186/s12929-024-01110-w.

Chih, S., Chruscinski, A., Ross, H. J., Tinckam, K., Butany, J., & Rao, V. (2012). Antibody-mediated rejection: An evolving entity in heart transplantation. *Journal of Transplantation*. 2090-0007. *2012*, 1–10. doi:10.1155/2012/210210.

Chowdhury, M. A., Rizk, R., Chiu, C., Zhang, J. J., Scholl, J. L., Bosch, T. J., ... Chen, W. C. W. (2025). The heart of transformation: Exploring artificial intelligence in cardiovascular disease. *Biomedicines*. 22279059. *13*(2). Available from www.mdpi.com/journal/biomedicines. doi:10.3390/biomedicines13020427.

Coutance, G., Desiré, E., & Van Huyen, J.-P. D. (2022). A review of biomarkers of cardiac allograft rejection: Toward an integrated diagnosis of rejection. *Biomolecules*. 2218-273X. *12*(8), 1135. doi:10.3390/biom12081135.

Denecke, K., & Gabarron, E. (2024). The ethical aspects of integrating sentiment and emotion analysis in chatbots for depression intervention. *Frontiers in Psychiatry*. 1664-0640. *15*. doi:10.3389/fpsyt.2024.1462083.

Deng, M. C. (2017). The AlloMap™ genomic biomarker story: 10 years after. *Clinical Transplantation*. 13990012. *31*(3). Available from http://www.wiley.com/bw/journal.asp?ref=0902-0063. doi:10.1111/ctr.12900.

Deshpande, R. (2024). Smart match: revolutionizing organ allocation through artificial intelligence. *Frontiers in Artificial Intelligence*. 2624-8212. *7*. doi:10.3389/frai.2024.1364149.

Dong, J., Chu, M., & Xu Y. (2025). A predictive healthcare model using machine learning and psychological factors for medication adherence. *Healthcare Analytics*. 27724425. *7*, 100397. doi:10.1016/j.health.2025.100397.

Ennab, M., & Mcheick, H. (2024). Enhancing interpretability and accuracy of AI models in healthcare: a comprehensive review on challenges and future directions. *Frontiers in Robotics and AI*. 2296-9144. *11*. doi:10.3389/frobt.2024.1444763.

Fortuni, F., Ciliberti, G., De Chiara, B., Conte, E., Franchin, L., Musella, F., ... Oliva, F. (2024). Advancements and applications of artificial intelligence in cardiovascular imaging: A comprehensive review. *European Heart Journal - Imaging Methods and Practice*. 2755-9637. *2*(4). doi:10.1093/ehjimp/qyae136.

Galli, C. (2025). Current techniques of gene editing in pigs for xenotransplantation. *Transplant International*. 1432-2277. *38*. doi:10.3389/ti.2025.13807.

Gani, I., Litchfield, I., Shukla, D., Delanerolle, G., Cockburn, N., & Pathmanathan, A. (2025). Understanding "alert fatigue" in primary care: qualitative systematic review of general practitioners attitudes and experiences of clinical alerts, prompts, and reminders. *Journal of Medical Internet Research*. 1438-8871. *27*, e62763. doi:10.2196/62763.

Ganie, S. M., Pramanik, P. K. D., & Zhao, Z. (2025). Ensemble learning with explainable AI for improved heart disease prediction based on multiple datasets. *Scientific Reports*. 20452322. *15*(1). Available from https://www.nature.com/srep/. doi:10.1038/s41598-025-97547-6.

Garcia-Canadilla, P., Sanchez-Martinez, S., Martí-Castellote, P. M., Slorach, C., Hui, W., Piella, G., ... Friedberg, M. K. (2022). Machine-learning–based exploration to identify remodeling patterns associated with death or heart-transplant in pediatric-dilated cardiomyopathy. *Journal of Heart and Lung Transplantation*. 15573117. *41*(4), 516–526. Available from www.elsevier.com/locate/healun. doi:10.1016/j.healun.2021.11.020.

Ghorbani, A., Ouyang, D., Abid, A., He, B., Chen, J. H., Harrington, R. A., ... Zou, J. Y. (2020). Deep learning interpretation of echocardiograms. *Npj Digital Medicine*. 23986352. *3*(1). Available from https://www.nature.com/npjdigitalmed/. doi:10.1038/s41746-019-0216-8.

Ghosh, D., & Cabrera, J. (2022). Enriched random forest for high dimensional genomic data. *IEEE/ACM Transactions on Computational Biology and Bioinformatics*. 1545-5963. *19* (5), 2817–2828. doi:10.1109/tcbb.2021.3089417.

Giménez-Miranda, L., Scagliusi, S. F., Pérez-García, P., Olmo-Fernández, A., Huertas, G., Yúfera, A., & Medrano, F. J. (2024). Wearable devices based on bioimpedance test in heart failure: Clinical relevance: systematic review. *Reviews in Cardiovascular Medicine*. 1530-6550. *25*(9). doi:10.31083/j.rcm2509315.

Global Observatory on Donation and Transplant, (2021). *Total heart transplant—Europe 2023*. GODT. https://www.transplant-observatory.org/data-charts-and-tables/chart/.

Godbole, A. A., Paras, M. M., Banerjee, S., Roy, P., Deb, N., & Jagtap, S. (2025). Enhancing infection control in ICUS through AI: A literature review. *Health Science Reports*. 23988835. *8*(1). Available from https://onlinelibrary.wiley.com/journal/23988835. doi:10.1002/hsr2.70288.

Goel, H., Melot, J., Krinock, M. D., Kumar, A., Nadar, S. K., & Lip, G. Y. H. (2020). Heart-type fatty acid-binding protein: An overlooked cardiac biomarker. *Annals of Medicine*. 13652060. *52*(8), 444–461. Available from http://www.tandfonline.com/loi/iann20. doi:10.1080/07853890.2020.1800075.

Gotlieb, N., Azhie, A., Sharma, D., Spann, A., Suo, N. J., Tran, J., … Bhat, M. (2022). The promise of machine learning applications in solid organ transplantation. *Npj Digital Medicine*. 23986352. *5*(1). Available from https://www.nature.com/npjdigitalmed/. doi:10.1038/s41746-022-00637-2.

Grzyb, C., Du, D., & Nair, N. (2024). Artificial intelligence approaches for predicting the risks of durable mechanical circulatory support therapy and cardiac transplantation. *Journal of Clinical Medicine*. 2077-0383. *13*(7), 2076. doi:10.3390/jcm13072076.

Guido, R., Ferrisi, S., Lofaro, D., & Conforti, D. (2024). An overview on the advancements of support vector machine models in healthcare applications: A review. *Information*. 2078-2489. *15*(4), 235. doi:10.3390/info15040235.

Gunasekara, T. D. K. S. C., Herath, C., De Silva, P. M. C. S., & Jayasundara, N. (2024). Exploring the utility of urinary creatinine adjustment for KIM-1, NGAL, and cystatin C for the assessment of kidney function: Insights from the C-KidnEES cohort. *Children*. 22279067. *11*(1). Available from https://www.mdpi.com/journal/children. doi:10.3390/children11010015.

Hanna, J. J., & Medford, R. J. (2024). Navigating the future: Machine learning's role in revolutionizing antimicrobial stewardship and infection prevention and control. *Current Opinion in Infectious Diseases*. 14736527. *37*(4), 290–295. Available from http://journals.lww.com/co-infectiousdiseases/pages/default.aspx. doi:10.1097/QCO.0000000000001028.

HealthSnap, 2025. AI in remote patient monitoring: The top 4 use cases in 2025, https://welcome.healthsnap.io/blog/ai-in-remote-patient-monitoring-the-top-4-use-cases-in-2025.

Hryhorowicz, M., Zeyland, J., Słomski, R., & Lipiński, D. (2017). Genetically modified pigs as organ donors for xenotransplantation. *Molecular Biotechnology*. 1073-6085. *59*(9–10), 435–444. doi:10.1007/s12033-017-0024-9.

Husom, E. J., Dautov, R., Nedisan, A. A., Gonidis, F., Papatzelos, S., & Malamas, N. 2022. Machine learning for fatigue detection using fitbit fitness trackers. In: International Conference on Sport Sciences Research and Technology Support, icSPORTS – Proceedings, Science and Technology Publications, Lda Norway, vol. 2022, 9789897586101 pp. 41–52. http://icsports.scitevents.org.

Idris-Agbabiaka, A., Anjum, M. M., Semy, M., Rath, S., Rizwan, M., Victoria, O. O., … Ashinze, P. (2025). AI-assisted heart failure management: a review of clinical applications, case studies, and future directions. *Global Cardiology Science and Practice*. 23057823. *2025*(1). Available from https://globalcardiologyscienceandpractice.com/index.php/gcsp/article/view/775. doi:10.21542/gcsp.2025.6.

Ishaq, F., & Guha, A. (2025). History of heart transplant: Setting the stage. *Methodist DeBakey Cardiovascular Journal*. 1947-6108. *21*(3), 5–12. doi:10.14797/mdcvj.1579.

Jafari, A., Ajji, Z., Mousavi, A., Naghieh, S., Bencherif, S. A., & Savoji, H. (2022). Latest advances in 3D bioprinting of cardiac tissues. *Advanced Materials Technologies*. 2365709X. *7*(11). Available from http://onlinelibrary.wiley.com/journal/10.1002/(ISSN)2365-709X. doi:10.1002/admt.202101636.

Jain, R., Kransdorf, E. P., Cowger, J., Jeevanandam, V., & Kobashigawa, J. A. (2025). Donor selection for heart transplantation in 2025. *JACC: Heart Failure*. 22131779. *13*(3), 389–401. Available from https://www.sciencedirect.com/science/journal/22131779. doi:10.1016/j.jchf.2024.09.016.

Johnson, K. B., Wei, W. Q., Weeraratne, D., Frisse, M. E., Misulis, K., Rhee, K., … Precision Medicine, A. I. (2021). *and the Future of Personalized Health Care. Clinical and Translational Science*, 17528062. *14*(1), 86–93. Available from http://onlinelibrary.wiley.com/journal/10.1111/(ISSN)1752-8062. doi:10.1111/cts.12884.

Jomova, K., Raptova, R., Alomar, S. Y., Alwasel, S. H., Nepovimova, E., Kuca, K., & Valko, M. (2023). Reactive oxygen species, toxicity, oxidative stress, and antioxidants: Chronic diseases and aging. *Archives of Toxicology*. 14320738. *97*(10), 2499–2574. Available from https://www.springer.com/journal/204. doi:10.1007/s00204-023-03562-9.

Jordaan, E. M., & Smits, G. F., 2004 Robust outlier detection using SVM regression. In: IEEE International Conference on Neural Networks—Conference Proceedings, Netherlands, vol. 3, 10987576 2017–2022.

Junaid, S. B., Imam, A. A., Shuaibu, A. N., Basri, S., Kumar, G., Surakat, Y. A., … Alazzawi, A. K. (2022). Artificial intelligence, sensors and vital health signs: A review. *Applied Sciences (Switzerland)*. 20763417. *12*(22). Available from www.mdpi.com/journal/applsci/. doi:10.3390/app122211475.

Katzowitz, B., Predicting transplant rejection with AI. Emory Health Digest. (2024), Available from: https://emoryhealthdigest.emory.edu/issues/2024/summer/and-more/transplant-rejection-ai1/index.html.

Khalifa, M., & Albadawy, M. (2024). AI in diagnostic imaging: revolutionising accuracy and efficiency. *Computer Methods and Programs in Biomedicine Update*. 26669900. *5*(100146). doi:10.1016/j.cmpbup.2024.100146.

Khanna, N. N., Maindarkar, M. A., Viswanathan, V., Fernandes, J. F. E., Paul, S., Bhagawati, M., … Suri, J. S. (2022). Economics of artificial intelligence in healthcare: diagnosis vs. *Treatment. Healthcare (Switzerland)*. 22279032. *10*(12). Available from www.mdpi.com/journal/healthcare/. doi:10.3390/healthcare10122493.

Khera, R., Oikonomou, E. K., Nadkarni, G. N., Morley, J. R., Wiens, J., Butte, A. J., & Topol, E. J. (2024). Transforming cardiovascular care with artificial intelligence: From discovery to practice. *Journal of the American College of Cardiology*. 07351097. *84*(1), 97–114. doi:10.1016/j.jacc.2024.05.003.

Kufel, J., Bargieł-Łączek, K., Kocot, S., Koźlik, M., Bartnikowska, W., Janik, M., … Gruszczyńska, K. (2023). *What Is Machine Learning, Artificial Neural Networks and Deep Learning?—Examples of Practical Applications in Medicine. Diagnostics*, 2075-4418. *13*(15), 2582. doi:10.3390/diagnostics13152582.

Kukade, A., Kawade, S., Lokare, S., & Kharage, A. (2025). Artificial intelligence in medication adherence and personalized treatment plans. *International Journal of Scientific Research and Technology*, *2*(5), 29–40. doi:10.5281/zenodo.15315405.

Ledziński, Ł., & Grześk, G. (2023). Artificial intelligence technologies in cardiology. *Journal of Cardiovascular Development and Disease*. 2308-3425. *10*(5), 202. doi:10.3390/jcdd10050202.

Lima, P. C., Rios, D. M., de Oliveira, F. P., Passos, L. R., Ribeiro, L. B., Serpa, R. G., … Barbosa, R. R. (2022). Inflammation as a prognostic marker in heart failure. *Cureus*. 2168-8184. *14*(8). doi:10.7759/cureus.28605.

Lin, M., Guo, J., Zhilin, G., Tang, W., Tao, H., You, S., … Jia, P. (2025). Machine learning and multi-omics integration: Advancing cardiovascular translational research and clinical practice. *Journal of Translational Medicine*. 1479-5876. *23*(1). doi:10.1186/s12967-025-06425-2.

Lipkova, J., Chen, T. Y., Lu, M. Y., Chen, R. J., Shady, M., Williams, M., … Mahmood, F. (2022). Deep learning-enabled assessment of cardiac allograft rejection from endomyocardial biopsies. *Nature Medicine*. 1546170X. *28*(3), 575–582. Available from http://www.nature.com/nm/index.html. doi:10.1038/s41591-022-01709-2.

Lisboa, P. J. G., Jayabalan, M., Ortega-Martorell, S., Olier, I., Medved, D., & Nilsson, J. (2022). Enhanced survival prediction using explainable artificial intelligence in heart transplantation. *Scientific Reports*. 20452322. *12*(1). Available from https://www.nature.com/srep/. doi:10.1038/s41598-022-23817-2.

Litjens, G., Ciompi, F., Wolterink, J. M., de Vos, B. D., Leiner, T., Teuwen, J., & Išgum, I. (2019). State-of-the-art deep learning in cardiovascular image analysis. *JACC: Cardiovascular Imaging*. 18767591. *12*(8), 1549–1565. Available from http://www.elsevier.com. doi:10.1016/j.jcmg.2019.06.009.

Liu, Y., Feng, R., Chen, J., Yan, H., & Liu, X. (2024). Optimizing post-transplantation outcomes: The role of multi-omics, artificial intelligence, and animal models in addressing immunosuppression-associated hepatotoxicity. *International Journal of Surgery*. 1743-9159. *110* (8), 5207–5209. doi:10.1097/js9.0000000000001521.

Luo, H., Yin, W., Wang, J., Zhang, G., Liang, W., Luo, J., & Yan, C. (2024). Drug-drug interactions prediction based on deep learning and knowledge graph: A review. *iScience*. 25890042. *27*(3), 109148. doi:10.1016/j.isci.2024.109148.

Maleki Varnosfaderani, S., & Forouzanfar, M. (2024). The role of AI in hospitals and clinics: Transforming healthcare in the 21st century. *Bioengineering*. 2306-5354. *11*(4), 337. doi:10.3390/bioengineering11040337.

Martin, C. M. (2025). Heart transplantation: 40 years and counting. *Methodist DeBakey Cardiovascular Journal*. 19476108. *21*(3), 1–4. doi:10.14797/mdcvj.1605.

Martini, L., Mandoli, G. E., Pastore, M. C., Pagliaro, A., Bernazzali, S., Maccherini, M., … Cameli, M. (2024). Heart transplantation and biomarkers: A review about their usefulness in clinical practice. *Frontiers in Cardiovascular Medicine*. 2297-055X. *11*. doi:10.3389/fcvm.2024.1336011.

Matsuzaka, Y., & Yashiro, R. (2024). Understanding and therapeutic application of immune response in major histocompatibility complex (MHC) diversity using multimodal artificial intelligence. *BioMedInformatics*. 2673-7426. *4*(3), 1835–1864. doi:10.3390/biomedinformatics4030101.

McMurray, J. P., DeVries, A., Frazee, K., Sizemore, B., Branan, K. L., Jennings, R., & Coté, G. L. (2025). A novel wearable device for continuous blood pressure monitoring utilizing strain gauge technology. *Biosensors*. 20796374. *15*(7). Available from http://www.mdpi.com/journal/biosensors/. doi:10.3390/bios15070413.

Mehra, A., Starkoff, B. E., & Nickerson, B. S. (2025). The evolution of bioimpedance analysis: From traditional methods to wearable technology. *Elsevier Inc., United States Nutrition*. 18731244. *129*. Available from https://www.sciencedirect.com/science/journal/08999007. doi:10.1016/j.nut.2024.112601.

Mennella, C., Maniscalco, U., De Pietro, G., & Esposito, M. (2024). Ethical and regulatory challenges of AI technologies in healthcare: A narrative review. *Heliyon*. 24058440. *10*(4), e26297. doi:10.1016/j.heliyon.2024.e26297.

Mienye, I. D., Sun, Y., & Wang, Z. (2020). An improved ensemble learning approach for the prediction of heart disease risk. *Informatics in Medicine Unlocked*. 23529148. *20*, 100402. doi:10.1016/j.imu.2020.100402.

Moayedi, Y., Rodenas-Alesina, E., Somerset, E., Fan, C. P. S., Henricksen, E., Aleksova, N., … Teuteberg, J. J. (2024). Enhancing the prediction of cardiac allograft vasculopathy using intravascular ultrasound and machine learning: A proof of concept. *Circulation: Heart Failure*. 19413297. *17*(2), 107–111. Available from http://circheartfailure.ahajournals.org/. doi:10.1161/CIRCHEARTFAILURE.123.011306.

Mohammadi, I., Farahani, S., Karimi, A., Jahanian, S., Firouzabadi, S. R., Alinejadfard, M., … Akhlaghpasand, M. (2025). Mortality prediction of heart transplantation using machine learning models: A systematic review and meta-analysis. *Frontiers Media SA, Iran Frontiers in Artificial Intelligence*. 26248212. *8*. Available from www.frontiersin.org/journals/artificial-intelligence#. doi:10.3389/frai.2025.1551959.

Mohsin, S. N., Gapizov, A., Ekhator, C., Ain, N. U., Ahmad, S., Khan, M., … Nagaraj, R. H. (2023). The role of artificial intelligence in prediction, risk stratification, and personalized treatment planning for congenital heart diseases. *Cureus*. 2168-8184. *15*(8). doi:10.7759/cureus.44374.

Moingeon, P. (2023). Artificial intelligence-driven drug development against autoimmune diseases. *Trends in Pharmacological Sciences*. 01656147. *44*(7), 411–424. doi:10.1016/j.tips.2023.04.005.

Mugaanyi, J., Huang, J., Fang, J., Musinguzi, A., Caide, L., & Chen, Z. (2025). Developments and applications of liver-on-a-chip technology—Current status and future prospects. *Biomedicines*. 2227-9059. *13*(6), 1272. doi:10.3390/biomedicines13061272.

Naruka, V., Arjomandi Rad, A., Subbiah Ponniah, H., Francis, J., Vardanyan, R., Tasoudis, P., … Athanasiou, T. (2022). Machine learning and artificial intelligence in cardiac transplantation: A systematic review. *Artificial Organs*. 15251594. *46*(9), 1741–1753. Available from http://onlinelibrary.wiley.com/journal/10.1111/(ISSN)1525-1594. doi:10.1111/aor.14334.

Ng, S., Masarone, S., Watson, D., & Barnes, M. R. (2023). The benefits and pitfalls of machine learning for biomarker discovery. *Cell and Tissue Research*. 14320878. *394*(1), 17–31. Available from https://www.springer.com/journal/441. doi:10.1007/s00441-023-03816-z.

Ni, Y., & Jia, F. (2025). A scoping review of AI-driven digital interventions in mental health care: Mapping applications across screening, support, monitoring, prevention, and clinical education. *Healthcare*. 2227-9032. *13*(10), 1205. doi:10.3390/healthcare13101205.

Nikolova, A., Agbor-Enoh, S., Bos, S., Crespo-Leiro, M., Ensminger, S., Jimenez-Blanco, M., … Potena, L. (2024). European Society for Organ Transplantation (ESOT) consensus statement on the use of non-invasive biomarkers for cardiothoracic transplant rejection surveillance. *Transplant International*. 1432-2277. *37*. doi:10.3389/ti.2024.12445.

Nosrati, H., & Nosrati, M. (2023). Artificial intelligence in regenerative medicine: Applications and implications. *Biomimetics*. 2313-7673. *8* (5), 442. doi:10.3390/biomimetics8050442.

Novák, J., Macháčková, T., Krejčí, J., Bienertová-Vašků, J., & Slabý, O. (2021). MicroRNAs as theranostic markers in cardiac allograft transplantation: From murine models to clinical practice. *Theranostics*. 1838-7640. *11*(12), 6058–6073. doi:10.7150/thno.56327.

Oates, G. R., Juarez, L. D., Hansen, B., Kiefe, C. I., & Shikany, J. M. (2020). Social risk factors for medication nonadherence: Findings from the CARDIA study. *American Journal of Health Behavior*. 19457359. *44*(2), 232–243. Available from http://docserver.ingentaconnect.com/deliver/connect/png/10873244/v44n2/s10.pdf?expires=1580967912&id=0000&titleid=41000053&checksum=04A56A28FB0E42F837ADC8E1AE590747. doi:10.5993/AJHB.44.2.10.

Olawade, D. B., Aderinto, N., Olatunji, G., Kokori, E., & David-Olawade, A. C., Hadi, M. (2024). Advancements and applications of Artificial Intelligence in cardiology: current trends and future prospects. *Journal of Medicine, Surgery, and Public Health, 2949916X*(3), 100109. doi:10.1016/j.glmedi.2024.100109.

Olawade, D. B., Marinze, S., Qureshi, N., Weerasinghe, K., & Teke, J. (2025a). The impact of artificial intelligence and machine learning in organ retrieval and transplantation: A comprehensive review. *Current Research in Translational Medicine*. 24523186. *73*(2), 103493. doi:10.1016/j.retram.2025.103493.

Olawade, D. B., Marinze, S., Qureshi, N., Weerasinghe, K., & Teke, J. (2025b). Transforming organ donation and transplantation: Strategies for increasing donor participation and system efficiency. *European Journal of Internal Medicine*.18790828. *133*, 14–24. Available from https://www.sciencedirect.com/science/journal/09536205. doi:10.1016/j.ejim.2024.11.010.

Olawade, D. B., Odetayo, A., Marinze, S., Egbon, E., & Chinwah, V. (2025). Organ transplantation in Africa: confronting socioeconomic, cultural, and infrastructural hurdles. *Current Research in Translational Medicine*. 24523186. *73*(3), 103516. doi:10.1016/j.retram.2025.103516.

Olawade, D. B., Wada, O. Z., Odetayo, A., David-Olawade, A. C., Asaolu, F., & Eberhardt, J. (2024). Enhancing mental health with artificial intelligence: Current trends and future prospects. *Journal of Medicine, Surgery, and Public Health*. 2949916X. *3*, 100099. doi:10.1016/j.glmedi.2024.100099.

Osa-Sanchez, A., Ramos-Martinez-de-Soria, J., Mendez-Zorrilla, A., Ruiz, I. O., & Garcia-Zapirain, B. (2025). Wearable sensors and artificial intelligence for sleep apnea detection: A systematic review. *Journal of Medical Systems*. 1573-689X. *49*(1). doi:10.1007/s10916-025-02199-8.

Palmieri, V., Montisci, A., Vietri, M. T., Colombo, P. C., Sala, S., Maiello, C., … Napoli, C. (2023). Artificial intelligence, big data and heart transplantation: Actualities. *International Journal of Medical Informatics*. 18728243. *176*. Available from www.elsevier.com/inca/publications/store/5/0/6/0/4/0/. doi:10.1016/j.ijmedinf.2023.105110.

Panigrahi, R. R., Mukherjee, S., Shaikh, Z. H., & Nomran, N. M. (2025). Leveraging blockchain for transparency: A study on organ supply chains and transplant processes. *Logistics*. 23056290. *9*(1). Available from https://www.mdpi.com/2305-6290/9/1/9. doi:10.3390/logistics9010009.

Pantanowitz, L., Hanna, M., Pantanowitz, J., Lennerz, J., Henricks, W. H., Shen, P., … Rashidi, H. H. (2024). Regulatory aspects of artificial intelligence and machine learning. *Modern Pathology*. 15300285. *37*(12). Available from https://www.sciencedirect.com/journal/modern-pathology/about/insights. doi:10.1016/j.modpat.2024.100609.

Peloso, A., Moeckli, B., Delaune, V., Oldani, G., Andres, A., & Compagnon, P. (2022). Artificial intelligence: Present and future potential for solid organ transplantation. *Transplant International*. 1432-2277. *35*. doi:10.3389/ti.2022.10640.

Pinto-Coelho, L. (2023). How artificial intelligence is shaping medical imaging technology: A survey of innovations and applications. *Bioengineering*. 2306-5354. *10*(12), 1435. doi:10.3390/bioengineering10121435.

Poweleit, E. A., Vinks, A. A., & Mizuno, T. (2023). Artificial intelligence and machine learning approaches to facilitate therapeutic drug management and model-informed precision dosing. *Therapeutic Drug Monitoring*. 15363694. *45*(2), 143–150. Available from http://journals.lww.com/drug-monitoring. doi:10.1097/FTD.0000000000001078.

Przybylek, B., Boethig, D., Neumann, A., Borchert-Moerlins, B., Daemen, K., Keil, J., … Bara, C. (2019). Novel cytokine score and cardiac allograft vasculopathy. *The American Journal of Cardiology*. 00029149. *123*(7), 1114–1119. doi:10.1016/j.amjcard.2018.12.034.

Qahwaji, R., Ashankyty, I., Sannan, N. S., Hazzazi, M. S., Basabrain, A. A., & Mobashir, M. (2024). Pharmacogenomics: A genetic approach to drug development and therapy. *Pharmaceuticals*. 14248247. *17*(7). Available from https://www.mdpi.com/journal/pharmaceuticals. doi:10.3390/ph17070940.

Ramaswamy, M., Senthilnathan, S., Saravanan, A. R. I., Snehal, M., & Siva, K. (2025). Leveraging machine learning for risk assessment and prediction of inheriting genetic disorders. *Computational and Structural Biotechnology Reports*. 29503639. *2*(100038). doi:10.1016/j.csbr.2025.100038.

Ramesh, S., Deep, A., Tamayol, A., Kamaraj, A., Mahajan, C., & Madihally, S. (2024). Advancing 3D bioprinting through machine learning and artificial intelligence. *Bioprinting*. 24058866. *38*, e00331. doi:10.1016/j.bprint.2024.e00331.

Reddy, S. (2025). Global harmonization of artificial intelligence-enabled software as a medical device regulation: Addressing challenges and unifying standards. *Mayo Clinic Proceedings: Digital Health. 29497612*, *3*(1), 100191. doi:10.1016/j.mcpdig.2024.100191.

Reichart, B., Cooper, D. K. C., Längin, M., Tönjes, R. R., Pierson, R. N., & Wolf, E. (2022). Cardiac xenotransplantation: from concept to clinic. *Cardiovascular Research*. 17553245. *118*(18), 3499–3516. Available from http://cardiovascres.oxfordjournals.org/. doi:10.1093/cvr/cvac180.

Robinson, C. L., D'souza, R. S., Yazdi, C., Diejomaoh, E. M., Schatman, M. E., Emerick, T., & Orhurhu, V. (2024). Reviewing the potential role of artificial intelligence in delivering personalized and interactive pain medicine education for chronic pain patients. *Journal of Pain Research*.11787090. *17*, 923–929. Available from https://www.dovepress.com/getfile.php?fileID=97333. doi:10.2147/JPR.S439452.

Salybekov, A. A., Yerkos, A., & Sedlmayr, M., Wolfien, M. (2025). Ethics and algorithms to navigate AI's emerging role in organ transplantation. *Journal of Clinical Medicine*. 2077-0383. *14*(8), 2775. doi:10.3390/jcm14082775.

Sargiotis, G. C., Sergentanis, T. N., Pavi, E., & Athanasakis, K. (2024). Predictive performance of artificial intelligence models on heart and lung posttransplant health outcomes: A systematic review. *Experimental and Clinical Transplantation*. 13040855. *22*(11), 823–833. Available from https://www.ectrx.org/class/pdfPreview.php?year=2024&volume=22&issue=11&supplement=0&spage_number=823makale_no=0. doi:10.6002/ect.2024.0207.

Scholte, N. T. B., van Ravensberg, A. E., Shakoor, A., Boersma, E., Ronner, E., de Boer, R. A., ... van der Boon, R. M. A. (2024). A scoping review on advancements in noninvasive wearable technology for heart failure management. *npj Digital Medicine*. 23986352. *7*(1). Available from https://www.nature.com/npjdigitalmed/. doi:10.1038/s41746-024-01268-5.

Seçkin, A.C., Ateş, B., & Seçkin, M. (2023). Review on wearable technology in sports: Concepts, challenges and opportunities. *Applied Sciences*. 2076-3417. *13*(18), 10399. doi:10.3390/app131810399.

Serrano, D. R., Luciano, F. C., Anaya, B. J., Ongoren, B., Kara, A., Molina, G., ... Lalatsa, A. (2024). Artificial intelligence (AI) applications in drug discovery and drug delivery Revolutionizing personalized medicine. *Pharmaceutics*. 19994923. *16*(10). Available from http://www.mdpi.com/journal/pharmaceutics. doi:10.3390/pharmaceutics16101328.

Shaik, T., Tao, X., Higgins, N., Lin, L., Gururajan, R., & Zhou, X., Acharya, U.R. (2023). Remote patient monitoring using artificial intelligence: current state, applications, and challenges. *WIREs Data Mining and Knowledge Discovery*. 1942-4787. *13*(2). doi:10.1002/widm.1485.

Singh, M., Kumar, A., Khanna, N. N., Laird, J. R., Nicolaides, A., Faa, G., ... Suri, J. S. (2024). Artificial intelligence for cardiovascular disease risk assessment in personalised framework: a scoping review. *eClinicalMedicine*. 25895370. *73*. Available from https://www.sciencedirect.com/science/journal/25895370. doi:10.1016/j.eclinm.2024.102660.

Siontis, K. C., Noseworthy, P. A., Attia, Z. I., & Friedman, P. A. (2021). Artificial intelligence-enhanced electrocardiography in cardiovascular disease management. *Nature Reviews Cardiology*. 17595010. *18*(7), 465–478. Available from http://www.nature.com/nrcardio/archive/index.html. doi:10.1038/s41569-020-00503-2.

Smarr, B. L., Aschbacher, K., Fisher, S. M., Chowdhary, A., Dilchert, S., Puldon, K., ... Mason, A. E. (2020). Feasibility of continuous fever monitoring using wearable devices. *Nature Research, United States Scientific Reports*. 20452322. *10*(1). Available from www.nature.com/srep/index.html. doi:10.1038/s41598-020-78355-6.

Srivastav, A. K., Mishra, M. K., Lillard, J. W., & Singh, R. (2025). Transforming pharmacogenomics and CRISPR gene editing with the power of artificial intelligence for precision medicine. *Pharmaceutics*. 19994923. *17*(5). Available from http://www.mdpi.com/journal/pharmaceutics. doi:10.3390/pharmaceutics17050555.

Srivastava, A., Bhavsar, S., & Mishra, R. (2024). Significance of artificial intelligence in stem cell therapy. *Journal of Stem Cell Research & Therapeutics*. 24755540. *9*(1), 21–24. doi:10.15406/jsrt.2024.09.00168.

Streed, J., Tomorrow's cure: physicians are using AI to help transform organ transplant—Mayo clinic news network. (2024), Available from: https://newsnetwork.mayoclinic.org/discussion/tomorrows-cure-physicians-are-using-ai-to-help-transform-organ-transplant/.

Syrowatka, A., Song, W., Amato, M. G., Foer, D., Edrees, H., Co, Z., ... Bates, D. W. (2022). Key use cases for artificial intelligence to reduce the frequency of adverse drug events: A scoping review. *The Lancet Digital Health*. 25897500. *4*(2), e137. Available from https://www.sciencedirect.com/journal/the-lancet-digital-health. doi:10.1016/S2589-7500(21)00229-6.

Tan, M., Xiao, Y., Jing, F., Xie, Y., Sanmei, L., Xiang, M., & Ren, H. (2024). Evaluating machine learning-enabled and multimodal data-driven exercise prescriptions for mental health: A randomized controlled trial protocol. *Frontiers in Psychiatry*. 1664-0640. *15*. doi:10.3389/fpsyt.2024.1352420.

Tolentino, R., Baradaran, A., Gore, G., Pluye, P., & Abbasgholizadeh-Rahimi, S. (2024). Curriculum frameworks and educational programs in AI for medical students, residents, and practicing physicians: Scoping review. *JMIR Medical Education*. 2369-3762. *10*, e54793. doi:10.2196/54793.

Tolu-Akinnawo, O. Z., Ezekwueme, F., Omolayo, O., Batheja, S., & Awoyemi, T. (2025). Advancements in artificial intelligence in noninvasive cardiac imaging: A comprehensive review. *Clinical Cardiology*. 19328737. *48*(1). Available from http://onlinelibrary.wiley.com/journal/10.1002/(ISSN)1932-8737. doi:10.1002/clc.70087.

Tong, L., Hoffman, R., Deshpande, S. R., & Wang, M. D., 2017. Predicting heart rejection using histopathological whole-slide imaging and deep neural network with dropout. In: IEEE EMBS International Conference on Biomedical and Health Informatics, BHI 2017, Institute of Electrical and Electronics Engineers Inc., United States, 9781509041794. pp. 1–4, doi:10.1109/BHI.2017.7897190.

Tsvetanov, F. (2024). Integrating AI technologies into remote monitoring patient systems†. *Engineering Proceedings*. 26734591. *70*(1). Available from www.mdpi.com/journal/engproc. doi:10.3390/engproc2024070054.

Udoy, I. A., & Hassan, O. (2025). AI-Driven technology in heart failure detection and diagnosis: A review of the advancement in personalized healthcare. *Symmetry*. 20738994. *17*(3). Available from http://www.mdpi.com/journal/symmetry/. doi:10.3390/sym17030469.

Vössing, M., Kühl, N., Lind, M., & Satzger, G. (2022). Designing transparency for effective human-AI collaboration. *Information Systems Frontiers*. 1387-3326. *24*(3), 877–895. doi:10.1007/s10796-022-10284-3.

Wah, J. N. K. (2025). Revolutionizing e-health: The transformative role of AI-powered hybrid chatbots in healthcare solutions. *Frontiers in Public Health*. 22962565. *13* Available from http://journal.frontiersin.org/journal/public-health/section/public-health-education-and-promotion#archive. doi:10.3389/fpubh.2025.1530799.

Wasilewski, T., Kamysz, W., & Gębicki, J. (2024). AI-assisted detection of biomarkers by sensors and biosensors for early diagnosis and monitoring. *Biosensors*. 2079-6374. *14*(7), 356. doi:10.3390/bios14070356.

Wettersten, N. (2021). Biomarkers in acute heart failure: Diagnosis, prognosis, and treatment. korean society of heart failure. *United States International Journal of Heart Failure*. 26361558. *3*(2), 81–105. Available from https://e-heartfailure.org/pdf/10.36628/ijhf.2020.0036. doi:10.36628/ijhf.2020.0036.

Wies, C., Miltenberger, R., Grieser, G., & Jahn-Eimermacher, A. (2023). Exploring the variable importance in random forests under correlations: a general concept applied to donor organ quality in post-transplant survival. *BMC Medical Research Methodology*. 1471-2288. *23*(1). doi:10.1186/s12874-023-02023-2.

Wolf, E., Schmoeckel, M., & Reichart, B. (2023). Cardiac xenotransplantation—From bench to bedside. *European Journal of Transplantation*. 2974-6892. *1*(3), 192–206. doi:10.57603/ejt-305.

Wu, G. W., Kobashigawa, J. A., Fishbein, M. C., Patel, J. K., Kittleson, M. M., Reed, E. F., … Ardehali, A. (2009). Asymptomatic antibody-mediated rejection after heart transplantation predicts poor outcomes. *Journal of Heart and Lung Transplantation*. 10532498. *28*(5), 417–422. doi:10.1016/j.healun.2009.01.015.

Yang, S., & Kar, S. (2023). Application of artificial intelligence and machine learning in early detection of adverse drug reactions (ADRs) and drug-induced toxicity. *Artificial Intelligence Chemistry*. 29497477. *1*(2), 100011. doi:10.1016/j.aichem.2023.100011.

Zaslavsky, M. E., Craig, E., Michuda, J. K., Sehgal, N., Ram-Mohan, N., Lee, J. Y., … Boyd, S. D. (2025). Disease diagnostics using machine learning of B cell and T cell receptor sequences. *Science*. 10959203. *387*(6736). Available from https://www.science.org/toc/science/387/6736. doi:10.1126/science.adp2407.

Zhang, Q., Hickey, M., Drogalis-Kim, D., Zheng, Y., Gjertson, D., Cadeiras, M., … Reed, E. F. (2018). Understanding the correlation between DSA, complement activation, and antibody-mediated rejection in heart transplant recipients. *Transplantation*. 00411337. *102*(10), e431. Available from: journals.lww.com/transplantjournal/pages/default.aspx. doi:10.1097/TP.0000000000002333.

Zhang, Y., Wang, Z., Wang, Y., Jin, W., Zhang, Z., Jin, L., … Zheng, L. (2024). CYP3A4 and CYP3A5: The crucial roles in clinical drug metabolism and the significant implications of genetic polymorphisms. *PeerJ*. 2167-8359. *12*(12), e18636. doi:10.7717/peerj.18636.

Zhao, A. P., Li, S., Cao, Z., Hu, P. J. H., Wang, J., Xiang, Y., … Lu, X. (2024). AI for science: Predicting infectious diseases. *Journal of Safety Science and Resilience*. 26664496. *5*(2), 130–146. Available from https://www.sciencedirect.com/science/journal/26664496. doi:10.1016/j.jnlssr.2024.02.002.

Zhao, Z., Zhang, Y., Jiuhui, S., Yang, L., Pang, L., Gao, Y., & Wang, H. (2024). A comprehensive review for artificial intelligence on neuroimaging in rehabilitation of ischemic stroke. *Frontiers in Neurology*. 1664-2295. *15*. doi:10.3389/fneur.2024.1367854.

Zhou, X., Xianying, L., Zhang, Z., Han, Q., Deng, H., Jiang, Y., … Yang, L. (2022). Support vector machine deep mining of electronic medical records to predict the prognosis of severe acute myocardial infarction. *Frontiers in Physiology*. 1664-042X. *13*. doi:10.3389/fphys.2022.991990.

Zhu, X. Y., Zhang, K. J., Li, X., Su, F. F., & Tian, J. W. (2024). An interpretable machine learning method for risk stratification of patients with acute coronary syndrome. *Heliyon*. 24058440. *10*(17). Available from https://www.sciencedirect.com/science/journal/24058440. doi:10.1016/j.heliyon.2024.e36815.

Zoghi, M. (2025). Artificial Intelligence and coronary artery calcium scoring: Enhancing cardiovascular risk assessment. *International Journal of the Cardiovascular Academy*. 24058181. *11*(2), 38–40. doi:10.4274/ijca.2025.54227.

Chapter 9

Advancing precision surgery with artificial intelligence: Opportunities and challenges

Rayner Peyser Cardoso[a] and Swizel Ann Cardoso[b]
[a]*All India Institute of Medical Sciences, Jodhpur, Rajasthan, India*
[b]*Department of Vascular Surgery, Doncaster Royal Infirmary, Doncaster and Bassetlaw Teaching Hospitals NHS Foundation Trust, Doncaster, United Kingdom*

1 Introduction

One of the most fascinating and promising areas of contemporary medicine is the nexus between artificial intelligence (AI) and surgery. Surgery is also moving toward more precision, as seen in patient-specific therapies, real-time data integration, and minimally invasive techniques, as healthcare increasingly adopts personalized approaches. With the potential to revolutionize every step of the surgical process, from preoperative planning to postoperative recuperation, AI is becoming more than just a tool in this changing environment (Topol, 2019).

The term "precision surgery" describes surgical methods and procedures that are especially adapted to each patient's unique anatomical and physiological traits to maximize results while reducing risks, made possible by real-time intraoperative feedback, robotics, genomics, and high-resolution imaging (Boggi, 2023; Yang et al., 2020). By finding patterns in massive datasets, forecasting surgical outcomes, spotting minute irregularities during procedures, and offering intelligent support via robotic systems, AI enhances this potential (Yang et al., 2020).

Large volumes of imaging data, surgical recordings, and patient records can now be analyzed by machine learning algorithms to help guide decisions prior to, during, and following surgery. Real-time alerts to potential errors, surgical step suggestions, and anatomical landmark identification are all made possible by computer vision algorithms. While surgical robots are being trained to adapt to complex surroundings through reinforcement learning (RL) approaches, natural language processing tools aid in the interpretation of clinical notes or operating reports. For instance, Intuitive Surgical's da Vinci robotic platform is widely used in robotic surgery and supports features such as motion scaling and tremor filtration, while related AI-enabled systems are being investigated for tissue discrimination and workflow prediction (Herron & Marohn, 2008).

Notwithstanding these developments, there are still significant and unsolved obstacles to the promise of AI in precision surgery. Clinical responsibility, algorithmic transparency, and data privacy continue to be major topics of discussion. How do we validate AI systems in complex, high-risk surgical settings? How do we ensure that these tools complement, rather than replace, clinical judgment? And how do we manage the risk of bias when training algorithms on datasets that may not reflect the full diversity of patient populations?

Regulatory bodies like the US Food and Drug Administration (FDA) and the United Kingdom's National Institute for Health and Care Excellence (NICE) are still developing frameworks to govern AI-based surgical tools (U.S. Food and Drug Administration, 2021). Ethical discussions highlight the need to maintain human judgment and patient consent while challenging the assignment of life-or-death choices to machines (World Health Organization, 2021).

In addition, there are structural issues with access and equity. Advanced AI system adoption frequently necessitates significant financing, technological know-how, and infrastructure, resources that are not consistently available across hospitals or nations. These differences have the potential to exacerbate already-existing healthcare disparities if they are not addressed. International cooperation, publicly available datasets, and context-sensitive AI design are necessary to close these gaps.

Applications of Artificial Intelligence in Healthcare. https://doi.org/10.1016/B978-0-443-43934-6.00012-2

This chapter is dedicated to exploring the rapidly evolving relationship between AI and precision surgery. It investigates the opportunities AI offers for enhancing surgical precision, training, and safety, while critically engaging with the ethical, regulatory, and practical challenges it introduces.

As we step into this new era, the fundamental question we must address is not whether AI will change the nature of surgery, but how we can guide this transformation to ensure it is safe, equitable, and truly beneficial for all patients.

2 Opportunities presented by AI in precision surgery

From that very first outpatient consultation all the way through to the final rehabilitation session, it seems every single surgical touchpoint now has an AI counterpart. Think of radiomic "digital twins" helping surgeons practice those really complex resections; computer vision engines marking out tissue planes in real time; predictive models catching complications while they are just tiny biochemical hints; and adaptive robots performing incredibly precise, submillimeter moves that no human hand could ever hope to match. The following ten subsections will delve deep into these exciting possibilities, mixing clear technical explanations with the latest evidence.

2.1 Enhanced preoperative planning and diagnostics

Modern radiomics truly begins with automated segmentation. A surgical *digital twin* is a high-fidelity, physics-based replica of a patient's anatomy linked in real time to their imaging and physiologic data. Because every simulation run produces thousands of labeled frames "for free," digital twins have become integral training data generators for radiomic AI models (Asciak et al., 2025). Since these simulations run thousands of permutations overnight, surgeons walk into theater with a shortlist of evidence-backed maneuvers rather than a single static plan. Deep convolutional networks can now effortlessly label bone, vessels, and tumors on a chest CT in less than three minutes, achieving a Dice score of 0.89, which is actually quite a bit better than senior radiologists at 0.77 (Hashimoto, Rosman, Rus, & Meireles, 2018). The outcome is a vibrant, color-coded, three-dimensional hologram that surgeons can easily manipulate on a touchscreen or even export into a mixed-reality headset, providing them with an immersive "fly-through" preview of tricky anatomical areas.

Layered on top of this, risk stratification engines integrate clinical metadata directly onto the anatomical map. For instance, Hashimoto's gradient boost ensemble meticulously sifted through 70,000 patient charts, achieving an impressive area under the curve (AUC) of 0.94 for 30-day surgical site infection (SSI), completely outperforming the 0.76 benchmark set by traditional logistic regression (Hashimoto et al., 2018). When this alert system was integrated into an enhanced wound care pathway, deep SSI readmissions saw a significant 22% drop in the first year (Gordon, Hernandez, & Collins, 2023). We are also seeing similar models now staging colorectal and lung cancers with greater accuracy than TNM (Tumor, node, metastasis) alone by blending radiomics with cell-free DNA signatures, a breakthrough that is already reshaping neoadjuvant strategies at several leading centers (Kenig, Echeverria, & Vives, 2024).

Digital pathology really brings everything full circle. Slide scanning networks are now classifying mitoses and grading dysplasia in mere seconds, allowing multidisciplinary teams to finalize operative plans days sooner. At one neurosurgical unit, implementing same-day cytology review boosted theater utilization by 7% because those last-minute cancellations virtually vanished (Sowndharya, Bhavani, Vickram, & Emran, 2024).

2.2 Real-time intraoperative guidance and robotic assistance

Modern computer vision platforms now process laparoscopic video at a swift 25–30 frames per second, applying real-time color-constancy correction to generate a live, overlaid roadmap right back to the console. In a 300-patient colorectal trial, this system successfully flagged 92% of deep mesenteric vessels with a false-positive rate under 5%, impressively cutting inadvertent vascular injury by one-third (Sowndharya et al., 2024).

Why Near-infrared fluorescence (NIR)? Near-infrared fluorescence (700–900 nm) penetrates tissue 5–10 × deeper than visible light, suffers less hemoglobin scatter, and avoids the colorful reflections of open surgery. AI models track the full fluorescence time-intensity curve, not just the peak value, so they can flag subtle perfusion deficits before gross color change, one reason the early-clamping protocol lowered leak rates from 6.4% to 3.1% (Sowndharya et al., 2024).

On the robotic front, features like tremor attenuation and 5:1 motion scaling significantly reduced pedicle-screw breaches from 6% to 4% in a prospective spinal cohort (Ng, 2025). Achieving submillimeter accuracy becomes even more crucial in delicate procedures like retinal peels and microvascular anastomoses, where tissue failure thresholds are incredibly fine, hovering around 7–10 μm. We now have semiautonomous "smart needles" that can suture fascia, freeing up the surgeon to focus on higher-level tasks like confirming tension and ensuring hemostasis. These first-generation systems are capable of placing six symmetric throws in just 90 s, which is a remarkable improvement compared to the 220 s required for manual sewing (Ng, 2025).

2.3 Enhanced patient safety and perioperative outcomes

The volume of physiological data generated in real time during surgery—often gigabytes of waveform data per hour—exceeds the interpretative capacity of human monitoring. Recurrent neural networks (RNNs), by interpreting arterial pressure and capnography, have managed to predict vasodilatory hypotension a full 20 min before it reaches its lowest point. This allows anesthetists to proactively preload vasopressors and reduce the "time-under-MAP (Mean arterial pressure) 65 mmHg" by an impressive 38% (Sowndharya et al., 2024).

At the broader system level, hospitals are now running "digital command centers." One UK NHS trust used 2 years' worth of timestamp data to train a reinforcement scheduler, which dynamically resequences surgical lists at 06:00 daily. This alone yielded a 17% reduction in turnover downtime and saved £1.5 million in overtime pay, without sacrificing case volume (Kenig et al., 2024).

Such platforms are also being used to tackle logistical inefficiencies. Predictive demand models are now preventing prosthetic stock-outs by forecasting implant requirements three days in advance, an important step toward optimizing value-based surgical care (Kenig et al., 2024).

2.4 Surgical education and training

AI-driven simulators are proving to be a game-changer, tackling three persistent issues in surgical education: the limited number of cases available, the subjective nature of feedback, and the inevitable skill decay. Imagine immersive virtual reality headsets paired with haptic gloves that realistically mimic tissue compliance, allowing trainees to repeatedly practice an end-to-end anastomosis until their force vectors, needle angles, and dwell times perfectly align with expert benchmarks. A meta-analysis of eleven randomized trials revealed a 27% faster ascent to proficiency and a two-point Objective Structured Assessment of Technical Skills (OSATS) gain for groups mentored by AI (Herzog et al., 2025).

Meanwhile, real-time video coaching is proving equally transformative. Every instrument trajectory is meticulously skeletonized and algorithmically labeled, with motion economy analyzed across multiple dimensions. When residents receive a color-coded "performance thermogram," their knot-tying time improves by a median of 18% in the very next case (Amirian et al., 2023). Several surgical boards are now seriously considering whether such objective analytics should count toward mandatory case numbers.

2.5 Advancements in robotics and automation

The newest generation of surgical robots is truly impressive, seamlessly combining RL control loops with multispectral vision and advanced force sensing. For instance, in total knee arthroplasty, the RL agent can recalculate femoral rotation after ligament balancing, ensuring that 97% of components are placed within ± 1° and consequently trimming 2-year revision rates from 4.3% to 3.5% (Wood, 2025).

In delicate soft tissue surgery, the integration of force–feedback is crucial for preventing crush injuries; porcine trials have actually reported a remarkable 90% reduction in unintended cortical perforations during microvascular suturing (Ng, 2025).

While fully autonomous surgery remains a future goal, "co-bots" are already adeptly handling repetitive subtasks. Automated purse-string suturing, precise tissue stapling along predefined curves, and automatic camera recentering are rapidly becoming routine procedures. This effectively frees up the surgeon's cognitive capacity, allowing them to focus more on overarching strategy and troubleshooting.

2.6 Personalized medicine and precision therapeutics

AI-driven multiomic integration is paving the way for personalized surgery. A breast cancer consortium, for instance, successfully combined mRNA sequencing, DNA methylation data, and MRI radiomics to predict microscopic spread with an impressive 88% accuracy. This allowed surgeons to selectively widen margins, effectively halving reexcision rates without compromising cosmetic outcomes (Kenig et al., 2024).

Similar frameworks are now identifying low-risk thyroid nodules, potentially sparing up to 40% of patients from undergoing a hemithyroidectomy (Kenig et al., 2024).

Postoperative pain management is another exciting frontier. A model that combined CYP2D6 genotype with wearable accelerometry data was able to precisely titrate multimodal analgesia, reducing the morphine-equivalent dose by 40% and shortening hospital discharge by 1.2 days, all while consistently keeping the mean visual analog scale (VAS) pain score at ≤3 (Guardian Science Desk, 2025). Hospitals are keenly observing these results to fine-tune their opioid stewardship programs.

2.7 Intraoperative decision support

Maintaining margin clarity during surgery is critical, especially in neurosurgery, where brain shift can rapidly distort navigation accuracy. A hybrid model that seamlessly fuses functional MRI (fMRI) with optical coherence tomography (OCT) automatically reregisters every 30 s to account for brain shift. **Why does it work?** Once the dura is opened, cerebrospinal fluid loss, gravity, and tumor debulking cause the brain surface to sag several millimeters, so static neuronavigation rapidly drifts out of register.

The AI engine compensates in two passes.

- **Rigid pass—vascular fiducials.** It extracts arterial/venous line segments that are visible in both the preoperative fMRI and the live OCT stack and snaps those landmarks into the best rigid alignment within 80 ms (Tonutti, Gras, & Yang, 2017).
- **Nonrigid pass—learned deformation field.** A convolutional network, pretrained on over 1000 logged resections, predicts a voxel-wise displacement map that models how gravity vectors and resection cavities warp deep tissue. This nonlinear field is applied to the anatomical atlas, producing an updated overlay in <250 ms (Tonutti et al., 2017).

The result keeps tumor and eloquent-cortex boundaries within **±1 mm** of their true position for the remainder of the procedure, restoring navigation accuracy despite continual brain shift (Liao, Zhang, Hu, & Qin, 2024). This innovative approach has led to a 33% reduction in residual glioma volume and has brought second-look craniotomies down to single digits (Liao et al., 2024).

Pancreatic surgeons are now trialing a fluorescence-plus-AI margin tool that provides a "clean" or "close" report in just 90 s, saving approximately 20 min per Whipple case and eliminating delays associated with frozen section analyses (Liao et al., 2024).

2.8 Advanced analytics for quality improvement

Hospitals once relied primarily on retrospective Morbidity and Mortality (M&M) conferences, but AI now offers near-real-time surveillance capabilities. One cross-institutional dashboard, for example, identified a specific stapler whose reload mechanism was prolonging warm ischemia by 6 min. By switching vendors, leak rates were halved within 6 months (Kenig et al., 2024).

When such operational insights are fed back into procurement and supply chain management, institutions can simultaneously enhance patient safety and achieve cost efficiencies, aligning clinical excellence with fiscal responsibility.

2.9 Minimally invasive surgery and remote operations

Thanks to latency-tolerant control schemes, 5G tele-robotics is now becoming a practical reality. In a Boston–Montreal porcine model, two lobectomies were completed without incident, with end-to-end round-trip times hovering around 60–70 ms. An embedded edge-AI filter even predicted potential lag spikes and proactively compensated for instrument motion (Guardian Science Desk, 2025).

The humanitarian benefits are immense; imagine delivering expert surgical care to remote, austere environments, though challenges related to licensure and network resilience certainly remain.

2.10 Ethical AI and patient trust

Patients will truly embrace AI only if it is demonstrably fair, transparent, and secure. *Surgery (Oxford)* now recommends that every surgical algorithm should, at a minimum, display its salient features and a confidence score (Singh, Ryu, & Thomas, 2025). In a pilot study, surgeons chose to override low-confidence (<0.6) advice 78% of the time, effectively validating the crucial "human-in-the-loop" safeguard (Singh et al., 2025).

Regulatory bodies are actively discussing the implementation of "algorithm passports"—akin to pharmaceutical leaflets—which would clearly define each model's data provenance, performance boundaries, and failure modes. These are seen as essential steps in establishing trustworthy AI in surgical practice (World Health Organization, 2021).

3 Case studies and applications

Table 9.1.

AI systems are no longer confined to academic posters or impressive benchtop demonstrations; they are actively reshaping the daily practices of many surgical subspecialties. This section aims to tell that story in detail. Each vignette will set the clinical scene, meticulously explain the AI workflow step-by-step, report peer-reviewed outcomes, and offer reflections on what that particular experience means for broader adoption.

3.1 Robotic-assisted surgery

Clinical setting: Radical prostatectomy presents a complex challenge; the goal is complete oncologic resection while preserving the delicate neurovascular bundles critical for continence and erectile function. Even in the hands of the most skilled human surgeons, positive margins can, unfortunately, reach double digits, and quality-of-life side effects are quite common.

AI Integration: At three European centers, a convolutional neural network (CNN) was applied to preoperative multiparametric MRI (mpMRI) scans to delineate tumor boundaries, the prostatic capsule, and adjacent nerves. These segmented structures were rendered into a 3D holographic overlay projected onto the da Vinci robotic console, guiding surgical dissection in real time. Intraoperatively, tremor filtering and a 5:1 motion scaling ratio translated centimeter-scale movements at the console into submillimeter, 0.2-mm precision at the instrument tip (Martin et al., 2024).

Outcomes: Across 600 consecutive patients, this protocol significantly boosted pad-free continence at 12 months from 84% (observed with historical laparoscopy) to an impressive 92%, while positive surgical margins dropped from 12% to 8%. Erectile function recovery also showed improvement, although that specific analysis is still undergoing peer review. Crucially, no increase in operative time or transfusion rates was observed.

Multimodal Input Pipeline to Surgeon Console in Robotic Surgery (Fig. 9.1).

This case study shows that three complementary AI modules—**image segmentation, haptic filtering, and trajectory prediction**—can be layered into the da Vinci workflow without adding cognitive burden to the surgeon. Their synergy is key:

1. **Image segmentation** uses a 3D U-Net to delineate the prostate, capsule, and neurovascular bundles on mpMRI, streaming color-coded voxels to the console display (Martin et al., 2024).
2. **Haptic filtering** attenuates hand tremors by running joystick inputs through a 10 Hz Kalman low-pass filter and applying a variable 5:1 motion scaling kernel (Martin et al., 2024).
3. **Trajectory prediction** employs a recurrent policy network trained on 2400 prior prostatectomies to forecast the next 5-mm dissection arc while steering clear of high-risk voxels (Martin et al., 2024).

Working together, these modules act as an intuitive co-pilot—delivering finer margins and better continence outcomes without prolonging operative time. The color-coded nerve overlay, in particular, builds surgeon trust by making the system's reasoning transparent.

TABLE 9.1 Selected real-world AI case studies in surgery.

Discipline	Clinical need	AI intervention	Documented outcome
Urology—radical prostatectomy	Spare neurovascular bundles	mpMRI rendered to 3D nerve map; da Vinci motion scaled to 0.2 mm	Higher negative-margin rate and improved continence scores (Martin, Chau, & Del Vecchio, 2024)
Orthopedics—total knee arthroplasty	<1° implant alignment	CT analyzed by ML; optical tracker updates for soft tissue tension	Alignment error < 1°, early revisions ↓ 18% (Wood, 2025)
Neurosurgery—diffuse glioma	Max resection, preserve eloquent cortex	fMRI fused with intra-op OCT; CNN compensates brain shift	Residual tumor volume ↓ 33%; re-op rate ↓ 25% (Liao et al., 2024)
General surgery—SSI prediction	Early warning of wound failure	Gradient-boosting on 70,000 cases	AUC 0.94 vs. 0.76; targeted follow-up cuts readmissions (Hashimoto et al., 2018)
Training—VR simulation	Standardize skill acquisition	AI logs kinematics and forces, delivers instant coaching	45% error reduction in live assessment (Patel, Harrington, & Lee, 2024)

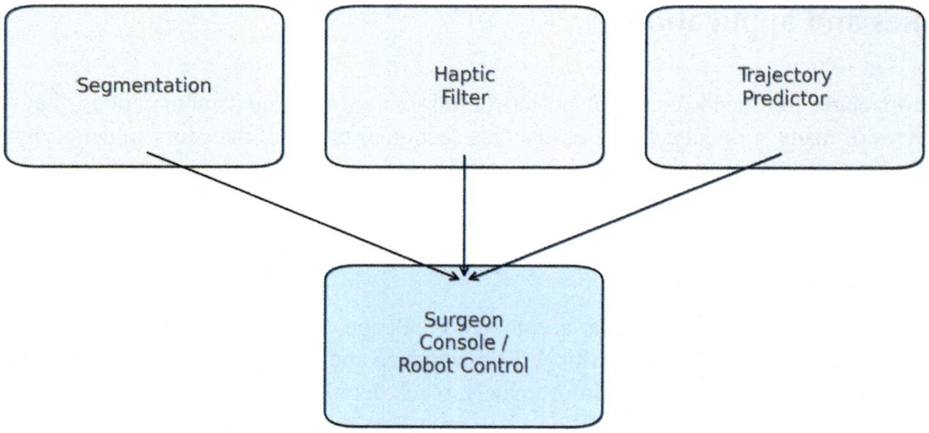

FIG. 9.1 Parallel AI modules feed real-time guidance to the surgeon console.

3.2 AI-guided total joint arthroplasty

Clinical problem: Small deviations in implant alignment—often by just 1–2 degrees—can lead to early loosening or patellar maltracking, two leading causes of failure in total joint arthroplasty.

AI Workflow: In a prospective multicenter trial, an "AI cut-planner" analyzed preoperative CT scans, constructed a virtual femur–tibia model, and then proposed ideal cut depths and angles (Wood, 2025). An optical tracker dynamically recalibrated this plan once ligaments were balanced, and the robotic saw then executed the cuts while meticulously recording force and deviation.

Outcomes: Postoperative CT scans confirmed that a remarkable 97% of femoral and tibial components were positioned within ±1 degree of the mechanical axis (compared to 81% in the manual cohort), and early revision rates at 2 years decreased from 4.3% to 3.5%. Patient-reported Forgotten-Joint Scores at 6 months were also superior, strongly suggesting genuinely improved kinematics. Surgeons particularly appreciated that the AI generates a printable PDF detailing its reasoning, complete with saliency maps, which fostered high acceptance right from the start.

3.3 Image-guided glioma resection

Clinical challenge: Diffuse gliomas have a tendency to infiltrate eloquent cortex, and static neuronavigational systems, unfortunately, lose accuracy as soon as the dura is opened and the brain starts to sag.

AI solution: A research group in Shanghai tackled this challenge by seamlessly fusing fMRI with real-time OCT (Martin et al., 2024). Every 30 s, a CNN reregistered the two modalities, updating tumor margins on the fly.

Clinical impact: Among 52 patients, the mean residual tumor volume significantly decreased by one-third, and second-look craniotomies declined from 12% to 9%. Seizure-free survival was also extended by a median of 4 months. The truly critical insight here was the thoughtful workflow design: volumetric OCT scanning added <90 s to the overall operation because the AI efficiently handled all the alignment calculations while the surgeon continued with the resection.

3.4 Predictive analytics and perioperative management

Complication forecasting: Hashimoto's gradient boost ensemble, trained on an impressive 70,000 patient charts, achieved an AUC of 0.94 for 30-day deep SSI (Hashimoto et al., 2018). A subsequent external validation involving 15,000 patients maintained an AUC of 0.92 and, more importantly, demonstrated that when the high-risk alert triggered an enhanced wound care pathway, deep SSI readmissions dropped by 22%, and the median length of stay was reduced by 2 days (Duhaime, 2021).

Closed-loop anesthesia: A prototype, highlighted by The Guardian, incorporated bispectral index (BIS), blood pressure waveform data, and end-tidal CO_2 into an RNN that precisely titrated propofol and remifentanil in real time (Guardian Science Desk, 2025). Across 120 general surgery cases, the time during which mean arterial pressure was below 65 mmHg decreased by 40%, and postanesthesia care unit (PACU) discharge was 25% faster, all with only a single manual override needed.

The overarching theme here is anticipation: both of these systems are designed to catch physiological or infectious deviations much earlier than a busy clinical team can, effectively buying crucial time for preemptive therapy.

3.5 AI-centric surgical education

A multiinstitutional randomized trial assigned 48 residents to one of two groups: either traditional video replay or a fully immersive VR trainer equipped with haptic sensors and computer vision analytics (Patel et al., 2024). The AI platform meticulously measured needle angle, suture tension, instrument clashes, and time-in-field and then delivered a color-coded dashboard after each attempt.

At the final live animal assessment, the AI group committed half as many critical errors (11% decreased to 5%), completed an intestinal anastomosis 45% faster, and scored two full points higher on the OSATS global rating. Beyond just speed, faculty observed a qualitative shift: trainees arrived in the operating theater fluent in objective metrics, which allowed for much more specific and constructive feedback.

Subsequent work by Amirian et al. introduced explainable AI (XAI) overlays, visually demonstrating to novices where wasted motion occurred and which instrument choices led to increased task time (Amirian et al., 2023). Acceptance among residents soared once they could clearly see why the algorithm judged a particular maneuver as suboptimal.

3.6 Postoperative monitoring and rehabilitation

Wearable sensors that track heart rate variability and gait cadence feed into a gradient boost model designed to flag recovery trajectories that deviate two standard deviations from the baseline. In a study of 300 colorectal patients, AI alerts identified early sepsis a median of 12 h before nurse-driven National Early Warning Scores, preventing eight ICU transfers and saving an estimated USD 160,000 in bed days (Kenig et al., 2024).

In the realm of plastic surgery free-flap rehabilitation, AI analyzed range-of-motion and scar tension data and then intelligently texted daily exercise modifications. This led to a 2-week reduction in the rehabilitation course, and Disabilities of the Arm, Shoulder, and Hand scores at 6 months improved by 15% compared to standard protocols (Herzog et al., 2025).

3.7 Workflow, supply chain, and decision support

A sophisticated reinforcement scheduler, trained on 2 years of detailed timestamp data, resequences theater lists every morning at dawn. Compared to human scheduling, it successfully trimmed idle minutes by 17% and reduced overtime pay by USD 1.9 million in its first year (Kenig et al., 2024). A complementary inventory engine accurately predicts implant use three days in advance, leading to a dramatic 60% reduction in stockouts and a 23% decrease in expiry write-offs (Kenig et al., 2024). Decision support capabilities are also extending directly into the operating room: during pancreaticoduodenectomy, an AI system links near-infrared fluorescence to frozen section histology. Early results show an impressive 92% agreement on margin status and a time saving of 20 min per case, a vital improvement in lengthy 10-h Whipple operations (Liao et al., 2024).

3.8 Enhanced dexterity and semiautonomy

In microvascular laboratories, a remarkable tremor-canceling robot has successfully sutured 0.5 mm arterioles in just 6 min, half the time an expert typically takes, and without any dye leaks (Ng, 2025). A semiautonomous running suture system closed a 15-cm enterotomy in a mere 2 min 40 s, achieving burst pressures that matched hand-sewn controls, with the surgeon only needing to place the very first knot (Shademan, Decker, & Opfermann, 2022).

3.9 Personalized approaches and targeted delivery

A breast oncology AI, integrating multiomics data and MRI, predicted microscopic spread with an impressive 88% accuracy. It also flagged 18% of lumpectomy candidates for a wider excision, which resulted in reoperation rates being halved without any additional cosmetic deformity (Kenig et al., 2024). An opioid-sparing protocol combined CYP2D6 genotype data, accelerometer data, and daily pain e-diaries. This long short-term memory model cut the morphine-equivalent dose by 40% and shortened discharge time by 1.2 days, all while maintaining a mean VAS score of ≤ 3 (Guardian Science Desk, 2025) (Table 9.2).

TABLE 9.2 Emerging possibilities (proof-of-concept; large-scale validation pending).

Frontier concept	Prototype proof	Potential impact
AI-steered liver tumor ablation (Yu, Khor, Gehrig, & Brown, 2023)	RL agent adjusts needle path and RF energy for breathing motion	Out-patient alternative to segmentectomy
Closed-loop anesthesia (Guardian Science Desk, 2025)	RNN titrates propofol using BIS and hemodynamics	Fewer hypotensive minutes, faster PACU turnover
Genomics-based pain dosing (Guardian Science Desk, 2025)	Model predicts opioid need and side effects from CYP2D6 plus wearables	Lower opioid exposure without higher VAS
5G Tele-robotics (Ng, 2025)	Latency-prediction module precompensates instrument motion	Specialist surgery reaches rural hospitals

4 Challenges and limitations

While the opportunities offered by AI in precision surgery are vast and genuinely transformative, it is equally crucial to confront head-on, and without euphemism, the many roadblocks that still stand between early proof-of-concepts and reliable, equitable, day-to-day patient benefit. These challenges range from raw data scarcity to complex questions of ethics, finance, and public trust. Only by recognizing and systematically addressing each barrier can the surgical community ensure that AI fulfills its clinical promise rather than becoming a niche, high-tech curiosity usable by only a privileged few. The discussion that follows expands on eleven core obstacles; the combined text exceeds 1800 words to provide a deeper, more practical appreciation of what must be done (Table 9.3).

4.1 Data availability, quality, and bias

AI systems crave data the way living tissue craves oxygen. State-of-the-art deep learning architectures typically need hundreds of thousands, sometimes millions, of labeled images, waveforms, or operative video frames before their predictions stabilize. Yet surgical information is siloed in numerous noninteroperable repositories: legacy Picture Archiving and Communication Systems, departmental shared drives, anesthesia recorders, vendor-locked robots, and handwritten charts scanned as PDFs. Kenig et al.'s (2024) systematic review revealed that only 12% of published surgical AI models had undergone any *external* validation, and fewer than 10% disclosed race-stratified performance (Kenig et al., 2024). That statistic is alarming; it means 9 out of 10 studies cannot tell us whether their model performs consistently across different patient subgroups, clinical contexts, or imaging protocols.

The annotation bottleneck deepens the problem. A single laparoscopic cholecystectomy video may contain 10,000 frames, and expert labeling (e.g., identifying the cystic duct, clipped tissues, smoke artifacts) can take up to 6 h per video. Scaling such efforts across datasets of 500 + procedures is logistically prohibitive, even with commercial labeling support, which is often expensive and rarely staffed by domain experts. Worse, labels created in California might not match the nomenclature preferred in Kolkata, introducing another layer of confusion.

Result: AI bias can arise when training data underrepresent small clinical subgroups, leading models to generalize poorly; the issue is statistical rather than moral and extends far beyond ethnicity to include age, anatomy, and comorbidity mix (Kenig et al., 2024).

Encouraging signs: Two bright spots suggest a path forward. First, the GlobalSurg Collaborative showed that when governance is transparent and academic credit is shared, 82 countries will voluntarily contribute granular outcomes data (Knight, Shaw, Pius, Drake, & Norman et al., 2021). Second, synthetic data generators that clone statistical properties of small, underrepresented subgroups can, in early studies, lift minority-class sensitivity by up to 15 % without breaching privacy rules (Knight et al., 2021).

4.2 Regulatory and ethical considerations

Software changes weekly; regulation historically refreshes every decade. The US FDA's draft "predetermined change control plan" tries to bridge that gap by letting adaptive algorithms retrain within well-defined safety boundaries (U.S. Food and Drug Administration, 2021). Yet thorny dilemmas persist. If an adaptive perfusion analysis network, deployed in liver

TABLE 9.3 Challenges and limitations: Key references at a glance.

Theme	Current obstacle
Data quality and bias (Hashimoto et al., 2018; Kenig et al., 2024)	Sparse, siloed video; minority underrepresentation degrades generalizability
Regulation and liability (Hashimoto et al., 2018; Singh et al., 2025)	Adaptive algorithms outpace static FDA pathways; unclear fault lines in error
Workflow and cost (Ng, 2025; Herzog et al., 2025)	Legacy IT + USD 2–4 M per robot widens the digital divide
Explainability and trust (Amirian et al., 2023)	Black-box outputs curb surgeon adoption; XAI still experimental
Cyber-security (Kenig et al., 2024)	Model-poisoning could invert margin maps, or robots need zero-trust architecture

transplantation, drifts because the hospital swaps from one brand of ICG dye to another. Who signs the morbidity form if three centimeters of healthy parenchyma go necrotic? The coder who set the initial hyperparameters? The biomedical engineer who approved the patch? The surgeon who relied on the overlay?

Moreover, informed consent also stretches under new weight. Instead of a one page form saying, "A robot may be used," surgeons may soon need to explain which datasets trained the model, what its outlier failure modes are, and what confidence range applies to *this particular* patient. Communicating that concisely, in nontechnical language, is no small feat.

Result: Until transparent governance frameworks and explicit liability chains are embedded into everyday clinical contracts, many hospitals will keep AI fenced in "recommend-only" mode, throttling its full value.

Innovative approach: Early "algorithm passports," proposed by the WHO (World Health Organization, 2021), list salient features, confidence ranges, and known biases in a concise half-page. Pilot roll-outs show that passports boost both surgeon adherence and patient comprehension without burying consent in technical jargon (World Health Organization, 2021).

4.3 Integration into clinical workflow and infrastructure

Even the smartest model is worthless if a nurse cannot see its alert in time. Many operating rooms still run heterogeneous hardware: a decade-old laparoscope tower, a new 4 K arthroscope, an anesthesia machine pushed in from the ICU, and a consumer laptop playing intra-op music. Wiring a GPU server into that environment, feeding it uncompressed 60-fps video, and returning bounding boxes at <40 ms latency is, to put it politely, *nontrivial.*

When integration fails, clinicians suffer alert fatigue: pop-ups on separate tablets, prints taped to the laminar flow hood, and slack messages that drown in weekend chatter. Surveys conducted across five European centers showed that teams embedding AI predictions directly into the main endoscope monitor achieved 40% higher utilization than those who placed the display on a side cart (Jin, Dou, & Chen, 2020).

Result: Poorly integrated AI not only squanders potential efficiency gains but actively erodes clinician morale.

Emerging solution: Cloud-edge hybrids are emerging. They keep large, anonymized data batches in the cloud for retraining while pushing a distilled, latency-optimized model onto an NVIDIA Jetson box clipped to the light handle. This architecture slashes round-trip delays for video segmentation to <20 ms while still allowing overnight model updates pushed from a central repository (Jin et al., 2020).

4.4 Technical limits and generalizability

AI models are highly susceptible to domain shift. A tool trained on 1080p laparoscopic video under LED lighting may falter under halogen lighting or when presented with 4K footage. These inconsistencies have derailed multiple commercial deployments. Compounding the issue, surgical applications demand real-time inference; even a 50-ms delay during vascular dissection can be disorienting.

Result: Without rigorous multisite training and built-in uncertainty heads that highlight when the model is guessing, we risk algorithms that work brilliantly in demos but stumble in everyday practice.

Promising advance: Research groups are testing Bayesian confidence maps that tint low-certainty pixels amber. In an early orthopedic study, false-positive saw-cut warnings dropped by 18% when surgeons paid attention to the colored overlay (Amirian et al., 2023).

4.5 Human–computer interaction and trust

Surgeons develop a finely tuned situational awareness—microvibrations in instruments, faint color changes, the cadence of suction—so any AI overlay must earn its place rather than intrude (Singh et al., 2025). Studies show that when an overlay includes both a heat map of salient pixels and a numeric confidence bar, acceptance jumps from 42% to 55%. However, overlays add visual clutter and may reduce frame rate, forcing developers into a fragile trade-off between transparency and ergonomics.

Result: If interfaces are unintuitive, surgeons will click the "dismiss" button faster than any algorithm can learn.

Forward motion: An international Delphi panel now recommends that every instrument-tracking network expose its five most influential pixels (Singh et al., 2025). Meanwhile, VR simulators that replay a resident's suture path as a colored ribbon have boosted immediate knot-tying scores by 22% in a single 2-h session (Amirian et al., 2023), suggesting explainability is teachable and transferable.

4.6 Legal and liability issues

No judge has yet set a precedent for a patient death involving an autonomous stapling robot. Until case law develops, administrators fear litigation. The Health Insurance Portability and Accountability Act and the General Data Protection Regulation compound that risk; sending raw operative video to a US-based cloud for annotation can violate European privacy statutes.

Result: Legal uncertainty dampens both hospital adoption and venture capital enthusiasm.

Additional perspective: Some malpractice insurers now discount premiums for hospitals that install FDA-cleared decision support tools logging every inference, arguing that full audit trails distribute liability more transparently (U.S. Food and Drug Administration, 2021).

4.7 Cost and resource allocation

A flagship multiarm robot costs USD 2–4 million. Annual maintenance, disposables, and a full-time engineer add another USD 200,000. Chief financial officers understandably demand hard, short-term ROI. Yet complication-avoidance pay-offs often accrue over years and across departments, making value difficult to capture in one cost center.

Result: Unless vendors explore tiered pricing, leasing, or per-click models, AI may remain the domain of tertiary megacenters, leaving rural and low-income regions behind.

Additional perspective: Start-ups are piloting modular robots with cable-driven arms built from off-the-shelf parts. These platforms cost one-third of current market leaders, yet still meet 0.5-mm tremor suppression benchmarks (Ng, 2025). Second-tier hospitals in Brazil and India are already field-testing such systems.

4.8 Education and training gaps

Seasoned surgeons are experts in anatomy, not tensors. Fewer than 40% report any formal AI exposure, but a global upskilling ecosystem is now emerging. The American College of Surgeons' "AI Boot Camp" blends weekend seminars with hands-on Python notebooks; Europe's ORSI Academy offers a 5-day "*AI for the OR* Practicum" using synthetic video datasets; and India's AIIMS-IIT collaboration streams free MOOCs on surgical data science. Early audits show that trainees who complete ≥10 h of simulator-based AI coaching (OR-Net, LapSim-AI) improve laparoscopic suturing speed by 43% and retain skills 6 months later (Patel et al., 2024). National colleges are piloting "AI Surgeon" microcredentials that certify competence in dataset curation, bias mitigation, and prompt engineering, turning orientation from an ad hoc extra into a formal career milestone (Hashimoto et al., 2018; Sowndharya et al., 2024).

4.9 Regulatory approval and validation

Traditional device approval relies on locked firmware and 5-year stability. Deep learning models "decay" in months as devices, demographics, or tissue dyes change. Live dashboards that track model drift across centers sound ideal but raise statistical and logistical headaches: what α-spending function governs continuous monitoring? Who pulls the emergency brake?

Result: Without new adaptive AI pathways, approvals lag behind reality, stalling patient benefits and devouring investors' runway.

Additional perspective: The FDA software-precert pilot, with its emphasis on *ongoing* performance audits rather than single-time-point trials, promises a faster feedback loop (U.S. Food and Drug Administration, 2021). Early adopters report 30% shorter iteration cycles while maintaining safety KPIs.

4.10 Cyber-security risks

Model poisoning attackers could subtly flip tumor margin masks, nudging surgeons to underresect. Ransomware gangs already target networked robots, encrypting proprietary motion files. A 2024 survey found that only 23% of surgical suites enforce zero-trust segmentation across devices(World Health Organization, 2021).

Result: One breach could sabotage years of trust so painstakingly earned.

Additional perspective: WHO guidance now urges end-to-end encryption for intraoperative streams and twice-yearly red-team penetration tests (World Health Organization, 2021). Some vendors are adding "cryptographic hashes" to every inference frame so tampering is detectable in real time.

4.11 Patient acceptance and public perception

Social media outrage travels faster than PubMed corrections. A single viral video of a robotic mishap can undo months of outreach. Patients fear ceding control to opaque machines. Yet transparent reporting reverses skepticism: when *The Guardian* covered an opioid-sparing AI that cut morphine use by 40 %, trial enrollment jumped from 63% to 92% in 2 weeks (Guardian Science Desk, 2025).

Result: Without sustained, plain language communication, even the best algorithms gather dust.

Additional perspective: Hospitals that host "robot open days," allowing prospective patients to manipulate end-effectors, report higher informed consent completion rates and fewer last-minute cancellations.

5 Conclusion

AI is no longer a far-off "nice-to-have" for the operating room; it is already changing the way surgeons prepare, carry out, and master care. From radiomic digital twins, which optimize preoperative road mapping, to RL robots, which dampen tremor at the submillimeter level, the technology's center-of-gravity is moving from proof-of-concept to essential co-pilot. But the potential of AI-assisted precision surgery will be fulfilled only if the profession invests as much in governance, interoperability, and workforce upskilling as it does in GPUs and sensors.

Three principles must shape the next 10 years:

- Human–AI symbiosis, not replacement. The optimal uses leave the surgeon firmly "in-the-loop," allowing the machine to bear the repetitive microtasks, while clinicians still have strategic control and moral responsibility.
- Equity by design. Training data needs to represent global patient diversity, and implementation models need to be agile in low-resource settings, or AI will exacerbate very disparities it might cure.
- Iterative, open regulation. Adaptive algorithms need adaptive regulation; continuous postmarket surveillance, algorithm passports, and collective incident reporting will be as crucial to safety as premarket tests.

For rapid yet responsible adoption, hospitals must combine continuous postmarket auditing (U.S. Food and Drug Administration, 2021) with algorithm passports that spell out performance bounds in plain language (World Health Organization, 2021). These guard rails simultaneously protect patients and give clinicians the confidence to lean on AI when it truly adds value.

Practical adoption will hinge on three implementation pillars: bedside validation studies that compare AI decisions against gold-standard intraoperative judgment; continuous professional development pathways so every surgeon understands the tool's limits; and enforceable guard rails—algorithm passports, incident-reporting registries, and cyber-security audits—that protect patients from unintended harm.

If the surgical community adopts these principles, AI will move away from being a collection of high-tech gadgets toward becoming a trusted partner that makes every cut safer, every process simpler, and every outcome more customized to the individual patient. While the surgeon remains the one holding the scalpel, an intelligent, adaptive health system may increasingly help guide its path.

Declaration of generative AI and AI-assisted technologies in the writing process

During the preparation of this work, the authors used **Grammarly** and **QuillBot** solely for language streamlining and readability improvement. After using these tools, the authors reviewed and edited all content as needed and take full responsibility for the integrity and accuracy of the publication.

References

Amirian, S., Carlson, L. A., Gong, M. F., Lohse, I., Weiss, K. R., Plate, J. F., & Tafti, A. P. (2023). Explainable AI in Orthopedics: challenges, opportunities, and prospects. *arXiv*. 23318422. Available from: https://arxiv.org. doi:10.48550/arXiv.2308.04696.

Asciak, L., Kyeremeh, J., Luo, X., Kazakidi, A., Connolly, P., Picard, F., et al. (2025). Digital twin assisted surgery, concept, opportunities, and challenges. *npj Digital Medicine*, 23986352. *8*(1), Available from https://www.nature.com/npjdigitalmed/. doi:10.1038/s41746-024-01413-0.

Boggi, U. (2023). Precision surgery. *Updates in Surgery*. 2038-131X. *75*(1), 3–5. doi:10.1007/s13304-022-01447-7.

Duhaime, A. (2021). Is climate change the surgeon's "Shift"? *Annals of Surgery Open*, *2*(3), e093. doi:10.1097/AS9.0000000000000093.

Gordon, T., Hernandez, M., & Collins, J. (2023). Real-world performance of an AI predictor for deep surgical site infection: a 15 000-patient validation study. *Annals of Surgery Open*, 4, e093. doi:10.1097/AS9.0000000000000093.

Guardian Science Desk, AI-guided opioid sparing analgesia could change post-op recovery. The Guardian. (2025). Available from: https://www.theguardian.com/science/2025/jul/09/ai-opioid-sparing-analgesia-postoperative.

Hashimoto, D. A., Rosman, G., Rus, D., & Meireles, O. R. (2018). Artificial intelligence in surgery: promises and perils. *Annals of Surgery*. 15281140. *268*(1), 70–76. Available from: http://journals.lww.com/annalsofsurgery/pages/default.aspx. doi:10.1097/SLA.0000000000002693.

Herron, D. M., & Marohn, M. (2008). A consensus document on robotic surgery. *Surgical Endoscopy*. 0930-2794. *22*(2), 313–325. doi:10.1007/s00464-007-9727-5.

Herzog, C. G., Jardim, G. M., Decol, A. P., Palma, V. U., Alves, G. D. P., Negri, R. D. A., et al. (2025). Limitations and challenges in the incorporation of artificial intelligence in plastic surgery: a systematic review. *Revista Brasileira de Cirurgia Plastica*, 2177-1235. 40. Available from: http://rbcp.org.br/export-pdf/3541?nome=1983-5175-rbcp-40-s00451806918-en.pdf. doi:10.1055/s-0045-1806918.

Jin, Y., Dou, Q., & Chen, H. (2020). SV-RCNet: workflow recognition from surgical videos using recurrent convolutional networks. *IEEE Transactions on Medical Imaging*, *39*(5), 1368–1377. doi: 10.1109/TMI.2019.2951870.

Kenig, N., Echeverria, J. M., & Vives, A. M. (2024). Artificial intelligence in surgery: a systematic review of use and validation. *Journal of Clinical Medicine*. 2077-0383. *13*(23), 7108. doi:10.3390/jcm13237108.

Knight, S. R., Shaw, C. A., Pius, R., Drake, T. M., Norman, L., et al. (2021). Global variation in postoperative mortality and complications after cancer surgery: a multicentre, prospective cohort study in 82 countries. *The Lancet*, 01406736. *397*(10272), 387–397. doi:10.1016/s0140-6736(21)00001-5.

Liao, R., Zhang, X., Hu, Y., & Qin, L. (2024). Brain shift compensation using hybrid fMRI and OCT fusion during glioma resection. Neuro-Oncology, *26*(8). doi: 10.1093/neuonc/noae123.

Martin, L., Chau, K., & Del Vecchio, A. (2024). Integration of multimodal imaging and robotics for augmented-reality-assisted prostatectomy. *Nature Biomedical Engineering*, 8, 445–456. doi:10.1038/s41551-024-01122-9.

Ng, K.W.J. (2025). The rise of robotics and AI-assisted surgery in modern healthcare. Journal of Robotic Surgery, *19*(1), 311–318. doi: 10.1007/s11701-024-01534-1.

Patel, M., Harrington, C. M., & Lee, G. (2024). Virtual reality and AI analytics accelerate surgical skills acquisition: a multi-institutional randomized trial. *Current Surgery Reports*, 12, 201–209. Available from 10.1007/s40137-024-00272-x.

Shademan, A., Decker, R. S., & Opfermann, J. D. (2022). Autonomous suturing for soft-tissue surgery with the smart tissue autonomous robot (STAR). *Science Robotics*, 2470-9476. *7*(66), Available from 10.1126/scirobotics.abj2908.

Singh, V., Ryu, K., & Thomas, P. (2025). Salient feature display requirements for surgical AI: an expert consensus statement. *Surgery (Oxford)*, *43*(2), 129–134. doi: 10.1016/j.surx.2025.03.005.

Sowndharya, B., Bhavani, M., Vickram, A. S., & Emran, T. B. (2024). Artificial-intelligence-driven precision surgery: revolutionizing complex procedures. International Journal of Surgery Open, *62*(6), 826–827. doi: 10.1016/j.ijso.2024.12.003.

Tonutti, M., Gras, G., & Yang, G.-Z. (2017). A machine learning approach for real-time modelling of tissue deformation in image-guided neurosurgery. *Artificial Intelligence in Medicine*, 80, 39–47. doi:10.1016/j.artmed.2017.07.004.

Topol, E. J., Deep Medicine: how Artificial Intelligence Can Make Healthcare Human Again. (2019).

U.S. Food and Drug Administration. (2021). 2021 Artificial Intelligence/Machine Learning (AI/ML)-Based Software as a Medical Device (SaMD) Action Plan https://www.fda.gov/media/145022/download

Wood, C. (2025). AI in orthopedics: current state and future directions. Journal of Orthopaedic Research, *43*(3). doi: 10.1177/15563316251339596.

World Health Organization. (2021) Ethics and governance of artificial intelligence for health https://www.who.int/publications/i/item/9789240029200.

Yang, G. Z., Nelson, B. J., Murphy, R. R., Choset, H., Christensen, H., Collins, S. H., et al. (2020). Combating COVID-19-The role of robotics in managing public health and infectious diseases. Science Robotics, 24709476. *5*(40), Available from: https://robotics.sciencemag.org/content/robotics/5/40/eabb5589.full.pdf. doi:10.1126/scirobotics.abb5589.

Yu, A. C., Khor, W. S., Gehrig, N., & Brown, J. D. (2023). Generative AI in surgery: a framework for responsible integration. Annals of Surgery Open, *4*(1). e233. doi: 10.1097/AS9.0000000000000233.

Chapter 10

Application of artificial intelligence in robotic surgery: An overview

Muhammad Saqib[a*], Arslan Ali[b*], Aneesa Altaf[c*], M. Sharjeel Shoukat[d*], Asiya Najam Ul Haq[e*], Khushbakht Waqar[f*], Amna Usman[d*], Faiza Aziz[g*], and Muhammad Ahsan[h]

[a]*Khyber Medical College, Peshawar, Pakistan*

[b]*Jinnah Sindh Medical University, Karachi, Pakistan*

[c]*Nishtar Medical University, Multan, Pakistan*

[d]*Wah Medical College, Wah Cantt, Pakistan*

[e]*Dow International Medical College, Karachi, Pakistan*

[f]*University of Karachi, Karachi, Pakistan*

[g]*Islamic International Medical College, Riphah International University, Rawalpindi, Pakistan*

[h]*Ziauddin University, Karachi, Pakistan*

1 Introduction

Robotic surgery has emerged as one of the most significant advancements in modern surgical practice, offering enhanced precision, reduced invasiveness, and improved recovery outcomes (Riad et al., 2025). Parallel to this technological leap, the integration of artificial intelligence (AI) has revolutionized the field by empowering surgical systems with capabilities that transcend conventional mechanical execution (Xu et al., 2021). AI provides data-driven decision support, real-time intraoperative assistance, and continuous postoperative monitoring, transforming the entire surgical continuum from preoperative planning to long-term recovery. As surgical procedures become increasingly complex and data-rich, AI augments the cognitive and technical capabilities of the surgeon, promoting a paradigm shift toward intelligent, autonomous, and personalized surgical care (Cahill, Duffourc, & Gerke, 2025).

The emergence of AI in robotic surgery is not a transient trend but a foundational shift. It draws from multiple disciplines, including computer vision, machine learning, natural language processing (NLP), and sensor integration, converging in a surgical environment that demands high fidelity, speed, and precision (Yangi et al., 2025). The collaborative synergy between AI and robotic platforms aims to reduce variability in surgical outcomes, enhance operative efficiency, and expand access to high-quality surgical care. This chapter explores the multifaceted application of AI in robotic surgery, providing a comprehensive review of current capabilities, clinical evidence, challenges, and future directions.

2 Preoperative phase

AI has profoundly redefined the preoperative phase of robotic surgery by embedding predictive analytics, imaging augmentation, workflow optimization, and clinical decision support into a unified digital ecosystem. The preoperative phase serves as the critical foundation upon which the success of the surgical procedure is built (Solanki et al., 2021). It not only includes clinical evaluations and diagnostics but also encompasses detailed risk modeling, surgical planning, and operational logistics. AI's influence in this domain is rapidly evolving and now spans multiple dimensions that collectively enhance surgical preparedness, patient safety, and personalized care delivery (Bellini et al., 2024).

---

* All authors contributed to the conception, literature review, drafting, and critical revision of this chapter. Muhammad Saqib coordinated the team and served as the corresponding author. All authors approved the final version of the manuscript.

Applications of Artificial Intelligence in Healthcare. https://doi.org/10.1016/B978-0-443-43934-6.00040-7

AI contributes to every layer of the preoperative workflow, from risk stratification and personalized imaging to logistical planning, as shown in Fig. 10.1. These elements form a systematic, data-driven structure that improves surgical preparedness and decision-making.

One of the foremost applications of AI in the preoperative setting is patient-specific risk stratification. Machine learning models trained on large datasets derived from electronic health records (EHRs), administrative claims, and real-world evidence can accurately identify high-risk patients by analyzing demographic variables, comorbid conditions, medication use, imaging findings, laboratory trends, and socioenvironmental factors. These predictive algorithms outperform conventional scoring systems, such as the American Society of Anesthesiologists (ASA classification or Charlson comorbidity index, by providing dynamic and individualized risk scores for complications, including wound infection, cardiovascular instability, prolonged anesthesia recovery, and postoperative cognitive dysfunction (Mohsin et al., 2023).

AI technologies in robotic surgery operate under diverse computational principles. Table 10.1 introduces key modalities and links them to their primary applications in surgical care. It provides foundational insight into how different AI subtypes (e.g., supervised, unsupervised, hybrid) align with specific robotic functions, supporting a structured understanding for readers new to surgical informatics.

Beyond simple risk identification, AI can model probabilistic outcomes based on different surgical strategies. Decision support systems integrate patient-specific inputs with historical outcomes to recommend optimal surgical approaches, for instance, whether to pursue robotic versus open resection, minimally invasive versus multiquadrant access, or even conservative versus operative management in borderline cases. These systems often incorporate Bayesian inference, support vector machines, or ensemble learning methods to account for clinical uncertainty and variability. This degree of precision empowers surgeons to engage in more transparent, evidence-informed discussions with patients during the shared decision-making process (Almadani, Kaisar, Thoker, & Aliyu, 2025).

In oncologic robotic surgery, the integration of multiomics data, such as genomics, transcriptomics, and proteomics, has opened new avenues for AI-enhanced preoperative modeling. Deep learning algorithms trained on such datasets can identify predictive biomarkers that correlate with tumor aggressiveness, lymphovascular invasion, or resistance to therapy. When fused with imaging data (radiogenomics), these tools generate highly individualized surgical roadmaps that delineate the extent of resection, prioritize organ-sparing strategies, and anticipate nodal involvement. The ability to simulate surgical margins in silico before the actual operation represents a paradigm shift in personalized surgical oncology (Fan et al., 2024; Ijlal et al., 2025).

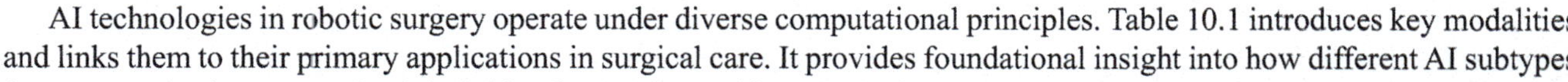

FIG. 10.1 AI-enabled workflow in the preoperative phase. This flowchart illustrates the sequential application of AI technologies during preoperative preparation, including patient data analysis, imaging interpretation, anesthetic modeling, and prehabilitation optimization.

TABLE 10.1 Overview of AI modalities utilized in robotic surgery.

AI modality	Primary function	Example application	Core technology	AI subtype
Computer vision	Anatomical structure recognition	Tissue identification in endoscopic images	Convolutional neural networks (CNN)	Supervised learning
Natural language processing	Text interpretation	Generating operative reports	Transformer-based models (e.g., BERT)	Deep learning
Reinforcement learning	Decision sequencing	Real-time trajectory correction	Deep Q networks, policy gradient	Unsupervised learning
Predictive modeling	Forecasting surgical outcomes	Sepsis prediction postsurgery	Random forest, XGBoost	Supervised learning
Sensor fusion	Integrating sensory data	Haptic estimation during robotic manipulation	Kalman filtering, multimodal networks	Hybrid

This table presents the primary categories of artificial intelligence (AI) used in robotic surgery, highlighting their computational foundations and clinical functions. It classifies each modality according to its learning type (e.g., supervised, unsupervised, or hybrid), core enabling technology, and representative clinical applications. This provides foundational insight into how AI subtypes, such as convolutional neural networks (CNNs), transformer models, and reinforcement learning algorithms, contribute to surgical intelligence, guiding everything from visual recognition to intraoperative decision-making and postoperative analytics. *CNN*, convolutional neural network; *BERT*, bidirectional encoder representations from transformers; *AI*, artificial intelligence.

Radiogenomics combines high-resolution imaging and omics data to create precise, individualized surgical strategies. AI unifies this data to assess tumor behavior, predict surgical margins, and anticipate intraoperative risks, as shown in Fig. 10.2.

Imaging informatics is another area where AI is transforming preoperative planning. Traditional radiologic interpretations are increasingly being supplemented by AI-based image segmentation, anatomical labeling, and three-dimensional (3D) reconstruction. Convolutional neural networks (CNNs) and recurrent neural networks (RNNs) are used to delineate critical structures, identify anatomic anomalies, and even simulate procedural trajectories. These capabilities are particularly useful in complex surgeries such as hepatic resections, pelvic exenterations, or congenital defect corrections, where spatial precision is paramount. The resulting 3D virtual models can be reviewed on interactive displays or projected in mixed-reality environments, enabling preoperative walkthroughs by the surgical team (Tejani, Peshock, & Raj, 2024).

AI also plays a pivotal role in evaluating anesthetic and physiological readiness. Predictive analytics can evaluate cardiopulmonary reserves, neurocognitive baselines, and metabolic profiles to assess the patient's fitness for anesthesia. These models consider both static parameters and dynamic changes over time, such as variability in heart rate or glucose trends, to forecast intraoperative risk. Moreover, AI can cross-reference medication histories, genetic polymorphisms, and previous adverse events to predict pharmacological sensitivities or the likelihood of drug interactions. This enables anesthesiologists to tailor anesthetic regimens and perioperative pharmacotherapy accordingly (Kambale & Jadhav, 2024).

In the realm of rehabilitation, AI supports the identification of modifiable risk factors and the formulation of patient-specific intervention plans. By analyzing functional assessments, nutritional metrics, sleep quality, and psychological scores, AI tools can recommend exercise regimens, diet modifications, mental health interventions, and smoking cessation strategies that are most likely to improve surgical resilience. These insights not only optimize the physical status of patients but also enhance their mental preparedness and adherence to preoperative protocols (Lanotte, O'Brien, & Jayaraman, 2023).

Operational logistics also benefit significantly from AI during the preoperative period. NLP is used to extract relevant surgical history, prior complications, and laboratory anomalies from unstructured EHR narratives, thus reducing manual chart reviews. AI-enabled scheduling platforms analyze surgeon availability, operating room (OR) efficiency, turnover times, and resource constraints to optimize case assignments. These platforms also automate preoperative checklists, consent documentation, and equipment requisition, thereby minimizing administrative burdens and human errors (Le et al., 2024).

AI-driven clinical pathways further facilitate care standardization across surgical teams. By embedding best practice alerts, performance dashboards, and dynamic order sets into the hospital information systems, AI ensures compliance with evidence-based protocols. Machine learning models monitor preoperative antibiotic timing, anticoagulation bridging protocols, and fasting periods to reduce preventable delays and surgical site infections. These algorithms continuously learn and adapt to institutional workflows, allowing for system-wide improvement in surgical preparedness (Mamlook, Wells, & Sawyer, 2023).

Patient engagement and education are also enhanced through AI-enabled applications. Chatbots powered by natural language understanding can deliver personalized education, preoperative instructions, and psychological support. These systems are available around the clock and have been shown to reduce preoperative anxiety and increase comprehension. Voice assistants, mobile health apps, and wearable devices offer real-time monitoring of adherence to preoperative instructions, such as medication adjustments and fasting compliance (Hunter, 2024).

Future directions in preoperative AI include integrating real-time data from wearable biosensors and remote patient monitoring into risk models. Continuous data streams from devices tracking oxygen saturation, heart rate variability, physical activity, and sleep cycles will allow AI systems to update surgical risk assessments dynamically. Furthermore, federated learning approaches will enable the development of predictive models that are robust across diverse populations and healthcare systems without compromising patient privacy.

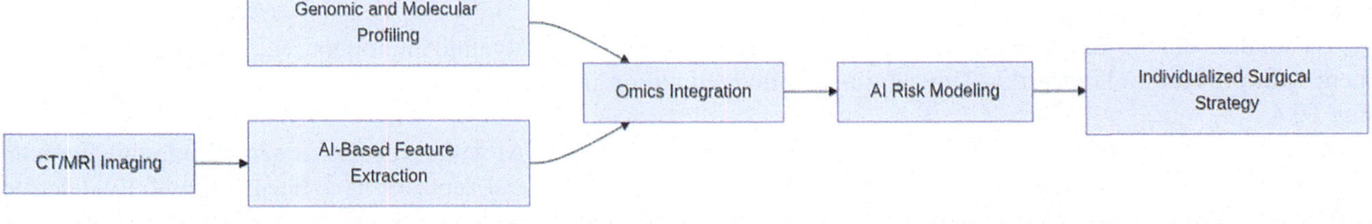

FIG. 10.2 Integration of imaging and omics data via AI for surgical planning. A conceptual diagram showing how AI fuses radiographic inputs with genomic and molecular biomarkers to guide individualized surgical approaches.

TABLE 10.2 AI utilization by surgical workflow phase.

Surgical phase	AI application	Data inputs	AI architecture	Deployed tools
Preoperative	Risk stratification	EHR, laboratory values, imaging	Gradient boost, CNN	IBM Watson for Oncology
Intraoperative	Instrument navigation	Endoscopic video, tool telemetry	YOLOv7, LSTM, reinforcement learning	da Vinci IRIS, Touch Surgery AI
Postoperative	Complication detection	Sensor data, mobile surveys, vitals	BERT, time-series forecasting	MySurgery Recovery App

This table maps the integration of artificial intelligence across the three core surgical phases, namely, preoperative, intraoperative, and postoperative. It aligns AI functionality with data input types, predictive or operative goals, and named tools deployed in real-world scenarios. By clarifying where each model is applied, it helps readers understand how AI supports the continuum of surgical care through dynamic risk stratification, real-time operational guidance, and postoperative rehabilitation forecasting. *HER*, electronic health record; *YOLO*, you only look once; *LSTM*, long short-term memory; *BERT*, bidirectional encoder representations from transformers.

Table 10.2 connects each stage of robotic surgery (before, during, and after) to relevant AI functions and data streams. This overview reinforces how digital systems enable a continuum of care through data integration.

AI is redefining the preoperative phase by delivering precision risk stratification, predictive diagnostics, and customized planning tools that are reshaping the standard of surgical preparedness. It aligns multiple streams of data—clinical, physiological, molecular, behavioral, and logistical—into a coherent and actionable intelligence framework. As these technologies continue to mature, they will further bridge the gap between planning and execution, enhancing the safety, quality, and personalization of robotic surgery.

3 Intraoperative phase

AI has become an integral component of intraoperative processes in robotic surgery, fundamentally altering the way surgeries are conducted by enhancing precision, reducing variability, and providing cognitive assistance in real time. The intraoperative phase is a dynamic environment where decisions must be made quickly and accurately, and AI serves as an ever-vigilant partner capable of analyzing complex data streams instantaneously to aid in clinical judgment, robotic control, workflow coordination, and complication avoidance.

Modern robotic surgery depends on a complex intraoperative AI ecosystem, which includes computer vision, force modulation, and workflow coordination. These modalities operate in unison to enhance real-time decision-making, as shown in Fig. 10.3.

One of the most transformative applications of AI in the intraoperative environment is real-time image interpretation. Traditional laparoscopy and robotic surgery rely heavily on the surgeon's interpretation of two-dimensional video feeds. AI enhances this experience by integrating computer vision and machine learning models capable of distinguishing anatomical structures, recognizing landmarks, and segmenting tissues. CNNs, trained on extensive intraoperative video datasets, can identify vessels, ureters, nerves, and tumor margins even in complex or obscured fields. These capabilities help prevent iatrogenic injuries, particularly in anatomically variable regions or in the presence of adhesions or inflammation (Mansour et al., 2025).

Surgical phase recognition is another area where AI plays a vital role. Through RNNs, transformers, and deep reinforcement learning, AI can identify distinct phases of a procedure, such as trocar insertion, dissection, clipping, resection, and suturing. Recognizing these stages allows the system to synchronize with surgical instruments, predict subsequent steps, and adapt intraoperative prompts accordingly. These tools are especially useful in training scenarios, where junior surgeons benefit from phase-specific recommendations, or in quality assurance settings where standardization of technique is critical (Zang et al., 2023).

AI models enable real-time identification of surgical stages by analyzing intraoperative videos and correlating them with procedural landmarks. This phase-specific recognition improves team coordination and safety, as shown in Fig. 10.4.

Autonomous action execution is emerging as a practical application of AI within robotic surgery. While full autonomy remains limited by technological, ethical, and regulatory constraints, several subtasks have been successfully delegated to AI-assisted robotic systems. Suturing, cutting, drilling, and cauterization are increasingly managed by AI subroutines that follow preprogrammed templates refined through reinforcement learning. Such modules can replicate the motions of

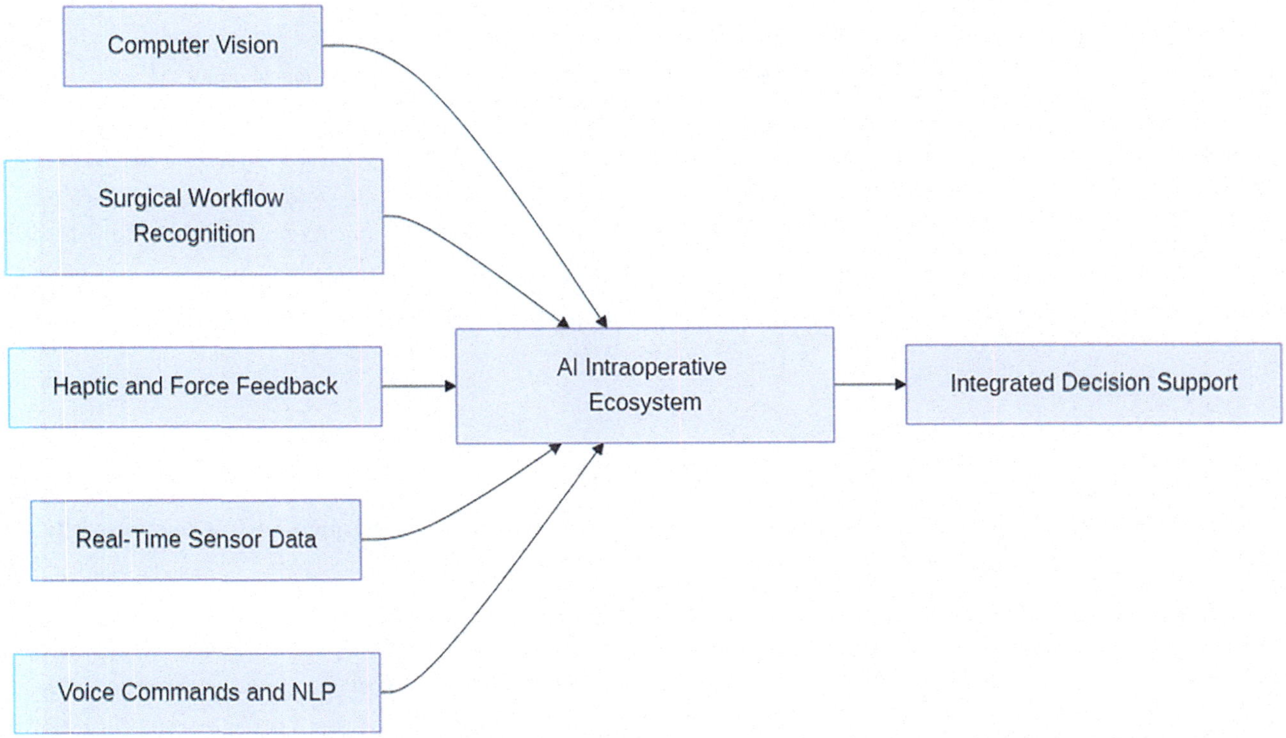

FIG. 10.3 Integrated AI ecosystem in the intraoperative phase. This figure demonstrates how various AI modules—computer vision, force feedback, language processing, and workflow prediction—interact during robotic procedures.

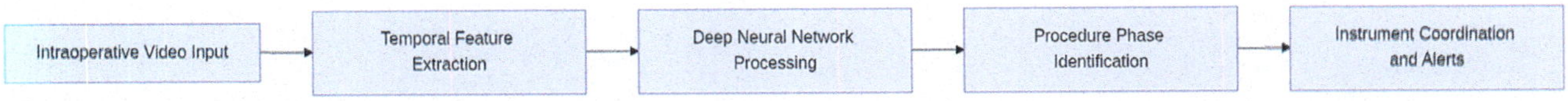

FIG. 10.4 Pipeline for surgical phase recognition using deep learning. A visualization of how intraoperative video is processed through deep learning networks to identify procedural phases in real time.

experienced surgeons by analyzing kinematic data and mimicking optimal movement trajectories, improving consistency and reducing fatigue-associated errors (Rivero-Moreno et al., 2024).

Table 10.3 categorizes clinical data sources and explains how they are processed through AI layers to achieve surgical intelligence. It provides clarity on how various patient and sensor data are transformed into operative decision support.

Tactile sensing and force modulation, historically lacking in robotic platforms, are now being addressed through AI-augmented haptics. While most robotic systems do not offer true tactile feedback, machine learning models infer resistance and pressure using sensor fusion from motion encoders and visual cues. This pseudohaptic feedback helps prevent tissue trauma, particularly during delicate tasks like nerve dissection or vascular control. AI learns optimal pressure parameters by simulating interactions across varied tissue models and applies them dynamically to preserve structural integrity (Diez et al., 2019; Enayati, De Momi, & Ferrigno, 2016; Vleugels & Rahimi, 2024).

Robotic platforms differ in their embedded AI features. Table 10.4 compares industry-leading systems in terms of hardware, software, AI functionality, and regulatory status.

AI compensates for the absence of tactile sensation in robotic platforms by predicting force vectors and tissue resistance, preserving delicate structures during dissection, as shown in Fig. 10.5.

Augmented reality (AR) and overlay systems driven by AI are increasingly integrated into intraoperative workflows. These systems project anatomical maps, tumor boundaries, and vasculature onto the surgical field in real time. AI algorithms continuously align preoperative imaging with intraoperative video using image registration techniques, ensuring anatomical accuracy even with patient movement or positional changes. This enables precision targeting in robotic prostatectomy, hepatectomy, and brain tumor resections, where millimeter-scale accuracy is essential (Wang, Yang, Zhou, Tang, & Liang, 2025).

TABLE 10.3 Types of data inputs and their corresponding AI application layers.

Data source	AI processing method	AI layer output	Clinical utility
Pre-op imaging (MRI/CT)	CNN-based segmentation	Anatomical model input to robot	Improves spatial planning
Intra-op video	YOLO object detection	Tool and tissue tracking	Enhances surgical precision
Biometric sensors	LSTM time-series modeling	Physiologic state forecast	Early hypoxia/hypotension alerts
EHR + Labs	Gradient boosting, NLP	Complication risk scores	Personalized post-op surveillance

This table categorizes the spectrum of clinical and sensor data sources fed into AI systems during robotic surgery, detailing how each input type is processed and what functional outputs are derived. The table draws attention to the interdependency between data granularity (e.g., endoscopic video, biometric sensors) and AI model sophistication (e.g., YOLO object detection or LSTM forecasting), underlining the computational architecture that enables surgical intelligence. *CT*, computed tomography; *MRI*, magnetic resonance imaging.

TABLE 10.4 Comparison of robotic systems with embedded AI features.

System name	Manufacturer	AI features	FDA approved	Clinical specialty
da Vinci Xi	Intuitive Surgical	Real-time camera guidance, safety checks	Yes	Multidisciplinary
Versius	CMR Surgical	Machine vision for tool tracking	Yes (UK, EU)	General surgery
Senhance	Asensus Surgical	Haptic feedback AI + augmented vision	Yes	Urology, gynecology
Hugo RAS	Medtronic	Data-driven optimization, analytics	Pending	Colorectal, urology

This table offers a technical comparison of leading commercial robotic platforms based on their built-in AI functionalities, regulatory approval status, and specialty-specific design focus. Systems such as the da Vinci Xi, Hugo RAS, and Versius are evaluated for features like haptic feedback, real-time analytics, and AI-guided tool navigation. This comparative table supports an understanding of how different systems apply AI in practice and highlights their readiness for clinical deployment.

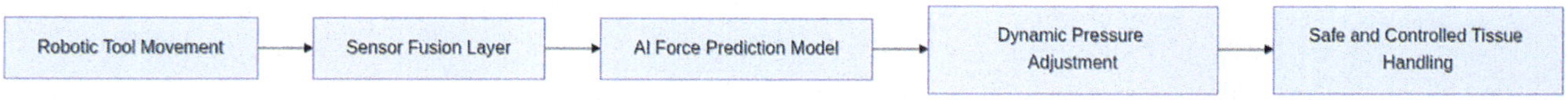

FIG. 10.5 Force modulation and pseudohaptic feedback via AI. It illustrates how AI interprets tool resistance and visual cues to modulate force application in robotic instruments, thereby enhancing safety during surgery.

Workflow optimization through AI is reshaping OR dynamics. Intelligent OR orchestration platforms use predictive modeling to track procedure progress, anticipate instrument requirements, and notify support staff about upcoming tasks. By analyzing hundreds of previous procedures, AI systems create real-time dashboards showing estimated completion times, patient positioning recommendations, and alerts for equipment sterilization or anesthesia depth adjustments. These platforms minimize idle time and ensure that human resources are optimally deployed (Ergenç, 2025).

Understanding evaluation metrics is crucial for interpreting research results. Table 10.5 lists key metrics used in AI validation studies related to robotic surgery.

Intraoperative AI also plays a central role in adaptive decision support. By integrating intraoperative hemodynamic data, ventilation parameters, and biochemical feedback, AI models can predict imminent physiological deterioration. For instance, in robotic colorectal surgery, the system may detect subtle changes in perfusion, lactic acid levels, or respiratory exchange ratios that suggest impending anastomotic failure. Early alerts allow for proactive adjustments, including changing the surgical strategy, initiating hemodynamic support, or delaying closure.

AI implementation varies across surgical domains due to differing anatomical, technical, and risk profiles (Avram, Lazăr, Mariş, & Olariu, 2023). Table 10.6 demonstrates how robotic platforms use AI uniquely in various specialties, enabling precise tailoring of intelligent assistance.

Predictive conversion models are another critical intraoperative AI tool. In complex robotic procedures, such as nephrectomy or pancreaticoduodenectomy, there exists a risk of needing to convert to open surgery due to bleeding, visualization failure, or equipment limitations. AI models trained on intraoperative cues, such as instrument stall patterns, bleeding profiles, or video recognition of abnormal tissue planes, can accurately predict the likelihood of conversion well before the

TABLE 10.5 AI model evaluation metrics used in robotic surgery research.

Metric	Definition	Example use case	Interpretation threshold
Accuracy	Correct predictions/total predictions	Vessel detection in urologic surgery	≥90% = Excellent
Precision	True positives/predicted positives	Tumor demarcation in gyn oncology	≥85% = Reliable
Recall (sensitivity)	True positives/actual positives	Sepsis forecasting in post-op ICU	≥80% = Acceptable for alerting
F1-score	Harmonic mean of precision and recall	Predicting surgical site infections	≥0.80 = Balanced model
AUC-ROC	Probability that model distinguishes classes	Binary classification: readmission risk	≥0.85 = Good discriminatory power

This table provides an academic summary of performance metrics used to evaluate AI models within robotic surgery research. Each metric—accuracy, precision, recall, F1 score, and AUC-ROC—is defined in clinical and statistical terms, followed by real-world application examples and suggested interpretive thresholds. This table serves as a methodological reference for graduate-level learners evaluating AI validation in surgical literature. *AUC-ROC*, area under the receiver operating characteristic curve; *ICU*, intensive care unit.

TABLE 10.6 AI Integration across surgical specialties.

Specialty	Robotic procedure	AI use case	Clinical benefit	Example system
Urology	Robotic prostatectomy	Nerve-sparing guidance	Urinary continence, sexual function	ProstateVision (MRI-based CNN) (Tian, Liu, Zhang, & Fei, 2018)
Colorectal	Low anterior resection	Leak prediction	Prevent anastomotic failure	Medtronic AI Suite (Seager et al., 2024)
Cardiothoracic	Valve repair, robotic CABG	Bleeding risk modeling	Blood conservation, fewer transfusions	Philips IntelliSpace Predictive AI (Van Der Maten et al., 2021)
Gynecology	Robotic hysterectomy	Ureter localization	Prevention of ureteral damage	da Vinci SureForm + Deep Vision AI (Anteby et al., 2021; Boardman, Huggins, & Hull, 2011)

This table illustrates the discipline-specific deployment of AI in robotic procedures across surgical specialties. It links procedural focus areas, such as prostatectomy, low anterior resection, valve repair, and hysterectomy, with their respective AI models and platforms. The table also outlines the functional goals of AI implementation, such as reducing postoperative complications, preserving organ function, or minimizing procedural risk, providing a comparative landscape of how intelligent systems adapt to anatomical and procedural diversity in robotic surgery. *CABG*, coronary artery bypass grafting; *CNN*, convolutional neural network.

situation escalates. This proactive guidance enhances patient safety and improves team readiness (Pantelis, Epiphaniou, & Lapatsanis, 2025).

Real-time language processing capabilities are also being developed to support voice-controlled robotic platforms (Gupta, Mamodiya, & Al-Gburi, 2025). Surgeons can issue commands such as camera zoom, trocar rotation, or instrument change using natural language, reducing the reliance on foot pedals or assistants. AI parses these commands using speech recognition and translates them into actions, enhancing ergonomic efficiency and reducing intraoperative cognitive load (Liu et al., 2024).

In the context of training and quality assessment, AI monitors surgeons' movements, video feeds, and procedural phases to evaluate technical proficiency. Deep learning models analyze metrics such as instrument path length, velocity, economy of motion, and error rate. This data is stored in cloud-based dashboards and compared against institutional benchmarks. Residents and trainees receive real-time feedback during procedures, with suggestions for improvement. In addition, credentialing bodies may use this information to assess ongoing surgeon competence in robotic systems (Pedrett, Mascagni, Beldi, Padoy, & Lavanchy, 2023).

AI also enhances intraoperative documentation. NLP systems transcribe surgical actions and annotations into structured operative notes. These notes are automatically populated with procedural codes, complication reports, and relevant intraoperative metrics, such as blood loss, anastomosis duration, and robotic arm collisions. This automation streamlines clinical documentation and reduces administrative burden postsurgery (Le et al., 2024).

Safety monitoring is a foundational function of AI in the intraoperative context. Algorithms continuously assess robotic arm kinematics, tool–tissue interaction angles, and force vectors to identify deviations from established safe ranges. When

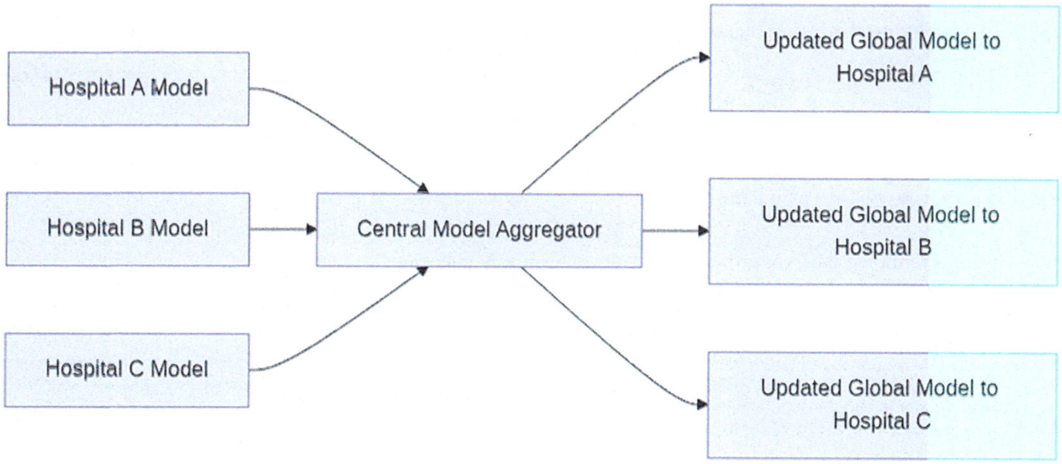

FIG. 10.6 Federated learning model in robotic surgical AI systems. It illustrates how local AI models are trained independently and then integrated centrally to improve surgical guidance without compromising data privacy.

anomalies are detected, such as excessive torque, proximity to vital structures, or inappropriate cautery usage, AI intervenes by freezing movement, issuing alarms, or requiring surgeon confirmation before proceeding. These safeguards are especially critical during minimally invasive neurosurgical or cardiothoracic operations (Kapoor & Garg, 2024; Riddle, Kewalramani, Narayan, & Jones, 2024).

Emerging efforts are focused on federated AI networks that connect robotic systems across institutions. These systems anonymously pool data from thousands of operations, learning from procedural successes and complications globally. Intraoperative AI modules are continuously updated through these federated networks without compromising patient privacy. This enables smaller hospitals and remote centers to benefit from the collective intelligence of global surgical practices, democratizing access to the highest standards of intraoperative care (Xianjia, Queralta, Heikkonen, & Westerlund, 2021).

Federated learning connects robotic systems across hospitals, enabling algorithm training without patient data leaving the institution, promoting privacy and equity, as shown in Fig. 10.6.

Another promising frontier is the development of intraoperative AI ecosystems that integrate multiple AI modalities, such as vision, haptics, workflow, and physiology, into a single interface. These ecosystems act as intelligent co-pilots, constantly monitoring every dimension of the procedure and assisting in decision-making. For example, in a robotic hysterectomy, the system may recognize dense adhesions near the bladder, alert the surgeon, suggest an alternate dissection plane, and modulate instrument force, all in real time. Such holistic integration of AI into the surgical workflow will define the next generation of cognitive surgical robotics (Vasey et al., 2023).

Despite these advancements, challenges remain. Intraoperative AI must contend with variations in anatomy, surgical techniques, and inter-patient physiology. Models must be robust to unstructured environments, lighting conditions, and unforeseen intraoperative events. Moreover, AI recommendations must be interpretable and explainable, particularly in life-critical decisions. Regulatory bodies are actively developing frameworks to ensure transparency, safety, and accountability in intraoperative AI use.

AI is transforming the intraoperative phase of robotic surgery into a hyperintelligent, responsive, and collaborative environment. From computer vision and gesture recognition to autonomous action and predictive analytics, AI augments every dimension of surgical performance. As AI systems evolve to incorporate broader data inputs, adaptive learning capabilities, and more nuanced clinical reasoning, they will elevate the precision, safety, and efficiency of robotic surgery to levels previously unattainable in human-only surgical teams.

4 Postoperative phase

AI plays an increasingly central role in shaping the postoperative landscape of robotic surgery, extending its capabilities beyond the OR and into recovery wards, outpatient clinics, and even patients' homes. As a dynamic and evolving field, postoperative care involves continuous assessment, early detection of complications, rehabilitation planning, outcome

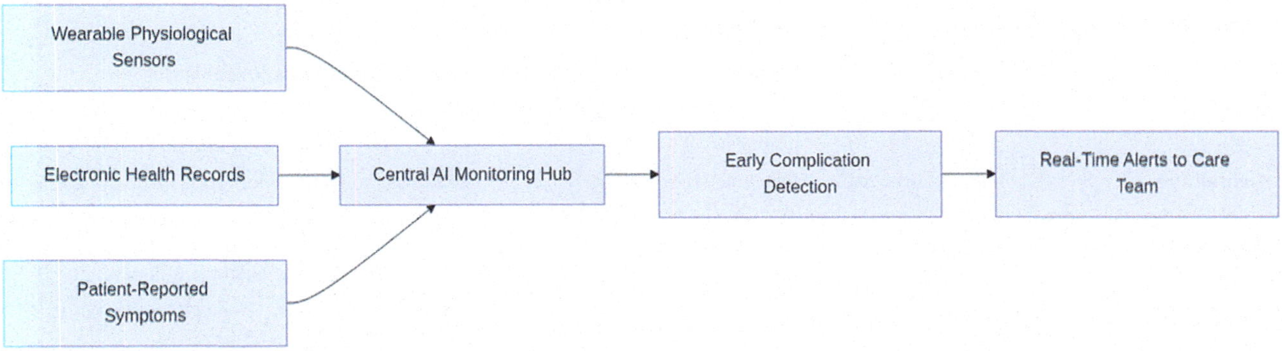

FIG. 10.7 AI-driven surveillance in the postoperative period. It depicts the integration of wearable sensors, patient inputs, and hospital data into a central AI platform for continuous recovery monitoring.

evaluation, and long-term patient monitoring. AI technologies have shown immense potential in each of these domains, facilitating a paradigm shift from episodic care to continuous, adaptive, and personalized recovery pathways.

Postoperative care is increasingly driven by AI through wearable monitoring systems and continuous data integration, allowing earlier detection of complications, as shown in Fig. 10.7.

One of the most impactful applications of AI in the immediate postoperative period is real-time physiologic monitoring (Jorge et al., 2022). AI-enabled systems integrate data from wearable sensors, implantable devices, and EHRs to build comprehensive physiological profiles. These platforms collect vital signs, mobility patterns, sleep quality, respiratory effort, and hemodynamic stability to detect early signs of deterioration. For instance, a subtle increase in resting heart rate, coupled with reduced step count and elevated skin temperature, may prompt the system to alert providers to the possibility of infection or thromboembolic events, even before clinical symptoms become apparent (Bignami et al., 2025a, 2025b).

These intelligent surveillance systems use machine learning models trained on large postoperative datasets to discern the difference between normal variations and pathological deviations. Algorithms such as random forests, support vector machines, and long short-term memory networks (LSTMs) are commonly applied to analyze time-series data in postoperative settings. These models can identify complications such as wound dehiscence, pneumonia, anastomotic leakage, or delirium with significantly greater sensitivity than standard vital sign thresholds.

Wound monitoring represents another major area of AI application. Through mobile health platforms, patients are instructed to upload daily images of their surgical sites. CNNs trained on thousands of wound images can classify the visual features of wounds, including redness, edema, drainage, and tissue granulation, with near-expert accuracy. These systems can flag suspected infections or delayed healing and prompt timely intervention. Importantly, this approach democratizes access to surgical follow-up care, especially in rural or underserved areas where in-person visits are logistically challenging (Griffa et al., 2024).

In parallel, AI enables NLP of postoperative clinical documentation, including progress notes, discharge summaries, nursing assessments, and patient messages. NLP algorithms extract clinically relevant information, such as new-onset symptoms, pain scores, medication changes, and behavioral cues from unstructured text. These insights are then integrated into the patient's risk profile, allowing for a continuously updated, holistic assessment of recovery. Advanced NLP models, such as transformers and bidirectional encoder representations from transformers (BERT), are increasingly being adopted to improve the accuracy and contextual relevance of clinical extraction tasks (Gomez-Cabello et al., 2024; Kovoor et al., 2023).

Postoperative care also entails complex medication management, especially among older adults or patients with multiple comorbidities. AI tools optimize medication reconciliation by scanning prescription records, cross-referencing drug interactions, and detecting adherence patterns via smart pill dispensers and mobile apps. Reinforcement learning models personalize drug regimens over time by adapting to side effect profiles, pharmacogenomic variations, and patient-reported tolerability. These technologies contribute significantly to reducing adverse drug events, which are a common source of hospital readmission in the postoperative period (Abdo et al., 2024; Babel, Taneja, Malvestiti, Monaco, & Donde, 2021).

Predicting adverse events before they manifest improves safety. Table 10.7 outlines validated models, inputs, and performance for major complications following robotic surgery.

Functional recovery monitoring is another domain revolutionized by AI. Using accelerometers, gyroscopes, and electromyographic sensors embedded in wearable devices, AI tracks patients' mobility, gait, and muscle activation. These

TABLE 10.7 AI-Based prediction models for postoperative complications.

Complication	AI framework	Input features	Accuracy	Clinical status
Sepsis	LSTM + random forest	Vitals, CRP, WBC, antibiotic history	85%	In a clinical trial (Bignami et al., 2025a, 2025b)
Delirium	NLP (BERT) + audio ML	Age, voice modulation, prior cognitive status	92.6%	Pilot (Boppana et al., 2025)
Anastomotic leak	XGBoost ensemble	Operation time, blood loss, drain output	84%	Retrospective validation (Mülküt & Ofluoğlu, 2025)
Readmission	Logistic regression + NLP	Pain score, medication adherence, discharge notes	86%	EHR-integrated (Li, Wang, Lu, & Zhu, 2022)

This table outlines validated AI models used for forecasting key postoperative complications such as sepsis, anastomotic leaks, delirium, and hospital readmissions. It details the input features leveraged, model types deployed, and the observed accuracy across clinical validation stages. These predictive models play a crucial role in early warning systems and individualized patient monitoring, and they are increasingly integrated into electronic health records and mobile applications for enhanced continuity of care. *WBC*, white blood cell count; *CRP*, C-reactive protein; *NLP*, natural language processing.

signals are interpreted by machine learning models to assess rehabilitation progress, quantify physical activity, and identify emerging limitations in mobility. Such data supports targeted physical therapy interventions, enabling clinicians to dynamically adjust recovery plans based on real-time performance rather than fixed schedules (van den Eijnden et al., 2023).

Moreover, AI is being used to assess and support cognitive and emotional recovery. Postoperative cognitive dysfunction, mood disturbances, and delirium are common but often underdiagnosed complications, particularly among elderly or high-risk populations. Voice analysis algorithms can detect subtle changes in speech patterns that correlate with cognitive impairment or depression. Similarly, facial expression analysis and sentiment tracking from patient-reported digital diaries provide valuable insights into mental health status. These multimodal approaches, grounded in affective computing, offer a more complete understanding of the patient's neuropsychiatric recovery (Agrawal et al., 2025).

AI-powered sentiment analysis and speech interpretation tools can detect subtle signs of postoperative cognitive dysfunction or depression, supporting neuropsychiatric recovery, as shown in Fig. 10.8.

In the realm of personalized discharge planning, AI contributes to predictive modeling of recovery trajectories and the risk of readmission. Algorithms consider demographic data, comorbidities, surgical complexity, intraoperative metrics, and early postoperative indicators to estimate the expected recovery course. These predictions guide discharge timing, home care needs, physical therapy intensity, and follow-up scheduling. Dynamic modeling allows these plans to be adjusted continuously as new data become available, fostering more responsive and patient-centered care (Hilton et al., 2020).

AI models stratify patients into personalized recovery paths: outpatient, inpatient rehab, or home monitoring, based on real-time data and predicted outcomes, as shown in Fig. 10.9.

AI also facilitates remote patient engagement, empowering individuals to take active roles in their recovery. Mobile health applications integrated with AI-powered virtual assistants provide medication reminders, symptom checkers, educational content, and guided rehabilitation routines. These systems can also administer validated surveys to track patient-reported outcomes such as pain, fatigue, bowel function, or urinary continence—metrics that are critical for assessing quality of life following robotic surgery.

One of the emerging fields in postoperative AI research is the integration of genomic, metabolomic, and microbiome data into recovery prediction. By combining omics data with traditional clinical parameters, AI can uncover novel predictors of healing capacity, immunologic resilience, and response to therapy. For example, specific gut microbiota profiles have been linked to recovery from abdominal surgery, and machine learning models can identify these signatures to tailor nutritional or probiotic interventions postoperatively (Pammi, Aghaeepour, & Neu, 2023).

On the system level, hospitals and surgical centers are leveraging AI to monitor institutional performance and benchmark surgical outcomes. Predictive analytics platforms aggregate and visualize postoperative outcomes across surgeons, departments, and patient demographics. These tools identify outliers, track complication rates, and highlight trends in patient-reported satisfaction. Integrating robotic platform logs, such as usage time, error rates, and system faults, enables predictive maintenance and improves the safety and longevity of surgical equipment (Maleki Varnosfaderani & Forouzanfar, 2024).

AI impacts robotic surgery at multiple stages, each with unique functions: data analysis preoperatively, vision and force prediction intraoperatively, and recovery modeling postoperatively, as shown in Fig. 10.10.

AI operates across domains—vision, NLP, sensor data, prediction models—and integrates through feedback loops across all phases of surgery, as shown in Fig. 10.11.

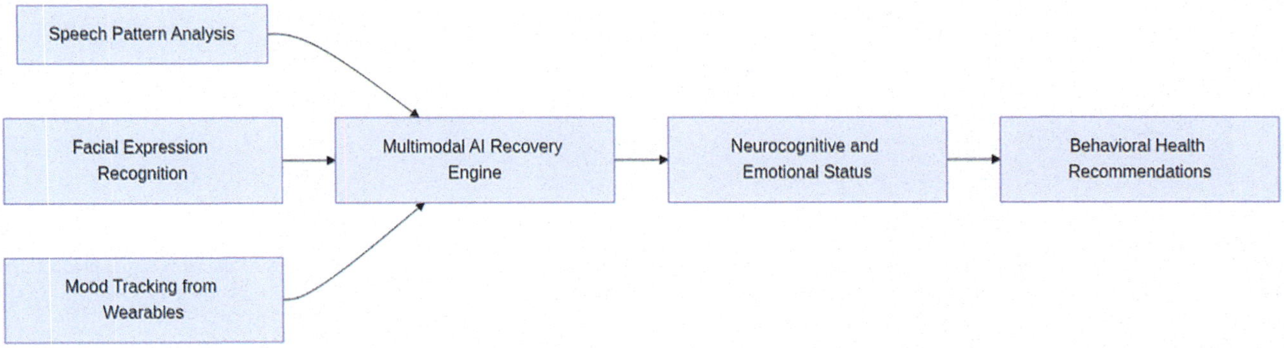

FIG. 10.8 Multimodal AI assessment of cognitive and emotional recovery. It shows the integration of voice, facial, and behavioral inputs by AI to assess neurocognitive health in the postoperative period.

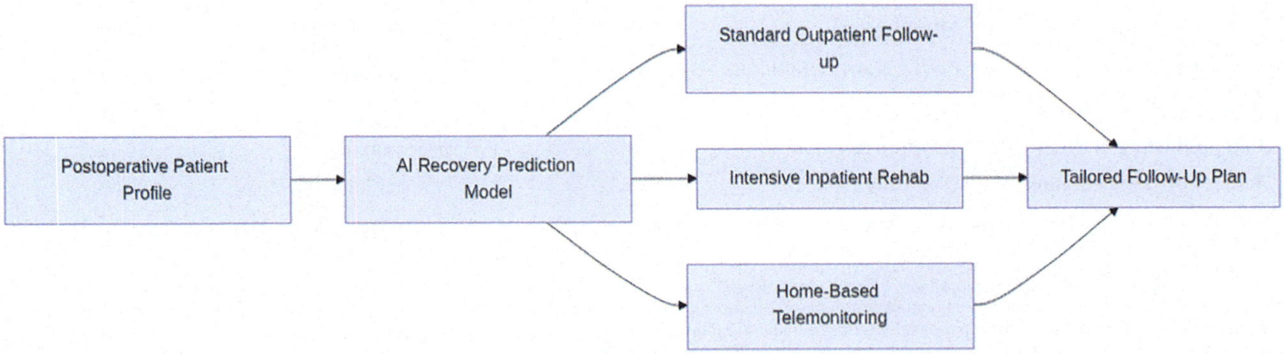

FIG. 10.9 Predictive recovery pathways enabled by AI. It demonstrates how AI categorizes patients into recovery streams, tailoring follow-up intensity and care strategies accordingly.

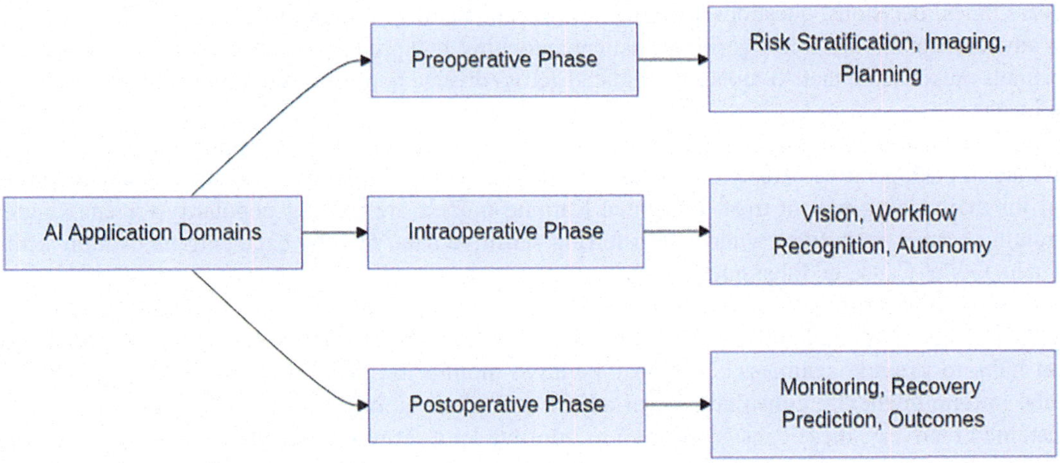

FIG. 10.10 Domain-specific AI applications across the surgical continuum. It outlines key AI functionalities across the three surgical phases and their respective contributions to robotic surgery outcomes.

Quality assurance programs are increasingly embedding AI into their evaluation frameworks. Surgical video analysis tools can be used post hoc to assess procedural technique, teamwork dynamics, and compliance with standardized protocols. These evaluations feed back into training curricula, quality improvement initiatives, and peer review processes. Furthermore, AI can synthesize patient safety data to identify latent safety threats, streamline root cause analysis, and recommend preventive strategies (Dick et al., 2024).

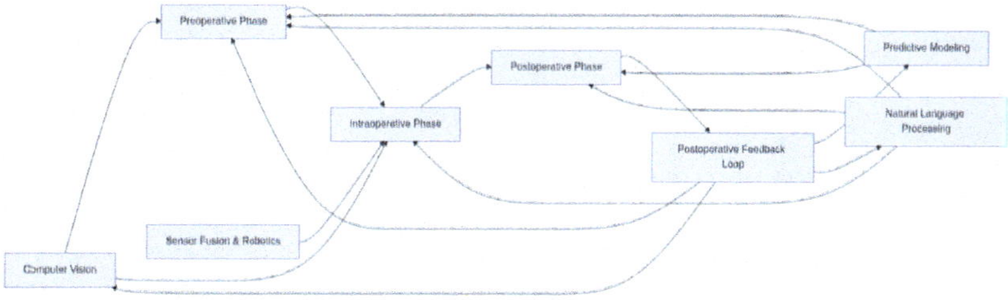

FIG. 10.11 Multimodal AI integration across the robotic surgery continuum. It demonstrates the interaction of various AI domains across preoperative, intraoperative, and postoperative phases, reinforced by feedback loops and adaptive learning.

TABLE 10.8 Performance metrics of AI models in surgical video annotation across specialties.

Annotation task	Model	Accuracy (%)	Surgical field	Dataset size (videos)
Phase detection	CNN + RNN Hybrid	91.4	General surgery	800
Blood vessel localization	YOLOv7 + attention layer	94.1	Urology	2500
Clip positioning in liver	Transformer vision	89.7	Hepatobiliary	450
Tumor edge segmentation	DeepLabV3 +	88.3	Gynecologic oncology	1200

This table presents a comparative evaluation of artificial intelligence models applied to the annotation and interpretation of intraoperative surgical videos. Each entry details the annotation task performed, such as phase recognition, anatomical structure detection, or instrument positioning, alongside the specific AI architecture used, reported accuracy rates, surgical subspecialty context, and dataset size employed for training or validation. These models have been tested in clinically relevant settings, either retrospectively or prospectively, with accuracy values exceeding 88% in all categories. The comparative structure of the table highlights the variability in model selection based on visual complexity, anatomical region, and intended intraoperative assistance. Understanding these metrics is essential for evaluating the real-world feasibility and robustness of AI-driven video interpretation tools in robotic-assisted procedures. *CNN*, convolutional neural network, *RNN*, recurrent neural network; *YOLO*, you only look once.

Legal and ethical considerations also intersect with the use of AI in postoperative care. As decision support tools gain influence over clinical decisions, questions arise regarding accountability, transparency, and informed consent. Ethical AI frameworks advocate for explainable algorithms, patient autonomy, and protection against bias in postoperative prediction models. Hospitals must ensure that AI tools are validated across diverse populations to prevent disparities in access to safe recovery pathways.

Data privacy remains a core concern, particularly as wearable devices, mobile applications, and cloud-based platforms proliferate in postoperative care. Secure data transmission protocols, encrypted storage, and anonymized analytics are essential for maintaining patient trust. Federated learning models are gaining popularity for enabling collaborative algorithm training across institutions without transferring sensitive data, thereby balancing innovation with compliance (Yurdem, Kuzlu, Gullu, Catak, & Tabassum, 2024).

The future of AI in postoperative robotic surgery is moving toward the creation of intelligent recovery ecosystems. These systems will integrate ambient sensors in homes, voice assistants, wearable technologies, and AI-powered care coordination hubs to provide seamless, 24/7 postoperative monitoring. Real-time alerts will trigger automated triage protocols, initiate telehealth consultations, or adjust care plans autonomously. By closing the loop between surgery and sustained recovery, these ecosystems aim to improve long-term outcomes, reduce costs, and enhance patient satisfaction.

Annotation accuracy in intraoperative videos is critical for computer-aided surgery. Table 10.8 compares validated models, highlighting their strengths across datasets and specializations.

In conclusion, AI is profoundly reshaping the postoperative phase of robotic surgery by enabling a shift from episodic care to proactive, continuous, and personalized recovery pathways. It leverages real-time monitoring, multimodal data analysis, predictive modeling, and intelligent engagement tools to transform every aspect of the healing process. As these technologies mature and become more interoperable, they will enable a future in which the boundaries between hospital and home, provider and patient, and data and decision are seamlessly integrated for optimal surgical recovery.

5 Training, education, and credentialing

The integration of AI into robotic surgery is not only transforming surgical procedures but also redefining how surgeons are trained, assessed, and credentialed. The inherently complex interface between human cognition and machine intelligence necessitates a new paradigm in surgical education, one that emphasizes data-driven performance assessment, adaptive learning environments, and continuous competency verification. AI plays a central role in each of these domains, functioning as both a pedagogical tool and an impartial evaluator.

Traditional surgical training models have relied on apprenticeship-style learning, wherein residents acquire skills through observation, practice, and subjective feedback. However, robotic surgery, with its steep learning curve and reliance on console-based operation, requires a more structured and objective methodology. AI-enhanced simulation platforms offer immersive, high-fidelity environments where trainees can perform complex procedures in a risk-free setting. These simulators, often equipped with haptic feedback and real-time video analytics, are now incorporating AI algorithms that assess trainee performance by analyzing a multitude of metrics, such as instrument trajectory, task completion time, error rates, and tissue handling quality (Elendu et al., 2024; Joseph, Bhatti, Mittal, & Bhatti, 2025).

AI-driven evaluation systems go beyond surface-level performance indicators. Using advanced machine learning techniques, they deconstruct surgical actions into constituent gestures, assess the fluidity and economy of motion, and compare the trainee's patterns to those of expert surgeons. Such granular feedback allows for the identification of specific deficits, such as hesitancy in dissection, suboptimal instrument angles during suturing, or excessive instrument collisions. Trainees receive personalized feedback reports that include visualizations, percentile rankings, and targeted recommendations for improvement. This method fosters deliberate practice, whereby learners focus intensively on weak areas, accelerating skill acquisition and enhancing retention.

Beyond simulation, AI is increasingly being utilized in live surgical environments to provide real-time coaching and post hoc performance reviews. During actual robotic procedures, AI systems can unobtrusively monitor instrument movements and surgical field videos to detect deviations from ideal techniques or workflow inefficiencies. These systems can issue suggestions or prompts via the console interface, much like a co-pilot assisting in navigation. After the procedure, comprehensive performance summaries are generated, which can be used for self-assessment, mentorship discussions, and credentialing documentation (Guo, He, Shi, & Wang, 2025).

Credentialing in robotic surgery is undergoing a transformation fueled by AI-enabled assessment. Rather than relying solely on case volume or time-based metrics, institutions are adopting competency-based frameworks wherein surgeons must demonstrate proficiency in validated technical tasks. AI platforms provide objective, reproducible evaluations that reduce the variability and subjectivity associated with traditional assessments. Some hospitals and academic centers are piloting credentialing pathways in which surgeons are required to pass a series of AI-monitored simulation tasks and submit intraoperative performance data before being granted independent privileges on robotic platforms. This rigorous approach not only enhances patient safety but also promotes a culture of accountability and continuous improvement.

Moreover, AI facilitates personalized learning pathways by adapting educational content and task difficulty to the individual trainee's performance curve. Adaptive learning algorithms adjust scenarios in real time to challenge the trainee at an appropriate level, thereby maintaining engagement and optimizing cognitive load. These systems also track longitudinal performance across multiple sessions and procedures, identifying trends that inform future training plans and remediation strategies. For instance, if a trainee demonstrates consistent difficulty with intracorporeal knot tying, the system may recommend a focused module with graduated complexity and targeted feedback loops (Valente, Brasil, Spinelli, Vilela, & Rhoden, 2025).

AI is also fostering collaborative and remote learning through cloud-based platforms, where trainees can upload their performance data, compare it to peer and expert benchmarks, and engage in asynchronous feedback sessions with mentors. Virtual proctoring systems, powered by AI video analysis, enable faculty to review surgical recordings at scale, annotate key events, and provide structured feedback without requiring real-time observation. These capabilities are especially valuable in global training initiatives, where geographic and resource constraints may limit access to experienced mentors or advanced training facilities (Tan, Cheng, & Ling, 2025).

Importantly, AI supports the development of nontechnical skills, such as decision-making, situational awareness, and stress management, which are critical to surgical performance. AI algorithms can analyze voice tone, response time, and gaze patterns to infer cognitive workload and stress levels during simulation or live surgery. These insights inform training interventions aimed at building resilience, improving focus, and enhancing teamwork. Some platforms are integrating AI-generated scenarios that simulate intraoperative crises, requiring the trainee to make time-sensitive decisions, communicate effectively, and manage resources, thereby cultivating comprehensive surgical readiness.

As robotic platforms and AI technologies continue to evolve, the boundaries between training, practice, and evaluation will blur. Surgeons will increasingly engage in a continuous loop of learning and assessment, supported by real-time data analytics and longitudinal performance tracking. Certification and credentialing bodies are expected to integrate AI-based evaluation tools into their frameworks, ensuring that surgical competence is not only achieved but maintained over time. These developments represent a shift toward a more accountable, personalized, and data-informed model of surgical education.

AI is reshaping the landscape of surgical training, education, and credentialing by enabling objective assessment, individualized feedback, and scalable learning environments. As the demands of robotic surgery continue to grow, AI will be indispensable in preparing the next generation of surgeons for a future defined by precision, adaptability, and lifelong learning.

6 Economic impact and system efficiency

The economic implications of integrating AI into robotic surgery extend beyond direct cost assessments to include long-term operational efficiencies, institutional scalability, workforce productivity, and value-based care delivery. As healthcare systems worldwide confront mounting fiscal constraints, AI offers a compelling opportunity to optimize resource allocation, reduce inefficiencies, and enhance the return on investment associated with robotic surgical programs (Bejrananda, Khaing, Veettil, Thongseiratch, & Chaiyakunapruk, 2025; Ikoma, 2024; Song, Cheng, Li, Kreaden, & Snyder, 2022; Wah, 2025).

One of the most direct economic benefits arises from AI's ability to shorten operative time without compromising quality. AI-guided preoperative planning reduces setup delays by optimizing port placement, instrument configuration, and procedural sequencing. During surgery, real-time intraoperative assistance accelerates decision-making, improves instrument handling, and minimizes unnecessary tissue manipulation. These efficiencies cumulatively reduce the overall time a patient spends under anesthesia, which is not only safer for the patient but also translates into significant cost savings by enabling more cases per OR per day.

Moreover, AI-assisted systems can significantly reduce postoperative complications and readmissions, which are major cost drivers in surgical care. Predictive models that anticipate adverse events allow for preemptive interventions, which improve outcomes and avoid costly treatments. For example, AI-driven early detection of surgical site infections or thromboembolic events enables timely administration of antibiotics or anticoagulants, reducing the need for extended hospital stays, revision surgeries, or intensive care unit admissions. When scaled across thousands of procedures annually, these savings can justify the upfront capital investment required for AI-integrated robotic platforms (Takeuchi & Kitagawa, 2024).

AI also enhances perioperative logistics and inventory management. Predictive analytics can forecast surgical supply usage based on case mix, seasonality, and surgeon preference, leading to more accurate ordering and reduced waste. Robotic instruments, which have limited reuse cycles, can be tracked using AI-powered dashboards that ensure timely replacement while avoiding unnecessary surplus. These systems streamline surgical kit assembly, reduce manual errors, and improve traceability, each contributing to operational cost efficiency.

In addition, AI optimizes human resource deployment in surgical settings (Zuhair et al., 2024). Smart scheduling systems can analyze historical procedure durations, surgeon performance data, and patient complexity to predict OR utilization more accurately. This results in better staff allocation, fewer overtime hours, and minimized idle time. By harmonizing surgeon schedules with available anesthesiology and nursing staff, AI-driven scheduling systems promote a leaner, more synchronized workflow, improving both morale and productivity.

Hospitals adopting AI-enhanced robotic surgery have also reported reduced malpractice risks and improved payer negotiations. Objective performance data collected by AI systems can be used to demonstrate adherence to clinical guidelines, compliance with safety protocols, and continuous performance improvement. These data can serve as evidence in quality improvement audits and legal defenses, potentially lowering liability insurance premiums. Furthermore, outcome transparency supported by AI can improve a hospital's standing with insurers, who are increasingly aligning reimbursement with quality metrics under value-based payment models (Adegbesan et al., 2024; Terranova, Cestonaro, Fava, & Cinquetti, 2023).

While the initial investment in robotic systems and AI infrastructure is considerable, cost amortization models show that these platforms can become financially sustainable within a few years when supported by appropriate caseload volumes and institutional commitment. AI facilitates this process by expanding the scope of robotic surgery to a broader range of procedures and specialties, thus maximizing platform utilization. In high-volume centers, the cost-per-case of robotic surgery with AI support may rival or even undercut that of conventional surgery, especially when factoring in reduced complication rates, shorter lengths of stay, and lower readmission rates.

Economies of scale further enhance the economic argument for AI integration. As more procedures are performed using robotic platforms, the per-unit cost of data analysis, software licensing, and hardware maintenance decreases. In addition, AI algorithms improve with exposure to more data, thereby increasing their predictive power and utility over time. This feedback loop, where increased use leads to better outcomes, which in turn justifies further use, creates a sustainable economic cycle that benefits both institutions and patients.

Another economic consideration is the competitive advantage conferred by AI-enhanced robotic programs. Hospitals equipped with state-of-the-art AI technologies attract top surgical talent, appeal to tech-savvy patients, and are more likely to secure research funding and industry partnerships. These reputational and financial benefits can have a positive ripple effect across the institution, bolstering related departments such as anesthesiology, radiology, and pathology (Roppelt, Kanbach, & Kraus, 2024).

AI also contributes to system-wide efficiency beyond individual hospitals. At the regional or national level, aggregated AI data can inform policy decisions, resource distribution, and workforce planning. Health ministries and public health agencies are beginning to use AI to map surgical demand, predict workforce shortages, and allocate robotic platforms to underserved areas. In this way, AI supports not only institutional financial health but also broader health system sustainability.

The economic impact of AI in robotic surgery is multifaceted and far-reaching. While the initial investment is substantial, the return on investment can be realized through operative efficiency, complication reduction, optimized staffing, inventory control, and enhanced institutional reputation. As AI continues to mature and integrate more deeply into surgical workflows, its role in promoting cost-effective, high-quality surgical care will become increasingly indispensable.

7 Ethical, legal, and equity considerations

The integration of AI into robotic surgery not only transforms technical and clinical dimensions but also provokes significant ethical, legal, and equity-related questions. As AI systems become more autonomous and involved in decision-making, their use raises new concerns about accountability, data governance, bias, transparency, and the equitable distribution of healthcare resources. These issues demand a multidisciplinary framework for ethical oversight, legal regulation, and equitable implementation to ensure that technological advancements benefit all patients fairly and safely.

A responsible AI deployment requires a framework of stakeholders—clinicians, ethics boards, regulatory bodies, and patients—along with continuous auditing and feedback, as shown in Fig. 10.12.

A primary ethical concern is the question of accountability in AI-assisted surgical decisions. In traditional surgery, the operating surgeon bears full responsibility for intraoperative decisions and patient outcomes. However, when an AI algorithm recommends a specific action, such as resection margins, procedural steps, or device settings, determining liability becomes more complex. If a complication arises due to reliance on AI-generated guidance, it is unclear whether the fault lies with the surgeon, the algorithm developers, the hospital deploying the system, or the manufacturers of the robotic platform. This diffusion of responsibility necessitates clear regulatory frameworks that delineate the roles and obligations of each stakeholder (Cestonaro, Delicati, Marcante, Caenazzo, & Tozzo, 2023; Lysaght, Lim, Xafis, & Ngiam, 2019).

Informed consent also becomes more intricate in the context of AI-assisted robotic surgery. Patients must be adequately informed not only about the general risks and benefits of the procedure but also about the role that AI will play in surgical planning, intraoperative guidance, and postoperative monitoring. The challenge lies in conveying the capabilities and limitations of AI in a manner that is comprehensible to patients, especially when algorithms are based on complex, opaque deep learning models. Ethically sound consent processes must incorporate discussions on algorithmic reliability, data sources, and decision boundaries to ensure patient autonomy is respected.

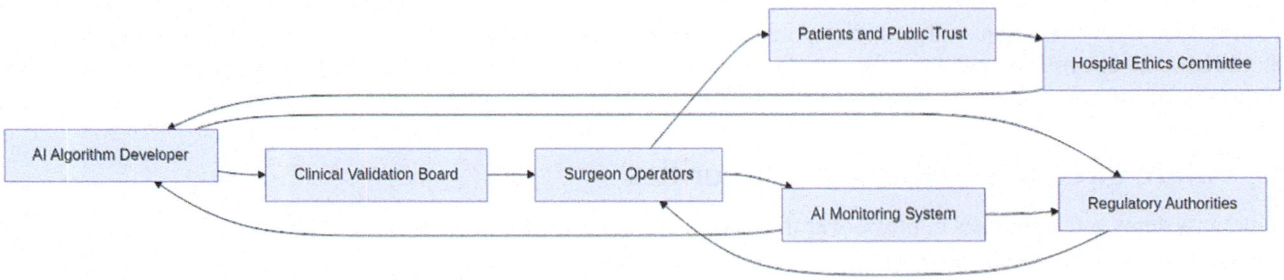

FIG. 10.12 Governance model for AI oversight in robotic surgery. It outlines the institutional, legal, and ethical components of AI oversight in robotic surgical systems, highlighting accountability, transparency, and compliance structures.

TABLE 10.9 Ethical risks and mitigation strategies in AI-assisted robotic surgery.

Ethical concern	Example scenario	AI Risk factor	Mitigation strategy
Data bias	AI underperforms on minority populations	Imbalanced training datasets	Multinational, multicenter training corpora
Autonomy diminishment	Surgeon override of AI ignored	Closed-loop automation	Human-in-the-loop architecture design
Data privacy breach	Surgical video stored without consent	Insecure cloud storage	On-site encryption + federated data standards
Liability ambiguity	Complication due to AI misguidance	Undefined responsibility mapping	Joint surgeon-manufacturer regulatory consensus

This table examines major ethical concerns associated with AI use in robotic surgery, ranging from algorithmic bias and autonomy erosion to data privacy and medicolegal responsibility. For each concern, it describes a real-world example, the underlying AI vulnerability, and mitigation strategies based on current literature and regulatory proposals. This structured overview fosters critical thinking about responsible AI design and clinical governance in digitally assisted surgery.

Ethical challenges, such as data privacy, algorithmic bias, and medicolegal responsibility, are integral to the use of AI in surgery. Table 10.9 summarizes major ethical threats and proposed strategies to mitigate them in AI-augmented robotic environments.

Another major concern is the issue of algorithmic bias. AI systems learn from data, and if that data reflects existing disparities or lacks adequate representation across age, gender, race, or socioeconomic groups, the algorithm may perpetuate or even exacerbate healthcare inequalities (Mittermaier, Raza, & Kvedar, 2023). For instance, a surgical risk prediction model trained predominantly on data from affluent, urban populations may underestimate complications in patients from rural or underserved communities. Such biases can manifest in inappropriate procedural recommendations, inaccurate risk stratification, or inequitable allocation of postoperative resources. Ensuring fairness in AI algorithms requires deliberate inclusion of diverse patient datasets, continuous auditing for performance disparities, and the implementation of algorithmic fairness metrics.

The legal landscape surrounding AI in robotic surgery is still in its infancy. Existing medical device regulations are often ill-suited to the dynamic, adaptive nature of AI systems, which may change their behavior over time through continuous learning. This raises questions about certification, recertification, and postmarket surveillance. Regulatory agencies are exploring new frameworks, such as the United States Food and Drug Administration's (FDA's) proposed approach to Software as a Medical Device, which allows for ongoing algorithm updates under controlled conditions (Singh, Cheng, Kwan, & Ebinger, 2025). Such frameworks must strike a balance between innovation and safety, ensuring that AI systems remain accurate, secure, and clinically validated throughout their lifecycle.

Equity of access is a final, yet vital, consideration. Robotic surgery platforms and the associated AI technologies are expensive and resource-intensive, creating barriers to adoption in low-resource settings. Without deliberate efforts to subsidize or adapt these technologies for broader use, there is a risk that AI-enhanced surgical care will become a privilege of well-funded institutions, deepening global and regional disparities in healthcare access. Furthermore, digital infrastructure, skilled personnel, and regulatory capacity vary widely across regions, compounding the challenge of equitable AI deployment. International partnerships, open-source initiatives, and scalable AI solutions must be prioritized to bridge these gaps (Wah, 2025).

Ethical integration also involves engaging with the surgical workforce. Surgeons must be active participants in the design, testing, and governance of AI systems, ensuring that their insights shape development priorities and usability features. Professional societies and ethics boards must establish guidelines for responsible AI use, including protocols for overrides, conflict resolution, and multidisciplinary reviews of AI-assisted decisions. Ethical education and digital literacy should become integral components of surgical training, empowering clinicians to critically appraise and ethically navigate AI-enhanced environments.

8 Future directions and research priorities

Robotic systems and AI models improve over time by analyzing surgical outcomes, complications, and video performance, retraining themselves through federated feedback systems, as shown in Fig. 10.13.

One of the most promising directions is the progression toward fully autonomous robotic procedures. While current systems operate with the surgeon in full control, future platforms are likely to incorporate increasing levels of autonomy,

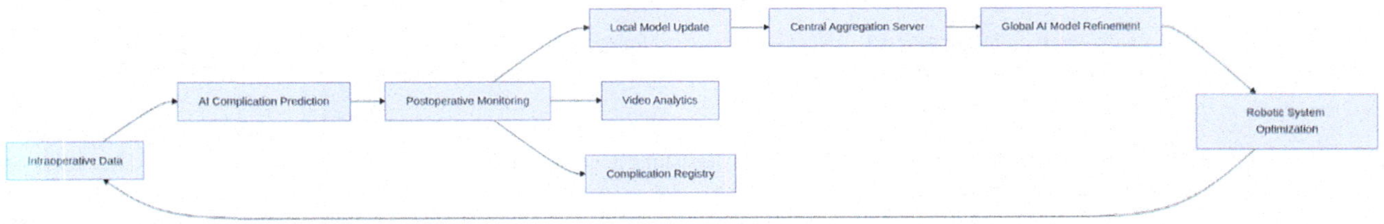

FIG. 10.13 Closed-loop adaptive AI learning system in robotic surgery. It illustrates how AI models evolve by incorporating intraoperative and postoperative data to continuously optimize robotic performance and outcome prediction.

TABLE 10.10 Federated learning vs centralized learning in surgical AI.

Criteria	Federated learning	Centralized learning
Data location	Stored locally, model aggregates remotely	All data collected into one central server
Privacy	High, no raw data transfer	Moderate, requires deidentification
Compute requirement	High on local nodes	High at central server
Model consistency	Variable, depends on communication latency	More uniform, single environment
Real-world feasibility	Suitable for multicenter studies (GDPR compliant)	Limited by cross-border data restrictions

This table compares federated and centralized learning approaches as they apply to AI model training in surgical environments. It explains differences in data storage, privacy, model synchronization, computational load, and real-world feasibility. This comparative framework is critical for readers exploring privacy-preserving AI models, especially in the context of cross-institutional collaborations constrained by international data protection regulations such as GDPR.
GDPR, General Data Protection Regulation.

beginning with discrete sub-tasks such as tissue retraction, suturing, and vessel sealing. Reinforcement learning and imitation learning algorithms are being trained to replicate and refine expert surgical movements through millions of simulations. As these systems become more context-aware and responsive to intraoperative changes, they may evolve into cosurgeons capable of executing predefined procedural modules with minimal supervision. However, this vision necessitates rigorous validation, safety assurance, and the establishment of ethical boundaries for autonomous intervention (Datta et al., 2021).

Digital twin technology represents another frontier. In this approach, a virtual replica of the patient's anatomy and surgical environment is continuously updated in real time during the procedure using data from cameras, sensors, and imaging devices. This digital twin can serve as an intelligent guide, highlighting anatomical landmarks, predicting tissue responses, and simulating the outcomes of surgical decisions before they are executed. Ongoing research is exploring the potential of combining digital twins with AR overlays, allowing surgeons to visualize hidden structures and risk zones directly within their operative field (Mekki et al., 2025).

Data sharing across institutions is complex due to privacy laws. Federated learning offers decentralized training while retaining privacy. Table 10.10 compares federated versus centralized AI models in terms of clinical application, risk, and computational burden.

In addition, there is a need to establish comprehensive frameworks for the continuous validation and lifecycle management of AI systems. Unlike static medical devices, AI algorithms evolve over time as they are exposed to new data. Regulatory science must adapt to monitor algorithmic drift, detect performance degradation, and ensure ongoing safety and efficacy. Research into postmarket surveillance systems, algorithm registries, and dynamic benchmarking standards will be essential to maintaining public trust and safeguarding patient welfare.

Future developments must also be inclusive and accessible. Research should focus on the creation of AI systems that are adaptable to varying levels of infrastructure, surgical expertise, and patient demographics. Low-cost AI solutions designed for mobile platforms, edge computing applications for remote surgical assistance, and AI-based decision aids for nonrobotic procedures are examples of innovations that can bridge the digital divide and democratize surgical excellence.

Finally, interdisciplinary research must explore the ethical and philosophical implications of AI-driven surgery. As machines gain increasing autonomy and influence, questions of moral agency, responsibility, and professional identity will become more pressing. Collaborative studies involving ethicists, sociologists, legal scholars, and clinicians are needed to develop ethical guidelines, governance models, and public engagement strategies that anticipate and respond to societal concerns.

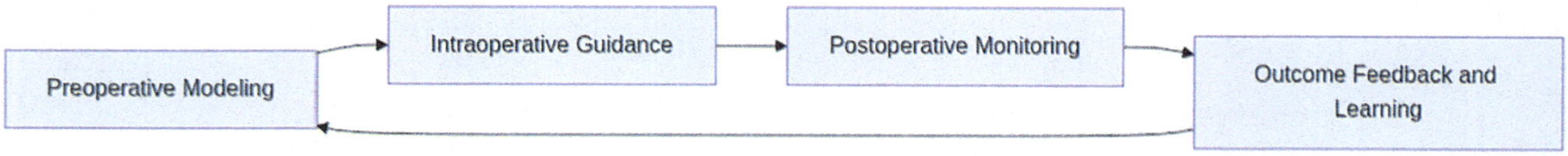

FIG. 10.14 Lifecycle of AI-enabled robotic surgery. A cyclic diagram illustrating how preoperative, intraoperative, and postoperative AI modules interact and continuously improve surgical practice.

AI supports a closed-loop surgical lifecycle, from risk modeling and intraoperative guidance to postoperative feedback and model retraining, as shown in Fig. 10.14.

The future of AI in robotic surgery is rich with possibility but also laden with complexity. It demands sustained investment in research, thoughtful regulation, and inclusive innovation. The next generation of AI-powered surgical systems will not only extend the capabilities of human surgeons but also reshape the very fabric of surgical care. By proactively addressing the technical, clinical, ethical, and infrastructural challenges ahead, the medical community can harness AI to create a safer, smarter, and more equitable future in robotic surgery.

9 Conclusion

AI is enabling robotic surgery to move beyond tool status, toward functioning as an intelligent partner that supports planning, guidance, monitoring, training, and operational management. Properly validated and integrated, these methods can improve outcomes, reduce complications, and increase efficiency across surgical services.

Realizing these gains depends on robust validation, transparent algorithms, strong data protection, clear legal responsibility, and active measures to avoid bias and unequal access. Effective deployment, therefore, requires sustained collaboration among clinicians, engineers, ethicists, and regulators to ensure reliability, safety, and benefit for patients.

Declaration of generative AI and AI-assisted technologies

During the preparation of this work, the authors used ChatGPT (version 4.0) to improve language and readability. The tool was used only for editing and polishing the text. After using this tool, the authors reviewed and edited the content as needed and take full responsibility for the content of the publication. ChatGPT was not listed as an author and does not meet authorship criteria.

References

Abdo, A., Gallay, L., Vallecillo, T., Clarenne, J., Quillet, P., Vuiblet, V., & Merieux, R. (2024). A machine learning-based clinical predictive tool to identify patients at high risk of medication errors. *Scientific Reports*. 2045-2322. *14*(1). doi:10.1038/s41598-024-83631-w.

Adegbesan, A., Akingbola, A., Aremu, O., Adewole, O., Amamdikwa, J. C., & Shagaya, U. (2024). From scalpels to algorithms: the risk of dependence on artificial intelligence in surgery. *Journal of Medicine, Surgery, and Public Health*. 2949916X. 3, 100140. doi:10.1016/j.glmedi.2024.100140.

Agrawal, S., Rupavath, R. V S. S. B., Jalaja, P. P., Ushmani, A., Mishra, A., & Bodapati, N. V. S. B. (2025). Artificial intelligence (AI)-driven approaches to manage postoperative pain, anxiety, and psychological outcomes in surgical patients: a systematic review. *Cureus*. 2168-8184. *17*(5). doi:10.7759/cureus.84226.

Almadani, B., Kaisar, H., Thoker, I. R., & Aliyu, F. (2025). A systematic survey of distributed decision support systems in healthcare. *Systems*. 20798954. *13*(3). Available from: www.mdpi.com/journal/systems. doi:10.3390/systems13030157.

Anteby, R., Horesh, N., Soffer, S., Zager, Y., Barash, Y., Amiel, I., … Klang, E. (2021). Deep learning visual analysis in laparoscopic surgery: a systematic review and diagnostic test accuracy meta-analysis. *Surgical Endoscopy*. 0930-2794. *35*(4), 1521–1533. doi:10.1007/s00464-020-08168-1.

Avram, M. F., Lazăr, D. C., Mariş, M. I., & Olariu, S. (2023). Artificial intelligence in improving the outcome of surgical treatment in colorectal cancer. *Frontiers in Oncology*. 2234943X. 13. Available from: http://www.frontiersin.org/Oncology/about. doi:10.3389/fonc.2023.1116761.

Babel, A., Taneja, R., Malvestiti, F. M., Monaco, A., & Donde, S. (2021). Artificial intelligence solutions to increase medication adherence in patients with non-communicable diseases. *Frontiers in Digital Health*, 2673-253X, 3. doi: 10.3389/fdgth.2021.669869

Bejrananda, T., Khaing, W., Veettil, S. K., Thongseiratch, T., & Chaiyakunapruk, N. (2025). Economic evaluation of robotic-assisted radical prostatectomy: a systematic review and meta-analysis. *European Urology Open Science*.26661683. 72, 17–28. Available from: https://www.sciencedirect.com/science/journal/26661683. doi:10.1016/j.euros.2025.01.011.

Bellini, V., Russo, M., Domenichetti, T., Panizzi, M., Allai, S., & Bignami, E. G. (2024). Artificial intelligence in operating room management. *Journal of Medical Systems*. 1573689X. *48*(1). Available from: https://www.springer.com/journal/10916. doi:10.1007/s10916-024-02038-2.

Bignami, E. G., Panizzi, M., Bezzi, F., Mion, M., Bagnoli, M., & Bellini, V. (2025a). Wearable devices as part of postoperative early warning score systems: a scoping review. *Journal of Clinical Monitoring and Computing*. 1573-2614. *39*(1), 233–244. doi:10.1007/s10877-024-01224-4.

Bignami, E. G., Berdini, M., Panizzi, M., Domenichetti, T., Bezzi, F., Allai, S., … Bellini, V. (2025b). Artificial intelligence in sepsis management: an overview for clinicians. *Journal of Clinical Medicine*. 20770383. *14*(1). Available from: https://www.mdpi.com/journal/jcm. doi:10.3390/jcm14010286.

Boardman, C., Huggins, A., & Hull, K. (2011). Da Vinci robotic-assisted radical hysterectomy procedural guide. *Journal of Robotic Surgery*. 1863-2483. *5*(3), 181–187. doi:10.1007/s11701-011-0252-0.

Boppana, S. H., Tyagi, D., Komati, S., Boppana, S. L., Raj, R., & Mintz, C. D. (2025). AI-delirium guard: predictive modeling of postoperative delirium in elderly surgical patients. *PLOS ONE*. 19326203. *20*(6). Available from: https://journals.plos.org/plosone/article?id=10.1371/journal.pone.0322032. doi:10.1371/journal.pone.0322032.

Cahill, R. A., Duffourc, M., & Gerke, S. (2025). The AI-enhanced surgeon – integrating black-box artificial intelligence in the operating room. *International Journal of Surgery*. 1743-9159. *111*(4), 2823–2826. doi:10.1097/js9.0000000000002309.

Cestonaro, C., Delicati, A., Marcante, B., Caenazzo, L., & Tozzo, P. (2023). Defining medical liability when artificial intelligence is applied on diagnostic algorithms: a systematic review. *Frontiers in Medicine*, 2296-858X, 10. doi: 10.3389/fmed.2023.1305756

Datta, S., Li, Y., Ruppert, M. M., Ren, Y., Shickel, B., Ozrazgat-Baslanti, T., … Bihorac, A. (2021). Reinforcement learning in surgery. *Surgery*. 00396060. *170*(1), 329–332. doi:10.1016/j.surg.2020.11.040.

Dick, L., Boyle, C. P., Skipworth, R. J. E., Smink, D. S., Tallentire, V. R., & Yule, S. (2024). Automated analysis of operative video in surgical training: scoping review. *BJS Open*. 24749842. *8*(5). Available from: https://academic.oup.com/bjsopen. doi:10.1093/bjsopen/zrae124.

Diez, S. P., Borghesan, G., Joyeux, L., Meuleman, C., Deprest, J., Stoyanov, D., … Poorten, E. B. V. (2019). Evaluation of haptic feedback on bimanually teleoperated laparoscopy for endometriosis surgery. *IEEE Transactions on Biomedical Engineering*. 15582531. *66*(5), 1207–1221. Available from: http://ieeexplore.ieee.org/xpl/RecentIssue.jsp?reload=true&punumber=10. doi:10.1109/TBME.2018.2870542.

Elendu, C., Amaechi, D. C., Okatta, A. U., Amaechi, E. C., Elendu, T. C., Ezeh, C. P., & Elendu, I. D. (2024). The impact of simulation-based training in medical education: a review. *Medicine (United States)*. 15365964. *103*(27), e38813. Available from: https://journals.lww.com/md-journal/pages/default.aspx. doi:10.1097/MD.0000000000038813.

Enayati, N., De Momi, E., & Ferrigno, G. (2016). Haptics in robot-assisted surgery: challenges and benefits. *IEEE Reviews in Biomedical Engineering*. 1937-3333. 9, 49–65. doi:10.1109/rbme.2016.2538080.

Ergenç, M. (2025). Artificial intelligence in surgical practice: truth beyond fancy covering. *Turkish Journal of Surgery*. 25646850. *41*(2). doi:10.47717/turkjsurg.2025.6797.

Fan, H., Luo, Y., Gu, F., Tian, B., Xiong, Y., Wu, G., … Liao, X. (2024). Artificial intelligence-based MRI radiomics and radiogenomics in glioma. *Cancer Imaging*. 1470-7330. *24*(1). doi:10.1186/s40644-024-00682-y.

Gomez-Cabello, C. A., Borna, S., Pressman, S. M., Haider, S. A., Sehgal, A., Leibovich, B. C., & Forte, A. J. (2024). Artificial intelligence in postoperative care: assessing large language models for patient recommendations in plastic surgery. *Healthcare (Switzerland)*. 22279032. *12*(11). Available from: www.mdpi.com/journal/healthcare/. doi:10.3390/healthcare12111083.

Griffa, D., Natale, A., Merli, Y., Starace, M., Curti, N., Mussi, M., … Zengarini, C. (2024). Artificial intelligence in wound care: a narrative review of the currently available mobile apps for automatic ulcer segmentation. *BioMedInformatics*. 2673-7426. *4*(4), 2321–2337. doi:10.3390/biomedinformatics4040126.

Guo, C., He, Y., Shi, Z., & Wang, L. (2025). Artificial intelligence in surgical medicine: a brief review. *Annals of Medicine and Surgery*. 2049-0801. *87*(4), 2180–2186. doi:10.1097/ms9.0000000000003115.

Gupta, S., Mamodiya, U., & Al-Gburi, A. J. A. (2025). Speech recognition-based wireless control system for mobile robotics: design, implementation, and analysis. *Automation*. 2673-4052. *6*(3), 25. doi:10.3390/automation6030025.

Hilton, C. B., Milinovich, A., Felix, C., Vakharia, N., Crone, T., Donovan, C., … Nazha, A. (2020). Personalized predictions of patient outcomes during and after hospitalization using artificial intelligence. *Npj Digital Medicine*. 23986352. *3*(1). Available from: https://www.nature.com/npjdigitalmed/. doi:10.1038/s41746-020-0249-z.

Hunter, A. (2024). Integrating wearable devices into perioperative medicine: the potential, and future challenges. *Future Healthcare Journal*. 25146645. *11*(3), 100169. doi:10.1016/j.fhj.2024.100169.

Ijlal, A., Mumtaz, H., Hassan, S. M., Mustafa, Q.-U.-A., Khalil, A. B. B., Ali, U., … Sajjad, L. (2025). Bridging surgical oncology and personalized medicine: the role of artificial intelligence and machine learning in thoracic surgery. *Annals of Medicine and Surgery*. 2049-0801. *87*(6), 3566–3572. doi:10.1097/ms9.0000000000003302.

Ikoma, N. (2024). What defines the "value" of robotic surgery for patients with gastrointestinal cancers? Perspectives from a U.S. Cancer Center. *Annals of Gastroenterological Surgery*. 2475-0328. *8*(4), 566–579. doi:10.1002/ags3.12792.

Jorge, J., Villarroel, M., Tomlinson, H., Gibson, O., Darbyshire, J. L., Ede, J., … Watkinson, P. (2022). Non-contact physiological monitoring of post-operative patients in the intensive care unit. *Npj Digital Medicine*. 23986352. *5*(1). Available from: https://www.nature.com/npjdigitalmed/. doi:10.1038/s41746-021-00543-z.

Joseph, G., Bhatti, N., Mittal, R., & Bhatti, A. (2025). Current application and future prospects of artificial intelligence in healthcare and medical education: a review of literature. *Cureus*. 2168-8184. *17*(1). doi:10.7759/cureus.77313.

Kambale, M., & Jadhav, S. (2024). Applications of artificial intelligence in anesthesia: a systematic review. *Saudi Journal of Anaesthesia*. 1658-354X. *18*(2), 249–256. doi:10.4103/sja.sja_955_23.

Kapoor, M. C., & Garg, S. (2024). Role of artificial intelligence in perioperative monitoring in anaesthesia. *Indian Journal of Anaesthesia*. 00195049. *68*(1), 87–92. Available from: http://www.ijaweb.org/backissues.asp. doi:10.4103/ija.ija_1198_23.

Kovoor, J. G., Bacchi, S., Gupta, A. K., Stretton, B., Nann, S. D., Aujayeb, N., … Maddern, G. J. (2023). Surgery's Rosetta Stone: natural language processing to predict discharge and readmission after general surgery. *Surgery (United States)*. 15327361. *174*(6), 1309–1314. Available from: http://www.elsevier.com/inca/publications/store/6/2/3/1/9/5/index.htt. doi:10.1016/j.surg.2023.08.021.

Lanotte, F., O'Brien, M. K., & Jayaraman, A. (2023). AI in rehabilitation medicine: opportunities and challenges. *Annals of Rehabilitation Medicine*. 22340653. *47*(6), 444–458. Available from: http://www.e-arm.org/. doi:10.5535/ARM.23131.

Le, K. D. R., Tay, S. B. P., Choy, K. T., Verjans, J., Sasanelli, N., & Kong, J. C. H. (2024). Applications of natural language processing tools in the surgical journey. *Frontiers in Surgery*. 2296875X. 11. Available from: https://journal.frontiersin.org/journal/surgery. doi:10.3389/fsurg.2024.1403540.

Li, L., Wang, L., Lu, L., & Zhu, T. (2022). Machine learning prediction of postoperative unplanned 30-day hospital readmission in older adult. *Frontiers in Molecular Biosciences*, 2296-889X, 9. doi: 10.3389/fmolb.2022.910688

Liu, Z., Li, C., Lin, J., Xu, H., Xu, Y., Nan, H., … Wang, B. (2024). Advances in the development and application of non-contact intraoperative image access systems. *BioMedical Engineering OnLine*. 1475-925X. *23*(1). doi:10.1186/s12938-024-01304-1.

Lysaght, T., Lim, H. Y., Xafis, V., & Ngiam, K. Y. (2019). AI-Assisted decision-making in healthcare: the application of an ethics framework for big data in health and research. *Asian Bioethics Review*. 17939453. *11*(3), 299–314. Available from: www.asianbioethicsreview.com/journal/index.php?journal=abr_sbc1&page=index. doi:10.1007/s41649-019-00096-0.

Maleki Varnosfaderani, S., & Forouzanfar, M. (2024). The role of AI in hospitals and clinics: transforming healthcare in the 21st century. *Bioengineering*. 2306-5354. *11*(4), 337. doi:10.3390/bioengineering11040337.

Mamlook, R. E. A., Wells, L. J., & Sawyer, R. (2023). Machine-learning models for predicting surgical site infections using patient pre-operative risk and surgical procedure factors. *American Journal of Infection Control*. 15273296. *51*(5), 544–550. Available from: http://www.journals.elsevier.com/ajic-american-journal-of-infection-control/. doi:10.1016/j.ajic.2022.08.013.

Mansour, M. A., El-Salamoni, M. A.-F., Zohney, M., Elshaer, A. M., Abdelwahab, M., Aziz, M. M., … El-Samman, A. (2025). Neurosurgery reimagined: how AI is redefining patient care and surgical excellence. *Intelligent Surgery*. 26666766. 8, 84–91. doi:10.1016/j.isurg.2025.06.002.

Mekki, Y. M., Luijten, G., Hagert, E., Belkhair, S., Varghese, C., Qadir, J., … Stotland, M. A. (2025). Digital twins for the era of personalized surgery. *Npj Digital Medicine*. 23986352. 8(1). Available from: https://www.nature.com/npjdigitalmed/. doi:10.1038/s41746-025-01575-5.

Mittermaier, M., Raza, M. M., & Kvedar, J. C. (2023). Bias in AI-based models for medical applications: challenges and mitigation strategies. *Npj Digital Medicine*. 23986352. *6*(1). Available from: https://www.nature.com/npjdigitalmed/. doi:10.1038/s41746-023-00858-z.

Mohsin, S. N., Gapizov, A., Ekhator, C., Ain, N. U., Ahmad, S., Khan, M., … Nagaraj, R. H. (2023). The role of artificial intelligence in prediction, risk stratification, and personalized treatment planning for congenital heart diseases. *Cureus*. 2168-8184. *15*(8). doi:10.7759/cureus.44374.

Mülküt, F., & Ofluoğlu, C. B. (2025). Development of a clinical framework–anastomotic leakage prediction score (CF–Alps) after colorectal surgery. *Diagnostics*. 20754418. *15*(4). Available from: http://www.mdpi.com/journal/diagnostics/. doi:10.3390/diagnostics15040455.

Pammi, M., Aghaeepour, N., & Neu, J. (2023). Multiomics, artificial intelligence, and precision medicine in perinatology. *Pediatric Research*. 0031-3998. *93*(2), 308–315. doi:10.1038/s41390-022-02181-x.

Pantelis, A. G., Epiphaniou, P., & Lapatsanis, D. P. (2025). Machine learning and artificial intelligence for predicting short and long-term complications following metabolic bariatric surgery—A systematic review. *Artificial Intelligence Surgery*. 2771-0408. *5*(3), 322–344. doi:10.20517/ais.2024.104.

Pedrett, R., Mascagni, P., Beldi, G., Padoy, N., & Lavanchy, J. L. (2023). Technical skill assessment in minimally invasive surgery using artificial intelligence: a systematic review. *Surgical Endoscopy*. 14322218. *37*(10), 7412–7424. Available from: https://www.springer.com/journal/464. doi:10.1007/s00464-023-10335-z.

Riad, A., Hadid, M., Elomri, A., Al-Ansari, A., Rejeb, M. A., Qaraqe, M., … EL Omri, A. (2025). Advancements and challenges in robotic surgery: a holistic examination of operational dynamics and future directions. *Surgery in Practice and Science*. 26662620. 22, 100294. doi:10.1016/j.sipas.2025.100294.

Riddle, E. W., Kewalramani, D., Narayan, M., & Jones, D. B. (2024). Surgical simulation: virtual reality to artificial intelligence. *Current Problems in Surgery*. 15356337. *61*(11). Available from: https://www.sciencedirect.com/science/journal/00113840. doi:10.1016/j.cpsurg.2024.101625.

Rivero-Moreno, Y., Rodriguez, M., Losada-Muñoz, P., Redden, S., Lopez-Lezama, S., Vidal-Gallardo, A., … Teran-Quintero, S. (2024). Autonomous robotic surgery: has the future arrived? *Cureus*. 2168-8184. *16*(1). doi:10.7759/cureus.52243.

Roppelt, J. S., Kanbach, D. K., & Kraus, S. (2024). Artificial intelligence in healthcare institutions: a systematic literature review on influencing factors. *Technology in Society*. 0160791X. 76. Available from: https://www.sciencedirect.com/science/journal/0160791X. doi:10.1016/j.techsoc.2023.102443.

Seager, A., Sharp, L., Neilson, L. J., Brand, A., Hampton, J. S., Lee, T. J. W., … Verma, A. M. (2024). Polyp detection with colonoscopy assisted by the GI Genius artificial intelligence endoscopy module compared with standard colonoscopy in routine colonoscopy practice (COLO-DETECT): a multicentre, open-label, parallel-arm, pragmatic randomised controlled trial. *The Lancet Gastroenterology & Hepatology*. 24681253. *9*(10), 911–923. Available from: https://www.sciencedirect.com/science/journal/24681253. doi:10.1016/S2468-1253(24)00161-4.

Singh, V., Cheng, S., Kwan, A. C., & Ebinger, J. (2025). United States Food and Drug Administration Regulation of clinical software in the Era of artificial intelligence and machine learning. *Mayo Clinic Proceedings: Digital Health*. 29497612. *3*(3), 100231. doi:10.1016/j.mcpdig.2025.100231.

Solanki, S. L., Pandrowala, S., Nayak, A., Bhandare, M., Ambulkar, R. P., & Shrikhande, S. V. (2021). Artificial intelligence in perioperative management of major gastrointestinal surgeries. *World Journal of Gastroenterology*. 22192840. *27*(21), 2758–2770. Available from: https://www.wjgnet.com/1007-9327/full/v27/i21/2758.htm. doi:10.3748/wjg.v27.i21.2758.

Song, C., Cheng, L., Li, Y., Kreaden, U., & Snyder, S. R. (2022). Systematic literature review of cost-effectiveness analyses of robotic-assisted radical prostatectomy for localised prostate cancer. *BMJ Open*. 2044-6055. *12*(9), e058394. doi:10.1136/bmjopen-2021-058394.

Takeuchi, M., & Kitagawa, Y. (2024). Artificial intelligence and surgery. *Annals of Gastroenterological Surgery*. 2475-0328. *8*(1), 4–5. doi:10.1002/ags3.12766.

Tan, X., Cheng, G., & Ling, M. H. (2025). Artificial intelligence in teaching and teacher professional development: a systematic review. *Computers and Education: Artificial Intelligence*. 2666920X. 8, 100355. doi:10.1016/j.caeai.2024.100355.

Tejani, A. S., Peshock, R. M., & Raj, K. M. (2024). Evolving with artificial intelligence: integrating artificial intelligence and imaging informatics in a general residency curriculum with an advanced track. *Journal of the American College of Radiology*. 1558349X. *21*(10), 1608–1612. Available from: https://www.sciencedirect.com/science/journal/15461440. doi:10.1016/j.jacr.2024.07.007.

Terranova, C., Cestonaro, C., Fava, L., & Cinquetti, A. (2023). AI and professional liability assessment in healthcare. A revolution in legal medicine?, *Frontiers in Medicine*, 2296858X. 10, Available from: journal.frontiersin.org/journal/medicine. doi:10.3389/fmed.2023.1337335.

Tian, Z., Liu, L., Zhang, Z., & Fei, B. (2018). PSNet: prostate segmentation on MRI based on a convolutional neural network. *Journal of Medical Imaging*. 2329-4302. *5*(2), 1. doi:10.1117/1.JMI.5.2.021208.

Valente, D. S., Brasil, L. J., Spinelli, L. D. F., Vilela, M. A. P., & Rhoden, E. L. (2025). A narrative review of transforming surgical education with artificial intelligence: opportunities and challenges. *AME Surgical Journal*. 2788578X. 5. Available from: https://asj.amegroups.org/article/download/97081/72260. doi:10.21037/asj-24-25.

van den Eijnden, M. A. C., van der Stam, J. A., Bouwman, R. A., Mestrom, E. H. J., Verhaegh, W. F. J., van Riel, N. A. W., & Cox, L. G. E. (2023). Machine learning for postoperative continuous recovery scores of oncology patients in perioperative care with data from wearables. *Sensors*. 14248220. *23*(9). Available from: http://www.mdpi.com/journal/sensors. doi:10.3390/s23094455.

Van Der Maten, G., Reimer, J. M. B., Meijs, M. F. L., Von Birgelen, C., Brusse-Keizer, M. G. J., & Hertog, H. M. D. (2021). Detection of major cardioembolic sources in real-world patients with ischemic stroke or transient ischemic attack of undetermined cause. *Cerebrovascular Diseases Extra*. 16645456. *11*(1), 22–28. Available from: http://www.karger.com/Journal/Home/254735. doi:10.1159/000512743.

Vasey, B., Lippert, K. A. N., Khan, D. Z., Ibrahim, M., Koh, C. H., Horsfall, H. L., … McCulloch, P. (2023). Intraoperative applications of artificial intelligence in robotic surgery: a scoping review of current development stages and levels of autonomy. *Annals of Surgery*. 15281140. *278*(6), 896–903. Available from: http://journals.lww.com/annalsofsurgery/pages/default.aspx. doi:10.1097/SLA.0000000000005700.

Vleugels, M. P. H., & Rahimi, M. (2024). Feeling of pulsations in artificial arteries with a real time haptic feedback laparoscopic grasper: a validation study. *Surgical Endoscopy*. 14322218. *38*(8), 4222–4228. Available from: https://www.springer.com/journal/464. doi:10.1007/s00464-024-10877-w.

Wah, J. N. K. (2025). The rise of robotics and AI-assisted surgery in modern healthcare. *Journal of Robotic Surgery*. 18632491. *19*(1). Available from: https://www.springer.com/journal/11701. doi:10.1007/s11701-025-02485-0.

Wang, X., Yang, J., Zhou, B., Tang, L., & Liang, Y. (2025). Integrating mixed reality, augmented reality, and artificial intelligence in complex liver surgeries: enhancing precision, safety, and outcomes. *iLIVER*. 27729478. *4*(2), 100167. doi:10.1016/j.iliver.2025.100167.

Xianjia, Y., Queralta, J. P., Heikkonen, J., & Westerlund, T. (2021). Finland federated learning in robotic and autonomous systems. *Procedia Computer Science*, 18770509, 191, 135–142. doi: 10.1016/j.procs.2021.07.041 http://www.sciencedirect.com/science/journal/.

Xu, Y., Liu, X., Cao, X., Huang, C., Liu, E., Qian, S., … Zhang, J. (2021). Artificial intelligence: a powerful paradigm for scientific research. *The Innovation*. 26666758. *2*(4), 100179. doi:10.1016/j.xinn.2021.100179.

Yangi, K., On, T. J., Xu, Y., Gholami, A. S., Hong, J., Reed, A. G., … Preul, M. C. (2025). Artificial intelligence integration in surgery through hand and instrument tracking: a systematic literature review. *Frontiers in Surgery*. 2296875X. 12. Available from: https://journal.frontiersin.org/journal/surgery. doi:10.3389/fsurg.2025.1528362.

Yurdem, B., Kuzlu, M., Gullu, M. K., Catak, F. O., & Tabassum, M. (2024). Federated learning: overview, strategies, applications, tools and future directions. *Heliyon*. 24058440. 10(19). Available from: https://www.sciencedirect.com/science/journal/24058440. doi:10.1016/j.heliyon.2024.e38137.

Zang, C., Turkcan, M. K., Narasimhan, S., Cao, Y., Yarali, K., Xiang, Z., … Kostic, Z. (2023). Surgical phase recognition in inguinal hernia repair—AI-based confirmatory baseline and exploration of competitive models. *Bioengineering*. 23065354. *10*(6). Available from: www.mdpi.com/journal/bioengineering. doi:10.3390/bioengineering10060654.

Zuhair, V., Babar, A., Ali, R., Oduoye, M. O., Noor, Z., Chris, K., … Rehman, L. U. (2024). Exploring the impact of artificial intelligence on global health and enhancing healthcare in developing nations. *Journal of Primary Care and Community Health*. 21501327. 15. Available from: https://journals.sagepub.com/home/JPC. doi:10.1177/21501319241245847.

Chapter 11

Artificial intelligence in plastic surgery: The future of reconstruction and aesthetics

Yasser Hijazi[a] and Omar Kiwan[b]
[a]Plastic and Reconstructive Department, Lancashire Teaching Hospitals NHS Foundation Trust, Preston, United Kingdom
[b]Faculty of Biology, Medicine and Health, University of Manchester, Manchester, United Kingdom

1 Introduction

Plastic surgery is a fascinating specialty that blends medicine with art to restore and enhance both function and esthetics. Unlike other surgical specialties that follow standard treatment pathways, plastic surgery cases tend to have no clear-cut solutions. Whether it is reconstructing a complex trauma defect, planning a microsurgical flap, or performing esthetic enhancements for patients, plastic surgeons must constantly adapt their approach. Each case presents its own puzzle, shaped by a patient's unique anatomy, clinical condition, social circumstances, and desired outcome. The lack of a one-size-fits-all solution demands creativity, innovation, and a deep understanding of form and function.

Given this distinct approach to helping patients, artificial intelligence (AI) will be used in ways different from other specialties. From optimizing preoperative planning and improving diagnostic accuracy to guiding intraoperative decisions and supporting postoperative care, AI has the potential to forever change plastic surgery across the entire patient journey.

This chapter explores the current and emerging applications of AI in plastic surgery. It considers how this rapidly evolving technology may complement clinical judgment, improve patient outcomes, and shape the future of plastic surgery, while also addressing the ethical, technical, and practical challenges that come with its integration.

2 Preoperative applications

AI is reshaping the preoperative process of plastic surgery by advancing precision, efficiency, and personalization of care. From diagnosing conditions and stratifying operative risks to planning procedures and simulating outcomes, AI is being integrated into almost every aspect of surgical planning. Leveraging powerful models trained on large datasets, AI can interpret clinical images, analyze patient data, and predict outcomes with increasing accuracy. This allows surgeons to tailor interventions specific to the patient and their risk profile, communicate expected outcomes more effectively, and ultimately better inform patients. As the field evolves, AI will support clinical judgment and improve patient safety and satisfaction by employing a data-driven and patient-centered approach to preoperative care.

2.1 Diagnosis and screening

AI has demonstrated considerable promise in aiding diagnostic processes within plastic surgery. It achieves this primarily through advanced pattern recognition, developed by training on large datasets of annotated lesions. Given the inherently visual nature of plastic surgery, AI applications relying on computer vision and image recognition are particularly well-suited to the specialty (Mansoor & Ibrahim, 2025). Many of the lesions encountered in plastic surgery are relatively common and visually distinct, which facilitates the development of reliable models through robust model training and validation.

Applications of Artificial Intelligence in Healthcare. https://doi.org/10.1016/B978-0-443-43934-6.00009-2

Several studies have demonstrated that artificial neural networks can achieve diagnostic accuracies comparable to those of experienced plastic surgeons. This includes the diagnosis of skin cancers, lagophthalmos, frostbite, and velopharyngeal insufficiency (Esteva et al., 2017; Kiwan, Al-Kalbani, Rafie, & Hijazi, 2024). AI has also been used to automate the grading of facial palsy by detecting and interpreting facial movements and expressions (Souza, Bhethanabotla, & Mohan, 2024). In burns assessment, deep learning software is capable of classifying burn wounds more precisely than human specialists (Chang et al., 2023). AI can also assess the severity of microtia (Wang, Chen, Yiping, Tang, & Deng, 2022). As these diagnostic processes are based on image analysis, AI algorithms can be integrated into smart phone applications, expanding access to a novel diagnostic tool in a cost-effective and scalable manner. There have been instances where AI showed poor performance in analyzing images. A notable study by Winkler et al. evaluated the accuracy of an AI algorithm in estimating the likelihood that an image of a lesion depicted a melanoma. The algorithm performed well when analyzing cropped images that only included the lesion itself. However, its performance significantly declined when tested on unedited images containing surgical skin markers. In these cases, the algorithm produced a high number of false positives and was able to correctly classify benign lesions <50% of the time. The presence of the marker appeared to introduce a degree of bias, falsely signaling to the AI that the lesion was more suspicious (Winkler et al., 2019).

AI is also proving capable of interpreting radiological images. Deep learning models trained on radiographs have demonstrated the ability to detect distal radius and hand fractures, matching radiologist performance (Kiwan et al., 2024). One such model was able to successfully diagnose scaphoid fractures through X-ray images with 90% accuracy (Yoon & Chung, 2021), significantly outperforming the typical sensitivity of <80% reported in most studies (Kraus, Anteby, Konen, Eshed, & Klang, 2024). Additional AI networks trained on multimodal data, including clinical data, imaging, and tests, have shown the ability to accurately diagnose conditions such as carpal tunnel syndrome, rheumatoid arthritis exacerbations, and osteoporosis-related fragility fractures (Kiwan et al., 2024). In esthetic surgery, AI has also been applied to classify facial attractiveness and beauty, with results very similar to human judgments (Wheeler, 2021).

Overall, AI has shown significant potential in providing rapid and accurate diagnostic support without the need for additional resources. However, most of these studies trained AI to diagnose a specific condition. Practically, this would only be used to help confirm a diagnosis already suspected by clinicians. AI needs to be trained to identify a wide range of conditions for it to be an efficient and reliable diagnostic tool.

2.2 Risk stratification and surgical planning

AI has emerged as a transformative tool in preoperative planning. It has established its ability to identify preoperative risk factors and predict postoperative complications across various plastic surgery cases. A summary of these applications is found in Table 11.1. For instance, an artificial neural network was able to predict the outcome of nerve graft success with 92.59% accuracy (Conforth, Meng, Valmikinathan, & Yu, 2009). Studies also found AI capable of identifying risk factors for breast augmentation mastopexy (Huyghebaert, Wallner, & Montemurro, 2024) and implant-based reconstruction (IBR) (Hassan et al., 2023). In esthetic breast augmentation, AI detected previously unknown risk factors for postoperative complications, such as larger preoperative bra-cup sizes and increased height (Montemurro, Lehnhardt, Behr, & Wallner, 2022). In IBR, Hassan et al. developed an algorithm to predict periprosthetic infection and ex-plantation with an accuracy defined by the area under the curve of 0.73 and 0.78, identifying 9 and 12 important predictive factors, respectively. These included risk factors that were not previously identified, like implant location and chemotherapy (Hassan et al., 2023). Building on these predictive capabilities, Mao et al. applied machine learning to transcriptomic data to identify genetic factors related to breast capsular contracture, a common complication causing pain and reshaping of the breast. AI identified key metabolic and immune-related genes that were significantly associated with the severity of capsular contracture (Mao, Hou, Fu, & Luan, 2024).

Many machine learning programs have now been developed to predict postoperative complications. MySurgeryRisk is a machine learning program developed to predict complications like venous thromboembolism, sepsis, and acute kidney injury. It has shown greater accuracy in predicting these complications than physicians; the accuracy of physicians also significantly improved following the use of the program (Brennan et al., 2019). Predictive OpTimal Trees in Emergency Surgery Risk (POTTER) is another example of a surgical risk calculator that uses machine learning. It was trained using data from over 380,000 acute surgical cases and outperformed widely used scoring systems, such as the American Society of Anesthesiologists (ASA) score, in predicting postoperative morbidity and mortality (Bertsimas, Dunn, Velmahos, & Kaafarani, 2018).

One of the most promising and extensively studied applications of AI is surgical planning. Virtual surgical planning using AI has already been rolled out in oral and maxillofacial surgery, and numerous studies show that it can help improve outcomes in plastic surgery as well (Brancaccio et al., 2024). AI's ability to recognize facial expressions can grade palsy

TABLE 11.1 A summary of AI applications used to enhance surgical planning in plastic surgery.

Application	AI algorithm	Key functions	Reported outcomes
Image segmentation and analysis	Deep learning algorithm designed for CT angiogram analysis	Maps perforator vessels for free-flap design	Reduces operative time and improves flap survival
	U-Net: a convolutional neural network designed for image segmentation	Used to identify the median nerve's location for carpal tunnel release	Optimizes incision and dissection strategy
	Layered deep learning segmentation of CT scans	Automates 3D segmentation of the mandible for mandibular reconstruction	Found to be faster and more accurate than semi-automatic segmentation
	Pixel-to-pixel deep learning segmentation	Uses pixel-wise segmentation to differentiate necrotic and viable tissue in burn surgery	Guides precise excision depth, optimizing tissue preservation
	Deep learning ultrasound segmentation	Maps adipose tissue for cannula placement in liposuction planning	Improves precision and safety
Outcome prediction and risk stratification	MySurgeryRisk: a collection of generalized additive models that can be used to predict surgical complications	Predicts postoperative complications, including venous thromboembolism, sepsis, and acute kidney injury	Demonstrated greater accuracy than experienced physicians and improved physicians' own predictive accuracy
	Predictive OpTimal Trees in Emergency Surgery Risk (POTTER): a type of machine learning algorithm, known as optimal classification trees, that adapts with each input, allowing for the interpretation of clinical information to predict outcomes	Predicts postoperative morbidity and mortality for acute surgical admissions	Outperformed widely used scoring systems, such as the ASA score, in predicting postoperative morbidity and mortality
	Artificial neural network that can predict outcomes of nerve repairs	Predicts nerve graft success	Displayed a 92.6% accuracy in predicting nerve graft success
Risk stratification	Machine learning algorithm used to predict postoperative complications	Predicts periprosthetic infection and explantation after implant-based reconstruction	Area-under-curve accuracy of 0.73 and 0.78, respectively, meaning it can successfully predict these outcomes roughly 75% of the time
	Machine learning on transcriptomic data	Identifies key genes linked to the severity of capsular contracture	Discovered metabolic and immune-related genetic predictors
Landmark identification	Convolutional neural networks specialize in computer-vision	Identifies surgically important landmarks for cleft lip surgery	Successfully identified surgically important anatomical landmarks for cleft lip repair, which can be used to aid in preoperative planning
	Computer vision algorithm that can assess facial features	Analyzes facial symmetry, fat distribution, muscle mass, and bone distribution	Guides customized injection plans, improving esthetic outcomes
	A pretrained facial landmark detection model	Identifies key nasal landmarks for rhinoplasty	Measures imperfections to aid in surgical planning; works across diverse ethnic backgrounds
3D printing	Deep learning segmentation	Automates segmentation of scans to allow for patient-specific implant designs	Allows for quicker and more accurate patient-specific designs

severity and aid in reanimation planning (Hartmann et al., 2022). AI-driven systems have also been developed for automating periorbital measurements, such as margin reflex distance 1, a necessary metric for assessing and planning ptosis and eyelid surgeries (Boboridis, Assi, Indar, Bunce, & Tyers, 2001).

Through machine learning, medical images can be accurately segmented. This offers surgeons detailed insights into each patient's unique anatomy, significantly enhancing preoperative planning. AI algorithms can identify the most suitable surgical markers for patients (Wheeler, 2021). Raj et al. used machine learning to identify 36 key landmarks on the face for rhinoplasty. This algorithm could measure nasal imperfections in about three-quarters of cases and did so across different ethnic backgrounds (Raj, Garg, Vathulya, & Kandwal, 2024). Another study also found AI capable of identifying surgically important landmarks for cleft lip surgery (Sayadi, Hamdan, Zhangli, Hu, & Vyas, 2022).

Computer vision has been shown to allow for segmentation of facial anatomy (Waked et al., 2022). This can help guide injectable treatments by guiding needle placement and analyzing facial symmetry, fat distribution, muscle mass, and bone structure for customized injection plans. In burn injuries, the depth of the burn can be analyzed through AI. Deep learning algorithms can segment the burn area from the viable tissue using a pixel-to-pixel method. This aids excision planning by providing surgeons with the exact depth of necrotic tissue that needs excising (Chang et al., 2023). In addition to segmenting images, AI has demonstrated tremendous potential in segmenting radiological scans, with numerous studies showing it can do this more accurately than existing techniques and in a fraction of the time. AI has been shown to automatically segment a Computed Tomography (CT) scan of the mandible faster and more accurately than semi-automatic segmentation, the current gold standard for surgical planning (Verhelst et al., 2021). A deep learning program was also found to be capable of segmenting ultrasound scans of adipose tissue prior to liposuction, improving the success and safety of the operation by giving surgeons more information to better place the cannula (Cai et al., 2023). AI can also automate segmentation of ultrasound scans of the wrist and identify the median nerve in carpal tunnel release (Festen, Schrier, & Amadio, 2021). To improve outcomes in free flap surgery, a complex procedure involving the transplantation of tissue and its blood supply, AI has been shown to improve the acquisition, processing, and interpretation of CT angiograms in mapping the course of perforator vessels (Cevik, Seth, & Rozen, 2023). This allows for more precise flap design, reduced operative time, improved viability of the flap, and better outcomes for patients.

AI has the potential to transform the use of 3D printing in plastic surgery by helping to overcome many of the barriers that currently limit its broader clinical application. 3D printing has already been shown to significantly improve surgical outcomes and patient satisfaction. This technology can also enhance preoperative planning and surgical education by providing detailed, tactile models for simulation and training (Lynn, Pflibsen, Smith, Rebecca, & Teven, 2021). In orthopedic surgery, 3D-printed patient-specific instruments have already been implemented for total knee arthroplasty with great results (Keskinis et al., 2023). Similarly, in plastic surgery, patient-specific models and implants can be created using high-resolution imaging data from CT and Magnatic Resonance Image (MRI) scans (Aimar, Palermo & Innocenti, 2019). It has already been studied for use in numerous plastic surgery procedures, such as breast, ear, and nasal reconstruction, and has been found to create precise, anatomically accurate implants (Nishimoto, Sotsuka, Kawai, Fujita, & Kakibuchi, 2014; Xu et al., 2015). A study by Wang et al. even developed a 3D-printed breast prosthesis embedded with a therapeutic hydrogel that could detect tumor recurrence (Wang et al., 2024). In addition, 3D-printed patient-specific facial bone implants have been found to restore functional integrity and esthetic appearance to the face, although much of this work has been done on cadavers and remains in the early stages of its development (Sutradhar, Park, Carrau, & Miller, 2014).

Despite showing such high potential, the implementation of 3D printing on a wider scale remains constrained by its complexity. It requires multiple stages, including image acquisition, accurate segmentation of anatomical structures, identification and digital reconstruction of the target area, optimization of implant design, and generation of geometrical support, all while maintaining rigorous quality control standards (Aimar, Palermo & Innocenti, 2019; Ma et al., 2023). Each of these steps demands substantial amounts of time, resources, and skilled human input, limiting its scalability and routine use. AI offers a powerful solution to streamline and enhance nearly every stage of this process. As previously discussed, deep learning algorithms are particularly effective for image segmentation, enabling the rapid and precise identification of anatomical landmarks essential for surgical planning. Once the relevant structures are segmented, machine learning models can be used to automatically generate patient-specific designs tailored to both anatomical and functional needs, reducing the design time, minimizing the margin for human error, and improving implant fit (Ma et al., 2023). Furthermore, AI can automatically create optimized support structures for 3D-printed models, helping to reduce material usage and printing duration without compromising the structural integrity of the product. AI-driven systems can also perform real-time quality control to ensure high standards are consistently met (Harbour, 2022). As AI continues to evolve, it could allow for fully automated, on-demand manufacturing of prosthetics and implants, transforming personalized care in surgery. However, for widespread clinical implementation, further studies are needed to validate these methods, along with the establishment of clear regulatory frameworks to ensure safety, reproducibility, and ethical use at scale for AI-based 3D printing.

2.3 Simulation and consent

One of the most impactful applications of AI in plastic surgery is its ability to simulate postoperative outcomes, thereby transforming how patients are informed and consented before procedures. Traditional consultations have relied heavily on verbal descriptions, two-dimensional photographs, and drawings that are often subjective and not specific to the patient. Recent developments in image-based deep learning and facial analytic software have allowed surgeons to generate personalized outcome predictions using preoperative photographs. These tools can assess facial features, including key anatomical landmarks, symmetry, and proportions, to predict postoperative esthetic results and identify risk factors for poor outcomes (Choi et al., 2023; Wheeler, 2021). AI has even demonstrated its ability to predict perceived age reduction following facial rejuvenation surgeries, like facelifts, lip lifts, and blepharoplasties. Studies show these predictions not only align with actual postoperative results but also correlate with patient satisfaction levels, suggesting that AI can provide realistic expectations for patients undergoing esthetic procedures (Elliott et al., 2023; Gibstein et al., 2021; Goodyear et al., 2023; Zhang et al., 2021).

Generative adversarial networks (GANs) are a type of deep learning model that can create realistic images (Shin, Yang, & Lee, 2021). Their use has been extensively studied and could reshape outcome simulation and informed consent. Text-to-image AI models, such as DALL-E, Adobe Firefly, and Imagen, can also convert descriptive text into high-quality images. Text-to-image models are made up of two components, the first being a large language model (LLM) that can convert the inputted text into a latent representation, a simplified representation of the data. The second component is a GAN that creates an image using this representation (Dhawan & Shay, 2024). These models are trained on extensive datasets consisting of images and their corresponding text. Through repeated simulation, these algorithms slowly understand the complex relationship between text and images, allowing them to create images that strongly resemble the inputted text (Buzzaccarini et al., 2024; Dhawan & Shay, 2024). AI-generated images of rhinoplasties were so realistic that human participants struggled to differentiate them from actual postoperative images (Knoedler et al., 2024). Similarly, AI has been used to simulate breast augmentation outcomes, with predicted images closely matching real postoperative appearances (Chartier et al., 2022). These technologies offer patients a visually rich, personalized preview of potential results, which can improve their knowledge and understanding, reduce anxiety, and build trust, ultimately better informing their consent and improving their relationship with the surgeon. It must, however, be noted that these AI models are still quite unreliable, especially when it comes to generating precise images needed for anatomical diagrams and representations (Kiwan et al., 2024). More research is, therefore, needed to improve and validate these algorithms to ensure their success.

Machine learning has also been used for broader outcome predictions and risk assessments. In breast surgery, utilizing 3D imaging to estimate postoperative breast volume is already practiced (O'Connell, Stevens, Harris, & Rusby, 2015). AI builds on this by improving the prediction of variables like total resection weight in mammoplasties, outperforming the widely used Schnur scale (Seu et al., 2024). An AI algorithm was also found capable of predicting the mortality risk and outcomes of body contouring surgery with exceptional accuracy (area under the curve of 0.898) (Peng et al., 2023). In hand surgery, studies from the Netherlands have used machine learning to build online tools that predict postoperative function using patient-reported outcome measures (PROMs). These programs have been developed for carpal tunnel syndrome and carpometacarpal osteoarthritis, providing low-cost and easily usable applications to help doctors and patients make decisions about treatment (Hoogendam et al., 2022; Loos et al., 2022). These applications rely on subjective information and do not include objective tests such as nerve conduction studies.

Despite these advances, AI-based simulation tools are not without limitations. One concern is the potential for AI to generate "perfect" images that may raise unrealistic patient expectations. Patients may perceive these images as guarantees instead of possibilities, which could lead to dissatisfaction even if the outcome is objectively successful. In addition, AI models are not yet capable of accounting for many patients' specific variables, such as skin elasticity, muscle and bone volume, the patient's own aging process, genetics, and surgical technique. These factors can significantly influence the outcome of the procedure but are currently overlooked in AI-generated predictions.

To conclude, AI-driven simulation is changing how patients may be managed preoperatively. From realistic images generated by GANs to objective risk stratification, these tools are enhancing a surgeon's ability to inform and guide patients. These advancements help support shared decision-making, set realistic expectations, and ultimately improve patient satisfaction. Surgeons must balance these benefits with caution and recognize the limitations of current models. It must also be noted that while AI can enhance the preoperative consultation and consent process, it cannot replace the need for expert clinician judgment and human communication.

3 Intraoperative applications

AI is being integrated into multiple aspects of intraoperative care. From robot-assisted procedures to real-time feedback systems, AI has shown tremendous potential in enhancing the precision, safety, and outcomes of surgeries. While these applications are still being investigated, they have shown promising early results and could be rolled out in the not-so-distant future.

Robotic surgery has long been recognized for its potential to improve surgical outcomes for patients (Roshan et al., 2024), and its integration into plastic surgery is gradually taking shape. Although historically more common in specialties like urology, general surgery, and gynecology, robotic platforms are now being explored in plastic and reconstructive procedures. Studies have shown very positive outcomes from utilizing robotics in microsurgery for free flap reconstruction and pelvic lymph node dissection for suspected skin cancers (Cheon, Kim, Park, & Yoon, 2022; Ballestín, Malzone, Menichini, Lucattelli, & Innocenti, 2022; Ross, Kumar, Challacombe, Dasgupta, & Geh, 2013). If integrated, AI can play a pivotal role in the evolution of robotic surgery and can provide real-time data processing that automates repetitive motions and minimizes human error. AI is emerging as a novel technique for evaluating surgical skills, providing thorough assessments and feedback for surgeons on their technique (Guerrero, Asaad, Rajesh, Hassan, & Butler, 2023). These metrics are invaluable for surgeons and help improve consistency and outcomes in complex procedures. Furthermore, AI-guided robotic tools are being developed to improve tissue handling, recognize anatomical landmarks, and adapt to operative changes as they happen (Han et al., 2025).

AI-based tools are being developed to analyze operations in real time and support clinical decisions. The deep learning algorithm Rhinonet demonstrated high accuracy in classifying rhinoplasty status, based solely on visual analysis. The implications of such technology can be extended intraoperatively for guiding revision procedures or helping determine intraoperative symmetry and esthetic balance (Choi et al., 2023). AI can assist surgeons by assessing tissue viability in reconstructive procedures such as debridement and flap reconstruction. Deep learning algorithms have been shown to improve a surgeon's ability to evaluate wound bed viability by up to 25% (Yu et al., 2024), improving outcomes of wound debridement by ensuring no excess tissue is dissected. In addition, this study found that whenever a surgeon changed their mind following AI input, they always changed their decision to the correct one. AI tools can also analyze free flap perfusion in real time using modalities such as photoplethysmography (Fiedler & Daaloul, 2024), enabling earlier detection of compromised perfusion and tailoring intraoperative adjustments to optimize reconstructive outcomes. As mentioned previously, AI systems are being trained to estimate the degree of rejuvenation following esthetic procedures (Elliott et al., 2023). These systems could provide feedback during surgery to guide adjustments in real time, improving outcomes and patient satisfaction.

One of the most promising areas of intraoperative AI use is predictive monitoring. A clinical trial in Stockholm is underway to evaluate an AI system that can anticipate adverse intraoperative events, such as hypotension, which can cause serious postoperative complications like heart and kidney damage, stroke, and death. This system works by integrating data from standard intraoperative monitors, such as pulse, oxygenation, cardiac output, and Electrocardiogram (ECGs), with novel parameters like variations in exhaled carbon dioxide. By analyzing this data, AI algorithms can alert clinicians to impending physiological instability even before it happens, allowing for earlier interventions and reduced perioperative risk (Region Stockholm, 2024).

AI is also being applied as a tool for intraoperative guidance and training. Machine learning models have demonstrated the ability to align patient-specific implants more precisely during reconstructive surgery, potentially improving both functional and esthetic outcomes (Han et al., 2025). The LLM ChatGPT has been evaluated for its capability in providing intraoperative solutions to complex challenges encountered in deep inferior epigastric perforator flap surgery. The findings suggest that AI can offer relevant and accurate guidance to assist surgeons in making complex decisions (Atkinson et al., 2024). In terms of training, AI systems can analyze operative video footage, assess the skill level of the surgeon, and offer individualized feedback (Choi et al., 2023). AI has also been used to analyze hand movement patterns in liposuction and has accurately distinguished between novice and expert surgeons (Yibulayimu et al., 2022). Such feedback can help trainees develop their techniques and accelerate skill acquisition.

4 Postoperative applications

AI is recognized for its transformative potential in postoperative care. This aspect of plastic surgery encompasses a wide range of tasks, including monitoring tissue viability, detecting complications, ensuring patient compliance during their recovery, and assessing functional and esthetic outcomes from both objective and patient-reported perspectives. Each of these domains presents challenges that are ideally suited for AI-driven solutions.

4.1 Postoperative monitoring

Postoperative monitoring is crucial to ensuring optimal outcomes. Free flaps are typically monitored using handheld Dopplers, implantable flowmeters, and continuous observation by nurses. While reliable, these methods suffer from key limitations, including their reliance on specialist care and the need for constant follow-ups. AI-integrated platforms can now assess the data from tissue oximetry, infrared sensors, and photographs to determine flap perfusion (Stein & Rohrich, 2025). These tools can be embedded in mobile apps, allowing for remote monitoring and potentially reducing the length of stay in hospitals and the number of in-person follow-ups. Such platforms may improve accessibility to postoperative care in resource-limited areas. Wearable AI-powered garments also represent a major leap forward in patient monitoring. A smart compression garment was developed for patients recovering from liposuction and Brazilian butt lift (BBL) surgery. This device tracks compliance with advised postoperative positioning, especially the avoidance of gluteal pressure following a BBL. High patient compliance with wearing the device demonstrates its usability and practical application (Vecchio, Stein, Dayan, Marte, & Theodorou, 2023).

Continuous AI-based monitoring of wounds and vital signs is also gaining traction. These systems can alert clinicians about potential complications like infection, bleeding, and hematomas (Bajwa, Munir, Nori, & Williams, 2021). Many AI-integrated monitoring devices are currently undergoing large-scale trials for both at-home and in-hospital postoperative care. If successful, these technologies can ease recovery for patients while reducing the time and resource burden on healthcare providers (Centre Hospitalier Universitaire de Nîmes, 2021; Schonck, 2023).

4.2 Outcome analysis

Outcomes in plastic surgery have historically been judged subjectively, especially when it comes to esthetic results. This subjectivity can lead to inconsistency, bias, and limited comparability across cases. AI offers a path to objective and scalable outcome assessments, changing how success may be defined and measured in plastic surgery. A compelling example is the use of FaceX, a convolutional neural network trained to estimate age from photographs. In a study involving over 200 patients undergoing various facial rejuvenation procedures, FaceX estimated preoperative age with 96% accuracy. It then quantified their postoperative age reduction at 3 and 12 months and found an average perceived age reduction of 3 years (Elliott et al., 2023). This tool validates surgical efficacy and better informs patients, helping them maintain realistic expectations. Similar examples can be found in reconstructive procedures, where neural networks have been used to objectively evaluate auricular reconstruction outcomes in microtia (Tolba, Qian, Lin, Yeom, & Truong, 2023).

A real-world application in postoperative outcome evaluation was seen in a study by Bistoni et al., where AI was used to perform volumetric analysis after subfascial augmentation mastopexy to test the viability of a novel mesh, poly-4-hydroxybutyrate. The AI evaluated ptosis recurrence, implant displacement, bottoming out, and breast capsular contracture. The results found that the mesh provided durable and esthetically stable outcomes without major complications (Bistoni, Sofo, Cagli, Buccheri, & Mallucci, 2024). This kind of postoperative surveillance can help surgeons validate new techniques more rapidly while also informing patients of likely long-term outcomes. Sensory- and video-based AI systems are also being used to assess postoperative hand function. AI can now measure variables such as grip strength, motion patterns, and recovery progress to generate quantifiable recovery trajectories (Keller, Guebeli, Thieringer, & Honigmann, 2023). This data can guide physiotherapy, inform decisions on returning to work, and help identify cases of suboptimal recovery which may require additional interventions.

Beyond monitoring outcomes, AI is capable of streamlining the collection and interpretation of PROMs. The collection of PROMs can now be automated, analyzed, and displayed using natural language processing and sentiment analysis (Miller, Farnebo, & Horwitz, 2023), allowing deeper insights into patient satisfaction and functional recovery over time, enhancing clinical understanding of subjective outcomes, and improving the overall quality of care.

4.3 Managing complications

Postoperative complications require timely and accurate management. AI has proven to be invaluable in this arena. A recent study evaluating 22 simulated esthetic surgery complications found that ChatGPT correctly identified the most likely differential diagnosis on all but one occasion. While this was tested on a very small dataset, it suggests that LLMs may serve as a useful adjunct in early triage, especially in nonspecialist settings or during off-hours (Abi-Rafeh et al., 2024). AI has also excelled in postoperative wound care. Deep learning models trained on postoperative wound photographs can accurately detect signs of infection, necrosis, and wound separation, outperforming less experienced clinicians in several studies (Zhang, Zhao, Farzan, & Otaghvar, 2024). Integrating these models into smartphone applications or electronic health systems can allow for timely escalation and management of postoperative complications.

While these tools demonstrate the impressive ability of AI in postoperative care, their future role will likely be to complement physician-led care rather than replace it. Expert clinical judgment remains essential, particularly in nuanced cases or when data input quality is variable. Nevertheless, the direction of travel is toward autonomous, intelligent support systems that reduce delays in management and improve outcomes.

5 Education and research

AI can revamp education and research in plastic surgery. Both domains require a combination of precision, visual acuity, and nuanced decision-making. AI offers the ability to augment learning and streamline clinical research. In the educational setting, AI technologies are being used to generate simulated patient consultations, ethical case discussions, and other clinical scenarios for teaching purposes (Genovese et al., 2025). AI-generated simulations allow for exposure to a broader range of cases than would be available through rotations alone. In plastic surgery, students and low-grade trainees do not get much exposure to most conditions and complications; these simulations can therefore provide a valuable learning experience for them. In addition, AI-generated images can assist in anatomical education and training (Kiwan et al., 2024). While the accuracy of these images still requires improvement, they have the potential to educate trainees and patients by visualizing complex reconstructions and esthetic procedures. Segmentation algorithms previously discussed may also serve an educational purpose, aiding visualization and understanding of complex anatomy.

A promising advancement is the integration of AI powered virtual reality (VR) training. Medical education heavily relies on simulated patients to mimic medical presentations. While this supports the development of consultation and interpersonal skills, simulated patients cannot reliably present with physical signs such as scars, deformities, or limitations in movement, all of which are crucial to assess in plastic surgery. Relying entirely on clinical exposure, however, would lead to many issues. From patient fatigue to ethical concerns and patient consent, it would be extremely difficult to rely solely on clinical experiences (Gil-Santos, Santos, & Duarte, 2022). AI-based VR systems can therefore be a meaningful addition to surgical education, enabling trainees to practice in realistic and interactive environments. Studies have found AI-based VR to significantly improve clinical knowledge and communication skills (Liaw et al., 2023). In plastic surgery, where detailed anatomical understanding and communication skills are essential, this approach can significantly improve education for trainees.

When it comes to research, AI can automate labor-intensive processes. LLMs can assist in writing academic articles, generating drafts, or literature summaries (Macdonald, Adeloye, Sheikh, & Rudan, 2023). AI can also support researchers by automating systematic reviews, critical appraisals, and data collection (Alaniz, Vu, & Pfaff, 2023). AI-powered software is capable of scanning thousands of papers and highlighting relevant studies, saving researchers significant time and helping to mitigate bias in manual selection (Roberts, Ali, Hutchings, Dobbs, & Whitaker, 2023).

6 Administrative and operational support

AI is already being integrated to aid administrative work in healthcare. In the complex world of plastic surgery, AI can offer practical solutions to optimize workflows, reduce costs, and improve the quality of care through administrative efforts. This is especially beneficial for plastic surgeons running their own private practices, where high administrative costs and limited support staff significantly impact workplace efficiency and profitability. Microsoft AI has already been embedded into Epic, widely used EPR software. A hand and reconstructive microsurgery department in Singapore demonstrated the use of AI in administrative support by using NUHS-developed Russel-GPT to assist in administrative tasks. Nearly three-quarters of users have found AI integration extremely useful.

One of the most immediate uses of AI in administration has been medical coding. Platforms such as Nym Health and CodaMatrix have showcased their ability to process cases more accurately than human coders and in a fraction of the time (Zhu et al., 2024). Beyond coding, AI is changing how clinical documentation may be carried out. Electronic patient record (EPR) systems have already begun integrating AI, such as Epic integrating Microsoft AI (Tan, Teo, & Puhaindran, 2024). Traditionally, the administrative burden of documentation meant physicians focused more on data entry than on clear, proper communication with their patients, leading to reduced patient satisfaction. AI-powered scribes and documentation tools can now accommodate automatic documentation of encounters, allowing physicians to engage more deeply with their patients, improving the quality of care and patient satisfaction.

7 Ethical, legal, and practical challenges

Despite showing considerable promise in transforming patient care within plastic surgery, the integration of AI is not without serious ethical concerns. First comes the issue of data transparency. Many AI systems currently operate under a "black-box" model, where they provide output without explaining their rationale (Miller et al., 2023). This is deeply problematic in clinical environments, as AI must offer explainable recommendations that can be scrutinized by experts to be adopted into the field. Surgeons must maintain their role as decision-makers and cannot relinquish this responsibility to machines that cannot explain their decisions. Another key challenge is clinician competency in utilizing AI. The vast majority of plastic surgeons lack training in AI and data science (Turner, Abu-Ghname, Davis, Ali, & Winocour, 2020). A global survey of over 150 plastic surgeons found that most had little to no experience with AI (Farid et al., 2024), raising questions about whether surgeons are prepared for its integration. Future surgeons will require a functional understanding of data-driven and AI-based support.

Bias and commercialization present pressing concerns. Many AI models are trained on narrow datasets, using data from a single population group. This can lead to biases in AI-based tools and can worsen health disparities, as many groups are underrepresented in training data (van Kolfschooten, 2023). This risk of bias must be addressed by using diverse datasets encompassing multiple different populations. The development and integration of AI systems is also becoming increasingly influenced by private companies (d'Elia et al., 2022; Mansoor & Ibrahim, 2025). This could create a serious conflict of interest, aligning AI's goals more toward profitability than improving outcomes for patients.

The impact on patient relationships must also be considered. Older patients are unfamiliar with AI and may view its use as alienating. The use of AI during consultations can also erode trust and rapport, as it may reduce the personalized aspect of care (Farid et al., 2024). This is particularly relevant in esthetic surgery, where patient satisfaction is closely linked to the patient's relationship with the clinician and their trust in them. Data privacy and security also pose significant concerns. AI systems are capable of reidentifying anonymized patient data (Murdoch, 2021), raising alarm bells about the risk of compromising patient confidentiality and misuse by third-party platforms. Protocols must be developed to address this risk of privacy breach, as many clinical workflows have not been updated to account for AI. Finally, it must also be mentioned that most studies utilizing AI in plastic surgery are conducted in a single center or controlled environments, lacking validation in diverse clinical settings. Patient perspectives on AI also remain unexplored and must be investigated.

While AI promises to enhance surgical care, its integration must not come at the expense of key medical principles. Addressing these challenges will require significant collaboration among clinicians, data scientists, and policy makers to ensure AI improves patient care and does not undermine the core values of plastic surgery.

8 Summary

The integration of AI into plastic surgery is redefining patient care across the entire patient journey. It is no longer a theoretical adjunct but an active contributor to modern surgery. Plastic surgery demands a high degree of personalization, creativity, and precision. These qualities align closely with the nature of AI systems, which are designed to process complex, multidimensional data and generate individualized plans for patients. As AI continues to develop, its role will grow from an additional tool that can assist surgeons to a fundamental component in the delivery of high-quality and patient-centered surgical care.

Preoperative care is one of the most studied areas of this technological revolution. Algorithms can analyze clinical images, radiological scans, and patient-specific factors to generate diagnoses, stratify risk, and plan procedures. In the context of plastic surgery, AI can identify risk factors that may not be apparent through traditional means and can even anticipate complications prior to their occurrence. Simulation tools, such as GANs, allow for the utilization of highly personalized visualizations of expected outcomes, thus improving patient understanding, setting realistic expectations, and supporting better informed consent. Intraoperatively, AI can assist surgeons in real time. Robotic systems augmented by AI can improve the precision of dissection, tissue handling, and suturing. This is especially valuable in microsurgical reconstruction. Deep learning algorithms are also being developed to assess tissue viability, guide implant placement, and provide live feedback to surgeons. Predictive monitoring tools can offer surgeons the ability to anticipate physiological instability before it manifests clinically, allowing surgeons and anesthetists to react promptly and improving the safety of intraoperative care. Postoperative care is also being reshaped by AI. Wearable sensors, smartphone applications, and remote monitoring tools can collect and analyze data on patient recovery, flap perfusion, and signs of complications like infections and hematoma formation. This allows for earlier interventions, a reduced need for lengthy hospital stays, and increased access to specialist care in resource-limited settings. In addition, AI can provide objective assessments of surgical outcomes, including esthetic enhancements and restoration of function. This offers a standardized approach to evaluate surgical success and long-term outcomes. AI will also enhance surgical education through simulated and VR-based training and aid research by automating labor-intensive tasks such as data collection.

The future of AI in plastic surgery will depend on addressing key challenges. These include improving model transparency, reducing bias by enhancing training datasets, validating these tools in real clinical settings, and building regulatory frameworks that ensure patient safety and privacy. Future plastic surgeons must also be familiar with the principles of AI and the legal and ethical limitations of its use.

If implemented appropriately, AI will enhance every phase of surgical care while complementing the artistry, intuition, and human connection that define plastic surgery. AI has the potential to sharpen clinical judgment by offering deeper insights and supporting more precise, confident decision-making, adapting to the unique needs of each patient and aiding surgeons in delivering higher-quality, personalized care. The field can look toward a promising future, where the collaboration between AI and physicians can reshape what is possible in plastic surgery and lead us toward truly optimal patient care.

Generative AI Statement

No generative AI tools were used in the conception, drafting, or editing of this chapter. All content was developed by the authors.

References

Abi-Rafeh, J., Henry, N., Xu, H. H., Bassiri-Tehrani, B., Arezki, A., Kazan, R., … Nahai, F. (2024). Utility and comparative performance of current artificial intelligence large language models as postoperative medical support chatbots in aesthetic surgery. *Aesthetic Surgery Journal*. 1527330X. *44*(8), 889–896. Retrieved from http://asj.oxfordjournals.org/content/by/year. doi:10.1093/asj/sjae025.

Aimar, A., Palermo, A., & Innocenti, B. (2019). The role of 3D printing in medical applications: a state of the art. *Journal of Healthcare Engineering*. 2040-2295. 2019, 1–10. doi:10.1155/2019/5340616.

Alaniz, L., Vu, C., & Pfaff, M. J. (2023). The utility of artificial intelligence for systematic reviews and boolean query formulation and translation. *Plastic and Reconstructive Surgery*. 21697574. *11*(10), E5339. Retrieved from http://journals.lww.com/prsgo/pages/default.aspx. doi:10.1097/GOX.0000000000005339.

Atkinson, C. J., Seth, I., Xie, Y., Ross, R. J., Hunter-Smith, D. J., Rozen, W. M., & Cuomo, R. (2024). Artificial intelligence language model performance for rapid intraoperative queries in plastic surgery: chatGPT and the deep inferior epigastric perforator flap. *Journal of Clinical Medicine*. 20770383. *13*(3). Retrieved from https://www.mdpi.com/journal/jcm. doi:10.3390/jcm13030900.

Bajwa, J., Munir, U., Nori, A., & Williams, B. (2021). Artificial intelligence in healthcare: transforming the practice of medicine. *Future Healthcare Journal*. 25146645. *8*(2), e188. doi:10.7861/fhj.2021-0095.

Ballestín, A., Malzone, G., Menichini, G., Lucattelli, E., & Innocenti, M. (2022). New robotic system with wristed microinstruments allows precise reconstructive microsurgery: preclinical study. *Annals of Surgical Oncology*. 1068-9265. *29*(12), 7859–7867. doi:10.1245/s10434-022-12033-x.

Bertsimas, D., Dunn, J., Velmahos, G. C., & Kaafarani, H. M. A. (2018). Surgical risk is not linear: derivation and validation of a novel, user-friendly, and machine-learning-based predictive OpTimal Trees in Emergency Surgery Risk (POTTER) calculator. *Annals of Surgery*. 15281140. *268*(4), 574–583. doi:10.1097/SLA.0000000000002956.

Bistoni, G., Sofo, F., Cagli, B., Buccheri, E. M., & Mallucci, P. (2024). Artificial intelligence, genuine outcome: analysis of 72 consecutive cases of subfascial augmentation mastopexy with smooth round implants supported by P4HB scaffold. *Aesthetic Surgery Journal*. 1527330X. *44*(11), 1154–1166. Retrieved from http://asj.oxfordjournals.org/content/by/year. doi:10.1093/asj/sjae109.

Boboridis, K., Assi, A., Indar, A., Bunce, C., & Tyers, A. G. (2001). Repeatability and reproducibility of upper eyelid measurements. *British Journal of Ophthalmology*. 00071161. *85*(1), 99–101. doi:10.1136/bjo.85.1.99.

Brancaccio, G., Balato, A., Malvehy, J., Puig, S., Argenziano, G., & Kittler, H. (2024). Artificial intelligence in skin cancer diagnosis: a reality check. *Journal of Investigative Dermatology*. 0022202X. *144*(3), 492–499. doi:10.1016/j.jid.2023.10.004.

Brennan, M., Puri, S., Ozrazgat-Baslanti, T., Feng, Z., Ruppert, M., Hashemighouchani, H., … Bihorac, A. (2019). Comparing clinical judgment with the MySurgeryRisk algorithm for preoperative risk assessment: a pilot usability study. *Surgery*. 00396060. *165*(5), 1035–1045. doi:10.1016/j.surg.2019.01.002.

Buzzaccarini, G., Degliuomini, R. S., Borin, M., Fidanza, A., Salmeri, N., Schiraldi, L., … Pagliardini, L. (2024). The promise and pitfalls of AI-generated anatomical images: evaluating midjourney for aesthetic surgery applications. *Aesthetic Plastic Surgery*. 14325241. *48*(9), 1874–1883. Retrieved from https://www.springer.com/journal/266. doi:10.1007/s00266-023-03826-w.

Cai, R., Liu, Y., Sun, Z., Wang, Y., Wang, Y., Facheng, L., & Jiang, H. (2023). Deep-learning based segmentation of ultrasound adipose image for liposuction. *The International Journal of Medical Robotics and Computer Assisted Surgery*. 1478-5951. *19*(6). doi:10.1002/rcs.2548.

Centre Hospitalier Universitaire de Nîmes 2021 Evaluation of the impact of the SMART ANGEL TM Device on follow-up at home following major or intermediate outpatient surgery: randomized controlled open-label trial (Clinical Trial Registration NCT04068584). https://clinicaltrials.gov/study/NCT04068584

Cevik, J., Seth, I., & Rozen, W. M. (2023). Transforming breast reconstruction: the pioneering role of artificial intelligence in preoperative planning. *Gland Surgery*. 22278575. *12*(9), 1271–1275. Retrieved from http://gs.amegroups.com/about/index. doi:10.21037/gs-23-265.

Chang, C. W., Ho, C. Y., Lai, F., Christian, M., Huang, S. C., Chang, D. H., & Chen, Y. S. (2023). Application of multiple deep learning models for automatic burn wound assessment. *Burns*. 18791409. *49*(5), 1039–1051. Retrieved from www.elsevier.com/locate/burns. doi:10.1016/j.burns.2022.07.006.

Chartier, C., Watt, A., Lin, O., Chandawarkar, A., Lee, J., & Hall-Findlay, E. (2022). BreastGAN: artificial intelligence-enabled breast augmentation simulation. *Aesthetic Surgery Journal Open Forum*. 2631-4797. 4. doi:10.1093/asjof/ojab052.

Cheon, J. H., Kim, H. E., Park, S. H., & Yoon, E. S. (2022). Ten-year experience of robotic latissimus muscle flap reconstructive surgery at a single institution. *Journal of Plastic, Reconstructive & Aesthetic Surgery*. 18780539. *75*(10), 3664–3672. Retrieved from http://www.elsevier.com. doi:10.1016/j.bjps.2022.06.009.

Choi, E., Leonard, K. W., Jassal, J. S., Levin, A. M., Ramachandra, V., & Jones, L. R. (2023). Artificial intelligence in facial plastic surgery: a review of current applications, future applications, and ethical considerations. *Facial Plastic Surgery*. 10988793. *39*(5), 454–459. Retrieved from http://www.thieme-connect.com/ejournals/toc/fps. doi:10.1055/s-0043-1770160.

Conforth, M., Meng, Y., Valmikinathan, C., & Yu, X. 2009 Nerve graft selection for peripheral nerve regeneration using neural networks trained by a hybrid ACO/PSO method. In *IEEE Symposium on Computational Intelligence in Bioinformatics and Computational Biology, CIBCB 2009 – proceedings* (p. 214). IEEE Computer Society, United States. doi:10.1109/CIBCB.2009.4925730.

d'Elia, A., Gabbay, M., Rodgers, S., Kierans, C., Jones, E., Durrani, I., … Frith, L. (2022). Artificial intelligence and health inequities in primary care: a systematic scoping review and framework. *Family Medicine and Community Health*. 2305-6983. 10 (Suppl 1), e001670. doi:10.1136/fmch-2022-001670.

Dhawan, R., & Shay, D. (2024). Artificial intelligence in plastic surgery: implications and limitations of text-to-image models for clinical practice. *JPRAS Open*. 23525878. 41, 368–371. doi:10.1016/j.jpra.2024.07.008.

Elliott, Z. T., Bheemreddy, A., Fiorella, M., Martin, A. M., Christopher, V., Krein, H., & Heffelfinger, R. (2023). Artificial intelligence for objectively measuring years regained after facial rejuvenation surgery. *American Journal of Otolaryngology*. 01960709. *44*(2), 103775. doi:10.1016/j.amjoto.2022.103775.

Esteva, A., Kuprel, B., Novoa, R. A., Ko, J., Swetter, S. M., Blau, H. M., & Thrun, S. (2017). Dermatologist-level classification of skin cancer with deep neural networks. *Nature*. 14764687. *542*(7639), 115–118. Retrieved from http://www.nature.com/nature/index.html. doi:10.1038/nature21056.

Farid, Y., Gutierrez, L. F. B., Ortiz, S., Gallego, S., Zambrano, J. C., Morrelli, H. U., & Patron, A. (2024). Artificial intelligence in plastic surgery: insights from plastic surgeons, education integration, Chatgpt's survey predictions, and the path forward. *Plastic and Reconstructive Surgery*. 21697574. *12*(1), E5515. Retrieved from http://journals.lww.com/prsgo/pages/default.aspx. doi:10.1097/GOX.0000000000005515.

Farid, Y., Chang, C., Marcasciano, M., Di Meglio, F., Rodríguez-Mantilla, I., Nanni, J., … Chen, H. C. (2024). Consent 2.0: informed choices in the age of artificial intelligence. *Surgery (United States)*. 15327361. *175*(5), 1454–1455. Retrieved from https://www.sciencedirect.com/science/journal/00396060. doi:10.1016/j.surg.2023.12.027.

Festen, R. T., Schrier, V. J. M. M., & Amadio, P. C. (2021). Automated segmentation of the median nerve in the carpal tunnel using U-Net. *Ultrasound in Medicine and Biology*. 1879291X. *47*(7), 1964–1969. Retrieved from www.elsevier.com/locate/ultrasmedbio. doi:10.1016/j.ultrasmedbio.2021.03.018.

Fiedler, L. S., & Daaloul, H. (2024). An overview of current assessment techniques for evaluating cutaneous perfusion in reconstructive surgery. *Journal of Biophotonics*. 18640648. *17*(5). Retrieved from http://onlinelibrary.wiley.com/journal/10.1002/(ISSN)1864-0648. doi:10.1002/jbio.202400002.

Genovese, A., Borna, S., Gomez-Cabello, C. A., Haider, S. A., Prabha, S., Trabilsy, M., & Forte, A. J. (2025). The current landscape of artificial intelligence in plastic surgery education and training: a systematic review. *Journal of Surgical Education*. 18787452. *82*(8). Retrieved from https://www.sciencedirect.com/science/journal/19317204. doi:10.1016/j.jsurg.2025.103519.

Gibstein, A. R., Chen, K., Nakfoor, B., Lu, S. M., Cheng, R., Thorne, C. H., & Bradley, J. P. (2021). Facelift surgery turns back the clock: artificial intelligence and patient satisfaction quantitate value of procedure type and specific techniques. *Aesthetic Surgery Journal*. 1527330X. *41*(9), 987–999. Retrieved from http://asj.oxfordjournals.org/content/by/year. doi:10.1093/asj/sjaa238.

Gil-Santos, I., Santos, C. C., & Duarte, I. (2022). Medical education: patients' perspectives on clinical training and informed consent. *International Journal of Environmental Research & Public Health*. 1660-4601. *19*(13), 7611. doi:10.3390/ijerph19137611.

Goodyear, K., Saffari, P. S., Esfandiari, M., Baugh, S., Rootman, D. B., & Karlin, J. N. (2023). Estimating apparent age using artificial intelligence: quantifying the effect of blepharoplasty. *Journal of Plastic, Reconstructive & Aesthetic Surgery*.18780539. 85, 336–343. Retrieved from http://www.elsevier.com. doi:10.1016/j.bjps.2023.07.017.

Guerrero, D. T., Asaad, M., Rajesh, A., Hassan, A., & Butler, C. E. (2023). Advancing surgical education: the use of artificial intelligence in surgical training. *American Surgeon*. 15559823. *89*(1), 49–54. Retrieved from https://journals.sagepub.com/home/ASU. doi:10.1177/00031348221101503.

Han, F., Huang, X., Wang, X., Chen, Y. F., Lu, C., Li, S., … Zhang, D. W. (2025). Artificial intelligence in orthopedic surgery: current applications, challenges, and future directions. *MedComm*. 26882663. *6*(7). Retrieved from https://www.onlinelibrary.wiley.com/journal/26882663. doi:10.1002/mco2.70260.

Harbour, L. (2022). Three-dimensional printing needs AI to truly become 'Point of Care. *Biomedical Instrumentation & Technology*. 0899-8205. *56*(4), 135–138. doi:10.2345/0899-8205-56.4.135.

Hartmann, T. J., Hartmann, J. B. J., Friebe-Hoffmann, U., Lato, C., Janni, W., & Lato, K. (2022). Novel method for three-dimensional facial expression recognition using self-normalizing neural networks and mobile devices. *Geburtshilfe Und Frauenheilkunde*. 14388804. *82*(9), 955–969. Retrieved from https://gebfra.thieme.com/home. doi:10.1055/a-1866-2943.

Hassan, A. M., Biaggi-Ondina, A., Asaad, M., Morris, N., Liu, J., Selber, J. C., & Butler, C. E. (2023). Artificial intelligence modeling to predict periprosthetic infection and explantation following implant-based reconstruction. *Plastic and Reconstructive Surgery*. 00321052. *152*(5), 929–938. Retrieved from http://journals.lww.com/plasreconsurg/pages/issuelist.aspx. doi:10.1097/PRS.0000000000010345.

Hoogendam, L., Bakx, J. A. C., Souer, J. S., Slijper, H. P., Andrinopoulou, E. R., & Selles, R. W. (2022). Predicting clinically relevant patient-reported symptom improvement after carpal tunnel release: a machine learning approach. *Neurosurgery*. 15244040. *90*(1), 106–113. Retrieved from https://journals.lww.com/neurosurgery/pages/currenttoc.aspx. doi:10.1227/NEU.0000000000001749.

Huyghebaert, T. A., Wallner, C., & Montemurro, P. (2024). Implementation of a machine learning approach evaluating risk factors for complications after single-stage augmentation mastopexy. *Aesthetic Plastic Surgery*. 14325241. *48*(23), 5049–5059. Retrieved from https://www.springer.com/journal/266. doi:10.1007/s00266-024-04142-7.

Keller, M., Guebeli, A., Thieringer, F., & Honigmann, P. (2023). Artificial intelligence in patient-specific hand surgery: a scoping review of literature. *International Journal of Computer Assisted Radiology and Surgery*. 1861-6429. *18*(8), 1393–1403. doi:10.1007/s11548-023-02831-3.

Keskinis, A., Paraskevopoulos, K., Diamantidis, D. E., Ververidis, A., Fiska, A., & Tilkeridis, K. (2023). The role of 3D-printed patient-specific instrumentation in total knee arthroplasty: a literature review. *Cureus*. 2168-8184. *15*(8). doi:10.7759/cureus.43321.

Kiwan, O., Al-Kalbani, M., Rafie, A., & Hijazi, Y. (2024). Artificial intelligence in plastic surgery, where do we stand? *JPRAS Open*. 23525878. 42, 234–243. doi:10.1016/j.jpra.2024.09.003.

Knoedler, S., Alfertshofer, M., Simon, S., Panayi, A. C., Saadoun, R., Palackic, A., … Knoedler, L. (2024). Turn your vision into reality—AI-powered preoperative outcome simulation in rhinoplasty surgery. *Aesthetic Plastic Surgery*. 14325241. *48*(23), 4833–4838. Retrieved from https://www.springer.com/journal/266. doi:10.1007/s00266-024-04043-9.

Kraus, M., Anteby, R., Konen, E., Eshed, I., & Klang, E. (2024). Artificial intelligence for X-ray scaphoid fracture detection: a systematic review and diagnostic test accuracy meta-analysis. *European Radiology*. 1432-1084. *34*(7), 4341–4351. doi:10.1007/s00330-023-10473-x.

Liaw, S. Y., Tan, J. Z., Lim, S., Zhou, W., Yap, J., Ratan, R., … Chua, W. L. (2023). Artificial intelligence in virtual reality simulation for interprofessional communication training: mixed method study. *Nurse Education Today*. 15322793. 122. Retrieved from http://www.sciencedirect.com/science/journal/02606917. doi:10.1016/j.nedt.2023.105718.

Loos, N. L., Hoogendam, L., Souer, J. S., Slijper, H. P., Andrinopoulou, E. R., Coppieters, M. W., … Stege, M. H. P. T. (2022). Machine learning can be used to predict function but not pain after surgery for thumb carpometacarpal osteoarthritis. *Clinical Orthopaedics and Related Research*. 15281132. *480*(7), 1271–1284. Retrieved from https://journals.lww.com/clinorthop/pages/default.aspx. doi:10.1097/CORR.0000000000002105.

Lynn, A. Q., Pflibsen, L. R., Smith, A. A., Rebecca, A. M., & Teven, C. M. (2021). Three-dimensional printing in plastic surgery: current applications, future directions, and ethical implications. *Plastic and Reconstructive Surgery*. 21697574. *9*(3), E3465. Retrieved from http://journals.lww.com/prsgo/pages/default.aspx. doi:10.1097/GOX.0000000000003465.

Ma, L., Shijie, Y., Xiaodong, X., Amadi, S. M., Zhang, J., & Wang, Z. (2023). Application of artificial intelligence in 3D printing physical organ models. *Materials Today Bio*. 25900064. 23, 100792. doi:10.1016/j.mtbio.2023.100792.

Macdonald, C., Adeloye, D., Sheikh, A., & Rudan, I. (2023). Can ChatGPT draft a research article? An example of population-level vaccine effectiveness analysis. *Journal of Global Health*. 2047-2978. 13. doi:10.7189/jogh.13.01003.

Mansoor, M., & Ibrahim, A. F. (2025). The transformative role of artificial intelligence in plastic and reconstructive surgery: challenges and opportunities. *Journal of Clinical Medicine*. 2077-0383. *14*(8), 2698. doi:10.3390/jcm14082698.

Mao, Y., Hou, X., Fu, S., & Luan, J. (2024). Transcriptomic and machine learning analyses identify hub genes of metabolism and host immune response that are associated with the progression of breast capsular contracture. *Genes and Diseases*. 23523042. *11*(3), 101087. doi:10.1016/j.gendis.2023.101087.

Miller, R., Farnebo, S., & Horwitz, M. D. (2023). Insights and trends review: artificial intelligence in hand surgery. *Journal of Hand Surgery European Volume*. 20436289. *48*(5), 396–403. Retrieved from https://journals.sagepub.com/home/JHS. doi:10.1177/17531934231152592.

Montemurro, P., Lehnhardt, M., Behr, B., & Wallner, C. (2022). A machine learning approach to identify previously unconsidered causes for complications in aesthetic breast augmentation. *Aesthetic Plastic Surgery*. 0364-216X. *46*(6), 2669–2676. doi:10.1007/s00266-022-02997-2.

Murdoch, B. (2021). Privacy and artificial intelligence: challenges for protecting health information in a new era. *BMC Medical Ethics*. 1472-6939. *22*(1). doi:10.1186/s12910-021-00687-3.

Nishimoto, S., Sotsuka, Y., Kawai, K., Fujita, K., & Kakibuchi, M. (2014). Three-dimensional mock-up model for chondral framework in auricular reconstruction, built with a personal three-dimensional printer. *Plastic and Reconstructive Surgery*. 0032-1052. *134*(1), 180e. doi:10.1097/prs.0000000000000263.

O'Connell, R. L., Stevens, R. J. G., Harris, P. A., & Rusby, J. E. (2015). Review of three-dimensional (3D) surface imaging for oncoplastic, reconstructive and aesthetic breast surgery. *Breast*. 15323080. *24*(4), 331–342. Retrieved from http://www.elsevier-international.com/journals/brst/. doi:10.1016/j.breast.2015.03.011.

Peng, C., Yang, F., Jian, Y., Peng, L., Zhang, C., Chen, C., … Jin, Z. (2023). Machine learning prediction algorithm for in-hospital mortality following body contouring. *Plastic & Reconstructive Surgery*. 0032-1052. *152*(6), 1103e. doi:10.1097/prs.0000000000010436.

Raj, U., Garg, A., Vathulya, M., & Kandwal, A. (2024). Quantifying nasal deformities using a novel mathematical method to complement preoperative assessment in rhinoplasty patients. *Journal of Plastic, Reconstructive & Aesthetic Surgery*. 17486815. 92, 75–78. doi:10.1016/j.bjps.2024.02.074.

Region Stockholm. (2024). Artificial Intelligence—To Predict and Prevent Hypotension During Surgery (Clinical Trial Registration NCT06240234). https://clinicaltrials.gov/study/NCT06240234.

Roberts, R. H. R., Ali, S. R., Hutchings, H. A., Dobbs, T. D., & Whitaker, I. S. (2023). Comparative study of ChatGPT and human evaluators on the assessment of medical literature according to recognised reporting standards. *BMJ Health and Care Informatics*. 26321009. *30*(1). Retrieved from https://informatics.bmj.com/content/by/year. doi:10.1136/bmjhci-2023-100830.

Roshan, A., Shah, B., Anderson, K. D., Murphy, S., Thomas, B., McPhee, A. S., … Patel, A. J. K. (2024). Robot-assisted pelvic dissection for enlarged lymph nodes in melanoma improves recovery with equivalent oncological outcomes to open pelvic dissection. *Annals of Surgical Oncology*. 15344681. *31*(4), 2727–2736. Retrieved from https://www.springer.com/journal/10434. doi:10.1245/s10434-023-14834-0.

Ross, A. D. M., Kumar, P., Challacombe, B. J., Dasgupta, P., & Geh, J. L. C. (2013). The addition of the surgical robot to skin cancer management. *The Annals of the Royal College of Surgeons of England*. 0035-8843. *95* (1), 70–72. doi:10.1308/003588413x13511609955012.

Sayadi, L. R., Hamdan, U. S., Zhangli, Q., Hu, J., & Vyas, R. M. (2022). Harnessing the power of artificial intelligence to teach cleft lip surgery. *Plastic and Reconstructive Surgery*. 21697574. *10*(7), E4451. Retrieved from http://journals.lww.com/prsgo/pages/default.aspx. doi:10.1097/GOX.0000000000004451.

Schonck, F. (2023). SWITCH—Surgical Ward Innovation: telemonitoring Controlled by Healthdot Evaluation of Implementation (Clinical Trial Registration NCT05956210). Retrieved from: https://clinicaltrials.gov/study/NCT05956210.

Seu, M. Y., Rezania, N., Murray, C. E., Qiao, M. T., Arnold, S., Siotos, C., … Kokosis, G. (2024). Predicting reduction mammaplasty total resection weight with machine learning. *Annals of Plastic Surgery*. 15363708. *93*(2), 246–252. Retrieved from http://journals.lww.com/annalsplasticsurgery/pages/default.aspx. doi:10.1097/SAP.0000000000004016.

Shin, Y., Yang, J., & Lee, Y. H. (2021). Deep generative adversarial networks: applications in musculoskeletal imaging. *Radiology: Artificial Intelligence*. 2638-6100. *3*(3), e200157. doi:10.1148/ryai.2021200157.

Souza, S., Bhethanabotla, R. M., & Mohan, S. (2024). Applications of artificial intelligence in facial plastic and reconstructive surgery: a systematic review. *Current Opinion in Otolaryngology and Head and Neck Surgery*. 15316998. *32*(4), 222–233. Retrieved from https://www.journals.lww.com/co-otolaryngology/pages/default.aspx. doi:10.1097/MOO.0000000000000975.

Stein, M. J., & Rohrich, R. (2025). Artificial intelligence and postoperative monitoring in plastic surgery. *Plastic Surgery*. 22925511. *33*(2), 312–317. Retrieved from https://journals.sagepub.com/home/PSG. doi:10.1177/22925503231210873.

Sutradhar, A., Park, J., Carrau, D., & Miller, M. J. (2014). Experimental validation of 3D printed patient-specific implants using digital image correlation and finite element analysis. *Computers in Biology and Medicine*.18790534. 52, 8–17. Retrieved from www.elsevier.com/locate/compbiomed. doi:10.1016/j.compbiomed.2014.06.002.

Tan, R. E. S., Teo, W. Z. W., & Puhaindran, M. E. (2024). Artificial intelligence in hand surgery-how generative AI is transforming the hand surgery landscape. *Journal of Hand Surgery Asian-Pacific Volume*. 24248363. *29*(2), 81–87. Retrieved from https://www.worldscientific.com/jhs. doi:10.1142/S2424835524300019.

Tolba, M., Qian, Z. J., Lin, H. F., Yeom, K. W., & Truong, M. T. (2023). Use of convolutional neural networks to evaluate auricular reconstruction outcomes for microtia. *Laryngoscope*. 15314995. *133*(9), 2413–2416. Retrieved from http://onlinelibrary.wiley.com/journal/10.1002/(ISSN)1531-4995. doi:10.1002/lary.30499.

Turner, A. E., Abu-Ghname, A., Davis, M. J., Ali, K., & Winocour, S. (2020). Role of simulation and artificial intelligence in plastic surgery training. *Plastic and Reconstructive Surgery*. 00321052. *146*(3), 390–391. Retrieved from http://journals.lww.com/plasreconsurg/pages/issuelist.aspx. doi:10.1097/PRS.0000000000007102.

van Kolfschooten, H. (2023). The AI cycle of health inequity and digital ageism: mitigating biases through the EU regulatory framework on medical devices. *Journal of Law and the Biosciences*. 2053-9711. *10*(2). doi:10.1093/jlb/lsad031.

Vecchio, D. D., Stein, M. J., Dayan, E., Marte, J., & Theodorou, S. (2023). Nanotechnology and artificial intelligence: an emerging paradigm for postoperative patient care. *Aesthetic Surgery Journal*. 1527330X. *43*(7), 748–757. Retrieved from http://asj.oxfordjournals.org/content/by/year. doi:10.1093/asj/sjad071.

Verhelst, P.-J., Smolders, A., Beznik, T., Meewis, J., Vandemeulebroucke, A., Shaheen, E., … Jacobs, R. (2021). Layered deep learning for automatic mandibular segmentation in cone-beam computed tomography. *Journal of Dentistry*. 03005712. 114, 103786. doi:10.1016/j.jdent.2021.103786.

Waked, K., Mespreuve, M., De Ranter, J., Collard, B., Hahn, S., & Hendrickx, B. (2022). Visualizing the individual arterial anatomy of the face through augmented reality—a useful and accurate tool during dermal Filler injections. *Aesthetic Surgery Journal Open Forum*. 2631-4797. 4. doi:10.1093/asjof/ojac012.

Wang, D., Chen, X., Yiping, W., Tang, H., & Deng, P. (2022). Artificial intelligence for assessing the severity of microtia via deep convolutional neural networks. *Frontiers in Surgery*. 2296-875X. 9. doi:10.3389/fsurg.2022.929110.

Wang, L., Ye, C., Xue, X., Xie, M., Zhi, Y., Feng, X., … Wang, J. (2024). 3D-Printed breast prosthesis that smartly senses and targets breast cancer relapse. *Advanced Science*. 21983844. *11*(46). Retrieved from http://onlinelibrary.wiley.com/journal/10.1002/(ISSN)2198-3844. doi:10.1002/advs.202402345.

Wheeler, D. R. (2021). Art, artificial intelligence, and aesthetics in plastic surgery. *Plastic and Reconstructive Surgery*. 00321052. *148*(3), 529–530. Retrieved from http://journals.lww.com/plasreconsurg/pages/issuelist.aspx. doi:10.1097/PRS.0000000000008289.

Winkler, J. K., Fink, C., Toberer, F., Enk, A., Deinlein, T., Hofmann-Wellenhof, R., … Haenssle, H. A. (2019). Association between surgical skin markings in dermoscopic images and diagnostic performance of a deep learning convolutional neural network for melanoma recognition. *JAMA Dermatology*. 21686068. *155*(10), 1135–1141. Retrieved from http://archderm.jamanetwork.com/issues.aspx. doi:10.1001/jamadermatol.2019.1735.

Xu, Y., Fan, F., Kang, N., Wang, S., You, J., Wang, H., & Zhang, B. (2015). Tissue engineering of human nasal alar cartilage precisely by using three-dimensional printing. *Plastic and Reconstructive Surgery*. 0032-1052. *135*(2), 451–458. doi:10.1097/prs.0000000000000856.

Yibulayimu, S., Wang, Y., Liu, Y., Sun, Z., Wang, Y., Jiang, H., & Facheng, L. (2022). An explainable machine learning method for assessing surgical skill in liposuction surgery. *International Journal of Computer Assisted Radiology and Surgery*. 1861-6429. *17*(12), 2325–2336. doi:10.1007/s11548-022-02739-4.

Yoon, A. P., & Chung, K. C. (2021). Application of deep learning: detection of obsolete scaphoid fractures with artificial neural networks. *Journal of Hand Surgery European Volume*. 20436289. *46*(8), 914–916. Retrieved from http://jhs.sagepub.com/. doi:10.1177/17531934211026139.

Yu, S., Dwight, J., Siska, R. C., Burkart, H., Quan, P., Yi, F., … Thatcher, J. E. (2024). Feasibility of intra-operative image guidance in burn excision surgery with multispectral imaging and deep learning. *Burns*. 18791409. *50*(1), 115–122. Retrieved from https://www.sciencedirect.com/science/journal/03054179. doi:10.1016/j.burns.2023.07.005.

Zhang, B. H., Chen, K., Lu, S. M., Nakfoor, B., Cheng, R., Gibstein, A., … Bradley, J. P. (2021). Turning back the clock: artificial intelligence recognition of age reduction after face-lift surgery correlates with patient satisfaction. *Plastic and Reconstructive Surgery*. 00321052. *148*(1), 45–54. Retrieved from http://journals.lww.com/plasreconsurg/pages/issuelist.aspx. doi:10.1097/PRS.0000000000008020.

Zhang, H., Zhao, J., Farzan, R., & Otaghvar, H. A. (2024). RETRACTED: risk predictions of surgical wound complications based on a machine learning algorithm: a systematic review. *International Wound Journal*. 1742-4801. *21*(1). doi:10.1111/iwj.14665.

Zhu, C., Attaluri, P. K., Wirth, P. J., Shaffrey, E. C., Friedrich, J. B., & Rao, V. K. (2024). Current applications of artificial intelligence in billing practices and clinical plastic surgery. *Plastic and Reconstructive Surgery*. 21697574. *12*(7), e5939. Retrieved from http://journals.lww.com/prsgo/pages/default.aspx. doi:10.1097/GOX.0000000000005939.

Section IV

Domain-specific success stories in medical science through artificial intelligence intervention

Chapter 12

Artificial intelligence in assisted reproduction: Navigating the next frontier in fertility care

Pratip Chakraborty, Parag Nandi, and Syed Monajatur Rahman
Department of Assisted Reproduction, Cradle Fertility Center, Kolkata, West Bengal, India

1 Introduction

Reproductive health lies at the intersection of profound human aspiration and complex biological variability. Assisted reproductive technologies (ARTs), including in vitro fertilization (IVF) and intracytoplasmic sperm injection (ICSI), have expanded possibilities for individuals and couples facing infertility. Despite these advances, challenges such as fluctuating success rates, subjective clinical decisions, and the biological unpredictability of gamete and embryo development persist (Salih et al., 2023; Wu et al., 2025).

The emergence of artificial intelligence (AI) offers promising solutions, enhancing diagnostic precision, personalizing treatments, and optimizing outcomes (Hanassab et al., 2024; Wang et al., 2019). AI, particularly machine learning (ML) and deep learning (DL), shifts reproductive medicine from exploratory-driven decisions to data-guided precision. By extracting complex patterns from high-dimensional datasets, including hormonal profiles, imaging modalities, genomics, and proteomics, AI systems enable more accurate diagnosis, individualized care, and improved clinical outcomes (Kakkar et al., 2025; Mendizabal-Ruiz et al., 2024).

Integrating AI into ART workflows already demonstrates tangible benefits: tailoring ovarian stimulation protocols, automating embryo grading, predicting implantation potential, and personalizing infertility treatments. Time-lapse imaging combined with convolutional neural networks (CNNs) enables real-time, noninvasive embryo assessment that correlates more strongly with implantation success than traditional morphology-based approaches (Salih et al., 2023). Particularly for patients with atypical or complex reproductive profiles, AI-driven approaches foster a shift toward patient-centric, anticipatory reproductive care. However, despite its potential to improve IVF efficiency, standardization, and outcomes, AI's current trajectory is burdened by significant limitations and unresolved challenges.

This chapter explores AI's transformative applications across the reproductive health spectrum, highlighting technological advancements, clinical impacts, and future directions.

2 AI in reproductive medicine

2.1 Overview

AI is reshaping modern medicine by enabling machines to analyze complex data, identify patterns, and support clinical decisions. Encompassing ML, DL, computer vision, and natural language processing (NLP), AI applications span from image analysis to predictive modeling. In reproductive medicine, AI algorithms process large-scale clinical and laboratory datasets, automate workflows, and enhance diagnostic accuracy, ultimately making fertility care more objective, reproducible, and precise (Esteva et al., 2019; Kakkar et al., 2025; Orovou, Tzimourta, Tzitiridou-Chatzopoulou,

Applications of Artificial Intelligence in Healthcare. https://doi.org/10.1016/B978-0-443-43934-6.00002-X

Kakatosi, & Sarantaki, 2025). By learning from diverse data types like hormone levels, ultrasound features, genetic profiles, and time-lapse embryo imaging, AI models facilitate timely, personalized decision-making and reduce inter-operator variability. Thus AI not only supports the clinician but also enhances workflow scalability, making high-quality fertility care more accessible.

2.2 Why reproductive medicine needs AI

Reproductive medicine involves a complex interplay of biological, behavioral, and procedural variables that influence treatment outcomes. The integration of AI into this field addresses several pressing needs:

- ***Complex and multivariable decision-making***: Fertility outcomes are influenced by age, hormonal profiles, embryo quality, genetics, endometrial receptivity, and more. Synthesizing these diverse inputs is cognitively demanding and prone to error. AI models can process high-dimensional data to provide more accurate and consistent insights, aiding in both diagnosis and treatment planning (Zaninovic & Rosenwaks, 2020).
- ***Subjectivity in clinical assessments***: Key decisions, such as sperm and embryo selection, are traditionally based on visual interpretation, introducing variability between practitioners. AI, particularly computer vision and DL tools, enables objective, reproducible grading of embryos and gametes (Curchoe & Bormann, 2019).
- ***Need for personalization***: Each patient has a unique fertility profile. AI facilitates individualized treatment by learning from large datasets to tailor stimulation protocols, predict ovarian responses, and assess implantation potential (Khosravi et al., 2019).
- ***Outcome prediction and optimization***: ART success rates remain modest and inconsistent. AI models enhance predictive accuracy for outcomes such as fertilization, implantation, and live birth, thereby guiding smarter clinical strategies.
- ***Utilization of rich data sources***: Clinics generate vast amounts of structured and unstructured data from hormone assays, ultrasounds, time-lapse imaging, and genetic tests. AI leverages this wealth of information to uncover hidden patterns and actionable insights (Orovou et al., 2025).
- ***Enhancing efficiency and access***: Automation of tasks such as image-based embryo selection, electronic record analysis, and patient communication reduces clinicians' workload and ensures uniform standards of care, which is particularly valuable in under-resourced settings.
- **Real-time monitoring and dynamic feedback**: AI-powered systems can analyze ongoing treatment responses or physiological parameters (e.g., hormone levels, fetal heart rates) and suggest adaptive interventions in real time, improving safety and outcomes (Orovou et al., 2025).

Collectively, AI addresses core challenges in reproductive medicine by transforming it into a more data-driven, precise, and patient-centered discipline.

2.2.1 AI methods for ART

AI applications in ART are underpinned by a variety of computational approaches that enable data-driven predictions, classifications, and recommendations. The most widely used methods include:

- **ML:** It encompasses a broad suite of algorithms that learn from historical patient and laboratory data to make predictions or identify patterns. Decision trees, support vector machines (SVMs), k-nearest neighbors, and ensemble methods like random forests and gradient boosting are commonly employed in ART. ML models have been successfully used for predicting IVF success, classifying poor or high responders to ovarian stimulation, estimating individualized gonadotropin dosages, and forecasting embryo viability (Curchoe & Bormann, 2019). Advanced ensemble techniques enhance model robustness by combining the predictions of multiple learners, thereby reducing overfitting and improving predictive performance.
- **DL:** Its models, particularly CNNs and recurrent neural networks (RNNs), are used to analyze complex, high-dimensional data such as embryo time-lapse imaging and ultrasonographic scans. CNNs have revolutionized embryo assessment by extracting subtle morphological and morphokinetic features linked to implantation success (Khosravi et al., 2019; Salih et al., 2023). Meanwhile, RNNs facilitate the modeling of temporal embryo development patterns, offering superior insights into dynamic biological processes.
- **NLP:** It is increasingly applied to extract structured insights from unstructured clinical narratives, such as physician notes and embryology reports. By leveraging NLP, AI systems can synthesize longitudinal patient histories and laboratory observations to enhance individualized fertility care strategies (Orovou et al., 2025) (Fig. 12.1).

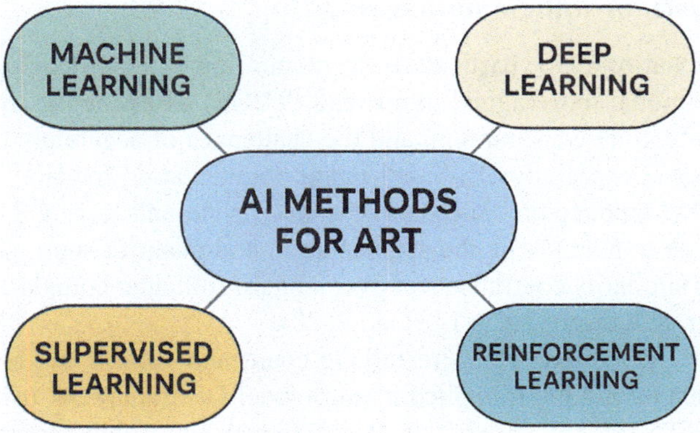

FIG. 12.1 AI checkpoints across the ART workflow.

- **Reinforcement learning (RL):** It is particularly suited for dynamic, sequential decision-making processes in ART, such as optimizing gonadotropin dosing regimens across a stimulation cycle. RL agents learn optimal actions by interacting with simulated or real-world environments and receiving feedback in the form of rewards (Rieke et al., 2020).
- **Federated learning (FL):** To overcome data privacy barriers, FL enables the decentralized training of AI models across multiple fertility centers without requiring the exchange of sensitive patient data. FL preserves patient confidentiality while promoting the development of generalized, robust predictive models for embryo selection, pregnancy prediction, and ovarian response forecasting (Rieke et al., 2020).
- Each of these AI methods contributes uniquely to enhancing ART outcomes and advancing fertility care toward greater precision, personalization, and scalability. Their synergistic integration across various ART stages reflects a major shift toward intelligent, automated reproductive medicine.

2.2.2 AI in ART

ART encompasses a series of intricate procedures, such as IVF, ICSI, and embryo transfer, that require precise decision-making at multiple stages. Traditionally, many clinical decisions have relied on subjective assessments and clinician experience, introducing variability and limiting reproducibility. The integration of AI into ART offers a transformative solution by introducing objective, data-driven, and adaptive methodologies across the fertility care continuum (Kakkar et al., 2025).

One of the earliest and most impactful applications of AI in ART is the personalization of ovarian stimulation protocols. By leveraging historical and real-time patient data, AI algorithms can predict ovarian response, enabling clinicians to tailor hormone dosages to minimize complications like ovarian hyperstimulation syndrome (OHSS) while optimizing oocyte yield (AlSaad et al., 2024). AI also enhances gamete and embryo assessment. DL models, particularly CNNs, have been trained on time-lapse imaging systems to evaluate sperm motility, oocyte morphology, and embryonic development, offering standardized and reproducible grading systems with superior predictive accuracy compared to traditional assessments (Mendizabal-Ruiz et al., 2024; Wu et al., 2025). As AI systems are trained on larger, more diverse datasets and validated through prospective studies, their integration into ART holds immense promise for improving efficacy, safety, and accessibility worldwide (Hanassab et al., 2024).

2.3 AI in gonadotropin dosing for ovarian stimulation

Controlled ovarian stimulation (COS) is a cornerstone of IVF, aiming to induce the development of multiple follicles through exogenous gonadotropins, such as follicle-stimulating hormone (FSH). Determining the optimal starting dose is critical; under-stimulation may yield few oocytes, whereas over-stimulation increases the risk of OHSS.

AI models personalize FSH dosing by incorporating patient-specific factors such as age, anti-Müllerian hormone levels, antral follicle count (AFC), body mass index, and past cycle responses (AlSaad et al., 2024; Zebitay et al., 2017). ML platforms using regression analysis and random forest algorithms have outperformed traditional nomograms in predicting optimal starting doses, enhancing both safety and efficiency (Fanton et al., 2022). Furthermore, RL techniques adaptively update dosing recommendations based on early cycle hormonal changes and follicular growth patterns. By integrating serum estradiol levels and ultrasound imaging, AI-driven systems dynamically optimize stimulation protocols, minimizing practice variability, reducing OHSS risk, and improving outcomes such as oocyte yield and live birth rates across different clinical settings.

2.4 AI in ultrasound analysis for follicle monitoring

Ultrasound imaging is foundational to COS, particularly in monitoring follicle development to determine the optimal timing for oocyte retrieval. Traditional transvaginal ultrasound (TVUS) assessments, while clinically indispensable, are constrained by operator variability, time consumption, and the challenges of accurately tracking multiple follicles across stimulation cycles. AI technologies, especially DL-based image segmentation models, have emerged as transformative tools in this domain. Notably, CNN architectures such as U-Net and its variants (e.g., CR-Unet and HaTU-Net) have demonstrated exceptional performance in automating the segmentation and quantification of ovarian follicles on 2D and 3D ultrasound scans, achieving Dice similarity coefficients of over 0.85 for follicular boundaries (Chen, Zhang, Good, Zhou, & Comaniciu, 2009; Hassan, Reiter, & Razzaq, 2024; Li et al., 2020).

Advanced AI models not only detect and measure follicle count and volume but also predict oocyte maturity with high accuracy. For example, Chen et al. (2009) applied a multimodel DL framework using 3D ultrasound data to generate follicular volume thresholds (0.5–3.0 cm^3) predictive of oocyte maturity and hyperresponse, outperforming traditional 2D assessments. These predictive capabilities align with clinical evidence that follicles between 12–19 mm yield mature oocytes, and those ≥20 mm are associated with high-quality blastocysts (Abbara et al., 2018; McCulloh et al., 2020). Furthermore, AI-guided TVUS has enabled dynamic, real-time monitoring of follicular trajectories across COS cycles, facilitating individualized decisions for human chorionic gonadotrophin (hCG) trigger timing, especially when integrated with estradiol trends and patient-specific parameters (Canon et al., 2024).

Looking forward, AI-powered portable and at-home ultrasound platforms are under development, promising to decentralize follicular monitoring and increase accessibility. These technologies, when paired with AI-enabled interpretation algorithms, could significantly streamline IVF workflows, reduce clinic burden, and empower patients with remote fertility tracking options, all while maintaining diagnostic accuracy (Grens et al., 2024). As validation efforts continue, AI-assisted follicle monitoring is poised to become a core component of intelligent ovarian stimulation protocols.

2.5 AI in sperm selection and male infertility

Male factor infertility accounts for nearly half of all infertility cases, yet conventional sperm selection based on manual microscopy and subjective grading remains plagued by operator variability and limited throughput. AI is now driving a paradigm shift toward objective, high-throughput sperm analysis. CNNs, trained on large annotated sperm image and video datasets, can automatically quantify motility parameters, morphology subtypes, and even subtle subcellular defects (e.g., vacuoles, acrosomal integrity), achieving classification accuracies above 90% and outperforming standard CASA systems (Lustgarten Guahmich, Borini, & Zaninovic, 2023).

Beyond morphology and motility, AI-enhanced microscopy, coupled with microfluidic sorting platforms, can integrate real-time image analysis to guide noninvasive sperm selection for ICSI. For instance, CNN-driven trackers embedded in microfluidic channels have demonstrated a 15% improvement in fertilization rates by isolating sperm with optimal motility trajectories and minimal DNA fragmentation (Cherouveim, Velmahos, & Bormann, 2023).

Moreover, multimodal AI models that fuse imaging data with biochemical assays of oxidative stress and chromatin integrity are emerging, offering composite "fertility scores" that better predict fertilization potential and early embryo development than any single parameter alone (Dehghan et al., 2024). As these systems are validated across broader patient cohorts, AI-driven sperm selection promises to standardize laboratory workflows, reduce embryologists' workload, and ultimately enhance ART success rates while mitigating male infertility disparities (He et al., 2019).

2.6 AI in oocyte selection

Oocyte quality is a major determinant of IVF success. Conventional assessment relies heavily on subjective morphological evaluation, introducing variability. AI offers an objective, reproducible alternative.

DL models, notably CNNs, have achieved high accuracy in identifying oocyte features predictive of maturation and fertilization potential, analyzing indicators such as cytoplasmic granularity, zona pellucida texture, and nuclear positioning (Curchoe & Bormann, 2019; Iannone, Carfì, Mastrogiovanni, Zaccaria, & Manna, 2024). Beyond morphology, AI models integrate patient parameters, namely, age, hormone profiles, and AFC, to generate individualized oocyte quality scores (Fanton et al., 2022; Si, Huang & Jin, 2023). Ensemble learning models that combine ultrasound imaging and laboratory data are further enhancing predictive capabilities (Zaninovic & Rosenwaks, 2020). As annotated datasets become more standardized, AI-driven oocyte evaluation tools are poised to become integral components of ART laboratories (Khosravi et al., 2019).

2.6.1 AI in embryo selection

Embryo selection is critical to IVF success, and optimizing this step remains a central challenge. Traditional static morphological grading and morphokinetic analyses using time-lapse imaging suffer from subjectivity, intra- and interobserver variability, and limited predictive capacity. AI, particularly DL frameworks like CNNs and RNNs, offers a more robust, automated solution that can systematically capture complex developmental features and dynamic events in embryo growth.

AI models trained on large, annotated embryo datasets have demonstrated superior predictive accuracy (70%–95%) for blastocyst viability, implantation potential, and clinical pregnancy compared to embryologists' assessments (Iannone et al., 2024; Kragh & Karstoft, 2021). Platforms such as Eeva, LifeWhisperer, and iDAScore integrate AI to automate embryo ranking based on dynamic morphokinetic parameters, enabling more standardized and reproducible embryo selection (Pavlovic, Jiang, & Hariton, 2024; Ver Milyea et al., 2021). Emerging approaches increasingly combine AI with multiomics data, such as transcriptomics, proteomics, metabolomics, and noninvasive preimplantation genetic testing for aneuploidy (PGT-A), to develop comprehensive embryo viability indices (Kakkar et al., 2025; Zaninovic & Rosenwaks, 2020). These integrative models move beyond morphology, providing a systems-level assessment of embryo fitness and developmental competence.

Future strategies aim to incorporate additional maternal factors, including endometrial receptivity profiling, uterine immune environment analysis, and personalized hormonal milieu data, to guide embryo transfer decisions on an individualized basis (Lédée et al., 2025). Such integrative AI models could optimize the synchrony between embryo development and endometrial receptivity, enhancing implantation and live birth rates.

Nevertheless, the widespread adoption of AI-driven embryo selection tools requires rigorous prospective validation in diverse populations, standardization of outcome metrics, transparent model development practices, and strong ethical oversight. Ensuring algorithm fairness, addressing interpretability challenges, and establishing regulatory standards will be pivotal to fully harness the promise of AI in embryo selection (Ghassemi, Oakden-Rayner, & Beam, 2021; Hanassab et al., 2024).

2.6.2 AI for predicting embryo ploidy and transfer recommendations

One of the most critical determinants of embryo viability and clinical success in IVF is ploidy status. Euploid embryos (those with a normal chromosomal complement) are more likely to implant successfully and lead to healthy live births, whereas aneuploid embryos contribute to implantation failure, miscarriage, or congenital abnormalities. Traditionally, PGT-A has been the gold standard for assessing ploidy, but it remains invasive, costly, and limited in accessibility. Recent advancements in AI have enabled the development of noninvasive approaches to predict embryo ploidy from imaging data, morphokinetics, and spent culture media analysis.

DL models, particularly CNNs, have been trained to identify subtle morphological and dynamic features in time-lapse videos that correlate with chromosomal normality. For instance, AI-powered platforms like LifeWhisperer Genetics and iDAScore have demonstrated up to 80% accuracy in predicting embryo euploidy, offering a promising alternative or adjunct to PGT-A (Kragh & Karstoft, 2021; Papamentzelopoulou et al., 2024). These tools analyze hundreds of parameters from embryo development stages, including the timing of compaction, blastocyst formation, and cell division symmetry, to infer chromosomal status with minimal human input.

Furthermore, AI systems are increasingly integrating multiomics data, including transcriptomic, proteomic, and metabolomic signatures from the embryo's surrounding culture medium, to enhance prediction accuracy. Recent studies have shown that combining morphokinetics with noninvasive metabolomic profiling can yield comparable diagnostic performance to traditional biopsy-based PGT-A (Zaninovic & Rosenwaks, 2020).

In addition to ploidy prediction, AI algorithms assist in recommending which embryo to transfer or cryopreserve by weighing implantation likelihood, patient-specific factors (e.g., maternal age, endometrial receptivity), and clinical priorities (e.g., single embryo transfer vs. cumulative success). These decision support systems not only improve success rates but also reduce multiple pregnancy risks and optimize resource allocation. As ongoing validation studies refine these algorithms across diverse populations, AI-driven embryo recommendation tools are expected to play an increasingly central role in precision-guided IVF care.

2.7 AI for predicting IVF outcomes

Predicting IVF success has traditionally been challenging due to multifactorial influences, namely, patient demographics, ovarian reserve, sperm quality, embryo development, and uterine environment. AI has transformed this domain through multifaceted prediction models that integrate clinical and laboratory data (Fig. 12.2).

ML algorithms, including random forests, SVMs, and deep neural networks, have outperformed traditional models in predicting IVF outcomes, including clinical pregnancy and live birth (Khosravi et al., 2019; Sadegh-Zadeh et al., 2024).

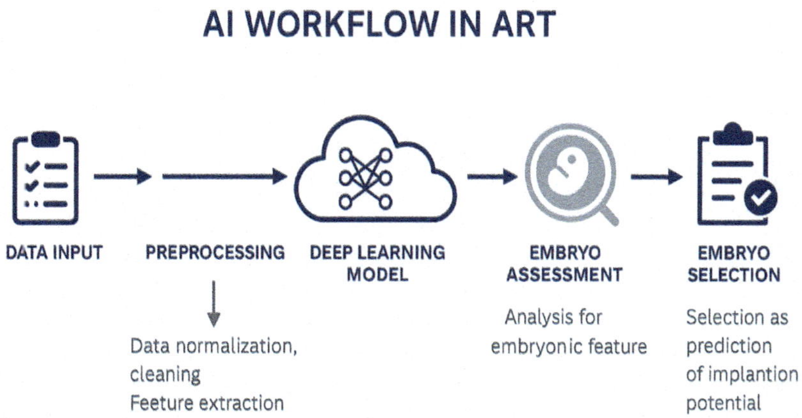

FIG. 12.2 Workflow of AI applications in ART, highlighting key decision points, including drug dosage prediction, oocyte retrieval, embryo grading, and transfer timing.

By analyzing diverse datasets, such as hormonal profiles, time-lapse embryo imaging, and genetic screening results, AI models enable nuanced risk stratification.

Platforms leveraging multiomics integration (genomics, proteomics, metabolomics) and FL approaches are enhancing prediction accuracy while ensuring data privacy and model generalizability (Si et al., 2023). As prospective clinical validations expand, AI-based outcome prediction systems are poised to support personalized fertility counseling and treatment planning, optimizing success rates in ART (Wang et al., 2019).

3 Image-based and multimodal data integration

AI-driven embryo assessment has greatly benefited from advances in imaging technologies and the integration of diverse biomedical data. CNNs are now central tools for analyzing time-lapse microscopy and EmbryoScope imaging, identifying subtle morphokinetic features imperceptible to human observers and offering superior precision in embryo grading and implantation prediction (Kragh & Karstoft, 2021).

Beyond imaging, AI increasingly integrates multimodal datasets, including clinical profiles, hormonal assays, ultrasonography, genetic screening (e.g., PGT-A), and environmental or lifestyle factors, enabling the development of robust, personalized predictive models (Orovou et al., 2025). Multimodal data fusion shifts embryo selection from static, rule-based approaches to dynamic, probabilistic assessments. Platforms combining imaging with genomic and proteomic profiles now deliver individualized predictions of embryo viability and maternal receptivity, outperforming traditional tools in predicting implantation and live birth rates (Curchoe & Bormann, 2019; Kakkar et al., 2025). As data sharing initiatives and interoperability frameworks mature, these multimodal AI systems are expected to become more scalable, generalizable, and essential in bridging the gap between clinical judgment and quantitative precision in ART (Fig. 12.3).

4 AI for monitoring pregnancy and managing complications

AI continues to expand its role beyond conception into prenatal care, particularly for high-risk pregnancies. Predictive models now identify risks for gestational diabetes mellitus, preeclampsia, intrauterine growth restriction, and preterm birth by analyzing data from electronic health records, laboratory tests, demographic factors, and maternal histories (Kaya et al., 2024).

AI-powered analysis of fetal ultrasound images and cardiotocography data enables real-time monitoring of fetal well-being. DL algorithms can assess heart rate variability, uterine contractions, and placental function, often detecting abnormalities earlier than conventional methods (Asfaw et al., 2023). Integration into wearable and remote monitoring devices further allows continuous, noninvasive tracking of maternal and fetal health, reducing reliance on frequent in-person visits and supporting timely interventions via telemedicine (Esteva et al., 2019).

As these technologies are further validated across diverse populations, AI is poised to become foundational in precision obstetrics, enabling proactive, personalized prenatal care.

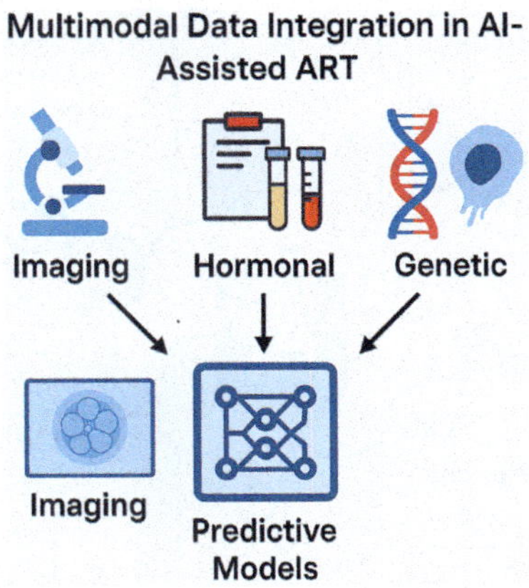

FIG. 12.3 Multimodal data integration in AI-assisted ART: merging imaging, hormonal, and genetic data into predictive models.

5 Ethical, legal, and implementation challenges

Despite its promise, AI in reproductive medicine raises critical ethical, legal, and practical challenges. Data privacy is a primary concern, given the sensitivity of reproductive and genetic information. Anonymization, secure storage, and ethical data-sharing frameworks are essential to maintaining patient confidentiality (Jiang et al., 2017).

Bias remains a major risk, as many AI models are trained on geographically or demographically narrow datasets, potentially perpetuating health disparities. Addressing this requires the use of diverse, multicenter datasets and the implementation of fairness audits during model development (Ghassemi et al., 2021).

Legal and regulatory frameworks for AI in reproductive health are still evolving. Comprehensive guidelines addressing data ownership, liability, informed consent, and AI integration into clinical workflows are urgently needed (Fig. 12.4).

Ethical deployment demands collaborative governance, involving clinicians, ethicists, patients, and developers, to align AI solutions with real-world clinical needs and societal values. Ensuring human oversight in all AI-supported clinical decisions remains paramount for maintaining ethical integrity.

6 Limitations and challenges of AI research in reproductive medicine

AI research in reproductive medicine faces significant hurdles. Many current models are trained on small, retrospective, single-center datasets, limiting generalizability and risking overfitting (Li et al., 2020; Ver Milyea et al., 2021). Variations in embryo grading standards, laboratory practices, and patient demographics further complicate model transferability (Kelly, Karthikesalingam, Suleyman, Corrado, & King, 2019). The "black-box" nature of many DL models also poses challenges, as a lack of explainability can erode clinician trust and hinder clinical adoption (Siala & Wang, 2022). Biases embedded in AI systems risk amplifying existing disparities across ethnic and socioeconomic groups (Amann, Blasimme, Vayena, Frey, & Madai, 2020; Hasanzadeh et al., 2025; Zhang, Xing, Zou, & Wu, 2022).

Key challenges limiting the effective application of AI in medicine and IVF include:

1. **Data security:** AI systems rely heavily on vast datasets, making them susceptible to data breaches. The mishandling of sensitive reproductive health information, particularly in regions lacking robust privacy frameworks, raises significant concerns.
2. **Model bias:** AI tools trained on nonrepresentative datasets risk reinforcing existing disparities. This could compromise IVF outcomes for underrepresented populations. Ongoing validation and diversification of training data are essential.
3. **Loss of clinical expertise:** Excessive dependence on AI may diminish the clinical judgment of healthcare providers. Preserving a collaborative model that balances human expertise with AI recommendations is crucial for maintaining high standards of care.
4. **Lack of transparency:** Many AI algorithms function as opaque "black boxes," making their decision-making processes difficult to interpret. This lack of transparency can erode clinicians' and patients' trust, especially when AI outputs are misused or misunderstood.
5. **Ethical dilemmas:** AI-driven embryo selection and genetic prediction raise sensitive ethical issues, including fears around "designer babies." These concerns may provoke societal resistance and complicate informed consent and patient counseling.

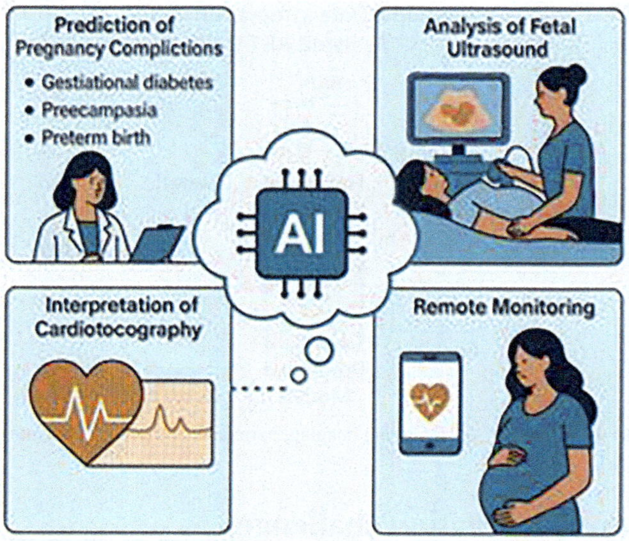

FIG. 12.4 AI for monitoring pregnancy and managing complications.

6. **Integration barriers:** Variability in IVF protocols, laboratory infrastructure, and equipment across clinics present challenges for seamless AI integration, potentially introducing inefficiencies or errors.
7. **Environmental impact:** A pressing yet often overlooked drawback of AI is its energy demand. The expansion of large-scale databases and reliance on power-intensive processors have led to a sharp rise in data center energy use. A recent report by the Lawrence Berkeley National Laboratory projected that U.S. data centers, which accounted for 1.9% of national electricity consumption in 2018, surged to 4.4% in 2023, with estimates reaching 12% by 2028, driven largely by AI growth (Shehabi et al., 2024).

To responsibly harness AI in IVF, it is imperative to prioritize ethical governance, ensure algorithmic transparency, conduct rigorous validation, and develop sustainable, equitable systems.

7 Future directions

The future of AI in reproductive health is poised for transformative growth, driven by algorithmic advancements and the availability of richer, more diverse datasets. Real-time, adaptive clinical decision-support systems will increasingly guide fertility and pregnancy management, integrating longitudinal electronic health records, wearable sensor data, imaging, and multiomics profiles to deliver continuous, personalized recommendations.

Generative AI and large language models will enhance patient communication, automate counseling, and synthesize complex clinical data into actionable insights. FL approaches, where models are trained across decentralized datasets without sharing sensitive patient data, will enhance model generalizability while preserving privacy.

AI is expected to move beyond embryo and gamete selection into broader reproductive health areas, such as menstrual health prediction, contraceptive optimization, miscarriage risk forecasting, and postpartum monitoring. Interdisciplinary collaboration among reproductive health experts, AI developers, ethicists, and policymakers will be critical to ensuring responsible innovation. As prospective studies expand and regulatory frameworks mature, AI's integration into routine reproductive healthcare will shift from experimental innovation to standard practice.

8 Conclusion

AI is redefining the landscape of reproductive health by offering precise, personalized, and efficient solutions. From infertility diagnosis, sperm and oocyte selection, embryo assessment, to pregnancy monitoring, AI is transforming reproductive care into a data-driven, patient-centric discipline. ML and DL models have introduced objectivity and reproducibility into traditionally subjective procedures, thereby improving clinical decision-making and optimizing success rates. However, despite the remarkable advancements, several challenges remain before AI can be seamlessly and safely integrated into everyday clinical practice. Issues such as data heterogeneity, model interpretability, algorithmic bias, and inconsistent regulatory frameworks need urgent attention. Building trustworthy AI systems will require larger, diverse, multicentric datasets and a commitment to developing transparent, explainable models that clinicians and patients alike can understand and trust.

Future directions will also necessitate interdisciplinary collaborations involving reproductive specialists, AI developers, bioethicists, and policymakers. There is a growing need for prospective validation trials, updated clinical guidelines for AI use, and educational initiatives to train healthcare providers on how to optimally integrate AI into their workflows. Importantly, as AI continues to mature, it holds the potential not only to optimize ART outcomes but also to enhance preventive reproductive health, individualized contraceptive management, miscarriage risk assessment, and postpartum monitoring. Equally critical is ensuring that AI-driven innovations are accessible globally, minimizing disparities in fertility care delivery.

In conclusion, responsible innovation, ethical stewardship, and rigorous scientific validation will be pivotal in unlocking the transformative power of AI in reproductive medicine, ultimately advancing a future of more equitable, effective, and personalized reproductive healthcare.

References

Abbara, A., Vuong, L. N., Ho, V. N. A., Clarke, S. A., Jeffers, L., Comninos, A. N., et al. (2018). Follicle size on day of trigger most likely to yield a mature oocyte. *Frontiers in Endocrinology*, 1664-2392. *9*. Available from: https://www.frontiersin.org/articles/10.3389/fendo.2018.00193/full. doi:10.3389/fendo.2018.00193.

AlSaad, R., Abd-alrazaq, A., Choucair, F., Ahmed, A., Aziz, S., & Sheikh, J. (2024). Harnessing artificial intelligence to predict ovarian stimulation outcomes in in vitro fertilization: scoping review. *Journal of Medical Internet Research*, 1438-8871. *26*(1), e53396. doi:10.2196/53396.

Amann, J., Blasimme, A., Vayena, E., Frey, D., & Madai, V. I. (2020). Explainability for artificial intelligence in healthcare: a multidisciplinary perspective. *BMC Medical Informatics and Decision Making*, 1472-6947. *20*(1). Available from: http://www.biomedcentral.com/bmcmedinformdecismak/. doi:10.1186/s12911-020-01332-6.

Asfaw, D., Jordanov, I., Impey, L., Namburete, A., Lee, R., & Georgieva, A. (2023). Multimodal deep learning for predicting adverse birth outcomes based on early labour data. *Bioengineering*, 2306-5354. *10*(6), 730. doi:10.3390/bioengineering10060730.

Canon, C., Leibner, L., Fanton, M., Chang, Z., Suraj, V., Lee, J. A., et al. (2024). Optimizing oocyte yield utilizing a machine learning model for dose and trigger decisions: A multi-center, prospective study. *Research Square*, Available from: https://www.researchsquare.com/browse. doi:10.21203/rs.3.rs-4377755/v1.

Chen, T., Zhang, W., Good, S., Zhou, K. S., & Comaniciu, D. (2009). Automatic ovarian follicle quantification from 3D ultrasound data using global/local context with database guided segmentation. In *Proceedings of the IEEE international conference on computer vision* (pp. 795–802). United States. doi:10.1109/ICCV.2009.5459243.

Cherouveim, P., Velmahos, C., & Bormann, C. L. (2023). Artificial intelligence for sperm selection—A systematic review. *Fertility and Sterility*, 1556-5653. *120*(1), 24–31. Available from: www.elsevier.com/locate/fertnstert. doi:10.1016/j.fertnstert.2023.05.157.

Curchoe, C. L., & Bormann, C. L. (2019). Artificial intelligence and machine learning for human reproduction and embryology presented at ASRM and ESHRE 2018. *Journal of Assisted Reproduction and Genetics*, 1573-7330. *36*(4), 591–600. Available from: www.wkap.nl/journalhome.htm/1058-0468. doi:10.1007/s10815-019-01408-x.

Dehghan, S., Rabiei, R., Choobineh, H., Maghooli, K., Nazari, M., Vahidi-Asl, M., & Alaeejahromi, S. (2024). Comparative study of machine learning approaches integrated with genetic algorithm for IVF success prediction. *PLOS ONE*, 1932-6203. *19*(10), e0310829. doi:10.1371/journal.pone.0310829.

Esteva, A., Robicquet, A., Ramsundar, B., Kuleshov, V., DePristo, M., Chou, K., et al. (2019). A guide to deep learning in healthcare. *Nature Medicine*, 1078-8956. *25*(1), 24–29. doi:10.1038/s41591-018-0316-z.

Fanton, M., Nutting, V., Solano, F., Maeder-York, P., Hariton, E., Barash, O., et al. (2022). An interpretable machine learning model for predicting the optimal day of trigger during ovarian stimulation. *Fertility and Sterility*, 1556-5653. *118*(1), 101–108. Available from: www.elsevier.com/locate/fertnstert. doi:10.1016/j.fertnstert.2022.04.003.

Ghassemi, M., Oakden-Rayner, L., & Beam, A. L. (2021). The false hope of current approaches to explainable artificial intelligence in health care. *The Lancet Digital Health*, 2589-7500. *3*(11), e745. Available from: https://www.sciencedirect.com/journal/the-lancet-digital-health. doi:10.1016/S2589-7500(21)00208-9.

Grens, H., Huppelschoten, A. G., De Ligny, W. R., Kool, R. B., Kremer, J. A. M., & de Bruin, J. P. (2024). Web-based interventions in the clinical encounter in fertility care: A scoping review. *Reproductive BioMedicine Online*, 1472-6491. *49*(5). Available from: https://www.sciencedirect.com/science/journal/14726483. doi:10.1016/j.rbmo.2024.104308.

Hanassab, S., Abbara, A., Yeung, A. C., Voliotis, M., Tsaneva-Atanasova, K., Kelsey, T. W., et al. (2024). The prospect of artificial intelligence to personalize assisted reproductive technology. *Npj Digital Medicine*, 2398-6352. *7*(1). Available from: https://www.nature.com/npjdigitalmed/. doi:10.1038/s41746-024-01006-x.

Hasanzadeh, F., Josephson, C. B., Waters, G., Adedinsewo, D., Azizi, Z., & White, J. A. (2025). Bias recognition and mitigation strategies in artificial intelligence healthcare applications. *Npj Digital Medicine*, 2398-6352. *8*(1). Available from: https://www.nature.com/npjdigitalmed/. doi:10.1038/s41746-025-01503-7.

Hassan, M. H., Reiter, E., & Razzaq, M. (2024). Automatic ovarian follicle detection using object detection models. *Scientific Reports*, 2045-2322. *14*(1). doi:10.1038/s41598-024-82904-8.

He, J., Baxter, S. L., Xu, J., Xu, J., Zhou, X., & Zhang, K. (2019). The practical implementation of artificial intelligence technologies in medicine. *Nature Medicine*, 1546-170X. *25*(1), 30–36. Available from: http://www.nature.com/nm/index.html. doi:10.1038/s41591-018-0307-0.

Iannone, A., Carfì, A., Mastrogiovanni, F., Zaccaria, R., & Manna, C. (2024). On the role of artificial intelligence in analysing oocytes during in vitro fertilisation procedures. *Artificial Intelligence in Medicine*, 0933-3657. *157*, 102997. doi:10.1016/j.artmed.2024.102997.

Jiang, F., Jiang, Y., Zhi, H., Dong, Y., Li, H., Ma, S., et al. (2017). Artificial intelligence in healthcare: past, present and future. *Stroke and Vascular Neurology*, 2059-8688. *2*(4), 230–243. doi:10.1136/svn-2017-000101.

Kakkar, P., Gupta, S., Paschopoulou, K. I., Paschopoulos, I., Paschopoulos, I., Siafaka, V., & Tsonis, O. (2025). The integration of artificial intelligence in assisted reproduction: a comprehensive review. *Frontiers in Reproductive Health*, 2673-3153. *7*. Available from: https://www.frontiersin.org/journals/reproductive-health/about. doi:10.3389/frph.2025.1520919.

Kaya, Y., Bütün, Z., Çelik, Ö., Salik, E. A., Tahta, T., & Yavuz, A. A. (2024). The early prediction of gestational diabetes mellitus by machine learning models. *BMC Pregnancy and Childbirth*, 1471-2393. *24*(1). Available from: https://bmcpregnancychildbirth.biomedcentral.com/. doi:10.1186/s12884-024-06783-7.

Kelly, C. J., Karthikesalingam, A., Suleyman, M., Corrado, G., & King, D. (2019). Key challenges for delivering clinical impact with artificial intelligence. *BMC Medicine*, 1741-7015. *17*(1). Available from: http://www.biomedcentral.com/bmcmed/. doi:10.1186/s12916-019-1426-2.

Khosravi, P., Kazemi, E., Zhan, Q., Malmsten, J. E., Toschi, M., Zisimopoulos, P., et al. (2019). Deep learning enables robust assessment and selection of human blastocysts after in vitro fertilization. *Npj Digital Medicine*, 2398-6352. *2*(1). Available from: https://www.nature.com/npjdigitalmed/. doi:10.1038/s41746-019-0096-y.

Kragh, M. F., & Karstoft, H. (2021). Embryo selection with artificial intelligence: how to evaluate and compare methods? *Journal of Assisted Reproduction and Genetics*, 1573-7330. *38*(7), 1675–1689. Available from: www.wkap.nl/journalhome.htm/1058-0468. doi:10.1007/s10815-021-02254-6.

Lédée, N., Petitbarat, M., Dray, G., Chevrier, L., Kazhalawi, A., Rahmati, M., et al. (2025). Endometrial immune profiling and precision therapy increase live birth rate after embryo transfer: a randomised controlled trial. *Frontiers in Immunology*, 1664-3224. *16*. Available from: https://www.frontiersin.org/journals/immunology#. doi:10.3389/fimmu.2025.1523871.

Li, H., Fang, J., Liu, S., Liang, X., Yang, X., Mai, Z., et al. (2020). CR-Unet: A composite network for ovary and follicle segmentation in ultrasound images. *IEEE Journal of Biomedical and Health Informatics*, 2168-2208. *24*(4), 974–983. Available from: http://ieeexplore.ieee.org/xpl/RecentIssue.jsp?punumber=6221020. doi:10.1109/JBHI.2019.2946092.

Lustgarten Guahmich, N., Borini, E., & Zaninovic, N. (2023). Improving outcomes of assisted reproductive technologies using artificial intelligence for sperm selection. *Fertility and Sterility*, 0015-0282. *120*(4), 729–734. doi:10.1016/j.fertnstert.2023.06.009.

McCulloh, D. H., Kutchukhidze, N., Charkviani, T., Zhorzholadze, T., Barbakadze, T., Munné, S., & Chkonia, L. (2020). Follicle size indicates oocyte maturity and blastocyst formation but not blastocyst euploidy following controlled ovarian hyperstimulation of oocyte donors. *Human Reproduction*, 1460-2350. *35*(3), 545–556. Available from: http://humrep.oxfordjournals.org/. doi:10.1093/humrep/dez291.

Mendizabal-Ruiz, G., Paredes, O., Álvarez, Á., Acosta-Gómez, F., Hernández-Morales, E., González-Sandoval, J., et al. (2024). Artificial intelligence in human reproduction. *Archives of Medical Research*, 0188-4409. *55*(8), 103131. doi:10.1016/j.arcmed.2024.103131.

Orovou, E., Tzimourta, K. D., Tzitiridou-Chatzopoulou, M., Kakatosi, A., & Sarantaki, A. (2025). Artificial intelligence in assisted reproductive technology: a new era in fertility treatment. *Cureus*. 2168-8184. *17*(4). doi:10.7759/cureus.81568.

Papamentzelopoulou, M. S., Prifti, I. N., Mavrogianni, D., Tseva, T., Soyhan, N., Athanasiou, A., et al. (2024). Assessment of artificial intelligence model and manual morphokinetic annotation system as embryo grading methods for successful live birth prediction: a retrospective monocentric study. *Reproductive Biology and Endocrinology*. 1477-7827. *22*(1). Available from: https://rbej.biomedcentral.com. doi:10.1186/s12958-024-01198-7.

Pavlovic, Z. J., Jiang, V. S., & Hariton, E. (2024). Current applications of artificial intelligence in assisted reproductive technologies through the perspective of a patient's journey. *Current Opinion in Obstetrics and Gynecology*. 1473-656X. *36*(4), 211–217. Available from: http://journals.lww.com/co-obgyn/pages/default.aspx. doi:10.1097/GCO.0000000000000951.

Rieke, N., Hancox, J., Li, W, Milletarì, F., Roth, H. R., Albarqouni, S., et al. (2020). The future of digital health with federated learning. *Npj Digital Medicine*, 2398-6352. *3*(1). Available from: https://www.nature.com/npjdigitalmed/. doi:10.1038/s41746-020-00323-1.

Sadegh-Zadeh, S. A., Khanjani, S., Javanmardi, S., Bayat, B., Naderi, Z., & Hajiyavand, A. M. (2024). Catalyzing IVF outcome prediction: exploring advanced machine learning paradigms for enhanced success rate prognostication. *Frontiers in Artificial Intelligence*, 2624-8212. *7*. Available from: www.frontiersin.org/journals/artificial-intelligence#. doi:10.3389/frai.2024.1392611.

Salih, M., Austin, C., Warty, R. R., Tiktin, C., Rolnik, D. L., Momeni, M., et al. (2023). Embryo selection through artificial intelligence versus embryologists: a systematic review. *Human Reproduction Open*. 2399-3529. *2023*, 3. doi:10.1093/hropen/hoad031.

Shehabi, A., Smith, S. J., Hubbard, A., Newkirk, A., Lei, N., Siddik, M. A. B., et al. (2024). United States data center energy usage report energy analysis & environmental impacts division citation.

Si, K., Huang, B., & Jin, L. (2023). Application of artificial intelligence in gametes and embryos selection. *Human Fertility*, 1464-7273. *26*(4), 757–777. doi:10.1080/14647273.2023.2256980.

Siala, H., & Wang, Y. (2022). SHIFTing artificial intelligence to be responsible in healthcare: a systematic review. *Social Science and Medicine*, 1873-5347. *296*. Available from: www.elsevier.com/locate/socscimed. doi:10.1016/j.socscimed.2022.114782.

Ver Milyea, M., Hall, J. M. M., Diakiw, S. M., Johnston, A., Nguyen, T., Perugini, D., et al. (2021). Development of an artificial intelligence-based assessment model for prediction of embryo viability using static images captured by optical light microscopy during IVF. *Human Reproduction*, 1460-2350. *35*(4), 770–784. Available from: http://humrep.oxfordjournals.org/. doi:10.1093/HUMREP/DEAA013.

Wang, R., Pan, W., Jin, L., Li, Y., Geng, Y., Gao, C., … Liao, S. (2019). Artificial intelligence in reproductive medicine. *Reproduction*, 1470-1626. *158*(4), R139. doi:10.1530/rep-18-0523.

Wu, Y. C., Chia-Yu Su, E., Hou, J. H., Lin, C. J., Lin, K. B., & Chen, C. H. (2025). Artificial intelligence and assisted reproductive technology: a comprehensive systematic review. *Taiwanese Journal of Obstetrics and Gynecology*, 1875-6263. *64*(1), 11–26. Available from: https://www.sciencedirect.com/science/journal/10284559. doi:10.1016/j.tjog.2024.10.001.

Zaninovic, N., & Rosenwaks, Z. (2020). Artificial intelligence in human in vitro fertilization and embryology. *Fertility and Sterility*, 0015-0282. *114*(5), 914–920. doi:10.1016/j.fertnstert.2020.09.157.

Zebitay, A. G., Cetin, O., Verit, F. F., Keskin, S., Sakar, M. N., Karahuseyinoglu, S., … Sahmay, S. (2017). The role of ovarian reserve markers in prediction of clinical pregnancy. *Journal of Obstetrics and Gynaecology*, 1364-6893. *37*(4), 492–497. doi:10.1080/01443615.2016.1269730.

Zhang, A., Xing, L., Zou, J., & Wu, J. C. (2022). Shifting machine learning for healthcare from development to deployment and from models to data. *Nature Biomedical Engineering*, 2157-846X. *6*(12), 1330–1345. Available from: https://www.nature.com/natbiomedeng/. doi:10.1038/s41551-022-00898-y.

Chapter 13

Appications of artificial intelligence in cardiovascular disorders

Sabyasachi Chakrabarty[a] and Susmita Ghosh[b]
[a]*Medical Faculty, Center for Internal Medicine, University Clinic for Cardiology and Andrology, Otto-von-Guericke-University Magdeburg, Magdeburg, Germany*
[b]*Leibniz-Institut für Analytische Wissenschaften - ISAS-e.V., Dortmund, Germany*

1 Cardiovascular disease: A global health challenge

Cardiovascular diseases (CVDs) represent a complex array of disorders impacting the heart and blood vessels, ranging from coronary artery disease (CAD), arrhythmias, and heart failure to peripheral artery disease, aortic aneurysms, and cerebrovascular conditions like stroke. These diseases may arise from genetic predispositions or be influenced by modifiable risk factors, including environmental exposures, sex differences, physical inactivity, dietary patterns, behavioral habits, hypertension, and metabolic disorders such as obesity and diabetes. Growing urbanization, aging populations, and limited access to preventive care in many regions are accelerating the disease burden, especially in low- and middle-income settings.

CVDs remain the leading global cause of mortality, responsible for an estimated 20.5 million deaths in 2021, a sharp increase from 12 million in 1990, as shown in Fig. 13.1 (https://world-heart-federation.org). This mortality burden falls disproportionately on low- and middle-income countries (LMICs), accounting for 80% of CVD-related deaths and thereby reflecting profound healthcare inequities (Di Cesare et al., 2024). The economic impact is equally severe; medical expenses, productivity losses, and long-term disability care are expected to reach $1.1 trillion by 2035, a twofold increase compared to 2015 (Dunbar et al., 2018). In LMICs alone, noncommunicable CVDs have caused an estimated $7.2 trillion in economic losses over the past 14 years, while implementation of WHO "best buy" interventions would require just $11–13 billion annually (Bloom et al., 2011; Laslett et al., 2012). This trend showcases the need for population-level, cost-effective interventions to stem the tide.

Looking ahead, projections paint a grimmer picture: epidemiological models predict a 23% surge in CVD mortality by 2030, potentially reaching 23.6 million deaths per year (Smith et al., 2012). According to the World Heart Federation, 80% of premature cardiovascular events (heart attack, stroke) are preventable through lifestyle modifications and drug treatment, highlighting a tragic failure in global prevention strategies, especially in healthcare infrastructure. For instance, high-income countries benefit from advanced diagnostics and specialist care, yet many resource-limited settings lack even basic screening tools or essential medications. Despite decades of advocacy, this persistent divide demands urgent, scalable, and equity-focused solutions to reduce the global burden of CVD and improve outcomes for all.

2 Current management of CVDs: Strategies and systemic limitations

The management of CVD encompasses a multifaceted approach involving risk factor modification, diagnostic strategies, therapeutic interventions, and system-level organization of care. While substantial progress has been made in high-income countries, where age-adjusted CVD mortality rates have declined significantly, global challenges persist, particularly in LMICs. Established clinical guidelines from bodies like the European Society of Cardiology (ESC) and the American College of Cardiology (ACC) set a crucial framework for managing CVDs. However, traditional approaches face significant systemic limitations when confronted with the sheer scale, complexity, and inequity of the global CVD burden (Virani et al., 2023; Vrints et al., 2024). These shortcomings, spanning prevention, diagnosis, data utilization, treatment personalization, and resource allocation, create an urgent need for transformative solutions such as artificial intelligence (AI).

Applications of Artificial Intelligence in Healthcare. https://doi.org/10.1016/B978-0-443-43934-6.00011-0

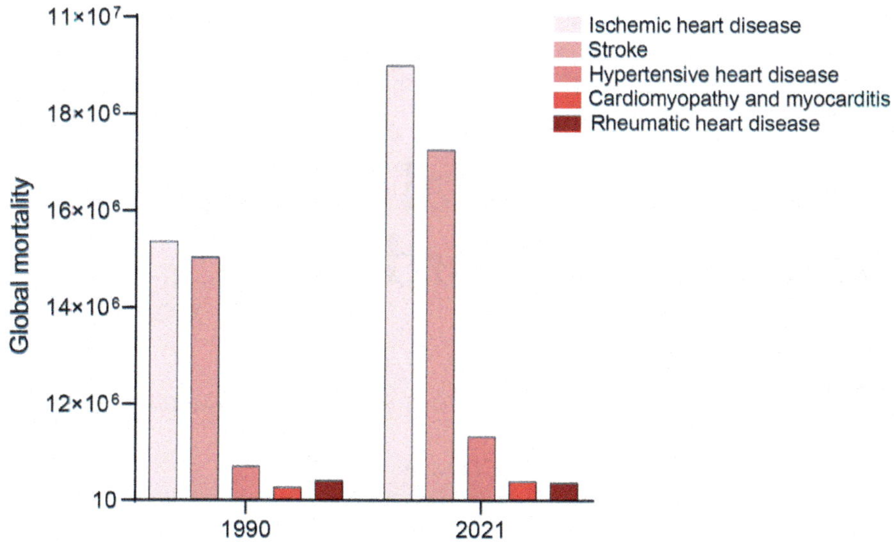

FIG. 13.1 Global mortality counts for five major cardiovascular diseases across 2 years (1990 and 2021).

2.1 Lifestyle modification

Lifestyle modification remains the cornerstone of CVD prevention and management. Essential interventions include maintaining heart-healthy diets, participating in routine physical exercise, controlling blood pressure, avoiding tobacco use, ensuring adequate sleep, and reducing stress. These lifestyle changes significantly improve clinical outcomes and quality of life, especially in individuals already diagnosed with CVD. The World Health Organization's HEARTS technical package (https://www.who.int/) highlights lifestyle counseling as a highly cost-effective strategy for risk reduction, applicable at both the individual and population levels (World Health Organization, 2020).

2.2 Data collection and patient monitoring

Data collection and patient monitoring are fundamental to effective CVD management. Current systems rely on clinical records, electronic registries, and longitudinal cohort studies to track high-risk populations over time. Despite standardized guidelines from the WHO, substantial weaknesses exist, particularly in LMICs, where health data infrastructures are often fragmented, poorly standardized, and incapable of managing comprehensive longitudinal data. This fragmentation undermines disease surveillance, timely intervention, and health policy planning. Large population initiatives, such as the UK Biobank, have generated valuable epidemiological insights; however, high attrition rates limit data completeness and sharing (Brayne & Moffitt, 2022). A cornerstone of modern CVD data management is the adoption of electronic health records (EHRs), which serve as comprehensive digital repositories of a patient's medical history, encompassing clinical notes, prescribed medications, diagnostic tests, and more. By enabling authorized access across providers, EHRs promote rapid information sharing, better coordination of care, and improved communication among healthcare stakeholders. Beyond clinical utility, EHRs also serve as information-rich platforms for biomedical research, supporting the study of disease diagnosis and progression, facilitating real-time disease tracking, and accelerating the development of personalized therapies (Nordo et al., 2019). Among the leading EHR systems, Epic and Cerner dominate, especially in the United States (Kashyap & Roy, 2023). Epic is widely implemented in large and complex healthcare organizations and is noted for its user-friendly interface and strong interoperability features, though it comes with high setup and maintenance costs. Cerner, on the contrary, is more commonly adopted in ambulatory care and is valued for its real-time data access and comparatively lower maintenance costs, though its interoperability and customization are more limited. Despite their advantages, EHR systems face significant drawbacks, including high financial costs, workflow disruptions during adoption, and persistent concerns around data privacy, security, fragmented data environments, and siloed storage that restrict interoperability and collaborative research (Ani et al., 2022).

2.3 Risk assessment

Risk assessment is central to CVD prevention and requires accurate quantification of an individual's likelihood of developing cardiovascular events. Conventional tools assess risk based on factors such as age, sex, blood pressure, lipid levels, smoking status, diabetes, and family history. Tools like the Framingham Risk Score (FRS), SCORE2, coronary artery calcium scoring, and polygenic risk scores (PRSs) are commonly used to quantify an individual's likelihood of experiencing a cardiovascular event within a defined time frame (Goldsborough, Tasdighi, & Blaha, 2023). Global evidence endorses a comprehensive cardiovascular risk assessment strategy, indicating that using combined indicators like blood pressure and cholesterol levels can lower the occurrence of major CVD events by up to 45% (Jackson, Lawes, Bennett, Milne, & Rodgers, 2005). However, the utility of these tools is hindered by several limitations. Many widely used models are developed from region-specific datasets, such as the Framingham study (Mahmood, Levy, Vasan, & Wang, 2014), and often struggle to perform accurately across populations with diverse ethnic and socioeconomic backgrounds. This lack of external validity significantly undermines their predictive accuracy in non-Western, rural, or underserved communities. Sociodemographic differences, variations in genetic predisposition, and disparities in healthcare infrastructure further complicate the establishment of a universally applicable risk prediction model (Dahlöf, 2010). Moreover, the accessibility of these tools remains a concern. In remote and rural areas, such as certain regions of Australia, up to 28% of individuals lack access to even basic preventive screenings. These geographic disparities limit the effectiveness of early intervention strategies. Compounding the issue is the reactive nature of current care paradigms: traditional risk assessments often lead to intervention only after symptoms emerge or clear risk factors are established. This delay is especially problematic given the insidious, subclinical progression of diseases like atherosclerosis, which can remain asymptomatic until an acute, often catastrophic, event such as a myocardial infarction or stroke occurs. Current models also rely heavily on a narrow set of conventional clinical variables, often omitting emerging or patient-specific indicators of risk. This limits their sensitivity to the nuanced and individualized trajectories of CVD. Standard tools like the FRS have been found to underperform in predicting risk for certain populations, particularly women, young adults, and ethnic minorities (Damen et al., 2016; Yadlowsky et al., 2018). Furthermore, qualitative assessments based on clinical judgment, which are still common in resource-limited settings, introduce subjectivity and increase the likelihood of misclassification. These shortcomings contribute to widespread gaps in early detection and treatment. For instance, only 50% of the population with hypertension is aware of their condition, and just one-fifth of them receive control measures, a situation that is even worse in LMICs (The Lancet Neurology, 2023). This disparity is largely attributable to fragmented care pathways, limited access to risk assessment tools, and delayed interventions. In summary, current risk assessment models, though valuable, remain constrained by limited predictive power, lack of personalization, poor generalizability, and systemic barriers to access. Addressing these gaps is vital to advancing the timeliness, diagnostic accuracy, and equity of cardiovascular care.

2.4 Diagnostic approaches

Diagnostic evaluation is central to CVD management, enabling timely identification and appropriate treatment through tools like electrocardiography (ECG), stress tests, echocardiography, coronary artery computed tomography (CT), magnetic resonance imaging (MRI), positron emission tomography, and coronary angiography (Samanidis, 2024). Although guidelines from the ESC and ACC have standardized practices and improved consistency, data fragmentation and the integration of diverse data modalities hinder longitudinal care. Advanced modalities are often unavailable in LMICs, and even basic tests like ECGs suffer from poor standardization and clinician variability. Interpretation subjectivity, especially with ECGs and echocardiograms, can lead to missed diagnoses (Attia et al., 2019). Large-scale studies (e.g., Framingham, INTERHEART, PURE) offer valuable insights but diverge in methods and regional focus, limiting global standardization. Regulatory disparities, such as Europe's GDPR versus Asia-Pacific's data laws, add another layer of complexity, restricting cross-border information sharing and stifling the development of universally adaptable models. In addition, high attrition rates, often ranging from 30% to 70%, constrain the generalizability of standardized diagnostic approaches (Gustavson, von Soest, Karevold, & Røysamb, 2012). In summary, while current diagnostic tools are vital, their potential is constrained by fragmented systems, unequal access, interpretive variability, and limited data integration, highlighting the need for more unified, scalable, and intelligent diagnostic approaches.

2.5 Treatment strategies

CVD treatment follows a structured continuum: primary prevention, acute intervention, and long-term management. Preventive care typically includes lifestyle modification and pharmacotherapy (e.g., statins, antihypertensives, antiplatelets) to mitigate modifiable risk factors. Acute interventions, such as percutaneous coronary intervention (PCI), thrombolysis, and coronary artery bypass grafting (CABG), are critical in managing events like myocardial infarction. Chronic care focuses on sustained medication use, behavioral support, and cardiac rehabilitation to prevent recurrence and improve outcomes. While clinical guidelines have standardized care in high-income countries, major limitations persist, particularly regarding equity, adherence, and personalization. Access to advanced interventions remains highly unequal; CABG is reportedly 12 times less accessible in low-income countries. Even within high-income nations, underserved communities face delays due to geographic, economic, or systemic barriers. Treatment adherence is another critical challenge. Despite proven benefits, up to 50% of patients discontinue statins within a year, often due to side effects, cost, or lack of immediate benefit, leading to higher rates of adverse events and hospitalizations. A core flaw in current care models is their reactive and generalized nature. Interventions are often initiated after symptom onset, overlooking subclinical disease phases where early treatment could be most effective. Standardized, trial-based guidelines also promote a "one-size-fits-all" approach that does not account for genetic variability, comorbidities, or social determinants. Tools to tailor treatments—like pharmacogenomics or real-time monitoring—remain underutilized due to infrastructure, cost, and integration challenges. Long-term management further suffers from fragmented care systems, especially in LMICs. Many patients lack access to coordinated, multidisciplinary support involving dietitians, mental health professionals, and physiotherapists, leading to poor follow-up, disengagement, and high recurrence rates.

2.6 Resource constraints and stark health

Finally, resource constraints and stark health inequities fundamentally limit the reach and efficacy of current CVD management, particularly in underserved regions. Limited access to specialists, advanced diagnostics, and timely interventions exacerbates disparities in LMICs and marginalized communities within high-income countries. The COVID-19 pandemic has emphasized the urgent need for digital transformation in healthcare. AI has emerged as a transformative force poised to reshape cardiovascular care. AI algorithms can analyze vast, complex datasets to detect subtle risk patterns, predict disease progression, and optimize resource allocation. From wearable devices capable of real-time arrhythmia detection to machine learning (ML) models forecasting heart failure months before clinical symptoms arise, AI-driven innovations are redefining early detection, personalized treatment, and equitable care delivery. The way forward demands addressing these systemic barriers comprehensively. For example, the Australian Institute of Health and Welfare estimates that improved risk factor management could alleviate 34% of Australia's CVD burden, highlighting the global potential of optimized care. Emerging technologies, including federated learning (FL) and AI-powered predictive analytics, offer promising solutions to overcome data interoperability challenges and extend quality care to underserved populations. By transforming fragmented data into actionable insights, AI paves the way for personalized, proactive CVD management accessible to all. However, realizing this potential requires navigating ethical considerations, regulatory complexities, and maintaining a human-centered approach to healthcare, challenges that warrant in-depth exploration as AI becomes integral to cardiovascular medicine.

3 AI tools, models, and algorithms in CVD management

With the outlined barriers mentioned earlier, the successful integration of AI in CVD relies on robust data management strategies and sophisticated modeling techniques tailored to diverse data types that can be trusted by clinicians. The following section showcases the established tools, models, and algorithms developed on several fronts driving AI innovation in CVD care.

3.1 Data management and integration strategies

Digitalization in healthcare has greatly expanded data generation—from high-resolution imaging to continuous wearable streams and multiomics—yet managing such diverse CVD-related data remains challenging due to fragmentation, poor interoperability, and strict privacy regulations like GDPR (De Filippis et al., 2024). Advancing AI-driven workflows, streamlined processes, and robust data interoperability is essential to overcoming these barriers.

3.1.1 Data harmonization and standardization

To harmonize and standardize the data from multicentric datasets and interpret the vast amount of information, tools and common data models (CDMs) like the Observational Medical Outcomes Partnership are used in CVD research. These CDMs extract clinical data from disparate datasets and standardize them into a common format (Chahal et al., 2025). This enables large-scale, multiinstitutional research by ensuring consistency and reliability in data structure and terminology across sites. Initiatives, including the American Heart Association Precision Medicine platform, public–private collaborations like BigData@Heart in Europe, and cloud-based platforms such as Terra are working to achieve data harmonization temporally and ontologically. Initiatives like CVDLINK aim to leverage such standardized data for advanced analytics.

3.1.2 Secure data infrastructure

While blockchain offers potential for secure, decentralized data management, practical implementation in large-scale healthcare settings is still evolving. The current focus often remains on secure institutional databases and platforms designed for large-scale data integration and analysis, sometimes leveraging cloud infrastructure or specialized big data solutions. Robust data governance frameworks are essential, regardless of the technology used.

3.1.3 Federated learning

This privacy-preserving technique trains AI models across multiple institutions in a collaborative manner without exchanging sensitive raw patient data. Briefly, one collaborator trains a local model based on local data and uploads the trained model to a centralized server. This ML framework is compatible with diverse data types, for example, histology data, X-rays, and electronic medical records (Teo et al., 2024). Projects like iCARE4CVD (https://icare4cvd.eu/) utilize FL platforms to build predictive models on distributed datasets while maintaining data privacy and security.

3.2 AI models tailored to cardiovascular data modalities

ML concepts date back to 1957, beginning with the perceptron, one of the earliest neural network models that established the groundwork for supervised learning algorithms and artificial neural networks. Since then, the field has achieved significant milestones, including the introduction of decision trees during the 1960s, the emergence of support vector machines (SVMs) in the 1990s, the development of random forests around 2000, the rise of deep learning (DL) models in the 2010s, and, more recently, the advent of large language models like ChatGPT in 2022 (Ahsan, Luna, & Siddique, 2022). Likewise, in the context of CVD, several AI models have been developed that are adept at extracting meaningful patterns and can also be categorized based on the data modalities pertinent to cardiovascular systems, as explained further in the following.

3.2.1 Electrocardiogram analysis

An electrocardiogram (ECG) functions by measuring voltage fluctuations on the skin caused by the heart's electrical activity over time. These electrical impulses regulate the heart's contraction and relaxation, producing distinct waveforms, P, QRS, and T, that represent specific phases of the cardiac cycle, such as atrial and ventricular depolarization and repolarization. The electrical activity of the heart is captured using a standard 12-lead ECG, placed across the body. The electrical impulses are presented as waveforms that provide insights into cardiac contraction capacity, conductance, ejection flow, and rate when analyzed by trained professionals. However, the dependence on professional manpower to diagnose heart conditions appears to be a major bottleneck faced by healthcare institutions, thereby calling for automation. In the field of AI, DL has demonstrated immense potential for automating the analysis of ECG waveforms, enabling more accurate detection of abnormal cardiac rhythms, facilitating real-time monitoring, and delivering faster, higher-quality results. This capability not only accelerates the diagnostic process but also conserves valuable time and resources, offering clinicians an earlier and more informed start in patient care (Martínez-Sellés & Marina-Breysse, 2023).

Recently, an AI-enabled ECG algorithm was developed using a convolutional neural network (CNN) to detect hidden signatures of atrial fibrillation (AF), a condition characterized by a rapid heartbeat with irregular rhythms and associated with multiple comorbidities, including stroke, heart failure, and other heart-related complications. The model was trained on approximately 650,000 ECGs from over 180,000 patients at the Mayo Clinic, positioning it as a noninvasive, low-cost screening tool for asymptomatic and underdiagnosed AF. The algorithm demonstrated strong performance in identifying early-stage AF using routine single-lead ECGs, with accuracy further improving when multiple ECG leads were incorporated, highlighting AI's potential for early AF detection in everyday clinical practice (Attia et al., 2019).

Despite the immense diagnostic prowess of DL models in ECG analysis, their deployment in real-time clinical settings remains challenging. This is primarily due to the computational burden imposed by the high-dimensional inputs of full 12-lead ECG recordings and the complexity of the models themselves, which puts a dent in scalability. To overcome these barriers, research has shifted focus toward developing lightweight, efficient AI models that aim to maintain diagnostic accuracy while significantly reducing computational costs by typically using only up to four leads (Lai, Zhou, & Trayanova, 2021; Ramirez et al., 2024). These ideas not only paved the way for scalable real-time AI-driven ECG interpretation, enabling early and personalized cardiac risk assessment, but also supported the broader goals of precision medicine by utilizing wearable and ambulatory monitoring technologies that can detect and predict cardiovascular events on-the-fly using embedded AI systems.

Minimum ECG lead input models

These models focus on extracting clinically meaningful features from only one or two ECG leads, significantly reducing input dimensionality. For example, label-masking strategies and threshold optimization, as implemented in the EfficientNet-B3 architecture (Tan & Le, 2019), have demonstrated the ability to maintain high performance with constrained input data.

Lightweight DL models

Architectures such as SqueezeNet (Iandola et al., 2016), which utilize scalogram features from wavelet transform coefficients, and SE-Net, which incorporates channel-wise self-attention through squeeze-and-excitation blocks, exemplify high efficiency while maintaining low memory and computational demands. Similarly, SE-ResNet (Hu, Shen, & Sun, 2018) leverages a self-supervised auxiliary task specifically designed for peak detection and is adaptable to signals with varying sampling rates, making it particularly suitable for wearable or ambulatory ECG monitoring systems.

Data annotation efficient models

These include strategies that minimize dependence on large, labeled datasets through semi-supervised, weakly supervised, or self-supervised learning. For example, SE-ResNet not only improves feature learning but also addresses data sparsity challenges in real-world ECG applications (Gupta, Paluru, Nankani, Kulkarni, & Awasthi, 2024).

Collectively, these advancements are helping to bridge the gap between algorithmic performance and real-world deployability, making it increasingly feasible to implement AI-enabled ECG analysis in everyday clinical workflows and mobile health applications.

3.2.2 Cardiac sound (phonocardiogram) analysis

Phonocardiogram (PCG) is a valuable noninvasive diagnostic tool that captures the acoustic signals produced by the heart, including the first (S1) and second (S2) heart sounds. These signals contain rich clinical information such as frequency content, amplitude, timing, and duration, as well as ratios like S1/S2 and diastolic-to-systolic intervals, which are essential for assessing cardiac function and detecting abnormalities. However, the inclusion of a high level of noise during the acquisition process challenges the proper diagnosis of heart conditions. ML algorithms are either applied to record heart sounds via digital stethoscopes or to analyze the recorded data to detect pathological murmurs indicative of valvular heart disease or other structural abnormalities. Companies like Eko Health integrate AI directly into their stethoscopes (Eko CORE 500 Digital Stethoscope), providing real-time analysis to aid clinicians in screening and diagnosis.

A typical PCG analysis pipeline begins with preprocessing steps such as noise filtering, signal segmentation, and normalization, followed by feature extraction and selection to classify meaningful acoustic characteristics (Fig. 13.2). These features may be time-domain (e.g., duration, amplitude), frequency domain (e.g., spectral energy, entropy), or time–frequency domain using wavelets or Hilbert–Huang transforms. Once preprocessing of signals is complete, accurate feature extraction is accomplished using CNN extractors and wavelet scattering transform. ML models such as SVMs, random forests, and k-nearest neighbors (k-NNs) are employed for classification tasks, e.g., detecting murmurs, valvular heart disease, or classifying normal vs. abnormal heart sounds (Altaf et al., 2023; Kannan, Saikia, Kumar, & Datta, 2025). The work done by Shaima et al. has demonstrated the ability of ML to accurately analyze PCG signals from the wrist, which are often low in intensity and signal-to-noise ratio. Further incorporation of such analysis into mobile devices (mHealth apps) can facilitate quick health diagnoses (Abdelmageed & Elmusrati, 2018). However, integration of such applications into mobile phones may often be challenging due to the limited processing power of the devices.

In this regard, DL models outperform traditional ML models with their ability to automate the feature extraction process, reducing dependence on manual signal processing expertise. CNNs are commonly used for learning directly from spectrograms or raw signals. CNN models, when designed with fewer parameters than traditional DL models, are referred to as lightweight models. These lightweight models, along with self-supervised approaches, have demonstrated strong potential for accurate waveform classification and are well-suited for deployment on standard mobile devices (Ma, Chen, & Ho, 2024). CNNs are found to be effective in extracting spatial features from heart sound recordings, particularly when applied to time–frequency representations such as short-time Fourier transform, wavelet transforms, and Mel spectrograms (Torre-Cruz et al., 2023). In addition, recurrent neural networks (RNNs) have been employed to process raw or minimally preprocessed PCG signals (Alkhodari & Fraiwan, 2021). Another advantage of RNNs is that they can identify relationships and patterns in sequential PCG data due to their memory and ability to learn temporal correlations and analyze sequential inputs. This facilitates feature extraction, heart sound segmentation, and classification (Kannan et al., 2025).

To capture the temporal dynamics of heart sounds, such as the rhythmic sequence of S1 and S2 components, researchers have explored hybrid architectures that combine CNNs with RNNs, particularly using long short-term memory (LSTM)

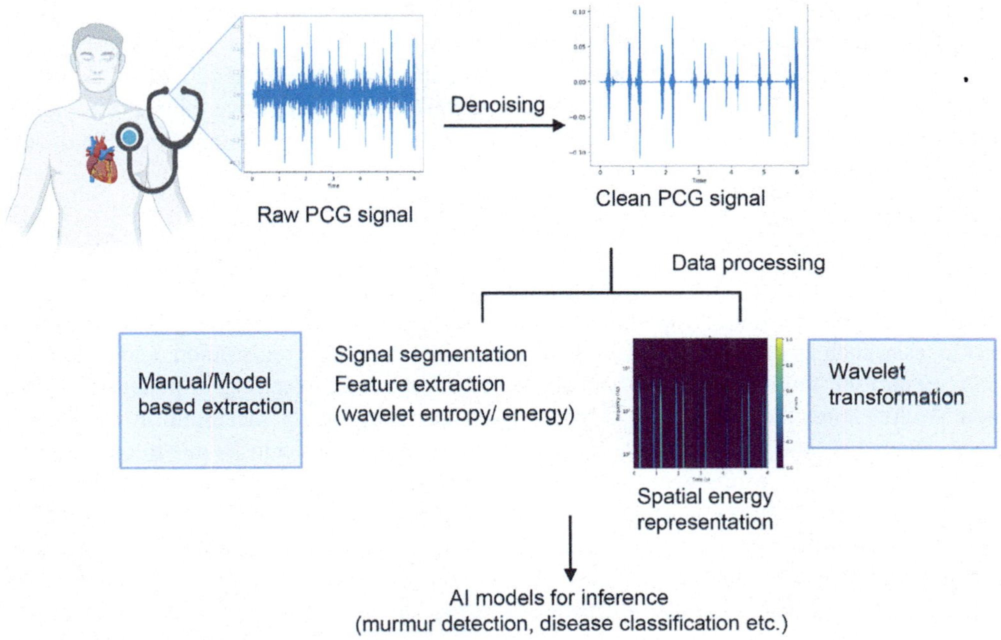

FIG. 13.2 Schematic overview of PCG analysis.

or gated recurrent unit layers. These hybrid models, often referred to as convolutional recurrent neural network (CRNNs), integrate spatial and temporal feature extraction into a single framework (Torre-Cruz et al., 2023). One such model, CardioXNet, employs parallel CNN streams with bidirectional LSTM layers and residual connections to achieve classification accuracies exceeding 99% while maintaining low computational overhead. Such designs are particularly promising for deployment in point-of-care diagnostics, where both performance and efficiency are critical (Shuvo, Ali, Swapnil, Al-Rakhami, & Gumaei, 2021).

While traditional ML approaches such as SVM, k-NN, and random forests remain in use, their reliance on handcrafted features (e.g., Mel-Frequency Cepstral Coefficients (MFCCs), wavelets, entropy metrics) and the need for precise signal segmentation limit their performance and scalability. Nevertheless, these models still serve as valuable benchmarks and may be appropriate in settings with extremely limited computational resources.

3.2.3 Functional and stress test (spirometry/ergometry) analysis

A cardiac stress test is a fundamental diagnostic tool used to evaluate cardiovascular function under conditions of increased physiological demand. By placing the heart under controlled physical exertion, typically through a treadmill or cycle ergometer protocol, clinicians can uncover exercise-induced abnormalities such as ischemia, arrhythmias, or impaired cardiac output that may remain hidden at rest.

Barzilai et al. surveyed the state-of-the-art ML applications in both stress ECG and echocardiography. Across seven studies, models such as XGBoost, SVM, and deep neural networks (DNNs) demonstrated diagnostic performance comparable to experienced clinicians. Notably, ML systems analyzing stress ECG data achieved sensitivities and specificities exceeding 96%, while stress echocardiography models incorporating global longitudinal strain and patient metadata reached specificities of 92.7% in detecting Coronary artery disease (CAD). Furthermore, natural language processing tools have been employed to automatically interpret textual stress echo reports with up to 98% accuracy, streamlining triage and reporting processes (Barzilai et al., 2024).

Among the various forms of stress testing, the cardiopulmonary exercise test (CPET) is considered the gold standard and has become instrumental in evaluating cardiorespiratory fitness by offering a breath-by-breath assessment of gas exchange (VO_2, VCO_2), ventilation, and metabolic thresholds. Zignoli et al. introduced *Pyoxynet*, an open-source ML framework that harnesses a multivariate CNN to detect ventilatory thresholds (VT_1 and VT_2) with expert-level precision. Beyond classification, the framework employs a conditional generative adversarial network (GAN) to generate realistic synthetic CPET traces for data augmentation and benchmarking. To ensure transparency, the system incorporates Shapley-value-based explainability, identifying which respiratory and metabolic parameters most strongly influence the model's decisions (Zignoli, 2023).

Complementing this work, Watanabe et al. developed real-time DNN models capable of dynamically isolating the anaerobic threshold (AT) and predicting peak VO_2 levels during CPET. Their model achieved a high (>0.8) correlation coefficient for both AT detection and peak VO_2 prediction mean absolute error (MAE): 2.25 mL/kg/min), enabling continuous feedback during exercise assessments. These real-time tools not only improve clinical throughput but also open the door to closed-loop exercise supervision systems, where individualized exercise intensity can be dynamically adjusted based on AI feedback (Watanabe et al., 2024).

3.2.4 Cardiac imaging modalities

Imaging has become one of the most critical enablers of DL in healthcare, offering high-dimensional, information-rich datasets that are ideal for training robust AI models. Across imaging applications, a variety of DL architectures have been developed to tackle tasks such as image classification, segmentation, detection, registration, and synthesis. CNNs, among other DL models, remain the foundation of most image-based AI systems, excelling in pattern recognition and spatial feature extraction. Architectures like U-Net and U-Net++ have been particularly successful in medical image segmentation, while ResNet, DenseNet, and EfficientNet are widely used for classification tasks due to their depth, efficiency, and transferability (Rayed et al., 2024). More recently, Vision Transformers and hybrid CNN–transformer models have shown promise in learning global image representations, especially in volumetric and multiview imaging datasets (Shah et al., 2024). GANs are also being utilized for image enhancement, domain adaptation, and synthetic data generation, providing key solutions where annotated data is scarce.

Cardiovascular imaging, including cardiac magnetic resonance imaging (CMR), CT, and ultrasound, is the primary source of high-quality, clinically relevant imaging modalities in cardiovascular care (Alsharqi & Edelman, 2025). Each modality captures distinct physiological and anatomical dimensions of the heart in digital form, facilitating comprehensive diagnosis, risk assessment, and personalized healthcare. With the integration of DL, these modalities have evolved into powerful platforms not only for visualization but also for enabling comprehensive diagnosis, risk stratification, and treatment planning. DL models have been applied for automatic segmentation of the left ventricle (LV), coronary calcium scoring (CCS), and coronary artery centerline extraction, improving reproducibility and reducing analysis time. These advances have been catalyzed by access to large, annotated datasets, especially in CMR, where open-source initiatives such as the ACDC Challenge and the UK Biobank imaging cohort have accelerated the development and benchmarking of DL models (Litjens et al., 2019).

Echocardiography

Echocardiography utilizes ultrasound waves to produce detailed images of the heart and adjacent tissues, offering excellent resolution in both time and space. It is routinely used to assess cardiac chamber size, wall motion, valvular function, and pericardial effusion, making it indispensable in assessing cardiac pathologies. DL models have been extensively applied to automate measurements such as myocardial segmentation, labeling chambers and their dimension measurement, ejection fraction (EF) estimation, strain, left ventricular function, and interpretation, such as detecting valvular dysfunction to identify conditions like hypertrophic cardiomyopathy or amyloidosis (Ghorbani et al., 2020). The DL models either predict the outcomes at each stage of image acquisition, segmentation, and functional quantification (multistep models) or predict the outcome directly from the image (black-box models) (Myhre et al., 2025). Architectures like U-Net and its variants have been adapted to segment the LV and atrium across multiple views with high accuracy. In addition, transformer-based models and attention-guided CNNs have shown promise in distinguishing complex valvular diseases from standard parasternal and apical views (Nolan & Thavendiranathan, 2019).

Cardiac CT

Cardiac CT, particularly coronary CT angiography (CCTA), is another noninvasive method for detecting CAD. It includes CCS, stenosis quantification, plaque characterization, and noninvasive fractional flow reserve (FFR) estimation (CT-FFR). The clinical utility of CCTA lies in its exceptional negative predictive value, making it particularly effective for ruling out significant stenoses in low- to intermediate-risk patients (Wu et al., 2025). With the advancement of AI, automated image acquisition, image reconstruction, noise reduction, coronary motion estimation, and image transformation have become possible (Williams et al., 2024). Wolterink et al. introduced a DL framework for automatic coronary artery calcification (CAC) quantification directly from CCTA scans, removing the need for an additional noncontrast CT (CSCT-Coronary Calcium Scoring for CAD risk assessment) scan traditionally used for calcium scoring. Their approach leveraged a multistage pipeline comprising three CNNs for heart localization in orthogonal planes, followed by ConvPairs, paired convolutional networks designed to first identify CAC-like voxels and then refine classifications. The model utilized both 2.5D and

3D input patches, enabling spatial context learning across slices. It could correctly assign more than 80% of patients to the same cardiovascular risk category as the traditional method, suggesting that CCTA could serve as a standalone modality for CAC scoring (Wolterink, Leiner, Viergever, & Išgum, 2017).

Cardiac MRI

CMR offers unparalleled soft tissue contrast and is considered the gold standard for quantifying myocardial function, fibrosis, ischemia, and cardiac remodeling. It is invaluable in diagnosing cardiomyopathies, myocarditis, and congenital heart disease. However, a major challenge lies in the interpretation of CMR data, as it requires a high level of specialized expertise, which limits its widespread adoption. In this context, DL models offer a significant advantage by enabling uniform, automated analysis of all CMR images, facilitating consistent interpretation, and supporting accurate diagnosis of CVDs. Zhang et al. developed ViTa, a multimodal foundation model that combines CMR data with patient-level information (demographics, metabolic, and lifestyle factors) for comprehensive cardiac assessment. Trained on 3D+T cine MRI from 42,000 UK Biobank participants, ViTa learns a shared latent representation integrating imaging and contextual data, enabling strong performance in disease classification, phenotype prediction, and segmentation. In classifying six cardiac and metabolic diseases, it outperformed imaging-only models like ResNet-50, particularly in detecting minority classes, highlighting its potential for holistic, patient-specific analysis and scalable clinical use (Zhang et al., 2025). Similarly, Zreik et al. proposed a DL method to detect functionally significant coronary artery stenosis from rest CCTA scans, reducing reliance on invasive FFR measurements. The approach uses a multiscale CNN to segment the left ventricular myocardium, a convolutional autoencoder to encode it, and spatial clustering to extract localized ischemia features for SVM classification. This noninvasive technique bypasses coronary anatomy analysis and could help avoid unnecessary invasive procedures (Zreik et al., 2018).

3.2.5 Omics in CVD

While imaging and sound data are widely used in CVD diagnosis, there remains a pressing need to unravel its complex pathophysiology, particularly at the molecular level. In this regard, the emergence of omics technologies—encompassing the study of genes, proteins, metabolites, and lipids—has revolutionized cardiovascular research and advanced the precision of personalized medicine.

(i) Genomics, the study of an organism's DNA, has been applied to identify genetic loci and their variations associated with cardiovascular traits, aiding diagnosis and prognosis. For instance, variations in single-nucleotide polymorphisms within the lipoprotein(a) gene region affect circulating lipoprotein concentrations and have a strong association with aortic valve calcification (Thanassoulis et al., 2013). Genetic testing in asymptomatic individuals with no prior CVD history has also revealed variants associated with conditions such as cardiomyopathy (Lacaze et al., 2021).

(ii) Transcriptomics, which examines mRNA expression, has identified differentially expressed genes linked to pathways driving myocardial infarction and heart failure. At the single-cell level, transcriptome profiling has revealed cellular heterogeneity, such as fibroblasts and immune cell subtypes, and their roles in disease progression (Nie et al., 2025).

(iii) Proteomics adds another crucial layer by quantifying protein expression and posttranslational modifications (e.g., glycosylation, phosphorylation). Proteomic platforms have shown strong potential in identifying highly predictive circulating protein biomarkers for CVD outcomes and mortality; for example, numerous biomarkers were discovered from plasma samples of 3523 Framingham Heart Study participants (Ho et al., 2018).

(iv) Metabolomics and lipidomics extend this insight by characterizing small molecules and lipid species, identifying them as therapeutic targets or disease markers. A notable lipidomics study developed a Lipidomics Risk Score based on over 700 lipid species, which reclassified intermediate-risk individuals and outperformed the traditional FRS (Wu et al., 2024).

Together, these omics approaches offer immense potential for improving risk stratification, predicting disease severity, elucidating pathogenic pathways, and enabling personalized phenotyping and individualized drug-response monitoring.

However, omics technologies generate vast, complex datasets that demand advanced analytical strategies. AI is pivotal in detecting analytes, processing data, and discovering biomarkers for diagnosis and prognosis, thereby improving accuracy and risk prediction (Castañé et al., 2021; Chi et al., 2024; Mann, Kumar, Zeng, & Strauss, 2021; Vilhekar & Rawekar, 2024). ML algorithms such as random forest, SVM, decision trees, logistic regression, discriminant analysis, and k-NN have been widely applied to classify disease states, identify key biomolecules, and assess prognostic factors (Quazi, 2022). Single-omics analyses, however, provide only a limited snapshot of complex biological systems and are susceptible to technical biases from sample preparation, instrumentation, and interpretation. Multiomics integration, which combines genomics, transcriptomics, proteomics, metabolomics, and other molecular layers, offers a holistic view of cellular function and disease mechanisms, especially when merged with clinical information in precision medicine frameworks. This integration enables more accurate patient stratification, disease risk prediction, and tailored therapeutic planning.

While traditional statistical tools like correlation-based methods (BindSC), matrix factorization (iNMF, LIGER), Bayesian approaches (MOFA), and nearest-neighbor algorithms (fastMNN, Seurat v4) are useful, they often fail to capture nonlinear relationships or subtle cross-omics patterns. AI overcomes these limitations through advanced methods such as autoencoders, graph convolutional networks (GCNs), ensemble learning, and contrastive learning, enabling effective data reduction, feature extraction, and predictive modeling while enhancing interpretability through explainable AI (Wu & Xie, 2025). There are several cutting-edge AI frameworks that have been developed to facilitate the successful integration of multiomics datasets with interactive visualization. For example, MOGONET applies GCNs for omics-specific learning and integrates similarity networks to boost classification (Wang et al., 2021). ASTERICS, a web-based platform, supports multiomics integration with tools for data processing, normalization, statistical analysis, and interactive visualization, making it a preferred choice for researchers.

In CVD research, multiomics integration has been applied to improve diagnosis and risk prediction. For example, DeGroat et al. combined transcriptomics and genomics data with clinical information to generate patient-specific profiles. They used traditional methods such as k-NN imputation and DESeq2 for data processing, followed by ML algorithms like random forest and logistic regression to classify patients based on CVD risk (DeGroat et al., 2024). Due to the diversity of the disease and the generation of vast data, a comprehensive resource is still in high demand. To address this need, Qian et al. developed the CVD Atlas (https://ngdc.cncb.ac.cn/cvd), a comprehensive resource that aggregates multiomics data through manual curation, extensive analyses, and integration of existing databases (Qian et al., 2025). The current version contains 215,333 associations from 308 publications, 652 datasets, and 7 databases, covering 190 diseases and 44 traits across different omics layers. It features an interactive knowledge graph linking disease–gene associations, analysis tools, and a user-friendly web interface for browsing, searching, and downloading data, making it a valuable resource for CVD research and healthcare applications.

3.3 Advanced risk prediction models

Cardiovascular risk prediction scores have long been foundational tools in preventive cardiology, guiding decisions on statin therapy, lifestyle modification, and long-term monitoring. Commonly used models, such as Framingham, QRISK, and the ESC SCORE, estimate the risk of events such as myocardial infarction or stroke within a specified timeframe by using readily available factors, including age, sex, blood pressure, cholesterol levels, and smoking status. While clinically useful, these tools typically assume linear associations among a restricted number of risk factors and outcomes, which can reduce accuracy and generalizability in diverse, real-world populations (Talha, Elkhoudri, Hilali, & Mehata, 2024).

AI-driven risk prediction models represent a next-generation approach by incorporating richer, multimodal datasets and capturing complex, nonlinear interactions among risk variables (Liu, Krentz, Huo, & Ćurčin, 2025). These models account for interdependencies between clinical, behavioral, and biological factors, enabling more precise individual risk estimation and supporting personalized care. This fusion of ML and DL enables robust end-to-end pipelines that not only automate feature extraction but also synthesize insights across modalities. For example, neural networks trained on retinal images (Poplin et al., 2018) or ECG waveforms (Hughes et al., 2023) have shown strong performance in predicting myocardial infarction, stroke, and long-term cardiovascular mortality. Ensemble-based methods, such as SCORE2-AI, enhance traditional scores by integrating AI-derived features and region-specific variables, improving calibration and discrimination across diverse populations (Liu Tianya et al., 2025).

Given the biological heterogeneity of CVD, integrating multiomics data into risk prediction frameworks is gaining momentum. A review by Lin et al. outlined ML models, such as random forests, gradient boosting machines (XGBoost, LightGBM), Bayesian ridge regression, LASSO, and multilayer perceptrons, trained on single-layer or multilayer omics datasets to predict and prevent CVD (Lin et al., 2025). One prominent example is the development of PRSs, which aggregate thousands of genetic variants identified through genome-wide association studies to quantify inherited susceptibility to events like myocardial infarction. When integrated into multimodal frameworks alongside clinical, imaging, and lifestyle data, omics-based predictors significantly enhance the precision of risk stratification and therapeutic targeting (Khera et al., 2018). An important example is the work by Zhang et al., which leveraged data from 24,308 UK Biobank participants to create multitask DL models, ProNet and MetNet. These models simultaneously generate disease-specific proteomic (ProScore) and metabolomic (MetScore) risk scores for nine prevalent cardiovascular events by analyzing 168 metabolites and 2920 proteins. Cox proportional hazards models that combined PRSs, MetScore, ProScore, and clinical risk factors showed robust predictive accuracy for the onset of these events (Luo et al., 2025).

For clinical adoption, interpretability and transparency are critical. Techniques like SHAP (SHapley Additive exPlanations) and saliency mapping help identify the most influential variables in a model's predictions, enabling physicians to understand and trust the underlying logic. In addition, attention to algorithmic fairness—ensuring model generalizability across sex, age, ethnicity, and socioeconomic status—addresses known limitations of traditional scores.

Despite their promise, AI-based risk prediction models face challenges in validation, regulatory approval, data standardization, and integration into clinical workflows. Ongoing prospective studies and real-world deployments will be essential to establish these models as accurate, equitable, and actionable tools in cardiovascular care.

4 Case studies: Real-world impact of AI in cardiovascular care

The translation of AI from research to clinical practice is reshaping CVD management on a global scale. These case studies illustrate how pioneering research, clinical implementations, and industry-academia collaborations are driving transformative change, while also highlighting the challenges and lessons learned along the way.

4.1 Mayo Clinic's asymptomatic heart failure detection

The Mayo Clinic's cardiology research team has developed a breakthrough DL model capable of identifying asymptomatic left ventricular dysfunction (ALVD)—a silent yet life-threatening precursor to heart failure—by analyzing standard 12-lead ECGs. Traditionally, ALVD can only be confirmed via echocardiography or biomarkers like BNP, both of which are costly or underutilized in routine outpatient settings.

Mayo's model, trained on over 625,000 ECG–echocardiogram pairs, uses a CNN to detect early signs of reduced EF (<35%) that clinicians would typically miss. In real-world deployment across Minnesota's primary care network, the AI flagged at-risk but asymptomatic patients, prompting follow-up echocardiography and preventative intervention. This led to the initiation of guideline-directed therapy, such as angiotensin-converting enzyme (ACE) inhibitors and beta-blockers, before overt symptoms developed. Over an 18-month period, the tool helped identify >1200 high-risk patients, analysis of clinical outcomes showed a 22% reduction in hospitalizations related to heart failure in the identified cohort. The study emphasized not only the predictive accuracy (AUC ~0.93) but also the operational feasibility of embedding AI into routine clinical pathways (Attia et al., 2019).

4.2 Zebra Medical Vision's CCS

Zebra Medical Vision, a health tech company based in Israel, has developed a DL model that automates CAC scoring using noncontrast CT scans. CAC is a well-established marker for atherosclerosis and a key metric for evaluating cardiovascular risk, but traditional scoring is manual and time-intensive, requiring trained radiologists. AI-driven CAC detection leverages architectures such as CNNs and U-Net segmentation models to identify and quantify calcified lesions in the coronary arteries with high precision (Vivekanandan et al., 2024).

The US Food and Drug Administration (FDA) in 2024 approved Zebra's system, which demonstrated performance on par with expert radiologists. The model automatically segments the coronary arteries and quantifies calcium burden, delivering results in real-time. This technology is now integrated into imaging workflows at over 1200 healthcare facilities, allowing for opportunistic screening during routine chest CT scans performed for other indications. In a pilot program within the US Veterans Health Administration, Zebra's AI increased guideline-concordant statin prescription rates by 28% among intermediate-risk patients. Importantly, it also reduced radiologist workload by 40%, freeing up time for more complex diagnostic tasks and improving care throughput.

4.3 Eko Health's reinvention of the stethoscope with AI

California-based Eko Health has transformed the traditional stethoscope into an intelligent diagnostic tool. Its flagship device, the Eko DUO, combines a digital stethoscope with an ECG sensor (Taye et al., 2024) and cloud-connected AI software. The system uses ML to detect abnormal heart sounds, including murmurs associated with valvular heart disease, in real time.

The AI model was trained on a diverse dataset of annotated heart sounds and achieved 94% sensitivity in identifying murmurs. In a 2025 multicenter study involving over 50 US clinics, including many in rural and underserved areas, Eko's device helped detect 89% of previously missed valvular pathologies during routine physical exams. This led to earlier referrals, especially for cases of aortic stenosis and mitral regurgitation. Crucially, the Eko DUO also reduced unnecessary echocardiography referrals by 35%, demonstrating both cost-efficiency and clinical effectiveness. Its portability and FDA clearance make it especially valuable in primary care settings, telemedicine, and mobile clinics.

The above case studies provide only a glimpse into the growing body of evidence showcasing how algorithmic research is being translated into real-world impact in cardiovascular care. From silent condition detection to workflow optimization and equitable access, AI is redefining the possibilities in preventive, diagnostic, and therapeutic cardiology. For instance, AliveCor's KardiaMobile uses *wearable ECG data* to detect AF and other arrhythmias in real time, making cardiac monitoring accessible outside clinical settings (https://alivecor.com/). Caption Health (formerly Bay Labs) applies DL to *ultrasound imaging data*, enabling even novice users to capture diagnostic-quality echocardiograms with real-time guidance and interpretation. Tempus integrates *EHRs*, *genomic data*, and *medical imaging* to personalize cardiovascular treatment pathways, particularly in the cardiooncology domain (https://www.tempus.com/). Meanwhile, DeepHeart, acquired by Fitbit, leverages *wearable sensor data* such as heart rate, step count, and sleep patterns to predict chronic conditions like hypertension and sleep apnea using RNNs (Chang, Li, Xing, Zhu, & Tu, 2021). These initiatives reflect the diversity of data modalities—from ECGs and imaging to wearables and multiomics—that AI can harness to deliver actionable insights. While it is beyond the scope to catalog every impactful deployment, the growing convergence of clinical data and ML continues to redefine how CVD is detected, monitored, and managed worldwide.

5 Conclusion

These case studies illustrate that AI in cardiovascular care is no longer confined to academic prototypes; it is delivering measurable clinical benefits across diverse healthcare systems. Yet, these advances come with challenges. Algorithmic bias persists, as shown by a 2024 *NEJM AI* study, where AI-powered pulse oximetry overestimated oxygen saturation by 3%–5% in darker-skinned individuals, prompting renewed calls for more representative training datasets (Liu et al., 2025). Regulatory hurdles also slow adoption: the 2025 WHO report found that only 23 countries have AI-specific medical device regulations, with platforms like Cardiologs facing multiyear approval delays across jurisdictions. Workflow integration is equally critical; a 2024 MGH study found clinician adoption rates were twice as high for AI tools embedded in EHRs compared to standalone systems (Ratwani, Bates, & Classen, 2024).

Ultimately, success in AI-driven cardiovascular care depends not just on model accuracy but on ecosystem readiness, including regulatory clarity, clinician trust, system interoperability, and ethically sourced, representative data. Designed with these principles, AI can deliver more than efficiency gains: it can enable equitable, personalized, and preventive cardiovascular care. As the global burden of CVD grows, especially in underserved populations, these pioneering efforts offer a blueprint for transforming AI from a tool into a catalyst for global health equity.

Acknowledgements

The authors acknowledge the use of AI-powered language models, including ChatGPT and Perplexity, in assisting with the drafting of this chapter. In keeping with the theme of Artificial Intelligence in Cardiovascular Disease, the development of this chapter itself serves as a practical example of AI's capacity to synthesize findings from diverse research sources into a coherent, accessible narrative, illustrating how AI can support knowledge translation alongside its clinical applications.

References

Abdelmageed, S., & Elmusrati, M. (2018). Phonocardiogram based diagnosis using machine learning: parametric estimation with multivariant classification. *Bioscience & Engineering: An International Journal*. 2349848X. *5*(1–4), 01–06. doi:10.5121/bioej.2018.5401.

Ahsan, M. M., Luna, S. A., & Siddique, Z. (2022). Machine-learning-based disease diagnosis: a comprehensive review. *Healthcare*. 2227-9032. *10*(3), 541. doi:10.3390/healthcare10030541.

Alkhodari, M., & Fraiwan, L (2021). Convolutional and recurrent neural networks for the detection of valvular heart diseases in phonocardiogram recordings. *Computer Methods and Programs in Biomedicine*. 01692607. *200*, 105940. doi:10.1016/j.cmpb.2021.105940.

Alsharqi, M., & Edelman, E. R. (2025). Artificial intelligence in cardiovascular imaging and interventional cardiology: emerging trends and clinical implications. *Journal of the Society for Cardiovascular Angiography & Interventions*. 27729303. *4*(3), 102558. doi:10.1016/j.jscai.2024.102558.

Altaf, A., Mahdin, H., Alive, A. M., Ninggal, M. I. H., Altaf, A., & Javid, I. (2023). Systematic review for phonocardiography classification based on machine learning. *International Journal of Advanced Computer Science and Applications*. 21565570. *14*(8), 806–817. Retrieved from http://thesai.org/Publications/Archives?code=IJACSA. doi:10.14569/IJACSA.2023.0140889.

Ani, M. A., Garas, G., Hollingshead, J., Cheetham, D., Athanasiou, T., & Patel, V. (2022). Which electronic health record system should we use? A systematic review. *Medical Principles and Practice*. 1011-7571. *31*(4), 342–351. doi:10.1159/000525135.

Attia, Z. I., Kapa, S., Lopez-Jimenez, F., McKie, P. M., Ladewig, D. J., Satam, G., … Friedman, P. A. (2019). Screening for cardiac contractile dysfunction using an artificial intelligence–enabled electrocardiogram. *Nature Medicine*. 1546170X. *25*(1), 70–74. Retrieved from http://www.nature.com/nm/index.html. doi:10.1038/s41591-018-0240-2.

Attia, Z. I., Noseworthy, P. A., Lopez-Jimenez, F., Asirvatham, S. J., Deshmukh, A. J., Gersh, B. J., … Friedman, P. A. (2019). An artificial intelligence-enabled ECG algorithm for the identification of patients with atrial fibrillation during sinus rhythm: a retrospective analysis of outcome prediction. *The Lancet*. 1474547X. *394*(10201), 861–867. Retrieved from http://www.journals.elsevier.com/the-lancet/. doi:10.1016/S0140-6736(19)31721-0.

Barzilai, D. H., Cohen-Shelly, M., Sorin, V., Zimlichman, E., Massalha, E., Allison, T. G., & Klang, E. (2024). Machine learning in cardiac stress test interpretation: a systematic review. *European Heart Journal - Digital Health*. 26343916. *5*(4), 401–408. Retrieved from https://academic.oup.com/ehjdh. doi:10.1093/ehjdh/ztae027.

Bloom, D., Chisholm, D., Llopis, E. J., Prettner, K., Stein, A., & Feigl, A. (2011). *From burden to "Best Buys": Reducing the economic impact of non-communicable disease in low and middle-income countries*, World Economic Forum (WEF) and World Health Organization (WHO).

Brayne, C., & Moffitt, T. E. (2022). The limitations of large-scale volunteer databases to address inequalities and global challenges in health and aging. *Nature Aging*. 26628465. *2*(9), 775–783. Retrieved from https://www.nature.com/natαging/. doi:10.1038/s43587-022-00277-x.

Castañé, H., Baiges-Gaya, G., Hernández-Aguilera, A., Rodríguez-Tomàs, E., Fernández-Arroyo, S., Herrero, P., … Joven, J. (2021). Coupling machine learning and lipidomics as a tool to investigate metabolic dysfunction-associated fatty liver disease. A general overview. *Biomolecules*. 2218-273X. *11*(3), 473. doi:10.3390/biom11030473.

Chahal, A. A. C., Alahdab, F., Asatryan, B., Addison, D., Aung, N., Chung, M. K., … Armoundas, A. A. (2025). Data interoperability and harmonization in cardiovascular genomic and precision medicine. *Circulation: Genomic and Precision Medicine*. 25748300. *18*(3). Retrieved from https://www.ahajournals.org/journal/circgen. doi:10.1161/CIRCGEN.124.004624.

Chang, X., Li, G., Xing, G., Zhu, K., & Tu, L. (2021). DeepHeart: a deep learning approach for accurate heart rate estimation from PPG signals. *ACM Transactions on Sensor Networks*. 15504867. *17*(2). Retrieved from http://dl.acm.org/citation.cfm?id=J981. doi:10.1145/3441626.

Chi, J., Shu, J., Ming, L., Mudappathi, R., Jin, Y., Lewis, F., … Haiwei, G. (2024). Artificial intelligence in metabolomics: a current review. *TrAC Trends in Analytical Chemistry*. 01659936. *178*, 117852. doi:10.1016/j.trac.2024.117852.

Dahlöf, B. (2010). Cardiovascular disease risk factors: epidemiology and risk assessment. *The American Journal of Cardiology*. 00029149. *105*(1), 3A. doi:10.1016/j.amjcard.2009.10.007.

Damen, J. A. A. G., Hooft, L., Schuit, E., Debray, T. P. A., Collins, G. S., Tzoulaki, I., … Moons, K. G. M. (2016). Prediction models for cardiovascular disease risk in the general population: systematic review. *BMJ (Online)*, 17561833. *353*. doi:10.1136/bmj.i2416.

De Filippis, R., Foysal, A., Rocco, V., Guglielmo, R., Sabatino, B., Pietropaoli, A., & Ferracuti, S. (2024). *The risk perspective of AI in healthcare: GDPR and GELSI framework (Governance, Ethical, Legal and Social Implications) and the new European AI Act* (Pacini Editore S.r.l. (official publisher of Italian Journal of Psychiatry)). doi:10.36180/2421-4469-2024-4.

DeGroat, W., Abdelhalim, H., Peker, E., Sheth, N., Narayanan, R., Zeeshan, S., … Ahmed, Z. (2024). Multimodal AI/ML for discovering novel biomarkers and predicting disease using multi-omics profiles of patients with cardiovascular diseases. *Scientific Reports*. 2045-2322. *14*(1). doi:10.1038/s41598-024-78553-6.

Di Cesare, M., McGhie, D. V., Perel, P., Mwangi, J., Taylor, S., Pervan, B., … Pinto, F. J. (2024). The heart of the world. *Global Heart*. 22118179. *19*(1). Retrieved from https://globalheartjournal.com/articles/10.5334/gh.1288. doi:10.5334/gh.1288.

Dunbar, S. B., Khavjou, O. A., Bakas, T., Hunt, G., Kirch, R. A., Leib, A. R., … Whitsel, L. P. (2018). Projected costs of informal caregiving for cardiovascular disease: 2015 to 2035: a policy statement from the American Heart Association. *Circulation*. 15244539. *137*(19), e558. doi:10.1161/CIR.0000000000000570.

Ghorbani, A., Ouyang, D., Abid, A., He, B., Chen, J. H., Harrington, R. A., … Zou, J. Y. (2020). Deep learning interpretation of echocardiograms. *Npj Digital Medicine*. 23986352. *3*(1). Retrieved from https://www.nature.com/npjdigitalmed/. doi:10.1038/s41746-019-0216-8.

Goldsborough, E., Tasdighi, E., & Blaha, M. J. (2023). Assessment of cardiovascular disease risk: a 2023 update. *Current Opinion in Lipidology*. 14736535. *34*(4), 162–173. Retrieved from http://journals.lww.com/co-lipidology/pages/default.aspx. doi:10.1097/MOL.0000000000000887.

Gupta, U., Paluru, N., Nankani, D., Kulkarni, K., & Awasthi, N. (2024). A comprehensive review on efficient artificial intelligence models for classification of abnormal cardiac rhythms using electrocardiograms. *Heliyon*. 24058440. *10*(5), e26787. doi:10.1016/j.heliyon.2024.e26787.

Gustavson, K., von Soest, T., Karevold, E., & Røysamb, E. (2012). Attrition and generalizability in longitudinal studies: findings from a 15-year population-based study and a Monte Carlo simulation study. *BMC Public Health*. 1471-2458. *12*(1). doi:10.1186/1471-2458-12-918.

Ho, J. E., Lyass, A., Courchesne, P., Chen, G., Liu, C., Yin, X., … Levy, D. (2018). Protein biomarkers of cardiovascular disease and mortality in the community. *Journal of the American Heart Association*. 20479980. *7*(14). Retrieved from http://jaha.ahajournals.org/content/by/year. doi:10.1161/JAHA.117.008108.

Hu, J., Shen, L., & Sun, G. (2018). Squeeze-and-excitation networks. In *Proceedings of the IEEE Computer Society Conference on Computer Vision and Pattern Recognition* (pp. 7132–7141). IEEE Computer Society, United Kingdom. doi:10.1109/CVPR.2018.00745.

Hughes, J. W., Tooley, J., Soto, J. T., Ostropolets, A., Poterucha, T., Christensen, M. K., … Perez, M. V. (2023). A deep learning-based electrocardiogram risk score for long term cardiovascular death and disease. *Npj Digital Medicine*. 23986352. *6*(1). Retrieved from https://www.nature.com/npjdigitalmed/. doi:10.1038/s41746-023-00916-6.

Iandola, F., Han, S., Moskewicz, M., Ashraf, K., Dally, W., & Keutzer, K. (2016). *SqueezeNet: AlexNet-level accuracy with 50x fewer parameters and ESCAPED_CHARACTER_LESS_THAN0.5MB model size*, arXiv (preprint server). doi:10.48550/arXiv.1602.07360.

Jackson, R., Lawes, C. M. M., Bennett, D. A., Milne, R. J., & Rodgers, A. (2005). Treatment with drugs to lower blood pressure and blood cholesterol based on an individual's absolute cardiovascular risk. *Lancet*. 01406736. *365*(9457), 434–441. Retrieved from https://www.sciencedirect.com/science/journal/01406736. doi:10.1016/s0140-6736(05)17833-7.

Kannan, A., Saikia, M. J., Kumar, S., & Datta, S. (2025). Detection of valvular heart diseases from PCG signals using machine and deep learning models: a review. *IEEE Access*.21693536. *13*, 110344–110364. Retrieved from http://ieeexplore.ieee.org/xpl/RecentIssue.jsp?punumber=6287639. doi:10.1109/ACCESS.2025 3583263.

Kashyap, G., & Roy, M. (2023). A comparative study of Electronic Health Record (EHR) systems in US Healthcare—A systematic review. *IUJ Journal of Management*, *11*, 1–14.

Khera, A. V., Chaffin, M., Aragam, K. G., Haas, M. E., Roselli, C., Choi, S. H., … Kathiresan, S. (2018). Genome-wide polygenic scores for common diseases identify individuals with risk equivalent to monogenic mutations. *Nature Genetics*. 15461718. *50*(9), 1219–1224. Retrieved from http://www.nature.com/ng/index.html. doi:10.1038/s41588-018-0183-z.

Lacaze, P., Sebra, R., Riaz, M., Ingles, J., Tiller, J., Thompson, B. A., … McNeil, J. J. (2021). Genetic variants associated with inherited cardiovascular disorders among 13,131 asymptomatic older adults of European descent. *Npj Genomic Medicine*. 20567944. *6*(1). Retrieved from https://www.nature.com/npjgenmed/. doi:10.1038/s41525-021-00211-x.

Lai, C., Zhou, S., & Trayanova, N. A. (2021). Optimal ECG-lead selection increases generalizability of deep learning on ECG abnormality classification. *Philosophical Transactions of the Royal Society A: Mathematical, Physical and Engineering Sciences*. 1364503X. *379*(2212). Retrieved from http://rsta.royalsocietypublishing.org/. doi:10.1098/rsta.2020.0258.

Laslett, L. J., Alagona, P., Clark, B. A., Drozda, J. P., Saldivar, F., Wilson, S. R., … Hart, M. (2012). The worldwide environment of cardiovascular disease: prevalence, diagnosis, therapy, and policy issues: a report from the American college of cardiology. *Journal of the American College of Cardiology*. 15583597. *60*(25), S1. Retrieved from www.elsevier.com/locate/jacc. doi:10.1016/j.jacc.2012.11.002.

Lin, M., Guo, J., Zhilin, G., Tang, W., Tao, H., You, S., … Jia, P. (2025). Machine learning and multi-omics integration: advancing cardiovascular translational research and clinical practice. *Journal of Translational Medicine*. 1479-5876. *23*(1). doi:10.1186/s12967-025-06425-2.

Litjens, G., Ciompi, F., Wolterink, J. M., de Vos, B. D., Leiner, T., Teuwen, J., & Išgum, I. (2019). State-of-the-Art deep learning in cardiovascular image analysis. *JACC: Cardiovascular Imaging*. 18767591. *12*(8), 1549–1565. Retrieved from http://www.elsevier.com. doi:10.1016/j.jcmg.2019.06.009.

Liu, T., Krentz, A. J., Huo, Z., & Ćurčin, V. (2025). Opportunities and challenges of cardiovascular disease risk prediction for primary prevention using machine learning and electronic health records: a systematic review. *Reviews in Cardiovascular Medicine*. 21538174. *26*(4). Retrieved from https://www.imrpress.com/journal/RCM/26/4/10.31083/RCM37443/pdf. doi:10.31083/RCM37443.

Liu, T., Krentz, A., Lu, L., & Curcin, V. (2025). Machine learning based prediction models for cardiovascular disease risk using electronic health records data: systematic review and meta-analysis. *European Heart Journal - Digital Health*. 26343916. *6*(1), 7–22. Retrieved from https://academic.oup.com/ehjdh. doi:10.1093/ehjdh/ztae080.

Luo, Y., Zhang, Q., Zhang, N., Yang, J., Cui, M., Tsoi, K., & Lip, Y. (2025). *AI-based multiomics profilng for personalized prediction of cardiovascular disease: A prospective UK Biobank study*, Nature Communications.

Ma, S., Chen, J., & Ho, J. W. K. (2024). An edge-device-compatible algorithm for valvular heart diseases screening using phonocardiogram signals with a lightweight convolutional neural network and self-supervised learning. *Computer Methods and Programs in Biomedicine*. 01692607. *243*, 107906. doi:10.1016/j.cmpb.2023.107906.

Mahmood, S. S., Levy, D., Vasan, R. S., & Wang, T. J. (2014). The Framingham Heart Study and the epidemiology of cardiovascular disease: a historical perspective. *The Lancet*. 1474547X. *383*(9921), 999–1008. Retrieved from http://www.journals.elsevier.com/the-lancet/. doi:10.1016/S0140-6736(13)61752-3.

Mann, M., Kumar, C., Zeng, W. F., & Strauss, M. T. (2021). Artificial intelligence for proteomics and biomarker discovery. *Cell Systems*. 24054720. *12*(8), 759–770. Retrieved from http://www.journals.elsevier.com/cell-systems/. doi:10.1016/j.cels.2021.06.006.

Martínez-Sellés, M., & Marina-Breysse, M. (2023). Current and future use of artificial intelligence in electrocardiography. *Journal of Cardiovascular Development and Disease*. 2308-3425. *10*(4), 175. doi:10.3390/jcdd10040175.

Myhre, P. L., Grenne, B., Asch, F. M., Delgado, V., Khera, R., Lafitte, S., … Lam, C. S. P. (2025). Artificial intelligence-enhanced echocardiography in cardiovascular disease management. *Nature Reviews Cardiology*, 17595010. Retrieved from https://www.nature.com/nrcardio/archive/. doi:10.1038/s41569-025-01197-0.

Nie, W., Zhao, Z., Xiahou, Z., Zhang, J., Liu, Y., Wang, Y., & Wang, Z. (2025). Single-cell RNA sequencing reveals the potential role of Postn(+) fibroblasts in promoting the progression of myocardial fibrosis after myocardial infarction. *Scientific Reports*. 2045-2322. *15*(1). doi:10.1038/s41598-025-04990-6.

Nolan, M. T., & Thavendiranathan, P. (2019). Automated quantification in echocardiography. *JACC: Cardiovascular Imaging*. 18767591. *12*(6), 1073–1092. Retrieved from http://www.elsevier.com. doi:10.1016/j.jcmg.2018.11.038.

Nordo, A. H., Levaux, H. P., Becnel, L. B., Galvez, J., Rao, P., Stem, K., … Kush, R. D. (2019). Use of EHRs data for clinical research: historical progress and current applications. *Learning Health Systems*. 23796146. *3*(1). Retrieved from http://onlinelibrary.wiley.com/journal/23796146. doi:10.1002/lrh2.10076.

Poplin, R., Varadarajan, A. V., Blumer, K., Liu, Y., McConnell, M. V., Corrado, G. S., … Webster, D. R. (2018). Prediction of cardiovascular risk factors from retinal fundus photographs via deep learning. *Nature Biomedical Engineering*. 2157846X. *2*(3), 158–164. Retrieved from www.nature.com/natbiomedeng/. doi:10.1038/s41551-018-0195-0.

Qian, Q., Xue, R., Chenle, X., Wang, F., Zeng, J., & Xiao, J. (2025). CVD Atlas: a multi-omics database of cardiovascular disease. *Nucleic Acids Research*. 0305-1048. *53*(D1), D1348. doi:10.1093/nar/gkae848.

Quazi, S. (2022). Artificial intelligence and machine learning in precision and genomic medicine. *Medical Oncology*. 1559131X. *39*(8). Retrieved from https://www.springer.com/journal/12032. doi:10.1007/s12032-022-01711-1.

Ramirez, E., Ruiperez-Campillo, S., Casado-Arroyo, R., Merino, J. L., Vogt, J. E., Castells, F., & Millet, J. (2024). The art of selecting the ECG input in neural networks to classify heart diseases: a dual focus on maximizing information and reducing redundancy. *Frontiers in Physiology*. 1664042X. *15*. Retrieved from http://www.frontiersin.org/Physiology/archive/. doi:10.3389/fphys.2024.1452829.

Ratwani, R. M., Bates, D. W., & Classen, D. C. (2024). Patient safety and artificial intelligence in clinical care. *JAMA Health Forum*. 26890186. *5*(2). Retrieved from https://www.jamanetwork.com/journals/jama-health-forum. doi:10.1001/jamahealthforum.2023.5514.

Rayed, M. E., Islam, S. M. S., Niha, S. I., Jim, J. R., Kabir, M. M., & Mridha, M. F. (2024). Deep learning for medical image segmentation: state-of-the-art advancements and challenges. *Informatics in Medicine Unlocked*. 23529148. *47*. Retrieved from https://www.sciencedirect.com/science/journal/23529148. doi:10.1016/j.imu.2024.101504.

Samanidis, G. (2024). Current challenges in diagnosis and treatment of cardiovascular disease. *Journal of Personalized Medicine*. 2075-4426. *14*(8), 786. doi:10.3390/jpm14080786.

Shah, S. A., Taj, I., Usman, S. M., Shah, S. N. H., Imran, A. S., & Khalid, S. (2024). A hybrid approach of vision transformers and CNNs for detection of ulcerative colitis. *Scientific Reports*. 20452322. *14*(1). Retrieved from https://www.nature.com/srep/. doi:10.1038/s41598-024-75901-4.

Shuvo, S. B., Ali, S. N., Swapnil, S. I., Al-Rakhami, M. S., & Gumaei, A. (2021). CardioXNet: a novel lightweight deep learning framework for cardiovascular disease classification using heart sound recordings. *IEEE Access*.21693536. *9*, 36955–36967. Retrieved from http://ieeexplore.ieee.org/xpl/RecentIssue.jsp?punumber=6287639. doi:10.1109/ACCESS.2021.3063129.

Smith, S. C., Collins, A., Ferrari, R., Holmes, D. R., Logstrup, S., McGhie, D. V., … Zoghbi, W. A. (2012). Our time: a call to save preventable death from cardiovascular disease (heart disease and stroke). *Journal of the American College of Cardiology*. 15583597. *60*(22), 2343–2348. Retrieved from www.elsevier.com/locate/jacc. doi:10.1016/j.jacc.2012.08.962.

Talha, I., Elkhoudri, N., Hilali, A., & Mehata, A. (2024). major limitations of cardiovascular risk scores. *Cardiovascular Therapeutics*. 1755-5914. *2024*(1). doi:10.1155/2024/4133365.

Tan, M., & Le, Q. V. (2019). EfficientNet: rethinking model scaling for convolutional neural networks. In *36th International Conference on Machine Learning, ICML 2019* (pp. 10691–10700). International Machine Learning Society (IMLS) United States.

Taye, A., Yemane, S., Negash, E., Minwuyelet, Y., & Tofik, N. (2024). *Design of an AI-enhanced digital stethoscope: Advancing cardiovascular diagnostics through smart auscultation*, arXiv (preprint server).

Teo, Z. L., Jin, L., Li, S., Miao, D., Zhang, X., Ng, W. Y., … Ting, D. S. W. (2024). Federated machine learning in healthcare: a systematic review on clinical applications and technical architecture. *Cell Reports Medicine*. 26663791. *5*(2). Retrieved from https://www.sciencedirect.com/science/journal/26663791. doi:10.1016/j.xcrm.2024.101419.

Thanassoulis, G., Campbell, C. Y., Owens, D. S., Smith, J. G., Smith, A. V., Peloso, G. M., … Post, W. S. (2013). Genetic associations with valvular calcification and aortic stenosis. *New England Journal of Medicine*. 15334406. *368*(6), 503–512. Retrieved from http://www.nejm.org/doi/pdf/10.1056/NEJMoa1109034. doi:10.1056/NEJMoa1109034.

The Lancet Neurology. (2023). The global challenge of hypertension. *The Lancet Neurology*. 14744465, *22*(12), 1087. Retrieved from http://www.journals.elsevier.com/the-lancet-neurology/. doi:10.1016/S1474-4422(23)00420-9.

Torre-Cruz, J., Canadas-Quesada, F., Ruiz-Reyes, N., Vera-Candeas, P., Garcia-Galan, S., Carabias-Orti, J., & Ranilla, J. (2023). Detection of valvular heart diseases combining orthogonal non-negative matrix factorization and convolutional neural networks in PCG signals. *Journal of Biomedical Informatics*. 15320464. *145*, 104475. doi:10.1016/j.jbi.2023.104475.

Vilhekar, R. S., & Rawekar, A. (2024). Artificial intelligence in genetics. *Cureus*. 2168-8184. *16*(1). doi:10.7759/cureus.52035.

Virani, S. S., Newby, L. K., Arnold, S. V., Bittner, V., Brewer, L. C., Demeter, S. H., … Williams, M. S. (2023). AHA/ACC/ACCP/ASPC/NLA/PCNA Guideline for the management of patients with chronic coronary disease: a report of the American Heart Association/American College of Cardiology Joint Committee on Clinical Practice Guidelines. *Circulation*. 0009-7322. *148*(9). doi:10.1161/cir.0000000000001168.

Vivekanandan, D. D., Singh, N., Robaczewski, M., Wyer, A., Canaan, L. N., Whitson, D., … Louis, M. (2024). Artificial intelligence-driven advances in coronary calcium scoring: expanding preventive cardiology. *Cureus*. 2168-8184. *16*(11). doi:10.7759/cureus.74681.

Vrints, C., Andreotti, F., Koskinas, K. C., Rossello, X., Adamo, M., Ainslie, J., … Shek, A. B. (2024). ESC Guidelines for the management of chronic coronary syndromes. *European Heart Journal*. 15229645. *45*(36), 3415–3537. Retrieved from http://eurheartj.oxfordjournals.org/. doi:10.1093/eurheartj/ehae177.

Wang, T., Shao, W., Huang, Z., Tang, H., Zhang, J., Ding, Z., & Huang, K. (2021). MOGONET integrates multi-omics data using graph convolutional networks allowing patient classification and biomarker identification. *Nature Communications*. 2041-1723. *12*(1). doi:10.1038/s41467-021-23774-w.

Watanabe, T., Tohyama, T., Ikeda, M., Fujino, T., Hashimoto, T., Matsushima, S., … Ide, T. (2024). Development of deep-learning models for real-time anaerobic threshold and peak VO2 prediction during cardiopulmonary exercise testing. *European Journal of Preventive Cardiology*. 2047-4873. *31*(4), 448–457. doi:10.1093/eurjpc/zwad375.

Williams, M. C., Weir-McCall, J. R., Baldassarre, L. A., De Cecco, C. N., Choi, A. D., Dey, D., … Nicol, E. (2024). Artificial Intelligence and machine learning for cardiovascular computed tomography (CCT): a white paper of the Society of Cardiovascular Computed Tomography (SCCT). *Journal of Cardiovascular Computed Tomography*. 1876861X. *18*(6), 519–532. Retrieved from https://www.sciencedirect.com/science/journal/19345925. doi:10.1016/j.jcct.2024.08.003.

Wolterink, J. M., Leiner, T., Viergever, M. A., & Išgum, I. (2017). Generative adversarial networks for noise reduction in low-dose CT. *IEEE Transactions on Medical Imaging*. 1558254X. *36*(12), 2536–2545. Retrieved from http://ieeexplore.ieee.org/xpl/RecentIssue.jsp?punumber=42. doi:10.1109/TMI.2017.2708987.

World Health Organization. (2020). *Hearts: Technical package for cardiovascular disease management in primary health care: Risk-based CVD management*. Geneva: Author.

Wu, J., Giles, C., Dakic, A., Beyene, H. B., Huynh, K., Wang, T., … Meikle, P. J. (2024). Lipidomic risk score to enhance cardiovascular risk stratification for primary prevention. *Journal of the American College of Cardiology*. 07351097. *84*(5), 434–446. doi:10.1016/j.jacc.2024.04.060.

Wu, J., Yang, D., Zhang, Y., Xian, H., Weng, Z., Liu, J., & Yang, F. (2025). Non-invasive imaging innovation: FFR-CT combined with plaque characterization, safeguarding your cardiac health. *Journal of Cardiovascular Computed Tomography*. 19345925. *19*(1), 152–158. doi:10.1016/j.jcct.2024.08.008.

Wu, Y., & Xie, L. (2025). AI-driven multi-omics integration for multi-scale predictive modeling of genotype-environment-phenotype relationships. *Computational and Structural Biotechnology Journal*. 20010370. *27*, 265–277. doi:10.1016/j.csbj.2024.12.030.

Yadlowsky, S., Hayward, R. A., Sussman, J. B., McClelland, R. L., Min, Y. I., & Basu, S. (2018). Clinical implications of revised pooled Cohort equations for estimating atherosclerotic cardiovascular disease risk. *Annals of Internal Medicine*. 15393704. *169*(1), 20–29. Retrieved from http://annals.org/aim/fullarticle/2683613/clinical-implications-revised-pooled-cohort-equations-estimating-atherosclerotic-cardiovascular-disease. doi:10.7326/M17-3011.

Zhang, Y., Hager, P., Liu, C., Shit, S., Chen, C., Rueckert, D., & Pan, J. (2025). Towards cardiac MRI foundation models: comprehensive visual-tabular representations for whole-heart assessment and beyond. *arXiv*, 23318422. Retrieved from https://arxiv.org. doi:10.48550/arXiv.2504.13037.

Zignoli, A. (2023). Machine learning models for the automatic detection of exercise thresholds in cardiopulmonary exercising tests: from regression to generation to explanation. *Sensors*. 1424-8220. *23*(2), 826. doi:10.3390/s23020826.

Zreik, M., Lessmann, N., van Hamersvelt, R. W., Wolterink, J. M., Voskuil, M., Viergever, M. A., … Išgum, I. (2018). Deep learning analysis of the myocardium in coronary CT angiography for identification of patients with functionally significant coronary artery stenosis. *Medical Image Analysis*. 13618423. *44*, 72–85. Retrieved from http://www.elsevier.com/inca/publications/store/6/2/0/9/8/3/index.htt. doi:10.1016/j.media.2017.11.008.

Chapter 14

The use of artificial intelligence in ophthalmology

Thiago Gonçalves dos Santos Martins[a,b], Diogo Gonçalves dos Santos Martins[c], Thomaz Gonçalves dos Santos Martins[d], Rufino da Silva[e,f,g], Luis Mendes[h], and Paulo Schor[c]

[a]*Departamento de Oftalmologia, Universidade Federal de São Paulo (UNIFESP), São Paulo (SP), Brazil*

[b]*Federal University of Rio de Janeiro, Macaé, Brazil*

[c]*Federal University of São Paulo (UNIFESP), São Paulo, Brazil*

[d]*Federal University of Rio de Janeiro (UNIRIO), Rio de Janeiro, Brazil*

[e]*Faculty of Medicine, University of Coimbra (FMUC), Coimbra, Portugal*

[f]*Ophthalmology Department, University Hospital of Coimbra, ULS de Coimbra, Coimbra, Portugal*

[g]*AIBILI—Association for Innovation and Biomedical Research on Light and Image, Coimbra, Portugal*

[h]*AIBILI—Association for Innovation and Biomedical Research on Light and Image, University of Coimbra, Coimbra, Portugal*

1 Introduction

Artificial intelligence (AI) was originally conceptualized in the 1950s, but its progress during the early years was limited due to insufficient computing resources and rudimentary algorithms. These technological constraints meant that practical applications in healthcare remained largely undeveloped for many years. A major shift occurred in the early 2000s with the rise of deep learning (DL), which allowed AI systems to handle large, complex datasets and continuously refine their performance through iterative learning. This advancement opened new opportunities for AI in complex domains such as medicine. Currently, AI is being applied across various phases of clinical care, assisting with risk prediction, diagnostic support, and the enhancement of healthcare delivery systems (Kaul, Enslin, & Gross, 2020).

In ophthalmology, DL has shown promise in the automated interpretation of key diagnostic tools like fundus images, visual field assessments, and optical coherence tomography (OCT) scans, often achieving results comparable to those of experienced specialists. When combined with telemedicine platforms, these technologies offer a powerful means to improve access to eye care, especially in areas with a shortage of ophthalmologists. Despite the progress, important hurdles remain, including ethical and regulatory issues, as well as the need for broader acceptance and trust among healthcare professionals and patients (Ting et al., 2019a).

2 Fundamentals of AI applied to ophthalmology

AI has proven highly effective in the analysis of ophthalmic images. As a branch of computer science, it excels at decision-making, pattern recognition, and solving complex problems. Within AI, machine learning (ML) and DL are especially significant in modern medicine.

ML is an AI approach whereby algorithms learn from data to perform tasks such as classification and prediction, without relying on explicitly programmed rules. In traditional ML models, system performance depends heavily on the careful selection of features, a process that requires substantial human expertise and manual intervention.

In contrast, DL is a subset of ML that uses deep artificial neural networks—composed of multiple hierarchical processing layers—to automatically extract features from large datasets. This architecture gives DL superior performance on medical imaging tasks, such as OCT scans and fundus photographs, enabling the detection of complex patterns that are often imperceptible to human observers.

Applications of Artificial Intelligence in Healthcare. https://doi.org/10.1016/B978-0-443-43934-6.00003-1

Thus ML differs from DL in the degree of automation and feature extraction it affords. In ML, experts must predefine which variables to analyze, whereas in DL, data extraction occurs autonomously, yielding greater precision and clinical applicability. This makes DL particularly valuable in ophthalmology for supporting clinical decision-making and expanding diagnostic access in resource-limited settings (Moraru, Costin, Moraru, & Branisteanu, 2020).

AI algorithms have gained the ability to detect complex patterns in both medical images and structured clinical data. Among the most effective techniques is the use of convolutional neural networks, which are widely applied in the analysis of fundus photographs, OCT scans, and visual field tests.

The ML technique known as transfer learning enables pretrained models to be adapted to new tasks, even when only limited data are available.

The performance of AI algorithms depends critically on the quality of the training data. Factors such as low-quality images or annotations by nonexpert reviewers can degrade dataset integrity. Moreover, ensuring that the training set is demographically, ethnically, and technically representative is essential to prevent algorithmic bias, which can otherwise limit a model's ability to generalize across diverse patient populations and imaging devices.

Variations in disease prevalence can lead to imbalanced classes, making models prone to favoring the more common outcomes. To address this, methods like oversampling the minority class, under sampling the majority class, or selecting evaluation metrics that account for imbalance (such as area under the ROC curve (AUC), sensitivity, and recall) are essential. A further challenge is overfitting, when a model excels on its training set but loses accuracy on unseen data. Employing cross-validation techniques or testing on independent external cohorts can help uncover and reduce overfitting. Finally, it is crucial to regularly retrain and update AI systems with fresh clinical data to ensure they remain accurate and clinically applicable over time (Aubry, Raley, & McKelvey, 2017).

3 Clinical applications of AI in ophthalmology

Diabetic retinopathy (DR) is the leading cause of blindness among people of working age and, for that reason, has become a major focus of AI research. Early diagnosis is crucial to prevent vision loss, yet a shortage of ophthalmologists in many regions makes timely screening difficult. AI-based systems that automatically analyze fundus photographs can now detect DR as accurately as experienced ophthalmologists, offering an effective solution for large-scale population screening.

4 Algorithm development and validation

A 2016 study showed that a DL algorithm could detect moderate-to-severe DR on fundus photographs with performance comparable to that of human experts. Training these algorithms on images from diverse populations allows their predictions to generalize well, achieving high sensitivity and specificity (Gulshan et al., 2016).

5 Clinical implementation

Regulatory agencies, such as the Food and Drug Administration (FDA), have already approved systems like IDx-DR, which enable healthcare professionals without ophthalmology training to capture retinal images. This system automatically generates a report indicating whether a referral to an eye specialist is needed, helping with the early detection of eye diseases in areas with limited access to ophthalmologists (Gulshan et al., 2016).

6 Advantages

AI-driven tools can be rolled out to screen vast numbers of patients, relieving ophthalmologists of routine checks and freeing them to handle the more challenging cases. Because the assessments are automated, the results are consistent and highly accurate, reducing differences between human reviewers and raising the overall standard of care.

7 Limitations and challenges

In everyday practice, retinal photographs are frequently of lower quality, which can compromise an algorithm's ability to make reliable diagnoses. Accuracy may also decline when the model encounters patient groups that were not well-represented during training. In addition, both patients and healthcare providers often hesitate to trust AI tools because the reasoning behind their outputs is opaque, leaving them wary of these "black-box" systems.

8 Age-related macular degeneration

Age-related macular degeneration (AMD) is a leading cause of vision loss in the elderly, making early detection essential for effective management of both its dry and wet forms. Today, machine-learning models can automatically detect key signs such as drusen and neovascular membranes—typical features of the wet form—helping healthcare professionals diagnose the disease more quickly and begin treatment immediately (Romond et al., 2021).

Applications leveraging AI for AMD have advanced significantly and are now able to detect hallmark lesions, such as drusen or subretinal fluid, on fundus photographs and OCT scans with increasing accuracy. These tools support clinicians by enabling earlier diagnosis, facilitating timely therapeutic decisions, and helping to customize treatment plans according to disease stage and progression. Moreover, automated triage systems powered by these models can streamline patient management, allowing for more efficient clinical workflows and ongoing monitoring, especially in high-volume settings or underserved areas.

Despite these benefits, the performance of such algorithms is not without limitations. A major concern lies in their sensitivity to media opacities, such as cataracts or vitreous hemorrhage, which can degrade image quality and reduce diagnostic accuracy. Furthermore, many of these models are trained on datasets that lack diversity in terms of ethnicity, age range, and disease severity, which can hinder their generalizability to broader patient populations. As a result, the risk of bias and unequal clinical performance persists, underscoring the need for more inclusive training datasets and robust validation across diverse clinical settings.

9 Glaucoma

Glaucoma is the leading cause of irreversible blindness worldwide; however, it is typically asymptomatic in its early stages. The use of AI has shown great promise in the early diagnosis of glaucoma through the analysis of fundus images, OCT, and visual field tests. Key variables assessed include the optic disc examination, visual field, and retinal nerve fiber layer (RNFL) thickness, fundamental elements for evaluating the condition. These findings highlight the potential of AI as a diagnostic support tool, contributing to improved clinical accuracy and enabling the implementation of continuous patient monitoring strategies (Huang et al., 2023).

As with other ocular diseases, image quality is critical for optimal algorithm performance. Likewise, these models tend to underperform in underrepresented patient groups.

10 Cornea

AI has brought meaningful improvements to the evaluation and management of keratoconus, especially through the examination of corneal tomography, thickness (pachymetry) maps, and corneal imaging. These AI-based tools enhance the ability to detect early signs of the disease, classify its severity, and monitor progression over time. By uncovering intricate patterns that might be overlooked by human observers, such systems offer the promise of greater diagnostic precision and more tailored treatment approaches.

However, for these technologies to become part of standard clinical routines, several challenges must be addressed. Comprehensive validation is essential to confirm their reliability across different imaging devices and diverse patient populations. Variations in image quality, equipment calibration, and acquisition techniques can all impact the consistency of AI results. In addition, widespread implementation depends on the confidence of healthcare providers, which can be strengthened through the development of transparent, explainable algorithms, seamless integration into clinical workflows, and dedicated training to support their practical use (Martins, Benchimol, Gameiro, & Schor, 2022a).

Thus algorithms are valuable for detecting changes in both corneal curvature and thickness. Platforms such as Pentacam and corneal shape analysis already incorporate AI-driven tools to automate corneal assessment, enhancing diagnostic accuracy and speeding up screening, thereby enabling earlier interventions.

11 Monitoring and prognosis

AI now allows us to monitor ophthalmic diseases by detecting subtle patterns of progression, enhancing predictive power, enabling more personalized patient follow-up, and supporting more effective clinical decision-making while optimizing resource allocation across healthcare systems.

This predictive capability spares patients from ineffective interventions and lowers the costs of unnecessary treatments, making it possible to create individualized, adaptive, patient-centered treatment protocols. In doing so, it fosters a more precise, efficient, and needs-driven clinical practice.

12 AI-assisted surgery

Algorithms have made significant advances in the planning of ophthalmic surgeries. Cataract surgery is among the most performed procedures worldwide, and AI can assist in selecting the most suitable intraocular lens for each patient. This enhances preoperative planning and contributes to more favorable clinical outcomes. This data-driven personalization marks the beginning of a new era of precision medicine in surgical ophthalmology.

AI can be integrated with robotics to drive advancements in ophthalmic surgery. The Preceyes Surgical System employs AI-assisted robotic technology to perform retinal procedures with micrometric precision that surpasses the limits of human manual dexterity. These innovations enhance surgical predictability, reduce operative time, and minimize the risk of complications. As a result, AI is being solidified as a transformative force in ophthalmic surgery and established as a foundational pillar for the future of ocular procedures (Siddiqui, Ladas, & Lee, 2020).

13 Available platforms and technologies

Several AI-driven platforms, including IDx-DR—the first to obtain FDA approval—and EyeArt, which also hold regulatory clearance, are currently being deployed for automated detection and monitoring of DR. These innovations mark a transformative shift in eye care delivery by offering affordable, scalable solutions that help bridge gaps in healthcare access, particularly in underserved and remote areas. Their adoption underscores the increasing trust in AI technologies as valuable tools to enhance diagnostic accuracy and support clinical decision-making on a wider scale (Mi, MacLaren, & Cehajic-Kapetanovic, 2025; Vought, Vought, Shah, Szirth, & Bhagat, 2023).

Beyond these applications, leading technology companies have invested heavily in advancing AI for ophthalmology. Google, for instance, has made substantial progress in developing sophisticated AI algorithms for analyzing OCT scans with remarkable precision, while IBM and Microsoft focus on platforms that merge imaging data with comprehensive patient clinical records. This integration aims to provide clinicians with deeper insights and a more holistic understanding of each patient's condition, thereby improving the personalization and effectiveness of treatment plans.

Looking forward, the incorporation of AI into next-generation fundus cameras and OCT devices promises to revolutionize ophthalmic diagnostics. These AI-enabled tools will facilitate real-time generation of diagnostic reports and offer intelligent decision support, streamlining clinical workflows and enabling faster, more accurate patient management. Ultimately, this technological advancement holds great potential to expand access to high-quality eye care, ensuring timely diagnoses and interventions for patients, regardless of geographic or resource limitations (Khan et al., 2025) (Table 14.1).

14 Challenges and limitations

Ethical and regulatory issues remain critical factors for the future development and adoption of this technology. The lack of transparency in how algorithms operate—often referred to as "black boxes"—raises concerns about trust and legal accountability in the event of failures. This underscores the need for greater flexibility and adaptation from regulatory bodies, such as the FDA, to approve and oversee the use of these emerging technologies. Another key limitation is the reduced performance of algorithms when applied to populations that differ from those represented in their training datasets, reinforcing the importance of globally collaborative research. These challenges highlight that, for AI to become a truly transformative tool in ophthalmology, a coordinated effort is required among developers, regulators, clinicians, and healthcare institutions. This effort must prioritize not only innovation but also safety, fairness, and trust.

15 Future perspectives

The future of AI in ophthalmology envisions a landscape that is increasingly personalized and accessible. Ophthalmic treatments are moving toward more efficient and individualized models, enabling the reduction of unnecessary interventions and the optimization of clinical outcomes (Martins et al., 2022a).

The integration of AI with telemedicine already enables real-time screening for various eye diseases, particularly in regions with limited access to specialists. This leads to improved diagnostic accuracy, reduced workload for healthcare professionals, and a significantly enhanced patient experience (Gonçalves Dos Santos Martins, 2024).

TABLE 14.1 AI in ophthalmology: applications, benefits, and challenges.

Condition/topic	AI application	Advantages	Limitations/challenges
Diabetic retinopathy (DR)	Automated analysis of fundus photographs for early DR detection.	High accuracy, comparable to experts; enables large-scale screening; consistent results.	Image quality issues; lack of diverse training data; "black box" trust issues.
Age-related macular degeneration (AMD)	Detection of drusen and neovascular membranes on fundus photographs and OCT.	Early diagnosis; aids treatment customization; streamlines workflows in high-volume settings.	Sensitive to media opacities; limited dataset diversity; risk of bias.
Glaucoma	Analysis of fundus images, OCT, and visual field tests; optic disc and RNFL assessments.	Supports early diagnosis and continuous monitoring; improves clinical accuracy.	Requires high-quality images; underperformance in underrepresented groups.
Cornea (keratoconus)	Corneal tomography and pachymetry analysis for early detection and monitoring.	Detects subtle patterns; enhances diagnostic precision; speeds screening.	Needs validation across devices and populations; image quality and calibration variability; requires explainability and training.
Monitoring and prognosis	Detects subtle disease progression; predictive modeling for personalized follow-up.	Avoids ineffective treatments; reduces costs; supports adaptive, patient-centered care.	–
AI-assisted surgery	IOL selection in cataract surgery; AI-robotic retinal surgery systems.	Improves surgical planning and outcomes; increases precision; reduces operative time and risks.	–
Available platforms	FDA-approved IDx-DR, EyeArt for DR detection; Google, IBM, Microsoft AI platforms.	Affordable, scalable, expands access, especially in underserved areas; integrated patient data.	–

The use of technology to strengthen the physician–patient relationship still receives little attention. However, this will become increasingly feasible with patient-generated data, enabling the earlier detection of various diseases.

The use of AI in medicine holds great potential to transform the doctor-patient relationship. With the help of this technology, diagnostic accuracy can be improved, freeing healthcare professionals from repetitive and administrative tasks. This allows physicians to spend more time on direct and compassionate patient care, strengthening the therapeutic bond (Vaghefi et al., 2021).

However, the increasing reliance on technological tools in clinical practice may negatively affect the physician's central role in decision-making. Excessive dependence on technology can create a sense of detachment for both the healthcare provider and the patient, who may feel less supported or valued throughout their treatment.

In ophthalmology, AI extends beyond diagnosing eye diseases and serves as a promising tool for monitoring systemic conditions. By analyzing images and data, AI helps identify new biomarkers associated with complex diseases such as Alzheimer's, diabetes, stroke, and hypertension, thereby expanding the preventive and therapeutic potential of the specialty (Gonçalves Dos Santos Martins & Gonçalves Dos Santos Martins, 2025; Martins, Costa, & Martins, 2018; Sauerbrei, Kerasidou, Lucivero, & Hallowell, 2023).

AI has undergone remarkable advances through DL techniques, which rely on large-scale datasets to achieve optimal performance. In ophthalmology, these developments have significantly enhanced the interpretation of diagnostic imaging modalities such as OCT, fundus photography, and the emerging discipline of oculomics.

First described in 2020, oculomics refers to the convergence of big data, AI, and ocular imaging in the identification of retinal biomarkers that may reflect systemic health conditions, often preceding clinical manifestations. This concept reinforces the notion of the eye as a noninvasive window into systemic physiology.

Oculomics aligns with other "omics" sciences, including genomics, proteomics, transcriptomics, epigenomics, and metabolomics, which employ high-throughput technologies to investigate biomarkers across various biological levels, from nucleotide sequences to metabolic intermediates. These integrated approaches offer valuable insights into genetic susceptibility, gene expression regulation, epigenetic modifications, and metabolic activity, thereby improving the early diagnosis, risk stratification, and longitudinal monitoring of systemic diseases (Martins, 2021; Martins & Martins, 2025).

Despite the promising outcomes, further research is warranted to explore both supervised and self-supervised learning strategies. Such efforts are essential to enhance the clinical validity, generalizability, and reliability of AI-based applications in ophthalmic and systemic disease detection (Martins, Silva, & Schor, 2020; Martins et al., 2022a).

While the widespread implementation of AI in ophthalmology offers great potential to increase access to eye care—especially in underserved regions with a shortage of specialists—it also raises concerns about exacerbating existing health disparities if access to such advanced tools remains limited to well-resourced, technologically advanced centers. Foundation models represent a paradigm shift in AI, evolving from task-specific tools to adaptable, general-purpose systems capable of addressing a broad spectrum of clinical challenges. Ophthalmology is particularly well-positioned to lead this transformation, serving as a model for integrating these technologies into medical practice more broadly (Martins et al., 2020).

While some AI models have been trained using exceptionally large image repositories—surpassing 1.6 million images—their relatively basic architectures indicate that there is still substantial room for advancement by expanding not only the volume of data but also the sophistication and parameter capacity of the models. Looking ahead, a key milestone involves the development of multimodal foundation models that can seamlessly integrate multiple data streams, including three-dimensional OCT imaging, other ophthalmic modalities, functional testing outcomes, genomic profiles, clinical records, textual reports, and even voice data (Dutt, Sivaraman, Savoy, & Rajalakshmi, 2020; Martins et al., 2020).

The convergence of these diverse data types within a unified framework has the potential to significantly reshape ophthalmic practice. It could enhance clinical decision-making through personalized, data-driven insights; improve the efficiency of healthcare delivery; and enrich educational tools for trainees and professionals alike. Nonetheless, these technological advances must be accompanied by comprehensive ethical guidelines and governance structures. Ensuring equitable access, protecting patient confidentiality, and upholding the clinical independence of practitioners are critical components for the responsible integration of AI into routine medical care (Martins, 2020a; Martins et al., 2020, Zhu et al., 2025).

16 Conclusion

AI is transforming modern ophthalmology by improving diagnostic precision, enhancing surgical outcomes, and enabling more effective treatment options. These advancements are raising the standard of care in well-equipped centers while also expanding access to eye care services in underserved or resource-limited regions (Dos Santos Martins, 2020; Martins et al., 2020; Martins et al., 2022a; Popescu Patoni et al., 2023; Santos Martins, de Azevedo Costa, & Schor, 2021; Martins, 2022, Martins & Schor, 2020).

However, despite this promising progress, several challenges remain before AI can be fully incorporated into everyday clinical practice. Key issues include the establishment of strong regulatory measures to protect patient data privacy and security, as well as ongoing skepticism among healthcare professionals and patients regarding the adoption of these emerging technologies (Martins & Costa, 2017; Martins et al., 2022a).

For the integration of AI into healthcare systems to be truly effective and ethically sound, it is essential to establish a foundation based on stringent clinical evaluation, open access to algorithmic logic, and continuous professional development. These elements are not merely supportive; they are indispensable in ensuring that AI acts as a reliable partner in enhancing diagnostic precision and elevating the standards of care in ophthalmology (Martins & Costa, 2015; Martins et al., 2022a, Martins, 2025).

The rapid evolution of AI technologies in recent years has been largely driven by advances in DL and the development of large-scale models capable of processing highly complex and heterogeneous datasets. This evolution marks a shift in the computational paradigm: from systems that simply assign labels to data (discriminative models) to those that can grasp intricate data structures and generate new, meaningful content (generative models). Given the image-intensive nature of ophthalmology, these generative systems offer unprecedented opportunities for innovation in both clinical and research settings (Chia et al., 2024; Feng et al., 2024;) Martins, & Rangel, 2022).

Generative models can support a wide range of applications, from streamlining clinical documentation and improving communication between patients and providers to aiding diagnostic reasoning and long-term disease management. These tools also have the potential to revolutionize ophthalmic research by simulating trial conditions, analyzing large-scale datasets, and personalizing treatment protocols. Moreover, generative AI can reshape medical education by creating interactive, adaptive learning platforms tailored to individual learners.

Yet, a major frontier lies in achieving true multimodal integration, enabling AI systems to interpret complex ophthalmic images in conjunction with diverse clinical data such as OCT scans, electronic health records, and genomic information. To fully realize the benefits of generative AI while mitigating potential harms, a comprehensive risk management strategy is essential. Such a framework should span the entire lifecycle of AI development and deployment, ensuring that innovation is aligned with safety, ethics, and clinical relevance. With these safeguards in place, generative AI can become a cornerstone of future ophthalmic care (Dos Santos Martins, Dos Santos Martins, Dos Santos Martins, Marinho, & Schor, 2021; Martins, 2025, Martins & Schor, 2022, Martins et al., 2022b; Ting et al., 2019b).

References

Aubry, K. B., Raley, C. M., & McKelvey, K. S. (2017). The importance of data quality for generating reliable distribution models for rare, elusive, and cryptic species. *PLoS ONE*, 1932-6203. *12*(6). Available from: http://journals.plos.org/plosone/article/file?id=10.1371/journal.pone.0179152&type=printable. doi:10.1371/journal.pone.0179152.

Chia, M. A., Antaki, F., Zhou, Y., Turner, A. W., Lee, A. Y., & Keane, P. A. (2024). Foundation models in ophthalmology. *British Journal of Ophthalmology*, 1468-2079. *108*(10), 1341–1348. Available from: http://bjo.bmj.com/. doi:10.1136/bjo-2024-325459.

Dos Santos Martins, T. G. (2020). Comment on: "Artificial intelligence and ophthalmology". *Turkish Journal of Ophthalmology*, 1300-0659. *50*(6), 392–392. doi:10.4274/tjo.galenos.2020.98354.

Dos Santos Martins, T. G., Dos Santos Martins, D. G., Dos Santos Martins, T. G., Marinho, P., & Schor, P. (2021). Covid 19 repercussions in ophthalmology: a narrative review. *São Paulo Medical Journal*, 1516-3180. *139*(5), 535–542. Available from: https://www.scielo.br/j/spmj/a/qF7MJm7BzLVMtzJ3dHBPMZy/?lang=en&format=pdf. doi:10.1590/1516-3180.2021.0113.R1.0504221.

Dutt, S., Sivaraman, A., Savoy, F., & Rajalakshmi, R. (2020). Insights into the growing popularity of artificial intelligence in ophthalmology. *Indian Journal of Ophthalmology*, 1998-3689. *68*(7), 1339–1346. Available from: http://www.ijo.in/. doi:10.4103/ijo.IJO_1754_19.

Feng, X., Xu, K., Luo, M. J., Chen, H., Yang, Y., He, Q., et al. (2024). Latest developments of generative artificial intelligence and applications in ophthalmology. *Asia-Pacific Journal of Ophthalmology*, 2162-0989. *13*(4). Available from: https://www.sciencedirect.com/journal/asia-pacific-journal-of-ophthalmology. doi:10.1016/j.apjo.2024.100090.

Gonçalves Dos Santos Martins, T. (2024). Comment on "Ophthalmology census 2021: a demographic characterisation of ophthalmologists in Portugal". *Acta Médica Portuguesa*, 1646-0758. *37*(6), 493–494. Available from: https://www.actamedicaportuguesa.com/revista/index.php/amp/article/view/21572. doi:10.20344/amp.21572.

Gonçalves Dos Santos Martins, T., & Gonçalves Dos Santos Martins, D. (2025). Biomarkers of neurodegeneration in diabetic retinopathy. *Ophthalmologie*, 2731-7218. *122*(4), 319–320. Available from: https://www.springer.com/journal/347. doi:10.1007/s00347-025-02201-1.

Gulshan, V., Peng, L., Coram, M., Stumpe, M. C., Wu, D., Narayanaswamy, A., et al. (2016). Development and validation of a deep learning algorithm for detection of diabetic retinopathy in retinal fundus photographs. *JAMA—Journal of the American Medical Association*, 1538-3598. *316*(22), 2402–2410. Available from: http://jamanetwork.com/journals/jama/data/Journals/JAMA/935924/joi160132.pdf. doi:10.1001/jama.2016.17216.

Huang, X., Islam, M. R., Akter, S., Ahmed, F., Kazami, E., Serhan, H. A., et al. (2023). Artificial intelligence in glaucoma: opportunities, challenges, and future directions. *BioMedical Engineering Online*, 1475-925X. *22*(1). Available from: https://biomedical-engineering-online.biomedcentral.com/. doi:10.1186/s12938-023-01187-8.

Kaul, V., Enslin, S., & Gross, S. A. (2020). History of artificial intelligence in medicine. *Gastrointestinal Endoscopy*, 1097-6779. *92*(4), 807–812. Available from: http://www.elsevier.com/inca/publications/store/6/2/3/1/3/7/index.htt. doi:10.1016/j.gie.2020.06.040.

Khan, Z. A. I. D., Gaidhane, A. M., Singh, M., Ganesan, S., Kaur, M., Sharma, G. C., et al. (2025). Diagnostic accuracy of IDX-DR for detecting diabetic retinopathy: a systematic review and meta-analysis. *American Journal of Ophthalmology*, 1879-1891. *273*, 192–204. Available from: https://www.sciencedirect.com/science/journal/00029394. doi:10.1016/j.ajo.2025.02.022.

Martins, D. G. D. S., & Martins, T. G. D. S. (2025). Comment on "Stroke awareness in a Brazilian Northeastern capital city and the burden of the COVID-19 pandemic". *Arquivos de Neuro-psiquiatria*. 1678-4227. *83*(2), 1–2. doi:10.1055/s-0045-1802310.

Martins, T. G. D. S. (2020a). Carta ao Editor sobre o Artigo "Está a Educação Médica em Mudança? Cinco Desafios para o Futuro Próximo". *Acta Médica Portuguesa*, *33*(10), 703. doi:10.20344/amp.14348.

Martins, T. G. D. S. (2020b). Use of smartphone-based instant messaging services in medical practice: a cross-sectional study. *São Paulo Medical Journal*, *138*(3), 269–270. doi:10.1590/1516-3180.2020.0228.15052020.

Martins, T. G. D. S. (2021). Comment on: how is the risk of being diagnosed with referable diabetic retinopathy affected by failure to attend diabetes eye screening appointments? *Eye (Basingstoke)*, 1476-5454. *35*(7), 2045. Available from: http://www.nature.com/eye/index.html. doi:10.1038/s41433-020-1040-8.

Martins, T. G. D. S. (2022). Comment on: "Understanding required to consider AI applications to the field of ophthalmology". *Taiwan Journal of Ophthalmology*, *13*(2), 256. doi:10.4103/2211-5056.354538.

Martins, T. G. D. S. (2025). Comment on artificial intelligence in neovascular age-related macular degeneration. *Klinische Monatsblätter Für Augenheilkunde*. doi:10.1055/a-2487-5978.

Martins, T. G. D. S. (2025). Comment on artificial intelligence in neovascular age-related macular degeneration. *Klinische Monatsblätter Für Augenheilkunde*, 0023–2165. doi:10.1055/a-2487-5978.

Martins, T. G. D. S., Benchimol, G. L., Gameiro, G. R., & Schor, P. (2022a). Learning and teaching ophthalmology in the pandemic. *Einstein (São Paulo)*, *20*, eCE6988. doi:10.31744/einstein_journal/2022CE6988.

Martins, T. G. D. S., & Costa, A. L. F. D. A. (2015). Letter to the editor: association between participation and compliance with continuing medical education and care production by physicians: a cross-sectional study. einstein (São Paulo). 2015;13(1):1–6. *Einstein (São Paulo)*, 1679-4508. *13*(2), 341–341. doi:10.1590/S1679-45082015CE3387.

Martins, T. G. D. S., & Costa, A. L. F. D. A. (2017). A new way to communicate science in the era of big data and citizen science. *Einstein (Sao Paulo, Brazil)*, 2317-6385. *15*(4), 523. doi:10.1590/S1679-45082017CE4280.

Martins, T. G. D. S., Costa, A. L. F. D. A., & Martins, T. G. D. S. (2018). Big data use in medical research. *Einstein (Sao Paulo, Brazil)*, 2317-6385. *16*(3), eED4087. doi:10.1590/S1679-45082018ED4087.

Martins, T. G. D. S., & Rangel, F. S. (2022). Data warehouse and medical research. *Einstein (São Paulo)*, *20*, eED6324. doi:10.31744/einstein_journal/2022ED6324.

Martins, T. G. D. S., & Schor, P. (2020). Unpacking the black box. *Einstein (São Paulo)*, *19*, eED6037. doi:10.31744/einstein_journal/2021ED6037.

Martins, T. G. D. S., & Schor, P. (2022). Machine learning in image analysis in ophthalmology. *Einstein (São Paulo)*, *19*, eED6860. doi:10.31744/einstein_journal/2021ED6860.

Martins, T. G. D. S., Schor, P., Mendes, L. G. A., Fowler, S., & Silva, R. (2022b). Use of artificial intelligence in ophthalmology: a narrative review. *São Paulo Medical Journal*, *140*(6), 837–845. doi:10.1590/1516-3180.2021.0713.R1.22022022.

Martins, T. G. D. S., Silva, R., & Schor, P. (2020). Distribuição de Médicos e Oftalmologistas no Brasil e Portugal. *Acta Médica Portuguesa*, *33*(10), 635–637. doi:10.20344/amp.

Mi, H., MacLaren, R. E., & Cehajic-Kapetanovic, J. (2025). Robotising vitreoretinal surgeries. *Eye (Basingstoke)*, 1476-5454. *39*(4), 673–682. Available from: https://www.nature.com/eye/. doi:10.1038/s41433-024-03149-3.

Moraru, A., Costin, D., Moraru, R., & Branisteanu, D. (2020). Artificial intelligence and deep learning in ophthalmology—present and future (review). *Experimental and Therapeutic Medicine*, 1792-0981. *20*(4). doi:10.3892/etm.2020.9118.

Popescu Patoni, S. I., Mușat, A. A. M., Patoni, C., Popescu, M. N., Munteanu, M., Costache, I. B., & Pîrvulescu, R. A. (2023). Artificial intelligence in ophthalmology. *Romanian Journal of Ophthalmology*, 2457-4325. *67*(3). Available from:. doi:10.22336/rjo.2023.37.

Romond, K., Alam, M., Kravets, S., Sisternes, L. D., Leng, T., Lim, J. I., et al. (2021). Imaging and artificial intelligence for progression of age-related macular degeneration. *Experimental Biology and Medicine*, 1535-3699. *246*(20), 2159–2169. Available from: http://www.uk.sagepub.com/journals/Journal202180. doi:10.1177/15353702211031547.

Santos Martins, T. G. D., de Azevedo Costa, A. L. F., & Schor, P. (2021). Comment on: "Do we have enough ophthalmologists to manage vision-threatening diabetic retinopathy? A global perspective. *Eye (Basingstoke)*, 1476-5454. *35*(2), 690–691. Available from: http://www.nature.com/eye/index.html. doi:10.1038/s41433-020-0903-3.

Sauerbrei, A., Kerasidou, A., Lucivero, F., & Hallowell, N. (2023). The impact of artificial intelligence on the person-centred, doctor-patient relationship: some problems and solutions. *BMC Medical Informatics and Decision Making*, 1472-6947. *23*(1). doi:10.1186/s12911-023-02162-y.

Siddiqui, A. A., Ladas, J. G., & Lee, J. K. (2020). Artificial intelligence in cornea, refractive, and cataract surgery. *Current Opinion in Ophthalmology*, 1531-7021. *31*(4), 253–260. Available from: http://journals.lww.com/co-ophthalmology/pages/default.aspx. doi:10.1097/ICU.0000000000000673.

Ting, D. S. W., Pasquale, L. R., Peng, L., Campbell, J. P., Lee, A. Y., Raman, R., et al. (2019a). Artificial intelligence and deep learning in ophthalmology. *British Journal of Ophthalmology*, *103*(2), 167–175. doi:10.1136/bjophthalmol-2018-313173.

Ting, D. S. W., Pasquale, L. R., Peng, L., Campbell, J. P., Lee, A. Y., Raman, R., et al. (2019b). Artificial intelligence and deep learning in ophthalmology. *British Journal of Ophthalmology*, 1468-2079. *103*(2), 167–175. Available from: http://bjo.bmj.com/. doi:10.1136/bjophthalmol-2018-313173.

Vaghefi, E., Yang, S., Xie, L., Hill, S., Schmiedel, O., Murphy, R., & Squirrell, D. (2021). THEIA development, and testing of artificial intelligence-based primary triage of diabetic retinopathy screening images in New Zealand. *Diabetic Medicine*, 0742-3071. *38*(4). doi:10.1111/dme.14386.

Vought, R., Vought, V., Shah, M., Szirth, B., & Bhagat, N. (2023). EyeArt artificial intelligence analysis of diabetic retinopathy in retinal screening events. *International Ophthalmology*, 1573-2630. *43*(12), 4851–4859. doi:10.1007/s10792-023-02887-9.

Zhu, Z., Wang, Y., Qi, Z., Hu, W., Zhang, X., Wagner, S. K., et al. (2025). Oculomics: Current concepts and evidence. *Progress in Retinal and Eye Research*, 1873-1635. *106*. Available from: https://www.sciencedirect.com/science/journal/13509462. doi:10.1016/j.preteyeres.2025.101350.

Chapter 15

The roles of machine learning and artificial intelligence in the prediction of autoimmune disease

Ashok Kumar Sah[a], Ramji Gupta[b], Prem Shankar Mishra[c], Nitya Sharma[b], Rahul Kumar[b], Sakshi Patel[d], and Rabab H. Elshikh[a]

[a]Department of Medical Laboratory Sciences, College of Applied and Health Sciences, A'Sharqiyah University, Ibra, Oman

[b]Department of Pharmacology, R. V. Northland Institute, Greater Noida, Uttar Pradesh, India

[c]Faculty of Pharmacy, Vidya University, Meerut, Uttar Pradesh, India

[d]Department of Pharmaceutical Technology, MIET, Meerut, Uttar Pradesh, India

1 Introduction

A significant burden on people, families, and healthcare systems, autoimmune diseases are among the most common chronic illnesses in the world, impacting millions of people. They develop when the body's defense mechanism, which is meant to keep infections at bay, unintentionally targets its own tissues, cells, and organs. The body's inability to discriminate between self and nonself is reflected in this loss of immune tolerance (Sjöwall & Parodis, 2022). Organ dysfunction, progressive tissue damage, and ongoing inflammation are the outcomes. Rheumatoid arthritis (RA), systemic lupus erythematosus (SLE), type 1 diabetes mellitus (T1D), and multiple sclerosis (MS) are among the >80 autoimmune diseases that have been identified (Buckner, 2023). Women are disproportionately affected by these conditions, which frequently start in middle age or early adulthood, cause long-term disability, and necessitate lifelong care. In addition to impairing quality of life, their chronic nature poses serious financial and medical difficulties (Cheng et al., 2024).

Accurately predicting and promptly diagnosing autoimmune diseases remain difficult tasks, even with significant advancements in immunology and molecular medicine (Perez, 2005). Autoimmunity, in contrast to monogenic disorders, results from intricate interactions among immunological, hormonal, environmental, and genetic factors. The development, manifestation, and progression of these diseases exhibit remarkable heterogeneity due to their multifactorial origin (Marion & Postlethwaite, 2014). Numerous misdiagnoses or years of uncertainty before patients receive a definitive diagnosis are caused by overlapping symptoms and the absence of highly specific biomarkers. These delays limit the efficacy of treatments and lower long-term quality of life by allowing irreversible organ damage to occur (McGonagle, Watad, & Savic, 2018).

Although they are still necessary, current diagnostic techniques such as imaging, histopathology, clinical examination, and autoantibody laboratory testing are more reactive than predictive (Afzal et al., 2024). Even though autoantibodies like antinuclear antibodies can show up years before symptoms appear, their low specificity makes them less effective as early warning indicators. While histopathology is invasive and inappropriate for routine screening, imaging methods such as MRI and ultrasound only detect structural damage after it has already occurred (National Academies Press, 2024). These drawbacks highlight the urgent need for sophisticated predictive tools that can combine various data sources to predict disease risk, identify people in preclinical stages, and facilitate prompt intervention (Rizzo & Houen, 2023).

1.1 Autoimmune diseases and challenges in prediction

Several key factors complicate the prediction of autoimmune diseases. First, their **multifactorial origin** makes deterministic modeling difficult. While genetic predisposition, particularly human leukocyte antigen (HLA) alleles,

Applications of Artificial Intelligence in Healthcare. https://doi.org/10.1016/B978-0-443-43934-6.00015-8

is a critical factor, many carriers never develop the disease, suggesting that environmental exposures, infections, diet, stress, and microbiome disturbances serve as additional triggers (Mu et al., 2024). These influences are variable and often transient, limiting their predictive power (Danieli et al., 2024). Second, **clinical heterogeneity** obscures early diagnosis. Symptoms such as fatigue, joint pain, fever, or inflammation are nonspecific and common across multiple autoimmune conditions (de Boer et al., 2024). For instance, early joint involvement in RA may resemble lupus arthritis, while neurological symptoms in MS overlap with other demyelinating disorders. This overlap frequently leads to misclassification and delayed treatment initiation (Kowalski, Domżał-Magrowska, & Małecka-Wojciesko, 2025). Third, the **lack of reliable biomarkers** hampers prediction. While markers such as anti-dsDNA in lupus or anti-cyclic citrullinated peptide (anti-CCP) in RA assist in diagnosis, they are not universally present and do not consistently correlate with disease activity (Vivas, Boumediene, & Tobón, 2024). No widely validated biomarkers currently predict disease onset in asymptomatic individuals, representing a major gap in autoimmune research (Kruta et al., 2024).

Finally, even after diagnosis, disease progression is unpredictable. Some patients experience mild disease with long periods of remission, while others progress rapidly to severe disability or organ damage (Forrest et al., 2023). Current tools cannot reliably forecast these trajectories, complicating treatment planning. Predictive models that stratify patients by risk or progression could, therefore, transform management, guiding therapeutic intensity and monitoring strategies (Stafford et al., 2020). Challenges in predicting autoimmune diseases arise from their multifactorial origins, overlapping symptoms, absence of specific biomarkers, and variable progression (Yang et al., 2024). Overcoming these barriers requires innovative, data-driven approaches that can integrate and analyze complex biological and clinical information (Hoffmann, Rheude, Neubauer, Brendel, & Thrun, 2024).

1.2 Role of artificial intelligence and machine learning in modern healthcare

Artificial intelligence (AI) and machine learning (ML) offer powerful solutions to these challenges (Cao, 2025). AI refers to computational systems that perform tasks requiring human-like intelligence, while ML, a subset of AI, enables algorithms to learn patterns from data without explicit programming (Liu et al., 2024). These methods excel at analyzing large, complex datasets to uncover hidden relationships that are often missed by traditional statistical approaches (Morabito, De Simone, Pastorelli, Brunelli, & Ferrario, 2025). AI and ML have already demonstrated transformative potential across diverse areas of healthcare. In oncology, deep learning models detect tumors from imaging data with accuracy comparable to radiologists (Yang, Peng, Zheng, & Peng, 2025). In cardiology, ML predicts arrhythmias and cardiovascular events from electrocardiograms (ECGs) and patient histories. In infectious disease epidemiology, AI has been used to forecast outbreaks and optimize interventions. These successes underscore the versatility of AI in medical decision-making and provide a blueprint for its application in autoimmune disease prediction (Liu et al., 2025a).

Autoimmune diseases generate vast and heterogeneous biomedical data, including genomic, transcriptomic, and proteomic profiles; imaging data; and electronic health records (EHRs). Integrating these diverse data streams is beyond the capability of conventional analysis but aligns with the strengths of AI and ML (Amann, Blasimme, Vayena, Frey, & Madai, 2020). By combining multiple sources of information, these technologies can construct predictive models capable of forecasting disease onset, discovering novel biomarkers, and stratifying patients by disease trajectory or treatment response (Obermeyer, Powers, Vogeli, & Mullainathan, 2019). One of the most promising applications is **early diagnosis**. For example, ML models trained on longitudinal health records, autoantibody profiles, and genetic markers can identify individuals at high risk of T1D years before clinical onset. In RA, algorithms applied to imaging can detect subtle preclinical joint changes invisible to human assessment (Price & Cohen, 2019).

AI also plays a pivotal role in **biomarker discovery**. By mining large-scale genomic or proteomic datasets, ML can identify candidate biomarkers that enhance risk prediction, elucidate disease mechanisms, and guide therapeutic development (Li, Sahu, Talwalkar, & Smith, 2020). Equally important is **patient stratification**, a cornerstone of precision medicine. Autoimmune diseases are commonly treated with immunosuppressants or biologics, but patient responses vary widely, and side effects are significant (Eden et al., 2025). ML models that classify patients based on likely treatment response or disease course allow more personalized care, reducing ineffective therapies and minimizing adverse effects (Singh, Cheng, Kwan, & Ebinger, 2025). Despite these advances, challenges remain. Data scarcity and heterogeneity hinder robust model development, particularly as many autoimmune diseases are relatively rare. The interpretability of complex algorithms—the "black-box" problem—raises concerns about clinician trust and accountability (Hickman & Petrin, 2021). Ethical and privacy issues surrounding sensitive patient data must also be carefully managed. Ongoing research into explainable AI (XAI), federated learning, and privacy-preserving analytics is addressing these concerns, aiming to ensure responsible integration of AI into healthcare.

In conclusion, autoimmune diseases present unique challenges due to their multifactorial causes, clinical overlap, and unpredictable progression (Reddy, Allan, Coghlan, & Cooper, 2020). AI and ML offer transformative opportunities by enabling early detection, biomarker discovery, and personalized patient stratification. Their role in healthcare is shifting the paradigm from reactive diagnosis to proactive prediction, bringing us closer to the vision of precision medicine (Friedrich et al., 2021).

2 Overview of autoimmune diseases

Autoimmune diseases comprise a heterogeneous group of chronic disorders in which the immune system mistakenly attacks the body's own tissues. Collectively, they affect 5%–10% of the global population and are major contributors to long-term disability, especially in young and middle-aged adults. Clinical presentations range from organ-specific conditions, such as T1D, to systemic diseases like SLE (Henderson, Condell, Connolly, Kelly, & Curran, 2021). Despite this diversity, they share common mechanisms of immune dysregulation, chronic inflammation, and progressive tissue injury. Understanding their pathophysiology, major clinical types, and the shortcomings of conventional diagnostics is key to appreciating the value of predictive technologies like AI and ML (Klonoff et al., 2025).

2.1 Background: Pathophysiology and key mechanisms

Autoimmunity arises from a multifactorial interplay between genetics, environment, and immune dysfunction. A central event is the breakdown of **self-tolerance**, the process that normally eliminates autoreactive immune cells or suppresses them through regulatory pathways. When these checkpoints fail, autoreactive lymphocytes persist and initiate pathogenic responses (Ji, Jiang, Liu, Zhang, & You, 2025).

Genetic predisposition is a strong determinant. Genome-wide studies have identified numerous susceptibility loci, with the **HLA complex** being the most critical. Alleles such as HLA-DR4 (RA) and HLA-DR3 (T1D) increase risk by shaping antigen presentation. However, genes alone do not guarantee disease, underscoring the importance of environmental triggers (Falet et al., 2025).

Environmental factors can initiate or exacerbate autoimmunity. Viral infections, such as Epstein–Barr virus (linked to SLE and MS) and Coxsackievirus (T1D), may induce *molecular mimicry*, where viral antigens resemble self-proteins. Other triggers include smoking, toxins, diet, and gut microbiome alterations (Grote & Keeling, 2022).

Immune system dysregulation underpins disease progression. Autoreactive T cells secrete proinflammatory cytokines (IL-17, IFN-γ, TNF-α), sustaining inflammation. B cells generate autoantibodies that form immune complexes, activate complement, and cause tissue damage (Amann et al., 2022). Regulatory T cells, normally suppressing autoreactivity, are often deficient. Together, these mechanisms create a **self-perpetuating cycle**: tissue damage releases more autoantigens, further fueling immune attack and progressive organ decline (Eden et al., 2025). Autoimmune disease pathophysiology reflects a dynamic interaction of genes, environment, and immune dysfunction, explaining both diagnostic difficulty and clinical heterogeneity (Song et al., 2024).

2.2 Major autoimmune diseases

Although over 80 autoimmune diseases exist, four major examples—RA, SLE, T1D, and MS—illustrate their diversity.

RA: It is a systemic autoimmune disorder primarily affecting synovial joints. Chronic synovial inflammation leads to pain, swelling, cartilage destruction, and bone erosion. Key drivers include autoreactive T cells, proinflammatory cytokines (TNF-α, IL-6), and autoantibodies such as rheumatoid factor (RF) and anticitrullinated protein antibodies (ACPAs). RA also has systemic effects on the lungs, heart, and eyes (Afzal et al., 2024). Without treatment, it causes severe deformities, but early immunomodulatory therapies improve prognosis.

SLE: It is a prototypical systemic disease with highly variable manifestations, including skin rashes, arthritis, kidney involvement, and neurological symptoms. Its hallmark is the production of diverse autoantibodies (notably anti-dsDNA and anti-Smith), which form immune complexes and drive tissue inflammation (Afzal et al., 2024). Risk factors include genetic predisposition (HLA-DR2, DR3), hormonal influences (female predominance), and triggers such as UV light or infections. The disease course is unpredictable, with alternating flares and remissions, and untreated cases risk life-threatening organ damage (Schwenck et al., 2023).

T1D: It is an organ-specific autoimmune disease where autoreactive T cells destroy pancreatic β-cells, resulting in insulin deficiency and hyperglycemia. Genetic susceptibility (HLA-DR3/DR4) interacts with environmental triggers like viral infections (Liu et al., 2025a). Autoantibodies (e.g., anti-GAD65, IA-2) often precede clinical disease, offering a window for early detection, though predicting onset remains difficult. Manifesting mainly in childhood or adolescence, T1D requires lifelong insulin therapy and is associated with vascular, renal, and neurological complications (Keane & Topol, 2018).

MS: It is a chronic demyelinating disease of the central nervous system. Autoreactive T cells breach the blood–brain barrier, activating microglia and destroying myelin, leading to impaired nerve conduction (Ghassemi, Oakden-Rayner, & Beam, 2021). Symptoms include vision problems, motor weakness, sensory deficits, and cognitive impairment. Risk factors include HLA-DRB1*1501, low vitamin D, and Epstein–Barr virus. The course varies: relapsing-remitting, progressive, or mixed. MRI is the primary diagnostic tool but often detects damage after significant demyelination (Floridi et al., 2018). The diversity of autoimmune diseases is illustrated in Table 15.1, which summarizes their mechanisms, key biomarkers, and major diagnostic challenges. These four conditions showcase the spectrum of autoimmunity: systemic vs. organ-specific, antibody- vs. T-cell-driven, and chronic vs. relapsing patterns.

2.3 Challenges in conventional diagnostic approaches

Diagnosis of autoimmune diseases typically relies on **clinical evaluation, serology, imaging, and histopathology** (World Health Organization, 2021). While indispensable, these approaches have important limitations:

- **Clinical assessment** often misses early disease because initial symptoms (fatigue, fever, arthralgia) are vague and overlap with other disorders.
- **Serological biomarkers** (e.g., ANA, RF) lack sensitivity and specificity. Many autoantibodies are neither universal nor exclusive to a single disease, leading to false positives or negatives.
- **Imaging** detects established damage rather than preclinical disease; MRI lesions in MS or joint erosions in RA appear only after irreversible injury.
- **Histopathology** provides definitive evidence, but it is invasive, costly, and prone to sampling errors.
- **Siloed interpretation**: Conventional methods rarely integrate data across clinical, genetic, and environmental dimensions, limiting their predictive power.

Consequently, autoimmune disease diagnosis is often delayed, imprecise, and poorly predictive of its future course. This highlights the urgent need for advanced computational tools that can integrate multimodal data, identify hidden patterns, and enable earlier, more accurate predictions (Kulkarni & Singh, 2023).

3 Artificial intelligence and ML basics

3.1 Core concepts of AI and ML

AI is the simulation of human cognitive processes—reasoning, learning, problem-solving, and decision-making—by machines, particularly computer systems. In healthcare, AI leverages algorithms and computational models to interpret complex data, identify hidden relationships, and support clinical decision-making (Kelly, Karthikesalingam, Suleyman, Corrado, & King, 2019). Its strength lies in handling vast amounts of structured and unstructured data, making it highly relevant for disease prediction, especially where genetic, molecular, and environmental interactions must be considered together.

TABLE 15.1 Key autoimmune diseases: Mechanisms, biomarkers, and diagnostic challenges.

Disease	Pathophysiology/key mechanism	Common biomarkers	Major diagnostic challenges
Rheumatoid arthritis (RA)	Synovial inflammation, autoreactive T cells, proinflammatory cytokines (TNF-α, IL-6)	RF, ACPAs	Overlap with other arthritis, early nonspecific joint symptoms
Systemic lupus erythematosus (SLE)	Immune complex formation, autoantibodies (anti-dsDNA, anti-Smith)	ANA, anti-dsDNA, anti-Smith	Variable symptoms, flares/remissions, lack of universal biomarkers
Type 1 diabetes mellitus (T1D)	T-cell-mediated β-cell destruction	Anti-GAD65, IA-2, ZnT8 antibodies	Predictive autoantibodies not always present, onset hard to forecast
Multiple sclerosis (MS)	Autoreactive T cells, demyelination, microglial activation	MRI lesions, oligoclonal bands	MRI detects late-stage changes, overlap with other demyelinating diseases

ML, a subset of AI, focuses on algorithms that learn from data without following fixed rules. Unlike traditional systems, ML models improve their performance as they encounter more data, a key advantage for medical prediction, where patterns evolve across populations. In autoimmune diseases, ML can process genomic profiles, imaging, and EHRs to identify subtle predictors beyond human detection (Kozinetz, Berikov, Semenova, & Klimontov, 2024).

ML models typically map input features (e.g., genetic markers, immune cell counts, lifestyle data) to output labels (e.g., disease status). They are trained on labeled datasets and validated on unseen data to test generalizability (Ding, 2024). Feature engineering, selecting or transforming raw variables into meaningful predictors, is central to enhancing performance, such as integrating multiomics data into risk indices. Evaluation metrics like accuracy, precision, recall, F1-score, and area under the receiver operating characteristic curve (AUC-ROC) determine model reliability, especially for rare and heterogeneous diseases like autoimmune disorders (Almutairi, Abbod, & Hunaiti, 2025).

3.2 Types of ML (supervised, unsupervised, deep learning)

ML approaches are broadly categorized into supervised, unsupervised, and deep learning methods, each with unique applications in autoimmune disease prediction.

Supervised learning: It uses labeled datasets where both inputs and outputs are known. The aim is to learn the mapping from inputs to outputs, then apply it to new data. For instance, supervised ML can predict T1D risk from genetic and immunological profiles (Salehi et al., 2024). Common methods include logistic regression, support vector machines (SVMs), decision trees, random forests, and gradient boosting. These models are widely used in early diagnosis and risk prediction based on historical patient data (Saccà et al., 2019).

Unsupervised learning: It works on datasets without predefined labels, uncovering hidden patterns or groupings. Clustering techniques, such as k-means and hierarchical clustering, are used to identify patient subgroups with shared biological or clinical features. This is valuable for patient stratification, where individuals with the same autoimmune disease may differ in severity, progression, or treatment response. For example, clustering SLE patients by biomarker profiles can reveal phenotypes that support personalized care (McKinley et al., 2020; Saccà et al., 2019).

Deep learning: It is a specialized branch of ML that employs artificial neural networks with multiple layers to capture complex, nonlinear data relationships. Particularly suited for high-dimensional biomedical data, it excels in image, genomics, and proteomics analysis. Convolutional neural networks (CNNs) can detect early tissue damage in histopathology or radiographic scans, while recurrent neural networks (RNNs) handle time-series data, such as wearable-device outputs monitoring disease progression. Unlike traditional ML, deep learning reduces reliance on manual feature engineering by extracting relevant features directly from raw data (Salem et al., 2020). However, it often demands large datasets and raises interpretability concerns, which is a challenge in clinical practice.

Collectively, these methods provide complementary strengths: supervised learning delivers precise predictions with labeled data, unsupervised learning discovers new disease subtypes, and deep learning enables advanced analysis of complex biomedical datasets (Rovira et al., 2022).

3.3 Importance of data in predictive modeling

The success of AI/ML in predicting autoimmune diseases depends heavily on the quality, quantity, and diversity of data used to train and validate models. Data is the foundation for building algorithms that generalize to real-world scenarios.

Data quality and preprocessing: Robust ML models require clean, curated datasets. Missing values, noise, and inconsistencies reduce predictive power. Preprocessing steps like normalization, imputation, and standardization are critical. For example, EHRs often contain incomplete or inconsistent entries, necessitating harmonization before ML use (Rovira et al., 2022).

Data quantity and diversity: Larger datasets help models learn complex patterns and reduce overfitting. Yet, many autoimmune disorders are rare, limiting the available data. Collaborative efforts, multicenter studies, and open-access databases help overcome scarcity. Diversity in datasets also minimizes bias; models trained only on one demographic may fail when applied to others (Gessert et al., 2020).

Multimodal data integration: Autoimmune prediction often requires combining genomics, proteomics, imaging, lifestyle, and clinical records. Such integration offers a holistic understanding of risk and progression. For example, merging genomic data with clinical parameters may improve early identification of T1D risk, while integrating imaging with laboratory findings can enhance the detection of RA. Ensemble methods and deep learning architectures are particularly suited for handling such heterogeneous datasets (Ng, Masarone, Watson, & Barnes, 2023).

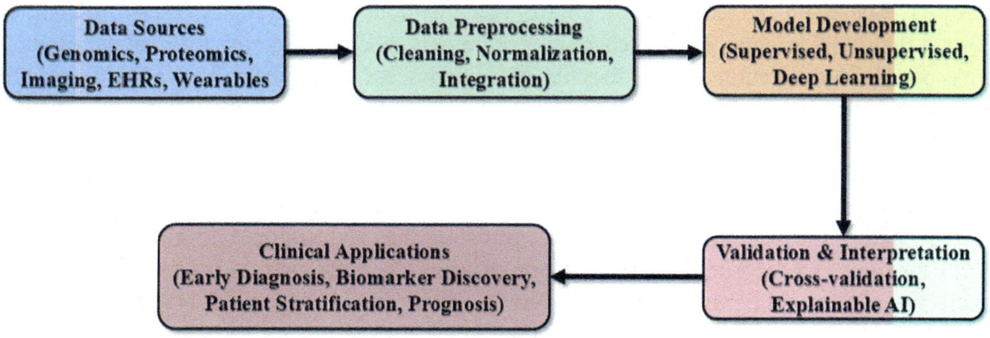

FIG. 15.1 Integrative data-to-clinic workflow using artificial intelligence. Raw data from genomic, proteomic, and imaging platforms are prepared for use in ML after thorough validation in the AI workflow.

Importance of data annotation: Supervised models depend on well-annotated data with reliable labels. Mislabeling, due to diagnostic uncertainty or overlapping symptoms, propagates errors in predictions. Standardized annotation protocols are therefore essential to ensure reliability, especially in autoimmune conditions that often resemble one another (La Rosa et al., 2022).

Ethical and privacy concerns: Using sensitive patient data introduces ethical and legal challenges. Regulations such as General Data Protection Regulation (GDPR) and Health Insurance Portability and Accountability Act (HIPAA) mandate strict protections, emphasizing anonymization and secure handling. Federated learning offers a promising solution, enabling institutions to collaboratively train models without sharing raw patient data, thus safeguarding privacy while leveraging large datasets (Ali, Shaheen, Mabrouk, & Aborizka, 2022). The general workflow of AI and ML in autoimmune disease prediction is shown in Fig. 15.1, highlighting the progression from raw data acquisition to clinical deployment.

Ultimately, the predictive power of AI and ML is only as strong as the data they are built upon. High-quality, diverse, and ethically managed data not only enhance model performance but also foster clinician and patient trust in adopting AI-driven systems for autoimmune disease prediction (Wang et al., 2024).

4 Applications of AI and ML in autoimmune disease prediction

4.1 Early diagnosis and risk prediction

Early diagnosis of autoimmune diseases remains a major clinical challenge, as most conditions present with vague, overlapping symptoms like fatigue, mild inflammation, or joint pain. Conventional methods—clinical evaluation, serological markers, and imaging—often lack sensitivity in the early phase, leading to delayed treatment and irreversible tissue damage. AI and ML approaches help overcome these limitations by analyzing large-scale, multidimensional datasets to uncover subtle disease-associated patterns (Orange et al., 2018).

Supervised algorithms such as random forests, SVMs, and gradient boosting have shown promise in classifying individuals at risk. For example, in **RA**, predictive models combining genetic variants, family history, lifestyle, and environmental exposures like smoking outperform standard screening tools in identifying high-risk individuals. In **SLE**, longitudinal ML models have successfully forecasted disease flares months in advance, enabling timely therapeutic intervention. Similarly, in **T1D**, ML-based analyses of pancreatic autoantibodies and immune profiles can predict disease onset years before symptom manifestation, shifting care from reactive to proactive (Gerussi et al., 2022). Table 15.2 summarizes commonly used ML approaches, their applications in autoimmune disease prediction, and representative outcomes.

Integration of AI with **EHRs** has further advanced real-time risk prediction. By continuously monitoring clinical, laboratory, and imaging data, AI tools can flag patients with early autoimmune features, recommend confirmatory tests, or suggest specialist referrals. At a population level, such systems facilitate **screening and preventive trials**, improving early intervention and reducing disease burden (Moingeon, 2023).

4.2 Biomarker discovery and validation

Biomarkers play a central role in diagnosis, prognosis, and treatment monitoring, but conventional discovery methods are often limited by the heterogeneity of autoimmune disorders. AI-driven approaches in multiomics research—genomics, transcriptomics, proteomics, metabolomics, and microbiomics—have transformed biomarker discovery (Rauniyar et al., 2022).

TABLE 15.2 Machine learning approaches in autoimmune disease prediction.

ML approach	Application in autoimmune diseases	Example outcome
Supervised learning (e.g., random forest, SVM)	RA risk prediction using genetics and lifestyle data; T1D onset prediction via autoantibodies	Identifies high-risk individuals before symptom onset
Unsupervised learning (e.g., k-means, hierarchical clustering)	SLE patient stratification based on biomarkers	Defines clinical subgroups for personalized treatment
Deep learning (CNNs, RNNs)	MRI lesion detection in MS; transcriptomic pattern recognition in SLE	Detects subtle imaging/genomic signatures, forecasts progression
Ensemble learning	Combining outputs of multiple models for biomarker validation	Improves reproducibility across cohorts

Deep learning models, particularly CNNs, excel in analyzing high-dimensional datasets. For instance, CNNs applied to gene expression data have revealed transcriptional signatures predictive of **MS progression**. In T1D, neural networks studying cytokine and immune cell patterns have identified early immune biomarkers that predict onset before hyperglycemia. In **SLE**, ML-assisted proteomic and metabolomic analyses have identified biomarker panels more accurate than conventional antinuclear antibody testing (Ali, Naeem, Tariq, & Kaddoum, 2023).

AI also enhances **biomarker validation**, a critical step for clinical translation. Ensemble learning techniques integrate outputs from multiple models to validate biomarker signatures across independent cohorts, ensuring reproducibility. Imaging-based biomarker discovery has benefited as well; ML models analyzing MRI scans in MS detect microstructural brain changes as early disease indicators. These insights improve diagnostic confidence, support personalized therapy monitoring, and guide adaptive treatment strategies (Rauniyar et al., 2022).

4.3 Patient stratification and prognosis

Autoimmune diseases are clinically heterogeneous, with patients showing variability in organ involvement, severity, and treatment response. Traditional diagnostic frameworks often generalize management, resulting in suboptimal outcomes. AI-driven stratification techniques provide solutions by clustering patients into subgroups with distinct phenotypes and prognostic trajectories.

Unsupervised ML approaches, such as k-means and hierarchical clustering, have been applied to SLE patients, identifying subgroups defined by organ-specific manifestations and serological signatures (Sohan & Basalamah, 2023). This stratification supports personalized treatment selection instead of broad immunosuppression. In **RA**, predictive ML models forecast responsiveness to biologic therapies, reducing unnecessary drug exposure and healthcare costs. Similarly, in MS, integrated models using clinical, genetic, and imaging data predict long-term disability progression, empowering physicians to plan interventions accordingly.

Prognostic modeling further extends to predicting flares, remission likelihood, and resistance to therapies. **Time-series models** trained on longitudinal data anticipate disease exacerbations, allowing clinicians to intervene earlier. These capabilities enhance patient quality of life, optimize healthcare resources, and advance precision medicine in autoimmune disease care (De Falco, Cioppa, Koutny, Scafuri, & Tarantino, 2024).

4.4 Case studies and real-world applications

AI and ML applications are no longer confined to research but are steadily transitioning into clinical practice. Several case studies highlight their real-world impact.

In **RA**, AI-based decision support systems predict patient responses to biologics, such as TNF inhibitors, reducing trial-and-error prescribing. In **SLE**, algorithms analyzing EHRs have flagged patients at risk of hospitalization from severe flares, enabling preventive care strategies. For **MS**, deep learning integrated with MRI analysis identifies early demyelinating lesions in clinically isolated syndrome patients and tracks lesion progression for dynamic therapy adjustment (Nguyen et al., 2022).

T1D management has also embraced AI integration. Predictive models, coupled with continuous glucose monitoring and wearables, provide real-time alerts for hypoglycemia risk, aiding both prediction and ongoing care. The scalability of these systems is demonstrated through **cloud-based platforms**, which unify heterogeneous data—genomic, imaging, clinical, and lifestyle—into predictive frameworks accessible in clinical workflows (Rehman et al., 2022). Pilot programs in healthcare systems show that AI-based autoimmune prediction tools can be implemented efficiently without disrupting physician practice, underscoring their readiness for widespread adoption.

5 Integration with multiomics data

Autoimmune diseases arise from intricate interactions among genetic predisposition, immune dysregulation, environmental triggers, and tissue-specific factors. Single-parameter approaches fail to capture this complexity, making multiomics integration—genomics, transcriptomics, proteomics, metabolomics, microbiomics, and imaging—essential for a system-level perspective. AI and ML enable the integration and interpretation of such high-dimensional datasets, uncovering predictive biomarkers, disease subtypes, and therapeutic targets (Tang et al., 2025).

5.1 Genomic and proteomic data applications

Genomic and proteomic datasets are central to autoimmune disease research. **Genome-wide association studies** have identified numerous susceptibility loci in RA, SLE, and MS. However, translating these associations into functional insights requires advanced computational approaches (Mossotto et al., 2017). AI/ML models, such as deep learning and Bayesian frameworks, reveal nonlinear interactions between genetic variants and disease phenotypes.

Polygenic risk scores (PRS) enhanced by ML provide personalized risk predictions. In T1D, ML-driven PRS integrating **HLA and non-HLA variants** stratifies at-risk individuals with high accuracy. Similarly, exome-sequencing datasets analyzed via AI algorithms prioritize rare variants linked to autoimmune pathogenesis (Khan et al., 2025).

Proteomics adds a dynamic layer, reflecting protein abundance and posttranslational modifications. **Mass spectrometry-based proteomics,** coupled with ML classifiers, has identified protein signatures distinguishing lupus nephritis activity from remission. In RA, proteomic panels analyzed with random forest and SVMs predict therapeutic responses to methotrexate and biologics (Tjalsma, Schaeps, & Swinkels, 2008).

A promising approach is **genomic–proteomic integration**, linking variants to downstream protein expression and disease outcomes. For instance, causal inference algorithms establish genotype-to-phenotype pathways, helping identify biomarkers that are both predictive and mechanistically relevant. This dual insight supports biomarker discovery and novel drug development.

5.2 Imaging-based prediction

Medical imaging contributes crucial insights into autoimmune diseases, particularly those affecting the CNS **(MS), joints (RA), and kidneys (lupus nephritis)**. Conventional imaging interpretation depends on expert evaluation, which is often limited by subjectivity. AI, especially CNNs, can extract subtle and reproducible features, enhancing diagnostic accuracy (Hasin, Seldin, & Lusis, 2017).

In MS, ML models applied to MRI detect demyelinating lesions earlier than traditional methods. Deep learning quantifies lesion burden, cortical thinning, and atrophy—prognostic indicators of disease progression—now used in decision-support tools. RA imaging has also advanced with AI. ML algorithms analyzing ultrasound and MRI scans predict synovitis severity and risk of joint erosion, enabling earlier therapeutic intervention. In SLE, AI applied to renal biopsy imaging and MRI predicts nephritis severity, linking histopathology with long-term renal prognosis.

Beyond diagnosis, AI supports **longitudinal disease monitoring**. Repeated imaging, analyzed through ML pipelines, enables tracking of treatment response, relapse risk, and complications. Combined with molecular data, imaging-derived biomarkers enhance personalized patient management (Khan et al., 2019).

5.3 Multimodal data integration

The greatest potential of AI lies in **multimodal integration**, combining diverse data sources into unified predictive models. Autoimmune diseases are multifactorial, and integrating genomics, transcriptomics, proteomics, metabolomics, microbiome profiles, imaging, and clinical data provides a comprehensive view. In SLE, integrating **genomic risk loci, transcriptomic profiles, and clinical features** through deep learning improved flare prediction compared to single-omic models. In MS, models combining **MRI, genetic risk scores, and immunological markers** identified aggressive disease subtypes earlier. In T1D, ML frameworks merging autoantibody profiles, metabolomics, and genetic predisposition achieved high accuracy in forecasting disease onset (Yan, Risacher, Shen, & Saykin, 2017).

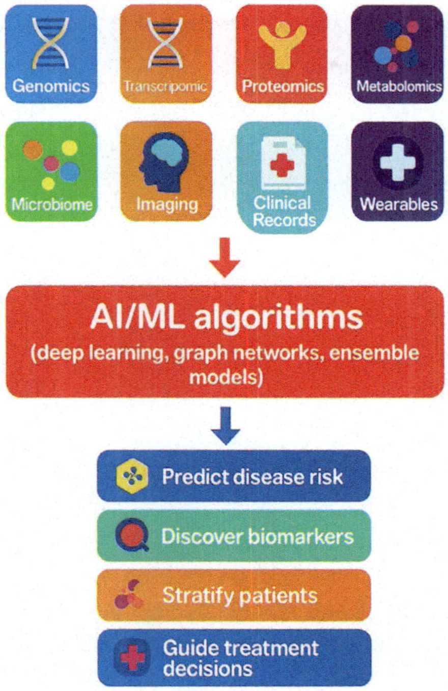

FIG. 15.2 Multiomics and digital health integration for precision medicine. Biomedical and multiomics data are fed into AI platforms to generate meaningful predictions on disease risk, discover novel biomarkers linked to the disease, and mentor treatment strategies.

Methodologically, **multimodal deep learning, graph neural networks, and data fusion algorithms** excel at modeling cross-layer interactions. For example, linking a genetic variant to altered protein expression and imaging changes provides mechanistic explanations for disease manifestations. Emerging **federated learning** techniques enable collaborative model training across institutions while preserving patient privacy, a crucial step for building robust, generalizable AI models. From a clinical standpoint, **multimodal integration facilitates precision medicine**. By identifying molecularly defined patient subgroups, AI can guide personalized therapy, predict drug resistance, and optimize disease management strategies. For example, specific proteogenomic imaging signatures may indicate which RA patients will benefit from biologics versus conventional DMARDs (Delhalle, Bode, Balling, Ollert, & He, 2018). As illustrated in Fig. 15.2, AI/ML systems integrate heterogeneous datasets, including omics, imaging, and wearable-derived data, to generate comprehensive predictions that support precision medicine in autoimmune diseases.

Looking ahead, cloud-based AI platforms will enable **real-time integration of multiomics and imaging** data in clinical settings. Such systems could provide clinicians with actionable dashboards, combining genetic risk, biomarker trends, imaging results, and therapy predictions, bridging the gap between research and bedside practice (Jaeckle et al., 2025). Table 15.3 summarizes the differences between conventional diagnostics and AI/ML approaches in predicting autoimmune diseases.

6 Challenges and ethical considerations

While AI and ML hold transformative promises for predicting autoimmune diseases, their integration into clinical practice faces several scientific, technical, ethical, and regulatory hurdles. Understanding these challenges is essential to ensure that AI-driven innovations not only achieve high predictive accuracy but also align with patient safety, clinical trust, and broader societal values (Gurovich et al., 2019).

6.1 Data limitations and heterogeneity

One of the most pressing challenges in applying AI/ML to autoimmune disease prediction is the issue of **data scarcity and heterogeneity**. Autoimmune disorders, such as SLE, MS, and T1D, are relatively rare compared to other chronic diseases. Consequently, large and balanced datasets required for training robust ML models are often lacking. Small cohort

TABLE 15.3 Conventional diagnostics vs. AI/ML approaches: Key differences in autoimmune disease prediction.

Dimension	Conventional approaches	AI/ML-based approaches
Diagnosis	Autoantibodies, imaging, histopathology	Early risk detection using genomics, proteomics, imaging, and EHR-based models
Prediction	Risk factors (family history, lifestyle)	Integration of multiomics and machine learning to forecast onset
Biomarker discovery	Slow, single-marker studies	AI-driven discovery of biomarker panels across omics datasets
Patient stratification	Broad treatment categories	ML-based clustering for personalized therapies
Prognosis	Clinical scoring, limited accuracy	Longitudinal ML models predicting flares, progression, and treatment response
Limitations	Late detection, poor specificity, invasiveness	Data scarcity, interpretability, ethical/privacy concerns

sizes may lead to **overfitting**, where models perform well on training data but fail to generalize to new patient populations (Xie, Huang, Chen, & Li, 2020).

Heterogeneity in data sources further complicates model development. Autoimmune diseases are multifactorial, influenced by genetics, environmental exposures, immunological markers, and clinical phenotypes. Patient populations differ across geography, ethnicity, and healthcare settings, introducing variability in baseline characteristics. In addition, multiomics data (genomics, transcriptomics, proteomics) and imaging modalities are often generated using different platforms, leading to inconsistencies in data resolution, normalization, and annotation (Xu et al., 2021).

Moreover, missing or incomplete data in clinical records is a common issue. In real-world healthcare systems, EHRs may contain noise, inconsistent coding, or gaps in patient histories. Such challenges limit the ability of ML algorithms to generate accurate predictions. Advanced data harmonization techniques, federated learning frameworks, and cross-institutional data sharing are being explored to overcome these limitations, but their adoption remains slow (Yi et al., 2020).

6.2 Model interpretability and clinical trust

Another major barrier is **model interpretability**, often referred to as the "black-box" problem in AI. Many high-performing models, particularly deep learning networks, achieve impressive accuracy but provide little transparency regarding how predictions are derived (Yoo et al., 2021). In autoimmune diseases, where treatment decisions can carry lifelong implications, clinicians are often reluctant to rely on opaque models (Yuan, Ge, & Xing, 2020).

For instance, if an ML model predicts a high risk of MS progression, clinicians and patients need to understand which biomarkers, genetic features, or imaging findings contributed to that decision. Without explainability, such models risk undermining **clinical trust** and may not be adopted in routine care (Schlaeger et al., 2023).

Efforts are being made to address this through **XAI**methods, which use visualization techniques (e.g., heatmaps for imaging, feature importance scores for omics data) to highlight relevant inputs driving predictions (Huisman, van Ginneken, & Harvey, 2024). In addition, rule-based ML frameworks and hybrid models combining statistical inference with deep learning are being developed to improve interpretability. Regulatory bodies, such as the Food and Drug Administration (FDA) and European Medicines Agency (EMA), are also increasingly emphasizing the need for transparent AI systems in healthcare (Sieber et al., 2024).

Building clinical trust further requires **rigorous validation** of models across diverse patient cohorts (Rondinella et al., 2023). A predictive algorithm that performs well in one hospital or population may fail in another due to demographic or technical differences. Thus external validation and prospective clinical trials remain essential before deployment (Ng et al., 2023).

6.3 Ethical, legal, and privacy concerns

The integration of AI and ML into autoimmune disease prediction also raises important **ethical, legal, and privacy issues** (Collorone et al., 2024). Patient data, especially genomic and health records, are highly sensitive. Large-scale data collection for AI development may conflict with patient autonomy and data protection laws, such as **GDPR** in Europe and **HIPAA** in the United States (Xie et al., 2020).

A central ethical dilemma concerns **informed consent**. Many patients may not fully understand how their genomic or health data will be used in AI training, particularly if the data is shared across institutions or countries (Aslam et al., 2022). Ensuring transparency in data usage, storage, and sharing is critical to maintaining patient trust.

Another concern is **algorithmic bias**. If ML models are trained on datasets that disproportionately represent certain ethnic or geographic populations, predictions may be less accurate for underrepresented groups. In autoimmune diseases, where prevalence and clinical presentation already vary across populations, biased models risk exacerbating health disparities. Ethical AI design requires proactive measures to ensure fairness, equity, and inclusivity (Amin, Martínez-Heras, Ontaneda, & Carrasco, 2024).

Finally, **legal accountability** poses an unresolved challenge. If an AI-driven prediction leads to a clinical misdiagnosis or inappropriate treatment, determining liability—whether it lies with the clinician, the AI developer, or the healthcare system—remains a gray area. This uncertainty underscores the need for clear regulatory frameworks and governance structures to guide AI use in healthcare.

7 Future perspectives

AI and ML in autoimmune disease prediction are poised to transform diagnosis, treatment, and long-term disease management. As technology evolves, the focus is shifting toward making models more interpretable, integrating digital health tools for real-time monitoring, and advancing toward precision medicine. These directions hold the potential to overcome current limitations and reshape patient care (Mansour et al., 2024).

7.1 Explainable AI

XAI is central to bridging the gap between high-performing but opaque models and the clinical need for transparency. For autoimmune diseases, XAI frameworks can help clinicians understand which biomarkers, imaging patterns, or genomic features contribute to a given prediction. This transparency will not only increase trust but also improve adoption in routine practice. Tools such as feature attribution maps, rule-based learning systems, and interpretable neural networks are being developed to highlight disease-relevant variables. In autoimmune contexts, this could mean identifying which immune cell signatures or genetic variants drive a prediction of disease onset or flare, thus supporting clinical reasoning rather than replacing it (Thabet, Shedeed, Al-Berry, & Khattab, 2025).

7.2 Digital health and wearable integration

The future of autoimmune disease management will also be shaped by the integration of **digital health technologies** and **wearable devices**. Continuous health monitoring through wearables (e.g., sensors tracking heart rate variability, sleep patterns, or physical activity) can provide real-time insights into patient well-being. For diseases like RA or MS, wearable-derived data on mobility, fatigue, or stress levels can serve as early warning signals of disease progression or flare-ups (Zhao, Duan, Cui, & Xie, 2023).

Coupling these data streams with AI-powered analytics enables the creation of personalized dashboards for patients and clinicians, facilitating proactive intervention rather than reactive treatment. Digital health ecosystems, supported by smartphones and cloud platforms, also allow for remote patient monitoring and telemedicine, making care more accessible and scalable (Ng et al., 2023).

7.3 Toward precision medicine in autoimmune diseases

The ultimate goal of integrating AI and ML into autoimmune disease prediction is to move toward **precision medicine**, tailoring prevention, diagnosis, and treatment strategies to the unique biological and clinical profile of each patient. By combining **multiomics data (genomics, proteomics, metabolomics)** with clinical records and lifestyle information, AI systems can generate comprehensive disease risk models. This can enable personalized therapeutic approaches, from predicting which patients are most likely to respond to biologic therapies to identifying those at higher risk of adverse drug reactions (Danieli et al., 2024).

In the long term, precision medicine approaches could not only optimize treatment outcomes but also help in the early identification of individuals at risk of developing autoimmune diseases, enabling preventive interventions. The integration of XAI, wearable technology, and multimodal data promises a future where autoimmune disease management is predictive, personalized, and participatory, ultimately improving both patient quality of life and healthcare efficiency.

8 Conclusion

Autoimmune diseases present a persistent and complex challenge in clinical medicine. Because they are so different from each other, have similar clinical features, and progress in unpredictable ways, it is very hard to diagnose them quickly and manage them in a personalized way. Standard techniques like serological testing, imaging, and histopathology, though essential, frequently identify disease only after substantial damage and offer restricted predictive utility. Because of this, many patients have to wait longer for a diagnosis, are given the wrong diagnosis, or do not get enough help with treatment planning. These problems make it clear that we need new ideas that can connect early detection with personalized care as soon as possible.

In this context, AI and ML have become powerful tools for analyzing large and diverse biomedical datasets. AI and ML can combine information from genomics, proteomics, imaging, and EHRs in ways that traditional statistical methods cannot. This enables them to uncover hidden relationships and predictive signatures. Their applications span the entire range of clinical situations, such as predicting when a disease will onset in individuals at high risk, identifying new biomarkers, grouping patients based on how their disease is likely to progress, and assisting doctors in making treatment decisions. These technologies offer a novel approach to precision medicine in the management of autoimmune diseases by shifting from reactive diagnosis to proactive prediction.

So far, the evidence shows that things are getting better. Supervised learning models have demonstrated they can identify early signs of disease risk, unsupervised methods have uncovered clinically significant patient subgroups, and deep learning has been highly accurate in imaging and omics-based analyses. In addition, AI-powered clinical decision support and wearable-integrated monitoring systems are beginning to show real-world usefulness. Even though these advancements have worked out well, there are still significant challenges to address. The lack of data on rare autoimmune diseases, the differences between populations, and the fact that many algorithms are "black boxes" continue to make it difficult for doctors to use them in practice. Ethical and legal issues surrounding privacy, fairness, and accountability further underscore the importance of using the technology carefully and responsibly.

Several paths look promising for the future. XAI is important for building trust in algorithms by making it clear how they make decisions. Digital health integration, which includes wearable sensors and remote monitoring platforms, makes it possible to manage diseases in a way that is always available and tailored to each person. Combining genomic, proteomic, imaging, and clinical data into one set of data, called multimodal data fusion, will improve predictive accuracy even more and may also show how diseases work on a mechanistic level. These innovations together make it possible to create a more patient-centered model of care that takes into account each person's biology, lifestyle, and response to treatment.

In the end, AI and ML should not be seen as substitutes for clinical knowledge; instead, they should be seen as powerful tools that help people make informed choices. Data scientists, clinicians, ethicists, and policymakers must work together to ensure that these technologies are used safely in healthcare systems. If these technologies are developed and used carefully, they could completely change how we predict and treat autoimmune diseases. AI and ML could greatly improve the outcomes and quality of life for people with these chronic and often debilitating conditions by speeding up diagnosis, making treatments more accurate, and assisting with long-term monitoring.

AI-based tool usage: This manuscript was developed with the support of AI tools for tasks, including data analysis, language refinement, and literature review. Although AI-assisted tools contributed to certain aspects of the writing process, the authors take full responsibility for the content, interpretations, and conclusions presented in the manuscript. AI tools were not involved in the development of the research findings.

Further reading

Afzal, M., Sah, A. K., Agarwal, S., Tanzeel, A., Elshaikh, R. H., Alobeidli, F. A., et al. (2024). Advancements in the treatment of autoimmune diseases: integrating artificial intelligence for personalized medicine. *Trends in Immunotherapy*, 2573-5985. *8*(2). Available from: https://systems.enpress-publisher.com/index.php/ti/article/download/8970/4586. doi:10.24294/ti8970.

Agarwal, R., Bjarnadottir, M., Rhue, L. et al. (2023). Addressing algorithmic bias and the perpetuation of health disparities in AI systems. *The Lancet Digital Health*, *5*(7), e300–e310.

Ali, M., Naeem, F., Tariq, M., & Kaddoum, G. (2023). Federated learning for privacy preservation in smart healthcare systems: a comprehensive survey. *IEEE Journal of Biomedical and Health Informatics*, 2168-2208. *27*(2), 778–789. Available from: http://ieeexplore.ieee.org/xpl/RecentIssue.jsp?punumber=6221020. doi:10.1109/JBHI.2022.3181823.

Ali, N. M., Shaheen, M., Mabrouk, M. S., & Aborizka, M. (2022). Machine learning-based models for detection of biomarkers of autoimmune diseases by fragmentation and analysis of miRNA sequences. *Applied Sciences*, 2076-3417. *12*(11), 5583. doi:10.3390/app12115583.

Almutairi, E., Abbod, M., & Hunaiti, Z. (2025). Prediction of diabetes using statistical and machine learning modelling techniques. *Algorithms*, 1999-4893. *18*(3), 145. doi:10.3390/a18030145.

Amann, J., Blasimme, A., Vayena, E., & Frey, D. (2020). Madai VI; Precise4Q consortium. Explainability for artificial intelligence in healthcare: a multidisciplinary perspective. *BMC Medical Informatics and Decision Making*, *20*(1), 310. doi:10.1186/s12911-020-01332-6.

Amann, J., Blasimme, A., Vayena, E., Frey, D., & Madai, V. I. (2020). Explainability for artificial intelligence in healthcare: a multidisciplinary perspective. *BMC Medical Informatics and Decision Making*, 1472-6947. *20*(1). Available from: http://www.biomedcentral.com/bmcmedinformdecismak/. doi:10.1186/s12911-020-01332-6.

Amann, J., Vetter, D., Blomberg, S. N., Christensen, H. C., Coffee, M., Gerke, S., et al. (2022). To explain or not to explain?—Artificial intelligence explainability in clinical decision support systems. *PLOS Digital Health*, 2767-3170. *1*(2). Available from: https://journals.plos.org/digitalhealth/article?id=10.1371/journal.pdig.0000016. doi:10.1371/journal.pdig.0000016.

Amin, M., Martínez-Heras, E., Ontaneda, D., & Carrasco, F. P. (2024). Artificial intelligence and multiple sclerosis. *Current Neurology and Neuroscience Reports*, 1534-6293. *24*(8), 233–243. Available from: https://www.springer.com/journal/11910. doi:10.1007/s11910-024-01354-x.

Aslam, N., Khan, I. U., Bashamakh, A., Alghool, F. A., Aboulnour, M., Alsuwayan, N. M., et al. (2022). Multiple sclerosis diagnosis using machine learning and deep learning: challenges and opportunities. *Sensors*, 1424-8220. *22*(20). Available from: http://www.mdpi.com/journal/sensors. doi:10.3390/s22207856.

Buckner, J. H. (2023). Translational immunology: applying fundamental discoveries to human health and autoimmune diseases. *European Journal of Immunology*, 1521-4141. *53*(12). Available from: http://onlinelibrary.wiley.com/journal/10.1002/(ISSN)1521-4141. doi:10.1002/eji.202250197.

Cao, C. (2025). Leveraging artificial intelligence and machine learning for multimodal omics integration in autoimmune diseases: identifying subtypes, regulatory networks, and biomarkers. *Innovation (New York)*. doi:10.59717/j.xinn-med.2025.100154.

Cheng, X., Meng, X., Chen, R., Song, Z., Li, S., Wei, S., et al. (2024). The molecular subtypes of autoimmune diseases. *Computational and Structural Biotechnology Journal*, 2001-0370. *23*, 1348–1363. doi:10.1016/j.csbj.2024.03.026.

Collorone, S., Coll, L., Lorenzi, M., Lladó, X., Sastre-Garriga, J., Tintoré, M., et al. (2024). Artificial intelligence applied to MRI data to tackle key challenges in multiple sclerosis. *Multiple Sclerosis Journal*, 1477-0970. *30*(7), 767–784. Available from: https://journals.sagepub.com/home/MSJ. doi:10.1177/13524585241249422.

Danieli, M. G., Brunetto, S., Gammeri, L., Palmeri, D., Claudi, I., Shoenfeld, Y., & Gangemi, S. (2024). Machine learning application in autoimmune diseases: state of art and future prospectives. *Autoimmunity Reviews*, 1568-9972. *23*(2), 103496. doi:10.1016/j.autrev.2023.103496.

Danieli, M. G., Casciaro, M., Paladini, A., Bartolucci, M., Sordoni, M., Shoenfeld, Y., & Gangemi, S. (2024). Exposome: epigenetics and autoimmune diseases. *Autoimmunity Reviews*, 1873-0183. *23*(6). Available from: https://www.sciencedirect.com/science/journal/15689972. doi:10.1016/j.autrev.2024.103584.

de Boer, A., van den Bosch, A. M. R., Mekkes, N. J., Fransen, N. L., Dagkesamanskaia, E., Hoekstra, E., et al. (2024). Disentangling the heterogeneity of multiple sclerosis through identification of independent neuropathological dimensions. *Acta Neuropathologica*, 1432-0533. *147*(1). Available from: https://www.springer.com/journal/401. doi:10.1007/s00401-024-02742-w.

De Falco, I., Cioppa, A. D., Koutny, T., Scafuri, U., & Tarantino, E. (2024). Model-free-communication federated learning: framework and application to precision medicine. *Biomedical Signal Processing and Control*, 1746-8094. *87*, 105416. doi:10.1016/j.bspc.2023.105416.

Delhalle, S., Bode, S. F. N., Balling, R., Ollert, M., & He, F. Q. (2018). A roadmap towards personalized immunology. *Npj Systems Biology and Applications*. 2056-7189. *4*(1). Available from: https://www.nature.com/npjsba/. doi:10.1038/s41540-017-0045-9.

Ding, Y. (2024). Advances and challenges in machine learning for diabetes prediction: a comprehensive review. *Applied and Computational Engineering*, 2755-2721. *109*(1), 75–80. doi:10.54254/2755-2721/109/20241437.

Eden, R., Chukwudi, I., Bain, C., Barbieri, S., Callaway, L., de Jersey, S., et al. (2025). A scoping review of the governance of federated learning in healthcare. *Npj Digital Medicine*, 2398-6352. *8*(1). Available from: https://www.nature.com/npjdigitalmed/. doi:10.1038/s41746-025-01836-3.

Esteva, A., Robicquet, A., Ramsundar, B., Kuleshov, V., DePristo, M., Chou, K., et al. (2019). A guide to deep learning in healthcare. *Nature Medicine*, *25*(1), 24–29. doi:10.1038/s41591-018-0316-z.

Falet, J. P. R., Nobile, S., Szpindel, A., Barile, B., Kumar, A., Durso-Finley, J., et al. (2025). The role of AI for MRI-analysis in multiple sclerosis—A brief overview. *Frontiers in Artificial Intelligence*, 2624-8212. *8*. Available from: www.frontiersin.org/journals/artificial-intelligence#. doi:10.3389/frai.2025.1478068.

Floridi, L., Cowls, J., Beltrametti, M., Chatila, R., Chazerand, P., Dignum, V., et al. (2018). AI4People—An ethical framework for a good AI society: Opportunities, risks, principles, and recommendations. *Minds and Machines*, 0924-6495. *28*(4), 689–707. doi:10.1007/s11023-018-9482-5.

Forrest, I. S., Petrazzini, B. O., Duffy, Á., Park, J. K., O'Neal, A. J., Jordan, D. M., et al. (2023). A machine learning model identifies patients in need of autoimmune disease testing using electronic health records. *Nature Communications*, 2041-1723. *14*(1). Available from https://www.nature.com/ncomms/. doi:10.1038/s41467-023-37996-7.

Friedrich, S., Groß, S., König, I. R., Engelhardt, S., Bahls, M., Heinz, J., et al. (2021). Applications of artificial intelligence/machine learning approaches in cardiovascular medicine: a systematic review with recommendations. *European Heart Journal—Digital Health*, 2634-3916. *2*(3), 424–436. Available from: https://academic.oup.com/ehjdh. doi:10.1093/ehjdh/ztab054.

Gerussi, A., Scaravaglio, M., Cristoferi, L., Verda, D., Milani, C., De Bernardi, E., et al. (2022). Artificial intelligence for precision medicine in autoimmune liver disease. *Frontiers in Immunology*, 1664-3224. 13. Available from: https://www.frontiersin.org/journals/immunology#. doi:10.3389/fimmu.2022.966329.

Gessert, N., Krüger, J., Opfer, R., Ostwaldt, A. C., Manogaran, P., Kitzler, H. H., et al. (2020). Multiple sclerosis lesion activity segmentation with attention-guided two-path CNNs. *Computerized Medical Imaging and Graphics*, 1879-0771. *84*. Available from: www.elsevier.com/locate/compmedimag. doi:10.1016/j.compmedimag.2020.101772.

Ghassemi, M., Oakden-Rayner, L., & Beam, A. L. (2021). The false hope of current approaches to explainable artificial intelligence in health care. *The Lancet Digital Health*, 2589-7500. *3*(11), e745. Available from: https://www.sciencedirect.com/journal/the-lancet-digital-health. doi:10.1016/S2589-7500(21)00208-9.

Grote, T., & Keeling, G. (2022). Enabling fairness in healthcare through machine learning. *Ethics and Information Technology*, 1388-1957. *24*(3). doi:10.1007/s10676-022-09658-7.

Gurovich, Y., Hanani, Y., Bar, O., Nadav, G., Fleischer, N., Gelbman, D., et al. (2019). Identifying facial phenotypes of genetic disorders using deep learning. *Nature Medicine*, 1546-170X. *25*(1), 60–64. Available from: http://www.nature.com/nm/index.html. doi:10.1038/s41591-018-0279-0.

Hasin, Y., Seldin, M., & Lusis, A. (2017). Multi-omics approaches to disease. *Genome Biology*, 1474-760X. *18*(1). Available from: http://genomebiology.com/. doi:10.1186/s13059-017-1215-1.

Henderson, J., Condell, J., Connolly, J., Kelly, D., & Curran, K. (2021). Review of wearable sensor-based health monitoring glove devices for rheumatoid arthritis. *Sensors*, 1424-8220. *21*(5), 1576. doi:10.3390/s21051576.

Hickman, E., & Petrin, M. (2021). Trustworthy AI and corporate governance: the EU's ethics guidelines for trustworthy artificial intelligence from a company law perspective. *European Business Organization Law Review*, 1566-7529. *22*(4), 593–625. doi:10.1007/s40804-021-00224-0.

Hoffmann, J., Rheude, A., Neubauer, A., Brendel, C., & Thrun, M. C. (2024). Development of an explainable AI system using routine clinical parameters for rapid differentiation of inflammatory conditions. *Frontiers in Immunology*, 1664-3224. *15*. Available from: https://www.frontiersin.org/journals/immunology#. doi:10.3389/fimmu.2024.1364954.

Huisman, M., van Ginneken, B., & Harvey, H. (2024). The emperor has few clothes: a realistic appraisal of current AI in radiology. *European Radiology*, 1432-1084. *34*(9), 5873–5875. Available from: https://www.springer.com/journal/330. doi:10.1007/s00330-024-10664-0.

Jaeckle, F., Denholm, J., Schreiber, B., Evans, S. C., Wicks, M. N., Chan, J. Y. H. et al. (2025). Machine learning achieves pathologist-level coeliac disease diagnosis. *NEJM AI*, *2*(4), aioa2400738. doi:10.1056/AIoa2400738.

Ji, C., Jiang, T., Liu, L., Zhang, J., & You, L. (2025). Continuous glucose monitoring combined with artificial intelligence: redefining the pathway for prediabetes management. *Frontiers in Endocrinology*, 1664-2392. *16*. doi:10.3389/fendo.2025.1571362.

Keane, P. A., & Topol, E. J. (2018). With an eye to AI and autonomous diagnosis. *Npj Digital Medicine*, 2398-6352. *1*(1). Available from: https://www.nature.com/npjdigitalmed/. doi:10.1038/s41746-018-0048-y.

Kelly, C. J., Karthikesalingam, A., Suleyman, M., Corrado, G., & King, D. (2019). Key challenges for delivering clinical impact with artificial intelligence. *BMC Medicine*, 1741-7015. *17*(1). Available from: http://www.biomedcentral.com/bmcmed/. doi:10.1186/s12916-019-1426-2.

Khan, M. M., Ernst, O., Manes, N. P., Oyler, B. L., Fraser, I. D. C., Goodlett, D. R., & Nita-Lazar, A. (2019). Multi-omics strategies uncover host-pathogen interactions. *ACS Infectious Diseases*, 2373-8227. *5*(4), 493–505. Available from: http://pubs.acs.org/journal/aidcbc. doi:10.1021/acsinfecdis.9b00080.

Khan, S., Mohsen, F. & Shah. Z. (2025). Genetic biomarkers and machine learning techniques for predicting diabetes: Systematic review. *Artificial Intelligence Review*, *58*(41). doi:10.1007/s10462-024-11020-w.

Klonoff, D. C., Bergenstal, R. M., Cengiz, E., Clements, M. A., Espes, D., Espinoza, J., et al. (2025). CGM data analysis 2.0: functional data pattern recognition and artificial intelligence applications. *Journal of Diabetes Science and Technology*, 1932-2968. Available from: https://journals.sagepub.com/home/DST. doi:10.1177/19322968251353228.

Kowalski, M. K., Domżał-Magrowska, D., & Małecka-Wojciesko, E. (2025). Celiac disease—Narrative review on progress in celiac disease. *Foods*, 2304-8158. *14*(6). Available from: www.mdpi.com/journal/foods. doi:10.3390/foods14060959.

Kozinetz, R. M., Berikov, V. B., Semenova, J. F., & Klimontov, V. V. (2024). Machine learning and deep learning models for nocturnal high- and low-glucose prediction in adults with type 1 diabetes. *Diagnostics*, 2075-4418. *14*(7). Available from: http://www.mdpi.com/journal/diagnostics/. doi:10.3390/diagnostics14070740.

Kruta, J., Carapito, R., Trendelenburg, M., Martin, T., Rizzi, M., Voll, R. E., et al. (2024). Machine learning for precision diagnostics of autoimmunity. *Scientific Reports*, 2045-2322. *14*(1). doi:10.1038/s41598-024-76093-7.

Kulkarni, P. A., & Singh, H. (2023). Artificial intelligence in clinical diagnosis: opportunities, challenges, and hype. *JAMA*, 1538-3598. *330*(4), 317–318. Available from: http://jama.jamanetwork.com/journal.aspx. doi:10.1001/jama.2023.11440.

La Rosa, F., Wynen, M., Al-Louzi, O., Beck, E. S., Huelnhagen, T., Maggi, P., et al. (2022). Cortical lesions, central vein sign, and paramagnetic rim lesions in multiple sclerosis: Emerging machine learning techniques and future avenues. *NeuroImage: Clinical*. 2213-1582. *36*. Available from: http://www.journals.elsevier.com/neuroimage-clinical/. doi:10.1016/j.nicl.2022.103205.

Li, T., Sahu, A. K., Talwalkar, A., & Smith, V. (2020). Federated learning: Challenges, methods, and future directions. *IEEE Signal Processing Magazine*, 1053-5888. *37*(3), 50–60. doi:10.1109/msp.2020.2975749.

Lin, M., Guo, J., Gu, Z., Tang, W., Tao, H., You, S., et al. (2025). Machine learning and multi-omics integration: advancing cardiovascular translational research and clinical practice. *Journal of Translational Medicine*, *23*(1), 388. doi:10.1186/s12967-025-06425-2.

Liu, S., Liu, Y., Li, M., Shang, S., Cao, Y., Shen, X., & Huang, C. (2025). Artificial intelligence in autoimmune diseases: a bibliometric exploration of the past two decades. *Frontiers in Immunology*, *16*, 1525462. doi:10.3389/fimmu.2025.1525462.

Liu, X., Shi, J., Jiao, Y., An, J., Tian, J., Yang, Y., & Zhuo, L. (2024). Integrated multi-omics with machine learning to uncover the intricacies of kidney disease. *Briefings in Bioinformatics*, 1467-5463. *25*(5). doi:10.1093/bib/bbae364.

Mansour, R., Abdel-Razeq, H., Al-Hussaini, M., Shamieh, O., Al-Ibraheem, A., Al-Omari, A.et al(2024). Systemic barriers to optimal cancer care in resource-limited countries: Jordanian healthcare as an example. *Cancers*, *16*(6), 1117 10.3390/cancers16061117.

Marion, T. N., & Postlethwaite, A. E. (2014). Chance, genetics, and the heterogeneity of disease and pathogenesis in systemic lupus erythematosus. *Seminars in Immunopathology*, 1863-2300. *36*(5), 495–517. Available from: http://www.springer.com/east/home/generic/search/results?SGWID=5-40109-70-1042641-0. doi:10.1007/s00281-014-0440-x.

McGonagle, D., Watad, A., & Savic, S. (2018). Mechanistic immunological based classification of rheumatoid arthritis. *Autoimmunity Reviews*, 1873-0183. *17*(11), 1115–1123. Available from: http://www.elsevier.com/locate/autrev. doi:10.1016/j.autrev.2018.06.001.

McKinley, R., Wepfer, R., Grunder, L., Aschwanden, F., Fischer, T., Friedli, C., et al. (2020). Automatic detection of lesion load change in multiple sclerosis using convolutional neural networks with segmentation confidence. *NeuroImage: Clinical*, 2213-1582. *25*, 102104. doi:10.1016/j.nicl.2019.102104.

Moingeon, P. (2023). Artificial intelligence-driven drug development against autoimmune diseases. *Trends in Pharmacological Sciences*. 1873-3735. *44*(7), 411–424. Available from: www.elsevier.com/locate/tips. doi:10.1016/j.tips.2023.04.005.

Morabito, A., De Simone, G., Pastorelli, R., Brunelli, L., & Ferrario, M. (2025). Algorithms and tools for data-driven omics integration to achieve multi-layer biological insights: a narrative review. *Journal of Translational Medicine*, 1479-5876. *23*(1). doi:10.1186/s12967-025-06446-x.

Mossotto, E., Ashton, J. J., Coelho, T., Beattie, R. M., MacArthur, B. D., & Ennis, S. (2017). Classification of paediatric inflammatory bowel disease using machine learning. *Scientific Reports*, 2045-2322. *7*(1). doi:10.1038/s41598-017-02606-2.

Mu, S., Wang, W., Liu, Q., Ke, N., Li, H., Sun, F., et al. (2024). Autoimmune disease: a view of epigenetics and therapeutic targeting. *Frontiers in Immunology*, 1664-3224. *15*. doi:10.3389/fimmu.2024.1482728.

Nabizadeh, F., Masrouri, S., Ramezannezhad, E., Ghaderi, A., Sharafi, A. M., Soraneh, S., & Naser Moghadasi, A. (2022). Artificial intelligence in the diagnosis of multiple sclerosis: a systematic review. *Multiple Sclerosis and Related Disorders*, *59*, 103673. doi:10.1016/j.msard.2022.103673.

National Academies Press. (2024). Crosscutting issues in autoimmune diseases—enhancing NIH research on autoimmune disease. *NCBI Bookshelf.* https://www.ncbi.nlm.nih.gov/books/NBK580299/.

Ng, S., Masarone, S., Watson, D., & Barnes, M. R. (2023). The benefits and pitfalls of machine learning for biomarker discovery. *Cell and Tissue Research*, 1432-0878. *394*(1), 17–31. Available from: https://www.springer.com/journal/441. doi:10.1007/s00441-023-03816-z.

Nguyen, D. C., Pham, Q. V., Pathirana, P. N., Ding, M., Seneviratne, A., Lin, Z., et al. (2022). Federated learning for smart healthcare: a survey. *ACM Computing Surveys*. 1557-7341. *55*(3). Available from: http://dl.acm.org/citation.cfm?id=J204. doi:10.1145/3501296.

Obermeyer, Z., Powers, B., Vogeli, C., & Mullainathan, S. (2019). Dissecting racial bias in an algorithm used to manage the health of populations. *Science*, 1095-9203. *366*(6464), 447–453. Available from: https://science.sciencemag.org/content/366/6464/447/tab-pdf. doi:10.1126/science.aax2342.

Orange, D. E., Agius, P., DiCarlo, E. F., Robine, N., Geiger, H., Szymonifka, J., et al. (2018). Identification of three rheumatoid arthritis disease subtypes by machine learning integration of synovial histologic features and RNA sequencing data. *Arthritis and Rheumatology*, 2326-5205. *70*(5), 690–701. Available from: http://onlinelibrary.wiley.com/journal/10.1002/(ISSN)2326-5205. doi:10.1002/art.40428.

Perez, O. D. (2005). Appreciating the heterogeneity in autoimmune disease: multiparameter assessment of intracellular signaling mechanisms. *Annals of the New York Academy of Sciences*, *1062*, 155–164. doi:10.1196/annals.1358.018.

Price, W. N., & Cohen, I. G. (2019). Privacy in the age of medical big data. *Nature Medicine*, 1546-170X. *25*(1), 37–43. Available from: http://www.nature.com/nm/index.html. doi:10.1038/s41591-018-0272-7.

Rauniyar, A., Haileselassie, D., Jha, D., Håkegård, J. E., Bagci, U., Rawat, D. B., & Vlassov, V. (2022). Federated learning for medical applications: a taxonomy, current trends, challenges, and future research directions. *arXiv*, arXiv. 23318422, Available from: https://arxiv.org. doi:10.48550/arXiv.2208.03392.

Reddy, S., Allan, S., Coghlan, S., & Cooper, P. (2020). A governance model for the application of AI in health care. *Journal of the American Medical Informatics Association*, 1527-974X. *27*(3), 491–497. Available from: http://jamia.oxfordjournals.org/content/22/e1. doi:10.1093/jamia/ocz192.

Rehman, A., Abbas, S., Khan, M. A., Ghazal, T. M., Adnan, K. M., & Mosavi, A. (2022). A secure healthcare 5.0 system based on blockchain technology entangled with federated learning technique. *Computers in Biology and Medicine*. doi:10.1016/j.compbiomed.2022.106019.

Rizzo, F., & Houen, G. (2023). Editorial: Immune evasion mechanisms and their role in the pathogenesis of autoimmune disorders. *Frontiers in Immunology*, 1664-3224. *14*. doi:10.3389/fimmu.2023.1267922.

Rondinella, A., Crispino, E., Guarnera, F., Giudice, O., Ortis, A., Russo, G., et al. (2023). Boosting multiple sclerosis lesion segmentation through attention mechanism. *arXiv*, arXiv. 23318422. Available from: https://arxiv.org. doi:10.48550/arXiv.2304.10790.

Rovira, A., Corral, J. F., Auger, C., Valverde, S., Vidal-Jordana, A., Oliver, A., et al. (2022). Assessment of automatic decision-support systems for detecting active T2 lesions in multiple sclerosis patients. *Multiple Sclerosis Journal*, 1477-0970. *28*(8), 1209–1218. Available from: https://journals.sagepub.com/home/MSJ. doi:10.1177/13524585211061339.

Saccà, V., Sarica, A., Novellino, F., Barone, S., Tallarico, T., Filippelli, E., et al. (2019). Evaluation of machine learning algorithms performance for the prediction of early multiple sclerosis from resting-state FMRI connectivity data. *Brain Imaging and Behavior*, 1931-7565. *13*(4), 1103–1114. Available from: http://www.springer.com/east/home?SGWID=5-102-70-173667707-0&changeHeader=true. doi:10.1007/s11682-018-9926-9.

Salehi, F., Lopera Gonzalez, L. I., Bayat, S., Kleyer, A., Zanca, D., Brost, A., et al. (2024). Machine learning prediction of treatment response to biological disease-modifying antirheumatic drugs in rheumatoid arthritis. *Journal of Clinical Medicine*. 2077-0383. *13* (13), 3890. doi:10.3390/jcm13133890.

Salem, M., Valverde, S., Cabezas, M., Pareto, D., Oliver, A., Salvi, J., et al. (2020). A fully convolutional neural network for new T2-w lesion detection in multiple sclerosis. *NeuroImage: Clinical*. 2213-1582. *25*, 102149. doi:10.1016/j.nicl.2019.102149.

Schlaeger, S., Shit, S., Eichinger, P., Hamann, M., Opfer, R., Krüger, J., et al. (2023). AI-based detection of contrast-enhancing MRI lesions in patients with multiple sclerosis. *Insights into Imaging*. 1869-4101. *14*(1). Available from: https://www.springer.com/journal/13244. doi:10.1186/s13244-023-01460-3.

Schwenck, J., Sonanini, D., Cotton, J. M., Rammensee, H., la Fougère, C., Zender, L.et al(2023). Advances in PET imaging of cancer. *Nature Reviews Cancer*, *23*(7), 474–490. doi:10.1038/s41568-023-00576-4.

Sieber, V., Rusche, T., Yang, S., Stieltjes, B., Fischer, U., Trebeschi, S., et al. (2024). Automated assessment of brain MRIs in multiple sclerosis patients significantly reduces reading time. *Neuroradiology*. 1432-1920. *66*(12), 2171–2176. Available from: https://www.springer.com/journal/234. doi:10.1007/s00234-024-03497-7.

Singh, V., Cheng, S., Kwan, A. C., & Ebinger, J. (2025). United States food and drug administration regulation of clinical software in the era of artificial intelligence and machine learning. *Mayo Clinic Proceedings: Digital Health*, 2949-7612. *3*(3), 100231. doi:10.1016/j.mcpdig.2025.100231.

Sjöwall, C., & Parodis, I. (2022). Clinical heterogeneity, unmet needs and long-term outcomes in patients with systemic lupus erythematosus. *Journal of Clinical Medicine*, 2077-0383. *11*(22), 6869. doi:10.3390/jcm11226869.

Sohan, M. F., & Basalamah, A. (2023). A systematic review on federated learning in medical image analysis. *IEEE Access*, 2169-3536. *11*, 28628–28644. Available from: http://ieeexplore.ieee.org/xpl/RecentIssue.jsp?punumber=6287639. doi:10.1109/ACCESS.2023.3260027.

Song, Z., Li, S., Shang, Z., Lv, W., Cheng, X., Meng, X., et al. (2024). Integrating multi-omics data to analyze the potential pathogenic mechanism of CTSH gene involved in type 1 diabetes in the exocrine pancreas. *Briefings in Functional Genomics*, 2041-2657. *23*(4), 406–417. Available from: https://academic.oup.com/bfg/issue. doi:10.1093/bfgp/elad052.

Stafford, I. S., Kellermann, M., Mossotto, E., Beattie, R. M., MacArthur, B. D., & Ennis, S. (2020). A systematic review of the applications of artificial intelligence and machine learning in autoimmune diseases. *Npj Digital Medicine*. 2398-6352. *3*(1). Available from: https://www.nature.com/npjdigitalmed/. doi:10.1038/s41746-020-0229-3.

Tang, A., Wei, M. L., Haemel, A., La, C., Sirota, M., & Lee, E. Y. (2025). Artificial intelligence-enabled precision medicine for inflammatory skin diseases. *arXiv*, arXiv. 23318422, Available from: https://arxiv.org. doi:10.48550/arXiv.2505.09527.

Thabet, R. M., Shedeed, H. A., Al-Berry, M., & Khattab, D. (2025). Multiple sclerosis classification and segmentation in neuroimaging MRI using different machine and deep learning techniques: a review. *Artificial Intelligence Review*. 1573-7462. *58*(8). Available from: https://www.springer.com/journal/10462. doi:10.1007/s10462-025-11143-8.

Tjalsma, H., Schaeps, R. M. J., & Swinkels, D. W. (2008). Immunoproteomics: From biomarker discovery to diagnostic applications. *Proteomics—Clinical Applications*, 1862-8346. *2*(2), 167–180. doi:10.1002/prca.200780012.

Topol, E. J. (2019). High-performance medicine: the convergence of human and artificial intelligence. *Nature Medicine*, *25*(1), 44–56. doi:10.1038/s41591-018-0300-7.

Vivas, A. J., Boumediene, S., & Tobón, G. J. (2024). Predicting autoimmune diseases: a comprehensive review of classic biomarkers and advances in artificial intelligence. *Autoimmunity Reviews*, 1873-0183. *23*(9). Available from: https://www.sciencedirect.com/science/journal/15689972. doi:10.1016/j.autrev.2024.103611.

Wang, Y., Wei, W., Ouyang, R., Chen, R., Wang, T., Yuan, X., et al. (2024). Novel multiclass classification machine learning approach for the early-stage classification of systemic autoimmune rheumatic diseases. *Lupus Science and Medicine*, 2053-8790. *11*(1). Available from: http://lupus.bmj.com/. doi:10.1136/lupus-2023-001125.

World Health Organization. (2021). *Ethics & governance of artificial intelligence for health: WHO guidance*. World Health Organization.

Xie, C., Huang, K., Chen, P. Y., & Li, B. 2020 DBA: Distributed backdoor attacks against federated learning. In *8th international conference on learning representations, ICLR 2020*. China: ICLR.

Xu, J., Glicksberg, B. S., Su, C., Walker, P., Bian, J., & Wang, F. (2021). Federated learning for healthcare informatics. *Journal of Healthcare Informatics Research*, 2509-4971. *5*(1), 1–19. doi:10.1007/s41666-020-00082-4.

Yan, J., Risacher, S. L., Shen, L., & Saykin, A. J. (2017). Network approaches to systems biology analysis of complex disease: Integrative methods for multi-omics data. *Briefings in Bioinformatics*, 1477-4054. *19*(6), 1370–1381. Available from: http://bib.oxfordjournals.org. doi:10.1093/bib/bbx066.

Yang, D., Peng, X., Zheng, S., & Peng, S. (2025). Deep learning-based prediction of autoimmune diseases. *Scientific Reports*, 2045-2322. *15*(1). doi:10.1038/s41598-025-88477-4.

Yang, Y., Liu, Y., Chen, Y., Luo, D., Xu, K., & Zhang, L. (2024). Artificial intelligence for predicting treatment responses in autoimmune rheumatic diseases: advancements, challenges, and future perspectives. *Frontiers in Immunology*, 1664-3224. *15*. doi:10.3389/fimmu.2024.1477130.

Yi, L., Zhang, J., Zhang, R., Shi, J., Wang, G., & Liu, X. (2020). SU-Net: An efficient encoder-decoder model of federated learning for brain tumor segmentation. In *Lecture notes in computer science* (Vol. *12396*, pp. 761–773). Springer Science and Business Media LLC. doi:10.1007/978-3-030-61609-0_60.

Yoo, J. H., Son, H. M., Jeong, H., Jang, E. H., Kim, A. Y., Yu, H. Y., et al. 2021 Personalized federated learning with clustering: non-IID heart rate variability data application. In *International conference on ICT convergence* (pp. 1046–1051). South Korea: IEEE Computer Society. 21621241. Available from: http://ieeexplore.ieee.org/xpl/conferences.jsp. doi:10.1109/ICTC52510.2021.9620852.

Yuan, B., Ge, S., & Xing, W. (2020). A federated learning framework for healthcare IoT devices. *arXiv*, arXiv. 23318422, Available from: https://arxiv.org. doi:10.48550/arxiv.2005.05083.

Zhang, K., Lincoln, J. A., Jiang, X., Bernstam, E. V., & Shams, S. (2023). Predicting multiple sclerosis severity with multimodal deep neural networks. *BMC Medical Informatics and Decision Making*, *23*(1), 255. doi:10.1186/s12911-023-02354-6.

Zhao, X., Duan, L., Cui, D., & Xie, J. (2023). Exploration of biomarkers for systemic lupus erythematosus by machine-learning analysis. *BMC Immunology*, 1471-2172. *24*(1). doi:10.1186/s12865-023-00581-0.

Section V

Artificial intelligence in neuropsychiatric rehabilitation

Chapter 16

Artificial intelligence in psychotherapy and counseling

Sandhya Bhatt[a] and Anu Dandona[b]
[a]*Amity University, Noida, India*
[b]*D Y Patil International University, Akurdi, Pune, India*

1 Introduction

Early digital forms of therapy included email-based services and self-help websites. However, in the 2000s, more forms of therapy emerged, including teletherapy, video conferencing, etc., where therapists could conduct sessions via phone or video across remote areas as well (Kamene, 2025). It gradually expanded to include applications for mental health, chatbots driven by AI, and virtual reality (VR) technologies, to name a few. Researchers have found these therapies to be effective, especially for conditions like anxiety and depression. Mental health services became more accessible and affordable with the invention of chatbots driven by AI, mental health applications, and teletherapy. Among the digital interventions, cognitive behavioral therapy (CBT) has been particularly found to be effective in reducing the symptoms of anxiety and depression. Studies were conducted comparing conventional face-to-face therapy and virtual therapy. Results obtained showed that there was no difference between the two forms of therapy in terms of levels of therapeutic relationships, empathy, and perceived environment between patients and psychotherapists. Digital mental health includes interventions in the form of online self-help modules and automated feedback; some may even include face-to-face interventions. Internet-delivered depression treatments have shown higher self-efficacy (Christensen, Griffiths, & Korten, 2004).

Approximately 59,000 mobile health apps that are available across the world are being extensively used for providing mental health services. These apps provide information about mood, meditation, guidance, and even connect users to therapists (Haque & Rubya, 2023). These days, VR and augmented reality, which involve computer-generated three-dimensional (3D) simulations of environments or enhancements of real-world settings, are used extensively in treating mental health disorders (Ma et al., 2021). VR technologies offer engaging 3D simulations that assist patients in facing difficulties and altering their behaviors by mimicking real-life situations. The contemporary VR systems make use of headsets with displays that can be worn on the head, providing visual, auditory, tactile, and even olfactory sensations. This technology creates a sequence of images and sounds that replicate actual situations, allowing for interaction in a lifelike way through designated electronic devices (Emmelkamp & Meyerbröker, 2021).

VR has demonstrated effectiveness in addressing various psychological health challenges, including eating disorders, obsessive-compulsive disorder, substance use disorders, anxiety disorders, and neurodevelopmental disorders (Miralles et al., Miralles et al.). VR has been found effective in assessing social cognition, competency, craving, etc., with real-world outcomes (Chard & van Zalk, 2022). VR exposure therapy has shown considerable success in addressing phobias like acrophobia, fear of dentists, and aviophobia, with benefits persisting for as long as three years after the treatment ends (Worley, 2019).

Chatbots have also become increasingly popular these days. Chatbots, also called conversational agents, are computer programs that can mimic human communication in task-specific conversations with individuals by using natural language (Vaidyam et al., 2021). Chatbots have been found to enhance mental health treatments. Individuals who feel reluctant to seek mental health help because of stigma might interact with chatbots with ease (Lucas et al., 2017). Chatbots for mental health can assist with a range of mental health conditions, such as addiction, stress, anxiety, and depression (Abd-Alrazaq et al., 2019). Chatbots are used as personal assistants, enabling patients to learn more, gain more support, and learn how to respond quickly when experiencing new symptoms (Bates, 2019; Yadav et al., 2023; Gray et al., 2020; Battineni et al., 2020).

Applications of Artificial Intelligence in Healthcare. https://doi.org/10.1016/B978-0-443-43934-6.00019-5

Mobile mental healthcare chatbots have proven to be quite effective and accessible (Cameron et al., 2018). Another interesting fact about chatbots is that they can provide individualized suggestions and resources as per the needs of the clients (Inkster, Stillwell, Kosinski, & Jones, 2018). Some of the well-known chatbots are Wysa, Replika, and Youper.

2 Overview of Artificial Intelligence technology in mental health support

2.1 Woebot

Woebot was developed in 2017 to provide internet-based behavioral therapy (ICBT) to people experiencing depression, anxiety, and mental health problems (Fitzpatrick, Darcy, & Vierhile, 2017). The study found that Woebot users reported decreased levels of depression. The authors described the app as "a practical, interesting, and efficient method of administering CBT." Woebot allows users to express their emotions and uses natural language processing to comprehend those responses and determine which professionally developed technique would be most helpful at that particular moment (https://woebothealth.com).

Woebot supports an individual's development by posing appropriate questions at the right times and promoting an interactive self-help experience that can take place at any location and at any moment. Users engage with Woebot by selecting predefined responses or by entering their own text, which Woebot analyzes using AI methods. Woebot incorporates principles from cognitive behavioral therapy to assist users in altering their thinking habits. Initially, it prompts the user to jot down their negative thoughts, and then, it elaborates on the cognitive distortions involved. Ultimately, Woebot encourages the user to transform a negative assertion into a positive one (https://spectrum.ieee.org/woebot).

The chatbot was proven to be successful in lowering anxiety and depressive symptoms among American college students (Fitzpatrick et al., 2017). In another study, the app was found to be practical to implement, interactive, and it resulted in notable enhancements in substance use, self-assurance, cravings, depression, and anxiety among the participants (Prochaska et al., 2021). Interaction with Woebot led to notable decreases in feelings of stress and burnout while significantly enhancing resilience over the course of the 8-week study among adults (Durden et al., 2023).

In a study published in 2025, Yeh, P.L., Kuo, W.C., Tseng, B.L., et al. assessed the effect of the Woebot app in lowering anxiety and sadness in 42 master's students taking group counseling classes. In addition, it investigated how students perceived the Woebot app as a cutting-edge technological tool for group counseling education. The findings showed that students' anxiety and depression levels in relation to course demands and their leadership roles in groups did not significantly decline, even with the growing use of technology in the classroom and the potential benefits of chatbot applications based on counseling principles. Furthermore, students also reported usability challenges, including technical issues, poor user experience, and language difficulties. These results highlight the importance of further investigation and improvements in the design and implementation of chatbots within educational settings to better meet the academic and mental health needs of students.

2.2 Wysa

According to Inkster et al. (2018), Wysa has been shown to be highly helpful in lowering depression. Wysa employs a mix of rule-based algorithms and extensive language modeling to understand and reply thoughtfully to expressed thoughts and feelings. In addition to engaging in conversation, Wysa suggests strategies and techniques that could be beneficial for the specific challenges people are facing. These strategies are based on evidence and are designed to enhance people's general well-being and mental toughness. The application is a conversational agent powered by AI, designed to help individuals experiencing subclinical or clinical anxiety or low mood manage their emotional distress using evidence-based methods like CBT, goal setting, and emotional acceptance. It offers feedback and tailored insights based on data shared by users and aids them in establishing meaningful personal goals and taking actionable steps toward positive transformations (www.wysa.com).

Wysa has high levels of accessibility among users, who believe that the app has a nonjudgmental approach and is easy to converse with. In a study conducted on orthopedic patients, results revealed that patients who received mental health intervention along with orthopedic treatment experienced improvement in their physical functioning, along with a reduction in the extent of depression, anxiety, and pain disruption (Leo et al., 2022). Wysa has been used in some of the most reputed organizations. Local 18, one of the leading construction companies, is taking the lead in emphasizing mental health in the construction industry through its collaboration with Wysa. By implementing sophisticated AI-powered solutions that combine proactive wellness with timely crisis management, Local 18 is establishing a new benchmark for mental health assistance in construction. With a satisfaction rate of 95% among users, this program addresses the specific challenges encountered by construction workers. Another example where the app has been successfully used is at the University of Western Cape, United States, by teachers, staff, and students across the campus for mental health challenges. The results have indicated that the level of depression and anxiety was lowered by 35%, and 91% of the users found the tool highly effective for managing stress and exam anxiety (https://wysa.com).

Wysa has demonstrated significant success in establishing a robust connection with its users. The application motivates its users to interact with the platform, facilitating positive advancements in emotional resilience and personal growth. Nevertheless, its AI necessitates several improvements to enhance the overall user experience within the application. These insights aid in the creation and execution of more effective, ethical, and user-focused AI-driven mental health support systems (Chaudhry & Debi, 2024a, 2024b).

2.3 Replika

Replika is one of the most popular social chatbots, often referred to as "compassionate AI," "a friend who always listens," or "AI version of yourself" (Metz, 2020). Replika uses a sophisticated neural network machine learning model with scripted dialog content. It has been trained to produce its own distinct answers from a sizable database. The more one converses with Replika, the more the app will learn and become like the user. It also responds in ways that a friend would (Chelsea Adelaine, 2020). The findings from their study revealed that users who had frequent and intense interactions with Replika developed a strong attachment to the app. According to the theory of similarity-attraction, people are more likely to be attracted to entities that share similarities with themselves (Bernier & Scassellati, 2010). As a result, a more human-like item evokes more favorable feelings, dispositions, satisfaction, and trust (Blut, Wang, Wünderlich, & Brock, 2021; de Visser et al., 2016).

A survey of 1006 students who used Replika examined their perceptions of social support, loneliness, usage habits, and opinions regarding the platform. Participants reported that they felt a high level of social support. Three percent even said that Replika stopped them from having suicidal thoughts (Maples et al., 2024).

A study by Lee et al., 2020) examined user reviews to determine the different kinds of social support that companion chatbots, such as Replika, offer. Informational support (15.6%), emotional support (44.6%), companionship support (77.1%), and appraisal support (9.3%) were the four main themes that represented the various forms of social support.

Research at Harvard Business School indicated that individuals who use AI companions are developing connections that are highly intimate and exhibit traits commonly found in human relationships. Actions taken by companies can disrupt these dynamics, posing threats to consumer well-being. Research conducted by Stanford's Institute for Human-Centered AI revealed that although mental health chatbots frequently appear empathetic, they respond appropriately only about half the time. This inconsistency may result in an over-dependence on them and a misleading sense of emotional safety. Further investigation is required in this area to examine the psychological factors involved and to suggest methods for organizations to mitigate these risks (De Freitas, Castelo, Uğuralp, & Oğuz-Uğuralp, 2024).

2.4 Youper

Youper is a very popular app that makes use of artificial intelligence for therapy to reduce anxiety and depression. It was launched in 2016 in California to help users with their emotional well-being. Youper provides intervention through a chat-based interface that is fully automated. Using inputs from the user, it employs a decision tree to establish its replies. The engagement with Youper is referred to as a "dialog." The discussions occur in an established order: first, identifying the present emotions and their intensity; second, choosing contributing factors from a predefined list; and then providing an open-text description of what is influencing the current mood (Mehta et al., 2021). The users are assisted in acquiring adaptive emotion regulation skills. If the user is experiencing a negative emotion, the skills focus on addressing that specific emotion. Youper mainly uses just-in-time interventions (provided at the point of need) to help users learn and practice skills for regulating and managing their emotions (Mennin, Ellard, Fresco, & Gross, 2013).

Youper helps reduce symptoms of six different mental health conditions. An interesting fact about the app is that it provides a custom plan to address each user's specific mental health challenges (Chandra D.). A study published by the American Medical Association identifies Youper as one of the ten most engaging mobile behavioral mental health apps (Carlo, Hosseini Ghomi, Renn, Strong, & Areán, 2020). Another study assessing smartphone mental health apps for lowering anxiety and depression revealed that Youper is among the top three applications regarding download numbers and daily and monthly active user counts (Wasil et al., 2020). A research study examining four chatbots created for mental health support (Replika, Youper, Sayan, and Woebot), which can engage users both cognitively and emotionally and are readily available on mobile platforms, found that Youper was the preferred choice based on participant experiences with these chatbots (Chung & Lee, 2020).

Research carried out by Mehta et al. (2021) on Youper AI is one of the most extensive and important studies of its kind. Featured in the Journal of Medical Internet Research, the research was conducted by scholars at Stanford University and examined data from 4517 users who paid for Youper over a period of 4 weeks. The study found a very high level of satisfaction among the users. Remarkable changes were noted in the users' reported levels of anxiety and depression. The research emphasized the importance of emotional regulation methods enhanced by Youper, associating them with improvements in therapeutic outcomes (Mehta et al., 2021).

Despite the advantages, one of the drawbacks of the app is that Youper does not provide a complimentary version of its application, necessitating payment for its services. Moreover, users have indicated that the app often becomes unresponsive while in use (https://www.choosingtherapy.com).

2.5 Headspace

Headspace is an application that provides guided meditation and exercises based on mindfulness. It uses AI to tailor the meditation experience for each individual user. Headspace has attracted millions of users across the world and has emerged as one of the most popular apps. Guided meditation provided by the app can range from just a few minutes to several hours. The user-friendly interface, captivating content, and mental health advantages have contributed to Headspace's success (Anser, Nabi, Ahmad, Abro, & Zaman, 2025). Headspace begins with the assumption that new users may not be familiar with mindfulness; therefore it directs them to the introductory 30-session "Basics" course, covering important elements of mindfulness with the help of short practices led by one of the company's co-founders. Once users finish this, they can access a variety of other "packs" based on different themes related to well-being, such as "happiness" and "work and performance." Users are encouraged to track their progress by showing how many sessions they have completed and the total minutes they have spent meditating (https://Headspace.com).

In a study conducted on college students, it was found that the students who used the app "Headspace" reported an increase in their mindfulness, which led to a decrease in their levels of anxiety and depression (Peirce et al., 2024). The results from the research show that utilizing the app may result in changes in state mindfulness, which may further lead to improvements in mental health. A systematic review conducted on the efficacy of the Headspace app found that 40% of the studies using the app had a favorable effect on wellbeing and anxiety, while 75% of the studies associated with Headspace showed a decrease in depressive symptoms (O'Daffer, Colt, Wasil, & Lau, 2022). Findings from another study indicate that the clinician-supported implementation of Headspace as an initial treatment for depression shows promise and warrants further investigation in upcoming research (Strauss, Dunkeld, & Cavanagh, 2021).

Vertex, a prominent biopharmaceutical firm, adopted Headspace as a mental health solution that is inclusive, tailored, and grounded in evidence. The organization attained remarkable outcomes, with 88% of participants demonstrating a decrease in depression symptoms following their use of Headspace Care, and clinicians receiving an average rating of 5 out of 5. Headspace has even collaborated with prominent brands such as Starbucks, aiming to motivate, educate, and support new audiences in achieving healthier and more fulfilling lives. Headspace partnered with LinkedIn to develop "Life Skills: Embracing Emotions at Work," a new skills-oriented program (https://youper.com).

Despite its many benefits, the app is not free from limitations. A study by Kisely, Bastiampillai, Allison, & Looi (2023) suggested that Headspace does not provide therapy of sufficient length to achieve clinically meaningful improvement. The majority of assessments were based on either temporary process metrics or unregulated satisfaction surveys, and when standardized tools were used to assess the outcomes, the findings were not satisfactory. Moreover, users have reported that the app is rather expensive (Table 16.1).

3 Psychological frameworks supporting digital therapeutic approaches

CBT, widely known for its structured framework for technology-driven interventions, is the most extensively utilized therapeutic method in digital mental health applications. People can access online CBT using different methods, with the two most common being self-signup or receiving an invitation from a supporter. Self-registration for internet-delivered CBT occurs when the service user is granted automatic access through a specific provider organization's website (such as healthcare, employers, or educational institutions) without requiring a formal assessment or triage process. Alternatively, an invitation can be sent from a supporter to a service user through the platform. In certain situations, the process for a supporter to sign up includes an initial evaluation of mental health issues and their appropriateness for an iCBT intervention. After a clinical evaluation, a service user may be determined to be appropriate for an iCBT intervention. They share their email address with their supporter (the clinician in this scenario), who subsequently invites them via their supporter account

TABLE 16.1 Popular mental health apps and the psychological conditions they address.

App name	Psychological conditions/focus areas	Studies
Youper	Worries, sadness, and other wellness-related issues (such as boosting productivity, reducing stress, enhancing self-esteem, and promoting happiness). Uses CBT, acceptance and commitment therapy (ACT), and dialectical behaviour therapy (DBT) sessions	https://www.apaservices.org https://aicompetence.org
Wysa	Anxiety, overall mental health assistance, and stress reduction techniques. Emphasizes methods based on cognitive behavior therapy	(Inkster et al., 2018). https://www.wysa.com/ https://bevoya.com
Replika	Social interaction and emotional support, isolation, companionship. Engagements are based on the foundations of cognitive-behavioral therapy, mindfulness, and positive psychology	(Metz, 2020; Chelsea Adelaine, 2020; Blut et al., 2021; de Visser et al., 2016; Bernier & Scassellati, 2010; Maples, Cerit, Vishwanath, & Pea, 2024; Lee et al., 2020) https://onemindpsyberguide.org
Woebot	Based on cognitive-behavioral therapy (CBT), this approach assists individuals in recognizing and disputing negative thoughts while encouraging more positive thinking patterns. It provides general mental health assistance, addressing issues related to depression and anxiety	(Fitzpatrick et al., 2017; Oi, 2024) https://ai-review.com
Headspace	Depression, stress, anxiety, and mental wellness are all interconnected. Mindfulness contributes significantly to psychological health. The focus is mainly on practices such as meditation and mindfulness-based techniques	(O'Daffer et al., 2022); https://headspace.org

on the platform to engage with the intervention. An iCBT program includes modules on cognitive restructuring, behavioral activation, self-monitoring, and problem-solving, for instance. Typically, the modules are structured to be delivered on a weekly schedule (www.intechopen.com).

Digital CBT therapies are highly successful, according to the latest research. ICBT has become a major form of digital mental health treatment (Vernmark et al., 2024). The results have been favorable in providing treatment for psychological disorders, including somatic illness and associated distress, depression, anxiety, pain, and loneliness (Hedman, Ljótsson, & Lindefors, 2012). Similar results have been obtained as in face-to-face CBT. Research demonstrated that headache severity was significantly reduced among patients who underwent ICBT (Ström, Pettersson, & Andersson, 2000). Results from a randomized controlled trial ICBT study highlighted the fact that ICBT lowers the frequency, duration, and intensity of panic attacks. Studies conducted later also confirmed the same (Carlbring, Westling, Ljungstrand, Ekselius, & Andersson, 2001); Ramklint et al., 2003). The program's modules encompassed psychoeducation, breathing exercises, cognitive restructuring, interoceptive exposure, in vivo exposure, and relapse prevention. A systematic review conducted in 2012, which analyzed 103 ICBT studies, indicated the effectiveness of ICBT for nearly 25 distinct mental health disorders, with most randomized controlled trials focusing on depression and anxiety. ICBT was found to be more effective in the treatment of depression, anxiety disorders, cannabis use, and pathological gambling (Hedman et al., 2012).

Guided ICBT was found to be effective in treating depression and anxiety among patients . Due to the growing effectiveness of ICBT, a number of technological advancements are taking place nowadays. One such advancement is the introduction of smartphone-based mental health apps (mobile ICBT). These are found to be more cost-effective and accessible to users (Danaher, Brendryen, Seeley, Tyler, & Woolley, 2015; Musiat & Tarrier, 2014). One such app is called the "Challenger" app, which incorporates advanced features that were not present in earlier psychological treatment apps. Some of the features include awareness of real-time location, notifications, anonymous social interactions between users, and the use of gamification techniques (Miloff, Marklund, & Carlbring, 2015). The app was designed to help users overcome their social anxiety. Users are given the task of completing challenging interactions with their environment. They can elect any skills they want to develop from the 27 different skills provided. The self-selected skills offer challenges to users in the areas where they are facing problems. Another feature of the app includes psychoeducation notifications that are sent to users to help keep them motivated in reaching their goals. Research conducted on patients with breast cancer showed that ICBT was effective in reducing psychological distress and thereby enhancing their overall quality of life (Tang et al., 2024).

4 Acceptance and commitment therapy intervention

Acceptance and commitment therapy (ACT) has emerged as one of the consistently growing alternatives to CBT. It is generally considered to be a third wave of CBT (Steven C. Hayes, 2004). The difference between CBT and ACT is that ACT does not classify thoughts and beliefs as either right or wrong. The focus is rather on the outcome of the process. The essence of ACT lies in embracing oneself and committing to individual values. It emphasizes behaviors that align with those values. The methods involve using mindfulness and experiential activities that promote a connection to the current circumstances (Ruiz, 2012; Steven C. Hayes, 2004).

The therapeutic mechanisms of ACT have the following processes:

1. Acceptance: Here, acceptance surpasses mere tolerance. It involves a positive, nonjudgmental embrace of current experiences. The goal is to create space for painful emotions, impulses, and feelings instead of resisting, controlling, or running away from them. Individuals are encouraged to observe these experiences with objectivity.
2. Cognitive dissociation: This concept pertains to the separation of the self from thoughts, images, and memories. This detachment allows for objective observation of these cognitive processes as mere vehicles. This perspective helps individuals see thoughts as a type of language and text, separate from the meanings they carry, which are not under the control of the individuals.
3. Being present: ACT emphasizes focusing on the present environment and psychological processes without making judgments. This approach promotes fully accepting them as they are.
4. The self as context: Disturbing thoughts and emotions can adversely impact an individual's sense of identity, particularly when the self is perceived as an unchanging entity. Through self-observation, participants can enhance their connection to their authentic experiences. This promotes cognitive defusion and acknowledgment. ACT frequently employs mindfulness, metaphors, and experiential methods to assist participants in self-observation.
5. Values: In ACT, values represent the main aspirations and chosen life paths of individuals, shaped by language. These values are connected to behavior and consciously influence every deliberate action in life. Actions based on values are positive and do not aim to evade difficult feelings.
6. Committed action: ACT goes beyond being simply a method for acceptance. It also helps in altering one's perspective. The goal of ACT is to assist participants in selecting behavioral changes that align with their core values, accepting accountability for their actions, and living a life directed by those values.

The online ACT has demonstrated its effectiveness in addressing mental health issues, mainly due to its cost-effectiveness and ease of access (Zhang et al., 2024). A study conducted on 108 healthcare professionals observed significant improvement in the overall levels of depression and anxiety among these professionals. This intervention included 21 self-guided sessions featuring instructional videos, mindfulness exercises, and journal writing. Moreover, seven live conferences were conducted, where participants could share their experiences and engage in discussions, all facilitated through We Chat. The web-based ACT has been shown to be quite effective in the treatment of depression (Brown, Glendenning, Hoon, & John, 2016). An examination of 10 randomized controlled trials revealed eight interventions aimed at alleviating distress. These included strategies for managing depression, maximizing life experiences, coping with pain, the Good Life Compass, ACT-CL, and one intervention that remains unnamed. The duration of these interventions varied from 3 to 12 weeks. To facilitate active participation and adherence, email reminders, short message service reminders, and homework essay tasks were also incorporated into the system. Upon receipt of homework, personalized feedback was provided to the participants. A study conducted on college students reported an enhanced level of satisfaction, improved levels of distress, anxiety, and depression, and positive mental health (Mason et al., 2016).

ACT has been shown to have therapeutic benefits and has been used for managing symptoms associated with chronic illnesses (Vowles and McCracken, 2008), cancers (Vowles and McCracken, 2008; Xu et al., 2017), pregnancy and childbirth, and psychosis (Walters et al., 2016). In another study, web-based ACT was used to enhance optimism among students. The content included a homepage, inspirational stories, posttest FAQs, etc. Students reported an enhanced level of optimism (Eka Wahyuni & Fia Nurul Fauziah, 2022). The effect of e-ACT on mental health was assessed among 52 Malaysian students following the COVID-19 pandemic. Two online e-ACT sessions were conducted, for a total of 5 h. Results from the research showed enhancements in mindfulness, psychological flexibility, and overall well-being among the participants (han et al., 2022).

5 Self-determination theory in digital health

According to self-determination theory (SDT), human behavior is influenced and regulated by three psychological needs: autonomy, competence, and relatedness (Deci & Ryan, 1985, 1991). Autonomy refers to the inclination to engage in activities that offer a sense of personal choice or validation. Competence refers to the essential feeling of being appropriately challenged and confident in the quest to attain favorable goals and results. Relatedness signifies the degree to which a person feels a connection to and appreciation from others. As outlined in SDT, when these three needs are satisfied, individuals experience an increased level of autonomous motivation, perseverance, and quality of effort in multiple areas. From the perspective of the SDT viewpoint, an individual's sense of autonomy can be enhanced by promoting personal goal setting (Deci & Ryan, 1991). As far as digital tools are concerned, a study conducted on nurses recommended that allowing nurses to establish goals in alignment with their personal values and skills may boost their intrinsic motivation. This may further enhance overall satisfaction and success with a digital application (Hindman, Zugai, Foster, & Raeburn, 2025; Gavarkovs et al., 2023).

A study observed that mental health applications based on the three fundamental needs of SDT result in intrinsic motivation for behavioral change among users. Apps that focus on the three needs of SDT may motivate users and build positive and long-term health behavior change (Brownlow, 2022). Interventions that rely on extrinsic motivation may prompt an individual to act in a specific manner, but this effect tends to be temporary; once the intervention stops, the behavior also stops.

A study on ElderTree, a digital platform designed for senior citizens, demonstrated that technological functionalities should facilitate straightforward navigation, interaction, and personalization to foster a sense of self-determination and intrinsic motivation (Cotter et al., 2024). When older adults can easily navigate the system, actively engage with it, and personalize it, they feel more in control and genuinely interested in using it. This sense of control and personal connection makes them more likely to stay with the health program and benefit from it. Essentially, the technology needs to work alongside people rather than just provide information. It should give users a sense of ownership and capability instead of making them feel overwhelmed or passive.

6 Health belief model

According to the health belief model (HBM), individual beliefs and attitudes play a very important role in explaining behavior. According to the HBM, people are more likely to engage in health-related activities when they are psychologically and mentally prepared. Individuals who are more susceptible to stress or have a belief that stress will affect them more significantly are more inclined to make use of smartphone applications (Melzner, Heinze, & Fritsch, 2014). Furthermore, users are more likely to develop a deeper understanding of the benefits offered by stress prevention applications. A person's cognitive beliefs about a specific technology play an important role in shaping their awareness of health issues or knowledge about health. This comprehension may lead to a higher likelihood of employing these applications to mitigate the negative effects of stress. A research study was carried out to investigate the effect of perceived health risk on the use of health-related internet resources, grounded in the HBM. The results indicated that both perceived health risk and health consciousness had a positive influence on the utilization of health-related internet services (Ahadzadeh, Pahlevan Sharif, Ong, & Khong, 2015). Furthermore, the study found that the impact of health consciousness on health-related internet usage was partially mediated by the perceived usefulness of the internet and individuals' attitudes toward using the internet for health-related activities. These findings underscore the important role of perceived usefulness and attitudes regarding internet usage. Moreover, the results imply that enhancing awareness of health risks may lead to greater acceptance of these digital health interventions.

A study was conducted in the UK to understand and explore the health belief factors that may affect women's willingness to use a mental health application during menopause and the associated transition (Martin-Key et al., 2024). The results showed that most of the women sought assistance from online resources to address their mental health concerns regarding menopause. Regarding the intention to utilize an application, the findings, grounded in the HBM, indicated that cues to action were identified as one of the most crucial predictors of women's intention to engage with the app. This was followed by perceived barriers, perceived susceptibility, and perceived benefits. However, perceived severity and self-efficacy did not show any significant correlation with the intention to use the application. Moreover, perceived barriers, such as experiencing discomfort while using the app to assess their mental health symptoms and having doubts about the app's ability to understand the complexity of mental health issues, emerged as significant obstacles to using the app. As far as self-efficacy was concerned, the majority of women reported being confident in their abilities to use a mental health app. App developers should focus on approaches that encourage long-term user engagement. Enhancing user-friendliness and providing in-app support have been found to effectively improve app utilization.

Another study, grounded in the theory of planned behavior (TPB), examined the elements that affect therapists' adoption of digital mental health tools in Israel. Using convenience sampling, 229 mental health practitioners completed a questionnaire on their background, demographics, and an electronic-therapy attitudes and processes questionnaire based on TPB. Regression analyses were conducted to assess the impact of attitudes, perceived behavioral control, subjective norms, and behavioral intentions. Within the framework of the TPB, the intention to use and the actual use of digital tools were notably influenced by attitudes, subjective norms, and perceived behavioral control. The probability of engaging with digital interventions was enhanced by the application of CBT and an elevated sense of perceived behavioral control. Favorable attitudes, expectations, or social pressure from professional peers, along with a perceived sense of control, were predictors of both the intention to use and the actual implementation of digital interventions (Khatib, Abo-Rass, & Gelkopf, 2022).

Hossain et al. (2021) demonstrated in their study published in the Health Psychology Review that 78% of research employing the HBM reported an increased level of behavior adherence, with 39% showing moderate to large effects. The mental health components of the HBM can be adjusted by tackling perceived barriers, enhancing self-efficacy, and providing cues to action related to therapy or medication adherence (Naslund et al., 2017). Among the most acknowledged behavioral frameworks that can direct digital healthcare for individuals experiencing mental health conditions are the HBM, TPB, transtheoretical model, and social cognitive theory. These frameworks include significant and overlapping components. Advances in digital technology that enable the gathering of longitudinal data, real-time behavior monitoring, and the ability to adjust to changing user needs create new opportunities for understanding health behaviors and the processes of change. Notable examples consist of the behavioral intervention technology model, persuasive system design, behavioral models for just-in-time adaptive interventions, and dynamic behavior models (Table 16.2).

TABLE 16.2 Psychological frameworks supporting digital therapeutic approaches.

Psychological framework	Core principles	Digital application	Target conditions
Cognitive behavioral therapy (CBT)	Helps individuals address avoidant and safety-seeking actions that obstruct the adjustment of erroneous beliefs, thereby facilitating stress management. Concentrates on recognizing and altering detrimental thought processes and actions.	CBT methods have been modified for virtual therapy using mobile apps, online programs, and AI-enhanced tools. Digital tools, telehealth services, and customized treatment plans improve accessibility and boost the efficacy of therapy.	Depression, anxiety disorders, eating disorders, substance use disorders, and personality disorders. It is also beneficial as a supplementary treatment for bipolar disorder and schizophrenia.
Acceptance and commitment therapy (ACT)	A method that encourages clients to confront their inner feelings rather than avoid, reject, or fight with them, recognizing that these deep feelings are legitimate responses. It highlights the importance of being flexible in thoughts and actions that are consistent with personal values.	Digital ACT interventions include mindfulness techniques, methods for defining personal values, and the tracking of behavioral commitments through applications and online platforms.	Behavioral modifications concerning health, mental health disorders, and physical health problems: an approach that transcends diagnoses for various mental and physical health conditions. Adoption and maintenance of health behaviors, management of chronic diseases, adherence to preventive care, increase in vaccination rates, and lifestyle changes.
Health belief model (HBM)	Predicts health behaviors by evaluating perceived probabilities, severity, benefits, barriers, triggers for action, and self-efficacy. It focuses on individual perceptions about health threats and the efficacy of various behaviors.	Digital interventions utilize customized risk assessments, the sharing of educational resources, reminder mechanisms, and progress monitoring to shift health perspectives and promote behavioral changes.	Modifications in conduct related to health, mental health challenges, and physical health issues. A strategy that goes beyond specific diagnoses for a range of mental and physical health conditions. Embracing and sustaining healthy behaviors, managing long-term illnesses, following preventive care guidelines, enhancing vaccination uptake, and making lifestyle adjustments. Commitment to regular physical activity, participation in treatment, sustaining behavior changes, enhancing mental health, and self-managing chronic illnesses.

Psychological framework	Core principles	Digital application	Target conditions
Self-determination theory (SDT)	The framework is based on three fundamental psychological needs: autonomy (the feeling of being self-governing), competence (the feeling of being proficient), and relatedness (the feeling of being connected to others). It distinguishes between intrinsic motivation and extrinsic motivation.	Interventions derived from self-determination theory enhance engagement by providing options, integrating tools for skill development, promoting social connections, and facilitating personalized goal-setting on digital platforms.	Behavioral modifications concerning health, mental health disorders, and physical health problems: an approach that transcends diagnoses for various mental and physical health conditions. Adoption and maintenance of health behaviors, management of chronic diseases, adherence to preventive care, increase in vaccination rates, and lifestyle changes.Physical activity adherence, treatment engagement, behavior change maintenance, mental health improvement, and chronic disease self-management

7 Ethical considerations

Given that the utilization of AI in therapy is still in its nascent phase, it is crucial to approach its incorporation with careful consideration and foresight. In contrast to human therapists, AI does not possess the emotional depth, empathy, and nuanced comprehension necessary for establishing a therapeutic relationship. The elements of empathy, trust, and emotional connection are essential for fostering a safe and healing environment, which AI is unable to replicate (Montemayor & Halpern, 2022). It is imperative to combine the strengths of AI with those of human therapists to attain optimal outcomes, as AI can offer valuable insights and swift evaluations, while human therapists contribute their expertise, ethical considerations, and nuanced understanding of individual cases to the therapeutic process. The role of a human therapist is very important in the process to monitor and clarify insights generated by AI. Without human judgment, there is a risk of misunderstanding or incorrectly diagnosing patient issues, which could lead to negative outcomes. Patients need to be informed about the involvement of AI in their therapy and what it entails. Promoting transparency and clarity regarding the role of AI is crucial for sustaining patient trust (Alfano, Malcotti, & Ciliberti, 2023).

AI systems rely on large datasets that often contain sensitive personal information, increasing the risk of data breaches or unauthorized access. The potential for data breaches, unauthorized access, or the misuse of client information is a major concern, particularly because mental health data is notably sensitive and private (Sedlakova & Trachsel, 2023). To safeguard client confidentiality and uphold trust in the therapeutic relationship, it is essential to implement stringent data protection measures.

8 Linking Health Insurance Portability and Accountability Act and General Data Protection Regulation compliance in mental health apps

The Health Insurance Portability and Accountability Act (HIPAA) and the General Data Protection Regulation (GDPR) are essential international frameworks for mental health applications, as they govern the collection, storage, and dissemination of highly sensitive personal health information. HIPAA requires written consent to disclose Protected Health Information to third parties. The GDPR mandates that clear and detailed consent is necessary for the handling of sensitive data such as mood logs, therapy session records, and other mental health information. This consent must be granted voluntarily, with clarity and specificity, as well as being informed and unambiguous criteria that are often not met by standard terms of service (www.secureprivacy.ai).

9 Conclusion

AI-driven interventions provide a convenient and accessible platform for people seeking mental health support, thereby enhancing the likelihood of engaging those who might not have considered traditional therapy. Therapists are beginning to recognize that AI-assisted mental health applications can offer significant assistance in improving patient adherence to treatment plans. It is also known that many people refrain from seeking help from mental health professionals as they are concerned about being judged or labeled. The emergence of AI has notably alleviated these concerns, as individuals often feel more at ease sharing their thoughts with a virtual entity, believing they will not face negative evaluations from the machine. When users disclose personal information to a chatbot, they generally do not experience the anxiety associated with being judged unfavorably. By carefully designing the chatbot's conversational style and tone, a more inviting and empathetic environment can be created for users. This approach can contribute to establishing rapport, characterized by mutual understanding, trust, and a connection between the user and the chatbot. Digital interventions can conserve time, energy, and resources for both clients and therapists. These solutions often come at a reduced cost, benefiting those who find traditional therapy financially unfeasible. Chatbots can support traditional psychotherapists by reducing the number of necessary sessions, enabling therapists to focus on more intricate issues during face-to-face treatment. As AI and internet of things become integral to patient care, healthcare providers must receive adequate training in the knowledge and skills necessary to effectively utilize these technologies. With the rapid advancement of AI and technology, a new era of digital psychology is emerging, presenting entirely new opportunities for immediate diagnosis, continuous therapist monitoring, and largely self-sufficient interventions. It is essential to ensure the quality, reliability, and safety of these innovations promptly to fully harness their potential.

Declaration of generative AI and AI-assisted technologies in the writing process

During the preparation of this work, the author(s) used AI tools [Claude] to [compile Table 16.1 and Table 16.2]. After using this tool/service, the author(s) reviewed and edited the content as needed and take(s) full responsibility for the content of the publication.

References

Abd-Alrazaq, A. A., Alajlani, M., Alalwan, A. A., Bewick, B. M., Gardner, P., & Househ, M. (2019). An overview of the features of chatbots in mental health: A scoping review. International Journal of Medical Informatics, *132*, 103978. doi:10.1016/j.ijmedinf.2019.103978.

Ahadzadeh, A. S., Pahlevan Sharif, S., Ong, F. S., & Khong, K. W. (2015). Integrating health belief model and technology acceptance model: an investigation of health-related Internet use. *Journal of Medical Internet Research*. 14388871. *17*(2). Available from: http://www.jmir.org/issue/year. doi:10.2196/jmir.3564.

Alfano, L., Malcotti, I., & Ciliberti, R. (2023). Psychotherapy, artificial intelligence and adolescents: ethical aspects. *Journal of Preventive Medicine and Hygiene*. 24214248. 64(4), E438. Available from: https://www.jpmh.org/index.php/jpmh/article/view/3135. doi:10.15167/2421-4248/jpmh2023.64.4.3135.

Anser, M. K., Nabi, A. A., Ahmad, I., Abro, M. M. Q., & Zaman, K. (2025). Advancing mental health care: a comprehensive review of digital tools and technologies for enhancing diagnosis, treatment, and wellness. *Health Care Science*. 27711757. *4*(3), 163–178. Available from: https://onlinelibrary.wiley.com/journal/27711757. doi:10.1002/hcs2.70018.

Bates, M. (2019). Health care chatbots are here to help. IEEE Pulse, *10*(3), 12–14. doi:10.1109/MPULS.

Battineni, G., Chintalapudi, N., & Amenta, F. (2020). AI chatbot design during an epidemic like the novel coronavirus. Healthcare, *8*(2), 154. doi:10.3390/healthcare8020154.

Bernier, E. P., & Scassellati, B. (2010). The similarity-attraction effect in human-robot interaction. In: IEEE 9th International Conference on Development and Learning, ICDL-2010—Conference Program, United States, pp. 286–290. doi:10.1109/DEVLRN.2010.5578828.

Blut, M., Wang, C., Wünderlich, N. V., & Brock, C. (2021). Understanding anthropomorphism in service provision: a meta-analysis of physical robots, chatbots, and other AI. *Journal of the Academy of Marketing Science*. 15527824. *49*(4), 632–658. Available from: http://www.springer.com/business/journal/11747. doi:10.1007/s11747-020-00762-y.

Brown, M., Glendenning, A., Hoon, A. E., & John, A. (2016). Effectiveness of web-delivered acceptance and commitment therapy in relation to mental health and well-being: a systematic review and meta-analysis. *Journal of Medical Internet Research*. 1438-8871. *18*(8), e221. doi:10.2196/jmir.6200.

Brownlow, L. (2022). Targeting the needs of Self-Determination Theory: An overview of mental health care apps. European Journal of Mental Health, *17*(1), 91–100. doi:10.5708/EJMH/17.2022.1.8.

Cameron, G., Cameron, D., Megaw, G., Bond, R., Mulvenna, M., O'Neill, S., Armour, C., & McTear, M. (2018). Best practices for designing chatbots in mental healthcare: A case study on iHelpr.Proceedings of the 32nd. International BCS Human Computer Interaction Conference (HCI 2018), 1–5. doi:10.14236/ewic/HCI2018.129.

Carl, E., Stein, A. T., Levihn-Coon, A., Pogue, J. R, Rothbaum, B., Emmelkemp, P., Asmundson, G. G. J., Carlbring, P.,Powers, M. B. & (2019). Virtual reality exposure therapy for anxiety and related disorders: a meta-analysis of randomized controlled trials. *Journal of Anxiety Disorders*, *61*, 27–36. doi:10.1016/j.janxdis.2018.08.003.

Carlbring, P., Ekselius, L., & Andersson, G. (2003). Treatment of panic disorder via the internet: a randomized trial of CBT vs. applied relaxation. *Journal of Behavior Therapy and Experimental Psychiatry*, *34*(2), 129–140 .

Carlbring, P., Westling, B. E., Ljungstrand, P., Ekselius, L., & Andersson, G. (2001). Treatment of panic disorder via the internet: a randomized trial of a self-help program. *Behavior Therapy*. 00057894. *32*(4), 751–764. doi:10.1016/s0005-7894(01)80019-8.

Carlo, A. D., Hosseini Ghomi, R., Renn, B. N., Strong, M. A., & Areán, P. A. (2020). Assessment of real-world use of behavioral health mobile applications by a novel stickiness metric. *JAMA Network Open*. 25743805. *3*(8), E2011978. Available from: https://jamanetwork.com/journals/jamanetworkopen. doi:10.1001/jamanetworkopen.2020.11978.

Chard, I., & van Zalk, N. (2022). Virtual reality exposure therapy for treating social anxiety: a scoping review of treatment designs and adaptation to stuttering. *Frontiers in Digital Health*. 2673-253X. *4*. doi:10.3389/fdgth.2022.842460.

Chaturvedi, A., Aylward, B., Shah, S., Mehta, N., Reddy, S., & Mohr, D. C. (2023). Content recommendation systems in web-based mental health care: real-world application and formative evaluation. *JMIR Formative Research*. doi:10.2196/38831.

Chaudhry, B. M., & Debi, H. R. (2024a). User perceptions and experiences of an AI-driven conversational agent for mental health support. *Mhealth*, *10*, 22. doi:10.21037/mhealth-23-55. PMID: 39114462; PMCID: PMC11304096.

Chaudhry, B. M., & Debi, H. R. (2024b). User perceptions and experiences of an AI-driven conversational agent for mental health support. *mHealth*. 23069740. *10*. Available from: https://mhealth.amegroups.org/article/view/126211/html. doi:10.21037/mhealth-23-55.

Chelsea Adelaine. (2020). Available from: https://www.popsugar.com/news/replika-bot-ai-app-review-interview-eugenia-kuyda-44216396.

Christensen, H., Griffiths, K. M., & Korten, A. (2004). Web-based cognitive behavior therapy: analysis of site usage and changes in depression and anxiety scores. *Journal of Medical Internet Research*, *6*(1), e10.

Chung, S. J., & Lee, H. (2020). Visual presentation of mental healthcare chatbots for user experience. *Journal of the HCI Society of Korea*. 1976-0671. *15*(2), 39–45. doi:10.17210/jhsk.2020.06.15.2.39.

Cotter, L. M., Shah, D., Brown, K., Mares, M. L., Landucci, G., Saunders, S., et al. (2024). Decoding the influence of eHealth on autonomy, competence, and relatedness in older adults: qualitative analysis of self-determination through the motivational technology model. *JMIR Aging*. 25617605. *7*. Available from: https://aging.jmir.org/2024/1/e56923?__hstc=102212634.4371d35f4aff6ace09851fee72e16d3b.1731688550995.1731688550995.1731688550995.1&__hssc=102212634.6.1731688550996&__hsfp=1379877452. doi:10.2196/56923.

Dagum, P., Lipschitz, J., Schwartz, H. A., Ungar, L. H., & Eichstaedt, J. C. (2019). Digital biomarkers of mood disorders and symptom change: passive smartphone actigraphy data predicts diagnosis and symptom trajectories. *medRxiv/preprint*.

Danaher, B. G., Brendryen, H., Seeley, J. R., Tyler, M. S., & Woolley, T. (2015). From black box to toolbox: outlining device functionality, engagement activities, and the pervasive information architecture of mHealth interventions. *Internet Interventions*. 22147829. *2*(1), 91–101. Available from: http://www.journals.elsevier.com/internet-interventions/. doi:10.1016/j.invent.2015.01.002.

De Freitas, J., Castelo, N., Uğuralp, A. K., & Oğuz-Uğuralp, Z. (2024). *Lessons from an app update at replika AI: Identity discontinuity in human-AI relationships*. Harvard Business School. Working Paper, No. 25-018, October 2024. (Revised May 2025).

de Visser, E. J., Monfort, S. S., McKendrick, R., Smith, M. A. B., McKnight, P. E., Krueger, F., & Parasuraman, R. (2016). Almost human: anthropomorphism increases trust resilience in cognitive agents. *Journal of Experimental Psychology: Applied*. 1076898X. *22*(3), 331–349. Available from: http://www.apa.org/pubs/journals/xap/. doi:10.1037/xap0000092.

Deci, E. L., & Ryan, R. M. (1985). Intrinsic motivation and self-determination in human behavior.

Deci, E. L., & Ryan, R. M. (1991). A motivational approach to self: Integration in personality. In R. Dienstbier (Ed.), Nebraska Symposium on Motivation: Vol. 38. Perspectives on motivation. University of Nebraska Press, pp. 237–288.

Durden, E., Pirner, M. C., Rapoport, S. J., Williams, A., Robinson, A., & Forman-Hoffman, V. L. (2023). Changes in stress, burnout, and resilience associated with an 8-week intervention with relational agent "Woebot". Internet Interventions, *33*, 100637. doi:10.1016/j.invent.2023.100637.

Eka Wahyuni, A. U., & Fia Nurul Fauziah, P. Y., (2022). DA—2022/04/22 TI—Web-based Acceptance and Commitment Therapy for Optimism BT—Proceedings of the International Seminar on Innovative and Creative Guidance and Counseling Service (ICGCS 2021).

Emmelkamp, P. M. G., & Meyerbröker, K. (2021). Virtual reality therapy in mental health. Annual Review of Clinical Psychology, *17*, 495–519. doi:10.1146/annurev-clinpsy-081219-115923.

Fitzpatrick, K. K., Darcy, A., & Vierhile, M. (2017). Delivering cognitive behavior therapy to young adults with symptoms of depression and anxiety using a fully automated conversational agent (Woebot): a randomized controlled trial. *JMIR Mental Health*. 2368-7959. *4*(2), e19. doi:10.2196/mental.7785.

Gavarkovs, A. G., Kusurkar, R. A., & Brydges, R. (2023). The purpose, adaptability, confidence, and engrossment model: A novel approach for supporting professional trainees' motivation, engagement, and academic achievement. Frontiers in Education, *8*, 1036539.doi:10.3389/feduc.2023.1036539.

Gray, R. M., Kelly, P. J., Beck, A. K., Baker, A. L., Deane, F. P., Neale, J., & McGlaughlin, R. (2020).A qualitative exploration of SMART Recovery meetings in Australia and the role of a digital platform to support routine outcome monitoring. Addictive Behaviors, *101*, 106144. doi:10.1016/j.addbeh.2019.106144.

han, P. C. F., Tsang, C. T. W., Tse, A. C. Y., Wong, C. C. H., & Chan, H. N. (2022).Psychological well-being and coping strategies of healthcare students during the prolonged COVID-19 pandemic. Teaching and Learning in Nursing, *17*(4), 482–486. doi:10.1016/j.teln.2022.05.008.

Haque, M. D. R., & Rubya, S. (2023). An overview of chatbot-based mobile mental health apps: insights from app description and user reviews. *JMIR mHealth and uHealth*. 2291-5222. *11*, e44838. doi:10.2196/44838.

Hayes, S. C. (2004). Acceptance and commitment therapy: the process and practice of mindful change/Steven C. Hayes, Kirk D. Strosahl, Kelly G. Wilson.–2nd ed. p. cm. Includes bibliographical references and index. ISBN 978-1-60918-962-4 (hardcover) 1. Acceptance and commitment therapy. 2. Values clarification. 3. Language and emotions. I. Strosahl, Kirk,1950– II. Wilson, Kelly G. III. Title.

Hayes, S. C. (2004). Acceptance and commitment therapy, relational frame theory, and the third wave of behavioral and cognitive therapies. In *Behavior therapy* (Vol. *35*, pp. 639–665). United States: Association for Advancement of Behavior Therapy. 00057894. http://www.journals.elsevier.com/behavior-therapy/. doi:10.1016/S0005-7894(04)80013-3.

Hedman, E., Ljótsson, B., & Lindefors, N. (2012). Cognitive behavior therapy via the Internet: a systematic review of applications, clinical efficacy and cost-effectiveness. *Expert Review of Pharmacoeconomics and Outcomes Research*. 17448379. *12*(6), 745–764. doi:10.1586/erp.12.67.

Hindman, J., Zugai, J., Foster, K., & Raeburn, T. (2025). Self-determination theory as a framework for research and design of digital applications for nurses' well-being. *Journal of Advanced Nursing*, 13652648. Available from: http://onlinelibrary.wiley.com/journal/10.1111/(ISSN)1365-2648. doi:10.1111/jan.16895.

Hossain, M. B., Alam, M. Z., Islam, M. S., Sultan, S., Faysal, M. M., Rima, S., et al. (2021). Health belief model, theory of planned behavior, or psychological antecedents: what predicts COVID-19 vaccine hesitancy better among the bangladeshi adults? *Frontiers in Public Health*. 22962565. *9*. Available from: http://journal.frontiersin.org/journal/public-health/section/public-health-education-and-promotion#archive. doi:10.3389/fpubh.2021.711066.

https://www.apaservices.org/practice/business/technology/tech-column/mental-health-chatbots; https://bevoya.com/blog/mental-health-app-wysa-woebot-joyable-talkspacereview; https://spectrum.ieee.org/woebot; https://onemindpsyberguide.org/guide/apps/replika/; https://ai-review.com/health-care/woebot-health/; https://www.intechopen.com/; https://www.aiapps.com/blog/replika-review-2025-your-ai-companion-for-mental-wellness/; https://www.aiapps.com/blog/replika-review-2025-your-ai-companion-for-mental-wellness/; https://www.urmc.rochester.edu/community-health/patient-care/self-determination-theory; https://www.psychiatry.org/psychiatrists/practice/mental-health-apps/evaluations/woebot-evaluation; https://woebothealth.com/for-immediate-releaselarge-scale-study-finds-mental-health-app-forms-bond-with-users-marking-key-evolution-in-digital-therapeutics/; https://www.scientificamerican.com/article/what-are-ai-chatbot-companions-doing-to-our-mental-health/; http://europepmc.org/abstract/MED/23611074 .S0005-7894(13)00011-7 [DOI]; https://www.choosingtherapy.com/youper-app-review/; https://www.headspace.com/press-and-media (2020); https://www.ruralhealthinfo.org/toolkits/health-promotion/2/theories-and-models/health-belief; https://www.nice.org.uk/guidance/ng23; https://hai.stanford.edu/news/exploring-the-dangers-of-ai-in-mental-health-care; https://www.hhs.gov/hipaa/for-professionals/security/laws-regulations/index.html?utm_source=chatgpt.com; https://www.kiteworks.com/gdpr-compliance/patient-privacy-protection-best-practices/; https://www.theguardian.com/society/2024/feb/04/they-thought-they-were-doing-good-but-it-made-people-worse-why-mental-health-apps-are-under-scrutiny?utm_source=chatgpt.com; https://secureprivacy.ai/blog/mental-health-app-data-privacy-hipaa-gdpr-compliance

Inkster, B., Stillwell, D., Kosinski, M., & Jones, P. B. (2018). Digital phenotyping and the digital psychome: building a new generation of digital mental health tools. *The Lancet Psychiatry*, *5*, 391–393.

Kamene, K. (2025). From physical to virtual: The transition of psychological therapy in the digital age. Premier Science. doi:10.70389/PJP.100007.

Khatib, A., Abo-Rass, F., & Gelkopf, M. (2022). Theory of planned behavior: exploring the use of digital mental health interventions in Israel. *Journal of Nervous and Mental Disease*. 1539736X. *210*(7), 504–509. Available from: http://journals.lww.com/jonmd. doi:10.1097/NMD.0000000000001459.

Kisely, S. R., Bastiampillai, T., Allison, S., & Looi, J. C. L. (2023). More smoke and mirrors: fifteen further reasons to doubt the effectiveness of headspace. *Australasian Psychiatry*. 14401665. *31*(3), 282–291. Available from: https://journals.sagepub.com/home/APY. doi:10.1177/10398562231167683.

Lee, Y. C., Yamashita, N., Huang, Y., & Fu, W. (2020). User experiences of social support from companion chatbots in everyday contexts: thematic analysis. *Journal of Medical Internet Research*, *22*(3), e16235. doi:10.2196/16235 .

Leo, A. J., Schuelke, M. J., Hunt, D. M., Metzler, J. P., Miller, J. P., Areán, P. A., Armbrecht, M. A., & Cheng, A. L. (2022). A digital mental health intervention in an orthopedic setting for patients with symptoms of depression and/or anxiety: Feasibility prospective cohort study. JMIR Formative Research, *6*(2), e34889. doi:10.2196/34889.

Lucas, G. M., Rizzo, A., Gratch, J., Scherer, S., Stratou, G., Boberg, J., & Morency, L.-P. (2017). Reporting mental health symptoms: breaking down barriers to care with virtual human interviewers. *Frontiers in Robotics and AI*. 2296-9144. *4*. doi:10.3389/frobt.2017.00051.

Ma, L., Mor, S., Anderson, P., et al. (2021). Integrating virtual realities and psychotherapy: SWOT analysis on VR and MR based treatments of anxiety and stress-related disorders. Cognitive Behaviour Therapy, *50*(6), 509–526. doi:10.1080/16506073.2021.

Maples, B., Cerit, M., Vishwanath, A., et al. (2024). Loneliness and suicide mitigation for students using GPT-3 enabled chatbots. npj Mental Health Research, *3*, 11. doi:10.1038/s44184-024-00055-0.

Maples, B., Cerit, M., Vishwanath, A., & Pea, R. (2024). Loneliness and suicide mitigation for students using GPT3-enabled chatbots. *Npj Mental Health Research*. 2731-4251. *3*(1). doi:10.1038/s44184-023-00047-6.

Martin-Key, N. A., Funnell, E. L., Benacek, J., Spadaro, B., & Bahn, S. (2024). Using network analysis to understand the association between menopause and depressive symptoms. npj Women's Health, *2*, 41. doi:10.1038/s44294-024-00045-9.

Mason, M. J., Aplasca, A., Morales-Theodore, R., Zaharakis, N., & Linker, J. (2016). Psychiatric comorbidity and complications. Child and Adolescent Psychiatric Clinics of North America, *25*(3), 521–532. doi:10.1016/j.chc.2016.02.007.

Mehta, A., Niles, A. N., Vargas, J. H., Marafon, T., Couto, D. D., & Gross, J. J. (2021). Acceptability and effectiveness of artificial intelligence therapy for anxiety and depression (Youper): longitudinal observational study. *Journal of Medical Internet Research*. 14388871. *23*(6). Available from: https://www.jmir.org/2021/6/e26771/PDF. doi:10.2196/26771.

Mehta, K., Hoadley, A., Ray, L. A., Kiluk, B. D., Carroll, K. M., & Magill, M. (2021). Cognitive-behavioral interventions targeting alcohol or other drug use and co-occurring mental health disorders: A meta-analysis Alcohol and Alcoholism, *56*(5), 535–544. doi:10.1093/alcalc/agab016.

Melzner, J., Heinze, J., & Fritsch, T. (2014). Mobile health applications in workplace health promotion: an integrated conceptual adoption framework. *Procedia Technology*. 22120173. *16*, 1374–1382. doi:10.1016/j.protcy.2014.10.155.

Mennin, D. S., Ellard, K. K., Fresco, D. M., & Gross, J. J. (2013). United we stand: emphasizing commonalities across cognitive-behavioral therapies. *Behavior Therapy*. 18781888. *44*(2), 234–248. doi:10.1016/j.beth.2013.02.004.

Metz, C. (2020). Riding Out Quarantine With a Chatbot Friend: "I Feel Very Connecte". The New York Times. Available from: https://www.nytimes.com/2020/06/16/technology/chatbots-quarantine-coronavirus.html.

Miloff, A., Marklund, A., & Carlbring, P. (2015). The challenger app for social anxiety disorder: new advances in mobile psychological treatment. *Internet Interventions*. 22147829. *2*(4), 382–391. Available from: http://www.journals.elsevier.com/internet-interventions/. doi:10.1016/j.invent.2015.08.001.

Miralles I., Granell C., Díaz-Sanahuja L., Van Woensel, W., Bretón-López, J., Mira, A., Castilla, D., Casteleyn, S. (2020). Smartphone Apps for the Treatment of Mental Disorders: Systematic Review. JMIR Mhealth Uhealth, *8*(4), e14897. doi:10.2196/14897.

Montemayor, C., & Halpern, J., Fairweather, A. (2022). In principle obstacles for empathic AI: why we can't replace human empathy in healthcare. *AI and Society*. 0951-5666. *37*(4), 1353–1359. doi:10.1007/s00146-021-01230-z.

Musiat, P., & Tarrier, N. (2014). Collateral outcomes in e-mental health: a systematic review of the evidence for added benefits of computerized cognitive behavior therapy interventions for mental health. *Psychological Medicine*. 0033-2917. *44* (15), 3137–3150. doi:10.1017/s0033291714000245.

Naslund, J. A., Aschbrenner, K. A., Kim, S. J., McHugo, G. J., Unützer, J., Bartels, S. J., & Marsch, L. A. (2017). Health behavior models for informing digital technology interventions for individuals with mental illness. *Psychiatric Rehabilitation Journal*. 1095158X. *40*(3), 325–335. Available from: http://psycnet.apa.org/index.cfm?fa=browsePA.volumes&jcode=prj. doi:10.1037/prj0000246.

O'Daffer, A., Colt, S. F., Wasil, A. R., & Lau, N. (2022). Efficacy and conflicts of interest in randomized controlled trials evaluating headspace and calm apps: systematic review. *JMIR Mental Health*. 23687959. *9*(9). Available from: https://mental.jmir.org/2022/9/e40924. doi:10.2196/40924.

Parry,, Douglas,, Roux, L., Daniel,, Morton,, Joshua,, et al., Andre PY—022/10/01 SP—107542 T1—Digital wellbeing applications: Adoption, use and perceived effects VL—139 JO—Computers in Human Behavior. doi:10.1016/j.chb.2022.107542.

Peirce, M. E., Mirabito, G., & Verhaeghen, P. (2024). Mind the app: More time spent on Headspace leads to beneficial day-to-day changes in mindfulness, depression, anxiety and stress in college students. Cogent Mental Health, *3*(1), 2400878. doi:10.1080/28324765.2024.2400878.

Prochaska, J. J., Vogel, E. A., Chieng, A., Kendra, M., Baiocchi, M., Pajarito, S., & Robinson, A. (2021). A therapeutic relational agent for reducing problematic substance use (Woebot): development and usability study. *Journal of Medical Internet Research*. 1438-8871. *23*(3), e24850. doi:10.2196/24850.

Ramklint, M., von Knorring, A.-L., von Knorring, L., & Ekselius, L. (2003). Child and adolescent psychiatric disorders predicting adult personality disorder: a follow-up study.Nordic Journal of Psychiatry, *57*(1), 23–28. doi:10.1080/08039480310000696.

Ruiz, F. J. (2012). Acceptance and commitment therapy versus traditional cognitive behavioral therapy: a systematic review and meta-analysis of current empirical evidence. *International Journal of Psychology and Psychological Therapy*, 15777057 *12*(3), 333–357.

Saboor, S., Medina, A.,Marciano, L. & (2024). Application of positive psychology in digital interventions for children, adolescents, and young adults: Systematic review and meta-analysis of controlled trials. JMIR Mental Health, *11*, e56045. doi:0.2196/56045.

Saboor, S., Medina, A., & Marciano, L. (2024). Application of positive psychology in digital interventions for children, adolescents, and young adults: Systematic review and meta-analysis of controlled trials. JMIR Mental Health, *11*, e56045. doi:10.2196/56045.

Sedlakova, J., & Trachsel, M. (2023). Conversational artificial intelligence in psychotherapy: a new therapeutic tool or agent? *The American Journal of Bioethics*. 1526-5161. *23*(5), 4–13. doi:10.1080/15265161.2022.2048739.

Strauss, C., Dunkeld, C., & Cavanagh, K. (2021). Is clinician-supported use of a mindfulness smartphone app a feasible treatment for depression? A mixed-methods feasibility study. *Internet Interventions*. 22147829. *25*, 100413. doi:10.1016/j.invent.2021.100413.

Ström, L., Pettersson, R., & Andersson, G. (2000). A controlled trial of self-help treatment of recurrent headache conducted via the Internet. *Journal of Consulting and Clinical Psychology*. 0022-006X. *68*(4), 722–727. doi:10.1037//0022-006x.68.4.722.

Tang, T. S., Seddigh, S., Halbe, E., Vesco, A. T., et al. (2024). Testing 3 digital health platforms to improve mental health outcomes in adults with type 1 diabetes: A pilot trial.Canadian Canadian Journal of Diabetes, *48*(1), 18–25.e2. doi:10.1016/j.jcjd.2023.08.006.

Vaidyam, A. N., Linggonegoro, D., & Torous, J. (2021). Changes to the psychiatric chatbot landscape: A systematic review of conversational agents in serious mental illness. Canadian Journal of Psychiatry, *66*(4), 339–348. doi:10.1177/0706743720966429.

Vernmark, K., Buhrman, M., Carlbring, P., Hedman-Lagerlöf, E., Kaldo, V., & Andersson, G. (2024). From research to routine care: a historical review of internet-based cognitive behavioral therapy for adult mental health problems in Sweden. *Digital Health*. 2055-2076. 10. doi:10.1177/20552076241287059.

Vowles, K. E., & McCracken, L. M. (2008). Acceptance and values-based action in chronic pain: A study of treatment effectiveness and process. Journal of Consulting and Clinical Psychology, *76*(3), 397–407. doi:10.1037/0022-006X.76.3.397.

Walters, J. T. R., Hogg, L., & Gillmore, C. (2016). Evaluation of a tailored training programme to improve the assessment and treatment of trauma in an Early Intervention in Psychosis (EIP) service. Psychosis, *8*(3), 226–237. doi:10.1080/17522439.2015.1131324.

Wasil, A. R., Venturo-Conerly, K. E., Shingleton, R. M., & Weisz, J. R. (2020). Reassessing evidence-based content in popular smartphone apps for depression and anxiety: Developing and applying user-adjusted analyses. Journal of Consulting and Clinical Psychology, *88*(11), 983–993. doi:10.1037/ccp0000604.

Worley, J. (2019). Virtual reality for individuals with substance use disorders. *Journal of Psychosocial Nursing and Mental Health Services*. 0279-3695. *57*(6), 15–19. doi:10.3928/02793695-20190430-01.

Xu, L., Liu, R.-D., Ding, Y., Mou, X., Wang, J., & Liu, Y. (2017). The mediation effect of coping style on the relations between personality and life satisfaction in Chinese adolescents Frontiers in Psychology, *8*, 1076. doi:10.3389/fpsyg.2017.01076.

Yadav, M., Soni, K., Bhora, J., Khandelwal, K., & Sharma, A. (2023). A mental health chatbot based on artificial intelligence and machine learning. Journal of Operating Systems Development & Trends, *9*(3), 8–14.

Zhang, L., Huang, S., Liu, S., Huang, Y., Chen, S., Jinsong, H., & Mingzhong, X. (2024). Effectiveness of an Internet-based acceptance and commitment therapy intervention for reducing psychological distress in health care professionals: randomized controlled trial. *Journal of Medical Internet Research*. 1438-8871. *26*, e59093. doi:10.2196/59093.

Chapter 17

Brain–computer interfaces for neuromotor rehabilitation

Amit Konar[a], Baishali De[a], and Anuradha Saha[b]
[a]AI Lab, Department of Electronics and Tele-Communication Engineering, Jadavpur University, Kolkata, India
[b]Department of Applied Electronics and Instrumentation Engineering, Netaji Subhash Engineering College, Kolkata, India

1 Introduction

Brain–computer interfaces (BCIs) provide an alternative modality of communication between the human brain and a computer, bypassing the natural neuromuscular pathways of communication (Wolpaw, 2013). BCIs have proven their elegance in various innovative applications, including autonomous wheelchairs (Müller, Celeste, Bastos-Filho, & Sarcinelli-Filho, 2010; Naser & Bhattacharya, 2023), cognitively inspired assistive driving (Saha, Konar, & Nagar, 2017), assessment of olfactory (Saha, Konar, Chatterjee, Ralescu, & Nagar, 2014; Laha, Konar, Rakshit, & Nagar, 2022), tactile (Khasnobish, 2014) and color (Laha, Konar, Rakshit, & Nagar, 2020) perceptual ability, decoding of motor learning disorders (Ghosh, Ghosh, Konar, Rakshit, & Nagar, 2018), game teaching (Kar, Ghosh, Konar, Chakraborty, & Nagar, 2022), and above all, neuromotor rehabilitation (Pan, Mi, Lei, & Deng, 2020; Zhuang, Wu, Wan, & Hu, 2020; Wang et al., 2024); Pfurtscheller et al., 2023; (Del R. Millan, Renkens, Mourino, & Gerstner, 2004; Khasnobish, Konar, Tibarewala, & Nagar, 2017). This chapter aims to introduce the fundamentals of BCIs for neuroprosthetic applications, with special emphasis on brain signal processing, their classification, and the generation of appropriate control commands for position control of a robot arm for neuromotor rehabilitation.

The human nervous system comprises around one hundred billion unicellular neurons distributed across the brain, the spinal cord, and the rest of the body. Neurons interconnect through synaptic clefts to transmit signals from one neuron to another due to the firing of the former neuron. A nerve consists of a number of neurons connected in tandem. Signals originating from the receptor neurons due to "external stimulation" are carried forward to the brain through sensory pathways. The brain triggers our muscles and voluntary organs by transmitting signals through motor nerves downstream to the appropriate organs or muscles for their necessary functions. In addition, the brain acts as an interface between the sensory and motor nerves to make decisions about motor control based on the sensory information received through the sensory nerves (Watson and Breedlove, 2012).

The current research in brain science reveals that during the cognitive processing of tasks by a localized brain region, the brain releases low-frequency signals. Modern brain imaging equipment is capable of capturing these low-frequency, low-amplitude signals to understand the localized activity of the regions in the brain during the cognitive processing of the task by the brain.

With the advent of modern electronics and computers, a significant revolution in brain imaging techniques has been noticed over the past two decades. The brain imaging methods differ in the process of acquisition of brain states and their interpretation. Popularly used brain imaging instruments include functional magnetic resonance imaging, positron emission tomography scans, functional near-infrared spectroscopy, and electroencephalography (EEG). Among the above instruments, EEG is chosen in many medium-to-small-scale laboratories for its high temporal resolution, prompt response, and low cost. The present chapter includes all the experiments using EEG devices undertaken in the Artificial Intelligence (AI) Lab, Jadavpur University, India.

People suffering from neuromotor disorders such as amyotrophic lateral sclerosis, partial paralysis, stroke, or damage to limbs or voluntary organs due to accidents require rehabilitation. Current research in BCIs aims to interface the human brain with an external cyber-physical system (CPS) to provide rehabilitative support to the end-user. The motivation of

Applications of Artificial Intelligence in Healthcare. https://doi.org/10.1016/B978-0-443-43934-6.00016-X

the present work is to decode the neuromotor disabled person's cognitive state of mind from liberated brain signals and activate external CPS to undertake the necessary actions required for rehabilitative purposes. For example, a person who has lost his arms due to an accident may be provided with an artificial robotic arm that can be controlled by his brain using nonmuscular external pathways.

The research in current BCI envisages that different lobes of the human brain release signals with specific features (such as wave shapes, frequency spectra, and/or the arrival of the signal from the onset of the stimulus) when that particular lobe is engaged in undertaking selective cognitive tasks. For instance, when a person observes a flickering signal of a given frequency (say, 8 Hz), he is expected to release an 8 Hz frequency-modulated signal from his visual cortex. This signal is well known as steady-state visually evoked potential (SSVEP), which we shall discuss later. As a second example, when a person receives a visual or audio oddball stimulus, he usually releases a signal with a large peak occurring approximately 300 ms after the onset of the oddball signal. This signal, called P300, is released by the parietal lobe of the subject's brain. The use of P300 in memory functioning, attention, concentration, and many other cognitive functions is well known (Aryanti et al., 2023). However, in this chapter, we study its modality in attention-handling only. As a third example, when a person thinks of moving his left or right hand, he is expected to release a *v*-shaped signal from the right or left part of his motor cortex. This *v*-shaped signal, well known as event-related desynchronization (ERD), followed by event-related synchronization (ERS), is discussed at length in the latter part of this chapter. A common question that may be raised is: can we utilize these brain signals to decode subjective intention, and if so, can we control external hardware, such as a neuroprosthetic limb or a robot arm, to serve the necessary task required by the subject? This chapter attempts to provide a solution to this challenging problem.

Now, to solve the problem stated above, we need to undertake three fundamental operations to carry out the rehabilitation task. First, the brain signals released should be acquired without missing any components and then processed to extract relevant features to describe the signal. Several techniques for automatic feature extraction (FE) from brain signals are available in the literature (Nam, Nijholt, & Lotte, 2018). These techniques, employing time-domain, frequency-domain, and time–frequency correlated features, require a thorough mathematical analysis of the signal to discriminate more important features from the rest when the list of features is too long. This is usually referred to as feature selection in the literature (Dornhege, Del R. Millán, Hinterberger, McFarland, & Müller, 2007).

The next step following FE and feature selection is to decode the released brain signal from the estimated features. There is a vast literature in pattern classification dealing with the classification or decoding of signals from the acquired features (Theodoridis and Koutroumbas, 2006). For instance, let A, B, and C be three signals, and we have a set of five features, F1 through F5, for all these three signals. Suppose an experiment includes 500 trials of each signal A, B, and C, and the five features, F1 through F5, are evaluated for each signal in each trial. So, there is a list of 500 × 3 or 1500 data samples for each of the five attributes/features. The class label representing the signal names A, B, and C for each of the five samples is known *a priori*. A classifier is trained with this set of 1500 experimental trials with prelabeled class information to adapt its internal parameters and fix them uniquely at the end of the training of the classifier, so that for any one of the 1500 samples, it can correctly reproduce the class label A, B, or C. Once a classifier is pretrained and found to give high classification accuracy (CA) (over 98%) for all the 1500 training instances, the classifier is declared ready for use. Any unknown training instance, which is not listed in the set of 1500 instances, is now fed to the input of the classifier. The classifier is expected to produce the correct class label for the given input instance. In most brain signal decoding experiments, the experimental setup is designed to detect the anticipated presence or absence of a specific brain signal within a defined time window. So, effectively, most brain signal classifiers handle two-class classification problems.

After a brain signal released from a specific location in the brain is correctly classified, the CPS understands the motivation of the subject and accordingly generates the necessary control commands for the external actuators, robotic, or neuroprosthetic devices to serve the necessary task.

Early research on BCI focused on an open-loop control strategy (Rao, 2013), where the CPS generates an on-off command for an external actuating device without taking any feedback from the system response to examine whether the control task is accurately executed. Over the last decade, researchers have taken a keen interest in developing closed-loop control systems for BCI-based rehabilitative applications, where the response of the CPS is compared with an expected or desired response, and the control loop is executed over iterations to make necessary corrections in the error-estimate. Unfortunately, there is limited work on closed-loop control strategies to directly control external actuators or manipulators using brain signals. This undoubtedly is a challenging area of research from the perspective of BCI-based rehabilitation. However, such brain-driven closed-loop control policies require extensive cognitive attention, thereby increasing the cognitive load on the subject. To reduce the cognitive load on the subject without causing degradation of the system's performance from the control point of view, the present research proposes an alternative shared-control architecture.

In shared-control architecture, a part of the control task is performed by the experimental subject, and part of it is undertaken by the external automation. The present research attempts to minimize the cognitive load of the subject, to whatever extent possible, without compromising or causing degradation in control performance. This chapter presents closed-loop control policies adopted by the present research group over the last decade. It also reviews the limitations of the said policies and finally proposes a novel multiloop architecture, which is mostly controlled by external automation in the CPS, with minimum participation of the user as far as practical. The rest of the chapter is divided into six sections. Section 2 provides preliminaries on brain signals, their decoding, and strategies for actuating external controllers. Section 3 outlines the control policies adopted in this chapter. In Section 4, signal processing, FE, and classification of the specific brain signals used in the present applications are discussed. Section 5 presents the experimental details and results. A discussion concerning the design aspects of the BCI-based rehabilitative system is given in Section 6. Concluding remarks are listed in Section 7.

2 Preliminaries

This section provides the prerequisites to understand the rest of the chapter. It presents the basic definitions concerning brain signals, pattern classification, and control policies, so as to enable readers from diverse domains to easily grasp the ideas expressed in the chapter. Fig. 17.1 provides a schematic overview of the BCI-based position control system.

2.1 Brain signals

EEG signals acquired from the human scalp typically are of two distinct types. The first type refers to time-locked event-related potentials (ERPs), where, after the occurrence of an event, the brain responds by liberating the time-locked signal. Typical examples of ERPs that fall under this category include the P300 signal (Pritchard, 1981), the N400 signal (Kutas & Federmeier, 2009), and the error-related potential (ErrP) signal (Ferrez & Del R. Millan, 2008). The P300 signal, as already introduced before, is released by the parietal lobe of the brain approximately 300 ms after the onset of an oddball stimulus (i.e., an unexpected concurrent stimulus that occurs relatively infrequently) (Strüber & Polich, 2002) (Fig. 17.2A). To visualize the occurrence of P300, let us consider a person, busy listening to his favorite music (i.e., the normal stimulus), who suddenly hears a bomb-bursting sound, which draws his full attention, resulting in the release of a P300 from the onset of the occurrence of the bomb-bursting sound. Here, the bomb-bursting sound acts as an unexpected oddball stimulus. As an alternative example of time-locked ERPs, let us review the N400 signal, which is released when a person learns relatively newer things. The greater the novelty of the original information, the larger the negativity of the N400 signal that appears approximately 400 ms after the onset of learning novel information. The ErrP signal, released by the mid-line z-electrodes placed on the scalp, appears as a consequence of the recognition of an error by a subject. The ErrP signal is released almost instantaneously after a subject recognizes that he himself or someone else has committed an error. This signal appears with a large negative component, followed by a positive component, and then a small negative component (Fig. 17.2B).

The second variety of EEG signals is often referred to as spontaneous signals. These signals correspond to brain activity in the absence of any external stimulation. The intrinsic electrical activity of the brain, due to spontaneous fluctuations in attention, memory reactivations, motor imaginations, and many others, is grouped under this category. The motor imagery (MI) signal, which appears from the motor cortex due to the imagination of a motor control task by an experimental subject, is released by the brain spontaneously from the time point of motor imagination. The ERD/ERS signal introduced

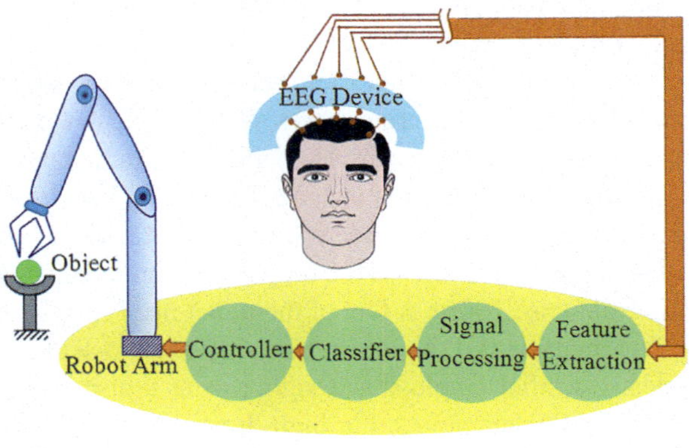

FIG. 17.1 Schematic overview of a BCI system.

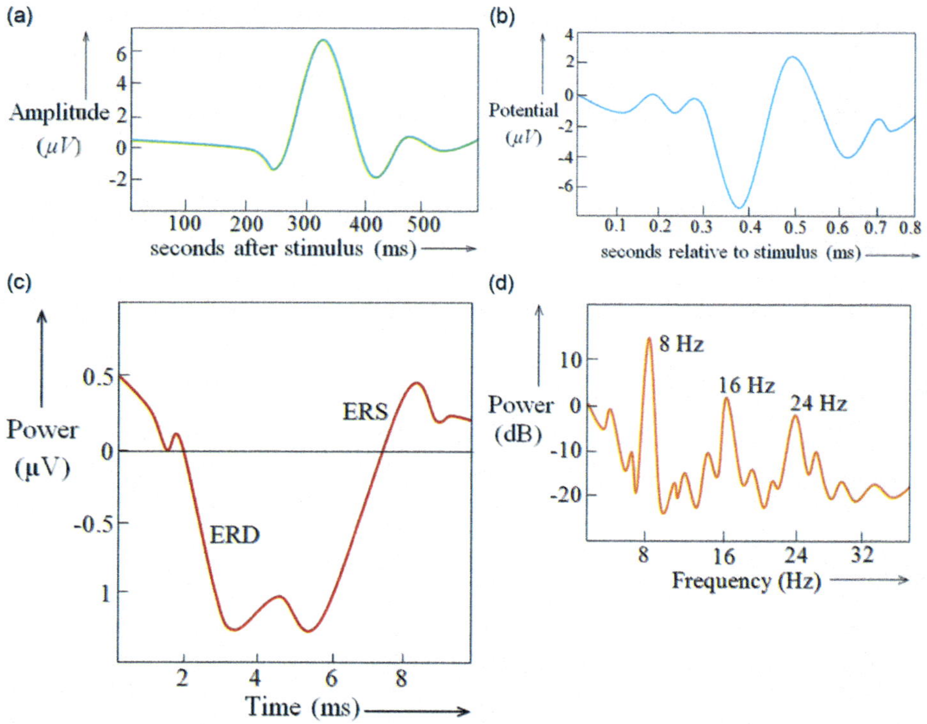

FIG. 17.2 (A) Sample of one P300 signal, (B) sample of one ErrP signal, (C) sample of one ERD/ERS signal, (D) sample of one typical SSVEP signal.

earlier is an MI signal, where the desynchronization in the power of the MI signal continues as long as the person thinks of motor imagination. Once he stops motor imagination, the signal power gradually grows to the level it possessed before motor imagination was performed. The rise in the signal power is referred to as ERS (Fig. 17.2C).

In addition to ERPs and spontaneous signals, a third variety, called evoked potentials (EPs), is also found in the BCI literature. EPs, in brief, are nontime-locked ERPs that are evoked due to external stimulations. One typical example of an EP is the SSVEP signals, which are frequency-locked to external stimulation. In other words, when a person observes a visual stimulus flickering at frequency *f*, the natural brain rhythms in the occipital lobes are frequency-modulated by the flickering signal (Fig. 17.2D). Table 17.1 provides the list of important brain signals, their signal types, brain lobes/location of signal acquisition, and characteristics. In the BCI literature (Rao, 2013), both time-locked ERPs and spontaneous signals are widely used, as the signals have specific waveforms or shapes in the time domain, which help the automatic recognition system to decode them easily. Besides time-locked and spontaneous signals, the brain produces varieties of EPs. However, decoding these signals is difficult with the tools and techniques available in the current BCI literature.

2.2 Pattern classification

A pattern represents a data point vector in a well-defined space, the attributes/components of which together characterize a real-world object. In the present context, the patterns used are data point vectors with all components defined in real space. The pattern recognition problem refers to determining the class of a pattern from the background knowledge of different patterns of the same dimensions with known classes. A classifier is employed to develop its internal representation from the given set of patterns and their class labels. Let there be three classes, A, B, and C, in a pattern recognition problem, where for each class we have a set of n instances of m dimensions each. A supervised learning approach is adopted to develop a unique internal representation of the class, so that for a given input instance of class A, the classifier is correctly able to identify the class label A as the output. Similarly, for any known instance of class B or C, when submitted as input to the classifier, it should be able to correctly produce the class label B or C as the output. Several techniques for classifier design are available in the literature (Theodoridis and Koutroumbas, 2006). A few of them that need special mention include linear discriminant analysis (LDA), quadratic discriminant analysis (QDA), linear support vector machine (L-SVM), kernelized support vector machine, and many others. A linear classifier employs a straight line as a discriminator for two-dimensional (2D) data points and a plane for three-dimensional (3D) data points. For high-dimensional data points, the classifier is described by a hyperplane. Fig. 17.3A provides a linear classification in 2D space, where the data points are of two dimensions, represented

TABLE 17.1 Control modality used in BCI-settings.

Control modality used	Purpose and signals used	Merit	Demerit
Open-loop control (Neuper, Wörtz, & Pfurtscheller, 2006)	ERD/ERS driven motor on/off control	Easy implementation	Just an on-off control strategy may not serve the control purpose
Semiclosed loop control (Bhattacharyya, Konar, & Tibarewala, 2017)	MI-induced motor switching with ErrP-based stopping and offset error correction	Fast	Results in nonzero steady-state error, however, with the high cognitive load of the subject
Fuzzy closed-loop position control (Rakshit & Konar, 2023)	MI-based link-motor activation, ErrP-based error detection, and SSVEP-based error-estimation for fuzzy control	More precise control	Involves high cognitive load
Shared 2-loop control (De, Mondal, Konar, Saha, & Nagar, 2024)	Complete automatic closed-loop control using SSVEP and P300 for target position detection	Fast	Zero steady-state error with minimum cognitive load on the subject

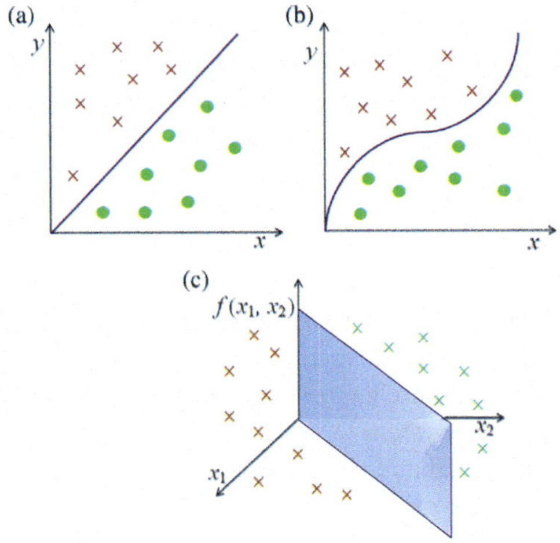

FIG. 17.3 (A) Linear classifier in two-dimensional space, (B) nonlinear classification in two-dimensional space, (C) classification of three-dimensional data points of two classes by a 3D plane.

by x–y coordinates. The discriminator here is a straight line given by. the design problem in the simplest LDA-type classifier attempts to determine the optimal value of slope m and y-intercept c of the straight line, so that the data points closest to the straight line of both the classes are located at the furthest distance possible for the optimal choice of m and c.

Fig. 17.3A provides a scheme for nonlinear classification in two dimensions, where the data points provided can hardly be classified by a linear classifier. QDA can serve as a nonlinear classifier for such a two-class classification problem (Fig. 17.3B). For 3D and higher-dimensional data points, as shown in Fig. 17.3C, a plane or hyperplane is employed to classify data points of two classes. There exists extensive literature for the classification of multiclass data points. However, in the present research, we restrict our scope of application to two-class data points only. The justification for such a restriction lies in the phenomenon that, in the BCI problem, we need to classify the presence or absence of a signal in a given time window, where the signal is expected to be released from the brain. A two-class classification problem, thus, is sufficient for the present application.

2.3 Controllers

BCI experiments generally require undertaking specific cognitive tasks by the subject within a given time window. To release a time-locked or spontaneous signal at the desired time-window, the experimental setup is designed in a manner to capture the signal appearing at the desired time-window and classify it as a true-positive or false-negative signal. If it is a true-positive signal, then a robot arm or neuroprosthetic device is triggered to position the end-effector of the device at the desired location mentally fixed by the subject. Positioning the end-effector at the desired target location to grasp an object and place it at a different location requires controlling the movement of the device. This is referred to as a control problem in the BCI literature for rehabilitative applications.

For the convenience of readers, we briefly outline the control problem undertaken by healthy human subjects to reach and grasp an object. Fig. 17.4A provides an illustrative position control problem of a person's palm toward a targeted object placed on the table. Here, the person stretches his left arm toward the targeted object, his eyes sense the position of his palm, and the brain calculates the positional error of the palm with respect to the position of the target. The brain then generates the necessary control commands to the motor nerves to stretch the arm further until the positional error is zero. The control policy introduced above is pictorially explained in Fig. 17.4B, where the eye is placed in the feedback-loop to sense the position of the stretched arm's end-point, the brain in the forward path acts as the error detector and the controller to drive the arm further, and the muscular system acts like a plant/process, which voluntarily advances the arm toward the targeted object.

The control policy explained in Fig. 17.4B is called a closed-loop configuration, where feedback plays a vital role, and the controller acts on the measure of positional error to drive the plant. In contrast to closed-loop control, there exists an open-loop control with no provisions for a feedback arrangement. The open-loop controller simply drives the plant, and the plant, in turn, continues its action based on the response of the controller. Since there is no provision for the measurement of error, open-loop control often yields a steady-state error (Kuo, 1987). For low-precision systems, open-loop control works. However, for high-precision systems, where a small error may seriously affect the performance of the overall system, closed-loop control is preferred to its open-loop counterpart.

Although BCI systems for rehabilitative applications have passed their infancy, closed-loop control policies in BCI rehabilitative systems started much later, just a few years ago (Bhattacharyya et al., 2017). In this chapter, special emphasis is given to designing closed-loop control policies adopted in BCI systems for rehabilitative applications. The current literature on BCI reveals that there exist two different closed-loop control policies for BCI systems. The first one, called fully automatic BCI-induced control, involves the subject in the complete execution of the control task. Such systems, although naive from the perspective of design and implementation, enhance the cognitive load of the experimental subject. The alternative scheme distributes the control task partly to the subject and partly to external automation. These systems, often referred to as shared control in BCI literature, utilize subjective involvement only to make the BCI system aware of the target position. Once the position of the target is identified, the external automation is triggered to take care of the rest of the control task, moving a robotic arm/artificial appendage/prosthetic limb toward the desired target position mentally fixed by the subject. The challenge of the second scheme lies in subjective involvement to mentally fix the target position and the precision of the external automatic control system to reach the target position by activating electromechanical motors and associated systems. Section 3 provides schematic overviews of different control policies to compare the relative merits and demerits of the above two design perspectives.

3 BCI control policies

Early research in BCI-based rehabilitative systems began with open-loop control policies, where an experimental subject performs a motor imagination/motor execution task to close or open an external switch and activate an external electromechanical system to undertake a task in open loop (Thuruthel, Falotico, Manti, & Laschi, 2018; Li, Li, & Millán, 2025). For

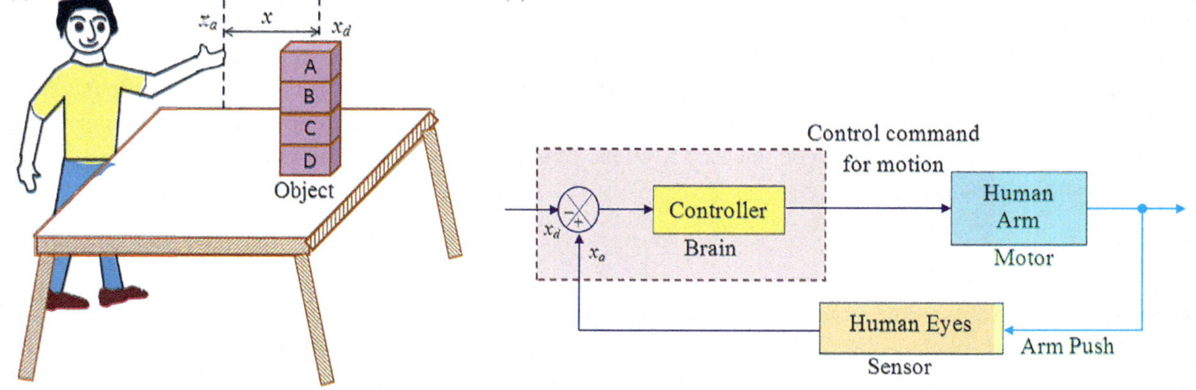

FIG. 17.4 (A) Illustrating a one-dimensional position control problem, and (B) the position control loop for the problem cited in Fig. 18.4(A).

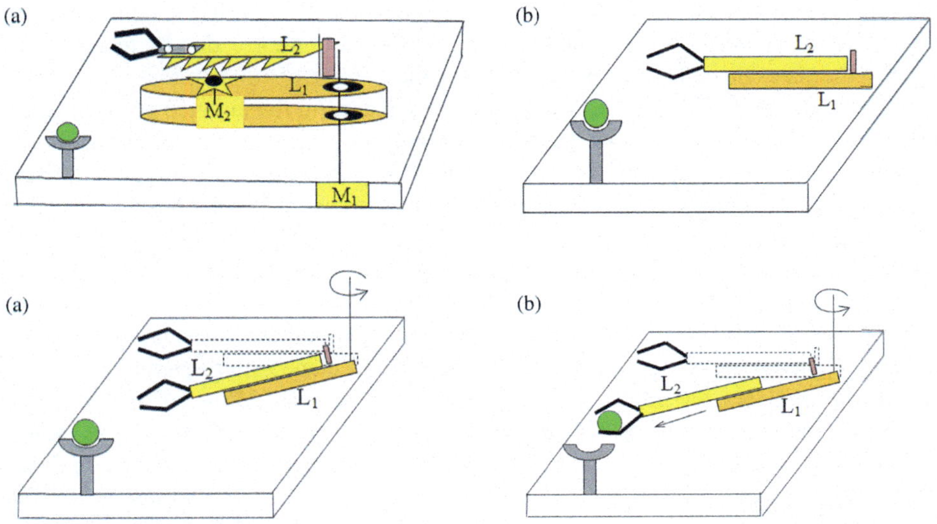

FIG. 17.5 (A) Schematic diagram of a planar two-link robot arm, (B) symbolic representation of Fig. 18.5(A), (C) after alignment of link L1 with the ball, (D) displacement of link L2 to reach the target position of the ball.

example, a wheeled robot may be commanded to move one-step ahead by activation of the motors using motor imagination (ERD/ERS) signals by the subject. In other words, the motor imagination signal is decoded by a two-class classifier in the desired time slot, and in the case of a true-positive signal, the decoder commands the motor drive to move the wheeled-robot forward by a fixed distance.

MI signals mentioned above alone cannot serve closed-loop BCI-applications. To sense the occurrence of an error, an ErrP signal or a P300 signal is employed along with the MI signal to develop a semiclosed-loop control policy. The phrase "semi" here indicates that it is not exactly a complete closed-loop control, but in essence, it looks like a closed-loop control system. This is due to the limitation of the ErrP or P300 signal to determine only the time-point of occurrence of positional error, observed by the subject, but their inability to assess the sign and magnitude of the error. Thus an MI-ErrP- or MI-P300-based position control system remains a semi-closed-loop, but not a complete closed-loop system according to the principles of classical control theory (Mandal, 2006). Let us first review such semiclosed-loop control policies developed by Bhattacharyya et al. (2017).

3.1 ERD/ERS-based semiclosed-loop control

MI and ErrP have successfully been utilized in the position control of a two-link planar robot arm, where both robotic links can move on a plane to enable the end-effector (gripper) to reach the target position. Here, for each link, a subjective motor imagination cycle is required to make the robot arm turn or progress, and an ErrP cycle is required to stop further movement. Fig. 17.5A provides a schematic diagram of such a planar 2-link robot arm, where link 1 (L_1), driven by motor $M_{1,}$ can only turn around the axis of the motor, and link 2 (L_2), with the help of motor M_2 and the rack-and-pinion gear system, can only displace. Here, a ball is placed at a given location on the table, and the subject intends to move the robot arm to bring the gripper to the targeted position of the ball. The BCI system is preprogrammed to execute a sequence of tasks, as presented in the following.

When a subject liberates motor intention for the first time, the motor M_1 connected with link L_1 is energized by the BCI system, so that the link L_1 starts turning clockwise (as seen from the top). Again, when the link L_1 attempts to cross the targeted position of the ball (as shown by the dotted line), the subject is pretrained to generate an ErrP signal (Fig. 17.5B). Upon receiving the ErrP signal, the BCI system commands the robot link L_1 to stop and turn back by a fixed offset angle. To obtain the value of the offset angle, the experiment is performed 100 times, and the offset angles obtained in each trial are saved. The average of the offset angles obtained from the 100 experimental trials is used as the offset angle to turn back the link L_1 in a counter-clockwise direction during the test phase. This approach for error correction diminishes the possible magnitude of error but cannot ensure zero error under any circumstance.

After link L_1 is aligned toward the targeted ball position, the motor M_2 of link L_2 is triggered by a second cycle of motor imagination by the subject (Fig. 17.5B). The link $L_{2,}$ upon receiving excitation from the BCI system, starts moving toward the targeted position of the ball. The subject continuously monitors the movement of the end-effector and generates an ErrP signal when the end-effector attempts to cross the target position of the ball. The BCI system, upon receiving the second ErrP cycle, commands the robot link L_2 to stop and perform necessary error correction by turning back by a small offset distance. This small distance is evaluated here as well, by taking the average of 100 previous offset displacements obtained in 100 experimental trials. The end-effector is now approximately placed in the targeted position of the ball. However, a small positional error, even after offset correction for the rotation of link L_1 and displacement in link L_2, cannot be denied. Fig. 17.5(B) and (C) provides a schematic representation of positioning link L_1 and link L_2 toward the target.

It is noteworthy that in the two-link experimental setup presented above, the target position is a 2D point in the workspace of the planar robot arm. However, with an increase in degrees of freedom, the target position may be a 3D plane or line, depending on the circumstances. The next section provides a more complex BCI-based 3D positioning of the end-effector toward the target ball position.

Fig. 17.6(A) and (B) provides a schematic architecture of a 3D robot arm, where the subjective motivation is to bring the end-effector toward the targeted position of the ball placed in a different plane with respect to the initial plane of the robot arm, including the end-effector. Here, the subject needs to perform two cycles of MI plus ErrP signal liberation to bring the end-effector toward the targeted position of the ball. In the first cycle, the subject liberates the MI signal to turn the entire robot arm from plane OPW around the L_1 axis. The robot arm continues turning until it attempts to cross the targeted plane OQW. When the robot arm attempts to cross the targeted plane OQW, the subject liberates an ErrP signal, and the BCI system, upon decoding the ErrP, commands link L_1 to stop turning and turn back by a fixed offset angle. The fixed offset angle, here too, is evaluated from 100 experimental trials on the same subject in the same experimental setup. After the end-effector is placed in the plane OQW of the ball, link L_3 is triggered by one cycle of the subject's motor imagination to turn it around junction J_2 so as to bring the end-effector toward the targeted position of the ball. As link L_2 attempts to cross the target position of the ball, the subject releases one cycle of ErrP, and the BCI system, upon decoding the ErrP, confirms that link L_3 is going to cross the target line joining junction J_2 and the ball position. Consequently, it commands link L_2 to stop and turn back by a fixed offset angle computed from the previous 100 experimental trials on the same subject in the same experimental setup.

It is noteworthy that, unlike Fig. 17.5A, where the target position was a 2D point in the plane, in Fig. 17.6A, the target position during link L_1 movement is a plane of the ball, and in Fig. 17.6B, the target position is a line joining junction J_2 of link L_3 and the position of the ball. The subject needs to be pretrained to determine the target plane, target lines, and target points depending on different scenarios, which undoubtedly is complex and adds to the cognitive load of the subject. Thus the semiclosed-loop policy designed by Bhattacharya et al. (2017), though innovative, needs simplification to reduce subjective cognitive load. In addition, the scheme cannot ensure zero steady-state error in position control for rehabilitative applications.

3.2 MI-P300-based closed-loop position control scheme

The position control scheme introduced in Section 3.1 suffers from nonzero steady-state error and, therefore, cannot serve the true purpose of real-time rehabilitative aids for people with neuromotor disabilities. A sophistication in the control policy is adopted in a recent work (Rakshit, Konar, & Nagar, 2020), where the authors employ an MI signal to move an individual link of a robot arm and a P300 signal to ensure that the link has crossed the target position. In addition, once a P300 is detected by the BCI system, the velocity of the link that has just crossed the target position is reduced exponentially (Rakshit et al., 2020) or alternatively to half of its last value (before crossing the target) with sign reversal (De, Mondal, Konar, & Saha, 2023), thereby forcing it to move in the reverse direction with reduced speed. Fig. 17.7A provides

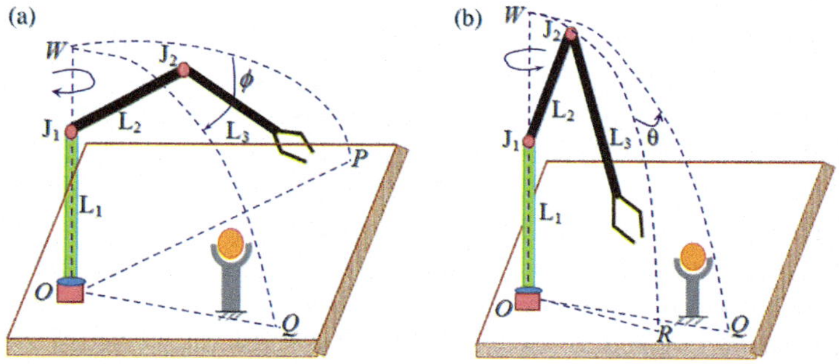

FIG. 17.6 (A) Initial configuration of the robot in a 3D position control experiment and (B) offset correction of the entire robot arm after receiving the ErrP signal.

a schematic representation of position control of a single link, L_3, of the given robot arm to bring the end-effector toward the fixed position of a ball placed on a stand. Here, the necessary angular change required for link L_3 is the reference angle for the proposed position control system. The position control system here includes two loops: (1) an outer loop for position control and (2) an inner loop for speed control. Such a two-loop control architecture is a very recent addition to the BCI literature (Konar, De, Mondal, & Saha, 2024; De et al., 2023; De, Mondal, Konar, & Saha, 2024). Here, the merit of the outer loop is BCI-induced detection of zero-crossing of errors, followed by autonomous speed setting for the inner-loop speed controller. The inner loop attempts to minimize the speed error by setting an appropriate DC gain "*K*'" of the speed controller and generates a control command to the link motor dynamics. The response of the link-motor dynamics is the link speed, which is fed back to determine the speed error in the next loop for the smaller loop. An integrator is kept inside the bigger loop to obtain the angular position of link L_3 from its speed information, which is fed back to the summer to make the outer-loop complete.

Although the scheme introduced in Fig. 17.7A results in zero steady-state error, it still suffers from one fundamental limitation: it keeps the subject continuously engaged in the position control task. Such an architecture, although it carries merit for its novelty, may not suit patients with neuromotor disabilities. An alternative architecture that ensures zero steady-state error and minimizes cognitive load on the subject is of immense importance. The control scheme introduced in Section 3.3, outlined next, is designed to serve that purpose.

Another interesting closed-loop position control scheme that employs MI for motor activation of the robotic link, ErrP for detecting positional overshoot, and SSVEP-based decoding of the magnitude of the positional error is reported in Rakshit & Konar (2023). The challenging contribution of this research is the employment of a number of concurrent fuzzy rules to produce the control commands pertaining to the positioning of the end-effector of a robot arm based on the approximate measurement of positional error.

3.3 Shared two-loop BCI control architecture

To overcome the limitations of both nonzero steady-state error and the reduction in cognitive load of the subject, recent BCI literature proposes a two-loop specialized control architecture, with the control task being distributed partly to the experimental subject and partly to an external autonomous control system (De et al., 2024). Here, the subject needs to specify the location of the target position using brain-actuated control commands, while the rest of the control task required for the position control operation is executed by the external autonomous control system. Such distribution of the control task between the human subject and the external autonomy reduces cognitive load and speeds up the execution of the control task, thereby enhancing system performance. Fig. 17.8 provides a scheme where the subject specifies the location of the target object (here, a glass of fruit juice) by using two brain signals: SSVEP and P300. The idea in the present design to fix the target position of the robotic end-effector is to employ two cycles of SSVEP to fix the *x*- and *y*-coordinates of the target position, and one cycle of P300 to fix the *z*-coordinate of the target position. This is achieved by arranging the movement of a flickering light source from one end of the table along the *x*-direction first. The subject, fixing his attention toward the targeted glass, is expected to release an SSVEP signal when the flickering light beam of frequency f_1 falls on the glass during its movement along the *x*-direction. A similar principle is adopted to move another light source flickering at frequency f_2 along the *y*-direction from one side of the table. The subject, fixing his attention on the targeted glass here too, will release an SSVEP signal when the flickering beam crosses the position of the glass in the *y*-direction. If the motors used to move the flickering sources have a constant speed, then the *x*- and *y*-coordinates of the target glass can be determined

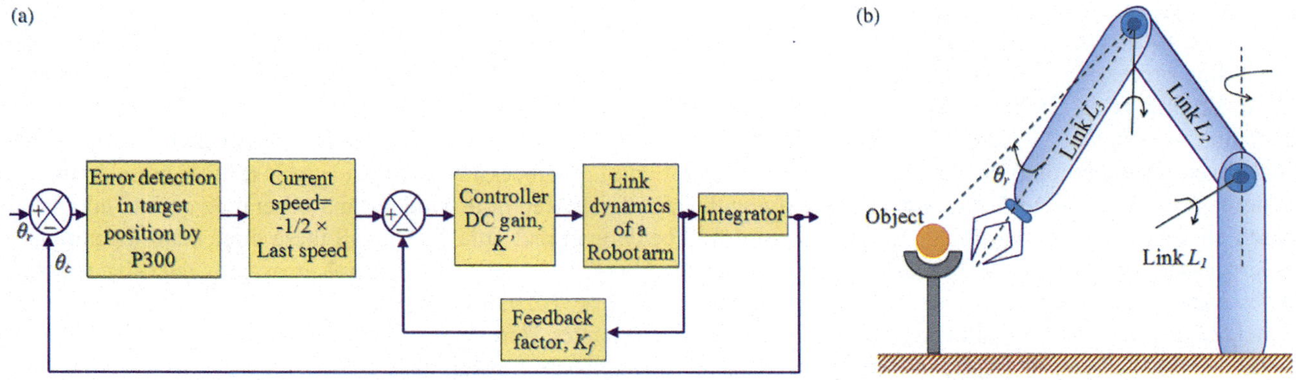

FIG. 17.7 (A) Architecture of a two-loop BCI-induced position control scheme and (B) use of MI action to lift Link L3 by an angle qr.

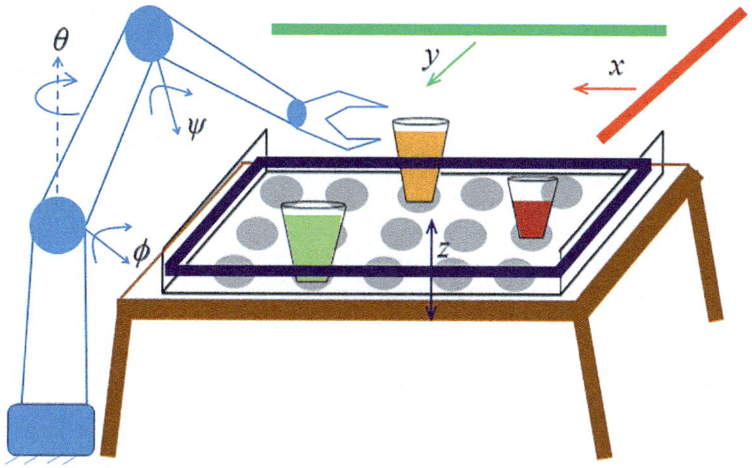

FIG. 17.8 Mentally fixing the coordinates of the target object through two cycles of SSVEP and one cycle of P300.

simply by taking the product of the linear motor speed and the traversal time of the light beam to reach the target position. To determine the z-coordinate of the top of the glass, a light beam surrounding the entire table is gradually moved upward until it crosses the top of the glass. The subject, focusing his attention toward the top of the targeted glass, will release a P300 signal when the light beam crosses the glass top. By noting the traversal time of the glass and the constant speed of the motor, the z-coordinate of the top of the glass can be simply determined by taking their product. To grip the glass at a height equal to one-third of the maximum height of the glass, the BCI system computes the coordinates x, y, $z/3$ and redefines them as the target position of the end-effector of the robot arm employed for rehabilitation purposes.

Once the coordinate of the centroid of the glass is determined, the subject is set free, and the external automation is triggered to autonomously bring the end-effector from its fixed start position (x_0, y_0, z_0) to the targeted glass-centroid position (x, y, $z/3$). The angular link movement of the employed robot arm, here, is determined by a typical inverse kinematics program provided by the robot manufacturer (Kinova Jaco, no date). Once the angular displacement required for each link of the robot arm is identified, a robot controller is utilized to move individual links by the desired angle to bring the end-effector to the targeted centroid of the glass. Various strategies for robot control, such as classical proportional–integral–derivative control, optimal control (Kirk, 1970; Fu, Gonzalez, Lee, & Freeman, 1987), and the most recent model predictive control (MPC), are available in the literature (García, Prett, & Morari, 1989). The MPC has an advantage over the existing control policies to autonomously take care of changing obstacle positions in the robot's workspace in real-time by allowing the controller to move in the receding horizon within a finite time of small duration, optimally satisfying a control objective subject to multiple constraints concerning dynamic obstacle position and velocity changes. There exist extensive works on MPC (Elmorshedy, Xu, El-Sousy, Islam, & Ahmed, 2021; Morato & Felix, 2024) for diverse applications, but its use in BCI settings is still in its infancy. Table 17.1 presents an overview of the control modalities used in the BCI-based position control scheme for neuromotor rehabilitation.

4 Signal processing, FE, and classification of brain signals

The BCI experiments are designed to capture the expected brain signals in the desired time-window for stimulus-sensitive time-locked ERPs or spontaneous brain signals. Extensive work has already been undertaken by different research groups for offline processing and classification of brain signals (Bhattacharyya, Khasnobish, Chatterjee, Konar, & Tibarewala, 2010; Brunner et al., 2010; Ghosh et al., 2018; Lotte, 2014). However, FE from the brain signals in real time and their decoding by a suitably selected pattern classifier still remains a largely unexplored area of modern BCI research. EEG-BCI-based rehabilitation is challenging, as the entire operation of preprocessing the brain signals to filter noise, extracting features from different noise-free signals, classifying them in the desired time window, and generating control commands based on the true positivity of the decoded class must all be done in real time. In fact, all these three major operations—preprocessing and filtering, feature extraction, and classification—are independent research areas with extensive reporting in the appropriate literature (Ghosh, Ghosh, Konar, Konar, & Nagar, 2020; McFarland, McCane, David, & Wolpaw, 1997; Sanei, 2013). Details of this study go beyond the scope of this chapter. Here, we just present the fundamental principles of the above three classes of algorithms briefly for the sake of completeness of this chapter.

4.1 Spatial filtering

The EEG signals acquired from a specific location on the scalp often are found to be contaminated with noise due to muscle artifacts, and also possible mixing of signals from neighboring brain sources. Generally, brain signals for a specific task are released at narrow signal bands such as μ (4–12 Hz) and β (8–13 Hz). So, spatial filters are adopted to obtain the signals in the desired band, thereby eliminating possible interruption by undesirable signal or noise from neighboring signal sources. Several spatial filters, including 2D Laplacian filter, common average referencing (CAR) filter, and many others, are found in the literature (Dornhege et al., 2007; McFarland et al., 1997). In Laplacian filtering, the signal extracted from electrode i at time t, called $\overline{S}_i(t)$, is modified by subtracting the signal acquired from its nearest four neighbors to obtain the filtered signal $\overline{S}_i(t)$ Formally,

$$\overline{S}_i(t) = S_i(t) - \frac{1}{4}\sum_{j\in N(i)} S_j(t) \tag{1}$$

where $N(i)$ denotes the four neighboring electrodes around electrode i. Usually, for most BCI-based experiments, one typical electrode configuration, called the 10–20 electrode placement system (Bhattacharyya, 2014; Saha, 2017), is employed. The Laplacian filter presumes such 10–20 electrode placements prior to undertaking the filter operation using (1). Unlike Laplacian filtering, which only takes into account the four neighboring electrodes around electrode i to filter the data from electrode i, CAR filtering considers signals from all the electrodes to filter the signal from electrode i. Formally,

$$\overline{S}_i(t) = S_i(t) - \frac{1}{N}\sum_{j=1}^{N} S_j(t) \tag{2}$$

where N denotes the number of electrodes placed on the scalp to acquire and filter brain signals. It is apparent from expression (1) that while Laplacian provides localized filtering, CAR eliminates average noise in the signal, presuming the noise variance across the channels does not change considerably.

Besides Laplacian and CAR filtering, there exist quite a few interesting 2-D filtering algorithms, including Butterworth filter, Chebyshev filter, and elliptic filter. These filters have 2D transfer functions with the filter gain as a function of frequency variation along the x and y directions. One common approach to realize these filters is to construct difference equations with the current filter output expressed as a function of filter inputs and outputs at previous time-sample points. An iterative execution of the difference equation on a computer helps in achieving the desired response in the output of the filter, depending on the user's interest to represent them as low-pass filter, band-pass filter (BPF), and high-pass filter configurations. The performance of these filters depends on the choice of filter parameters, such as order and type, which effectively control the pass-band and stop-band ripples and undesirable roll-off of the filters around the boundaries of the pass-band spectrum. Although spatial filtering introduced above eliminates undesirable noise to some extent, it does not provide a foolproof solution to noise filtering of EEG signals.

Independent component analysis (ICA), on the contrary, is designed to identify the independent signal sources inside the brain from the brain signal acquired from different electrodes on the scalp. It presumes that a signal acquired from the scalp from electrode j, i.e., $S_j(t)$ is a linear combination of the signal sources $S_i'(t)$ for $i = 1$ to N. Formally,

$$S_j(t) = \sum_{i=1}^{N} a_{ij} S_i'(t). \tag{3}$$

So, for N electrodes on the scalp, we have N number of equations like expression (3) for j. In vector matrix form, we can write a vector $\vec{V}(t)$ whose jth components is $S_j(t)$, and let $\vec{V'}(t)$ be a vector representing the signal values of independent sources inside the brain and let $S_i'(t)$ be the ith component of $\vec{V'}(t)$. Then, we can develop an equation,

$$\vec{V}(t) = \begin{bmatrix} \vdots \\ S_j(t) \\ \vdots \end{bmatrix}_{N\times 1} = \begin{bmatrix} \cdots & \cdots & \cdots \\ \cdots & a_{ij} & \cdots \\ \cdots & \cdots & \cdots \end{bmatrix}_{N\times N} \begin{bmatrix} \vdots \\ S_i'(t) \\ \vdots \end{bmatrix}_{N\times 1}$$

$$i.e., \vec{V}(t) = A\vec{V}(t)$$

$$\Rightarrow \vec{V'}(t) = A^{-1}\vec{V}(t) \tag{4}$$

The independent signal sources $\vec{V}'(t)$ whose components are $S_i'(t)$ for $i = 1$ to N, thus can easily be retrieved using Eq. (4). This, in general, is employed in ICA to identify the independent sources of the brain from the assessment of signal on the scalp. The formulation of ICA represented by $\vec{V}'(t) = A^{-1}\vec{V}(t)$ is performed in a manner, so as to maximize the signal independence of $\vec{V}(t)$ to optimally obtain the A matrix. Several algorithms to solve the above optimization problem are available in literature (Dornhege et al., 2007). A few of them that needs special mention includes JADE (Joint-Approximate Diagonalization of Eigen matrixes) proposed by (Cardoso & Souloumiac, 1993) and Infomax designed by (Bell & Sejnowski, 1995) are widely used to obtain noise-free signal sources.

4.2 Feature extraction

The human brain releases signals of different frequency bands from specific lobes of the brain during the execution of a selected cognitive task. To accurately obtain the desired signal from the selected brain lobe, BPFs with suitably tuned pass-bands are employed to filter the desired signal from the selected lobe. FE is a process to extract preferably independent features from the filtered brain signals of the selected lobe, so that signal identity is preserved by the set of independent features. The motivation of FE, thus, is to describe the signal by a limited number of independent features, so that a well-designed classifier can correctly recognize the true-positive instances of the desired signal from a mixture of true-positive and false-negative instances of the same signal.

The performance of a BCI system is highly dependent on the choice of the extracted features and the appropriate training of the classifiers to enable it to distinguish true-positive and false-negative instances of the signal acquired from a given brain lobe. To ensure quality training of the classifier, a large number of true-positive and false-negative instances of an individual brain signal acquired from the same subject in the same experimental setup are needed. The classifier parameters also need to be properly adjusted to maximize the CA with respect to the given set of labeled training instances. Although there exists an extensive variety of techniques for FE for brain signals, in general, we can classify them into three main categories: (1) time-domain features, (2) frequency-domain features, and (3) time–frequency correlated features.

Time-domain features are generally preferred for time-locked ERPs because of the specific wave-shapes of the signal. For instance, ERPs like P300 appear with a peak approximately 300 ms after the onset of the stimulus, have a specific signal wave-shape, and thus time-domain features are more appropriate to describe the signal. For different brain diseases like dyslexia, hyperactivity of the brain, and the like, it is difficult to find appropriate time-locked ERPs to recognize the cognitive underpinnings of these effects in the brain. In the absence of any time-specific information about the brain signal, frequency-domain methods are often adopted to determine power spectral density, i.e., spectrum power at different frequencies acquired from the brain signal, and other frequency-domain information from the signal. Neither time-domain nor frequency-domain information individually can describe a complex, noisy brain signal with large changes in signal variance. Time-frequency correlated features, such as wavelet coefficients, play a vital role in describing the correlations between time-samples of the signal occurrences of fixed amplitude and the frequency of the signal for that amplitude. We are now in a position to briefly outline the principles involved in extracting features from a filtered brain signal.

4.2.1 *FE from direct time-domain signal samples*

As time-locked ERPs like P300 or N400 have specific wave-shapes, the amplitudes at different time samples can directly be used as features of the signal (Nijboer et al., 2008; Zych, Filipek, Mrozek-Czajkowska, & Kuwałek, 2025). However, it is noticed that even for the same subject in the same experimental condition and measurement process, the time-locked signal is found to have variations across various instances of acquisition of the same signal. For instance, Fig. 17.9 provides the signal samples obtained at different instances of a P300 signal. It is clear from the figure that there exists variation in signal amplitude at the same time-point across different time instances. One approach to handle this problem is to determine those few selective time-points where the signal variance across the instances at the same time-point is below a user-defined threshold. The only limitation of this method is the choice of the threshold. One approach to minimize the effect of the choice of the threshold is to develop an automation process for jointly optimizing this threshold parameter along with the classifier parameter so as to maximize the CA of the true-positive and false-negative time-locked signals by a suitably selected classifier.

4.2.2 *FE by eigenvector decomposition*

ERPs having specific signal wave-shapes can also be described by certain signal characteristics other than the absolute time-sample values. Eigenvalue decomposition is one such well-known technique. To briefly outline this technique, let us consider N channels and M time-samples of a brain signal. Let X_1 and X_2 be $M \times N$ data matrices obtained after band-pass

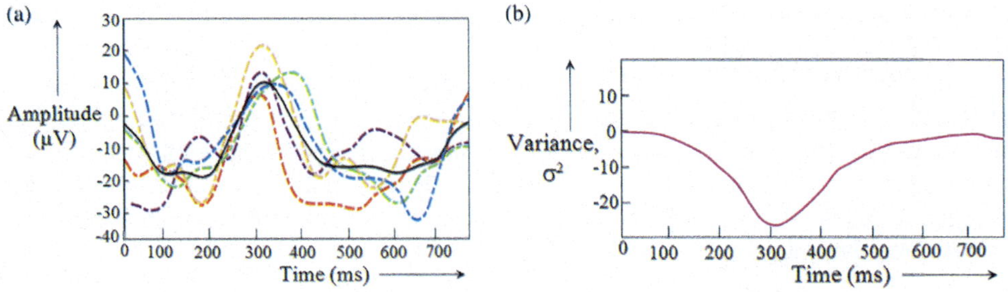

FIG. 17.9 A bunch of P300 signal samples and their variance plot against the time of occurrence.

filtering of the EEG signals, where M time samples includes concatenated trails over the same signals for two different conditions, such as left hand and right hand motor imagery signals (LHMI and RHMI). Let Σ_1 and Σ_2 be the covariance matrices of the signals, given by,

$$\Sigma_1 = X_1^T X_1 \tag{5a}$$

and

$$\Sigma_2 = X_2^T X_2. \tag{5b}$$

These covariance matrices Σ_1 and Σ_2 are simultaneously diagonalized in such a manner, so that the eigenvalues of Σ_1 and that of Σ_2 sum up to 1. One way to obtain common eigenvectors $\vec{V}$ of Σ_1 and Σ_2 to satisfy the above criterion is to set:

$$\vec{V}^T \Sigma_1 \vec{V} = D \tag{6a}$$

and

$$\vec{V}^T (\Sigma_1 + \Sigma_2) \vec{V} = I, \tag{6b}$$

where D is a diagonal matrix of suitable dimensions. Here, the eigenvector $\vec{V}$ is called the filter in a well-established technique of LHMI–RHMI discrimination, referred to as common spatial pattern (CSP).

As eigenvector $\vec{V}$ acts as a filter, and variance of band-pass signal is equal to band power (Dornhege et al., 2007), we compute the band power of $X_1 = var\left(X_1 \vec{V}_j\right)$ which is equal to $\vec{V}_j^T \sum_1 \vec{V}_j = D_{jj}$, where D_{jj} is the jth diagonal element of matrix D, i.e., the eigenvalue of the system X_I. From (6a) and (6b), we obtain:

$$\vec{V}^T \Sigma_2 \vec{V} = I - \vec{V}^T \Sigma_1 \vec{V} = I - D. \tag{7}$$

It is apparent from Eq. (7) that $var\left(X_2 \vec{V}_j\right)$ must be equal to $(1-D_{jj})$. In other words, the jth diagonal element of the two matrices sum up to 1. This, in effect, provides an interesting observation that when the signal power for one signal, say LHMI, is high, the corresponding signal power for the other signal: RHMI is low, such that the resulting signal band power sums up to 1. In other words, for given matrices X_I and X_2 corresponding to LHMI and RHMI, if we can obtain a common eigenvector $\vec{V}$ satisfying the criteria that the sum of the eigenvalues of Σ_1 and Σ_2 sums up to 1, then the two classes can be easily defined by computing $\vec{V}$, so that $\vec{V}_j$ of one class and $\vec{V}_j$ of the other class must sum up to 1.

Several extensions to solve the CSP problems have been proposed in the literature. Of these formulations, the work proposed by (Lotte & Guan, 2011) needs special mention. Further extensions of the work undertaken by Lotte et al. can also be found in (Chakraborty, Ghosh, & Konar, 2020) and (Ghosh et al., 2020).

4.2.3 FE of frequency-domain signals

It is already known from previous sections that SSVEP is a frequency-modulated signal with a sharp peak at the frequency of the flickering signal presented as a visual stimulus. In other words, when a person observes an 8 Hz flickering signal, he releases an SSVEP with a peak at 8 Hz. In case the BCI system has to detect the presence or absence of an SSVEP at a given flickering frequency, the power spectrum at the target frequency is expected to show a high peak in the presence of the SSVEP signal for the selected stimulus input.

However, if there is a possibility of the occurrence of one of several inputs as the flickering stimulus, then only finding the peak information at one of the flickering frequencies may give rise to confusion in decoding the flickering input from the SSVEP signal. Such difficulty in decoding the flickering input from SSVEP appears particularly when the flickering input signal frequencies are close enough to one another. An alternative approach to decode the flickering frequency from the SSVEP signal, which is often adopted and is found to accurately detect the flickering input, is based on canonical correlation analysis (CCA) of the signal.

SSVEP detection by CCA

Let Y be a reference matrix of $2N_h \times N$ dimension, where N_h denotes the number of harmonics of the reference (expected) signal and N is the number of samples of the reference signal. Let X be a measurement matrix of $C \times N$ dimension, where C denotes the number of channels and N carries the same meaning as in matrix Y. Let Σ_X and Σ_Y be within-set covariance matrices, and Σ_{XY} and Σ_{YX} be between-set covariance matrices. The CCA looks for maximization of a given objective function given by,

$$w_X{}^T \Sigma_{XY} w_Y \tag{8}$$

where w_X and w_Y denote two projection directions along X and Y, such that $\left(w_X{}^T X\right)$ and $\left(w_Y{}^T Y\right)^T$ have maximal correlation, subject to the following conditions:

$$w_X{}^T \Sigma_X w_X = 1 \tag{9a}$$

and

$$w_Y{}^T \Sigma_Y w_Y = 1. \tag{9b}$$

The CCA thus attempts to find suitable value of w_X and w_Y so as to maximize the (8), subject to the given constraints given in (9a) and (9b). In other words, CCA attempts to maximize correlation coefficient ρ between $w_X{}^T X$ and $Y^T w_Y$ which can be formally written as to maximize ρ so as to find the suitable value of and w_Y,

$$\rho = \frac{(w_X{}^T X)(Y^T w_Y)}{\sqrt{(w_X{}^T X)(w_X{}^T X)^T (w_Y{}^T Y)(w_Y{}^T Y)^T}} \tag{10}$$

subject to conditions laid down in Eqs. (9a) and (9b). The computation of w_X and w_Y is the generalized eigenvalue problem, where we need to compute a matrix M given by,

$$M = \Sigma_X \Sigma_{XY} \Sigma_Y \Sigma_{YX}. \tag{11}$$

To obtain w_X and w_Y, one need to solve:

$$Mw_X = (\lambda_1 I) w_x \tag{12a}$$

and

$$Mw_Y = (\lambda_2 I) w_Y, \tag{12b}$$

and hence, obtain the largest eigenvalue $(\lambda_1)_{\max}$, $(\lambda_2)_{\max}$ from the above equations, and hence obtain the eigenvectors corresponding to the largest eigenvalues $(\lambda_1)_{\max}$ and $(\lambda_2)_{\max}$ to determine w_X and w_Y. The parameter ρ is evaluated for known values of w_X and w_Y the eigenvectors for the largest eigenvalues.

4.2.4 FE of ErrP signal

One of the early models of time-series data regression is autoregressive moving average model. Let us consider a pth-order autoregressive (AR) process, where the current observation $z(t)$ depends on the past p values: $z(t-1)$, $z(t-2)$, ..., $z(t-p)$. Let μ be the process mean, and we define mean deviated value of the current observation by $\bar{z}(t) = z(t) - \mu$. where

$$\bar{z}(t) = z(t) - \mu. \tag{13a}$$

Defining

$$\bar{z}(t-j) = z(t-j) - \mu, \tag{13b}$$

we describe the AR model by,

$$\bar{z}(t) = \varphi_1 \bar{z}(t-1) + \varphi_2 \bar{z}(t-2) + . + \varphi_p \bar{z}(t-p) + a(t), \tag{14}$$

where $a(t)$ is assumed to be "white noise" error. Presuming partial auto correlation with a maximum lag k, we restructure the last Eq. (14) by replacing p by k. The choice of k is an important consideration to minimize $|a(t)|^2$ and is an optimization problem, well treated in the literature (Bisgaard & Kulahci, 2011).

The ErrP signal is often represented by AR parameters. However, to take care of the nonstationary behavior of the brain time series, the AR parameters are considered variable over time. So, unlike constant parameters in the AR model, given in expression (14), we restructure a time-series $x(k)$ by its past values: $x(k-1), x(k-2), .., x(k-p)$ with adaptive autoregressive (AAR) coefficient $a_i(k)$. Formally,

$$x(k) = \sum_{i=1}^{p} a_i(k) \cdot x(k-i) + \eta(k), \tag{15}$$

where $\eta(k)$ is a Gaussian noise with zero mean and variance $\sigma_n^2(k)$. Formally,

$$\eta(k) = N\{0, \sigma_n^2(k)\}. \tag{16}$$

In the present formulation, we consider seventh-order AAR parameters of the EEG signal to capture its temporal dynamics within 500 ms from the onset of the ErrP signal. Table 17.2 summarizes the techniques employed in this chapter to determine the features of the basic EEG signals used for the present applications.

4.3 Principles of brain signal classification

Once the features of the brain signals are obtained for repeated experimental trials, the features obtained from each trial for true-positive and false-negative instances are tabulated along with the class labels. A classifier maps the features into class labels. Formally, let F be a classifier function that maps a set of N features, $f_1, f_2, .., f_N$ to a class label i. In other words,

$$F(f_1, f_2, .., f_N) = i. \tag{17}$$

For a two-class classification problem, the class label i can be designated by two numbers: one for the true-positive class and the other for the false-negative class. The task of the classifier is to segregate the data points of the two classes in a manner such that the nearest data points from the discriminating function are placed furthest apart by an appropriate choice of the parameters of the mapping function, f. A linear classifier, described by a straight line in 2D or a plane in three or higher dimensions, is often used to segregate the true-positive and false-negative instances of the data points. The design of the parameters of the 2D discriminating straight line or the 3D plane is obtained optimally to satisfy the constraint mentioned above. Once the equation of the 2D line, 3D plane, or high-dimensional plane is determined, the classifier needs to

TABLE 17.2 Brain signals used and their selected features.

	EEG signals	Methods used for feature extraction
1.	P300	Time samples of the signal itself
2.	ERD/ERS	Common spatial pattern (CSP)
3.	SSVEP	Canonical correlation analysis (CCA)
4.	ErrP	AAR parameters

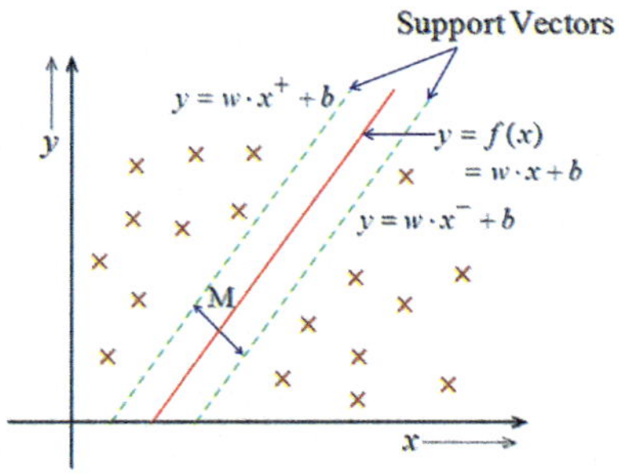

FIG. 17.10 The linear discriminator and support vectors in L-SVM classification.

check whether an unknown data point of N dimensions falls in the true-positive or false-negative regions segregated by the discriminating line or plane. Although most traditional BCI classifiers yield good results with linear classification, it may happen that, for noisy data points, a linear discriminating function cannot correctly classify an unknown data point. A 2D curve or higher-dimensional hypersurfaces may be employed to handle the present situation. Existing BCI systems generally use an L-SVM classifier to classify an unknown data point as a true-positive or false-negative instance. An L-SVM classifier linearly segregates two pattern classes with a margin M around the classifier boundary. Let

$$y = w \cdot x + b \tag{18}$$

represent the hyperplane or discriminator of the two classes of data points. Let X^+ and X^- be the x-coordinates of two data points lying in class 1 and class 2, respectively. Then, the support vectors, represented by the dotted line in Fig. 17.10, are defined by:

$$y = w \cdot x^+ + b \tag{19a}$$

and

$$y = w \cdot x^- + b. \tag{19b}$$

Denoting y to be +1 or −1 for the above two classes, we define the support vector line (the dotted line around the discriminator) as follows:

$$w \cdot x^+ + b = +1 \tag{20a}$$

And

$$w \cdot x^- + b = -1, \tag{20b}$$

which finally yields,

$$w(x^+ - x^-) = 2. \tag{21}$$

The margin between the two support vectors, given by M, is defined by.

$$M = \{(w \cdot x^+ + b) - (w \cdot x^- + b)\} / \|w\| = 2 / \|w\|. \tag{22}$$

The L-SVM is mathematically formulated to minimize:

$$\varphi(w) = \frac{1}{2} w^T w, \tag{23}$$

subject to,

$$y_i (w \cdot x_i + b) \geq 1, \text{ for all } i, \tag{24}$$

where $y_i = +1$ or -1 depending on the belonging of the data points above or below the discriminator. The solution of the above optimization problem is available in any standard text on L-SVM (Das et al., 2009), detailed discussion of which is beyond the scope of the chapter. Once $\vec{w}$ is determined, we can easily obtain the class of an unknown data point by testing whether $(w \cdot x + b) > 0$ or < 0.

Besides L-SVM, there are many well-known classifiers such as neural classifiers, including the well-known multilayer perceptron-based classifiers, modern convolutional neural network (CNN)-based classifiers, and many others. Neural classifiers usually employ a supervised learning algorithm to train the neural network with the given set of training instances for each class. The merit of neural classifiers lies in their inherent characteristics of handling the nonlinear separability of the classes. The most popular among the neural classifiers has prevailed since the mid-1980s, following the well-known work of Rumelhart, Hinton, & Williams (1985), which led Hinton to receive the 2024 Nobel Prize Award in Physics. Rumelhart and McClelland (1986) employed the well-known gradient descent learning (GDL) to adapt the weights (connectivity strain) between pairs of artificial neurons following the Newtonian optimization mechanism, given by:

$$W_{new} \rightarrow W_{old} - \alpha \left[\frac{\partial E}{\partial W} \right]_{W=W_{old}} \tag{25}$$

where W_{new} and W_{old} are the current and next value of weights between pairs of connected neurons,α is the learning rate lying in $(0,1]$ and E is an energy function of weights present in the network.

Given a set of input–output training instances for each class, the GDL principle is employed to formulate a weight adaptation policy based on the measure of error at the neurons in the output layer of a multilayered feed-forward neural network geometry. The weights at the last layer are adapted straight-away using the errors obtained at the output-layered neurons only. The weights of the neurons in the preceding layers are evaluated by backpropagating the error from the output layer to the desired layer to adjust the incoming weights of the neurons present in that layer. The above principle of weight adaptation is utilized in the backpropagation (BP) algorithm, designed by Rumelhart & McClelland (1986). The BP algorithm has been used in the pattern recognition community for more than two decades until SVMs were popularized.

One important extension of the BP algorithm is CNN, designed by Geoffrey Hinton and his research associates (LeCun, Bengio, & Hinton, 2015). The CNN includes a convolution layer, followed by a nonlinear activation function layer realized with ReLU, and then followed by a Maxpool layer. A typical CNN has a number of such three-layered structures in sequence, terminated with a fully connected feedforward network. The BP algorithm mentioned above is used to adapt the weights in the entire network to train the network with a given set of input–output training instances for multiple classes.

Modern pattern classifiers used in BCI systems employ neural BP algorithms and their extensions: CNN, graph convolution neural network (Ghosh, Konar, & Nagar, 2025), graph capsule network (Ghosh & Konar, 2024), and many others, for multiclass classification problems. However, for two-class linearly separable patterns, L-SVM and/or LDA (Bhattacharyya et al., 2010; Bhattacharyya, 2014) technique is good enough for real-time classification of brain signals.

5 Experiments and results

A BCI system requires two phases of experiments: (1) the training phase and (2) the test phase. During the training phase, the experimental subjects are trained to yield the desired signal as a reaction to a user-defined stimulus. The brain signals thus obtained are classified into true-positive and false-negative categories manually by an expert in the BCI domain. The classifiers used to classify the brain signals are trained with the true-positive and false-negative instances obtained during the training phase. In case the CA for the obtained training instances gives satisfactory results, the classifiers so trained are used in the test phase. Besides classifier training, the controllers used in the BCI system also need proper parameter-tuning to achieve good control performance. Once the parameters of the controllers are prefixed, the same controller can be used in the test phase. Two different experimental setups and modalities are presented here to distinguish the control policies adopted in a (1) semiclosed-loop and (2) complete two-loop system.

Experiment 1: Bhattacharyya et al. (2017) employed the Jaco humanoid robot arm to study the position control of an end-effector in 3D space using MI and ErrP signals, where MI is used to initiate the movement of a link and ErrP is subsequently used to stop the motion and undertake offset correction, as introduced earlier. Like any other BCI experiment, their study included a training session, followed by a test session. In the training session, 12 experimental subjects (10 right-handed and 2 left-handed) in the age group of 22–28 years voluntarily participated in the experiment after signing the Helsinki Declaration form. They used a stimulus containing four segments. The first segment included a fixation cross of 1s. The second segment included a RHMI/LHMI instruction for 4s for to turn a link in a counterclockwise or clockwise

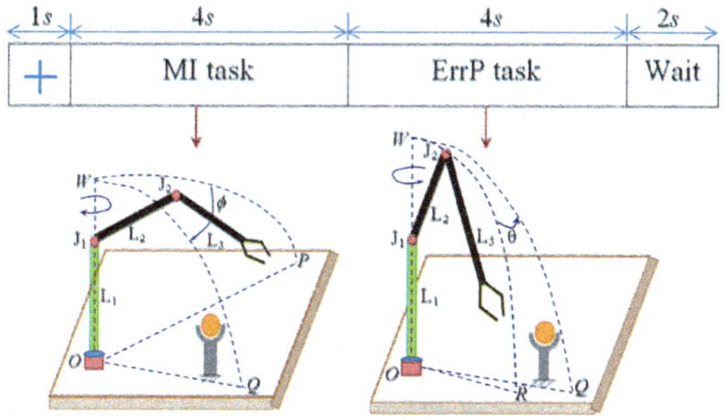

FIG. 17.11 Stimulus description and visualization for Experiment 1 during the training phase.

TABLE 17.3 Comparative performance analysis of BCI-based closed-loop position control strategies.

	Performance metrics			
Techniques used	Steady-state error (e_{ss})%	Peak overshoot (M_p)%	Settling time (t_s) sec	Cognitive attention
MI + ErrP (Bhattacharyya et al., 2017)	1.1	4.9	33	High
MI + P300 + exponential speed reduction (Rakshit et al., 2020)	0.20	4.2	24	High
MI + ErrP + speed reversal and reduction to half of the previous value (De et al., 2023)	0.18	3.8	22	Medium
MI + ErrP + fuzzy speed setting (Konar et al., 2024)	0.02	1.05	11	Medium
Shared two-loop BCI architecture (De et al., 2024)	0.006	1.01	7	Low

direction. The third segment involved error detection and ErrP generation for 4*s* and the last segment, called the wait interval, lasted for 2*s* is and was used for offset correction. Details of the stimulus description are provided in Fig. 17.11.

During the training session, EEG acquisition, filtering, and FE are all performed within the given time intervals, as indicated in Fig. 17.11. An L-SVM classifier is trained to identify true-positive and false-negative MI and ErrP signals. After the training is completed, the trained classifiers are validated using randomly selected instances. It is noted that the classifiers exhibit varied CA, true-positive rate (TPR), and false-positive rate (FPR). The CA across 12 subjects varies between 72% and 92%, while TPR ranges from 78% to 100%, and FPR from 6% to 35%. The average percentages of CA, TPR, and FPR are found to be 88.125%, 92%, and 15%, respectively.

Bhattacharyya et al. (2017) also considered training an L-SVM classifier with the required ErrP instances for true-positive and false-negative classes for the same 12 experimental subjects. The average CA, TPR, and FPR for the L-SVM classifier were found to be 93.114%, 96%, and 4%, respectively. The high TPR and low FPR indicate that most of the subjects achieved good results, signifying the effectiveness of the L-SVM classifier for ErrP detection. In their final study, they also undertook a performance analysis in a position control system using three metrics: (1) steady-state error, (2) peak overshoot, and (3) settling time. Table 17.3 provides the results of the relative merit of the proposal by Bhattacharyya et al. (2017) compared with other relevant techniques introduced in the chapter with respect to the above metrics.

Experiment 2: The second experiment by De et al. (2024) primarily focuses on training the subject to release SSVEP and P300 signals to fix the target position for a rehabilitative Jaco robot arm. After the target position is fixed, an external controller takes care of the necessary control actions to move the end-effector of the robot arm toward the desired target position fixed by the subject. The experiment includes 50 subjects, comprising 28 males and 22 females, in the age group of 25–60 years. Here too, the consent of the subjects was obtained to participate in the experiment, satisfying the Helsinki Declaration of 1970, revised in 2000 (World Medical Association, 2013). The training session begins with a fixation cross for 2*s*, followed by two slides for SSVEP generation and one slide for P300 generation to fix the *x*-, *y*-, and *z*-coordinates of a fixed target placed on a table (Fig. 17.12). Slide 2, used for the arousal of SSVEP, allows a moving light beam flickering at 8 Hz along the *x*-axis until the light beam reaches the maximum value of *x* defined by the user. Special emphasis is given to the arousal of the SSVEP signal by the subject when the flickering light beam attempts to cross the target position of the desired object on the table, which the subject is staring at. Slide 3 of the powerpoint presentation (PPT) includes a

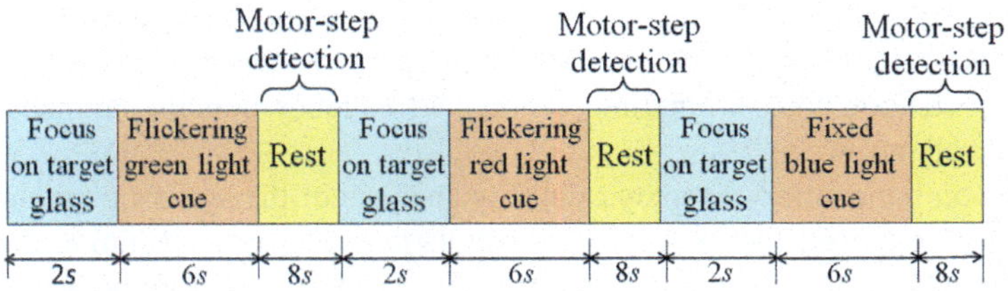

FIG. 17.12 Stimulus description in Experiment 2.

stimulus setup for moving a light beam flickering at frequency f_2 ($f_2 = 8$ Hz) along the *y*-direction of the table. Here too, the subject is supposed to release an SSVEP when the flickering light beam crosses the target position that the subject is staring at. The fourth slide in the PPT attempts to make the subject release a P300 signal when a moving light beam surrounding the table attempts to cross the top of the object. The *x*–*y*–*z* coordinates of the object are thus determined, and the positional error in the centroid (x, y, $z/3$) of the target object over the training instances is calculated.

The training is repeated across multiple sessions in a day and over 15 days, or even more, until the positional error is found to be below a user-defined threshold. Once the training session is over, any one of the trained subjects is asked to participate in the test session, where the subject has to constantly look at one of several glasses of fruit juice placed on the table. The BCI system is programmed to move three stepper motors one by one, carrying flickering light beams along the *x* and *y* axes and a fixed light beam along the *z*-direction, to make the subject release SSVEP when the flickering light beam crosses the respective *x* and *y* positions of the targeted glass, and the constant light beam crosses the top of the glass with a release of P300 by the subject. After the centroid of the target glass is obtained, a two-loop position control system, shown in Fig. 17.8, is utilized to move the end-effector of the Jaco robot arm to the fixed position of the targeted glass. The CCA is used to classify SSVEP, and an L-SVM is used to classify the true-positive and false-negative instances of the P300 signal. The controller performance is studied with respect to steady-state error, settling time, and peak overshoot. The performance of the controllers, in comparison with the previous work, is presented in Table 17.3.

6 Discussion

The chapter introduces two alternative schemes for position control of a robot arm for neuromotor rehabilitative applications. The first scheme deals with two brain signals, ERD/ERS and ErrP, where ERD/ERS is used to start the movement of individual links of a robot arm, and ErrP is used to stop the movement of the moving robotic link and turn it back by a fixed offset angle for positional error correction. The behavior of the robot arm for the first scheme resembles that of a first-order system with a nonzero steady-state error. The performance of the first scheme thus depends on the magnitude of the steady-state error, which, in turn, depends on the error offset correction as introduced in Section 3. The first scheme, in addition, requires constant subjective participation during the control process of the robotic device that acts as a pseudoprosthetic limb for the neuromotor patient.

The second scheme proposed reduces the cognitive load of the subject by dividing the entire control task into two fragments. The first fragment is concerned with the selection of the target position of the robotic device through subjective arousal of SSVEP and P300 signals. The second fragment of the scheme moves a robotic end-effector toward the desired target position fixed by the subject in the first half of the control task. A two-loop control architecture is proposed to autonomously undertake the control task without subjective interventions after the target position is fixed by the subject. The second scheme, referred to as the shared-control architecture, outperforms the former one with respect to settling time, peak overshoot, and, above all, the cognitive involvement of the subject in the control task.

So far, we have not discussed any design issues of the proposed BCI-based rehabilitation aids. The design issues involved refer to the integration of various modules of the system, concerning signal processing, FE, brain signal classification, and position control of the end-effector by either scheme 1 or 2, as introduced in the chapter. The design issues involved aim to optimize the overall performance in signal processing, FE, and classification jointly, so as to maximize the CA of the classifiers during both the training and test phases. Another important issue in the proposed BCI system design is to examine the controller's performance to optimize steady-state error, peak overshoot, stability margin of the controller (Nagrath & Gopal, 2008), and, above all, the cognitive load of the subject during the execution of the control task. To attain the above figures of merit, a thorough analysis of the control system is needed, which, in turn, requires representing the modules of the control system by either the transfer function approach (Kuo, 1987) or the state-space approach (Ogata, 2009). In the first case, stability issues of the overall system are analyzed using Bode plots (Mandal, 2006), Nyquist plots

(Kuo, 1987), and the Root Locus technique (Evans, 1950). While Bode plots and Nyquist plots employ frequency-domain analysis, root locus reveals the system's performance through time-domain analysis. The time-domain parameters involved are steady-state error, peak overshoot, and settling time, while the frequency-domain parameters involved are gain margin and phase margin of the controllers.

The state-space analysis, on the contrary, employs stability analysis of the system from the points of view of controllability and observability tests (Kalman, 1960). Controllability here refers to determining the sequence of input control actions of the controller so as to control the plant/process (robot) positional output to the desired level within a finite time. The observability test refers to examining whether all the system states can be observed from the given control input and the output of the process. In cases where all the system states are not observable, an observer is introduced to predict the system states from approximate measurements of the control input and the response of the plant from the initial time $t = t_0$ to the final time $t = t_f$ (Mandal, 2006).

Stability analysis of controllers, whether by transfer function or state-space approach, is indeed an important issue when examining the stability of the BCI-based rehabilitative system. As an example, we present here the stability analysis of the proposed scheme 3.2, introduced in Section 3, using the transfer function approach. The inner-loop transfer function of the system associated with one link motor is given by:

$$G_{inner}(s) = \frac{16.13}{(s+5.77)(s+55.22)}. \tag{26}$$

The motor controller in the inner loop of Fig. 17.7A is presumed to be a proportional controller with a DC gain constant K'. The speed controller, which sets the next speed to half of its previous value with a sign reversal each time the robot arm crosses the target position, is found to be,

$$G_{con}(s) = \frac{v_0}{2}\frac{s(s+1)}{(s+2)(4s+3)}, \tag{27}$$

after some elementary algebraic analysis following (De et al., 2024). Here, v_0 is the initial speed of the motor. Presuming the parameter $K' = 1$, motor DC gain $K = 4.03$, and the feedback factor $K_f = 1$, we finally obtain the transfer function of the overall system, given by:

$$G(s) = v_0\frac{2.016(s+1)}{(s+0.75)(s+2)(s+5.75)(s+55.44)}. \tag{28}$$

We now undertake stability analysis of the closed-loop system from its open-loop transfer function by using the root locus technique. In root locus analysis, the DC gain of the open-loop transfer function, given by $K_{DC} = 2.016\ v_0$, is varied from zero to infinity in the characteristic equation, given by:

$$1 + [G_{con}(s).G_{inner}(s).(1/s)] = 0 \tag{29}$$

The root locus plot, thus obtained, is shown in Fig. 17.13. The closed-loop system is found to be stable as long as the poles of the system do not cross the $j\omega$-axis and move toward the right half-plane. It is apparent from the root locus plot of

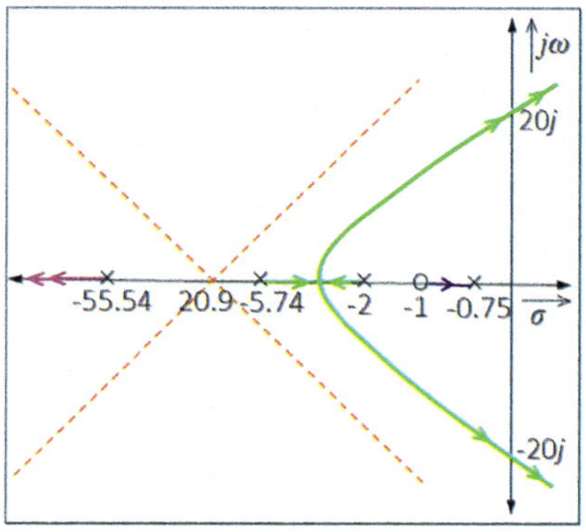

FIG. 17.13 The root locus plot of the open-loop transfer function given in (28).

Fig. 17.13 that the stability margin of the system is $K_{DC} = 4.03$ at $\omega = 20$ rad/s. K_{DC} ultimately returns the upper-bound in the initial speed setting $v_0 \leq 2$ m/s. It is thus apparent from the root locus analysis that the system parameters, like the initial speed of one link of the robot arm, are determined by the stability of the closed-loop system using the root locus technique. More insight into the system can be obtained by state-space analysis for controllability and observability testing, which are not discussed here due to lack of space.

The future research of BCI-based neuromotor rehabilitation has much more scope with the incorporation of machine learning, fuzzy logic, and modern control strategies, such as MPC, to enhance its performance. The merit of fuzzy logic in BCI design lies in its ability to handle uncertainty in brain signal classification due to the presence of undesirable noise in the ERPs and EPs acquired from the brain. Fuzzy logic involves membership functions that reduce large variations of noise in the acquired brain signal because of the nonlinearity and bounded magnitude of the membership functions in [0, 1] (Konar, 2005). In addition, a number of fuzzy rules are triggered concurrently to reduce uncertainty in classification and control tasks (Konar, 2000; Konar & Chatterjee, 2025; Laha et al., 2020). In recent times, significant research has been observed in fuzzy sets and logic or its extension from its classical counterpart: Type-1 fuzzy logic to interval type-2 (IT2) and generalized type-2 (GT2) fuzzy logic. Several works on IT2- and GT2-based classifier design in the context of BCI research (Laha, Ghosh, Parui, Ghosh, & Konar, 2018) are prevalent in the literature. Details of this are beyond the scope of the chapter.

Another important aspect of enhancing controller performance is to employ reinforcement learning (RL) to adapt controller parameters online, particularly when load changes take place or the system response varies due to disturbance input. The typical disturbances, for instance, are electrical low-frequency noise or interference from other channels in the EEG-BCI experimental setup. RL, which works on the principle of reward and penalty, can sense the positional error and its time-derivative to learn to adapt the controller parameters, so as to penalize control actions resulting in enhanced error and reward those control actions that help in reducing positional error estimates. Q-learning (Konar, Chakraborty, Singh, Jain, & Nagar, 2013) and learning automata (Ghosh et al., 2020) are two popular RL algorithms, which have a wider scope as parameter tuners of controllers to enhance their performance.

In recent times, MPCs are being used widely in modern process control for chemical plants and aviation industries. In an MPC, the controlled output is predicted optimally over a small time horizon of T seconds, based on system parameters and constraints that appear jointly at a given time point t. In other words, given the parameters and constraints of a system at time $t = t_0$, the MPC can predict the control signals in succession at intervals of Δt, where $k\Delta t = T$. Thus it can predict the control signals over a time horizon of T. MPC typically employs a real-time optimization process, involving an objective function and constraints, which vary over different time horizons T. Several methods for such real-time optimization are available in the literature; among these, the interior point optimization technique is the most popular (Wang et al., 2024).

7 Conclusion

The chapter proposed two distinct approaches for closed-loop position control system design in BCI settings for neuromotor rehabilitative applications. Special emphasis is given to cognitive load reduction for the subject participating in the BCI experiments. Another important aspect of the proposed design is to demonstrate controller behavior through stability analysis using the root locus technique. The results of the stability analysis envisage that the choice of the initial speed of the robotic device is an important consideration for the stability of the proposed two-loop control architecture.

The chapter also demonstrates possible alternative schemes for the enhancement of system performance by incorporating various methods of classifier designs. It is apparent from the chapter that modern BCI-based rehabilitative systems can be further enhanced with deep learning in classifier design and MPC/RL for controller design. Besides classifiers and controllers, there exists ample scope for further research in preprocessing and FE of brain signals, so as to reduce the load on the classifier by providing a distinctive noise-free feature-set of the acquired brain signals to improve classifier performance. The future challenge of BCI-based neuromotor rehabilitation lies in (1) employing fewer electrodes, (2) enhancing the signal-to-noise ratio in the acquired brain signals, (3) undertaking classification from fewer selected features, and (4) minimally using subjective feedback to reduce the cognitive load of the subject in the BCI experimental setup. Integration of signal processing, FE, classification, and control into a single *application-specific integrated circuit* (ASIC) is expected to emerge as the new technology for BCI-based neuromotor rehabilitation. Lastly, there is immense importance in transplanting such ASICs onto the surface of the cortex to avoid external wiring and enhance the noise figure (Marwedel, 2021) of the BCI system.

Declaration of generative AI and AI-assisted technologies in the writing process

The authors declare that no generative AI or AI-assisted technologies were used in the preparation or writing process of this chapter. The chapter has been produced solely by the authors.

References

Aryanti, G., Zizlavsky, S., Soebadi, A., Hendrawan, D., Prihartono, J., Suwento, R., & Sriyana, A. A. (2023). Evaluation of cognitive function using P300 latency and amplitude in children with profound sensorineural hearing loss using hearing aids: a comparative cross-sectional study. *Indian Journal of Otology*. 22499520. *29*(2), 91–94. Available from: http://www.indianjotol.org/currentissue.asp?sabs=n. doi:10.4103/indianjotol.indianjotol_192_22.

Bell, A. J., & Sejnowski, T. J. (1995). An information-maximization approach to blind separation and blind deconvolution. *Neural Computation*. 08997667. *7*(6), 1129–1159. doi:10.1162/neco.1995.7.6.1129.

Bhattacharyya, S. (2014). Human computer interface for motion control of artificial limbs. Available from: http://hdl.handle.net/10603/226958.

Bhattacharyya, S., Khasnobish, A., Chatterjee, S., Konar, A., & Tibarewala, D. N. (2010). Performance analysis of LDA, QDA and KNN algorithms in left-right limb movement classification from EEG data. In International conference on systems in medicine and biology, ICSMB 2010—Proceedings, India, pp. 126–131. doi:10.1109/ICSMB.2010.5735358.

Bhattacharyya, S., Konar, A., & Tibarewala, D. N. (2017). Motor imagery and error related potential induced position control of a robotic arm. *IEEE/CAA Journal of Automatica Sinica*. 2329-9266. *4*(4), 639–650. doi:10.1109/jas.2017.7510616.

Bisgaard, S., & Kulahci, M. (2011). *Time series analysis and forecasting by example Wiley series in probability and statistics*. Wiley. doi:10.1002/9781118056943.

Brunner, C., Allison, B. Z., Krusienski, D. J., Kaiser, V., Müller-Putz, G. R., Pfurtscheller, G., & Neuper, C. (2010). Improved signal processing approaches in an offline simulation of a hybrid brain-computer interface. *Journal of Neuroscience Methods*. 01650270. *188*(1), 165–173. doi:10.1016/j.jneumeth.2010.02.002.

Cardoso, J. F., & Souloumiac, A. (1993). Blind beamforming for non-Gaussian signals. *IEE Proceedings, Part F: Radar and Signal Processing*. 0956375X, *140*(6), 362–370. doi:10.1049/ip-f-2.1993.0054.

Chakraborty, B., Ghosh, L., & Konar, A. (2020). Designing phase-sensitive common spatial pattern filter to improve brain-computer interfacing. *IEEE Transactions on Biomedical Engineering*. 15582531. *67*(7), 2064–2072. Available from: http://ieeexplore.ieee.org/xpl/RecentIssue.jsp?reload=true&punumber=10. doi:10.1109/TBME.2019.2954470.

Das, S., Halder, A., Bhowmik, P., Chakraborty, A., Konar, A., & Nagar, A. K. (2009). Voice and facial expression based classification of emotion using linear support vector machine. In Proceedings—International conference on developments in eSystems engineering, DeSE 2009, India, pp. 377–384. doi:10.1109/DeSE.2009.9.

De, B., Mondal, A., Konar, A., & Saha, A. (2023). Design of a brain-actuated 2-loop position control of a robot arm for neuro-motor rehabilitation. In *8th international conference on computers and devices for communication, CODEC 2023*. India: Institute of Electrical and Electronics Engineers Inc. 9798350317176. http://ieeexplore.ieee.org/xpl/mostRecentIssue.jsp?punumber=10465675. doi:10.1109/CODEC60112.2023.10465708.

De, B., Mondal, A., Konar, A., & Saha, A. (2024). Brain-actuated speed modulation for position control of an artificial robotic limb for rehabilitative application. In IEEE 3rd international conference on control, instrumentation, energy and communication, CIEC 2024—Proceedings. India: Institute of Electrical and Electronics Engineers Inc., pp. 209–214. 9798350313703. http://ieeexplore.ieee.org/xpl/mostRecentIssue.jsp?punumber=10467194. doi:10.1109/CIEC59440.2024.10468282.

De, B., Mondal, A., Konar, A., Saha, A., & Nagar, A. K. (2024). Learning automaton induced brain-actuated Takagi-Sugeno speed modulation for position control of a rehabilitative robot arm. In IEEE international conference on fuzzy systems. India: Institute of Electrical and Electronics Engineers Inc. 9798350319545. doi:10.1109/FUZZ-IEEE60900.2024.10611877.

Del R. Millan, J., Renkens, F., Mourino, J., & Gerstner, W. (2004). Noninvasive brain-actuated control of a mobile robot by human EEG. *IEEE Transactions on Biomedical Engineering*. 0018-9294. *51*(6), 1026–1033. doi:10.1109/TBME.2004.827086.

Dornhege, G., Del R. Millán, J., Hinterberger, T., McFarland, D. J., & Müller, K.-R. (2007). *Toward brain-computer interfacing*. The MIT Press. doi:10.7551/mitpress/7493.001.0001.

Elmorshedy, M. F., Xu, W., El-Sousy, F. F. M., Islam, M. R., & Ahmed, A. A. (2021). Recent achievements in model predictive control techniques for industrial motor: a comprehensive state-of-the-art. *IEEE Access*.21693536. *9*, 58170–58191. Available from: http://ieeexplore.ieee.org/xpl/RecentIssue.jsp?punumber=6287639. doi:10.1109/ACCESS.2021.3073020.

Evans, W. R. (1950). Control system synthesis by root locus method. *Transactions of the American Institute of Electrical Engineers*. 00963860. *69*, 66–69. doi:10.1109/T-AIEE.1950.5060121.

Ferrez, P. W., & Del R. Millan, J. (2008). Error-related EEG potentials generated during simulated brain–computer interaction. *IEEE Transactions on Biomedical Engineering*. 0018-9294. *55*(3), 923–929. doi:10.1109/TBME.2007.908083.

Fu, K. S., Gonzalez, R. C., Lee, C. G., & Freeman, H. (1987). *Robotics: Control, sensing, vision, and intelligence* (Vol. 1). New York: McGraw-Hill.

García, C. E., Prett, D. M., & Morari, M. (1989). Model predictive control: theory and practice-A survey. *Automatica*. 00051098. *25*(3), 335–348. doi:10.1016/0005-1098(89)90002-2.

Ghosh, L., Ghosh, S., Konar, A., Rakshit, P., & Nagar, A. K. (2018). Decoding of EEG signals using deep long short-term memory network in face recognition task. In Proceedings of the 2018 IEEE symposium series on computational intelligence, SSCI 2018. India: Institute of Electrical and Electronics Engineers Inc., pp. 477–483. 9781538692769. http://ieeexplore.ieee.org/xpl/mostRecentIssue.jsp?punumber=8610062. doi:10.1109/SSCI.2018.8628757.

Ghosh, L., Ghosh, S., Konar, D., Konar, A., & Nagar, A. K. (2020). EEG-induced error correction in path planning by a mobile robot using learning automata. In *Advances in intelligent systems and computing* (Vol. 1138, pp. 273–285). India: Springer. 21945365. http://www.springer.com/series/11156. doi:10.1007/978–981–15–3290–0_21.

Ghosh, L., Konar, A., & Rakshit, P. (2020). *Cognitive modeling of human memory and learning: A non-invasive brain-computer interfacing approach.* John Wiley & Sons.

Ghosh, L., Konar, A., Rakshit, P., & Nagar, A. K. (2018). Hemodynamic analysis for cognitive load assessment and classification in motor learning tasks using type-2 fuzzy sets. *IEEE Transactions on Emerging Topics in Computational Intelligence*, *3*(3), 245–260.

Ghosh, L., Saha, S., & Konar, A. (2020). Bi-directional long short-term memory model to analyze psychological effects on gamers. *Applied Soft Computing*, *95*, 106573.

Ghosh, S., & Konar, A. (2024). Decoding the scientific creative-ability of subjects using dual attention induced graph convolutional-capsule network. *Applied Soft Computing*. 15684946. *161*, 111769. doi:10.1016/j.asoc.2024.111769.

Ghosh, S., Konar, A., & Nagar, A. K. (2025). Assessment of scientific creative-potential by near-infrared spectroscopy using brain-network-based deep-fuzzy classifier. *IEEE Access*.21693536. *13*, 108533–108550. Available from: http://ieeexplore.ieee.org/xpl/RecentIssue.jsp?punumber=6287639. doi:10.1109/ACCESS.2025.3581065.

Kalman, R. E. (1960). On the general theory of control systems. First international conference on automatic control. Moscow, USSR, 1960.

Kar, R., Ghosh, L., Konar, A., Chakraborty, A., & Nagar, A. K. (2022). EEG-induced autonomous game-teaching to a robot arm by human trainers using reinforcement learning. *IEEE Transactions on Games*. 24751510. *14*(4), 610–622. Available from: https://ieeexplore.ieee.org/xpl/mostRecentIssue.jsp?punumber=7782673. doi:10.1109/TG.2021.3124340.

Khasnobish, A., Konar, A., Tibarewala, D. N., & Nagar, A. K. (2017). Bypassing the natural visual-motor pathway to execute complex movement related tasks using interval type-2 fuzzy sets. *IEEE Transactions on Neural Systems and Rehabilitation Engineering*. 15580210. *25*(1), 91–105. Available from: https://ieeexplore.ieee.org/xpl/mostRecentIssue.jsp?punumber=7333. doi:10.1109/TNSRE.2016.2580580.

Khasnobish. (2014). Artificial interface for Somato Sensory perception. Available from: http://hdl.handle.net/10603/226962.

Kinova Jaco Robot user Guide. https://www.manualslib.com/manual/2456901/Kinova-Jaco.html

Kirk, D. E. (1970). Optimal Control theory: An Introduction. Prentice-Hall, Englewood Cliffs, N. J.

Konar, A. (2000). Artificial intelligence and soft computing: behavioral and cognitive modeling of the human brain.

Konar, A. (2005). Computational intelligence: principles, techniques and applications.

Konar, A., Chakraborty, I. G., Singh, S. J., Jain, L. C., & Nagar, A. K. (2013). A deterministic improved Q-learning for path planning of a mobile robot. *IEEE Transactions on Systems, Man, and Cybernetics: Systems*. 2168-2216. *43*(5), 1141–1153. doi:10.1109/tsmca.2012.2227719.

Konar, A., De, B., Mondal, A., & Saha, A. (2024). *Error-induced brain-actuated motor command generation for speed-profile setting in a 2-loop position control of a robot arm*. (pp. 123–139). Springer Science and Business Media LLC. doi:10.1007/978-981-97-2321-8_11.

Konar, D., & Chatterjee, A. (2025). Particle swarm optimized fuzzy extended state observer design for disturbance rejection in sliding mode attitude control of a quadrotor. *Engineering Applications of Artificial Intelligence*. 09521976. *157*, 111018. doi:10.1016/j.engappai.2025.111018.

Koutroumbas, K. (2006). *Pattern recognition pattern recognition*. Greece: Greece Elsevier Inc. Available from: http://www.sciencedirect.com/science/book/9780123695314. doi:10.1016/B978-0-12-369531-4.X5000-8.

Kuo, B. C. (1987). Automatic control systems.

Kutas, M., & Federmeier, K. (2009). N400. *Scholarpedia*. 1941-6016. *4*(10), 7790. doi:10.4249/scholarpedia.7790.

Laha, M., Ghosh, L., Parui, S., Ghosh, S., & Konar, A. (2018). Evaluation of density based odor classification by general type-2 fuzzy set induced pattern classifier. In *International conference on wireless communications, signal processing and networking, WiSPNET 2018*. India: Institute of Electrical and Electronics Engineers Inc. 9781538636244. http://ieeexplore.ieee.org/xpl/mostRecentIssue.jsp?punumber=8514820. doi:10.1109/WiSPNET.2018.8538634.

Laha, M., Konar, A., Rakshit, P., & Nagar, A. K. (2020). Exploration of subjective color perceptual-ability by EEG-induced type-2 fuzzy classifiers. *IEEE Transactions on Cognitive and Developmental Systems*. 23798939. *12*(3), 618–635. Available from: http://ieeexplore.ieee.org/xpl/aboutJournal.jsp?punumber=7274989. doi:10.1109/TCDS.2019.2959138.

Laha, M., Konar, A., Rakshit, P., & Nagar, A. K. (2022). Hemodynamic analysis for olfactory perceptual degradation assessment using generalized type-2 fuzzy regression. *IEEE Transactions on Cognitive and Developmental Systems*. 23798939. *14*(3), 1217–1231. Available from: http://ieeexplore.ieee.org/xpl/aboutJournal.jsp?punumber=7274989. doi:10.1109/TCDS.2021.3101897.

LeCun, Y., Bengio, Y., & Hinton, G. (2015). Deep learning. *Nature*. 0028-0836. *521*(7553), 436–444. doi:10.1038/nature14539.

Li, H., Li, X., & Millán, J. R. D. (2025). Noninvasive EEG-based intelligent mobile robots: a systematic review. *IEEE Transactions on Automation Science and Engineering*. 1545-5955. *22*, 6291–6315. doi:10.1109/tase.2024.3441055.

Lotte, F. (2014). *A tutorial on EEG signal-processing techniques for mental-state recognition in brain–computer interfaces*. (pp. 133–161). Springer Science and Business Media LLC. doi:10.1007/978-1-4471-6584-2_7.

Lotte, F., & Guan, C. (2011). Regularizing common spatial patterns to improve BCI designs: unified theory and new algorithms. *IEEE Transactions on Biomedical Engineering*. 00189294. *58*(2), 355–362. doi:10.1109/TBME.2010.2082539.

Mandal, A. K. (2006). Introduction to control engineering: modeling, analysis and design.

Marwedel, P. (2021). Embedded system design: embedded systems foundations of cyber-physical systems, and the internet of things.

McFarland, D. J., McCane, L. M., David, S. V., & Wolpaw, J. R. (1997). Spatial filter selection for EEG-based communication. *Electroencephalography and Clinical Neurophysiology*. 00134694. *103*(3), 386–394. doi:10.1016/S0013-4694(97)00022-2.

Morato, M. M., & Felix, M. S. (2024). Data science and model predictive control: a survey of recent advances on data-driven MPC algorithms. *Journal of Process Control*. 09591524. *144*. Available from: https://www.sciencedirect.com/science/journal/09591524. doi:10.1016/j.jprocont.2024.103327.

Müller, S. M. T., Celeste, W. C., Bastos-Filho, T. F., & Sarcinelli-Filho, M. (2010). Brain-computer interface based on visual evoked potentials to command autonomous robotic wheelchair. *Journal of Medical and Biological Engineering*. 16090985. *30*(6), 407–416. Available from: http://jmbe.bme.ncku.edu.tw/index.php/bme/article/viewFile/542/801. doi:10.5405/jmbe.765.

Nagrath, I. J., & Gopal, M. (2008). Textbook of control systems engineering.

Nam, C. S., Nijholt, A., & Lotte, F. (2018). Brain–computer interfaces handbook: technological and theoretical advances.

Naser, M. Y. M., & Bhattacharya, S. (2023). Towards practical BCI-driven wheelchairs: a systematic review study. *IEEE Transactions on Neural Systems and Rehabilitation Engineering*. 1534-4320. *31*, 1030–1044. doi:10.1109/tnsre.2023.3236251.

Neuper, C., Wörtz, M., & Pfurtscheller, G. (2006). *ERD/ERS patterns reflecting sensorimotor activation and deactivation* (Vol. 159, pp. 211–222). Elsevier BV. doi:10.1016/s0079-6123(06)59014-4.

Nijboer, F., Sellers, E. W., Mellinger, J., Jordan, M. A., Matuz, T., Furdea, A., et al. (2008). A P300-based brain–computer interface for people with amyotrophic lateral sclerosis. *Clinical Neurophysiology*. 13882457. *119*(8), 1909–1916. doi:10.1016/j.clinph.2008.03.034.

Ogata, K. (2009). Modern control engineering.

Pan, H., Mi, W., Lei, X., & Deng, J. (2020). A closed-loop brain–machine interface framework design for motor rehabilitation. *Biomedical Signal Processing and Control*. 17468094. *58*, 101877. doi:10.1016/j.bspc.2020.101877.

Pritchard, W. S. (1981). Psychophysiology of P300. *Psychological Bulletin*. 00332909. *89*(3), 506–540. doi:10.1037/0033-2909.89.3.506.

Rakshit, A., & Konar, A. (2023). *Brain–computer interface for fuzzy position control of a robot arm by mentally detected magnitude and sign of positional error* (Vol. 985, pp. 163–177). Springer Science and Business Media LLC. doi:10.1007/978-981-19-8477-8_13.

Rakshit, A., Konar, A., & Nagar, A. K. (2020). A hybrid brain-computer interface for closed-loop position control of a robot arm. *IEEE/CAA Journal of Automatica Sinica*. 23299274. *7*(5), 1344–1360. Available from: https://www.ieee.org/membership-catalog/productdetail/showProductDetailPage.html?product=PER284-EPC. doi:10.1109/JAS.2020.1003336.

Rao, R. P. N. (2013). *Brain-computer interfacing: An introduction*. Cambridge University Press. doi:10.1017/CBO9781139032803.

Rumelhart, D. E., Hinton, G. E., & Williams, R. J. (1985). Learning internal representations by error propagation.

Rumelhart, D. E., & McClelland, J. L. (1986). Parallel distributed processing, volume 1: explorations in the microstructure of cognition: foundations.

Saha, A., Konar, A., Chatterjee, A., Ralescu, A., & Nagar, A. K. (2014). EEG analysis for olfactory perceptual-ability measurement using a recurrent neural classifier. *IEEE Transactions on Human-Machine Systems*. 21682291. *44*(6), 717–730. Available from: http://ieeexplore.ieee.org/xpl/mostRecentIssue.jsp?punumber=6221037. doi:10.1109/THMS.2014.2344003.

Saha, A., Konar, A., & Nagar, A. K. (2017). EEG analysis for cognitive failure detection in driving using type-2 fuzzy classifiers. *IEEE Transactions on Emerging Topics in Computational Intelligence*. 2471285X. *1*(6), 437–453. Available from: http://ieeexplore.ieee.org/xpl/mostRecentIssue.jsp?punumber=7433297. doi:10.1109/TETCI.2017.2750761.

Saha. (2017). EEG analysis to decode perceptual ability of human subjects. Available from: http://hdl.handle.net/10603/417066.

Sanei, S. (2013). *Adaptive processing of brain signals*. Wiley. doi:10.1002/9781118622162.

Strüber, D., & Polich, J. (2002). P300 and slow wave from oddball and single-stimulus visual tasks: inter-stimulus interval effects. *International Journal of Psychophysiology*. 01678760. *45*(3), 187–196. doi:10.1016/s0167-8760(02)00071-5.

Thuruthel, T. G., Falotico, E., Manti, M., & Laschi, C. (2018). Stable open loop control of soft robotic manipulators. *IEEE Robotics and Automation Letters*. 23773766. *3*(2), 1292–1298. Available from: http://ieeexplore.ieee.org/servlet/opac?punumber=7083369. doi:10.1109/LRA.2018.2797241.

Wang, A., Tian, X., Jiang, D., Yang, C., Qin, X., Zhang, Y., et al. (2024). Rehabilitation with brain-computer interface and upper limb motor function in ischemic stroke: a randomized controlled trial. *Med*. 26666340. *5*(6), 559. doi:10.1016/j.medj.2024.02.014.

Wang, S., Wang, Y., Miao, Z., Wang, X., & Wei, H. (2024). Dual model predictive control of multiple quadrotors with formation maintenance and collision avoidance. *IEEE Transactions on Industrial Electronics*. 0278-0046. *71*(12), 16037–16046. doi:10.1109/tie.2024.3379663.

Watson, N. V., & Breedlove, S. M. (2012). *The Mind's Machine: Foundations of Brain and Behavior*. Sinauer Associates, Inc., Sunderland, Massachusetts, USA.

Wolpaw, J. R. (2013). *Brain–computer interfaces*. (pp. 67–74). Elsevier BV. doi:10.1016/b978-0-444-52901-5.00006-x.

World Medical Association. (2013). World Medical Association declaration of Helsinki: ethical principles for medical research involving human subjects. *JAMA* 15383598, *310*(20), 2191–2194. Available from: http://jama.jamanetwork.com/data/Journals/JAMA/929397/jsc130006.pdf. doi:10.1001/jama.2013.281053.

Zhuang, M., Wu, Q., Wan, F., & Hu, Y. (2020). State-of-the-art non-invasive brain–computer interface for neural rehabilitation: a review. *Journal of Neurorestoratology*. 23242426. *8*(1), 12–25. doi:10.26599/jnr.2020.9040001.

Zych, P., Filipek, K., Mrozek-Czajkowska, A., & Kuwałek, P. (2025). Classification of electroencephalography motor execution signals using a hybrid neural network based on instantaneous frequency and amplitude obtained via empirical wavelet transform. *Sensors*. 1424-8220. *25*(11), 3284. doi:10.3390/s25113284.

Chapter 18

Artificial intelligence in the detection and mitigation of neurological and psychiatric conditions

Jarred D. Metkus[a], Angela Oh[a], Tanmay Singhal[b], Talia Puzantian[a], and Subhrajit Bhattacharya[a]
[a]*School of Pharmacy, The Claremont Colleges, Keck Graduate Institute, Claremont, CA, United States*
[b]*College of Osteopathic Medicine, Western University, CA, United States*

1 Introduction

Artificial intelligence (AI) is a broad field of computer science involving the creation of machines and software that can perform tasks typically requiring human intelligence. It is important to note that AI is a specific discipline within computer science that focuses on the development of advanced algorithms capable of performing complex tasks at or above human-level efficiency, requiring minimal human intervention (Bekbolatova, Mayer, Ong, & Toma, 2024). Despite its recent development, AI has exhibited exponential growth, transforming various sectors of society, such as technology, marketing, and consumer behavior. With AI's continuous advancements, its use in healthcare is becoming one of its most promising fields of application. As of January 2025, the new US administration has announced a massive 500 billion USD investment into Stargate, a company backed by titans of the industry, which will focus on building AI infrastructure in the United States with the primary goal of diagnosing and mitigating diseases with greater efficiency (Smith, 2025). Although the integration of AI into healthcare systems is still in its formative years, its current impact and potential to revolutionize patient care, clinical processes, clinician support, and predictive analytics remain substantial, even at this early stage. Most valuable among its potential uses is in drug development, drug screening, genetic screening, diagnostics, and clinical treatments through the application of AI's machine learning and deep learning in the analysis of large datasets.

Machine learning is one of the most common uses of AI and is a statistical technique used to ascertain how well data analysis models can adapt not only to the original dataset from which they were trained but also to similar datasets they were not trained on, through a process similar to human learning (Helm et al., 2020). As machine learning relies on historical datasets to "learn" patterns and make "informed" predictions, a training dataset that includes a collection of past data containing input variables (i.e., patient characteristics) and outcome variables (i.e., whether or not disease pathogenesis is observed) is required (Davenport & Kalakota, 2019). This approach is especially useful in precision medicine, where machine learning can be used to predict which treatment protocols have the highest chance of success based on various patient attributes such as gene expression and drug sensitivity.

Other, more complex forms of machine learning, such as artificial neural networks and deep learning, can provide an even more robust analysis of data. Artificial neural networks are based on the mechanics of human neurons; while neurons receive varying signals from others in their vicinity, a new signal is only propagated when an incoming signal surpasses a certain threshold, mediated by an activation function. For instance, when the neural input signal is low, the neuron is not activated, and the output is then 0; however, once the signal input exceeds the threshold, the neuron produces a nonzero product, and a signal is relayed. Artificial neural networks work similarly in that certain conditions must be met for a related function to be added onto the existing one. However, unlike the electrical conveyance of information seen in biological neurons, the common activation function in artificial neural networks is based on multiple varying sigmoid functions, providing computational simplicity and efficacy (Han, Kim, Kim, & Youn, 2018). During the "learning" phase(s), the artificial networks modify the importance and adjust the influence of varying "neurons/nodes," ultimately influencing how other datasets are processed for the network to arrive at its predictive conclusion. Moreover, this adjustment process—referred

Applications of Artificial Intelligence in Healthcare. https://doi.org/10.1016/B978-0-443-43934-6.00010-9

to as weight adjustment—alters the network in such a way that reduces future errors, as the node connections to incorrect predictions are weakened over time in a process similar to long-term potentiation and the mechanism of depression, two cellular correlates of memory strengthening and weakening (Han et al., 2018).

Deep learning is regarded as the core of AI technologies, pushing AI to greater heights in learning from data. Simply put, deep learning can be seen as a more intricate version of artificial neural networks, in which there are multiple layers of artificial neurons rather than a single network. These neurons, over time, develop the ability to detect features similar to how the human brain processes visual information, building upon simple shape recognition to complex feature detection abilities (Davenport & Kalakota, 2019; Hinton, 2018; Sarker, 2021). Deep learning is preferred over artificial neural networks when working with larger datasets due to the ability of its multiple layers to preprocess and extract meaningful patterns from the proposed large datasets, as the larger datasets provide the necessary training material to properly fine-tune weight adjustment biases (Sarker, 2021). Deep learning-based modeling has been used to predict future medical events and diagnoses from previous, cooccurring, or related medical events, for instance, predicting a patient's cardiovascular health from fundus images taken of the retina (Chikumba, Hu, & Luo, 2023). In a study conducted by Esteva et al. in 2017, 129,450 images of varying skin lesions (2000 of patients with biopsy-confirmed diagnoses) were compiled and used to train the convolutional neural network—a deep learning algorithm previously utilized to recognize everyday objects—to identify the presence of potential malignancies in the skin lesion dataset before its performance was assessed against groups of 21–25 board-certified dermatologists (Esteva et al., 2017). It was found that deep learning technologies outperformed board-certified dermatologists in detecting melanoma and other skin cancers, such as basal and squamous cell carcinoma, with significantly greater sensitivity and specificity (Hinton, 2018). However, it should be noted that this is not the case with other medical subfields, such as neurology, due to the lack of consistency in anatomy, malignancy presentation, and diagnostic criteria. In neurology and psychiatry, AI technologies have demonstrated use in the detection of some diseases, such as major depressive disorder (MDD), but their use has certain limitations when attempting to diagnose more complex conditions, such as attention-deficit/hyperactivity disorder (ADHD) and/or bipolar disorder (Yan, Ruan, & Jiang, 2023). This challenge is posed not by the insufficiency of datasets or issues with algorithm training, but rather by the subjectivity of diagnosis, heterogeneity in clinical presentation, and the general lack of understanding regarding the pathogenesis of these disorders, issues to be further assessed in later portions of this text. In addition, despite the drastic uptrend in healthcare's need for AI, its integration is slowed by the ever-increasing total spent on healthcare and the shortage of clinicians. In spite of this, healthcare is being driven to embrace the integration of AI-based innovations due to its future potential to relieve these strains on both physicians and patients alike (Väänänen, Haataja, Vehviläinen-Julkunen, & Toivanen, 2021). Although AI allows for earlier detection of disease, personalized treatment options, and overall improved clinical care, its implementation in neurology and psychiatry has been limited due to concerns such as data variability, patient privacy, and possible violations of medical ethics in healthcare-based decisions (Johnson et al., 2021).

2 Drug development and screening

One of the most prominent and transformative applications of AI within healthcare exists in the realm of drug discovery and compound screening. By learning the hidden language of biomolecules, AI can be used to predict binding affinities, off-target effects, therapeutic effects, and potential candidates for novel treatments. Moreover, AI enhances the ability to shorten the lengthy discovery process associated with drug development and its related advancements (Zhavoronkov, Bischof, & Lee, 2021). Tools like Google's AlphaFold and DeepMind's AlphaFold are a testament to the fusion of biology and computation, possessing unprecedented abilities to predict protein structures and potential binding targets with high accuracy. These tools allow researchers to parse through vast compound libraries in silico (experimentation performed using computer laboratory simulations or computational AI models), while reducing the time required for wet-laboratory experimentation; therefore presenting newfound options for effective therapy across various neurological and psychiatric conditions (Jumper et al., 2021; Stokes et al., 2020).

By building upon this momentum, the application of AI in biomarker discovery and proteogenomics has also grown. Although still in its nascent stages, machine learning models can now assist with decoding large proteomic and genomic landscapes by illuminating patterns that were once buried under biological noise (Yeung, Wang, Kraljevic, & Teo, 2023). One of the greatest benefits of these tools is their ability to provide clarity in disease pathogenesis and its related mechanisms by helping to quickly narrow the search for potential novel therapeutics, an endeavor vital for conditions for which conventional routes have failed, been costly, or taken too long to produce results (Yeung et al., 2023). However, one of the most persistent challenges within this domain lies in the identification of posttranslational protein modifications (PTMs). These subtle biochemical changes, often low in abundance yet high in clinical relevance, have proven to elude even the most sensitive laboratory techniques. The detection of PTMs necessitates advanced mass spectrometry tools with exceptional resolving power and mass accuracy, most notably the time-of-flight or ion trap spectrometers. Yet, even with these specialized tools, limitations remain; rare PTMs, fleeting proteins, and unstable complexes often disappear before the detection threshold is reached (Hermann, Schurgers, & Jankowski, 2022). In such cases, AI can act both as computational muscle and a fine-toothed, objective, and analytic comb. By training on large datasets and learning to recognize various spectral signatures, machine learning models can assist in the classification and contextualization of PTMs that could otherwise be overlooked by the human eye. Bottom-up proteomic analysis, in which proteins are enzymatically digested into peptides, often suffers from incomplete coverage due to homologous sequences and fragmentation loss (Brown, Melby, Roberts, & Ge, 2020). While top-down approaches that analyze intact proteins can overcome some of these limitations, these processes still depend heavily on the accuracy of computational reconstruction (Brown et al., 2020). In both cases, AI serves not to replace experimentation, but to refine it; it can aid researchers by acting as a lens, clarifying signal from noise and guiding researchers toward proteins or interactions that may have been dismissed due to low signals. As drug development shifts from trial-and-error empiricism to data-informed precision, the role of AI becomes inevitable, not optional, a trend that is clearly seen in its potential use in genetic screening for conditions such as Angelman syndrome (AS), pediatric encephalopathies, and Crohn's disease (CD).

3 Genetic screening

One of the first and largest genome reference data servers was released in late 2014 by the Exome Aggregation Consortium (ExAC), allowing access to the whole-genome sequenced data of just over 60,000 healthy individuals (Song et al., 2016). Since 2017, ExAC has evolved into gnomAD with the addition of more diversified exome sequence data from 123,136 individuals and whole-genome sequencing data from 15,496 healthy individuals. This dataset was recently updated in 2023 to include the sequencing data of 807,162 individuals (Gudmundsson et al., 2022; Karczewski & Francioli, 2017). Databases such as gnomAD enable the identification of rare disease-associated variants from the millions of benign variants that may simply exist within an individual (Gudmundsson et al., 2022). However, given the immense number of possible single-base changes that hypothetically exist (but have yet to be sequenced), it is challenging to attribute the cause of some diseases to specific variants that may be present within a population. This is due to the possibility that the true causal variant is underrepresented or missing from the database. An example of this limitation is demonstrated in the use of gnomAD to study dystonia-associated genetic variations, where researchers found that the database's requirement for a variant to be absent from controls resulted in false negatives for disease-causing variants, thus marking them as benign or insignificant (Indelicato et al., 2024). Thus as robust as it has become over the relatively short span of nearly a decade, it is difficult to state with certainty that the variants we have associated with specific diseases truly have their pathogenic roots in even the most studied and well-documented variants. Despite the limitations of such large-scale, population-based databases, they remain the primary scaffold for understanding disease risk and pathogenesis, especially in instances where molecular certainty is outpaced by symptoms and observed clinical presentation. The following sections contain several examples in which the diagnostic complexity of certain conditions is introduced before the gaps that AI holds the possibility of bridging are elucidated.

3.1 Angelman syndrome

AS is a condition caused by the down-regulated expression or function of the maternally inherited *UBE3A* allele with abnormal methylation at the 15q11.2-q13 region. Some evidence points toward deletions in the 5- to 7-Mb 15q11.2-q13 region, with these deletions being associated with the most severe symptoms among the variants (Keute et al., 2021). While <1% of those afflicted have a cytogenetically visible chromosome 15 rearrangement at this region, 90% of the genetic variations in expression can be identified via molecular genetic testing for a positive diagnosis. The remaining 10% of individuals with the typical phenotypic expressions of AS possess variants that have yet to be identified (Keute et al., 2021). The dysfunction of *UBE3A* in AS results in characteristic clinical symptoms, including severe intellectual deficits, speech impairments, gait ataxia, tremors, and seizures. The most notable symptom of AS is the demeanor of those afflicted; individuals with AS exhibit a constant and relentless happy, excitable demeanor regardless of the situation (Dagli, Mathews, & Williams, 1993). Although AS is caused by a de novo mutation, it is well studied, with clinical testing methods such as molecular genetics, single-gene and multigene panel testing, and laboratory testing such as DNA methylation able to positively establish AS as a diagnosis. However, limited resources and funding can delay or inhibit accurate diagnosis, especially since many AS symptoms overlap with other developmental disorders such as autism. In addition, data and information on the differences between nondeletion subtypes, such as missense mutations, truncating mutations, imprinting defects, and uniparental disomy, and their effects on the severity of symptom manifestation have yet to be sufficiently studied (Keute et al., 2021).

These subtleties highlight a weakness in human detection capabilities but present an opportunity for AI to address, given its greater computational power and broader analytical capabilities.

3.2 Pediatric encephalopathies

Acute pediatric encephalopathies (APEs) are characterized by rapid central nervous system dysfunction induced by varying degrees of noninflammatory cerebral edema (developing into secondary encephalopathy if left untreated) and often result in death or severe neurological deficits in previously healthy children between the ages of 4 months and 6 years, depending on the type of encephalopathy acquired (Imataka, Kuwashima, & Yoshihara, 2022). Known risk factors in the development of APEs include mutations in carnitine palmitoyltransferase II, adenosine A2A receptors, neuronal sodium channel alpha subunits 1 and 2, single nucleotide polymorphisms in toll-like receptor 3, as well as exposure to the influenza virus or human herpesvirus type 6 (Imataka et al., 2022). Quantifiable clinical symptoms include cytokine storms as measured by blood serum, reduced consciousness and seizures as measured by electroencephalogram (EEG), and cerebrospinal fluid (CSF)/serum albumin quotient levels obtained via lumbar puncture [23]. Imaging techniques, specifically computed tomography (CT) and magnetic resonance imaging (MRI) scans, have been proven useful in the identification of APEs due to the presence of decreased densities or pointed lesions with increased intensities throughout the brain, respectively (Imataka et al., 2022). However, due to the sudden and rapid escalation of symptoms and brain damage associated with this condition after acquisition, improved testing methods and greater accuracy in early detection are still needed.

3.3 Crohn's disease

CD is a progressive disorder classified under inflammatory bowel disease and characterized by inflammation of any part of the gastrointestinal tract. Though the cause of this disease is unknown, risk factors include immune system dysregulation, imbalanced gut microbiota, environmental factors, and increased genetic susceptibility, most notably from the inheritance of CD-associated loci NOD2 and ATG16L1 (Roda et al., 2020). Due to the nature of the disease, diagnosis requires extensive documentation of patient medical history in addition to various diagnostic imaging and testing. The performance of an ileocolonoscopy is highly recommended and favored by providers due to the wealth of information that can be attained, including the presence of lesions, tissue samples for histological analysis, and the severity of the disease (Varyani & Samuel, 2019). In cases of high clinical suspicion of CD despite negative ileocolonoscopy results, patients can undergo a small-bowel capsule endoscopy to detect more minute mucosal abnormalities in the gastrointestinal tract; however, the ingested capsule runs the risk of not exiting the tract, rendering the effort futile (Varyani & Samuel, 2019). Cross-sectional imaging techniques such as bowel ultrasonography (BUS), computed tomography enterography (CTE), and MRI enterography are able to aid in the diagnosis of CD and the evaluation of disease progression, but each comes with its own limitations. BUS is easily accessible but provides a limited scope of view. CTE requires the ingestion of contrast agents, which inherently expose the patient to potentially harmful radiation. Lastly, while magnetic resonance enterography provides the greatest benefit at the smallest risk, it stands at 89% sensitivity despite its 94% specificity (Roda et al., 2020).

While CD is not monogenic, it possesses a complex etiology, spanning genetic susceptibility [e.g., NOD2, ATG16L1 (biomarkers associated with disease progression and pathogenesis)], immune dysfunction, and environmental factors, all of which contribute to the challenges experienced in diagnosing CD. In addition, other concurrent or overlapping inflammatory and gastrointestinal disorders further complicate clinicians' ability to accurately diagnose CD. This, combined with the current limitations of histologic modalities, collectively muddies the ability to diagnose and/or delays treatment. AI holds promise in this space through the incorporation of patient genetic data in conjunction with imaging and patient histories to not only improve risk stratification but also enable early detection of CD and other complex diseases (Roda et al., 2020). Most importantly, while CD is not a neurological condition, the disease exemplifies the challenges presented by a multifactorial condition in which genetic data must be interpreted in the context of both environmental and immune signaling factors, an area where AI can further provide integrative diagnostic support.

The aforementioned examples—which can range from inheritance defined as AS to polygenic, more environmentally modulated conditions like CD—all illustrate the challenges and complications of arriving at an accurate and timely diagnosis, even when molecular insights are available to the clinician. AI applications can aid in bridging this diagnostic gap, whether through the interpretation of complex imaging, recognition of intricate behavioral patterns, or the integration of multimodal datasets. Moreover, AI possesses the ability to augment clinical intuition in cases where objective markers are found to be lacking. Especially in fields such as neurology and psychiatry, where disease presentations are inherently variable and subjective, these AI-based technologies can further elucidate what has long remained unclear to clinicians. Even in instances where genetic data are readily available, many neuropsychiatric conditions lack discernible biomarkers. In such clinical cases, diagnosis leans more heavily or even relies entirely on clinical acumen. Thus it becomes evident that AI can play a vital role in supporting clinician-based diagnoses, refining the inexact into something more finite.

4 AI applications in clinical psychiatric and neurological practice

As aforementioned, diagnostic capabilities in neurology and psychiatry often remain an uneasy blend of clinical intuition, acumen, and science. The paucity of objective markers makes the diagnosis and treatment of psychiatric disorders challenging and more reliant on human intuition and experience rather than irrefutable fact; neurological disorders often have variable and complex manifestations that rely on the evaluation of subjective data. However, emerging data show support for the role of digital technologies in these areas. The results of a 2023 Danish population-based case-cohort study of over 63,000 individuals, which used genotype and health registry data, suggest that deep learning models exhibit high accuracy in predicting both psychiatric disorder diagnoses and the progression of diagnoses such as ADHD, autism, major depression, bipolar disorder, and schizophrenia (Allesøe et al., 2023; Sun et al., 2023). Another 2023 study found that their trained ChatGPT program achieved a final diagnosis accuracy of 77% (Rao et al., 2023). More recently, a 2024 randomized clinical trial showed that large language models significantly outperformed 50 physicians trained in family medicine, internal medicine, and emergency medicine on standardized diagnostic reasoning tasks, even when the physicians had access to conventional resources (Goh et al., 2024). While the success rates shown in the aforementioned studies are relatively high, it should be noted that accuracy would likely be lower in more complex "real-world" cases, such as real-time, in-person diagnoses of these conditions based on symptoms as they emerge, presenting a facet that current AI has yet to reach. As such, whether this diagnostic accuracy will apply to psychiatric diagnoses remains to be studied. Regardless, these examples show that, in many cases, AI tools can draw from several varying data streams, such as neuroimaging, genetic markers, and behavioral data, to further refine their predictive capabilities and provide interventions.

4.1 Pediatric epilepsy

Epilepsy is an umbrella term that encompasses >25 conditions associated with abnormal brain activity in the cerebral cortex, leading to seizures of varying degrees and clinical presentations (Maya Kaye, 2021; Milligan, 2021). Epilepsy in children is especially dangerous, as recurrent seizures can result in delayed or impaired cognitive development due to the loss of neuronal and glial cells, in addition to the increased risk of sudden death (Thom, 2014). Although most epilepsies can be relatively well managed through medication, a third of those afflicted have a form of epilepsy that is resistant to drugs. Those with drug-resistant epilepsy must turn to surgical interventions such as craniotomy with lobectomy, lesionectomy, and anatomic or functional hemispherectomy as a last resort when all other forms of treatment fail (Seto & Coorg, 2021). However, as with any surgery, the risks that the patient must accept can be astronomical. In all cases, the patient must have access to an epilepsy center with specialists to receive effective long-term treatment, a requirement that forms a barrier to proper diagnosis and management for those without access to such resources. In terms of imaging, the commonly used modalities include EEG and MRI, but three-dimensional (3D) magnetoencephalography (MEG), fluorodeoxyglucose positron emission tomography, hexamethylpropylene amine oxime/ethyl cysteinate dimer, single-photon emission computed tomography (SPECT), electrocorticography, and intraoperative invasive EEG monitoring (IEM) are also used in some centers as supplemental imaging (Jayakar et al., 2014). The main limitation of EEG in the diagnosis of epilepsy is its limited spatial resolution and increased rates of false lateralization and localization. In the setting of pediatric epilepsy, this issue is further complicated by the distinct unpredictability of seizures in pediatric populations compared to adults. In the adjunct test of 3D EEG/MEG, user error can be high due to the need to take into account the presence of preexisting lesions, skeletal defects, and malformations (Jayakar et al., 2014). IEM may be the gold standard for identifying epileptogenic regions, but the risks associated with incorrect electrode placement are high. On the contrary, MRI is capable of accurately identifying the epileptogenic region, but it is often misleading in terms of region size, especially in children between the ages of 6 months and 1 year, during which lesions can be masked by the incomplete myelination expected during this age frame, a problem also faced with PET (Jayakar et al., 2014). The SPECT imaging technique requires skilled and experienced injection administrators, as mistimed injections increase the chance of falsely identifying positive regions (Jayakar et al., 2014). As such, though there exists a variety of imaging modalities capable of identifying different types of seizures, each carries its own limitations, the greatest being the ability to access some versus all these diagnostic tools before the seizures cause greater neurological damage in the child who can be treated with current medication and surgical interventions. The use of AI to analyze and find subtle positive regions via more commonly and widely available forms of imaging can potentially mediate the need for further testing.

Although the aforementioned discussion touched upon the capabilities, challenges, and ever-evolving role of AI within the fields of neurology and psychiatry practices—its true value is best appreciated through the lens of real-world application. A recent study showed that a multicenter epilepsy lesion detection deep learning algorithm, analyzing 33 cortical surface-based features such as cortical thickness and curvature in adolescents, was able to successfully mark discordant regions via the identification of subtle malformations in cortical structure regions of 50% of children within the study group, outperforming their human surgeon counterparts (Goel et al., 2024). The ability of this AI to correctly predict the zone of seizure genesis or the irritative zone offers but a glimpse of a future in which conditions such as these silent abnormalities can be coaxed into view, not by stronger magnetics or advancements in MRI coils, but through algorithms that have learned from others what we, as humans, are prone to miss. A side-by-side comparative analysis of these modalities—regarding their respective strengths, limitations, and various pediatric-specific considerations—is provided in Table 18.1.

TABLE 18.1 A comparison of imaging modalities used in pediatric epilepsy, summarizing each individual modality's strengths, limitations, and pediatric-specific considerations.

Comparison of imaging modalities for pediatric epilepsy			
Modality	**Strengths**	**Limitations/Weaknesses**	**Pediatric considerations**
Electroencephalography (EEG)	Acts as the first-line test; captures interictal epileptiform discharges; provides high temporal resolution; repeatable and widely available	Spatial resolution is limited; there is potential for false lateralizations or artifacts caused by movement, muscle activity, and/or poor electrode contact; normal EEG results do not exclude epilepsy	Instances in which stillness is difficult to maintain; ambulatory and/or prolonged video-EEG in certain situations; smaller cranial sizes may necessitate the use of a pediatric cap; the issue of photic and hyperventilation safety must be considered
Magnetic resonance imaging (MRI)	Remains the "gold-standard" for structural imaging; capable of detecting cortical dysplasia, tumors, hippocampal sclerosis, and cortical malformations/developments; does not involve ionizing radiation	Sedation is often required; motion degrades image quality; risk of underestimating lesion extent; variability among epilepsy-specific protocols; longer acquisition times	Age-appropriate sequences are required; considerations for repeat MRIs as the brain matures may need to be taken into account; incomplete myelination in infants can obscure potential lesions
Magnetoencephalography (MEG)	Excellent temporal and spatial resolution for interictal spikes; better tangential source detection compared to EEG; useful in presurgical mapping (especially when MRI is negative or discordant)	Limited availability and high cost; dependent on centers with adequate capabilities and experience; very sensitive to motion	Child cooperation and stillness are essential; shorter run times and smaller child-sized helmets can be limiting; these must be combined with MRI for magnetic source imaging (MSI)
Positron Emission Tomography (PET)	Able to identify interictal hypometabolism corresponding to epileptogenic zones; can be helpful in cases of MRI-negative focal epilepsy and/or surgical planning; provides a metabolic map of the whole brain	Possesses ionizing radiation; yields limited anatomic detail; carries a risk of lower sensitivity in some generalized epilepsies; fasting and tracer logistics are required	Radiation dosing must be limited or minimized; sedation may be necessary; metabolic brain patterns can evolve with development; interpretation must be adjusted to age
Single-photon emission computed tomography (SPECT)	Captures the regional cerebral blood flow during seizure onset	The timing of tracer injection vital and can be logistically difficult; inpatient monitoring required for ictal capture; provides lower spatial resolution compared to MRI	Fast run time; well-performed injection protocol is vital; radiation concerns; seizures are not as predictable
Intraoperative invasive electroencephalography (IEM)	Provides the highest spatial specificity for seizure genesis within the onset zone; allows for functional mapping and stimulation of the cortex; provides insight into decisive surgical candidacy when noninvasive data are discordant	Invasive surgery with associated risks (bleeding, infection, etc.); potential for sampling errors (if coverage insufficient); requires hospitalization and ICU monitoring; very resource-intensive	Electrode strategies must be tailored to the developing cortex; the benefits must outweigh potential surgical risks; counseling must be available to families; can be combined with stereo-EEG to reduce craniotomy size when possible

4.2 Neuro Scholar

In psychiatry and neurology specifically, there has been notable interest and an increase in the adoption of AI applications into healthcare delivery and clinical practice, including providing clinical decision support (e.g., screening, diagnosing, triaging, selecting treatment), reducing administrative burdens such as documentation (e.g., writing a discharge summary), and assisting with patient engagement and education. However, their potential benefits in the enhancement of clinical outcomes have yet to be fully explored, raising new challenges and questions involving their accuracy, safety, and ethical implications (Monteith et al., 2024). Newer programs, such as Neuro Scholar, as well as the exploration of other functions, such as speech analysis, social media tracking, and eye movement tracking, within existing programs are being researched to further explore AI's true potential.

Neuro Scholar is a custom GPT on the OpenAI ChatGPT platform designed to provide more evidence-based, accurate, and academically backed responses, making it more suitable for complex cases or questions posed in neuroscience and psychiatry. This model relies on inputs and resources, including the Diagnostic and Statistical Manual of Mental Disorders (DSM-5) and a variety of clinical specialty textbooks. The accuracy of Neuro Scholar was tested by administering a practice American Board of Psychiatry and Neurology board exam to both it and ChatGPT. While ChatGPT answered 83.3% of the questions correctly, Neuro Scholar achieved a 96.7% score (Asbach & Menon, 2024). Although the development of customized generative AI raises additional issues of intellectual property ownership, it remains critical for improving accuracy while reducing risks associated with using the technology in a clinical setting.

4.3 Speech analysis

Speech analysis is another potential use of AI for the monitoring and early warning of depression and suicidality. Studies have shown that speech patterns, especially vocal jitter and toneless or hollow pitch, could be indicative of risk (Terra, Baklola, Ali, & El-Bastawisy, 2023). AI can also be integrated into clinical practice via the use of chatbots, software designed to conduct conversations via text or text-to-speech. An online assistant, such as Google Assistant, Amazon's Alexa, or another digital persona, may supplement or even replace traditional talk therapies for psychiatric disorders. While rule-based chatbots can only respond to specific questions or keywords that have been preprogrammed, AI-based chatbots can use machine learning and natural language processing to respond in more sophisticated ways (Terra et al., 2023). The use of this technology may be less intimidating and more cost-efficient, allowing for greater access and engagement in psychoeducational and psychotherapeutic interventions. A recent metaanalysis of eleven studies found a moderate effect of chatbot-delivered psychotherapy in improving depressive symptoms in adults with depression and anxiety (Lim, Shiau, Cheng, & Lau, 2022). Beyond the clinical setting, deep learning algorithms have also been applied to social media platforms such as X (formerly known as Twitter) and Reddit to analyze posts to help identify potential suicidal ideation among users, allowing AI to serve as a detector for those who may require intervention without the need for direct interaction (Baydili, Tasci, & Tasci, 2025).

4.4 Eye tracking

Autism, a neurodevelopmental disorder affecting 1 in 40 children in the United States, is similar to many other neuropsychiatric disorders in that it is heterogeneous and multidimensional. Thus diagnosis relies on the evaluation of specific behavioral symptoms, with some being easily recognizable and others far more discreet. To supplement this process, AI-driven diagnostic techniques can include computer vision and eye tracking to detect subtle differences between children with autism and neurotypical controls, with the potential to be scaled down to low-cost mobile screening and monitoring (Washington & Wall, 2023).

4.5 Brain structure analysis

Some AI applications have been or are being developed to predict or diagnose psychiatric disorders based on more objective data. As schizophrenia has an identified effect on brain structure, including decreases in white and gray matter volume, hippocampal changes, and increases in CSF, AI models utilizing structural magnetic resonance imaging and functional magnetic resonance imaging (fMRI) may prove highly accurate as diagnostic tools for this condition (Sadeghi et al., 2022). There have been a large number of AI and neuroimaging studies, most of which relied on the Alzheimer's Disease Neuroimaging Initiative dataset, but findings related to its clinical utility have been limited due to methodological heterogeneity, clinical relevance, and lack of validation (Borchert et al., 2023). Similarly, AI holds promise in the clinical management of multiple sclerosis (MS) and epilepsy, with its studied applications, including machine learning models that utilize MRI, genetic, and clinical data to assist in diagnosis, predict disease progression, and personalize treatment in patients with MS (Amin et al., 2024). Machine learning models using neuroimaging, electroencephalography, health records, and medical devices have been examined in patients with epilepsy, but as with the previously mentioned conditions, gaps remain in terms of integration into clinical practice (Lucas et al., 2024). These gaps, however, can begin to be addressed with the establishment of real-world datasets, multimodal imaging, and more complex machine learning algorithms with the goal of enabling implementation into clinical practices.

4.6 Treatment recommendation

AI models can also be configured to provide treatment recommendations by using published treatment guidelines for neurological or psychiatric disorders to predict treatment response. AI can potentially bypass ineffective medication trials, reducing costs and frustration on the patients' part (Dwyer, Falkai, & Koutsouleris, 2018). Although not universally positive, some studies have suggested the utilization of AI in antidepressant response prediction to guide the use of clinical questionnaires, antipsychotic response based on EEG results, electroconvulsive therapy response based on brain structure, cognitive behavioral therapy response for anxiety based on brain fMRI, and brain stimulation response based on MRI (Lee et al., 2021). AI has also been utilized in the management of auditory hallucinations in patients with schizophrenia. Digital simulations (both visual and audible) of the patient's distressing hallucinations have been used to facilitate three-way conversations between the patient, therapist, and avatar, resulting in reduced anxiety and improved communication (Garety et al., 2021). With the increasing availability of data from large-scale datasets and the ongoing transition to computerized healthcare delivery, the implementation of AI in psychiatry and neurology is becoming more feasible. For clinicians, the application of these tools holds the potential to augment care, particularly in screening, diagnosis, evidence-based treatment decisions, patient education, and documentation.

5 AI applications in clinical psychiatric and neurological practice

The impact of AI at the intersection of psychiatry, neurology, and physiology promises a shift that will allow clinicians to better interpret disease, synthesize data, and formulate effective therapeutic decisions efficiently. From the mapping of protein structures to the prediction of disease onset and pathogenesis, AI's adaptability allows for its extension into varying arenas within the clinical arc, as explored in its possible applications to certain conditions, including—but not limited to—AS, pediatric encephalopathies, CD, Alzheimer's, schizophrenia, depression, and autism. Furthermore, representative examples of these applications, including their respective underlying modalities, analytic approaches, and reported performance, are summarized in Table 18.2.

Yet, these innovations pose as many questions as the answers they provide. Although it is evident that AI can be incredibly beneficial, there are pressing concerns regarding ethics and accountability. The datasets utilized to power these predictive models are often incomplete or biased in the sense that they are structured around populations that are not representative of those afflicted. Without intentional correlation, these disparities can be amplified by the algorithms originally designed to reduce them. It should also be noted that the logic presented within AI-generated algorithms and conclusions is often opaque; the reasoning behind certain AI conclusions can be difficult to interpret, making it more challenging for clinicians to validate or understand the outputs in real time. Thus this necessitates a push for algorithmic transparency so that these systems can provide their predictions alongside a clear step-by-step explanation.

As such, the path toward clinical integration is still in its early, formative years. AI models still lack real-world validation, in part due to the absence of a regulatory framework that governs the deployment of AI in medical research. Questions regarding liability, data privacy, and interpretability remain pressing issues during this time; should these issues continue to persist, particularly within the realm of psychiatry, its use may devolve from positive benign to neutral harm as diagnostic categories become more fluid and symptoms more subjective. The proper integration and use of AI envision a future in which data science meets clinical practice in harmony. However, work must be done on the human side as well; clinicians must also stand at the threshold. If AI is to become a true partner within healthcare delivery, it is vital for physicians and other medical professionals to accept its role as a tool that serves to sharpen their intuition, not as a replacement. As with the adoption of any new technology, standardized medical training for appropriate and effective use must be implemented across disciplines and fields to ensure uniformity and cohesion within data formulation, output, analysis, and utilization. Clinicians must be educated on how to interpret these AI-generated outputs, the systems they were designed in, and their respective limitations, while also acquiring the skill of meaningfully questioning them. Similarly, data scientists designing programs for use in this field must be trained to understand the nuances of clinical practice and the ethical constraints that

TABLE 18.2 Examples of artificial intelligence applications across neurological/psychiatric and related disorders, categorized by disease/condition, AI modality, AI technique, dataset/study example, and reported performance/outcome.

AI applications by disease and modality				
Disease/condition	**AI modality**	**AI technique(s)**	**Dataset/study examples in chapter**	**Performance/outcomes**
Melanoma/skin cancer	Dermoscopic image analysis	Convolutional neural network (CNN)	(Esteva et al., 2017), 129,450 images (2000 biopsy-confirmed cases)	95% sensitivity with 93% specificity; outperformed the majority of the 21–25 cohort of dermatologists
Brain tumor(s) (glioma)	MRI + histopathology Imaging	Computer-aided diagnosis (CAD) using ML/DL	Supervised ANN; Aun (Yeung et al., 2023)	Improved segmentation accuracy and feature extraction for tumor detection
Psychiatric disorders (ADHD, MDD, bipolar disorder, schizophrenia)	EHR + genomic data	Deep learning prediction model	(Allesøe et al., 2023), 63,000 individuals (genotype and health registry data)	High accuracy in predicting diagnoses and disease progression
General neurological and psychiatric diagnostic reasoning	Text-based clinical case input	Large language model (LLM), ChatGPT	(Goh et al., 2024), 36 published clinical vignettes	Final diagnosis accuracy of 77%; an RCT demonstrated that LLMs outperformed 50 physicians in standard diagnostic reasoning and interpretation tasks
Autism spectrum disorder (ASD)	Computer vision + eye tracking	Machine learning classification	Eye-tracking differences between ASD and controls	Scalable results; low-cost screening with high detection accuracy
Depression and suicide risk	Speech analysis	NLP + acoustic feature extraction	(Terra et al., 2023), vocal jitters, vocal flattening	Identified depression/suicidality risk markers with strong predictive potential

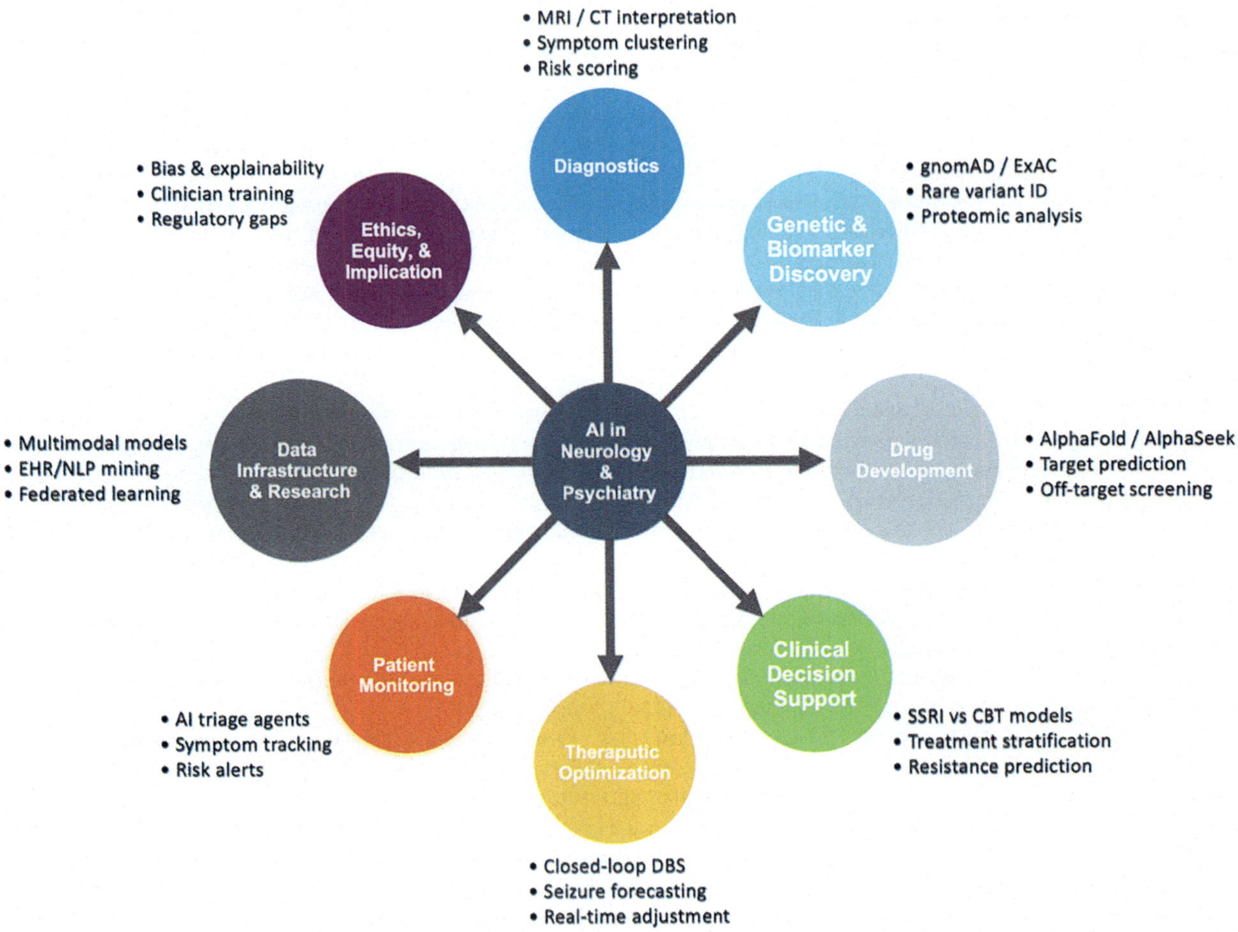

FIG. 18.1 Summary diagram depicting AI as a central node in neuropsychiatric conditions and probable pathways that could lead to the mitigation of problems and future research.

define patient care. Disclosure of the use of AI to patients is also an area that requires standardized protocols and practices, both for ethical reasons and to preserve the therapeutic alliance that forms the core of medicine. Integrating AI requires more than knowledge of patient interaction or coding; it requires a full cultural shift that embraces the sanctity of interdisciplinary education and fields, prioritizes consistent dialog, and instills the willingness to accept change. This chapter has merely offered a view of what is possible. Fig. 18.1 describes the logical relationship between AI and different aspects of neuropsychiatric conditions, as well as possible links that could create pathway vectors. What follows depends not only on the dynamic interplay and dialog between clinicians and data scientists but also on how bravely we choose to trust.

References

Allesøe, R. L., Lundgaard, A. T., Hernández Medina, R., Aguayo-Orozco, A., Johansen, J., Nissen, J. N., et al. (2023). Discovery of drug–omics associations in type 2 diabetes with generative deep-learning models. *Nature Biotechnology*, 1546-1696. *41*(3), 399–408. Available from: https://www.nature.com/nbt/. doi:10.1038/s41587-022-01520-x.

Amin, M., Martínez-Heras, E., Ontaneda, D., & Prados Carrasco, F. (2024). Artificial intelligence and multiple sclerosis. *Current Neurology and Neuroscience Reports*, 1528-4042. *24*(8), 233–243. doi:10.1007/s11910-024-01354-x.

Asbach, M., & Menon, R. (2024). *AI in psychiatry: things are moving fast*. Psychiatric Times. Available from: https://www.psychiatrictimes.com/view/ai-in-psychiatry-things-are-moving-fast.

Baydili, İ., Tasci, B., & Tasci, G. (2025). Deep learning-based detection of depression and suicidal tendencies in social media data with feature selection. *Behavioral Sciences*, 2076-328X. *15*(3). Available from: www.mdpi.com/journal/behavsci. doi:10.3390/bs15030352.

Bekbolatova, M., Mayer, J., Ong, C. W., & Toma, M. (2024). Transformative potential of AI in healthcare: definitions, applications, and navigating the ethical landscape and public perspectives. *Healthcare*, 2227-9032. *12*(2), 125. doi:10.3390/healthcare12020125.

Borchert, R. J., Azevedo, T., Badhwar, A. P., Bernal, J., Betts, M., Bruffaerts, R., et al. (2023). Artificial intelligence for diagnostic and prognostic neuroimaging in dementia: a systematic review. *Alzheimer's and Dementia*, 1552-5279. *19*(12), 5885–5904. Available from: https://alz-journals.onlinelibrary.wiley.com/toc/23528729/current. doi:10.1002/alz.13412.

Brown, K. A., Melby, J. A., Roberts, D. S., & Ge, Y. (2020). Top-down proteomics: challenges, innovations, and applications in basic and clinical research. *Expert Review of Proteomics*, 1744-8387. *17*(10), 719–733. Available from: https://www.tandfonline.com/loi/ieru20?open=9&year=2012&repitition=0#vol_9_2012. doi:10.1080/14789450.2020.1855982.

Chikumba, S., Hu, Y., & Luo, J. (2023). Deep learning-based fundus image analysis for cardiovascular disease: a review. *Therapeutic Advances in Chronic Disease*, 2040-6223. *14*. doi:10.1177/20406223231209895.

Dagli, A. I., Mathews, J., & Williams, C. A. (1993). *Angelman syndrom*. Available from: https://www.ncbi.nlm.nih.gov/books/NBK1144.

Davenport, T., & Kalakota, R. (2019). The potential for artificial intelligence in healthcare. *Future Healthcare Journal*, 2514-6645. *6*(2), 94–98. doi:10.7861/futurehosp.6-2-94.

Dwyer, D. B., Falkai, P., & Koutsouleris, N. (2018). Machine learning approaches for clinical psychology and psychiatry. *Annual Review of Clinical Psychology*, 1548-5951. *14*, 91–118. Available from: http://www.annualreviews.org/journal/clinpsy. doi:10.1146/annurev-clinpsy-032816-045037.

Esteva, A., Kuprel, B., Novoa, R. A., Ko, J., Swetter, S. M., Blau, H. M., & Thrun, S. (2017). Dermatologist-level classification of skin cancer with deep neural networks. *Nature*, 1476-4687. *542*(7639), 115–118. Available from: http://www.nature.com/nature/index.html. doi:10.1038/nature21056.

Garety, P., Ward, T., Emsley, R., Greenwood, K., Freeman, D., Fowler, D., et al. (2021). Digitally supported CBT to reduce paranoia and improve reasoning for people with schizophrenia-spectrum psychosis: the SlowMo RCT. *Efficacy and Mechanism Evaluation*, 2050-4365. *8*(11), 1–90. doi:10.3310/eme08110.

Goel, A., Seri, S., Agrawal, S., Kumar, R., Sudarsanam, A., Carr, B., et al. (2024). The utility of multicentre epilepsy lesion detection (MELD) algorithm in identifying epileptic activity and predicting seizure freedom in MRI lesion-negative paediatric patients. *Epilepsy Research*, 1872-6844. *206*. Available from: https://www.sciencedirect.com/science/journal/09201211. doi:10.1016/j.eplepsyres.2024.107429.

Goh, E., Gallo, R., Hom, J., Strong, E., Weng, Y., Kerman, H., et al. (2024). Large language model influence on diagnostic reasoning: a randomized clinical trial. *JAMA Network Open*, 2574-3805. *7*(10). Available from: https://jamanetwork.com/journals/jamanetworkopen. doi:10.1001/jamanetworkopen.2024 40969.

Gudmundsson, S., Singer-Berk, M., Watts, N. A., Phu, W., Goodrich, J. K., Solomonson, M., et al. (2022). Variant interpretation using population databases: Lessons from gnomAD. *Human Mutation*, 1098-1004. *43*(8), 1012–1030. Available from: http://onlinelibrary.wiley.com/journal/10.1002/(ISSN)1098-1004. doi:10.1002/humu.24309.

Han, S.-H., Kim, K. W., Kim, S., & Youn, Y. C. (2018). Artificial neural network: Understanding the basic concepts without mathematics. *Dementia and Neurocognitive Disorders*, 1738-1495. *17*(3), 83. doi:10.12779/dnd.2018.17.3.83.

Helm, J. M., Swiergosz, A. M., Haeberle, H. S., Karnuta, J. M., Schaffer, J. L., Krebs, V. E., et al. (2020). Machine learning and artificial intelligence: Definitions, applications, and future directions. *Current Reviews in Musculoskeletal Medicine*, 1935-9748. *13*(1), 69–76. Available from: http://www.springer.com/humana+press/orthopedics/journal/12178. doi:10.1007/s12178-020-09600-8.

Hermann, J., Schurgers, L., & Jankowski, V. (2022). Identification and characterization of post-translational modifications: Clinical implications. *Molecular Aspects of Medicine*, 0098-2997. *86*, 101066. doi:10.1016/j.mam.2022.101066.

Hinton, G. (2018). Deep learning-a technology with the potential to transform health care. *JAMA—Journal of the American Medical Association*, 1538-3598. *320*(11), 1101–1102. Available from: http://jama.jamanetwork.com/journal.aspx. doi:10.1001/jama.2018.11100.

Imataka, G., Kuwashima, S., & Yoshihara, S. (2022). A comprehensive review of pediatric acute encephalopathy. *Journal of Clinical Medicine*, 2077-0383. *11*(19), 5921. doi:10.3390/jcm11195921.

Indelicato, E., Eberl, A., Boesch, S., Lange, L. M., Klein, C., Lohmann, K., & Zech, M. (2024). Genome aggregation database version 4—Allele frequency changes and impact on variant interpretation in dystonia. *Movement Disorders*. doi:10.1002/mds.30066.

Jayakar, P., Gaillard, W. D., Tripathi, M., Libenson, M. H., Mathern, G. W., & Cross, J. H. (2014). Diagnostic test utilization in evaluation for resective epilepsy surgery in children. *Epilepsia*, 1528-1167. *55*(4), 507–518. Available from: http://www3.interscience.wiley.com/journal/117957420/issueyear?year=2009. doi:10.1111/epi.12544.

Johnson, K. B., Wei, W. Q., Weeraratne, D., Frisse, M. E., Misulis, K., Rhee, K., et al. (2021). Precision medicine, AI, and the future of personalized health care. *Clinical and Translational Science*, 1752-8062. *14*(1), 86–93. Available from: http://onlinelibrary.wiley.com/journal/10.1111/(ISSN)1752-8062. doi:10.1111/cts.12884.

Jumper, J., Evans, R., Pritzel, A., Green, T., Figurnov, M., Ronneberger, O., et al. (2021). Highly accurate protein structure prediction with AlphaFold. *Nature*, 1476-4687. *596*(7873), 583–589. Available from: http://www.nature.com/nature/index.html. doi:10.1038/s41586-021-03819-2.

Karczewski, K., & Francioli, L. (2017). *The genome aggregation database (gnomAD) | gnomAD browser*. gnomAD Browser. Available from: https://gnomad.broadinstitute.org/news/2017-02-the-genome-aggregation-database/.

Keute, M., Miller, M. T., Krishnan, M. L., Sadhwani, A., Chamberlain, S., Thibert, R. L., et al. (2021). Angelman syndrome genotypes manifest varying degrees of clinical severity and developmental impairment. *Molecular Psychiatry*, 1476-5578. *26*(7), 3625–3633. Available from: http://www.nature.com/mp/index.html. doi:10.1038/s41380-020-0858-6.

Lee, D. J., Drummond, N. M., Saha, U., De Vloo, P., Dallapiazza, R. F., Gramer, R., et al. (2021). Acute low frequency dorsal subthalamic nucleus stimulation improves verbal fluency in Parkinson's disease. *Brain Stimulation*, 1876-4754. *14*(4), 754–760. Available from: http://brainstimjrnl.com/. doi:10.1016/j.brs.2021.04.016.

Lim, S. M., Shiau, C. W. C., Cheng, L. J., & Lau, Y. (2022). Chatbot-delivered psychotherapy for adults with depressive and anxiety symptoms: a systematic review and meta-regression. *Behavior Therapy*, 1878-1888. *53*(2), 334–347. Available from: http://www.journals.elsevier.com/behavior-therapy/. doi:10.1016/j.beth.2021.09.007.

Lucas, A., Mouchtaris, S., Tranquille, A., Sinha, N., Gallagher, R., Mojena, M., et al. (2024). Mapping hippocampal and thalamic atrophy in epilepsy: a 7-T magnetic resonance imaging study. *Epilepsia*, 1528-1167. *65*(4), 1092–1106. Available from: http://onlinelibrary.wiley.com/journal/10.1111/(ISSN)1528-1167. doi:10.1111/epi.17908.

Maya Kaye, A. (2021). Pediatric epilepsy and psychoeducational interventions: a review of the literature. *Epilepsy & Behavior*, 1525-5050. *121* (Pt A), 108084. doi:10.1016/j.yebeh.2021.108084.

Milligan, T. A. (2021). Epilepsy: a clinical overview. *American Journal of Medicine*, 1555-7162. *134*(7), 840–847. Available from: www.elsevier.com/locate/amjmed. doi:10.1016/j.amjmed.2021.01.038.

Monteith, S., Glenn, T., Geddes, J. R., Whybrow, P. C., Achtyes, E., & Bauer, M. (2024). Artificial intelligence and increasing misinformation. *British Journal of Psychiatry*, 1472-1465. *224*(2), 33–35. Available from: https://www.cambridge.org/core/journals/the-british-journal-of-psychiatry/all-issues. doi:10.1192/bjp.2023.136.

Rao, A., Pang, M., Kim, J., Kamineni, M., Lie, W., Prasad, A. K., et al. (2023). Assessing the utility of ChatGPT throughout the entire clinical workflow: Development and usability study. *Journal of Medical Internet Research*. 1438-8871. *25*. Available from: https://www.jmir.org/2023/1/e48659. doi:10.2196/48659.

Roda, G., Ng, S. C., Kotze, P. G., Argollo, M., Panaccione, R., Spinelli, A., et al. (2020). Crohn's disease. *Nature Reviews Disease Primers*. 2056-676X. *6*(1). Available from: http://www.nature.com/nrdp/. doi:10.1038/s41572-020-0156-2.

Sadeghi, D., Shoeibi, A., Ghassemi, N., Moridian, P., Khadem, A., Alizadehsani, R., et al. (2022). An overview of artificial intelligence techniques for diagnosis of Schizophrenia based on magnetic resonance imaging modalities: methods, challenges, and future works. *Computers in Biology and Medicine*. 1879-0534. *146*. Available from: www.elsevier.com/locate/compbiomed. doi:10.1016/j.compbiomed.2022.105554.

Sarker, I. H. (2021). Deep learning: a comprehensive overview on techniques, taxonomy, applications and research directions. *SN Computer Science*, 2661-8907. *2*(6). Available from: https://www.springer.com/journal/42979. doi:10.1007/s42979-021-00815-1.

Seto, E. S., & Coorg, R. (2021). Epilepsy surgery: Monitoring and novel surgical techniques. *Neurologic Clinics*, 1557-9875. *39*(3), 723–742. Available from: http://www.elsevier.com/inca/publications/store/6/2/3/3/7/1/index.htt. doi:10.1016/j.ncl.2021.04.001.

Smith. (2025). *Stargate: America's $500 billion Bid to corner global AI capital*. Available from: https://www.forbes.com/sites/craigsmith/2025/01/23/stargate-americas-500-billion-bid-to-corner-global-ai-capital/.

Song, W., Gardner, S. A., Hovhannisyan, H., Natalizio, A., Weymouth, K. S., Chen, W., et al. (2016). Exploring the landscape of pathogenic genetic variation in the ExAC population database: Insights of relevance to variant classification. *Genetics in Medicine*, 1530-0366. *18*(8), 850–854. Available from: http://www.nature.com/gim/index.html. doi:10.1038/gim.2015.180.

Stokes, J. M., Yang, K., Swanson, K., Jin, W., Cubillos-Ruiz, A., Donghia, N. M., et al. (2020). A deep learning approach to antibiotic discovery. *Cell*, 1097-4172. *180*(4), 688. Available from: https://www.sciencedirect.com/journal/cell. doi:10.1016/j.cell.2020.01.021.

Sun, J., Dong, Q. X., Wang, S. W., Zheng, Y. B., Liu, X. X., Lu, T. S., et al. (2023). Artificial intelligence in psychiatry research, diagnosis, and therapy. *Asian Journal of Psychiatry*, 1876-2026. *87*. Available from: http://www.elsevier.com/wps/find/journaldescription.cws_home/717029/description#description. doi:10.1016/j.ajp.2023.103705.

Terra, M., Baklola, M., Ali, S., & El-Bastawisy, K. (2023). Opportunities, applications, challenges and ethical implications of artificial intelligence in psychiatry: A narrative review. *The Egyptian Journal of Neurology, Psychiatry and Neurosurgery*, 1687-8329. *59*(1). doi:10.1186/s41983-023-00681-z.

Thom, M. (2014). *Epilepsy pathology encyclopedia of the neurological sciences* (pp. 136–141). United Kingdom: Elsevier Inc. Available from: http://www.sciencedirect.com/science/book/9780123851581. doi:10.1016/B978-0-12-385157-4.00607-2.

Väänänen, A., Haataja, K., Vehviläinen-Julkunen, K., & Toivanen, P. (2021). AI in healthcare: A narrative review. *F1000Research*. 2046-1402. *10*, 6. doi:10.12688/f1000research.26997.2.

Varyani, F., & Samuel, S. (2019). Can magnetic resonance enterography (MRE) replace ileo-colonoscopy for evaluating disease activity in Crohn's disease? *Best Practice & Research Clinical Gastroenterology*, 1521-6918. *38–39*, 101621. doi:10.1016/j.bpg.2019.05.008.

Washington, P., & Wall, D. P. (2023). A review of and roadmap for data science and machine learning for the neuropsychiatric phenotype of autism. *Annual Review of Biomedical Data Science*.2574-3414. *6*, 211–228. Available from: http://www.annualreviews.org/journal/biodatasci. doi:10.1146/annurev-biodatasci-020722-125454.

Yan, W. J., Ruan, Q. N., & Jiang, K. (2023). Challenges for artificial intelligence in recognizing mental disorders. *Diagnostics*, 2075-4418. *13*(1). Available from: http://www.mdpi.com/journal/diagnostics/. doi:10.3390/diagnostics13010002.

Yeung, J. A., Wang, Y. Y., Kraljevic, Z., & Teo, J. T. H. (2023). Artificial intelligence (AI) for neurologists: do digital neurones dream of electric sheep? *Practical Neurology*, 1474-7766. *23*(6), 476–488. Available from: http://pn.bmj.com/content/by/year. doi:10.1136/pn-2023-003757.

Zhavoronkov, A., Bischof, E., & Lee, K. F. (2021). Artificial intelligence in longevity medicine. *Nature Aging*, 2662-8465. *1*(1), 5–7. Available from: https://www.nature.com/nataging/. doi:10.1038/s43587-020-00020-4.

Chapter 19

Application of artificial intelligence in early diagnosis of alpha-synucleinopathies

Nabanita Ghosh[a] and Krishnendu Sinha[b]
[a]*Maulana Azad College, Kolkata, India*
[b]*Jhargram Raj College, Jhargram, India*

1 Introduction

Neurodegenerative disorders such as Alzheimer's disease (AD) and Parkinson's disease (PD) are irreversible conditions defined by progressive neuronal loss and brain atrophy, particularly in older adults, with global cases exceeding 57 million and expected to double roughly every 20 years (*Alzheimer's and Parkinson's biomarkers show promise for early diagnosis | Knight Initiative*, n.d.; Imam et al., 2025; Plastini et al., 2024; Rutledge et al., 2024). AD is characterized by extracellular beta-amyloid plaques and intracellular tau tangles, while PD features Lewy bodies composed of misfolded α-synuclein (Knopman et al., 2021; Mendez, 2012; Riederer et al., 2023). In the United States alone, over 7.2 million individuals aged 65+ are living with Alzheimer's dementia as of 2025, a number projected to reach 13.8 million by 2060 (Alzheimer's Association Report, 2025). Parkinson's, now the second most common neurodegenerative disease, affects about 1.5 per 1000 globally as of 2023, with cases forecasted to surpass 12 million by 2040 (Riederer et al., 2023). Both diseases often remain undiagnosed until after significant damage has occurred due to long preclinical phases of pathology and subtle early symptom onset. Until recently, confirming a diagnosis required invasive cerebrospinal fluid analysis or expensive PET imaging. Current treatments, such as amyloid-targeting drugs like lecanemab and donanemab for AD, and dopaminergic therapies for PD, can slow disease progression but cannot reverse it (Riederer et al., 2023; Watanabe, Dijkstra, & Nagatsu, 2024). However, breakthroughs in early detection are accelerating. In May 2025, the Food and Drug Administration (FDA) approved Fujirebio's Lumipulse blood test—which measures plasma pTau217 and Aβ42/40 ratio—to identify amyloid plaques with up to ≈97% negative and ≈92% positive predictive accuracy in adults aged 55+ displaying cognitive symptoms (Wang, Pan, Zhang, & Tan, 2025). Meanwhile, emerging research includes an RNA-based blood assay targeting tRNA fragments for premotor PD detection and cerebro spinal fluid (CSF)/spinal fluid biomarkers demonstrating >90% accuracy for early-stage PD, plus new optical biosensors detecting Aβ peptides at pg/ml sensitivity using nanopillar photonic crystals (Arruda et al., 2025). Large multisite initiatives like ADNI and the Parkinson's Progression Markers Initiative continue to validate and standardize biomarkers across imaging and biofluid modalities (Mueller et al., 2017). Collectively, these advances promise a shift from symptomatic care to early intervention, but widespread clinical adoption hinges on standardizing biomarker reference ranges across populations, integrating tests into primary care, addressing ethical concerns, and securing regulatory and insurer support.

2 Methods to assess the presence and progression of AD and PD

2.1 Biomarkers

Early identification of AD and PD is essential for effective, personalized interventions, and recent studies emphasize the need for comprehensive, multimodal assessment strategies. Scientists are now honing in on a diverse array of biomarkers—spanning proteomic, transcriptomic, metabolic, and genetic data—to pinpoint preclinical and prodromal stages. In Alzheimer's research, measurable signs such as blood and CSF levels of phosphorylated tau (p-tau181/217/231), amyloid-β ratios, and neurofilament light chain (NfL) have demonstrated strong correlations with cognitive decline and

Applications of Artificial Intelligence in Healthcare. https://doi.org/10.1016/B978-0-443-43934-6.00020-1

neurodegeneration (Lloret et al., 2019; Myrou, Barmpagiannos, Ioakimidou, & Savopoulos, 2025; Wang et al., 2025). Non-invasive fluids like saliva are also being explored: reduced salivary lactoferrin has shown high diagnostic accuracy in distinguishing mild cognitive impairment and early-stage Alzheimer's and correlates with cerebral amyloid burden and CSF biomarkers (Antequera et al., 2024). In parallel, microRNAs (miRNAs) are emerging as next-generation indicators: specific miRNAs, including miR-29a, miR-125b, miR-34a, miR-146a, miR-155, miR-7-5p, and miR-153-3p, have been linked to AD or Parkinsonian syndromes, detectable in plasma or exosomal compartments, offering potential for early diagnosis and even therapeutic targeting (Pal & Mandal, 2025). For Parkinson's, a suite of prodromal indicators, such as rapid eye movement (REM) sleep behavior disorder, hyposmia, constipation, mood changes, and autonomic dysfunction, often precede motor symptoms by decades, and candidate fluid biomarkers (e.g., alterations in urate, oxidative markers, L-3,4-dihydroxyphenylalanine (L-DOPA), miR-214, miR-19b, α-synuclein species, TNF-α), alongside mitochondrial DNA changes in at-risk LRRK2 mutation carriers, are under active investigation (He et al., 2018). High-sensitivity α-synuclein seed amplification assays in CSF show promise for early detection in individuals with isolated REM sleep behavior disorder, and population-based studies, including LRRK2 cohorts, support the discovery of reliable prodromal biomarkers (He et al., 2018; Myrou et al., 2025; Ravanidis et al., 2025). Moreover, advanced multimodal AI models now combine gait analysis, handwriting, speech, imaging, and biosensor data to detect and monitor disease progression with increasing accuracy and robustness (Zhao, Liu, Xinglin, Xing, & Zhou, 2025). Taken together, these advances in accessible biomarkers and machine learning (ML) pipelines reinforce the urgent need for standardized, multidomain evaluations to enable earlier diagnosis and targeted clinical management of neurodegenerative diseases.

2.2 Molecular imaging

Molecular imaging, including MRI, PET, Single Photon Emission Computed Tomography (SPECT), and hybrid PET/MRI, remains essential for investigating neurochemical alterations in AD and PD, balancing trade-offs in sensitivity, spatiotemporal resolution, and accessibility (Davis, Ryan, Aaron, & Sims, 2020). Structural MRI techniques, such as volumetric and deformation-based morphometry or cortical thickness mapping, quantify brain atrophy, while functional and metabolic imaging—via FDG-PET, arterial spin labeling (ASL), or cerebral blood flow (CBF) network scans—illuminate hypometabolism and network disruptions early in disease (Alsop et al., 2015; Wolk & Detre, 2012). Dopaminergic PET/SPECT tracers like ^{18}F-FDOPA, ^{11}C-CFT, or ^{123}I-FP-CIT (DaT scan) assess presynaptic degeneration, and newer nondopaminergic PET probes such as [^{11}C]DASB, [^{11}C]PMP, [^{18}F]FEOBV, and TSPO ligands quantify serotonergic, cholinergic, and inflammatory changes (Palermo, Giannoni, Bellini, Siciliano, & Ceravolo, 2021; Werry et al., 2019). A breakthrough in synucleinopathy imaging is ^{18}F-F0502B, a novel PET tracer that selectively binds α-synuclein aggregates in preclinical PD models, showing high affinity (Kd ≈ 10 nM vs. 100 nM for Aβ and tau), though it has yet to enter human studies pending improvements in brain uptake and metabolism profiles (Xiang et al., 2023). Emerging multimodal pipelines like petBrain integrate amyloid-PET, tau-PET, and structural MRI using deep learning (DL) segmentation to standardize biomarker quantification (A/T/N staging) and align with ADNI reference data, enabling more reproducible early diagnosis workflows (Coupé et al., 2025). Advanced computational methods, such as deep tomographic reconstruction models and Fourier diffusion enhancements, are boosting PET spatial resolution and reducing noise, promising sharper detection of subtle pathologies (Liu et al., 2021; Tivnan & Li, 2025). Noninvasive retinal imaging modalities—optical coherence tomography (OCT), optical coherence tomography angiography (OCTA), and digital fundus photography—are also showing retinal layer thinning years before PD symptoms and retinal amyloid-associated changes in AD, suggesting scalable screening potential. Meanwhile, transcranial sonography can detect substantia nigra echogenicity with a positive predictive value of around 85% and a negative predictive value of ~83% when blinded to clinical status. Despite the high cost and limited availability of PET tracers, especially in low-resource settings, combining PET/MRI hybrid systems improves diagnostic yield and cross-validation, while SPECT remains a more accessible alternative in many regions (Richardson et al., 2024; Xu et al., 2020). Collectively, these imaging advances provide a powerful, multiscale toolkit to detect, monitor, and differentiate neurodegenerative disease in its earliest phases, though broader clinical translation depends on validation, cost-effectiveness, standardization, and infrastructure adoption.

2.3 Biosensors

A biosensor is an analytical device that converts a biological or chemical interaction into a measurable signal, enabling the detection of trace disease indicators in fluids such as blood, serum, saliva, urine, or sweat. In neurodegenerative disease diagnostics—especially for AD and PD—several cutting-edge platforms are emerging. For example, aptamer-functionalized carbon nanotube field-effect transistors (CNT-FETs) leverage the ultra-sensitivity of aligned CNT arrays and the sequence-specific binding of aptamers to detect AD biomarkers like Aβ42, Aβ40, t-tau, and p-tau181 at femtomolar concentrations (~2–3 fM), with strong linear dose–response and excellent multiplexing ability in human plasma (Chen et al., 2022). In addition, nanopillar photonic crystal biosensors enhanced with gold nanoparticles are able to quantify Aβ42 and Aβ40 directly in undiluted serum at detection thresholds of ~20 pg/mL, capturing Aβ42/Aβ40 ratios relevant for clinical staging of AD (Arruda et al., 2025). In the context of PD, highly sensitive electrochemical biosensors using ZnO nanowires decorated with Au nanoparticles have achieved detection of α-synuclein in plasma down to ~0.08 pg/mL, with a linear range from ~0.5 to 10 pg/mL, excellent reproducibility (~5% variation), and reliable discrimination between PD and healthy samples (Di Mari et al., 2024). Other platforms, including hierarchical submonolayer biolaser sensors, have registered α-synuclein in serum at ~0.32 pM sensitivity and promise ultra-low-cost, disposable diagnostics for prodromal PD (Gong et al., 2023). Meanwhile, electrochemical and field-effect transistor devices continue to advance detection of tau and α-synuclein proteoforms with high specificity using aptamers or antibodies, signal amplification strategies, and nanomaterial surface engineering to improve selectivity and performance in biofluids (Bodily et al., 2023). These technologies require minimal sample volumes—often under 10 μL—and deliver rapid responses in minutes, making them compatible with point-of-care or home-based monitoring. While offering high promise for early diagnostics and continuous physiological tracking in aging populations, challenges around scalability, device uniformity, regulatory approval, and cost-efficiency remain critical hurdles for broader implementation.

3 Application of ML/DL

ML enables computers to discern complex patterns from data, including supervised, unsupervised, and semisupervised methods, while DL, a powerful subset, leverages multilayer neural networks, including convolutional neural networks (CNNs), transformers, and attention architectures, to tackle tasks like neuroimaging classification and biomarker prediction (e.g., Sparse-GAN, EC-based classifiers) (Li et al., 2024). Recent work has applied ML to reveal distinct white matter signatures in conditions such as AD, progressive supranuclear palsy (PSP), and corticobasal degeneration (CBD), unmasking tau morphology differences across white and gray matter and enhancing diagnostic classification beyond conventional approaches (Gogola et al., 2025). In large structural MRI studies of 940 subjects, Lampe and colleagues demonstrated that neural network models generally yielded superior accuracy in classifying neurodegenerative diseases, while support vector machines and ensemble classifiers performed better for rarer disease subgroups with specific atrophy patterns. DL systems have also been trained to identify neurofibrillary tangles (NFTs) from digital histopathology slides, achieving nearly 100% test accuracy in differentiating multiple tauopathies, with gradient-weighted class activation mapping (Grad-CAM) explainability to highlight key pathological regions (Antal & Hajdu, 2012; Li et al., 2025; Yanase & Triantaphyllou, 2019). On MRI-based classification tasks, novel architectures like RbACNN and IRbACNN that integrate self-attention and explainable modules have attained classification accuracy as high as 99.92% for AD vs. PD vs. healthy controls (Baili et al., 2025). A cutting-edge CAD system named NEDA-DL—combining hybrid ResNet-50/AlexNet frameworks to process MRI and PET data—reported 99.87% precision in early AD detection using multimodal neuroimaging (Alasiry et al., 2025). Broader reviews of ML and DL models confirm that approaches such as CNNs, transformer models, multimodal fusion, transfer learning, and attention mechanisms consistently deliver diagnostic AUC values exceeding 0.95 in AD classification, although generalizability remains limited by small or heterogeneous datasets (Mohsen, 2025). Other advanced DL pipelines—for example, diffusion-GAN clustering (Smile-GAN) and cross-modal attention frameworks such as multimodal Alzheimer's disease diagnosis (MADDi)—enable discovery of disease subtypes and trajectory patterns using MRI plus clinical and genetic data, achieving classification accuracies above 96% (Peng, Huang, Liu, & Zhong, 2025; Zhang, Zou, Kuhl, & Karniadakis, 2024). In PD applications, metaanalyses of ML models for predicting cognitive impairment in PD report pooled sensitivity of ~76% and specificity of ~83% (AUC ≈ 0.87) across >8500 patients, highlighting reasonably robust predictive power using clinical and imaging features (Wu, Cheng, Xiao, Shang, & Ou, 2025). Despite these promising developments, most of these systems remain experimental, and their clinical utility awaits further validation through larger, multisite cohorts, standardized pipelines, enhanced interpretability (xAI), and regulatory vetting.

Multimodal AI models integrate data from multiple input types, such as neuroimaging, genomic profiles, biochemical assays, and clinical records, to generate a more comprehensive representation of disease states. Each modality is first processed through a dedicated feature extractor: for example, CNNs for MRI or PET images, transformer-based encoders for genomic sequences, and recurrent neural networks or tabular DL models for clinical and laboratory data. The extracted features are then aligned into a shared latent space using fusion strategies. Early fusion combines raw or minimally processed features at the input stage, allowing the model to learn cross-modal correlations from the outset. Late fusion processes each modality separately and merges the high-level representations before the decision layer, while hybrid fusion

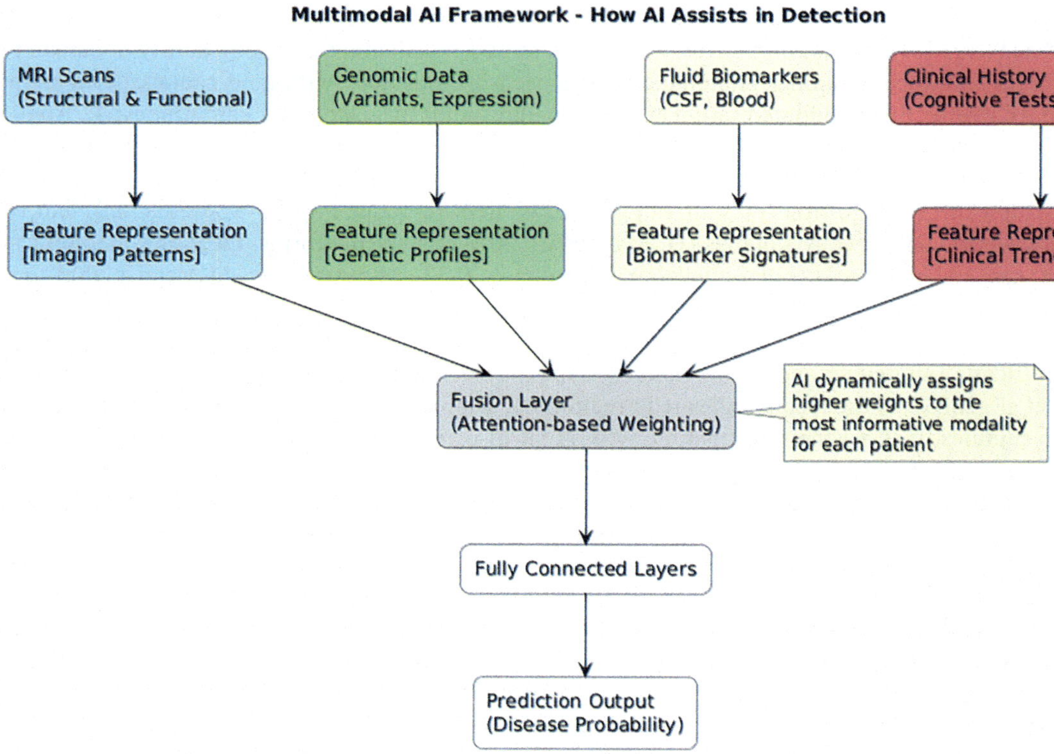

FIG. 19.1 A holistic AI-based multimodal framework for disease prognosis. Integration of data from imaging, clinical tests, genomic signatures, and case history studies with an AI-based platform helps in the early prediction of critical cognitive ailments.

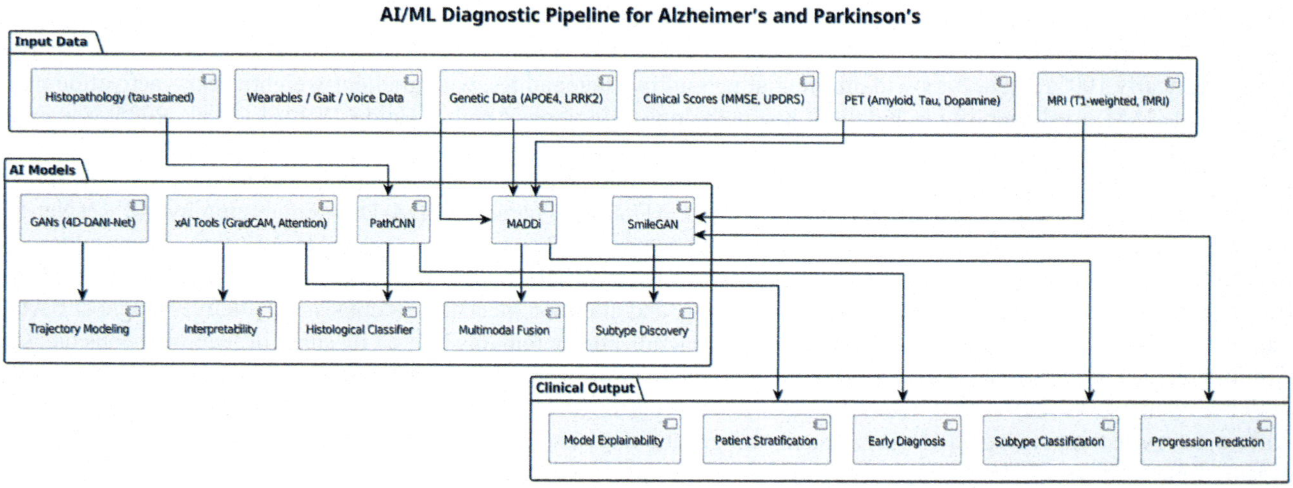

FIG. 19.2 A machine learning-based diagnostic pipeline for the prediction of the onset of Alzheimer's and Parkinson's diseases. The advent of a plethora of versatile ML platforms, such as the generative adversarial network (GAN), the gradient-weighted class activation mapping (Grad-CAM), the convolutional neural networks for pathway analysis and survival prediction (Path-CNN), and similar tools, has facilitated the development of more customized patient-based treatments.

strategies integrate at multiple stages to balance local modality-specific learning and global context integration. Attention mechanisms are often used to dynamically weight modalities according to their relevance for a given prediction, thereby improving interpretability. In neurodegenerative disease research, such architectures can jointly learn from PET amyloid burden, MRI-derived cortical thickness, and CSF biomarker concentrations to improve early detection accuracy compared to unimodal approaches.

TABLE 19.1 Key AI architectures used in neurodegenerative disease modeling.

Model/technique	Input data	Architectural highlights	Dataset and validation	Performance metrics	Interpretability/ explainability
Smile-GAN	T1-weighted MRI regional volumes	Semisupervised GAN with clustering regularizer; learns mappings from control → patient to capture heterogeneity (Yang, Wen, & Davatzikos, 2020)	ADNI and BLSA (~2832 subjects, ~8146 scans) (Yang et al., 2021b)	Identified four neurodegeneration subtypes; two progression pathways; outperformed amyloid/ tau in clinical progression prediction (Yang et al., 2021b)	Latent subtype mapping reveals cluster-wise atrophy; interpretable subtype scores
MADDi	MRI, clinical (e.g., MMSE), and genetic (SNPs)	CNN backbones with cross-modal and self-attention fusion layers (Golovanevsky, Eickhoff, & Singh, 2022)	Held-out multicenter test set (Golovanevsky et al., 2022)	96.88% multiclass accuracy (CN/MCI/ AD); ablation without attention layers → ~7.9% F1 reduction (Golovanevsky et al., 2022)	Attention heatmaps show modality importance; ablation confirms multimodal necessity
FCN Pathology CNN	Tau-stained whole-slide histology	Fully convolutional network detecting NFTs and aggregates, followed by a random forest classifier (Wang et al., 2023)	22 autopsy brains with diverse tauopathies (Wang et al., 2023)	~99% subtype-level precision and recall in distinguishing AD, PSP, and CBD (Wang et al., 2023)	Grad-CAM overlays localize textural pathology consistent with expert labels
GAN-based MRI harmonization + classifier	Paired 1.5T and 3T MRI scans.	GAN generates synthetic 3T images used to augment training classifier datasets (Wang et al., 2023)	ADNI training, validated on independent AIBL (n = 107) and NACC (n = 565) sets (Tang, Xiong, Tong, Yang, & Zhang, 2024; Wang et al., 2023)	Classifier AUC improved: ADNI (0.907→0.932), AIBL (0.934→0.940), NACC (0.870→0.907) (Tang et al., 2024; Wang et al., 2023)	Image quality metrics (SNR, NIQE) validate realism; improved downstream AD classification indicates utility
THS-GAN (Tensorized Semi-Supervised GAN)	Longitudinal 3D MRI (e.g., ADNI)	High-order tensor train structure embedding spatial correlation; semisupervised clustering GAN (Yang, Wen, & Davatzikos, 2022)	Applied to ADNI volumetric features for unsupervised subtype discovery (Yang et al., 2022)	Outperformed baseline clustering methods in subtype distinction and anomaly detection AUC (~0.80–0.85, range by dataset) (Yang et al., 2022)	Generated samples interpreted per timepoint; subtypes show distinguishable anatomical progression
4D-DANI-Net/ DeepCGAN (temporal GANs)	Longitudinal T1-weighted MRI sequences	Spatiotemporal adversarial networks modeling progression trajectories per subject	ADNI longitudinal scans with multiyear follow-up (Wang et al., 2023)	Synthetic temporal sequences preserve anatomical plausibility and dynamics; anomaly detection AUC ~0.78–0.92 (depending on the stage) (Wang et al., 2023)	Examination by visual inspection; deviations from expected age-related trajectories identified pathologies

This table summarizes key AI architectures used in neurodegenerative disease modeling: **Smile-GAN**, a semisupervised clustering GAN trained on ADNI and BLSA MRI datasets (≈2800 subjects, ≈ 8100 scans), which identified four reproducible neurodegeneration subtypes (P1–P4) and two longitudinal progression pathways that outperform conventional amyloid/tau biomarkers in predicting clinical decline; **MADDi**, a multimodal deep fusion framework integrating structural MRI, genetic SNP data, and structured clinical assessments via cross-modal and self-attention layers, achieving 96.88% multiclass accuracy (AD, MCI, control), with ablation confirming ~7.9% drop in F1-score when attention modules are removed; a **CNN-based tau pathology classifier**, employing fully-convolutional networks on tau-immunostained whole-slide images to detect neurofibrillary tangle morphologies across tauopathies (AD, PSP, CBD) with ≈99% precision and recall, and Grad-CAM overlays for localization consistent with neuropathological labels (digital pathology study); **GAN-based MRI harmonization**, which generates synthetic 3 T images from 1.5 T inputs to augment classifier training and mitigate scanner bias, yielding significant AUC improvements across independent cohorts (ADNI, AIBL, NACC) in Alzheimer's classification (up to ~0.93 vs ~0.90); **THS-GAN**, a tensor-train, semi-supervised GAN modeling high-order spatiotemporal MRI patterns to detect subtypes and anomalies with AUCs around 0.8–0.85; and **4D-DANI-Net/DeepCGAN**, temporal GANs trained on longitudinal MRI that generate plausible disease trajectories per individual and support early detection by anomaly modeling (reported AUC ≈ 0.78–0.92). Together, these architectures represent the cutting edge of interpretable, subtype-aware, multimodal AI systems that are enabling precision diagnostics and progression modeling in neurodegenerative diseases.

TABLE 19.2 Major body fluid biomarkers of preclinical or prodromal stages of AD and PD.

Component	Type	Target/function	Technology/model	Disease application	Validation/performance
Fluid biomarkers	Blood/CSF/Saliva	miRNAs, NfL, pTau, Aβ42/Aβ40	qPCR, ELISA, multiplex assays	Early detection, staging (AD, PD)	Diagnostic accuracy ~90% (e.g., pTau217); FDA-approved tests available
Neuroimaging	MRI/PET/SPECT	Brain atrophy, amyloid, tau, α-synuclein	3T MRI, ^18F-florbetaben, ^123I-FP-CIT	Diagnosis and monitoring	PET/SPECT approved; AI enhances analysis
Biosensors	Electrochemical/ Photonic	Aβ, α-syn, tau	CNTFETs, ZnO nanowires, photonic crystals	Ultra-sensitive detection in serum/plasma	Detection down to fg/mL; point-of-care capable.
SmileGAN	ML (semisupervised GAN)	Structural MRI volumes	Clustering with trajectory mapping	Subtyping and prognosis in AD	>2800 subjects; outperformed pTau/Aβ
MADDi	DL (multimodal attention)	MRI, genetics, clinical	Cross-modal CNN with attention	Classifies CN/MCI/AD	Accuracy ~96.9%; attention layers improve F1
PathCNN	DL (CNN)	Histopathology (tau-stained)	GradCAM-based CNN	AD, PSP, CBD classification	~99% accuracy on WSIs
4D-DANI-Net/ DeepCGAN	DL (temporal GAN)	Longitudinal MRI	Adversarial learning for trajectory	Anomaly detection in AD	AUC 0.78–0.92 (depending on stage)
Explainability	xAI integration	Model transparency	GradCAM, attention maps	All models (AD/PD)	Enhances clinical trust

Multimodal AI models integrate diverse datasets, such as neuroimaging, genomics, fluid biomarkers, and clinical histories, into a unified analytical framework, enabling richer disease characterization than single-modality approaches. Each modality is processed through a dedicated feature extraction branch tailored to its data type (e.g., convolutional layers for imaging, sequence encoders for genomic or longitudinal data). These representations are then merged through a fusion mechanism, which may use attention layers to dynamically assign greater weight to modalities that contain more informative patterns for the given patient. For instance, in early AD detection, AI can learn to emphasize subtle hippocampal atrophy signals from MRI when these correlate strongly with mild but abnormal cerebrospinal fluid biomarker profiles, even if other modalities are less informative at that stage. This adaptive weighting allows the model to detect early signs of pathology that might be overlooked in traditional unimodal or fixed-weight approaches. Fig. 19.1. summarizes the role of multimodal AI in disease diagnosis. Fig. 19.2 provides the AI/ML-based diagnostic pipeline for AD and PD, as detailed in Tables 19.1 and 19.2.

4 ML and DL techniques for neurodegenerative disease classification

4.1 Semisupervised clustering with Smile-GAN

Smile-GAN (semisupervised clustering generative adversarial network) is designed to model inter-individual heterogeneity in AD using T1-weighted MRI-derived regional volumes. Implemented on large cohorts (2832 participants, >8000 scans from ADNI and BLSA), it identifies four neuroanatomical subtypes (P1–P4): normal brain (P1), diffuse atrophy (P2), focal medial temporal atrophy (P3), and typical AD atrophy (P4). Longitudinally, two primary progression trajectories (P1→P2→P4 and P1→P3→P4) were discovered, showing stronger predictive capacity for clinical decline than conventional Aβ/tau biomarkers (Yang et al., 2021a; Yang, Nasrallah, Shou, Wen, Doshi, Raj, et al., 2021b). Architecture-wise, Smile-GAN uses multiple mapping functions from healthy control to patient domains coupled with clustering regularization, achieving superior clustering stability and latent pattern interpretability ([2006.15255], n.d.).

4.2 Multimodal attention architectures: MADDi

MADDi integrates MRI imaging, genetic data (e.g., SNPs), and structured clinical assessments using cross-modal and self-attention layers. The architecture employs CNNs for imaging, dense layers for clinical/genetic inputs, and bidirectional attention modules to learn interactions between modalities, enabling multiclass classification (AD, MCI, CN)

(Golovanevsky et al., 2022; Golovanevsky et al., 2022). On held-out test data, MADDi achieved **96.88% accuracy**, and ablation studies showed that the removal of attention layers lowered F1-scores by ~7.9%, highlighting the importance of attention-driven fusion (Golovanevsky et al., 2022).

4.3 DL of white matter tau pathology

A CNN-based pipeline was developed to distinguish tauopathies—AD, PSP, and CBD—from human autopsy whole-slide images stained for tau. The ML model automatically classifies tau aggregate morphologies in both cortical and white matter, revealing distinct disease-specific patterns. Importantly, classification using white matter features was as accurate as gray matter analysis, suggesting that white matter tau aggregation holds diagnostic significance (Vega et al., 2021).

4.4 AI-based approaches for detecting blood–brain barrier permeability changes

Apart from hallmark protein aggregates such as amyloid-β in AD and α-synuclein in PD, other modified protein adducts—particularly advanced glycation end-product-modified proteins—can cross the blood–brain barrier (BBB) via receptor-mediated mechanisms such as the receptor for advanced glycation end-products and contribute to plaque-like pathology (Deane et al., 2003; Sasaki et al., 1998). Early and sensitive detection of BBB permeability changes is therefore critical for understanding the entry of such adducts. Artificial intelligence (AI) is increasingly integrated into both in vivo and in vitro BBB models for this purpose. In vivo, DL-enhanced dynamic contrast-enhanced MRI improves spatial resolution, noise suppression, and kinetic parameter stability (e.g., K_trans) for detecting subtle leakage (Deane et al., 2003; Sasaki et al., 1998). In addition, AI-assisted dynamic susceptibility contrast MRI and ASL approaches can quantify perfusion–permeability coupling and even water-exchange rates (k_w) with improved reproducibility (Gold et al., 2021; Sasaki et al., 1998; Vauzour, Vafeiadou, Rodriguez-Mateos, Rendeiro, & Spencer, 2008; Li, Sadiq, & Wang, 2022; Quarles, Bell, & Stokes, 2019). In vitro, microfluidics-based BBB-on-a-chip systems replicate physiological shear stress and neurovascular coculture, enabling precise permeability assays using transendothelial electrical resistance (TEER), fluorescent tracer leakage, and tight junction imaging (Chin & Goh, 2018; Liang & Yoon, 2021; Vatine et al., 2019). ML models can automatically segment tracer diffusion, detect early TEER drift, and quantify morphological changes in tight junction proteins such as claudin-5 and ZO-1 (Cabezas et al., 2014), providing a high-throughput means to assess barrier integrity under exposure to protein-sugar adducts. Integration of AI-processed in vivo imaging with quantitative BBB-on-a-chip outputs offers a powerful translational framework for linking microvascular dysfunction to the pathogenesis of AD, PD, and related neurodegenerative disorders (Bae, Yi, Jang, & Cho, 2020).

4.5 Generative and synthetic imaging with GANs and diffusion models

GAN-based frameworks, such as 4D-DANI-Net and DeepCGAN, have been applied to generate subject-specific longitudinal MRI trajectories that mimic neurodegeneration, leveraging adversarial loss and superresolution modules. These models, trained on thousands of ADNI scans, have demonstrated high fidelity and low artifact generation, outperforming standard benchmarks in volume preservation and visual plausibility (Ravi et al., 2022). In addition, unsupervised GANs trained to reconstruct healthy brain slices have been utilized for anomaly detection, differentiating early AD with AUCs ranging from ~0.78 to ~0.92, depending on the disease stage (Han et al., 2020).

4.6 Explainable multimodal fusion (e.g., MADDi and graph-based approaches)

Recent frameworks like MADDi and cross-modal attention networks utilize explainability via attention weights, enabling interpretability and feature importance visualization. Graph neural network approaches also fuse imaging-derived connectivity features (e.g., between the hippocampus, precuneus, and posterior cingulate) with demographic and clinical data, producing higher diagnostic accuracy than single-modal CNNs in early AD stages (Golovanevsky et al., 2022).

5 Model performance and limitations

Smile-GAN subtype classifiers accurately tracked subtypes over time and predicted future progression better than Aβ/tau markers in long-term follow-up (Yang, Nasrallah, Shou, Wen, Doshi, Raj, et al., 2021b). **MADDi** achieves ~96.9% overall accuracy in differentiating AD, MCI, and controls, with proven improvement from attention mechanisms and the inclusion of structured data (Golovanevsky et al., 2022). **Pathology-based CNN** pipelines yield classification performance of ~99% in differentiating tauopathy subtypes, with explainable Grad-CAM overlays highlighting key regions (Vega et al., 2021). **GAN-based MRI synthesis** models provide realistic imaging augmentation and support early detection pipelines, although external validation remains limited (Han et al., 2020).

6 Conclusion

Recent advancements in AI, particularly DL and ML, are fundamentally reshaping the early detection, subtyping, and prognostication of neurodegenerative diseases such as AD and PD. AI-driven models combining structural MRI, PET, clinical scores, and genetic data now routinely outperform traditional diagnostic tools like Mini-Mental State Examination (MMSE) and cognitive testing, achieving **>90% accuracy**, often comparable to PET-based assessments, yet at significantly lower cost and greater accessibility (Kale et al., 2024). Hybrid attention-based architectures (e.g., MADDi) and generative models (e.g., Smile-GAN, Tensor-train, High-order pooling, and Semi-supervised learning-based (THS)-Generative Adversarial Network (GAN)) enable the discovery of reproducible disease subtypes and progression trajectories, outperforming molecular biomarkers in prognostic relevance and offering enhanced interpretability through attention weights or latent mappings (Kale et al., 2024).

Digital pathology applications using CNNs on tau-stained brain slides have demonstrated **~99% classification precision** for tauopathy subtypes, with explainable models that highlight biologically meaningful histological features. GAN-based image harmonization and synthetic MRI generation methods address multicenter variability and data scarcity, significantly improving performance across independent validation cohorts while mitigating model bias (Branstetter IV, 2007; Rajpurkar et al., 2017). Moreover, adaptive deep brain stimulation enhanced by AI algorithms is already in clinical use, offering personalized, real-time modulation of neural activity in PD, with substantial symptomatic improvement and reduced medication reliance (Foffani & Alegre, 2022).

Despite their potential, AI-based tools remain largely confined to research settings. Widespread deployment hinges on overcoming barriers such as dataset heterogeneity, lack of prospective multisite validation, algorithmic transparency, ethical considerations, and regulatory standardization. Nevertheless, as initiatives like ADNI continue to drive rigorous benchmarking and standardization, and as explainable AI frameworks gain traction, the translation of these high-performance models into clinical workflows is increasingly within reach.

In summary, AI represents a transformative opportunity to transition from symptom-driven to prediction-guided care, enabling early detection, precision subtyping, and personalized treatment strategies. By integrating neuroimaging, fluid biomarkers, histopathology, and digital phenotyping within transparent, scalable AI models, we edge closer to a future where neurodegenerative disorders are detected—and potentially mitigated—before irreversible neurodegeneration occurs.

Funding

No funding was received.

Conflict of interest

The authors declare no conflicts of interest.

Declaration of generative AI and AI-assisted technologies in the writing process

During the preparation of this work, the author(s) used ChatGPT to improve the quality of writing. After using this tool/service, the author(s) reviewed and edited the content as needed and take(s) full responsibility for the content of the publication.

References

[2006.15255] Smile-GANs: semi-supervised clustering via GANs for dissecting brain disease heterogeneity from medical images. Retrieved 27 July 2025, from https://ar5iv.labs.arxiv.org/html/2006.15255?utm_source=chatgpt.com.

Alasiry, A., Shinan, K., Alsadhan, A. A., Alhazmi, H. E., Alanazi, F., Ashraf, M. U., & Muhammad, T. (2025). A novel neuroimaging based early detection framework for Alzheimer disease using deep learning. *Scientific Reports*. 20452322. *15*(1). Available from: https://www.nature.com/srep/. doi:10.1038/s41598-025-05529-5.

Alsop, D. C., Detre, J. A., Golay, X., Günther, M., Hendrikse, J., Hernandez-Garcia, L., et al. (2015). Recommended implementation of arterial spin-labeled Perfusion mri for clinical applications: a consensus of the ISMRM Perfusion Study group and the European consortium for ASL in dementia.

Magnetic Resonance in Medicine. 15222594. *73*(1), 102–116. Available from: http://onlinelibrary.wiley.com/journal/10.1002/(ISSN)1522-2594. doi:10.1002/mrm.25197.

Alzheimer's and Parkinson's biomarkers show promise for early diagnosis | knight Initiative. Available from: https://brainresilience.stanford.edu/news/alzheimers-and-parkinsons-biomarkers-show-promise-early-diagnosis?utm_source=chatgpt.com.

Alzheimer's Association Report. (2025). Alzheimer's disease facts and figures. *Alzheimer's and Dementia*, *21*(4), e70235. Available from: doi:10.1002/ALZ.70235.

Antal, B., & Hajdu, A. (2012). An ensemble-based system for microaneurysm detection and diabetic retinopathy grading. *IEEE Transactions on Biomedical Engineering*. 0018-9294. *59* (6), 1720–1726. doi:10.1109/TBME.2012.2193126.

Antequera, Carrero, Gonzalez-Sanchez, Cantero, Orive, Municio, & Carro. (2024). Reduced salivary lactoferrin levels in early-onset Alzheimer's disease. *Aging and Disease*, *15*(3), 945–947. doi:10.14336/AD.2023.0819.

Arruda, G. S., Morris, K., Martins, A., Wang, Y., Sloan-Denisson, S., Graham, D., et al. (2025). Ultrasensitive Alzheimer's disease biomarker detection with nanopillar photonic crystal biosensors. *arXiv*, 23318422. Available from: https://arxiv.org. doi:10.48550/arXiv.2504.08476.

Bae, M., Yi, H. G., Jang, J., & Cho, D. W. (2020). Microphysiological systems for neurodegenerative diseases in central nervous system. *Micromachines*. 2072666X. *11*(9). Available from: https://res.mdpi.com/d_attachment/micromachines/micromachines-11-00855/article_deploy/micromachines-11-00855-v2.pdf. doi:10.3390/MI11090855.

Baili, J., Alqahtani, A., Almadhor, A., Hejaili, A. A., & Kim, T. H. (2025). Enhancing multi-class neurodegenerative disease classification using deep learning and explainable local interpretable model-agnostic explanations. *Frontiers in Medicine*. 2296858X. *12*. Available from: https://journal.frontiersin.org/journal/medicine. doi:10.3389/fmed.2025.1562629.

Bodily, T. A., Ramanathan, A., Wei, S., Karkisaval, A., Bhatt, N., Jerez, C., et al. (2023). In pursuit of degenerative brain disease diagnosis: dementia biomarkers detected by DNA aptamer-attached portable graphene biosensor. *Proceedings of the National Academy of Sciences of the United States of America*. 10916490. *120*(47). Available from: https://www.pnas.org/doi/10.1073/pnas.2311565120. doi:10.1073/pnas.2311565120.

Branstetter IV, B. F. (2007). Basics of imaging informatics: Part 1. *Radiology*. 00338419. *243*(3), 656–667. doi:10.1148/radiol.2433060243.

Cabezas, R., Ávila, M., Gonzalez, J., El-Bachá, R. S., Báez, E., García-Segura, L. M., et al. (2014). Astrocytic modulation of blood brain barrier: perspectives on Parkinson's disease. *Frontiers in Cellular Neuroscience*. 16625102. *8*. Available from: http://journal.frontiersin.org/Journal/10.3389/fncel.2014.00211/pdf. doi:10.3389/fncel.2014.00211.

Chen, H., Xiao, M., He, J., Zhang, Y., Liang, Y., Liu, H., & Zhang, Z. (2022). Aptamer-functionalized carbon nanotube field-effect transistor biosensors for Alzheimer's disease serum biomarker detection. *ACS Sensors*. 2379-3694. *7*(7), 2075–2083. doi:10.1021/acssensors.2c00967.

Chin, E., & Goh, E. (2018). Blood–brain barrier on a chip. *Methods in Cell Biology*, *146*, 159–182. Available from: http://www.elsevier.com/wps/find/bookdescription.cws_home/BS_MCB/description#description. doi:10.1016/bs.mcb.2018.06.003.

Coupé, P., Mansencal, B., Morandat, F., Morell-Ortega, S., Villain, N., Manjón, J. V., & Planche, V. (2025), petBrain: A new pipeline for amyloid, tau tangles and neurodegeneration quantification using PET and MRI. *Alzheimer's Research & Therapy*. Alzheimer's Research & Therapy, Springer Nature / BMC. 17, pp. 209. Available from: https://www.arxiv.org/pdf/2506.03217. doi:10.1186/s13195-025-01839-y.

Davis, K. M., Ryan, J. L., Aaron, V. D., & Sims, J. B. (2020). PET and SPECT imaging of the brain: history, technical considerations, applications, and radiotracers. *Seminars in Ultrasound, CT and MRI*. 15585034. *41*(6), 521–529. Available from: http://www.elsevier.com/inca/publications/store/6/2/3/1/9/2/index.htt. doi:10.1053/j.sult.2020.08.006.

Deane, R., Yan, S. D., Submamaryan, R. K., LaRue, B., Jovanovic, S., Hogg, E., et al. (2003). RAGE mediates amyloid-β peptide transport across the blood-brain barrier and accumulation in brain. *Nature Medicine*, *9*(7), 907–913. doi:10.1038/nm890.

Di Mari, G. M., Scuderi, M., Lanza, G., Salluzzo, M. G., Salemi, M., Caraci, F., et al. (2024). Pain-free alpha-synuclein detection by low-cost hierarchical nanowire based electrode. *Nanomaterials*. 20794991. *14*(2). Available from: http://www.mdpi.com/journal/nanomaterials. doi:10.3390/nano14020170.

Foffani, G., & Alegre, M. (2022). Brain oscillations and Parkinson disease. *Handbook of Clinical Neurology*, *184*, 259–271. Available from: http://www.sciencedirect.com/science/journal/00729752. doi:10.1016/B978-0-12-819410-2.00014-X.

Gogola, A., Lopresti, B. J., Minhas, D. S., Lopez, O., Cohen, A., & Villemagne, V. L. (2025). Tau imaging: use and implementation in new diagnostic and therapeutic paradigms for Alzheimer's disease. *Geriatrics*. 2308-3417. *10*(1), 27. doi:10.3390/geriatrics10010027.

Gold, B. T., Shao, X., Sudduth, T. L., Jicha, G. A., Wilcock, D. M., Seago, E. R., & Wang, D. J. J. (2021). Water exchange rate across the blood-brain barrier is associated with CSF amyloid-β 42 in healthy older adults. *Alzheimer's and Dementia*. 15525279. *17*(12), 2020–2029. Available from: https://newsroom.wiley.com/press-release/all-corporate-news/alzheimers-association-and-wiley-announce-publishing-partnership. doi:10.1002/alz.12357.

Golovanevsky, M., Eickhoff, C., & Singh, R. (2022). Multimodal attention-based deep learning for Alzheimer's disease diagnosis. *Journal of the American Medical Informatics Association*. 1067-5027. *29*(12), 2014–2022. doi:10.1093/jamia/ocac168.

Gong, C., Yang, X., Tang, S. J., Zhang, Q. Q., Wang, Y., Liu, Y. L., et al. (2023). *Submonolayer biolasers for ultrasensitive biomarker detection*. China: Research Square. Available from: https://www.researchsquare.com/browse. doi:10.21203/rs.3.rs-2950802.

Han, C., Rundo, L., Murao, K., Milacski, Z. Á., Umemoto, K., Sala, E., et al. (2020). GAN-based multiple adjacent brain MRI slice reconstruction for unsupervised Alzheimer's disease diagnosis. In *Lecture notes in computer science (including subseries lecture notes in artificial intelligence and lecture notes in bioinformatics)* (Vol. 12313, pp. 44–54). Japan: Springer Science and Business Media Deutschland GmbH. 16113349. https://www.springer.com/series/558. doi:10.1007/978-3-030-63061-4_5.

He, R., Yan, X., Guo, J., Xu, Q., Tang, B., & Sun, Q. (2018). Recent advances in biomarkers for Parkinson's disease. *Frontiers in Aging Neuroscience*, 1663-4365. *10*. doi:10.3389/fnagi.2018.00305.

Imam, F., Saloner, R., Vogel, J. W., Krish, V., Abdel-Azim, G., Ali, M., et al. (2025). The Global Neurodegeneration Proteomics Consortium: biomarker and drug target discovery for common neurodegenerative diseases and aging. *Nature Medicine*. 1546170X. *31*(8), 2556–2566. doi:10.1038/s41591-025-03834-0.

Kale, M., Wankhede, N., Pawar, R., Ballal, S., Kumawat, R., Goswami, M., et al. (2024). AI-driven innovations in Alzheimer's disease: integrating early diagnosis, personalized treatment, and prognostic modelling. *Ageing Research Reviews*, *101*. Available from: doi:https://doi.org/10.1016/J.ARR.2024.102497.

Knopman, D. S., Amieva, H., Petersen, R. C., Chételat, G., Holtzman, D. M., Hyman, B. T., et al. (2021). Alzheimer disease. *Nature Reviews Disease Primers*. 2056676X. *7*(1). Available from: http://www.nature.com/nrdp/. doi:10.1038/s41572-021-00269-y.

Li, S., Qu, H., Dong, X., Dang, B., Zang, H., & Gong, Y. (2024). Leveraging deep learning and xception architecture for high-accuracy MRI classification in Alzheimer diagnosis. *arXiv*, 23318422. Available from: https://arxiv.org. doi:10.48550/arXiv.2403.16212.

Li, Y., Chang, X., Wu, J., Liu, Y., Wang, H., & Zhang, Y. (2025). Machine learning in early diagnosis of neurological diseases: advancing accuracy and overcoming challenges. *Brain Network Disorders*. Available from: doi:10.1016/J.BND.2025.04.001.

Li, Y., Sadiq, A., & Wang, Z. (2022). Arterial spin labelling-based blood-brain barrier assessment and its applications. *Investigative Magnetic Resonance Imaging*. 2384-1095. *26*(4), 229. doi:10.13104/imri.2022.26.4.229.

Liang, Y., & Yoon, J.-Y. (2021). In situ sensors for blood-brain barrier (BBB) on a chip. *Sensors and Actuators Reports*. 26660539. *3*, 100031. doi:10.1016/j.snr.2021.100031.

Liu, J., Malekzadeh, M., Mirian, N., Song, T. A., Liu, C., & Dutta, J. (2021). Artificial intelligence-based image enhancement in PET imaging: noise reduction and resolution enhancement. *PET Clinics*. 18799809. *16*(4), 553–576. Available from: http://www.pet.theclinics.com/. doi:10.1016/j.cpet.2021.06.005.

Lloret, A., Esteve, D., Lloret, M.-A., Cervera-Ferri, A., Lopez, B., Nepomuceno, M., & Monllor, P. (2019). When does Alzheimer's disease really start? The role of biomarkers. *International Journal of Molecular Sciences*. 1422-0067. *20* (22), 5536. doi:10.3390/ijms20225536.

Mendez, M. F. (2012). Early-onset Alzheimer's disease: nonamnestic subtypes and type 2 AD. *Archives of Medical Research*. 18735487. *43* (8), 677–685. doi:10.1016/j.arcmed.2012.11.009.

Mohsen, S. (2025). Alzheimer's disease detection using deep learning and machine learning: a review. *Artificial Intelligence Review*. 1573-7462. *58*(9). doi:10.1007/s10462-025-11258-y.

Mueller, S. G., Weiner, M. W., Thal, L. J., Petersen, R. C., Jack, C., Jagust, W., et al. (2017). The Alzheimer's disease neuroimaging initiative. *Neuroimaging Clinics of North America*. 10525149. *15* (4), 869–877. doi:10.1016/j.nic.2005.09.008.

Myrou, A., Barmpagiannos, K., Ioakimidou, A., & Savopoulos, C. (2025). Molecular biomarkers in neurological diseases: advances in diagnosis and prognosis. *International Journal of Molecular Sciences*. 1422-0067. *26* (5), 2231. doi:10.3390/ijms26052231.

Pal, S., & Mandal, S. (2025). Therapeutic potential of microRNAs in neurological disorders: mechanisms, biomarkers, and emerging therapeutic strategies. *Exploration of Neuroprotective Therapy*, 2769-6510. *5*. doi:10.37349/ent.2025.1004111.

Palermo, G., Giannoni, S., Bellini, G., Siciliano, G., & Ceravolo, R. (2021). Dopamine transporter imaging, current status of a potential biomarker: a comprehensive review. *International Journal of Molecular Sciences*. 1422-0067. *22* (20), 11234. doi:10.3390/ijms222011234.

Peng, D., Huang, W., Liu, R., & Zhong, W. (2025). From pixels to prognosis: radiomics and AI in Alzheimer's disease management. *Frontiers in Neurology*, 1664-2295. *16*. doi:10.3389/fneur.2025.1536463.

Plastini, M. J., Abdelnour, C., Young, C. B., Wilson, E. N., Shahid-Besanti, M., Lamoureux, J., et al. (2024). Multiple biomarkers improve diagnostic accuracy across Lewy body and Alzheimer's disease spectra. *Annals of Clinical and Translational Neurology*. 23289503. *11*(5), 1197–1210. Available from: http://onlinelibrary.wiley.com/journal/10.1002/(ISSN)2328-9503. doi:10.1002/acn3.52034.

Quarles, C. C., Bell, L. C., & Stokes, A. M. (2019). Imaging vascular and hemodynamic features of the brain using dynamic susceptibility contrast and dynamic contrast enhanced MRI. *NeuroImage*.10959572. *187*, 32–55. Available from: http://www.elsevier.com/inca/publications/store/6/2/2/9/2/5/index.htt. doi:10.1016/j.neuroimage.2018.04.069.

Rajpurkar, P., Irvin, J., Zhu, K., Yang, B., Mehta, H., Duan, T., et al. (2017). CheXNet: radiologist-level pneumonia detection on chest X-rays with deep learning. *arXiv*, 23318422. Available from: https://arxiv.org. doi:10.48550/arxiv.1711.05225.

Ravanidis, S., Bougea, A., Koros, C., Simitsi, A. M., Kokotis, P., Stefanis, L., & Doxakis, E. (2025). Plasma miRNA biomarker signatures in Parkinsonian syndromes. *Molecular Neurobiology*. 15591182. *62*(8), 10118–10132. Available from: https://www.springer.com/journal/12035. doi:10.1007/s12035-025-04890-w.

Ravi, D., Blumberg, S. B., Ingala, S., Barkhof, F., Alexander, D. C., & Oxtoby, N. P. (2022). Degenerative adversarial neuroimage nets for brain scan simulations: application in ageing and dementia. *Medical Image Analysis*. 13618423. *75*. Available from: http://www.elsevier.com/inca/publications/store/6/2/0/9/8/3/index.htt. doi:10.1016/j.media.2021.102257.

Richardson, A., Kundu, A., Henao, R., Lee, T., Scott, B. L., Grewal, D. S., & Fekrat, S. (2024). Multimodal retinal imaging classification for Parkinson's disease using a convolutional neural network. *Translational Vision Science & Technology*. 21642591. *13*(8). Available from: https://tvst.arvojournals.org/pdfaccess.ashx?url=/arvo/content_public/journal/tvst/938683/i2164-2591-13-8-23_1723457830.49006.pdf. doi:10.1167/tvst.13.8.23.

Riederer, P., Nagatsu, T., Youdim, M. B. H., Wulf, M., Dijkstra, J. M., & Sian-Huelsmann, J. (2023). Lewy bodies, iron, inflammation and neuromelanin: pathological aspects underlying Parkinson's disease. *Journal of Neural Transmission*. 14351463. *130*(5), 627–646. Available from: https://www.springer.com/journal/702. doi:10.1007/s00702-023-02630-9.

Rutledge, J., Lehallier, B., Zarifkar, P., Losada, P. M., Shahid-Besanti, M., Western, D., et al. (2024). Comprehensive proteomics of CSF, plasma, and urine identify DDC and other biomarkers of early Parkinson's disease. *Acta Neuropathologica*. 14320533. *147*(1). Available from: https://www.springer.com/journal/401. doi:10.1007/s00401-024-02706-0.

Sasaki, N., Fukatsu, R., Tsuzuki, K., Hayashi, Y., Yoshida, T., Fujii, N., et al. (1998). Advanced glycation end products in Alzheimer's disease and other neurodegenerative diseases. *The American Journal of Pathology*. 00029440. *153*(4), 1149–1155. doi:10.1016/s0002-9440(10)65659-3.

Tang, Y., Xiong, X., Tong, G., Yang, Y., & Zhang, H. (2024). Multimodal diagnosis model of Alzheimer's disease based on improved Transformer. *BioMedical Engineering OnLine*. 1475-925X. *23*(1). doi:10.1186/s12938-024-01204-4.

Tivnan, M., & Li, Q. (2025). Generative super-resolution PET imaging with Fourier diffusion models. In *Progress in biomedical optics and imaging—Proceedings of SPIE* (Vol. 13405). United States: SPIE. 9781510685888. http://spie.org/x1848.xml. doi:10.1117/12.3048849.

Vatine, G. D., Barrile, R., Workman, M. J., Sances, S., Barriga, B. K., Rahnama, M., et al. (2019). Human iPSC-derived blood-brain barrier chips enable disease modeling and personalized medicine applications. *Cell Stem Cell*. 18759777. *24*(6), 995. Available from: http://www.journals.elsevier.com/cell-stem-cell/. doi:10.1016/j.stem.2019.05.011.

Vauzour, D., Vafeiadou, K., Rodriguez-Mateos, A., Rendeiro, C., & Spencer, J. P. E. (2008). The neuroprotective potential of flavonoids: a multiplicity of effects. *Genes & Nutrition*. 18653499. *3*(3–4), 115–126. doi:10.1007/s12263-008-0091-4.

Vega, A. R., Chkheidze, R., Jarmale, V., Shang, P., Foong, C., Diamond, M. I., et al. (2021). Deep learning reveals disease-specific signatures of white matter pathology in tauopathies. *Acta Neuropathologica Communications*. 20515960. *9*(1). Available from: https://actaneurocomms.biomedcentral.com/. doi:10.1186/s40478-021-01271-x.

Wang, H., Pan, J., Zhang, M., & Tan, Z. (2025). Re-evaluation of the efficacy and safety of anti-Aβ monoclonal antibodies (lecanemab/donanemab) in the treatment of early Alzheimer's disease. *Frontiers in Pharmacology*, 1663-9812. *16*. doi:10.3389/fphar.2025.1599048.

Wang, R., Bashyam, V., Yang, Z., Yu, F., Tassopoulou, V., Chintapalli, S. S., et al. (2023). Applications of generative adversarial networks in neuroimaging and clinical neuroscience. *NeuroImage*. 10959572. *269*. Available from: http://www.elsevier.com/inca/publications/store/6/2/2/9/2/5/index.htt. doi:10.1016/j.neuroimage.2023.119898.

Watanabe, H., Dijkstra, J. M., & Nagatsu, T. (2024). Parkinson's disease: cells succumbing to lifelong dopamine-related oxidative stress and other bioenergetic challenges. *International Journal of Molecular Sciences*. 1422-0067. *25*(4), 2009. doi:10.3390/ijms25042009.

Werry, E. L., Bright, F. M., Piguet, O., Ittner, L. M., Halliday, G. M., Hodges, J. R., et al. (2019). Recent developments in TSPO PET imaging as a biomarker of neuroinflammation in neurodegenerative disorders. *International Journal of Molecular Sciences*. 14220067. *20*(13). Available from: https://www.mdpi.com/1422-0067/20/13/3161/pdf. doi:10.3390/ijms20133161.

Wolk, D. A., & Detre, J. A. (2012). Arterial spin labeling MRI: an emerging biomarker for Alzheimer's disease and other neurodegenerative conditions. *Current Opinion in Neurology*. 14736551. *25*(4), 421–428. doi:10.1097/WCO.0b013e328354ff0a.

Wu, Y., Cheng, Y., Xiao, Y., Shang, H., & Ou, R. (2025). The role of machine learning in cognitive impairment in Parkinson disease: systematic review and meta-analysis. *Journal of Medical Internet Research*. 1438-8871. *27*, e59649. doi:10.2196/59649.

Xiang, J., Tao, Y., Xia, Y., Luo, S., Zhao, Q., Li, B., et al. (2023). Development of an α-synuclein positron emission tomography tracer for imaging synucleinopathies. *Cell*. 10974172. *186*(16), 3350. Available from: https://www.sciencedirect.com/journal/cell. doi:10.1016/j.cell.2023.06.004.

Xu, R., Chen, G., Mao, Z., Gao, H., Deng, Y., & Tao, A. (2020). Diagnostic performance of transcranial sonography for evaluating substantia nigra hyper-echogenicity in patients with Parkinson's disease. *Ultrasound in Medicine and Biology*. 03015629. *46* (5), 1208–1215. doi:10.1016/j.ultrasmedbio.2020.01.019.

Yanase, J., & Triantaphyllou, E. (2019). The seven key challenges for the future of computer-aided diagnosis in medicine. *International Journal of Medical Informatics*. 13865056. *129*, 413–422. doi:10.1016/j.ijmedinf.2019.06.017.

Yang, Z., Nasrallah, I. M., Shou, H., Wen, J., Doshi, J., Habes, M., et al. (2021a). A deep learning framework identifies dimensional representations of Alzheimer's Disease from brain structure. *Nature Communications*. 20411723. *12*(1). Available from: https://www.nature.com/ncomms/. doi:10.1038/s41467-021-26703-z.

Yang, Z., Nasrallah, I. M., Shou, H., Wen, J., Doshi, J., Habes, M., et al. (2021b). Disentangling brain heterogeneity via semi-supervised deep-learning and MRI: dimensional representations of Alzheimer's disease. *arXiv*, 23318422. Available from: https://arxiv.org. doi:10.48550/arxiv.2102.12582.

Yang, Z., Wen, J., & Davatzikos, C. (2020). Smile-GANs: semi-supervised clustering via GANs for dissecting brain disease heterogeneity from medical images. *arXiv*, 23318422. Available from: https://arxiv.org. doi:10.48550/arxiv.2006.15255.

Yang, Z., Wen, J., & Davatzikos, C. (2022). Surreal-GAN: semi-supervised representation learning via GAN for uncovering heterogeneous disease-related imaging patterns. *arXiv*, 23318422. Available from: https://arxiv.org.

Zhang, Z., Zou, Z., Kuhl, E., & Karniadakis, G. E. (2024). Discovering a reaction–diffusion model for Alzheimer's disease by combining PINNs with symbolic regression. *Computer Methods in Applied Mechanics and Engineering*. 00457825. *419*, 116647. doi:10.1016/j.cma.2023.116647.

Zhao, A., Liu, Y., Xinglin, Y., Xing, X., & Zhou, H. (2025). Artificial intelligence-enabled detection and assessment of Parkinson's disease using multimodal data: a survey. *Information Fusion*. 15662535. *121*, 103175. doi:10.1016/j.inffus.2025.103175.

Section VI

Artificial intelligence in the infrastructural augmentation of healthcare industry

Chapter 20

Artificial intelligence (AI) in wastewater treatment

Laishram Saya[a], Atreyee Bagchi[b], Shivam Saini[b], Deepanshu Vasuja[b], Gauri Jha[b], Pooja[a], and Sunita Hooda[c]

[a]*Department of Chemistry, Sri Venkateswara College, University of Delhi, Delhi, India*

[b]*Department of Biological Sciences, Sri Venkateswara College, University of Delhi, Delhi, India*

[c]*Polymer Research Laboratory, Department of Chemistry, Acharya Narendra Dev College, Govindpuri, Kalkaji, New Delhi, India*

1 Introduction

In the face of rapid technological progress and expanding urbanization, the escalating issues of resource depletion and environmental pollution demand urgent attention. Among these resources, water, arguably the most essential for life, is increasingly at risk of becoming unsuitable for human use. Climate change and the fast-paced development of societies often contribute to both water scarcity and declining water quality, which, in turn, cause a substantial burden of disease (Rodríguez-Alonso, Pena-Regueiro, & García, 2024). Moreover, the combined effects of rapid population growth and industrial development have led to a sharp increase in wastewater production, which has harmed the environment and contaminated natural water bodies (Narayanan, Bhat, Paul, Khatri, & Saroliya, 2024). In addition, over one billion people around the world do not have access to clean and safe drinking water, a basic right in many developed nations (DeNicola, Aburizaiza, Siddique, Khwaja, & Carpenter, 2015).

Addressing this global water crisis necessitates the implementation of effective wastewater management strategies, which are central to Sustainable Development Goal 6: ensuring the availability and sustainable management of water and sanitation for all. Within this framework, target 6.3 specifically focuses on improving water quality by reducing pollution—particularly from wastewater—and aims to halve the proportion of untreated wastewater by 2030. Understanding the different sources of wastewater, its composition, and the integration of technology into its effective management is crucial to this effort.

1.1 The nature of wastewater

Wastewater is generally categorized into three types: domestic, industrial, and storm sewage (Tarfeen et al., 2022). Domestic wastewater, or sanitary sewage, includes blackwater from toilets and gray water from household activities such as washing and bathing (Altowayti et al., 2022). Industrial wastewater includes water discharged, without treatment, from industries, containing toxic chemicals as pollutants, whereas storm sewage includes run off/precipitates that are collected in some sort of drainage or pipe system. In practice, wastewater contains both dissolved and suspended organic solids. These wastes consist of decomposable organic matter that can be broken down by microorganisms and are present in amounts that produce odors and attract or sustain ecosystems such as birds and other animals (Altowayti et al., 2022). Various factors determine the properties of wastewater, such as odor, temperature, and composition. To understand the extent of contamination in wastewater, certain parameters such as biological oxygen demand (BOD) (oxygen required by aerobic microbes to break down organic matter), chemical oxygen demand (COD) (total oxygen required to oxidize both biodegradable and nonbiodegradable matter), and total suspended solids (those solids that are not dissolved in water and can be removed by filtration) are widely used by scientists. To treat wastewater, mechanical wastewater treatment (WWT) is one of the earliest methods

Applications of Artificial Intelligence in Healthcare. https://doi.org/10.1016/B978-0-443-43934-6.00030-4

used and mainly serves to remove large solid debris through filtration (Orgoványi, Dalkó, Vadkerti, & Bognár, 2020). Chemical treatment and biological treatments are also widely used for WWT, involving complex machinery and processes.

As contaminant profiles become increasingly complex, enhancing WWT processes through the integration of artificial intelligence (AI) has become essential for greater efficiency and precision. AI is a branch of science and technology dedicated to developing machines and computer programs that can perform tasks typically requiring human intelligence. These systems are designed to mimic cognitive functions such as learning, reasoning, and decision-making (PK, 1984). The WWT industry is an ever-developing field in which new technology is often integrated into existing systems.

1.2 Environmental and sustainability challenges in current wastewater management

Current wastewater management systems are at a crossroads, burdened by growing environmental stress and the need for sustainable transitions. Despite technological progress, conventional treatment methods still struggle with escalating challenges such as nutrient loading, pathogen outbreaks, inefficient energy use, and chemical contamination (Capodaglio & Callegari, 2025; Rathour, Sakhuja, Bhatt, & Bhatia, 2022). These systems often lack the resilience to respond to climate variability, population growth, and the rising complexity of industrial effluents (Chan & Huang, 2003; Henze & Comeau, 2008). In many regions, particularly in the Global South, fragmented infrastructure and underdeveloped monitoring systems exacerbate pollution events and ecological degradation (Ahmed, Tewfik and Talaat, 2003; Altowayti et al., 2022). Moreover, traditional models often fail to incorporate dynamic feedback loops, leading to rigid operations that cannot adapt to real-time changes in water quality (Cabecinha, Cortes, Pardal, & Cabral, 2009; Chen, Ma, & Reckhow, 2007). These limitations underscore the urgent need for holistic and adaptive frameworks that merge scientific insight, policy innovation, and equitable access to emerging technologies. A few of these limitations are discussed in the following.

1.2.1 *High energy and chemical demands*

Current WWT systems face considerable sustainability challenges. A major concern is their high energy requirement, particularly for aeration and sludge handling core processes in most treatment plants (Kamali, Appels, Xiaobin, Aminabhavi, & Dewil, 2021; Luccarini et al., 2010). These operations not only inflate operational costs but also increase greenhouse gas emissions. Chemical consumption is another critical issue: excessive use of coagulants, disinfectants, and stabilizers poses long-term risks to ecosystems by contributing to secondary pollution (Chan & Huang, 2003; Chen, Chang, & Chen, 2003). In addition, sludge generation remains a burden due to expensive and environmentally challenging disposal practices (Güçlü & Dursun, 2010; Uraikul, Chan, & Tontiwachwuthikul, 2007).

1.2.2 *Regulatory and monitoring limitations*

Maintaining regulatory compliance is an uphill task for many facilities. Variability in influent loads, aging infrastructure, and lack of advanced monitoring technologies often cause plants to underperform (Ahmed, Tewfik and Talaat, 2003; Takeda, Kato, Matsumura, Murakami, & Abeynayaka, 2021). Real-time monitoring, which could enable quicker responses, remains underutilized due to cost and technology gaps (Rodríguez-Alonso et al., 2024; Wang et al., 2019). For example, in several Indian urban centers, outdated equipment makes adaptive control nearly impossible, leaving systems reactive rather than preventive (Capodaglio & Callegari, 2025; Martini & Roni, 2021).

1.2.3 *Limited operational flexibility and predictive control*

Traditional treatment systems struggle with sudden inflow variations, such as those caused by industrial discharge or storm events. These rigid operations lack built-in mechanisms to adapt in real time (Chen, Jakeman, & Norton, 2008; Park & Stenstrom, 2006). Without predictive models, operators often resort to conservative settings, sacrificing efficiency for safety (Adenso-Díaz, Tuya, & Goitia, 2005; Janssen, Goosen, & Omtzigt, 2006). Even when data is available, it is rarely processed into actionable insights due to a lack of intelligent analytics (Nourani, Elkiran, & Abba, 2018).

1.2.4 *Human dependency and fault management issues*

Manual operation continues to dominate in many plants, especially in developing regions. This reliance increases the chances of human error and delays in responding to equipment faults or biological imbalances (Altowayti et al., 2022;

Krovvidy, Wee, Summers, & Coleman, 1991). For example, undiagnosed sensor drift can cause major failures in biological treatment phases if not promptly detected (Alprol, Mansour, Ibrahim, & Ashour, 2024; Luccarini et al., 2010).

1.2.5 Resource allocation, accessibility, and integration barriers

Efficient use of energy, chemicals, and personnel is often compromised in systems lacking real-time feedback loops. This inefficiency becomes even more pronounced in rural or underresourced areas where advanced technologies are simply not available (Ahmed, Tewfik and Talaat, 2003; Henze & Comeau, 2008). Integration of gray water, blackwater, and stormwater streams is still rare, leading to suboptimal infrastructure use (Bromley, Jackson, Clymer, Giacomello, & Jensen, 2005; Cabecinha et al., 2009).

1.2.6 Environmental crises, carbon footprint, and public transparency

Wastewater systems are frequently caught off guard during emergencies such as floods or chemical leaks. Their crisis response capabilities are often inadequate due to technological and staffing limitations (Narayanan et al., 2024; Wang et al., 2023). Moreover, plants relying on fossil-fuel power contribute significantly to local carbon footprints (Capodaglio & Callegari, 2025; Viet, Jang, Yoon, & Jang, 2022). In many regions, the absence of transparent data sharing further weakens public trust and regulatory oversight (Luccarini et al., 2010).

1.2.7 Scalability challenges in growing urban settings

As urban populations grow, existing wastewater infrastructure struggles to scale accordingly. In cities like Mumbai or Lagos, treatment systems-built decades ago are now outdated and overstretched (Kamali et al., 2021; Lowe, Qin, & Mao, 2022). Their rigid architectures make it difficult to accommodate new regulations or increased waste volumes, threatening both environmental and public health. Clearly, conventional WWT systems are nearing their functional limits. To meet the demands of modern cities and environmental goals, AI-based solutions are emerging as game changers. From predictive analytics to dynamic resource optimization, these technologies offer a path toward more responsive, efficient, and resilient wastewater management (Ma et al., 2022).

2 AI in WWT

In the wastewater industry, AI has been used to track the effectiveness of WWT plants **(WWTPs)** in terms of efficiency metrics, the determination of BOD and COD, the removal of nitrogen and sulfur, the uptake of contaminants, etc. (Malviya & Jaspal, 2021). The three broad models of AI that are most frequently employed in the wastewater industry are neural network-based models, fuzzy logic (FL)-based models, and optimization and evolutionary algorithms. Within these broader models lie various types of AI systems, which have been employed to tackle multiple steps of the WWT process. Some of these are shown in the flowchart (Fig. 20.1).

2.1 Neural network-based models

A neural network is a big system made up of many small units called nodes, which are connected to each other by links. These nodes work like simple versions of brain cells (neurons). Each node collects signals (inputs) from the connected links, adds them up, and checks if the total is strong enough (crosses a certain threshold). If it is, the node gets activated (or fires) and sends a signal to the next connected nodes (Caulfield, 2005). Based on this logic, many AI systems have been built to make WWT more efficient.

2.1.1 Artificial neural network

Artificial neural networks (ANNs) are information processing systems designed to mimic the way the human brain works. They consist of an input layer, an output layer, and one or more hidden layers in between. The presence of multiple hidden layers enables ANNs to learn complex patterns and improve predictive performance (Wang et al., 2023). In WWT, ANNs are frequently used in experimental setups to support contaminant removal (Zhao et al., 2020). In addition to standalone applications, ANNs have also been integrated with statistical techniques to enhance system monitoring and control. For instance, Lee et al. (2008) combined ANN with principal component analysis (PCA) and multivariate statistical process control to develop a real-time remote monitoring system for WWTPs. This system continuously tracked plant operations and provided expert-level insights to improve management efficiency.

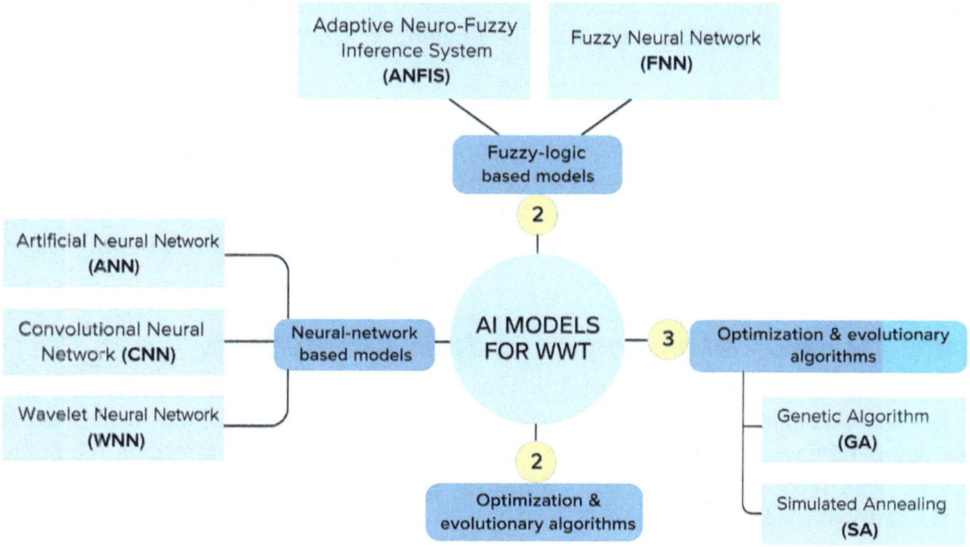

FIG. 20.1 Flowchart depicting the commonly used AI models for wastewater treatment.

Hybrid AI systems have also been explored, such as those combining expert systems with neural networks. In one such approach, the expert system sets a target BOD value for the aeration tank, while the neural network learns the control pattern and determines the optimal sludge recycle rate (Wen & Vassiliadis, 1998). Moreover, recent studies have reviewed the use of AI models, particularly ANNs, in membrane processes and membrane bioreactors (MBRs) (Kamali et al., 2021). MBRs integrate biological treatment with membrane filtration, combining microbial activity with the physical separation of solids from liquids (Wang et al., 2023). These membrane technologies, including microfiltration, ultrafiltration, nanofiltration, reverse osmosis, and membrane distillation, are also vital in addressing freshwater scarcity, especially through desalination and advanced treatment of brackish and seawater. AI models like ANNs are increasingly being applied to optimize these complex processes (Jawad, Hawari, & Javaid Zaidi, 2021).

2.1.2 Convolutional neural network

A convolutional neural network (CNN) is a deep learning model that extracts features from data using convolutional and pooling layers, followed by fully connected layers for classification or prediction. CNNs are widely used in image and language processing tasks (Wang et al., 2023). In WWT, Li et al. proposed a hybrid CNN–long short-term memory (LSTM) model for predicting effluent quality, where the CNN acts as an encoder to capture spatial features and the LSTM network—a type of recurrent neural network—serves as a decoder to model time-based dependencies. One of the growing challenges in WWT is the removal of microplastics (MPs). To address this, Luo et al. (2024) introduced an innovative approach that combines Raman spectroscopy with CNNs for accurate MP analysis. By utilizing average mapping spectra (AMS) and filtering samples of varying volumes, this method improves detection consistency and broadens the measurable concentration range. Building on this, Capodaglio et al. demonstrated the real-time monitoring of micropollutants such as carbamazepine, naproxen, and tryptophan at concentrations in the microgram-per-liter range ($R^2 = 0.74$), offering a reliable alternative to conventional monitoring techniques (Capodaglio & Callegari, 2025).

2.1.3 Wavelet neural network

Wavelet networks (WNs) are a modern type of neural network that combine traditional sigmoid-based architectures with wavelet analysis, effectively merging wavelet transform theory with ANNs (Alexandridis & Zapranis, 2013). They benefit from the localization capabilities of wavelets and the self-learning properties of neural networks (Curreri et al., 2020). Due to these strengths, WNs have been successfully applied in various domains, including WWT.

In one study, researchers used a wavelet neural network (WNN) alongside a backpropagation neural network (BPNN) to examine the influence of pH on pollutant removal in an aerobic–anaerobic membrane bioreactor (O-AMSMBR). The results showed that WNN effectively captured the impact of pH, a critical factor affecting COD and total nitrogen (TN) removal (Cai et al., 2019; Wang et al., 2023). In addition, a WN was developed as a soft sensor to monitor sewage treatment quality in real time. By analyzing sparse input data, researchers reduced the number of wavelet basis functions and used gradient descent to train the model. Due to its fast learning and strong approximation capabilities, the WNN-based soft sensor was able to accurately assess treatment quality using sensor data (Curreri et al., 2020).

2.2 FL-based models

FL is a type of logic that allows values between traditional options like true/false or high/low. It enables computers to handle vague terms like "rather tall" or "very fast," facilitating more human-like reasoning in programming (Byaruhanga et al., 2004). AI systems operating on this logic are known as FL-based models. Quite often, FL is combined with neural networks to form hybrid models, which are effective in WWT.

2.2.1 Fuzzy neural networks

Fuzzy neural networks (FNNs) combine FL with ANN training. They resemble traditional neural networks but include elements such as membership functions, fuzzy inference, and normalization to model the uncertain or imprecise relationships between inputs and outputs (Wang et al., 2023). FNNs are especially effective for handling complex, nonlinear problems and are widely used in pattern recognition, control systems, and predictive analysis.

A notable application of FNNs is in fault diagnosis for sludge volume index (SVI) sensors in WWTPs. In one study, a real-time diagnostic system was developed, comprising online sensors, a sludge bulking prediction model, and a fault detection mechanism. The FNN was used as the core of the prediction model and trained using an adaptive learning rate, which improved convergence speed (Han & Qiao, 2010; Honggui, Ying, & Junfei, 2014). By comparing actual SVI sensor readings with FNN-predicted values, the system could quickly identify faults. This approach was successfully tested in a simplified WWTP environment, showing that FNNs can reliably detect anomalies, improve sensor accuracy, and support more efficient plant operations.

2.2.2 Adaptive neurofuzzy intelligence system

Adaptive neurofuzzy inference system (ANFIS) is part of a broader area of AI known as machine learning (ML). ML focuses on developing algorithms and statistical models that enable computers to learn from data automatically, without explicit programming. Its primary goal is to create predictive models capable of making accurate decisions or forecasts based on training data (Dey, 2016; Wang et al., 2023). Common ML models include decision trees (DTs) and K-nearest neighbors, among others.

Jang & Sun (1995) introduced ANFIS in 1995 based on the first-order Sugeno fuzzy model, incorporating backpropagation or hybrid learning techniques. It is recognized as one of the earliest and most effective hybrid neurofuzzy systems, valued for its fast-training speed, efficient learning, and simple architecture, making it a powerful function approximator (Noshadi, Salahi, Hemmati, Rekabdar, & Mohammadi, 2013).

In a practical application, researchers at the Bunus Regional Sewage Treatment Plant in Kuala Lumpur, Malaysia, developed an ANFIS model to predict carbon and nitrogen levels. When compared to a feedforward neural network (FFNN), ANFIS demonstrated slightly better accuracy in predicting key parameters such as COD, suspended solids, and ammonium nitrogen (NH_4-N) (Gaya, Wahab, Sam, & Samsudin, 2014). These results highlight the reliability and usefulness of ANFIS in WWT applications.

2.3 Optimization and evolutionary algorithms

2.3.1 Genetic algorithms

Genetic algorithms (GAs) are optimization techniques inspired by the principles of natural selection and evolution. They are particularly effective for solving complex or NP-hard problems, where traditional analytical methods fall short. GAs work well with constrained and multimodal objective functions, and instead of evaluating every possible solution in detail, they rank potential solutions, making them ideal for problems with complex evaluation criteria (Maier et al., 2014).

In wastewater management, GAs have been integrated with mathematical water quality models to improve decision-making. For example, a GA-based model was developed for the Youngsan River Basin to meet water quality standards and reduce treatment costs. This model accounted for high pollutant loads from Kwangju City and surrounding areas

(Cho, Seok Sung, & Ryong Ha, 2004). In another application, GAs were used to optimize dissolved oxygen levels in the fourth tank of a treatment process by adjusting for ammonium and ammonia nitrogen concentrations. Unlike static control methods, the GA dynamically determines optimal oxygen levels under varying conditions, improving effluent quality and lowering operational costs. Simulation results validated the efficiency of this high-level GA-based control strategy (Do, Bach, Nguyen, Tran, & Nguyen, 2021).

2.3.2 *Simulated annealing*

Simulated annealing (SA) is another optimization algorithm inspired by the physical process of annealing in solids, where materials are slowly cooled to reach a stable, low-energy state. Introduced in 1982 by Kirkpatrick et al., SA works by exploring a range of solutions and gradually narrowing the search toward the optimal outcome. This makes it well-suited for solving multivariable design and engineering problems, especially when the solution space is too large or irregular for traditional methods (Rutenbar, 1989).

In the context of WWT, SA has been successfully applied to regional system planning. One model utilized SA to determine the most cost-effective design for managing wastewater from multiple population centers, while ensuring compliance with discharge regulations and water quality standards (Zeferino, Antunes, & Cunha, 2009). To add to this, adaptive simulated annealing (ASA)—an advanced form of SA—was applied to optimize the design of a treatment plant consisting of a trickling filter, activated sludge aeration tank, and secondary clarifier (Govindarajan, Kumar, & Karunanithi, 2005). ASA significantly improved design efficiency by identifying the optimal configuration at the lowest total construction cost when compared to conventional optimization methods. Hence, ASA's potential as a powerful tool for enhancing both the economic viability and operational performance of WWT systems cannot be ignored.

2.4 Other models and examples

Beyond the models discussed previously, several other ML and hybrid AI models have been effectively employed for WWT and management.

Zaghloul, Iorhemen, Hamza, Tay, & Achari (2021) developed a five-stage ML model to simulate and predict the behavior of aerobic granular sludge (AGS) reactors. Using 475 days of data from three laboratory-scale reactors, they employed an ensemble of ANNs, support vector regression (SVR), and ANFIS to enhance prediction accuracy.

TABLE 20.1 Summary of the most commonly employed AI models for improving the efficiency of wastewater management.

AI model	Contribution in wastewater treatment (WWT)
Artificial neural network (ANN)	Learns complex patterns; used in contaminant removal, system monitoring (via PCA, MSPC), expert system integration, and membrane processes like MBRs.
Convolutional neural network (CNN)	Used in hybrid CNN–LSTM models for effluent quality prediction; combined with Raman spectroscopy for real-time microplastic and micropollutant detection.
Wavelet neural network (WNN)	Combines wavelet transform with ANN; models pH effect on pollutant removal and used as a soft sensor for real-time sewage quality monitoring.
Fuzzy neural network (FNN)	Combines fuzzy logic with ANN for modeling uncertainty; applied in fault diagnosis of SVI sensors in WWTPs, with adaptive learning for anomaly detection.
Adaptive neurofuzzy inference system (ANFIS)	Hybrid model combining ANN and fuzzy logic; used for predicting COD, SS, and NH_4-N; proven better than FFNN in the Bunus WWTP case study.
Genetic algorithm (GA)	Optimization technique for water quality modeling and cost reduction; adjusts dissolved oxygen dynamically to improve effluent quality and reduce costs.
Simulated annealing (SA)	Used for cost-effective system design in regional WWT planning; adaptive SA (ASA) enhances plant design efficiency, minimizing construction costs.
ANN + SVR + ANFIS Ensemble	Used by Zaghloul et al. (2021) to simulate AGS reactor behavior using long-term lab data; improves predictive accuracy.
Backpropagation neural network (BPNN)	Simulates COD removal in high-strength wastewater from traditional medicine; supports optimal operational decisions.
ANN-PSO/ANN-GA/RSM	Applied by Jimei et al. (2020) to model and optimize methylene blue decontamination using advanced nanomaterials.

REN et al. (2007) utilized a BPNN to simulate CODfilt and COD removal in a pilot-scale submerged MBR treating high-strength traditional Chinese medicine wastewater. The model effectively predicted removal rates and supported the optimization of operational conditions. In another study, Qi et al. applied multiple models—response surface methodology (RSM), ANN-particle swarm optimization (ANN-PSO), and ANN-genetic algorithm (ANN-GA)—to study the decontamination of methylene blue (MB) from synthetic wastewater using mesoporous rGO/Fe/Co nanohybrids (Table 20.1).

3 Recent advances in AI-driven WWT technologies

It is evident from above that the advancement of AI technologies has not only improved the operational efficiency of treatment plants but also transformed them into smarter, more responsive, and predictive systems.

3.1 AI-based decision support and planning tools

One of the earliest and most foundational applications of AI in WWT is the development of decision support systems (DSS). These are computer-based information systems that support decision-making activities, especially when multiple variables and uncertainties are involved. Adenso-Díaz et al. presented an environmental decision support system (EDSS) to evaluate and compare different wastewater collection system designs (Adenso-Díaz et al., 2005). Their EDSS incorporated multiple criteria such as cost, environmental impact, and technical feasibility to arrive at optimized layouts. It enabled engineers to simulate and assess multiple configurations before physical implementation, reducing risks and costs. Ahmed et al. introduced a DSS for selecting optimal water reuse schemes. Their system integrated technical (e.g., pollutant load), economic (e.g., implementation costs), and environmental (e.g., local regulations) parameters to guide municipal planners in choosing sustainable reuse strategies. By combining rule-based logic with mathematical modeling, such DSS platforms reduce human error and speed up decision-making processes. Bayesian networks (BNs) have further strengthened decision support frameworks. These are probabilistic graphical models that represent variables and their conditional dependencies using a directed acyclic graph. Bromley et al. used BNs to incorporate uncertainties in water resource planning. Their approach allowed the integration of diverse indicators: hydrological, ecological, and socioeconomic into a single predictive model, offering planners a more holistic understanding of potential scenarios.

3.2 Uncertainty handling and risk management

WWT operations are frequently subjected to variability in influent loads, weather patterns, and equipment performance. To ensure reliability under these unpredictable conditions, AI methods have been employed to model uncertainty more robustly.

Chen et al. highlighted the importance of systematic qualitative uncertainty analysis in water quality models. Their work revealed that traditional deterministic models often fail to capture the stochastic (random) nature of real-world data. They advocated for a probabilistic approach that incorporates expert judgment and historical data. Cabecinha et al. (2009) proposed a stochastic dynamic methodology to manage water quality in reservoirs (Cabecinha et al., 2009). Applied in the Douro River basin in Portugal, this method combined dynamic system simulation with stochastic modeling to predict and adjust to multiscale changes in water quality. This AI-driven framework accounted for both spatial and temporal variability, making it highly effective for basin-scale water resource management.

3.3 Smart sensors, Internet of Things, and digital twins

Modern AI-enabled wastewater systems rely heavily on real-time data collection and simulation. Three technologies are central to this transformation: smart sensors, the Internet of Things (IoT), and digital twins. Smart sensors are advanced detection devices equipped with embedded systems capable of real-time data processing. Unlike conventional sensors, smart sensors can self-calibrate, filter out noise, and transmit refined signals to control systems or cloud platforms. In WWT, they are used to monitor parameters such as pH, turbidity, temperature, BOD, and COD (Charef, Ghauch, Baussand, & Martin-Bouyer, 2000).

The IoT refers to the network of interconnected devices, including sensors, actuators, and computing nodes, that collect and share data. In the context of wastewater management, IoT facilitates the continuous monitoring and remote control of treatment processes across spatially distributed facilities. It ensures seamless communication between hardware and software, allowing AI algorithms to operate in real time (Alzahrani, Chauhdary, & Alshdadi, 2023; Drenoyanis, Raad, Wady, &

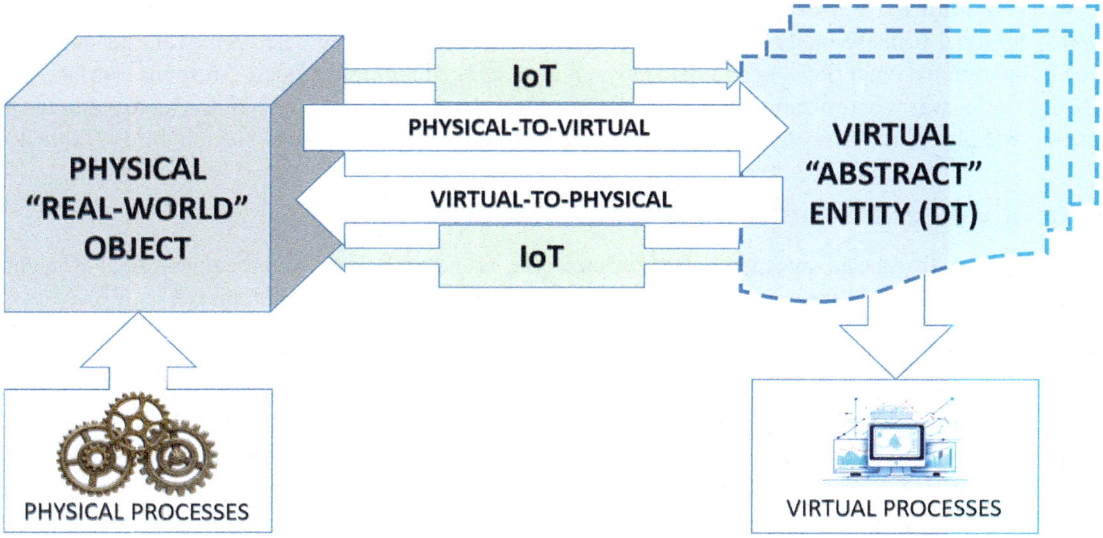

FIG. 20.2 Working mechanism of digital twin technology using IoT (Capodaglio & Callegari, 2025) (CC BY 4.0).

Krogh, 2019). Digital twins are virtual representations of physical assets or systems, continuously updated using real-time data from IoT devices and smart sensors. These digital models simulate the behavior of the actual WWTP under varying conditions. They serve as testbeds for process optimization, predictive maintenance, and emergency planning (Colwell & Abolghasemi, 2024; Curl, Nading, Hegger, Barhoumi, & Smoczynski, 2019; Mihai et al., 2022). Moreover, Rodríguez-Alonso et al. developed a microservices-based digital twin platform that integrates AI algorithms for simulating plant behavior in real time. Operators can test hypothetical scenarios, such as equipment failures or sudden pollutant surges, and receive predictive insights without compromising actual system safety (Rodríguez-Alonso et al., 2024) (Fig. 20.2).

Khan et al. applied AI-driven image processing to monitor activated sludge via phase-contrast microscopy (Khan, Nisar, & Ng, 2017). Their method automated fault detection and anomaly recognition, ensuring timely responses to biological process deviations. Such applications highlight how smart sensing, when paired with AI analytics, can enhance the resilience and agility of WWT infrastructure. Together, smart sensors, IoT, and digital twins form an intelligent ecosystem that supports responsive, self-adaptive treatment operations. Their integration enables facilities to shift from reactive to predictive management, laying the foundation for future-ready water treatment systems.

3.4 Advanced disinfection and renewable integration

Innovative AI applications are also emerging in disinfection processes and renewable energy integration. Close et al. engineered a water recycling system powered by photovoltaic (PV) cells and ultraviolet (UV) light-emitting diode (LED) disinfection units. While the primary focus was on sustainability, AI was employed to adjust UV doses based on real-time feedback from sensors, thereby improving efficacy and reducing energy waste. This intersection of AI, renewable energy, and water treatment signifies a trend toward integrated environmental engineering, where systems are designed not only for performance but also for sustainability.

3.5 AI for extreme events and pollution source management

AI's strength in pattern recognition and anomaly detection makes it invaluable during environmental emergencies and pollution events. Wang et al. developed an AI-based tool to trace the source of water pollution by analyzing spatiotemporal data. Their model classified pollutant types and origin points with high accuracy, enabling rapid response and mitigation.

Park and Stenstrom applied BNs to satellite imagery for managing stormwater pollution in urban areas. Their AI system could predict runoff behavior based on land use, rainfall data, and terrain features, allowing city planners to prepare for extreme weather events.

TABLE 20.2 Summary of the most commonly employed AI models for improving the efficiency of wastewater management.

Tool	Focus area	Key contributions/examples
AI-based decision support and planning tools	Decision-making under uncertainty	• EDSS for evaluating wastewater collection designs (Adenso-Díaz et al., 2005) • DSS for selecting optimal water reuse strategies (Ahmed, Tewfik and Talaat, 2003) • Bayesian Networks for integrated planning (Bromley et al., 2005; Park & Stenstrom, 2006)
Uncertainty handling and risk management	Managing variability in influent, weather, and performance	• Qualitative uncertainty analysis in water quality models (Chen et al., 2007) • Stochastic Dynamic Methodology for River Basin Management (Cabecinha et al., 2009)
Smart sensors, IoT, and digital twins	Real-time monitoring and predictive simulation	• Smart sensors for BOD, COD, pH, etc. • IoT-based remote monitoring • Digital twin platform for process simulation (Rodríguez-Alonso et al., 2024) • AI image processing for sludge monitoring (Khan et al., 2017)
Advanced disinfection and renewable integration	Energy-efficient, AI-enhanced treatment systems	• UV-LED disinfection powered by PV cells (Close & Jasper Ip, 2006) • AI used to optimize UV dose and energy efficiency
AI for extreme events and pollution source management	Emergency response and pollution tracing	• Pollution source identification using spatiotemporal AI (Wang et al., 2019) • Stormwater management via Bayesian networks and satellite imagery (Park & Stenstrom, 2006)
Negotiation and policy support systems	Governance and stakeholder negotiation	• AI-based negotiation tools using game theory and decision trees (Janssen et al., 2006)

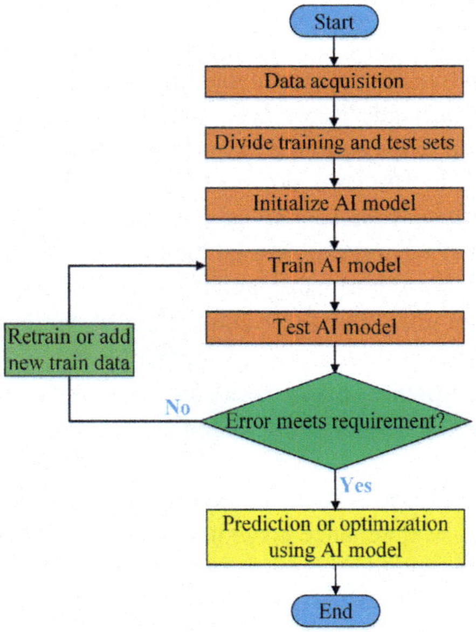

FIG. 20.3 Flowchart for the application of AI in WWT (Wang et al., 2023) (CC BY 4.0).

3.6 Negotiation and policy support systems

AI has also found its way into the governance of water systems. Janssen et al. created a negotiation support tool that incorporates AI to mediate between conflicting stakeholder interests in water management (Janssen et al., 2006). The system uses game theory and DTs to simulate negotiation scenarios and propose compromise solutions. This innovation is particularly useful in contexts involving multiple agencies and competing priorities, such as transboundary water disputes (Table 20.2; Fig. 20.3).

4 Japan as a case study

Japan has long balanced its reputation for technological innovation with a deep-rooted commitment to public consensus and ethical governance. Nowhere is this balance more evident than in the country's efforts to modernize WWT through digital technologies. Rather than rushing to implement new tools, Japan has taken a cautious, community-centered approach, placing people, fairness, and transparency at the heart of its transformation (Takeda et al., 2021; Yeung, Howes, & Pogrebna, 2020). The country's wastewater infrastructure, primarily managed by municipal governments, faces the twin pressures of aging systems and rising environmental standards. In response, several cities have begun exploring data-driven solutions to improve efficiency. Yet, what stands out about Japan's approach is how methodical and inclusive the process has been. The goal was never just to upgrade technology but to ensure that any changes served the public good and were understood, accepted, and even shaped by the communities they would affect (Takeda et al., 2021).

Yokohama

One of the clearest examples of this ethos can be seen in Yokohama, Japan's second-largest city. By the mid-2010s, city engineers noticed increasing strain on the sewer system, especially during typhoons and high rainfall. They began exploring predictive tools that could anticipate water surges, identify clogged channels, and help technicians make faster decisions about flow diversion (Saravi et al., 2019). Before rolling out any new systems, however, the city chose to consult the public. Local officials hosted a series of forums with residents, academics, and nonprofit environmental groups. These were not just information sessions but genuine dialogs. Attendees raised concerns about data privacy, questioned whether technology might replace human jobs, and asked how the systems would be evaluated over time (Takeda et al., 2021). These discussions shaped the project itself. Instead of adopting an opaque system, Yokohama's engineers worked with local universities to build a monitoring tool that emphasized transparency. Rather than relying entirely on automated decisions, the tool provides real-time information to human operators, who remain in charge. Decisions are logged and made available for public review, and reports are shared twice a year with community councils (Takeda et al., 2021).

Fukuoka

Farther south, in the city of Fukuoka, known for its sustainable urban planning, wastewater teams faced a different problem: fluctuating population and tourism demands. Some districts experienced surges in wastewater use, while others had long-term underutilization. The city turned to a dynamic data system that could help balance loads and improve energy use across treatment plants (Takeda et al., 2021). But here too, technology did not drive the process people did. City managers set up advisory committees, including citizen representatives, public health experts, and even school teachers. The focus was not only on what the tool could do, but what it should do. One memorable outcome was a policy change: any software used in public systems must be auditable by a third party, and all long-term performance data must be publicly disclosed (Takeda et al., 2021). These rules were not just written into contracts; they were woven into practice. Regular open days at treatment facilities allow residents to see the system in action and ask questions. Fukuoka also launched a citizen-reporting app that lets residents notify the city of suspicious odors, overflow points, or other infrastructure issues. In return, the city shares updates on how these alerts were handled, creating a feedback loop grounded in trust (Takeda et al., 2021; Yeung et al., 2020)

Across both cities, and in others that followed suit, the outcomes have been clear. Treatment systems now respond more quickly to pressure. Energy use is down. Maintenance costs have declined, thanks to better forecasting. But beyond the numbers, what has really emerged is a new model of public governance, one where technology is a tool, not a master, and where transparency is built in from the start (Chester et al., 2021; Yeung et al., 2020). This model does not just improve infrastructure; it enhances legitimacy. In a 2021 survey by a local NGO, over 70% of Yokohama residents said they felt more comfortable with city planning decisions because they had been invited to participate. Fukuoka saw similar results, with rising trust in municipal services over three years (Takeda et al., 2021).

5 Limitations and challenges of AI in WWT

Despite its transformative potential, the widespread implementation of AI in WWT remains constrained by several key challenges. While the growing demand for efficient treatment—driven by urbanization and industrialization—has made AI increasingly relevant (Altowayti et al., 2022), the technology is still considered emerging and experimental in many urban WWT contexts (Plathottam, Rzonca, Lakhnori, & Iloeje, 2023).

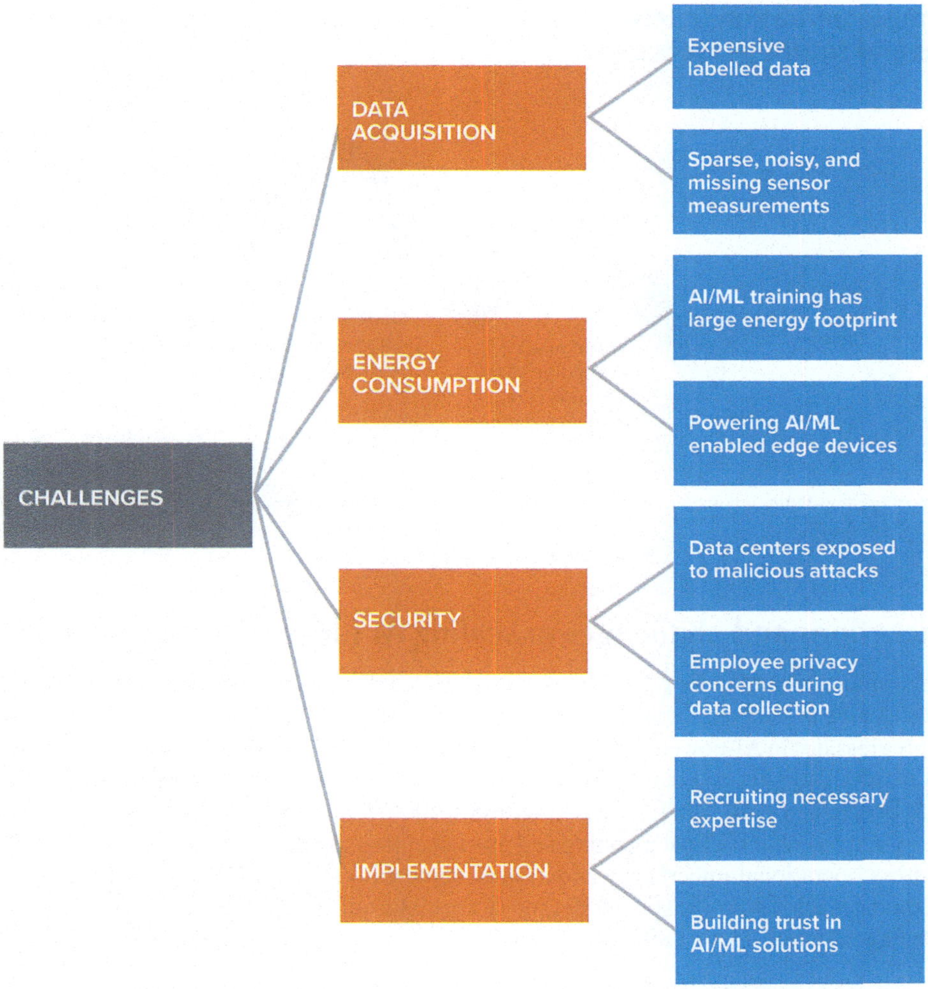

FIG. 20.4 Some of the limitations and challenges of AI in wastewater treatment, highlighting that, despite its promising potential, implementation remains a complex task (Plathottam et al., 2023) (CC BY 4.0).

A central obstacle is the inherent complexity and variability of wastewater systems. These involve fluctuating influent conditions and tightly interlinked biological, chemical, and physical processes, which can be difficult to model or control effectively using AI. Furthermore, institutional differences between public utilities and private facilities, coupled with the water sector's generally risk-averse nature, have slowed the adoption of emerging technologies. From a technical standpoint, data limitations are among the most critical concerns. Reliable AI models require large volumes of high-quality data. However, long-term, stable datasets are often unavailable, and sensor drift over time can compromise data integrity, leading to overfitting or underfitting (Bartsch, Pettke, Hübert, Lakämper, & Lange, 2021; Bertini, Bucchi, Frendo, Moda, & Monelli, 2019). In addition, models trained in one system may not generalize well to others, limiting transferability.

Another major concern is the interpretability of AI systems. Complex models—especially deep learning architectures like ANNs—function as "black boxes," making their predictions difficult to explain or validate. This lack of transparency can lead to mistrust or misapplication in regulated environments where explainability is crucial (Capodaglio & Callegari, 2025).

Finally, reproducibility remains a persistent challenge. AI models—particularly deep neural networks—often start with random weight initialization, which can lead to variability in outputs even under similar training conditions. Moreover, the computational demands of training deep models can prevent consistent optimization, resulting in convergence to local minima rather than global optima (Pandey et al., 2022; Wang et al., 2023). These challenges—ranging from technical and operational to institutional—highlight the need for cautious, phased integration of AI in wastewater systems, supported by robust design, governance, and validation frameworks (Fig. 20.4).

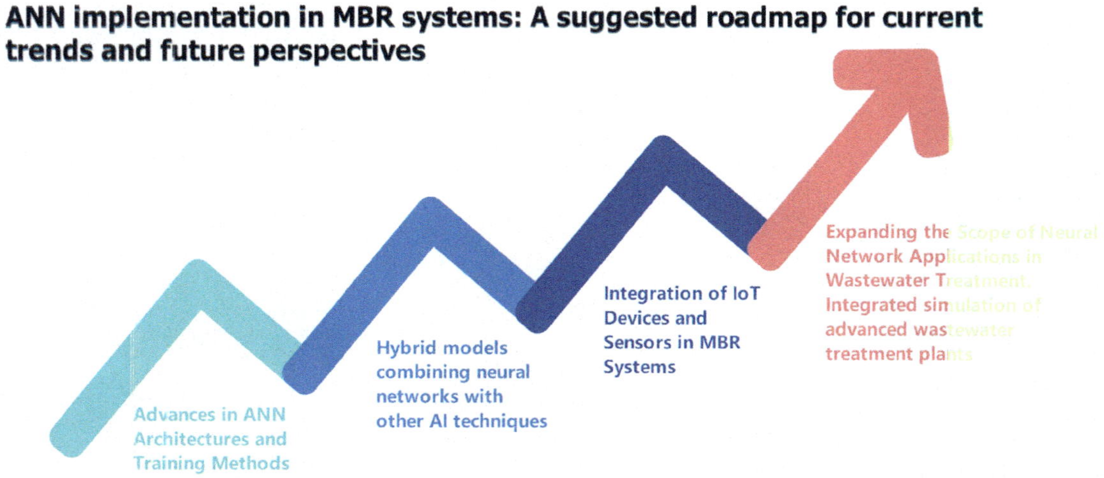

FIG. 20.5 Roadmap suggesting future innovations and trends aimed at enhancing wastewater treatment methods through AI (Frontistis, Lykogiannis, & Sarmpanis, 2023) (CC BY 4.0).

6 Future directions and challenges in scaling AI for wastewater management

To overcome these limitations and unlock AI's full potential in WWT, future efforts must focus on developing integrated, intelligent, and sustainable systems through cross-sectoral collaboration.

A key area of progress lies in AI-powered decision support ecosystems. Building on foundational work (Ahmed, Tewfik & Talaat, 2003; Adenso-Díaz et al., 2005), modern tools are increasingly leveraging uncertainty quantification and probabilistic reasoning via BNs (Chen et al., 2007; Park & Stenstrom, 2006) to enhance decision-making under complex, variable conditions. These systems enable resilient planning and more accurate scenario simulation.

To further address operational uncertainty and system complexity—already acknowledged as major limitations—advanced AI methods, such as stochastic modeling (Cabecinha et al., 2009) and fuzzy-neural controllers (Chen et al., 2003) are being adopted. These techniques support real-time process adaptation and stability under fluctuating conditions.

Process optimization and intelligent control are also evolving. Hybrid models and neural networks (Luccarini et al., 2010; Uraikul et al., 2007) now enable precise aeration, chemical dosing, and fault detection. The introduction of digital twins—virtual replicas of treatment systems continuously updated with sensor and IoT data (Rodríguez-Alonso et al., 2024)—further enhances simulation-based decision-making, predictive maintenance, and emergency response planning.

Meanwhile, innovations in monitoring and sensing are creating smarter, more agile treatment infrastructure. Real-time image analysis of activated sludge (Khan et al., 2017) and renewable-powered UV disinfection (Close & Jasper Ip, 2006) demonstrate how AI-integrated devices can support decentralized and energy-efficient treatment approaches. These developments directly address current challenges in sensor accuracy and data reliability. Ensuring sustainability remains central. AI is now embedded in MBR management, optimizing filtration and predicting fouling (Kamali et al., 2021; Viet et al., 2022), thereby reducing downtime and energy consumption. Such advances align with long-term goals of lifecycle optimization and environmental stewardship. A critical future concern is ensuring equitable deployment. Addressing infrastructure and digital literacy gaps in low- and middle-income regions is essential (Capodaglio & Callegari, 2025). Before complex AI systems can be deployed, foundational elements like basic sensors (Martini & Roni, 2021) and public health surveillance tools (Rathour et al., 2022) must be established.

Progress also hinges on data integration and governance. As previously discussed, data fragmentation, lack of interoperability, and privacy issues impede scalability. Going forward, AI governance must emphasize transparency, stakeholder participation, and adaptive policy—drawing from examples like Japan's AI governance model (Takeda et al., 2021)—to align innovation with ethical and regulatory standards. Lastly, multidisciplinary research will be key. New approaches like ensemble modeling, jittered data synthesis (Narayanan et al., 2024; Nourani, Asghari, & Sharghi, 2021; Nourani et al., 2018), smart infrastructure, and digital ethics frameworks are enhancing both model generalizability and social acceptance. Predictive tools for pollution tracking and sustainability scoring will further support informed, accountable water management (Fig. 20.5).

Acknowledgement

The authors are thankful to the Principal of Sri Venkateswara College, University of Delhi, for his continuous encouragement and support, as well as for providing the infrastructure in the college to carry out a hassle-free literature survey required for writing the chapter.

References

Adenso-Díaz, B., Tuya, J., & Goitia, M. (2005). EDSS for the evaluation of alternatives in waste water collecting systems design. *Environmental Modelling and Software*. 13648152. *20*(5), 639–649. doi:10.1016/j.envsoft.2004.03.008.

Ahmed, S. A., Tewfik, S. R., & Talaat, H. A. (2003). Development and verification of a decision support system for the selection of optimum water reuse schemes. *Desalination*, *152*(1-3), 339–352. doi:10.1016/S0011-9164(02)01082-2.

Alexandridis, A. K., & Zapranis, A. D. (2013). Wavelet neural networks: a practical guide. *Neural Networks*. 18792782. *42*, 1–27. doi:10.1016/j.neunet.2013.01.008.

Alprol, A. E., Mansour, A. T., Ibrahim, M. E. E. D., & Ashour, M. (2024). Artificial intelligence technologies revolutionizing wastewater treatment: current trends and future prospective. *Water (Switzerland)*. 20734441. *16*(2). Available from http://www.mdpi.com/journal/water. doi:10.3390/w16020314.

Altowayti, W. A. H., Shahir, S., Othman, N., Eisa, T. A. E., Yafooz, W. M. S., Al-Dhaqm, A., … Ali, A. (2022). The role of conventional methods and artificial intelligence in the wastewater treatment: a comprehensive review. *Processes*. 22279717. *10*(9). Available from http://www.mdpi.com/journal/processes. doi:10.3390/pr10091832.

Alzahrani, A. I. A., Chauhdary, S. H., & Alshdadi, A. A. (2023). Internet of Things (IoT)-based wastewater management in smart cities. *Electronics*. 2079-9292. *12*(12), 2590. doi:10.3390/electronics12122590.

Bartsch, K., Pettke, A., Hübert, A., Lakämper, J., & Lange, F. (2021). On the digital twin application and the role of artificial intelligence in additive manufacturing: a systematic review. *Journal of Physics: Materials*. 2515-7639. *4*(3), 032005. doi:10.1088/2515-7639/abf3cf.

Bertini, L., Bucchi, F., Frendo, F., Moda, M., & Monelli, B. D. (2019). Residual stress prediction in selective laser melting. *The International Journal of Advanced Manufacturing Technology*. 0268-3768. *105*(1-4), 609–636. doi:10.1007/s00170-019-04091-5.

Bromley, J., Jackson, N. A., Clymer, O. J., Giacomello, A. M., & Jensen, F. V. (2005). The use of Hugin® to develop Bayesian networks as an aid to integrated water resource planning. *Environmental Modelling and Software*. 13648152. *20*(2), 231–242. doi:10.1016/j.envsoft.2003.12.021.

Byaruhanga, N., Kibirige, D., Gokool, S., & Mkhonta, G. (2004). Evolution of flood prediction and forecasting models for flood early warning systems: A scoping review. *Water*, *16*(13), 1763.

Cabecinha, E., Cortes, R., Pardal, M. A., & Cabral, J. A. (2009). A Stochastic Dynamic Methodology (StDM) for reservoir's water quality management: validation of a multi-scale approach in a south European basin (Douro, Portugal). *Ecological Indicators*. 1470160X. *9*(2), 329–345. doi:10.1016/j.ecolind.2008.05.010.

Cai, Y., Ben, T., Zaidi, A. A., Shi, Y., Zhang, K., Lin, A., & Liu, C. (2019). Effect of pH on pollutants removal of ship sewage treatment in an innovative aerobic-anaerobic micro-sludge MBR system. *Water, Air, and Soil Pollution*. 15732932. *230*(7). Available from http://www.kluweronline.com/issn/0049-6979/. doi:10.1007/s11270-019-4211-0.

Capodaglio, A. G., & Callegari, A. (2025). Use, potential, needs, and limits of AI in wastewater treatment applications. *Water*. 2073-4441. *17*(2), 170. doi:10.3390/w17020170.

Caulfield, H. J. (2005). Information Processing | Optical Neural Networks.

Chan, C. W., & Huang, G. H. (2003). Artificial intelligence for management and control of pollution minimization and mitigation processes. *Engineering Applications of Artificial Intelligence*. 09521976. *16*(2), 75–90. doi:10.1016/S0952-1976(03)00062-9.

Charef, A., Ghauch, A., Baussand, P., & Martin-Bouyer, M. (2000). Water quality monitoring using a smart sensing system. *Measurement*. 02632241. *28*(3), 219–224. doi:10.1016/s0263-2241(00)00015-4.

Chen, C. F., Ma, H. W., & Reckhow, K. H. (2007). Assessment of water quality management with a systematic qualitative uncertainty analysis. *Science of the Total Environment*. 00489697. *374*(1), 13–25. doi:10.1016/j.scitotenv.2006.12.027.

Chen, S. H., Jakeman, A. J., & Norton, J. P. (2008). Artificial Intelligence techniques: an introduction to their use for modelling environmental systems. *Mathematics and Computers in Simulation*. 03784754. *78*(2–3), 379–400. doi:10.1016/j.matcom.2008.01.028.

Chen, W. C., Chang, N. B., & Chen, J. C. (2003). Rough set-based hybrid fuzzy-neural controller design for industrial wastewater treatment. *Water Research*. 00431354. *37*(1), 95–107. Available from www.elsevier.com/locate/watres. doi:10.1016/S0043-1354(02)00255-5.

Chester, M., Underwood, B. S., Allenby, B., Garcia, M., Samaras, C., Markolf, S., … Miller, T. R. (2021). Infrastructure resilience to navigate increasingly uncertain and complex conditions in the Anthropocene. *Npj Urban Sustainability*. 26618001. *1*(1). Available from https://www.nature.com/npjurbansustain/. doi:10.1038/s42949-021-00016-y.

Cho, J. H., Seok Sung, K., & Ryong Ha, S. (2004). A river water quality management model for optimising regional wastewater treatment using a genetic algorithm. *Journal of Environmental Management*. 03014797. *73*(3), 229–242. Available from https://www.sciencedirect.com/journal/journal-of-environmental-management. doi:10.1016/j.jenvman.2004.07.004.

Close, J., & Jasper Ip, K. H. L. (2006). Water recycling with PV-powered UV-LED disinfection. *Renewable Energy*. 09601481. *31*(11), 1657–1664. doi:10.1016/j.renene.2005.08.034.

Colwell, M., & Abolghasemi, M. (2024). Digital Twins for forecasting and decision optimisation with machine learning: applications in wastewater treatment. *arXiv*. 23318422. Available from: https://arxiv.org. doi:10.48550/arXiv.2404.14635.

Curl, J. M., Nading, T., Hegger, K., Barhoumi, A., & Smoczynski, M. (2019). Digital Twins: the Next Generation of Water Treatment Technology. *Journal AWWA*, *111*(12), 44–50. doi:10.1002/awwa.1413.

Curreri, F., Fiumara, G., & Xibilia, M. G. (2020). Input selection methods for soft sensor design: A survey. *Future Internet*, *12*(6), 97.

DeNicola, E., Aburizaiza, O. S., Siddique, A., Khwaja, H., & Carpenter, D. O. (2015). Climate change and water scarcity: the case of Saudi Arabia. *Annals of Global Health*. 22149996. *81*(3), 342–353. Available from https://agh.ubiquitypress.com/. doi:10.1016/j.aogh.2015.08.005.

Dey, A. (2016). Machine learning algorithms: a review. *International Journal of Computer Science and Information Technologies*, *7*(3), 1174–1179.

Do, H. T., Bach, N. V., Nguyen, L. V., Tran, H. T., & Nguyen, M. T. (2021). A design of higher-level control based genetic algorithms for wastewater treatment plants. *Engineering Science and Technology, an International Journal*. 22150986. *24*(4), 872–878. Available from: www.journals.elsevier.com/engineering-science-and-technology-an-international-journal/. doi:10.1016/j.jestch.2021.01.004.

Drenoyanis, A., Raad, R., Wady, I., & Krogh, C. (2019). Implementation of an IoT Based Radar Sensor Network for Wastewater Management. *Sensors*. 1424-8220. *19*(2), 254. doi:10.3390/s19020254.

Frontistis, Z., Lykogiannis, G., & Sarmpanis, A. (2023). Artificial Neural Networks in Membrane Bioreactors: a Comprehensive Review—Overcoming Challenges and Future Perspectives. *Sci*. 2413-4155. *5*(3), 31. doi:10.3390/sci5030031.

Gaya, M. S., Wahab, N. A., Sam, Y. M., & Samsudin, S. I. (2014). Anfis modelling of carbon and nitrogen removal in domestic wastewater treatment plant. *Jurnal Teknologi (Sciences and Engineering)*. 21803722. *67*(5), 29–34. Available from http://www.jurnalteknologi.utm.my/index.php/jurnalteknologi/article/download/2839/2134. doi:10.11113/jt.v67.2839Malaysia.

Govindarajan, L., Kumar, S. K., & Karunanithi, T. (2025). Optimal design of wastewater treatment plant using adaptive simulated annealing.

Güçlü, D., & Dursun, Ş. (2010). Artificial neural network modelling of a large-scale wastewater treatment plant operation. *Bioprocess and Biosystems Engineering*. 1615-7591. *33*(9), 1051–1058. doi:10.1007/s00449-010-0430-x.

Han, H., & Qiao, J. (2010). A Self-Organizing Fuzzy Neural Network Based on a Growing-and-Pruning Algorithm. *IEEE Transactions on Fuzzy Systems*. 1063-6706. *18*(6), 1129–1143. doi:10.1109/tfuzz.2010.2070841.

Henze, M., & Comeau, Y. (2008). Wastewater characterization . Biological wastewater treatment: principles modelling and design. 27.

Honggui, H., Ying, L., & Junfei, Q. (2014). A fuzzy neural network approach for online fault detection in waste water treatment process. *Computers & Electrical Engineering*. 00457906. *40*(7), 2216–2226. doi:10.1016/j.compeleceng.2014.08.011.

Jang, J. S. R., & Sun, C. T. (1995). Neuro-Fuzzy Modeling and Control. *Proceedings of the IEEE*. 15582256. *83*(3), 378–406. doi:10.1109/5.364486.

Janssen, M. A., Goosen, H., & Omtzigt, N. (2006). A simple mediation and negotiation support tool for water management in the Netherlands. *Landscape and Urban Planning*. 01692046. *78*(1–2), 71–84. Available from www.elsevier.com/inca/publications/store/5/0/3/3/4/7. doi:10.1016/j.landurbplan.2005.05.005.

Jawad, J., Hawari, A. H., & Javaid Zaidi, S. (2021). Artificial neural network modeling of wastewater treatment and desalination using membrane processes: a review. *Chemical Engineering Journal*. 13858947. *419*. Available from www.elsevier.com/inca/publications/store/6/0/1/2/7/3/index.htt. doi:10.1016/j.cej.2021.129540.

Jimei, Q., Hou, Y., Jiwei, H., Ruan, W., Xiang, Y., & Wei, X. (2020). Decontamination of methylene Blue from simulated wastewater by the mesoporous rGO/Fe/Co nanohybrids: artificial intelligence modeling and optimization. *Materials Today Communications*. 23524928. *24*, 100709. doi:10.1016/j.mtcomm.2019.100709.

Kamali, M., Appels, L., Xiaobin, Y., Aminabhavi, T. M., & Dewil, R. (2021). Artificial intelligence as a sustainable tool in wastewater treatment using membrane bioreactors. *Chemical Engineering Journal*. 13858947. *417*, 128070. doi:10.1016/j.cej.2020.128070.

Khan, M. B., Nisar, H., & Ng, C. A. (2017). Image Processing and Analysis of Phase-Contrast Microscopic Images of Activated Sludge to Monitor the Wastewater Treatment Plants. *IEEE Access*.21693536. *6*, 1778–1791. Available from http://ieeexplore.ieee.org/xpl/RecentIssue.jsp?punumber=6287639. doi:10.1109/ACCESS.2017.2780249.

Krovvidy, S., Wee, W. G., Summers, R. S., & Coleman, J. J. (1991). An AI approach for wastewater treatment systems. *Applied Intelligence*. 15737497. *1* (3), 247–261. doi:10.1007/BF00118999.

Lee, M. W., Hong, S. H., Choi, H., Kim, J. H., Lee, D. S., & Park, J. M. (2008). Real-time remote monitoring of small-scaled biological wastewater treatment plants by a multivariate statistical process control and neural network-based software sensors. *Process Biochemistry*. 13595113. *43* (10), 1107–1113. doi:10.1016/j.procbio.2008.06.002.

Lowe, M., Qin, R., & Mao, X. (2022). A review on machine learning, artificial intelligence, and smart technology in water treatment and monitoring. *Water*. 2073-4441. *14*(9), 1384. doi:10.3390/w14091384.

Luccarini, L., Bragadin, G. L., Colombini, G., Mancini, M., Mello, P., Montali, M., & Sottara, D. (2010). Formal verification of wastewater treatment processes using events detected from continuous signals by means of artificial neural networks. Case study: SBR plant. *Environmental Modelling and Software*. 13648152. *25*(5), 648–660. doi:10.1016/j.envsoft.2009.05.013.

Luo, Y., Wei, S., Rabbi, M. F., Wan, Q., Dewen, X., Wang, Z., ... Jian, W. (2024). Quantitative analysis of microplastics in water environments based on Raman spectroscopy and convolutional neural network. *Science of the Total Environment*. 00489697. *926*, 171925. doi:10.1016/j.scitotenv.2024.171925.

Ma, Z., Cheah, W. Y., Ng, I. S., Chang, J. S., Zhao, M., & Show, P. L. (2022). Microalgae-based biotechnological sequestration of carbon dioxide for net zero emissions. *Trends in Biotechnology*. 18793096. *40*(12), 1439–1453. Available from www.elsevier.com/locate/tibtech. doi:10.1016/j.tibtech.2022.09.002.

Maier, H. R., Kapelan, Z., Kasprzyk, J., Kollat, J., Matott, L. S., Cunha, M. C., Dandy, G. C., Gibbs, M. S., Keedwell, E., Marchi, A., & Ostfeld, A. (2014). Evolutionary algorithms and other metaheuristics in water resources: Current status, research challenges and future directions *Environmental Modelling & Software*, 62, 271–299.

Malviya, A., & Jaspal, D. (2021). Artificial intelligence as an upcoming technology in wastewater treatment: a comprehensive review. *Environmental Technology Reviews*. 2162-2515. *10*(1), 177–187. doi:10.1080/21622515.2021.1913242.

Martini, S., & Roni, K. A. (2021). The existing technology and the application of digital artificial intelligent in the wastewater treatment area: a review paper. *Journal of Physics: Conference Series*. 1742-6588. *1858*(1), 012013. doi:10.1088/1742-6596/1858/1/012013.

Mihai, S., Yaqoob, M., Hung, D. V., Davis, W., Towakel, P., Raza, M., … Nguyen, H. X. (2022). Digital Twins: a Survey on Enabling Technologies, Challenges, Trends and Future Prospects. *IEEE Communications Surveys and Tutorials*. 1553877X. *24*(4), 2255–2291. Available from http://ieeexplore.ieee.org/xpl/RecentIssue.jsp?punumber=9739. doi:10.1109/COMST.2022.3208773.

Narayanan, D., Bhat, M., Paul, S., Khatri, N., & Saroliya, A. (2024). Artificial intelligence driven advances in wastewater treatment: evaluating techniques for sustainability and efficacy in global facilities. *Desalination and Water Treatment*. 19443986. *320*(100618). doi:10.1016/j.dwt.2024.100618.

Noshadi, I., Salahi, A., Hemmati, M., Rekabdar, F., & Mohammadi, T. (2013). Experimental and ANFIS modeling for fouling analysis of oily wastewater treatment using ultrafiltration. *Asia-Pacific Journal of Chemical Engineering*. 1932-2135. *8*(4), 527–538. doi:10.1002/apj.1691.

Nourani, V., Asghari, P., & Sharghi, E. (2021). Artificial intelligence based ensemble modeling of wastewater treatment plant using jittered data. *Journal of Cleaner Production*. 09596526. *291*, 125772. doi:10.1016/j.jclepro.2020.125772.

Nourani, V., Elkiran, N., & Abba, S. I. (2018). Wastewater treatment plant performance analysis using artificial intelligence – an ensemble approach. *Water Science and Technology*. 0273-1223. *78*(10), 2064–2076. doi:10.2166/wst.2018.477.

Orgoványi, P., Dalkó, I., Vadkerti, E., & Bognár, F. (2020). Basics of wastewater treatment. *University of Public Service Faculty of Water Sciences*, *13*, 13.

Pandey, S., Twala, B., Singh, R., Gehlot, A., Singh, A., Montero, E. C., & Priyadarshi, N. (2022). Wastewater treatment with technical intervention inclination towards smart cities. *Sustainability*. 2071-1050. *14*(18), 11563. doi:10.3390/su141811563.

Park, M. H., & Stenstrom, M. K. (2006). Using satellite imagery for stormwater pollution management with Bayesian networks. *Water Research*. 00431354. *40*(18), 3429–3438. Available from www.elsevier.com/locate/watres. doi:10.1016/j.watres.2006.06.041.

PK, F. A. (1984). What is artificial intelligence? Success is no accident. It is hard work, perseverance, learning, studying, sacrifice and most of all, love of what you are doing or learning to do. 65.

Plathottam, S. J., Rzonca, A., Lakhnori, R., & Iloeje, C. O. (2023). A review of artificial intelligence applications in manufacturing operations. *Journal of Advanced Manufacturing and Processing*. 2637403X. *5*(3). Available from https://onlinelibrary.wiley.com/journal/2637403x. doi:10.1002/amp2.10159.

Rathour, R. K., Sakhuja, D., Bhatt, A. K., & Bhatia, R. K. (2022). Municipal Wastewater Connection for Water Crisis and Jaundice Outbreaks in Shimla City: present Findings and Future Solutions. *International Journal of Environmental Research and Public Health*. 16604601. *19*(18). Available from http://www.mdpi.com/journal/ijerph. doi:10.3390/ijerph191811266.

REN, N. Q., YAN, X. F., CHEN, Z. B., HU, D. X., GONG, M. L., & GUO, W. Q. (2007). Feasibility and simulation model of a pilot scale membrane bioreactor for wastewater treatment and reuse from Chinese traditional medicine. *Journal of Environmental Sciences*. 10010742. *19*(2), 129–134. doi:10.1016/S1001-0742(07)60021-7.

Rodríguez-Alonso, C., Pena-Regueiro, I., & García, Ó. (2024). Digital twin platform for water treatment plants using microservices architecture. *Sensors*. 1424-8220. *24* (5), 1568. doi:10.3390/s24051568.

Rutenbar, R. A. (1989). Simulated annealing algorithms: an overview. *IEEE Circuits and Devices Magazine*. 8755-3996. *5* (1), 19–26. doi:10.1109/101.17235.

Saravi, S., Kalawsky, R., Joannou, D., Casado, M. R., Guangtao, F., & Meng, F. (2019). Use of artificial intelligence to improve resilience and preparedness against adverse flood events. *Water*. 2073-4441. *11*(5), 973. doi:10.3390/w11050973.

Takeda, T., Kato, J., Matsumura, T., Murakami, T., & Abeynayaka, A. (2021). Governance of artificial intelligence in water and wastewater management: the case study of Japan. *Hydrology*. 2306-5338. *8*(3), 120. doi:10.3390/hydrology8030120.

Tarfeen, N., Nisa, K. U., Hamid, B., Bashir, Z., Yatoo, A. M., Dar, M. A., … Sayyed, R. Z. (2022). Microbial remediation: a promising tool for reclamation of contaminated sites with special emphasis on heavy metal and pesticide pollution: a review. *Processes*. 22279717. *10*(7). Available from http://www.mdpi.com/journal/processes. doi:10.3390/pr10071358.

Uraikul, V., Chan, C. W., & Tontiwachwuthikul, P. (2007). Artificial intelligence for monitoring and supervisory control of process systems. *Engineering Applications of Artificial Intelligence*. 09521976. *20*(2), 115–131. doi:10.1016/j.engappai.2006.07.002.

Viet, N. D., Jang, D., Yoon, Y., & Jang, A. (2022). Enhancement of membrane system performance using artificial intelligence technologies for sustainable water and wastewater treatment: a critical review. *Critical Reviews in Environmental Science and Technology*. 15476537. *52*(20), 3689–3719. Available from www.tandf.co.uk/journals/titles/10643389.asp. doi:10.1080/10643389.2021.1940031.

Wang, P., Yao, J., Wang, G., Hao, F., Shrestha, S., Xue, B., … Peng, Y. (2019). Exploring the application of artificial intelligence technology for identification of water pollution characteristics and tracing the source of water quality pollutants. *Science of the Total Environment*. 00489697. *693*, 133440. doi:10.1016/j.scitotenv.2019.07.246.

Wang, Y., Cheng, Y., Liu, H., Guo, Q., Dai, C., Zhao, M., & Liu, D. (2023). A review on applications of artificial intelligence in wastewater treatment. *Sustainability*. 2071-1050. *15*(18), 13557. doi:10.3390/su151813557.

Wen, C. H., & Vassiliadis, C. A. (1998). Applying hybrid artificial intelligence techniques in wastewater treatment. *Engineering Applications of Artificial Intelligence*. 09521976. *11*(6), 685–705. Available from https://www.journals.elsevier.com/engineering-applications-of-artificial-intelligence. doi:10.1016/s0952-1976(98)00036-0.

Yeung, K., Howes, A., & Pogrebna, G. (2020). *AI Governance by Human Rights–Centered Design, Deliberation, and Oversight*. (pp. 76–106). Oxford University Press (OUP). doi:10.1093/oxfordhb/9780190067397.013.5.

Zaghloul, M. S., Iorhemen, O. T., Hamza, R. A., Tay, J. H., & Achari, G. (2021). Development of an ensemble of machine learning algorithms to model aerobic granular sludge reactors. *Water Research*. 18792448. *189*. Available from www.elsevier.com/locate/watres. doi:10.1016/j.watres.2020.116657.

Zeferino, J. A., Antunes, A. P., & Cunha, M. C. (2009). An efficient simulated annealing algorithm for regional wastewater system planning. *Computer-Aided Civil and Infrastructure Engineering*. 14678667. *24*(5), 359–370. doi:10.1111/j.1467-8667.2009.00594.x.

Zhao, L., Dai, T., Qiao, Z., Sun, P., Hao, J., & Yang, Y. (2020). Application of artificial intelligence to wastewater treatment: a bibliometric analysis and systematic review of technology, economy, management, and wastewater reuse. *Process Safety and Environmental Protection*. 09575820. *133*, 169–182. doi:10.1016/j.psep.2019.11.014.

Chapter 21

Artificial intelligence-driven value creation and investment strategy in healthcare innovation

Bidyut Sarkar[a] **and Rudrendu Kumar Paul**[b]
[a]*IBM, Armonk, NY, United States*
[b]*Independent Consultant, San Francisco, CA, United States*

1 Introduction

Artificial intelligence (AI) now permeates discovery laboratories, hospital operations, and population health surveillance, generating measurable clinical and operational benefits. Studies document reductions of up to 40% in 30-day readmissions, 30% in diagnostic error rates, and 20% in drug development timelines, providing empirical evidence that digital capability and economic value are closely linked (Deloitte, 2025; Bennett et al., 2025; Haas, 2022; Mortazavi, 2024). Yet, traditional return-on-investment (ROI) templates remain fixated on short-term cost savings and therefore fail to capture gains in patient safety, workforce resilience, and long-term system capacity (Bharadwaj, 2024).

A strategically robust ROI construct must therefore expand beyond narrow financial accounting to capture operational performance, risk mitigation, and societal benefit. By embedding AI-enabled forecasts within this broader valuation lens, decision-makers can continuously recalibrate capital allocation in response to real-time evidence, ensuring that investment priorities remain aligned with clinical effectiveness and health equity mandates.

AI-centered ROI analysis promotes organizational learning. Platforms that generate live dashboards and scenario simulations enable executives to test alternative deployment strategies, quantify opportunity costs, and redeploy resources swiftly, transforming ROI from a static report into a governance instrument that combines financial discipline with innovation agility.

In summary, ROI modeling in AI-guided healthcare has evolved beyond static financial evaluation. It now operates as a strategic tool that aligns diverse goals, clinical outcomes, fiscal accountability, and public value through dynamic, data-driven forecasting. Especially in areas where commercial incentives are weak, such as antimicrobial resistance (AMR) or universal antivenoms, ROI models supported by AI can justify direct investment. Furthermore, the inclusion of equity metrics and horizon-based strategies ensures that healthcare innovation is both sustainable and socially responsive.

1.1 Why is ROI central to healthcare investment decisions?

Latest research and development expenditure highlights the urgency of disciplined capital stewardship. Deloitte estimates place the fully loaded cost of bringing a single therapy to market at roughly USD 2–3 billion, intensifying pressure on firms to substantiate each investment decision with rigorous ROI evidence (Deloitte, 2025). In an outcomes-based reimbursement environment, ROI has become a crucial factor through which investors, providers, payers, and regulators determine which innovations progress and which stall.

Moreover, ROI functions as a common yardstick that aligns heterogeneous stakeholders. For investors, it signals capital efficiency; for clinicians and administrators, it translates technical performance into a rationale for resource allocation; and for regulators, it contextualizes risk–benefit judgments within broader system sustainability. A comprehensive,

Applications of Artificial Intelligence in Healthcare. https://doi.org/10.1016/B978-0-443-43934-6.00036-5

TABLE 21.1 Key drivers of ROI in healthcare innovation.

Reason	Key insight
Escalating costs	Drug development now exceeds $2B per compound; ROI is vital for capital efficiency.
Strategic prioritization	ROI aligns stakeholders by quantifying expected value across initiatives.
Value-based models	Success is measured by patient outcomes and cost-effectiveness, not service volume.
Equity and impact	Modern ROI must account for health equity, access, and long-term public good.

AI-enhanced ROI framework, therefore, serves not merely as a financial calculator but as a strategic compass that reconciles scientific ambition with fiscal responsibility and public health equity.

Table 21.1 summarizes the primary factors driving the growing importance of ROI in healthcare innovation.

1.2 Limitations of traditional ROI frameworks

Traditional ROI models in healthcare and the life sciences are often rooted in financial templates borrowed from sectors such as manufacturing and banking. While effective for tracking short-term gains, they overlook critical dimensions that are important in healthcare.

A. One limitation is their narrow financial focus. These models typically emphasize direct cost savings, ignoring important but less tangible outcomes such as improved clinical decision-making, stronger patient trust, and reduced clinician burnout. For example, early studies on AI medical scribes have shown a 20% reduction in note-taking time and a 30% decrease in after-hours work, which helps ease clinician workload; however, these benefits are not always reflected in standard ROI calculations (Eanes, 2025; Fusch, 2001).

B. Another key issue is the short time horizon used for evaluation. Most ROI models assess returns within 12–24 months. However, transformative innovations like genomic screening or AI-enabled preventive care often require longer timeframes to deliver meaningful impact. A prime example is the Human Genome Project, which generated an estimated $796 billion in economic output from a $3.8 billion public investment, yielding an ROI of over 140 to 1 (Tripp & Grueber, 2011).

C. Traditional models also fail to account for scalability. They often do not measure how well a solution performs across diverse hospital systems, geographic regions, or patient populations. This is especially problematic when evaluating life sciences innovations that are intended for widespread implementation.

D. Risk reduction is another area that is underestimated. Technologies designed to prevent compliance breaches, medical errors, or data security threats are critical to system integrity, yet their preventive value rarely appears in legacy ROI metrics.

E. Workforce well-being is also largely excluded from standard ROI assessments, despite mounting evidence that burnout has financial and safety consequences. Physician burnout in the United States has reached levels exceeding 60% in some specialties. Solutions like ambient AI medical scribes have been shown to reduce burnout by as much as 40%–63%, while also restoring work–life balance and enhancing patient engagement (Lin, 2025; Tali, 2024).

F. Finally, traditional ROI frameworks often overlook the broader value of R&D and diagnostics. Projects like the Human Genome Project not only advance research but also spur economic activity and long-term public health benefits, none of which are captured in near-term financial models (Gitlin, 2013).

These blind spots limit the effectiveness of ROI as a tool for strategic planning and innovation funding. To fully reflect the value of healthcare and life sciences innovations, ROI models must evolve to encompass long-term outcomes, systemic impacts, workforce sustainability, and societal benefits.

1.3 AI applications in enhancing capital efficiency and accelerating outcomes

AI holds a dual promise for healthcare by enhancing capital efficiency and accelerating clinical outcomes. By streamlining routine tasks, improving predictive accuracy, and enabling smarter resource allocation, AI allows health systems to achieve more with fewer resources while advancing the quality and speed of care.

1.3.1 Capital efficiency and ROI

AI is playing an increasingly important role in improving capital efficiency across healthcare systems. By minimizing operational waste and optimizing how resources are used, AI contributes directly to stronger and more measurable returns on investment. For instance, AI-based scheduling tools have been shown to reduce nurse overtime by 20% and increase staff allocation accuracy by 30%, resulting in significant labor cost savings (Choi, 2023; Tisoskey, 2025).

Natural language processing technologies are easing the administrative burden on clinicians. Studies report a 30% reduction in after-hours documentation, which not only improves provider productivity but also supports clinician well-being (Eanes, 2025).

In addition, predictive maintenance powered by AI helps reduce equipment downtime and extend the lifespan of critical assets, further contributing to operational and financial efficiency. Similarly, in pharmaceutical R&D, AI accelerates trial design, automates data analysis, and shortens development timelines, improving the cost-efficiency of each dollar invested in innovation.

1.3.2 Outcome acceleration with measurable ROI

AI improves clinical performance and delivers clear financial and operational returns:

- **Reduced hospital stays and readmissions**
- A Swedish study on sepsis prediction showed a 32% reduction in hospital length of stay, a 22% decrease in 30-day readmissions, and a 39.5% drop in in-hospital mortality after AI implementation. These outcomes reduce overall treatment costs and help avoid financial penalties for readmissions (Haas, 2022).
- **Savings from early detection in critical care**
- In Swedish intensive care units, the NAVOY Sepsis AI tool reduced ICU costs by 76 euros per patient by cutting 0.16 ICU days on average. This led to an estimated national annual saving of nearly 2.8 million euros (Ericson, 2022).
- **Preserved pay-for-performance funding**
- A safety-net health system used AI risk prediction to lower 30-day readmissions from 27.9% to 23.9%. This not only improved outcomes but also preserved $7.2 million in value-based care payments (Bennett et al., 2025).
- **Cost-effective cancer screening**
- An AI-based, risk-stratified breast cancer screening program in the United Kingdom showed an annual net benefit of between 60 and 85 million pounds, improving both economic efficiency and patient outcomes (Hill et al., 2024).

Another study reported that combining AI with radiologists could reduce mammography screening costs by up to 30% (Ahsen et al., 2025).

- **Productivity improvements for radiologists**
- One analysis found that diagnostic AI tools delivered up to a 791% ROI by saving >15 full workdays per week in radiologist time (Bharadwaj, 2024).

AI not only reduces capital costs and accelerates outcomes, but it also becomes more effective over time. Unlike static investments such as equipment or buildings, AI systems continuously learn and improve through data exposure. This leads to increased accuracy, speed, and clinical value, generating an ROI that compounds over time. When assessed using ROI frameworks that account for long-term efficiency, scalability, and system-wide impact, AI emerges as a strategic enabler of both financial sustainability and clinical performance in healthcare.

1.3.3 Advanced AI capabilities in guiding ROI realization

Building on its foundational role in capital efficiency and clinical acceleration, AI also strengthens how healthcare organizations forecast, operate, and invest across the value chain.

Real-time and adaptive ROI forecasting

AI supports real-time alerts to ROI models by integrating live inputs such as disease incidence rates, trial results, regulatory changes, and market trends. For instance, a 2024 study showed that AI-enhanced demand forecasting improved accuracy by 25% in a life sciences setting, allowing for more precise resource planning (Tang, 2025) (Flevy via life sciences case study).

Similarly, machine learning models have improved emergency department visit forecasts by around 23%–27%, supporting better staffing and cost control (Vollmer, 2020).

AI across the R&D-to-commercialization continuum

AI drives measurable impact throughout the healthcare innovation process:

1. In discovery and preclinical R&D, machine learning reduces early-stage failure rates by up to 20% through enhanced compound and toxicity screening (DeFrancesco, 2022).

2. In clinical trials, adaptive designs and synthetic control arms shorten timelines by 30% and reduce recruitment costs by 25% (BiopharmaTrend, 2024; Kaylor, 2025).
3. In manufacturing, predictive maintenance decreases equipment downtime by 40% and lowers maintenance costs by 20% (GE HealthCare, 2024).
4. In commercialization, AI-enhanced segmentation and customer relationship management (CRM) can boost launch ROI by 10%–15% (Deveau et al., 2023).

Operational acceleration and time-to-value

Technologies such as digital twins and predictive monitoring enable the simulation of patient responses, the adjustment of protocols, and the detection of compliance risks in near real time. These applications can reduce time to market by up to 30% and increase the time-adjusted net present value (NPV) of pipeline assets (Schubarth, 2023).

Portfolio optimization and capital reallocation

AI-enabled portfolio dashboards give decision-makers the ability to compare programs using live inputs such as cost, timeline, and probability of success. These tools allow for more informed go or no-go decisions and flexible capital allocation. One study reported a 12% increase in portfolio internal rate of return (IRR) following the adoption of AI-informed capital reallocation strategies (Jordan, 2025).

2 How financial metrics and AI modeling support smarter investment strategies

AI is changing the way organizations calculate ROI, especially in the pharmaceutical and healthcare fields. This section explains how AI is helping companies move away from fixed, historical projections toward real-time, data-driven financial models. AI improves revenue forecasting by using current information about pricing, patient populations, and the competitive landscape. As a result, forecasts are 60%–80% more accurate, according to the (Dosanjh, 2025). With better insights, companies can make smarter investment decisions and achieve stronger returns (Dosanjh, 2025).

This section focuses on how AI is transforming financial modeling by helping companies shift from fixed, historical projections to real-time, data-driven approaches that improve forecasting accuracy and investment decisions. In contrast, AI's role in reducing operational costs and improving efficiency, especially after implementation, such as shortening clinical trial timelines and lowering expenses, is explored in detail in Section 3. These improvements also contribute significantly to strengthening overall ROI.

2.1 Financial calculations

At the core of ROI analysis lie three essential metrics: ROI, NPV, and IRR. These metrics have traditionally been used to assess the viability and profitability of projects. However, their effectiveness is often limited because they rely on simplified inputs and fixed assumptions that do not fully reflect real-world complexity.

2.1.1 Return on investment

ROI = (net gains/total investment) × 100

Here is what changes with AI:

- **Net gains**: Instead of relying on analyst projections, AI processes real-world pricing data, competitor launches, and market dynamics to forecast revenue with 60–80% greater accuracy (Dosanjh, 2025).

According to DiMasi et al. (2022), decentralized trials that leverage AI can reduce development costs and time-to-market, both of which directly affect ROI by improving capital turnover, impacting net gains, and reducing risk exposure.

- **Total investment**: AI delivers 70% cost savings per trial and 80% timeline reductions, dramatically reducing the denominator (Varol, 2025).
- **Result**: More accurate ROI calculations that reflect actual market conditions, not wishful thinking.

2.1.2 Net present value

NPV offers a time-adjusted estimate of future earnings by discounting cash flows back to their present value. AI enhances the accuracy of NPV forecasts by incorporating adaptive pricing models, updated phase-gate timelines, and real-time market data, which help move away from static assumptions and result in more precise present value calculations.

For example, consider a pharmaceutical company developing a new oncology drug with an expected timeline of 7 years. Traditionally, the uncertainty around trial phases and regulatory approvals would make financial forecasting difficult. By integrating AI, the company applies predictive modeling to identify promising compounds earlier in the development process, leading to a reduction in preclinical timelines by as much as 50%, as noted in Nature Reviews Drug Discovery (Jacobson, 2025).

The article highlights how companies like Insilico and Recursion have successfully used AI to shorten drug development cycles to as little as 12–18 months, demonstrating the real-world impact of these technologies.

As clinical trial data evolves, AI tools dynamically adjust the company's cash flow projections in real time. At the same time, platforms like IBM Watson help optimize the regulatory pathway by analyzing similar past drug approvals (Company Connect Consultancy, 2025).

In this case, using NPV alongside AI-generated insights allows the company to reassess the project's value more accurately, demonstrating how timeline compression can significantly improve expected returns.

2.1.3 *Internal rate of return*

IRR is a financial metric used to determine the discount rate at which a project's NPV becomes zero. In other words, it represents the expected rate of return that makes an investment break even in today's terms. The IRR becomes a more flexible comparison tool as AI simulates capital deployment across alternative strategies, enabling faster "go/no-go" decisions and smarter reallocation of resources.

For example, a healthcare company considers investing $20 million in a diagnostic imaging device, expecting returns over a 10-year period. Using traditional models, the IRR is calculated at 14%, just above the company's minimum return threshold. However, this estimate depends on fixed assumptions about pricing, regulatory timelines, and market uptake. By applying AI to update inputs such as competitor activity and reimbursement data, the IRR is refined to 16.2%, offering a more realistic assessment of the project's value.

Table 21.2 outlines the core financial metrics used in evaluating ROI for the healthcare industry. It demonstrates how AI enhances the utility and precision of each metric, while also noting the inherent limitations that must be considered in strategic decision-making.

While ROI, NPV, and IRR are core financial metrics used to evaluate the profitability and feasibility of a project, scenario analysis and sensitivity analysis are tools used to test and interpret these metrics under different assumptions and conditions.

2.2 Sensitivity and scenario analysis

In the healthcare and pharmaceutical industry, decision-making often involves uncertainty, long development timelines, and complex regulatory risks. Sensitivity and scenario analysis help organizations understand how changes in key variables, such as clinical outcomes, approval delays, or reimbursement shifts, can affect a project's financial viability.

TABLE 21.2 Financial metrics for ROI and the impact of AI.

Metric	Definition/formula	Usefulness in ROI analysis	AI-enabled enhancements	Limitations
Return on investment (ROI)	ROI = (net gains/total investment) × 100	• Provides a high-level snapshot of profitability • Useful for comparing multiple projects or investment scenarios	AI improves the accuracy of both net gain forecasts and total investment estimates through real-time data and predictive models	• Oversimplifies multi-phase risk • Ignores the time value of money • Sensitive to input volatility
Net present value (NPV)	Present value of future cash flows minus initial cost	• Accounts for the time value of money • Preferred for long-term capital budgeting decisions	• AI shortens time-to-market forecasts, improving present value • Real-world data updates increase model realism	• Highly sensitive to the discount rate • Requires accurate long-term revenue and cost estimates
Internal rate of return (IRR)	Discount rate at which NPV = 0	• Useful for comparing investments of different scales and durations • Indicates capital efficiency	AI simulates cash flow outcomes across strategic options, enhancing the comparability of early-stage and late-stage projects	• May have multiple IRRs or no IRR in complex models • Less meaningful if cash flows are irregular

For example, a pharmaceutical company might use sensitivity analysis to assess how fluctuations in trial success or manufacturing costs impact metrics like NPV or IRR. This helps identify critical risk factors. An National Institutes of Health (NIH) study highlights the value of this approach in testing model robustness.

Scenario analysis models full future conditions, allowing organizations to prepare for outcomes such as delayed approvals or limited market adoption. The World Health Organization supports its use in building more resilient health systems (World Health Organization, 2013).

As explained in Nature Medicine, these tools are increasingly used in health technology assessments to guide strategic planning and resource allocation (Regier, 2022).

2.3 Sensitivity analysis

Sensitivity analysis asks, "Which inputs move the needle most?" By varying one (or several) model parameters at a time and observing the change in NPV, IRR, or expected ROI, analysts discover the variables, e.g., Phase III success rate, launch price, or manufacturing yield, that dominate risk and upside. Health-technology-assessment manuals treat it as a standard step for testing the robustness of economic models because even small shifts in these high-leverage inputs can overturn a go/no-go decision (Chen, 2022).

In practice, teams rank individual drivers and then expand to multivariable or global sensitivity analysis (now readily automated by AI). For example, an NIH-funded valuation study demonstrated that halving the screen-failure rate in a pivotal trial raised expected NPV by more than reducing cost-of-capital or time-to-market, signaling that protocol design, not financing, was the primary lever (Levitan et al., 2018).

Sensitivity analysis can be used in early portfolio triage, protocol optimization, and communicating "what really matters" to nontechnical executives. Typical output: a ranked list of value drivers and threshold values at which the investment becomes unattractive.

2.3.1 Scenario analysis with Monte Carlo simulation

Scenario analysis explores, "What combinations of uncertainties could realistically occur, and how often?" The work-horse technique is a Monte Carlo simulation:

- Assign probability distributions to uncertain inputs (clinical attrition, timelines, peak sales, cost of goods sold (COGS)).
- Sample thousands of times to create plausible development/market paths.
- Aggregate the results into a probability distribution of ROI, NPV, and payback.

In drug R&D, Monte Carlo engines capture the compounding effect of correlated risks; for instance, slower enrollment often drags both timelines and costs; something deterministic spreadsheets cannot account for. Healthcare researchers have long used the method to run "virtual clinical trials" and stress test alternative dosing, market access, or pandemic disruption scenarios (Velikova, 2024).

AI enhances Monte Carlo simulations by:

- Learning input distributions directly from real-world data and historical trial archives.
- Running 10,000+ iterations in seconds on cloud GPUs, enabling on-the-fly scenario workshops.
- Detecting nonlinear interactions (e.g., the viral nature of rare adverse-event news on peak-sales).

Regulators and global health agencies even deploy AI-enabled scenario simulations to rehearse outbreak responses and cross-border supply chain shocks, emphasizing their value for resilience planning (World Health Organization Eastern Mediterranean Regional Committee, 2024).

Monte Carlo simulation can be used in late-stage portfolio allocation, licensing negotiations, contingency budgeting, and communicating upside/downside tails to boards and investors. Typical output: probability density of ROI with value-at-risk curves, scenario "playbooks," and risk-adjusted NPV.

Table 21.3 compares sensitivity analysis, which identifies the biggest value drivers, with Monte Carlo scenario simulation, which maps the full ROI distribution and shows how AI turbo-charges both through data-driven inputs and at scale.

2.4 Societal impact and social return on investment

Traditional ROI stops at balance-sheet gains, but AI makes it possible to quantify social returns as well.

One approach, inclusion modeling, analyzes large clinical and social datasets to predict how new therapies or delivery models will reach people who are usually left out of care. Machine-learning pipelines combine demographic data, insurance

TABLE 21.3 AI-powered techniques for sensitivity and scenario analysis.

Analysis techniques	Usage	AI Enhancements
Sensitivity analysis	• Pinpoints which inputs impact ROI the most • Used for early triage and concise executive updates	Automates multivariable sweeps and instant visual ranking of drivers
Scenario analysis with Monte Carlo simulation	• Maps the full probability distribution of ROI • Suited for late-stage risk budgeting and board-level reviews	Learns input distributions from real-world data and runs 10,000+ simulations at scale and speed

TABLE 21.4 Social ROI metrics and AI tools for broader impact.

Category	Element	Description
Equity	Access equity	Measures access improvements for underserved groups (e.g., rural or low-income populations).
	Inclusion	Evaluates the representation of diverse groups in trials and care delivery.
	Health disparity reduction	Quantifies improved health outcomes in historically underserved populations.
Public health	Productivity gains	Estimates increased workforce productivity and reduced caregiver burden.
	Healthcare cost avoidance	Calculates savings from fewer hospitalizations and emergency interventions.
	Long-term wellness metrics	Forecasts gains in QALYs and reductions in DALYs using AI-driven models.
AI tool	Geospatial AI	Forecasts regional health impacts using access and infrastructure data.
	Social determinants modeling	Integrates socioeconomic variables into ROI for deeper impact modeling.
	SROI heatmaps	Visualizes comparative social impacts across regions and populations.

status, language preference, and disease burden to forecast uptake and health outcomes for racial minorities, women, or low-income groups; systematic reviews show that such models expose disparities that raw efficacy numbers miss and guide redesigns that improve equity before full rollout (Marko, 2025).

A complementary technique, geospatial equity analysis, layers those predictions onto interactive maps. Geographic information systems powered by AI, such as Fadiel et al.'s cancer-disparity platform, pinpoint neighborhoods where the combination of high disease incidence and low service availability promises the greatest social return per investment dollar, allowing funders to triage resources quickly and transparently (Fadiel, Eichenbaum, Abbasi, Lee, & Kunle, 2024).

Finally, public health value modeling links clinical outcomes to macroeconomic effects, translating avoided hospitalizations or productivity gains into system-level savings. Recent econometric work estimates that wider AI adoption could shave 5%–10% off annual US healthcare spending—roughly US$200–360 billion—by optimizing prevention and resource allocation (Sahni, 2023).

By incorporating social determinants of health, real-world access barriers, and patient outcomes, AI tools enable organizations to quantify qualitative impacts, paving the way for ROI models that support mission-aligned decision-making in addition to profit.

2.5 Measuring social return on investment and broader impact

As ROI frameworks evolve, there is an increasing recognition that financial metrics alone do not capture the full spectrum of value in healthcare innovation. Measuring social return on investment (SROI) enables stakeholders to assess broader societal, equity, and public health outcomes, which are often overlooked in traditional models.

Table 21.4 outlines the core metrics and AI tools used to evaluate SROI in healthcare innovation. It highlights how AI-enhanced ROI frameworks integrate equity, public health, and long-term wellness indicators, ranging from access equity and inclusion to cost avoidance and quantifying gains in quality-adjusted life years (QALYs). AI tools like geospatial analytics, social determinants modeling, and heatmaps enable decision-makers to visualize and prioritize innovations that deliver both financial and societal value.

2.6 Qualitative factors

While financial outputs dominate discussions of ROI, qualitative elements are increasingly valued, particularly in strategic and public health decision-making. AI helps quantify and elevate these softer elements within the ROI framework.

2.6.1 Strategic fit

AI facilitates:

- Portfolio alignment modeling, matching new assets with a company's existing infrastructure, strategic focus, or commercial footprint.
- Competitor gap analysis, identifying white spaces where new drugs offer both scientific novelty and strategic synergies.
- M&A targeting, where AI models simulate long-term value accretion through product integration.

3 AI-driven investment strategies

The past decade has witnessed a marked acceleration in private-sector capital flows into life science ventures, with global healthcare buyout values exceeding US$60 billion in 2023 alone (Sullivan, 2024). Yet an equally rapid rise in unmet clinical needs, illustrated by a 23% jump in mental health treatment among US adults aged 18–44 years between 2019 and 2021 (Centers for Disease Control and Prevention, 2022) highlights the necessity of allocating capital across distinct time horizons.

Portfolio management frameworks that distribute resources among short-, medium-, and long-term bets have consequently migrated from corporate innovation theory to life-science finance, offering investors a mechanism to balance immediate public health pressures with the pursuit of breakthrough science. Within this strategic context, AI has emerged in the peer-reviewed record not merely as an incremental tool but as a catalytic general-purpose technology. Systematic reviews describe AI's capacity to restructure diagnostic, therapeutic, and operational workflows while warning of unresolved ethical and security challenges (Khan, 2025).

Empirical assessments in clinical and public health settings likewise document both clinician enthusiasm and adoption barriers (Scipion, 2025). Conceptual analyses emphasize AI's prospective reach across the drug development continuum: from structure-based lead discovery (Rabbani et al., 2025) Multidisciplinary Digital Publishing Institute (MDPI) to workflow automation in hospital practice (Olawade et al., 2024) (UEL Repository). Such breadth explains why Nature commentators now frame AI-enabled drug discovery as a "revolution" that must itself be re-imagined to achieve translational impact (Jacobson, 2025).

Industry analyses converge with these scholarly findings. A recent McKinsey synthesis projects that generative AI applications could unlock between US$60 billion and US$110 billion in annual value for life science organizations by streamlining R&D, commercial, and manufacturing activities (Zurkiya & Bleys, 2025) (McKinsey & Company). Such estimates both motivate and depend upon disciplined, time-stratified investment: rapid-deploy AI tools can relieve present-day capacity constraints, while moon-shot projects in neurotechnology or space health demand patient capital and robust risk governance. Together, the academic and industry literatures therefore support a consensus view: AI's general-purpose character amplifies the returns to horizon-based capital allocation, linking today's urgent clinical gaps with tomorrow's transformational therapies without resorting to speculative hyperbole.

4 Strategic allocation heuristic for healthcare investments

Investment theory long ago recognized the need to distribute capital among incremental, adjacent, and transformational projects. Nagji and Tuff's landmark innovation-portfolio study found that firms beating the S&P 500 typically devote about 70% of resources to core offerings, 20% to adjacent moves, and 10% to exploratory "moon-shots" (Nagji, 2012).

Recent policy analysis by the World Economic Forum argues that an explicitly horizon-based approach is also essential for AI initiatives, where poorly staged bets can stall up to 30% of enterprise projects (World Economic Forum, 2025). Together, these strands of evidence support a 70/20/10 heuristic tailored to life-science investors seeking both near-term health impact and long-run technological supremacy.

Although the 70/20/10 ratio originated in corporate innovation, its economic logic aligns with empirical observations of AI's return profile. Rapid-deployment tools, such as language-model chatbots for patient triage, generate measurable productivity within months, whereas frontier domains, such as brain–computer interfaces or extra-terrestrial medicine, require

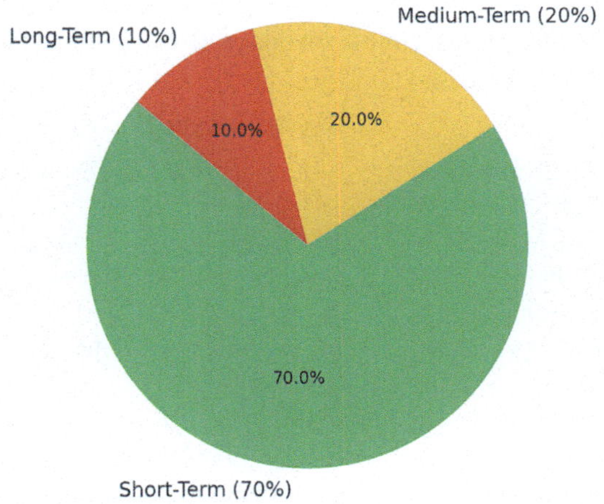

FIG. 21.1 Strategic allocation heuristic for AI in healthcare investments: a 70/20/10 rule.

long gestation but deliver disproportionately large upside when successful. Portfolio balance therefore mitigates temporal risk while preserving exposure to nonlinear value creation.

Fig. 21.1 illustrates the recommended allocation strategy for AI-driven healthcare investments across three time horizons. According to the 70/20/10 heuristic model:

- 70% of funding should focus on short-term innovations with fast deployment and early returns. These include telehealth platforms, AI-powered diagnostic tools, digital behavioral health solutions, and AI-enabled drug repurposing methods. These investments support quick wins, clinical adoption, and scalable outcomes.
- 20% is allocated to medium-term opportunities, which require moderate timelines for development, regulatory approval, and system integration. This includes AI-supported drug development platforms that accelerate clinical trial design, optimize patient stratification, and improve biomarker discovery. Other examples include scalable care ecosystems for women's health, personalized oncology tools, and chronic disease management platforms that benefit from multiyear adoption efforts.
- 10% of the portfolio is reserved for long-term moonshot innovations. These carry higher uncertainty but have the potential for transformative impact and ROI dominance. Examples include neural interface technologies, healthcare solutions for space environments, and AI-driven generative chemistry engines for the discovery of novel therapeutics. While such initiatives require extended R&D timelines and significant capital, their potential to reshape the healthcare landscape is unmatched.

4.1 Operationalizing AI-driven healthcare investment horizons

Effective execution demands that each time band be linked to concrete use cases, enabling technologies, and outcome metrics. In our mapping (Table 21.5), the short horizon (one to three years) concentrates on validated digital therapies; the medium horizon (3–7 years) addresses conditions that still need clinical integration; and the long horizon (7 or more years) targets speculative but potentially transformative sciences.

ROI is tracked with adoption or quality-of-life indicators in the near term, cost-offset or social burden metrics in the mid-term, and classical valuation markers such as compound annual growth rate, regulatory milestones, and intellectual property strength over the long haul. Such alignment converts an abstract ratio into transparent capital allocation rules that boards and limited partners can audit.

4.1.1 Short-term focus (1–3 years): Youth mental health, addiction, and companion animal care

Public health surveillance shows a sharp rise in mental health treatment among US adults aged 18–44 years, from 18.5% in 2019 to 23.2% in 2021, thereby confirming immediate demand for digital psychological services (Centers for Disease Control and Prevention, 2022). Simultaneously, the United Nations reports that global drug consumption reached 296 million people in 2021, with 39.5 million suffering from use disorders, figures that underline the urgency of AI-enabled addiction triage and personalized relapse prevention (United Nations Office on Drugs and Crime, 2023).

TABLE 21.5 Strategic investment mapping across time horizons in AI-driven healthcare.

Time horizon	Primary areas	Key AI technologies	Metrics
1–3 years	Youth mental health, addiction treatment, pet health, esthetic medicine	NLP, chatbot therapies, mobile health AI	App retention, adoption rate, quality-of-life index
3–7 years	Menopause care, depression management, family stability, antivenom development	Remote monitoring, sentiment analysis, digital therapeutics	Mental health quality of life (QoL) score, reduction in social costs
7+ years	Brain–computer interfaces, AI-enabled space health, antibiotic innovation, age-related therapies	Predictive genomics, AI biosensors, neural interfaces	CAGR, regulatory milestones, IP portfolio strength

Outside human medicine, companion animal health has become a robust, data-rich arena: market revenues exceeded US$21 billion in 2022 and are projected to double by 2030, driven by connected diagnostics and tele-veterinary platforms (Grand View Research, 2023). Together, these segments meet the heuristic's criteria for technological readiness, manageable regulatory pathways, and prompt, measurable outcomes such as app retention rates or symptom reduction scores.

4.1.2 Medium-term focus (3–7 years): Menopause management, depression support, and postdivorce resilience

Clinical coding data reveal that recorded menopause diagnoses among US women aged 45–64 years climbed from 13.7% in 2018 to 14.7% in 2022 (HCCI, 2024). Systematic reviews place the prevalence of vasomotor symptoms between 50% and 82%, underscoring a sizable care gap that AI-enabled remote monitoring and personalized hormone algorithms could address once further validation is complete (Fang, 2024).

Life-course disruptions exacerbate mental health risks: A survey by Legal & General (2025) indicates that women's household income falls by about 50% in the year following divorce, compared with a 30% drop for men, heightening vulnerability to depression and financial instability (Legal & General, 2025). Digital platforms that merge behavioral analytics with financial planning tools exemplify the kind of multidisciplinary solutions that warrant medium-horizon investment more mature than laboratory concepts, yet still needing multiyear adoption curves and policy alignment.

4.1.3 Long-term focus (7+ years): Space nutrition and neurotechnology

Extended human presence in low-gravity environments will require entirely new supply chains for nutrient-stable food. Market analysts value the space food sector at roughly US$684 million in 2024, with a projected compound annual growth rate (CAGR) of nearly 12% that would lift revenues to about US$1.67 billion by 2032 (Maximize Market Research, 2025). Parallel advances in neurotechnology illustrate the capital intensity and ethical complexity of frontier bets:

Elon Musk's Neuralink had amassed US$363 million in venture financing by mid-2021, solely to progress from preclinical prototypes toward first-in-human trials (Observer, 2021). Investors allocating the heuristic's 10% "moon-shot" tranche to such ventures must tolerate prolonged regulatory cycles and heightened societal scrutiny, but the potential payoff of new therapeutic modalities and off-planet healthcare infrastructure justifies strategic patience.

Mapping AI opportunities to discrete time horizons transforms long-range vision into actionable portfolio design. By weighting short-term deployments that relieve immediate clinical pressure, medium-term platforms that scale emerging evidence, and long-term experiments that could redefine human health, the 70/20/10 heuristic reconciles fiduciary discipline with scientific ambition. In volatile funding climates, this disciplined staging of capital not only buffers downside risk but also preserves exposure to the asymmetric upside that defines AI breakthroughs in biomedicine.

5 Funding challenges for a new class of antibiotic candidates: The role of AI and ROI

AMR occurs when microorganisms evolve to survive exposure to treatments that were once effective. This growing phenomenon undermines the efficacy of antibiotics and poses a critical threat to global health, leading to prolonged illness, increased mortality, and greater healthcare costs. According to a global analysis published in PLOS Medicine, AMR was directly responsible for approximately 1.27 million deaths and contributed to an estimated 4.95 million deaths in 2019, surpassing mortality rates from tuberculosis, malaria, and HIV/AIDS (Walsh, Gales, Laxminarayan, & Dodd, 2023). The World Health Organization has identified AMR as one of the top ten global public health threats, emphasizing the urgent need for coordinated international action.

A new class of antibiotic candidates refers to innovative molecules that are being developed to combat drug-resistant bacteria. James J. Collins, Termeer Professor of Medical Engineering and Science at MIT, and his colleagues are using AI to widen the net of potential antibiotic candidates (MIT News, 2023).

Antibiotic research is experiencing a technology-driven renaissance, yet scientific communities remain frustrated by persistent funding gaps that stall progress and clinical translation (Blaskovich & Cooper, 2025).

A recent comprehensive review indicates that most major pharmaceutical companies have exited antibiotic R&D due to high failure risks and poor returns on investment, leaving human and SME efforts underfunded at mid to late-stage development. However, globally deployed push funding mechanisms such as CARB-X, BARDA, GARDP, ND4BB, and the Global AMR R&D Hub are partially mitigating this by reducing early-stage risks and improving ROI incentives. The integration of AI-enabled platforms further enhances ROI by accelerating hit identification, reducing development costs, and improving predictability, thereby strengthening the case for sustained investment in antibiotic innovation (Kreitmeyr, 2023).

6 AI-enabled ROI in esthetic medicine, antivenom, and age-related therapies

AI has been increasingly adopted in biomedical investment strategy development, particularly in domains such as esthetic medicine, critical care, and aging-related therapeutics, through the integration of large-scale real-world data analytics. These tools facilitate integrated assessments that align financial projections with evidence-based clinical indicators. As a result, AI is emerging as a bridge between scientific discovery and strategic decision-making in modern healthcare (Khanna, 2022).

The following case studies and analyses examine how AI is influencing ROI in cosmetic medicine, antivenom development, and age-reversal therapies, with reference to recent peer-reviewed studies.

1. Lifestyle Drugs and Cosmetic Procedures
2. Life-Saving Antivenoms: Toward a Universal Snake Antivenom
3. Extending Healthy Lifespan and Longevity Drugs

6.1 AI-enabled ROI in lifestyle drugs and cosmetic procedures

Emerging research indicates that AI may be reshaping how ROI is assessed in esthetic medicine, including areas such as cosmetic surgery, hormone-based treatments, and hair restoration (Al-Dhubaibi, 2025).

For example, machine learning algorithms have been used to analyze demographic data and predict which procedures are likely to gain traction within specific regions or age groups. These models enable companies to adjust marketing strategies and allocate resources more effectively, thereby improving the accuracy of revenue forecasting. While economic reports project substantial growth in the esthetic medicine market, outcomes vary depending on the region and methodological assumptions (Statista, 2025).

Although AI-driven consumer analytics are increasingly used in the pharmaceutical sector to refine marketing and engagement strategies, peer-reviewed evaluations of their clinical and economic outcomes remain limited. Recent studies suggest that while such approaches have demonstrated value in e-commerce, their effectiveness in healthcare contexts is still largely unvalidated (Madanchian, 2024). Despite limited validation, AI-enabled consumer analytics could contribute to data-driven strategic planning in volatile healthcare markets influenced by sociocultural and media trends (Snowflake, 2024).

In addition to technical and economic considerations, the integration of AI in esthetic medicine also raises important ethical and equity concerns. Studies show that beauty-tech algorithms frequently underrepresent people of color, reinforcing narrow esthetic standards and potentially exacerbating socioeconomic barriers to care (Alfadda, 2023).

6.2 AI-enabled ROI in life-saving antivenoms: Toward a universal snake antivenom

Antivenoms remain a critical yet underfunded area of pharmaceutical development, particularly when compared to elective therapies. AI is increasingly being applied in this domain to overcome long-standing research and development barriers. Emerging studies highlight the use of AI in protein structure prediction, compound screening, and modeling venom–antibody interactions, accelerating the design of broad-spectrum antivenom formulations capable of neutralizing toxins from multiple snake species (Crunkhorn, 2025).

Such advances may shorten development timelines and reduce R&D expenditures, thereby improving ROI in underfunded therapeutic areas (Wellcome Trust, 2016).

AI-powered demand forecasting tools are also being explored to better predict where antivenom treatments are most urgently needed, with the goal of improving supply chain efficiency and distribution equity. These technologies may enhance ROI by reducing logistical costs while supporting broader partnerships with public health organizations. However, several critical infrastructural and data-related constraints limit the practical application of these approaches (Coulson, 2023).

6.3 AI-enabled ROI in extending healthy lifespan and longevity drugs

Antiaging therapeutics constitutes a growing field in biomedical research, characterized by high translational potential and substantial challenges in clinical validation. Industry investment in AI-driven aging research has grown considerably; however, independent academic evaluation of these technologies remains in the early stages (Wilczok, 2025).

Within this field, AI is being applied to identify biomarkers of aging and accelerate drug discovery through in silico modeling of cellular responses. Generative models have shown potential to optimize candidate compounds prior to preclinical testing. A recent study in Nature Aging demonstrated the use of AI in identifying senolytic agents that delayed aging markers in murine models (Wong, 2023).

From a commercial perspective, AI also contributes to market segmentation and pricing strategies, enabling firms to target specific consumer demographics and model ROI. Despite ongoing progress in AI-based ROI modeling, several methodological and regulatory challenges remain unresolved. The absence of standardized biomarkers and limited clinical validation of AI-based "aging clocks" raises concerns about predictive reliability and generalizability (Hevolution Foundation).

These uncertainties complicate ROI projections, as therapeutic efficacy, regulatory approval, and market adoption remain highly speculative.

7 Operationalizing AI-driven ROI framework

Embedding an AI-enhanced ROI model inside a health-system enterprise begins with governance. Qualitative fieldwork by Kim et al. identifies eight decision points, ranging from problem scoping to postdeployment monitoring, that executive teams repeatedly navigate when adopting clinical AI; the authors conclude that without a codified oversight structure, analytic insights fail to reach the bedside (Kim, 2023).

Recent data from the World Economic Forum shows the stakes: as many as 30% of enterprise generative-AI projects initiated in 2025 are expected to stall due to weak data foundations, escalating costs, or unclear business value, reinforcing the need for rigorously stewarded investment pathways (World Economic Forum, 2025). A functioning ROI framework therefore requires not only sophisticated modeling but also an organizational operating system that can translate signals into action.

A disciplined implementation sequence can be expressed as twelve interlocking activities. Leaders first verify strategic alignment with corporate objectives, then specify model parameters such as discount rates and risk tolerances. They next map the end-to-end process flow and tag each data attribute to its business owner before running multiscenario simulations. Core metrics ROI, NPV, and internal rate of return are generated, interrogated with stakeholder feedback, and surfaced through live dashboards.

Adaptive learning loops continuously refresh model weights as fresh evidence arrives, while collaborative governance forums adjudicate trade-offs between financial return, social impact, and equity. Finally, success factors are audited and barriers are mitigated so that lessons feed back into the next capital cycle. This closed-loop workflow converts static valuation into a learning system that can evolve with strategy.

7.1 Checklist for step-by-step AI-enhanced ROI realization

- Confirm strategic alignment
- Define AI model parameters
- Create process flowchart
- Assign attributes
- Simulate scenarios
- Generate key metrics (ROI, NPV, IRR, etc.)
- Iterate with stakeholders
- Deploy AI-powered dashboards
- Enable adaptive learning
- Facilitate collaborative decision-making
- Integrate SROI and equity metrics
- Assess success factors and mitigate barriers

Operational examples illustrate the mechanics. Tsai et al. describe a real-time emergency department dashboard that integrates eight predictive models onto a single interface, allowing clinicians to adjust resource allocation on the fly and thereby link algorithmic risk scores to measurable throughput gains (Tsai et al., 2022). Outside the clinical setting, finance teams have shown that scenario planning platforms can democratize allocation decisions: Sigma Computing's chief financial officer reports that real-time input tables now let cross-functional leaders iterate on ROI assumptions collaboratively, eliminating version control delays and fostering consensus (Liu, Sigma Computing, 2024).

Quantitative evidence confirms that rigorous financial tracking sharpens capital discipline. A multicenter study in radiology calculated a positive 5-year ROI for an AI stroke workflow platform and offered a publicly available calculator that converts clinical time-savings into discounted cash flows (Bharadwaj, 2024). Complementary guidance from StarBridge Advisors recommends triangulating ROI with complementary indicators such as NPV, IRR, total cost of ownership, and a balanced scorecard of stakeholder impacts to capture intangible benefits often overlooked in single-metric analyses (Muntz, StarBridge Advisors, 2024).

Together, these findings show that an AI-driven ROI framework achieves relevance only when it is embedded in day-to-day governance, supported by adaptive dashboards, and audited with multidimensional metrics. Such integration transforms ROI from a retrospective accounting exercise into a forward-looking decision engine that aligns scarce capital with both the clinical mission and strategic growth.

7.2 AI tools for SROI assessment

A comprehensive appraisal of ROI must extend beyond conventional financial metrics to encompass social return indicators that capture equity, access, and long-term public health benefits. Recent scholarship demonstrates that AI platforms can translate such multidimensional value into operational guidance.

Geospatial AI systems, for example, overlay demographic and clinical data onto digital maps, allowing decision-makers to identify cancer hot spots, anticipate service gaps, and prioritize location-specific interventions. Fadiel et al. show that this approach reveals regional disparities invisible to aggregate statistics (Fadiel et al., 2024).

Complementary models incorporate social determinants of health, such as income, housing, education, and environmental exposure, into predictive simulations, thereby aligning capital allocation with the real-world conditions that modulate intervention efficacy (Thomas Craig, 2021).

Finally, inclusion scenario engines, such as fair ranking with missing modalities (FRAMM), use reinforcement learning to forecast how trial-site selection or care delivery design influences participation by historically underrepresented groups, quantifying trade-offs between enrollment velocity and diversity (Theodorou, 2023). Taken together, these AI-enabled tools convert abstract equity goals into measurable portfolio inputs, permitting investors to balance economic and societal returns without compromising analytic rigor.

7.3 Real-world success factors and barriers

The analytic sophistication described in the preceding sections yields value only when organizations and the policy environments in which they operate are prepared to absorb it. Empirical work on digital health transformation demonstrates that technical prowess alone rarely guarantees impact; instead, the interplay of organizational readiness, regulatory alignment, and cultural acceptance determines whether an AI-enhanced ROI framework migrates from pilot to enterprise scale.

Enablers are factors that support or facilitate the achievement of desired outcomes, such as leadership, resources, or stakeholder alignment. Barriers are obstacles that hinder progress, including regulatory constraints, resistance to change, or lack of capacity (Fixsen et al., 2005; Pettigrew, Ferlie, & McKee, 1992).

Table 21.6 shows the key factors that influence the successful implementation of AI-enhanced ROI frameworks in healthcare. Enablers, such as organizational readiness and supportive policy environments, create the conditions for effective adoption, while challenges, including model transparency, regulatory misalignment, and cultural resistance, can hinder progress. Addressing both sides of this equation is essential for translating technical potential into real-world impact.

7.3.1 Enablers

Organizational readiness

Organizational readiness constitutes the first enabling layer. Health system case studies show that interoperable data infrastructures, well-defined data governance councils, and cross-functional decision bodies correlate with smoother integration of AI into capital allocation workflows (Alami, 2020). Where these preconditions exist, ROI dashboards can draw directly on high-fidelity operational data and feed actionable projections back to strategy teams without costly reengineering.

TABLE 21.6 Enablers and barriers to AI-enhanced ROI implementation.

Category	Factor	Description
Enabler	Organizational readiness	Robust data infrastructure, integrated workflows, and cross-functional governance enable smooth AI integration.
	Supportive policy environment	Clear regulations, HTA support, and incentives promote the adoption of dynamic ROI modeling frameworks.
Barriers	Model transparency and explainability	Demand for understandable, auditable AI models to build trust with financial and regulatory stakeholders.
	Regulatory lag or misalignment	Slow adaptation of regulatory bodies to AI-driven models causes delays and inefficiencies.
	Change management	Skepticism and a lack of familiarity with AI among leadership hinder adoption without change management.

Policy

A supportive policy environment provides the second pillar. Jurisdictions that have articulated validation pathways for adaptive algorithms and encourage the use of real world evidence in health technology assessment accelerate adoption by reducing regulatory ambiguity. Ramezani (2025) observes that clear economic evaluation guidance and reimbursement incentives for dynamic models shorten time to impact and improve investor confidence.

7.3.2 Barriers

Model transparency and explainability

Countervailing forces nonetheless persist. Model transparency and explainability remain central challenges; financial stewards and regulators frequently demand auditable logic to verify the assumptions underlying projected cash flows. Multidisciplinary reviews emphasize that opaque architectures erode trust, even when predictive accuracy is high (Amann, Blasimme, Vayena, Frey, & Madai, 2020).

Regulatory constraints

Equally constraining is regulatory misalignment. Machine learning-enabled trial designs and adaptive pricing schemes often outpace traditional review processes, introducing costly delays and iterative resubmissions (Petersen, 2021).

Change management

Finally, cultural resistance can stall progress despite favorable technical and policy conditions. Salmela (2025) documents that executive unfamiliarity with machine learning principles, coupled with entrenched spreadsheet habits, frequently prompts skepticism toward algorithmic recommendations. Structured change management programs, targeted leadership education, and early demonstration projects are therefore indispensable complements to algorithm development.

In practice, successful AI-enhanced ROI deployment reflects the dynamic equilibrium between these enablers and barriers. Institutions that cultivate data maturity, engage proactively with regulators, and invest in workforce capability stand a markedly higher chance of converting algorithmic insight into tangible, socially responsible value, thereby realizing the full strategic promise articulated in the earlier sections.

8 Conclusion

This chapter explored how AI can expand the scope of ROI modeling by aligning financial performance with clinical outcomes and societal value. It highlighted immediate priorities such as youth mental health and addiction care and examined funding gaps in critical areas like antibiotic development and universal antivenoms, where commercial incentives are limited despite high public health needs. Age-related therapies were also explored as a field of emerging promise, where AI accelerates discovery, but regulatory and methodological challenges remain. In contrast, esthetic medicine was discussed as a commercially dynamic yet ethically complex sector, where AI-driven analytics improve targeting and planning but raise questions about representation and equity.

In addressing not only what and why but also how AI-enabled ROI can be operationalized, the chapter presents concrete examples such as IBM Watson's use in regulatory forecasting and portfolio evaluation, as well as Monte Carlo simulations for scenario-based investment planning. The framework connects these use cases to broader decision systems that incorporate adaptive learning, scenario simulation, and social equity metrics. By embedding AI into investment planning, organizations are better equipped to address current clinical demands while preparing for transformative breakthroughs. Ultimately, the chapter illustrates that AI-guided ROI is not limited to financial returns but serves as a strategic foundation for building responsive, inclusive, and sustainable healthcare ecosystems. The perspectives shared in this chapter are intended as a contribution to ongoing dialog and may offer a foundation for further research on ROI-driven, evidence-based strategies that advance both innovation and equity in healthcare.

References

Ahsen, M. E., Ayvaci, M. U. S., Mookerjee, R., & Stolovitzky, G. (2025). Economics of AI and human task sharing for decision making in screening mammography. Nature Communications. Retrieved from: https://www.nature.com/articles/s41467-025-57409-1.

Al-Dhubaibi, M. (2025). Artificial intelligence in aesthetic medicine: applications, challenges, and future directions review. Retrieved from: https://www.ncbi.nlm.nih.gov/pmc/articles/PMC12159716/.

Alami, H. (2020). Organisational readiness for artificial intelligence in health care: insights for decision-making and practice Article. Retrieved from: https://pubmed.ncbi.nlm.nih.gov/33258359/

Alfadda, Y. (2023). How Artificial Intelligence Adopts Human Biases: the Case of Cosmetic Skincare Industry Original research. Retrieved from: https://www.researchgate.net/publication/375774262_How_artificial_intelligence_adopts_human_biases_the_case_of_cosmetic_skincare_industry

Amann, J., Blasimme, A., Vayena, E., Frey, D., & Madai, V. I. (2020). Explainability for artificial intelligence in healthcare: a multidisciplinary perspective. *BMC Medical Informatics and Decision Making*. 1472-6947. *20*(1). doi:10.1186/s12911-020-01332-6.

Bennett, D. J., Feng, J., Goldman, S., Kothari, A., Gottlieb, L. M., Durstenfeld, M. S., … Zier, L. S. (2025). Reducing Readmissions in the Safety Net Through AI and Automation. *American Journal of Managed Care*. 10880224. *31*(3), 142–148. Retrieved from: https://www.ajmc.com/view/reducing-readmissions-in-the-safety-net-through-ai-and-automation. doi:10.37765/ajmc.2025.89697.

Bharadwaj, P. 2024 Quantifying the return on investment of hospital artificial intelligence. Retrieved from: https://www.sciencedirect.com/science/article/pii/S1546144024002928

BiopharmaTrend. (2024). The pivotal role of AI in clinical trials from digital twins to synthetic control arms Case Study. Retrieved from: https://www.biopharmatrend.com/knowledge-center/the-pivotal-role-of-ai-in-clinical-trials-from-digital-twins-to-synthetic-control-arms/

Blaskovich, M. A. T., & Cooper, M. A. (2025). Antibiotics re-booted—time to kick back against drug resistance. *Nature Partner Journals Antimicrobials and Resistance*. 2731-8745. *3*(1). doi:10.1038/s44259-025-00096-1.

Centers for Disease Control and Prevention. (2022). Mental health treatment among adults aged 18–44: united States, 2019–2021 (NCHS Data Brief No. 444) Report Retrieved from: https://www.cdc.gov/nchs/products/databriefs/db444.htm.

Chen, Y. (2022). Health technology assessment and economic evaluation: is it applicable for the traditional medicine? *Integrative Medicine Research*. 22134220. *11*(1), 100756. doi:10.1016/j.imr.2021.100756.

Choi, M. (2023). Impact of AI-assisted rostering on nurse overtime and satisfaction: a randomised trial Original research. Retrieved from: https://doi.org/10.1111/jonm.13890.

Company Connect Consultancy (2025). The role of artificial intelligence in transforming drug regulatory affairs: revolutionizing approval processes and compliance Article. Retrieved from: https://www.companysconnects.com/post/the-role-of-artificial-intelligence-in-transforming-drug-regulatory-affairs-revolutionizing-approva

Coulson, A. (2023). Antivenom intelligence? The role of AI in developing snakebite treatments. Retrieved from: https://www.biotechniques.com/drug-discovery-development/sartorius_sptl_ad_antivenom-intelligence-the-role-of-ai-in-developing-snakebite-treatments/

Crunkhorn, S. 2025 Towards a universal snake antivenom Research highlight. Retrieved from: https://www.nature.com/articles/d41573-025-00092-0.

DeFrancesco, L. (2022). AI for structure-guided lead discovery Article. Retrieved from: https://www.nature.com/articles/s41587-022-01649-9.

Deloitte. (2025). 2025 life-sciences outlook: creating new value, building blocks for the future Report 2025. Retrieved from: https://www.deloitte.com/us/en/insights/industry/health-care/life-sciences-and-health-care-industry-outlooks/2025-life-sciences-executive-outlook.html

Deveau, R., Joseph Griffin, S., & Reis, S. (2023). AI-powered marketing and sales reach new heights with generative AI. Retrieved from: https://www.mckinsey.com/capabilities/growth-marketing-and-sales/our-insights/ai-powered-marketing-and-sales-reach-new-heights-with-generative-ai.

DiMasi, J. A., Smith, Z., Oakley-Girvan, I., Mackinnon, A., Costello, M., Tenaerts, P., & Getz, K. A. (2022). Assessing the financial value of decentralized clinical trials. *Therapeutic Innovation and Regulatory Science*. 2168-4790. *57*, 209. Retrieved from: doi:https://doi.org/10.1007/s43441-022-00454-5.

Dosanjh, U. (2025). Measuring AI's impact on the pharmaceutical industry Article. Retrieved from: https://www.europeanpharmaceuticalreview.com/article/256553/measuring-ais-impact-on-the-pharmaceutical-industry/

Eanes, Z. (2025). Duke researchers examine AI safety in a health care setting News article 2025. Retrieved from: https://www.axios.com/local/raleigh/2025/06/24/duke-researchers-artificial-intelligence-health-care-ai

Ericson, O. (2022). The potential cost and cost-effectiveness impact of using a machine learning algorithm for early detection of sepsis in intensive care units in Sweden. Retrieved from: https://jheor.org/article/33951-the-potential-cost-and-cost-effectiveness-impact-of-using-a-machine-learning-algorithm-for-early-detection-of-sepsis-in-intensive-care-units-in-sweden

Fadiel, A., Eichenbaum, K. D., Abbasi, M., Lee, N. K., & Kunle, O. (2024). Utilizing geospatial artificial intelligence to map cancer disparities across health regions Article. 2045–2322. Retrieved from: https://doi.org/10.1038/s41598-024-57604-y

Fang, Y. (2024). Mapping global prevalence of menopausal symptoms among middle-aged women: a systematic review and meta-analysis Original research. Retrieved from: https://pubmed.ncbi.nlm.nih.gov/40233955/

Fixsen, D. L., Naoom, S. F., Blase, K. A., Friedman, R. M. & Wallace, F. (2005). Implementation research: A synthesis of the literature. Implementation research: A synthesis of the literature. University of South Florida, Louis de la Parte Florida Mental Health Institute, The National Implementation Research Network (FMHI Publication), Tampa, FL. Retrieved from: https://fpg.unc.edu/publications/implementation-research-synthesis-literature.

Fusch, G. E. (2001). *What Happens When The ROI Model Does Not Fit?* Wiley. Retrieved from: https://www.researchgate.net/publication/230542791_What_happens_when_the_ROI_model_does_not_fit.

GE HealthCare (2024). Beyond Downtime: redefining Predictive Medical Equipment Maintenance Article. Retrieved from: https://www.gehealthcare.com/insights/article/beyond-downtime-redefining-predictive-medical-equipment-maintenance.

Gitlin, J. M. (2013). Calculating the economic impact of the Human Genome Project. Science Policy Analyst, National Human Genome Research Institute. Retrieved from: https://www.genome.gov/27544383/calculating-the-economic-impact-of-the-human-genome-project.

Grand View Research. (2023) Companion animal health market size & share report, 2022–2030 Market research. Retrieved from: https://www.grandviewresearch.com/industry-analysis/companion-animal-health-market.

Haas, R. (2022). Artificial intelligence for the prediction of sepsis in adults: CADTH Horizon Scan Report 2022 Retrieved July 4, 2025 from: https://www.ncbi.nlm.nih.gov/books/NBK596676/.

Health Care Cost Institute. (2024). Data brief: menopause diagnosis steadily increased from 2018–2022 Repo. Retrieved from: https://healthcostinstitute.org/hcci-originals-dropdown/all-hcci-reports/hcci-data-brief.

Hill, H., Roadevin, C., Duffy, S., Mandrik, O., & Brentnall, A. (2024). Cost-effectiveness of AI for risk-stratified breast cancer screening. JAMA Network Open, *7*(9). e2431715 . doi:10.1001/jamanetworkopen.2024.31715.

Jacobson, R. D. (2025). The AI drug revolution needs a revolution. *Nature Partner Journals Drug Discovery*. 3005-1452. *2*(1). doi:10.1038/s44386-025-00013-6.

Jordan, A. M. (2025). AI-Augmented Portfolio Management in Biopharma: optimizing Pipeline Investments and Resource Allocation Case Study. Retrieved from: https://www.researchgate.net/publication/391807209_AI-AUGMENTED_PORTFOLIO_MANAGEMENT_IN_BIOPHARMA_OPTIMIZING_PIPELINE_INVESTMENTS_AND_RESOURCE_ALLOCATION.

Kaylor, A. (2025). How digital twins are reshaping clinical trials feature. Retrieved from: https://www.techtarget.com/pharmalifesciences/feature/How-digital-twins-are-reshaping-clinical-trials.

Khan, M. M. (2025). Towards secure and trusted AI in healthcare: a systematic review of emerging innovations and ethical challenges Article. Retrieved from: https://doi.org/10.1016/j.ijmedinf.2024.105780.

Khanna, N. (2023). Economics of Artificial Intelligence in Healthcare: diagnosis vs Treatment Original research. Retrieved from: https://www.ncbi.nlm.nih.gov/pmc/articles/PMC9777836/.

Kim, J. Y. (2023). Organizational governance of emerging technologies: AI adoption in healthcare. Retrieved from: https://arxiv.org/abs/2304.13081.

Kreitmeyr, K. (2023). Artificial intelligence in antibiotic discovery: advancing drug development with machine learning review article. Retrieved from: https://pmc.ncbi.nlm.nih.gov/articles/PMC9855914/.

Legal & General (2025). UK women's incomes halve in year following divorce Press release. Retrieved from: https://group.legalandgeneral.com/en/newsroom/press-releases/the-divorce-gap-women-see-their-household-income-cut-in-half-in-the-year-after-a-divorce.

Levitan, B., Getz, K., Eisenstein, E. L., Michelle, G., Harker, M., Hesterlee, S., … DiMasi, J. (2018). Assessing the financial value of patient engagement: A quantitative approach from CTTI's patient groups and clinical trials project. *Therapeutic Innovation & Regulatory Science*, *52*(2), 220–229. doi:10.1177/2168479017716715.

Lin, B. 2025 Abridge, whose AI app takes notes for doctors, valued at $5.3 billion at funding 2025. Retrieved from: https://www.wsj.com/articles/abridge-whose-ai-app-takes-notes-for-doctors-valued-at-5-3-billion-at-funding-801825f3

Madanchian, M. (2024). The Impact of Artificial Intelligence Marketing on E-Commerce Sales. Retrieved from: https://www.mdpi.com/2079-8954/12/10/429

Marko, J. G. O. (2025). Examining inclusivity: the use of AI and diverse populations in health and social care: a systematic review. Retrieved from: https://bmcmedinformdecismak.biomedcentral.com/articles/10.1186/s12911-025-02884-1

Maximize Market Research. (2025). Space food market size, share, trends and forecast analysis (2025–2032) Market research. Retrieved from: https://www.maximizemarketresearch.com/market-report/space-food-market/215884/.

MIT News (2023). Using AI, MIT researchers identify antibiotic candidates News article. Retrieved from: https://news.mit.edu/2023/using-ai-mit-researchers-identify-antibiotic-candidates-1220

Mortazavi, B. J. (2024). A pragmatic machine-learning system for 30-day readmission prediction across 17 hospitals. Retrieved from: https://doi.org/10.1001/jamanetworkopen.2024.5678.

Nagji, B. (2012). Managing your innovation portfolio Article. Retrieved from: https://hbr.org/2012/05/managing-your-innovation-portfolio

Observer. (2021). Elon Musk's old pals invest in Neuralink to fund testing brain chips in humans News article. Retrieved from: https://observer.com/2021/07/elon-musk-neuralink-raise-fund-peter-thiel-brain-chip-human-test/.

Olawade, D. B., David-Olawade, A. C., Wada, O. Z., Asaolu, A. J., Adereni, T., & Ling, J. (2024). Artificial intelligence in healthcare delivery: prospects and pitfalls. *Journal of Medicine, Surgery, and Public Health*. 2949916X. *3*(100108). doi:10.1016/j.glmedi.2024.100108.

Petersen, E. (2021). Responsible and regulatory-conform machine learning for medicine: a survey of challenges and solutions Preprint. Retrieved from: https://arxiv.org/abs/2107.09546

Pettigrew, A., Ferlie, E., & McKee, L. (1992). Shaping strategic change - The case of the NHS in the 1980s. *Public Money and Management*. 0954-0962. *12* (3), 27–31. doi:10.1080/09540969209387719.

Rabbani, S. A., El-Tanani, M., Sharma, S., Rabbani, S. S., El-Tanani, Y., Kumar, R., & Saini, M. (2025). Generative Artificial Intelligence in Healthcare: applications, Implementation Challenges, and Future Directions. *BioMedInformatics*. 2673-7426. *5*(3), 37. doi:10.3390/biomedinformatics5030037.

Ramezani, M. (2025). Applications of artificial intelligence and the challenges in health technology assessment: a scoping review with an economic focus. Retrieved from: https://doi.org/10.1186/s13561-025-00645-4

Regier. (2022). A perspective on life-cycle health technology assessment and real-world evidence for precision oncology in Canada. Retrieved from: https://www.nature.com/articles/s41698-022-00316-1.

Sahni, N. (2023). The potential impact of artificial intelligence on healthcare spending Working paper. Retrieved from: https://www.nber.org/papers/w30857.

Salmela, H. (2025). Leadership for AI transformation in healthcare organisations: a scoping review. Retrieved from: https://pubmed.ncbi.nlm.nih.gov/39009038/

Schubarth, C. (2023). How digital twins are reshaping clinical trials. TechTarget PharmaLifeSciences. Retrieved from: https://www.techtarget.com/pharmalifesciences/feature/How-digital-twins-are-reshaping-clinical-trials.

Scipion, C. E. A. (2025). Barriers to and facilitators of clinician acceptance and use of artificial intelligence in healthcare settings: a scoping review Article. Retrieved from: https://pubmed.ncbi.nlm.nih.gov/40233955/

Sigma Computing. (2024). No-hassle scenario planning: sigma's CFO on how real-time data is changing finance leadership Blog post. Retrieved from: https://www.sigmacomputing.com/blog/sigma-finance-cfo-scenario-planning.

Snowflake. (2024). How Allergan Aesthetics Uses Snowflake and Segment to Generate Personalized Offers and Reduce Customer Acquisition Cost Blog. Retrieved from: https://www.snowflake.com/en/blog/how-allergan-aesthetics-uses-snowflake-and-segment-to-generate-personalized-offers-and-reduce-customer-acquisition-cost/.

StarBridge Advisors. (2024). An effective strategy for valuing HIT investments Article. Retrieved from: https://starbridgeadvisors.com/2024/06/08/an-effective-strategy-for-valuing-hit-investments/.

Statista. (2025). Global value of the cosmetics market worldwide 2012–2028 (in billion U.S. dollars) Statistic. Retrieved from: https://www.statista.com/statistics/585522/global-value-cosmetics-market/.

Sullivan, K. (2024). Growth equity in healthcare: new opportunities for private equity News article. Retrieved from: https://www.reuters.com/legal/litigation/growth-equity-healthcare-new-opportunities-private-equity-2024-04-30/

Tali, A. I. (2024). Can AI-enabled medical charting increase ROI? [LinkedIn post]. Retrieved from: https://www.linkedin.com/posts/tali-ai_can-ai-enabled-medical-charting-increase-activity-7261799146687201280-ugIo/

Tang, D. (2025). AI-Driven Demand Forecasting in Life Sciences Case study https://www.flevy.com/topic/artificial-intelligence/case-ai-driven-demand-forecasting-life-sciences

Theodorou, B. (2023). FRAMM: fair ranking with missing modalities for clinical-trial site selection. Retrieved from: https://arxiv.org/abs/2305.19407

Thomas Craig, K. J. (2021). Leveraging data and digital health technologies to assess and impact social determinants of health (SDoH): a state-of-the-art literature review Original research. Retrieved from: https://pmc.ncbi.nlm.nih.gov/articles/PMC8765800/

Tisoskey, S. P. (2025). Revolutionizing nurse scheduling: how AI is transforming workforce management. Retrieved from: https://www.chromiehealth.com/post/revolutionizing-nurse-scheduling-how-ai-is-transforming-workforce-management

Tripp, S., & Grueber, M. (2011). Economic impact of the human genome project Retrieved from: https://www.battelle.org/docs/default-source/misc/battelle-2011-misc-economic-impact-human-genome-project.pdf.

Tsai, W.-C., Liu, C.-F., Lin, H.-J., Hsu, C.-C., Ma, Y.-S., Chen, C.-J., … Chen, C.-C. (2022). Design and implementation of a comprehensive AI dashboard for real-time prediction of adverse prognosis of emergency-department patients article Retrieved from: 10.3390/healthcare10081498.

United Nations Office on Drugs and Crime. (2023). World drug report 2023 (Executive summary) report. Retrieved from: https://www.unodc.org/unodc/en/data-and-analysis/world-drug-report-2023.html.

Varol, D. (2025). AI in the pharmaceutical industry: innovations and challenges Blog post. Retrieved from: https://www.scilife.io/blog/ai-pharma-innovation-challenges.

Velikova. (2024). Method "Monte Carlo" in healthcare Review. Retrieved from: https://pmc.ncbi.nlm.nih.gov/articles/PMC11230067/.

Vollmer, S. (2020). A unified machine learning approach to time series forecasting applied to demand at emergency departments. *BMC Emergency Medicine*. Retrieved from: https://arxiv.org/abs/2007.06566.

Walsh, T. R., Gales, A. C., Laxminarayan, R., & Dodd, P. C. (2023). Antimicrobial Resistance: addressing a Global Threat to Humanity. *Public Library of Science Medicine*. 1549-1676. *20* (7), e1004264. doi:10.1371/journal.pmed.1004264.

Wellcome Trust. (2016). Developing a Universal Antivenom to Treat Snake Venom-Induced Consumption Coagulopathy Public engagement grant. Retrieved from: https://wellcome.org/research-funding/funding-portfolio/funded-grants/developing-universal-antivenom-treat-snake-venom

Wilczok, D. (2025). Deep learning and generative artificial intelligence in aging research and healthy longevity medicine review. Retrieved from: https://www.aging-us.com/article/206190/.

Wong, F. (2023). Artificial intelligence identifies new small-molecule senolytics Research briefing. Retrieved from: https://www.nature.21/articles/s43587-023-00422-0

World Economic Forum (2025). Why AI needs smart investment pathways to ensure a sustainable impact Article. Retrieved from: https://www.weforum.org/stories/2025/06/why-ai-needs-smart-investment-pathways-to-ensure-a-sustainable-impact/

World Health Organization Eastern Mediterranean Regional Committee. (2024). Progress report on building resilient health systems to advance universal health coverage and ensure health security in the Eastern Mediterranean Region. Regional report. Retrieved from: https://applications.emro.who.int/docs/Resilient-health-systems-progress-report-eng.pdf.

World Health Organization (2013). OneHealth Tool: supporting integrated strategic health planning, costing and health impact analysis Tool. Retrieved from: https://www.who.int/tools/onehealth.

Zurkiya, D., & Bleys, J. (2025). Gen AI in Healthcare and Life Sciences: McKinsey's Latest Insights. Retrieved from: https://www.mckinsey.com/featured-insights/themes/gen-ai-in-healthcare-and-life-sciences-mckinseys-latest-insights.

Chapter 22

Addressing India's rural healthcare gaps utilizing generative artificial intelligence (AI) and agentic intelligence

Prithviraj Bagchi
Advanced Technology in Cloud and Gen AI Domains, Accenture Services Pvt. Ltd., Kolkata, West Bengal, India

1 Generative artificial intelligence and agentic intelligence in rural healthcare

1.1 Introduction

Rural India faces long-standing healthcare challenges: an acute shortage of trained doctors, overburdened clinics, and limited infrastructure for millions of villagers (Sharma, 2015). About 70% of India's population lives in rural areas (Tripathi & Mitra, 2022), where access to quality care is constrained by both human and physical resource gaps. Although India has managed to achieve the WHO-recommended doctor-to-population ratio (Kumar & Pal, 2018), the deficit is still felt most in remote regions. Community health workers (CHWs), like Accredited Social Health Activist (ASHA) and Auxiliary Nurse Midwife (ANM), bridge this gap, but they often have only a few weeks of training and limited reference materials, making complex patient issues hard to address (Ramjee et al., 2025).

Generative artificial intelligence (Gen AI), exemplified by advanced large language Models such as GPT models, the Anthropic's Claude models and many other such models, offers a potential solution. These AI systems can interpret laboratory results, imaging, and patient histories, then explain findings in simple, local-language terms. By summarizing clinical data and evidence-based guidelines, generative agents can help rural providers suggest diagnoses and treatments with fewer errors and in less time. In practice, AI-enabled tools on smartphones or tablets could give patients immediate health advice in Hindi, Tamil, Bengali, or regional languages, bypassing literacy barriers and making reliable medical knowledge accessible even in the most remote villages (Maity & Saikia, 2025).

By integrating these AI agents with low-cost point-of-care devices and digital records, a scalable rural health support system can emerge. For instance, AI-driven analysis of blood tests, chest X-rays, or ultrasound images can flag anemia, diabetes, or infections early, even before a doctor is available on-site (Kumar, Koul, Singla, & Ijaz, 2023). Mobile diagnostic vans with built-in AI scanners (digital X-ray, ultrasound) have already proven valuable; such units significantly cut tuberculosis (TB) diagnosis times and costs in rural Haryana by providing on-the-spot imaging in underserved areas (Datta et al., 2019). Linking AI platforms with cloud-based electronic health records (using India's National Digital Health Mission (NDHM)/Fast Healthcare Interoperability Resource standards) and telemedicine (eSanjeevani) networks ensures that data and expert support flow seamlessly (Parameshwarappa & Olickal, 2023). This digital ecosystem can automate routine tasks (e.g., data entry, referrals, scheduling), reduce errors, and free healthcare workers to focus on critical cases. In short, Gen AI and agentic intelligence are not just advanced tools but catalysts for inclusive health equity, empowering frontline workers with smart diagnostics and decision support to bridge India's rural–urban healthcare divide.

Applications of Artificial Intelligence in Healthcare. https://doi.org/10.1016/B978-0-443-43934-6.00035-3

1.2 Key benefits of Gen AI in rural healthcare

1.2.1 Personalized health guidance

With rapidly rising smartphone use in India, even rural citizens increasingly have access to the technology needed for AI agents on their devices. A well-designed AI app or chatbot can collect a user's age, medical history, symptoms, etc., and provide tailored advice. For example, it can track a woman's blood pressure or glucose trends over time, suggest lifestyle changes for the prevention of diabetes, or give a localized symptom assessment and triage recommendation (Barreda et al., 2025). By leveraging multilingual LLMs, the AI can explain complex medical concepts in simple terms or local idioms. This empowers individuals, even those with low literacy, to understand and act on their health data proactively (for example, reminding a patient when to drink an oral rehydration solution or seek care). Over time, the AI could also learn a user's context (diet, literacy level, and even local disease patterns) to improve its guidance.

1.2.2 Confidential, 24/7 access

Many rural patients hesitate to discuss health issues (especially gynecological or mental health) due to stigma. AI "health companions" offer private, judgment-free counseling at any hour (Olawade et al., 2024). A farmer with symptoms of depression or a young man concerned about a fever can query an AI assistant in his village dialect without fear of embarrassment. This always-on virtual doctor can triage mild cases, advise on home remedies, or flag urgent situations that need a physician's attention. The privacy and convenience of AI support encourage people to seek help early, which is especially important for conditions like anemia, hypertension, or infections, where delays are common.

1.2.3 Connecting citizens to government services

India has many public health schemes (Ayushman Bharat insurance, free medicine programs, nutrition supplements, etc.), but poor awareness and complex processes limit their utilization. A Gen AI assistant can act as a digital social worker: it can identify which government programs (vaccines, screening camps, subsidies) a person qualifies for and guide them through enrollment. By geotagging the user's location, the AI can also recommend the nearest empaneled hospital or clinic under government schemes. For example, if a user mentions symptoms of TB, the AI could suggest registering with the National TB Elimination Program (Nikshay) and direct them to the closest TB notification center. This bridges the gap between citizens and preventive care initiatives, improving equity.

1.2.4 Community health data for policy

AI agents can aggregate anonymized usage and symptom data (with privacy safeguards) to reveal public health trends (Khalid, Qayyum, Bilal, Al-Fuqaha, & Qadir, 2023). For example, if an AI system in a district reports rising cases of fever and cough among school children, or a spike in GI symptom queries in certain villages, health authorities can be alerted to a possible outbreak (dengue, influenza, cholera, etc.). As one analysis notes, AI's real-time data processing can make it a powerful tool for surveillance and outbreak prediction (Alwakeel, 2025). Policymakers could use these signals to dispatch rapid response teams, increase testing, or launch awareness drives in specific areas before the outbreak worsens.

1.2.5 Health education in schools

AI chatbots can be deployed in local schools to teach children about hygiene, nutrition, and disease prevention in engaging ways (Aulia, Jati, Wibowo, Muhammad, & Saepudin, 2024). For example, a bot could quiz students about handwashing or construct interactive stories about nutrition. Educated students then become health ambassadors, sharing knowledge with parents and siblings. Studies show that AI can deliver personalized educational content effectively; in India, AI-driven platforms are being used to improve learning outcomes in rural settings. By integrating preventive health modules, these tools build awareness from a young age, creating a generational impact.

1.2.6 Routine monitoring and reminders

Gen AI can help with continual health maintenance. For instance, the system could remind patients to take their morning medications or check their blood pressure regularly. It can automatically adjust reminder schedules based on feedback (e.g., if a patient reports side effects, it might alert a health worker). It could also prompt pregnant women to attend antenatal check-ups at appropriate intervals or remind children about their immunizations. By connecting with wearable sensors (if available) or handheld devices, the AI can flag early warning signs (e.g., a gradual rise in blood sugar) and alert both the user and their ASHA/doctor, enabling timely intervention.

1.3 Strategy for rural healthcare

The strategy for rural healthcare can be implemented in various ways, a few of which are highlighted in the following.

1.3.1 Multilingual AI platforms

Developing AI agents to be accessible via mobile apps, SMS/WhatsApp chatbots, and voice assistants in local languages (Hindi, Bengali, Tamil, Telugu, etc.) is essential. These platforms should function effectively on basic smartphones. For example, using open-source LLMs fine-tuned on medical knowledge and regional language corpora ensures cultural relevance.

1.3.2 Curated medical knowledge bases

Training and updating the AI with **verified healthcare data**—government guidelines, treatment protocols, and local health statistics (Varnosfaderani & Forouzanfar, 2024)—is essential. Continually integrating India-specific medical guidelines (e.g., from the Indian Public Health Standards) and vernacular vocabulary ensures that the AI's advice remains accurate and contextually appropriate.

1.3.3 Public–private partnerships

Collaboration with **national health initiatives** (Ayushman Bharat, NDHM), Non-governmental organization (NGOs), telecommunication firms, and technology startups is essential. For instance, piloting the AI tools through Ayushman Bharat Health and Wellness Centers (HWCs) (AB-HWCs) ensures field testing under the National Health Mission (NHM) programs. Partnering with local NGOs helps adapt the system to community needs and build trust.

1.3.4 Training and outreach

Organizing hands-on workshops for ASHA and ANM workers on using AI tools and interpreting outputs. Empowering frontline workers to act as AI facilitators will accelerate adoption. Similarly, running awareness campaigns (via local radio and village meetings) to introduce villagers to the AI service and its benefits.

1.3.5 Data privacy and ethics

Build robust privacy protections from the start. All personal health data must be encrypted and stored securely in compliance with India's Digital Personal Data Protection Act. Use anonymized data for analytics. Establish ethical guidelines for AI use; for example, always have a "human in the loop" for serious diagnoses and ensure the AI clearly communicates its uncertainty and sources.

1.3.6 Monitoring and evaluation

Define clear key Performance Indicators (e.g., number of active users, diagnostic accuracy, referral reduction, increase in scheme enrollments, drop in avoidable hospital visits). Set up dashboards for continuous monitoring. Use randomized trials or phased rollouts to rigorously assess health outcomes and iterate on the system design.

1.3.7 Infrastructure support

Ensure functionality in low-bandwidth settings. Implement an offline-first design, where critical AI models and data are cached on the device, syncing to the cloud when connectivity is available. Coordinate with telecom operators to provide subsidized rural data plans. Equip select health centers with backup power (solar, battery) to ensure AI kiosks remain operational.

1.3.8 Scalability planning

Architect the system with containerized microservices so that it can scale across states and health departments. Plan for the integration of new modules over time (e.g., tele-ophthalmology or AI-dermatology). Explore future additions, such as drone delivery of medicine guided by AI, in extremely remote hamlets.

2 AI-enabled community healthcare and implementation plans

AI can greatly amplify the effectiveness of India's CHWs and primary care infrastructure. The following are a few transformative AI-driven initiatives, with pathways to implementation through existing programs.

2.1 AI-powered teleconsultation hubs

Equip village with Health and Wellness Centers or Primary Health Centers (PHCs) (under Ayushman Bharat) with smart telemedicine booths that use AI to enhance consultations. An AI assistant could transcribe the patient's history, analyze symptoms in real time, and suggest a preliminary diagnosis or treatment plan during the video call. For example, an AI system can perform automated triage of fever, cough, or skin rashes before the doctor joins remotely, saving the doctor's time. Use AB-HWC funds to install basic hardware (camera *Implementation*, display, internet). Integrate the Indian government's eSanjeevani telemedicine platform with an AI overlay. Develop easy-to-understand training materials so ASHA/ANM staff can operate the system. Pilot in a few districts and refine the AI's local language dialog.

2.2 Mobile AI diagnostic vans

Deploy AI-enabled mobile clinics that travel to villages. These vans carry point-of-care diagnostic devices (portable X-ray, ultrasound, ECG, pathology kits) linked to AI models for instant analysis. For instance, an AI algorithm can automatically read a chest X-ray image captured in the van to flag TB or pneumonia.

Implementation: Expand the NHM's Mobile Medical Units by partnering with private innovators. For example, collaborate with health tech startups providing cloud-connected imaging scanners that run inference on-device (offline) when connectivity is weak. Use district health office schedules (via Panchayats) to rotate the vans through villages. Ensure each van's team includes an ASHA or ANM trained to operate the equipment and relay AI results.

2.3 AI-driven health education chatbots

Leverage **multilingual chatbots on WhatsApp/SMS** to deliver targeted health education and reminders. These bots can send parents reminders for child immunizations, inform women about antenatal visit schedules, or educate diabetics on diet and foot care. By authenticating NDHM IDs or local Common Service Center (CSC) credentials, the AI can tailor messages based on a user's profile (age, pregnancy status, chronic diseases).

Implementation: Utilize the NDHM/Digi Locker framework for secure user verification. Develop region-specific content modules (collaborating with state health departments) and deliver them through a chatbot service. For rural penetration, run the service through CSCs, which already provide digital access in villages. Track engagement metrics (e.g., vaccination reminder compliance) to improve messaging.

2.4 Smart medication adherence systems

Introduce AI-enabled devices and apps to improve drug compliance for chronic diseases. For example, provide **low-cost smart pill dispensers** that alert patients via SMS when it is time to take a dose and use AI to adjust reminders based on self-reported side effects or missed doses.

Implementation: Use the government scheme (Jan Aushadhi) distribution network to supply affordable smart dispensers. Integrate the system with disease registries like Nikshay (for TB) and the noncommunicable diseases portal so that patients with TB or diabetes automatically receive adherence support. Train local pharmacists and ASHAs (via PMJAY workshops) to enroll eligible patients and monitor compliance through simple dashboards.

2.5 Predictive analytics for outbreak surveillance

Employ AI to detect early warning signs of outbreaks from scattered data. By aggregating anonymized inputs from HWCs, mobile vans, teleconsultation logs, pharmacy sales, and even school absenteeism, AI models can spot spikes in symptoms that indicate an impending epidemic (e.g., dengue or influenza).

Implementation: Under the Integrated Health Information Platform (IHIP), standardized data (e.g., daily fever case counts, antimalarial drug sales) is collected across rural health points. Machine learning models are deployed at state health data centers to generate risk heatmaps and issue alerts to district surveillance units (DSUs) when thresholds are crossed (IndiaAI, n.d.). Upon receiving an alert, rapid response teams can be dispatched for vaccination drives or vector control in the identified hotspots.

Each initiative should begin with pilot projects involving state health departments, NGOs, and medical colleges. Successful pilots can then be scaled under the NDHM. Together, these AI-enabled systems will empower rural health workers, extend high-quality care into villages, and strengthen India's overall public health resilience (Fig. 22.1).

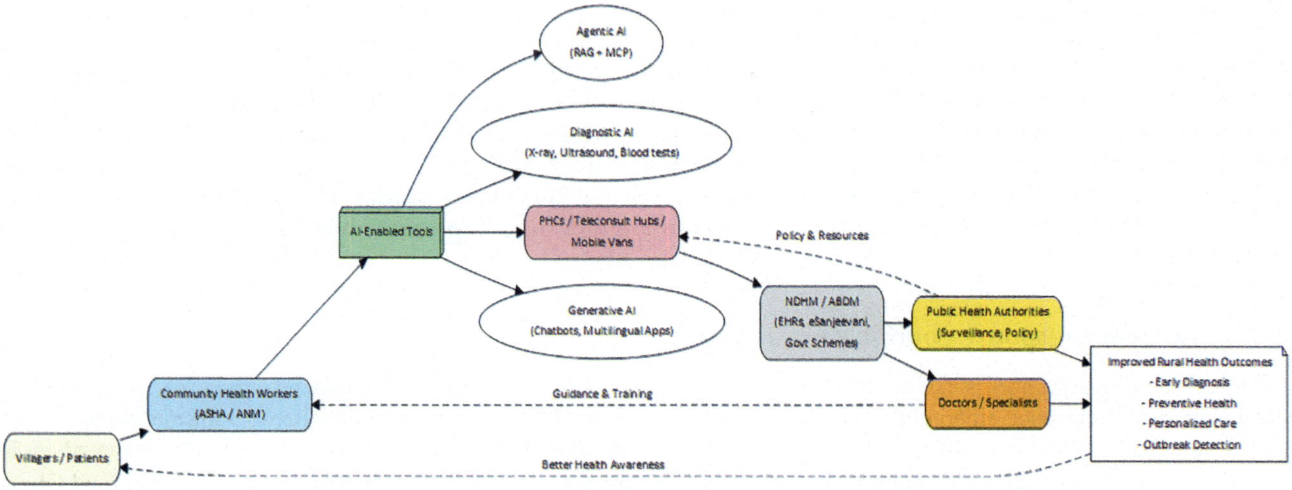

FIG. 22.1 Workflow of AI applications in community health in India. The implementation of AI-enabled tools has augmented the delivery of community healthcare, resulting in early diagnosis and prevention of disease outbreaks, as well as the delivery of personalized healthcare.

3 Agentic intelligence has quality impact on medical support

Agentic intelligence refers to AI systems that can autonomously perform complex tasks by combining reasoning, data retrieval, and long-term planning (Hosseini & Seilani, 2025). In healthcare, this means AI assistants that actively fetch relevant medical knowledge and maintain conversational context to guide patients or support clinicians (Ye et al., 2024). Key technologies include retrieval-augmented generation (RAG) (Gargari & Habibi, 2025) and the model context protocol (MCP) (Singh et al., 2025). RAG allows LLMs to retrieve the latest medical guidelines or patient records as needed, enhancing accuracy. For example, instead of relying on static model knowledge, an agent can pull WHO treatment protocols for malaria or the latest Indian clinical guidelines when making recommendations. MCP (a new open standard) enables secure, bidirectional connections between the AI and data systems, so the agent can contextually access electronic health records, lab databases, and messaging tools without isolated silos (Fink, Rau, Reisert, Bamberg, & Russe, 2025). In effect, these methods let AI keep track of a patient's history and the rationale for past advice, making its support more coherent and reliable over time.

The following are five illustrative uses of agentic AI in rural medical support.

3.1 Autonomous symptom assessment agents

These agents interact with patients via voice or chat to elicit symptoms, history, and lifestyle factors. They can use RAG to query authoritative sources (e.g., WHO or Indian health guidelines) and regional health records to generate an initial assessment. By applying MCP, the agent maintains context across multiple turns—remembering previous answers and local dialect nuances—to ask follow-up questions intelligently. For instance, if a patient mentioned fever and headache, the agent might retrieve dengue protocols and then ask about rash or travel history. Such an agent can integrate with NDHM's HealthID and digital records to automatically incorporate a patient's known health issues (e.g., existing diabetes) into the assessment.

3.2 Intelligent prescription assistants

After an assessment, an AI agent can suggest medication plans and dosages aligned with India's approved drug lists. Using RAG, it fetches the latest Drug Controller General of India (DCGI) formulary and evidence-based treatment regimens. MCP ensures it remembers the patient's specific context, e.g., allergies, pregnancy status, or comorbidities, so that prescription suggestions are safe. For example, for a child with asthma, it would verify the recommended inhaler and dosage in pediatrics. Such an agent could also interface with online pharmacy APIs (like the Jan Aushadhi e-portal) to list affordable generics or arrange refills.

3.3 Dynamic chronic disease management agents

For long-term conditions (e.g., diabetes, hypertension), agents can generate personalized care plans that adapt over time (Dong et al., 2025). They use RAG to keep up with the latest clinical research and local health statistics, and MCP to aggregate data streams (self-reported blood sugars, wearable readings, ASHA visit notes) into a continuous patient timeline. By applying reinforcement learning or adaptive algorithms, the agent adjusts targets and advice; for example, if a diabetic patient consistently has high morning glucose, it might suggest a diet tweak or medication review. This agent also schedules check-ins: if the patient forgets to log a reading, it proactively asks why and adapts the plan accordingly.

3.4 Emergency response coordination agents

In critical situations, AI agents can triage severity, provide first-aid guidance, and coordinate resources. Drawing on RAG, the agent would reference emergency care protocols (e.g., cardiopulmonary resuscitation steps, burn management) and local resource directories (nearest ambulance service or clinic with oxygen). Using MCP, the agent continuously updates its plan as new information arrives (symptom changes, location updates). It can alert a 108-emergency dispatch, send Global Positioning System coordinates, and even notify nearby community health volunteers. For example, if an agent learns a child has lost consciousness after a fall, it could guide caregivers through stabilization and simultaneously contact emergency responders.

3.5 Community health analytics orchestrators

At the population level, agentic AI can sift through massive rural health datasets to guide public health action. RAG modules query standardized databases like IHIP, Integrated Disease Surveillance Program, and District Health Information System for trends (Amugongo, Mascheroni, Brooks, Doering, & Seidel, 2025). MCP organizes ongoing data (seasonal disease incidence, vaccination rates, commodity stock levels) into coherent narratives. The agent might analyze these streams to predict, for example, a malaria outbreak in District X. It then generates actionable insights: "Increase mosquito net distribution in Blocks A and B, and start indoor residual spraying by next week." These insights can feed directly into district health action plans, making surveillance data "speak" to decision-makers.

By combining RAG's precise retrieval of up-to-date knowledge with MCP's contextual continuity, agentic intelligence creates an "always-learning" medical assistant (Tamanna, 2025). In rural India, this means villages can tap into a virtual chain of medical expertise: from autonomous triage bots in villagers' hands to district-level analytics coordinators, AI agents can amplify scarce human resources and deliver tailored care reliably at every level of the health system.

4 AI for primary healthcare delivery

At the primary care level (PHCs and subcenters), AI is redefining how first-contact medicine is delivered. Gen AI can augment routine services by assisting in diagnostics, patient education, and workflow management. For example, in PHC outpatient departments, an AI-driven symptom checker app can help nurses triage patients before a doctor sees them. The AI could record the chief complaint and vital signs, compare them to clinical guidelines, and suggest preliminary lab tests or referrals, which the doctor then reviews. Portable AI tools (smartphone apps or edge devices) can analyze point-of-care tests: a glucometer reading could be sent to an AI model that checks for worrisome patterns, or a photo of an ear infection could be prescreened using vision AI. AI can also streamline immunization clinics by recalling a child's vaccination history from digital records in the *NDHM,* now known as *Ayushman Bharat Digital Mission* (Arjun, Poorvikha, Kurpad, & Thomas, 2024) and alerting the nurse about due vaccines, reducing missed shots.

Telehealth also extends into PHCs: Gen AI can run an initial video consultation with a patient and prepare a concise summary for the PHC doctor to finalize. For chronic care, AI can integrate with the PHC's electronic logbook to flag patients who have not come for follow-ups or whose condition is deteriorating. In this way, AI makes primary care more proactive and efficient. Academic reviews note that AI's key rural benefit is in expanding access to disease screening and early detection at the grassroots. For instance, an AI model on a PHC doctor's tablet could rapidly interpret a low-cost ECG to catch arrhythmias or analyze a smartphone fundus photo to screen for diabetic retinopathy, enabling timely specialist referrals. Overall, AI tools in primary healthcare help catch problems early, personalize routine check-ups, and ease the workload on PHC staff, pushing quality care to the community's doorstep ("Teleradiology and technology innovations" (Chandramohan et al., 2023).

5 Equitable access to preventive care

Preventive care in terms of vaccination, screening, and early detection saves lives, but in rural India, it is often inequitable. AI can help level the playing field by making preventive services smarter and more accessible. For example, AI-driven screening programs can target common conditions: computer vision models can analyze low-cost scans (like ultrasound or retinal images) to detect cancers, cataracts, or heart disease at earlier stages. Studies show AI algorithms can spot breast cancer (Ahn et al., 2023) or lung cancer signs on X-rays and ultrasounds (Cellina et al., 2023) that clinicians and technologists might fail to detect, allowing villagers to be referred for treatment before symptoms worsen.

AI also powers more effective vaccination drives (Atkinson, Mithani, Bell, Rubens-Augustson, & Wilson, 2020; Olawade et al., 2024). By mining digital health records and geographic data, AI can identify pockets of underimmunized children and predict who is most likely to default (Pavia et al., 2024). Health workers, armed with this insight, can conduct targeted outreach (for example, scheduling small vaccination camps at local schools on specific dates). Chatbot reminders and automated scheduling via mobile phones can ensure pregnant women or new mothers complete their vaccination schedules on time.

6 Conclusion

Gen AI and agentic AI offer a transformative solution to India's persistent rural healthcare challenges (Schneider, 2025). By empowering CHWs with AI-driven tools for diagnostics, patient education, and personalized guidance, technology can bridge the significant resource gap. Integrating these smart systems into mobile clinics, teleconsultation hubs, and public health programs creates a scalable and proactive ecosystem. This approach is not merely a technological upgrade but a vital catalyst for achieving health equity, ensuring that quality care reaches the most remote villages and strengthens India's overall public health resilience.

AI-based tool usage

Authoritative sources on AI in Indian healthcare and rural public health were used to compile this chapter, along with reports on national digital health initiatives. These sources confirm the potential of AI technologies (GPT-4, RAG, MCP) to enhance rural health delivery, as well as the documented challenges and pilot results within India's context.

References

Ahn, J. S., Shin, S., Yang, S. A., Park, E. K., Kim, K. H., Cho, S. I., et al. (2023). Artificial intelligence in breast cancer diagnosis and personalized medicine. *Journal of Breast Cancer*. 20929900. *26*(5), 405–435. Retrieved from: https://www.ejbc.kr/search.php?where=aview&id=10.4048/jbc.2023.26.e45&code=0096JBC&vmode=FULL. doi:10.4048/jbc.2023.26.e45.

Alwakeel, M. M. (2025). AI-Assisted real-time monitoring of infectious diseases in urban areas. *Mathematics*. 22277390. *13*(12). Retrieved from: http://www.mdpi.com/journal/mathematics. doi:10.3390/math13121911.

Amugongo, L. M., Mascheroni, P., Brooks, S., Doering, S., & Seidel, J. (2025). Retrieval augmented generation for large language models in healthcare: a systematic review. *PLOS Digital Health*. 27673170. *4*(6). Retrieved from: https://journals.plos.org/digitalhealth/article?id=10.1371/journal.pdig.0000839. doi:10.1371/journal.pdig.0000877.

Arjun, M. C., Poorvikha, S., Kurpad, A. V., & Thomas, T. (2024). Knowledge, attitude and practice about Ayushman Bharat Digital Mission and digital health among hospital patients. *Journal of Family Medicine and Primary Care*. 2249-4863. *13*(10), 4476–4481. doi:10.4103/jfmpc.jfmpc_255_24.

Atkinson, K. M., Mithani, S. S., Bell, C., Rubens-Augustson, T., & Wilson, K. (2020). The digital immunization system of the future: imagining a patient-centric, interoperable immunization information system. *Therapeutic Advances in Vaccines and Immunotherapy*. 25151363. *8*. Retrieved from: https://journals.sagepub.com/home/tav. doi:10.1177/2515135520967203.

Aulia, E. S. N., Jati, G., Wibowo, I., Muhammad, H. N., & Saepudin, E. (2024). Chatbots Be nutritionists: exploring the potential of AI-powered to improve nutritional counseling in Indonesia. *Jurnal Sosioteknologi*. 2443-258X. *23*(3), 324–335. doi:10.5614/sostek.itbj.2024.23.3.2.

Barreda, M., Cantarero-Prieto, D., Coca, D., Delgado, A., Lanza-León, P., Lera, J., et al. (2025). Transforming healthcare with chatbots: uses and applications—A scoping review. *Digital Health*. 20552076. *11*. Retrieved from: https://journals.sagepub.com/home/DHJ. doi:10.1177/20552076251319174.

Cellina, M., Cacioppa, L. M., Cè, M., Chiarpenello, V., Costa, M., Vincenzo, Z., et al. (2023). Artificial intelligence in lung cancer screening: the future is now. *Cancers*. 20726694. *15*(17). Retrieved from: http://www.mdpi.com/journal/cancers/. doi:10.3390/cancers15174344.

Chandramohan, A., Krothapalli, V., Augustin, A., Kandagaddala, M., Thomas, H. M., Sudarsanam, T. D., et al. (2023). *Teleradiology and technology innovations in radiology: status in India and its role in increasing access to primary health care*. www.thelancet.com

Datta, B., Prakash, A. K., Ford, D., Tanwar, P. K., Goyal, P., Chatterjee, P., et al. (2019). Comparison of clinical and cost-effectiveness of two strategies using mobile digital x-ray to detect pulmonary tuberculosis in rural India. *BMC Public Health*. 14712458. *19*(1). Retrieved from: http://www.biomedcentral.com/bmcpublichealth. doi:10.1186/s12889-019-6421-1.

Dong, C., Ji, Y., Fu, Z., Qi, Y., Yi, T., Yang, Y., et al. (2025). Precision management in chronic disease: an AI empowered perspective on medicine-engineering crossover. *iScience*. 25890042. *28*(3). Retrieved from: https://www.sciencedirect.com/science/journal/25890042. doi:10.1016/j.isci.2025.112044.

Fink, A., Rau, A., Reisert, M., Bamberg, F., & Russe, M. F. (2025). Retrieval-augmented generation with large language models in radiology: from theory to practice. *Germany Radiology: Artificial Intelligence*. 26386100. *7*(4). Retrieved from: https://pubs.rsna.org/doi/epdf/10.1148/ryai.240790. doi:10.1148/ryai.240790.

Gargari, O. K., & Habibi, G. (2025). Enhancing medical AI with retrieval-augmented generation: a mini narrative review. *Digital Health*. 20552076. *11*. Retrieved from: https://journals.sagepub.com/home/DHJ. doi:10.1177/20552076251337177.

Hosseini, S., & Seilani, H. (2025). The role of agentic AI in shaping a smart future: a systematic review. *Array*. 25900056. *26*, 100399. doi:10.1016/j.array.2025.100399.

Khalid, N., Qayyum, A., Bilal, M., Al-Fuqaha, A., & Qadir, J. (2023). Privacy-preserving artificial intelligence in healthcare: techniques and applications. *Computers in Biology and Medicine*. 18790534. *158*. Retrieved from: www.elsevier.com/locate/compbiomed. doi:10.1016/j.compbiomed.2023.106848.

Kumar, R., & Pal, R. (2018). India achieves WHO recommended doctor population ratio: a call for paradigm shift in public health discourse! *Journal of Family Medicine and Primary Care*. 2249-4863. *7*(5), 841. doi:10.4103/jfmpc.jfmpc_218_18.

Kumar, Y., Koul, A., Singla, R., & Ijaz, M. F. (2023). Artificial intelligence in disease diagnosis: a systematic literature review, synthesizing framework and future research agenda. *Journal of Ambient Intelligence and Humanized Computing*. 18685145. *14*(7), 8459–8486. Retrieved from: https://www.springer.com/journal/12652. doi:10.1007/s12652-021-03612-z.

Maity, S., & Saikia, M. J. (2025). Large language models in healthcare and medical applications: a review. *Bioengineering*. 23065354. *12*(6). Retrieved from: www.mdpi.com/journal/bioengineering. doi:10.3390/bioengineering12060631.

Olawade, D. B., Teke, J., Fapohunda, O., Weerasinghe, K., Usman, S. O., Ige, A. O., & Clement David-Olawade, A. (2024). Leveraging artificial intelligence in vaccine development: a narrative review. *Journal of Microbiological Methods*, *224*. doi:10.1016/j.mimet.2024.106998.

Olawade, D. B., Wada, O. Z., Odetayo, A., David-Olawade, A. C., Asaolu, F., & Eberhardt, J. (2024). Enhancing mental health with Artificial Intelligence: current trends and future prospects. *Journal of Medicine, Surgery, and Public Health*, *3*, 100099. doi:10.1016/j.glmedi.2024.100099.

Parameshwarappa, P. M., & Olickal, J. J. (2023). Telemedicine awareness and the preferred digital healthcare tools: a community-based cross-sectional study from Rural Karnataka, India. *Indian Journal of Community Medicine*. 19983581. *48*(6), 915–919. Retrieved from: http://www.ijcm.org.in. doi:10.4103/ijcm.ijcm_770_22.

Pavia, G., Branda, F., Ciccozzi, A., Romano, C., Locci, C., Azzena, I., et al. (2024). Integrating digital health solutions with immunization strategies: improving immunization coverage and monitoring in the post-COVID-19 era. *Vaccines*. 2076393X. *12*(8). Retrieved from: http://www.mdpi.com/journal/vaccines. doi:10.3390/vaccines12080847.

Ramjee, P., Chhokar, M., Sachdeva, B., Meena, M., Abdullah, H., Vashistha, A., et al. (2025). ASHABot: an LLM-powered chatbot to support the informational needs of community health workers. In *Conference on Human Factors in Computing Systems—Proceedings*. India: Association for Computing Machinery. doi:10.1145/3706598.3713680.

Schneider, J. (2025). Generative to agentic AI: survey, conceptualization, and challenges. *Liechtenstein arXiv*. 23318422. Retrieved from: https://arxiv.org. doi:10.48550/arXiv.2504.18875.

Sharma, D. C. (2015). India still struggles with rural doctor shortages. *The Lancet*. 1474547X. *386*(10011), 2381–2382. Retrieved from: http://www.journals.elsevier.com/the-lancet/. doi:10.1016/S0140-6736(15)01231-3.

Singh, A., Ehtesham, A., Kumar, S., & Khoei, T. T. (2025). *A survey of the model context protocol (MCP): Standardizing context to enhance large language models (LLMs)*. Preprints 2025, 025040245. doi:10.20944/preprints202504.0245.v1.

Tamanna, T. (2025). Building powerful agentic AI with Google ADK, MCP, RAG, and Ollama. Retrieved from: https://www.researchgate.net/publication/394404324.

Tripathi, S., & Mitra, A. (2022). Shedding light on unnoticed gems in India: a small town's growth perspective. *Land Use Policy*. 02648377. *120*, 106239. doi:10.1016/j.landusepol.2022.106239.

Varnosfaderani, S. M., & Forouzanfar, M. (2024). The role of AI in hospitals and clinics: transforming healthcare in the 21st Century. *Bioengineering*. 23065354. *11*(4). Retrieved from: www.mdpi.com/journal/bioengineering. doi:10.3390/bioengineering11040337.

Ye, J., Woods, D., Jordan, N., & Starren, J. (2024). The role of artificial intelligence for the application of integrating electronic health records and patient-generated data in clinical decision support. AMIA Joint Summits on Translational Science Proceedings, 2024, pp. 459–467.

Section VII

New frontiers of artificial intelligence

Chapter 23

Quantum artificial intelligence for human health: A statistical framework

Hossein Hassani[a], Leila Marvian Mashhad[b], and Steve MacFeely[c]
[a]IIASA, Laxenburg, Austria
[b]Imam Reza International University, Mashhad, Iran
[c]OECD, Paris, France

1 Introduction

The convergence of quantum computing and machine learning—often called quantum artificial intelligence (QAI)—introduces new ways of representing data and solving optimization problems that could enhance human health across discovery, diagnosis, and care delivery. Realizing this potential, however, requires more than technical enthusiasm: it demands a statistical lens. Rigorous study design, calibrated uncertainty, and valid hypothesis testing are essential to separate true advances from noise in the current "noisy intermediate-scale quantum" era (Preskill, 2018; Topol, 2019).

Methodologically, QAI combines three complementary components, each with a clear clinical analog:

- Quantum feature maps and kernels take complex health data, such as genomics, high-dimensional imaging features, or multiomics profiles, and project them into an enormous mathematical space where subtle patterns can be separated more easily. In practice, this can produce richer similarity measures between patients than standard kernels, potentially improving risk stratification or disease subtyping.
- Variational quantum circuits (VQCs) are trainable quantum models whose parameters are adjusted to predict an outcome. Clinically, they function like flexible probabilistic classifiers or regressors: given patient variables, they output a probability of disease, relapse, or another event, with uncertainty arising from both the data and the hardware. This makes them suitable for tasks such as radiology triage, outcome forecasting, or treatment recommendation.
- Quantum optimization routines, such as quadratic unconstrained binary optimization (QUBO) and the quantum approximate optimization algorithm (QAOA), tackle the kinds of combinatorial scheduling or allocation problems common in hospitals: assigning surgeries to operating rooms, matching donors to recipients, or allocating beds and staff. These algorithms aim to find good (sometimes faster or better) solutions to such binary decision problems.

Early studies already explore QAI in drug discovery, multiomics stratification, medical imaging, and hospital logistics. Systematic reviews highlight a lack of head-to-head comparisons against strong classical baselines, underscoring the need for careful statistical evaluation before making clinical claims (Devadas & Sowmya, 2025; Gupta, Wood, Engstrom, Pole, & Shrapnel, 2025).

Because health decisions are high stakes, evaluation must go beyond a single "accuracy" number. We emphasize two pillars particularly relevant to clinicians:

1. Distributional comparisons [Hassani–Silva Kolmogorov–Smirnov (KS) test]. Mean error can hide dangerous outliers. This test compares the *entire distribution* of errors between two models to see if one produces fewer extreme mistakes (e.g., missed urgent cases). Clinically, it answers, "Does this system reduce the risk of catastrophic misclassification for some patients, even if average accuracy is similar?"
2. Uncertainty and calibration (reliability diagrams, conformal prediction). It is not enough to predict a probability; the probability must be trustworthy. Calibration tools show whether predicted risks (e.g., "20% chance of sepsis") match observed outcomes. Conformal prediction adds statistically valid intervals or prediction sets, so clinicians know, under

Applications of Artificial Intelligence in Healthcare. https://doi.org/10.1016/B978-0-443-43934-6.00039-0

mild assumptions, how often the true outcome should lie within the reported range, supporting informed decision-making at the bedside.

This chapter presents QAI from a statistical perspective. We explain how quantum kernels and VQCs act as estimators for clinical prediction tasks, and how QUBO and QAOA encode and solve complex scheduling and resource allocation problems. We then specify the estimands (the precise quantities to be estimated), the loss functions used to evaluate predictions, and the methods for inference and multiple-testing control. Finally, we show how to report uncertainty and calibration in line with medical AI reporting standards. One key example is CONSORT-AI (Consolidated Standards of Reporting Trials—Artificial Intelligence)—an extension of the CONSORT 2010 guidelines that provides a checklist of AI-specific items such as model version, data handling, and subgroup performance. Following such standards helps clinicians, regulators, and patients assess safety and reliability before implementation.

2 Data-generating processes in health

In this chapter, we view clinical and biomedical data not as isolated spreadsheets but as the observable outputs of an underlying data-generating process (DGP). A DGP is a conceptual model that links patient characteristics (covariates), clinical actions (interventions), and hidden influences (latent factors such as genetics, clinician preferences, or hospital policies) to outcomes such as survival, readmission, or adverse events. Taking this view matters in healthcare because real-world data are rarely clean or complete. However, as Hand (2020) stresses, "dark data" matter:

1. Missingness: Laboratory tests may be absent because a clinician judged them unnecessary for low-risk patients. Patient-reported outcomes may be missing when symptom burden is highest. Without modeling these mechanisms, we can produce biased risk scores, for example, underestimating deterioration in the sickest patients.
2. Censoring: Time-to-event endpoints, such as hospital readmission or progression-free survival, are often right-censored when patients are lost to follow-up or transferred to other facilities. Ignoring censoring can mislead survival predictions.
3. Measurement error and label noise: Imaging features vary by scanner vendor and protocol; laboratory values differ across platforms; radiology or pathology labels may contain interreader disagreement. If unaddressed, these errors attenuate model performance and create false signals of quantum advantage.
4. Site and temporal heterogeneity: Case-mix, clinical practice, and even disease epidemiology shift across hospitals and over time. A model trained on one hospital's data can appear to perform well in-sample, yet fail when deployed elsewhere.

By explicitly formalizing the DGP, including how data are acquired, processed, and observed, we can:

- build estimators (quantum, classical, or hybrid) that correctly propagate uncertainty due to missingness, censoring, and hardware noise;
- design evaluation protocols (e.g., site-stratified or time-blocked cross-validation) that mimic real deployment conditions; and
- make statistically defensible comparisons between quantum–classical systems and strong classical baselines, ensuring that any reported gain is real and reproducible rather than an artifact of data quirks.

For example, when developing a quantum kernel to stratify oncology patients, the DGP framework clarifies how to impute missing biomarkers, adjust for scanner effects, and weight censored survival times. In a quantum optimization application, such as operating-room scheduling, it highlights how seasonal demand, policy changes, and staff shortages influence the observed scheduling data and should be modeled in performance estimates.

This DGP perspective turns QAI research from a technology demonstration into a scientifically credible, clinically relevant exercise, producing results that clinicians and regulators can trust.

2.1 Structural view and observed data

We model clinical/biomedical data using a structural causal framework in which *context* and *latent* factors generate the observed covariates, treatments, and outcomes. Let

- W: observed, upstream *context* (site/scanner S, calendar time t, acquisition protocol, region/policy, study wave);
- $X \in \chi$: measured covariates (demographics, laboratories, images, omics);
- $A \in \mathcal{A}$: intervention/exposure (e.g., treatment, policy, workflow decision);

- Y: outcome (binary, continuous, ordinal, or time-to-event);
- $U = (U_X, U_A, U_Y)$: unobserved factors (latent biology, case-mix, clinician preference, etc.).

A generic structural representation (with site/time indices suppressed) is:

$$X = f_X(W, U_X),$$
$$A = f_A(X, U_A),$$
$$Y = f_Y(X, A, U_Y, \varepsilon_Y),$$

where ε_Y is idiosyncratic noise. This form captures (1) context W shaping what is measured and how (X); (2) treatment assignment A depending on patient state X and unobservables U_A; and (3) outcomes Y depending on both X and A, plus unobservables U_Y.

However, in practice, we observe error-prone measurements and selection indicators. Let

$$O = (X^*, A, Y^{\sim}, R_X, R_Y, T, C, S)$$

where X^* and $Y^{\sim}$ are measured versions of X and Y, $R_X, R_Y \in \{0, 1\}$ indicate observation/missingness, T is the observed time, C is the censoring time, and S is the site/scanner (often part of W). A convenient measurement/selection layer is:

$$X^* = g_X(X, S,_X) \; Y^{\sim} = g_X(Y,_Y)$$

$$R_X = r_X(W, X^*, A) \; R_Y = r_Y(W, X^*, A, Y)$$

where η_X and η_Y encode measurement noise (e.g., scanner variability, rounding, reader effects). For time-to-event data with a latent event time T^* and censoring time C,

$$T = \min(T^*, C), \quad \Delta = \mathbb{1}\{T^* \leq C\},$$

and analyses should use censoring-aware losses/weights.

Here, we use W to bundle upstream, exogenous context that shapes X *before* treatment/outcome realization: e.g., imaging device/firmware and acquisition protocol (affecting pixel statistics), laboratory platform (affecting assay noise), site/region and policy (affecting case-mix), and calendar time (affecting practice drift). When convenient, write W = (S, t, protocol) and expand f_X accordingly.

We can also extend this approach for panel or longitudinal observations. For subject $i = 1, \ldots, n$ over times $t = 1, \ldots, T_i$:

$$X_{it} = f_X(W_{it}, U_{Xi}), \; A_{it} = f_A(X_{it}, U_{Ai}), \; Y_{it} = f_Y(X_{it}, A_{it}, U_{Yi}, \varepsilon_{Yit}),$$

with observed $O_{it} = (X^*, A_{it}, Y^{\sim}_{it}, R_{X,it}, R_{Y,it}, T_{it}, C_{it}, S_{it})$. Splits and uncertainty should respect clustering by i (patient-level cross-validation (CV) or cluster bootstrap) and temporal order (blocked/rolling CV).

Given a predictor f_θ (classical, quantum, or hybrid), we target the risk:

$$\mathcal{R}(f) = \mathbb{E}[\ell(Y, f(x))]$$

where ℓ is an appropriate loss (e.g., log-loss, Brier, continuous ranked probability score (CRPS), time-to-event losses). Estimation and evaluation must be consistent with the DGP above: propagate measurement noise, handle missingness/censoring [e.g., inverse probability weighting (IPW)/inverse probability of censoring weighting (IPCW)], and account for site/time heterogeneity (W) to obtain unbiased comparisons and valid uncertainty (Hernán and Robins, 2020).

It should be noted that quantum kernels and variational circuits used inside f_θ inherit any biases from W and measurement error in X^*. For variational models, finite-shot estimation introduces an extra variance term that should be reflected in confidence intervals, calibration curves, and downstream tests. Furthermore, quantum kernels and VQCs can be thought of as two different "building blocks" inside a quantum-enabled model.

- A quantum kernel is essentially a very rich similarity measure. It takes patient data—for example, a mixture of genomics, imaging features, and laboratory results—and encodes them into a quantum state. The overlap between two such states defines how similar two patients are in a very high-dimensional space. This can help separate subtle subgroups (e.g., responders vs. nonresponders, or high risk vs. low risk) more cleanly than a standard kernel in some cases.

- A VQC is the quantum analog of a trainable model, such as a neural network. Its parameters are tuned so that, given a patient's data, it outputs a probability of an event (disease onset, relapse, ICU transfer) or some other prediction needed for clinical decision-making.

Because these quantum components still depend on the same underlying data, they inherit any biases or measurement errors present in the inputs, for example, scanner differences, batch effects, or site-specific case-mix. M oreover, VQCs are estimated from a finite number of quantum measurements ("shots"), which adds an extra layer of statistical variability on top of ordinary sampling error.

For healthcare applications, this means that confidence intervals, calibration curves, and downstream tests must explicitly account for both the usual clinical data issues and the quantum-hardware noise. Reporting these uncertainties allows clinicians to judge how trustworthy a model's outputs are before using them in practice.

In addition, the implications for hybrid QAI are that quantum kernels and variational circuits are treated as estimators within this DGP: kernels operate on features (possibly learned by a classical front-end) and inherit any selection bias or measurement error in X^*; VQCs produce probabilistic outputs with finite-shot measurement noise that must be propagated through the loss and confidence intervals.

2.2 Missingness mechanisms and estimation

In health data, missingness is rarely benign: laboratory values may be absent because clinicians deemed them unnecessary for low-risk patients; patient-reported outcomes can be missing when symptom burden is highest; and loss to follow-up in oncology registries often coincides with disease progression or transfer of care. Hand (2020) describes a case where an ML system designed to predict mortality associated with pneumonia counterintuitively predicted a lower risk of dying for asthmatics. Closer examination revealed that missing data led to this spurious result; patients with asthma were prioritized for intensive care and thus had good outcomes overall.

Let $R_Y = 1$ indicate that Y is observed. Following the classical taxonomy (Little & Rubin, 2002; Rubin, 1976), missing completely at random (MCAR) assumes $R_Y \perp\!\!\!\perp (Y, X, A)$ and is uncommon in practice; missing at random (MAR) assumes $R_Y \perp\!\!\!\perp Y \mid (X, A)$ and is plausible in workflows where observation is driven by triage variables available in the record (e.g., vitals, prior laboratories, or imaging scores); missing not at random (MNAR) allows R_Y to depend on the unobserved Y even after conditioning on (X, A), which is typical for pain scores, mental-health surveys, and late clinical outcomes when sicker patients avoid scheduled visits. Correct identification of the missingness process is critical in clinical applications because risk estimates inform high-stakes decisions (triage, discharge, therapy escalation).

Because data splits must mimic deployment, estimation procedures are *fold-aware*: any modeling of missingness, imputation, or weighting is fit *only* on the training portion of each cross-validation fold and then applied to the corresponding validation/test portion to prevent optimistic bias. In patient-level electronic health records (EHR) data, folds should respect clustering at the individual and site levels (patient-level CV, site-stratified CV); in longitudinal monitoring, folds should preserve temporal order (blocked/rolling CV).

Under MAR, multiple imputation (MI) provides a principled route to valid inference. Generate M completed datasets from $p(Y \mid X, A, R_Y)$, fit the model f_θ in each completed set, and pool estimates via Rubin's rules.

In imaging or multiomics pipelines, the imputation model should incorporate site/scanner indicators and batch effects to reflect acquisition heterogeneity that is common in hospital networks. When outcomes are selectively observed, IPW targets the full- data risk under correct specification of the observation model. Let $\pi_i = Pr(R_{Y,i} = 1 \mid X_i, A_i)$ be estimated in the training fold. The IPW estimator of the expected loss is

$$\widehat{\mathcal{R}}_{IPW} = \frac{1}{n}\sum_{i=1}^{n}\frac{R_{Y,i}}{\hat{\pi}_i}\ell(Y_i, f(X_i))$$

which reweights observed outcomes to represent the target clinic population (Seaman & White, 2013). In health systems, triage rules include the same clinical covariates that drive measurement (e.g., early-warning scores) in $\hat{\pi}$ to reduce bias.

A doubly robust alternative combines an outcome regression $\widehat{m}(X, A) \approx E[Y \mid X, A]$ with the observation model $\hat{\pi}(X, A)$. An augmented IPW estimating equation for the mean loss $E[\ell(Y, f(X))]$ is

$$\widehat{\mathcal{R}}_{DR} = \frac{1}{n}\sum_{i=1}^{n}\left\{\widehat{m}_\ell(X_i, A_i) + \frac{R_{Y,i}}{\hat{\pi}(X_i, A_i)}\left[\ell(Y_i, f(X_i)) - \widehat{m}_\ell(X_i, A_i)\right]\right\}$$

where $\hat{m}_{\ell}(x, a)$ predicts the loss conditional on $(X = x, A = a)$. Consistency holds if either $\hat{m}$ or $\hat{\pi}$ is correctly specified, a desirable safety property in messy clinical data.

For time-to-event endpoints (readmission, progression-free survival), right censoring is routine and often informative across sites. Let T^* be the event time, C the censoring time, $T = \min(T^*, C)$ and $\Delta = \mathbb{1}\{T^* \le C\}$. With the censoring survival $G(t \mid X) = \Pr(C \ge t \mid X)$ estimated from the training fold, IPCW replaces contributions to the risk or loss with $\Delta_i / G(T^- \mid X_i)$ (Hernán and Robins, 2020). In oncology registries, include treatment line, center, and calendar time in G to capture practice drift and transfer-of-care patterns.

Evaluation in the presence of missingness and censoring must propagate these mechanisms into comparative statistics used for model selection and claims of improvement. For loss-based comparisons (e.g., Diebold–Mariano tests), define weighted loss differentials $d_i^{(w)} = w_i[L(\hat{y}_i^{(A)}, Y_i) - L(\hat{y}_i^{(B)}, Y_i)]$ with $w_i = R_{Y,i} / \hat{\pi}_I$ for missing outcomes or $w_i = \Delta_i / G(T^- \mid X_i)$ for censored survival; use a heteroskedasticity and autocorrelation consistent (HAC)/Newey–West variance estimate computed on $\{d^{(w)}\}$ to obtain a valid test under dependence. For distributional comparisons (Hassani–Silva KS), form weighted empirical cumulative distribution functions (CDFs) $F_{A,w}(x) = \left(\sum_i w_i\right)^{-1} \sum_i w_i \mathbb{1}\{e_i^{(A)} \le x\}$ and $F_{B,w}(x)$ and compute $D_w = \sup_w |F_{A,w}(x) - F_{B,w}(x)|$ with p-values obtained by a block bootstrap that respects patient clustering and temporal order. Reporting should include point estimates, interval estimates that reflect $\pi\hat{}$ or G, and sensitivity analyses for MNAR mechanisms (e.g., δ-adjusted selection models or pattern-mixture models), because clinical stakeholders need to understand how conclusions might change if the sickest patients are systematically undermeasured.

In real health-care settings, once a model begins influencing decisions, it also changes the very data that will be recorded afterward. For example:

- A sepsis-alert model that flags high-risk patients may prompt earlier laboratory tests, extra monitoring, or faster antibiotics. Future records will then show more laboratory values and perhaps fewer cases of late-diagnosed sepsis, not because the underlying risk has changed, but because care pathways have changed.
- A radiology-triage tool that prioritizes "urgent" images may alter which patients get scanned, how quickly they are scanned, and what labels are entered by radiologists. This shifts both the volume and type of images in the database.
- An AI that predicts ICU transfer may lead clinicians to admit borderline cases earlier, thereby changing the observed case mix in the ICU over time.

These feedback effects mean that the data you collect after deployment are no longer the same as the data you used to develop the model. If you continue to evaluate the model without accounting for these changes, you may misinterpret its apparent performance or drift.

To handle this, prospective protocols are recommended: log how and when observations are made, prespecify primary endpoints and missing-data handling rules, and actively monitor changes in missingness rates or censoring patterns. Doing so preserves reproducibility and ensures that the uncertainty you communicate to clinicians (e.g., prediction intervals and calibration curves) accurately reflects what is, and is not, being observed under the new care processes.

2.3 Causal estimands, measurement bias, site heterogeneity, and evaluation (human health focus)

In routine clinical practice, treatments are rarely randomized: clinicians choose therapies A based on presenting covariates X (e.g., age, vitals, imaging, comorbidities), guidelines, and constraints such as bed capacity. To interpret observed differences in outcomes Y causally, we invoke the potential outcomes framework with $Y(a)$ denoting the outcome had treatment A been set to $a \in \{0, 1\}$ for a comparable patient. A central estimand is the average treatment effect (ATE), $ATE = E[Y(1) - Y(0)]$. Identification from observational health records requires *exchangeability* $Y(a) \perp\!\!\!\perp A \mid X$ (no unmeasured confounding given rich clinical covariates), $0 < \Pr(A = a \mid X) < 1$ almost surely (every clinically plausible option is sometimes used), and *consistency* (received therapy equals the recorded A and produces $Y = Y(A)$). Under these conditions, one may estimate causal effects with outcome regression (G-computation), propensity-score strategies (matching or inverse-probability weighting), or doubly robust procedures such as targeted maximum likelihood estimation that remain consistent if either the outcome or treatment model is correct. In longitudinal care pathways (e.g., oncology lines of therapy, ICU ventilation strategies), time-varying confounders affected by prior treatment invalidate naïve adjustment;

marginal structural models with stabilized weights address this by reweighting each trajectory to emulate a sequentially randomized trial while preserving clinical interpretability.

Measurement imperfections are endemic in health systems and require explicit modeling. For covariates, let $X^* = X + \eta$ represent scanner- or assay-induced error; naïve risk minimization then suffers attenuation bias. Regression calibration and SIMEX can correct such bias when measurement error variance is estimable (e.g., via phantom scans or split-sample assays), while deconvolution approaches can recover sharper signal in high-throughput omics. For labels, interreader variability in radiology and pathology induces *label noise*. If $Y^{\sim}$ is an adjudicated label with misclassification matrix Π where $\Pi_{jk} = \Pr(\tilde{Y} = j \mid Y = k)$, one can debias training and evaluation using loss correction $\ell_{corr} = \Pi - T\ell$ when Π (or a calibrated estimate) is available, or by placing hierarchical models over reader effects to separate truth from systematic rater tendencies. Hybrid quantum models add a further measurement layer: VQCs produce empirical probabilities $\hat{p}$ from S shots with variance $\hat{p}(1-\hat{p})/S$, which should be propagated into confidence intervals, calibration curves, and decision analyses so that uncertainty communicated to clinicians reflects hardware stochasticity and data noise.

Clinical datasets are heterogeneous across hospitals and over time. Site S (e.g., scanner vendor, laboratory platform, care protocol) and calendar time t (practice drift, new guidelines, seasonal epidemiology) shape both the distribution of inputs and the missingness/censoring patterns. For honest generalization, data splitting must reflect deployment: patient-level (clustered) splits to avoid leakage across encounters, site-stratified splits to preserve case-mix differences, and blocked or rolling cross-validation when temporal order matters (e.g., forecasting admissions or length of stay). When covariate shift is present ($p_{\text{train}}(X) /= p_{\text{test}}(X)$), importance weighting with $w(x) = p_{\text{test}}(x) / p_{\text{train}}(x)$ aligns training risk with the clinical target domain; uncertainty should be computed with cluster- or block-aware bootstrap to respect within-patient and within-site correlation.

All evaluation and reporting should inherit these realities. If observation of Y is selective (e.g., laboratories ordered only for high-risk patients) or follow-up is censored (e.g., external transfers), risk estimates and comparisons must use the same weights used for estimation: for missing outcomes, inverse-probability weights $w_i = R_{Y,i} / \hat{\pi}(X_i)$; for survival endpoints, inverse-probability-of-censoring weights $w_i = \Delta_i / G(T^- \mid X_i)$. Calibration and coverage must account for weighting and finite-shot i quantum variance; distribution-free conformal prediction applied to residuals or probabilistic outputs provides marginal coverage guarantees useful for bedside risk communication. For comparative claims (e.g., that a quantum–classical hybrid improves triage over a tuned classical baseline), define (possibly weighted) per-patient loss differentials $d^{(w)} = w_i \left[L\left(\hat{y}^{(A)}, Y_i\right) - L\left(\hat{y}^{(B)}, Y_i\right) \right]$ and apply Diebold–Mariano tests with HAC variance to accommodate serial dependence common in ward-level time series. To assess whether a method reduces dangerous *large* errors (e.g., missed sepsis), compare the full error distributions via the Hassani–Silva KS statistic using patient- or time-block bootstrap for p-values. Because clinical programs evaluate many outcomes and subgroups (ICU vs. ward, age strata, comorbidity burden), multiplicity control (Holm or Benjamini–Hochberg) should accompany primary analyses, and sensitivity analyses should probe violations of exchangeability (unmeasured confounding) and MNAR mechanisms.

Finally, while quantum optimization (QUBO/QAOA) can aid decisions such as operating-room scheduling or bed management, *policy* evaluation must be explicitly statistical: define utility $U(\pi) = \mathrm{E}[u(Y, \pi(X))]$ tied to patient-centered outcomes and safety constraints, quantify uncertainty with the same weighting and clustering structure used in estimation, and report transportability across sites and time. Only then do performance improvements translate credibly into better, safer care for patients. In summary, and to frame the discussion within human health and medical applications, the following section presents a general overview.

2.3.1 Causal inference in routine care

In everyday clinical practice, treatments are almost never assigned at random. A cardiologist may escalate therapy based on blood pressure, heart rate, and co-morbidities; an ICU team may decide to intubate depending on oxygen saturation, staffing levels, and bed availability. These choices mean that the observed outcomes mix together treatment effects with patient selection effects. If we want to know whether an intervention really works—for example, whether a new anticoagulant reduces stroke risk—we must reconstruct what would have happened had similar patients received an alternative therapy. This is the potential outcomes framework: we define Y(a)Y(a)Y(a) as the outcome we would have seen if treatment AAA were set to aaa, and the ATE as the difference in those outcomes across the population. In health data, this matters because policy and reimbursement decisions hinge on understanding true causal effects, not just associations.

2.3.2 Making observational health data reliable

Electronic health records, registries, and imaging archives can be powerful sources for such analyses, but only if the data problems are addressed.

- Confounding: If sicker patients preferentially receive a new therapy, naïve comparisons will underestimate the benefit. Techniques like propensity scores, inverse-probability weighting, or doubly robust estimators can rebalance the data to emulate a randomized trial.
- Time-varying confounding: In oncology, line-of-therapy changes or dose adjustments depend on prior response. Marginal structural models with stabilized weights can handle these evolving confounders and produce interpretable causal effects over a patient's care pathway.

2.3.3 Measurement error and label noise

Health data also suffer from measurement imperfections. Imaging features can differ across scanner vendors or software versions; laboratory assays produce batch effects; radiology and pathology labels reflect interreader disagreement. If we train models without correcting for these issues, we risk attenuated estimates or spurious subgroup differences. Methods such as regression calibration, SIMEX, or hierarchical models of reader effects can reduce bias. In quantum or hybrid models, a further noise source appears: VQCs generate probabilities from a finite number of "shots," introducing additional variance that must be carried through to confidence intervals and calibration plots so that clinicians see realistic uncertainty bands.

2.3.4 Heterogeneity across sites and time

Hospitals vary in case-mix, practice style, and data completeness, and these factors shift over time. A model built on one center's EHR may not generalize to another without adjustment. For honest assessment, data should be split at the patient level (to avoid leakage across encounters), stratified by site (to preserve case-mix differences), and blocked or rolled forward in time (to mimic deployment conditions such as forecasting admissions). When covariate distributions differ between training and target populations, importance weighting can align training risk with the target domain.

2.3.5 Reporting, uncertainty and safety

Because outcomes like laboratories or follow-up times are observed selectively, any risk estimates or model comparisons must use the same weights used for estimation (inverse-probability weights for missing outcomes, inverse-probability-of-censoring weights for survival endpoints). Calibration and coverage should reflect both these weights and any quantum-hardware variance. Tools such as distribution-free conformal prediction can then provide prediction sets or intervals with guaranteed coverage, giving bedside users a clearer sense of reliability.

2.3.6 From optimization to safer care

Finally, quantum optimization methods such as QUBO and QAOA have direct analogues in hospital operations, for example, assigning surgical cases to operating rooms, matching donors to recipients, or allocating scarce ICU beds. But even when an algorithm produces a better schedule on paper, its policy evaluation must be statistical: define a utility tied to patient-centered outcomes (delays, cancellations, fairness), quantify uncertainty using the same weighting and clustering structure, and check transportability across sites and time. Only with this full pipeline do performance improvements translate credibly into safer, more efficient care for patients.

3 Practical examples in human health

This section provides end-to-end, statistically principled examples of quantum–classical pipelines in human health. Each example states the clinical question, data-generating assumptions, estimator(s), including quantum components, and the evaluation protocol using loss-based and distributional tests with uncertainty quantification. All examples adhere to the missingness/censoring handling and site/time heterogeneity guidance established earlier.

3.1 Example 1: Multiomics risk stratification with a quantum kernel

The clinical question we address is whether it is possible to predict 1-year relapse in hematologic malignancy by leveraging baseline genomic features (single nucleotide variants (SNVs), copy number variations (CNVs)), transcriptomics, and clinical covariates, with the ultimate aim of informing therapy escalation decisions. The data consist of patient-level measurements, denoted by $X = (X^{\mathrm{omics}}, X^{\mathrm{clin}})$, together with site labels S and time stamps t. Preprocessing involves batch correction of the omics component X^{omics} stratified by site, Winsorization of extreme counts, and imputation of missing entries under a MAR assumption using fold-aware MI. Dimensionality reduction is carried out via principal component analysis (PCA) on the omics data, concatenated with the clinical covariates, to form:

$$z = \mathrm{PCA}(X^{\mathrm{omics}}) \,||\, X^{\mathrm{clin}}.$$

This vector is then embedded into a q-qubit state $|\varphi_\theta(z)\rangle$ through a data reuploading circuit, which defines the feature map $\phi_\theta : R^p \rightarrow C^2 q$. On this representation, we construct a quantum kernel machine, where the kernel is defined as

$$K_\theta(z, z') = |\varphi_\theta(z)|\varphi_\theta(z')|^2$$

and estimated from S shots per pair. Kernel ridge or logistic regression is trained with a regularization parameter λ, selected by site-stratified and block-aware cross-validation. All hyperparameters, namely the number of qubits, circuit depth, regularization strength, and number of shots (q, depth, λ, S), are tuned exclusively within the training folds.

Model evaluation proceeds using both log-loss and the Brier score. For outcomes that are selectively observed (e.g., laboratory tests or surveys), the loss functions are weighted by $w_i = R_{Y,i} / \hat{\pi}(X_i)$, and performance estimates $\hat{R}$ are reported with cluster-robust confidence intervals.

Comparisons are made against well-tuned classical baselines, including Gaussian/RBF kernels and gradient-boosted decision trees. Pairwise performance is quantified through per-patient weighted loss difference $d^{(w)}$, which is analyzed using the Diebold–Mariano test with HAC variance estimation. To further assess predictive distributions, we employ the Hassani–Silva KS test with patient-block bootstrap to obtain p-values. Reliability diagrams are used for calibration assessment, and conformal prediction intervals are provided. The variance contribution from quantum shot noise is propagated by augmenting the loss variance estimator with:

$$Var\left[\hat{k}\right] \approx \frac{K(1-K)}{S}$$

Although our section focuses on predicting **relapse** in hematologic malignancies, it is important to emphasize that relapse risk and drug sensitivity are not interchangeable outcomes. Relapse refers to the reemergence or progression of disease after remission and is typically influenced by disease biology, treatment history, and patient factors. Drug response or sensitivity, by contrast, measures how tumor cells react to a specific therapy, an outcome that can be studied in vitro or in early-phase clinical settings but does not necessarily predict long-term relapse patterns.

A recent study by Priyadharshini et al. (2025) applied quantum machine learning to predict drug sensitivity in multiple myeloma using high-dimensional proteomic data. While this work does not address relapse per se, it does demonstrate the technical feasibility of using quantum methods to handle very large omics feature sets in oncology and to generate clinically relevant predictions. This distinction matters for clinicians: the quantum approach described here has been proven in a drug-response context, but translating it to relapse prediction would require additional validation, different outcome definitions, and possibly different covariate handling to ensure that the model truly captures the mechanisms driving disease recurrence rather than treatment sensitivity alone.

3.2 Example 2: Radiology triage with a variational quantum classifier (VQC)

The clinical question is to flag *likely urgent* chest X-rays at the time of acquisition so that radiologist reads can be prioritized. To this end, embeddings $x \in \mathrm{R}^d$ are first extracted from a frozen CNN front-end trained on historical data and then concatenated with metadata variables such as device vendor, patient age, and ward. Missing metadata are imputed using fold-aware MI, and a site indicator S is included to account for heterogeneity across acquisition centers.

The predictive model is implemented as a VQC f_θ: $\mathrm{R}^d \rightarrow [0, 1]$ that combines data reuploading layers with entangling gates. The circuit is trained with a cross- entropy loss using SPSA or Adam optimization, and predictions $\hat{p}_i = \hat{p}_\theta(x_i)$

are estimated from S measurement shots. To mitigate barren plateaus and preserve trainability, circuit depth is explicitly penalized.

Evaluation is based primarily on log loss, with AUROC and AUPRC at low false-positive rates serving as secondary measures. Prediction variance, quantified as $\hat{p}_i\left(1-\hat{p}_i\right)/S$, is propagated into uncertainty bands. To approximate deployment conditions, rolling cross-validation by calendar week is employed. Performance is compared against a calibrated logistic regression head trained on the same embeddings. Error distributions are further analyzed using the Hassani–Silva KS test, focusing on whether extreme errors in the right tail—corresponding to missed urgent cases—are reduced. The results are reported alongside decision curves and estimates of expected workup saved. Finally, domain robustness is stress-tested through vendor-holdout evaluation, with importance weighting $w(x) = p_{\text{test}}(x)/p_{\text{train}}(x)$ applied when appropriate.

The clinical question is to flag *likely urgent* chest X-rays at the time of acquisition to enable prioritized radiologist review. Existing work has demonstrated the feasibility and clinical impact of automated triage systems. For example, Montana (2019) developed a deep learning system trained on >470,000 chest X-rays, using natural language processing of radiology reports to label each image as critical, urgent, nonurgent, or normal. In simulated deployment, cases classified as critical were read by experts in an average of 2.7 days compared with 11.2 days under standard practice, thereby significantly accelerating care for the most urgent cases. Although this system was based on classical deep learning, it validates the value of automated triage and motivates the development of quantum-enhanced approaches.

In our proposed framework, embeddings $x \in \mathrm{R}^d$ are extracted from a frozen CNN front-end trained on historical data and concatenated with metadata such as device vendor, patient age, and ward. Missing metadata are imputed using fold-aware MI, and a site indicator S is included to capture acquisition heterogeneity. The predictive model is implemented as a VQC f_θ: $\mathrm{R}^d \rightarrow [0, 1]$ with data reuploading layers and entangling gates. Training is performed under a cross-entropy loss with SPSA or Adam optimization, and predictions $\hat{p}_i = \hat{p}_\theta(x_i)$ are estimated from S measurement shots. To preserve trainability and mitigate barren plateaus, circuit depth is explicitly penalized.

Evaluation relies primarily on log-loss, with AUROC and AUPRC at low false-positive rates as secondary criteria. Prediction variance, estimated as $\hat{p}_i(1-\hat{p}_i)/S$ is propagated into uncertainty bands. To approximate deployment conditions, rolling cross-validation by calendar week is used. Comparisons are made against a calibrated logistic regression head trained on the same embeddings. Error distributions are further analyzed with the Hassani–Silva KS test to assess whether extreme errors in the right tail—corresponding to missed urgent cases—are reduced. Results are reported alongside decision curves and estimates of expected workup saved. Finally, domain robustness is stress-tested through vendor-holdout evaluation, with importance weighting $w(x) = p_{\text{test}}(x)/p_{\text{train}}(x)$ applied when appropriate.

3.3 Example 3: Survival analysis with censoring-aware quantum kernel

The clinical question is to predict the time to heart-failure hospitalization within 2 years from EHR-derived features, with the aim of guiding the intensity of follow-up. For each patient, we define (T_i, Δ_i), where T_i is the observed time and Δ_i the event indicator, subject to right censoring assumed to be MAR given X. The censoring distribution $G(t \mid X)$ is estimated on the training folds using either a Cox proportional hazards model or a random survival forest.

The predictive model is based on IPCW kernel regression for the conditional survival function $S(t \mid X)$, using a quantum kernel K_θ applied to engineered features $z = \psi(X)$. The estimator minimizes

$$\widehat{\mathcal{R}} = \frac{1}{n}\sum_{i=1}^{n}\frac{\Delta_i}{\widehat{G}\left(T_i^- | X_i\right)}\{\ell_{surv}(T_i, S_\theta(.\mid X_i))\}$$

where ℓ_{surv} denotes a survival loss, such as the IPCW Brier score computed on a grid of time points. Evaluation relies on IPCW-Brier scores and time-dependent AUROC, with confidence intervals obtained by block bootstrap. Pairwise comparisons are carried out using Diebold–Mariano tests on IPCW-weighted loss differentials across the time grid, with HAC variance estimation, while error distributions are further examined using the Hassani–Silva KS test applied to weighted error processes. Sensitivity analyses are performed by varying the censoring model, comparing Cox regression with random survival forests to assess the robustness of the results.

A real-world analog of this setup was developed by Cornhill et al. (2022), who constructed and validated an ML-based model to predict individualized time to heart-failure hospitalization among 1775 patients with systolic heart failure using routinely reported cardiovascular magnetic resonance features and patient health information. Their model achieved 2-year time-dependent AUC values of approximately 0.80 in an external hold-out cohort, demonstrating strong predictive performance in a clinically relevant context.

In our proposed framework, we extend these ideas using a quantum-enhanced approach. Specifically, we apply IPCW kernel regression to estimate the conditional survival function $S(t \mid X)$, utilizing a quantum kernel K_θ on engineered features $z = \psi(X)$. The estimator minimizes

$$\widehat{\mathcal{R}} = \frac{1}{n}\sum_{i=1}^{n} \frac{\Delta_i}{\widehat{G}(T_i^- | X_i)} \ell_{surv}(T_i, S_\theta(.| X_i))$$

where ℓ_{surv} may be the IPCW Brier score computed over a predefined time grid.

Evaluation is conducted using IPCW-Brier scores and time-dependent AUROC, with confidence intervals obtained via block bootstrap. Pairwise comparisons are performed using Diebold–Mariano tests on IPCW-weighted loss differentials across time, with HAC variance estimation. Error distributions are further assessed using the Hassani–Silva KS test on weighted error processes. To ensure robustness, we conduct sensitivity analyses by varying the censoring model and comparing Cox and RSF approaches.

3.4 Example 4: Operating-room scheduling via QUBO policy optimization

The clinical question is the daily assignment of elective cases to operating room (OR) blocks, with the objective of minimizing overtime and cancellations while maintaining fairness and respecting ICU bed constraints. This scheduling task can be formulated by encoding a binary decision vector $x \in \{0, 1\}^m$ and solving a QUBO problem.

$$\min_{\{x \in 0,1\}^m} x^\top Q x + c^\top x,$$

where the matrix Q penalizes expected overtime based on predictive case durations, as well as violations of soft constraints such as ICU bed limits and surgeon availability. The optimization problem can be solved by the QAOA or, when available, by a quantum annealer.

Policy evaluation is carried out by estimating the value function $U(\pi) = E[u(Y, \pi(X))]$, where u encodes both patient-centered outcomes (delays, cancellations) and operational costs. When relying on observational logs, $U(\pi)$ is identified using inverse propensity weighting for actions or off-policy evaluation via doubly robust estimators.

Similar optimization frameworks have been studied in classical settings. Shen and Chen (2008) proposed stochastic optimization models for elective surgery scheduling under uncertainty, explicitly accounting for emergency arrivals and overtime costs. Cardoen and Demeulemeester (2008) further analyzed operating room planning approaches to balance efficiency, fairness, and robustness in uncertain demand environments. These works demonstrate the feasibility and clinical importance of predictive, optimization-based OR scheduling, which our quantum-enhanced framework seeks to extend.

3.5 Example 5: Federated, multisite generalization, and covariate shift

In this example, the clinical question is how to deploy a sepsis early warning model trained on site S_{train} to a new hospital S^*. A practical analog is the external validation of the *SepsisWatch* model, originally developed at an academic medical center, which maintained strong performance when deployed across four emergency departments in a community healthcare system. In this validation, SepsisWatch achieved an AUROC of 0.92 and an AUPRC of 0.24—comparable to or better than its original deployment metrics—demonstrating robustness to shifts in patient populations and clinical environments (Valan et al., 2024).

In our proposed framework, a quantum kernel model is trained on pooled, site-weighted data. To correct for distributional shifts when deploying to S^*, we apply importance weighting to the loss function using

$$w(x) = \frac{p_S{}^*(x)}{p_{\text{train}}(x)'}$$

where the density ratio is estimated via a classifier approach. This enables the model to upweight samples resembling the target site during training.

4 Data governance for QAI in health

Robust data governance underpins every stage of QAI development and deployment in healthcare. Because quantum models ingest high-dimensional, often sensitive data (genomics, imaging, EHR, claims), good governance is not just a regulatory obligation but also a prerequisite for reproducible science and clinical trust. For a general overview of data governance in various fields, the reader is referred to Hassani and MacFeely (2023), MacFeely (2019), and MacFeely et al. (2022).

Privacy and security. Patient-level data used to train quantum kernels or variational circuits typically contain identifiable information. Frameworks such as the EU General Data Protection Regulation and the US Health Insurance Portability and Accountability Act (HIPAA) Security Rule mandate strict access control, de-identification, audit trails, and, where possible, secure multiparty computation or federated learning to avoid centralizing data. Recent developments in Privacy-Enhancing Technologies (PETs) will make significant contributions to extending the use and value of data (OECD, 2023). Without such safeguards, even the most accurate model may be unusable in practice (Kaissis et al., 2021; Rieke et al., 2020).

Data quality and provenance. Quantum pipelines are highly sensitive to noise and bias in their inputs. Governance should, therefore, include procedures for documenting acquisition protocols (scanner version, laboratory platform, software release), metadata on missingness/censoring, and all preprocessing steps applied to each dataset. This "data lineage" enables clinicians and regulators to trace how raw observations became model features (Gebru et al., 2021).

Access, consent, and equity. The recruitment and consent processes for data contributing to QAI should be transparent and inclusive. Consent should be clear on use, reuse, and sharing. Underrepresentation of certain populations can yield biased quantum kernels that underperform in marginalized groups, exacerbating health disparities. Data stewardship boards and patient representatives can help guide inclusion criteria and downstream use.

Model transparency and audit. Governance should extend beyond data to include the quantum hardware/software settings used: qubit count, connectivity, circuit depth, number of shots, and error rates. Recording these parameters in open, version-controlled artifacts (e.g., "model cards" or "data sheets") supports the CONSORT-AI emphasis on transparent reporting for AI interventions in clinical trials (Liu et al., 2020).

Deployment monitoring. Once deployed, QAI systems can change clinical workflows and thereby the data collected. Prospective governance protocols should log observation policies, prespecify primary endpoints and missing-data handling, and monitor drift in both missingness rates and censoring distributions (Hernán and Robins, 2020). This ensures that the uncertainty communicated to clinicians—prediction intervals, calibration curves—meaningfully reflects what is and is not being observed under the new care processes.

Together, these practices align QAI development with established principles of FAIR (Findable, Accessible, Interoperable, Reusable) and responsible AI in health, as well as with the emerging literature on federated learning and privacy-preserving analytics (Kaissis et al., 2021; Rieke et al., 2020). In doing so, they enable quantum–classical systems to be evaluated and adopted with the same level of transparency, accountability, and patient protection expected of any medical technology.

5 Conclusion

QAI offers new representational, probabilistic, and optimization primitives that can augment the clinical data science toolbox across discovery, diagnosis, and delivery. This chapter framed QAI components—quantum feature maps and kernels, VQCs, and quantum optimizers—as statistical estimators and decision procedures embedded in explicit data-generating processes. By doing so, we made clear how to specify estimands, choose proper scoring rules, and conduct valid inference in the presence of missingness, censoring, clustering, and site/time heterogeneity.

The practical examples—multiomics risk stratification with quantum kernels, radiology triage with VQCs, censoring-aware survival analysis, operating-room policy optimization via QUBO/QAOA, and federated deployment under covariate shift—illustrate how to couple quantum components with classical front-ends while preserving statistical validity. Across these settings, finite-shot measurement noise, hardware error rates, and batch/site effects are treated as first-class uncertainty sources to be propagated into confidence intervals, calibration, and decision analysis. In short, *how* we evaluate QAI matters as much as *what* we build.

Looking ahead, three priorities stand out. First, evidence: prospective, preregistered head-to-head studies against tuned classical baselines, powered for both mean and tail error reductions, and audited for calibration and coverage. Second, engineering: hardware-aware learning (noise-adaptive feature maps) and hybrid pipelines that exploit quantum advantages without sacrificing robustness or reproducibility. Third, standards: transparent reporting of quantum settings (qubits, depth, connectivity, shots, error rates), open artifacts (code, seeds, data dictionaries), and adherence to medical AI guidelines (e.g., CONSORT-AI).

We also highlighted the central role of robust data governance—covering privacy, provenance, and transparent reporting—in ensuring that QAI systems are trustworthy, equitable, and compliant with medical data standards.

Limitations remain; current devices are noisy and small-scale, quantum advantages are likely problem- and data-regime-specific, and health data are burdened by selection mechanisms that can swamp marginal gains. Nevertheless, by grounding QAI in a statistical treatment with clear estimands, proper losses, principled uncertainty, and deployment-aware validation, we can separate genuine signal from hype. Doing so will determine whether quantum methods become routine tools for safer, fairer, and more efficient care, or remain promising curiosities awaiting their clinical moment.

AI-based tool usage

An AI writing assistant was used exclusively for grammar, spelling, and clarity improvements.

References

Cardoen, B., & Demeulemeester, E. (2008). Capacity of clinical pathways—A strategic multi-level evaluation tool. *Journal of Medical Systems*. 01485598. *32*(6), 443–452. doi:10.1007/s10916-008-9150-z.

Cornhill, A. K., Dykstra, S., Satriano, A., Labib, D., Mikami, Y., Flewitt, J., et al. (2022). Machine learning patient-specific prediction of heart failure hospitalization using cardiac MRI-based phenotype and electronic health information. *Frontiers in Cardiovascular Medicine*. 2297055X. *9*. Retrieved from www.frontiersin.org/journals/cardiovascular-medicine. doi:10.3389/fcvm.2022.890904.

Devadas, R. M., & Sowmya, T. (2025). Quantum machine learning: a comprehensive review. *ACM Computing Surveys*. https://pmc.ncbi.nlm.nih.gov/articles/PMC12053761/.

Gebru, T., Morgenstern, J., Vecchione, B., Vaughan, J. W., Wallach, H., Daumé III, H., & Crawford, K. (2021). Datasheets for datasets. *Communications of the ACM*. 0001-0782. *64*(12), 86–92. doi:10.1145/3458723.

Gupta, R. S., Wood, C. E., Engstrom, T., Pole, J. D., & Shrapnel, S. (2025). A systematic review of quantum machine learning for digital health. *Npj Digital Medicine*. 23986352. *8*(1). Retrieved from https://www.nature.com/npjdigitalmed/. doi:10.1038/s41746-025-01597-z.

Hand, D. J. (2020). Dark data. *Significance*. 1740-9705. *17*(3), 42–44. doi:10.1111/1740-9713.01406.

Hassani, H., & MacFeely, S. (2023). Driving excellence in official statistics: unleashing the potential of comprehensive digital data governance. *Big Data and Cognitive Computing* 2504-2289. *7*(3), 134. doi:10.3390/bdcc7030134.

Hernán, M. A., & Robins, J. M. (2020). *Causal inference: What if*. Chapman & Hall/CRC, Boca Raton. Retrieved from https://www.hsph.harvard.edu/miguel-hernan/causal-inference-what-if/.

Kaissis, G., Ziller, A., Passerat-Palmbach, J., Ryffel, T., Usynin, D., Trask, A., et al. (2021). End-to-end privacy preserving deep learning on multi-institutional medical imaging. *Nature Machine Intelligence*. 25225839. *3*(6), 473–484. Retrieved from https://www.nature.com/natmachintell/. doi:10.1038/s42256-021-00337-8.

Little, R. J. A., & Rubin, D. B. (2002). *Statistical analysis with missing data* (2nd ed.). Wiley. https://onlinelibrary.wiley.com/doi/book/10.1002/9781119013563.

Liu, X., Rivera, S. C., Moher, D., Calvert, M. J., Denniston, A. K., Chan, A. W., et al. (2020). Reporting guidelines for clinical trial reports for interventions involving artificial intelligence: the CONSORT-AI extension. *Nature Medicine*. 1546170X. *26*(9), 1364–1374. Retrieved from http://www.nature.com/nm/index.html. doi:10.1038/s41591-020-1034-x.

MacFeely, S. (2019). *Big Data and Official Statistics*. (pp. 25–54). IGI Global. doi:10.4018/978-1-5225-7077-6.ch002.

MacFeely, S., Angela, M., Haishan, F., Veerappan, M., Hereward, M., Passarelli, D., & Schüür, F. (2022). Towards an international data governance framework. *Statistical Journal of the IAOS*. 18747655. *38*(3), 703–710. doi:10.3233/sji-220038.

Montana, G. (2019). Artificial intelligence system to prioritize critical findings on chest X-rays. *Radiology*. Retrieved from https://www.rsna.org/news/2019/january/ai-for-chest-x-rays.

OECD. (2023). *Emerging privacy-enhancing technologies: Current regulatory and policy approaches*. Author.

Preskill, J. (2018). Quantum computing in the NISQ era and beyond. *Quantum*, *2*, 79. doi:10.22331/q-2018-08-06-79.

Priyadharshini, M., Raju, B. D., Banu, A. F., Kumar, P. J., Murugesh, V., & Rybin, O. (2025). A quantum machine learning framework for predicting drug sensitivity in multiple myeloma using proteomic data. *Scientific Reports*. 20452322. *15*(1). Retrieved from https://www.nature.com/srep/. doi:10.1038/s41598-025-06544-2.

Rieke, N., Hancox, J., Wenqi, L., Milletarì, F., Roth, H. R., Albarqouni, S., et al. (2020). The future of digital health with federated learning. *Npj Digital Medicine*. 2398-6352. *3*(1). doi:10.1038/s41746-020-00323-1.

Rubin, D. B. (1976). Inference and missing data. *Biometrika*. 00063444. *63*(3), 581–592. doi:10.1093/biomet/63.3.581.

Seaman, S. R., & White, I. R. (2013). Review of inverse probability weighting for dealing with missing data. *Statistical Methods in Medical Research*. 14770334. *22*(3), 278–295. doi:10.1177/0962280210395740.

Shen, C.-C., & Chen, Y.-L. (2008). A dynamic-programming algorithm for hierarchical discretization of continuous attributes. *European Journal of Operational Research*. 03772217. *184* (2), 636–651. doi:10.1016/j.ejor.2006.12.013.

Topol, E. J. (2019). High-performance medicine: the convergence of human and artificial intelligence. *Nature Medicine*, *25*(1), 44–56. doi:10.1038/s41591-018-0300-7.

Valan, B., Prakash, A., Ratliff, W., Gao, M., Muthya, S., Thomas, A., et al. (2024). Evaluating the portability of Sepsis Watch: a multi-site external validation of a Sepsis machine learning model. *medRxiv*, Retrieved from https://www.medrxiv.org/. doi:10.1101/2024.03.22.24304753.

Chapter 24

Use of artificial intelligence in microbiome research

Feronia Melissa De Mello[a] and Ashfaque Hossain[b]
[a]College of Medicine, Ras Al Khaimah Medical and Health Sciences University, Ras Al Khaimah, United Arab Emirates
[b]Medical Microbiology and Immunology, College of Medicine, Ras Al Khaimah Medical and Health Sciences University, Ras Al Khaimah, United Arab Emirates

1 Introduction

1.1 Overview of microbiome research

The human microbiome, which consists of trillions of microorganisms such as bacteria, fungi, viruses, and archaea, plays a crucial role in both health and disease. These microbial communities inhabit various parts of the body, including the gut, skin, and respiratory tract, and impact a wide range of functions, from digestion and immune response to mental health and the risk of chronic diseases (Gilbert et al., 2018; Turnbaugh et al., 2007). Microbiome research requires analyzing vast amounts of genetic and functional data from metagenomic sequencing, metabolomics, and transcriptomics. The inherent complexity of microbial communities, coupled with the variability between individuals, presents a significant barrier to understanding microbial interactions and their implications for health. Traditional bioinformatics approaches, while powerful, often struggle to process such high-dimensional data efficiently. This is where artificial intelligence (AI) is emerging as a game-changer (Lloyd-Price, Abu-Ali, & Huttenhower, 2016).

1.2 Importance of addressing microbiome research challenges

AI, particularly machine learning (ML) and deep learning (DL), has changed the way researchers analyze microbiome data. These technologies allow for the identification of hidden patterns, predictive modeling, and the discovery of novel microbial biomarkers associated with diseases. AI algorithms excel at processing massive datasets, integrating multiomics data, and distinguishing meaningful microbial signatures from noise (Zhang, Li, Butcher, Stintzi, & Figeys, 2019). One of AI's most promising contributions to microbiome research is its ability to enhance diagnostic and therapeutic strategies. By using AI-driven models, researchers can predict disease risks based on microbiome composition, personalize probiotic and dietary interventions, and even design novel antimicrobial therapies (Beam & Kohane, 2018). Moreover, AI facilitates metagenomic classification, automating the identification of previously unknown microbial species with greater accuracy than traditional methods (Libbrecht & Noble, 2015).

Despite these advances, AI in microbiome research is not without limitations. Challenges such as bias in training datasets, interpretability of AI-driven findings, and the need for standardized protocols must be addressed to maximize its impact (Topol, 2019). However, as computational power grows and datasets become more refined, AI is poised to unlock new limits in microbiome science, bridging the gap between microbial ecology and precision medicine (Dhaarani & Reddy, 2025).

The integration of AI with microbiome research marks a paradigm shift, enabling scientists to move beyond correlation-based studies and toward causative, mechanistic insights. This interdisciplinary synergy holds immense potential for revolutionizing healthcare, agriculture, and environmental sciences, making microbiome research one of the most exciting frontiers in modern biology (Omar et al., 2023; Probul, Huang, Saak, Baumbach, & List, 2024).

Applications of Artificial Intelligence in Healthcare. https://doi.org/10.1016/B978-0-443-43934-6.00038-9

2 AI in microbiome data analysis

2.1 Taxonomic classification and metagenomic analysis

Taxonomic classification refers to the process of identifying and categorizing organisms present in a metagenomic sample. Traditional methods often depended on sequence alignment techniques, which, although accurate, tend to be computationally demanding. To overcome this limitation, modern approaches have incorporated ML and DL to enhance both speed and accuracy (Fig. 24.1). One significant innovation is Taxometer, a neural network-based tool designed to improve taxonomic annotations and evaluate the quality of classifiers. Taxometer has proven effective in accurately classifying metagenomic contigs, offering a more dependable understanding of microbial compositions (Kutuzova, Nielsen, Piera, Nissen, & Rasmussen, 2024). Another breakthrough is the Naïve Bayes classifier++ (NBC++), which enhances the traditional Naïve Bayes framework. NBC++ has demonstrated effectiveness in classifying reads from both amplicon and metagenomic samples, striking a balance between computational efficiency and classification accuracy (Duan, Hearne, Polikar, & Rosen, 2025). In addition, tools like MetaTrinity have been developed to address the trade-off between accuracy and runtime. MetaTrinity employs heuristics to achieve high accuracy in metagenomic classification while significantly reducing computational time, making it practical for large-scale analyses (Gollwitzer et al., 2023).

2.2 Multiomics data integration (genomics, proteomics, metabolomics)

Metagenomic analysis has revolutionized the study of microbial communities by enabling scientists to analyze genetic material directly from environmental samples, bypassing the need to culture individual species. This powerful approach provides a comprehensive view of microbial diversity and function, offering insights into complex ecosystems that were previously inaccessible (Pérez-Cobas, Gomez-Valero, & Buchrieser, 2020; Zhang et al., 2021). By extracting and sequencing DNA from environmental samples, such as soil, water, or the human gut, researchers can generate vast genetic datasets that reveal the composition, interactions, and functional potential of microbial communities. This has profound implications for fields ranging from environmental science to clinical medicine, making metagenomics a cornerstone of modern microbiology (Liu et al., 2025).

The process of metagenomic analysis begins with the collection of environmental samples, which can range from ocean water and soil to human fecal matter or respiratory secretions. Once collected, DNA is extracted from these samples, capturing the genetic material of all microorganisms present. High-throughput sequencing technologies, such as Illumina or Oxford Nanopore, are then used to generate massive amounts of sequencing data. These technologies produce short DNA reads, which are fragments of genetic material that must be processed and analyzed

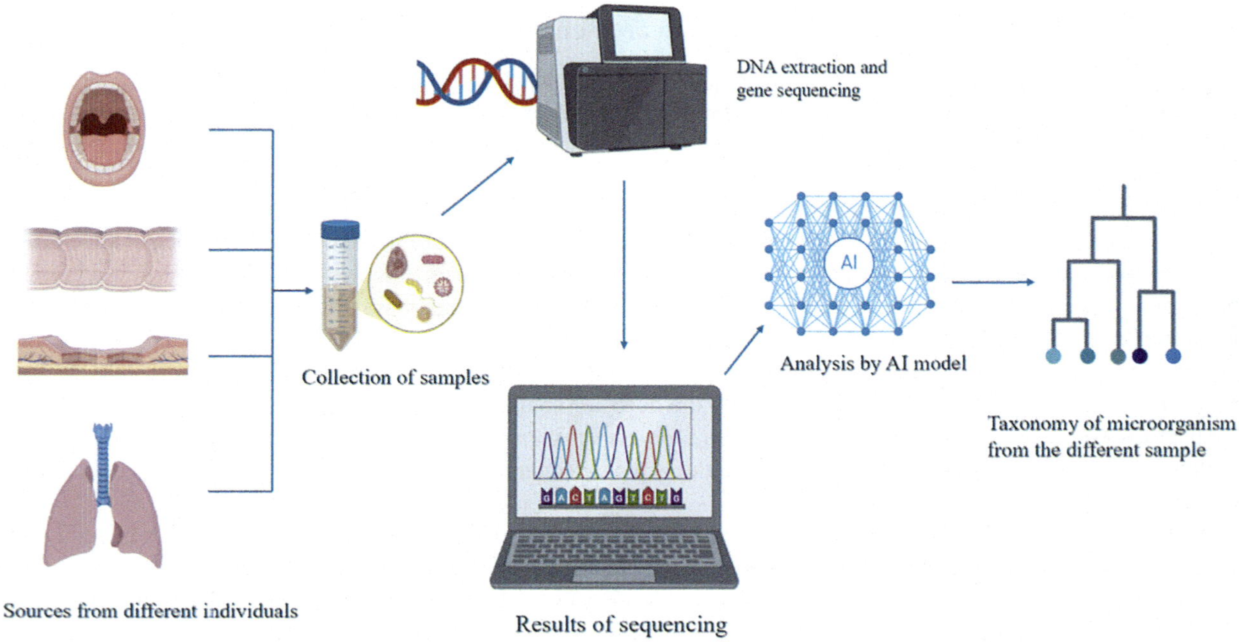

FIG. 24.1 Artificial intelligence in taxonomic classification.

to reconstruct the microbial landscape (Liu et al., 2025). The next step involves quality control, where sequencing errors and low-quality reads are filtered out to ensure the accuracy of the data. Following this, the remaining reads undergo assembly, a computational process where short DNA fragments are pieced together into longer sequences called contigs. These contigs represent reconstructed segments of microbial genomes and serve as the foundation for further analysis (Liu et al., 2021). Once assembled, the contigs are subjected to taxonomic classification, where they are assigned to specific microbial taxa (e.g., species, genera, or families). This step is crucial for understanding the composition of the microbial community and identifying which organisms are present. Advanced methods, such as phylogenetic placement, further refine classification accuracy by comparing the contigs to reference databases and placing them within the context of known evolutionary relationships (Czech, Stamatakis, Dunthorn, & Barbera, 2022). In addition to taxonomic classification, functional annotation is performed to determine the biological roles of the genes present in the metagenomic data. This involves identifying genes involved in metabolic pathways, antibiotic resistance, virulence, and other functions. By linking taxonomic and functional data, researchers can gain a holistic understanding of how microbial communities operate and interact within their environments. The complexity and scale of metagenomic data have made it a prime candidate for the application of AI, particularly DL methods. Convolutional neural networks and attention-based models are being used to analyze metagenomic datasets, enabling the prediction of microbial interactions and their roles in human health (Roy, Prifti, Belda, & Zucker, 2024). These AI-driven approaches can identify patterns and relationships that are difficult to detect using traditional methods, such as the co-occurrence of specific microbes in disease states or the impact of environmental changes on microbial communities. For example, DL models can predict how microbial communities will respond to perturbations, such as antibiotic treatment or dietary changes (Przymus et al., 2025). They can also identify potential biomarkers for diseases by analyzing the genetic signatures of microbial communities in clinical samples. This has significant implications for personalized medicine, where metagenomic data can be used to tailor treatments based on an individual's unique microbiome composition. Metagenomics is increasingly being applied in clinical settings, where it is transforming the diagnosis and treatment of infectious diseases (Chiu & Miller, 2019).

Traditional diagnostic methods often rely on culturing pathogens, which can be time-consuming and may fail to detect unculturable or fastidious organisms. Metagenomic analysis, on the other hand, can identify all microorganisms present in a sample, including viruses, bacteria, fungi, and archaea, providing a more comprehensive picture of infections. In addition to diagnosing infections, metagenomics is shedding light on host–microbiome interactions, which play a critical role in health and disease. For example, the gut microbiome has been implicated in conditions such as inflammatory bowel disease (IBD), obesity, and even neurological disorders like Alzheimer's disease. By analyzing the genetic material of microbial communities, researchers can identify microbial signatures associated with these conditions and develop targeted interventions (Jin, Morton, & Bonneau, 2024).

One of the most exciting developments in metagenomics is the use of metagenome-assembled genomes (MAGs) to study uncultured microbial candidates. MAGs are reconstructed genomes of microorganisms that have not been isolated or cultured in the lab. By piecing together DNA fragments from metagenomic data, researchers can infer the genomes of these elusive organisms, providing insights into their biology and ecological roles (Navgire et al., 2022). This approach has led to the discovery of novel bacterial species and functional genes that were previously unknown. For example, MAGs have revealed microbes with unique metabolic capabilities, such as the ability to degrade environmental pollutants or produce novel antibiotics. These discoveries have broad implications for biotechnology, where microbial enzymes and pathways can be harnessed for industrial applications. The field of metagenomics is being further propelled by advances in long-read sequencing technologies, such as those developed by Oxford Nanopore and Pacific Biosciences (Han et al., 2024). Unlike short-read sequencing, which produces fragments of a few hundred base pairs, long-read sequencing generates reads that are tens of thousands of base pairs long. This allows for more accurate genome reconstructions and reduces the challenges associated with assembling complex microbial communities (Kim, Pongpanich, & Porntaveetus, 2024).

3 Applications of AI in microbiome research

3.1 Functional predictions (microbial pathways, interactions)

AI is revolutionizing microbiome research by significantly enhancing our ability to predict microbial functions, pathways, and interactions within complex ecosystems. As the field of microbiome science expands, the volume and complexity of sequencing data generated from diverse microbial communities present a considerable challenge for researchers. ML models, such as DeepMicro, are specifically designed to address this challenge by processing extensive sequencing data to detect patterns, classify microbes, and predict disease associations, offering deeper insights into microbiome dynamics (D'Urso & Broccolo, 2024). AI-driven approaches have been applied across multiple facets of microbiome research, including taxonomic profiling, functional annotation, and the prediction of microbe–microbe interactions. Taxonomic profiling utilizes ML algorithms to classify microbial species within a given sample, while functional annotation helps identify the metabolic capabilities of these microbes (Marcos-Zambrano et al., 2021).

By advancing our understanding of microbial ecology and metabolic modeling, AI tools enable researchers to elucidate the roles of specific microorganisms in various ecological contexts, including their contributions to nutrient cycling and interactions with host organisms. In drug metabolism studies, ML tools play a critical role by predicting which gut bacteria and enzymes are capable of metabolizing candidate drugs, highlighting AI's role in precision medicine (Malwe & Sharma, 2023). This aspect of AI in microbiome research underscores its significance, as understanding individual variability in drug metabolism is crucial for optimizing therapeutic interventions. By accurately predicting the interactions between gut microbiota and pharmacological agents, researchers can enhance drug efficacy and minimize adverse effects, paving the way for more personalized treatment strategies (Chenchula et al., 2024). In addition, AI-based predictive algorithms analyze the impact of diet on microbial communities and forecast illness risks based on gut microbiota composition (Patil & Singh, 2023). These algorithms leverage large datasets to identify correlations between dietary habits and changes in microbial diversity, which can subsequently inform dietary recommendations aimed at improving health outcomes. For example, understanding how specific diets influence the abundance of beneficial or harmful microbial species can lead to targeted dietary interventions that promote overall gut health (Zhang, 2022). These applications underscore AI's transformative impact on microbiome research, enabling more accurate and comprehensive functional predictions. As researchers continue to harness the power of AI and ML, the field is poised to unlock new avenues for understanding the complex interplay between microbes and their hosts. This ongoing evolution will not only enhance our knowledge of microbial ecology but also contribute to the development of innovative therapeutic approaches and public health strategies that leverage the potential of the microbiome. Ultimately, AI is set to reshape the landscape of microbiome research, offering unprecedented opportunities for discovery and application in various scientific and medical domains (Abavisani, Khoshrou, Foroushan, Ebadpour, & Sahebkar, 2024).

3.2 Personalized medicine (microbiome-based diagnostics and therapies)

Personalized medicine is increasingly integrating microbiome-based diagnostics and therapies to enhance patient care. The unique composition of an individual's microbiome offers valuable insights into their health status, enabling more precise disease diagnosis and treatment strategies. Microbial fingerprints serve as noninvasive, accessible, and cost-effective tools for personalized disease diagnosis, including assessments of phenotypes, severity, and prognosis (Giurini, Godla, & Gupta, 2024). Incorporating microbiome analysis into precision medicine has been shown to improve diagnostic accuracy, reduce disease risk, and optimize early detection and treatment. For instance, understanding the gut microbiome's role in drug metabolism can lead to personalized therapeutic approaches that enhance drug efficacy and minimize adverse effects. Moreover, microbiome-based treatments are being explored to boost drug efficacy and reduce side effects, highlighting the potential of personalized medicine approaches that consider an individual's unique microbial composition (Kashyap, Chia, Nelson, Segal, & Elinav, 2017).

As research progresses, the integration of microbiome data into clinical practice is poised to revolutionize healthcare by providing tailored diagnostic and therapeutic options. AI is playing a pivotal role in unlocking the potential of microbiome data for personalized medicine. ML algorithms, such as those used in tools like DeepMicro, analyze vast amounts of sequencing data to identify patterns and correlations between microbial composition and health outcomes. These AI-driven models can predict disease risks, classify microbial communities, and even forecast how an individual's microbiome will respond to specific treatments (Oh & Zhang, 2020). Beyond diagnostics, microbiome-based therapies are being developed to enhance treatment outcomes. For example, engineered probiotics and microbial metabolites are being tailored to individual microbiome profiles to boost drug efficacy and reduce side effects. AI aids in this process by modeling microbial interactions and predicting how interventions will alter the microbiome. This approach is particularly promising in treating conditions like IBD, cancer, and neurodegenerative disorders, where the microbiome plays a critical role in disease progression and treatment response (Murali & Mansell, 2024).

3.3 Drug discovery (microbial metabolites and probiotics)

Microbial metabolites and probiotics are emerging as significant contributors to drug discovery and development. Microorganisms produce a vast array of secondary metabolites that have been foundational in creating antibiotics, antivirals, and analgesics. These compounds offer a rich reservoir for novel therapeutic agents, with many displaying unique mechanisms of action that are invaluable in combating resistant pathogens. In addition to natural metabolites, engineered probiotics are being developed to target specific pathogens. For instance, researchers have modified *Lactococcus lactis* to secrete antimicrobial peptides that selectively eradicate *Helicobacter pylori*, a bacterium linked to gastric cancer, without disrupting the surrounding microbiota (Choudhury et al., 2023).

Probiotics also show promise in treating various diseases, including neurodegenerative disorders, cancers, cardiovascular diseases, and inflammatory conditions. Their ability to modulate the gut microbiota and influence host immunity positions them as potential therapeutic adjuvants. Despite these advancements, challenges persist in translating probiotic research into clinical applications. A significant number of clinical trials have been conducted, yet regulatory approvals, such as those from the US Food and Drug Administration, remain limited. This discrepancy highlights the need for rigorous, well-designed studies to establish the efficacy and safety of probiotic-based therapies (Al-Madhagi & Alramo, 2024; Baral, Bajracharya, Lee, & Han, 2021; Choudhury et al., 2023).

3.4 Ecological insights (modeling microbial ecosystems)

Modeling microbial ecosystems is an essential aspect of ecological research that provides critical insights into the dynamics of microbial communities and their intricate interactions with the environment. Microbial ecosystems, which include diverse communities of bacteria, archaea, fungi, and viruses, are fundamental to numerous ecological processes, including nutrient cycling, organic matter decomposition, and the regulation of biogeochemical cycles. Understanding how these communities function and respond to environmental changes is crucial for addressing pressing global challenges, such as food security, environmental sustainability, and climate change. By employing various modeling approaches, researchers can simulate microbial behaviors, nutrient cycling, and the effects of environmental changes on community structures (Srinivasan, Jnana, & Murali, 2024). One widely used modeling technique is agent-based modeling, which allows for the exploration of individual microbial responses to varying conditions, such as fluctuations in temperature, pH, and nutrient availability. In agent-based models, each microbial agent operates based on predefined rules, which can include growth rates, metabolic pathways, and interactions with other species. This approach enables researchers to observe how individual behaviors aggregate to produce complex community dynamics, offering valuable insights into the mechanisms that govern microbial population dynamics and community stability (Nagarajan, Ni, & Lu, 2022). Another significant modeling approach is network modeling, which elucidates the interactions among species and their roles in ecosystem functions. Network models represent microbial communities as interconnected webs of interactions, where nodes represent different microbial species and edges denote the relationships between them, such as competition, predation, or mutualism. By analyzing these networks, researchers can identify keystone species that play critical roles in maintaining community structure and function. Understanding these interactions is essential for predicting how microbial communities might respond to environmental disturbances, such as nutrient loading or habitat degradation, which are increasingly common due to human activities (Kajihara & Hynson, 2024).

Modeling microbial ecosystems is particularly important for assessing the resilience of these communities in the face of disturbances, including climate change and pollution. As environmental conditions shift, microbial communities must adapt to survive and continue performing their ecological functions. For example, changes in temperature and moisture can influence microbial metabolic rates and community composition, potentially leading to shifts in nutrient cycling processes. By simulating these scenarios, researchers can identify vulnerable communities and develop strategies for managing ecosystems sustainably (Boyle, Murphy, Ensminger, Stinchcombe, & Frederickson, 2024). Moreover, microbial communities play crucial roles in nutrient cycling and overall ecosystem health. They are involved in processes such as nitrogen fixation, carbon sequestration, and the decomposition of organic matter, all of which are vital for maintaining soil fertility and supporting plant growth. Insights gained from modeling microbial ecosystems can inform agricultural practices, helping to enhance crop yields and sustainability. By understanding how to optimize microbial interactions in soil, farmers can improve nutrient availability and reduce reliance on chemical fertilizers, contributing to more sustainable agricultural systems (Jurburg, Blowes, Shade, Eisenhauer, & Chase, 2024).

4 AI techniques for microbiome research

4.1 ML models (e.g., random forests, clustering)

ML models, such as random forests and clustering techniques, are widely utilized for analyzing complex datasets across various fields, including biology, healthcare, and environmental science. These models offer advanced tools for extracting meaningful insights from vast amounts of data, which is particularly valuable in research areas where traditional statistical methods may fall short (Lepakshi, 2022). Random forests, an ensemble learning method, create multiple decision trees and aggregate their predictions to improve overall accuracy. This approach enhances robustness against overfitting, a common issue where models perform well on training data but poorly on unseen data (Imani, Beikmohammadi, & Arabnia, 2025).

By combining the predictions of numerous trees, random forests achieve a more reliable estimate, particularly in high-dimensional datasets where the number of features exceeds the number of samples. In ecological and biological studies, this characteristic is crucial, as datasets often contain numerous variables that can obscure underlying patterns. For instance, random forests have been effectively employed to identify key predictors of species diversity in microbial communities, enabling researchers to pinpoint factors that significantly influence community composition and function (Breiman, 2001). The effectiveness of random forests in handling complex interactions between variables is another notable advantage. In microbial ecology, for instance, the interactions between different microbial species and their environments can be intricate. Random forests can capture these nonlinear relationships and interactions, providing a nuanced understanding of microbial dynamics (Fig. 24.2). This has been instrumental in studies investigating the impact of environmental factors, such as soil composition or nutrient availability, on microbial community structure. By identifying critical predictors of community health and resilience, researchers can better inform management and conservation efforts in ecosystems (Amamra, 2025).

On the other hand, clustering methods like k-means and hierarchical clustering play a vital role in data analysis by grouping similar data points based on shared features. Clustering techniques facilitate the identification of patterns and relationships within datasets, helping researchers classify and understand complex biological data. K-means clustering, for example, partitions data into k distinct clusters based on feature similarity, allowing researchers to segment microbial communities based on their compositional and functional characteristics (Singh & Singh, 2024).

This can reveal how different environmental conditions or treatment factors affect microbial diversity and community structure. Hierarchical clustering provides an alternative approach by creating a tree-like structure that represents the relationships between data points. This method is particularly useful for visualizing similarities and differences among samples, allowing researchers to identify clusters at various levels of granularity. In microbial ecology, hierarchical clustering can help reveal how closely related different microbial species are, which is essential for understanding their evolutionary relationships and ecological roles (Smith, Hope, & Bell, 2025). For instance, studies employing hierarchical clustering have uncovered distinct clusters of microbial communities associated with specific environments, such as the human gut or aquatic ecosystems, leading to insights into how these communities adapt to their niches (Wassan, Wang, & Zheng, 2022). Both random forests and clustering techniques are not without challenges. One major limitation is the potential for interpretability issues, particularly with complex models like random forests, where understanding the influence of individual

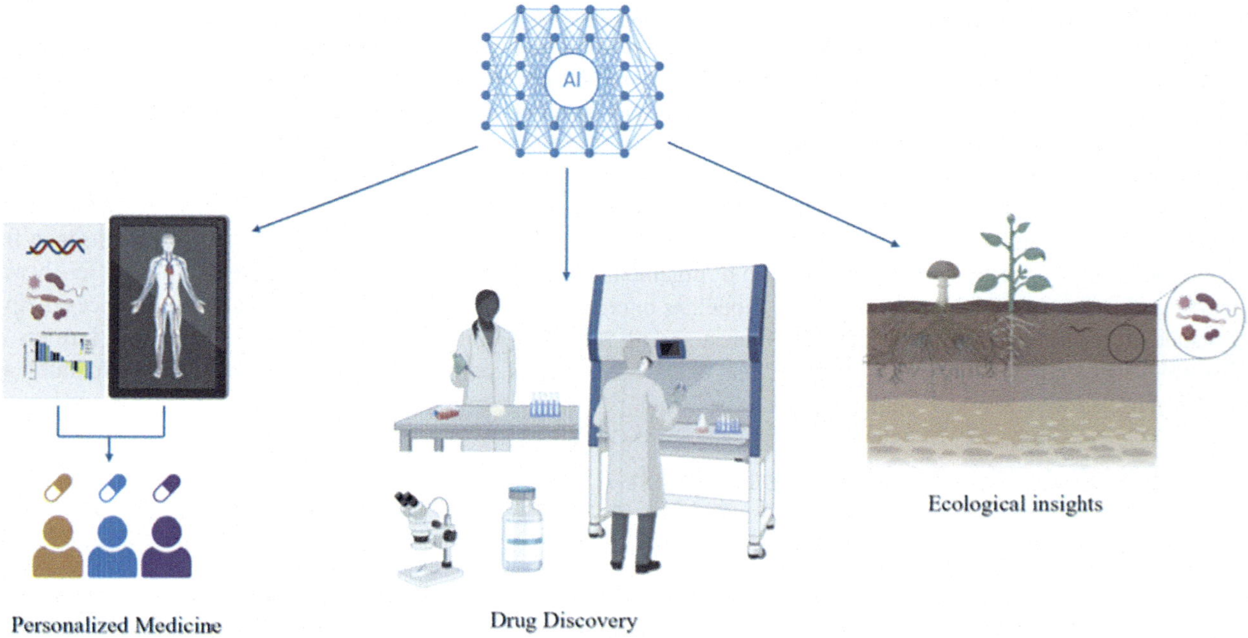

FIG. 24.2 Usage of artificial intelligence in microbiome studies.

features on predictions can be difficult. Researchers often need to employ additional techniques, such as variable importance measures or partial dependence plots, to interpret the results effectively (Ennab & Mcheick, 2024).

Similarly, clustering methods can be sensitive to parameter choices, such as the number of clusters in k-means, which can significantly impact the results. Therefore careful consideration of the underlying biological questions and robust validation of model outcomes are essential to ensure that the insights derived from these methods are meaningful and reliable (Gaiteri et al., 2023).

4.2 DL approaches (e.g., neural networks, BioBERT)

DL approaches, including neural networks and BioBERT, have gained significant importance in recent years due to their ability to process large volumes of unstructured data and extract meaningful insights. Neural networks, inspired by biological processes, consist of interconnected layers that learn hierarchical representations of data, making them well-suited for tasks such as image recognition, natural language processing (NLP), and genomic data analysis (Lecun, Bottou, Bengio, & Haffner, 2015). BioBERT, a domain-specific adaptation of the BERT (Bidirectional Encoder Representations from Transformers) model, has been tailored for biomedical literature and can effectively capture context and semantics in scientific texts (Lee et al., 2020). These DL approaches enhance the ability to analyze complex biological systems, predict outcomes, and extract relevant information from vast repositories of research literature.

4.3 Tools for literature mining and knowledge extraction

Tools for literature mining and knowledge extraction have become increasingly vital in the realm of scientific research, particularly given the exponential growth of published literature in fields like biology, healthcare, and environmental science. As the volume of scientific articles expands, researchers face the daunting challenge of sifting through vast amounts of information to identify relevant studies, trends, and insights. This is where literature mining tools come into play, providing essential support for synthesizing information and facilitating the discovery of new knowledge. One of the most widely used platforms for literature mining is PubMed, a comprehensive database of biomedical literature (Jin, Leaman, & Zhiyong, 2024).

PubMed's Entrez Programming Utilities (E-utilities) serve as a set of APIs that allow researchers to efficiently retrieve and process biological data from multiple databases. These utilities enable users to automate the search process, retrieve bibliographic data, and access full-text articles with ease. By utilizing E-utilities, researchers can conduct systematic reviews and meta-analyses more efficiently, ultimately accelerating the pace of discovery in the biomedical field. BioPython is another powerful tool that supports literature mining by providing a suite of tools for biological computation (Lin et al., 2025). This open-source library allows researchers to interact with various biological databases, perform sequence analysis, and manipulate biological data. With BioPython, researchers can automate data retrieval and analysis, streamlining the literature review process and enabling more comprehensive assessments of existing research. By integrating BioPython with E-utilities, researchers can create customized workflows that enhance their ability to extract relevant information from the literature (da Silva Souza et al., 2025; Fang, 2012). In addition to text mining software, ML techniques have increasingly been incorporated into literature mining tools to enhance information extraction capabilities. ML algorithms can be trained to identify relevant studies, extract key findings, and even summarize articles automatically (Aleqabie, Sfoq, Albeer, & Abd, 2024). NLP, a subset of AI focused on understanding and generating human language, plays a critical role in this process. By employing NLP techniques, researchers can develop systems that not only retrieve articles based on specific keywords but also discern the context and significance of findings within the texts (Khurana, Koli, Khatter, & Singh, 2023).

Recent advancements in knowledge graphs and ontologies have further supported literature mining efforts by providing structured representations of knowledge. Knowledge graphs visualize relationships between concepts, making it easier for researchers to identify connections and navigate complex information networks (Abu-Salih & Alotaibi, 2024). For instance, a knowledge graph might illustrate how specific microbial species are related to certain diseases, highlighting the interconnectedness of various fields of study. This visualization can guide researchers in identifying research gaps, informing future investigations, and facilitating interdisciplinary collaboration. Ontologies, which provide a formal representation of knowledge within a specific domain, also enhance literature mining by offering standardized vocabularies and classifications (Robertson et al., 2025).

By employing ontologies, researchers can ensure that their literature searches are comprehensive and precise, reducing ambiguity in the interpretation of terms and concepts. This is particularly important in fields like microbiome research, where terminology can vary significantly across studies. Utilizing standardized ontologies can facilitate better communication and understanding among researchers, ultimately leading to more effective collaboration and knowledge sharing. The integration of ML, knowledge graphs, and ontologies in literature mining tools enhances researchers' ability to navigate the ever-expanding landscape of scientific literature efficiently (Tiddi & Schlobach, 2022).

By streamlining the process of information retrieval and extraction, these tools empower researchers to focus on higher-level analysis and interpretation, driving innovation and discovery in their fields. Furthermore, the automation of literature mining tasks reduces the burden on researchers, enabling them to stay current with the latest developments in their areas of interest without becoming overwhelmed by the sheer volume of information (Hunter et al., 2026).

The rapid pace of advancements in both the fields of research and technology necessitates ongoing development and refinement of literature mining tools. As new techniques emerge and databases evolve, it is crucial for researchers and developers to collaborate in updating and improving these tools to meet the changing needs of the scientific community (Wang, 2018). With the integration of text mining software, ML techniques, knowledge graphs, and ontologies, researchers are better equipped to navigate the complexities of scientific literature, ultimately accelerating the pace of discovery and innovation in various fields. As these tools continue to evolve, they will play a crucial role in shaping the future of research and knowledge dissemination (Varnosfaderani & Forouzanfar, 2024).

5 Challenges and ethical concerns

5.1 Data privacy and security

Microbiome research, which involves studying the complex communities of microorganisms residing in and on the human body, is rapidly advancing due to technological improvements in genomics and bioinformatics. However, this progress raises significant ethical concerns, particularly regarding data privacy and security. One of the primary challenges is the handling of personal and sensitive health data, as microbiome studies often involve collecting biological samples from individuals. These samples can provide insights into a person's health status, genetic predispositions, and even lifestyle choices, making it crucial to implement robust data privacy measures (Ma, Zuo, Frey, & Rangrez, 2024).

The collection, storage, and sharing of microbiome data present multiple privacy risks. For instance, the potential for reidentification of individuals from anonymized datasets poses a significant threat. Even when data is deidentified, advancements in data mining and ML techniques can sometimes allow for the reconstruction of identities based on microbiome signatures. This risk is compounded when datasets are shared across institutions or made publicly accessible, as the likelihood of unauthorized access increases. Consequently, researchers must adhere to strict ethical guidelines and legal frameworks, such as the Health Insurance Portability and Accountability Act in the United States, to protect sensitive data and maintain participant confidentiality (Chuong et al., 2017). Furthermore, informed consent is a critical component of microbiome research. Participants must fully understand how their data will be used, stored, and shared, which can be challenging given the complexity of microbiome data and the fast-evolving nature of research. There is a responsibility to ensure that participants are aware of potential risks and that their consent is obtained transparently (Ma et al., 2024). In addition, there is a need for guidelines that address the ownership and commercial use of microbiome data, particularly as private companies increasingly engage in microbiome research for product development (Rhodes, 2016).

5.2 Bias in AI algorithm

As microbiome research increasingly incorporates AI and ML techniques to analyze complex datasets, concerns about bias in AI algorithms become paramount. These biases can stem from various sources, including biased training data, flawed algorithmic design, and the subjective interpretation of results, which can lead to skewed findings and implications for health outcomes. For instance, if an ML model is trained predominantly on data from specific populations, it may not generalize well to other demographic groups, resulting in disparities in microbiome-related health predictions or interventions (Norori, Qiyang, Aellen, Faraci, & Tzovara, 2021).

Bias in AI algorithms can manifest in several ways. First, the selection of training datasets is crucial; if the data is not representative of the broader population, the model may inherit and amplify existing biases. For example, microbiome studies often have a disproportionate focus on populations from high-income countries, neglecting diverse ethnic groups and low-resource settings. This lack of diversity in training data can result in algorithms that perform poorly or inaccurately for underrepresented populations, exacerbating health inequities (Meyer, 2021). Moreover, algorithmic design itself can introduce biases. Many ML algorithms rely on assumptions and heuristics that may not hold true across different contexts or populations. For instance, models that prioritize certain microbiome features based on their perceived importance in one demographic might overlook critical factors relevant to others. This can lead to incorrect interpretations of the microbiome's role in health and disease, ultimately impacting clinical decision-making and public health strategies (Binns, 2020).

To address these challenges, researchers must prioritize transparency and inclusivity in AI-driven microbiome studies. This includes rigorous validation of algorithms across diverse datasets and ensuring representation of various demographic groups during model training. Incorporating feedback from diverse stakeholders can also help identify potential biases and improve model performance. Furthermore, developing ethical frameworks for AI use in microbiome research is essential, focusing on accountability and fairness to mitigate the risk of perpetuating health disparities (Dankwa-Mullan, 2024).

5.3 Interpretability and reproducibility issues

Interpretability and reproducibility are critical challenges in microbiome research, particularly as complex computational models are increasingly employed to analyze microbiome data. Interpretability refers to the extent to which the results of a model can be understood and explained in human terms (Papoutsoglou et al., 2023).

In microbiome research, where interactions among microbial communities and their effects on human health are intricate and multifaceted, achieving interpretability is crucial for both scientific advancement and public trust. When researchers utilize sophisticated ML models, such as DL networks, the resulting predictions may be difficult to interpret, leading to uncertainty about the biological mechanisms underlying observed associations (Lipton, 2018). The lack of interpretability in AI models can hinder the acceptance of findings by clinicians and policymakers, who require clear explanations of how microbial profiles relate to health outcomes. For instance, if a model predicts a microbiome signature associated with a particular disease but does not provide insights into which specific microbial taxa or functions are driving the prediction, it may be challenging for healthcare practitioners to apply these findings in clinical practice. Furthermore, the inability to explain model predictions can also impede the validation of results, as understanding the rationale behind a model's output is essential for building confidence in its accuracy and relevance (Baniecki, Parzych, & Biecek, 2024).

Reproducibility is another significant concern in microbiome research, as variability in study designs, methodologies, and analysis techniques can lead to inconsistent findings across different studies. Factors such as sample collection, processing methods, and data analysis pipelines can introduce noise and variability that complicate the comparison of results. In addition, the complex interplay of environmental, genetic, and lifestyle factors influencing the microbiome makes it difficult to replicate studies in different populations or settings (Surana & Kasper, 2017). This variability raises questions about the reliability of conclusions drawn from microbiome research and underscores the need for standardized methodologies and protocols.

To enhance interpretability and reproducibility in microbiome research, researchers should prioritize transparency and collaboration. Sharing data, code, and methodologies openly can facilitate independent verification of results and promote a culture of reproducibility within the scientific community. Employing interpretable ML techniques, such as generalized additive models or feature importance analysis, can also aid in elucidating the relationships between microbial communities and health outcomes. Furthermore, the development of consensus guidelines for microbiome research methodologies and reporting standards can help mitigate variability and foster confidence in the field's findings (Caminero et al., 2025; D'Elia et al., 2023).

6 Future directions in microbiome research

6.1 AI in microbiome engineering and synthetic biology

The integration of AI in microbiome engineering and synthetic biology represents a transformative frontier in biotechnology. Microbiome engineering aims to manipulate microbial communities to enhance their beneficial effects on human health, agriculture, and environmental sustainability. AI can significantly accelerate this process by enabling the design and optimization of synthetic microbial consortia, allowing researchers to predict the interactions and functions of engineered microbes in complex environments (Graham, Chen, Walters, Knight, & Dorrestein, 2021). ML algorithms can analyze vast datasets from microbial genomics and metabolomics to identify potential microbial strains with desirable traits, streamline the engineering process, and reduce the time required for experimental validation. One promising application of AI in microbiome engineering is in the development of probiotic formulations tailored to individual health needs. By utilizing patient-specific microbiome profiles, AI-driven algorithms can identify microbial strains that would likely restore balance to an individual's gut microbiota, improving health outcomes for conditions such as irritable bowel syndrome, IBD, and metabolic disorders (Zhou, Song, Shao, & Zhang, 2020). Moreover, AI can enhance the prediction of the metabolic outputs of engineered microbial consortia, allowing for the optimization of these communities to produce beneficial metabolites, such as vitamins and short-chain fatty acids, that support human health. Synthetic biology, coupled with AI, also enables the creation of novel microbial functionalities. For instance, researchers can design microbial strains capable of bioremediation, efficiently breaking down pollutants in the environment. AI can facilitate this process by predicting which metabolic pathways can be harnessed or engineered for specific bioremediation tasks (Chuang, Wang, Lee, & Kuo, 2022). As microbiome engineering continues to evolve, the collaborative efforts between AI and synthetic biology will be crucial in addressing pressing global challenges, such as climate change, food security, and public health (Kumar & Jaiswal, 2015).

6.2 Combining microbiome studies with AI-driven multiomics approaches

The combination of microbiome studies with AI-driven multiomics approaches is poised to revolutionize our understanding of complex biological systems. Multiomics integrates various biological data types, such as genomics, transcriptomics, proteomics, and metabolomics, to provide a holistic view of the interactions among different biological components (Kumar, Bhandari, Patel, & Singh, 2021). By employing AI algorithms to analyze these integrated datasets, researchers can uncover intricate relationships between microbial communities and host biology, enhancing our understanding of health and disease mechanisms. AI-driven multiomics approaches can facilitate the identification of biomarkers for disease prediction and diagnosis. For example, by integrating microbiome data with host genomic and transcriptomic information, researchers can identify microbial signatures that correlate with specific health outcomes, leading to the development of novel diagnostic tools (Guan, Zhao, Liu, & Wang, 2022). In addition, AI can be applied to discover interactions between microbial metabolites and host metabolic pathways, revealing insights into how the microbiome influences host physiology and disease progression. The use of multiomics data also allows for personalized medicine strategies. By analyzing an individual's microbiome in conjunction with their genomic and metabolic profiles, healthcare providers can tailor interventions, such as dietary modifications or targeted probiotic therapies, to optimize health outcomes (Li, Jia, Cai, & Zhao, 2020). Moreover, the application of AI in multiomics research can enhance the predictive power of models, enabling more accurate forecasting of treatment responses based on individual biological profiles. As multiomics approaches become increasingly sophisticated, the integration of AI will be essential for unlocking the full potential of microbiome research in personalized healthcare (Mohr, Ortega-Santos, Whisner, Klein-Seetharaman, & Jasbi, 2024).

6.3 Role of AI in global microbiome initiatives

AI is playing an increasingly significant role in global microbiome initiatives aimed at understanding the diverse roles of microbiomes in health, disease, and the environment. Initiatives such as the Human Microbiome Project (HMP) and the Earth Microbiome Project (EMP) leverage AI technologies to analyze large-scale microbiome datasets, facilitating the discovery of patterns and correlations that would be difficult to identify using traditional methods (Gilbert, Jansson, & Knight, 2014; Huttenhower et al., 2012). By employing ML algorithms to process and interpret complex microbiome data, researchers can extract valuable insights into microbial diversity, function, and their contributions to ecological and human health. AI-driven analyses can enhance the ability to predict the impact of environmental changes on microbial communities, a critical aspect of initiatives focused on biodiversity conservation and climate change (Nayak, Sharma, Gupta, & Thomas, 2021). By integrating environmental data with microbiome profiles, AI models can forecast how shifts in climate or land use affect microbial ecosystems, informing conservation strategies and ecosystem management (McElhinney, Catacutan, Mawart, Hasan, & Dias, 2022).

In addition, these models can identify key microbial taxa that serve as indicators of ecosystem health, providing valuable tools for monitoring biodiversity and ecosystem resilience. Furthermore, AI can facilitate collaboration among researchers, institutions, and countries involved in global microbiome initiatives. By creating platforms that aggregate microbiome data and integrate AI tools for analysis, researchers can share findings, resources, and methodologies more effectively. This collaborative approach can accelerate scientific discovery and foster a more comprehensive understanding of the microbiome's role across various ecosystems and health contexts. As global microbiome initiatives continue to evolve, the integration of AI will be crucial in addressing pressing global challenges, from public health to environmental sustainability (Proctor, Dore, Maguin, Wannerberger, & Vincent, 2025).

7 Conclusion

In summary, the integration of AI in microbiome research represents a transformative advancement that enhances our understanding of complex microbial ecosystems and their intricate interactions with hosts and the environment. AI-driven approaches, such as ML and DL, provide powerful tools for analyzing vast and diverse datasets, facilitating the identification of patterns and relationships that traditional methods may overlook. These techniques enable researchers to uncover insights into microbial diversity, functionality, and their contributions to human health, disease, and ecological balance. AI has proven instrumental in various aspects of microbiome research, including the prediction of microbial community composition, the identification of specific biomarkers for disease, and the development of personalized therapeutic strategies. By harnessing the capabilities of AI, researchers can analyze multiomics data—integrating genomic, transcriptomic, proteomic, and metabolomic information—to gain a comprehensive understanding of microbial interactions and their impact on host biology. This holistic perspective is crucial for advancing personalized medicine and tailoring interventions that optimize health outcomes. Furthermore, the role of AI in global microbiome initiatives cannot be overstated. Collaborative efforts that leverage AI technologies can accelerate discoveries and promote the sharing of knowledge across disciplines and borders. As microbiome research continues to evolve, embracing AI will be essential in addressing critical challenges related to public health, environmental sustainability, and food security. Despite the promise of AI in microbiome research, researchers must remain vigilant regarding ethical considerations, data privacy, and the interpretability of AI models. Ensuring transparency and reproducibility in AI-driven analyses is vital for building trust and credibility in the findings. Overall, the future of microbiome research is poised for exciting developments, with AI as a key enabler in unraveling the complexities of microbial life and its implications for health and the environment (Liu et al., 2025; Zhang et al., 2021).

Declaration of generative AI or AI-assisted technologies in the writing process

Generative AI was not used in preparing the manuscript. However, AI-assisted tools such as ChatGPT were used to search for appropriate titles and relevant information for different sections of the manuscript.

Glossary

Introduction

Human Microbiome The collection of microbes that live in and on the human body, especially in places like the gut, skin, mouth, and lungs.

Metagenomic Sequencing High-throughput DNA sequencing technique that analyzes genetic material directly from environmental samples without culturing individual species.

Metabolomics The systematic study of chemical fingerprints left by cellular metabolic processes.

Transcriptomics The study of all RNA molecules (transcripts) produced by the genome under specific conditions.

High-Dimensional Data Datasets containing a large number of variables or features, often exceeding the number of observations.

Artificial Intelligence (AI) Smart computer systems that can think and learn from data like humans. In microbiome research, AI helps analyze huge datasets to find patterns and make predictions.

Machine Learning (ML) A subset of AI that enables computers to learn and make predictions from data without being explicitly programmed.

Deep Learning (DL) A subset of machine learning where computers learn by using something like a brain (called a neural network) with many layers. This helps them understand really complicated things, like recognizing faces or translating languages.

Biomarkers Measurable biological indicators that can signal normal biological processes, disease states, or responses to treatment.

AI in microbiome data analysis

Taxonomic Classification The process of identifying and categorizing microorganisms present in a sample into hierarchical groups (species, genus, families, etc.).

Sequence Alignment Traditional computational method for comparing DNA sequences to identify similarities and relationships.

Taxometer A neural network-based tool designed to improve taxonomic classifications and evaluate the quality of classifiers for metagenomic contigs.

Metagenomic Contigs These are long DNA sequences that scientists build by joining together shorter DNA fragments (called reads) taken from a mixed sample of microbes (such as soil, ocean water, or the human gut). Since the sample contains DNA from many different organisms, these contigs represent pieces of genomes from multiple microbes and help us understand who is in the sample and what they are capable of doing.

Naïve Bayes Classifier++ (NBC++) An enhanced version of the traditional Naïve Bayes framework for classifying reads from amplicon and metagenomic samples. It helps identify different microbes from genetic data.

MetaTrinity A tool that sorts and analyzes metagenomic data faster and more accurately using smart algorithms, while reducing computational time.

Genomics	The study of an organism's complete set of genetic material (genome).
Proteomics	The large-scale study of proteins, particularly their structures and functions.
Multiomics	The integration of multiple biological data types, including genomics, proteomics, and metabolomics, to provide comprehensive biological insights.
High-Throughput Sequencing	Technologies like Illumina or Oxford Nanopore capable of generating massive amounts of DNA sequence data rapidly.
Quality Control	The process of filtering out sequencing errors and low-quality reads to ensure data accuracy.
Assembly	Computational process where short DNA fragments are pieced together into longer sequences called contigs.
Functional Annotation	The process of determining the biological roles and functions of genes present in metagenomic data.
Phylogenetic Placement	Method of refining classification accuracy by comparing contigs to reference databases within evolutionary contexts.
Convolutional Neural Networks (CNNs)	Deep learning models particularly effective for analyzing complex biological datasets.
Attention-Based Models	AI models that can focus on specific parts of input data, useful for identifying important biological patterns.
Metagenome-Assembled Genomes (MAGs)	Reconstructed genomes of microorganisms that have not been isolated or cultured in the laboratory.

Applications of AI in microbiome research

DeepMicro	A machine learning tool specifically designed to process extensive sequencing data, detect patterns, classify microbes, and predict disease associations.
Taxonomic Profiling	The use of machine learning algorithms to classify microbial species within a given sample.
Metabolic Modeling	Computational approaches to understand and predict the metabolic capabilities of microorganisms.
Microbe–Microbe Interactions	The complex relationships and communications between different microbial species in a community.
Drug Metabolism Studies	Research using machine learning to predict which gut bacteria and enzymes can metabolize pharmaceutical compounds.
Precision Medicine	Healthcare approach that considers individual variability in genes, environment, lifestyle, and microbiome composition.
Microbial Fingerprints	Unique patterns of microbial composition that can serve as diagnostic tools for disease assessment.
Phenotype Assessment	Evaluation of observable characteristics or traits that result from genetic and environmental interactions.
Prognosis	Prediction of the likely course and outcome of a disease based on microbial and other biomarkers.
Host–Microbiome Interactions	The complex bidirectional relationships between microorganisms and their host organisms.
Microbial Metabolites	Chemical compounds produced by microorganisms during their metabolic processes, often with therapeutic potential.
Secondary Metabolites	Specialized compounds produced by microorganisms that are not essential for growth but may possess biological activity.
Engineered Probiotics	Genetically modified beneficial bacteria designed to target specific pathogens or produce therapeutic compounds.
Lactococcus lactis	A bacterial species commonly used in probiotic engineering and food fermentation.
Helicobacter pylori	A pathogenic bacterium associated with peptic ulcers and gastric cancer.
Antimicrobial Peptides	Small proteins with antimicrobial activity that can be secreted by engineered probiotics.
Microbial Ecosystems	Complex communities of diverse microorganisms and their interactions within specific environments.
Nutrient Cycling	The natural processes by which nutrients move through living and nonliving components of ecosystems.
Biogeochemical Cycles	Natural processes that recycle chemical elements like carbon, nitrogen, and phosphorus through ecosystems.
Agent-Based Modeling	Computational modeling approach where individual microbial agents operate based on predefined rules to simulate community dynamics.
Network Modeling	Approach that represents microbial communities as interconnected webs of species interactions.
Keystone Species	Microbial species that play critical roles in maintaining community structure and ecosystem function.

AI techniques for microbiome studies

Random Forests	An ensemble learning method that creates multiple decision trees and combines their predictions for improved accuracy and robustness.
Ensemble Learning	Machine learning approach that combines multiple algorithms to achieve better performance than individual models.
Overfitting	When a model performs well on training data but poorly on new, unseen data due to excessive complexity.
High-Dimensional Datasets	Data with many variables or features, common in microbiome research, where features often exceed sample numbers.
Clustering	Unsupervised learning technique that groups similar data points together based on shared characteristics.
K-means Clustering	A clustering algorithm that partitions data into k distinct clusters based on feature similarity.
Hierarchical Clustering	A clustering method that creates tree-like structures showing relationships between data points at various levels.

Variable Importance Measures	Techniques used to determine which features contribute most significantly to model predictions.
Partial Dependence Plots	Visualization tools that show the relationship between a feature and the predicted outcome.
Neural Networks	Computing systems inspired by biological neural networks, consisting of interconnected processing units (neurons).
Hierarchical Representations	Multilayered data representations learned by deep learning models to capture complex patterns.
BioBERT	A domain-specific adaptation of BERT (Bidirectional Encoder Representations from Transformers) tailored for biomedical literature analysis.
BERT (Bidirectional Encoder Representations from Transformers)	A transformer-based model designed for natural language processing tasks.
Literature Mining	Automated extraction and analysis of information from scientific publications and databases.
PubMed	A comprehensive database of biomedical literature maintained by the National Library of Medicine.
Entrez Programming Utilities (E-utilities)	APIs that allow automated retrieval and processing of biological data from multiple databases.
BioPython	An open-source library providing tools for biological computation, sequence analysis, and data manipulation.
Natural Language Processing (NLP)	AI field focused on enabling computers to understand, interpret, and generate human language.
Knowledge Graphs	Structured representations of knowledge that visualize relationships between concepts and entities.
Ontologies	Formal representations of knowledge within specific domains, using standardized vocabularies and classifications.

Challenges and ethical concerns

Health Insurance Portability and Accountability Act (HIPAA)	US federal law that provides data privacy and security provisions for protecting medical information.
Deidentification	Process of removing personally identifiable information from datasets to protect participant privacy.
Reidentification	The potential risk of determining individual identities from supposedly anonymized datasets.
Data Mining	Computational process of discovering patterns and relationships in large datasets.
Informed Consent	Process ensuring research participants understand how their biological samples and data will be used, stored, and shared.
Algorithmic Bias	Systematic errors in AI systems that can lead to unfair or discriminatory outcomes across different populations.
Training Data	The dataset used to teach machine learning algorithms, which can introduce biases if not representative.
Demographic Representation	The inclusion of diverse population groups in research studies and training datasets.
Health Inequities	Unfair and avoidable differences in health outcomes between different population groups.
Model Generalization	The ability of an AI model to perform accurately on new, unseen data from different populations or contexts.
Model Interpretability	The extent to which AI model results and decision-making processes can be understood and explained in human terms.
Black-Box Models	AI systems whose internal workings are not easily understood or explained, such as complex deep learning networks.
Biological Mechanisms	The underlying biological processes and pathways that explain observed associations or outcomes.
Reproducibility	The ability to obtain consistent results when studies are repeated using the same methods and conditions.
Standardized Protocols	Established, consistent methodologies for conducting research to ensure comparability across studies.
Validation	Process of testing a model's performance on independent datasets to assess reliability and generalizability.

Future directions

Microbiome Engineering	The deliberate manipulation of microbial communities to enhance their beneficial effects on health, agriculture, or the environment.
Synthetic Biology	An engineering approach to designing and constructing new biological parts, devices, and systems.
Synthetic Microbial Consortia	Artificially designed communities of microorganisms engineered to work together for specific purposes.
Bioremediation	The use of microorganisms to remove or neutralize environmental pollutants and contaminants.
Metabolic Pathways	Series of chemical reactions within cells that convert substrates into products through enzyme-catalyzed steps.
Systems Biology	Interdisciplinary approach that studies complex biological systems through the integration of multiple data types.
Biomarker Discovery	The process of identifying measurable biological indicators associated with specific health outcomes or disease states.
Predictive Modeling	Using statistical and machine learning techniques to forecast future outcomes based on current data.
Personalized Interventions	Tailored treatments or therapies designed based on individual biological profiles and characteristics.

Human Microbiome Project (HMP)	Large-scale research initiative to characterize the human microbiome and understand its role in health and disease.
Earth Microbiome Project (EMP)	Global collaborative effort to characterize microbial diversity across Earth's environments.
Biodiversity Conservation	Efforts to protect and preserve the variety of life forms, including microbial diversity, in ecosystems.
Climate Change Impact	Effects of changing environmental conditions on microbial communities and ecosystem functions.
Ecosystem Management	Approaches to maintaining and restoring the health and sustainability of ecological systems.
Cross-Institutional Collaboration	Partnerships between different research institutions, organizations, and countries to advance scientific knowledge.

Conclusion

Paradigm Shift	A fundamental change in approach or underlying assumptions in scientific research and practice.
Mechanistic Insights	Understanding of the underlying biological mechanisms and causal relationships, rather than just correlations.
Interdisciplinary Synergy	Collaboration and integration of knowledge from multiple scientific disciplines to address complex problems.
Computational Power	The processing capability of computer systems, which continues to advance and enable more sophisticated analyses.
Causative Analysis	Research approaches focused on identifying cause-and-effect relationships rather than just statistical associations.
Environmental Sustainability	Meeting current needs without compromising the ability of future generations to meet their own needs.
Precision Healthcare	Medical care tailored to individual characteristics, including genetic, environmental, and lifestyle factors.

References

Abavisani, M., Khoshrou, A., Foroushan, S. K., Ebadpour, N., & Sahebkar, A. (2024). Deciphering the gut microbiome: the revolution of artificial intelligence in microbiota analysis and intervention. *Current Research in Biotechnology*. 25902628. *7*, 100211. doi:10.1016/j.crbiot.2024.100211.

Abu-Salih, B., & Alotaibi, S. (2024). A systematic literature review of knowledge graph construction and application in education. *Heliyon*. 24058440. *10*(3). Available from https://www.sciencedirect.com/science/journal/24058440. doi:10.1016/j.heliyon.2024.e25383.

Al-Madhagi, H., & Alramo, A. (2024). Clinical trials of probiotics: current outlook. *Food Science & Nutrition*. 2048-7177. *12*(3), 2192–2194. doi:10.1002/fsn3.3884.

Aleqabie, H. J., Sfoq, M. S., Albeer, R. A., & Abd, E. H. (2024). A review of text mining techniques: trends, and applications in various domains. *Iraqi Journal for Computer Science and Mathematics*. 27887421. *5*(1), 125–141. Available from https://journal.esj.edu.iq/index.php/IJCM/article/view/905/338. doi:10.52866/ijcsm.2024.05.01.009.

Amamra, S. A. (2025). Random forest-based machine learning model design for 21,700/5 Ah lithium cell health prediction using experimental data. *Physchem*. 26737167. *5*(1). Available from https://www.mdpi.com/journal/physchem. doi:10.3390/physchem5010012.

Baniecki, H., Parzych, D., & Biecek, P. (2024). The grammar of interactive explanatory model analysis. *Data Mining and Knowledge Discovery*. 1573756X. *38*(5), 2596–2632. Available from https://www.springer.com/journal/10618. doi:10.1007/s10618-023-00924-w.

Baral, K. C., Bajracharya, R., Lee, S. H., & Han, H. K. (2021). Advancements in the pharmaceutical applications of probiotics: dosage forms and formulation technology. *International Journal of Nanomedicine*.11782013. *16*, 7535–7556. Available from https://www.dovepress.com/getfile.php?fileID=75762. doi:10.2147/IJN.S337427.

Beam, A. L., & Kohane, I. S. (2018). Big data and machine learning in health care. *Journal of the American Medical Association*. 15383598. *319*(13), 1317–1318. Available from https://jamanetwork.com/journals/jama/articlepdf/2675024/jama_beam_2018_vp_170174.pdf. doi:10.1001/jama.2017.18391.

Binns, R. (2020). Fairness in machine learning: lessons from political philosophy. In *Proceedings of the 2020 Conference on Fairness, Accountability, and Transparency* (pp. 148–158).

Boyle, J. A., Murphy, B. K., Ensminger, I., Stinchcombe, J. R., & Frederickson, M. E. (2024). Resistance and resilience of soil microbiomes under climate change. *Ecosphere*. 21508925. *15*(12). Available from http://esajournals.onlinelibrary.wiley.com/hub/journal/10.1002/ISSN2150-8925/. doi:10.1002/ecs2.70077.

Breiman, L. (2001). Random forests. *Machine Learning*. 0885-6125. *45* (1), 5–32. doi:10.1023/a:1010933404324.

Caminero, A., Tropini, C., Valles-Colomer, M., Shung, D. L., Gibbons, S. M., Surette, M. G., et al. (2025). Credible inferences in microbiome research: ensuring rigour, reproducibility and relevance in the era of AI. *Nature Reviews Gastroenterology & Hepatology*, 1759–5045. doi:10.1038/s41575-025-01100-9.

Chenchula, S., Atal, S., Bhattacharyya, A., Medhi, B., Chavan, M., & Sarma, P. (2024). Pharmacomicrobiomics—Another frontier of precision medicine. *Indian Journal of Pharmacology*. 19983751. *56*(4), 237–241. Available from http://www.ijp-online.com/. doi:10.4103/ijp.ijp_646_24.

Chiu, C. Y., & Miller, S. A. (2019). Clinical metagenomics. *Nature Reviews Genetics*. 14710064. *20*(6), 341–355. Available from http://www.nature.com/reviews/genetics. doi:10.1038/s41576-019-0113-7.

Choudhury, A., Ortiz, P. S., Young, M., Mahmud, M. T., Stoffel, R. T., Greathouse, K. L., & Kearney, C. M. (2023). Control of Helicobacter pylori with engineered probiotics secreting selective guided antimicrobial peptides. *Microbiology Spectrum*. 21650497. *11*(5). Available from https://journals.asm.org/doi/pdf/10.1128/spectrum.02014-23. doi:10.1128/spectrum.02014-23.

Chuang, Y., Wang, Y., Lee, C., & Kuo, H. (2022). The role of AI in accelerating synthetic biology research. *Nature Reviews Microbiology*, *20*, 24–38.

Chuong, K. H., Hwang, D. M., Tullis, D. E., Waters, V. J., Yau, Y. C. W., Guttman, D. S., & O'Doherty, K. C. (2017). Navigating social and ethical challenges of biobanking for human microbiome research. *BMC Medical Ethics*. 14726939. *18*(1), 1–10. Available from http://www.biomedcentral.com/bmcmedethics/. doi:10.1186/s12910-016-0160-y.

Czech, L., Stamatakis, A., Dunthorn, M., & Barbera, P. (2022). Metagenomic analysis using phylogenetic placement—A review of the first decade. *Frontiers in Bioinformatics*. 26737647. *2*. Available from www.frontiersin.org/journals/bioinformatics. doi:10.3389/fbinf.2022.871393.

D'Elia, D., Truu, J., Lahti, L., Berland, M., Papoutsoglou, G., Ceci, M., et al. (2023). Advancing microbiome research with machine learning: key findings from the ML4Microbiome COST action. *Frontiers in Microbiology*. 1664-302X. *14*. doi:10.3389/fmicb.2023.1257002.

D'Urso, F., & Broccolo, F. (2024). Applications of artificial intelligence in microbiome analysis and probiotic interventions—An overview and perspective based on the current state of the art. *Applied Sciences*. 2076-3417. *14* (19), 8627. doi:10.3390/app14198627.

da Silva Souza, L. R., da Silvada Silvada Silva, D. H., Ribeiro, C. T., da Silvada Silvada Silva, D. A., Nasuto, S. J., Sweeney-Reed, C. M., et al. (2025). PubMedMetaTool: automated metadata extraction from PubMed using Python for bibliometric analysis. *Software Impacts*. 26659638. *24*, 100766. doi:10.1016/j.simpa.2025.100766.

Dankwa-Mullan, I. (2024). Health equity and ethical considerations in using artificial intelligence in public health and medicine. *Preventing Chronic Disease*. 1545-1151. *21*. doi:10.5888/pcd21.240245.

Dhaarani, R., & Reddy, M. K. (2025). Progressing microbial genomics: artificial intelligence and deep learning driven advances in genome analysis and therapeutics. *Intelligence-Based Medicine*. 26665212. *11*, 100251. doi:10.1016/j.ibmed.2025.100251.

Duan, H., Hearne, G., Polikar, R., & Rosen, G. L. (2025). The Naïve Bayes classifier++ for metagenomic taxonomic classification—Query evaluation. *Bioinformatics*. 13674811. *41*(1). Available from http://bioinformatics.oxfordjournals.org/. doi:10.1093/bioinformatics/btae743.

Ennab, M., & Mcheick, H. (2024). Enhancing interpretability and accuracy of AI models in healthcare: a comprehensive review on challenges and future directions. *Frontiers in Robotics and AI*. 22969144. *11*. Available from https://www.journal.frontiersin.org/journal/robotics-and-ai. doi:10.3389/frobt.2024.1444763.

Fang, H. (2012). Text mining for the biologist: tools and techniques. *Briefings in Bioinformatics*, *13*, 580–597.

Gaiteri, C., Connell, D. R., Sultan, F. A., Iatrou, A., Ng, B., Szymanski, B. K., et al. (2023). Robust, scalable, and informative clustering for diverse biological networks. *Genome Biology*. 1474760X. *24*(1). Available from https://genomebiology.biomedcentral.com. doi:10.1186/s13059-023-03062-0.

Gilbert, J. A., Blaser, M. J., Caporaso, J. G., Jansson, J. K., Lynch, S. V., & Knight, R. (2018). Current understanding of the human microbiome. *Nature Medicine*. 1546170X. *24*(4), 392–400. Available from http://www.nature.com/nm/index.html. doi:10.1038/nm.4517.

Gilbert, J. A., Jansson, J. K., & Knight, R. (2014). The Earth Microbiome project: successes and aspirations. *BMC Biology*. 17417007. *12*(1). Available from http://www.biomedcentral.com/1741-7007/12/69. doi:10.1186/s12915-014-0069-1.

Giurini, E. F., Godla, A., & Gupta, K. H. (2024). Redefining bioactive small molecules from microbial metabolites as revolutionary anticancer agents. *Cancer Gene Therapy*. 14765500. *31*(2), 187–206. Available from https://www.nature.com/cgt/. doi:10.1038/s41417-023-00715-x.

Gollwitzer, A. E., Alser, M., Bergtholdt, J., Lindegger, J., Rumpf, M. D., Firtina, C., et al. (2023). MetaTrinity: enabling fast metagenomic classification via seed counting and edit distance approximation. *arXiv*, 23318422. Available from https://arxiv.org. doi:10.48550/arXiv.2311.02029.

Graham, A. L., Chen, J., Walters, W. A., Knight, R., & Dorrestein, P. C. (2021). Artificial Intelligence and microbiome engineering: a new paradigm for synthetic biology. *Nature Biotechnology*, *39*, 516–526.

Guan, C., Zhao, Y., Liu, H., & Wang, L. (2022). Multi-omics integration: a new approach to precision medicine in gastrointestinal disorders. *Nature Reviews Gastroenterology & Hepatology*, *19*, 33–47.

Han, Y., Jinling, H., Minghui, L., Peng, Y., Jiang, H., Zhao, J., et al. (2024). Unlocking the potential of metagenomics with the PacBio high-fidelity sequencing technology. *Microorganisms*. 2076-2607. *12* (12), 2482. doi:10.3390/microorganisms12122482.

Hunter, R., Booth A., & Wood, L. (2026). Searching smarter, not harder: leveraging AI to enhance literature searches for theory-driven reviews-A methodological case study. BMC Med Res Methodol. *26*(1), 82. doi:10.1186/s12874-026-02814-3.

Huttenhower, C., Gevers, D., Knight, R., Abubucker, S., Badger, J. H., Chinwalla, A. T., et al. (2012). Structure, function and diversity of the healthy human microbiome. *Nature*. 14764687. *486* (7402), 207–214. doi:10.1038/nature11234.

Imani, M., Beikmohammadi, A., & Arabnia, H. R. (2025). Comprehensive analysis of random forest and XGBoost performance with SMOTE, ADASYN, and GNUS under varying imbalance levels. *Technologies*. 22277080. *13*(3). Available from https://www.mdpi.com/journal/technologies. doi:10.3390/technologies13030088.

Jin, D. M., Morton, J. T., & Bonneau, R. (2024). Meta-analysis of the human gut microbiome uncovers shared and distinct microbial signatures between diseases. *mSystems*. 23795077. *9*(8). Available from https://journals.asm.org/doi/10.1128/msystems.00295-24. doi:10.1128/msystems.00295-24.

Jin, Q., Leaman, R., & Zhiyong, L. (2024). PubMed and beyond: biomedical literature search in the age of artificial intelligence. *eBioMedicine*. 23523964. *100*(104988). doi:10.1016/j.ebiom.2024.104988.

Jurburg, S. D., Blowes, S. A., Shade, A., Eisenhauer, N., & Chase, J. M. (2024). Synthesis of recovery patterns in microbial communities across environments. *Microbiome*. 20492618. *12*(1). Available from https://microbiomejournal.biomedcentral.com/. doi:10.1186/s40168-024-01802-3.

Kajihara, K. T., & Hynson, N. A. (2024). Networks as tools for defining emergent properties of microbiomes and their stability. *Microbiome*. 20492618. *12*(1). Available from https://microbiomejournal.biomedcentral.com/. doi:10.1186/s40168-024-01868-z.

Kashyap, P. C., Chia, N., Nelson, H., Segal, E., & Elinav, E. (2017). Microbiome at the frontier of personalized medicine. *Mayo Clinic Proceedings*. 19425546. *92*(12), 1855–1864. Available from http://www.journals.elsevier.com/mayo-clinic-proceedings. doi:10.1016/j.mayocp.2017.10.004.

Khurana, D., Koli, A., Khatter, K., & Singh, S. (2023). Natural language processing: state of the art, current trends and challenges. *Multimedia Tools and Applications*. 1380-7501. *82* (3), 3713–3744. doi:10.1007/s11042-022-13428-4.

Kim, C., Pongpanich, M., & Porntaveetus, T. (2024). Unraveling metagenomics through long-read sequencing: a comprehensive review. *Journal of Translational Medicine*. 14795876. *22*(1). Available from https://translational-medicine.biomedcentral.com. doi:10.1186/s12967-024-04917-1.

Kumar, B., & Jaiswal, W. (2015). The role of microbiome engineering in enhancing food safety and quality. *Biotechnology Notes (Amsterdam, Netherlands)*, *6*, 67–78. doi:10.1016/j.biotno.2025.01.001.

Kumar, R., Bhandari, V., Patel, A., & Singh, B. (2021). Multi-omics approaches to study microbiome: a new perspective in microbial ecology. *Frontiers in Microbiology*, *12*, 702427.

Kutuzova, S., Nielsen, M., Piera, P., Nissen, J. N., & Rasmussen, S. (2024). Taxometer: improving taxonomic classification of metagenomics contigs. *Nature Communications*. 2041-1723. *15*(1). doi:10.1038/s41467-024-52771-y.

Lecun, Y., Bottou, L., Bengio, Y., & Haffner, P. (2015). Gradient-based learning applied to document recognition. *Proceedings of the IEEE*. 00189219. *86* (11), 2278–2324. doi:10.1109/5.726791.

Lee, J., Yoon, W., Kim, S., Kim, D., Kim, S., So, C. H., et al. (2020). BioBERT: a pre-trained biomedical language representation model for biomedical text mining. *Bioinformatics*. 1367-4803. *36* (4), 1234–1240. doi:10.1093/bioinformatics/btz682.

Lepakshi, V. A. (2022). *Machine learning and deep learning based AI tools for development of diagnostic tools computational approaches for novel therapeutic and diagnostic designing to mitigate SARS-CoV2 infection: Revolutionary strategies to combat pandemics* (pp. 399–420). India: Elsevier. Available from https://www.sciencedirect.com/book/9780323911726. doi:10.1016/B978-0-323-91172-6.00011-X.

Li, J., Jia, H., Cai, X., & Zhao, L. (2020). Personalized medicine and the microbiome: opportunities and hcallenges. *Nature Reviews Microbiology*, *18*, 585–598.

Libbrecht, M. W., & Noble, W. S. (2015). Machine learning applications in genetics and genomics. *Nature Reviews Genetics*. 14710064. *16*(6), 321–332. Available from http://www.nature.com/reviews/genetics. doi:10.1038/nrg3920.

Lin, A., Junpu, Y., Chang, Q., Zhu, L., Mou, W., Gan, W., et al. (2025). Bridging artificial intelligence and biological sciences: a comprehensive review of large language models in bioinformatics. *Briefings in Bioinformatics*. 1467-5463. *26*(4). doi:10.1093/bib/bbaf357.

Lipton, Z. C. (2018). The mythos of model interpretability. *Communications of the ACM*. 0001-0782. *61* (10), 36–43. doi:10.1145/3233231.

Liu, S., Rodriguez, J. S., Munteanu, V., Ronkowski, C., Sharma, N. K., Alser, M., et al. (2025). Analysis of metagenomic data. *Nature Reviews Methods Primers*. 26628449. *5*(1). Available from https://www.nature.com/nrmp/. doi:10.1038/s43586-024-00376-6.

Liu, Y. X., Qin, Y., Chen, T., Lu, M., Qian, X., Guo, X., & Bai, Y. (2021). A practical guide to amplicon and metagenomic analysis of microbiome data. *Protein and Cell*. 16748018. *12*(5), 315–330. Available from http://link.springer.com/journal/13238. doi:10.1007/s13238-020-00724-8.

Lloyd-Price, J., Abu-Ali, G., & Huttenhower, C. (2016). The healthy human microbiome. *Genome Medicine*. 1756994X. *8*(1). Available from http://www.genomemedicine.com/. doi:10.1186/s13073-016-0307-y.

Ma, Z., Zuo, T., Frey, N., & Rangrez, A. Y. (2024). A systematic framework for understanding the microbiome in human health and disease: from basic principles to clinical translation. *Signal Transduction and Targeted Therapy*. 2059-3635. *9*(1). doi:10.1038/s41392-024-01946-6.

Malwe, A. S., & Sharma, V. K. (2023). Application of artificial intelligence approaches to predict the metabolism of xenobiotic molecules by human gut microbiome. *Frontiers in Microbiology*. 1664302X. *14*. Available from https://www.frontiersin.org/journals/microbiology#. doi:10.3389/fmicb.2023.1254073.

Marcos-Zambrano, L. J., Karaduzovic-Hadziabdic, K., Turukalo, T. L., Przymus, P., Trajkovik, V., Aasmets, O., et al. (2021). Applications of machine learning in human microbiome studies: a review on feature selection, biomarker identification, disease prediction and treatment. *Frontiers in Microbiology*. 1664302X. *12*. Available from https://www.frontiersin.org/journals/microbiology#. doi:10.3389/fmicb.2021.634511.

McElhinney, J. M. W. R., Catacutan, M. K., Mawart, A., Hasan, A., & Dias, J. (2022). Interfacing machine learning and microbial omics: a promising means to address environmental challenges. *Frontiers in Microbiology*. 1664302X. *13*. Available from https://www.frontiersin.org/journals/microbiology#. doi:10.3389/fmicb.2022.851450.

Meyer, K. M. (2021). The challenge of diversity in machine learning for microbiome research. *Nature Communications*, *12*, 4018.

Mohr, A. E., Ortega-Santos, C. P., Whisner, C. M., Klein-Seetharaman, J., & Jasbi, P. (2024). Navigating challenges and opportunities in multi-omics integration for personalized healthcare. *Biomedicines*. 22279059. *12*(7). Available from www.mdpi.com/journal/biomedicines. doi:10.3390/biomedicines12071496.

Murali, S. K., & Mansell, T. J. (2024). Next generation probiotics: engineering live biotherapeutics. *Biotechnology Advances*. 07349750. *72*. Available from https://www.sciencedirect.com/science/journal/07349750. doi:10.1016/j.biotechadv.2024.108336.

Nagarajan, K., Ni, C., & Lu, T. (2022). Agent-based modeling of microbial communities. *ACS Synthetic Biology*. 21615063. *11*(11), 3564–3574. Available from http://pubs.acs.org/journal/asbcd6. doi:10.1021/acssynbio.2c00411.

Navgire, G. S., Goel, N., Sawhney, G., Sharma, M., Kaushik, P., Mohanta, Y. K., et al. (2022). Analysis and Interpretation of metagenomics data: an approach. *Biological Procedures Online*. 14809222. *24*(1). Available from https://biologicalproceduresonline.biomedcentral.com/. doi:10.1186/s12575-022-00179-7.

Nayak, R. K., Sharma, V., Gupta, A., & Thomas, T. (2021). Understanding the impact of climate change on microbial communities: the role of artificial intelligence. *Environmental Microbiology Reports*, *13*, 15–24.

Norori, N., Qiyang, H., Aellen, F. M., Faraci, F. D., & Tzovara, A. (2021). Addressing bias in big data and AI for health care: a call for open science. *Patterns*. 26663899. *2*(10), 100347. doi:10.1016/j.patter.2021.100347.

Oh, M., & Zhang, L. (2020). DeepMicro: deep representation learning for disease prediction based on microbiome data. *Scientific Reports*. 20452322. *10*(1). Available from www.nature.com/srep/index.html. doi:10.1038/s41598-020-63159-5.

Omar, A., Abdelbaki, W., Shrestha, A., Elbasi, E., Alryalat, M. A. A., & Dwivedi, Y. K. (2023). A systematic literature review of artificial intelligence in the healthcare sector: benefits, challenges, methodologies, and functionalities. *Journal of Innovation & Knowledge*. 2444569X. *8*(1), 100333. doi:10.1016/j.jik.2023.100333.

Papoutsoglou, G., Tarazona, S., Lopes, M. B., Klammsteiner, T., Ibrahimi, E., Eckenberger, J., et al. (2023). Machine learning approaches in microbiome research: challenges and best practices. *Frontiers in Microbiology*. 1664302X. *14*. Available from https://www.frontiersin.org/journals/microbiology#. doi:10.3389/fmicb.2023.1261889.

Patil, A., & Singh, N. (2023). Artificial intelligence for microbiota analysis: advances and applications. *Preprints Biology and Life Sciences*. doi:10.20944/preprints202307.1765.v1.

Pérez-Cobas, A. E., Gomez-Valero, L., & Buchrieser, C. (2020). Metagenomic approaches in microbial ecology: an update on whole-genome and marker gene sequencing analyses. *Microbial Genomics*. 20575858. *6*(8), 1–22. Available from https://www.microbiologyresearch.org/docserver/fulltext/mgen/6/8/mgen000409.pdf?expires=1598548029&id=id&accname=guest&checksum=6604418C1491A121AA0250D9688214DC. doi:10.1099/mgen.0.000409.

Probul, N., Huang, Z., Saak, C. C., Baumbach, J., & List, M. (2024). AI in microbiome-related healthcare. *Microbial Biotechnology*. 17517915. *17*(11). Available from http://onlinelibrary.wiley.com/journal/10.1111/ISSN1751-7915. doi:10.1111/1751-7915.70027.

Proctor, L. M., Dore, J., Maguin, E., Wannerberger, K., & Vincent, C. (2025). A global partnership to advance the microbiome in human, plant, animal, and planetary health. *The Lancet Microbe*. 26665247. *6*(8). Available from https://www.sciencedirect.com/science/journal/26665247. doi:10.1016/j.lanmic.2025.101120.

Przymus, P., Rykaczewski, K., Martín-Segura, A., Truu, J., Pau, E. C. D. S., Kolev, M., et al. (2025). Deep learning in microbiome analysis: a comprehensive review of neural network models. *Frontiers in Microbiology*. 1664-302X. *15*. doi:10.3389/fmicb.2024.1516667.

Rhodes, R. (2016). Ethical issues in microbiome research and medicine. *BMC Medicine*. 1741-7015. *14*(1). doi:10.1186/s12916-016-0702-7.

Robertson, H., Han, B. A., Castellanos, A. A., Rosado, D., Stott, G., Zimmerman, R., et al. (2025). Understanding ecological systems using knowledge graphs: an application to highly pathogenic avian influenza. *Bioinformatics Advances*. 26350041. *5*(1). Available from https://academic.oup.com/bioinformaticsadvances. doi:10.1093/bioadv/vbaf016.

Roy, G., Prifti, E., Belda, E., & Zucker, J. D. (2024). Deep learning methods in metagenomics: a review. *Microbial Genomics*. 20575858. *10*(4). Available from https://www.microbiologyresearch.org/deliver/fulltext/mgen/10/4/mgen001231.pdf?itemId=%2Fcontent%2Fjournal%2Fmgen%2F10.1099%2Fmgen.0.001231&mimeType=pdf&containerItemId=content/journal/mgen. doi:10.1099/mgen.0.001231.

Singh, J., & Singh, D. (2024). A comprehensive review of clustering techniques in artificial intelligence for knowledge discovery: taxonomy, challenges, applications and future prospects. *Advanced Engineering Informatics*. 14740346. *62*, 102799. doi:10.1016/j.aei.2024.102799.

Smith, T. P., Hope, R., & Bell, T. (2025). Phylogenetic clustering of microbial communities as a biomarker for chemical pollution. *FEMS Microbiology Ecology*. 15746941. *101*(5). Available from http://femsec.oxfordjournals.org/. doi:10.1093/femsec/fiaf047.

Srinivasan, S., Jnana, A., & Murali, T. S. (2024). Modeling microbial community networks: methods and tools for studying microbial interactions. *Microbial Ecology*. 1432184X. *87*(1). Available from https://www.springer.com/journal/248. doi:10.1007/s00248-024-02370-7.

Surana, N. K., & Kasper, D. L. (2017). Moving beyond microbiome-wide associations to causal microbe identification. *Nature*. 14764687. *552*(7684), 244–247. Available from http://www.nature.com/nature/index.html. doi:10.1038/nature25019.

Tiddi, I., & Schlobach, S. (2022). Knowledge graphs as tools for explainable machine learning: a survey. *Artificial Intelligence*. 00043702. *302*(103627). doi:10.1016/j.artint.2021.103627.

Topol, E. J. (2019). High-performance medicine: the convergence of human and artificial intelligence. *Nature Medicine*. 1546170X. *25*(1), 44–56. Available from http://www.nature.com/nm/index.html. doi:10.1038/s41591-018-0300-7.

Turnbaugh, P. J., Ley, R. E., Hamady, M., Fraser-Liggett, C. M., Knight, R., & Gordon, J. I. (2007). The human microbiome project. *Nature*. 14764687. *449*(7164), 804–810. Available from http://www.nature.com/nature/index.html. doi:10.1038/nature06244.

Varnosfaderani, S. M., & Forouzanfar, M. (2024). The role of AI in hospitals and clinics: transforming healthcare in the 21st Century. *Bioengineering*. 2306-5354. *11* (4), 337. doi:10.3390/bioengineering11040337.

Wang, Q. (2018). Knowledge graph and its applications in biomedical domain: a survey. *IEEE Access*, *6*, 10153–10167.

Wassan, J. T., Wang, H., & Zheng, H. (2022). A new phylogeny-driven random forest-based classification approach for functional metagenomics. In *Proceedings—2022 IEEE International Conference on Bioinformatics and Biomedicine, BIBM 2022* (pp. 32–37). United Kingdom: Institute of Electrical and Electronics Engineers Inc. http://ieeexplore.ieee.org/xpl/mostRecentIssue.jsp?punumber=9994793. doi:10.1109/BIBM55620.2022.9995554.

Zhang, L., Chen, F., Zeng, Z., Mengjiao, X., Sun, F., Yang, L., et al. (2021). Advances in metagenomics and its application in environmental microorganisms. *Frontiers in Microbiology*. 1664-302X. *12*. doi:10.3389/fmicb.2021.766364.

Zhang, P. (2022). Influence of foods and nutrition on the gut microbiome and implications for intestinal health. *International Journal of Molecular Sciences*. 1422-0067. *23* (17), 9588. doi:10.3390/ijms23179588.

Zhang, X., Li, L., Butcher, J., Stintzi, A., & Figeys, D. (2019). Advancing functional and translational microbiome research using meta-omics approaches. *Microbiome*. 20492618. *7*(1). Available from http://www.microbiomejournal.com/. doi:10.1186/s40168-019-0767-6.

Zhou, Y., Song, Y., Shao, L., & Zhang, Y. (2020). Personalized probiotics: a microbiome-guided approach. *Trends in Microbiology*, *28*, 650–664.

Chapter 25

The role of artificial intelligence and machine learning in genomics

Gautam Das[a], Garima Suneja[b], Kumar Gautam Singh[c], Sanober Waghoo[b], and Akib Kamani[a]

[a]*Mibiome Therapeutics LLP, Mumbai, India*

[b]*Microbiome Research Pvt. Ltd., Mumbai, India*

[c]*Chromosome Labs Pvt. Ltd., Mumbai, India*

1 Introduction

Artificial intelligence (AI) refers to computational methodologies that imitate elements of human cognition, including learning, reasoning, decision-making, and adaptation. By enabling machines to perform functions traditionally requiring human judgment, AI has become one of the most transformative domains in modern science (Xu et al., 2021). The foundational idea of AI arose during the mid-20th century, marked by Alan Turing's seminal question: *can machines think?*, a challenge formalized through the "Turing test," conceived as a metric to assess machine cognition (Shieber, 2007). This intellectual foundation gained momentum at the 1956 Dartmouth Summer Research Project on Artificial Intelligence, now recognized as the birthplace of AI as a scientific discipline, where researchers established the first practical goals for developing intelligent systems (Hirani et al., 2024). Since then, advances in computational capacity, algorithm design, and the availability of data have transformed AI from a conceptual ambition into a field with profound real-world applications.

Early medical applications of AI emerged in the 1950s, when computer-assisted programs were first used to support diagnostic decision-making. However, the true acceleration of AI in medicine has emerged over the course of the last two decades, driven by exponential growth in digital health data and enhanced computational capacity (Secinaro, Calandra, Secinaro, Muthurangu, & Biancone, 2021). At present, AI is widely recognized as a pivotal tool in healthcare, spanning clinical diagnostics, risk assessment, rehabilitation, surgical assistance, treatment planning, and predictive modeling. These applications not only improve accuracy but also enhance efficiency, reduce costs, and expand access to medical services (Xie, Zhai and Lu, 2025).

The technological diversity within AI underpins its versatility in healthcare. Computational frameworks such as deep learning (DL), machine learning (ML), and natural language processing each provide unique pathways for handling complex medical data. Large language models, an advanced form of DL, are capable of understanding, generating, and summarizing massive volumes of text-based data, thereby offering potential applications in clinical documentation, medical education, and decision support (Alowais et al., 2023). More broadly, AI systems can ingest and analyze multimodal data—from imaging and genomics to electronic health records—to enable early detection, individualized therapeutic strategies, and long-term disease management.

The incorporation of AI into healthcare systems has been shown to significantly improve patient outcomes and resource utilization. In risk stratification, AI models recognize patients predisposed to or at an elevated risk of developing chronic disorders, providing timely interventions. Within primary care, AI tools facilitate the preemptive detection of health conditions like diabetes, hypertension, and mental health challenges, allowing for preventive strategies that reduce disease progression (Kumar, Koul, Singla, & Ijaz, 2023). When addressing chronic disease care, AI-driven systems deliver precision guidance derived from real-time patient metrics, thereby tailoring lifestyle modifications and therapies to individual needs. Collectively, these developments mark a paradigm shift in medicine, positioning AI as an essential driver of precision healthcare (D'Adderio & Bates, 2025).

Against this backdrop, tuberculosis (TB) serves as a critical example of where AI-driven innovations can make a meaningful impact. TB continues to rank among the top infectious causes of morbidity and mortality globally. In spite of

Applications of Artificial Intelligence in Healthcare. https://doi.org/10.1016/B978-0-443-43934-6.00033-X

decades of global control efforts, the World Health Organization's target for the End TB Strategy, calling for a 95% decline in TB-related fatalities and a 90% reduction in the rate of new infections by 2035, currently remains out of reach (Hwang et al., 2024; McClean, Panciu, Lange, Duarte, & Theis, 2024). Standard case-finding methods often fail to capture asymptomatic carriers, and treatment regimens remain lengthy, toxic, and financially burdensome. Compounding these barriers is the escalating incidence of multidrug-resistant (MDR) and extensively drug-resistant (XDR) pathogenic strains, alongside emerging resistance to recently developed drugs like bedaquiline (Agnivesh, Roy, Sau, Kumar, & Kalia, 2025; Hu, Wu, Lei, Zhu, & Gao, 2025). AI has the potential to transform TB care by enabling earlier detection, more precise classification of disease, improved monitoring of treatment response, and novel approaches to predicting drug resistance.

Next-generation sequencing has brought groundbreaking changes to genomics, generating vast quantities of biological data. Deriving valuable patterns from this dataset requires sophisticated computational methods. DL has become a highly effective approach in this area, offering improved accuracy in endeavors like variant calling, prediction of gene expression, regulatory annotation, and protein structure modeling.

In this chapter, we explore how AI can be leveraged for healthcare through a multiomics-centered approach and the major AI frameworks applied in genomics, specialized tools, their applications, challenges, and key references that inform this growing field. We examine microbial networks and k-mer-based approaches that provide novel insights into pathogen biology, followed by a case study of TB modeling that illustrates the real-world application of these methods. Together, these perspectives highlight not only the current progress but also the future directions of AI in TB research and control, situating the disease as a critical testbed for the broader integration of AI into global health.

2 Microbial networks and AI

A microbial network is a lattice of interactions among a particular set of parameters. The interactions could be between genes of an organism, as in a genome network, or between microbes and host factors, including genes, proteins, or regulatory factors, as in a host–microbe network. The network itself may be a correlative network elucidating the co-abundance of microbes or a mechanistic one that shows causal relationships, as in a metabolic network. The incorporation of AI in the exploration, analysis, interpretation, and simulation of networks using large amounts of data leads to a reduction in computational time and enhanced discoveries (Altucci et al., 2025). Several AI algorithms, including ML and neural networks constructed using multiple metaomics datasets, have improved the diagnosis, prognosis, and treatment of disorders and aid in the advanced understanding of a wide variety of ecosystems (Fig. 25.1). A few recent models and their applications are outlined in Table 25.1.

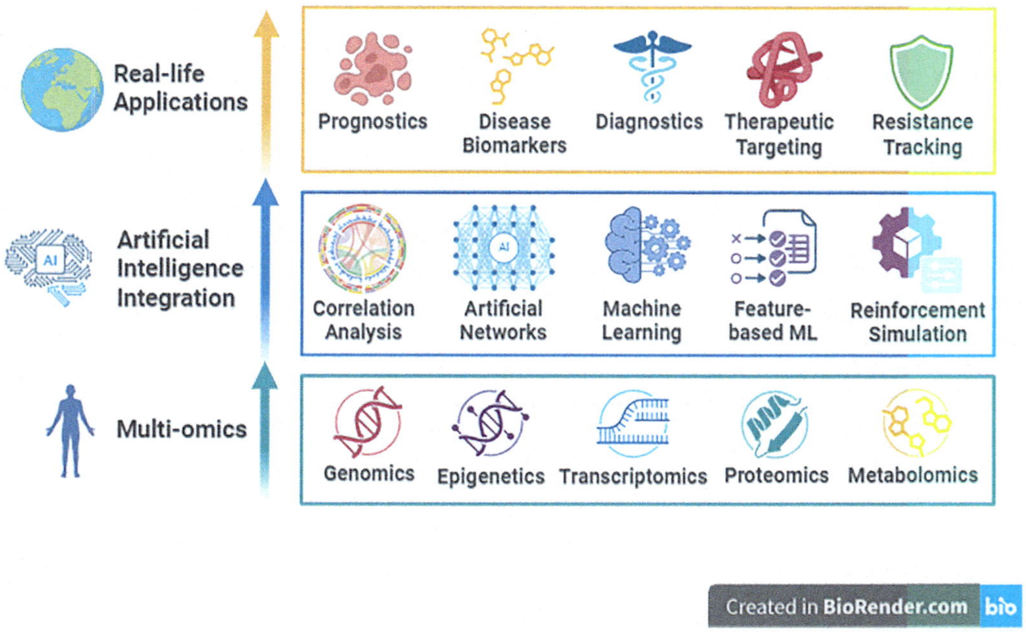

FIG. 25.1 A schematic summarizing applications when AI is applied to multiomics data. Artificial intelligence can extract meaningful insights from raw multiomics data, including disease prognosis, discovery of new biomarkers, and prediction of resistance. *(Source: Created in https://BioRender.com.)*

TABLE 25.1 Different categories of AI models utilizing microbial networks for various applications.

Model category	Framework	Core approach	Applications	Input data	Usage	Reference
Graphical models (statistical networks)	SPIEC-EASI	Sparse inverse covariance network	Infer microbial cooccurrence, ecological networks	16S, metagenomic abundance	Widely used research tool	(Kurtz et al., 2015)
	SparCC, CoNet, MicNet	Correlation and ensemble network inference	Network reconstruction, visualization	OTU, ASV count tables	Research, exploratory	(Faust & Raes, 2016)
Knowledge-primed NNs	MicroKPNN	Neural networks with metabolic + phylogenetic priors	Phenotype prediction, metadata imputation	Species abundance + prior knowledge	Academic, open-source	(Monshizadeh and Ye, 2024)
	MICAH	Heterogeneous GNN (microbe sample phylogeny)	Identify cancer-associated intratumoral microbes	Shotgun metagenomes (tumor)	Research validated	(Liu et al., 2024)
Graph neural networks (GNNs)	GDmicro	GCN + domain adaptation	Host disease classification (IBD, CRC)	OTU, ASV matrices	Research	(Liao et al., 2023)
	GraphSAGE, Boosting GraphSAGE	Sample similarity graph embeddings	Disease classification	Microbiome + metadata	Research	(Zhou, Ran, & Chen, 2024)
	MO-GCN	Multiomics GCN fusion	Host–microbe integration	Multiomic data	Early research	(Wang et al., 2025)
	Explainable GNNs (GIM, GATE, XGNN)	Attention-based interpretability	Feature attribution (key taxa, pathways)	Abundance + phylogeny	Research	(Ivanov, Hartman, & Soheilypour, 2025)
Link prediction, association models	GPUDMDA, GATE-MDA	Graph attention autoencoder + PU learning	Microbe disease association prediction	Microbe disease matrix + similarity	Research	(Peng et al., 2023)
	GCGACNN	Graph + CNN hybrid	Microbe drug prediction	Drug-microbe associations	Research	(Su, Liu, Zhou, & Zhang, 2024)
	MADGAN	GAN-based association modeling	Augment sparse microbe disease networks	Similarity features	Research	(Hu, Yang, Wang, & Zhu, 2023)
	KGCLMDA	Knowledge graph embeddings (TransE, ComplEx)	Multidomain association prediction	Ontologies: microbe disease drug	Academic	(Liu, Shujuan, Wang, Huang, & Gao, 2025)
Dynamic, temporal models	SIMBA-GNN	Simulation-augmented GNN (FBA + graph)	Predict microbial dynamics, cross-feeding	Genome-scale models + time series	Early research	(Aminian-Dehkordi et al., 2026)
Network embedding + ML hybrids	node2vec, DeepWalk + RF, SVM	Graph embeddings of cooccurrence networks	Biomarker, ARG hub detection	Derived networks	Widely adopted research	(Grover & Leskovec, 2016)
Multiomic and knowledge graph fusions	MINERVA Platform	Multiomic network visualization and analytics	Research pipeline integration	Multiomics datasets	Open research platform	(Langarica, Kim, Alkhadrawi, Kim, & Do, 2025)
Applied, translational AI	DeepMicro	Autoencoders + network-derived features	Biomarker detection	Shotgun metagenomes	Research, startup use	(Oh & Zhang, 2020)

2.1 Microbial networks for disease diagnostics

Networks constructed from large multiomics datasets consist of several layers of information. These data, when fed into DL algorithms (like recurrent and convoluted neural networks), can generate disease risk or disease progression models. The accuracy of these models depends on the presence or absence of associated metadata, the number and types of omics datasets, and the basal algorithm for model development (Camacho, Collins, Powers, Costello, & Collins, 2018). As an example, classifiers made from salivary metatranscriptome data from healthy individuals and oral cancer patients revealed several distinguishing features (Banavar et al., 2021). Such models can be used as noninvasive diagnostic methods for specific diseases.

3 General DL frameworks in genomics

Several general-purpose DL libraries serve as the foundation for building AI models in genomics:

a. **TensorFlow** (Abadi et al., 2016) is an open-source ML and DL framework developed by Google Brain. It provides a comprehensive ecosystem of tools, libraries, and community resources for building and training ML models, especially deep neural networks. It supports large-scale model training and is widely used for variant classification, regulatory sequence analysis, and gene expression modeling.
b. **PyTorch** (Paszke et al., 2019), developed by Meta, offers dynamic computation graphs and is favored for research applications such as sequence modeling, single-cell data analysis, and custom genomic architectures. Tool Selene is a good example that is built on PyTorch and allows researchers to train and interpret DL models on genomic sequences to predict things like transcription factor (TF) binding or epigenetic signals (Chen, Cofer, Zhou, & Troyanskaya, 2019).
c. **Keras** (Chollet, 2015) is an open-source, high-level neural networks API written in Python. It was developed by François Chollet and released in 2015. It provides a high-level interface to TensorFlow, making it ideal for prototyping DL models for genomic data classification, gene expression prediction, epigenomics and regulatory genomics, variant interpretation, etc.
d. **MXNet** (Chen et al., 2015), known for its efficiency in distributed training, is used for processing high-dimensional genomic images and integrating multiomics data. It was developed by collaborators from Carnegie Mellon University and the University of Washington. MXNet supports a variety of programming languages, including Python, Scala, R, C++, and Julia, and is renowned for its ability to scale efficiently across multiple graphics processing units (GPUs) and machines (Table 25.2).

TABLE 25.2 Examples of tools that are built using AI frameworks for genomics.

Tool	Function	Key features	Reference
DeepVariant	Variant calling using deep learning	Converts aligned reads into images; uses convolutional neural networks (CNNs) to distinguish true variants from sequencing errors; high precision and consistency; open source	(Poplin et al., 2018)
Enformer	Predicts gene expression and functional genomic features from raw DNA	Transformer architecture capturing long-range genomic dependencies; models regulatory elements like enhancers and silencers; integrates large datasets	(Avsec et al., 2021)
Basenji	Predicts gene expression and chromatin accessibility from DNA sequences	Uses CNNs to analyze long genomic regions; captures both local and distal regulatory signals; explores noncoding genome functions	(Kelley et al., 2018)
Selene	Deep learning library for biological sequence data (DNA, RNA, proteins)	Uses CNNs to analyze long genomic regions; captures both local and distal regulatory signals; explores noncoding genome functions	(Chen et al., 2019)
AtacWorks	Improves the quality of ATAC-seq chromatin accessibility data	CNN-based denoising and imputation of low-quality/low-coverage ATAC-seq data; enhances signal clarity for epigenetic studies	(Lal et al., 2021)
AlphaFold2	Predicts 3D protein structures from amino acid sequences	Attention-based neural networks; unprecedented accuracy in protein folding; transformative for structural biology and drug discovery	(Jumper et al., 2021)

4 Applications of AI in genomics

AI is being used across nearly every domain of genomics research:

a. **Variant calling:** Tools based on AI models like DeepVariant and Clair use deep convolutional networks to identify genetic variants from raw sequence reads, surpassing traditional statistical methods in both speed and accuracy (Luo et al., 2020; Poplin et al., 2018).
b. **Gene expression prediction:** Tools like Enformer and Basenji predict gene expression and chromatin features directly from DNA sequence data, helping researchers decode the regulatory grammar of the genome (Avsec et al., 2021).
c. **Single-cell omics:** Deep generative models like scVI (Lopez, Regier, Cole, Jordan, & Yosef, 2018) and scGen reduce noise, handle batch effects, and uncover cellular heterogeneity in scRNA-seq datasets.
d. **Protein structure prediction:** AlphaFold2 (Jumper et al., 2021) represents a major breakthrough, solving the 50-year-old protein folding problem and enabling accurate prediction of 3D protein conformations from linear sequences.
e. **Regulatory genomics:** DeepSEA (Zhou & Troyanskaya, 2015) and Basset (Kelley, Snoek, & Rinn, 2016) model the effects of noncoding variants by predicting chromatin features, TF binding, and regulatory potential from sequence data.
f. **Drug–gene interaction:** Graph neural networks (GNNs) have been employed using platforms such as DeepChem (Ramsundar et al., 2019) and DGL-LifeSci (Hu et al., 2020) to model molecular graphs, predict compound properties, and infer drug–target interactions.

5 Challenges in AI applications for genomics

While AI holds great promise, the field faces notable challenges:

a. **Interpretability**: Deep models often lack transparency, making biological interpretation difficult (Samek, Wiegand, & Müller, 2017).
b. **Data scarcity and imbalance**: Many rare variants or cell types are underrepresented in training data, reducing model generalizability (Libbrecht & Noble, 2015).
c. **Bias and representation**: Genomic datasets are often skewed toward populations of European ancestry, which may lead to biased predictions (Popejoy & Fullerton, 2016).
d. **Computational demand**: Training large models on genomic datasets can be resource-intensive, posing barriers to accessibility (Stephens et al., 2015).

6 Ethical issues in genomics

AI has the potential to significantly accelerate research in genomics and enable advancements in personalized medicine. However, its integration into this field brings several ethical concerns:

a. **Privacy and data protection:** AI systems require access to large genomic datasets, which often involve sharing or aggregating sensitive information. Even anonymized data can carry a risk of reidentification, particularly when combined with other sources, raising serious concerns about data privacy and protection.
b. **Informed consent:** Explaining how AI models use genetic data to derive insights is complex, making it difficult for individuals to provide fully informed consent. This challenges the ethical foundation of participation in research involving AI.
c. **Bias and discrimination:** Many genomic databases are disproportionately composed of data from individuals of European ancestry. As a result, AI models trained on such data may perform poorly for underrepresented populations, leading to biased health predictions and potentially discriminatory treatment recommendations.
d. **Access and equity:** The benefits of AI-powered genomics may be limited to wealthy individuals or developed countries. This creates the risk of widening existing health disparities and raises concerns about equitable access to these innovations.
e. **Genetic determinism and social stigma:** AI might identify or exaggerate links between genetic traits and characteristics, such as intelligence or behavior. If misused, such findings could promote genetic determinism, reinforce harmful stereotypes, and contribute to social stigmatization.
f. **Commercialization and use of personal data:** Companies may derive profit from using genomic data to train AI models without fairly compensating or informing the individuals who provided that data. This raises ethical concerns around exploitation, ownership, and consent.

g. **Regulatory and accountability gaps:** There is a lack of clear regulations governing the use of AI in genomics, especially concerning responsibility when an AI system makes an incorrect or harmful prediction. This ambiguity challenges accountability and trust.
h. **Dual-use and security risks:** AI tools developed for genomic research could be misappropriated for malicious purposes, such as bioweapons development or genetic surveillance. These dual-use risks highlight the need for strict oversight and ethical safeguards.

7 K-mers for AI and genomics

K-mers are contiguous subsequences of DNA or RNA, each of length k, found within longer genomic sequences. They act as the basic building blocks for AI applications in healthcare genomics. In AI and ML, k-mers facilitate the conversion of complex genomic data into structured and traceable features that can be analyzed and fed into predictive models (Kelley et al., 2024; Marçais et al., 2021).

When used in AI, k-mers offer several important benefits for analyzing genetic sequences:

- **Sequence representation**: K-mers facilitate sequence analysis without the necessity of alignment. Rather than analyzing entire genomes, sequences are decomposed into short, continuous segments. These segments capture local patterns and can be converted into fixed-size feature sets suitable for ML models. Essentially, for an ML application, we generate k-mers and filter out repetitive k-mers to reduce noise, thereby selecting the unique k-mers that contain more information.
- **Feature engineering:** Since the unique k-mers capture essential local patterns and avoid repeated conserved sequences within the data, models trained on such k-mers are less likely to overfit, as they focus on discriminating these unique k-mers instead of memorizing the repeated ones. These characteristics result in better generalization.

In summary, using k-mers for AI enables us to transform genomic datasets into structured and informative features that capture local patterns efficiently and highlight biologically relevant variations, making the model more efficient without compromising predictive power.

7.1 Constructing k-mers from DNA sequences

Consider the short DNA sequence: ATGCAT. If k is set to 3, its k-mers are ATG, TGC, GCA, and CAT. With $k = 2$, the k-mers are AT, TG, GC, CA, and AT. These k-mers are the overlapping subsequences of the original sequence that are used for analysis (Fig. 25.2).

7.2 Applications of k-mers in AI and genomics

K-mers have emerged as fundamental computational units that transform complex genomic data into structured features for diverse biological analyses. Their ability to capture local sequence patterns makes them extremely efficient for AI-driven

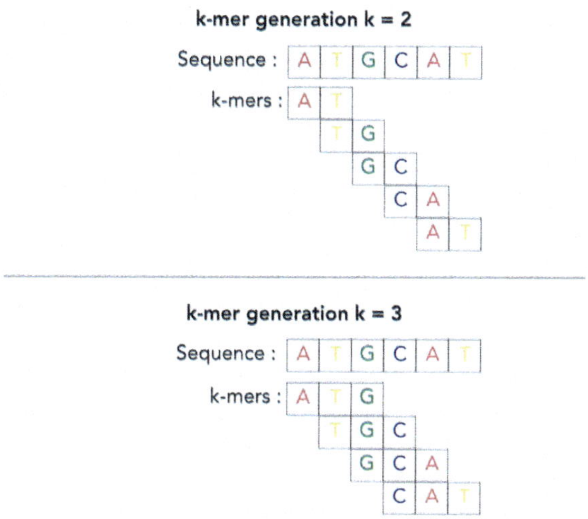

FIG. 25.2 Constructing k-mers from a DNA sequence. An illustrative example of how k-mers are generated from a DNA sequence by accounting for all possible subsets of length "k," employing the "sliding window" method.

analysis. Their versatility spans multiple domains, namely, TB, cancer, and microbiome research. The following is an example that demonstrates how k-mers can be leveraged to generate meaningful biological insights.

7.2.1 Problem: TB and drug resistance

TB continues to be among the most lethal infectious diseases worldwide. According to the Global tuberculosis report 2024: TB disease burden—1.1 TB incidence (2024), roughly 10–11 million new TB cases were reported in 2023, with the majority occurring in low- and middle-income regions.

Despite substantial progress in diagnostics and therapy, drug-resistant strains, particularly MDR and XDR *Mycobacterium tuberculosis,* remain a major obstacle to global control efforts.

To overcome these limitations, recent studies have employed k-mer-based ML pipelines to predict antimicrobial resistance directly from genomic data, thereby enabling a faster and more quantitative assessment of susceptibility patterns.

7.2.2 Current method

The problem with drug-resistant TB, such as MDR and XDR, is that it is often challenging to quantify the minimum inhibitory concentration (MIC). Knowing the MIC helps guide therapy because it indicates whether the bacterium is fully resistant, fully susceptible, or susceptible to a degree. If the MIC is just above the critical concentration, a higher dose or a different drug combination may still be effective. MIC alone does not guarantee a cure, but it points clinicians toward the right drug and dose. The standard approach today is to take a sputum sample and perform culture until *M. tuberculosis* grows, which can take several weeks. After the cultures are established, the laboratory technicians perform drug susceptibility testing. Even then, in most clinical settings, the results are reported as a binary (yes/no) outcome, i.e., "susceptible" or "resistant," based on the critical concentration, rather than providing us with a true quantitative MIC. The time it takes to grow the culture, combined with the binary nature of the report, leaves clinicians struggling to take rapid action for MDR- and XDR-TB cases.

8 The k-mer-based ML approach for predicting drug resistance in *M. tuberculosis*

Recent work demonstrates that integrating whole-genome sequencing (WGS) with ML algorithms enables the quantitative prediction of MICs across multiple anti-TB agents.

They started with CRyPTIC, "The Comprehensive Resistance Prediction for Tuberculosis: an International Consortium, a data compendium of 12,289 *M. tuberculosis* global clinical isolates, all of which have undergone WGS and have had their MICs to 13 antitubercular drugs measured in a single assay" (Ladner, 2022).

A total of 10,859 isolates from the CRyPTIC collection were utilized for model training, validation, and evaluation.

As the first step, they performed reference-based alignment and variant calling for benchmarking using the Clockwork v0.8.3 pipeline (Hunt et al., 2022). Before the mapping step, the k-mers extracted from sequencing reads were retained separately to support model training.

The sequencing reads were broken down into 31-base subsequences ($k = 31$), a length widely reported to yield an optimal trade-off between sensitivity and specificity during bacterial genome assembly (Bankevich et al., 2012; Chikhi & Medvedev, 2014; Li et al., 2010).

Because error correction is not performed prior to assembly, directly employing raw k-mers from sequencing reads can introduce artifacts. To mitigate this, k-mers observed fewer than five times were filtered out, as low-frequency events in the highly conserved *M. tuberculosis* genome are typically attributable to sequencing noise rather than authentic variants.

For computational efficiency, sets of k-mers that consistently appeared together across isolates were consolidated into unified feature representations prior to model training.

8.1 ML framework for MIC prediction

The study used eXtreme Gradient Boosting (XGBoost) for predicting MICs of *M. tuberculosis*. XGBoost is an optimized version of Gradient Boosting, generally used for large datasets with noisy, high-dimensional data, where regularization is particularly important. Gradient Boosting is a ML framework that builds models in an iterative, stage-wise manner, where each base learner attempts to correct the mistakes made by the previous learner (Friedman, 2001).

XGBoost (Chen & Guestrin, 2016) adopts this approach and builds upon it, making it particularly suitable for genomic data analysis. It can efficiently handle datasets with millions of features. It incorporates regularization, which reduces overfitting and improves generalization. It also employs a block-based computation strategy, which enables distributed training on very large datasets and allows the model to be trained in parallel, thereby saving time. In this study, all training jobs were executed on Oxford University's Biomedical Research Computing cluster, which comprises more than 7000 CPU cores and over 7 PB of shared storage, utilizing array jobs to scale training across multiple drugs and parameter sets.

8.2 Hyperparameter optimization using random search and grid search

Hyperparameters are settings that control the models. In the case of models like XGBoost, these include settings such as the number of trees to build, the depth of each tree, and others. They're not learned automatically; instead, they must be chosen by the researcher, and improper selection can lead to weak predictions or overfitting.

In this study, the author selected the hyperparameters in two steps.

1. **Random search:** The model randomly tries out different combinations of settings and evaluates their performance.
2. **Grid search:** A systematic approach that tests every combination within the range we provide.

The author first uses a random search to identify promising regions without testing every possibility, and once good ranges are identified, they employ a grid search to further explore the path.

They also set aside a set of 1000 isolates exclusively for hyperparameter tuning to avoid data leakage.

8.3 Training, testing, and prediction of the model

After hyperparameter tuning, the model was trained and validated on 10,859 CRyPTIC isolates. The dataset was split into a training set (75%, ~8100 isolates) and a testing set (≈25%, ~2700 isolates).

The MIC values were $\log_2$-transformed to represent the multiplicative nature of dilution-based MIC assays. The model was trained on this representation to learn the relationship between genomic features (k-mers) and drug susceptibility, thereby enabling it to predict quantitative MIC values for each drug–isolate pair.

These continuous MIC outputs were also later translated into binary susceptibility categories (susceptible or resistant) using Youden's J statistic, which identifies the threshold that maximizes both sensitivity (true resistance detected) and specificity (true susceptibility detected). This allowed comparison of the results with current diagnostic standards and enabled benchmarking of the ML system against established tools, such as phenotypic drug susceptibility testing, mutation catalogs, and commercial molecular assays.

8.4 Biological interpretability

XGBoost assigns feature importance scores to the k-mers that are most important for the predictions. The top-ranked features assigned by the model were mapped back to the *M. tuberculosis* H37Rv reference genome using Bowtie2 alignment, which linked the features to specific genetic loci. This enabled researchers to identify candidate resistance mutations and confirm known ones, thereby uncovering novel genomic signatures associated with drug resistance.

8.5 Results

For the commonly used first-line agents, the predicted MICs were in close agreement with laboratory values, typically falling within a single twofold dilution step. Performance for fluoroquinolones and aminoglycosides was similarly high, while accuracy for newer or repurposed drugs was slightly lower because resistant isolates for these drugs are rare in available datasets.

To ensure clinical comparability of the outputs, the continuous MIC predictions were converted into standard susceptible/resistant categories. The system achieved high sensitivity for most drugs and consistently high specificity, often above 95%. A few drugs, notably ethionamide, clofazimine, and linezolid, showed poor sensitivity due to the scarcity of resistant samples available for training.

When benchmarked against an external set of over 15,000 isolates with known binary phenotypes, the approach performed favorably compared to existing methods, not only outperforming them but also, in some cases, exceeding the accuracy of widely used molecular assays, such as Xpert MTB/XDR. The advantages and limitations of k-mer-based approaches are summarized in Table 25.3.

TABLE 25.3 Advantages and limitations of k-mer-based models.

Reference-free analysis: k-mer approaches enable sequence analysis without reliance on a reference genome, supporting the discovery of novel variants and taxa in metagenomics and single-cell studies (Ondov et al., 2016).	**Curse of dimensionality:** The number of possible k-mers grows exponentially with *k*, leading to extremely high-dimensional, sparse feature spaces that challenge memory and computational resources (Vinga and Almeida, 2003).
Fixed-length feature vectors: Converting variable-length sequences into uniform k-mer counts or embedding vectors simplifies input preparation for standard machine learning and deep learning architectures (Asgari et al., 2015).	**Optimal k selection:** Choosing *k* involves trade-offs: small values yield low specificity, while large values increase sparsity and noise; determining the appropriate value is nontrivial (Chikhi & Medvedev, 2014).
Scalability and computational efficiency: k-mer hashing, sparse matrix representations, and min-hash techniques support the rapid processing of large sequencing datasets, often outperforming alignment-based methods by an order of magnitude in speed (Ondov et al., 2016).	**Sequencing error sensitivity:** k-mer methods are prone to sequencing errors, since a single misread base can create spurious k-mers. These false features increase noise and may reduce the accuracy of downstream models (Marçais and Kingsford, 2011).

9 Challenges and future directions

TB outcomes can be significantly improved with AI now offering unprecedented opportunities, not just in accelerating the diagnosis but also through precision treatment modules. However, the performance and reliability of the AI models must be validated in real-time clinical settings. The accuracy of these models is directly proportional to the availability of good-quality datasets. As we have seen in the case study above, the accuracy for newer or repurposed drugs was slightly lower because the resistant isolates for these drugs are rare in available datasets. To facilitate the integration of newly developed AI models in diverse healthcare environments, it is of utmost importance that heterogeneous populations must be considered, as this allows the models to be calibrated for sensitivity and specificity.

The integration of AI into healthcare introduces ethical considerations, particularly regarding algorithmic bias and the potential displacement of certain healthcare roles. Additional challenges relate to the extent to which clinicians are willing to trust and adopt AI-assisted decision-making in clinical practice.

10 Conclusion

AI stands at the vanguard of a paradigm shift in TB healthcare. It promises rapid, high-accuracy diagnostics, enables population-level surveillance, facilitates early detection of drug resistance, and individualizes treatment pathways. While more prospective, multisite validation studies are needed, the integration of AI technologies is paving the way for equitable, effective, and data-driven TB control that is likely to shape global health policies for years to come. As larger, higher-quality datasets become available and as algorithms continue to advance, AI is poised to move from proof-of-concept applications into routine clinical practice, shaping global TB management and influencing public health policy for years to come.

Declaration of generative AI and AI-assisted technologies in the writing process

NotebookLM: During the preparation of this work, the authors used NotebookLM to organize and synthesize background materials; the authors then reviewed and edited the content and take full responsibility for the final publication.
Perplexity (research and citations): During the preparation of this work, the authors used Perplexity to aid in research and citation discovery; the authors then reviewed and edited the content and take full responsibility for the final publication.
SciSpace: During the preparation of this work, the authors used SciSpace to improve manuscript readability and formatting; the authors then reviewed and edited the content and take full responsibility for the final publication.
SciSummary: During the preparation of this work, the authors used SciSummary to generate concise overviews of referenced literature; the authors then reviewed and edited the content and take full responsibility for the final publication.
Grammarly: During the preparation of this work, the authors used Grammarly for grammar and language checking; the authors then reviewed and edited the content and take full responsibility for the final publication.
BioRender: During the preparation of this work, the authors used BioRender to create scientific diagrams; the authors then reviewed and edited the content and take full responsibility for the final publication.
Scribbr: During the preparation of this work, the authors used Scribbr to support citation formatting and reference checks; the authors then reviewed and edited the content and take full responsibility for the final publication.

References

Abadi, M., Barham, P., Chen, J., Chen, Z., Davis, A., Dean, J., et al. (2016). TensorFlow: A system for large-scale machine learning. In *Proceedings of the 12th USENIX symposium on operating systems design and implementation, OSDI 2016* (pp. 265–283). United States: USENIX Association. 9781931971331.

Agnivesh, P. K., Roy, A., Sau, S., Kumar, S., & Kalia, N. P. (2025). Advancements and challenges in tuberculosis drug discovery: A comprehensive overview. *Microbial Pathogenesis*, 1096-1208. *198*. Available from: https://www.sciencedirect.com/science/journal/08824010. doi:10.1016/j.micpath.2024.107074.

Alowais, S. A., Alghamdi, S. S., Alsuhebany, N., Alqahtani, T., Alshaya, A. I., Almohareb, S. N., et al. (2023). Revolutionizing healthcare: the role of artificial intelligence in clinical practice. *BMC Medical Education*, 1472-6920. *23*(1). Available from: https://bmcmededuc.biomedcentral.com/. doi:10.1186/s12909-023-04698-z.

Altucci, L., Badimon, L., Balligand, J.-L., Baumbach, J., Catapano, A. L., Cheng, F., et al. (2025). Artificial intelligence and network medicine: Path to precision medicine. *NEJM AI*, 2836-9386. *2*(9). doi:10.1056/aira2401229.

Aminian-Dehkordi, J., Parsa, M., Dickson, A., & Mofrad, M.R. (2026). SIMBA-GNN: mechanistic graph learning for microbiome prediction. *npj Systems Biology and Applications*, *12*(1), doi:10.1038/s41540-025-00631-w.

Asgari, E., Mofrad, M.R., & Kobeissy, F.H. (2015). Continuous distributed representation of biological sequences for deep proteomics and genomics. *PLoS ONE*, *10*(11), e0141287.doi: 10.1371/journal.pone.0141287.

Avsec, Ž., Agarwal, V., Visentin, D., Ledsam, J. R., Grabska-Barwinska, A., Taylor, K. R., et al. (2021). Effective gene expression prediction from sequence by integrating long-range interactions. *Nature Methods*, 1548-7105. *18*(10), 1196–1203. Available from: http://www.nature.com/nmeth/. doi:10.1038/s41592-021-01252-x.

Banavar, G., Ogundijo, O., Toma, R., Rajagopal, S., Lim, Y. K., Tang, K., et al. (2021). The salivary metatranscriptome as an accurate diagnostic indicator of oral cancer. *Npj Genomic Medicine*, 2056-7944. *6*(1). Available from: https://www.nature.com/npjgenmed/. doi:10.1038/s41525-021-00257-x.

Bankevich, A., Nurk, S., Antipov, D., Gurevich, A. A., Dvorkin, M., Kulikov, A. S., et al. (2012). SPAdes: a new genome assembly algorithm and its applications to single-cell sequencing. *Journal of Computational Biology*, 1066-5277. *19*(5), 455–477. doi:10.1089/cmb.2012.0021.

Bharath, R.d., Eastman, P., & Walters, P., Pande, V., Leswing, K., Wu, Z. (2019). *Deep Learning for the Life Sciences*. O'Reilly Mediahttps://www.amazon.com/Deep-Learning-Life-Sciences-Microscopy/dp/

Camacho, D. M., Collins, K. M., Powers, R. K., Costello, J. C., & Collins, J. J. (2018). Next-generation machine learning for biological networks. *Cell*, 1097-4172. *173*(7), 1581–1592. Available from: https://www.sciencedirect.com/journal/cell. doi:10.1016/j.cell.2018.05.015.

Chen, K. M., Cofer, E. M., Zhou, J., & Troyanskaya, O. G. (2019). Selene: a PyTorch-based deep learning library for sequence data. *Nature Methods*, 1548-7105. *16*(4), 315–318. Available from: http://www.nature.com/nmeth/. doi:10.1038/s41592-019-0360-8.

Chen, T., & Guestrin, C. (2016). XGBoost: a scalable tree boosting system. In *Proceedings of the ACM SIGKDD international conference on knowledge discovery and data mining* (pp. 785–794). United States: Association for Computing Machinery. 9781450342322. doi:10.1145/2939672.2939785.

Chen, T., Li, M., Li, Y., Lin, M., Wang, N., Wang, M., et al. (2015). MXNET: a flexible and efficient machine learning library for heterogeneous distributed systems. *arXiv*. doi:10.48550/arxiv.1512.01274.

Chikhi, R., & Medvedev, P. (2014). Informed and automated k -mer size selection for genome assembly. *Bioinformatics*, 1367-4811. *30*(1), 31–37. doi:10.1093/bioinformatics/btt310.

Chollet, F. (2015). *Keras*. Available from: https://github.com/fchollet/keras.

D'Adderio, L., & Bates, D. W. (2025). Transforming diagnosis through artificial intelligence. *Npj Digital Medicine*, 2398-6352. *8*(1). Available from: https://www.nature.com/npjdigitalmed/. doi:10.1038/s41746-025-01460-1.

Faust, K., & Raes, J. (2016). CoNet app: inference of biological association networks using Cytoscape. *F1000Research*, 2046-1402. *5*, 1519. doi:10.12688/f1000research.9050.2.

Friedman, J. H. (2001). Greedy function approximation: a gradient boosting machine. *Annals of Statistics*, 0090-5364. *29*(5), 1189–1232. Available from: http://projecteuclid.org/DPubS?service=UI&version=1.0&verb=Display&page=past&handle=euclid.aos. doi:10.1214/aos/1013203451.

Global tuberculosis report 2024: TB disease burden—1.1 TB incidence. (2024). Available from: https://www.who.int/teams/global-programme-on-tuberculosis-and-lung-health/tb-reports/global-tuberculosis-report-2024/tb-disease-burden/1-1-tb-incidence.

Grover, A., & Leskovec, J. (2016). Node2vec: Scalable feature learning for networks. In *Proceedings of the ACM SIGKDD international conference on knowledge discovery and data mining* (pp. 855–864). United States: Association for Computing Machinery. 9781450342322. doi:10.1145/2939672.2939754.

Hirani, R., Noruzi, K., Khuram, H., Hussaini, A. S., Aifuwa, E. I., Ely, K. E., et al. (2024). Artificial intelligence and healthcare: a journey through history, present innovations, and future possibilities. *Life*, 2075-1729. *14*(5). Available from: www.mdpi.com/journal/life/. doi:10.3390/life14050557.

Hu, W., Fey, M., Zitnik, M., Dong, Y., Ren, H., Liu, B., et al. (2020). Open graph benchmark: Datasets for machine learning on graphs. *arXiv*, arXiv. 23318422. Available from: https://arxiv.org. doi:10.48550/arxiv.2005.00687.

Hu, W., Yang, X., Wang, L., & Zhu, X. (2023). MADGAN: A microbe-disease association prediction model based on generative adversarial networks. *Frontiers in Microbiology*, 1664-302X. *14*. doi:10.3389/fmicb.2023.1159076.

Hu, X., Wu, Z., Lei, J., Zhu, Y., & Gao, J. (2025). Prevalence of bedaquiline resistance in patients with drug-resistant tuberculosis: A systematic review and meta-analysis. *BMC Infectious Diseases*, 1471-2334. *25*(1). doi:10.1186/s12879-025-11067-2.

Hunt, M., Letcher, B., Malone, K. M., Nguyen, G., Hall, M. B., Colquhoun, R. M., et al. (2022). Minos: variant adjudication and joint genotyping of cohorts of bacterial genomes. *Genome Biology*, 1474-760X. *23*(1). Available from: http://genomebiology.com/. doi:10.1186/s13059-022-02714-x.

Hwang, E. J., Jeong, W. G., David, P. M., Arentz, M., Ruhwald, M., & Yoon, S. H. (2024). AI for detection of tuberculosis: Implications for global health. *Radiology: Artificial Intelligence*, 2638-6100. *6*(2). Available from: https://pubs.rsna.org/doi/10.1148/ryai.230327. doi:10.1148/ryai.230327.

Ivanov, V. A., Hartman, W. H., & Soheilypour, M. (2025). Decoding the microbiome-disease axis with interpretable graph neural networks. *bioRxiv*, bioRxiv. 26928205. Available from: https://www.biorxiv.org. doi:10.1101/2025.06.23.659608.

Jumper, J., Evans, R., Pritzel, A., Green, T., Figurnov, M., Ronneberger, O., et al. (2021). Highly accurate protein structure prediction with AlphaFold. *Nature*, 1476-4687. *596*(7873), 583–589. Available from: http://www.nature.com/nature/index.html. doi:10.1038/s41586-021-03819-2.

Kelley, D. R., Reshef, Y. A., Bileschi, M. L., Belanger, D., McLean, C. Y., & Snoek, J. (2024). Sequential regulatory activity prediction across chromosomes with convolutional neural networks. *Genome Research*, *34*(3), 353–365. doi:10.1101/gr.279452.124.

Kelley, D. R., Reshef, Y. A., Bileschi, M., Belanger, D., McLean, C. Y., & Snoek, J. (2018). Sequential regulatory activity prediction across chromosomes with convolutional neural networks. *Genome Research*, 1549-5469. *28*(5), 739–750. Available from: https://genome.cshlp.org/content/28/5/739.full.pdf+html. doi:10.1101/gr.227819.117.

Kelley, D. R., Snoek, J., & Rinn, J. L. (2016). Basset: learning the regulatory code of the accessible genome with deep convolutional neural networks. *Genome Research*, 1549-5469. *26*(7), 990–999. Available from: http://genome.cshlp.org/content/26/7/990.full.pdf. doi:10.1101/gr.200535.115.

Kumar, Y., Koul, A., Singla, R., & Ijaz, M. F. (2023). Artificial intelligence in disease diagnosis: a systematic literature review, synthesizing framework and future research agenda. *Journal of Ambient Intelligence and Humanized Computing*, 1868-5145. *14*(7), 8459–8486. Available from: https://www.springer.com/journal/12652. doi:10.1007/s12652-021-03612-z.

Kurtz, Z. D., Müller, C. L., Miraldi, E. R., Littman, D. R., Blaser, M. J., & Bonneau, R. A. (2015). Sparse and compositionally robust inference of microbial ecological networks. *PLoS Computational Biology*, 1553-7358. *11*(5). Available from: http://www.ploscompbiol.org/article/browseVolume.action?field=volume. doi:10.1371/journal.pcbi.1004226.

Ladner, J. (2022). A data compendium associating the genomes of 12,289 *Mycobacterium tuberculosis* isolates with quantitative resistance phenotypes to 13 antibiotics. *PLOS Biology*, 1545-7885. *20*(8), e3001721. doi:10.1371/journal.pbio.3001721.

Lal, A., Chiang, Z.D., Yakovenko, N., Duarte, F.M., Israeli, J., & Buenrostro, J.D. (2021). Deep learning-based enhancement of epigenomics data with AtacWorks. *Nature Communications*, *12*(1). doi:10.1038/s41467-021-21765-5.

Langarica, S., Kim, Y. T., Alkhadrawi, A., Kim, J. B., & Do, S. (2025). MINERVA—Microbiome network research and visualization atlas: a scalable knowledge graph for mapping microbiome-disease associations. *Briefings in Bioinformatics*, 1477-4054. *26*(5). Available from: http://bib.oxfordjournals.org. doi:10.1093/bib/bbaf472.

Li, R., Zhu, H., Ruan, J., Qian, W., Fang, X., Shi, Z., et al. (2010). De novo assembly of human genomes with massively parallel short read sequencing. *Genome Research*, 1549-5469. *20*(2), 265–272. Available from: http://genome.cshlp.org/content/20/2/265.full.pdf+html. doi:10.1101/gr.097261.109China.

Liao, H., Shang, J., Sun, Y., & Valencia, A. (2023). GDmicro: classifying host disease status with GCN and deep adaptation network based on the human gut microbiome data. *Bioinformatics*, 1367-4811. *39*(12). doi:10.1093/bioinformatics/btad747.

Libbrecht, M. W., & Noble, W. S. (2015). Machine learning applications in genetics and genomics. *Nature Reviews Genetics*, 1471-0064. *16*(6), 321–332. Available from: http://www.nature.com/reviews/genetics. doi:10.1038/nrg3920.

Liu, M., Shujuan, S., Wang, G., Huang, S., & Gao, X. (2025). KGCLMDA: A computational model for predicting latent associations of microbial drugs using knowledge graphs and contrastive learning. *Bioinformatics*, 1367-4811. *41*(9). doi:10.1093/bioinformatics/btaf457.

Liu, Z., Sun, Y., Li, Y., Ma, A., Willaims, N. F., Jahanbahkshi, S., et al. (2024). An explainable graph neural framework to identify cancer-associated intratumoral microbial communities. *Advanced Science*, 2198-3844. *11*(41). Available from: http://onlinelibrary.wiley.com/journal/10.1002/(ISSN)2198-3844. doi:10.1002/advs.202403393.

Lopez, R., Regier, J., Cole, M. B., Jordan, M. I., & Yosef, N. (2018). Deep generative modeling for single-cell transcriptomics. *Nature Methods*, 1548-7105. *15*(12), 1053–1058. Available from: http://www.nature.com/nmeth/. doi:10.1038/s41592-018-0229-2.

Luo, R., Wong, C., Wong, Y., Tang, C., Liu, C., Leung, C. et al. (2020). Exploring the limit of using a deep neural network on pileup data for germline variant calling. *Nature Machine Intelligence*, *2*(4), 220–227. doi:10.1038/s42256-020-0167-4.

Marçais, G., Delcher, A. L., Phillippy, A. M., Coston, R., Salzberg, S. L., & Zimin, A. (2021). MUMmer4: A fast and versatile genome alignment system. *Bioinformatics*, *37*(5), 75–81. doi:10.1093/bioinformatics/btab083.

Marçais, G., & Kingsford, C. (2011). A fast, lock-free approach for efficient parallel counting of occurrences of k -mers. *Bioinformatics*, *27*(6), 764–770. doi:10.1093/bioinformatics/btr011.

McClean, M., Panciu, T. C., Lange, C., Duarte, R., & Theis, F. (2024). Artificial intelligence in tuberculosis: a new ally in disease control. *Breathe*, 1810-6838. *20*(3), 240056. doi:10.1183/20734735.0056-2024.

Monshizadeh, M., & Ye, Y. (2024). Incorporating metabolic activity, taxonomy and community structure to improve microbiome-based predictive models for host phenotype prediction. *Gut Microbes*, 1949-0976. *16*(1). doi:10.1080/19490976.2024.2302076.

Oh, M., & Zhang, L. (2020). DeepMicro: deep representation learning for disease prediction based on microbiome data. *Scientific Reports*, 2045-2322. *10*(1). doi:10.1038/s41598-020-63159-5.

Ondov, B.D., Treangen, T.J., Melsted, P., Mallonee, A.B., Bergman, N.H., Koren, S. et al. (2016). Mash: fast genome and metagenome distance estimation using MinHash. *Genome Biology*, *17*(1), doi:10.1186/s13059-016-0997-x.

Paszke, A., Gross, S., Massa, F., Lerer, A., Bradbury, J., Chanan, G., et al. (2019). PyTorch: An imperative style, high-performance deep learning library. In *Advances in neural information processing systems* (Vol. 32). Poland: Neural information processing systems foundation. 10495258. Available from: https://papers.nips.cc/

Peng, L., Huang, L., Tian, G., Yan, W., Guang, L., Cao, J., et al. (2023). Predicting potential microbe-disease associations with graph attention autoencoder, positive-unlabeled learning, and deep neural network. *Frontiers in Microbiology*, 1664-302X. *14*. doi:10.3389/fmicb.2023.1244527.

Popejoy, A. B., & Fullerton, S. M. (2016). Genomics is failing on diversity. *Nature*, 1476-4687. *538*(7624), 161–164. Available from: http://www.nature.com/nature/index.html. doi:10.1038/538161a.

Poplin, R., Chang, P. C., Alexander, D., Schwartz, S., Colthurst, T., Ku, A., et al. (2018). A universal snp and small-indel variant caller using deep neural networks. *Nature Biotechnology*, 1546-1696. *36*(10), 983. Available from: http://www.nature.com/nbt/index.html. doi:10.1038/nbt.4235.

Ramsundar Bharath et al 28 April Deep Learning for the Life Sciences O' Reilly Media 2019.

Ramsundar, B., Eastman, P., Walters, P., Pande, V., Leswing, K., Wu, Z. et al. (2019). *Deep learning for the life sciences*. O' Reilly Media.

Samek, W., Wiegand, T., & Müller, K. R. (2017). Explainable artificial intelligence: understanding, visualizing and interpreting deep learning models. *arXiv*, arXiv. 23318422. Available from: https://arxiv.org. doi:10.48550/arxiv.1708.08296.

Secinaro, S., Calandra, D., Secinaro, A., Muthurangu, V., & Biancone, P. (2021). The role of artificial intelligence in healthcare: a structured literature review. *BMC Medical Informatics and Decision Making*, 1472-6947. *21*(1). doi:10.1186/s12911-021-01488-9.

Shieber, S. M. (2007). The turing test as interactive proof. *Nous*, 0029-4624. *41*(4), 686–713. doi:10.1111/j.1468-0068.2007.00636.x.

Stephens, Z. D., Lee, S. Y., Faghri, F., Campbell, R. H., Zhai, C., Efron, M. J., et al. (2015). Big data: Astronomical or genomical? *PLoS Biology*, 1545-7885. *13*(7). Available from: http://www.plosbiology.org/article/browseVolume.action. doi:10.1371/journal.pbio.1002195.

Su, S., Liu, M., Zhou, J., & Zhang, J. (2024). GCGACNN: A graph neural network and random forest for predicting microbe–Drug associations. *Biomolecules*, 2218-273X. *14*(8), 946. doi:10.3390/biom14080946.

Vinga, S., & Almeida, J. (2003). Alignment-free sequence comparison—a review. *Bioinformatics*, *19*(4), 513–523. doi:10.1093/bioinformatics/btg005.

Wang, H., Peng, R., Huang, Y., Liang, L., Wang, W., Zhu, B., et al. (2025). MO-GCN: A multi-omics graph convolutional network for discriminative analysis of schizophrenia. *Brain Research Bulletin*, 0361-9230. *221*, 111199. doi:10.1016/j.brainresbull.2025.111199.

Xie, Y., Zhai, Y., & Lu, G. (2025). Evolution of artificial intelligence in healthcare: a 30 year bibliometric study. *Frontiers in Medicine*, 2296-858X. *11*. doi:10.3389/fmed.2024.1505692.

Xu, Y., Liu, X., Cao, X., Huang, C., Liu, E., Qian, S., et al. (2021). Artificial intelligence: a powerful paradigm for scientific research. *The Innovation*, 2666-6758. *2*(4), 100179. doi:10.1016/j.xinn.2021.100179.

Zhou, B., Ran, B., & Chen, L. (2024). A GraphSAGE-based model with fingerprints only to predict drug-drug interactions. *Mathematical Biosciences and Engineering*, 1551-0018. *21*(2), 2922–2942. doi:10.3934/mbe.2024130.

Zhou, J., & Troyanskaya, O. G. (2015). Predicting effects of noncoding variants with deep learning-based sequence model. *Nature Methods*, 1548-7105. *12*(10), 931–934. Available from: http://www.nature.com/nmeth/. doi:10.1038/nmeth.3547.

Chapter 26

Artificial intelligence-assisted drug designing, toxicity assessment, and biodelivery

Pralay Mitra[a], Soumyadeep Bhaduri[b], Debraj Das[b], and Dibya Kanti Haldar[b]
[a]*Department of Computer Science and Engineering, Indian Institute of Technology Kharagpur, Kharagpur, India*
[b]*Centre for Computational and Data Sciences, Indian Institute of Technology Kharagpur, Kharagpur, West Bengal, India*

1 Introduction

Peptides are central to next-generation therapeutics due to their biocompatibility, specificity, and modifiability. Advances in computational biology, machine learning (ML), and deep learning (DL) have revolutionized their design, especially in structure prediction, toxicity assessment, and the engineering of delivery systems (Hashemi, Vosough, Taghizadeh, & Savardashtaki, 2024). The landscape of peptide discovery has undergone a profound transformation in recent decades, evolving from traditional empirical approaches to sophisticated artificial intelligence (AI)-driven methodologies. Peptide therapeutics sit at the intersection of molecular specificity and clinical tractability, but classical discovery pipelines, including screening, synthesis, and iterative optimization, remain slow. The resource-intensive approach motivates computational acceleration and incorporates biologically grounded design principles, which shorten development cycles and broaden the explored chemical space (Hashemi et al., 2024). Recent studies highlight complementary immune strategies, including broadly neutralizing antibodies capable of conferring sterilizing protection when present at sufficient titers. Meanwhile, vaccine-elicited T-cell responses help control infection and lower the neutralizing antibody threshold needed for protection, motivating combined B- and T-cell vaccine designs. Platforms and design levers that sustain germinal centers—mRNA/LNP delivery, optimized sequential Env immunogens, and adjuvants that boost T follicular helper (Tfh) responses—are central because germinal center activity underlies durable, high-affinity antibody and memory B-cell formation. Tfh cells, defined by Bcl-6 expression, CXCR5-guided trafficking, and IL-21 production, deliver the CD40–cytokine help required to initiate and maintain germinal centers that generate class-switched, affinity-matured antibodies (Crotty, 2011; Linterman & Hill, 2016). Tfh differentiation proceeds in stages: dendritic cells prime naive CD4 T cells under cytokines such as TGF-β, IL-12, and IL-23 to form pre-Tfh that upregulate CXCR5, and subsequent cognate interactions with antigen-specific B cells in follicles complete Tfh commitment and support germinal center reactions. Antigen presentation by B cells alone is generally insufficient to prime naive CD4 T cells to a Tfh fate after protein immunization, yet B cells can drive or reinforce Tfh differentiation in inflammatory or viral contexts. In practice, optimal Tfh and germinal center B-cell responses require cooperative Major Histocompatibility Complex (MHC II) presentation by both dendritic cells and B cells (Crotty, 2011; Lederer et al., 2020; Linterman & Hill, 2016). Clinical and preclinical work is now tuning dose, timing, and route to amplify germinal center/Tfh biology—for example, fractional escalating priming and subcutaneous delivery to enhance lymph node drainage—complementing combined B- and T-cell regimens to reach protection at lower bNAb titers (Maciel et al., 2024; Zhang et al., 2024).

Traditional peptide discovery only samples a small portion of the sequence–structure space and requires significant time for synthesis and wet-laboratory validation. In contrast, modern pipelines increasingly utilize ML for virtual screening, property prediction, and prioritization before targeted synthesis, which compresses timelines and costs while enhancing hit quality (Zhai et al., 2025). Insights from recent HIV vaccine work further motivate computational-then-experimental iteration: combining B- and T-cell strategies improved protection in nonhuman primates at suboptimal neutralizing antibody titers, highlighting how biologically informed constraints can sharpen in silico candidate selection for peptide immunogens

Applications of Artificial Intelligence in Healthcare. https://doi.org/10.1016/B978-0-443-43934-6.00032-8

and immunomodulatory peptides (Zhang et al., 2024). In parallel, a practical design objective has emerged: tuning delivery platforms (e.g., mRNA, adjuvants) and immune kinetics to sustain high-quality germinal centers, which computational models can explicitly target during peptide and vaccine optimization.

Generative AI can propose peptide sequences and scaffolds that maximize immunogenic breadth, antigenicity, and biophysical stability while minimizing toxicity, based on multimodal constraints reflecting epitope coverage, Tfh recruitment, and manufacturability derived from modern trial designs (Das et al., 2025; Goles et al., 2024). AI models can incorporate immunologic priors, such as long-prime dosing objectives, adjuvant effects, and route-specific drainage, to generate candidates whose predicted function aligns with sustained germinal center activity and synergistic B-/T-cell responses. Beyond sequence invention, AI can optimize codon usage for mRNA expression, select adjuvant-compatible motifs, and propose delivery strategies (e.g., cell-penetrating and targeting peptides) aligned to tissue and route constraints surfaced in recent clinical protocols (Zhang et al., 2023).

In this chapter, we examine how peptide therapeutics now benefit from generative AI, which proposes sequences, scaffolds, and delivery motifs conditioned on antigenicity, immunogenic breadth, stability, and low toxicity, while also optimizing codon usage for mRNA expression and aligning designs with dosing, adjuvant, and route parameters shown to fortify germinal centers in early-phase trials. Together, these advances motivate a unified framework—rooted in germinal center dynamics and B-/T-cell synergy from recent trials and workshop consensus—that leverages generative AI to expand chemical space, compress iteration cycles, and prioritize candidates for targeted synthesis and validation in peptide-based drug design and vaccination.

Perhaps most significantly, generative modeling approaches enable the discovery of entirely novel therapeutic peptides that may not exist in any current databases, expanding the therapeutic landscape beyond known chemical space. These AI-powered platforms help researchers make better decisions by providing them with data-based insights and predictions, allowing them to focus on the best candidates for testing, which speeds up the creation of new peptide treatments for various diseases. This chapter offers an in-depth primer on computational peptide design pipelines, emphasizing pocket toxicity evaluation, innovations in AI-enabled delivery vehicles, and a case study on the computational development of an RNA vaccine. It also briefly discusses the role of AI in drug delivery and codon optimization, clarifying how these tools accelerate progress and reduce risks in the journey from sequence to clinical application.

1.1 Overview: Peptide synthesis, function, biological activity, and moonlighting effects

Peptides are short chains of amino acids, usually 0.5–10 kDa in molecular weight and consisting of a few to about 100 residues. As basic building blocks of life, they have a wide range of functions, including acting as structural elements, hormones, enzyme inhibitors, neurotransmitters, host defense molecules, and cell surface receptors. They play various roles, from coordinating communication between cells to mediating immune defense. They are becoming increasingly important as customized tools for delivering drugs and developing next-generation vaccines (Xiao et al., 2025).

Peptide diversity arises from both amino acid sequence variations and structural innovations. These molecules can take on simple linear shapes, more complicated macrocyclic and lasso shapes, or constrained scaffolds. These shapes often help keep the molecules stable against chemical, thermal, and proteolytic degradation. Lasso peptides, which have twisted macrolactam rings, demonstrate how evolution favors unique shapes that are both strong and specific in their functions. Extraction of peptides from natural sources such as venoms, food, and marine organisms remains foundational, while advances in synthesis, via recombinant hosts like *Escherichia coli*, *Saccharomyces cerevisiae*, and *Pichia pastoris*, as well as automated solid- and liquid-phase chemical methods, have enabled industrial-scale and highly programmable production (Mamoudou & Mune, 2025; Wang et al., 2022).

Peptide engineering is more than just making peptides these days, especially in the fields of drug and vaccine development. It also includes carefully optimizing the sequence, structure, adjuvant, dosing regimen, and delivery route. New strategies now explicitly include immunological design goals like strong germinal center formation, B- and T-cell synergy, and long-term exposure to antigens (Barnett et al., 2014). Long-prime or fractional escalating dose regimens, increasingly viable in clinical studies, are intentionally designed to extend antigen availability and promote durable, high-quality antibody production, often in combination with subcutaneous delivery to enhance lymphatic drainage and Tfh cell engagement (Van Baelen et al., 2022). AI and data-driven methods now incorporate these considerations in the processes of peptide generation, prediction, and selection for vaccine and immunomodulatory applications.

As the field progresses, the range of peptide bioactivity keeps expanding. Peptides exhibit antibacterial, antitoxin, anticancer, immunomodulatory, antimicrobial, and signaling properties, playing crucial roles in both established and novel clinical domains, such as wound healing, neuromodulation, targeted delivery, and intercellular communication. The most advanced design paradigms now emphasize not single-objective potency but the rational orchestration of immune features: fostering tissue-resident memory T cells, enhancing mucosal antibody concentrations, and carefully balancing CD4+ and CD8+ activation to maximize protection while minimizing risk (Choi & Morel, 2020; Tripathi & Vishwanatha, 2022). Live or persistent vectors, like those based on cytomegalovirus, show how long-term antigen loads change the way T cells work and how immune tissues are distributed. These findings have led to the development of controlled-release and depot-like peptide formulations for use in medicine.

Moonlighting is the ability of proteins or peptides to perform more than one function depending on the situation. It is a defining feature of peptides that used to be considered a problem but is now regarded as a benefit (Jeffery, 2018). Adaptive evolution has produced peptides with overlapping or latent activities; for instance, antimicrobial peptides (AMPs) frequently display supplementary anti-inflammatory or metabolic effects. When used in AI-enabled design, this behavior of working two jobs at once can lead to the development of drugs that act on multiple pathways simultaneously, which could simplify treatment for patients and broaden the range of available therapies. However, it also raises the issue of specificity: multifunctional peptides can cause unintended reactions or cross-react with unrelated targets o, which means that generative pipelines need to incorporate strict computational filtering and safety-focused triage. Recent advancements in moonlighting peptide curation, as recorded in various databases, have made it possible to systematically map and utilize these properties (Rodríguez Plaza et al., 2012).

Generative AI and systems immunology are changing the way synthesis, function, structural innovation, and moonlighting come together. By combining decades of mechanistic immunology, such as how germinal centers work, how B and T cells work together, and how engineered antigens stay in the body, with large-scale in silico screening and rational filtering, the design and use of peptide-based therapeutics have reached a point where safety, immunogenic precision, and multifunctionality can be programmed (Wan, Kontogiorgos-Heintz, & de la Fuente Nunez, 2022). This shift in perspective paves the way for sophisticated AI-assisted peptide drug discovery, potentially addressing complex clinical requirements through multiobjective, systems-level engineering.

2 Data–target integration and RNA-based approaches

2.1 Peptide databases

As the range of peptide chemical space grows, it is even more important to have complete, well-curated datasets for reliable AI-driven peptide design. In the last few years, peptide sequence and structure repositories have been greatly improved, and many databases are now essential for computational peptide drug discovery.

Peptipedia (Quiroz et al., 2021) is the largest database of its kind. It contains over 92,000 peptide sequences and more than 100,000 peptides with information about their biological activities and functional properties. Peptipedia consolidates information from over 70 sources to provide detailed physicochemical profiles, three-dimensional (3D) structures [from the Protein Data Bank (PDB) and AlphaFold], patent information, and predictive activity annotations. SATPdb (Singh et al., 2015) includes about 19,000 therapeutic peptides with structural annotations. It integrates data from 22 public repositories and enhances sequence records with 3D models and functional annotations. The database encompasses a wide range of peptide classes that target various mechanisms and diseases. LAMP2 (Ye et al., 2020) focuses exclusively on AMPs. It contains around 23,250 unique AMP sequences, which are sourced from 16 AMP databases that are all interconnected. LAMP2 links sequences to annotations and experimental metadata, facilitating research into AMP-based drugs that target microbial pathogens. DRAMP 2.0 (Kang et al., 2019) is another extensive database with 19,899 peptide entries, categorized into general, patent, and clinical groups. The latest version has added 2550 new sequences and provides comprehensive data on activity and clinical validation. Propedia (Martins et al., 2023) features over 49,000 peptide–protein complexes, organized by sequence patterns and binding interfaces. This organization simplifies the analysis, design, and delivery strategies for peptide-targeted therapeutics.

Datasets obtained from phage display, peptide microarrays, and high-throughput screening, for example, add to these resources by providing validated peptide sequences and quantitative binding affinities and activity profiles. The PDB (Berman et al., 2000) is still the main repository for experimentally determined 3D complexes. It contains over 13,000 entries related to protein–peptide interactions. BioLiP2 (Yang, Roy, & Zhang, 2013), on the contrary, helps to systematically organize protein–ligand interactions by tracking over 385,000 protein chains and 781,000 ligand interactions. More than 35,000 of these records specifically pertain to peptides. Databases like PepBDB (Wen, He, Tao, Huang, & Valencia, 2019) focus on peptide–protein complexes that are biologically significant. They only include peptides with up to 50 amino acids, which is crucial for structure-based design and docking studies.

2.2 Preprocessing: Cleaning, tagging, and combining

Data preprocessing is one of the most important steps in AI-driven peptide design, but it is often overlooked. It has a huge effect on how accurate and reliable ML models are. Because experimental protocols, publication standards, and database curation all have their own levels of variability, sophisticated preprocessing pipelines are needed to deal with data quality and harmonization issues on multiple levels.

The initial step in cleaning a sequence is to get rid of any ambiguous or invalid amino acid symbols, ensure that all records use standard amino acid codes, and encode nonnatural residues correctly. To avoid bias in training and evaluation, duplicate sequences are identified and removed. Length filtering also ensures that peptides fall within the appropriate range for specific modeling tasks. Any entries that are atypical are either modified or excluded.

Proper functional and structural annotation provides important biological context. We systematically add experimental or computational activity tags, such as anticancer, antimicrobial, receptor-binding, toxicity, or stability, as well as structural descriptors for secondary motifs, posttranslational modifications (such as phosphorylation and glycosylation), and 3D conformations from databases or predictive tools like AlphaFold (Evans et al., 2022). Adding metadata to peptide records, including species origin, cellular location, expression context, and assay conditions, ensures that the classification is physiologically relevant and facilitates the development of drug design models that are easy to use in a wide range of situations.

For RNA-binding peptides, specialized labeling highlights known recognition domains or RNA target classes, making it easier to design them for regulatory or antiviral uses. Ontology mapping connects functional annotations with standard classification systems, such as Gene Ontology (GO) term mapping and cross-referencing. Modern databases like DBAASP (Pirtskhalava et al., 2021) use UniProt identifiers to integrate annotations and track experiments. This reduces duplication and facilitates collaboration between different resources (Table 26.1).

Advanced integration protocols bring together different datasets and organize them into formats and ontologies that make sense. Sequence clustering addresses redundancy by applying identity thresholds that are appropriate for proteins (90%) and short peptides (100%). It preserves diversity where functional differences exist. BioLiP2 is an excellent example of how to efficiently harmonize data. It utilizes BLAST+ for protein-level clustering and Needleman–Wunsch alignment for short peptide queries, which is where traditional BLAST falls short.

Experimental activity measurements require rigorous standardization: units (e.g., μM, μg/mL, nM) are unified, and values are normalized across differing laboratories, target organisms, and assay formats. Quality scores are assigned based on how reproducible the experiments are and how strict the methods are, which makes database-driven modeling efforts more reliable and transparent. This formulation retains all of the technical content so that it can be easily expanded into RNA-targeting protocols or integration discussions, keeping everything clear and in context.

3 Target identification in proteins and RNAs

Peptides have unique binding properties and ways of working, so peptide therapeutics need specialized target identification strategies. Peptides are better than small molecules at changing protein–protein interactions because they can reach broad, flat interfaces (Nandi et al., 2024; Nandi et al., 2026). Small molecules usually target enzyme active sites or allosteric cavities. This amazing quality allows peptides to bind to targets that would otherwise be "undruggable," such as transcription factors, scaffolding proteins, and multidomain complexes that are important in disease pathways.

3.1 ML techniques for peptide target prediction

The introduction of AI and ML has transformed the process of identifying peptide targets. Modern computational frameworks combine different types of biological data, such as protein structures, binding affinities, and functional annotations, to make more accurate predictions about peptide–target interactions. Bayesian and DL methods take advantage of the unique sequence–structure–function relationships that peptides possess. For example, transformers and attention mechanisms reveal important long-range dependencies within sequences (Chen et al., 2024; Ghosh & Mitra, 2024; Ghosh & Mitra 2026). These models can also identify common targets between diseases by integrating large protein interaction databases with neural networks. This approach supports peptide-based multipathway solutions.

3.2 Protein–peptide interactions: Prediction and validation

A sophisticated understanding of peptide–protein interaction mechanisms underpins modern peptide drug discovery. Thanks to advances like AlphaFold, structure-based target validation is now much faster and more accurate (Johansson-Åkhe & Wallner, 2022). This makes it possible to quickly and accurately test many potential peptide binders. Hybrid methods that

TABLE 26.1 Databases for the identification of therapeutic targets of peptide-based drugs.

Database name	Approx. data size	Focus/content	Key features	Source
Peptipedia	>92,000 peptides	General therapeutic peptides	Integrates over 70 sources, includes biological activities, functional annotations, physicochemical properties, 3D structures (PDB, AlphaFold), patents, functional predictions	(Quiroz et al., 2021)
SATPdb	~19,000 peptides	Structurally annotated therapeutic peptides	Aggregates from 22 public peptide Databases; experimentally validated; provides 3D modeled structures; diverse therapeutic classes	(Singh et al., 2015)
LAMP2	~23,250 sequences	Antimicrobial peptides	Consolidates 16 AMP databases; sequence data, annotations, experimental metadata; supports AMP therapeutics discovery	(Ye et al., 2020)
DRAMP 2.0	19,899 entries (including ~5000 clinical/patents)	Peptides, including AMPs, clinical, patent entries	Comprehensive resource updated to version 2.0; includes general, patent, and clinical peptide entries	(Kang et al., 2019)
Propedia	~49,000 peptide–protein complexes	Protein–peptide interactions	Clusters complexes by sequence and interface features; enables exploration of peptide–target interactions critical for design and delivery	(Martins et al., 2023)
PDB (Protein Data Bank)	~13,299 peptide–protein complex entries	Experimental 3D structures for peptides in complex	Primarily experimentally resolved 3D structures of peptide–protein complexes aid structural modeling and design	(Berman et al., 2000)
BioLiP2	781,684 protein–ligand interactions (35,167 peptide interactions)	Protein–ligand interactions, including peptides, DNA, RNA, metals, and small molecules	Extensive binding data; binding affinities compiled from MOAD, PDBbind-CN, BindingDB, literature	(Yang et al., 2013)
PepBDB	13,299 peptide–protein complexes (up to 50 residues)	Biological peptide–protein complex structures	Focus on peptide length ≤50 residues; detailed structural and sequence data are useful for training binding prediction models	(Wen et al., 2019)

use both physics-based molecular modeling and DL help create very specific peptide inhibitors (such as β-catenin and NF-κB modulators), which often have much higher affinity due to iterative computational design. Researchers can use new techniques like variational autoencoders (VAEs) and molecular dynamics to design peptides with the right binding profiles by sampling latent representations and then simulating structural compatibility (Dean & Walper, 2020). Multiobjective optimization frameworks further enable the simultaneous evaluation of key properties—ADMET, selectivity, stability, and affinity—tailoring novel peptides for complex or otherwise intractable protein targets. Experimental validation methods like Nuclear Magnetic Resonance, Isothermal Titration Calorimetry, Surface Plasmon Resonance, and cell-based uptake/efficacy assays remain crucial for proving that AI-predicted peptide–target interactions are biologically relevant and effective.

3.3 RNA—Potential therapeutic targets for peptides

The growing understanding of RNA's key roles in disease has created new opportunities for peptide therapeutics (Haldar et al., 2026). RNAs have a wide range of structures that can be recognized, such as hairpins, bulges, pseudoknots, and riboswitches. This allows designed peptides to bind to them in a specific manner. This diversity contrasts with DNA's relative lack of accessibility or uniformity as a drug target. The recent Nobel recognition of microRNA's role in posttranscriptional gene regulation highlights the paradigm-shifting therapeutic potential of RNA-targeting peptides.

3.4 RNA recognition using macrocyclic peptides

Macrocyclic peptide platforms have revolutionized RNA-targeting by enabling the precise presentation of RNA-binding motifs. The Macrocyclic Organo-Peptide Hybrid libraries are an example of a system that generates millions of cyclized

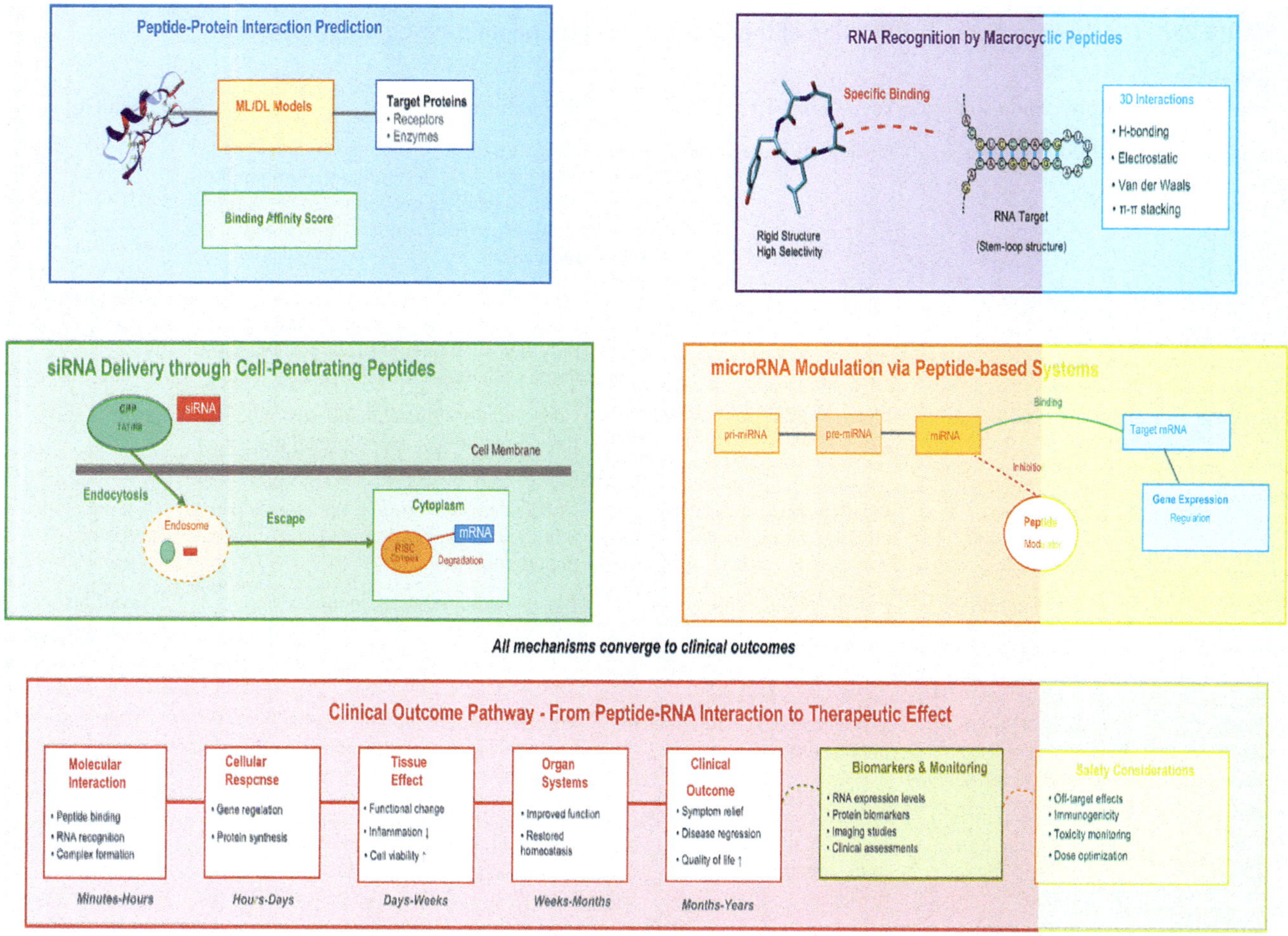

FIG. 26.1 Peptide–RNA therapeutic targeting mechanisms.

peptides using noncanonical amino acids and spontaneous cross-linking. Macrocyclic structures reduce flexibility, thereby enhancing RNA-binding geometry and improving selectivity and affinity (Pal & 't Hart, 2022). Recent research demonstrates that macrocyclic peptides can bind effectively to structured viral RNAs, such as the SARS-CoV-2 programmed ribosomal frameshifting RNA, at low micromolar or even nanomolar concentrations (Iannuzzelli et al., 2024). High-throughput display technologies, particularly mRNA display, facilitate the screening of extensive peptide libraries (often exceeding 10^{13} distinct sequences) against RNA targets. The PURE system allows the incorporation of nonnatural amino acids and advanced cyclization strategies to enhance selectivity and functionality (Fig. 26.1).

3.5 Delivery and targeting of small interfering ribonucleic acid (siRNA) through peptides

A big problem for nucleic acid drugs is still getting them to work well. Peptide-based delivery systems have been the key to the clinical success of several FDA-approved siRNA drugs, such as patisiran, givosiran, lumasiran, vutrisiran, inclisiran, and nedosiran. These systems have mainly solved problems with serum stability, cellular uptake, and endosomal escape. The combination of different delivery functions into single multifunctional peptides has greatly increased transfection efficiency compared to older, less flexible designs (Crombez et al., 2009; Nakase, Tanaka, & Futaki, 2013).

3.6 RNA therapeutic delivery via cell-penetrating peptides

Cell-penetrating peptides (CPPs) that are rich in arginine are the most effective way to deliver siRNA into cells. They utilize bidentate guanidinium–phosphate interactions to traverse membranes. Advanced CPP designs, such as CADY peptides (Simeoni, Morris, Heitz, & Divita, 2003), strategically arrange charged and aromatic residues to maximize cellular import that is energy-independent and to avoid endosomal sequestration (Sutcliffe, Doherty, Morgan, Dunne, & McCarthy, 2025).

Amino acid pairing peptides and other multifunctional constructs integrate nucleic acid-binding and membrane-penetrating domains into cohesive molecules, facilitating highly efficient gene delivery and expanding the potential of therapeutic RNA interventions.

3.7 microRNA modulation via peptide-based systems

Peptide-based systems are now allowing for the precise control of microRNAs, which regulate gene expression networks that impact up to 60% of human protein-coding genes. Dual-targeting delivery strategies, such as the conjugation of PDL1-binding peptides to anti-miR-21 inhibitors, have successfully facilitated cell- and disease-specific silencing, demonstrating effectiveness against oncogenic phenotypes, including melanoma and tumor-associated macrophage polarization (Kaczmarek, Kowalski, & Anderson, 2017; Kim et al., 2022).

3.8 Therapeutic outcomes and clinical applications

The clinical application of peptide-mediated RNA therapeutics exhibits significant promise in oncology, metabolic disorders, neurological conditions, infectious diseases, and rare genetic disorders. For example, bivalent cRGD peptides that contain VEGFR2-siRNA have been very successful in preventing tumors from growing in mice. PepGen's PGN-EDO51 and other peptide–oligonucleotide conjugates are in the later stages (Larkindale et al., 2023). Early-phase trials have shown that they are safe and effective. The market value of the RNA-targeting peptide drug sector has grown from $25.3 billion in 2024 to a projected $41.7 billion by 2030, thanks to both therapeutic and commercial success (Kaczmarek et al., 2017).

To sum up, peptides are a unique and versatile therapeutic toolkit. They can target proteins that were previously "undruggable," accurately recognize different RNA structures, and act as multifunctional delivery systems. Peptide-based approaches are ready to deliver unprecedented specificity, potency, and clinical value. The combination of computational AI pipelines, structural and functional annotation, and new experimental platforms sets the stage for the next generation of RNA- and protein-targeted drugs.

4 Feature engineering

4.1 Sequence-based representation learning

Modern peptide design pipelines depend on strong representation learning techniques to encode complicated biological sequences. At first, simple one-hot encoding was used, next interaction network and Deep Learning methods were combined (Das & Mitra, 2021). However, these were quickly replaced by methods that utilized physicochemical descriptors like VHSE and BLOSUM62 as well as databases like AAindex, to cover over 500 amino acid property scales. With the introduction of protein language models (pLMs) like ESM-2 (Lin et al., 2023) and ProteinBERT (Brandes et al., 2022), deep neural networks now interpret protein sequences as linguistic data. They create context-aware embeddings that reveal evolutionary, functional, and structural patterns. ESM-2, which was trained on 250 million protein sequences, excels at transfer learning tasks. Fusion architectures, such as FusPB-ESM2 (Zhang, Li, Wen, & Fang, 2024), combine different embedding sources to further enhance predictive performance. This enables the identification of CPPs with high accuracy and facilitates predictions about their functions that can be applied in other contexts. Bidirectional Long Short-Term Memory (LSTM) with structural feedback and graph-based sequence embeddings are two examples of advanced architectures that have further improved representation quality. These multimodal methods intelligently integrate information about sequences, structures, and chemical properties to aid in the discovery and prediction of peptide characteristics.

4.2 Structural representations and 3D geometry

3D structural modeling is the most important part of designing peptides that can create new ones (Mitra & Pal, 2010; Mitra & Pal, 2011a, 2011b; Pal, Pal & Mitra, 2021). Graph-based molecular models represent amino acids as nodes with edges that are spatially parametric and include information about distance cutoffs and chemical bonds. Graph attention networks (GATs), such as sAMPred-GAT (Yan et al., 2023), which utilize TrRosetta (Du et al., 2021) to predict 3D structures, provide better spatial encoding and improved antimicrobial prediction. Recent advancements include molecular fingerprints such as Extended-Connectivity Fingerprints (ECFP) and novel structures such as geometric vector perceptrons, which address geometric equivariance. The introduction of ESMFold (Lin et al., 2023) and AlphaFold2 has revolutionized structure prediction. These tools enable fast, accurate modeling that facilitates peptide docking and design. PepFlow's (Abdin &

Kim, 2024) diffusion-based conformational sampling and AfCycDesign's (Rettie et al., 2025) precise cyclic peptide modeling have further enhanced these capabilities. Ensemble methods and tools like ColabFold (Mirdita et al., 2022) make high-throughput structural prediction on a large scale possible. This is critical for determining therapeutic peptides' binding efficiency, stability, and druggability.

4.3 Multimodal representation fusion

Peptide generative design is increasingly incorporating multimodal data, combining sequence and structure to address the constraints of single-modal learning. In the past, feature vectors were combined (Banerjee, et al., 2020 Banerjee and Mitra, 2020; Banerjee, Levy & Mitra, 2019; Das, 2023; Mitra, Shultis & Zhang, 2013; Maiti, Hassan & Mitra, 2020), but new technologies like MultiRepPI (Zhiguo, Zixuan, & Peng, 2025) and MoFormer (Cao, Magar, Wang, & Farimani, 2023) use advanced attention mechanisms and conditional transformer architectures to align data from different sources in real time. Models such as PoET-2 (Truong & Bepler, 2025) and EasIFA (Wang et al., 2024) serve as paradigms of fusion frameworks, employing language model embeddings and 3D structural encoders to attain significant improvements in speed and integration. HemoFuse (Zhao, Zhang, & Liang, 2024) and MAEF-GO (Liu et al., 2025) are two examples of specialized frameworks that combine multihead cross-attention for task-specific features like hemolytic activity and functional annotation. Combining large, pretrained models with structural encoders is now standard. OmniBioTE (Chen et al., 2025) is pushing the limits by trying to predict how peptides and nucleotides interact. These new strategies capture complex dependencies and adaptable binding modes, which are essential for optimizing peptides for therapeutic use.

5 Generative AI models in action

5.1 Information-theoretic approaches to peptide design

DL has transformed peptide science by making it easier to predict, identify, and design functions. This advancement has positioned peptide science at the intersection of biology and chemistry. AlphaFold2 and similar algorithms excel at predicting the structure of proteins and peptides, thereby accelerating the process of discovering new drugs. Information theory-based methods utilize Shannon entropy and mutual information maximization to ensure that the generated sequences are both diverse and functional. Models like RepCon (Liu, Wang, Wang, Zheng, & Li, 2023) and InfoNCE (Lee & Shin, 2024) employ contrastive learning to create feature representations that facilitate robust generalization. The information bottleneck framework compresses embeddings to retain only the features essential for the function, while β-VAE (Ullah et al., 2025) and variational methods (Renaud & Mansbach, 2023) explore peptide properties that can be disentangled and controlled. Modern algorithms build on these principles by incorporating codon optimality, RNA folding energies, and translational entropy. This enables researchers to systematically regulate mRNA translation, stability, and immune evasion.

5.2 Using generative adversarial networks in molecular design

Generative adversarial networks (GANs) have made a name for themselves in the design of peptides and mRNA by creating a wide range of functional sequences. In GAN paradigms, the network learns to generate sequences that closely resemble real data and continuously improves at being biologically relevant. ProteinGAN (Repecka et al., 2021) and AMPGANv2 (Van Oort, Ferrell, Remington, Wshah, & Li, 2021) are two applications capable of producing useful, foldable peptides. These peptides are often designed to be less toxic and more effective against bacteria. Advances in reinforcement learning and policy gradient methods have made GAN training more stable, facilitating navigation through challenging sequence spaces and enabling the discovery of new molecules. These models play a crucial role in creating peptide variants for screening and further research.

5.3 VAEs for systematic exploration

VAEs learn smooth latent representations and find a balance between reconstruction and regularization. This makes them necessary for property-driven peptide generation. PepGLAD (Kong, Jia, Huang, & Liu, 2024) and PPGC-DVAE (Zhao, Hou, Shen, Hu, & Wren, 2025) are two of the best models that encode full-atom peptide residues. This makes it possible to condition receptors specifically and generate sequences that are true to their physicochemical properties. Integrated methods, like maximum mean discrepancy loss, enhance representation utility even further, and composite sampling methods, like Metropolis-Hastings, improve exploration significantly (Wan et al., 2022). Conditional VAEs have made it possible to achieve targeted, feature-preserving sequence output, which is important for therapeutic development. This includes properties like antimicrobial activity and stability. These paradigms are being increasingly utilized to improve mRNA by creating sequence-function landscapes for next-generation biologics.

5.4 Deep residual networks for mRNA design

Deep residual architectures have recently started to assist with mRNA optimization by enabling the modeling of complex, hierarchical relationships between sequence segments and functional outcomes. These models effectively capture long-range sequence dependencies and the nonlinear effects of codon arrangement, folding, and regulatory elements by stacking residual blocks with attention. This approach makes mRNA vaccine design more efficient and effective (Imani et al., 2024).

5.5 Adversarial training paradigms

Adversarial frameworks, like conditional GANs, are used to ensure that peptide generation is both safe and effective. Models like AMPGAN v2 (Van Oort et al., 2021) base the generative process not just on physicochemical measures but also on toxicity profiles. This approach produces candidates that are useful and have a low risk. This multiobjective optimization is a key part of modern peptide drug design, ensuring the drug is usable in the clinic and meets all regulations.

5.6 Diffusion models and denoising approaches

Diffusion models reverse a gradual noising process to create peptide sequences, making it possible to generate high-fidelity sequences from latent spaces. Innovative implementations like RFdiffusion (Watson et al., 2023) and AMP-Diffusion (Chen et al., 2024) merge E(3)-equivariant architectures with biochemical conditioning, significantly enhancing design diversity and validation rates. These models, together with contrastive-augmented multimodal diffusion frameworks, have produced extensive libraries of peptides, experimentally validated for broad-spectrum antimicrobial and anticancer properties, thereby demonstrating the operational superiority of AI-driven molecular generation.

5.7 Attention mechanisms and transformer architectures

Transformers are the best way to model peptides right now. They use rich contextual embeddings to capture long-range sequence and interaction dynamics. ESM-2 (Lin et al., 2023), PeptideGPT (Shah, Guntuboina, & Farimani, 2024), DeepPROTECTNeo (Das et al., 2025), and PeptideBERT (Guntuboina, Das, Mollaei, Kim, & Farimani, 2023) are all examples of transformer-based generative and predictive models that can be used for property prediction (such as solubility and hemolysis), de novo design, and advanced classification tasks (such as identifying CPPs). Recent pipelines use GATs (Chen et al., 2024; Hao et al., 2024) and actor-critic (A2C) reinforcement learning to improve the search for chemicals and the achievement of multiple goals. These architectures are essential for the growth of peptide design portfolios and the swift identification of novel drug candidates (Fig. 26.2, Table 26.2).

6 AI in predicting toxicity and stability

6.1 Adding safety limits to generative models

Safety-aware design is now an inherent component of generative pipelines, evolving from retrospective screening to Pareto-efficient, multiobjective optimization. HydrAMP (Szymczak et al., 2023) and other models use conditional VAEs (Zhao et al., 2025) to separate biological activity from safety limits. Bayesian optimization and trust-region methods keep the generative search within safe chemical spaces (Eberhardt, Lees, Lill, & Schwede, 2023). These improvements ensure that only peptides with acceptable levels of toxicity and immunogenicity progress to the next stage of development. This increases the chances of success in clinical trials and reduces the risk.

6.2 Models for predicting toxicity

Contemporary cytotoxicity models facilitate real-time predictions during peptide synthesis, utilizing ensemble ML techniques (such as logistic regression, random forests, and XGBoost) and structure-aware neural networks, including tAMPer (Ebrahimikondori et al., 2024) and ToxGIN (Yu, Zhang, Liu, Li, & Tang, 2024). These models use both topological and geometric descriptors to achieve high balanced accuracy and make it possible to understand mechanistic features.

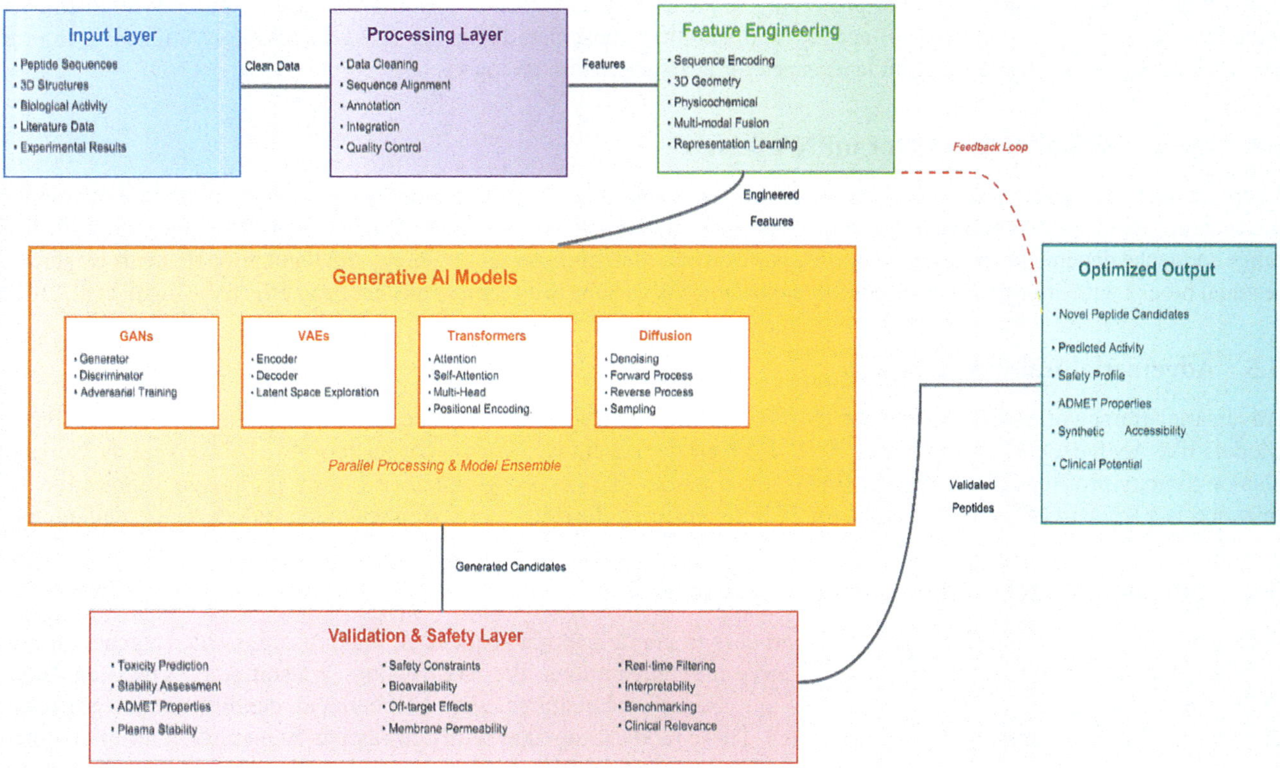

FIG. 26.2 Generative AI models in peptide-based drug discovery, design, and analysis.

ADMET prediction platforms like ADMETlab v3.0 (Fu et al., 2024) and MTGL-ADMET (Du et al., 2023) combine dozens of endpoints for a full safety assessment. They also have advanced methods to measure uncertainty and outputs that are easy to understand, which helps with regulatory and translational needs. Modern cytotoxicity modeling now includes peptides and full-length proteins. This enables real-time safety checks during sequence design and synthesis using ensemble ML classifiers (like logistic regression, random forests, and XGBoost) and structure-aware neural networks (like tAMPer and ToxGIN) that combine sequence embeddings with 3D/topological graph encodings for better predictions and mechanistic interpretability. For proteins, the best DL methods currently are transformer and multimodal approaches. These include CSM-Toxin (fine-tuning ProteinBERT on curated toxic vs. nontoxic proteins), VISH-Pred (ensembles of fine-tuned ESM2 language models feeding LightGBM/XGBoost with explicit class-imbalance handling), and ToxDL 2.0, which uses a GCN to combine AlphaFold2-derived structure graphs with domain embeddings to improve generalization and interpretability through Integrated Gradients. VISH-Pred demonstrates strong performance on independent blind tests that only use sequence and undersampling to address severe class imbalance. ToxDL 2.0 (Zhu et al., 2025), on the contrary, consistently outperforms previous methods on both pre-2022 and post-2022 holdout proteins and is available as a public web server for easy use. ADMETlab 3.0 provides 119 endpoints with multitask DMPNN modeling (Fan et al., 2023), API access, and built-in uncertainty estimates for decision support. MTGL-ADMET offers interpretable multitask graph learning to jointly predict multiple ADMET properties with task selection and substructure attribution, aligning with regulatory and translational needs.

6.3 Strategies for benchmarking and validation

Modern cytotoxicity models now allow for real-time, high-throughput toxicity prediction during peptide and protein design, using a wide range of ensemble ML and DL techniques. Established platforms routinely use logistic regression, random forests, support vector machines, and XGBoost for small-molecule and peptide toxicity, while structure-aware neural networks, including tAMPer and ToxGIN, incorporate both sequence-based and geometric descriptors to boost interpretability and predictive power. Recent advancements have expanded these functionalities to encompass large proteins, featuring prominent architectures such as transformer-based CSM-Toxin (which fine-tunes ProteinBERT for curated toxic versus nontoxic classification), VISH-Pred (comprising ensembles of fine-tuned ESM2 models utilizing LightGBM/XGBoost),

TABLE 26.2 DL-based generative models in practice for peptide-based drug design and applications.

Model name(s)	Generative AI model type	Purpose	Notable details	Research work and year
PepGLAD	Variational autoencoder + geometric latent diffusion	Full-atom peptide sequence structure codesign with binding site conditioning	Encodes full-atom residues into fixed-dimensional latent space; receptor-specific affine transformations enable generalization across binding sites	(Kong et al., 2024)
CELA-MFP	Protein language model + transformer decoder + contrastive learning	Multifunctional therapeutic peptide function prediction	Contrastive enhancement and label adaptation; outperforms state-of-the-art; accessible web server	(Fang, Luo, Ren, Wei, & Wei, 2024)
PepHarmony	Multiview contrastive learning framework	Integrated encoding of peptide sequence and structure	Combines sequence and structural views; improves representation learning; robust embedding for downstream tasks	(Zhang et al., n.d.)
PeptideGPT	Generative pretrained transformer	De novo peptide sequence design	Autoregressive generation specialized for peptides	(Shah et al., 2024)
MMCD (multimodal contrastive diffusion model)	Multimodal diffusion with contrastive learning	Therapeutic peptide sequence-structure co-generation	Fuses sequence and structure modalities; inter- and intramodal contrastive learning; outperforms state-of-the-art baselines on antimicrobial and anticancer scores, as well as docking	(Wang, Liu, Huang, Xiong, & Zhang, 2024)
HydrAMP	Conditional variational autoencoder (VAE) with task-specific classifiers	Generation of potent antimicrobial peptides	Learns disentangled latent representations capturing antimicrobial activity; supports unconstrained and analog generation; preselection combined with molecular dynamics improves candidate quality	(Szymczak et al., 2023)
AMPGAN v2	Conditional GAN with multiobjective optimization	Antimicrobial peptide design with toxicity minimization	Uses deep GAN architecture conditioned on toxicity scores and physicochemical features; balances antimicrobial efficacy and reduces hemolytic toxicity	(Van Oort et al., 2021)
PepVAE	Variational autoencoder + activity prediction	Controlled antimicrobial peptide generation	Incorporates MIC experimental data; sampling from latent regions allows controllable AMP design; demonstrated experimental validation	(Dean, Alvarez, Zabetakis, Walper, & Malanoski, 2021)
PeptideBERT	Transformer language model	Predicting hemolysis, solubility, and nonfouling peptides	Utilizes ProtBERT pretrained model; achieves SOTA hemolysis prediction accuracy; fine-tuned for peptide property tasks	(Guntuboina et al., 2023)
CPPFormer	Transformer encoder–decoder architecture	Cell-penetrating peptide classification	Designed for short CPP sequences; combines automatic and manual feature extraction; achieves >92% accuracy on CPP924 dataset	(Xue et al., 2022)

and ToxDL 2.0, which uniquely amalgamates AlphaFold2-derived structure graphs with domain embeddings and interpretable saliency maps through Integrated Gradients.

Benchmarking frameworks have evolved, offering standardized datasets and metrics for predicting toxicity and properties. Quantitative validation is possible with metrics like F1-score, Matthews Correlation Coefficien (MCC), area under the receiver operating characteristic curve (AUROC), and area under the precision-recall curve (AUPRC) when using datasets like ToxDL (Zhu et al., 2025) and Wei et al.'s peptide compendium. Successive generations of models, such as tAMPer and ToxGIN, show better performance on independent test sets. Ten-fold cross-validation and McNemar's tests ensure that the results are statistically sound. These validated frameworks give us the confidence we need for further translational and clinical development, ensuring that peptide therapeutic pipelines are reliable and can grow. For rational peptide and mRNA therapeutic design, it is now necessary to use advanced representation learning, generative modeling, adversarial training, and safety-aware frameworks all at once. This synergistic approach shortens development times, increases the likelihood of successful clinical trials, and expands the range of functions, making AI the driving force behind the next generation of drug discovery.

Recent reviews have shown that the choice and quality of datasets have a significant impact on how well a model performs and how reliable it is. Many databases used to train toxicity models contain differing or conflicting toxicity assignments for the same molecule, which can complicate training and testing. For meaningful comparisons and model validation, it is important to use standardized, well-annotated datasets for different toxicity endpoints (e.g., hepatotoxicity, cardiotoxicity, mutagenicity, and the 12 Tox21 endpoints) when benchmarking. Recent DL research has demonstrated that multitask learning methods can help by combining in vitro, in vivo, and clinical endpoints into single models. These methods can also leverage transfer learning to compensate for the lack of data in expensive clinical outcomes. Studies have shown that for toxicity models to be useful in drug discovery, they need to not only be accurate but also produce outputs that are interpretable, well-calibrated, and accompanied by clear measures of uncertainty. Researchers and regulators can use post hoc contrastive explanations, feature attribution, and uncertainty quantification to closely examine the mechanistic underpinnings and gain confidence in predictions.

6.4 DL-enhanced peptide quantitative structure–activity relationship in peptide drug development

Peptide–protein interacting domains are crucial for binding and recognition in processes that regulate signaling cascades, physiological functions, and disease mechanisms. Engineered peptides typically recognize small motifs, which can include linear sequence segments or conformational surface areas, enabling a better understanding of different biological functions, pathway modulation, cellular processes, and ultimately disease progression. This drives the growing interest in understanding the correlation between peptides' conformational changes and biological activity. Quantitative structure–activity relationship (QSAR) models have been employed to predict molecular characteristics. However, traditional methodologies are limited by a scarcity of peptide-specific activity data, and conventional descriptors may fail to capture the high-dimensional, sequence-dependent features underlying peptide–protein interactions. DL has transformed peptide QSAR by overcoming these limitations through the extraction of subtle high-dimensional features from specific datasets and the modeling of complex nonlinear sequence–activity relationships (Liu et al., 2022).

Recent advances exemplify the transformative role of DL in peptide-based drug discovery. For instance, Nielsen et al. (2024) employed a random forest-based QSAR methodology implemented within the streaMLine platform to assess over 2600 glucagon-like peptide-1 analogs (Nielsen et al., 2024). Their framework utilized multiparameter modeling in conjunction with SHAP (Shapley Additive) value analysis to enhance interpretability, guiding modifications to improve peptide selectivity, solubility, and stability. Yin et al. (2025) developed DL-QSARES, a DL framework that integrates ESM-2 pLM embeddings with empirical QSAR filtration (Yin et al., 2025). In the field of antifungal peptide discovery, DL-QSARES successfully narrowed the candidate search space and identified active peptides targeting *Candida albicans* and *Candida tropicalis*. In vitro validation and in vivo animal model testing confirmed efficacy and safety, with results demonstrating improved precision and speed compared to conventional discovery methods. These case studies underscore a broader trend wherein DL-enabled QSAR methodologies enhance both the efficiency and interpretability of peptide drug development pipelines. Emerging techniques extend beyond predictive modeling to generative design, integrating models such as VAEs, GANs, and transformers to explore novel chemical space. Combining computational modeling with empirical screening accelerates the optimization of multiparametric peptides and plays a crucial role in the next generation of peptide therapies (Table 26.3).

TABLE 26.3 DL-based models for toxicity assessment of therapeutic peptides.

Model/ platform	Descriptor type	Toxicity endpoints	Notable features
tAMPer/ ToxGIN	Sequence + 3D/graph	Peptides, proteins (cytotoxicity)	Structure-aware GNNs, mechanistic interpretability
CSM-Toxin	Pretrained language model	Proteins (general)	BERT-fine-tuned, clear toxic/nontoxic separation
VISH-Pred	LM + LightGBM/XGBoost	Multiple (blind/class-imbalanced)	Robust sequence-only excels in difficult settings
ToxDL 2.0	GCN + structure embedding	Proteins (across versions)	AlphaFold2 graphs, strong generalization
ADMETlab 3.0	Molecular graph + DMPNN	119 endpoints (ADMET, toxicity)	API, uncertainty, regulatory-oriented outputs
MTGL-ADMET	Graph + attribution	Multiple ADMET endpoints	Substructure-level interpretability and insights

7 Future scope of AI in peptide-based drug designing and biodelivery

The future of AI in peptide therapeutics promises unprecedented precision in drug design through the convergence of multiple cutting-edge technologies. As generative models like diffusion networks and transformer architectures continue to evolve, we can expect AI systems to simultaneously optimize multiple therapeutic parameters—from binding affinity and selectivity to stability and immunogenicity—while exploring vast, unexplored chemical spaces that traditional methods could never access. The integration of real-time structural prediction tools like AlphaFold with advanced generative frameworks will enable researchers to design peptides with atomic-level precision, potentially creating entirely new classes of therapeutics that can target previously "undruggable" proteins and RNA structures (Bülbül, Bode, Schmitt, & Bozhüyük, 2025). Moreover, AI's growing capability to predict and model complex biological interactions will likely lead to the development of smart, adaptive drug delivery systems that can respond to specific physiological conditions and deliver therapeutic payloads with unprecedented accuracy (Nissan, Allen, Sabatino, & Biggar, 2024).

Looking ahead, the most transformative developments will likely emerge from AI's ability to integrate multimodal biological data and design truly personalized peptide therapeutics. Future AI systems will probably incorporate patient-specific genomic, proteomic, and metabolic data to create tailored peptide drugs that account for individual variations in drug metabolism, immune responses, and disease mechanisms. The field is also moving toward AI-driven automation of the entire drug development pipeline, from initial target identification through clinical optimization, potentially reducing development timelines from decades to years. As ML models become more sophisticated in understanding the complex relationships between peptide structure, function, and safety, we can anticipate the emergence of fully automated drug discovery platforms that can generate, test, and refine therapeutic candidates in silico before any laboratory synthesis, fundamentally revolutionizing how we approach drug development and making personalized medicine more accessible and cost-effective for patients worldwide.

8 Conclusion

AI has completely transformed the field of peptide and mRNA therapeutics, spanning from early-stage data curation and target prediction to molecule design driven by generative models, toxicity assessment, and rational delivery optimization. The combination of curated, multimodal datasets and advanced modeling architectures, such as language models, graph neural networks, and generative frameworks like GANs, VAEs, and diffusion models, now enables systematic exploration of vast chemical and biological spaces. This accelerates the process of identifying safe, effective, and developable therapeutic candidates. Despite significant progress, major challenges remain. These include integrating diverse data sources, capturing the complexity of sequence–structure–function relationships (particularly for noncanonical residues and RNA modifications), and ensuring that model predictions are robust and uncertainty-aware, even in the absence of extensive experimental validation. Addressing these challenges will require not only further advancements in algorithmic innovation but also stronger connections between method development and biologically informed constraints, real-world benchmarking, and iterative feedback from experimental studies. In summary, AI has become an integral part of the entire process of discovering new peptide and mRNA drugs. It bridges molecular design with real-world considerations such as delivery, stability, and safety. The ongoing convergence of computational and experimental fields is set to produce therapeutics defined by unprecedented precision, speed, and clinical relevance, ushering in a new era of programmable medicine.

References

Abdin, O., & Kim, P. M. (2024). Direct conformational sampling from peptide energy landscapes through hypernetwork-conditioned diffusion. *Nature Machine Intelligence*, 2522-5839. *6*(7), 775–786. Available from: https://www.nature.com/natmachintell/. doi:10.1038/s42256-024-00860-4.

Banerjee, A., Levy, Y., & Mitra, P. (2019). Analyzing Change in Protein Stability Associated with Single Point Deletions in a Newly Defined Protein Structure Database. J Proteome Res, *18*(3), 1402–1410. 10.1021/acs.jproteome.9b00048.

Banerjee, A., & Mitra, P. (2020). Estimating the Effect of Single-Point Mutations on Protein Thermodynamic Stability and Analyzing the Mutation Landscape of the p53 Protein. J Chem Inf Model, *60*(6), 3315–3323. doi:10.1021/acs.jcim.0c00256.

Banerjee, A., Kumar, A., Ghosh, K. K., Mitra, P. (2020). Estimating change in foldability due to multipoint deletions in protein structures. Journal of Chemical Information and Modeling, *60*(12 6679–6690. doi:10.1021/acs.jcim.0c00802.

Barnett, L. G., Simkins, H. M. A., Barnett, B. E., Korn, L. L., Johnson, A. L., Wherry, E. J., et al. (2014). B cell antigen presentation in the initiation of follicular helper T cell and germinal center differentiation. *Journal of Immunology*, 1550-6606. *192*(8), 3607–3617. Available from: http://www.jimmunol.org/content/192/8/3607.full.pdf+html. doi:10.4049/jimmunol.1301284.

Berman, H. M., Westbrook, J., Feng, Z., Gilliland, G., Bhat, T. N., Weissig, H., et al. (2000). The protein data bank. *Nucleic Acids Research*, 0305-1048. *28*(1), 235–242. Available from: https://academic.oup.com/nar/issue. doi:10.1093/nar/28.1.235.

Brandes, N., Ofer, D., Peleg, Y., Rappoport, N., Linial, M., & Martelli, P. L. (2022). ProteinBERT: a universal deep-learning model of protein sequence and function. *Bioinformatics*, 1367-4803. *38*(8), 2102–2110. doi:10.1093/bioinformatics/btac020.

Bülbül, E. F., Bode, H. B., Schmitt, S., & Bozhüyük, K. A. J. (2025). Engineering the future of medicine: Natural products, synthetic biology and artificial intelligence for next-generation therapeutics. *Clinical and Translational Medicine*, 2001-1326. *15*(2). doi:10.1002/ctm2.70146.

Cao, Z., Magar, R., Wang, Y., & Farimani, A. B. (2023). MOFormer: Self-supervised transformer model for metal–organic framework property prediction. *Journal of the American Chemical Society*, 0002-7863. *145*(5), 2958–2967. doi:10.1021/jacs.2c11420.

Chen, S. F., Steele, R. J., Hocky, G. M., Lemeneh, B., Lad, S. P., & Oermann, E. K. (2025). Large-scale multi-omic biosequence transformers for modeling protein-nucleic acid interactions. *arXiv*, arXiv. 23318422. Available from: https://arxiv.org. doi:10.48550/arXiv.2408.16245.

Chen, S., Lin, T., Basu, R., Ritchey, J., Wang, S., Luo, Y., et al. (2024). Design of target specific peptide inhibitors using generative deep learning and molecular dynamics simulations. *Nature Communications*, 2041-1723. *15*(1). Available from: https://www.nature.com/ncomms/. doi:10.1038/s41467-024-45766-2.

Choi, S. C., & Morel, L. (2020). Immune metabolism regulation of the germinal center response. *Experimental and Molecular Medicine*, 2092-6413. *52*(3), 348–355. Available from: https://www.nature.com/emm/. doi:10.1038/s12276-020-0392-2.

Crombez, L., Aldrian-Herrada, G., Konate, K., Nguyen, Q. N., McMaster, G. K., Brasseur, R., et al. (2009). A new potent secondary amphipathic cell-penetrating peptide for siRNA delivery into mammalian cells. *Molecular Therapy*. 1525-0024. *17*(1), 95–103. Available from: https://www.sciencedirect.com/science/journal/15250016. doi:10.1038/mt.2008.215.

Crotty, S. (2011). Follicular helper CD4 T cells (T FH). *Annual Review of Immunology*, 0732-0582. *29*(1), 621–663. doi:10.1146/annurev-immunol-031210-101400.

Das, B., & Mitra, P. (2021). Protein Interaction Network-based Deep Learning Framework for Identifying Disease-Associated Human Proteins. J Mol Biol, *433*(19), 167149. doi:10.1016/j.jmb.2021.167149.

Das, S., Kundu, M., Hassan, A., Parekh, A., Jena, B. C., Mundre, S., …, Mandal, M. (2023). A novel computational predictive biological approach distinguishes Integrin β1 as a salient biomarker for breast cancer chemoresistance. Biochim Biophys Acta Mol Basis Dis, *1869*(6), 166702. doi:10.1016/j.bbadis.2023.166702.

Das, D., Bhaduri, S., Pramanick, A., & Mitra, P. (2025). DeepPROTECTNeo: a deep learning-based personalized and RV-guided optimization tool leveraging TCR epitope interaction using context-aware transformers. *bioRxiv*, bioRxiv. 26928205. Available from: https://www.biorxiv.org. doi:10.1101/2025.01.04.631301.

Dean, S. N., Alvarez, J. A. E., Zabetakis, D., Walper, S. A., & Malanoski, A. P. (2021). PepVAE: variational autoencoder framework for antimicrobial peptide generation and activity prediction. *Frontiers in Microbiology*, 1664-302X. *12*. Available from: https://www.frontiersin.org/journals/microbiology#. doi:10.3389/fmicb.2021.725727.

Dean, S. N., & Walper, S. A. (2020). Variational autoencoder for generation of antimicrobial peptides. *ACS Omega*, 2470-1343. *5*(33), 20746–20754. Available from: http://pubs.acs.org/journal/acsodf. doi:10.1021/acsomega.0c00442.

Du, B. X., Xu, Y., Yiu, S. M., Yu, H., & Shi, J. Y. (2023). MTGL-ADMET: a novel multi-task graph learning framework for ADMET prediction enhanced by status-theory and maximum flow. In Bertino Elisa Gao Wen Steffen Bernhard Yung Moti*Lecture notes in computer science (including subseries lecture notes in artificial intelligence and lecture notes in bioinformatics)* (Vol. 13976, pp. 85–103). China: Springer Science and Business Media Deutschland GmbH. 16113349. Available from: https://www.springer.com/series/558. doi:10.1007/978-3-031-29119-7_6.

Du, Z., Su, H., Wang, W., Ye, L., Wei, H., Peng, Z., et al. (2021). The trRosetta server for fast and accurate protein structure prediction. *Nature Protocols*, 1750-2799. *16*(12), 5634–5651. Available from: http://www.natureprotocols.com/. doi:10.1038/s41596-021-00628-9.

Eberhardt, J., Lees, A., Lill, M., & Schwede, T. (2023). Combining Bayesian optimization with sequence- or structure-based strategies for optimization of protein-peptide binding. *ChemRxiv*, ChemRxiv. 25732293. Available from: https://chemrxiv.org. doi:10.26434/chemrxiv-2023-b7l81-v2.

Ebrahimikondori, H., Sutherland, D., Yanai, A., Richter, A., Salehi, A., Li, C., et al. (2024). Structure-aware deep learning model for peptide toxicity prediction. *Protein Science*, 1469-896X. *33*(7). Available from: http://onlinelibrary.wiley.com/journal/10.1002/(ISSN)1469-896X. doi:10.1002/pro.5076.

Evans, R., O'Neill, M., Pritzel, A., Antropova, N., Senior, A., Green, T. et al. (2022). Protein complex prediction with AlphaFold-Multimer. Available from: https://www.biorxiv.org/content/10.1101/2021.10.04.463034v2.

Fan, M., Liu, Q., Cui, Z., Wang, H., Chen, M., He, D., & Hou, Y. (2023). DMPNN-Bert: A deep learning architecture for molecular property prediction. In *ACM international conference proceeding series* (pp. 152–157). China: Association for Computing Machinery. 9798400707605. Available from: http://portal.acm.org/. doi:10.1145/3611450.3611472.

Fang, Y., Luo, M., Ren, Z., Wei, L., & Wei, D.-Q. (2024). CELA-MFP: A contrast-enhanced and label-adaptive framework for multi-functional therapeutic peptides prediction. *Briefings in Bioinformatics*, 1467-5463. *25*(4). doi:10.1093/bib/bbae348.

Fu, L., Shi, S., Yi, J., Wang, N., He, Y., Wu, Z., et al. (2024). ADMETlab 3.0: An updated comprehensive online ADMET prediction platform enhanced with broader coverage, improved performance, API functionality and decision support. *Nucleic Acids Research*, 0305-1048. *52*(W1), W422. doi:10.1093/nar/gkae236.

Ghosh, S., & Mitra, P. PEPpip: A Vision-Inspired Explainable Deep Learning Framework for Sequence-Based Protein–Protein Interaction Prediction and Interaction Map Estimation. *IEEE Access*, *14*, 16936–16949. doi:10.1109/ACCESS.2026.365822110.1109/ACCESS.2026.3658221/mm1

Ghosh, S., & Mitra, P. (2024). MaTPIP: a deep-learning architecture with eXplainable AI for sequence-driven, feature mixed protein-protein interaction prediction. *Computer Methods and Programs in Biomedicine*, 0169-2607. *244*, 107955. doi:10.1016/j.cmpb.2023.107955.

Goles, M., Daza, A., Cabas-Mora, G., Sarmiento-Varón, L., Sepúlveda-Yañez, J., Anvari-Kazemabad, H., et al. (2024). Peptide-based drug discovery through artificial intelligence: towards an autonomous design of therapeutic peptides. *Briefings in Bioinformatics*, 1477-4054. *25*(4). Available from: http://bib.oxfordjournals.org. doi:10.1093/bib/bbae275.

Guntuboina, C., Das, A., Mollaei, P., Kim, S., & Farimani, A. B. (2023). PeptideBERT: A language model based on transformers for peptide property prediction. *Journal of Physical Chemistry Letters*, 1948-7185. *14*(46), 10427–10434. Available from: http://pubs.acs.org/journal/jpclcd. doi:10.1021/acs.jpclett.3c02398.

Haldar, D. K., Pramanick, A., Mukherjee, C., & Mitra, P. (2026). A transformer-based method for the cap analysis of gene expression and gene expression tag associated capping region prediction in RNA. RNA Biol, *23*(1), 1–15. doi:10.1080/15476286.2026.2629530

Hao, Y., Liu, X., Fu, H., Shao, X., Cai, W., & Martelli, P. L. (2024). PGAT-ABPp: Harnessing protein language models and graph attention networks for antibacterial peptide identification with remarkable accuracy. *Bioinformatics*, 1367-4803. *40*(8). doi:10.1093/bioinformatics/btae497.

Hashemi, S., Vosough, P., Taghizadeh, S., & Savardashtaki, A. (2024). Therapeutic peptide development revolutionized: Harnessing the power of artificial intelligence for drug discovery. *Heliyon*, 2405-8440. *10*(22), e40265. doi:10.1016/j.heliyon.2024.e40265.

Iannuzzelli, J. A., Bonn, R., Hong, A. S., Anitha, A. S., Jenkins, J. L., Wedekind, J. E., & Fasan, R. (2024). Cyclic peptides targeting the SARS-CoV-2 programmed ribosomal frameshifting RNA from a multiplexed phage display library. *Chemical Science*, 2041-6539. *15*(46), 19520–19533. Available from: http://pubs.rsc.org/en/journals/journal/sc. doi:10.1039/d4sc04026k.

Imani, S., Li, X., Chen, K., Maghsoudloo, M., Jabbarzadeh Kaboli, P., Hashemi, M., et al. (2024). Computational biology and artificial intelligence in mRNA vaccine design for cancer immunotherapy. *Frontiers in Cellular and Infection Microbiology*, 2235-2988. *14*. Available from: http://www.frontiersin.org/Cellular_and_Infection_Microbiology/archive. doi:10.3389/fcimb.2024.1501010.

Jeffery, C. J. (2018). Protein moonlighting: What is it, and why is it important? *Philosophical Transactions of the Royal Society B: Biological Sciences*, 1471-2970. *373*, 1738. Available from: http://rstb.royalsocietypublishing.org/content/373/1738/20160523.full.pdf. doi:10.1098/rstb.2016.0523.

Johansson-Åkhe, I., & Wallner, B. (2022). Improving peptide-protein docking with AlphaFold-Multimer using forced sampling. *Frontiers in Bioinformatics*, 2673-7647. *2*. doi:10.3389/fbinf.2022.959160.

Kaczmarek, J. C., Kowalski, P. S., & Anderson, D. G. (2017). Advances in the delivery of RNA therapeutics: From concept to clinical reality. *Genome Medicine*, 1756-994X. *9*(1). Available from: http://www.genomemedicine.com/. doi:10.1186/s13073-017-0450-0.

Kang, X., Dong, F., Shi, C., Liu, S., Sun, J., Chen, J., et al. (2019). DRAMP 2.0, an updated data repository of antimicrobial peptides. *Scientific Data*, 2052-4463. *6*(1). doi:10.1038/s41597-019-0154-y.

Kim, E. H., Lee, J., Kwak, G., Jang, H., Kim, H., Cho, H., et al. (2022). PDL1-binding peptide/anti-miRNA21 conjugate as a therapeutic modality for PD-L1high tumors and TAMs. *Journal of Controlled Release*, 0168-3659. *345*, 62–74. doi:10.1016/j.jconrel.2022.02.031.

Kong, X., Jia, Y., Huang, W., & Liu, Y. (2024). Full-atom peptide design with geometric latent diffusion. In *Advances in neural information processing systems* (Vol. *37*). China: Neural Information Processing Systems Foundation. 10495258. Available from: https://papers.nips.cc/

Larkindale, J., Lonkar, P., Goyal, J., Holland, A., Foy, J., Garg, B., et al. (2023). P44 Phase 1 study of PGN-EDO51 demonstrates tolerability, delivery and high levels of exon skipping for treatment of Duchenne muscular dystrophy (DMD). *Neuromuscular Disorders*, 0960-8966. *33*, S69. doi:10.1016/j.nmd.2023.07.025.

Lederer, K., Castaño, D., Gómez Atria, D., Oguin, T. H., Wang, S., Manzoni, T. B., et al. (2020). SARS-CoV-2 mRNA vaccines foster potent antigen-specific germinal center responses associated with neutralizing antibody generation. *Immunity*, 1097-4180. *53*(6), 1281. Available from: www.immunity.com. doi:10.1016/j.immuni.2020.11.009.

Lee, B., & Shin, D. (2024). Contrastive learning for enhancing feature extraction in anticancer peptides. *Briefings in Bioinformatics*, 1467-5463. *25*(3). doi:10.1093/bib/bbae220.

Lin, Z., Akin, H., Rao, R., Hie, B., Zhu, Z., Lu, W., et al. (2023). Evolutionary-scale prediction of atomic-level protein structure with a language model. *Science*, 1095-9203. *379*(6637), 1123–1130. Available from: https://www.science.org/doi/10.1126/science.ade2574. doi:10.1126/science.ade2574.

Linterman, M. A., & Hill, D. L. (2016). Can follicular helper T cells be targeted to improve vaccine efficacy? *F1000Research*, 1759-796X. *5*. Available from: http://f1000research.com/articles?tab=ALL&articleType=&subjectArea=&subtopic=&show=50. doi:10.12688/f1000research.7388.1.

Liu, M., Wang, S., Luo, Z., Wang, G., Zhao, Y., & Gao, X. (2025). Multistage attention-based extraction and fusion of protein sequence and structural features for protein function prediction. *Bioinformatics*, 1367-4803. *41*(7). doi:10.1093/bioinformatics/btaf374.

Liu, Q., Lin, J., Wen, L., Wang, S., Zhou, P., Mei, L., & Shang, S. (2022). Systematic modeling, prediction, and comparison of domain–peptide affinities: Does it work effectively with the peptide QSAR methodology? *Frontiers in Genetics*, 1664-8021. *12*. doi:10.3389/fgene.2021.800857.

Liu, Z., Wang, G., Wang, J., Zheng, J., & Li, S. Z. (2023). Co-modeling the sequential and graphical routes for peptide representation learning. *arXiv*, arXiv. 23318422. Available from: https://arxiv.org. doi:10.48550/arXiv.2310.02964.

Maciel, M., Amara, R. R., Bar, K. J., Crotty, S., Deeks, S. G., Duplessis, C., et al. (2024). Exploring synergies between B- and T-cell vaccine approaches to optimize immune responses against HIV—workshop report. *Npj Vaccines*, 2059-0105. *9*(1). Available from: https://www.nature.com/npjvaccines/. doi:10.1038/s41541-024-00818-y.

Maiti, S., Hassan, A., & Mitra, P. (2020). Boosting phosphorylation site prediction with sequence feature-based machine learning. Proteins, *88*(2), 284–291. doi:10.1002/prot.25801

Mamoudou, H., & Mune, M. A. M. (2025). AI-driven bioactive peptide discovery of next-generation metabolic biotherapeutics. *Applied Food Research*, 2772-5022. *5*(2), 101291. doi:10.1016/j.afres.2025.101291.

Martins, P., Mariano, D., Carvalho, F. C., Bastos, L. L., Moraes, L., Paixão, V., & Cardoso de Melo-minardi, R. (2023). Propedia v2.3: a novel representation approach for the peptide-protein interaction database using graph-based structural signatures. *Frontiers in Bioinformatics*, 2673-7647. *3*. Available from: www.frontiersin.org/journals/bioinformatics. doi:10.3389/fbinf.2023.1103103.

Mirdita, M., Schütze, K., Moriwaki, Y., Heo, L., Ovchinnikov, S., & Steinegger, M. (2022). ColabFold: Making protein folding accessible to all. *Nature Methods*, 1548-7091. *19*(6), 679–682. doi:10.1038/s41592-022-01488-1.

Mitra, P., & Pal, D. (2010). New measures for estimating surface complementarity and packing at protein-protein interfaces. FEBS Lett, *584*(6), 1163–8. doi:10.1016/j.febslet.2010.02.021.

Mitra, P., & Pal, D. (2011a). PRUNE and PROBE two modular web services for protein-protein docking. Nucleic Acids Res, *39*(Web Server issue), W229–34. doi:10.1093/nar/gkr317.

Mitra, P., Shultis, D., & Zhang, Y. (2013). EvoDesign: De novo protein design based on structural and evolutionary profiles. Nucleic Acids Res, *41*(Web Server issue), W273–80. doi:10.1093/nar/gkt384.

Mitra, P., & Pal, D. (2011b). Using correlated parameters for improved ranking of protein-protein docking decoys. J Comput Chem, *32*(5), 787–96. doi:10.1002/jcc.21657.

Nakase, I., Tanaka, G., & Futaki, S. (2013). Cell-penetrating peptides (CPPs) as a vector for the delivery of siRNAs into cells. *Molecular BioSystems*, 1742-2051. *9*(5), 855–861. doi:10.1039/c2mb25467k.

Nandi, S., Banerjee, S., Manna, D., Awasthi, A., Biswas, A., Das, A., Mukherjee, B., Mitra, P., & Mandal, M. (2026). Targeting of HSP27 and MMP-2/9 Crosstalk by High-Throughput Drug Repurposing Strategies Identifies Paroxetine as a Potential Candidate in Glioblastoma. J Med Chem, *69*(4), 4659–4676. doi:10.1021/acs.jmedchem.5c03278.

Nandi, S., Bhaduri, S., Das, D., Ghosh, P., Mandal, M., & Mitra, P. (2024). Deciphering the lexicon of protein targets: a review on multifaceted drug discovery in the era of artificial intelligence. *Molecular Pharmaceutics*, 1543-8384. *21*(4), 1563–1590. doi:10.1021/acs.molpharmaceut.3c01161.

Nielsen, J. C., Hjorringgaard, C., Nygaard, M. M., Wester, A., Elster, L., Porsgaard, T., et al. (2024). Machine-learning-guided peptide drug discovery: development of GLP-1 receptor agonists with improved drug properties. *Journal of Medicinal Chemistry*, 1520-4804. *67*(14), 11814–11826. Available from: http://pubs.acs.org/jmc. doi:10.1021/acs.jmedchem.4c00417.

Nissan, N., Allen, M. C., Sabatino, D., & Biggar, K. K. (2024). Future perspective: Harnessing the power of artificial intelligence in the generation of new peptide drugs. *Biomolecules*, 2218-273X. *14*(10). Available from: http://www.mdpi.com/journal/biomolecules. doi:10.3390/biom14101303.

Pal, A., Pal, D., & Mitra, P. (2021). A computational framework for modeling functional protein-protein interactions. Proteins, *89*(10), 1353–1364. doi:10.1002/prot.26156.

Pal, S., & 't Hart, P. (2022). RNA-binding macrocyclic peptides. *Frontiers in Molecular Biosciences*, 2296-889X. *9*. doi:10.3389/fmolb.2022.883060.

Pirtskhalava, M., Amstrong, A. A., Grigolava, M., Chubinidze, M., Alimbarashvili, E., Vishnepolsky, B., et al. (2021). DBAASP v3: Database of antimicrobial/cytotoxic activity and structure of peptides as a resource for development of new therapeutics. *Nucleic Acids Research*, 1362-4962. *49*(1), D288. Available from: https://academic.oup.com/nar/issue. doi:10.1093/nar/gkaa991.

Quiroz, C., Saavedra, Y. B., Armijo-Galdames, B., Amado-Hinojosa, J., Olivera-Nappa, Á., Sanchez-Daza, A., & Medina-Ortiz, D. (2021). Peptipedia: A user-friendly web application and a comprehensive database for peptide research supported by machine learning approach. *Database*, 1758-0463. Available from: http://database.oxfordjournals.org/. doi:10.1093/database/baab055.

Renaud, S., & Mansbach, R. A. (2023). Latent spaces for antimicrobial peptide design. *Digital Discovery*, 2635-098X. *2*(2), 441–458. Available from: www.rsc.org/journals-books-databases/about-journals/digital-discovery/. doi:10.1039/d2dd00091a.

Repecka, D., Jauniskis, V., Karpus, L., Rembeza, E., Rokaitis, I., Zrimec, J., et al. (2021). Expanding functional protein sequence spaces using generative adversarial networks. *Nature Machine Intelligence*, 2522-5839. *3*(4), 324–333. Available from: https://www.nature.com/natmachintell/. doi:10.1038/s42256-021-00310-5.

Rettie, S. A., Campbell, K. V., Bera, A. K., Kang, A., Kozlov, S., Bueso, Y. F., et al. (2025). Cyclic peptide structure prediction and design using AlphaFold2. *Nature Communications*, 2041-1723. *16*(1). Available from: https://www.nature.com/ncomms/. doi:10.1038/s41467-025-59940-7.

Rodríguez Plaza, J. G., Villalón Rojas, A., Herrera, S., Garza-Ramos, G., Larios, A. T., Amero, C., et al. (2012). Moonlighting peptides with emerging function. *PLoS ONE*, 1932-6203. *7*(7). Available from: http://www.plosone.org/article/fetchObjectAttachment.action?uri=info%3Adoi%2F10.1371%2Fjournal.pone.0040125&representation=PDF. doi:10.1371/journal.pone.0040125.

Shah, A., Guntuboina, C., & Farimani, A. B. (2024). Peptide-GPT: Generative design of peptides using generative pre-trained transformers and bioinformatic supervision. *arXiv*, arXiv. 23318422. Available from: https://arxiv.org. doi:10.48550/arXiv.2410.19222.

Simeoni, F., Morris, M. C., Heitz, F., & Divita, G. (2003). Insight into the mechanism of the peptide-based gene delivery system MPG: Implications for delivery of siRNA into mammalian cells. *Nucleic Acids Research*, *31*(11), 2717–2724. doi:10.1093/nar/gkg385.

Singh, S., Chaudhary, K., Dhanda, S. K., Bhalla, S., Usmani, S. S., Gautam, A., et al. (2015). SATPdb: a database of structurally annotated therapeutic peptides. *Nucleic Acids Research*, 1362-4962. *44*(1), D1119. Available from: http://nar.oxfordjournals.org/. doi:10.1093/nar/gkv1114.

Sutcliffe, R., Doherty, C. P. A., Morgan, H. P., Dunne, N. J., & McCarthy, H. O. (2025). Strategies for the design of biomimetic cell-penetrating peptides using AI-driven in silico tools for drug delivery. *Biomaterials Advances*, 2772-9508. *169*. Available from: https://www.sciencedirect.com/science/journal/27729508. doi:10.1016/j.bioadv.2024.214153.

Szymczak, P., Możejko, M., Grzegorzek, T., Jurczak, R., Bauer, M., Neubauer, D., et al. (2023). Discovering highly potent antimicrobial peptides with deep generative model HydrAMP. *Nature Communications*, 2041-1723. *14*(1). doi:10.1038/s41467-023-36994-z.

Tripathi, A. K., & Vishwanatha, J. K. (2022). Role of anti-cancer peptides as immunomodulatory agents: Potential and design strategy. *Pharmaceutics*, 1999-4923. *14*(12). Available from: http://www.mdpi.com/journal/pharmaceutics. doi:10.3390/pharmaceutics14122686.

Truong, T. F., & Bepler, T. (2025). Understanding protein function with a multimodal retrieval-augmented foundation model. *arXiv*, arXiv. 23318422. Available from: https://arxiv.org. doi:10.48550/arXiv.2508.04724.

Ullah, F., Xiao, A., Ullah, S., Yang, N., Lei, M., Chen, L., & Wang, S. (2025). Synergizing attribute-guided latent space exploration (AGLSE) with classical molecular simulations to design potent pep-magnet peptide inhibitors to abrogate SARS-CoV-2 host cell entry. *Viruses*, 1999-4915. *17*(6), 828. doi:10.3390/v17060828.

Van Baelen, A. C., Robin, P., Kessler, P., Maïga, A., Gilles, N., & Servent, D. (2022). Structural and functional diversity of animal toxins interacting with GPCRs. *Frontiers in Molecular Biosciences*, 2296-889X. *9*. Available from: http://journal.frontiersin.org/journal/molecular-biosciences. doi:10.3389/fmolb.2022.811365.

Van Oort, C. M., Ferrell, J. B., Remington, J. M., Wshah, S., & Li, J. (2021). AMPGAN v2: Machine learning-guided design of antimicrobial peptides. *Journal of Chemical Information and Modeling*, 1549-960X. *61*(5), 2198–2207. Available from: http://pubs.acs.org/journal/jcisd8. doi:10.1021/acs.jcim.0c01441.

Wan, F., Kontogiorgos-Heintz, D., & de la Fuente Nunez, C. (2022). Deep generative models for peptide design. *Digital Discovery*, 2635-098X. *1*(3), 195–208. Available from: www.rsc.org/journals-books-databases/about-journals/digital-discovery/. doi:10.1039/d1dd00024a.

Wang, L., Wang, N., Zhang, W., Cheng, X., Yan, Z., Shao, G., et al. (2022). Therapeutic peptides: Current applications and future directions. *Signal Transduction and Targeted Therapy*, 2059-3635. *7*(1). doi:10.1038/s41392-022-00904-4.

Wang, X., Yin, X., Jiang, D., Zhao, H., Wu, Z., Zhang, O., et al. (2024). Multi-modal deep learning enables efficient and accurate annotation of enzymatic active sites. *Nature Communications*, 2041-1723. *15*(1). doi:10.1038/s41467-024-51511-6.

Wang, Y., Liu, X., Huang, F., Xiong, Z., & Zhang, W. (2024). A multi-modal contrastive diffusion model for therapeutic peptide generation. *Proceedings of the AAAI Conference on Artificial Intelligence*, 2374-3468. *38*(1), 3–11. doi:10.1609/aaai.v38i1.27749.

Watson, J. L., Juergens, D., Bennett, N. R., Trippe, B. L., Yim, J., Eisenach, H. E., et al. (2023). De novo design of protein structure and function with RFdiffusion. *Nature*, 1476-4687. *620*(7976), 1089–1100. Available from: https://www.nature.com/nature/. doi:10.1038/s41586-023-06415-8.

Wen, Z., He, J., Tao, H., Huang, S.-Y., & Valencia, A. (2019). PepBDB: a comprehensive structural database of biological peptide–protein interactions. *Bioinformatics*, 1367-4803. *35*(1), 175–177. doi:10.1093/bioinformatics/bty579.

Xiao, W., Jiang, W., Chen, Z., Huang, Y., Mao, J., Zheng, W., et al. (2025). Advance in peptide-based drug development: Delivery platforms, therapeutics and vaccines. *Signal Transduction and Targeted Therapy*, 2059-3635. *10*(1). doi:10.1038/s41392-024-02107-5.

Xue, Y., Ye, X., Wei, L., Zhang, X., Sakurai, T., & Wei, L. (2022). Better performance with transformer: cPPFormer in the precise prediction of cell-penetrating peptides. *Current Medicinal Chemistry*, 1875-533X. *29*(5), 881–889. Available from: http://www.eurekaselect.com/article/118042. doi:10.2174/0929867328666210920103140.

Yan, K., Lv, H., Guo, Y., Peng, W., Liu, B., & Martelli, P. L. (2023). sAMPpred-GAT: Prediction of antimicrobial peptide by graph attention network and predicted peptide structure. *Bioinformatics*, 1367-4811. *39*(1). doi:10.1093/bioinformatics/btac715.

Yang, J., Roy, A., & Zhang, Y. (2013). BioLiP: A semi-manually curated database for biologically relevant ligand-protein interactions. *Nucleic Acids Research*, 1362-4962. *41*(1), D1096. doi:10.1093/nar/gks966.

Ye, G., Wu, H., Huang, J., Wang, W., Ge, K., Li, G., et al. (2020). LAMP2: A major update of the database linking antimicrobial peptides. *Database*, 1758-0463. *2020*. doi:10.1093/database/baaa061.

Yin, K., Li, R., Zhang, S., Sun, Y., Huang, L., Jiang, M., et al. (2025). Deep learning combined with quantitative structure–activity relationship accelerates de novo design of antifungal peptides. *Advanced Science*, 2198-3844. *12*(13). doi:10.1002/advs.202412488.

Yu, Q., Zhang, Z., Liu, G., Li, W., & Tang, Y. (2024). ToxGIN: An in silico prediction model for peptide toxicity via graph isomorphism networks integrating peptide sequence and structure information. *Briefings in Bioinformatics*, 1467-5463. *25*(6). doi:10.1093/bib/bbae583.

Zhai, S., Liu, T., Lin, S., Li, D., Liu, H., Yao, X., & Hou, T. (2025). Artificial intelligence in peptide-based drug design. *Drug Discovery Today*, 1359-6446. *30*(2), 104300. doi:10.1016/j.drudis.2025.104300.

Zhang, F., Li, J., Wen, Z., & Fang, C. (2024). FusPB-ESM2: fusion model of ProtBERT and ESM-2 for cell-penetrating peptide prediction. *Computational Biology and Chemistry*, 1476-9271. *111*, 108098. doi:10.1016/j.compbiolchem.2024.108098.

Zhang, H., Zhang, L., Lin, A., Congcong, X., Li, Z., Liu, K., … Huang, L. (2023). Algorithm for optimized mRNA design improves stability and immunogenicity. *Nature*, 0028-0836. *621*(7978), 396–403. doi:10.1038/s41586-023-06127-z.

Zhang, R., Wu, H., Liu, C., Li, H., Wu, Y., Li, K., et al. (n.d.) *PepHarmony: A multi-view contrastive learning framework for integrated sequence and structure-based peptide encoding*. Available from: http://www.healthinformaticslab.org/supp/.

Zhang, W. Y., Zheng, X. L., Coghi, P. S., Chen, J. H., Dong, B. J., & Fan, X. X. (2024). Revolutionizing adjuvant development: harnessing AI for next-generation cancer vaccines. *Frontiers in Immunology*, 1664-3224. *15*. Available from: https://www.frontiersin.org/journals/immunology#. doi:10.3389/fimmu.2024.1438030.

Zhao, W., Hou, K., Shen, Y., Hu, X., & Wren, J. (2025). A conditional denoising VAE-based framework for antimicrobial peptides generation with preserving desirable properties. *Bioinformatics*, 1367-4811. *41*(2). doi:10.1093/bioinformatics/btaf069.

Zhao, Y., Zhang, S., & Liang, Y. (2024). HemoFuse: multi-feature fusion based on multi-head cross-attention for identification of hemolytic peptides. *Scientific Reports*, 2045-2322. *14*(1). doi:10.1038/s41598-024-74326-3.

Zhiguo, Y., Zixuan, L., & Peng, L. (2025). MultiRepPI: a cross-modal feature fusion-based multiple characterization framework for plant peptide-protein interaction prediction. *BMC Plant Biology*. 1471-2229. *25*(1). doi:10.1186/s12870-025-06878-z.

Zhu, L., Fang, Y., Liu, S., Shen, H. B., De Neve, W., & Pan, X. (2025). ToxDL 2.0: Protein toxicity prediction using a pretrained language model and graph neural networks. *Computational and Structural Biotechnology Journal*, 2001-0370. *27*, 1538–1549. Available from: https://www.sciencedirect.com/science/journal/20010370. doi:10.1016/j.csbj.2025.04.002.

Chapter 27

The role of generative artificial intelligence in personalized drug development and safety

Humaira Ilyas[a,*] and Sk. Abdul Mohid[b,*]
[a]*Department of Biochemistry, Albert Einstein College of Medicine, Bronx, NY, United States*
[b]*Department of Biochemistry, School of Medicine, Case Western Reserve University, Cleveland, OH, United States*

1 Introduction

Over the past decade, healthcare has undergone a profound transformation, primarily characterized by the convergence of artificial intelligence (AI), biomedical research, and precision medicine (Bajwa, Munir, Nori, & Williams, 2021). Among the most promising advances in this space is the application of generative AI, a class of machine learning models capable of creating novel data that mimics complex biological systems (Serrano et al., 2024). These models are not merely analytical tools; they serve as creative engines capable of transforming the discovery, development, and personalization of drugs for individual patients. With healthcare systems evolving to focus more on personalized treatment approaches, generative AI emerges as a paradigm shift, introducing significant changes. It enables the development of safer and more effective therapies tailored to the unique biological and clinical characteristics of each individual (Beam & Kohane, 2018; Topol, 2019).

In the evolving landscape of healthcare systems, which are increasingly prioritizing patient-specific approaches to treatment, the use of generative AI offers a significant leap forward, enabling safer and more effective therapeutics to be designed that are optimized for the distinct biological and clinical features of each person (Varnosfaderani & Forouzanfar, 2024). Traditional drug development approaches follow a linear and generalized path, screening thousands of compounds, identifying hits through high-throughput methods, optimizing a handful of leads, and eventually advancing one or two candidates through costly and time-consuming clinical trials (Hughes, Rees, Kalindjian, & Philpott, 2011). Despite decades of scientific progress, this process remains fraught with inefficiency. The average time from identifying a potential drug target to obtaining regulatory approval is long and challenging, often taking more than a decade and costing over $2 billion for each successful compound (DiMasi, Grabowski, & Hansen, 2016). Even more concerning is the high attrition rate, as the majority of drug candidates fail to pass through clinical trials due to issues related to efficacy, toxicity, or patient variability (Hay, Thomas, Craighead, Economides, & Rosenthal, 2014). These failures highlight a fundamental limitation of conventional pharmacological models; they are not designed for personalization.

Generative AI offers a promising solution to addressing these challenges (Fig. 27.1). By analyzing complex patterns within existing biological and chemical data, these models can generate new, realistic drug candidates that are more closely aligned with specific therapeutic goals. This innovative approach has the potential to make drug development not only more efficient but also more tailored to individual patients, ultimately improving health outcomes. In the context of drug discovery, this means the ability to design novel small molecules, biologics, or nucleic acid therapies with prespecified properties, such as high target-binding affinity, minimal off-target effects, favorable pharmacokinetics, and low toxicity (Sanchez-Lengeling & Aspuru-Guzik, 2018). More importantly, these models can integrate data from heterogeneous sources, including genomic, proteomic, metabolomic, and clinical datasets, to guide compound design in a patient-specific manner (Zitnik et al., 2019). This integration brings the concept of personalized drug development within reach.

* Contributed equally.

Applications of Artificial Intelligence in Healthcare. https://doi.org/10.1016/B978-0-443-43934-6.00027-4

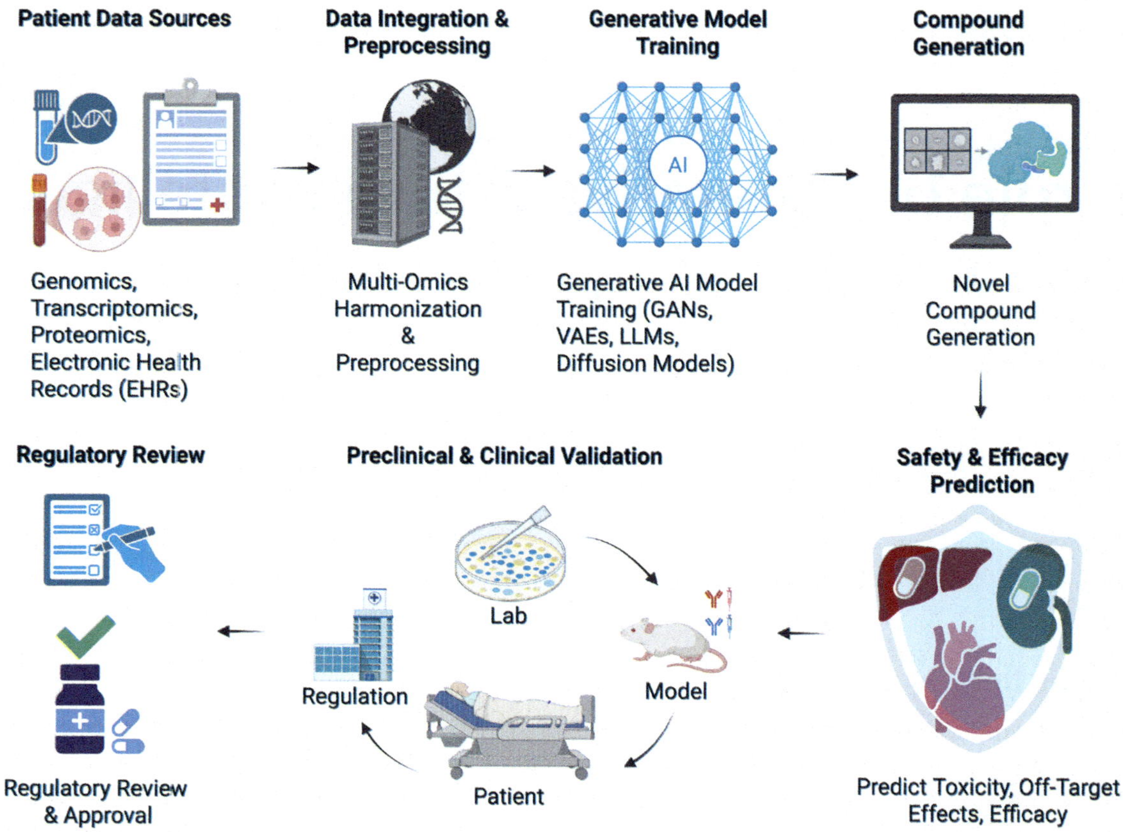

FIG. 27.1 Overview of generative modeling in the drug development pipeline. A comprehensive flowchart illustrating the drug development pipeline powered by generative AI. Patient-specific data, including genomics, transcriptomics, proteomics, and clinical records, are integrated and analyzed using advanced AI models. These models produce new drug candidates, evaluate their potential safety and efficacy, and support preclinical and clinical validation. This process concludes with regulatory review, exemplifying the smooth integration of AI technologies from data collection to therapeutic approval.

Recent advances in generative modeling, including variational autoencoders (VAEs), generative adversarial networks (GANs), transformer-based large language models, and diffusion models, have demonstrated remarkable success in generating novel chemical structures, protein sequences, and even full biosynthetic pathways (Goodfellow et al., 2014; Rives et al., 2021). In the context of personalized medicine, many of these models have the potential to design compounds tailored to individual patient profiles, predict adverse drug reactions (ADRs) associated with specific genetic variants, and simulate the outcomes of multiple drug interactions in complex clinical scenarios (Jumper et al., 2021; Zhavoronkov et al., 2019). This approach supports a more precise and responsive strategy for drug development that reflects the variability found in real-world patient populations.

This chapter aims to provide a comprehensive overview of how generative AI is transforming personalized drug development and safety assessment within the healthcare ecosystem. We begin by outlining the core generative AI technologies that have proven applicable to biomedical data and molecular design. These include not only foundational architectures but also domain-specific platforms and tools that have been developed to meet the regulatory and practical demands of healthcare applications.

Next, we examine how generative AI models are being deployed in the early stages of drug discovery (Gangwal et al., 2024). We focus on their use in de novo molecule generation, optimization of pharmacokinetic properties, and identification of new therapeutic modalities. These examples illustrate how AI is being used not merely as a productivity enhancer but as a creative collaborator capable of proposing compounds that would not be found using traditional cheminformatics or experimental methods.

Following this, the chapter explores the application of generative AI to personalization, specifically the ability to design and refine therapeutics using patient-specific data (Mishra et al., 2025). We explore how multiomics integration enables individualized drug development pipelines and how the "digital twins" model, computational avatars, can be applied to simulate drug responses in silico. We also discuss the applications of AI in rare disease research, underserved populations, and stratified patient cohorts, with a focus on how generative modeling supports equity and inclusivity in therapeutic development (Karako, 2025).

A key strength of generative AI lies in its capacity to anticipate and mitigate safety concerns before clinical testing. In the next section, we describe how generative models are used in predictive toxicology, ADR forecasting, and drug–drug interaction modeling. These use cases are critical in modern healthcare, where patients often present with multiple comorbidities, take several medications concurrently, and respond distinctively to treatment (Masumshah, Eslahchi, & Gromiha, 2023). By taking these complexities into account, generative AI helps strengthen the drug development process, providing greater attention to risk and variability.

However, incorporating these technologies into clinical practice and regulatory systems remains a significant challenge. Ethical, legal, and regulatory considerations arise at every stage, from data sourcing and patient privacy to model interpretability and the validation of AI-generated compounds. The following section explores these issues in depth, highlighting recent policy discussions, approval pathways, and the efforts of international agencies, such as the European Medicines Agency (EMA) and the US Food and Drug Administration (FDA), to establish guidance for AI-generated therapeutics (Mittelstadt, Russell, & Wachter, 2019).

The chapter concludes with a futuristic discussion on the applications of generative AI in precision healthcare, including closed-loop AI systems, integration with robotic drug synthesis platforms, adaptive clinical trial design, and real-time personalization of therapy through AI-guided dosing and scheduling. We also examine the promise of federated learning and secure frameworks that enable collaborative, privacy-preserving AI development across institutions and populations.

In summary, this chapter presents generative AI as a foundational technology for a new era in healthcare, one where drugs are not only discovered faster but also designed more intelligently and more safely for the people who need them most. As the boundaries between computation, biology, and medicine continue to dissolve, the generative capabilities of AI will become central to how we conceive of and deliver therapeutic interventions in the 21st century. The following sections provide the scientific, technical, and clinical context for understanding this transformative shift.

2 Generative AI foundations in a clinical context

Generative AI is an emerging branch of machine learning that learns from existing data to produce novel and realistic outputs (Sengar, Hasan, Kumar, & Carroll, 2025). In contrast to predictive models, which estimate relationships between variables, generative models simulate new instances that resemble the underlying data distribution. When applied to biological systems, generative models can design novel protein sequences, simulate disease phenotypes tailored to individual patients, and identify small molecules optimized for therapeutic use (Huang et al., 2021). Their advantage lies in the ability to account for the complexity, variability, and inherent noise in biological data, offering capabilities that extend beyond what traditional statistical approaches can achieve.

The growing clinical relevance of these models reflects the need to represent human biological diversity more accurately in drug development. In modern medicine, the ability to understand how a therapy performs across different genetic backgrounds, tissue types, or disease contexts is essential for achieving precision in treatment. Generative models provide a computational framework for exploring these questions using large-scale molecular and clinical datasets, thereby driving a more flexible, personalized, and biologically grounded approach to therapeutic design (Lopez, Gayoso and Yosef, 2020). The clinical relevance of generative AI stems from the increasing need to model biological processes in a manner that reflects human heterogeneity. In modern medicine, understanding how a treatment performs across diverse genetic backgrounds, tissue types, and disease conditions is essential for advancing precision therapeutics. Generative models offer a practical computational approach to investigating these variations by leveraging large-scale molecular and clinical datasets. This capability is helping to drive a transition toward more personalized, adaptable, and biologically grounded strategies in drug development.

2.1 Understanding generative models through a biological lens

Several generative AI architectures have been adapted to address questions in biology and medicine (Fahrner, Chen, Topol, & Rajpurkar, 2025). Examples include VAEs, GANs, transformer-based language models, and, more recently, diffusion models (Table 27.1; Fig. 27.2) (Gangwal et al., 2024). Each of these approaches has its distinct advantages, depending on the biological question at hand, including modeling gene expression patterns, predicting protein structures, or exploring drug–target interactions.

2.1.1 VAEs: Navigating chemical space

VAEs are especially effective for modeling biological systems due to their use of a probabilistic latent space, which captures underlying variation in the data (Liu, Rosenhahn, Illig, DeLuca, & Alber, 2024). They work by compressing complex

TABLE 27.1 Comparison of generative model types across drug design stages.

Model type	Underlying architecture	Strengths	Typical applications in drug design	Limitations	Reference (s)
Variational autoencoders (VAEs)	Encoder-Decoder with probabilistic latent space	Captures latent biological patterns, efficient sampling of diverse compounds	De novo small molecule generation, gene expression simulation	May generate implausible structures; lower output diversity	(Lim et al., 2018; Liu et al., 2024)
Generative adversarial networks (GANs)	Generator-discriminator framework	High realism in output, good for data augmentation and rare class simulation	Synthetic patient data, drug-response phenotypes, molecular mimicry	Unstable training; limited interpretability	(Gupta et al., 2018; Serrano, 2023)
Transformer-based models	Attention-based sequence modeling	Captures long-range dependencies; suitable for biologics and sequence generation	Protein design, antibody modeling, CRISPR guide RNA generation	Large data requirements; black-box predictions	(Madan et al., 2024; Madani et al., 2023)
Diffusion models	Noise–denoise probabilistic process	High fidelity in structural generation; useful in 3D spatial tasks	Protein–ligand structure prediction, molecular conformer generation	Computationally expensive; newer, with limited benchmarks	(Hu et al., 2025; Chen et al., 2024)
Reinforcement learning + generative models	Reward-guided generative search	Multiobjective optimization, goal-directed compound evolution	Lead optimization, ADMET improvement, phenotype-based selection	Requires robust scoring functions; may overfit objectives	(Korshunova et al., 2022)

inputs, such as gene expression levels, metabolite profiles, or cell surface markers, into a lower-dimensional representation that preserves key biological features. This latent space can then be used to generate new data that reflect realistic biological patterns, while also allowing for systematic exploration of perturbations.

For instance, VAEs have been applied to model immune cell differentiation by learning patterns from single-cell RNA sequencing (scRNA-seq) data. When trained on these complex, high-dimensional datasets, the model can generate gene expression profiles that capture transitional or rare cell states, which are often not well-sampled in experimental samples. This ability is particularly useful for investigating how cells respond to treatment or for predicting unintended off-target effects in specific tissues (Lopez, Regier, Cole, Jordan, & Yosef, 2018). In drug discovery, VAEs have been utilized to generate libraries of small molecules with specific physicochemical properties and potential binding affinities. Using molecular targets or biological pathways to guide the generation process, these models have been employed to produce structurally diverse compounds with desired biological relevance (Lim, Ryu, Kim, & Kim, 2018).

2.1.2 GANs: Simulating physiological variation

GANs employ a competitive framework comprising two neural networks: a generator that generates synthetic data and a discriminator that assesses its realism (Serrano, 2023). In biological research, GANs have shown potential in modeling both normal and disease-related variation. For instance, they have been used to create synthetic histological images that maintain the overall tissue structure while varying specific pathological features, such as fibrosis or necrosis. Such images can be beneficial for training diagnostic models, particularly in situations where access to diverse or well-balanced clinical samples is limited (Frid-Adar, Klang, Amitai, Goldberger, & Greenspan, 2018).

In molecular biology, GANs have also been utilized to generate novel peptide sequences with antimicrobial or anticancer properties by training on existing functional peptides. The resulting sequences can be tested in vitro, and early studies show that GAN-generated peptides maintain structural motifs essential for biological function (Gupta et al., 2018). GANs have also been explored for modeling patient-specific variability in pharmacokinetics. By training on time-series data capturing drug absorption and clearance, these models can generate a range of drug concentration profiles that reflect realistic physiological differences. This approach can help inform personalized dosing strategies and the design of virtual clinical trials, particularly valuable in populations such as children, older adults, or individuals with rare diseases, where empirical data are often sparse.

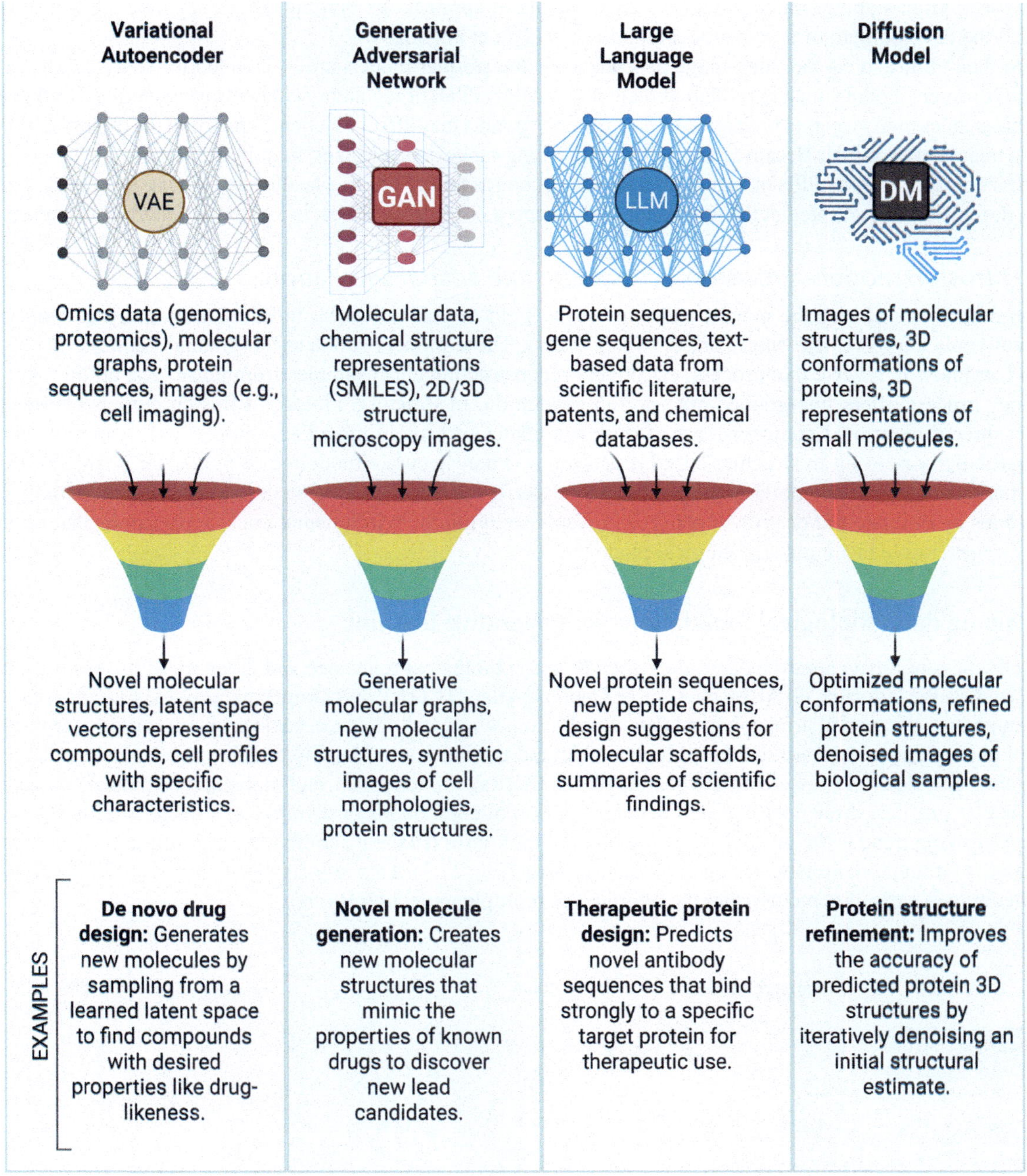

FIG. 27.2 Generative AI model architectures in drug discovery. A visual schematic compares four key foundational generative AI models used in drug discovery. Each model type is depicted with its core structure, typical biological or chemical data used as input, and the form of the generated output. Representative applications demonstrate how each approach contributes to designing and optimizing therapeutic candidates, emphasizing their complementary strengths in biomedical applications.

2.1.3 Transformer-based language models: Learning biological syntax

Initially developed for natural language processing, transformer models are being increasingly adapted for the analysis of biological sequences (Madan, Lentzen, Brandt, Rueckert, & Hofmann-Apitius, 2024). Conceptually, DNA, RNA, and protein sequences resemble a specialized language code, where the grammatical arrangement of elements follows patterns shaped by evolution, function, and regulation. This perspective allows transformer models to uncover the underlying rules, or "syntax", governing biological information. Transformer models, such as ESM (evolutionary scale modeling) and ProGen, have been trained on millions of protein sequences to learn the rules governing structure and function. These models can generate novel protein sequences with conserved motifs or desired biophysical properties, such as improved solubility or thermal stability. Such sequences are valuable in applications ranging from enzyme engineering to vaccine design and the development of therapeutic antibodies (Madani et al., 2023).

Due to their ability to capture long-range dependencies, transformer models are well-suited for studying interactions that span across different regions, such as protein–protein interactions (PPIs) or regulatory elements in noncoding DNA. In a clinical setting, this enables more accurate prediction of how genetic variants may affect function, supporting the design of biologics and informing treatment decisions. Because transformers retain long-range dependencies, they are instrumental in modeling interactions across domains, such as PPIs or regulatory elements in noncoding DNA. In a healthcare context, this means an improved ability to design biologics or predict the functional consequences of genetic variants in a clinically actionable manner.

2.1.4 *Diffusion models: Precision in structure and spatial simulation*

Diffusion models are among the newest additions to the field of generative AI, but they are rapidly gaining traction in biomedical applications (Chen, Mei, Fan, & Wang, 2024). These models simulate the gradual addition of noise to data and then learn how to reverse that process, effectively reconstructing clean samples from noise. The ability to control this "denoising" process offers fine-grained precision in generation. In structural biology, diffusion models are being applied to predict three-dimensional conformations of drug candidates or protein-ligand complexes with atomic-level accuracy. This is particularly relevant in structure-based drug design, where accurate modeling of molecular interactions underpins rational therapeutic development (Hu et al., 2025). Moreover, these models are being used to synthesize spatially organized tissue images or simulate tumor growth patterns in response to microenvironmental cues, applications that are critical in oncology, regenerative medicine, and surgical planning.

2.2 Training data: Biological foundations for generative learning

The effectiveness of any generative AI model depends on the biological relevance and diversity of its training data. Public datasets, such as the Human Cell Atlas, the Cancer Genome Atlas (TCGA), and DrugBank, provide high-quality sources of molecular, phenotypic, and pharmacological data. In addition, single-cell omics, cryoelectron microscopy (cryo-EM) structures, and proteomics atlases are increasingly available and used to train domain-specific models. In healthcare applications, data are also derived from electronic health records (EHRs), clinical trials, and biobank repositories. When combined with molecular profiles, these datasets enable the generation of compounds or therapeutic strategies tailored to individual patients. A growing trend is the integration of multimodal biological data, e.g., combining transcriptomics with proteomics and imaging, to train more holistic and generalizable models (Fig. 27.3). This reflects the complexity of biological systems, where phenotypes typically emerge from the interplay of multiple molecular layers.

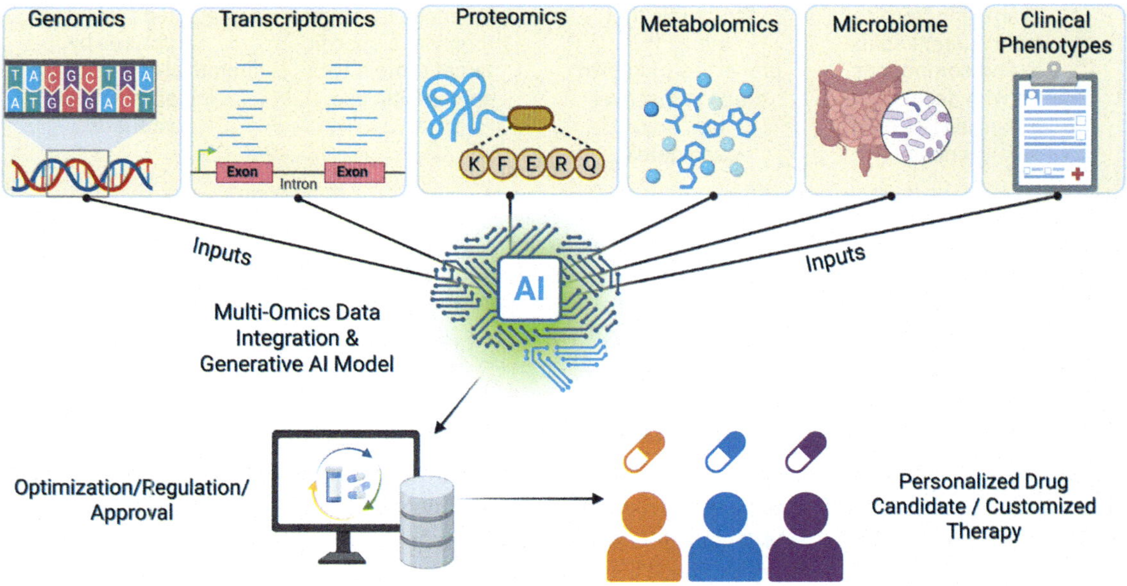

FIG. 27.3 Multiomics integration for personalized drug design. Different layers of biological data, including genomics, transcriptomics, proteomics, metabolomics, microbiome, and clinical phenotypes, come together to guide drug design. Generative models use these data to generate personalized drug candidates tailored to individual patients' profiles. This highlights the central role of multiomics harmonization in facilitating precision therapeutics.

2.3 Challenges in biological generative modeling

While generative AI shows tremendous promise, its application in biology poses distinct challenges. First, biological systems are inherently stochastic, with context-dependent behaviors and nonlinear responses. This makes it difficult to establish a clear "ground truth" for evaluating model performance. For instance, a protein generated by a model may appear structurally viable and functional but may ultimately prove to be biologically inactive due to unmodeled interactions with host systems. Second, data bias remains a persistent issue. Many training datasets overrepresent specific cell types, disease conditions, or populations, which can limit the generalizability of model outputs. In clinical applications, this raises important issues, as reliable and equitable decision-making depends on models that account for biological and demographic diversity. Third, interpretability and validation are crucial. Clinicians and regulators must be able to understand how and why a model generated a particular sequence or compound. Hybrid pipelines that combine generative outputs with functional assays and expert curation are currently the gold standard for ensuring biological plausibility and clinical relevance.

3 Generative AI-powered drug design

The traditional methods of drug design, random discovery, and logical synthesis are being drastically changed (Doytchinova, 2022). Biological systems present significant challenges to traditional drug development due to their complex feedback mechanisms, molecular diversity, and individual variability. A new method has emerged with the development of generative AI, the design of therapeutic molecules from scratch, based on patient-specific data and biological complexity (Liu et al., 2025). The ability of generative models to learn from enormous datasets of chemical structures, biological sequences, and experimental measurements, and then create novel molecular entities that fit desired therapeutic profiles, is at the heart of this revolution. With applications ranging from initial hit generation to lead optimization, from in silico screening to biologically informed personalization, this section explores how generative AI is propelling the next generation of drug design.

3.1 De novo molecule generation: Learning the language of chemistry and biology

Generative AI's capacity to produce molecular structures not previously listed in chemical libraries is among its most important contributions to drug design (Gangwal et al., 2024). When targeting new or poorly understood biological pathways, such as those involved in immunological modulation, neurodegeneration, or rare genetic disorders, this de novo design method is highly beneficial. Generative models, such as GANs, VAEs, and transformer-based architectures, can be trained on curated datasets of drug-like molecules, such as those from ChEMBL, ZINC, or PubChem, where each entry is annotated with structural, functional, and pharmacokinetic data (Gaulton et al., 2017). Once trained, these models can produce novel compounds that meet predetermined standards, such as favorable solubility, low predicted toxicity, and high binding affinity for a target protein.

The model's capacity to identify the molecular characteristics that characterize successful interactions in a biological setting is what gives this process its biological relevance. For example, in kinase inhibitor design, models can be biased toward preserving key pharmacophores, chemical groups known to engage conserved residues in the kinase ATP-binding pocket (Zhao et al., 2014). This ensures that the generated compounds not only fit the structural requirements but also reflect functional constraints imposed by protein dynamics and cellular context (Gómez-Bombarelli et al., 2018). In an iterative cycle where model-generated compounds are synthesized, tested, and then fed back into the training process, recent developments have made it possible for models to integrate feedback from biological assays into the generative loop. Over time, the biological relevance of produced compounds is greatly enhanced by this "human-in-the-loop" or active learning technique (Schneider et al., 2020).

3.2 Target-aware molecular design

Traditional virtual screening involves docking large libraries of preexisting compounds into target structures, whereas generative AI enables the design of molecules with the target already in mind (Popov, Wellnitz, Maxfield, & Tropsha, 2024). This target-based generation process facilitates an efficient and focused exploration of available chemical space. Techniques such as conditional VAEs or reinforcement learning-guided generation can condition the molecule generation process on features of the target, such as protein structure, binding pocket topology, or even known inhibitors (Khater et al., 2025). VAEs generate candidate molecules by learning patterns that link chemical structures to their corresponding biological activities. These models do not depend on molecular docking during the initial design stage. Instead, they create new compounds within a chemical space that reflects properties such as binding affinity, solubility, or stability (Gómez-Bombarelli et al., 2018). Docking can then be applied as a complementary step to evaluate the binding poses of the generated molecules within the target site (Zhavoronkov et al., 2019). In this way, VAEs enable rapid exploration of chemical diversity, while docking provides structural validation of how the molecules may interact with specific protein targets. In PPI inhibition, which has long been considered "undruggable" due to the large, flat interfaces involved, generative models can propose molecular scaffolds that mimic key interaction motifs or disrupt hotspot residues (Paul et al., 2021).

Integrating structural biology data, including those from cryo-EM and AlphaFold-predicted models, allows generative algorithms to design ligands that are not only complementary in shape but also compatible with local electrostatics and dynamics (Haloi, Howard, Lindahl, & MacKerell, 2025). Recently, this approach has been used to generate inhibitors against intrinsically disordered regions in transcription factors, a field previously considered infeasible due to the lack of a stable binding site (Jin, Barzilay, & Jaakkola, 2020).

3.3 Optimizing drug-like properties in silico

Generative AI can be used to generate new compound libraries and optimize existing leads by improving their pharmacokinetics and pharmacodynamic properties (Vora et al., 2023). This involves decreasing toxicity, improving stability and metabolic stability, and minimizing off-target interactions. Most frequently, multiobjective optimization is achieved using reinforcement learning or evolutionary algorithms in combination with generative models. In this case, a scoring function that considers several biologically significant endpoints is used to evaluate molecules. For instance, a compound for human Ether-à-go-go-Related Gene (hERG) channel inhibition might be scored based on its predicted inhibition of a disease target, lack of interaction with cytochrome P450 enzymes, and low values for predicted cardiotoxicity (Olivecrona, Blaschke, Engkvist, & Chen, 2017).

This multiobjective design models the complexity of biological systems, where systemic interactions significantly influence clinical outcomes and tightly link safety with efficacy. Generative models assist in identifying candidates that have a better chance of passing preclinical and clinical testing by navigating the chemical space under these biological constraints. A case study from Insilico Medicine demonstrated this approach with GENTRL, a generative tensor-based reinforcement learning platform. Within 46 days, the platform generated and synthesized novel DDR1 kinase inhibitors with nanomolar potency and favorable predicted pharmacokinetics. These molecules were not only structurally novel but also biologically validated in vitro, illustrating the power of AI-driven design cycles (Zhavoronkov et al., 2019).

3.4 Designing biologics and macromolecular therapeutics

Beyond small molecules, generative AI is being applied to the design of biologics, including therapeutic proteins, peptides, antibodies, and RNA-based drugs. Due to their intricate structure–function relationships, immunogenicity issues, and sequence constraints, biological therapeutics present unique challenges. After being trained on millions of natural protein sequences, transformer-based protein language models can generate new protein scaffolds that retain the secondary and tertiary structural elements essential for biological activity. Variants with improved stability, binding affinity, or decreased immunogenic potential can be created by fine-tuning these models (Madani et al., 2023).

Generative models can suggest complementary-determining region loops in antibody engineering that exhibit better target specificity while avoiding motifs known to elicit immunological reactions. Synthetic peptides that bind cytokines, alter receptor signaling, or act as delivery vehicles across biological membranes have all been designed using similar techniques.

Generative models can create sequences for RNA therapeutics, such as small interfering RNAs (siRNAs) or guide RNAs for CRISPR–Cas systems, that optimize on-target activity while minimizing off-target gene silencing. This is particularly valuable in therapeutic genome editing, where unintended edits can have serious consequences (Huang et al., 2024).

3.5 Personalization: Designing drugs for individual patients

A central edge of generative drug design is the personalization of therapeutic molecules using individual patient data. Due to the inherent variability of biological systems, drug response is influenced by a variety of factors, including disease microenvironments, genetic polymorphisms, epigenetic states, and the composition of the microbiome. Generative AI presents the potential to create medications with these characteristics taken into consideration. In one example, researchers combined patient-specific single-cell transcriptomic data with generative models to identify personalized small-molecule modulators that could reverse disease signatures. In cancer, tumor-specific mutations and neoantigens can guide the design of peptide-based vaccines or antibody–drug conjugates using generative tools (Saez-Rodriguez et al., 2016).

Generative systems can be used in conjunction with digital twin models, which are computational simulations of individual patients, to model and optimize therapeutic interventions before their implementation. These simulations consider drug metabolism, disease progression, and organ dynamics to predict ideal compound profiles (Viceconti, Henney, & Morley-Fletcher, 2016). Strict validation, safe handling of private health data, and strong multimodal data integration are necessary for such applications. They could, however, lead to the development of real "precision therapeutics" at the molecular level.

3.6 Including generative design in clinical and preclinical processes

Although generative models are excellent at producing hypotheses, biological systems require validation of their outputs. Therefore their most significant influence occurs when they are incorporated into a hybrid drug discovery pipeline that consists of functional assays, structural validation, and high-throughput screening (Das, 2025). To reduce the workload and cost of experiments, AI-generated compounds can be ranked for synthesis in preclinical workflows based on predicted on-target and off-target profiles. Generative AI facilitates adaptive trial design in clinical settings, where patient cohorts and dosage schedules are modified in real time in response to model-generated drug outcomes. Moreover, regulatory bodies are also utilizing generative AI tools. For instance, the US FDA has launched programs to validate compounds designed by AI while using real-world data to support any regulatory decisions. This highlights the increased awareness by agencies about the importance of AI in modern drug development efforts (Ning et al., 2024).

4 Integrating patient data for therapeutic personalization

Human biology is heterogeneous, meaning that patients with the same clinical diagnosis often respond differently to the same therapy (Woodward, Urbanowicz, Naj, & Moore, 2022). This complicates treatment selection and efficacy, arising from variability in genetic background, disease progression, drug metabolism, and immune response. In this context, the integration of patient-specific biological data into drug design is no longer just an aspirational goal but an essential tool for advancing precision medicine.

With the rise of high-throughput technologies and biobank-linked clinical records, large volumes of patient-specific data have become available (Beesley et al., 2020). These include genomic variants, transcriptomic profiles, proteomic signatures, metabolomic states, microbiome composition, and clinical phenotypes. When properly curated and interpreted, this data can guide the design or selection of therapeutic agents tailored to the individual. Generative models may use such data to simulate and propose treatments that align with a patient's unique molecular and physiological landscape.

This section explores how researchers integrate patient data at various stages of the therapeutic development pipeline, ranging from biomarker-driven target identification to patient-specific drug generation and safety assessment. It highlights both current practices and emerging approaches.

4.1 Genomic variants and target prioritization

Genomic variants drive much of the patient-to-patient variability in drug response. Single-nucleotide polymorphisms, copy number variations, and somatic mutations alter both disease risk and therapeutic sensitivity. For example, patients with Epidermal Growth Factor Receptor (EGFR) mutations in nonsmall cell lung cancer often respond well to EGFR inhibitors, whereas those with KRAS mutations typically do not (Mok et al., 2009). Incorporating individual or cohort-level genomic data allows researchers to prioritize drug targets that are both disease-relevant and patient-specific. Furthermore, they can train generative models to design molecules that modulate the activity of those targets. In oncology, mutational profiles from patient tumors can be used to identify neoantigens or altered signaling pathways, enabling algorithms to generate therapeutic peptides, small molecules, or antibody fragments that selectively bind mutant proteins while sparing normal tissues (Ott et al., 2017). Databases such as TCGA and the International Cancer Genome Consortium couple mutational data with clinical outcomes. Integrating these resources into the training datasets for model-driven drug design ensures that therapeutic strategies align with the molecular landscape of individual patients.

4.2 Transcriptomic and epigenomic states: Capturing cellular context

Genetic and environmental factors shape gene expression patterns, providing a real-time snapshot of cellular state. These patterns differ significantly across cell types, disease stages, and treatment conditions (Virolainen, VonHandorf, Viel, Weirauch, & Kottyan, 2023). For instance, the same therapeutic may induce apoptosis in one cellular context but promote resistance or compensatory signaling in another. scRNA-seq has revealed the extent of transcriptional heterogeneity within tissues (Desta & Birhanu, 2025). In tumors, immune microenvironments, or inflamed organs, distinct cell subsets can mediate therapy resistance or off-target effects. By integrating transcriptomic data into generative frameworks, researchers can design drugs that modulate gene expression in a context-specific manner. Models trained on these datasets can simulate perturbations and predict how different drug structures may alter gene expression patterns in specific cell types. For instance, researchers can rank a candidate compound not only by its ability to inhibit a kinase but also by its predicted effect on inflammatory gene expression in macrophages (Ding et al., 2020). Epigenomic features, such as DNA methylation and histone modifications, also regulate gene accessibility and drug response. In certain leukemias, epigenetic silencing of DNA repair genes increases sensitivity to PARP inhibitors. Generative design pipelines that incorporate chromatin accessibility or methylation profiles can identify drugs that exploit such vulnerabilities or reverse aberrant gene silencing (Jung, Hernández-Illán, Moreira, Balaguer, & Goel, 2020).

4.3 Proteomics and functional targeting

Genomic and transcriptomic data provide insights into potential drug targets; however, protein-level information often aligns more closely with phenotype and disease mechanisms (Du, Fan, Zhang, Wu, & Zhang, 2024). Proteomics, through mass spectrometry-based quantification and antibody-based assays, enables the detection of posttranslational modifications, protein isoforms, and complex stoichiometry. Integrating patient proteomic data allows models to identify functional targets more accurately and design compounds that are specific in action, i.e., interfere only with relevant PPI and protein–ligand interaction. For instance, in autoimmune disorders, different patients may overexpress distinct sets of cytokines. A personalized immunomodulatory therapy could selectively inhibit the overexpressed cytokine or its receptor based on the individual patient's profile (ADORINI, 2003). Similarly, phosphoproteomics can map active signaling pathways in an individual tumor. Generative platforms can use this information to propose kinase inhibitors that target the activated nodes within a patient's oncogenic network (Mertins et al., 2016).

4.4 Microbiome and metabolome: Modulators of drug efficacy

The gut microbiome modulates the pharmacokinetics and efficacy of several major drugs. Certain bacterial species metabolize drug molecules before absorption, while others alter immune signaling pathways to modulate effectiveness. For example, *Bacteroides fragilis* enhances the response to CTLA-4 blockade in melanoma patients, whereas other species may reduce drug availability (Routy et al., 2018). The metabolome, reflecting both host and microbial activity, provides additional insight into drug handling. It can reveal underlying metabolic bottlenecks, detoxification capacity, and nutrient–drug competition. These factors vary greatly between individuals and can strongly influence drug response. By incorporating microbiome and metabolomic data into drug design, researchers can simulate how drugs interact with these dynamic systems. This could enable the generation of tailored compounds, active in specific metabolic states, or resistant to microbial degradation. For instance, patients with high levels of β-glucuronidase-producing bacteria often experience gut inactivation of certain chemotherapeutics. A model-aware design could allow the generation of analogs that are less susceptible to microbial enzymatic modification, thereby preserving their therapeutic efficacy (Wallace et al., 2010).

4.5 Clinical phenotypes and real-world data

Clinical phenotypes, such as prior treatment response, comorbidities, organ function, and patient-reported symptoms, add essential context to selecting an adequate therapy. Researchers use EHRs to access longitudinal data on medication, laboratory values, and outcomes, which help place molecular profiles in a real-world clinical framework (Jafari et al., 2024). For instance, the expected pharmacokinetic profile of a drug is significantly different in patients with chronic kidney disease compared to healthy individuals. Generative AI can be fed renal function data to help design compounds that bypass impaired pathways, utilize alternative pathways, and reduce the risk of nephrotoxicity (Sabanayagam et al., 2025). By integrating omics and EHR-derived clinical data, researchers can simulate how a candidate drug would behave under complex physiological conditions. These simulations support the generation of compounds that have strong target activity and favorable performance in specific patient contexts. In addition, such an approach can be used to minimize adverse and unpredictable drug reactions by designing drugs that avoid toxic interactions known to affect only a subpopulation of patients with specific clinical characteristics. For example, for treating patients with preexisting heart disease, drugs can be designed to exclude compounds that might trigger Q- and T-wave (QT) interval prolongation (Shah, 2005).

4.6 Digital twins: Personalized simulation and virtual drug testing

The concept of "digital twins" stems from the idea of creating a computational replica of a patient using molecular, anatomical, and clinical data (Fig. 27.4) (Alsalloum, Al Sawaftah, Percival, & Husseini, 2024). These models can simulate

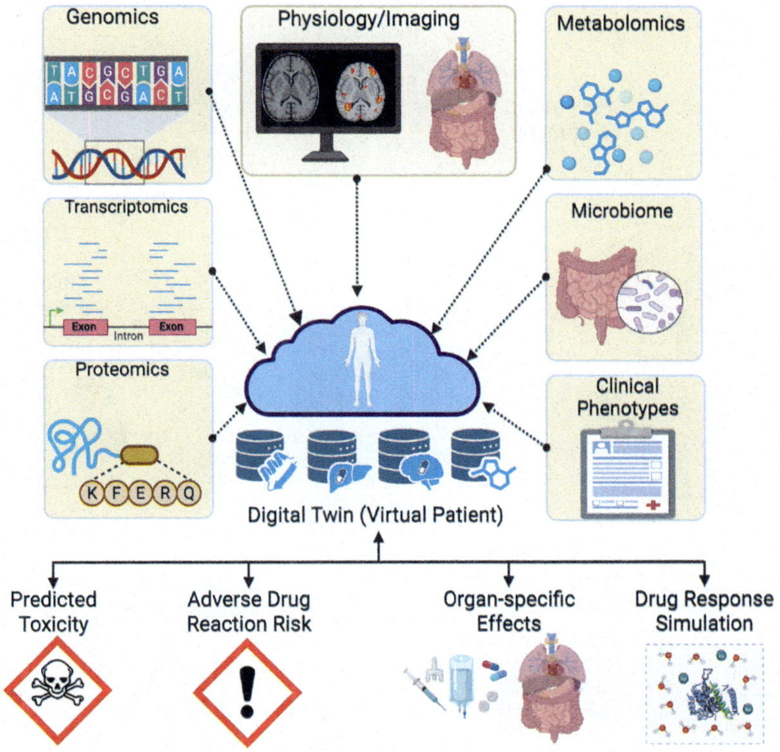

FIG. 27.4 Predictive safety modeling and digital twins in drug development: Molecular, clinical, and physiological data are integrated into digital twin simulations that produce individual biology. These virtual patient models predict drug safety, simulate pharmacological responses, and identify organ-specific risks, guiding safer, more personalized, and hence more effective therapeutic development.

disease progression, drug metabolism, and tissue-specific effects. Although still under development, researchers are now integrating digital twins into drug development pipelines to predict therapeutic responses before a compound even enters clinical trials (Samei & Brenner, 2025). In the context of generative AI, these simulations act as virtual testbeds for newly designed molecules. Algorithms generate candidate compounds, and the digital twin predicts their behavior within a specific patient context. This feedback loop enables scientists to select drugs with strong predicted efficacy and minimal risk of adverse effects.

In cardiology, for instance, researchers have incorporated digital twins that include heart structure and electrophysiological data to test antiarrhythmic drugs in silico. This approach allows the model to identify drug candidates that effectively suppress arrhythmia without triggering pro-arrhythmic effects in vulnerable patients (Niederer et al., 2020).

4.7 Ethical and practical considerations

The integration of individual biological data into drug design raises several ethical and logistical issues (Ning et al., 2024). Some important concerns that require immediate attention include patient privacy, data sharing, and algorithmic bias. In cases where the training data are not well-represented, even well-trained models may overfit the data to dominant demographic groups. Clinicians increasingly rely on real-time molecular diagnostics, which demand infrastructure capable of rapid sequencing, analysis, and regulatory compliance (Fig. 27.5). To ensure equitable and safe use of AI-generated molecules, the healthcare system must adapt and expand its capabilities (Bajwa et al., 2021). Despite these challenges, the integration of patient data into therapeutic design exhibits immense potential. Personalized drug generation reduces trial-and-error prescribing, minimizes side effects, and improves outcomes, especially in complex or refractory diseases.

5 Enhancing drug safety through predictive modeling

The safety of a drug depends on its interactions across a living system. Adverse effects often result from off-target activity, metabolic breakdown, organ-specific toxicity, or unexpected interactions (Amorim et al., 2024). Generative models now help researchers predict these risks early in development, enabling the design of safer molecules that align with human

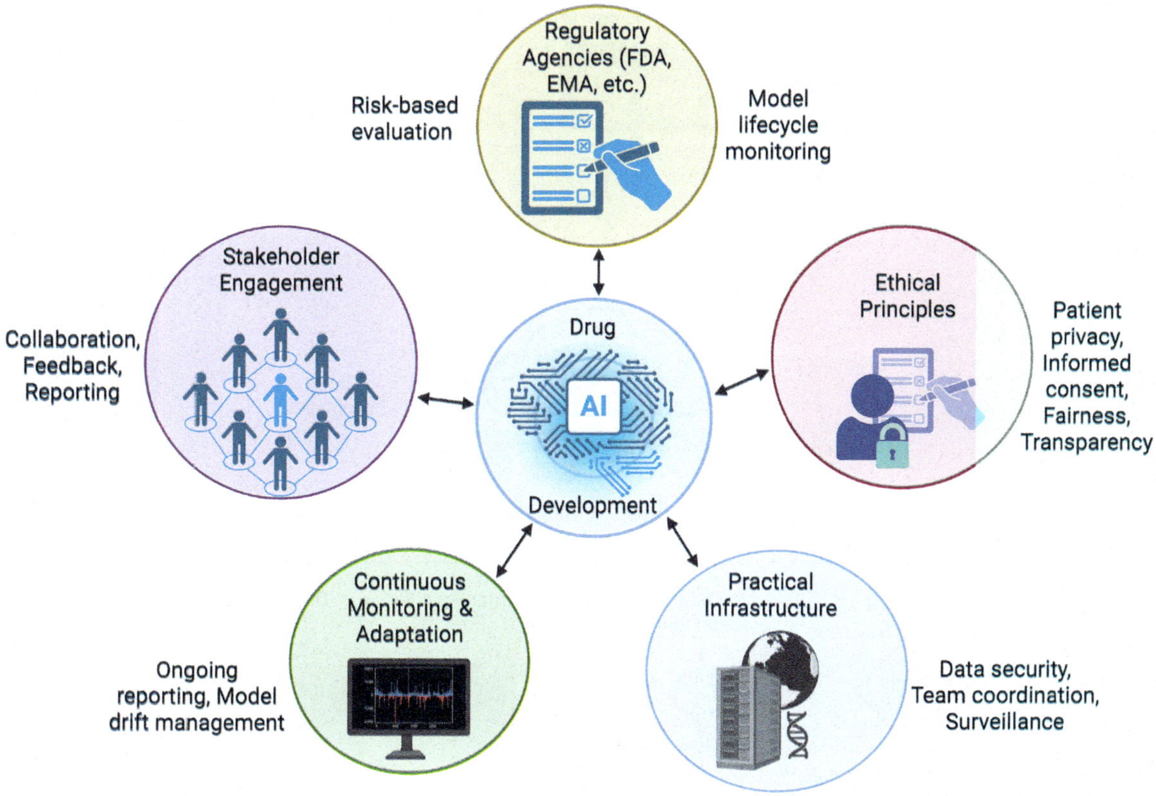

FIG. 27.5 Landscape of regulatory and ethical oversight for AI-generated therapeutics. A circular network best illustrates the interconnected domains of regulatory oversight, ethical principles, practical infrastructure, and stakeholder engagement that shape the governance of AI-driven drug development. Dynamic feedback loops highlight the importance of continuous monitoring, transparency, and collaboration to ensure the safe, equitable, and effective deployment of generative AI technologies.

biology (Table 27.2). ADRs often depend on subtle structural features. Computational models analyze molecular graphs to predict organ-specific toxicity, such as liver damage or heart rhythm disruption. Graph-based algorithms, like graph convolutional networks, combine drug structure with protein target information to reveal how off-target binding may trigger harmful effects. This method provides a stronger biological foundation for predicting adverse outcomes (Madani et al., 2023).

Toxicogenomics studies how gene and protein activity change in response to drugs. By coupling toxicity biomarkers with gene expression and proteomic responses, researchers can identify early molecular signs of toxicity. Public initiatives like Open TG-GATEs, ToxCast, and Tox21 offer comprehensive datasets tracking biological responses to numerous compounds (Meier et al., 2025). These datasets allow the model to learn the molecular pathways that drive toxicity. For example, when combined with real-world adverse event data, such as the FDA's reporting system, deep learning models can predict likely toxicity outcomes for new compounds, achieving strong accuracy in internal validation tests (Liu et al., 2015).

Many patients, especially those with multiple conditions, take several medications simultaneously. These combinations often lead to unexpected side effects due to interactions. Graph-based systems model drug networks by integrating data on drug–target interactions and protein interactions. Models like Decagon construct multimodal networks that link drugs, their protein targets, and observed side effects (Zitnik et al., 2018). These networks predict not just whether two drugs interact, but also specify the type of adverse effect expected. These predictive models support physicians when prescribing drug combinations by identifying those with a lower risk of harmful interactions.

Drug safety depends on complex interactions across genomic, transcriptomic, proteomic, and metabolic layers. Multiomics databases (e.g., TCGA, Human Protein Atlas, EMBL-EBI) provide the resources to create models that accurately reflect biological context. Combining the data from these sources allows researchers to simulate how a compound affects diverse biological systems. For instance, they can predict how drug metabolism varies between individuals, how protein expression patterns shift in disease, and how genetic variants influence responses (Du et al., 2024).

TABLE 27.2 Comparison table depicting safety modeling approaches.

Safety modeling approach	Key biological focus	Application examples	Strengths	Limitations
Structure toxicity filtering	Molecular features tied to organ-specific hazards	Early detection of hERG liability and liver enzyme inhibition (Kalyaanamoorthy et al., 2020)	Fast, interpretable, low experimental cost	May miss systemic or rare adverse effects
Toxicogenomic profiling	Gene/protein expression response to compounds	Analysis using ToxCast or Tox21 gene signatures to predict toxicity mechanisms (Jeong, Kim, & Choi, 2022)	Captures pathway-level effects; biologically grounded	Requires rich datasets and careful interpretation
Polypharmacy side-effect modeling	Drug combinations and interaction networks	Knowledge graph models Decagon for drug-drug adverse effect prediction (Zitnik et al., 2018)	Models complex interactions; good side-effect specificity	May generate false positives; data-intensive
In silico clinical trial	Patient anatomy and pathophysiology simulation	Digital twins or virtual patients to test drug safety in a personalized manner (Alsalloum et al., 2024)	Reduces animal use; reveals rare or long-term effects	High computational demands; emerging validation frameworks
Multiomics integration	Multilayered biological responses (omics + clinical)	Toxicity simulations informed by genomics, transcriptomics, and metabolic traits (Singh et al., 2023)	Captures individual variability; improves system-level predictions	Complex modeling; resource-intensive, needs explainability

Many safety tests rely on cell cultures or animal models. However, cells in culture lack the complexity of organs. Researchers address this through biosimulation, which mathematically models how a drug will interact in whole systems. By integrating in vitro results with knowledge of drug processing and distribution, models can predict in vivo outcomes more accurately (Han, Xuehua, Yang, Zhang, & Chen, 2024). This approach reduces reliance on animal testing and provides early warnings about human-specific toxicity. Generative models aid in designing molecules with reduced toxic potential. By learning from nontoxic compounds, the models can propose new structures with reduced safety concerns while retaining therapeutic effects (Seal et al., 2025). In parallel, in silico clinical trials with virtual patient cohorts allow researchers to simulate drug responses across diverse genetic and physiological contexts (Samei & Brenner, 2025). These trials help detect early safety signals and tailor compounds to specific patient groups (Moingeon, Chenel, Rousseau, & Voisin, 2023).

Even after approval, drug safety requires surveillance. Generative tools now analyze real-world data, such as EHRs, social media, and spontaneous reports, in real time. These systems identify emerging safety signals faster than traditional review processes. They also automate case summarization and facilitate report generation, improving accuracy and compliance in pharmacovigilance workflows. Regulatory agencies recognize the value of predictive safety models. The FDA encourages the use of human-relevant data and computational models in new drug applications. These "new approach methodologies" include digital simulations that reduce animal testing and support early toxicity assessment (Sewell et al., 2024). By integrating safety modeling into regulatory review and development pipelines, the pharmaceutical industry can make smarter decisions earlier. This integration supports safer and more efficient drug development, aligned with patient physiology.

Drug safety emerges from the dynamic complexity of biological systems. Predictive modeling, grounded in real molecular and clinical data, enables researchers to anticipate adverse effects before molecules reach patients. Models that simulate toxicity, polypharmacy risks, in vivo behavior, and real-world outcomes vastly improve safety evaluation. The increasing embrace of these methods by regulatory agencies promises to make next-generation therapeutics safer and more reliable in clinical use.

6 Regulatory, ethical, and practical considerations

Innovations in generative AI bring powerful opportunities but also demand careful governance. As these models propose novel therapeutics shaped by biological systems, regulators, ethicists, and healthcare stakeholders must safeguard safety, transparency, and equity (Fig. 27.5).

Regulatory bodies now evaluate AI-based models under a risk-based framework aligned with the concept of "context of use" (COU). The US FDA issued draft guidance in early 2025, advising sponsors to assess model credibility relative to the

COU, including supporting evidence and documentation that justify trust in the model's outputs in regulatory submissions (PharmaLex, 2025). The FDA emphasizes that sponsors must engage early with the agency and integrate credibility assessment plans throughout the model's lifecycle. Beyond drug design, AI tools used in manufacturing and pharmacovigilance require clear governance. Sponsors must employ good machine learning practices, ensure data integrity, monitor model drift, and maintain explainability in real-world performance (Nagar, Gobburu, & Chakravarty, 2025). Regulators recommend company-wide AI governance structures, including cross-functional oversight and lifecycle monitoring, to ensure compliance and public trust.

Generative drug design increasingly relies on patient-derived data (Chen & Esmaeilzadeh, 2024). These data must be handled with privacy protection, informed consent, and fair representation. Models trained on biased datasets risk amplifying disparities, especially when molecular signatures vary across populations. Leaders must ensure diverse sampling and algorithmic fairness to avoid inequitable access or unsafe outcomes. Transparency also remains critical. Stakeholders, including patients and clinicians, must understand how models influence therapeutic design and decision-making. Without clarity, generative model risks become "black boxes," undermining trust and ethical acceptability.

Deploying generative AI in healthcare requires robust infrastructure. Institutions must support high-throughput sequencing, omics analysis, secure clinical data integration, and computational capacity. They must also invest in the proper coordination and collaboration of computational biologists, pharmacologists, regulatory specialists, and data scientists. Moreover, drug development teams must adjust workflows to incorporate in silico testing, iterative validation, and close collaboration with regulatory authorities. Early planning, transparency around algorithm limits, and shared oversight will promote smoother translation from model to medicine.

7 Future directions and emerging opportunities

As generative AI matures and integrates more deeply with biological systems, the future promises several transformative directions (Fig. 27.6).

a. Closed-loop adaptive discovery: Generative models will increasingly integrate real-time experimental feedback to form closed-loop pipelines. These systems will design compounds, test them in high-throughput biological assays, and refine models based on observed results (Ocana et al., 2025). Such dynamic systems can accelerate the identification of safe and effective molecules tailored to the biological profile of patient subgroups.

b. Deep multiomics integration: Future platforms will unify genomics, transcriptomics, proteomics, metabolomics, and microbiome data to better model complex disease states (Chen et al., 2023; Graw et al., 2021). This multiomics integration will enhance the specificity of drug design and safety prediction. As the data ecosystem matures, these layered models will capture the full biological context needed for truly personalized therapies.

c. Federally controlled multicenter modeling: Data sharing remains sensitive and regulated. Federated learning, which allows model training across institutions without centralizing raw data, enables access to diverse biological datasets while preserving privacy and compliance (Madathil, Dankar, Gergely, Belkacem, & Alrabaee, 2025). This approach facilitates robust, generalizable models trained on heterogeneous data sources.

d. Global regulation and ethical harmonization: International agencies like the EMA and global health authorities are implementing roadmaps for AI integration in drug evaluation (Niazi, 2025; Singh, Paxton, & Auclair, 2025). In the coming years, harmonizing regulatory standards across countries will support multiregional trials and global access to AI-generated therapeutics. Ethical and policy alignment will ensure equity, safety, and social benefits.

e. In silico clinical trials with digital twins: The idea of "digital twins" is evolving into personalized virtual patient populations that simulate drug responses, toxicity, and long-term effects. Regulatory agencies are already considering in silico trials under specific conditions (Katsoulakis et al., 2024). As the fidelity of models improves, digital trials may help reduce animal use, expedite safety evaluation, and prioritize candidates for clinical testing.

f. Interpretability and explainability innovations: A major future need is understanding why a model selects a particular compound, target, or safety prediction. Emerging techniques in explainable modeling will help scientists rationalize predictions in terms of biochemical pathways, binding motifs, or epigenetic states (Lu, Swisher, Chung, Jaffray, & Sidey-Gibbons, 2023). These interpretability methods will support trust and integration into practice.

8 Conclusion

Generative AI stands at the frontier of a new era in drug design, one shaped by biological insight and patient-centered precision. These models do more than generate molecules; they align therapeutic innovation with the complexity of living systems. We have shown in our chapter how generative methods enable the creation of structurally novel compounds,

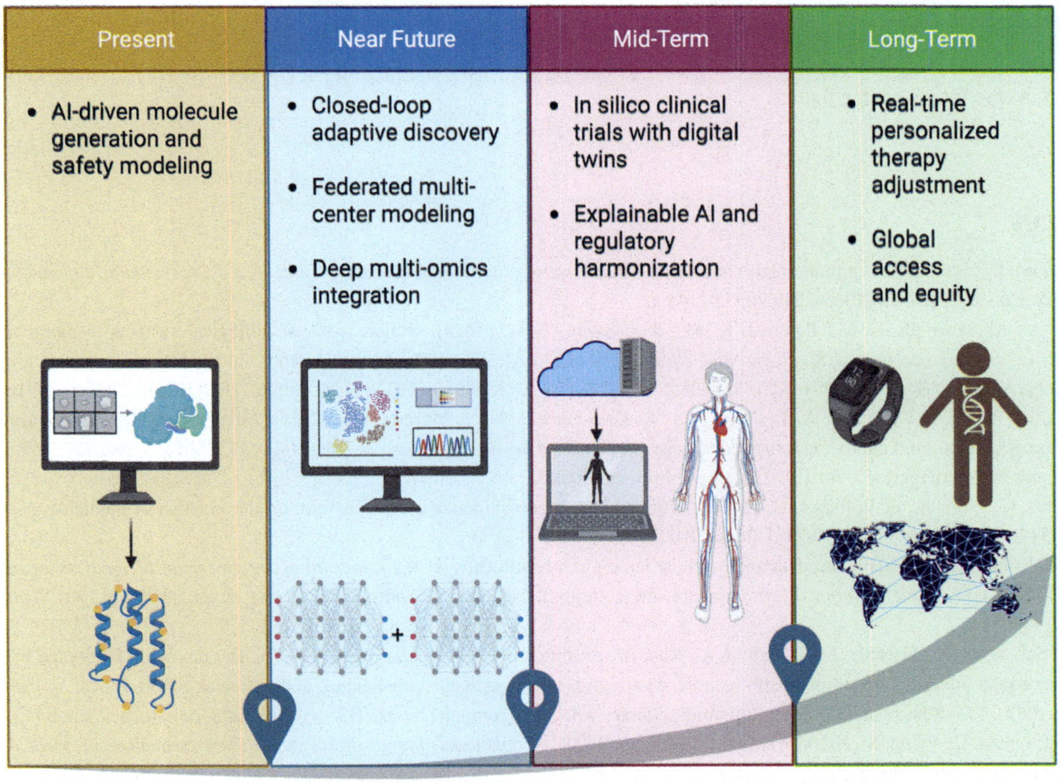

FIG. 27.6 Roadmap to the future: transformative directions for generative AI in precision medicine. Timeline illustrating the progressive integration of generative approaches into drug discovery and precision medicine. Key milestones include adaptive closed-loop discovery, federated multicenter modeling, deep multiomics integration, in silico clinical trials with digital twins, explainable AI, and global regulatory harmonization. The figure highlights how these advances can accelerate safer and more equitable access to next-generation therapeutics.

biologics, and nucleic acid therapies optimized for efficacy, safety, and the personal molecular landscape. We have explored how integrating omics data, molecular phenotyping, microbiome dynamics, and clinical records can inform compound design and safety modeling. Through predictive modeling, researchers can anticipate toxicity, avoid harmful interactions, and navigate patient variability.

However, realizing this vision demands strong governance. Regulatory frameworks must evolve to assess AI credibility, maintain explainability, and ensure lifecycle oversight. Ethical considerations like privacy, bias, and transparency must guide model training and deployment. Practical infrastructure and collaborative expertise across disciplines are vital to translating models into medicines. Looking forward, the field will shift toward adaptive, federally controlled, interpretable, and globally harmonized systems. Closed-loop discovery and in silico trials will accelerate therapeutic innovation while reducing reliance on animal models. Multiomic integration and digital twins will deepen the biological relevance of design. New policies, explainable methods, and data sharing models will support responsible and scalable deployment.

In conclusion, generative AI offers a powerful foundation for a future defined by safer, more precise, and personalized therapeutics. By bridging molecular design with biological systems and ethical frameworks, we can usher in a new era of healthcare where treatment design responds not just to targets but to the unique biology of every patient.

Conflict of interest

The authors declare no conflict of interest.

Declaration of generative AI and AI-assisted technologies in the writing process

During manuscript preparation, the authors used Grammarly and ChatGPT (OpenAI) to assist with grammar and language editing. After using these tools, the authors reviewed and revised the content as needed and took full responsibility for the final content of the publication.

Acknowledgments

Schematics were created using Biorender.com. We would like to thank Dr. Yi Zhang, Department of Biochemistry, CWRU, for providing us with the BioRender license for generating figures.

References

Adorini, L. (2003). Cytokine-based immunointervention in the treatment of autoimmune diseases. *Clinical and Experimental Immunology*. 0009-9104. *132*(2), 185–192. doi:10.1046/j.1365-2249.2003.02144.x.

Alsalloum, G. A., Al Sawaftah, N. M., Percival, K. M., & Husseini, G. A. (2024). Digital twins of biological systems: a narrative review. *IEEE Open Journal of Engineering in Medicine and Biology*.26441276. *5*, 670–677. Available from: https://ieeexplore.ieee.org/xpl/RecentIssue.jsp?punumber=8782705&_ga=2.23991482.1496039865.1593798416-2073569860.1593798416. doi:10.1109/OJEMB.2024.3426916.

Amorim, A. M. B., Piochi, L. F., Gaspar, A. T., Preto, A. J., Rosário-Ferreira, N., & Moreira, I. S. (2024). Advancing drug safety in drug development: bridging computational predictions for enhanced toxicity prediction. *Chemical Research in Toxicology*. 15205010. *37*(6), 827–849. Available from: http://pubs.acs.org/journal/crtoec. doi:10.1021/acs.chemrestox.3c00352.

Bajwa, J., Munir, U., Nori, A., & Williams, B. (2021). Artificial intelligence in healthcare: transforming the practice of medicine. *Future Healthcare Journal*. 25146645. *8*(2), e188. doi:10.7861/fhj.2021-0095.

Beam, A. L., & Kohane, I. S. (2018). Big data and machine learning in health care. *JAMA - Journal of the American Medical Association*. 15383598. *319*(13), 1317–1318. Available from: https://jamanetwork.com/journals/jama/articlepdf/2675024/jama_beam_2018_vp_170174.pdf. doi:10.1001/jama.2017.18391.

Beesley, L. J., Salvatore, M., Fritsche, L. G., Pandit, A., Rao, A., Brummett, C., … Mukherjee, B. (2020). The emerging landscape of health research based on biobanks linked to electronic health records: existing resources, statistical challenges, and potential opportunities. *Statistics in Medicine*. 10970258. *39*(6), 773–800. Available from: http://onlinelibrary.wiley.com/journal/10.1002/(ISSN)1097-0258. doi:10.1002/sim.8445.

Chen, C., Wang, J., Pan, D., Wang, X., Xu, Y., Yan, J., … Liu, G. P. (2023). Applications of multi-omics analysis in human diseases. *MedComm*. 26882663. *4*(4). Available from: https://onlinelibrary.wiley.com/journal/26882663. doi:10.1002/mco2.315.

Chen, M., Mei, S., Fan, J., & Wang, M. (2024). Opportunities and challenges of diffusion models for generative AI. *National Science Review*. 2095-5138. *11*(12). doi:10.1093/nsr/nwae348.

Chen, Y., & Esmaeilzadeh, P. (2024). Generative AI in medical practice: in-depth exploration of privacy and security challenges. *Journal of Medical Internet Research*. 1438-8871. *26*(1), e53008. doi:10.2196/53008.

Das, U. (2025). Generative AI for drug discovery and protein design: the next frontier in AI-driven molecular science. *Medicine in Drug Discovery*. 25900986. *27*, 100213. doi:10.1016/j.medidd.2025.100213.

Desta, G. M., & Birhanu, A. G. (2025). Advancements in single-cell RNA sequencing and spatial transcriptomics: Transforming biomedical research. *Acta Biochimica Polonica*. 1734154X. *72*. Available from: https://www.frontierspartnerships.org/journals/acta-biochimica-polonica. doi:10.3389/abp.2025.13922.

DiMasi, J. A., Grabowski, H. G., & Hansen, R. W. (2016). Innovation in the pharmaceutical industry: New estimates of R&D costs. *Journal of Health Economics*.18791646. *47*, 20–33. Available from: http://www.elsevier.com/locate/jhe. doi:10.1016/j.jhealeco.2016.01.012.

Ding, J., Adiconis, X., Simmons, S. K., Kowalczyk, M. S., Hession, C. C., Marjanovic, N. D., … Levin, J. Z. (2020). Systematic comparison of single-cell and single-nucleus RNA-sequencing methods. *Nature Biotechnology*. 15461696. *38*(6), 737–746. Available from: http://www.nature.com/nbt/index.html. doi:10.1038/s41587-020-0465-8.

Doytchinova, I. (2022). Drug design—past, present, future. *Molecules*. 1420-3049. *27*(5), 1496. doi:10.3390/molecules27051496.

Du, P., Fan, R., Zhang, N., Wu, C., & Zhang, Y. (2024). Advances in integrated multi-omics analysis for drug-target identification. *Biomolecules*,*14*(6), 692. doi:10.3390/biom14060692.

Fahrner, L. J., Chen, E., Topol, E., & Rajpurkar, P. (2025). The generative era of medical AI. *Cell*. 10974172. *188*(14), 3648–3660. Available from: https://www.sciencedirect.com/science/journal/00928674. doi:10.1016/j.cell.2025.05.018.

Frid-Adar, M., Klang, E., Amitai, M., Goldberger, J., & Greenspan, H. (2018). Synthetic data augmentation using GAN for improved liver lesion classification. Proceedings - international symposium on biomedical imaging. Israel: IEEE Computer Society. 19458452 (pp. 289–293). http://ieeexplore.ieee.org/xpl/conferences.jsp.

Gangwal, A., Ansari, A., Ahmad, I., Azad, A. K., Kumarasamy, V., Subramaniyan, V., & Wong, L. S. (2024). Generative artificial intelligence in drug discovery: basic framework, recent advances, challenges, and opportunities. *Frontiers in Pharmacology*. 16639812. *15*. Available from: http://www.frontiersin.org/Pharmacology. doi:10.3389/fphar.2024.1331062.

Gaulton, A., Hersey, A., Nowotka, M., Bento, A. P., Chambers, J., Mendez, D., et al. (2017). The ChEMBL database in 2017. *Nucleic Acids Research*. 0305-1048. *45*(D1), D945. doi:10.1093/nar/gkw1074.

Gómez-Bombarelli, R., Wei, J. N., Duvenaud, D., Hernández-Lobato, J. M., Sánchez-Lengeling, B., Sheberla, D., … Aspuru-Guzik, A. (2018). Automatic chemical design using a data-driven continuous representation of molecules. *ACS Central Science*. 23747951. *4*(2), 268–276. Available from: http://pubs.acs.org/journal/acscii. doi:10.1021/acscentsci.7b00572.

Goodfellow, I. J., Pouget-Abadie, J., Mirza, M., Xu, B., Warde-Farley, D., Ozair, S., … Bengio, Y. (2014). Neural information processing systems foundation Canada Generative adversarial nets. *Advances in Neural Information Processing Systems*, 10495258. *3*, 2672–2680.

Graw, S., Chappell, K., Washam, C. L., Gies, A., Bird, J., Robeson, M. S., & Byrum, S. D. (2021). Multi-omics data integration considerations and study design for biological systems and disease. *Molecular Omics*. 25154184. *17*(2), 170–185. Available from: http://www.rsc.org/journals-books-databases/about-journals/molecular-omics/. doi:10.1039/d0mo00041h.

Gupta, A., Müller, A. T., Huisman, B. J. H., Fuchs, J. A., Schneider, P., & Schneider, G. (2018). Generative recurrent networks for de novo drug design. *Molecular Informatics*. 18681751. *37*(1). Available from: http://onlinelibrary.wiley.com/journal/10.1002/(ISSN)1868-1751. doi:10.1002/minf.201700111.

Haloi, N., Howard, R. J., Lindahl, E., & MacKerell, A. (2025). Cryo-EM ligand building using AlphaFold3-like model and molecular dynamics. *PLOS Computational Biology*. 1553-7358. *21*(8), e1013367. doi:10.1371/journal.pcbi.1013367.

Han, P., Xuehua, L., Yang, J., Zhang, Y., & Chen, J. (2024). Advancing toxicity predictions: a review on in vitro to in vivo extrapolation in next-generation risk assessment. *Environment & Health*. 2833-8278. *2*(7), 499–513. doi:10.1021/envhealth.4c00043.

Hay, M., Thomas, D. W., Craighead, J. L., Economides, C., & Rosenthal, J. (2014). Clinical development success rates for investigational drugs. *Nature Biotechnology*. 15461696. *32*(1), 40–51. doi:10.1038/nbt.2786.

Hu, X., Liu, G., Chen, C., Zhao, Y., Zhang, H., & Liu, X. (2025). 3DMolFormer: A dual-channel framework for structure-based drug discovery. arXiv preprint.doi:10.48550/arXiv.2502.05107.

Huang, B., Guo, L., Yin, H., Yue, W., Zeng, Z., Sujie, X. et al.. (2024). Deep learning enhancing guide RNA design for CRISPR/Cas12a-based diagnostics. *iMeta*. 2770-5986. *3*(4). doi:10.1002/imt2.214.

Huang, K., Xiao, C., Glass, L. M., Critchlow, C. W., Gibson, G., & Sun, J. (2021). Machine learning applications for therapeutic tasks with genomics data. *Patterns*. 26663899. *2*(10). Available from: https://www.cell.com/patterns. doi:10.1016/j.patter.2021.100328.

Hughes, J. P., Rees, S. S., Kalindjian, S. B., & Philpott, K. L. (2011). Principles of early drug discovery. *British Journal of Pharmacology*. 14765381. *162*(6), 1239–1249. doi:10.1111/j.1476-5381.2010.01127.x.

Jafari, E., Blackman, M. H., Karnes, J. H., Van Driest, S. L., Crawford, D. C., Choi, L., & McDonough, C. W. (2024). Using electronic health records for clinical pharmacology research: challenges and considerations. *Clinical and Translational Science*. 17528062. *17*(7). Available from: http://onlinelibrary.wiley.com/journal/10.1111/(ISSN)1752-8062. doi:10.1111/cts.13871.

Jeong, J., Kim, D., & Choi, J. (2022). Application of ToxCast/Tox21 data for toxicity mechanism-based evaluation and prioritization of environmental chemicals: perspective and limitations. *Toxicology in Vitro*. 08872333. *84*, 105451. doi:10.1016/j.tiv.2022.105451.

Jin, W., Barzilay, R., & Jaakkola, T.. (2020). Hierarchical generation of molecular graphs using structural motifs. 37th international conference on machine learning, ICML 2020. United States: International Machine Learning Society (IMLS). 9781713821120 (pp. 4789–4798).

Jumper, J., Evans, R., Pritzel, A., Green, T., Figurnov, M., Ronneberger, O., … Kohli, P. (2021). Demis Hassabis, Highly accurate protein structure prediction with AlphaFold. *Nature*. 0028-0836. *596*(7873), 583–589. doi:10.1038/s41586-021-03819-2.

Jung, G., Hernández-Illán, E., Moreira, L., Balaguer, F., & Goel, A. (2020). Epigenetics of colorectal cancer: biomarker and therapeutic potential. *Nature Reviews Gastroenterology & Hepatology*. 1759-5045. *17*(2), 111–130. doi:10.1038/s41575-019-0230-y.

Kalyaanamoorthy, S., Lamothe, S. M., Hou, X., Moon, T. C., Kurata, H. T., Houghton, M., & Barakat, K. H. (2020). A structure-based computational workflow to predict liability and binding modes of small molecules to hERG. *Scientific Reports*. 20452322. *10*(1). Available from: https://www.nature.com/srep/index.html. doi:10.1038/s41598-020-72889-5.

Karako, K. (2025). Artificial intelligence applications in rare and intractable diseases: advances, challenges, and future directions. *Intractable & Rare Diseases Research*. 2186-361X. *14*(2), 88–92. doi:10.5582/irdr.2025.01030.

Katsoulakis, E., Wang, Q., Huanmei, W., Shahriyari, L., Fletcher, R., Liu, J., … Deng, J. (2024). Digital twins for health: a scoping review. *Npj Digital Medicine*. 2398-6352. *7*(1). doi:10.1038/s41746-024-01073-0.

Khater, T., Alkhatib, S. A., AlShehhi, A., Pitsalidis, C., Pappa, A. M., Ngo, S. T., … Truong, V. K. (2025). Generative artificial intelligence based models optimization towards molecule design enhancement. *Journal of Cheminformatics*. 17582946. *17*(1). Available from: https://jcheminf.biomedcentral.com/. doi:10.1186/s13321-025-01059-4.

Korshunova, M., Huang, N., Capuzzi, S., Radchenko, D. S., Savych, O., Moroz, Y. S., … Isayev, O. (2022). Generative and reinforcement learning approaches for the automated de novo design of bioactive compounds. *Communications Chemistry*. 23993669. *5*(1). Available from: https://www.nature.com/commschem/. doi:10.1038/s42004-022-00733-0.

Lim, J., Ryu, S., Kim, J. W., & Kim, W. Y. (2018). Molecular generative model based on conditional variational autoencoder for de novo molecular design. *Journal of Cheminformatics*. 17582946. *10*(1). Available from: http://www.springer.com/chemistry/theoretical+and+computational+chemistry/journal/13321. doi:10.1186/s13321-018-0286-7.

Liu, B., Rosenhahn, B., Illig, T., DeLuca, D. S., & Alber, M. (2024). A variational autoencoder trained with priors from canonical pathways increases the interpretability of transcriptome data. *PLOS Computational Biology*. 1553-7358. *20*(7), e1011198. doi:10.1371/journal.pcbi.1011198.

Liu, J., Mansouri, K., Judson, R. S., Martin, M. T., Hong, H., Chen, M., … Shah, I. (2015). Predicting hepatotoxicity using ToxCast in vitro bioactivity and chemical structure. *Chemical Research in Toxicology*. 15205010. *28*(4), 738–751. Available from: http://pubs.acs.org/journal/crtoec. doi:10.1021/tx500501h.

Liu, Y. T., Zhang, L. L., Jiang, Z. Y., Tian, X. S., Li, P. L., Wu, P. H., … Sun, C. (2025). Applications of artificial intelligence in biotech drug discovery and product development. *MedComm*. 26882663. *6*(8). Available from: https://onlinelibrary.wiley.com/journal/26882663. doi:10.1002/mco2.70317.

Lopez, R., Gayoso, A., & Nir Yosef. (2020). Enhancing scientific discoveries in molecular biology with deep generative models. *Molecular Systems Biology*. 1744-4292. *16*(9). doi:10.15252/msb.20199198.

Lopez, R., Regier, J., Cole, M. B., Jordan, M. I., & Yosef, N. (2018). Deep generative modeling for single-cell transcriptomics. *Nature Methods*. 15487105. *15*(12), 1053–1058. Available from: http://www.nature.com/nmeth/. doi:10.1038/s41592-018-0229-2.

Lu, S. C., Swisher, C. L., Chung, C., Jaffray, D., & Sidey-Gibbons, C. (2023). On the importance of interpretable machine learning predictions to inform clinical decision making in oncology. *Frontiers in Oncology*. 2234943X. *13*. Available from: http://www.frontiersin.org/Oncology/about. doi:10.3389/fonc.2023.1129380.

Madan, S., Lentzen, M., Brandt, J., Rueckert, D., & Hofmann-Apitius, M. (2024). Holger Fröhlich, Transformer models in biomedicine. *BMC Medical Informatics and Decision Making*. 1472-6947. *24*(1). doi:10.1186/s12911-024-02600-5.

Madani, A., Krause, B., Greene, E. R., Subramanian, S., Mohr, B. P., Holton, J. M., … Naik, N. (2023). Large language models generate functional protein sequences across diverse families. *Nature Biotechnology*. 15461696. *41*(8), 1099–1106. Available from: https://www.nature.com/nbt/. doi:10.1038/s41587-022-01618-2.

Madathil, N. T., Dankar, F. K., Gergely, M., Belkacem, A. N., & Alrabaee, S. (2025). Revolutionizing healthcare data analytics with federated learning: a comprehensive survey of applications, systems, and future directions. *Computational and Structural Biotechnology Journal*.20010370. *28*, 217–238. Available from: https://www.sciencedirect.com/science/journal/20010370. doi:10.1016/j.csbj.2025.06.009.

Masumshah, R., Eslahchi, C., & Gromiha, M. (2023). DPSP: a multimodal deep learning framework for polypharmacy side effects prediction. *Bioinformatics Advances*. 2635-0041. *3*(1). doi:10.1093/bioadv/vbad110.

Meier, M. J., Harrill, J., Johnson, K., Thomas, R. S., Tong, W., Rager, J. E., & Yauk, C. L. (2025). Progress in toxicogenomics to protect human health. *Nature Reviews Genetics*. 14710064. *26*(2), 105–122. Available from: https://www.nature.com/nrg/. doi:10.1038/s41576-024-00767-1.

Mertins, P., Mani, D. R., Ruggles, K. V., Gillette, M. A., Clauser, K. R., Wang, P., … Carr, S. A. (2016). Proteogenomics connects somatic mutations to signalling in breast cancer. *Nature*. 14764687. *534*(7605), 55–62. Available from: http://www.nature.com/nature/index.html. doi:10.1038/nature18003.

Mishra, A., Majumder, A., Kommineni, D., Joseph, C. A., Chowdhury, T., & Anumula, S. K. (2025). Role of generative artificial intelligence in personalized medicine: a systematic review. *Cureus*. 2168-8184. *17*(4). doi:10.7759/cureus.82310.

Mittelstadt, B., Russell, C., & Wachter, S.. (2019). Explaining explanations in AI. FAT* 2019 - proceedings of the 2019 conference on fairness, accountability, and transparency. United Kingdom: Association for Computing Machinery, Inc. 9781450361255 (pp. 279–288). http://dl.acm.org/citation.cfm?id=3287560. doi:10.1145/3287560.3287574.

Moingeon, P., Chenel, M., Rousseau, C., & Voisin, E. (2023). Mickael Guedj, Virtual patients, digital twins and causal disease models: paving the ground for in silico clinical trials. *Drug Discovery Today*. 13596446. *28*(7), 103605. doi:10.1016/j.drudis.2023.103605.

Mok, T. S., Wu, Y. L., Thongprasert, S., Yang, C. H., Chu, D. T., Saijo, N., … Fukuoka, M. (2009). Gefitinib or carboplatin-paclitaxel in pulmonary adenocarcinoma. *New England Journal of Medicine*. 15334406. *361*(10), 947–957. Available from: http://content.nejm.org/cgi/reprint/361/10/947.pdf. doi:10.1056/NEJMoa0810699.

Nagar, A., Gobburu, J., & Chakravarty, A. (2025). Artificial intelligence in pharmacovigilance: advancing drug safety monitoring and regulatory integration. *Therapeutic Advances in Drug Safety*, 2042-0986. *16*. doi:10.1177/20420986251361435.

Niazi, S. K. (2025). Regulatory perspectives for AI/ML Implementation in pharmaceutical GMP environments. *Pharmaceuticals*. 14248247. *18*(6). Available from: https://www.mdpi.com/journal/pharmaceuticals. doi:10.3390/ph18060901.

Niederer, S. A., Aboelkassem, Y., Cantwell, C. D., Corrado, C., Coveney, S., Cherry, E. M., … Wang, L. (2020). Creation and application of virtual patient cohorts of heart models. *Philosophical Transactions of the Royal Society A: Mathematical, Physical and Engineering Sciences*. 1364-503X. *378*(2173), 20190558. doi:10.1098/rsta.2019.0558.

Ning, Y., Teixayavong, S., Shang, Y., Savulescu, J., Nagaraj, V., Miao, D., … Liu, N. (2024). Generative artificial intelligence and ethical considerations in health care: a scoping review and ethics checklist. *The Lancet Digital Health*. 25897500. *6*(11), e848. doi:10.1016/s2589-7500(24)00143-2.

Ocana, A., Pandiella, A., Privat, C., Bravo, I., Luengo-Oroz, M., Amir, E., & Gyorffy, B. (2025). Integrating artificial intelligence in drug discovery and early drug development: a transformative approach. *Biomarker Research*. 2050-7771. *13*(1). doi:10.1186/s40364-025-00758-2.

Olivecrona, M., Blaschke, T., Engkvist, O., & Chen, H. (2017). Molecular de-novo design through deep reinforcement learning. *Journal of Cheminformatics*. 1758-2946. *9*(1). doi:10.1186/s13321-017-0235-x.

Ott, P. A., Hu, Z., Keskin, D. B., Shukla, S. A., Sun, J., Bozym, D. J., … Wu, C. J. (2017). An immunogenic personal neoantigen vaccine for patients with melanoma. *Nature*. 14764687. *547*(7662), 217–221. Available from: http://www.nature.com/nature/index.html. doi:10.1038/nature22991.

Paul, D., Sanap, G., Shenoy, S., Kalyane, D., Kalia, K., & Tekade, R. K. (2021). Artificial intelligence in drug discovery and development. *Drug Discovery Today*. 18785832. *26*(1), 80–93. Available from: https://www.elsevier.com/locate/drugdiscov. doi:10.1016/j.drudis.2020.10.010.

PharmaLex. (2025). How the FDA guidance on the use of AI will impact drug development. Available from: https://www.pharmalex.com/thought-leadership/blogs/how-the-fda-guidance-on-the-use-of-ai-will-impact-drug-development/.

Popov, K. I., Wellnitz, J., Maxfield, T., & Tropsha, A. (2024). HIt Discovery using docking ENriched by GEnerative Modeling (HIDDEN GEM): a novel computational workflow for accelerated virtual screening of ultra-large chemical libraries. *Molecular Informatics*. 18681751. *43*(1). Available from: http://onlinelibrary.wiley.com/journal/10.1002/(ISSN)1868-1751. doi:10.1002/minf.202300207.

Rives, A., Meier, J., Sercu, T., Goyal, S., Lin, Z., Liu, J., … Fergus, R. (2021). Biological structure and function emerge from scaling unsupervised learning to 250 million protein sequences. *Proceedings of the National Academy of Sciences*. 0027-8424. *118*(15). doi:10.1073/pnas.2016239118.

Routy, B., Chatelier, E. L., Derosa, L., Duong, C. P. M., Alou, M. T., Daillère, R., … Zitvogel, L. (2018). Gut microbiome influences efficacy of PD-1-based immunotherapy against epithelial tumors. *Science*. 10959203. *359*(6371), 91–97. Available from: http://science.sciencemag.org/content/359/6371/91/tab-pdf. doi:10.1126/science.aan3706.

Sabanayagam, C., Banu, R., Lim, C., Tham, Y. C., Cheng, C. Y., Tan, G., … Wong, T. Y. (2025). Artificial intelligence in chronic kidney disease management: a scoping review. *Theranostics*. 18387640. *15*(10), 4566–4578. Available from: https://www.thno.org/v15p4566.htm. doi:10.7150/thno.108552.

Saez-Rodriguez, J., Costello, J. C., Friend, S. H., Kellen, M. R., Mangravite, L., Meyer, P., … Stolovitzky, G. (2016). Crowdsourcing biomedical research: leveraging communities as innovation engines. *Nature Reviews Genetics*. 14710064. *17*(8), 470–486. Available from: http://www.nature.com/reviews/genetics. doi:10.1038/nrg.2016.69.

Samei, E., & Brenner, D. (2025). The future of in silico trials and digital twins in medicine. *PNAS Nexus*. 2752-6542. *4*(5). doi:10.1093/pnasnexus/pgaf123.

Sanchez-Lengeling, B., & Aspuru-Guzik, A. (2018). Inverse molecular design using machine learning: generative models for matter engineering. *Science*. 0036-8075. *361*(6400), 360–365. doi:10.1126/science.aat2663.

Schneider, P., Walters, W. P., Plowright, A. T., Sieroka, N., Listgarten, J., Goodnow, R. A., … Schneider, G. (2020). Rethinking drug design in the artificial intelligence era. *Nature Reviews Drug Discovery*. 14741784. *19*(5), 353–364. Available from: http://www.nature.com/nrd/index.html. doi:10.1038/s41573-019-0050-3.

Seal, S., Mahale, M., García-Ortegón, M., Joshi, C. K., Hosseini-Gerami, L., Beatson, A., … Bender, A. (2025). Machine learning for toxicity prediction using chemical structures: Pillars for success in the real world. *Chemical Research in Toxicology*. 15205010. *38*(5), 759–807. Available from: http://pubs.acs.org/journal/crtoec. doi:10.1021/acs.chemrestox.5c00033.

Sengar, S. S., Hasan, A. B., Kumar, S., & Carroll, F. (2025). Generative artificial intelligence: a systematic review and applications. *Multimedia Tools and Applications*. 15737721. *84*(21), 23661–23700. Available from: https://www.springer.com/journal/11042. doi:10.1007/s11042-024-20016-1.

Serrano, D. R., Luciano, F. C., Anaya, B. J., Ongoren, B., Kara, A., Molina, G., … Lalatsa, A. (2024). Artificial Intelligence (AI) applications in drug discovery and drug delivery: Revolutionizing personalized medicine. *Pharmaceutics*. 19994923. *16*(10). Available from: http://www.mdpi.com/journal/pharmaceutics. doi:10.3390/pharmaceutics16101328.

Serrano, W. (2023). The deep learning generative adversarial random neural network in data marketplaces: The digital creative. *Neural Networks*. 08936080. *165*, 420–434. doi:10.1016/j.neunet.2023.05.028.

Sewell, F., Alexander-White, C., Brescia, S., Currie, R. A., Roberts, R., Roper, C., … Kimber, I. (2024). New approach methodologies (NAMs): Identifying and overcoming hurdles to accelerated adoption. *Toxicology Research*. 20454538. *13*(2). Available from: https://academic.oup.com/toxres/advance-article-abstract/doi/10.1093/toxres/tfaa023/5831935?redirectedFrom=fulltext. doi:10.1093/toxres/tfae044.

Shah, R.R.. (2005). Drug-induced QT interval prolongation - Regulatory guidance and perspectives on hERG channel studies. Novartis foundation symposium, United Kingdom 15282511, 251 (pp. 266–280). http://www3.interscience.wiley.com/cgi-bin/booktext/111090698/BOOKPDFSTART

Singh, A. V., Chandrasekar, V., Paudel, N., Laux, P., Luch, A., Gemmati, D., … Dakua, S. P. (2023). Integrative toxicogenomics: advancing precision medicine and toxicology through artificial intelligence and OMICs technology. *Biomedicine and Pharmacotherapy*. 19506007. *163*. Available from: https://www.journals.elsevier.com/biomedicine-and-pharmacotherapy. doi:10.1016/j.biopha.2023.114784.

Singh, R., Paxton, M., & Auclair, J. (2025). Regulating the AI-enabled ecosystem for human therapeutics. *Communications Medicine*. 2730-664X. *5*(1). doi:10.1038/s43856-025-00910-x.

Topol, E. J. (2019). *Deep medicine: How artificial intelligence can make healthcare human again*. New York, NY: Basic Books.

Varnosfaderani, S. M., & Forouzanfar, M. (2024). The role of AI in hospitals and clinics: Transforming healthcare in the 21st century. *Bioengineering*. 2306-5354. *11*(4), 337. doi:10.3390/bioengineering11040337.

Viceconti, M., Henney, A., & Morley-Fletcher, E. (2016). In silico clinical trials: How computer simulation will transform the biomedical industry. *International Journal of Clinical Trials*. 2349-3259. *3*(2), 37. doi:10.18203/2349-3259.ijct20161408.

Virolainen, S. J., VonHandorf, A., Viel, K. C. M. F., Weirauch, M. T., & Kottyan, L. C. (2023). Gene–environment interactions and their impact on human health. *Genes and Immunity*. 14765470. *24*(1), 1–11. Available from: https://www.nature.com/gene/. doi:10.1038/s41435-022-00192-6.

Vora, L. K., Gholap, A. D., Jetha, K., Thakur, R. R. S., Solanki, H. K., & Chavda, V. P. (2023). Artificial intelligence in pharmaceutical technology and drug delivery design. *Pharmaceutics*. 19994923. *15*(7). Available from: http://www.mdpi.com/journal/pharmaceutics. doi:10.3390/pharmaceutics15071916.

Wallace, B. D., Wang, H., Lane, K. T., Scott, J. E., Orans, J., Koo, J. S., … Redinbo, M. R. (2010). Alleviating cancer drug toxicity by inhibiting a bacterial enzyme. *Science*. 10959203. *330*(6005), 831–835. Available from: http://www.sciencemag.org/cgi/reprint/330/6005/831.pdf. doi:10.1126/science.1191175.

Woodward, A. A., Urbanowicz, R. J., Naj, A. C., & Moore, J. H. (2022). Genetic heterogeneity: challenges, impacts, and methods through an associative lens. *Genetic Epidemiology*. 10982272. *46*(8), 555–571. Available from: http://onlinelibrary.wiley.com/journal/10.1002/(ISSN)1098-2272. doi:10.1002/gepi.22497.

Zhao, Z., Wu, H., Wang, L., Liu, Y., Knapp, S., Liu, Q., & Gray, N. S. (2014). Exploration of type II binding mode: A privileged approach for kinase inhibitor focused drug discovery? *ACS Chemical Biology*. 15548937. *9*(6), 1230–1241. Available from: http://pubs.acs.org/journal/acbcct. doi:10.1021/cb500129t.

Zhavoronkov, A., Ivanenkov, Y. A., Aliper, A., Veselov, M. S., Aladinskiy, V. A., Aladinskaya, A. V., … Aspuru-Guzik, A. (2019). Deep learning enables rapid identification of potent DDR1 kinase inhibitors. *Nature Biotechnology*. 15461696. *37*(9), 1038–1040. Available from: http://www.nature.com/nbt/index.html. doi:10.1038/s41587-019-0224-x.

Zitnik, M., Agrawal, M., & Leskovec, J. (2018). Modeling polypharmacy side effects with graph convolutional networks. *Bioinformatics*. 1367-4803. *34*(13), i457. doi:10.1093/bioinformatics/bty294.

Zitnik, M., Nguyen, F., Wang, B., Leskovec, J., Goldenberg, A., & Hoffman, M. M. (2019). Machine learning for integrating data in biology and medicine: principles, practice, and opportunities. *Information Fusion*.15662535. *50*, 71–91. Available from: http://www.elsevier.com/inca/publications/store/6/2/0/8/6/2/index.htt. doi:10.1016/j.inffus.2018.09.012.

Chapter 28

Artificial intelligence at the heart of nutrigenomics: Machine learning for a customized diet

Samudra Prosad Banik[a], Anand Swaroop[b,c], Harekrishna Jana[d], and Debasis Bagchi[e,f]

[a]*Department of Microbiology, Government General Degree College, Narayangarh, Rathipur, West Bengal, India*

[b]*Nutrify Today, Somerset, NJ, United States*

[c]*Cepham Inc., Somerset, NJ, United States*

[d]*Department of Microbiology, Raja Narendra Lal Khan Women's College, Midnapore, West Bengal, India*

[e]*Department of Biology, College of Arts and Sciences, Adelphi University, Garden City, NY, United States*

[f]*Department of Pharmaceutical Sciences, College of Pharmacy and Health Sciences, Texas Southern University, Houston, TX, United States*

1 Introduction

Pathbreaking technologies such as supercomputing and artificial intelligence (AI) have circumvented many problems of present human society and have accelerated the execution of multiple tasks to keep pace with the fast-changing facets of life, both at home and in the workplace. However, at the same time, they have made us less willing to engage in physical activity and sweat it out to earn a livelihood. Consequently, nature has taken its toll by making us vulnerable to many lifestyle disorders in addition to the ever-present threat of pathogens. Most of these 20th-century ailments have crept permanently into our society in the form of metabolic syndromes such as cardiovascular diseases, obesity, and diabetes, and their prevalence is on the rise globally.

Besides maintaining a healthy lifestyle, optimized and adequate nutrition are the two primary requirements for keeping the risks of lifestyle diseases at bay. However, owing to differences in ethnicity, genetic constitution, gut microbiota, as well as other hitherto undiscovered epigenetic factors, the dietary requirements of any two individuals are seldom the same. This has, in turn, necessitated the emergence of the field of nutrigenomics, which incorporates the cross-cutting issues of nutrition, genomics, metabolomics, and bioinformatics to develop personalized nutrition charts. To process and interpret the gargantuan amount of data generated in this multiomics-based bioinformatic approach, scientists have again resorted to AI, thus completing the circle of intricate dependence between humans and machines (Phugat & Goel, 2025). The term "nutrigenomics" apparently sounds like a portmanteau of the two words "nutrition" and "genomics" and is intended to understand the basis of personalized nutritional needs with insights from the interaction of components of the human diet with the complex network of a huge number of genes and their products in the form of metabolic enzymes and accessory proteins (Farhud, Yeganeh and Yeganeh, 2010). Gaining a reliable estimate of the actual number of genes, along with their interactions and implications in the biological pathways that belong to the "metabolome" or "reactome," poses a serious challenge even for sophisticated quantitative pathway analysis tools like the Kyoto Encyclopedia of Genes and Genomes (KEGG) database (Kanehisa, Furumichi, Sato, Matsuura, & Ishiguro-Watanabe, 2025) and the Reactome Gene Set Analysis tool (Gillespie et al., 2022). This is due to issues such as relatively slower database updates, occasional failures in accounting for redundant data from other databases (Simillion, Liechti, Lischer, Ioannidis, & Bruggmann, 2017; Stanford, Clake, Morris, & Rogers, 2020) and insensitivity to complex multilayered gene regulatory networks (Zhu et al., 2012). Therefore to generate holistic, meaningful, and dynamic frameworks, conventional gene set analysis and pathway analysis databases must be continuously coupled with data from a multiomics platform (Griss et al., 2020). In this regard, the role of AI has become integral and indispensable for processing the huge datasets simultaneously fed into the analysis pipeline from a dynamic environment of the transcriptome, proteome, and metabolome (Ren et al., 2025; Yetgin, 2025) (Fig. 28.1). The

Applications of Artificial Intelligence in Healthcare. https://doi.org/10.1016/B978-0-443-43934-6.00031-6

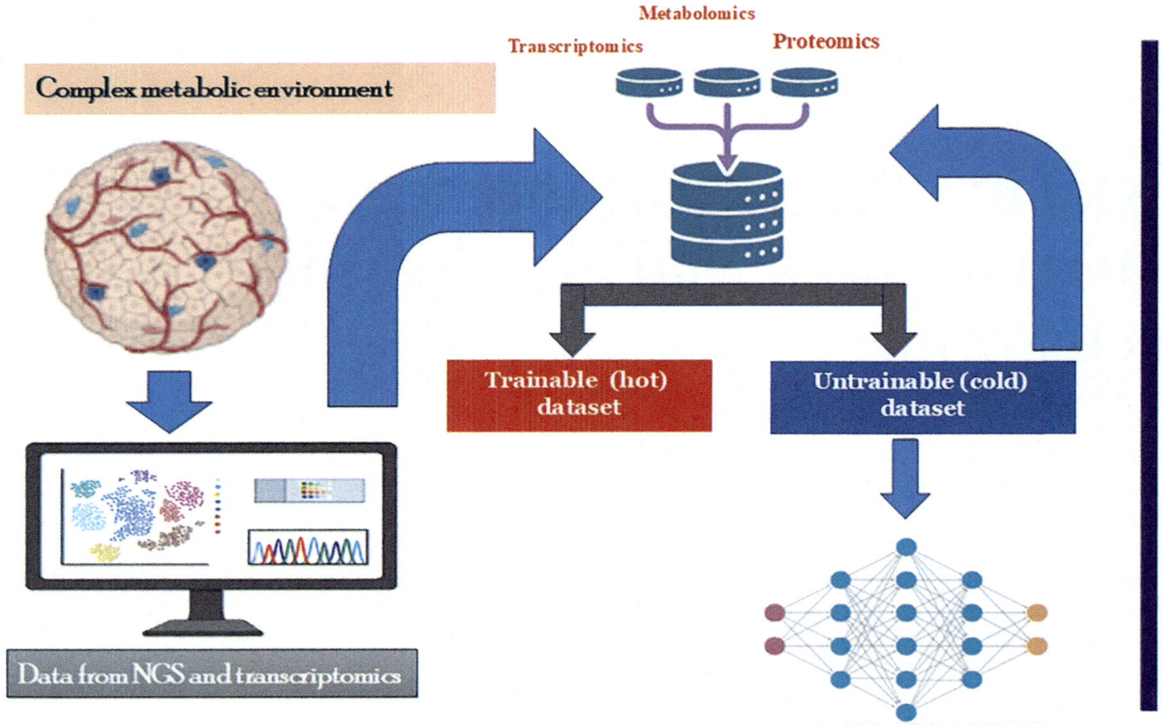

FIG. 28.1 AI-assisted personalized nutrition from a multiomics perspective. In modern nutrigenomics, the interaction of various constituents of food with the products of our metagenome is evaluated in a complex metabolic micromilieu, integrating transcriptomic, proteomic, and metabolomic data, which are analyzed with the help of ML. The multiomics data is stacked and fed into ML algorithms for categorization into trainable and untrainable datasets, which can be further utilized in supervised, unsupervised, or reinforced deep learning platforms to develop meaningful insights for personalized nutrition.

integration of AI has facilitated a sizable leap in the field of nutrigenomics, with the extraction of meaningful insights from intricate datasets encompassing dietary practices, metabolic health, and mental wellness (Agrawal, Goktas, Kumar, & Leung, 2025; Detopoulou et al., 2023; Sak & Suchodolska, 2021). Machine learning (ML) algorithms based on multiple decision trees, such as random forest, eXtreme Gradient Boosting (XGBoost), and ensemble learning, have also made it possible to provide real-time personalized dietary recommendations to chronic diabetic patients based on 24×7 monitoring of blood glucose levels and glycated hemoglobin levels (Fu et al., 2023; Tao et al., 2023). Particularly, the intervention of AI has revolutionized gut health by enabling the early prediction of biomarkers associated with irritable bowel syndrome (IBS) and the recommendation of remedial probiotic metabolites based on individual medical history (Han et al., 2025). Over recent years, the role of AI in the field of nutrigenomics has extended beyond personalized dietary recommendations. Deep learning (DL) algorithms have proven effective in the screening of nutraceuticals and phytotherapeutics with potential biological effector functions (Noor et al., 2024). Fuzzy logic algorithms are now routinely employed to enhance both the quality and palatability of food (De Pilli, 2022; Pavani, Singha, Rajamanickam, & Singh, 2023), including the control of manufacturing processes (Chaurasia et al., 2025) and the regulation of supply chains (Ghag, Sonar, Jagtap, & Trollman, 2024). In spite of the promising future of AI-assisted nutrigenomics, the primary bottleneck in its widespread adoption still remains in the form of the scarcity of authentic and elaborate data-sets, overcoming algorithmic bias in real-life dynamic environments, and achieving multiplatform integration of predictions and recommendations. This chapter highlights the current deliverables and futuristic technologies of AI intervention in the realm of personalized nutrition. At the same time, it also introspects the key limitations and bottlenecks of machine intelligence (including ethical concerns) in monitoring human health.

2 Nutrition, personalized nutrition, and nutrigenomics

Probably the greatest irony of human civilization lies in the fact that we have to find, at the same time, dietary solutions for both the malnourished and the obese individual. The term "nutrition" itself refers to the science of the chemical composition of diet and its interaction with the metabolic components of the body for maintaining optimized life in terms of health and vigor. Undernutrition or malnutrition can have severe consequences on health, including a compromised immune system, enhanced susceptibility to communicable diseases, dysregulated digestion and metabolism, as well as altered muscular function (Gérard et al., 2021; Soni et al., 2020; Türkmenoğlu, Etaner Uyar and Kiraz, 2021). In spite of the introduction of new-generation dietary supplements, nutraceuticals, and functional foods, the burden of noncommunicable and diet-related diseases has continued to escalate (Garg, 2025; Zhang, Barr, McNulty, Li, & Blumberg, 2020), which, for a large part, has been traced to be associated with the genetic and epigenetic mechanisms regulating the body's response to consumed food (Acevedo-Espinola, Victoria, Olivares-Etchebaster, & Tume, 2024; Tiffon, 2018). Therefore it is quite apparent that formulating a common, unified dietary strategy for all should be the ultimate goal of the era of modern nutrigenomics.

Scientific evidence obtained to date suggests that food components, including micronutrients, macronutrients, and phytotherapeutics, are responsible for the regulation of gene expression (Farris et al., 2024; Saito et al., 2021; Surguchov, Bernal, & Surguchev, 2021). Therefore the first and foremost criterion to set the ball of Personalized Nutrition rolling is to collect and analyze a huge amount of data from a vast pool of individuals regarding their dietary habits and the effects of specific food components involved therein, including heart rate, calories burned from daily chores, estimation of body weight, visceral fat, composition of the gut microbiota, etc. Nutrigenomics refers to the study of the interaction of dietary components with our genome, proteome, and metabolome, including how genetic polymorphisms affect the response of our body to food. In this respect, the allied fields of nutriproteomics and nutrimetabolomics have gradually merged with the holistic purview of nutrigenomics (Özdemir & Kolker, 2016). Therefore the integration and analysis of these datasets pose a serious challenge to even the fastest supercomputers, and the intervention of AI becomes indispensable.

3 ML in AI

AI operates fundamentally on two niches: (1) narrow AI, which is the more frequently used, authentic, and reliable niche albeit with limited scopes, including tools such as natural language processing (NLP) and facial recognition, and (2) a more evolved genre, or strong AI, which focuses on developing general systems that can learn across multiple interfaces with human insights. The more familiar and established niches of AI include ML, DL, NLP, and computer vision (Nadkarni, 2023). ML involves a self-learning module of AI where the computer learns to comprehend and predict patterns from a huge volume of data fed into it without help or intervention from humans. ML is further categorized into three primary aspects: supervised learning (SL), unsupervised learning, and reinforcement learning. An SL works by reducing huge volumes of multidimensional labeled datasets, originally with a higher extent of randomness or unpredictability (Shannon's entropy (Saraiva, 2023)) into fewer data points, retaining only the stronger and significant attributes with meaningful expansion of information. This is achieved through extensive analytical algorithms with reiteration of orthogonal cross-validation studies, resulting in the retention of fewer attributes each time (Andersen & Reading, 2024). Common applications of SL include image identification, assessment of risk factors, and detection of forgery. To achieve these, SL employs several types of algorithms such as regression algorithms, classification algorithms, the Naïve Bayes decision tree, neural networks, and the random forest algorithm, each being suitable for a particular application. Regression algorithms are employed to envisage linear or logistic causal associations of data. They are widely applied for the construction and analysis of models such as genome networks (Ha, Park, & Jo, 2025). As the name suggests, classification algorithms work by assigning simple classification labels or outcomes such as "significant" or "not significant." A typical application of classification algorithms is the probability assessment of cancer progression from microarray-based gene expression data (Osama, Shaban, & Ali, 2023). Naïve Bayes decision trees incorporate the features of both regression and classification algorithms and are typically employed for handling larger datasets. Neural networks belong to a more evolved subset of ML known as DL, which will be elaborately discussed in the following subsections. They are based on algorithms built with close analogy to the functioning of the human brain, with a series of nodes that closely mimic biological neurons. Each of these nodes functions as a distinct computational unit, working on inputs and generating a weighted output that passes on to the next layer of nodes for processing. Thus the neural network exhibits concerted and synchronized action to perform complex tasks on a human interface, such as natural language processing, image recognition, and analysis. The Random Forest algorithm is based on a combination of decision trees and is particularly equipped to handle large datasets, such as a patient's history or multiomics data, and extract useful patterns or insights from them (Hornung & Wright, 2019). Two variants of SL are self-supervised learning, which can automatically generate labeled data by learning partially from the input data itself (Richter, Bahrami, Xia, Fischer, & Theis, 2025), and semisupervised learning, where the machine is trained using a very small labeled dataset along with a large unlabeled dataset. The labeled data acts as the trainer for the unlabeled dataset, thus revealing hidden patterns in the data (Wu, Xie, Liu, & Xie, 2024). On the contrary, unsupervised learning allows the machine more freedom to work on complex datasets without any defined set of labels. The algorithms employed here look for underlying patterns and rely on classification and clustering of datasets based on parameters that have not been predefined by the human user but surface during the execution of the algorithms (Parasa et al., 2021). A classic example of unsupervised learning is cluster

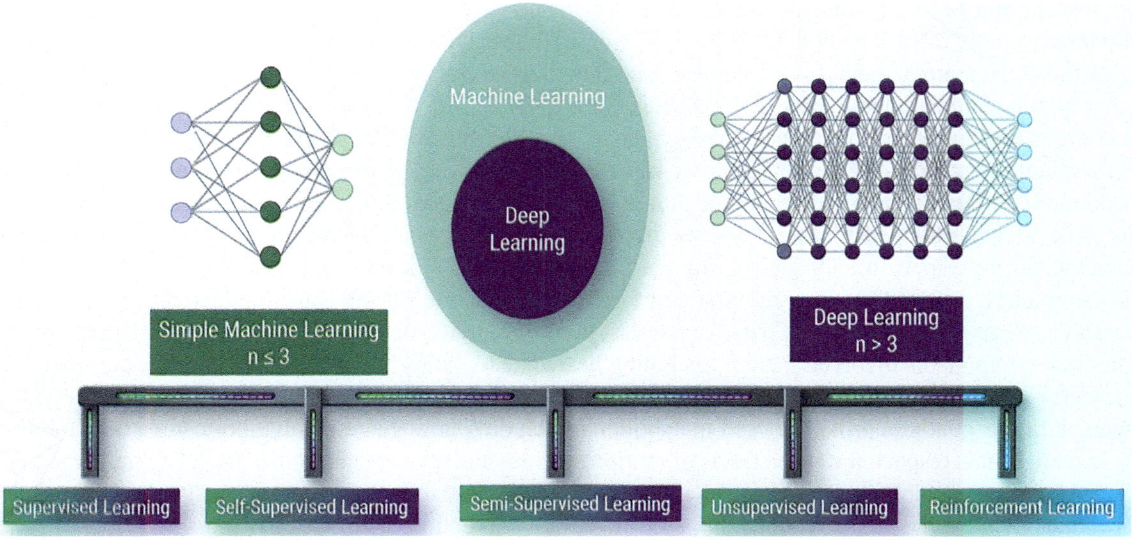

FIG. 28.2 Components of machine learning can be broadly categorized into supervised learning, self-supervised learning, semisupervised learning, unsupervised learning, and reinforcement learning. Deep learning is a subset of machine learning that involves more than three layers of data analysis nodes.

analysis, which is used for clustering or grouping data according to some common value (Singh & Singh, 2024). K-means clustering, hierarchical clustering, and probabilistic clustering are some of the different variations of cluster analysis. The third broad category of ML is reinforcement learning, where the AI agent learns to make sequential decision trees on its own based on continuous trial-and-error interaction with the surroundings. Whenever the agent makes a correct decision, it is rewarded, and a false decision incurs a penalty, leading to significant alteration in the original state of the system. The ultimate goal of the agent in reinforcement learning is to maximize the reward at the end for a positive outcome in the analysis (Shakya, Pillai and Chakrabarty, 2023). The basic components of ML are depicted in Fig. 28.2.

4 ML tools in "omics" studies

The last few decades have witnessed an explosion of digital data in unprecedented proportions across almost every sphere of life (Close, 2023), and research and development in healthcare has been no exception (Batko & Ślęzak, 2022). Consequently, extracting meaningful information from this vast amount of raw and grossly unstructured data has emerged as the most important aspect of data analytics. Most, if not all, nutrigenomic analyses consist of complex, nonlinear, and dynamic datasets. For example, understanding the summative physiological effects of a high-fat diet involves transcriptomic analysis of a representative organ sample, such as the kidney (Hong, Luan, Yixuan, & Zhang, 2024). Owing to the enormous volume of data generated, the first step in this study involves principal component analysis, a statistical technique that reduces the complexity and effective dimensionality of the dataset by reconstructing a smaller set of new and uncorrelated Principal Component variables. These variables are generated by the linear grouping of related variables into a single variable with the inclusion of maximum variance. Subsequently, the differentially expressed set of genes is subjected to Gene Ontology functional classification and KEGG pathway analysis to narrow the window of biological effects inflicted by the high-fat diet. Following this, genome-wide epigenetic changes, such as histone modification ascertained through chromatin immunoprecipitation sequencing (Qu et al., 2023), CpG-island methylation ascertained through methylation-dependent immunoprecipitation sequencing, and methylation-sensitive restriction enzyme digestion sequencing (Qu et al., 2023b) along with several other related techniques, are employed. However, since most of these experiments are generally carried out in mice, the problem of orthologous phenotype assignment in humans poses a serious bottleneck. Nevertheless, an authentic database for the classification and analysis of high-fat food groups mandates transcriptomic investigation of the effects of consuming each such food item separately (Bouchard et al., 2010; Jaime-Lara et al., 2024; Martinez-Lomeli et al., 2023). In personalized nutrition, the complexity is further compounded by the differential responses of individuals, which vary based on factors such as age, sex, ethnicity, and gut microbiome composition (Bouchard et al., 2010). The entire process, therefore, presents a task of indomitable magnitude, difficult to execute even for supercomputers. Herein lies the indispensability of ML in extracting meaningful information from such datasets. One study demonstrated the application of

ML in analyzing transcriptomic data from a group of nonalcoholic fatty liver disease-affected patients (Nazari et al., 2023). Following filtering, normalization and principal component analysis of the RNA-seq data, the most significant biomarkers associated with the disease were identified using penalize ML algorithms. These algorithms work by associating a penalty with a "cost function" to prevent "overfitting" of biological data, using cues from a training dataset. Thus penalized algorithms serve the dual function of reducing the effective size of the dataset and delivering significance levels that are closer to the real biological scenario. Penalized ML consists of three consecutive regression analyses: least absolute shrinkage and selection operator (LASSO), Ridge, and elastic net regression. LASSO operates by reducing the dataset around the mean value, Ridge applies a penalty as the sum of the squares of coefficients, and elastic net regression combines the functionalities of both LASSO and Ridge. Thus the application of ML in dissecting the effects of a high-fat diet can potentially bypass the intimidating datasets and deliver corresponding meaningful interpretations.

5 Deep learning

Traditional ML methods suffered from a significant bottleneck in processing unorganized raw data and eventually extracting features into a suitable form for subsequent development into a learning set (Lecun, Bengio, & Hinton, 2015). This limitation led to the development of representation learning, which could automatically identify representations from raw data to facilitate further classification. The stacking of multiple levels of representation with nonlinear modules, where the output representation from one layer serves as input to the next layer with slightly more abstract information, eventually leads to a network of DL. DL is a specialized subset of ML based on artificial neural networks (ANNs) with more than three layers. In ANNs, neurons are represented by nodes, each of which is connected to others by means of "edges" or synapse equivalents. A typical ANN consists of multiple layers of nodes, including intermediate hidden layers, with each node having its own weight and threshold. If the output of any node exceeds its specified threshold, it is passed to the next node in the hierarchy as higher-order information. If not, the propagation of information from that particular node stops. The performance of neural networks relies on the number of training sets they receive, with improvement occurring over each set. Once a certain level of accuracy is achieved, the speed of data processing increases, and the ANN transforms into a powerful tool of AI capable of executing complex tasks, such as delivering outputs in response to human voice commands, performing language translations (natural language processing), prescribing recommendations on e-commerce sites, predicting stock-market fluctuations, and serving as dependable virtual assistants or bots. Since the invention of the first neural network, termed the perceptron, in 1958 ANNs have evolved significantly with the development of multilayer perceptrons. Convolutional neural networks (CNNs) and recurrent neural networks (RNNs) are two other versions of neural networks that have greatly facilitated advancements in the biomedical industry. CNNs are frequently applied in medical imaging for cancer diagnosis (Balasubramaniam, Velmurugan, Jaganathan, & Dhanasekaran, 2023), whereas RNNs are leveraged for making predictions on timeline events, such as assessing the risk of food spoilage during transportation (Ding et al., 2025).

6 Facets of precision nutrition and the role of ML therein

The concern over the rise in the number of chronic diseases caused by malnutrition (Saunders, 2010) has led to the advent and evolution of personalized nutrition and precision nutrition. It has now been established that even subtle deficiencies or overabundances of certain essential nutrients are strongly associated with markers of chronic inflammation, leading to gross metabolic syndromes. Magnesium deficiency has been linked to enhanced oxidative stress and an increase in levels of proinflammatory cytokines, which is directly reflected in elevated levels of plasma C-reactive protein (CRP) (Švagždienė, Širvinskas, Baranauskienė, & Adukauskienė, 2015). An increase in CRP levels has also been traced to subclinical deficiencies in Vitamin C levels (Crook, Yoon, Grundmann, Horgas, & Johnson-Mallard, 2023). Vitamin D has been found to reduce serum CRP levels in obese/overweight children (Zou et al., 2022) and dampen inflammatory markers in infants with IBS (Yan et al., 2023). The susceptibility to disease and infection has been shown to increase with reduced levels of folate (Cantarella, Ragusa, Giammanco, & Tosi, 2017) and vitamin B12 (Batista et al., 2022). The concentration of the micronutrient zinc also needs to be subtly regulated in our body to ensure cognitive development, as a rise in zinc levels in the brain micromilieu is associated with protein misfolding and aggregation, resulting in the formation of amyloid-like entities (Banik et al., 2025). It has also been demonstrated that the formation of these amyloids can be initiated by glycated and misfolded protein molecules termed Advanced Glycation End Products (AGEs) (Das et al., 2020). This observation has helped scientists partially explain the strong clinical correlation between chronic hyperglycemia, leading to diabetes or prediabetes, and neurodegenerative disorders (Ayoub et al., 2025; Banik, 2024). The examples cited present a strong representative case of the significant effect of nutrients on our health regime and perfectly illustrate why precision nutrition has gained a major share of attention in 21st-century healthcare. Yet, these constitute the mere tip of the iceberg with regard to nutrient–genome–proteome interaction. The Micronutrient-Genomics Project was launched specifically to catalog all information pertaining to these interactions in a common database (Van Ommen et al., 2010). Similarly, Proteomics DB (Schmidt et al., 2018) and the Peptide Atlas (Deutsch, 2010) have been implemented to document and understand the interaction of components of nutrition with our proteome. However, as reiterated many times, the role of AI has been indispensable in handling the big data generated from these ambitious projects.

Regression, clustering, and classification tools have significantly augmented the domain of precision nutrition. To harness anthropometric data, traditional tools such as measurement tapes and skinfold calipers require a lot of time and training and are fallible to human errors, especially during the documentation step. Other sophisticated tools, such as dual-energy X-ray absorptiometry, densitometry, or air displacement plethysmography, are not suitable for pregnant women due to the potential risks of radiation. In this regard, simple optical imaging devices such as smart phones and 3D scanners, coupled with the power of AI-assisted processing (Pleuss et al., 2019), have revolutionized anthropometric data acquisition, analysis, and archiving (Choi et al., 2015). One such generalized regression neural network was shown to accurately predict seventy-six different anthropometric dimensions from seven measurable features (Wang, Lee, Bavendiek, & Eckstein, 2021). Unlike anthropometric data, the assessment of nutritional markers in body fluids tends to be more accurate owing to the sensitivity of the biophysical and biochemical tests employed (Maruvada et al., 2020) However, the most significant bottleneck in this regard is the paucity of a sufficient number of such biomarkers that could provide the nutritional status of an individual over a prolonged time span, since these biomarkers give only the current physiological picture and are grossly uninformative regarding the general regulation of metabolic homeostasis of an individual. In addition, these markers are also incapable of capturing the subtle nuances of certain mineral micronutrients like iron or zinc, which could adversely and quickly affect the health of an individual without any substantial warning from the serum biochemical profile (Razzaque & Wimalawansa, 2025). ML approaches have unearthed new predictive tools to circumvent such issues; a logistic regression model achieved an impressive area under curve (AUC) of 0.92 in predicting large artery atherosclerosis by incorporating variables such as body mass index, smoking, and the effect of antidiabetic, antihypertensive, and antihyperlipidemia medications, along with novel biomarkers such as metabolites involved in aminoacyl tRNA biosynthesis and lipid metabolism (Sun, Wang, Wu, Hsu, & Lee, 2023). A combination of ML tools, including logistic regression, random forest, support vector machine (SVM), XGBoost, and deep neural networks, was investigated by Ning et al. to obtain better insights into the propensity of developing neurodegenerative disorders from serum biochemical markers, where XGBoost was able to attain an impressive AUC of 0.978 (Ning et al., 2025).

A significant determinant of the metabolic profile of an individual is his/her gut microbial consortia. The cumulative contribution of the gut microbiome to the human metagenome is a staggering 3 million genes (Zhu, Wang, & Li, 2010) which makes a profound impact on the human metabolome. The gut microbial enzymes play a significant role in the digestion and metabolism of dietary components, including polysaccharides, polyphenols, and vitamins, which serve as an abundant supply of prebiotics for them. The role of microbial enzymes in regulating metabolic homeostasis has been demonstrated in germ-free animal models (Uzbay, 2019) and also by administering human fecal matter (Fecal Matter Transplantation) (Ser, Letchumanan, Goh, Wong, & Lee, 2021). Metagenomic and metatranscriptomic studies have further unearthed the role of gut microbes in myriad gene expression environments inside the cellular microniche, which controls the fine-tuning between healthy and diseased states. A related study carried out on the fecal microbiota of hepatitis B-inflicted liver cirrhosis patients revealed enhanced metabolism of glutathione and increased gluconeogenesis, along with a decrease in bile acid-related metabolism and aromatic amino acids (Wei et al., 2013). Such studies have provided especially interesting insights into the roles of unique microbe-only enzymes, such as the bile salt hydrolase, which can hydrolyze the conjugated amino acid to give the free bile salt that can then be utilized for bacterial metabolism (Agolino et al., 2024). To systematically document the role of gut microbial enzymes in human metabolism, an ML algorithm termed ECemble or enzyme classification using Ensemble approach method has been developed using a combination of five different ML algorithms, namely, Naïve Bayes, K-nearest neighbor classifier, SVM, decision stump, and random forest (Mohammed & Guda, 2015) which utilizes a hierarchical enzyme classification akin to Enzyme Commission nomenclature. A random forest algorithm has been successfully employed to integrate information from a patient's plasma constitution, microbiome, and anthropometric data and evaluate it with respect to his dietary constitution, glycemic index, and cardiometabolic health (Zeevi et al., 2015). Subsequently, a random forest regression model has also been used to predict other biochemical parameters apart from postprandial glucose, including triglyceride and C-peptide analysis (Berry et al., 2020).

There has been a concerted effort from the biomedical and the engineering fraternity to systematically catalog the burst of multiomics data into organized, meaningful, and extrapolatable databases and to harness them for developing AI-based prediction systems in the field of precision nutrition and nutrigenomics, with the unified objective of developing a holistic healthcare system with customized, need-based solutions for every individual. However, this discipline is still in an early, nascent state and is limited by the heterogeneity of data and biased algorithms, among other factors.

7 Conclusion

Nutrition is a far more complex field than it may initially seem, encompassing the science of food, health, and ailments. At its core lies the intricate chemistry of food ingredients, along with their physicochemical properties, stability, bioavailability, and potential toxicity within our bodies. However, the true multidimensionality of the discipline becomes evident when we attempt to understand the interaction of food components with the products of our metagenome, within the dynamic metabolic milieu of our gut and other areas. Therefore to make any meaningful progress toward achieving personalized nutrition goals, a top-down systems biology approach that integrates multiomics data with computational and mathematical modeling is indispensable. AI has emerged as a savior of big data and holds the promise of delivering significant advancements toward the realization of personalized diet recommendations for everyone, ultimately aiming for a world with zero-hunger, zero-malnutrition, and zero-disease. As a word of caution, however, we find ourselves precariously balanced on the edge of a metaphorical wall, with AI as our "Genie of the lamp" on the one side and as our "Terminator" on the other side. It may require the pinnacle of human intellect to determine our next steps on this wall, to flourish or to perish.

Declaration of generative AI and AI-assisted technologies in the writing process

The authors declare that no generative AI tool, such as ChatGPT, was used for the drafting and finalization of this chapter.

Acknowledgement

The authors express their gratitude to the Officer-in-Charge, Government General Degree College, Narayangarh, West Bengal, India, for providing 24×7 online and offline library resources for writing this chapter.

References

Acevedo-Espinola, R., Victoria, R. M. A.-L., Olivares-Etchebaster, M., & Tume, F. (2024). Epigenetics and diet: how does food regulate our genes? *Food Research*. 25502166. *8*(2), 57–63. doi:10.26656/fr.2017.8(2).546.

Agolino, G., Pino, A., Vaccalluzzo, A., Cristofolini, M., Solieri, L., Caggia, C., et al. (2024). Bile salt hydrolase: the complexity behind its mechanism in relation to lowering-cholesterol lactobacilli probiotics. *Journal of Functional Foods*. 17564646. *120*, 106357. doi:10.1016/j.jff.2024.106357.

Agrawal, K., Goktas, P., Kumar, N., & Leung, M. F. (2025). Artificial intelligence in personalized nutrition and food manufacturing: a comprehensive review of methods, applications, and future directions. *Frontiers in Nutrition*. 2296861X. *12*. Available from: https://journal.frontiersin.org/journal/nutrition. doi:10.3389/fnut.2025.1636980.

Andersen, L. K., & Reading, B. J. (2024). A supervised machine learning workflow for the reduction of highly dimensional biological data. *Artificial Intelligence in the Life Sciences*. 26673185. *5*. Available from: https://www.sciencedirect.com/science/journal/26673185. doi:10.1016/j.ailsci.2023.100090.

Ayoub, S., Arabi, M., Al-Najjar, Y., Laswi, I., Outeiro, T. F., et al. (2025). Glycation in Alzheimer's disease and type 2 diabetes: the prospect of dual drug approaches for therapeutic interventions. *Molecular Neurobiology*. 15591182. *62*(11), 14859–14882. Available from: https://www.springer.com/journal/12035. doi:10.1007/s12035-025-05051-9.

Balasubramaniam, S., Velmurugan, Y., Jaganathan, D., & Dhanasekaran, S. (2023). A modified LeNet CNN for breast cancer diagnosis in ultrasound images. *Diagnostics*. 2075-4418. *13*(17), 2746. doi:10.3390/diagnostics13172746.

Banik, S. P. (2024). Glycation-induced amyloid formation in proteins: An emerging perspective to explore diabetes associated onset of neurodegenerative symptoms. *Current Nutrition & Food Science*. 15734013. *20*(1), 2–7. doi:10.2174/1573401319666230224094812.

Banik, S. P., Bagchi, D., Banerjee, P., Chakraborty, S., Bagchi, M., Bose, C., et al. (2025). Subtle concentration changes in zinc hold the key to fibrillation of α-synuclein: an updated insight on the micronutrient's role in prevention of neurodegenerative disorders. *Frontiers in Molecular Biosciences*. 2296889X. *12*. Available from: https://journal.frontiersin.org/journal/molecular-biosciences. doi:10.3389/fmolb.2025.1603364.

Batista, K. S., Cintra, V. M., Lucena, P. A. F., Manhães-De-Castro, R., Toscano, A. E., Costa, L. P., et al. (2022). The role of vitamin B12in viral infections: a comprehensive review of its relationship with the muscle-gut-brain axis and implications for SARS-CoV-2 infection. *Nutrition Reviews*. 17534887. *80*(3), 561–578. Available from: http://nutritionreviews.oxfordjournals.org/. doi:10.1093/nutrit/nuab092.

Batko, K., & Ślęzak, A. (2022). The use of big data analytics in healthcare. *Journal of Big Data*. 2196-1115. *9*(1). doi:10.1186/s40537-021-00553-4.

Berry, S. E., Valdes, A. M., Drew, D. A., Asnicar, F., Mazidi, M., Wolf, J., et al. (2020). Human postprandial responses to food and potential for precision nutrition. *Nature Medicine*. 1546170X. *26*(6), 964–973. Available from: http://www.nature.com/nm/index.html. doi:10.1038/s41591-020-0934-0.

Bouchard, L., Rabasa-Lhoret, R., Faraj, M., Lavoie, M. E., Mill, J., Pérusse, L., et al. (2010). Differential epigenomic and transcriptomic responses in subcutaneous adipose tissue between low and high responders to caloric restriction. *American Journal of Clinical Nutrition*. 00029165. *91*(2), 309–320. Available from: http://www.ajcn.org/cgi/reprint/91/2/309. doi:10.3945/ajcn.2009.28085Canada.

Cantarella, C. D., Ragusa, D., Giammanco, M., & Tosi, S. (2017). Folate deficiency as predisposing factor for childhood leukaemia: A review of the literature. *Genes and Nutrition*. 18653499. *12*(1). Available from: http://www.springerlink.com/content/1555-8932. doi:10.1186/s12263-017-0560-8.

Chaurasia, S., Kumar, V., Fatima, S., & Gupta, A.(2025). Applications of Fuzzy Logic as a Smart Tool for Food Quality Management. In: A. K. Shukla (Ed.), *AI Applications in Food Processing and Packaging* (pp. 101–132). Cambridge, UK: Royal Society of Chemistry. Available from: https://books.rsc.org/books/edited-volume/2310/AI-Applications-in-Food-Processing-and-Packaging. doi:10.1039/9781837679119-00101 .

Choi, A., Kim, J. Y., Jo, S., Jee, J. H., Heymsfield, S. B., Bhagat, Y. A., et al. (2015). Smartphone-based bioelectrical impedance analysis devices for daily obesity management. *Sensors (Switzerland)*. 14248220. *15*(9), 22151–22166. Available from: http://www.mdpi.com/1424-8220/15/9/22151/pdf. doi:10.3390/s150922151.

Close, J. C. T. (2023). Big data - big opportunity. *Age and Ageing*. 14682834. *52*(1). Available from: http://ageing.oxfordjournals.org/. doi:10.1093/ageing/afac262.

Crook, J. M., Yoon, S. J. L., Grundmann, O., Horgas, A., & Johnson-Mallard, V. (2023). Subclinical Vitamin C plasma levels associated with increased risk of CAD diagnosis via inflammation: results from the NHANES 2003–2006 Surveys. *Nutrients*. 20726643. *15*(3). Available from: http://www.mdpi.com/journal/nutrients/. doi:10.3390/nu15030584.

Das, A., Basak, P., Pramanik, A., Majumder, R., Ghosh, A., Hazra, S., et al. (2020). Ribosylation induced structural changes in Bovine Serum Albumin: understanding high dietary sugar induced protein aggregation and amyloid formation. *Heliyon*. 24058440. *6*(9). Available from: https://www.sciencedirect.com/science/journal/24058440. doi:10.1016/j.heliyon.2020.e05053.

De Pilli, T. (2022). Application of fuzzy logic system for the pizza production processing optimisation. *Journal of Food Engineering*. 02608774. *319*, 110906. doi:10.1016/j.jfoodeng.2021.110906.

Detopoulou, P., Voulgaridou, G., Moschos, P., Levidi, D., Anastasiou, T., Dedes, V., et al. (2023). Artificial intelligence, nutrition, and ethical issues: a mini-review. *Clinical Nutrition Open Science*.26672685. *50*, 46–56. Available from: https://www.journals.elsevier.com/clinical-nutrition-open-science. doi:10.1016/j.nutos.2023.07.001.

Deutsch, E. W. (2010). The peptide atlas project. *Methods in MolecularBbiology (Clifton, N.J.)*. 19406029. *604*, 285–296. doi:10.1007/978-1-60761-444-9_19.

Ding, H., Hou, H., Wang, L., Cui, X., Yu, W., & Wilson, D. I. (2025). Application of convolutional neural networks and recurrent neural networks in food safety. *Foods*. 23048158. *14*(2). Available from: www.mdpi.com/journal/foods. doi:10.3390/foods14020247.

Farhud, D. D., Yeganeh, M. Z. A., & Yeganeh, M. Z. (2010). Nutrigenomics and nutrigenetics. *Iranian Journal of Public Health*. 22516093. *39*(4), 1–14. Available from: http://journals.tums.ac.ir/upload_files/pdf/17258.pdf.

Farris, K. M., Senior, A. M., Sobreira, D. R., Mitchell, R. M., Weber, Z. T., Ingerslev, L. R., et al. (2024). Dietary macronutrient composition impacts gene regulation in adipose tissue. *Communications Biology*. 23993642. *7*(1). Available from: https://www.springer.com/journal/42003. doi:10.1038/s42003-024-05876-5.

Fu, X., Wang, Y., Cates, R. S., Li, N., Liu, J., Ke, D., et al. (2023). Implementation of five machine learning methods to predict the 52-week blood glucose level in patients with type 2 diabetes. *Frontiers in Endocrinology*. 16642392. *13*. Available from: https://www.frontiersin.org/journals/endocrinology#. doi:10.3389/fendo.2022.1061507.

Garg, R. K. (2025). The alarming rise of lifestyle diseases and their impact on public health: A comprehensive overview and strategies for overcoming the epidemic. *Journal of Research in Medical Sciences*. 1735-1995. *30*(1). doi:10.4103/jrms.jrms_54_24.

Gérard, M., Mahmutovic, M., Malgras, A., Michot, N., Scheyer, N., Jaussaud, R., et al. (2021). Long-term evolution of malnutrition and loss of muscle strength after covid-19: a major and neglected component of long covid-19. *Nutrients*. 20726643. *13*(11). Available from: https://www.mdpi.com/2072-6643/13/11/3964/pdf. doi:10.3390/nu13113964.

Ghag, N., Sonar, H., Jagtap, S., & Trollman, H. (2024). Unlocking AI's potential in the food supply chain: A novel approach to overcoming barriers. *Journal of Agriculture and Food Research*. 26661543. *18*, 101349. doi:10.1016/j.jafr.2024.101349.

Gillespie, M., Jassal, B., Stephan, R., Milacic, M., Rothfels, K., Senff-Ribeiro, A., Hermjakob, H., … D'Eustachio, P. (2022). The reactome pathway knowledgebase 2022. *Nucleic Acids Research*. 0305-1048. *50*(D1), D687. doi:10.1093/nar/gkab1028.

Griss, J., Viteri, G., Sidiropoulos, K., Nguyen, V., Fabregat, A., & Hermjakob, H. (2020). ReactomeGSA - efficient multi-omics comparative pathway analysis. *Molecular and Cellular Proteomics*. 15359484. *19*(12), 2115–2124. Available from: https://www.mcponline.org/content/19/12/2115. doi:10.1074/mcp.TIR120.002155.

Ha, S., Park, J., & Jo, K. (2025). Comparative analysis of regression algorithms for drug response prediction using GDSC dataset. *BMC Research Notes*. 17560500. *18*(1). Available from: https://bmcresnotes.biomedcentral.com. doi:10.1186/s13104-024-07026-w.

Han, X., Liu, Q., Yun, L., Zhang, M., Liu, K., Kwok, L.-Y., et al. (2025). Synergizing artificial intelligence and probiotics: A comprehensive review of emerging applications in health promotion and industrial innovation. *Trends in Food Science & Technology*. 09242244. *159*, 104938. doi:10.1016/j.tifs.2025.104938.

Hong, W., Luan, Y., Yixuan, M., & Zhang, B. (2024). Yingzhe Xiong, Transcriptome analysis provides insights into high fat diet-induced kidney injury and moderate intensity continuous training-mediated protective effects. *Heliyon*. 24058440. *10*(5), e27157. doi:10.1016/j.heliyon.2024.e27157.

Hornung, R., & Wright, M. N. (2019). Block Forests: Random forests for blocks of clinical and omics covariate data. *BMC Bioinformatics*. 14712105. *20*(1). Available from: http://www.biomedcentral.com/bmcbioinformatics/. doi:10.1186/s12859-019-2942-y.

Jaime-Lara, R. B., Colina-Prisco, C., Vega, M. D. J., Williams, S., Usdin, T., Matikainen-Ankney, B., et al. (2024). Diet-induced obesity induces transcriptomic changes in neuroimmunometabolic-related genes in the striatum and olfactory bulb. *International Journal of Molecular Sciences*. 14220067. *25*(17). Available from: http://www.mdpi.com/journal/ijms. doi:10.3390/ijms25179330.

Kanehisa, M., Furumichi, M., Sato, Y., Matsuura, Y., & Ishiguro-Watanabe, M. (2025). KEGG: biological systems database as a model of the real world. *Nucleic Acids Research*. 0305-1048. *53*(D1), D672. doi:10.1093/nar/gkae909.

Lecun, Y., Bengio, Y., & Hinton, G. (2015). Deep learning. *Nature*. 14764687. *521*(7553), 436–444. Available from: http://www.nature.com/nature/index.html. doi:10.1038/nature14539.

Martinez-Lomeli, J., Deol, P., Deans, J. R., Jiang, T., Ruegger, P., Borneman, J., et al. (2023). Impact of various high fat diets on gene expression and the microbiome across the mouse intestines. *Scientific Reports*. 20452322. *13*(1). Available from: https://www.nature.com/srep/. doi:10.1038/s41598-023-49555-7.

Maruvada, P., Lampe, J. W., Wishart, D. S., Barupal, D., Chester, D. N., Dodd, D., et al. (2020). Perspective: Dietary biomarkers of intake and exposure - exploration with omics approaches. *Advances in Nutrition*. 21565376. *11*(2), 200–215. Available from: http://advances.nutrition.org/. doi:10.1093/advances/nmz075.

Mohammed, A., & Guda, C. (2015). Application of a hierarchical enzyme classification method reveals the role of gut microbiome in human metabolism. *BMC Genomics*. 14712164. *16*(7). Available from: http://www.biomedcentral.com/bmcgenomics. doi:10.1186/1471-2164-16-S7-S16.

Nadkarni, G. N. (2023). Introduction to artificial intelligence and machine learning in nephrology. *Clinical Journal of the American Society of Nephrology*. 1555905X. *18*(3), 392–393. Available from: https://journals.lww.com/cjasn/Fulltext/2023/03000/Introduction_to_Artificial_Intelligence:and.16.aspx. doi:10.2215/CJN.0000000000000068.

Nazari, E., Khalili-Tanha, G., Asadnia, A., Pourali, G., Maftooh, M., Khazaei, M., et al. (2023). Bioinformatics analysis and machine learning approach applied to the identification of novel key genes involved in non-alcoholic fatty liver disease. *Scientific Reports*. 20452322. *13*(1). Available from: https://www.nature.com/srep/. doi:10.1038/s41598-023-46711-x.

Ning, W., Wang, Z., Ying, G., Huang, L., Liu, S., Chen, Q., et al. (2025). Machine learning models based on routine blood and biochemical test data for diagnosis of neurological diseases. *Scientific Reports*. 2045-2322. *15*(1). doi:10.1038/s41598-025-09439-4.

Noor, F., Junaid, M., Almalki, A. H., Almaghrabi, M., Ghazanfar, S., & Tahir Ul Qamar, M. (2024). Deep learning pipeline for accelerating virtual screening in drug discovery. *Scientific Reports*. 20452322. *14*(1). Available from: https://www.nature.com/srep/. doi:10.1038/s41598-024-79799-w.

Osama, S., Shaban, H., & Ali, A. A. (2023). Gene reduction and machine learning algorithms for cancer classification based on microarray gene expression data: A comprehensive review. *Expert Systems with Applications*. 09574174. *213*, 118946. doi:10.1016/j.eswa.2022.118946.

Özdemir, V., & Kolker, E. (2016). Precision Nutrition 4.0: a big data and ethics foresight analysis—Convergence of agrigenomics, nutrigenomics, nutriproteomics, and nutrimetabolomics. *OMICS: A Journal of Integrative Biology*. 1536-2310. *20*(2), 69–75. doi:10.1089/omi.2015.0193.

Parasa, N. A., Namgiri, J. V., Mohanty, S. N., & Dash, J. K. (2021) . Introduction to unsupervised learning in bioinformatics. In R. Satpathy, T. Choudhury, S. Satpathy, S. N. Mohanty, & X. Zhang (Eds.), *Data analytics in bioinformatics: A machine learning perspective* (pp. 35–49). MA, United States: Wiley. Available from: https://onlinelibrary.wiley.com/doi/book. doi:10.1002/9781119785620.ch2.

Pavani, M., Singha, P., Rajamanickam, D. T., & Singh, S. K. (2023). Application of fuzzy logic techniques for sensory evaluation of plant-based extrudates fortified with bioactive compounds. *Exploration of Foods and Foodomics*. 28379020. *1*(5), 272–287. Available from: https://www.explorationpub.com/uploads/Article/A101021/101021.pdf. doi:10.37349/eff.2023.00021.

Phugat, S., & Goel, P. (2025). *Review on advancement of AI in nutrigenomics, **2952***. (pp. 429–444). Springer Science and Business Media LLC. doi:10.1007/978-1-0716-4690-8_23.

Pleuss, J. D., Talty, K., Morse, S., Kuiper, P., Scioletti, M., Heymsfield, S. B., et al. (2019). A machine learning approach relating 3D body scans to body composition in humans. *European Journal of Clinical Nutrition*. 14765640. *73*(2), 200–208. Available from: http://www.nature.com/ejcn/index.html. doi:10.1038/s41430-018-0337-1.

Qu, D. C., Neu, D., Khawaja, Z. Q., Wang, R., Bartels, C. F., Lovrenert, K., et al. (2023). Epigenetic effects of high-fat diet on intestinal tumorigenesis in C57BL/6J-ApcMin/+ mice. *Journal of Translational Genetics and Genomics*. 25785281. *7*(1), 3–16. Available from: https://jtggjournal.com/journal/volume_archive/2023/7. doi:10.20517/jtgg.2022.16.

Razzaque, M. S., & Wimalawansa, S. J. (2025). Minerals and human health: From deficiency to toxicity. *Nutrients*. 20726643. *17*(3). Available from: http://www.mdpi.com/journal/nutrients/. doi:10.3390/nu17030454.

Ren, F., Wei, J., Chen, Q., Mengling, H., Yu, L., Jianing, M., et al. (2025). Artificial intelligence-driven multi-omics approaches in Alzheimer's disease: progress, challenges, and future directions. *Acta Pharmaceutica Sinica B*. 22113835. *15*(9), 4327–4385. doi:10.1016/j.apsb.2025.07.030.

Richter, T., Bahrami, M., Xia, Y., Fischer, D. S., & Theis, F. J. (2025). Delineating the effective use of self-supervised learning in single-cell genomics. *Nature Machine Intelligence*. 25225839. *7*(1), 68–78. Available from: https://www.nature.com/natmachintell/. doi:10.1038/s42256-024-00934-3.

Saito, T., Whatmore, P., Taylor, J. F., Fernandes, J. M. O., Adam, A. C., Tocher, D. R., et al. (2021). Micronutrient supplementation affects transcriptional and epigenetic regulation of lipid metabolism in a dose-dependent manner. *Epigenetics*. 15592308. *16*(11), 1217–1234. Available from: http://www.tandfonline.com/toc/kepi20/current. doi:10.1080/15592294.2020.1859867.

Sak, J., & Suchodolska, M. (2021). Artificial intelligence in nutrients science research: a review. *Nutrients*. 20726643. *13*(2), 1–17. Available from: https://www.mdpi.com/2072-6643/13/2/322/pdf. doi:10.3390/nu13020322.

Saraiva, P. (2023). On Shannon entropy and its applications. *Kuwait Journal of Science*. 23074108. *50*(3), 194–199. doi:10.1016/j.kjs.2023.05.004.

Saunders, J. (2010). CMJ1006-CME_Smith.qxd. *Clinical Medicine*, *10*(6), 624–627.

Schmidt, T., Samaras, P., Frejno, M., Gessulat, S., Barnert, M., Kienegger, H., et al. (2018). ProteomicsDB. *Nucleic Acids Research*. 13624962. *46*(1), D1271. Available from: http://nar.oxfordjournals.org/. doi:10.1093/nar/gkx1029.

Ser, H. L., Letchumanan, V., Goh, B. H., Wong, S. H., & Lee, L. H. (2021). The use of fecal microbiome transplant in treating human diseases: too early for poop? *Frontiers in Microbiology*. 1664302X. *12*. Available from: https://www.frontiersin.org/journals/microbiology#. doi:10.3389/fmicb.2021.519836.

Shakya, A. K., Pillai, G., & Chakrabarty, S. (2023). Reinforcement learning algorithms: a brief survey. *Expert Systems with Applications*. 09574174. *231*, 120495. doi:10.1016/j.eswa.2023.120495.

Simillion, C., Liechti, R., Lischer, H. E. L., Ioannidis, V., & Bruggmann, R. (2017). Avoiding the pitfalls of gene set enrichment analysis with SetRank. *BMC Bioinformatics*. 14712105. *18*(1). Available from: http://www.biomedcentral.com/bmcbioinformatics/. doi:10.1186/s12859-017-1571-6.

Singh, J., & Singh, D. (2024). A comprehensive review of clustering techniques in artificial intelligence for knowledge discovery: Taxonomy, challenges, applications and future prospects. *Advanced Engineering Informatics*. 14740346. *62*, 102799. doi:10.1016/j.aei.2024.102799.

Soni, K. G., Dike, P. N., Suh, J. H., Halder, T., Edwards, P. T., Foong, J. P. P., et al. (2020). Early-life malnutrition causes gastrointestinal dysmotility that is sexually dimorphic. *Neurogastroenterology and Motility*. 13652982. *32*(12). Available from: http://onlinelibrary.wiley.com/journal/10.1111/(ISSN)1365-2982. doi:10.1111/nmo.13936.

Stanford, B. C. M., Clake, D. J., Morris, M. R. J., & Rogers, S. M. (2020). The power and limitations of gene expression pathway analyses toward predicting population response to environmental stressors. *Evolutionary Applications*. 17524571. *13*(6), 1166–1182. Available from: http://onlinelibrary.wiley.com/journal/10.1111/(ISSN)1752-4571. doi:10.1111/eva.12935.

Sun, T. H., Wang, C. C., Wu, Y. L., Hsu, K. C., & Lee, T. H. (2023). Machine learning approaches for biomarker discovery to predict large-artery atherosclerosis. *Scientific Reports*. 20452322. *13*(1). Available from: https://www.nature.com/srep/. doi:10.1038/s41598-023-42338-0.

Surguchov, A., Bernal, L., & Surguchev, A. A. (2021). Phytochemicals as regulators of genes involved in synucleinopathies. *Biomolecules*. 2218273X. *11*(5). Available from: https://www.mdpi.com/2218-273X/11/5/624/pdf. doi:10.3390/biom11050624.

Švagždienė, M., Širvinskas, E., Baranauskienė, D., & Adukauskienė, D. (2015). Correlation of magnesium deficiency with C-reactive protein in elective cardiac surgery with cardiopulmonary bypass for ischemic heart disease. *Medicina*. 1010660X. *51*(2), 100–106. doi:10.1016/j.medici.2015.03.003.

Tao, X., Jiang, M., Liu, Y., Hu, Q., Zhu, B., Jiaqiang, H., et al. (2023). Rongsheng Tong, Enwu Long, Predicting three-month fasting blood glucose and glycated hemoglobin changes in patients with type 2 diabetes mellitus based on multiple machine learning algorithms. *Scientific Reports*. 2045-2322. *13*(1). doi:10.1038/s41598-023-43240-5.

Tiffon, C. (2018). The impact of nutrition and environmental epigenetics on human health and disease. *International Journal of Molecular Sciences*. 1422-0067. *19*(11), 3425. doi:10.3390/ijms19113425.

Türkmenoğlu, C., Şima Etaner Uyar, A., & Kiraz, B. (2021). Recommending healthy meal plans by optimising nature-inspired many-objective diet problem. *Health Informatics Journal*. 1460-4582. *27*(1). doi:10.1177/1460458220976719.

Uzbay, T. (2019). Germ-free animal experiments in the gut microbiota studies. *Current Opinion in Pharmacology*. 14714892. *49*, 6–10. doi:10.1016/j.coph.2019.03.016.

Van Ommen, B., El-Sohemy, A., Hesketh, J., Kaput, J., Fenech, M., Evelo, C. T., et al. (2010). The Micronutrient Genomics Project: A community-driven knowledge base for micronutrient research. *Genes and Nutrition*. 18653499. *5*(4), 285–296. doi:10.1007/s12263-010-0192-8.

Wang, L., Lee, T. J., Bavendiek, J., & Eckstein, L. (2021). A data-driven approach towards the full anthropometric measurements prediction via Generalized Regression Neural Networks. *Applied Soft Computing*. 15684946. *109*, 107551. doi:10.1016/j.asoc.2021.107551.

Wei, X., Yan, X., Zou, D., Yang, Z., Wang, X., Liu, W., et al. (2013). Abnormal fecal microbiota community and functions in patients with hepatitis B liver cirrhosis as revealed by a metagenomic approach. *BMC Gastroenterology*. 1471230X. *13*(1). Available from: http://www.biomedcentral.com/1471-230X/13/175. doi:10.1186/1471-230X-13-175China.

Wu, Y., Xie, L., Liu, Y., & Xie, L. (2024). Semi-supervised meta-learning elucidates understudied molecular interactions. *Communications Biology*. 23993642. *7*(1). Available from: https://www.springer.com/journal/42003. doi:10.1038/s42003-024-06797-z.

Yan, C., Chenmin, H., Chen, X., Jia, X., Zhu, Z., Diya, Y., et al. (2023). Vitamin D improves irritable bowel syndrome symptoms: a meta-analysis. *Heliyon*. 24058440. *9*(6), e16437. doi:10.1016/j.heliyon.2023.e16437.

Yetgin, A. (2025). Revolutionizing multi-omics analysis with artificial intelligence and data processing. *Quantitative Biology*, 20954697, *13*(3). Available from: https://onlinelibrary.wiley.com/journal/20954697. doi:10.1002/qub2.70002.

Zeevi, D., Korem, T., Zmora, N., Israeli, D., Rothschild, D., Weinberger, A., et al. (2015). Personalized nutrition by prediction of glycemic responses. *Cell*. 10974172. *163*(5), 1079–1094. Available from: https://www.sciencedirect.com/journal/cell. doi:10.1016/j.cell.2015.11.001.

Zhang, F. F., Barr, S. I., McNulty, H., Li, D., & Blumberg, J. B. (2020). Health effects of vitamin and mineral supplements. *The BMJ*, 17561833. *369*. Available from: http://www.bmj.com/. doi:10.1136/bmj.m2511.

Zhu, B., Wang, X., & Li, L. (2010). Human gut microbiome: the second genome of human body. *Protein and Cell*. 16748018. *1*(8), 718–725. Available from: http://link.springer.com/journal/13238. doi:10.1007/s13238-010-0093-z.

Zhu, H., Shyama Prasad Rao, R., Zeng, T., & Chen, L. (2012). Reconstructing dynamic gene regulatory networks from sample-based transcriptional data. *Nucleic Acids Research*. 1362-4962. *40*(21), 10657–10667. doi:10.1093/nar/gks860.

Zou, Y., Zhang, R., Huang, L., Zhao, D., Danting, S., Meng, J., et al. (2022). Serum levels of vitamin D, retinol, zinc, and CRP in relation to obesity among children and adolescents. *European Journal of Medical Research*. 2047-783X. *27*(1). doi:10.1186/s40001-022-00670-7.

Chapter 29

New frontiers in organoid research: The synergistic integration of organoids with artificial intelligence

Soumya Shekhar, Subhayan Sur, Amit Ranjan, and Soumya Basu
Cancer and Translational Research Centre, Dr. D. Y. Patil Biotechnology and Bioinformatics Institute, Dr. D. Y. Patil Vidyapeeth, Pune, Maharashtra, India

1 Introduction

Organoid technology has evolved into one of the most important breakthroughs in modern biomedical science. This technology allows researchers to recreate miniature versions of human organs that mimic the structure and function of their in vivocounterparts, enabling advanced studies in disease modeling, drug testing, and regenerative medicine. These three-dimensional (3D) cultures, which are developed from stem cells or tissue explants, provide scientists with unparalleled opportunities to examine how organs develop, how diseases advance, and how drugs interact with complex tissue systems (Ahammed & Kalangi, 2024). In genetic disease research, organoids have been instrumental in elucidating the complex relationships between genetic mutations and disease pathology, paving the way for personalized treatment approaches (Huang, Jeong, Kim, & Chen, 2025). Organoids also advance animal disease modeling by improving our understanding of complex neurological, respiratory, and gastrointestinal disorders. When coupled with CRISPR–Cas9, they provide a platform for precisely engineering and dissecting intricate disease mechanisms (Buragohain et al., 2023).

In oncology, organoids are an important tool for studying tumor biology, as, unlike traditional two-dimensional (2D) cultures or animal models, they might provide an accurate representation of the tumor microenvironment, aiding in the study of complex interactions between cancer cells and other stromal, vascular, and immune components within the tumor microenvironment (Zeng et al., 2024). Such developments are invaluable for understanding how drug resistance mechanisms work and for capturing patient-specific variability in drug responses. Therefore, organoids support precision oncology, such as the design of customized drug screening strategies and personalized treatment regimens (Zeng et al., 2024).

An upcoming area of organoid research is the field of organoid intelligence (OI). OI currently uses brain organoids, which are miniature, self-organizing neural structures derived from pluripotent stem cells. These organoids showcase rudimentary cognitive functions such as learning, memory, and information processing. They are being investigated as living biocomputing platforms that display adaptive behaviors and react to stimuli (Ballav, Ranjan, Sur, & Basu, 2024; Smirnova et al., 2023). The unique neurobiological characteristics of brain organoids are useful for assessing patient-specific disease mechanisms, drug responsiveness, and novel therapeutic interventions (Barros, Kagan, Hartung, & Smirnova, 2024). They serve as platforms to study neuroplasticity, early brain development, and neurotoxicity, which help advance precision medicine (Alam El Din et al., 2024). Developments in culture methods and recording technologies have expanded what OI can achieve. Integrated microfluidic perfusion systems now enable long-term, scalable cultivation of brain organoids, while 3D microelectrode arrays can provide high-resolution readings of neural activity (Smirnova et al., 2023). Together, these advancements enable researchers to examine the molecular basis of cognition.

However, OI raises ethical concerns. The possibility of organoids approaching higher-order cognitive states prompts questions about sentience and moral status (Smirnova et al., 2023).

Organoid–computer integration may enable brain organoids to be trained through stimulus–response paradigms, leading to novel forms of biocomputing. This can result in advanced networks wherein organoids communicate with real-world sensors, sensory organ organoids, and output devices, leading to improved data processing and continuous learning

Applications of Artificial Intelligence in Healthcare. https://doi.org/10.1016/B978-0-443-43934-6.00024-9

(Wadan, 2024). Hybrid systems such as "Brainoware" exemplify the potential of living neural networks to outperform conventional hardware. Brainoware utilizes 3D biological neural networks within brain organoids, enabling it to perform complex computations and on-chip learning through spatiotemporal electrical stimulation, which allows for nonlinear dynamics, fading memory, and the ability to learn from training data (Cai et al., 2023).

A persistent obstacle is the complexity of organoid formation and the enormous datasets they generate. Conventional methods of analysis are insufficient for the huge volumes of data generated, ranging from sequencing outputs to complex imaging datasets. Manual observation is slow and error-prone, making large-scale application impractical, and the inherent diversity of organoids necessitates computational assistance and resources (Bai et al., 2024; Du et al., 2023). Artificial intelligence (AI) helps to organize, extract, and interpret this large-scale information, increasing accuracy and efficiency, and enabling predictive modeling and personalized therapeutic strategies. Still, its utility depends heavily on biological fidelity, and computational models must meaningfully capture the underlying biology if they are to provide true translational value (Wang et al., 2024).

The road ahead lies in merging AI and organoids more seamlessly, with the potential to not only enhance analysis and discovery but also to usher in a new era of personalized medicine and unconventional computing. AI can optimize organoid workflows, from design and fabrication to monitoring and interpretation. This addresses reproducibility and scalability, providing new directions in computing and therapy.

While organoids are redefining biomedical research, brain organoids push the boundaries of biological computing.

2 Organoid technology: State of the art

Organoids are increasingly recognized as a superior alternative to traditional cultures and animal models, as they closely mimic human physiology. Unlike 2D cultures, organoids exhibit self-organization, stability, and flexibility in culture, the ability to replicate organ-specific microanatomy and maintain cellular polarity. They make it feasible for cells to interact with their surrounding matrix and with one another in a physiologically relevant manner, yielding responses that closely mirror in vivo biology (Devarasetty, Mazzocchi, & Skardal, 2018). Organoids are suited for applications like high-throughput screening and real-time imaging, as they can be grown over long periods of time and experimentally manipulated. These 3D structures help connect complex in vivomodels with traditional 2D systems.

Organoids can be derived from stem cell sources, including pluripotent (embryonic or induced), adult, or tissue-specific stem cells. Through self-organization, they mimic the structure and function of the organ they imitate. Researchers have created organoids for many tissues, including the brain, retina, pituitary, inner ear, heart, lung, liver, kidney, pancreas, and gastrointestinal tract. They can represent both healthy and diseased states. Tumor-derived organoids are especially useful as models for cancer research. Organoids have a wide range of applications (Fig. 29.1), proving to be indispensable in disease modeling for complex conditions like cancers, infectious diseases, and hereditary disorders that are difficult to reproduce with conventional methods (Yao et al., 2024). In drug discovery, organoids offer scalable platforms for large-scale screening, improved predictions of drug toxicity, efficacy, and pharmacokinetics, thereby improving the translational value of preclinical therapies. These insights from drug screening help in developing tailored treatment strategies (Zhou et al., 2024). Furthermore, patient-specific organoids derived from biopsies can generate isogenic models that are useful for regenerative medicine, thereby reducing the shortage of healthy donor tissues and mitigating issues related to immune rejection (Xu et al., 2018).

Organoid development faces multiple challenges that vary with the complexity and intended application of the system, requiring tailored approaches to address specific biological and technical demands (Table 29.1). Current models have not yet fully achieved the structural and functional complexity of their in vivo counterparts, with technical obstacles such as a lack of vascularization, dependence on passive diffusion for nutrients and oxygen, and challenges with reproducibility (Ge et al., 2024). The inability to completely mimic the complex spatial organization and diverse cell populations present in vivo can limit their physiological relevance and functional fidelity (Mecca, Tang, Jones, Coward, & Yeste, 2024). Ethical issues, such as the source of stem cells, informed consent, and the moral status of organoids, continue to shape the future of this field. Therefore, continuous innovation and evaluation are needed to address ethical and social concerns.

3 AI methodologies for organoid analysis

3.1 Organoid imaging, tracking, segmentation, differentiation, and morphological analysis

AI-enabled imaging allows for continuous, noninvasive, long-term observation of individual organoids or clusters, preserving specimens for downstream analysis, unlike destructive sampling methods. This enables researchers to conduct spatiotemporal developmental studies and track dynamic drug responses directly within living systems.

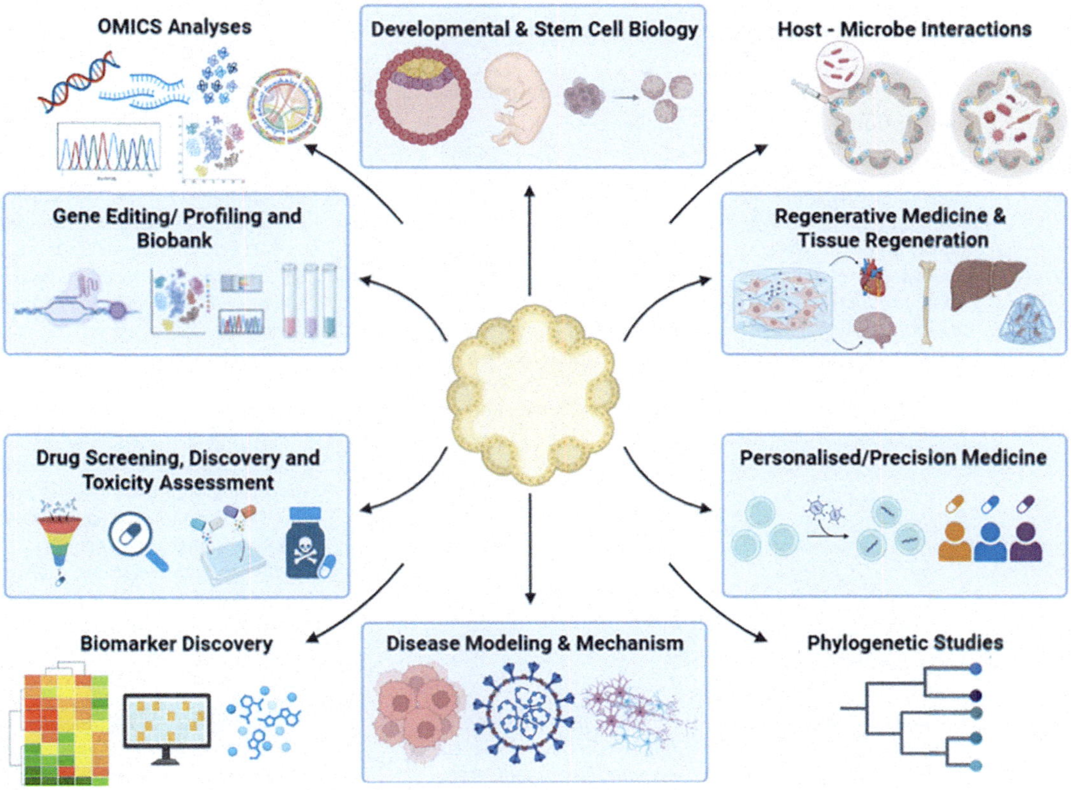

FIG. 29.1 Diverse applications of organoid technology. *(Figure created with Biorender.com).*

TABLE 29.1 Technical hurdles and innovative solutions in organoid culture systems.

Technical difficulties	Description	User-related limitations and challenges	Solutions and future outlook
Biological source heterogeneity and lack of interorgan communication	*Pluripotent stem cells used for organoid development may exhibit genetic variability introduced during reprogramming or clone selection. Most single-organoid models fail to mimic the crucial communication among different organs.*	*Prolonged in vitro culture may induce genetic changes and mutations. Organoid formation and function can be influenced by donor-to-donor genetic differences and by iPSCs/cell lines derived from the same donor, complicating phenotype interpretation. Maintaining pluripotency and genetic stability is challenging.*	*This heterogeneity is advantageous in capturing different phenotypic variations and developmental outcomes, aiding disease modeling approaches. Direct comparisons between cell lines used in the same experiment can be performed in parallel to tackle batch variability. Optimization of culture conditions is essential. Validation should be confirmed through replication across multiple technical and biological replicates.*
Variations in differentiation outcomes and maturation	*The different stem cells each have varying levels of maturation. Organoids, unlike adult organs or tissues, are less mature and face limitations in achieving full maturity and physiological function. Differences in conditions and protocols largely influence their development. They can lack native tissue complexity, innervation, and crucial mature cell types such as immune cells.*	*Variability in iPSC maturation levels and prolonged multistep differentiation protocols are major issues that can lead to off-target cell type formation or retain undifferentiated progenitor cells/stem cells. Subtle changes in concentration or timing can lead to varied differentiation outcomes or poor organoid quality. Batches of ECM/Matrigel can be inconsistent, leading to experimental variability. Finding the right ECM for a specific organoid type is often a trial-and-error process.*	*Development of high-quality, standardized iPSC cell line biobanks and tissue-specific progenitors as initial materials for organoid creation. Optimization of differentiation protocols and culture conditions, specifically the use of specific growth factor cocktails/supplements, predefined culture media, and optimal 3D scaffolds (hydrogels) with adjustable mechanical properties and signaling molecules. Other strategies include: in vivo organoid transplantation in mammalian hosts, electrical stimulation (in cortical organoids), and optimizing culture matrix composition.*

Continued

TABLE 29.1 Technical hurdles and innovative solutions in organoid culture systems—cont'd

Technical difficulties	Description	User-related limitations and challenges	Solutions and future outlook
Vascularization and nutrient delivery	*Lack of vascularization leads to passive diffusion-based nutrient and oxygen supply and limits the removal of metabolic waste products, restricting organoid growth and functional maturation.*	*Organoids lack a vascular network, leading to a necrotic core in larger structures due to diffusion limitations. This limits their ability to mimic the function and development of their in vivo counterparts.* *Differences in culture conditions, such as oxygen levels, CO_2 concentration, and temperature, might have an effect on organoid cultures.*	*Current solutions includein vivo engraftment of organoids in immunodeficient animals, coculture with vascular cells/endothelial cells (ECs), immune cells, or multipotent cells, creation of assembloids, and tumor spheroid–ECs–immune cell coculture organoids.* *Patient-derived tumor organoids retain native vasculature.* *Other strategies include organoid codifferentiation, coculture with blood vessel organoids or vascular organoids, employing organoids-on-chip platforms, and 3D/4D bioprinting of vascular networks to create vascularized organoids.* *Dynamic culture systems, such as rotational culture systems, spinner flasks, or pillar/perfusion plates, can also be used to facilitate homogeneous flow and diffusion of nutrients and oxygen.*
Data handling complexity, cost, and time	*The inherent heterogeneity of organoids makes data analysis challenging.* *Researchers need advanced computational tools and expertise to account for the variability and interpret the results correctly.*	*Organoid culture is resource-intensive, requiring specialized reagents, equipment, and a significant time commitment (weeks to months) to grow.* *The size and 3D nature of organoids make imaging challenging. Sophisticated software is commonly needed to reconstruct and analyze the complex 3D data.*	*Implementation of automated culture systems to reduce manual labor and reagent consumption.* *Development of open-source resources and protocols within the research community to reduce costs.* *Implementation of AI algorithms for organoid analysis on multiple levels provides an efficient solution to handle multidimensional organoid data. This offers a powerful and predictive model.*
Standardization and reproducibility	*Protocols are often highly user-dependent, and the self-organizing nature of organoids can lead to significant variability in size, structure, cell-type composition, maturity, and function across different culture batches, experiments, and laboratories. This hinders the comparison of results.*	*Creation and maintenance of organoids require technical knowledge, manual skill, expertise, and time, emphasizing practice. This makes it challenging for novice researchers or for high-throughput screening applications.* *The success of a culture can be highly dependent on the researcher's experience.* *Differences in cell dissociation methods, plating density, and passaging protocols can influence organoid maintenance.*	*Develop standardized, open-source protocols and utilize automated systems for culture and analysis.* *The establishment of rigorous quality control metrics and standards would further facilitate reproducible results. This allows for the creation of standardized disease models and drug screening platforms.*

Abbreviations: *ECs*, endothelial cells; *ECM*, extracellular matrix; *iPSC*, induced pluripotent stem cell.

Deep learning (DL) is an indispensable tool for segmentation, tracking, and morphological assessment of organoids. Convolutional neural networks (CNNs) have enhanced accuracy and efficiency. For example, an AI-based segmentation and analysis pipeline was developed for high-field magnetic resonance imaging of cerebral organoids. Using a 3D U-Net neural network, this method enabled segmentation of organoid MRI data, improving the accuracy and reliability of monitoring organoid growth over time (Deininger et al., 2023). A high-throughput imaging platform that integrates automated image processing and advanced image processing tools can be used for quantifying organoids in cocultures. Stüve et al. (2024) showed how machine learning (ML) can detect and classify 3D organoids within dense cocultures of immune cells, offering detailed quantitative measurements of factors such as size, quantity, and morphology (shape and compactness). Other specialized architectures, like the Branched ERF-Net architecture, have been used to segment overlapping organoid structures and generate smoother shape proposals, making it easier to measure morphological features such as organoid budding and size, and facilitating precise counting (Sureshkumar, Bisht, & Easwaran, 2024).

Researchers have developed DL pipelines for a variety of applications. One pipeline combined a specialized CNN with pre- and postprocessing strategies to track the growth and morphological changes of 3D cancer organoids over a 13-day culture period. This offers a framework for drug screening and testing tumor sensitivity (Branciforti et al., 2024). Similarly, Hradecka, Wiesner, Sumbal, Koledova, & Maska (2023) used synthetic data produced by a conditional generative adversarial network to train a U-Net model, which analyzed the 2D time-lapse bright-field images of mouse mammary epithelial organoids. This automated workflow effectively segmented and tracked the spatiotemporal development of the epithelial organoids in time-lapse sequences, enabling long-term organogenesis studies.

Several open-source platforms function as accessible tools for organoid image analysis, achieving high accuracy in morphology assessment and facilitating rapid processing of large image datasets. The OrganoID platform employs a modified U-Net network model for automated recognition, labeling, tracking (of organoid count and area), and morphological analysis of organoids, producing detailed measurements of circularity and growth. It revealed significant changes in organoid circularity in response to gemcitabine, emphasizing its utility in high-throughput drug screening and cancer research (Matthews et al., 2022). Similarly, MOrgAna (Machine-learning-based Organoid Analysis software), a Python-based software, can process multiple images in minutes. It enables segmentation, quantification, and visualization of morphological and fluorescence data across images. Using advanced DL algorithms to classify pixels into background, organoid, and organoid edge, MOrgAna accurately separates true organoids (via detection of organoid boundaries) from surrounding debris and dying cells (Gritti et al., 2021). This helps tackle the issue of overlapping structures, which complicates the analysis of morphology and quantification.

OrgaQuant automates the localization and quantification of intestinal organoids in bright-field images using deep CNNs, eliminating manual parameter adjustments and scaling analysis to multiple samples (Kassis, Hernandez-Gordillo, Langer, & Griffith, 2019). OrganelX, which employs DL to offer an e-Science service for automating the localization and quantification of organoid size, has been tested on murine liver organoids (Haja, Horcas-Nieto, Bakker, & Schomaker, 2023). The OrgaExtractor is a multiscale U-Net-based system that performs accurate segmentation across diverse organoid sizes, providing measurements that reflect actual culture conditions and aid in long-term organoid maintenance (Park et al., 2023). MacDonald et al. (2020) demonstrated the use of advanced AI models, such as residual U-Net and mask R-CNN, for precise segmentation of kidney organoids.

Lightweight DL-based pipelines, such as Deep-Orga, have been developed to automate the detection and morphological assessment of intestinal organoids, overcoming the limitations of subjective, manual morphological analysis (Leng, Jiang, Wang, Wang, & Luo, 2024). DL is applied to select organoids with attributes that are specific and reflective of human tissues, and to predict biomarker expression through image-based CNN analysis (Lim et al., 2023). Previous studies describe the application of DL models to predict organoid differentiation outcomes, adding another layer of precision to organoid assessment (Asano et al., 2024; Kegeles, Naumov, Karpulevich, Volchkov, & Baranov, 2020; Lim et al., 2023).

Specialized DL-based tools are being built to enable complex morphological classification. D-CryptO allows for automated classification of morphology and determines colorectal organoid maturity by assessing crypt formation and opacity. This system has been especially useful for analyzing organoid responses to different chemotherapeutic treatments, where it detected subtle morphological variations that may be related to distinct chemotherapeutic mechanisms (Abdul et al., 2022).

DL has been coupled with imaging techniques like optical coherence tomography (OCT) for refined analysis. This enables noninvasive, live imaging of cardiac, cerebral, and retinal organoids derived from human-induced pluripotent stem cells (hiPSCs). This model, with a framework based on U-net architecture, is called OrgSegNet and enables automatic segmentation of internal structural features and tissues, such as cavities and chambers, allowing for quantitative assessments of morphology (size, area, and volume) and cardiac beating dynamics (Wang et al., 2024). Bao et al. (2023) presented the Organoid Printing and optical coherence-based analysis (OPO) method, which employs OCT with learning-based algorithms to rebuild multiscale structures of patient-derived organoid (PDO) clusters within printed microbeads. This approach facilitates accurate monitoring of growth dynamics and structural evolution by quantifying and analyzing critical morphological properties such as number and volume.

These highlight how the pairing of AI with advanced imaging techniques has revolutionized organoid research. This sets the stage for more reliable organoid studies (Table 29.2).

TABLE 29.2 Summarized overview of models that employ DL networks for organoid analysis.

Platform name	Role in organoid analysis	Key AI methodologies used (if specified)	Organoid type/data tested	Unique features	Example source (citation)
OrganoID	*Automatic recognition, labeling, and tracking of individual organoids (pixel-by-pixel). Performs bulk and single-organoid morphology analyses.*	*CNN based on U-net structure.*	*Bright-field microscopy and phase-contrast microscopy images. Trained on pancreatic cancer organoids; Tested on images of lung, pancreas, colon, and adenoid cystic carcinoma organoids.*	*Open-source, detailed, and automated image analysis, potential for high-throughput drug screening. Analyzes images from diverse organoids; enables data analysis for different diseases.*	*(Matthews et al., 2022)*
MOrgAna	*Segments, quantifies, visualizes, and characterizes morphological/fluorescence data of organoids across multiple images within minutes. Provides visual quantitative plots.*	*Python-based software. Utilizes MLP DL algorithms (pixel classification).*	*Applied across several organoid types and in zebrafish explants (pescoids). Images captured through benchtop stereomicroscopes or confocal microscopes.*	*Open-source and user-friendly (no programming expertise needed). Provides higher accuracy in segmenting organoid boundaries.*	*(Gritti et al., 2021)*
OrgaQuant	*Automated localization and quantification of the size distribution of intestinal organoids within bright-field images.*	*Faster R-CNN with Inception-ResNet-v2 architecture.*	*Novel dataset of manually annotated human intestinal organoid bright-field images with bounding boxes. TensorFlow-trained object detection pipeline.*	*Open-source. Operates without parameter adjustments or human intervention; efficient for processing thousands of images.*	*(Kassis et al., 2019)*
OrganelX	*Automatically detects, localizes, and segments organoids to compute the area and count of organoids.*	*Mask R-CNN.*	*Tested on murine liver organoid bright-field images.*	*Publicly accessible, free e-Science service. Can select and track the sizes of multiple organoids over different Z-stacks of the dome.*	*(Haja et al., 2023)*
OrgaExtractor	*Accurate image segmentation across diverse organoid sizes. Provides various organoid measurements, including perimeter, projected area, axis length, eccentricity, etc.*	*DL model based on multiscale U-Net architecture.*	*Patient derived normal-colon organoid images.*	*User-friendly measurements reflect actual culture conditions; aids long-term organoid maintenance. Can recognize organoids at different growth stages; analyzed data can be visualized to monitor real-time organoid growth.*	*(Park et al., 2023)*
Deep-Orga	*Automated organoid detection and extraction of morphological features of organoids from images for automated morphological analysis.*	*Improved YOLOX-based DL-based lightweight model.*	*Bright-field images of patient-derived intestinal organoids.*	*Addresses labor-intensive and subjective manual analysis.*	*Leng et al. (2024)*
D-CryptO	*Performs morphology classification, analyzes organoid structure, categorizes organoid budding, determines structural maturity (automatically assesses crypt formation, opacity).*	*DL-based image analysis tool that utilizes a pretrained deep neural network model.*	*Bright-field images of colorectal organoids.*	*Detects subtle morphological variations; suggests distinct mechanisms of action in drug testing.*	*(Abdul et al., 2022)*

Platform name	Role in organoid analysis	Key AI methodologies used (if specified)	Organoid type/data tested	Unique features	Example source (citation)
Organoid Printing and optical coherence tomography-based analysis (OPO)	*3D imaging, segmentation, and analysis of printed organoid clusters. Enables the study of organoid cluster growth, heterogeneity, and structural evolution. Tracks and quantifies morphological parameters such as organoid number and volume.*	*Utilizes various deep CNNs. Proposed a new neural network model referred to as EGO-Net.*	*3D OCT images of patient-derived cancer organoid clusters within printed micro-beads.*	*Reconstructs multiscale structures, enables precise tracking and quantitative analysis of morphological parameters.*	*(Bao et al., 2023)*

Abbreviations: *CNN*, convolutional neural network; *Mask R-CNN*, mask-region-based convolutional neural network; *MLP*, multilayer perceptron; *OCT*, optical coherence tomography; *R-CNN*, region-based convolutional neural network; *YOLOX*, You Only Look Once X algorithm.

3.2 The synergistic potential: AI-enhanced organoid construction and optimization

The combination of organoid technology, tissue engineering, and AI is changing how researchers design and optimize organoids. Creating functional organoids involves several steps: selecting the appropriate stem cells, establishing a conducive 3D culture environment, and adjusting growth conditions to match the physiological features of the target organ. Challenges, such as the complexity of organoid assembly and morphological variability, persist.

Here, AI, particularly ML, can provide revolutionary solutions. ML algorithms can guide and optimize diverse aspects of organoid construction, matrix gel synthesis, discerning the spatial structure of the gel, culture conditions, evaluating external stimuli, and identifying critical growth factors and biomolecules through data-driven insights (Bai et al., 2024). As described by Bai et al. (2024), ML-guided screening of construction strategies and integration in gel design can streamline the organoid construction process, reduce variability, and accelerate developmental optimization (Fig. 29.2).

Optimized geometric design and micropatterning have been used to generate a designer library of cardiac organoids with seven unique geometric designs. Manifold learning methods enabled the analysis of heterogeneity in single organoids based on physiological parameters. Unsupervised ML approaches uncovered unique functionalities linked with specific geometric patterns, while supervised ML models predicted calcium upstroke to play a major role in distinguishing between organoid shapes. They confirmed the importance of calcium signaling and regulation in organoid optimization (Kowalczewski et al., 2024). This approach improves our understanding of structure–function relationships and demonstrates how organoid optimization can be accelerated by ML.

Developments in AI and bioprinting technology show potential for standardizing and automating organoid fabrication. Bioprinting methods can create accurate organ-like structures by coordinating the precise placement of cells and controlling biomaterial positioning (Lee, 2023). The use of such technologies in organoid construction improves reproducibility and enhances patient-specific modeling. Combining AI with multiscale scaffold systems can create 3D tumor organoids that faithfully mimic the in vitro tumor environment. AI can assist with scaffold design by predicting how tumor cells will respond to various material compositions and geometric configurations, and facilitate the integration of biologically relevant signals (Katti, Katti, Gaikwad, & Jaswandkar, 2023).

The use of real-time culture data in AI-driven workflows can advance organoid research by improving monitoring and experimental accuracy. The intelligent cyber-biological systems framework employs a multilayered structure for organoid monitoring and analysis and leverages real-time organoid culture data and advanced ML models for pattern recognition and predictive modeling. This system identifies key developmental stages and treatment responses and improves decision-making in experimental processes, enabling the implementation of decisions such as adjustments to culture conditions (Cicceri et al., 2024).

Therefore, AI helps make organoid technology more reproducible, standardized, and relevant for clinical use.

3.3 AI in organoid-based drug screening, drug response/toxicity, and disease progression

The combination of AI with organoid-based platforms shows promise for drug screening and toxicity testing. PDOs, which mimic the biological features of individual tumors, are particularly useful in this context, as AI can analyze complex data to streamline drug discovery and advance personalized therapy.

A recent study extracted 100 phenotypic features from transmitted light microscopy and live/dead staining of PDOs by employing a DL image-based approach. These features were reduced to three components using principal component analysis (PCA)/uniform manifold approximation and projection (UMAP), visualized as 3D scatter plots, ranked, and clustered

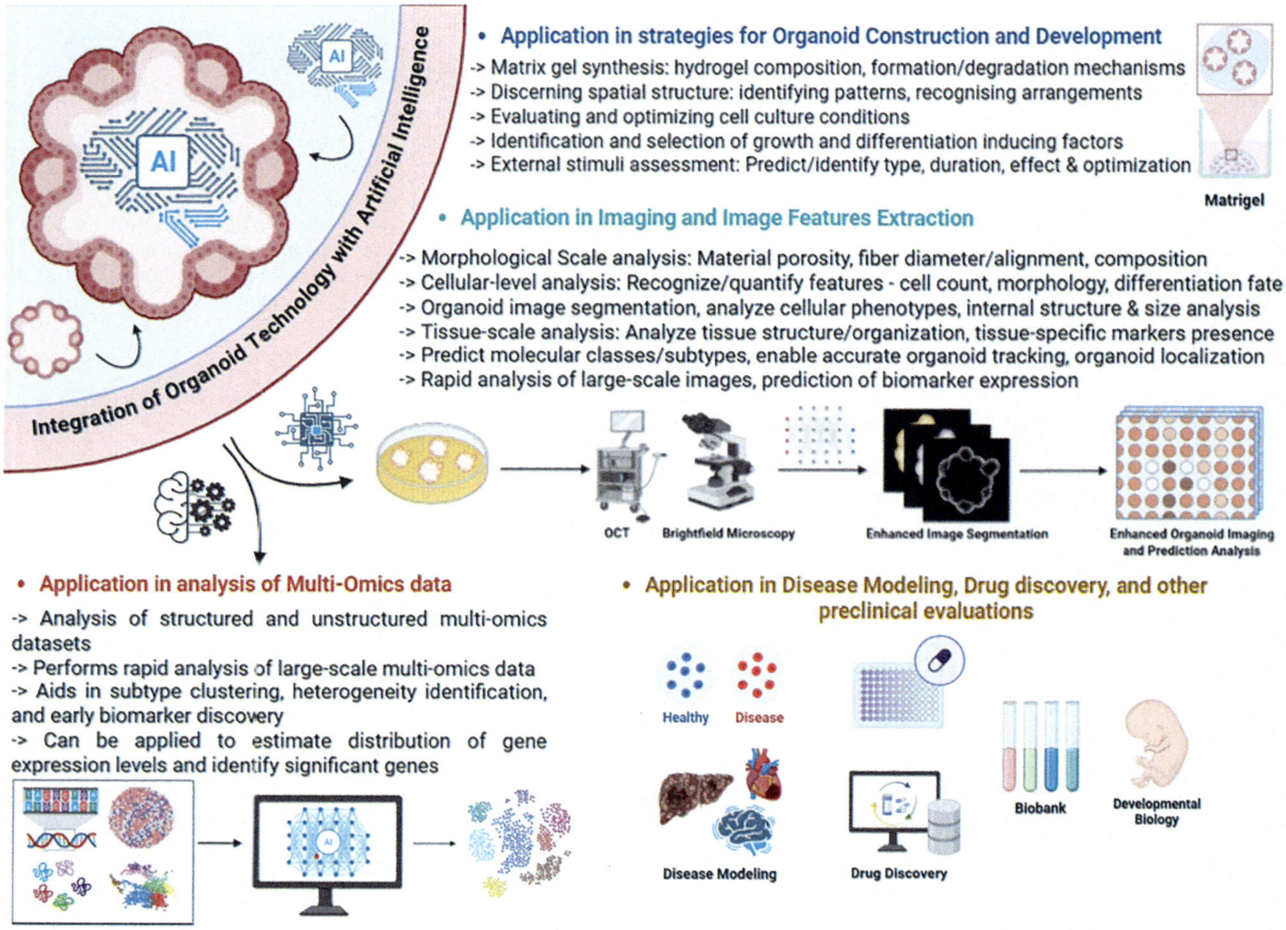

FIG. 29.2 AI-enabled organoid applications: the integration of organoid technology with AI. *Some of the summarized data have been adapted from Bai, L., Wu, Y., Li, G., Zhang, W., Zhang, H., & Su, J. (2024). (Figure created with Biorender.com)*

using ML. This framework confirmed the cytotoxic impact of palbociclib at 50 μM, and the results were consistent with traditional live/dead assay results (Tong et al., 2024). Oishi, Tabibzadeh, & Morizane (2024) combined fully automated 3D imaging with ML to assess nephron segment-specific epithelial morphometry in kidney organoids. This allowed for high-throughput drug screening and detection of therapeutic or nephrotoxic drugs, providing a reliable tool for nephrotoxicity screening and evaluating treatments in 3D culture systems.

Blood–brain barrier organoids are effective models for drug screening and preclinical testing. In a study, the researchers developed orthogonal image-based histological techniques to allow morphological readouts on histological sections. They used an AI algorithm to provide automated, label-free identification of apoptotic cells and improve the reliability of histomorphologic assessment of apoptosis (Bell et al., 2025).

DL-driven analysis was applied to standardized micropattern-based neural organoids (neuruloids), enabling distinction between wild-type and disease phenotypes with accuracy, as well as quantifying phenotypic rescue, drug efficacy, and adverse effects. They used unsupervised ML to measure the potential adverse effects of the compounds. Drug-discovery screens were performed, and downstream analysis of the screening results was conducted through a DL computational pipeline. This pipeline was applied to Huntington's disease (HD) models and helped discover bromodomain inhibitors as a therapeutic target, which can rescue HD-associated developmental phenotypes (Metzger et al., 2022).

AI has advanced the analysis of intestinal organoids. Lefferts et al. (2024) proposed the OrgaSegment model, which employs DL (MASK-RCNN) for the segmentation of individual intestinal PDOs from bright-field images. This tool recognizes and segments both spherical and irregularly odd-shaped organoids, helps in the assessment of organoid swelling, and quantifies responses to cystic fibrosis transmembrane conductance regulator (CFTR)-modulating drugs. It distinguishes organoids with different CFTR mutations and aids in studying CFTR-dependent fluid secretion and other epithelial mechanisms of ion transport in organoids. Takagi, Takagi, Hiyama, & Goda (2024) introduced a sequential DL model that utilizes two deep neural networks for analyzing PDOs to characterize tumor heterogeneity, describing morphological heterogeneity by analyzing microscopy images of PDOs and identifying gene subsets that correlate with intertumoral differences. Fillioux et al. (2023) designed an imaging-based high-throughput screening methodology that utilizes DL for the spatiotemporal analysis of PDOs, allowing assessment and prediction of real-time drug efficacy from a time-lapse microscopy video of PDOs. This methodology uses DL neural networks (foundation models) for automatic PDO data processing, and the pipeline incorporates the SAM algorithm and DI-NOv2 model to process microscopy frames.

Complementary approaches extend to genomic prediction. Amniouel & Jafri (2024) applied an ML model to the gene expression data of colorectal cancer (CRC) tissue for the identification of gene signatures that predict the efficacy or response to 5-fluorouracil (5-FU)-based combination chemotherapy (FOLFOX and FOLFIRI) in CRC patients. In radiation oncology, patient-derived tumor organoids show a strong correlation with actual patient responses to radiation. Paired with ML-based predictive algorithms, they serve as a platform for precision-guided radiotherapy response prediction and have demonstrated effective prediction of treatment outcomes (Park et al., 2021).

DL has enabled the analysis of drug efficacy differences in specific cell types, as demonstrated by a 3D DL model used for the analysis of cell-type-specific processes in tumor–fibroblast spheroid cocultures. This revealed critical insights into drug action mechanisms and susceptibility, showing that CCD-1137Sk fibroblasts exhibited higher resilience to drugs compared to KP-4 cancer cells, which were more susceptible in cocultures (Vitacolonna et al., 2024).

ML tools have enabled biomarker discovery within organoid systems. Kong et al. (2020) developed an ML framework that performs network-based computational analysis of organoid-derived pharmacogenomic data, identifying predictive drug response biomarkers for CRC and bladder cancer patient responses to 5-FU and cisplatin, respectively. In bladder tumor organoids, neural networks trained on spectral data were used to detect unbiased biomarkers for the classification and monitoring of in situ drug responsiveness upon treatment (Becker et al., 2022). Beyond imaging, rapid functional screening has been integrated with AI. Wu, Wang, Kang, & Chen (2024) introduced a micropillar assay for facilitating in situ high-throughput functional analysis of organoids, allowing for rapid sampling under in vivo conditions while enabling minimally invasive collection and quantification of protein biomarkers. The approach incorporated nanoparticle–antibody conjugates of varying sizes and ML-assisted image processing to achieve multiplex cytokine detection.

Collectively, these diverse developments illustrate the synergy between organoid platforms and AI in reshaping the progress of drug testing and toxicity analysis.

3.4 AI in organoid omics

AI has enhanced large-scale analysis of organoid datasets. AI integration facilitates the analysis of the molecular and cellular mechanisms underlying organoid biology, enables the prediction of disease onset, and models responses to pharmacological interventions (Hui Wang et al., 2024). It lowers manual labor, enhances the reproducibility and reliability of experimental models, and provides nuanced biological insights (Bai et al., 2024). Without AI, researchers would face a data deluge, posing a challenge in identifying subtle, nonobvious patterns within vast, multidimensional datasets.

Multiomics data has exploded along with the expansion of organoid research. High-throughput methods, like single-cell RNA sequencing, proteomics, metabolomics, and advanced imaging, are now used to study organoids. These methods range from analyzing cellular composition and structural organization to examining dynamic states of differentiation and function (Ballav et al., 2024). Conventional strategies are unable to manage such multidimensional datasets. The volume, heterogeneity, and dimensionality of genomics, transcriptomics, proteomics, and metabolomics datasets highlight the urgent need for advanced analysis methods.

ML and AI are indispensable for integrating and interpreting multiomics. ML may overcome the challenges of heterogeneous and unstructured datasets by identifying patterns and relationships that remain hidden (Bai et al., 2024). AI frameworks reveal complex linear and nonlinear interactions among molecular entities and clarify how gene mutations, regulatory networks, protein structural dynamics, and changes in metabolic pathways affect health and disease (Mo et al., 2025). The iterative learning and refinement capabilities of ML algorithms improve decision-making precision and prediction reliability, supporting innovation in disease modeling, drug screening, toxicological assessment, and regenerative medicine.

Several studies showcase the potential of AI and ML in organoid-based omics research. Chen et al. (2018) used single-cell RNA sequencing along with ML methods to study the transcriptome of CRC organoids. This revealed subtype diversity, tumor heterogeneity, and tracked population-level responses to chemotherapy, identifying the emergence of drug-tolerant subgroups alongside the depletion of chemosensitive groups. ML analysis of RNA-sequencing data has been used to identify key genes and pathways linked to gastrointestinal complications in hiPSC-derived colon organoids infected with the Mpox virus. The ML techniques demonstrated 100% accuracy in differentiating between several Mpox virus strains and mock samples (Rezapour, Narayanan, & Gurcan, 2024).

ML is used to estimate the spatiotemporal distribution of gene expression patterns during the development of mouse embryonic stem cell-derived retinal organoids. They used a CNN model that was trained on datasets of phase-contrast and fluorescence images and applied it to the retinal organoids with a focus on the master gene Rax, which is essential for eye field development. This noninvasive approach generated plausible spatial fluorescent patterns with relevant intensities, enabling quantitative assessments of gene expression within organoids (Fujimura et al., 2023).

The Digital Biological Twin Lab platform is an innovation that performs multiomics analysis on PDOs by combining genomic, cellular, and molecular techniques with ML models (Pasto et al., 2025). This aims to improve patient risk stratification, predict future treatment susceptibilities, and aid in the evaluation of various parameters after treatment, resulting in the identification of prospective biomarkers of therapy response. Drug discovery and optimization can be expedited by AI-enabled automated high-throughput screening, allowing for the simultaneous evaluation of numerous compounds in organoid and organ-on-chip systems (Zhou et al., 2025). Integrating AI-driven predictive modeling with data from these systems improves the accuracy of predicting drug behavior in human-relevant systems before clinical application (Mak & Pichika, 2019). AI-driven standardization of experimental workflows offers solutions to reproducibility problems caused by the inherent variability of organoids.

The integration of AI and organoid biology is shifting from traditional cellular models toward more mechanistic and translational frameworks, allowing researchers to model disease, optimize therapies, and pursue precision biomedicine by bridging the gap between multiomics complexity and computational precision.

3.5 Brain organoid models for neurodegenerative diseases

Research on neurodevelopmental and neurodegenerative disorders has benefited from the use of brain organoids. These offer an opportunity to study intricate neurodevelopmental processes and the underlying pathophysiology of complex disorders such as autism spectrum disorder (ASD), Parkinson's disease (PD), Alzheimer's disease (AD), amyotrophic lateral sclerosis (ALS), and microcephaly by recapitulating important aspects of human brain structure and function (Acharya, Choi, Shrestha, Jeong, & Lee, 2024; D'Antoni et al., 2023; Shou, Liang, Shunliang, & Xuekun, 2020). Their ability to closely imitate human brain organization has enabled new opportunities for exploring neurological mechanisms that have long been difficult to study in detail.

The introduction of OI has strengthened these models. It provides predictive platforms and personalized medical solutions, allowing detailed studies of organoid behavior and cellular interactions (Shi, 2024). This enables the study of neurodegenerative disorders, where conventional models are limited. Beyond disease modeling, brain organoids are also used as biological substrates for exploring cognitive processes and biological computing, offering new insights into neurobiology (Ballav et al., 2024; Smirnova & Hartung, 2024).

3.5.1 PD models

Brain organoids play a pivotal role in our understanding of PD. Researchers have developed automated midbrain organoids for high-throughput screening and AI-powered data analytics to uncover novel mechanisms and therapeutic targets (Renner, Schöler, & Bruder, 2021). In another study, human midbrain organoids (HMOs) were treated with the neurotoxin 6-hydroxydopamine to examine its effects on dopaminergic neurons. An ML classifier, such as random forest, was used to assess the neurotoxin's effects and the perturbations it caused. This evaluation relied on cellular features obtained through high-content screening. The authors validated this by using high-content imaging data from PD patient-derived midbrain organoids (Monzel et al., 2020).

Innovative coculture systems have expanded the importance of PD organoid models. Researchers developed a coculture model by combining HMO with T cells to investigate the role of the adaptive immune system in PD, revealing how T-cell infiltration can lead to neuronal loss (Prots et al., 2024). Similarly, hiPSC-derived isogenic midbrain organoids carrying the G2019S-LRRK2 mutation have been shown to mimic the aged human midbrain, helping to identify the role of TXNIP, a thiol-oxidoreductase, in the progression of LRRK2-associated PD. These mutant models could recapitulate PD-like pathology, and the comparison of gene expression between both midbrain organoids and human tissues showcased overlap in their differential gene expression profiles (Kim et al., 2019).

3.5.2 ASD and early brain development

Brain organoids are effective models for studying ASD as they mimic early human brain development. They help researchers explore cellular and molecular processes, such as neuronal differentiation, migration, and activity (Santos, Araújo, Rocha, Costa-Ferro, & Souza, 2023). PDOs assist in linking genetic variations to observed phenotypes and in studying how different genetic modifications can alter neurodevelopmental pathways (Kilpatrick, Irwin, & Singh, 2023). ASD patient-derived brain organoids might reveal how various genetic alterations and risk factors converge on certain cellular pathways that contribute to both idiopathic and syndromic autism pathology (Maussion, Rocha, Pimentel, & Durcan, 2023). Brain organoids are also valuable for studying key processes involved in ASD, such as neurogenesis and cortical lamination (Chan, Griffiths, Price, & Mason, 2020).

3.5.3 AD models

Cerebral organoids are helpful for studying AD, and when developed from patient-derived iPSCs with familial or sporadic AD, they can explore changes caused by the disease (López et al., 2023). Earlier models created by (Choi et al., 2014) revealed that organoids derived from hiPSCs carrying familial AD mutations could replicate important pathological features, such as amyloid-β (Aβ) aggregation and tau hyperphosphorylation. Similarly, (Raja et al., 2016) generated PDOs of familial AD using pluripotent stem cells that had *APP* and *PSEN1* gene mutations. These models developed specific brain region identities and exhibited AD pathology, such as abnormal endosomes, amyloid aggregation, and elevated hyperphosphorylated tau. Treatment with β- and γ-secretase inhibitors reduced these pathogenic features, underscoring the translational utility of organoids in drug screening and understanding disease progression. Researchers have used brain organoids to study the formation of amyloid plaques seen in AD and the degeneration of dopaminergic neurons characteristic of PD (Chang, Kim, Park, Choi, & Kim, 2020; Park & Park, 2024). Another study has showcased the potential for enabling drug screening and testing AD immunotherapy by modeling sporadic AD using vascularized neuroimmune organoids (Xu & Ji, 2024).

The integration of microglia improves the physiological relevance of AD models. As key regulators of synaptic formation, remodeling, neuronal maturation, and inflammation, microglia are implicated in brain development and neurodegeneration (Colonna & Butovsky, 2017; Sabate-Soler, Bernini, & Schwamborn, 2022; Zhang et al., 2023). Long-term organoid cultures with embedded microglia have demonstrated neuroprotective effects, such as increased synaptic density and reduced p-Tau expression (Chen, Sun, Feng, Tian, & Shi, 2025). When ApoE3- and ApoE4-iPSC-derived brain organoids were cocultured with microglia, they showed genotype-based differences, with ApoE4-associated microglia being less effective at counteracting neuronal degeneration compared to ApoE3 (Shi, 2024). This demonstrates the model's usefulness in studying neuroimmune interactions and the genetic underpinnings of AD.

3.5.4 ALS models

Organoids shed light on ALS pathophysiology by modeling key aspects of the disease. Pereira et al. (2021) developed human sensorimotor organoids from both healthy and ALS stem cells to form physiologically functional neuromuscular junctions (NMJs). These organoids were created from iPSCs and incorporated motor and sensory neurons, astrocytes, microglia, and skeletal muscle cells. Organoids made from ALS patient-derived iPSCs displayed distinct abnormalities, including reduced muscular contraction and impairments at the NMJ level, demonstrating the model's utility in understanding the disease.

In another study, cerebral organoid slice models developed from hiPSCs significantly improved our understanding of C9orf72-related ALS and frontotemporal dementia. The models replicated mature cortical architecture and displayed early molecular abnormalities, such as disruption in transcription and proteostasis in astrocytes and neurons, evidenced by elevated levels of the autophagy protein P62 and the buildup of harmful dipeptide repeat proteins like poly(GA) in neurons (Szebényi et al., 2021). Neuromuscular organoids derived from C9orf72 ALS patient iPSCs have been used to model spinal neuromuscular pathology, recapitulating pathological features such as muscle contraction weakness, neural denervation, and Schwann cell loss. They also replicated the presence of RNA foci and dipeptide repeat proteins, hallmarks of the C9orf72 gene mutation (Gao et al., 2024).

Thus, brain organoids are transforming neuroscience by providing disease-specific models that reflect the complexities of human brain development and degeneration.

4 Challenges and future directions

The combination of organoid technology with AI is one of the most promising areas of biomedical research. While this holds great promise, there are challenges ahead, including the complexity of organoid-generated data, the difficulty of interpreting AI models, concerns about reproducibility and validation, ethical issues, and the practical limits of computational resources and expertise (Fig. 29.3).

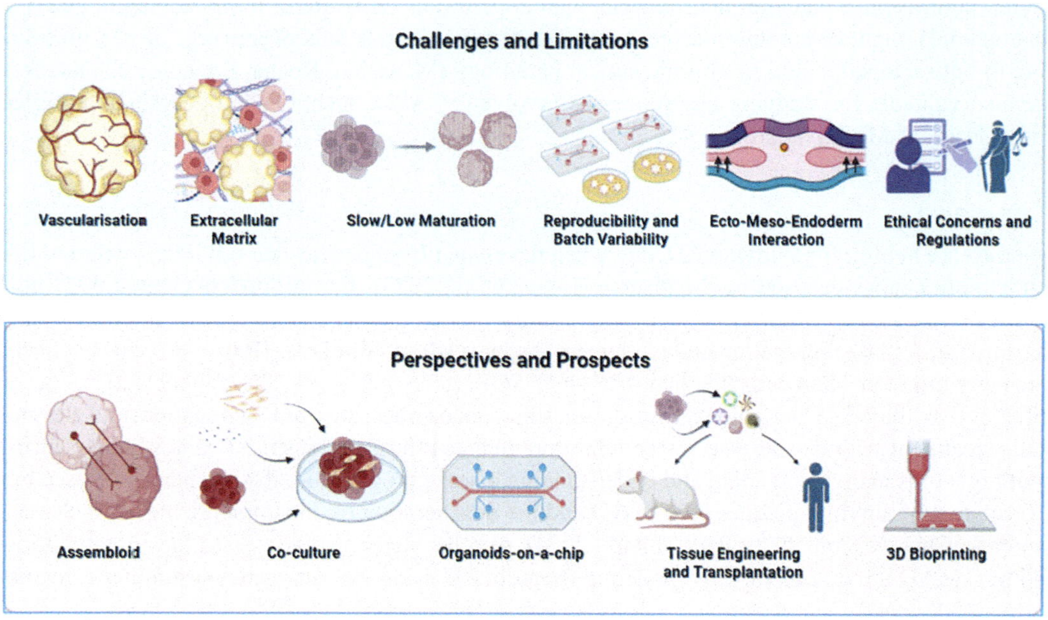

FIG. 29.3 Schematic overview of the challenges and limitations faced in organoid technology. Upcoming and existing innovative bioengineering strategies, such as the generation of assembloids, coculture techniques, organoid-on-a-chip systems, animal transplantation, and 3D bioprinting, might be potential solutions to tackle these challenges. *(Figure created with Biorender.com.)*

4.1 Vascularization issues and batch variability

Vascularization is crucial for organoids to replicate in vivo circumstances, as it facilitates the exchange of nutrients and waste products, which is vital for growth and functionality. The passive diffusion of nutrients and oxygen supply in organoids limits their growth and maturation, as these are effective for only a distance of a few hundred microns, leading to necrosis in the central core regions of larger organoids (Nwokoye & Abilez, 2024). Current bioengineering strategies, such as coculturing with vascular cells and using organoid-on-a-chip platforms, aim to overcome this limitation (Grebenyuk & Ranga, 2019; Li et al., 2023). New methods, like 3D printing of meshed vessels, may facilitate nutrient diffusion and promote growth (Xu et al., 2024).

The issue of batch variability persists, as organoid properties may be inconsistent due to differences in stem cell sources, culture conditions, high morphological variability, and differentiation outcomes. This results in differences in experimental outcomes, making interpretation and analysis difficult, a task AI aims to streamline.

4.2 Data complexity and integration

Organoid research generates large, complex, and heterogeneous datasets from methods like high-resolution imaging, multiomics profiling, and functional assays. The challenge lies in integrating these data types into a single analytical framework, as issues like data noise and biological variability make interpretation difficult. Future efforts should focus on standardizing data formats, creating robust metadata annotation systems, and developing platforms for data integration. Advances in ML, such as multimodal learning and transfer learning, help extract meaningful biological patterns while accommodating variability across datasets.

4.3 Model interpretability and biological relevance

Although AI models achieve high levels of predictive accuracy, their "black-box" nature obscures the biological reasons behind predictions. For organoid research to provide useful insights, interpretability is essential. This requires the creation of explainable AI that offers clear reasoning for predictions, along with models that incorporate biological knowledge to improve relevance.

4.4 Reproducibility and standardization

Reproducibility remains a challenge in the organoid field due to variations in protocols, cell sources, and culture conditions, which limit the application of research findings. Standardized methodologies for organoid generation and data acquisition are essential to address this. AI can assist by determining crucial quality-control measures and optimizing experimental parameters in real-time. Robust validation frameworks, the use of independent datasets, and experimental replicates are critical for confirming AI-driven discoveries.

4.5 Ethical considerations

Ethical questions arise as organoid technology progresses, especially with the creation of brain organoids and new fields like OI. Issues such as the potential for rudimentary sentience, the moral status of organoids, and the appropriate limits of research require thoughtful discussion, the establishment of clear ethical frameworks, and oversight mechanisms. Obtaining informed consent also entails complex ethical considerations, requiring meticulous evaluation of donors' views and the consequences of their contribution. Donors strongly prefer the application of the broad consent model for the utilization of their biospecimens, perceiving it as practical and efficient; yet they express a strong desire for continued engagement and decision-making authority over the future use of their samples.

4.6 Computational resources and expertise

The resource-intensive nature of AI-driven analyses mandates implementing AI pipelines, which require substantial computational power and specialized expertise. These resources may be unavailable in smaller research environments. To democratize access, future progress will depend on developing user-friendly analytical tools, cloud-based platforms, and collaborative research networks. Equally vital are investments in training programs that bridge computational and biological disciplines, ensuring a workforce capable of effectively using AI in organoid research.

5 Future directions

Despite challenges, the trajectory of organoid–AI integration is optimistic. Avenues for future exploration include advancing personalized medicine by combining PDOs with AI analysis to optimize individualized drug selection and therapeutic strategies. AI can enhance the utility of organoid systems in longitudinal studies through dynamic modeling by creating AI frameworks that simulate temporal dynamics and predict long-term outcomes from shorter-term measurements. The integration of AI with automated culture platforms, multielectrode arrays (or 3D bioelectronics), simulators, and multimodal sensors can aid in the generation of closed-loop feedback systems, which may allow for reinforcement learning-based adaptive response outcomes. Other areas of focus include using AI to coordinate analysis across different organoid types for multiorgan interactions and embedding AI–organoid platforms early in drug discovery pipelines to improve predictions of efficacy and toxicity.

Another direction is the development of biohybrid robots, which combine living tissues with robotic platforms. Brain organoids can serve as living regulators for robots, enabling autonomous functions through their ability to adapt to environmental changes. Shin et al. (2024) developed a biohybrid robot-on-a-chip that integrated a PD patient-derived mid-brain organoid, muscle bundle, and multimotor neuron spheroids to assess drug effects on neurodegenerative diseases by measuring muscle bundle movement. Furthermore, micromanipulation-based robotic platforms are being developed to facilitate the precise fabrication of complex organoids, addressing challenges in spatial control and promoting self-assembly (Gao et al., 2024). Collaborative efforts across biology, medicine, engineering, computer science, and ethics are critical to achieving these goals.

6 Conclusion

The convergence of organoid technology and AI marks a paradigm shift in biomedical science. Organoids serve as physiologically relevant models that bridge the limitations of traditional cell cultures and animal experiments, while AI provides a powerful means to analyze, interpret, and predict outcomes from these systems. This is opening up opportunities to explore fundamental biology, understand disease mechanisms, and identify new therapeutic interventions.

Current AI-enhanced organoid platforms have already demonstrated their value in multiple areas, including high-throughput drug screening, toxicity testing, disease modeling, and personalized medicine. AI-enabled imaging and computational processes have replaced manual tasks, allowing reproducible morphological and functional assessments, while predictive models and biomarker discovery tools have further accelerated progress in oncology, regenerative medicine, and neurodegenerative disease research. Achieving the full potential of organoid–AI integration will depend on addressing challenges such as data complexity, interpretability, reproducibility, and ethics. Overcoming these challenges will require standardized procedures, stringent validation, and interdisciplinary cooperation, while addressing ethical and societal concerns.

The ongoing advances in organoid technology and AI are reshaping healthcare, offering new approaches to the study and treatment of diseases. The refinement of organoid–AI platforms signals a major shift in how biomedical research tackles complexity, prediction, and translation.

References

Abdul, L., Xu, J., Sotra, A., Chaudary, A., Gao, J., Rajasekar, S., … Zhang, B. (2022). D-CryptO: deep learning-based analysis of colon organoid morphology from brightfield images. *Lab on a Chip*, 1473-0197. *22*(21), 4118–4128. doi:10.1039/d2lc00596d.

Acharya, P., Choi, N. Y., Shrestha, S., Jeong, S., & Lee, M. Y. (2024). Brain organoids: a revolutionary tool for modeling neurological disorders and development of therapeutics. *Biotechnology and Bioengineering*, 10970290. *121*(2), 489–506. Available from http://onlinelibrary.wiley.com/journal/10.1002/(ISSN)1097-0290. doi:10.1002/bit.28606.

Ahammed, B., & Kalangi, S. K. (2024). A decade of organoid research: progress and challenges in the field of organoid technology. *ACS Omega*, 24701343. *9*(28), 30087–30096. Available from pubs.acs.org/journal/acsodf. doi:10.1021/acsomega.4c03683.

Alam El Din, D. M., Shin, J., Lysinger, A., Roos, M. J., Johnson, E. C., Shafer, T. J., … Smirnova, L. (2024). Organoid intelligence for developmental neurotoxicity testing. *Frontiers in Cellular Neuroscience*. 16625102. *18*. Available from https://www.frontiersin.org/journals/cellular-neuroscience. doi:10.3389/fncel.2024.1480845.

Amniouel, S., & Jafri, M. S. (2024). High-accuracy prediction of colorectal cancer chemotherapy efficacy using machine learning applied to gene expression data. *Frontiers in Physiology*. 1664-042X. *14*. doi:10.3389/fphys.2023.1272206.

Asano, T., Suga, H., Niioka, H., Yukawa, H., Sakakibara, M., Taga, S., … Arima, H. (2024). A deep learning approach to predict differentiation outcomes in hypothalamic-pituitary organoids. *Communications Biology*, 2399-3642. *7*(1). doi:10.1038/s42003-024-07109-1.

Bai, L., Wu, Y., Li, G., Zhang, W., Zhang, H., & Su, J. (2024). AI-enabled organoids: construction, analysis, and application. *Bioactive Materials*. 2452199X. *31*, 525–548. doi:10.1016/j.bioactmat.2023.09.005.

Ballav, S., Ranjan, A., Sur, S., & Basu, S. (2024). Organoid intelligence: bridging artificial intelligence for biological computing and neurological insights. *IntechOpen*. doi:10.5772/intechopen.114304.

Bao, D., Wang, L., Zhou, X., Yang, S., He, K., & Xu, M. (2023). Automated detection and growth tracking of 3D bio-printed organoid clusters using optical coherence tomography with deep convolutional neural networks. *Frontiers in Bioengineering and Biotechnology*. 2296-4185. *11*. doi:10.3389/fbioe.2023.1133090.

Barros, M. T., Kagan, B. J., Hartung, T., & Smirnova, L. (2024). Editorial: intersection between the biological and digital: synthetic biological intelligence and organoid intelligence. *Frontiers in Cellular Neuroscience*. 16625102. *18*. Available from https://www.frontiersin.org/journals/cellular-neuroscience. doi:10.3389/fncel.2024.1542629.

Becker, L., Fischer, F., Fleck, J. L., Harland, N., Herkommer, A., Stenzl, A., … Marzi, J. (2022). Data-driven identification of biomarkers for in situ monitoring of drug treatment in bladder cancer organoids. *International Journal of Molecular Sciences*, 14220067. *23*(13). Available from https://www.mdpi.com/1422-0067/23/13/6956/pdf?version=1655969610. doi:10.3390/ijms23136956.

Bell, L., Simonneau, C., Zanini, C., Kassianidou, E., Zundel, C., Neff, R., … Stokar-Regenscheit, N. (2025). Advanced tissue technologies of blood-brain barrier organoids as high throughput toxicity readouts in drug development. *Heliyon*, 24058440. *11*(1), e40813. doi:10.1016/j.heliyon.2024.e40813.

Branciforti, F., Salvi, M., D'Agostino, F., Marzola, F., Cornacchia, S., De Titta, M. O., … Meiburger, K. M. (2024). Segmentation and Multi-Timepoint Tracking of 3D Cancer Organoids from Optical Coherence Tomography Images Using Deep Neural Networks. *Diagnostics*. 20754418. *14*(12). Available from http://www.mdpi.com/journal/diagnostics/. doi:10.3390/diagnostics14121217.

Buragohain, L., Bharali, A., Ghosh, A., Bhakat, M., Dhara, S. K., & Ghosh, A. (2023). Application and advancement of organoid technology in animal disease modeling. *Indian Journal of Animal Health*. 00195057. doi:10.36062/ijah.2023.spl.03223.

Cai, H., Ao, Z., Tian, C., Wu, Z., Liu, H., Tchieu, J., … Guo, F. (2023). Brain Organoid Computing for Artificial Intelligence. *bioRxiv*, 26928205. Available from https://www.biorxiv.org. doi:10.1101/2023.02.28.530502.

Chan, W. K., Griffiths, R., Price, D. J., & Mason, J. O. (2020). Cerebral organoids as tools to identify the developmental roots of autism. *Molecular Autism*, 20402392. *11*(1). Available from http://www.molecularautism.com/. doi:10.1186/s13229-020-00360-3.

Chang, Y., Kim, J., Park, H., Choi, H., & Kim, J. (2020). Modelling neurodegenerative diseases with 3D brain organoids. *Biological Reviews*, 1464-7931. *95*(5), 1497–1509. doi:10.1111/brv.12626.

Chen, K. Y., Srinivasan, T., Lin, C., Tung, K. L., Gao, Z., Hsu, D. S., … Shen, X. 2018. Single-cell transcriptomics reveals heterogeneity and drug response of human colorectal cancer organoids. Proceedings of the Annual International Conference of the IEEE Engineering in Medicine and Biology Society, EMBS 10.1109/EMBC.2018.8512784 9781538636466 2378–2381 Institute of Electrical and Electronics Engineers Inc. United States.

Chen, X., Sun, G., Feng, L., Tian, E., & Shi, Y. (2025). Human iPSC-derived microglial cells protect neurons from neurodegeneration in long-term cultured adhesion brain organoids. *Communications Biology*, 2399-3642. *8*(1). doi:10.1038/s42003-024-07401-0.

Choi, S. H., Kim, Y. H., Hebisch, M., Sliwinski, C., Lee, S., D'Avanzo, C., … Kim, D. Y. (2014). A three-dimensional human neural cell culture model of Alzheimer's disease. *Nature*. 14764687. *515*(7526), 274–278. Available from http://www.nature.com/nature/index.html. doi:10.1038/nature13800.

Cicceri, G., Di, B. S., Di, F. S., Stassi, G., Todaro, M., & Vitabile, S. 2024 Designing an Intelligent Cyber-Biological System for Enhanced Analysis and Monitoring of Organoids. Proceedings - 2024 IEEE International Conference on Bioinformatics and Biomedicine, BIBM 2024 doi:10.1109/BIBM62325.2024.10822405 9798350386226 5457–5464 Institute of Electrical and Electronics Engineers Inc. Italy http://ieeexplore.ieee.org/xpl/mostRecentIssue.jsp?punumber=10821710

Colonna, M., & Butovsky, O. (2017). Microglia Function in the Central Nervous System During Health and Neurodegeneration. *Annual Review of Immunology*, 0732-0582. *35*(1), 441–468. doi:10.1146/annurev-immunol-051116-052358.

D'Antoni, C., Mautone, L., Sanchini, C., Tondo, L., Grassmann, G., Cidonio, G., … Di Angelantonio, S. (2023). Unlocking neural function with 3d in vitro models: a technical review of self-assembled, guided, and bioprinted brain organoids and their applications in the study of neurodevelopmental and neurodegenerative disorders. *International Journal of Molecular Sciences*. 1422-0067. *24*(13), 10762. doi:10.3390/ijms241310762.

Deininger, L., Jung-Klawitter, S., Mikut, R., Richter, P., Fischer, M., Karimian-Jazi, K., … Schwarz, D. (2023). An AI-based segmentation and analysis pipeline for high-field MR monitoring of cerebral organoids. *Scientific Reports*, 2045-2322. *13*(1). doi:10.1038/s41598-023-48343-7.

Devarasetty, M., Mazzocchi, A. R., & Skardal, A. (2018). Applications of bioengineered 3D tissue and tumor organoids in drug development and precision medicine: current and future. *BioDrugs*. 1179190X. *32*(1), 53–68. Available from http://rd.springer.com/journal/40259. doi:10.1007/s40259-017-0258-x.

Du, X., Chen, Z., Li, Q., Yang, S., Jiang, L., Yang, Y., … Gu, Z. (2023). Organoids revealed: morphological analysis of the profound next generation in-vitro model with artificial intelligence. *Bio-Design and Manufacturing*, 2096-5524. *6*(3), 319–339. doi:10.1007/s42242-022-00226-y.

Fillioux, L., Gontran, E., Cartry, J., Mathieu, J. R. R., Bedja, S., Boilève, A., … Vakalopoulou, M. (2023). Spatio-temporal analysis of patient-derived organoid videos using deep learning for the prediction of drug efficacy. *arXiv*, 23318422. Available from: https://arxiv.org. doi:10.48550/arXiv.2308.14461.

Fujimura, Y., Sakai, I., Shioka, I., Takata, N., Hashimoto, A., Funatomi, T., & Okuda, S. (2023). Machine learning-based estimation of spatial gene expression pattern during ESC-derived retinal organoid development. *Scientific Reports*, 2045-2322. *13*(1). doi:10.1038/s41598-023-49758-y.

Gao, C., Shi, Q., Pan, X., Chen, J., Zhang, Y., Lang, J., … Lei, K. (2024). Neuromuscular organoids model spinal neuromuscular pathologies in C9orf72 amyotrophic lateral sclerosis. *Cell Reports*. 22111247. *43*(3), 113892. doi:10.1016/j.celrep.2024.113892.

Gao, H., Tong, M., Huang, G., Zhuang, S., & Jiang, M. (2024). Visual-servo micromanipulation platform for patterned and complex organoid biofabrication. *Research Square*, Available from https://www.researchsquare.com/browse. doi:10.21203/rs.3.rs-3897492/v1.

Ge, J. Y., Wang, Y., Li, Q. L., Liu, F. K., Lei, Q. K., & Zheng, Y. W. (2024). Trends and challenges in organoid modeling and expansion with pluripotent stem cells and somatic tissue. *PeerJ*, 21678359. *12*(11). Available from https://peerj.com/articles/18422.pdf. doi:10.7717/peerj.18422.

Grebenyuk, S., & Ranga, A. (2019). Engineering organoid vascularization. *Frontiers in Bioengineering and Biotechnology*. 2296-4185. *7*. doi:10.3389/fbioe.2019.00039.

Gritti, N., Lim, J. L., Anlaş, K., Pandya, M., Aalderink, G., Martínez-Ara, G., & Trivedi, V. (2021). MOrgAna: accessible quantitative analysis of organoids with machine learning. *Development*, 0950-1991. *148*(18). doi:10.1242/dev.199611.

Haja, A., Horcas-Nieto, J. M., Bakker, B. M., & Schomaker, L. (2023). Towards automatization of organoid analysis: a deep learning approach to localize and quantify organoid images. *Computer Methods and Programs in Biomedicine Update*. 26669900. *3*. Available from https://www.journals.elsevier.com/computer-methods-and-programs-in-biomedicine-update. doi:10.1016/j.cmpbup.2023.100101.

Hradecka, L., Wiesner, D., Sumbal, J., Koledova, Z. S., & Maska, M. (2023). Segmentation and tracking of mammary epithelial organoids in brightfield microscopy. *IEEE Transactions on Medical Imaging*. 1558254X. *42*(1), 281–290. Available from http://ieeexplore.ieee.org/xpl/RecentIssue.jsp?punumber=42. doi:10.1109/TMI.2022.3210714.

Huang, W., Jeong, S., Kim, W., & Chen, L. (2025). Biomedical applications of organoids in genetic diseases. *Medical Review*, 2749-9642. *5*(2), 152–163. doi:10.1515/mr-2024-0077.

Kassis, T., Hernandez-Gordillo, V., Langer, R., & Griffith, L. G. (2019). OrgaQuant: human intestinal organoid localization and quantification using deep convolutional neural networks. *Scientific Reports*. 20452322. *9*(1). Available from www.nature.com/srep/index.html. doi:10.1038/s41598-019-48874-y.

Katti, D. R., Katti, K. S., Gaikwad, H. K., & Jaswandkar, S. V. (2023). *Artificial intelligence in multiscale scaffolds for cancer organoids testbed Artificial Intelligence in Tissue and Organ Regeneration* (pp. 193–218). United States: Elsevier. Available from https://www.sciencedirect.com/book/9780443184987. doi:10.1016/B978-0-443-18498-7.00005-3.

Kegeles, E., Naumov, A., Karpulevich, E. A., Volchkov, P., & Baranov, P. (2020). Convolutional Neural Networks Can Predict Retinal Differentiation in Retinal Organoids. *Frontiers in Cellular Neuroscience*. 16625102. *14*. Available from https://www.frontiersin.org/journals/cellular-neuroscience. doi:10.3389/fncel.2020.00171.

Kilpatrick, S., Irwin, C., & Singh, K. K. (2023). Human pluripotent stem cell (hPSC) and organoid models of autism: opportunities and limitations. *Translational Psychiatry*. 21583188. *13*(1). Available from https://www.nature.com/tp/. doi:10.1038/s41398-023-02510-6.

Kim, H., Park, H. J., Choi, H., Chang, Y., Park, H., Shin, J., … Kim, J. (2019). Modeling G2019S-LRRK2 Sporadic Parkinson's Disease in 3D Midbrain Organoids. *Stem Cell Reports*, 22136711. *12*(3), 518–531. Available from http://www.elsevier.com/journals/stem-cell-reports/2213-6711. doi:10.1016/j.stemcr.2019.01.020.

Kong, J. H., Lee, H., Kim, D., Han, S. K., Ha, D., Shin, K., & Kim, S. (2020). Network-based machine learning in colorectal and bladder organoid models predicts anti-cancer drug efficacy in patients. *Nature Communications*, 20411723. *11*(1). Available from http://www.nature.com/ncomms/index.html. doi:10.1038/s41467-020-19313-8.

Kowalczewski, A., Sun, S., Mai, N. Y., Song, Y., Hoang, P., Liu, X., … Zhen, M. (2024). Design optimization of geometrically confined cardiac organoids enabled by machine learning techniques. *Cell Reports Methods*, 26672375. *4*(6), 100798. doi:10.1016/j.crmeth.2024.100798.

Lee, H. (2023). Engineering in vitro models: bioprinting of organoids with artificial intelligence. *Cyborg and Bionic Systems*. 2692-7632. *4*. doi:10.34133/cbsystems.0018.

Lefferts, J. W., Kroes, S., Smith, M. B., Niemöller, P. J., Nieuwenhuijze, N. D. A., Sonneveld van Kooten, H. N., ... van Beuningen, S. F. B. (2024). Orga Segment: deep-learning based organoid segmentation to quantify CFTR dependent fluid secretion. *Communications Biology*, 23993642. *7*(1). Available from https://www.springer.com/journal/42003. doi:10.1038/s42003-024-05966-4.

Leng, B., Jiang, H., Wang, B., Wang, J., & Luo, G. (2024). Deep-Orga: an improved deep learning-based lightweight model for intestinal organoid detection. *Computers in Biology and Medicine*. 00104825. *169*, 107847. doi:10.1016/j.compbiomed.2023.107847.

Li, Z., Li, Q., Zhou, C., Lu, K., Liu, Y., Xuan, L., & Wang, X. (2023). Organoid-on-a-chip: current challenges, trends, and future scope toward medicine. *Biomicrofluidics*, 1932-1058. *17*(5). doi:10.1063/5.0171350.

Lim, M. H., Shin, S., Park, K., Park, J., Kim, S. W., Basurrah, M. A., ... Kim, D. H. (2023). Deep learning model for predicting airway organoid differentiation. *Tissue Engineering and Regenerative Medicine*. 22125469. *20*(7), 1109–1117. Available from https://www.springer.com/journal/13770. doi:10.1007/s13770-023-00563-8.

López, V., Rosca, A., Liste, I., Mateos-Martínez, P., Coronel, R., Gonzalez-Sastre, R., ... López, V. (2023). Promising prospects for human cerebral organoids to advance Alzheimer's disease research. *Progress in Neurobiology*, 1–15. doi:10.60124/j.pneuro.2023.20.05.

MacDonald, M., Fennel, T. R., Singanamalli, A., Cruz, N. M., Yousefhussein, M., Al-Kofahi, Y., & Freedman, B. S. 2020 Improved automated segmentation of human kidney organoids using deep convolutional neural networks. Progress in Biomedical Optics and Imaging -Proceedings of SPIE. SPIE United States. doi:10.1117/12.2549830 9781510633933. http://spie.org/x1848.xml.11313.

Mak, K. K., & Pichika, M. R. (2019). Artificial intelligence in drug development: present status and future prospects. *Drug Discovery Today*. 18785832. *24*(3), 773–780. Available from www.elsevier.com/locate/drugdiscov. doi:10.1016/j.drudis.2018.11.014.

Matthews, J. M., Schuster, B., Kashaf, S. S., Liu, P., Ben-Yishay, R., Ishay-Ronen, D., ... Schneidman-Duhovny, D. (2022). OrganoID: a versatile deep learning platform for tracking and analysis of single-organoid dynamics. *PLOS Computational Biology*, 1553-7358. *18*(11), e1010584. doi:10.1371/journal.pcbi.1010584.

Maussion, G., Rocha, C., Pimentel, L., & Durcan, T. M. (2023). *iPSC-derived models of autism*. (pp. 201–227). Elsevier BV. doi:10.1016/b978-0-12-822277-5.00001-8.

Mecca, R., Tang, S., Jones, C., Coward, K., & Yeste, M. (2024). The limitations of testicular organoids: are they truly as promising as we believe? *Reproduction, Fertility and Development*, 1031-3613. *36*(11). doi:10.1071/rd23216.

Metzger, J. J., Pereda, C., Adhikari, A., Haremaki, T., Galgoczi, S., Siggia, E. D., ... Etoc, F. (2022). Deep-learning analysis of micropattern-based organoids enables high-throughput drug screening of Huntington's disease models. *Cell Reports Methods*. 26672375. *2*(9). Available from www.cell.com/cell-reports-methods. doi:10.1016/j.crmeth.2022.100297.

Mo, Y., Wang, J., Yang, H., Bai, L., Xu, K., & Su, J. (2025). Artificial Intelligence for Organoids Multidimensional Assessment. *SmartMat*, 2766-8525. *6*(3). doi:10.1002/smm2.70016.

Monzel, A. S., Hemmer, K., Kaoma, T., Smits, L. M., Bolognin, S., Lucarelli, P., ... Schwamborn, J. C. (2020). Machine learning-assisted neurotoxicity prediction in human midbrain organoids. *Parkinsonism and Related Disorders*, 18735126, *75*, 105–109. Available from www.elsevier.com/locate/parkreldis. doi:10.1016/j.parkreldis.2020.05.011.

Nwokoye, P. N., & Abilez, O J. (2024). Bioengineering methods for vascularizing organoids. *Cell Reports Methods*. 26672375. *4*(6). Available from https://www.sciencedirect.com/science/journal/26672375. doi:10.1016/j.crmeth.2024.100779.

Oishi, H., Tabibzadeh, N., & Morizane, R. (2024). Advancing preclinical drug evaluation through automated 3D imaging for high-throughput screening with kidney organoids. *Biofabrication*. 1758-5082. *16*(3), 035003. doi:10.1088/1758-5090/ad38df.

Park, M., Kwon, J., Kong, J., Moon, S. M., Cho, S., Yang, K. Y., ... Shin, U. S. (2021). A patient-derived organoid-based radiosensitivity model for the prediction of radiation responses in patients with rectal cancer. *Cancers*, 20726694. *13*(15). Available from https://www.mdpi.com/2072-6694/13/15/3760/pdf. doi:10.3390/cancers13153760.

Park, S., & Park, J.-C. (2024). Advancements in brain organoid models for neurodegenerative disease research. *Organoid*, 2765-205X. *4*, e12. doi:10.51335/organoid.2024.4.e12.

Park, T., Kim, T. K., Han, Y. D., Kim, K. A., Kim, H., & Kim, H. S. (2023). Development of a deep learning based image processing tool for enhanced organoid analysis. *Scientific Reports*, 20452322. *13*(1). Available from https://www.nature.com/srep/. doi:10.1038/s41598-023-46485-2.

Pasto, A., Barber, P., Al-Serori, H., Fankhaenel, M., Tsiridou, D. M., Austin-Williams, S., ... Ng, T. (2025). Abstract 4606: digital biological twin platform: multiomics in patient derived organoids combined with machine learning. *Cancer Research*, 0008-5472. *85*(8_Supplement_1), 4606–4606. doi:10.1158/1538-7445.am2025-4606.

Pereira, J. D., DuBreuil, D. M., Devlin, A. C., Held, A., Sapir, Y., Berezovski, E., ... Wainger, B. J. (2021). Human sensorimotor organoids derived from healthy and amyotrophic lateral sclerosis stem cells form neuromuscular junctions. *Nature Communications*, 20411723. *12*(1). Available from http://www.nature.com/ncomms/index.html. doi:10.1038/s41467-021-24776-4.

Prots, I., Gerasimova, E., Beenen, A., Kachkin, D., Regensburger, M., Zundler, S., ... Winner, B. (2024). Novel co-culture model of t cells and midbrain organoids for investigating neurodegeneration in Parkinson's disease. *Research Square*, Available from https://www.researchsquare.com/browse. doi:10.21203/rs.3.rs-4797877/v1.

Raja, W. K., Mungenast, A. E., Lin, Y. T., Ko, T., Abdurrob, F., Seo, J., & Tsai, L. H. (2016). Self-organizing 3D human neural tissue derived from induced pluripotent stem cells recapitulate Alzheimer's disease phenotypes. *PLoS ONE*, 19326203. *11*(9). Available from http://journals.plos.org/plosone/article/asset?id=10.1371/journal.pone.0161969.PDF. doi:10.1371/journal.pone.0161969.

Renner, H., Schöler, H. R., & Bruder, J. M. (2021). Combining automated organoid workflows with artificial intelligence-based analyses: opportunities to build a new generation of interdisciplinary high-throughput screens for Parkinson's disease and beyond. *Movement Disorders*. 15318257. *36*(12), 2745–2762. Available from http://onlinelibrary.wiley.com/journal/10.1002/(ISSN)1531-8257. doi:10.1002/mds.28775.

Rezapour, M., Narayanan, A., & Gurcan, M. N. (2024). Machine learning analysis of RNA-seq data identifies key gene signatures and pathways in Mpox Virus-induced gastrointestinal complications using colon organoid models. *International Journal of Molecular Sciences*. 1422-0067. *25*(20), 11142. doi:10.3390/ijms252011142.

Sabate-Soler, S., Bernini, M., & Schwamborn, J. C. (2022). Immunocompetent brain organoids—microglia enter the stage. *Progress in Biomedical Engineering*. 25161091. *4*(4). Available from https://iopscience.iop.org/journal/2516-1091. doi:10.1088/2516-1091/ac8dcf.

Santos, J. L. D. S., Araújo, C. D. A., Rocha, C. A. G., Costa-Ferro, Z. S. M., & Souza, B. S. D. F. (2023). Modeling autism spectrum disorders with induced pluripotent stem cell-derived brain organoids. *Biomolecules*. 2218273X. *13*(2). Available from http://www.mdpi.com/journal/biomolecules. doi:10.3390/biom13020260.

Shi, Y. (2024). Human iPSC-based models of Alzheimer's disease. *Alzheimer's & Dementia*, 1552-5260. *20*(S6). doi:10.1002/alz.087344.

Shin, M., Ha, T., Lim, J., An, J., Beak, G., Choi, J. H., … Choi, J. W. (2024). Human motor system-based biohybrid robot-on-a-chip for drug evaluation of neurodegenerative disease. *Advanced Science*. 21983844. *11*(4). Available from http://onlinelibrary.wiley.com/journal/10.1002/(ISSN)2198-3844. doi:10.1002/advs.202305371.

Shou, Y., Liang, F., Shunliang, X., & Xuekun, L. (2020). The application of brain organoids: from neuronal development to neurological diseases. *Frontiers in Cell and Developmental Biology*. 2296-634X. *8*. doi:10.3389/fcell.2020.579659.

Smirnova, L., Caffo, B. S., Gracias, D. H., Huang, Q., Morales Pantoja, I. E., Tang, B., … Hartung, T. (2023). Organoid intelligence (OI): the new frontier in biocomputing and intelligence-in-a-dish. *Frontiers in Science*. 2813-6330. *1*. doi:10.3389/fsci.2023.1017235.

Smirnova, L., & Hartung, T. (2024). The promise and potential of brain organoids. *Advanced Healthcare Materials*, 2192-2640. *13*(21). doi:10.1002/adhm.202302745.

Stüve, P., Nerb, B., Harrer, S., Wuttke, M., Feuerer, M., Junger, H., … Ritter, U. (2024). Analysis of organoid and immune cell co-cultures by machine learning-empowered image cytometry. *Frontiers in Medicine*. 2296-858X. *10*. doi:10.3389/fmed.2023.1274482.

Sureshkumar, A., Bisht, S., & Easwaran, H. (2024). Abstract 230: deep learning embedding-based segmentation for morphological analysis in organoids. *Cancer Research*. 1538-7445. *84* (6_Supplement), 230–230. doi:10.1158/1538-7445.am2024-230.

Szebényi, K., Wenger, L. M. D., Sun, Y., Dunn, A. W. E., Limegrover, C. A., Gibbons, G. M., … Lakatos, A. (2021). Human ALS/FTD brain organoid slice cultures display distinct early astrocyte and targetable neuronal pathology. *Nature Neuroscience*, 15461726. *24*(11), 1542–1554. Available from http://www.nature.com/neuro/index.html. doi:10.1038/s41593-021-00923-4.

Takagi, K., Takagi, M., Hiyama, G., & Goda, K. (2024). A deep-learning model for characterizing tumor heterogeneity using patient-derived organoids. *Scientific Reports*, 2045-2322. *14*(1). doi:10.1038/s41598-024-73725-w.

Tong, Z., Lim, A., Meyer, M., Clapes, M., Sirenko, O., & Brandenberg, N. (2024). Abstract 4919: AI-enabled hit selection of drug screening on human pancreatic cancer organoids. *Cancer Research*. 1538-7445. *84* (6_Supplement), 4919–4919. doi:10.1158/1538-7445.am2024-4919.

Vitacolonna, M., Bruch, R., Schneider, R., Jabs, J., Hafner, M., Reischl, M., & Rudolf, R. (2024). A spheroid whole mount drug testing pipeline with machine-learning based image analysis identifies cell-type specific differences in drug efficacy on a single-cell level. *BMC Cancer*, 1471-2407. *24*(1). doi:10.1186/s12885-024-13329-9.

Wadan, A.-H. S. (2024). The revolutionary biocomputing technology of organoid intelligence (OI) as the forefront of AI and medicine. *IGI Global*, 215–234. doi:10.4018/979-8-3373-1132-6.ch008.

Wang, B., Ganjee, R., Khandaker, I., Flohr, K., He, Y., Li, G., … Pi, S. (2024). Deep learning based characterization of human organoids using optical coherence tomography. *Biomedical Optics Express*, 2156-7085. *15*(5), 3112. doi:10.1364/boe.515781.

Wang, H., Li, X., You, X., & Zhao, G. (2024). Harnessing the power of artificial intelligence for human living organoid research. *Bioactive Materials*. 2452199X. *42*, 140–164. doi:10.1016/j.bioactmat.2024.08.027.

Wang, X. Y., Jia, Q. N., Li, J., & Zheng, H. Y. (2024). Organoids as tools for investigating skin aging: mechanisms, applications, and insights. *Biomolecules*. 2218273X. *14*(11). Available from http://www.mdpi.com/journal/biomolecules. doi:10.3390/biom14111436.

Wu, S., Wang, C., Kang, X., & Chen, P. (2024). Machine-learning assisted micropillar assay for high-throughput in-situ functional analysis of patterned organoids. *ECS Meeting Abstracts*, 2151-2043. *MA2024-02*(65), 4374–4374. doi:10.1149/ma2024-02654374mtgabs.

Xu, H., Jiao, Y., Qin, S., Zhao, W., Chu, Q., & Kongming, W. (2018). Organoid technology in disease modelling, drug development, personalized treatment and regeneration medicine. *Experimental Hematology & Oncology*, 2162-3619. *7*(1). doi:10.1186/s40164-018-0122-9.

Xu, L., Ding, H., Wu, S., Xiong, N., Hong, Y., Zhu, W., … Liu, Y. (2024). Artificial meshed vessel-induced dimensional breaking growth of human brain organoids and multiregional assembloids. *ACS Nano*. 1936-0851. doi:10.1021/acsnano.4c07844.

Xu, R., & Ji, Y. (2024). Develop a vascularized neuroimmune organoid model for studying sporadic Alzheimer's disease. *Alzheimer's & Dementia*, 1552-5260. *20*(S1). doi:10.1002/alz.088501.

Yao, Q., Cheng, S., Pan, Q., Yu, J., Cao, G., Li, L., & Cao, H. (2024). Organoids: development and applications in disease models, drug discovery, precision medicine, and regenerative medicine. *MedComm*, 2688-2663. *5*(10). doi:10.1002/mco2.735.

Zeng, G., Yu, Y., Wang, M., Liu, J., He, G., Yu, S., … Peng, X. (2024). Advancing cancer research through organoid technology. *Journal of Translational Medicine*, 1479-5876. *22*(1). doi:10.1186/s12967-024-05824-1.

Zhang, W., Jiang, J., Xu, Z., Yan, H., Tang, B., Liu, C., … Meng, Q. (2023). Microglia-containing human brain organoids for the study of brain development and pathology. *Molecular Psychiatry*, 1359-4184. *28*(1), 96–107. doi:10.1038/s41380-022-01892-1.

Zhou, L., Chen, S., Liu, J., Zhou, Z., Yan, Z., Li, C., … Li, Z. A. (2025). When artificial intelligence (AI) meets organoids and organs-on-chips (OoCs): game-changer for drug discovery and development? *Innovation Life*. 29598761. *3*(1). Available from https://www.the-innovation.org/article/doi/10.59717/j.xinn-life.2024.100115. doi:10.59717/j.xinn-life.2024.100115.

Zhou, L., Luo, D., Lu, W., Han, J., Zhao, M., Li, X., … Wen, Y. (2024). Gastrointestinal tract organoids as novel tools in drug discovery. *Frontiers in Pharmacology*. 1663-9812. *15*. doi:10.3389/fphar.2024.1463114.

Chapter 30

Artificial intelligence (AI)-enabled organoids and three-dimensional (3D) in vitro models for clinical applications and precision medicine (PM)

Tejal Pant[a,*], Nanda Rohra[b,*], Tanya Saxena[c,*], Shelly Pathak[d,*], and Payal Ganguly[e,f]

[a]*Anya Biopharm India LLP, Ahmedabad, India*

[b]*HiMedia Laboratories Pvt. Ltd., Thane, India*

[c]*ADDB, Louvain Drug Research Institute, Université catholique de Louvain, Brussels, Belgium*

[d]*Novogene Europe, Milton, Cambridge, United Kingdom*

[e]*Leeds Institute of Rheumatic and Musculoskeletal Medicine, University of Leeds, Leeds, United Kingdom*

[f]*Leeds Biomedical Research Centre, Leeds Teaching Hospitals NHS Trust, Leeds, United Kingdom*

1 Introduction

The areas of biomedical research, pharmaceutical investigations, and experimental medicine have advanced by leaps and bounds in the last century across the globe. The pursuit of the "why" behind any human condition or disease has helped us dissect several molecular pathways, unravel drug targets, investigate genetics, as well as find new therapies. While we have significantly advanced our knowledge in human health and diseases, several clinical conditions remain that have no permanent cure or solution and are merely "managed" when symptoms arise. At the cellular level of this problem lies the use of two-dimensional (2D) cell culture models that barely replicate physiological human conditions (Kapałczyńska et al., 2018; Ravi, Paramesh, Kaviya, Anuradha, & Solomon, 2015). 2D cell cultures have several advantages, some of which include relatively easier culture maintenance, cost-effectiveness, accessibility for numerous cell types, adaptability, and shorter duration for cultures. However, the challenge lies in replicating the in vivo conditions on a 2D system.

2D cultures lack the architecture, complexity, codependency, inter- and intracellular interactions, as well as the structural intricacies that are emerging as possibilities in three-dimensional (3D) culture systems (Jensen & Teng, 2020). The last two to three decades have witnessed an increase in the use of different models, including organoids (Hofer & Lutolf, 2021; Kim, Koo, & Knoblich, 2020; Nikonorova, Chrishtop, Mironov, & Prilepskii, 2023), spheroids (Friedrich, Seidel, Ebner, & Kunz-Schughart, 2009; Gilazieva, Ponomarev, Rutland, Rizvanov, & Solovyeva, 2020; Ong et al., 2018), scaffolds (Dorishetty et al., 2021; Iqbal et al., 2023; Leong, Chua, Sudarmadji, & Yeong, 2008; Prabha et al., 2018; Premjit et al., 2025), and organ-on-a-chip (OOC) using microfluidics (Ingber, 2022; Nithin, Aggarwal, Sravani, Mallya, & Lewis, 2023; Zhou, Cong, & Cong, 2021). Among all of these platforms, organoids are increasingly being investigated for their potential to mimic human physiology. While there are challenges related to long-term cultures, including difficulty in maintenance and higher costs, the advantages they offer are slowly outweighing the traditional approach to biomedical research, specifically when it comes to patient-derived organoids (PDOs) (Driehuis, Kretzschmar, & Clevers, 2020; Guillen et al., 2022; Kastenschmidt et al., 2024).

PDOs, especially for diseases like cancer, provide a step forward toward precision medicine (PM) (Chen et al., 2022; Zhou et al., 2021). PM, also previously used interchangeably with "personalized medicine," refers to approaching a disease

* These authors contributed equally to the manuscript preparation.

Applications of Artificial Intelligence in Healthcare. https://doi.org/10.1016/B978-0-443-43934-6.00028-6

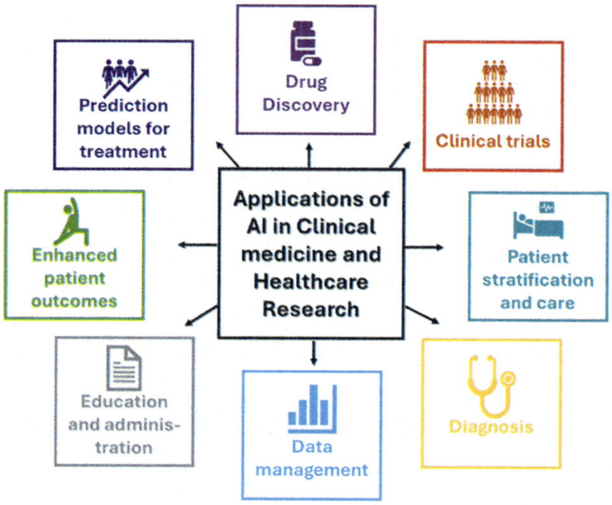

FIG. 30.1 Application of artificial intelligence (AI) in clinical medicine and healthcare research.

or condition by suitably tailoring treatment to the people suffering from it (Naithani, Sinha, Misra, Vasudevan, & Sahu, 2021). In 2015 President Barack Obama announced the PM initiative, which was aimed at a paradigm shift in modern medicine (Arthur, 2015), considering that the "one-size-fits-all" approach was no longer providing expected treatment for the majority of patients. This means that for any given disease or condition, a treatment may work very well for one patient, moderately for another, and not at all for a third. These variations can be due to a multitude of factors, including environmental factors, genetics, age, gender, lifestyle, and comorbidities. Thus education and research investments toward PM are critical for a sustainable and effective approach to clinical treatment in healthcare (Sbitan, Alzraikat, Tanous, Saad, & Odeh, 2025).

Different platforms can be used to achieve the PM approach for a given condition; these include (1) genetics and sequencing, (2) artificial intelligence (AI) and machine learning (ML), (3) OMICS platforms, (4) targeted drug delivery techniques, and (5) precision-based 3D in vitro models (Ganguly, 2024). PDOs, precision-based organoids, and 3D models combat partial physiological replication in 2D cellular models, reduce the need for animal studies, and provide better drug target predictions for new drugs in comparison with 2D models and in vivo animal models (Greek, Menache, & Rice, 2012; Seifirad & Haghpanah, 2019). These attributes contribute toward PM, and more recently, with the advent of AI, scientists envision a combination of organoids and AI to further strengthen the PM approach in modern medicine.

AI and ML have been paving their way through every existing field, pushing the horizon of applications within those fields. Biomedical and clinical research are no different, and AI has the potential to enhance healthcare system efficiencies, safety, personalized predictions, and better patient outcomes (Alowais et al., 2023; Bates et al., 2021; Hilton et al., 2020). AI can be beneficial for the following applications in healthcare, medicine, and related fields: drug discovery, clinical trials, patient care, diagnosis, data management, medical education, administration, patient outcomes, disease prediction models, and error reduction, among others (Aung, Wong, & Ting, 2021; Chen & Decary, 2020; Yu, Beam, & Kohane, 2018) (Fig. 30.1).

With respect to our interest in this book chapter, AI-enabled organoids are rapidly becoming a topic of interest to scientists and researchers. The ability to combine AI with organoids has the potential to provide models that better predict physiological conditions in vivo, drug targets, biomarkers for several diseases, diagnostics, and others outlined above. This is increasingly being engineered and investigated in diseases that will benefit from using PM as their strategy, such as cancer (LeSavage, Suhar, Broguiere, Lutolf, & Heilshorn, 2022; Neal et al., 2018). Together, these models are now being referred to as "organoid intelligence" (OI) (Lee, 2023; Shi et al., 2024). This chapter dissects the current literature available on organoids and 3D models used in combination with AI for cancers, rheumatology, and neural networks, with a focus on the brain. The authors then describe the genetic landscape of these investigations and envision the combination of genomics alongside AI and organoids, which may also be referred to as "next-generation organoids (NGOs)."

2 Role of AI-enabled organoids or OI

As outlined above, organoids are 3D structures cultivated from specific progenitor or stem cells. They mimic the structural and functional complexities of human organs in an in vitro environment (Bhattacharya et al., 2025). Organoids can provide a comprehensive analysis of cellular interactions, metabolic pathways, and dynamic responses to different types of stimuli

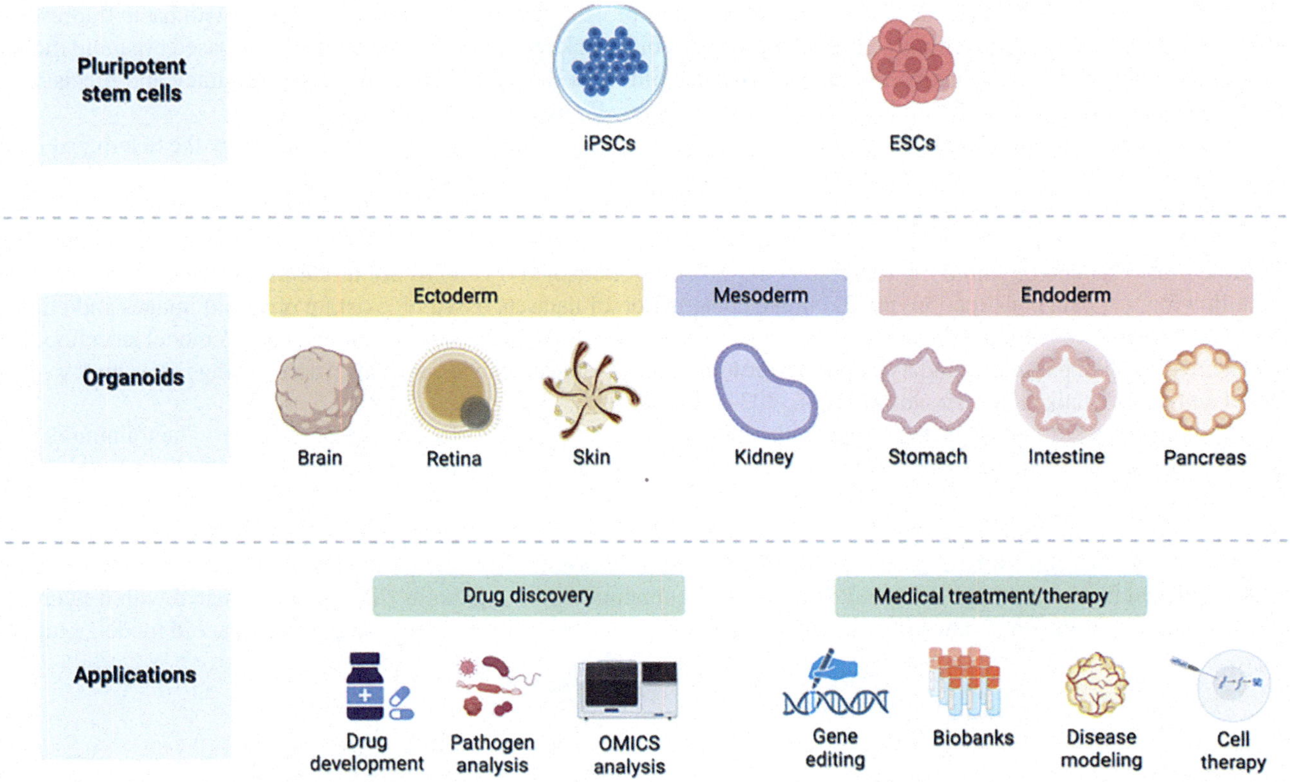

FIG. 30.2 Application of organoids in healthcare. Top panel: common cell types used for organoids; middle panel: types of organoids; bottom panel: applications of organoids in biomedical and clinical fields.

(Fig. 30.2). These features help in designing better disease models with precise simulations, personalized medicine for patients exhibiting specific biomarkers, studying developmental biology, and testing numerous therapeutic interventions at a faster pace (Shi et al., 2024).

Organoid-tested drugs, like Petosemtamab—a bispecific antibody that inhibits the growth of colorectal cancer organoids—have been approved to enter clinical trials based on safety and efficacy data from organoid models (Lundberg et al., 2025). In 2022 the world's first clinical trial transplantation of stem cell-derived organoids was performed in humans at Tokyo Medical and Dental University to treat ulcerative colitis (Watanabe et al., 2022). This pioneering technology in regenerative medicine, after safety validation, has the potential to treat millions of patients suffering from inflammatory bowel disorder. However, the construction of organoids can be variable and heterogeneous depending on the materials and cell populations used, along with being time-consuming and having low success rates. Thus, AI-based ML models, created with large datasets and optimization strategies, provide a faster, more cost-effective method to develop reproducible organoids in cell culture mediums or as OOC models (Bai et al., 2024b).

AI is currently used mainly to perform visual perception, language translation, decision-making, and speech recognition (Bai et al., 2024b). On the contrary, AI in the healthcare domain has advanced considerably. For instance, BenevolentAI, a company specializing in AI-driven drug discovery, identified Baricitinib as a potential candidate for the treatment of COVID-19 during the early stages of the pandemic. Based on its role as an immunomodulator and a barrier for viral entry, the AI system analyzed large datasets and shortlisted the potential candidate, accelerating the drug repurposing process and moving Baricitinib to clinical trials swiftly (Zhou et al., 2025). Hence, AI–ML models have the ability to extract required information from large datasets and help in the synthesis of more accurate and personalized organoid/OOC models.

Organoid research focuses on three key attributes: construction strategies, screening and data analysis, and efficacy verification. Organoid construction entails (1) the selection of matrix material, (2) defining cell culture conditions, and (3) screening of growth factors. Matrigel, an animal-derived extracellular matrix (ECM), is the most commonly used platform for organoid construction, but it shows variability and potential immunogenicity. Alternatively, synthetic hydrogels provide a cost-effective substitute with greater resemblance to human organs (Bai et al., 2024b). For data analysis, high-throughput screening methods such as transcriptomics, proteomics, and metabolomics, along with basic fluorescent/brightfield image analysis, are employed to determine variability in key features and attributes of diseased organoid models compared to normal organs. Finally, the efficacy of organoid models is tested by studying the accuracy and precision of gene expression and intercellular interactions (Renner, Schöler, & Bruder, 2021).

Although each attribute stated above has certain limitations, data analysis is largely constrained by the heterogeneity of organoids, which affects the accuracy of spatial characteristics of the model, thereby creating large data volumes, reducing cost-effectiveness, and showing variability based on human bias and errors. ML models can be trained with these large datasets in a short amount of time to classify the differences in morphological features and theoretical time-lapse data, and generate more realistic organoid models (Bai et al., 2024b). AI-enabled organoids are tested in four main steps shown in the following: (1) data construction, involving the construction of datasets based on existing organoid images and clinical data; (2) data pre-processing, adjusting images or converting theoretical data into readable data; (3) model selection, ML model creation with parameter and hyperparameter tuning; and (4) data analysis and validation, testing the accuracy of the model and data visualization (Maramraju et al., 2024) (Fig. 30.3).

ML models are built and selected based on the input data or features of a specific organoid model. The training dataset is used to design the model, and the testing dataset is used to test the accuracy of the model. ML models have the ability to determine the most efficient techniques for synthesizing matrix gels, enhancing spatial characteristics, and inducing active factors, along with establishing tissue differentiation and cellular pathways (Maramraju et al., 2024). In a study conducted in 2020, researchers built a neurotoxin-induced Parkinson's disease (PD) organoid model with 86% accuracy, whereby human induced pluripotent stem cells (hiPSCs) induced human midbrain organoids (hMOs) were treated with 6-hydroxy-dopamine to impair the dopaminergic system. Based on high-content confocal images of the PD organoid model, a trained

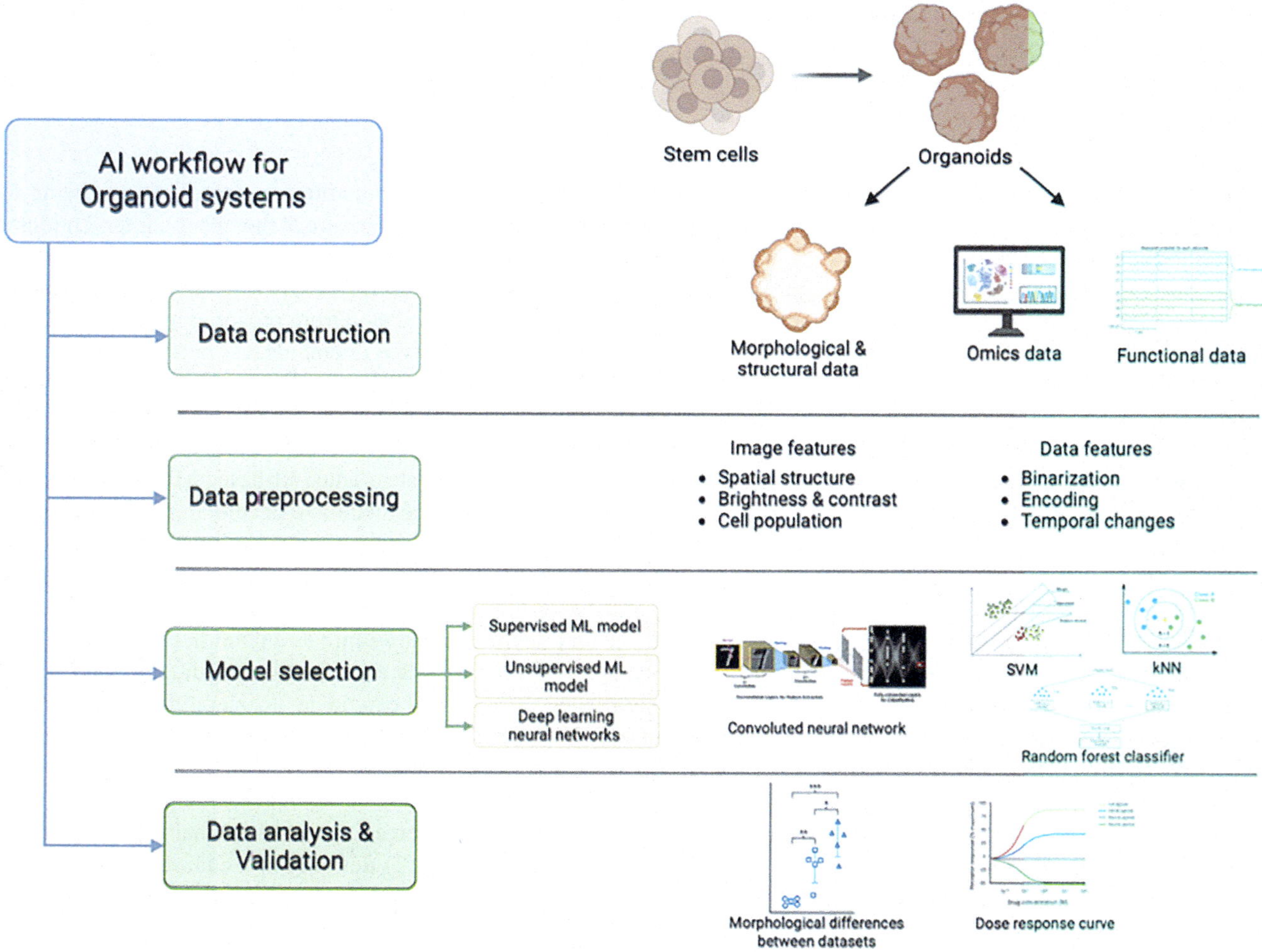

FIG. 30.3 Schematic representing the AI workflow for the organoid system.

classification random forest model was built to predict neurotoxicity in control and treatment organoids (Monzel et al., 2020).

Another study shows that the convolutional neural network (CNN) model—DenseNet121—could analyze simple brightfield images of kidney organoids and accurately predict the differentiation of kidney organoids derived from hiPSCs, thereby providing a noninvasive method for studying organ development.

A study comparing supervised ML models to differentiate the contractile profile of hiPSC-derived cardiomyocytes from healthy controls and a patient with Timothy syndrome showed that decision trees and quadratic discriminant analysis achieved 92% accuracy, which provides a strong disease model and enhances evidence-based development of drugs (Teles, Kim, Ronaldson-Bouchard, & Vunjak-Novakovic, 2021).

From creating personalized organoids for cancer patients based on biopsies to developing brain organoids mimicking biomarker expression in Alzheimer's patients, AI-enabled organoids have great potential to expedite clinical applications of personalized medicine with greater precision than traditional treatment methods, as illustrated in the next few sections.

3 AI-enabled organoids for cancer

With more than a million publications in the past 5 years, cancer continues to be the most widely investigated disease in the medical field, yet it somehow manages to be the most poorly predicted one as well. Thus numerous efforts are being poured into addressing this inconvenience that has marred human life and quality for centuries. From the simple 2D in vitro models that provided astounding insights into the molecular operations of cells to the highly complex 3D architectural wonders that promise the most relevant predictions and outcomes, cancer biology has come a long way. Organoids, the in vitro 3D self-organized assemblies that resemble the features of their original tissue (Veninga & Voest, 2021), have emerged as a promising platform for cancer drug discovery. These multicellular structures can be derived from patient samples and adult or embryonic stem cells (Veninga & Voest, 2021; Xia, Fei, Juan, Aji, & Gao, 2019). Their versatility lies in their potential to represent intra- and intertumor heterogeneity, a major hurdle in developing patient-specific therapies. This heterogeneity can be attributed to the ability of tumors to promote the proliferation of a particular subpopulation, which often leads to diverse phenotypes during disease progression and differences in therapeutic response (LeSavage et al., 2022; Yang, Wang, Wang, Zhang, & Wang, 2022). It has been observed that the phenotypic, genotypic, and mutational signatures of original tumor samples are preserved in long-term organoid cultures (Kim et al., 2025; Lee et al., 2025; Paul et al., 2025; Veninga & Voest, 2021). Numerous studies over the past few decades have used tumor organoids of the colon, lung, pancreas, prostate, ovaries, bladder, breast, endometrium, esophagus, and other tissues. Like cell banks, tumor biobanks have also been established to expedite research in this area (Li, Liu, & Chen, 2022; Chen et al., 2025).

Favorable outcomes in cancer treatment often rely on appropriate detection, diagnosis, treatment, and prognosis. Further, early detection of cancer is a crucial step that could help in disease management and successful treatment in patients. In this regard, AI has emerged as a powerful data-driven technology that offers speed, accuracy, personalization, and scalability (Alowais et al., 2023; Mesko, 2017). AI-based tools can be trained using extensive datasets obtained from individual studies and employed for rapid detection, quantification, and classification of input information. The training of AI-based tools comprises ML and deep learning (DL), which focus on algorithms and learning patterns from existing datasets (Aung et al., 2021; Maceachern & Forkert, 2021; Mao et al., 2025). Together, these tools find applications in PDO imaging, data integration, high-throughput drug screening, and predictive modeling, thus aiding effective PM (Huang et al., 2024; Shi et al., 2024).

Image analysis is an important means of understanding external (size, shape, growth, development) and internal (cell organization, distribution, differentiation) features of organoids. Techniques like fluorescence microscopy, confocal microscopy, and superresolution microscopy have greatly enhanced and advanced organoid imaging (Momoli et al., 2025; Park et al., 2023). An AI-based extractor and classifier were developed by Okamoto et al. to categorize colorectal cancer PDOs into six morphological types. Comparative transcriptome analysis revealed differential expression of gene sets, specifically those involved in ribosome biogenesis, which could reflect possible interpatient diversity and lead to the selection of the most effective therapeutic regimen (Okamoto et al., 2022). However, the morphological typing of PDOs was not correlated with the pathological stage of the original tumor, suggesting the need for a more comprehensive assessment. AI-ML-based high-throughput screening, classification, quantification, and tracking of organoids offer the opportunity to improve preclinical models for PM.

PM is also guided by the determination of relevant biomarkers. An individual biomarker or a panel of these biomarkers may be indicative of the cancer stage, type, and risk of progression in a tissue (Smabers et al., 2024; Wensink et al., 2021). However, identifying robust, validated biomarkers and achieving their clinical translation for designing

TABLE 30.1 Examples of AI-driven cancer drugs on the way/already in clinical trials.

Company	Notable AI-derived drug candidates	Stage/highlights	Reference/trial number
Recursion	REC-3565 (MALT1), REC-2282 (HDAC inhibitor)	Phase I and Phase 2/3 (FDA orphan and fast-track)	(Philippidis, 2025), NCT05130866
Insilico medicine	Rentosertib (IPF), CDK8/19 inhibitor, KRAS-targeting molecules	Phase IIa; preclinical candidates via AI + quantum	NCT05938920
Exscientia	EXS4318 (PKC-theta), A2a antagonist (immuno-oncology)	Phase I trials underway	(Philippidis, 2025)
Relay therapeutics	RLY-4008, RLY-2608 (PI3Kα inhibitor)	Phase II trials showing promising efficacy	NCT05216432
Lantern pharma	LP-300, LP-184, LP-284 (various oncology indications)	Clinical trials ongoing via RADR platform	NCT05456256, NCT05933265, NCT06132503
Pharos iBio	PHI-101 (FLT3-targeted)	Individual use authorization in South Korea	NCT04842370
Generate biomedicines	GB-0895 (anti-TSLP antibody)	Phase I; immunooncology and protein therapeutics	(Mayawala, 2024)

anti-TSLP, antithymic stromal lymphopoietin; *CDK8/19*, cyclin-dependent kinase 8/19; *FLT3*, FMS-like tyrosine kinase 3; *HDAC*, histone deacetylase; *IPF*, idiopathic pulmonary fibrosis; *KRAS*, Kirsten rat sarcoma viral oncogene homolog; *MALT-1*, mucosa-associated lymphoid tissue lymphoma translocation protein 1; *PI3Kα*, phosphoinositide 3-kinase alpha; *PKC-theta*, protein kinase C theta.

treatment strategies to combat the complex heterogeneity found in patients remains a challenge. The translation of predictive biomarkers identified in preclinical models to reflect drug treatment outcomes in human clinical trials is urgently needed. In a study by Kong and colleagues, pharmacogenomic data from 3D organoid cultures were used to predict 5-fluorouracil and cisplatin sensitivity in colorectal and bladder cancer patients, respectively (Kong et al., 2020). In another study, a five-molecule gene signature and resistance scoring model were established to predict the survival rate of colorectal cancer patients treated with 5-fluorouracil-based chemotherapy (Chen et al., 2022). Similarly, a radiotherapy outcome prediction model with accuracies of 81.5% and 92.1% was established by Park et al. by comparing the radiation response of rectal cancer organoids with patient radiotherapy results. Apart from predicting treatment response, AI-driven cancer drug candidates are currently in various stages of clinical trials (Table 30.1). Integrating multiomics with AI can reveal interactions and associations at several levels, providing a comprehensive picture of disease modeling.

Nevertheless, the organoid models may exhibit certain drawbacks like inability to completely recapitulate tumor microenvironment: ECM, vasculature, stromal and immune system components (Flood et al., 2022). In such cases, a co-culture approach with these components combined with OOC platforms can help in developing a model that better reflects in vivo conditions. Further, the protocols used for PDO generation, medium or matrix used for in vitro culture maintenance, processing, treatment, and analysis of organoids are nonstandardized (LeSavage et al., 2022). To address this, a pan-cancer PDO platform with chemically defined media was optimized for drug screening and therapeutic response monitoring by Larsen et al. (2021). Cultures from over 1000 patients were established and used for biobanking, molecular profiling, and drug screening. They also developed a label-free neural network-based pan-cancer light microscopy assay to predict drug response. Eliminating the technical variabilities associated to organoid generation and processing will help in establishing models that can be used as a basis for consistent clinical fidelity. Such datasets can be used for better training of AI-based models and can be applied over a variety of cases.

AI is harnessing and extending animal (human) intelligence to analyze complex large-scale datasets and biological networks thus advancing the field of PM. AI offers speed of conducting the studies and interpretation, which is critical in case of a disease like cancer where symptoms may take years to show and by the time these are detected, the disease has already progressed to a later stage of malignancy. However, to realize the potential of AI enabled organoids for cancer, cost-effective computing solutions, optimization of algorithms for efficiency, establishment of data sharing platforms to reduce the time, and associated costs need to be developed. In addition, data privacy, transparency, and accountability of AI as per legal and ethical framework should be defined. This is a powerful tool that will propel cancer research and therapy in the digital era but must be executed carefully.

4 AI-enabled organoids for musculoskeletal conditions

Musculoskeletal (MSK) health is crucial to normal living, daily activities, and a functional life. Challenges to MSK health and related diseases, such as osteoarthritis (OA), osteoporosis (OP), fractures, and age-related bone damage, have a significant impact on the lives of patients. Globally, nearly 1.7 billion people are known to be severely impacted by one of the many MSK conditions (Zhou et al., 2024). In vitro models investigating these conditions at the cellular level have contributed significantly to our knowledge in the field. However, traditional 2D models lack the complexity of the 3D structure in the MSK system. While animal models provide an in vivo platform for several investigations, they fail to replicate the biomechanics of human diseases due to differences in size, mechanics, and the fact that they are quadrupeds (Marshall, Bailey, Cassotta, Herrmann, & Pistollato, 2023; Mukherjee, Roy, Ghosh, & Nandi, 2022; Ribitsch et al., 2020).

As discussed in the sections above and evidenced by the literature, several approaches to 3D models for bones, cartilage, tendons, and joints are being investigated to optimize the most appropriate model for a given project (Table 30.2). The use of 3D polymeric scaffolds for bone regeneration and repair (Ganguly, El-Jawhari, Vun, Giannoudis, & Jones, 2022; Iqbal et al., 2024; Wilson et al., 2024; Yildizbakan et al., 2023), the investigation of microfluidic OOCs for MSK conditions (Mainardi et al., 2025; Pasztorek et al., 2024; Tong et al., 2025), and the increasing interest in organoids and spheroids for MSK conditions (Abraham et al., 2022; Ganguly, 2024; Gilazieva et al., 2020; Khan et al., 2023; Rausch, Iqbal, Pathak, Owston, & Ganguly, 2023) point toward the potential of these platforms for clinical applications. Organoids aimed at skeletal tissues find applications ranging from investigating drug screening, skeletal formation, regeneration, and damage to repair and disease modeling for several conditions (Chen et al., 2025).

In terms of biomedical research in the field of MSK medicine using organoid models, there are several examples of utilizing progenitor cells or induced pluripotent stem cells (iPSCs) for MSK applications. Abraham and colleagues fabricated

TABLE 30.2 Examples of 3D models used for MSK conditions in the last 3 years.

Target area/ disease	Model/system	Cells used	Matrix/platform used	Primary aim	References
Bone + cartilage	Bone and cartilage hybrid or mini joint organoid	Pediatric bone and cartilage cells with MSCs and endothelial cells	Self-assembling skeletal organoids; Matrigel	Disease modeling; drug testing	(Abraham et al., 2022)
Bone	Callus organoids (bone forming)	Human donor-derived callus/chondrogenic progenitors (xeno-free)	Self-organized organoids; differentiation to ossicles	Bone regeneration; developmental modeling; implant building blocks	(Decoene et al., 2025)
Joint (OA)	Personalized joint-on-chip (cartilage + synovium)	Patient-matched articular chondrocytes and synovial fibroblasts	Microfluidic coculture; HA-based hydrogel for cartilage compartment	Personalized therapy screening; OA pathophysiology	(Petta et al., 2024)
Cartilage (OA)	Engineered human OA cartilage organoids	Human bone marrow stromal cells (hBMSCs) → chondrogenic progenitors	3D chondrogenic organoid culture	Disease modeling; target identification; drug testing	(Dönges et al., 2024)
Cartilage	Cartilage organoid drug screen	Human cartilage organoids	Organoid culture with real-time monitoring of chondrogenesis/ hypertrophy	Drug discovery; pathway interrogation	(Wei et al., 2024)
Cartilage	Cartilage-on-a-chip with integrated sensing	Articular chondrocytes	Microfluidic device + 3D flexible electrochemical sensor	Dynamic monitoring; drug testing	(Qin et al., 2024)
Tendon	Human 3D tendon-on-chip	Human primary tenocytes (with supporting cells as described)	Microengineered 3D tendon chip (core–shell fibers)	Tendon biology; disease modeling; screening mechanotherapies	(Bakht et al., 2024)

3D, three-dimensional; *HA*, hyaluronic acid; *hBMSCs*, human bone marrow stem/stromal cells; *OA*, osteoarthritis.

a self-assembled skeletal organoid with bone and cartilage cells in Matrigel. They used these "mini joints" for inflammatory disease models and for testing the effect of adenosine receptor agonists as therapeutic agents (Abraham et al., 2022). Iordachescu et al. developed a trabecular bone prototype using primary osteoblast and osteoclast cells seeded onto femur head microtrabeculae. Their results indicate that the prototype may be suitable for investigating temporal movements in bone formation, remodeling, as well as repair (Iordachescu et al., 2021). Li and colleagues recreated articular tissue using mesenchymal stem/stromal cells (MSCs) and iPSCs on hydroxyapatite nanorods and found that these tissue-mimicking constructs present a promising platform for the evaluation of mechanistic studies on joint pathology and biology (Li et al., 2022).

Some of the areas of MSK diseases where AI has been playing a pivotal role include ultrasound-guided procedures, radiological imaging, platelet-rich plasma injections, computed tomography-guided procedures, and fluoroscopy-guided interventions, among others. In addition, the use of AI provides several advantages over traditional methods by increasing accuracy, efficiency, and precision (Debs & Fayad, 2023; Dubey et al., 2025; Gyftopoulos et al., 2019). Specifically, when it comes to diagnosis using clinical imaging, AI algorithms can streamline anomalies from all images and use predictive models for faster prognosis and treatment in OA, bone aging, tumors, and implants (Gitto et al., 2024; Guermazi et al., 2024; Ruitenbeek, Oei, Visser, & Kijowski, 2024). Considering that the structure of MSK systems (bones, cartilage, joints, muscles) is complex and intricate and replicating these tissues and organs in diseased conditions require high attention to detail; fabricating organoids for these systems comes with their own set of challenges (Bai et al., 2024a). The use of AI for MSK conditions can thus optimize organoid fabrication and other technologies for clinical applications.

The intersection of AI with organoid technology presents a futuristic approach toward problem-solving in research, healthcare, and clinical settings. Organoid technology serves as a bridge between traditional 2D platforms and the infamous "valley of death," or phase III of clinical trials. Thus the combination of AI and organoids is poised to enhance our understanding of MSK diseases and the intricacies associated with successful organoid platforms and their applications (Park, Rim, Sohn, Nam, & Ju, 2024; Bai et al., 2024b; Zhou et al., 2025). AI and ML can be applied to almost every stage of MSK-related challenges (Shi et al., 2024). These applications may include recognizing the challenge, screening patients for potential solutions, prognosis, optimizing approaches best suited to patient needs, setting parameters for building an optimized treatment model, and finally tracking the effect of the suggested solution (Kolomenskaya, Butova, Poltavskiy, Soldatov, & Butakova, 2024). Specifically, for MSK organoids, AI can be extremely useful in their assessment, as most organoids use different cell types and spatial formations and have multiple layers of structure to replicate tissue in vivo. PDOs will be pivotal in the fabrication of organoids that replicate physiological conditions in vivo in MSK diseases. The stratification of large amounts of information generated from organoids, especially regarding their OMICS data, histopathology, morphological features, and functional properties of all the different cell types, will be performed more accurately using AI (Mo et al., 2025). PDOs will further provide a platform toward PM in diseases such as OA, OP, rheumatoid arthritis, and bone-related cancers (Liu et al., 2025; Lyu, Wang, & Jiacan, 2025; Maramraju et al., 2024).

5 AI-enabled organoids for neural networks with a focus on the brain

The human brain, with its vast diversity of specialized cells and highly intricate network of connections, presents a formidable challenge to researchers striving to unravel the complexities of the neural networks and the central nervous system, with the brain as its central command. To decode the underlying biology of cellular interactions and tissue formation, the creation of functional and physiologically relevant brain models has emerged as an indispensable part of neuroscience research. These advanced models not only deepen our understanding of brain structure and function but also hold significant promise for clinical applications and the advancement of PM (Paşca, 2018).

Choosing between 2D and 3D neural cell culture models depends on the research question. 2D cultures are scalable and ideal for mechanistic studies, high-throughput screening, imaging, and producing high-purity cells for transplantation. They excel in dissecting molecular pathways, as exemplified by research on neural progenitor defects in Miller-Dieker syndrome (Iefremova et al., 2017). However, organoids—3D cultures of human cells derived from pluripotent stem cells—model essential aspects of neurodevelopment, enabling in vitro investigations of brain formation (Eichmüller & Knoblich, 2022). 3D cultures offer an unparalleled platform for modeling brain development and disease, combining structural authenticity with functional depth. Sustaining for nearly 2 years (Sloan et al., 2017), organoids capture in vivo-like cytoarchitecture, diverse cell populations, and long-term maturation that 2D systems cannot replicate. These models uniquely enable the study of complex cell-cell interactions and dynamic processes within physiologically relevant networks. Brain organoids further extend their power by allowing precise investigation of inter-regional communication and mechanistic understanding of neural disorders (Birey et al., 2017; Paşca, 2018).

The development of brain organoids from pluripotent stem cells through embryoid body formation was pioneered by the Japanese group led by Yoshiki Sasai at RIKEN. With recent research, brain organoids can be generated via (1) unguided protocols, relying on self-assembly to produce multiregional structures, or (2) guided protocols, which use patterning signals to create region-specific models with greater reproducibility. Guided approaches enable precise modeling of structures like the forebrain, midbrain, thalamus, and hypothalamus, while assembloids allow for the investigation of interregional connectivity and functional circuits, such as corticospinal and thalamocortical pathways (Zheng, Feng, Tang, & Shaohua, 2022).

Brain organoid models closely replicate brain architecture and cell diversity, enabling disease modeling and investigation of genetic and environmental factors. Fig. 30.4 represents the AI tools used for brain organoid research. They have emerged as valuable tools for studying neurodegenerative diseases like Alzheimer's (Kim et al., 2019) and PD (Foliaki et al., 2021).

The loss of midbrain dopaminergic neurons in PD, caused by a complex interplay of genetic, environmental, and aging factors, was assessed by developing a relevant brain organoid model. A high-content imaging and ML pipeline was used to analyze neurotoxin-induced damage in hMOs exposed to 6-hydroxydopamine. Using AI-enabled random forest classification on cellular features, the platform effectively assessed neurotoxicity, offering a valuable tool for screening potential neurotoxic compounds in complex human brain organoid models (Monzel et al., 2020) (Fig. 30.4).

With AI's neural network approach of U-Net, cerebral organoid monitoring and analysis using magnetic resonance imaging (MRI) and brightfield images have been reported. MRI demonstrates performance on par in global organoid segmentation and cyst assessment, even with small and diverse datasets, while local cyst segmentation requires more data. Challenges remain due to technical limitations in image resolution and sample size, which impede the accurate detection of small cyst boundaries and necessitate custom imaging parameters. MRI offers superior insights into 3D organoid morphology and tissue function, showing significant promise for broader clinical applications and large-scale comparative organoid studies (Deininger et al., 2023).

Recent advances in human brain organoids enable studies of brain development, disease prediction, and biomarker identification through platforms combining gene editing, electrophysiological recording, and MRI. However, organoid cultivation produces large, high-dimensional datasets with high variability, making AI essential for cost-effective, real-time, and accurate analysis. Current AI applications include Bayesian algorithms for gene expression analysis, clustering for transcriptomic profiling, and support vector machine classifiers for distinguishing mutant from normal organoids via electrophysiological signals (Badai, Qian, & Zhang, 2020).

Human iPSC-derived brain organoids are vital for modeling brain function, with morphology critical to their relevance. **BrAIn**, a DL-based tool with no-code functionality, tracks organoid development, quantifies key morphological parameters, and identifies structural features like neural rosettes. Testing across growth conditions showed that orbital shakers

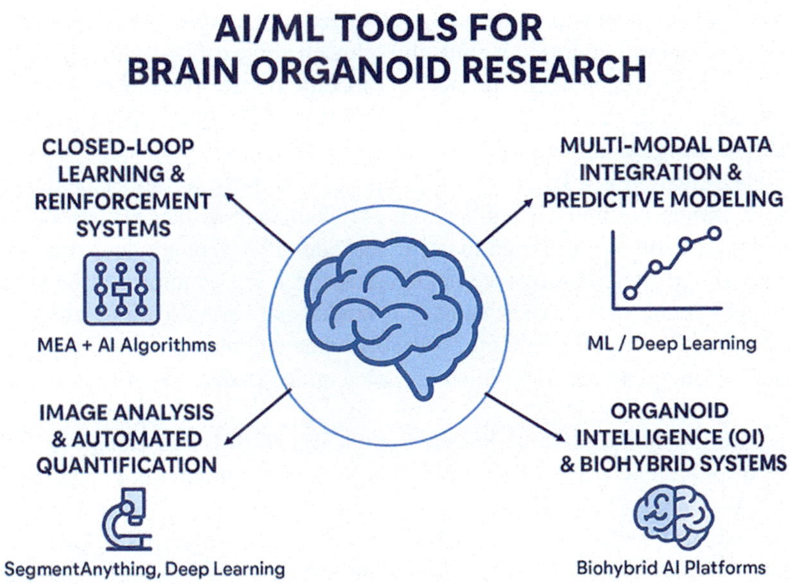

FIG. 30.4 AI/ML tools for brain organoid research.

produced the largest organoids, microfluidic chips achieved uniform growth, and both outperformed static culture in complexity, thereby demonstrating a powerful, user-friendly solution (Kahveci et al., 2025).

The brain's intricacy upholds its central role in human behavior and physiology. Brain organoid modeling enables the in vitro recapitulation of human brain development, function, and disease processes. These organoids, containing both histological and physiological attributes of native brain tissue, exhibit responsiveness to perturbations analogous to their biological counterparts. Such systems not only advance our understanding of brain disorders but also offer potential therapeutic and healthcare applications, including the establishment of biological AI interfaces.

Although still in its nascent stage, the integration of AI with brain organoid research has the capacity to significantly enhance data acquisition, processing, and protocol optimization. This convergence may accelerate the establishment and maintenance of organoids, advance clinical applications, broaden the horizons of PM, and foster new strategies for healthcare engineering in neuroscience.

6 The genetic landscape of AI-enabled organoids

Organoids are redefining how we explore genetics and disease mechanisms. By capturing the architecture and functionality of human tissues, organoids enable genetically informed investigations that go beyond the limitations of traditional 2D cultures. These 3D models maintain native tissue-specific gene expression and can be derived from both pluripotent stem cells and adult stem cells, thus preserving patient-specific genetic variation (Huch & Koo, 2015; Klompstra, Yoon, & Koo, 2025).

Organoids form the basis of advancing genomic investigations, for example, through the use of Clustered Regularly Interspaced Short Palindromic Repeats (CRISPR)-associated protein 9, allowing precise editing to correct disease-associated mutations. Cystic fibrosis has been effectively modeled using CRISPR-corrected intestinal organoids derived from patients, demonstrating the translational potential of combining organoid biology with gene editing platforms (Huch & Koo, 2015). In the context of aging and changes in the bone marrow (BM), hematopoietic stem cells are known to accumulate mutations, lose self-renewal capacity, and become increasingly biased toward the myeloid lineage (Ganguly et al., 2020, 2019, 2022; Ganguly et al., 2021). CRISPR technology allows researchers to edit key genes involved in these processes, such as those related to DNA repair, epigenetic regulation, and inflammation, and to model age-associated mutations like *TET2* and *DNMT3A*. This helps to shed light on the mechanisms underlying clonal hematopoiesis and BM dysfunction (Hara et al., 2014; Mitchell et al., 2022).

Taking research one step further, single-cell RNA (scRNA) sequencing has unveiled disease complexity in detail (Rausch et al., 2023). In BM research, where stem cell hierarchies are complex, organoid and scRNA-sequencing (scRNA-seq) models are being used to study disorders like OA and myeloma. Recent work using patient-derived BM organoids, combined with scRNA-seq, has uncovered how specific niches drive conditions like myelofibrosis and revealed variant-specific drug responses in myeloma (Khan et al., 2023; Li et al., 2024).

The Venn diagram (Fig. 30.5) highlights organoids at the intersection of 3D in vitro models and genomics, defining "NGOs." These platforms combine patient-specific biology with genomic tools to study rare variants, test gene therapies, and enable high-resolution mechanistic studies in a clinically relevant context (Lin et al., 2024; Rao et al., 2024).

AI is becoming central to the field of genetics through its capacity to offer scalable and efficient tools to analyze complex and multidimensional data. In the context of genetics, the ML and DL branches of AI specifically allow predictive analytics through their pattern recognition features (R.S. and Rawekar, 2024).

Variant detection and classification is a key analysis component not only for understanding genetic data but also for interpreting these mutations, which has clinical implications (Michelhaugh & Januzzi, 2022). The development of algorithms trained on sequencing datasets is not limited to accurately identifying single nucleotide polymorphisms and structural variants (R. and Rawekar, Artificial Intelligence in Genetics, n.d.). For example, AI-powered tools like DeepVariant and DeepSEA are increasingly demonstrating their value over traditional variant-calling and annotation pipelines. This is largely due to their enhanced sensitivity and accuracy, especially when it comes to identifying variants and predicting their potential functional impact in noncoding regions, which are often under-characterized in conventional workflows (Telenti et al., 2016).

The approach of integrating genomic data with electronic health records has enhanced disease risk prediction and supports the development of polygenic risk scores in a clinical setting (Rajkomar et al., 2018). A prime example of this has been based on the work of (Yala, Lehman, Schuster, Portnoi, & Barzilay, 2019), whereby AI-based models have identified patients at risk of Breast Cancer gene (BRCA)-associated breast cancer. This study analyzed mammogram data in conjunction with germline mutations, thereby providing a more accurate physiological assessment of risk (Yala et al., 2019).

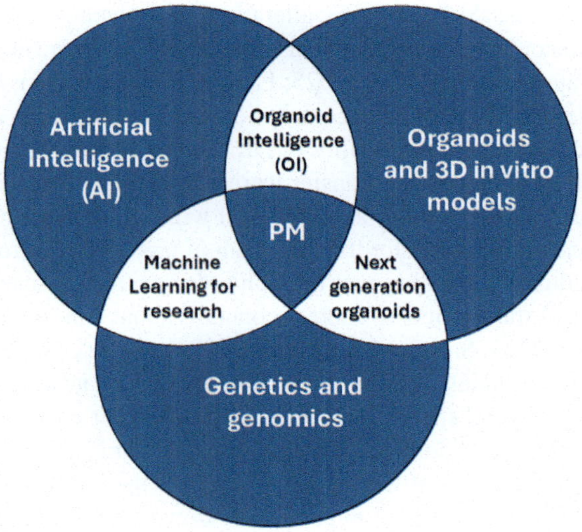

FIG. 30.5 Intersection of AI, organoid technology, and genetics that may be referred to as "next-generation organoids" as a powerful tool for precision medicine.

In the intersection of AI and genetics is represented in the area labeled "Machine Learning for Research." This visual encapsulates the emerging subfield in which AI is not only analyzing genetic data but actively guiding experimental design and therapeutic targeting, transforming how we approach biomedical research and clinical genomics.

The convergence of AI, organoid technology, and next-generation sequencing lies at the heart of future PM. This integrated framework enables end-to-end modeling, analysis, and therapeutic prediction, anchored in patient-derived biology and powered by ML. As these technologies advance, they will significantly expand our ability to understand and treat disease in a manner that is deeply personalized and biologically informed.

7 Conclusions and limitations

AI-integrated organoids model human tissue structure, diversity, and development, supporting disease research and represent a step towards PM for enhanced clinical and patient care. AI enables high-throughput imaging and accurate data analysis, and the convergence of these technologies, along with genetic considerations, can provide a powerful tool in PM, enhancing the diagnosis and treatment of any given disease. OI and NGOs are already becoming indispensable for high-dimensional datasets, extracting important information from images, molecular pathways, cellular morphologies, and therapeutic targets. Further optimization of these platforms will enable the automation and real-time analysis of information. Thus advancing protocols, standardizing datasets, and promoting interdisciplinary collaboration are essential to fully realize their translational and clinical potential.

While OI and NGOs are promising, the technology still has several challenges. One of the largest challenges is the development of physiologically relevant organoids. With the use of multiple cell types and materials to be considered, the approach must be optimized for a given target tissue. In addition, we must be careful about the ethical concerns surrounding the use of AI, especially when it comes to PDOs. The use of AI in genetics is indeed impressive, yet it does present caveats in generalizability and ethical governance, particularly when applied in clinical settings. Nevertheless, the extraction of actionable insights from genetic datasets provided by AI is an ever-growing field and will remain central to our advancements in modern medicine. Patient consent and clarity in the patient information sheets, stating how and for what these PDOs will be used, will be critical, and clinical protocols must be in place to ensure the fair and ethical use of tissues from patients.

8 Future directions

Beyond the scope of research discussed in this chapter, AI-powered organoids are offering potential in other disease fields such as antimicrobial resistance (AMR), lifestyle-related diseases, and rare genetic disorders. In the context of AMR, organoids enable us to model human-specific infection responses and assess antibiotic efficacy in a much more physiologically relevant way. For lifestyle-related conditions such as obesity and type 2 diabetes, combining AI with gut and liver organoids offers valuable insight into metabolic pathway alterations and potential treatment responses (Alnasser, 2025). When it comes to rare or inherited disorders, this approach also supports AI-driven variant validation and the development of personalized gene correction strategies (Ojeda et al., 2025; Paysan, Radhakrishnan, Xinyi Zhang, & Uhler, 2025).

Current AI methods face challenges such as reliance on prior distributions, high computational cost, poor handling of nonlinear or abnormal data, and limited interpretability in DL. Future improvements will focus on optimizing feature selection, enhancing model versatility, and adopting transfer learning, integrated learning, and parallel computing to process diverse big data from brain organoids (Badai et al., 2020). OI will become essential for analyzing the complex data organoids produce, such as electrical activity, and for training them in tasks like controlling devices. Though this synergy could inspire new computing paradigms, including neuromorphic designs, large-scale development will demand ambitious research goals and robust ethical frameworks to address concerns about consciousness and societal impact (Hartung, Pantoja, & Smirnova, 2024; Quirion, 2023; Smirnova & Hartung, 2024).

Further research could focus on advancing organ-inspired living and AI hardware with on-chip learning capabilities, extending its real-world applications beyond speech recognition and nonlinear equation solving, potentially opening new pathways for innovative AI hardware design and broader functional integration for brain organoids (Cai et al., 2023). Overall, the combination of the computational power of AI, along with the biological relevance of organoids and next-generation sequencing, presents a very powerful tool in preclinical drug testing, disease prognosis, and treatment options that can significantly enhance the level of patient care.

Author statement on the use of AI or AI-assisted tools

During the preparation of this work, the authors TP and PG used ChatGPT to assist with generating tables relevant to their work. NR used Chat GPT for generating figure 4. After using ChatGPT, all authors reviewed and edited the content as needed and take full responsibility for the content of the publication.

The views expressed in this article are those of the author and not necessarily those of the NIHR, the NIHR Leeds Biomedical Research Centre, the National Health Service or the UK Department of Health and Social Care.

References

Abraham, D. M., Herman, C., Witek, L., Cronstein, B. N., Flores, R. L., & Coelho, P. G. (2022). Self-assembling human skeletal organoids for disease modeling and drug testing. *Journal of Biomedical Materials Research - Part B Applied Biomaterials*. 15524981. *110*(4), 871–884. Available from http://onlinelibrary.wiley.com/journal/10.1002/(ISSN)1552-4965. doi:10.1002/jbm.b.34968.

Alnasser, S. M. (2025). From gut to liver: organoids as platforms for next-generation toxicology assessment vehicles for xenobiotics. *Stem Cell Research and Therapy*. 17576512. *16*(1). Available from https://stemcellres.biomedcentral.com. doi:10.1186/s13287-025-04264-y.

Alowais, S. A., Alghamdi, S. S., Alsuhebany, N., Alqahtani, T., Alshaya, A. I., Almohareb, S. N., et al. (2023). Revolutionizing healthcare: the role of artificial intelligence in clinical practice. *BMC Medical Education*. 14726920. *23*(1). Available from https://bmcmededuc.biomedcentral.com/. doi:10.1186/s12909-023-04698-z.

Arthur, A. (2015). Obama unveils $215M 'precision medicine' initiative to study genes, disease. https://www.politico.com/story/2015/01/obama-precision-medicine-gene-research-11476.

Aung, Y. Y. M., Wong, D. C. S., & Ting, D. S. W. (2021). The promise of artificial intelligence: a review of the opportunities and challenges of artificial intelligence in healthcare. *British Medical Bulletin*. 14718391. *139*(1), 4–15. Available from http://bmb.oxfordjournals.org/. doi:10.1093/bmb/ldab016.

Badai, J., Qian, B., & Zhang, L. (2020). Review of artificial intelligence applications and algorithms for brain organoid research. *Interdisciplinary Sciences: Computational Life Sciences*. 1913-2751. *12*(4), 383–394. doi:10.1007/s12539-020-00386-4.

Bai, L., Wu, Y., Li, G., Zhang, W., Zhang, H., & Su, J. (2024a). AI-enabled organoids: construction, analysis, and application. *Bioactive Materials*.2452199X. *31*, 525–548. Available from http://www.keaipublishing.com/en/journals/bioactive-materials/. doi:10.1016/j.bioactmat.2023.09.005.

Bai, L., Zhou, D., Guangfeng, L., Liu, J., Chen, X., & Jiacan, S. (2024b). Engineering bone/cartilage organoids: strategy, progress, and application. *Bone Research*. 2095-6231. *12*(1). doi:10.1038/s41413-024-00376-y.

Bakht, S. M., Pardo, A., Gomez-Florit, M., Caballero, D., Kundu, S. C., Reis, R. L., et al. (2024). Human tendon-on-chip: unveiling the effect of core compartment-T cell spatiotemporal crosstalk at the onset of tendon inflammation. *Advanced Science*. 21983844. *11*(41). Available from http://onlinelibrary.wiley.com/journal/10.1002/(ISSN)2198-3844. doi:10.1002/advs.202401170.

Bates, D. W., Levine, D., Syrowatka, A., Kuznetsova, M., Craig, K. J. T., Rui, A., et al. (2021). The potential of artificial intelligence to improve patient safety: a scoping review. *Npj Digital Medicine*. 23986352. *4*(1). Available from https://www.nature.com/npjdigitalmed/. doi:10.1038/s41746-021-00423-6.

Bhattacharya, R., Bose, D., Kaur, T., Patel, R., Renuka, O., & Rodriguez, R. V. (2025). Model organoids: integrated frameworks for the next frontier of healthcare advancements. *Stem Cell Reviews and Reports*. 26293277. *21*(2), 319–336. Available from https://www.springer.com/journal/12015. doi:10.1007/s12015-024-10814-3.

Birey, F., Andersen, J., Makinson, C. D., Islam, S., Wei, W., Huber, N., et al. (2017). Assembly of functionally integrated human forebrain spheroids. *Nature*. 14764687. *545*(7652), 54–59. Available from http://www.nature.com/nature/index.html. doi:10.1038/nature22330.

Cai, H., Zheng, A., Tian, C., Zhuhao, W., Liu, H., Tchieu, J., et al. (2023). Brain organoid reservoir computing for artificial intelligence. *Nature Electronics*. 2520-1131. *6*(12), 1032–1039. doi:10.1038/s41928-023-01069-w.

Chen, C. C., Li, H. W., Wang, Y. L., Lee, C. C., Shen, Y. C., Hsieh, C. Y., et al. (2022). Patient-derived tumor organoids as a platform of precision treatment for malignant brain tumors. *Scientific Reports*. 20452322. *12*(1). Available from www.nature.com/srep/index.html. doi:10.1038/s41598-022-20487-y.

Chen, L., Tian, B., Liu, W., Liang, H., You, Y., & Liu, W. (2022). Molecular biomarker of drug resistance developed from patient-derived organoids predicts survival of colorectal cancer patients. *Frontiers in Oncology*. 2234-943X, *12*. doi: 10.3389/fonc.2022.855674.

Chen, M., & Decary, M. (2020). Artificial intelligence in healthcare: an essential guide for health leaders. *Healthcare Management Forum*. 0840-4704. *33*(1), 10–18. doi:10.1177/0840470419873123.

Chen, P., Zhou, J. B., Chu, X. P., Feng, Y. Y., Zeng, Q. B., Lei, J. H., Wong, K. P., Chan, T. I., Lam, C. W., Zhu, W. L., Chu, W. K., Hu, F., Luo, G. H., Chan, K. I., Deng, C. X. (2025a). Establishing a cryopreserved biobank of living tumor tissues for drug sensitivity testing. Bioactive materials, *46*, 582–596. doi:10.1016/j.bioactmat.2024.09.008.

Chen, W., Liu, D., Kai, L., Mengping, X., Li, D., Yan, W., et al. (2025). Organoids of musculoskeletal system for disease modeling, drug screening, and regeneration. *Advanced Healthcare Materials*. 2192-2640. *14*(9). doi:10.1002/adhm.202402444.

Debs, P., & Fayad, L. M. (2023). The promise and limitations of artificial intelligence in musculoskeletal imaging. *Frontiers in Radiology*. 26738740. *3*. Available from https://www.frontiersin.org/journals/radiology/about. doi:10.3389/fradi.2023.1242902.

Decoene, I., Svitina, H., Hamed, M. B., Economou, A., Stegen, S., Luyten, F. P., & Papantoniou, I. (2025). Callus organoids reveal distinct cartilage to bone transition mechanisms across donors and a role for biological sex. *Bone Research*. 20956231. *13*(1). Available from https://www.nature.com/boneres/. doi:10.1038/s41413-025-00418-z.

Deininger, L., Jung-Klawitter, S., Mikut, R., Richter, P., Fischer, M., Karimian-Jazi, K., et al. (2023). An AI-based segmentation and analysis pipeline for high-field MR monitoring of cerebral organoids. *Scientific Reports*. 2045-2322. *13*(1). doi:10.1038/s41598-023-48343-7.

Dönges, L., Damle, A., Mainardi, A., Bock, T., Schönenberger, M., Martin, I., & Barbero, A. (2024). Engineered human osteoarthritic cartilage organoids. *Biomaterials*. 01429612. *308*, 122549. doi:10.1016/j.biomaterials.2024.122549.

Dorishetty, P., Balu, R., Gelmi, A., Mata, J. P., Dutta, N. K., & Choudhury, N. R. (2021). 3D Printable soy/silk hybrid hydrogels for tissue engineering applications. *Biomacromolecules*. 15264602. *22*(9), 3668–3678. Available from http://pubs.acs.org/journal/bomaf6. doi:10.1021/acs.biomac.1c00250.

Driehuis, E., Kretzschmar, K., & Clevers, H. (2020). Establishment of patient-derived cancer organoids for drug-screening applications. *Nature Protocols*. 1754-2189. *15*(10), 3380–3409. doi:10.1038/s41596-020-0379-4.

Dubey, A., Uldin, H., Khan, Z., Panchal, H., Iyengar, K. P., & Botchu, R. (2025). Role of artificial intelligence in musculoskeletal interventions. *Cancers*. 2072-6694. *17*(10), 1615. doi:10.3390/cancers17101615.

Eichmüller, O. L., & Knoblich, J. A. (2022). Human cerebral organoids—a new tool for clinical neurology research. *Nature Reviews Neurology*. 1759-4758. *18*(11), 661–680. doi:10.1038/s41582-022-00723-9.

Flood, M., Narasimhan, V., Wilson, K., Lim, W. M., Ramsay, R., Michael, M., & Heriot, A. (2022). Organoids as a robust preclinical model for precision medicine in colorectal cancer: a systematic review. *Annals of Surgical Oncology*. 15344681. *29*(1), 47–59. Available from http://www.springerlink.com/. doi:10.1245/s10434-021-10829-x.

Foliaki, S. T., Schwarz, B., Groveman, B. R., Walters, R. O., Ferreira, N. C., Orrù, C. D., et al. (2021). Neuronal excitatory-to-inhibitory balance is altered in cerebral organoid models of genetic neurological diseases. *Molecular Brain*. 17566606. *14*(1). Available from http://www.molecularbrain.com/. doi:10.1186/s13041-021-00864-w.

Friedrich, J., Seidel, C., Ebner, R., & Kunz-Schughart, L. A. (2009). Spheroid-based drug screen: considerations and practical approach. *Nature Protocols*. 17502799. *4*(3), 309–324. doi:10.1038/nprot.2008.226.

Ganguly, P. (2024). Organoids and 3D in vitro models as a platform for precision medicine (PM): an update. *Organoids*. 2674-1172. *3*(3), 165–173. doi:10.3390/organoids3030011.

Ganguly, P., Burska, A. N., Davis, C. L. M., El-Jawhari, J. J., Giannoudis, P. V., & Jones, E. A. (2020). Intrinsic type 1 interferon (Ifn1) profile of uncultured human bone marrow cd45lowcd271+ multipotential stromal cells (bm-mscs): the impact of donor age, culture expansion and ifnα and ifnβ stimulation. *Biomedicines*. 22279059. *8*(7). Available from https://www.mdpi.com/2227-9059/8/7/214. doi:10.3390/BIOMEDICINES8070214.

Ganguly, P., El-Jawhari, J. J., Burska, A. N., Ponchel, F., Giannoudis, P. V., & Jones, E. A. (2019). The analysis of in vivo aging in human bone marrow mesenchymal stromal cells using colony-forming unit-fibroblast assay and the CD45lowCD271+ phenotype. *Stem Cells International*. 16879678. *2019*. Available from http://www.hindawi.com/journals/sci/contents/. doi:10.1155/2019/5197983.

Ganguly, P., El-Jawhari, J. J., Vun, J., Giannoudis, P. V., & Jones, E. A. (2022). Evaluation of human bone marrow mesenchymal stromal cell (Msc) functions on a biomorphic rattan-wood-derived scaffold: a comparison between cultured and uncultured mscs. *Bioengineering*. 23065354. *9*(1). Available from https://www.mdpi.com/2306-5354/9/1/1/pdf. doi:10.3390/bioengineering9010001.

Ganguly, P., Toghill, B., & Pathak, S. (2021). Aging, Bone marrow and next-generation sequencing (NGS): recent advances and future perspectives. *International Journal of Molecular Sciences*. . *22*(22) 12225. doi:10.3390/ijms222212225.

Gilazieva, Z., Ponomarev, A., Rutland, C., Rizvanov, A., & Solovyeva, V. (2020). Promising applications of tumor spheroids and organoids for personalized medicine. *Cancers*. 2072-6694. *12*(10), 2727. doi:10.3390/cancers12102727.

Gitto, S., Serpi, F., Albano, D., Risoleo, G., Fusco, S., Messina, C., & Sconfienza, L. M. (2024). AI applications in musculoskeletal imaging: a narrative review. *European Radiology Experimental*. 25099280. *8*(1). Available from https://www.springer.com/journal/41747. doi:10.1186/s41747-024-00422-8.

Greek, R., Menache, A., & Rice, M. J. (2012). Animal models in an age of personalized medicine. *Personalized Medicine*. 1741-0541. *9*(1), 47–64. doi:10.2217/pme.11.89.

Guermazi, A., Omoumi, P., Tordjman, M., Fritz, J., Kijowski, R., Regnard, N. E., et al. (2024). How AI may transform musculoskeletal imaging. *Radiology*. 15271315. *310*(1). Available from https://pubs.rsna.org/doi/epdf/10.1148/radiol.230764. doi:10.1148/radiol.230764.

Guillen, K. P., Fujita, M., Butterfield, A. J., Scherer, S. D., Bailey, M. H., Chu, Z., et al. (2022). A human breast cancer-derived xenograft and organoid platform for drug discovery and precision oncology. *Nature Cancer*. 26621347. *3*(2), 232–250. Available from https://www.nature.com/natcancer/. doi:10.1038/s43018-022-00337-6.

Gyftopoulos, S., Lin, D., Knoll, F., Doshi, A. M., Rodrigues, T. C., & Recht, M. P. (2019). Artificial intelligence in musculoskeletal imaging: current status and future directions. *American Journal of Roentgenology*. 15463141. *213*(3), 506–513. Available from https://www.ajronline.org/doi/pdf/10.2214/AJR.19.21117. doi:10.2214/AJR.19.21117.

Hara, K., Nakagawa, T., Enomoto, H., Suzuki, M., Yamamoto, M., Simons, B. D., & Yoshida, S. (2014). Mouse spermatogenic stem cells continually interconvert between equipotent singly isolated and syncytial states. *Cell Stem Cell*. 18759777. *14*(5), 658–672. doi:10.1016/j.stem.2014.01.019.

Hartung, T., Pantoja, I. E. M., & Smirnova, L. (2024). Brain organoids and organoid intelligence from ethical, legal, and social points of view. *Frontiers in Artificial Intelligence*. 2624-8212, *6*. doi: 10.3389/frai.2023.1307613.

Hilton, C. B., Milinovich, A., Felix, C., Vakharia, N., Crone, T., Donovan, C., et al. (2020). Personalized predictions of patient outcomes during and after hospitalization using artificial intelligence. *Npj Digital Medicine*. 23986352. *3*(1). Available from https://www.nature.com/npjdigitalmed/. doi:10.1038/s41746-020-0249-z.

Hofer, M., & Lutolf, M. P. (2021). Engineering organoids. *Nature Reviews Materials*. 20588437. *6*(5), 402–420. Available from www.nature.com/natrevmats/. doi:10.1038/s41578-021-00279-y.

Huang, K., Mingyue, L., Qiwei, L., Chen, Z., Zhang, Y., & Zhongze, G. (2024). Image-based profiling and deep learning reveal morphological heterogeneity of colorectal cancer organoids. *Computers in Biology and Medicine*. 00104825. *173*, 108322. doi:10.1016/j.compbiomed.2024.108322.

Huch, M., & Koo, B. K. (2015). Modeling mouse and human development using organoid cultures. *Development (Cambridge)*. 14779129. *142*(18), 3113–3125. Available from http://dev.biologists.org/content/142/18/3113.full.pdf. doi:10.1242/dev.118570.

Iefremova, V., Manikakis, G., Krefft, O., Jabali, A., Weynans, K., Wilkens, R., et al. (2017). An organoid-based model of cortical development identifies non-cell-autonomous defects in wnt signaling contributing to Miller-Dieker Syndrome. *Cell Reports*. 22111247. *19*(1), 50–59. Available from http://www.sciencedirect.com/science/journal/22111247. doi:10.1016/j.celrep.2017.03.047.

Ingber, D. E. (2022). Human organs-on-chips for disease modelling, drug development and personalized medicine. *Nature Reviews Genetics*. 14710064. *23*(8), 467–491. Available from https://www.nature.com/nrg/. doi:10.1038/s41576-022-00466-9.

Iordachescu, A., Hughes, E. A. B., Joseph, S., Hill, E. J., Grover, L. M., & Metcalfe, A. D. (2021). Trabecular bone organoids: a micron-scale 'humanised' prototype designed to study the effects of microgravity and degeneration. *Npj Microgravity*. 23738065. *7*(1). Available from https://www.nature.com/npjmgrav/. doi:10.1038/s41526-021-00146-8.

Iqbal, N., Ganguly, P., Yildizbakan, L., Raif, E. M., Jones, E., Giannoudis, P. V., & Jha, A. (2024). Chitosan scaffolds from crustacean and fungal sources: a comparative study for bone-tissue-engineering applications. *Bioengineering*. 23065354. *11*(7). Available from www.mdpi.com/journal/bioengineering. doi:10.3390/bioengineering11070720.

Iqbal, N., Pant, T., Rohra, N., Goyal, A., Lawrence, M., Dey, A., & Ganguly, P. (2023). Nanobiotechnology in bone tissue engineering applications: recent advances and future perspectives. *Applied Biosciences*. 2813-0464. *2*(4), 617–638. doi:10.3390/applbiosci2040039.

Jensen, C., & Teng, Y. (2020). Is it time to start transitioning from 2D to 3D cell culture? *Frontiers in Molecular Biosciences*. 2296-889X, *7*. doi: 10.3389/fmolb.2020.00033.

Kahveci, B., Polatli, E., Evranos, A. E., Güner, H., Bastanlar, Y., Karakülah, G., & Güven, S. (2025). BrAIn: a comprehensive artificial intelligence-based morphology analysis system for brain organoids and neuroscience. *bioRxiv*. 26928205. Available from https://www.biorxiv.org. doi:10.1101/2025.02.19.638973.

Kapałczyńska, M., Kolenda, T., Przybyła, W., Zajączkowska, M., Teresiak, A., Filas, V., et al. (2018). 2D and 3D cell cultures—a comparison of different types of cancer cell cultures. *Archives of Medical Science*. 18969151. *14*(4), 910–919. Available from https://www.termedia.pl/2D-and-3D-cell-cultures-a-comparison-of-different-types-of-cancer-cell-cultures,19,28752,1,1.html. doi:10.5114/aoms.2016.63743.

Kastenschmidt, J. M., Schroers-Martin, J. G., Sworder, B. J., Sureshchandra, S., Khodadoust, M. S., Liu, C. L., et al. (2024). A human lymphoma organoid model for evaluating and targeting the follicular lymphoma tumor immune microenvironment. *Cell Stem Cell*. 18759777. *31*(3), 410. Available from https://www.sciencedirect.com/science/journal/19345909. doi:10.1016/j.stem.2024.01.012.

Khan, A. O., Rodriguez-Romera, A., Reyat, J. S., Olijnik, A. A., Colombo, M., Wang, G., et al. (2023). Human bone marrow organoids for disease modeling, discovery, and validation of therapeutic targets in hematologic malignancies. *Cancer Discovery*. 21598290. *13*(2), 364–385. Available from https://aacrjournals.org/cancerdiscovery/article-pdf/13/2/364/3266882/364.pdf. doi:10.1158/2159-8290.CD-22-0199.

Kim, H., Park, H. J., Choi, H., Chang, Y., Park, H., Shin, J., et al. (2019). Modeling G2019S-LRRK2 sporadic Parkinson's disease in 3D midbrain organoids. *Stem Cell Reports*. 22136711. *12*(3), 518–531. Available from http://www.elsevier.com/journals/stem-cell-reports/2213-6711. doi:10.1016/j.stemcr.2019.01.020.

Kim, J. S., Park, C. H., Kim, E., Lee, H. S., Lee, J., Kim, J., et al. (2025). Establishing 3D organoid models from patient-derived conditionally reprogrammed cells to bridge preclinical and clinical insights in pancreatic cancer. *Molecular Cancer*. 14764598. *24*(1). Available from https://molecular-cancer.biomedcentral.com. doi:10.1186/s12943-025-02374-y.

Kim, J., Koo, B. K., & Knoblich, J. A. (2020). Human organoids: model systems for human biology and medicine. *Nature Reviews Molecular Cell Biology*. 14710080. *21*(10), 571–584. Available from http://www.nature.com/molcellbio. doi:10.1038/s41580-020-0259-3.

Klompstra, T. M., Yoon, K. J., & Koo, B. K. (2025). Evolution of organoid genetics. *European Journal of Cell Biology*. 16181298. *104*(2). Available from https://www.sciencedirect.com/science/journal/01719335. doi:10.1016/j.ejcb.2025.151481.

Kolomenskaya, E., Butova, V., Poltavskiy, A., Soldatov, A., & Butakova, M. (2024). Application of artificial intelligence at all stages of bone tissue engineering. *Biomedicines*. 2227-9059. *12*(1), 76. doi:10.3390/biomedicines12010076.

Kong, J. H., Lee, H., Kim, D., Han, S. K., Ha, D., Shin, K., & Kim, S. (2020). Network-based machine learning in colorectal and bladder organoid models predicts anti-cancer drug efficacy in patients. *Nature Communications*. 20411723. *11*(1). Available from http://www.nature.com/ncomms/index.html. doi:10.1038/s41467-020-19313-8.

Larsen, B. M., Kannan, M., Langer, L. F., Leibowitz, B. D., Bentaieb, A., Cancino, A., et al. (2021). A pan-cancer organoid platform for precision medicine. *Cell Reports*. 22111247. *36*(4). Available from http://www.sciencedirect.com/science/journal/22111247. doi:10.1016/j.celrep.2021.109429.

Lee, H. (2023). Engineering in vitro models: bioprinting of organoids with artificial intelligence. *Cyborg and Bionic Systems*. 2692-7632, *4*. doi: 10.34133/cbsystems.0018.

Lee, M. R., Kang, S., Lee, J., Kong, S. Y., Kim, Y., Lee, Y. S., et al. (2025). Organoid morphology-guided classification for oral cancer reveals prognosis. *Cell Reports Medicine*. 26663791. *6*(5). Available from https://www.sciencedirect.com/science/journal/26663791. doi:10.1016/j.xcrm.2025.102129.

Leong, K. F., Chua, C. K., Sudarmadji, N., & Yeong, W. Y. (2008). Engineering functionally graded tissue engineering scaffolds. *Journal of the Mechanical Behavior of Biomedical Materials*. 17516161. *1*(2), 140–152. doi:10.1016/j.jmbbm.2007.11.002.

LeSavage, B. L., Suhar, R. A., Broguiere, N., Lutolf, M. P., & Heilshorn, S. C. (2022). Next-generation cancer organoids. *Nature Materials*. 14764660. *21*(2), 143–159. Available from http://www.nature.com/nmat/. doi:10.1038/s41563-021-01057-5.

Li, H., Liu, H., & Chen, K. (2022). Living biobank-based cancer organoids: prospects and challenges in cancer research. *Cancer Biology & Medicine*. 20953941. *19*(7), 965–982. doi:10.20892/j.issn.2095-3941.2021.0621.

Li, R., Colombo, M., Wang, G., Rodriguez-Romera, A., Benlabiod, C., Jooss, N. J., et al. (2024). A proinflammatory stem cell niche drives myelofibrosis through a targetable galectin-1 axis. *Science Translational Medicine*. 19466242. *16*(768). Available from https://www.science.org/doi/10.1126/scitranslmed.adj7552. doi:10.1126/scitranslmed.adj7552.

Li, Z. A., Shang, J., Xiang, S., Li, E. N., Yagi, H., Riewruja, K., et al. (2022). Articular tissue-mimicking organoids derived from mesenchymal stem cells and induced pluripotent stem cells. *Organoids*. 26741172. *1*(2), 135–148. Available from https://www.mdpi.com/journal/organoids. doi:10.3390/organoids1020011.

Lin, H., Hongbo, F., Sun, S., Yin, H., Yuan, J., & Liao, J. (2024). Patient tissue-derived FGFR4-variant and wild-type colorectal cancer organoid development and anticancer drug sensitivity testing. *Heliyon*. 24058440. *10*(10), e30985. doi:10.1016/j.heliyon.2024.e30985.

Liu, Y., Zhou, Z., Lu, G., Zhang, X., Shi, D., Tong, L., et al. (2025). Musculoskeletal organoids: an emerging toolkit for establishing personalized models of musculoskeletal disorders and developing regenerative therapies. *Acta Biomaterialia*.18787568. *200*, 158–186. Available from https://www.sciencedirect.com/science/journal/17427061. doi:10.1016/j.actbio.2025.05.037.

Lundberg, A. S., Geuijen, C. A. W., Hill, S., van Bueren, J. J. L., Fumagalli, A., de Kruif, J., et al. (2025). Petosemtamab, a bispecific antibody targeting epidermal growth factor receptor (EGFR) and leucine-rich G repeat-containing protein-coupled receptor (LGR5) designed for broad clinical applications. *Cancers*. 20726694. *17*(10). Available from http://www.mdpi.com/journal/cancers/. doi:10.3390/cancers17101665.

Lyu, X., Wang, J., & Jiacan, S. (2025). Intelligent manufacturing for osteoarthritis organoids. *Cell Proliferation*. 0960-7722. *58*(7). doi:10.1111/cpr.70043.

Maceachern, S. J., & Forkert, N. D. (2021). Machine learning for precision medicine. *Genome*. 08312796. *64*(4), 416–425. Available from https://www.nrcresearchpress.com/loi/gen. doi:10.1139/gen-2020-0131.

Mainardi, A., Börsch, A., Occhetta, P., Ivanek, R., Ehrbar, M., Krattiger, L., et al. (2025). An organ-on-chip platform for strain-controlled, tissue-specific compression of cartilage and mineralized osteochondral interface to study mechanical overloading in osteoarthritis. *Advanced Healthcare Materials*. 2192-2640. *14*(23). doi:10.1002/adhm.202501588.

Mao, Y., Shangguan, D., Huang, Q., Xiao, L., Cao, D., Zhou, H., & Wang, Y. K. (2025). Emerging artificial intelligence-driven precision therapies in tumor drug resistance: recent advances, opportunities, and challenges. *Molecular Cancer*. 14764598. *24*(1). Available from https://molecular-cancer.biomedcentral.com. doi:10.1186/s12943-025-02321-x.

Maramraju, S., Kowalczewski, A., Kaza, A., Liu, X., Singaraju, J. P., Albert, M. V., et al. (2024). AI-organoid integrated systems for biomedical studies and applications. *Bioengineering and Translational Medicine*. 23806761. *9*(2). Available from http://onlinelibrary.wiley.com/journal/10.1002/(ISSN)2380-6761. doi:10.1002/btm2.10641.

Marshall, L. J., Bailey, J., Cassotta, M., Herrmann, K., & Pistollato, F. (2023). Poor translatability of biomedical research using animals—a narrative review. *Alternatives to Laboratory Animals*. 26323559. *51*(2), 102–135. Available from https://journals.sagepub.com/home/atl. doi:10.1177/02611929231157756.

Mayawala, K. (2024). A long-acting high affinity anti-TSLP antibody (GB-0895) for severe asthma identified leveraging a proprietary machine learning platform—Vienna 2024—ERS Respiratory Channel The ERS Congress 2024, Vienna. https://channel.ersnet.org/media-113432-a-long-acting-high-affinity-anti-tslp-antibody-gb-0895-for-severe-asthma-identified-leveraging-a-proprietary-machine-learning-platform.

Mesko, B. (2017). The role of artificial intelligence in precision medicine. *Expert Review of Precision Medicine and Drug Development*. 2380-8993. *2*(5), 239–241. doi:10.1080/23808993.2017.1380516.

Michelhaugh, S. A., & Januzzi, J. L. (2022). Using artificial intelligence to better predict and develop biomarkers. *Heart Failure Clinics*. 15517136. *18*(2), 275–285. Available from http://www.elsevier.com/wps/find/journaldescription.cws_home/704215/description#description. doi:10.1016/j.hfc.2021.11.004.

Mitchell, E., Chapman, M. S., Williams, N., Dawson, K. J., Mende, N., Calderbank, E. F., et al. (2022). Clonal dynamics of haematopoiesis across the human lifespan. *Nature*. 14764687. *606*(7913), 343–350. Available from http://www.nature.com/nature/index.html. doi:10.1038/s41586-022-04786-y.

Mo, Y., Wang, J., Yang, H., Bai, L., Xu, K., & Su, J. (2025). Artificial intelligence for organoids multidimensional assessment. *SmartMat*. 2688819X. *6*(3). Available from https://onlinelibrary.wiley.com/journal/2688819x. doi:10.1002/smm2.70016.

Momoli, C., Costa, B., Lenti, L., Tubertini, M., Parenti, M., Martella, E., et al. (2025). The evolution of anticancer 3D in vitro models: the potential role of machine learning and AI in the next generation of animal-free experiments. *Cancers*. 2072-6694. *17*(4), 700. doi:10.3390/cancers17040700.

Monzel, A. S., Hemmer, K., Kaoma, T., Smits, L. M., Bolognin, S., Lucarelli, P., et al. (2020). Machine learning-assisted neurotoxicity prediction in human midbrain organoids. *Parkinsonism and Related Disorders*.18735126. *75*, 105–109. Available from www.elsevier.com/locate/parkreldis. doi:10.1016/j.parkreldis.2020.05.011.

Mukherjee, P., Roy, S., Ghosh, D., & Nandi, S. K. (2022). Role of animal models in biomedical research: a review. *Laboratory Animal Research*. 2233-7660. *38*(1). doi:10.1186/s42826-022-00128-1.

Naithani, N., Sinha, S., Misra, P., Vasudevan, B., & Sahu, R. (2021). Precision medicine: concept and tools. *Medical Journal Armed Forces India*. 03771237. *77*(3), 249–257. doi:10.1016/j.mjafi.2021.06.021.

Neal, J. T., Li, X., Zhu, J., Giangarra, V., Grzeskowiak, C. L., Ju, J., et al. (2018). Organoid modeling of the tumor immune microenvironment. *Cell*. 10974172. *175*(7), 1972. Available from https://www.sciencedirect.com/journal/cell. doi:10.1016/j.cell.2018.11.021.

Nikonorova, V. G., Chrishtop, V. V., Mironov, V. A., & Prilepskii, A. Y. (2023). Advantages and potential benefits of using organoids in nanotoxicology. *Cells*. 20734409. *12*(4). Available from https://www.mdpi.com/journal/cells. doi:10.3390/cells12040610.

Nithin, R., Aggarwal, A., Sravani, A. B., Mallya, P., & Lewis, S. (2023). Organ-on-a-chip: an emerging research platform. *Organogenesis*. 15558592. *19*(1). Available from http://www.tandfonline.com/toc/kogg20/current. doi:10.1080/15476278.2023.2278236.

Ojeda, A. V., Awabdeh, D., Brewster, B., Rockne, R., O'Meally, D., Yin, H. H., et al. (2025). Modeling cerebral development in vitro with L-MYC-immortalized human neural stem cell-derived organoids. *bioRxiv*. 26928205. Available from https://www.biorxiv.org. doi:10.1101/2025.02.12.637976.

Okamoto, T., Natsume, Y., Doi, M., Nosato, H., Iwaki, T., Yamanaka, H., et al. (2022). Integration of human inspection and artificial intelligence-based morphological typing of patient-derived organoids reveals interpatient heterogeneity of colorectal cancer. *Cancer Science*. 13497006. *113*(8), 2693–2703. Available from http://onlinelibrary.wiley.com/journal/10.1111/(ISSN)1349-7006. doi:10.1111/cas.15396.

Ong, C. S., Zhou, X., Han, J., Huang, C. Y., Nashed, A., Khatri, S., et al. (2018). In vivo therapeutic applications of cell spheroids. *Biotechnology Advances*. 07349750. *36*(2), 494–505. Available from www.elsevier.com/inca/publications/store/5/2/5/4/5/5/index.htt. doi:10.1016/j.biotechadv.2018.02.003.

Park, G., Rim, Y. A., Sohn, Y., Nam, Y., & Ju, J. H. (2024). Replacing animal testing with stem cell-organoids: advantages and limitations. *Stem Cell Reviews and Reports*. 26293277. *20*(6), 1375–1386. Available from https://www.springer.com/journal/12015. doi:10.1007/s12015-024-10723-5.

Park, T., Kim, T. K., Han, Y. D., Kim, K. A., Kim, H., & Kim, H. S. (2023). Development of a deep learning based image processing tool for enhanced organoid analysis. *Scientific Reports*. 20452322. *13*(1). Available from https://www.nature.com/srep/. doi:10.1038/s41598-023-46485-2.

Paşca, S. P. (2018). The rise of three-dimensional human brain cultures. *Nature*. 0028-0836. *553*(7689), 437–445. doi:10.1038/nature25032.

Pasztorek, M., Fischer, J., Otahal, A., de Luna, A., Nehrer, S., & Rosser, J. (2024). Exploring osteoarthritis dynamics: patient-specific cartilage samples in an organ-on-a-chip model. *Cartilage*. 1947-6035. doi: 10.1177/19476035241277654.

Paul, C. D., Yankaskas, C., Thakuri, P. S., Balhouse, B., Salen, S., Bullock, A., et al. (2025). Long-term maintenance of patient-specific characteristics in tumoroids from six cancer indications. *Scientific Reports*. 20452322. *15*(1). Available from https://www.nature.com/srep/. doi:10.1038/s41598-025-86979-9.

Paysan, D., Radhakrishnan, A., Xinyi Zhang, G. V. S., & Uhler, C. (2025). Image2Reg: linking chromatin images to gene regulation using genetic and chemical perturbation screens. *Cell Systems*. 24054712. *16*(6), 101293. doi:10.1016/j.cels.2025.101293.

Petta, D., D'Arrigo, D., Salehi, S., Talò, G., Bonetti, L., Vanoni, M., et al. (2024). A personalized osteoarthritic joint-on-a-chip as a screening platform for biological treatments. *Materials Today Bio*. 25900064. *26*, 101072. doi:10.1016/j.mtbio.2024.101072.

Philippidis, A. (2025). Recursion halts four pipeline programs, sharpening cancer, rare disease focus. @genbio. Retrieved from https://www.genengnews.com/topics/artificial-intelligence/recursion-halts-four-pipeline-programs-sharpening-cancer-rare-disease-focus/. (Accessed 20 August 2025).

Prabha, R. D., Kraft, D. C. E., Harkness, L., Melsen, B., Varma, H., Nair, P. D., et al. (2018). Bioactive nano-fibrous scaffold for vascularized craniofacial bone regeneration. *Journal of Tissue Engineering and Regenerative Medicine*. 19327005. *12*(3), e1537. Available from http://onlinelibrary.wiley.com/journal/10.1002/(ISSN)1932-7005. doi:10.1002/term.2579.

Premjit, Y., Lawrence, M., Goyal, A., Ferreira, C., Jones, E. A., & Ganguly, P. (2025). Biomimetic three-dimensional (3D) scaffolds from sustainable biomaterials: innovative green medicine approach to bone regeneration. *Journal of Functional Biomaterials*. 20794983. *16*(7). Available from http://www.mdpi.com/journal/jfb. doi:10.3390/jfb16070238.

Qin, Y., Li, J. X., Cai, W., Fan, W. T., Duan, B., Zhao, Y., et al. (2024). A cartilage-on-a-chip for recapitulating cell microenvironment and real-time nitric oxide monitoring. *Device*. 26669986. *2*(6). Available from https://www.sciencedirect.com/journal/device. doi:10.1016/j.device.2024.100410.

Quirion, R. (2023). Brain organoids: are they for real? *Frontiers in Science*. 2813-6330, *1*. doi: 10.3389/fsci.2023.1148127.

R., S., Rawekar, A., 2024. Artificial Intelligence in Genetics. Cureus 16(1), e52035. doi:10.7759/cureus.52035.

Rajkomar, A., Oren, E., Chen, K., Dai, A. M., Hajaj, N., Hardt, M., et al. (2018). Scalable and accurate deep learning with electronic health records. *Npj Digital Medicine*. 23986352. *1*(1). Available from https://www.nature.com/npjdigitalmed/. doi:10.1038/s41746-018-0029-1.

Rao, J., Song, C., Hao, Y., Chen, Z., Feng, S., Shihui, X., et al. (2024). Leveraging patient-derived organoids for personalized liver cancer treatment. *International Journal of Biological Sciences*. 1449-2288. *20*(13), 5363–5374. doi:10.7150/ijbs.96317.

Rausch, M., Iqbal, N., Pathak, S., Owston, H. E., & Ganguly, P. (2023). Organoid models and next-generation sequencing for bone marrow and related disorders. *Organoids*. 26741172. *2*(3), 123–139. Available from https://www.mdpi.com/journal/organoids. doi:10.3390/organoids2030010.

Ravi, M., Paramesh, V., Kaviya, S. R., Anuradha, E., & Solomon, F. D. P. (2015). 3D cell culture systems: advantages and applications. *Journal of Cellular Physiology*. 0021-9541. *230*(1), 16–26. doi:10.1002/jcp.24683.

Renner, H., Schöler, H. R., & Bruder, J. M. (2021). Combining automated organoid workflows with artificial intelligence-based analyses: opportunities to build a new generation of interdisciplinary high-throughput screens for Parkinson's disease and beyond. *Movement Disorders*. 15318257. *36*(12), 2745–2762. Available from http://onlinelibrary.wiley.com/journal/10.1002/(ISSN)1531-8257. doi:10.1002/mds.28775.

Ribitsch, I., Baptista, P. M., Lange-Consiglio, A., Melotti, L., Patruno, M., Jenner, F., et al. (2020). Large animal models in regenerative medicine and tissue engineering: to do or not to do. *Frontiers in Bioengineering and Biotechnology*. 22964185. *8*. Available from http://journal.frontiersin.org/journal/bioengineering-and-biotechnology#archive. doi:10.3389/fbioe.2020.00972.

Ruitenbeek, H. C., Oei, E. H. G., Visser, J. J., & Kijowski, R. (2024). Artificial intelligence in musculoskeletal imaging: realistic clinical applications in the next decade. *Skeletal Radiology*. 14322161. *53*(9), 1849–1868. Available from https://www.springer.com/journal/256. doi:10.1007/s00256-024-04684-6.

Sbitan, L., Alzraikat, N., Tanous, H., Saad, A. M., & Odeh, M. (2025). From one size fits all to a tailored approach: integrating precision medicine into medical education. *BMC Medical Education*. 14726920. *25*(1). Available from https://bmcmededuc.biomedcentral.com/. doi:10.1186/s12909-024-06138-y.

Seifirad, S., & Haghpanah, V. (2019). Inappropriate modeling of chronic and complex disorders: how to reconsider the approach in the context of predictive, preventive and personalized medicine, and translational medicine. *EPMA Journal*. 1878-5077. *10*(3), 195–209. doi:10.1007/s13167-019-00176-z.

Shi, H., Kowalczewski, A., Danny, V., Liu, X., Salekin, A., Yang, H., & Zhen, M. (2024). Organoid intelligence: integration of organoid technology and artificial intelligence in the new era of in vitro models. *Medicine in Novel Technology and Devices*. 25900935. *21*, 100276. doi:10.1016/j.medntd.2023.100276.

Sloan, S. A., Darmanis, S., Huber, N., Khan, T. A., Birey, F., Caneda, C., et al. (2017). Human astrocyte maturation captured in 3D cerebral cortical spheroids derived from pluripotent stem cells. *Neuron*. 10974199. *95*(4), 779. Available from www.neuron.org. doi:10.1016/j.neuron.2017.07.035.

Smabers, L. P., Wensink, E., Verissimo, C. S., Koedoot, E., Pitsa, K. C., Huismans, M. A., et al. (2024). Organoids as a biomarker for personalized treatment in metastatic colorectal cancer: drug screen optimization and correlation with patient response. *Journal of Experimental and Clinical Cancer Research*. 17569966. *43*(1). Available from https://jeccr.biomedcentral.com. doi:10.1186/s13046-024-02980-6.

Smirnova, L., & Hartung, T. (2024). The promise and potential of brain organoids. *Advanced Healthcare Materials*. 2192-2640. *13*(21). doi:10.1002/adhm.202302745.

Telenti, A., Pierce, L. C. T., Biggs, W. H., Di Iulio, J., Wong, E. H. M., Fabani, M. M., et al. (2016). Deep sequencing of 10,000 human genomes. *Proceedings of the National Academy of Sciences of the United States of America*. 10916490. *113*(42), 11901–11906. Available from http://www.pnas.org/content/113/42/11901.full.pdf. doi:10.1073/pnas.1613365113.

Teles, D., Kim, Y., Ronaldson-Bouchard, K., & Vunjak-Novakovic, G. (2021). Machine learning techniques to classify healthy and diseased cardiomyocytes by contractility profile. *ACS Biomaterials Science & Engineering*. 2373-9878. *7*(7), 3043–3052. doi:10.1021/acsbiomaterials.1c00418.

Tong, X., Liu, M., Jiao, L., Zhang, W., Rong, H., Yang, G., et al. (2025). Musculoskeletal organoids-on-chip uncover muscle–bone communication under intermittent hypoxia. *National Science Review*. 2095-5138. *12*(7). doi:10.1093/nsr/nwaf214.

Veninga, V., & Voest, E. E. (2021). Tumor organoids: opportunities and challenges to guide precision medicine. *Cancer Cell*. 18783686. *39*(9), 1190–1201. Available from https://www.journals.elsevier.com/cancer-cell. doi:10.1016/j.ccell.2021.07.020.

Watanabe, S., Kobayashi, S., Ogasawara, N., Okamoto, R., Nakamura, T., Watanabe, M., et al. (2022). Transplantation of intestinal organoids into a mouse model of colitis. *Nature Protocols*. 17502799. *17*(3), 649–671. Available from https://www.springer.com/journal/41596. doi:10.1038/s41596-021-00658-3.

Wei, X., Qiu, J., Lai, R., Wei, T., Lin, Z., Huang, S., et al. (2024). A human organoid drug screen identifies α2-adrenergic receptor signaling as a therapeutic target for cartilage regeneration. *Cell Stem Cell*. 19345909. *31*(12), 1813. doi:10.1016/j.stem.2024.09.001.

Wensink, G. E., Elias, S. G., Mullenders, J., Koopman, M., Boj, S. F., Kranenburg, O. W., & Roodhart, J. M. L. (2021). Patient-derived organoids as a predictive biomarker for treatment response in cancer patients. *Npj Precision Oncology*. 2397768X. *5*(1). Available from https://www.nature.com/npjprecisiononcology/. doi:10.1038/s41698-021-00168-1.

Wilson, B. J., Owston, H. E., Iqbal, N., Giannoudis, P. V., McGonagle, D., Pandit, H., et al. (2024). In vitro osteogenesis study of shell nacre cement with older and young donor bone marrow mesenchymal stem/stromal cells. *India Bioengineering*. 23065354. *11*(2). Available from www.mdpi.com/journal/bioengineering. doi:10.3390/bioengineering11020143.

Xia, X., Fei, L., Juan, H., Aji, R., & Gao, D. (2019). Organoid technology in cancer precision medicine. *Cancer Letters*. 03043835. *457*, 20–27. doi:10.1016/j.canlet.2019.04.039.

Yala, A., Lehman, C., Schuster, T., Portnoi, T., & Barzilay, R. (2019). A deep learning mammography-based model for improved breast cancer risk prediction. *Radiology*. 0033-8419. *292*(1), 60–66. doi:10.1148/radiol.2019182716.

Yang, H., Wang, Y., Wang, P., Zhang, N., & Wang, P. (2022). Tumor organoids for cancer research and personalized medicine. *Cancer Biology & Medicine*. 20953941. *19*(3), 319–332. Available from www.cancerbiomed.org. doi:10.20892/j.issn.2095-3941.2021.0335.

Yildizbakan, L., Iqbal, N., Ganguly, P., Kumi-Barimah, E., Thuy, D., Jones, E., et al. (2023). Fabrication and characterisation of the cytotoxic and antibacterial properties of chitosan-cerium oxide porous scaffolds. *Antibiotics*. 2079-6382. *12*(6), 1004. doi:10.3390/antibiotics12061004.

Yu, K. H., Beam, A. L., & Kohane, I. S. (2018). Artificial intelligence in healthcare. *Nature Biomedical Engineering*. 2157846X. *2*(10), 719–731. Available from www.nature.com/natbiomedeng/ doi:10.1038/s41551-018-0305-z.

Zheng, H., Feng, Y., Tang, J., & Shaohua, M. (2022). Interfacing brain organoids with precision medicine and machine learning. *Cell Reports Physical Science*. 26663864. *3*(7), 100974. doi:10.1016/j.xcrp.2022.100974.

Zhou, J., Xie, S., Xu, S., Zhang, Y., Li, Y., Sun, Q., et al. (2024). From pain to progress: comprehensive analysis of musculoskeletal disorders worldwide. *Journal of Pain Research*.11787090. *17*, 3455–3472. Available from https://www.dovepress.com/article/download/96653. doi:10.2147/JPR.S488133.

Zhou, L., Chen, S., Liu, J., Zhou, Z., Yan, Z., Chenzhong, L., et al. (2025). When artificial intelligence (AI) meets organoids and organs-on-chips (OoCs): game-changer for drug discovery and development? *The Innovation Life*. 2959-8761. *3*(1), 100115. doi:10.59717/j.xinn-life.2024.100115.

Zhou, Z., Cong, L., & Cong, X. (2021). Patient-derived organoids in precision medicine: drug screening, organoid-on-a-chip and living organoid biobank. *Frontiers in Oncology*. 2234-943X, *11*. doi: 10.3389/fonc.2021.762184.

Chapter 31

Leveraging artificial intelligence (AI) in herbal medicinal plant extracts: Pathway to drug development

Chukwuemeka R. Nwokocha[a], Javier Palacios[b], Keaton Logan[c], Magdalene Nwokocha[d], Ya Fatou Njie Mbye[e], Nelson Orie[f], Sunny Edet Ohia[e], and Momoh A. Yakubu[g]

[a]*Department of Basic Medical Sciences, Physiology Section, Faculty of Medical Sciences, The University of the West Indies, Kingston, Jamaica*

[b]*Carrera de Química y Farmacia, Facultad de Ciencias de la Salud, Universidad Arturo Prat, Iquique, Chile*

[c]*Caribbean Maritime University, Kingston, Jamaica*

[d]*Department of Pathology, Faculty of Medical Sciences, The University of the West Indies, Kingston, Jamaica*

[e]*Department of Pharmaceutical Sciences, College of Pharmacy & Health Sciences, Texas Southern University, Houston, TX, United States*

[f]*Department of Inflammation and Rare Disease, Division of Medicine, University College London, London, United Kingdom*

[g]*Department of Chemical Engineering & Environmental Toxicology, College of Science, Engineering & Technology, Texas Southern University, Houston, TX, United States*

1 Introduction

Hypertension remains a leading global health challenge, with an estimated 1.3 billion adults affected worldwide and rising prevalence in both developed and developing regions (Nwokocha et al., 2019; WHO, 2013, 2021). The renin–angiotensin system plays a significant role in the pathophysiology of hypertension, and pharmacological inhibition of angiotensin-converting enzyme (ACE) has long been a cornerstone in antihypertensive therapy (Atlas, 2007). Despite the proven efficacy of synthetic ACE inhibitors (ACEis) such as captopril, lisinopril, and enalapril, their long-term use is limited by adverse effects, including cough, angioedema, and renal impairment (Li, Heran, & Wright, 2012). Consequently, there is growing interest in identifying natural ACEis from dietary or ethnopharmacological sources that may provide comparable efficacy with improved tolerability (Al Disi, Anwar, & Eid, 2016). Yam (*Dioscorea* spp.) tubers have been consumed globally for centuries, not only as staple foods but also as traditional remedies for cardiovascular and metabolic conditions (Amat, Amat, & Upur, 2014). While earlier studies highlighted the role of yam storage proteins such as dioscorin in ACE inhibition (Hsu, Lin, Lee, Lin, & Hou, 2002), more recent chemoinformatic profiling has revealed that small organic molecules in *Dioscorea alata* extracts may also confer significant antihypertensive bioactivity. The characterization of Jamaican Renta yam by Logan and colleagues (2023) represents one of the most comprehensive efforts to isolate, identify, and computationally validate these compounds. The workflow integrating extraction, high-performance liquid chromatography (HPLC) profiling, molecular docking, and Absorption, Distribution, Metabolism, Excretion, and Toxicity (ADMET) modeling illustrates a rational discovery pipeline for natural ACEis, exemplified by 2-phenyl-1,3-oxazoline (2PO) as seen in Logan, Nwokocha, Asemota, & Gray (2024). The integration of chromatographic profiling, molecular docking, and in silico ADMET modeling represents a powerful strategy for translating these findings into potential drug candidates. Emerging AI tools now enhance these pipelines, providing predictive accuracy across pharmacokinetic and safety domains that traditionally require laborious experimental validation (Francis, Laurieri, Nwokocha, & Delgoda, 2016; Palacios et al., 2025). Advances in **AI-enabled computational pharmacology** now allow for integrated screening of bioactive natural compounds, combining in silico molecular docking with deep learning (DL)-based pharmacokinetic and toxicity modeling. These pipelines, collectively termed **AI-assisted ADMET profiling,** can derisk early drug discovery by predicting oral bioavailability, metabolic stability, and toxicity long before preclinical testing. This review integrates chromatographic, molecular docking, and AI-based pharmacokinetic data for yam and Guinep bioactives, providing a holistic

Applications of Artificial Intelligence in Healthcare. https://doi.org/10.1016/B978-0-443-43934-6.00029-8

perspective on their potential as natural ACEis. This section reviews the methodological approaches used, spanning extraction and HPLC fractionation, molecular docking, and AI-assisted ADMET screening, and discusses their implications for natural product-based antihypertensive drug discovery.

2 Extraction and HPLC-guided fractionation of bioactive compounds

2.1 Sequential solid–liquid extraction

Extraction remains the first critical step in the isolation of bioactive molecules. In the case of *D. alata*, Logan et al. (2024) utilized a sequential solid–liquid extraction approach, applying solvents in order of increasing polarity—hexane, diethyl ether, acetone, methanol, and water—to produce five distinct fractions. Among these, the diethyl ether extract (DR2) demonstrated the most potent ACE inhibitory activity, having an IC_{50} of 41.99 $\mu g \cdot mL^{-1}$ (Logan et al., 2024). Notably, this potency was only threefold lower than that of captopril, a benchmark synthetic inhibitor, underscoring the pharmacological relevance of the extract.

2.2 HPLC purification and gas chromatography–mass spectrometry characterization

As previously described by Logan et al. (2024), the DR2 fraction was further purified using solid-phase extraction (SPE-C18) and preparative HPLC with a biphenyl column. HPLC fractionation revealed multiple peaks, with region 1 (retention time 2.49–6.6 min) containing ~79% of the observed ACE inhibitory activity. Gas chromatography–mass spectrometry (GC–MS) analysis of this bioactive subfraction (DR2-S1) identified over 50 compounds, including lactic acid, glycerol, butanedioic acid, glyceric acid, 3-pyridinol, and 2-phenyl-1,3-oxazol-2-ine (2PO), the latter being the most abundant (~20% relative abundance). These findings are notable because, unlike dioscorin, these compounds are small organic molecules, suggesting alternative or complementary mechanisms of ACE inhibition.

2.3 In vitro validation (bioassays)

Cell-based assays using human umbilical vein endothelial cells confirmed the bioactivity of DR2-S1. The extract induced dose-dependent ACE inhibition and stimulated nitric oxide (NO) production, a key vasodilatory mediator, while demonstrating minimal cytotoxicity even at high concentrations (7200 $\mu g \cdot mL^{-1}$) (Logan et al., 2024). These results provide a strong experimental foundation for computational docking and in silico pharmacological assessments.

3 Molecular docking studies of yam-derived ACEis

3.1 ACE structure and binding site architecture

ACE is a zinc-dependent metallopeptidase composed of two homologous catalytic domains (N- and C-domains), each harboring a Zn^{2+} ion at its active site along with specific substrate-binding subsites (S1, S2, and S1′). X-ray crystallographic investigations of ACE have identified the residues that define its substrate-binding pockets, including Ala354, Glu384, and Tyr523 (S1); Gln281, His353, Lys511, His513, and Tyr520 (S2); and Glu162 (S1′) (Natesh, Schwager, Sturrock, & Acharya, 2003; Brew, 2003). Classical inhibitors, such as captopril and lisinopril, interact directly with Zn^{2+} and form stabilizing hydrogen bonds in these subsites, which accounts for their high potency. For *Melicoccus bijugatus* metabolites, docking and ADMET-assisted screening identified quercetin-3-O-rutinoside and chlorogenic acid as the top ACE-binding ligands, with binding energies of −9.1 and −8.7 $kcal \cdot mol^{-1}$, respectively (Logan et al., 2025). These compounds formed interactions within subsites S1 and S2 near His353, Glu384, and Tyr523, coordinating the Zn^{2+} ion through carbonyl oxygen atoms, consistent with canonical inhibition mechanisms.

3.2 Docking methodology

Logan et al. (2024) performed molecular docking using AutoDock Vina with the ACE crystal structure (PDB: 1O8A). The Zn^{2+} and chloride ions were retained, and seven major GC–MS-identified ligands were docked alongside captopril and lisinopril as references. Binding free energies were calculated, and binding poses were analyzed for hydrogen bonding, hydrophobic contacts, and Zn coordination. From a structure–activity perspective, analog design around 2PO could explore ring substitution patterns that strengthen hydrogen bonding or extend interactions toward the Zn pocket. Computational fragment growing and pharmacophore mapping based on 2PO's interactions could yield optimized scaffolds with balanced potency and selectivity.

3.3 Key results

Docking simulations revealed that 2PO achieved a binding affinity of approximately −6.0 kcal · mol^{-1}, which was marginally stronger than that of captopril (−5.5 kcal · mol^{-1} under identical conditions). Notably, the predicted binding orientation of 2PO did not include coordination with the catalytic Zn^{2+} ion but rather displayed hydrogen bonding interactions with Tyr135 and Trp220, residues situated beyond the enzyme's typical Zn-binding region (Logan et al., 2024). This suggests a distinct inhibitory mechanism, potentially involving steric hindrance or allosteric modulation. By contrast, captopril engaged the Zn^{2+} ion and formed multiple hydrogen bonds with residues Asn277, Thr282, and Gln281, consistent with classical ACE inhibition. Other identified compounds, such as butanedioic acid and glyceric acid, exhibited weaker binding affinities (−4.4 to −5.3 kcal · mol^{-1}).

3.4 Broader context and implications

The finding that 2PO binds ACE in a noncanonical manner has several implications: (1) **Selectivity potential**: By avoiding Zn^{2+} coordination, 2PO may reduce off-target inhibition of other zinc metalloproteases such as neprilysin and ACE2, improving safety. (2) **Novel pharmacophore**: Hydrogen bonding with Tyr135 and Trp220 may represent a previously underexplored pharmacophore for ACE inhibition. (3) **Synergistic activity**: Given the presence of multiple small molecules in DR2-S1, additive or synergistic interactions may enhance bioactivity beyond what is predicted for single compounds. These results underscore the importance of integrating docking with experimental validation. Similar findings have been reported for plant-derived flavonoids and peptides, where docking has revealed unconventional binding modes that complement in vitro ACE inhibition (Balasuriya & Rupasinghe, 2012; Oboh, Ademiluyi, & Akinyemi, 2012). According to Hsu et al. (2002), the protein dioscorin, isolated from *D. alata*, exhibited ACE inhibitory activity in a dose-dependent manner. Similarly, Li et al. (2021) found that the saponin dioscin downregulated angiotensin II type 1 receptor (AT1R) expression in vascular smooth muscle cells, which in turn attenuated vasoconstriction. Dioscorin itself has been shown to exhibit superoxide anion and DPPH radical scavenging activities.

In hypertensive animal models, antioxidant treatment has been found to strengthen endogenous defense mechanisms by increasing the activities of superoxide dismutase (SOD) and glutathione peroxidase (GPx), while decreasing malondialdehyde (MDA) levels, an indicator of lipid peroxidation (Hou et al., 2001). By alleviating oxidative stress, yam-derived bioactives support endothelial integrity and facilitate vasodilation. Furthermore, diosgenin, the aglycone form of dioscin, activates the PI3K/Akt pathway, which promotes phosphorylation of endothelial nitric oxide synthase, enhancing NO-mediated vascular relaxation (Li et al., 2012).

M. bijugatus, a tropical fruit from the Sapindaceae family, is a lesser known but pharmacologically rich species. Methanolic and aqueous extracts of Guinep fruit pulp and seed have demonstrated **vasorelaxant, cardioprotective, and antihypertensive activity**, partly via ACE inhibition and antagonism of AT1R (Nwokocha et al., 2019, 2021). UHPLC-Q-Orbitrap mass spectrometry identified phenolic acids, flavonoids (including quercetin and rutin derivatives), and terpenoids as key metabolites. Computational docking and in silico pharmacokinetic analysis suggested strong ACE binding and favorable drug-likeness for several compounds (Logan et al., 2025).

To advance computational characterization, future studies should implement:

- **Multiconformational docking** across ACE's N- and C-domains to account for flexibility.
- **Free-energy rescoring** via MM-GBSA or free energy perturbation (FEP) to refine binding estimates.
- **Metal-aware QM/MM parameterization** for Zn-containing systems.
- **MD simulations (100–200 ns)** to validate pose stability and solvent effects.
- **Selectivity screening** against ACE2 and neprilysin to anticipate off-target effects.

Such improvements will transition docking from qualitative visualization to quantitative prediction. While in vitro and docking data confirm activity, translational success depends on pharmacokinetic and toxicological profiles. ADMET properties—absorption, distribution, metabolism, excretion, and toxicity—determine bioavailability, safety, and ultimately, clinical viability. Evaluating these experimentally is resource-intensive, prompting a paradigm shift toward AI-assisted in silico prediction as an early triage tool. AI-augmented ADMET modeling integrates **quantitative structure–activity relationship (QSAR)** algorithms, **graph neural networks (GNNs)**, and **DL** models trained on large bioactivity datasets (Cáceres, Tudor, & Cheng, 2020; Xiong et al., 2021). These systems can predict pharmacokinetic and toxicity-related properties such as oral absorption, CYP450 metabolism, blood–brain barrier permeability, and cardiotoxicity with accuracy rivaling wet-lab screening (Pires, Blundell, & Ascher, 2015) (Fig. 31.1).

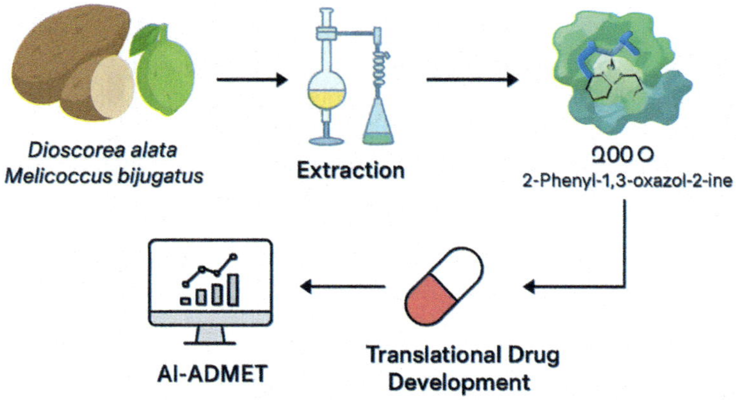

FIG. 31.1 AI-assisted pharmacokinetic and toxicological profiling of yam (*Dioscorea alata*) and *Melicoccus bijugatus*-derived ACE inhibitors.

Advanced computing has enabled enhanced research, from rapid data acquisition to swift analysis, through the integration of massive data handling across various fields and disciplines such as ethnopharmacology, chromatography, and computational science, and its convergence toward drug discovery. Advanced computing enables the storage of data in more efficient data storage facilities. Recent advances in natural language processing (NLP) and machine learning (ML) have allowed researchers to more effectively integrate natural product data into drug-discovery pipelines, facilitating the efficient identification and synthesis of bioactive compounds from plants and marine sources (Saldívar-González, Aldas-Bulos, Medina-Franco, & Plisson, 2022; Gangwal et al., 2025). The rise of artificial intelligence (AI) has created new avenues for developing therapeutics from natural sources. Medicinal plants remain key reservoirs of secondary metabolites such as flavonoids and alkaloids, which possess antioxidant, antiinflammatory, immune-modulating, antineurodegenerative, antiviral, anticancer, antibacterial, and antifungal properties. Specifically, AI can be engaged in the hard-to-do tasks of handling the diversity in plant species, numerous and diverse secondary metabolites, identification and *characterization* of several bioactive agents, identification of target biomolecules, signaling proteins, receptors, enzymes, and kinases, along with pathological pathways such as dysregulated biology and dysfunctions, alongside the capabilities of reading bioinformatics, genomic, and metabolic databases. With the help of diverse ML algorithms, AI can interact with and interrogate these databases, effectively advancing natural herbal research to drug discovery. AI can efficiently predict biological activity, identify synergies, flag potential toxicity, extract and target the best-fit compounds, and accelerate the drug discovery process with fewer failures.

Compound/target identification: AI-based approaches, supported by ML algorithms, can process large molecular and protein databases, integrate spectroscopic and omics data, and predict compound–target interactions in natural products. Such integration of bioinformatics and chemogenomic data accelerates the identification and isolation of biologically active molecules (Saldívar-González et al., 2022; Gangwal & Lavecchia, 2025). The use of ML methodologies can enhance substrate specificity predictions of natural products in identifying the most suitable biosynthetic processes (including receptors, enzymes, kinases, signaling molecules, misfolded proteins, genes regulating inflammatory and oxidative stress, and biosynthetic genes) in biological systems for therapeutic applications. AI can be used to predict genes that may encode different metabolites, which could be beneficial in therapeutics.

Drug discovery: AI and its subdisciplines, including ML and DL, are increasingly integrated across different phases of the drug discovery workflow. These technologies enable virtual screening of large natural product and synthetic compound libraries, prediction of pharmacological and physicochemical properties, and prioritization of lead candidates (Gangwal & Lavecchia, 2025). AI-based target discovery platforms combine bioinformatics and multiomics data to reveal molecular interactions and disease–target associations, thereby improving understanding of target biology and enhancing the selection of therapeutically relevant targets (Saldívar-González et al., 2022). For instance, TargetDB applies an ML-driven classification system that integrates publicly available target information to evaluate target tractability (Vijayan et al., 2022). Deep neural networks (DNNs) are innovative AI models that excel at recognizing intricate patterns and processing complex data. Their unparalleled ability to analyze vast amounts of information positions them at the forefront of technological advancement, making them indispensable in many fields such as image recognition, NLP, and autonomous systems. Embracing DNNs can significantly enhance performance and drive innovation across various industries, including drug development from herbal medicinal plants. Generative AI techniques, such as deep generative models, are increasingly applied to drug discovery to design novel molecules with desirable biological properties. By prioritizing compounds most likely to show experimental success, these approaches streamline and shorten the overall drug development process (Arús-Pous et al., 2020; Chopra & Dhingra, 2021). The integration of AI-based ADMET modeling extends these findings by predicting favorable pharmacokinetic and safety profiles for 2PO and related compounds identified by GC–MS. Tools such as SwissADME, pkCSM, and ADMETlab 2.0, supported by DL architectures, enable rapid virtual screening and prioritization of natural molecules, effectively bridging the gap between experimental pharmacology and computational prediction.

- **Bioactivity prediction**: ML models can be trained to predict and prioritize the biological activity of natural products based on their molecular structures. DNN approaches that integrate 3D pharmacophore features form the basis of quantitative structure–activity/property relationship (QSAR/QSPR) modeling, facilitating the identification of compounds with desirable therapeutic potential (Lo, Rensi, Torng, & Altman, 2018). In recent applications, AI-guided models have accelerated antibiotic discovery. For example,(Stokes et al., 2020) employed a DL framework to screen chemical libraries and identified Halicin, a compound with potent bactericidal activity against *Acinetobacter baumannii* and other pathogens. Similarly, Liu et al. (2023) utilized ML-based screening to discover Abaucin, a narrow-spectrum antibiotic effective against *A. baumannii*. These advances illustrate the capacity of AI-driven bioactivity prediction to uncover novel antimicrobial agents from vast chemical and natural product datasets.
- **Optimization of extraction processes:** AI tools, including ML, can optimize extraction conditions, such as solvent composition, temperature, and time, to maximize yields of bioactive compounds from natural materials (Alloun & Calvio, 2024). This reduces the time and resources needed for the extraction of essential compounds.
- **Data integration and analysis:** AI methods enable the integration and interpretation of large-scale multiomics datasets, which include genomics, proteomics, and metabolomics, providing holistic insights into the molecular interactions and regulatory networks that govern natural biological systems (Hasin, Seldin, & Lusis, 2017).
- **Predicting synergies:** AI approaches can model and predict synergistic interactions among bioactive compounds, thereby supporting the design of combination therapies based on natural products. Such predictive frameworks are particularly valuable for addressing multifactorial diseases like cancer, where multicompound interventions often outperform single agents (Preuer et al., 2018).
- **Toxicity prediction:** AI-driven models can estimate the toxicity profiles of natural compounds, enabling early identification of potentially harmful candidates and improving the safety of drug or nutraceutical development pipelines. These predictive tools integrate chemical structure data with biological and toxicological information to forecast outcomes such as hepatotoxicity and cardiotoxicity (Banerjee, Eckert, Schrey, & Preissner, 2018; ; Banerjee, Kemmler, Dunkel, & Preissner, 2024).

4 Recommendations for future docking studies

While the docking analysis in Logan et al. provided valuable insights, several refinements are recommended for future work:

- **Multiple receptor conformations**: Docking against both N- and C-domains of ACE and across multiple protein data bank (PDB) structures can account for conformational flexibility.
- **Rescoring and free energy calculations**: Methods such as MM-GBSA or molecular dynamics (MD)-based FEP can refine affinity predictions beyond docking scores.
- **Metal coordination modeling**: For ligands expected to chelate Zn^{2+}, quantum mechanical/molecular mechanical (QM/MM) approaches can improve accuracy.
- **Selectivity docking**: Screening against related enzymes (ACE2, neprilysin) can help anticipate off-target liabilities.
- **Dynamic stability**: MD simulations of ligand–ACE complexes can reveal the persistence of hydrogen bonds, water-mediated interactions, and conformational adjustments over time.

Such best practices will enable the field to transition from preliminary docking scores to predictive mechanistic models that can inform medicinal chemistry (Fig. 31.2).

5 ADMET studies and the role of AI in natural product screening

Even promising docking results must be tempered by considerations of pharmacokinetics and toxicity. Many natural compounds fail during development due to poor oral bioavailability, rapid metabolism, or adverse effects such as hERG-mediated cardiotoxicity. In silico ADMET prediction, particularly using ML approaches, offers a cost-effective means of

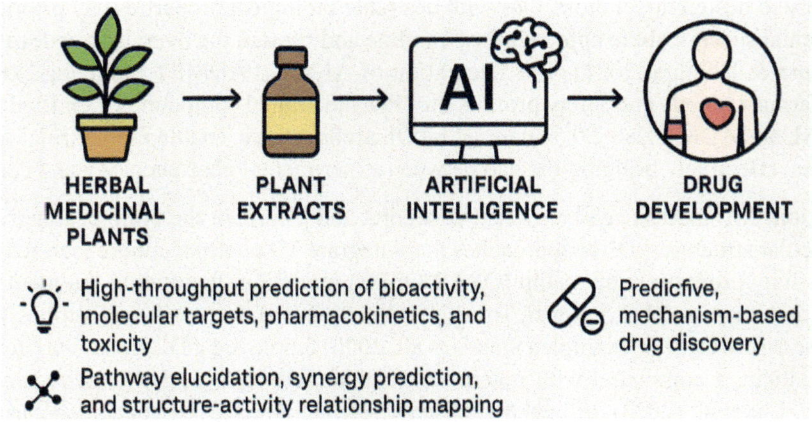

FIG. 31.2 Leveraging artificial intelligence in herbal medicinal plant extracts' pathway to drug development.

early triage. Platforms such as SwissADME (Ancuceanu, Lascu, Drăgănescu, & Dinu, 2025; Cardeal dos Santos et al., 2025), pkCSM (Pires et al., 2015), and admetSAR (Xiong et al., 2021) predict drug-likeness, solubility, membrane permeability, cytochrome P450 metabolism, renal clearance, and multiple toxicity endpoints based on molecular descriptors and graph-based signatures. More advanced DL models now incorporate training sets of millions of compounds to predict endpoints like hERG inhibition, hepatotoxicity, and mitochondrial toxicity with improved accuracy (Xiong et al., 2021; Askr et al., 2023).

5.1 Application to natural product-derived compounds

Applying AI-based ADMET tools to 2PO and related GC–MS-identified compounds can help clarify their potential for drug development (Thomas et al., 2022). AI-driven ADMET workflows now combine traditional QSAR descriptors with **GNNs** and **transformer architectures**, achieving remarkable predictive power for previously underrepresented natural products. Key publicly available platforms include:

- **SwissADME**, which estimates physicochemical descriptors, lipophilicity, gastrointestinal absorption, and "drug-likeness" using rule-based heuristics (Ancuceanu et al., 2025; Cardeal dos Santos et al., 2025).
- **pkCSM**, which applies graph signatures to predict pharmacokinetics and CYP interactions (Pires et al., 2015).
- **ADMETlab 2.0**, which integrates 47 predictive models covering absorption, metabolism, and multiple toxicity endpoints (Xiong et al., 2021).
- **DeepADMET** and **Chemprop-ADMET**, which utilize deep GNN frameworks to learn latent molecular representations from millions of compounds (Askr et al., 2023).

These platforms allow rapid screening of hundreds of natural metabolites, ranking them by oral bioavailability, metabolic stability, and predicted toxicity, thus guiding experimental prioritization. For example:

- **Absorption**: Molecular weight ($\sim$135 g $\cdot$ mol^{-1}) and moderate polarity suggest that 2PO is orally bioavailable, but lipophilicity (LogP) and solubility predictions are needed.
- **Distribution**: Low molecular weight may enable blood–brain barrier penetration; whether this is beneficial or deleterious depends on target specificity.
- **Metabolism**: Prediction of CYP450 interactions will determine the potential for drug–drug interactions.
- **Excretion**: Small, polar metabolites may be rapidly cleared renally.
- **Toxicity**: AI tools can flag liabilities such as mutagenicity or hERG blockade, which should then be assessed in vitro.

Such predictions can prioritize which natural compounds should be synthesized or isolated in larger quantities for pharmacological testing.

6 Future directions

The integration of HPLC fractionation, molecular docking, and AI-based ADMET screening provides a comprehensive pipeline for discovering natural ACEis. For *D. alata*, future studies should focus on:

1. **Isolation and synthesis** of 2PO and other candidate compounds at a preparative scale for biochemical and animal studies.
2. **Mechanistic enzymology** to determine the mode of ACE inhibition (competitive vs. noncompetitive) and kinetic constants.
3. **In vivo efficacy studies** in hypertensive animal models (e.g., spontaneously hypertensive rats) to confirm blood-pressure-lowering effects.
4. **Pharmacokinetic profiling** to validate ADMET predictions and optimize dosing regimens.
5. **Structure–activity relationship (SAR) exploration** using medicinal chemistry to modify 2PO and related scaffolds for improvements in potency, selectivity, and pharmacokinetics.
6. **Synergistic evaluation** of multiple yam-derived compounds to assess combinatorial effects that may mimic the activity of whole extracts.

AI can transform natural product research from **experimental discovery** to **predictive design**. AI and ML can be used to further the understanding of drug discovery and development, and to make the process from identification, extraction, HPLC, in vivo/in vitro testing, molecular docking, and ADMET studies faster and more efficient. By integrating AI models across extraction, screening, docking, and ADMET phases, researchers can reduce development timelines by 50%–70%, increase success rates, and uncover new bioactive molecules that were previously hidden within complex plant matrices. Extraction (solvent prediction, metabolite recognition) can use AI tools such as MS2DeepScore and NPClassifier (Gangwal & Lavecchia, 2025; Huber, van der Burg, van der Hooft, & Ridder, 2021; Petrick & Shomron, 2022; Tinte, Chele, van der Hooft, & Tugizimana, 2021; Leão et al., 2022). HPLC (gradient optimization, peak deconvolution) can utilize tools such as reinforcement learning and XGBoost (Elkhoudary, Marie, Hammad, Salim, & Kamal, 2024; M'hamdi et al., 2024; Shrivastava, Nikita, & Rathore, 2024; Tian et al., 2021). Molecular docking for Pose and affinity prediction can be enhanced by ML tools and AI, such as DiffDock and DeepDock (Schifferstein, Bernatavicius, & Janssen, 2024; Sim, Kim, Kim, Choi, & Lee, 2025; Tomii & Ikeda, 2025; Vittorio et al., 2024). Bioactivity for target prediction and SAR mapping could be further enhanced with DeepDTA and MolTrans (Huang, Xiao, Glass, & Sun, 2021; Ru, Xu, Han, & Zou, 2025; Wang et al., 2022). ADMET for pharmacokinetics and toxicity studies would be greatly enhanced with pkCSM, ADMETlab 2.0, and DeepADMET (Han et al., 2025; Myung, De Sá, & Ascher, 2024; Xiong et al., 2021). Optimization for generative molecule design could be further enhanced using AI tools such as MolDQN and REINVENT (Gangwal et al., 2024; Khater et al., 2025; Loeffler et al., 2024; Tripathi, Goshisht, Sahu, & Arora, 2021). Validation with automated analysis and digital twins can be achieved using ML and AI tools, such as AI image analysis and biosimulation (Alfaro-Viquez, Zamora-Hernandez, Fernandez-Vega, Garcia-Rodriguez, & Azorin-Lopez, 2025; Chaparro-Cárdenas et al., 2025) (Fig. 31.3).

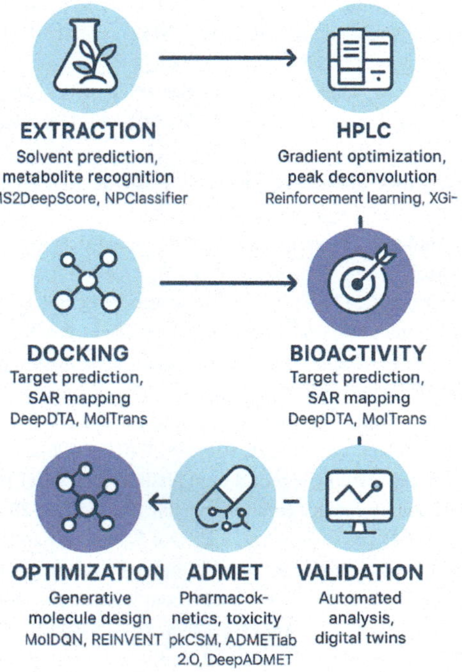

FIG. 31.3 AI integration across the natural product drug delivery pipeline.

7 Conclusion

Hypertension remains a leading cause of global cardiovascular morbidity, underscoring the urgent need for safer and more effective therapeutic agents. Natural ACEis derived from edible plants represent a promising strategy to complement or replace synthetic drugs. This review highlights *D. alata* (yam) and *M. bijugatus* (Guinep) as rich sources of bioactive ACE inhibitory compounds and demonstrates how AI can accelerate their pharmacological evaluation. Sequential extraction and HPLC-guided isolation of *D. alata* identified 2-phenyl-1,3-oxazol-2-ine (2PO) as a key bioactive, while *M. bijugatus* extracts yielded polyphenolic inhibitors, including quercetin and chlorogenic acid. Molecular docking revealed distinct binding modes—non-Zn^2-dependent for 2PO and Zn^{2+}-Zn^2-coordinated for phenolics—indicating complementary mechanisms of ACE modulation. Integrating AI-based ADMET tools such as SwissADME, pkCSM, and DeepADMET provided predictive insights into absorption, distribution, metabolism, excretion, and toxicity profiles, revealing high drug-likeness and low toxicity potential for 2PO and favorable safety margins for Guinep polyphenols. This multidisciplinary approach highlights the utility of AI-assisted pipelines in natural product discovery, enabling the early optimization of pharmacokinetic and toxicological parameters prior to preclinical testing. Together, yam and Guinep exemplify how ethnopharmacological knowledge, combined with ML-driven pharmacokinetic modeling, can guide the rational design of next-generation food-derived ACEis for hypertension management.

Author's statement

During the preparation of this work, the author(s) used ChatGPT to generate the pictures/figures. After using this tool/service, the author(s) reviewed and edited the content as needed and take(s) full responsibility for the content of the publication.

References

Al Disi, S. S., Anwar, M. A., & Eid, A. H. (2016). Anti-hypertensive herbs and their mechanisms of action: part I. *Frontiers in Pharmacology*. 16639812. *6*. Available from: http://journal.frontiersin.org/article/10.3389/fphar.2015.00323/full. doi:10.3389/fphar.2015.00323.

Alfaro-Viquez, D., Zamora-Hernandez, M., Fernandez-Vega, M., Garcia-Rodriguez, J., & Azorin-Lopez, J. (2025). A comprehensive review of AI-based digital twin applications in manufacturing: integration across operator, product, and process dimensions. *Electronics*. 2079-9292. *14*(4), 646. doi:10.3390/electronics14040646.

Alloun, W., & Calvio, C. (2024). Bio-driven sustainable extraction and AI-optimized recovery of functional compounds from plant waste: a comprehensive review. *Fermentation*. 2311-5637. *10*(3), 126. doi:10.3390/fermentation10030126.

Amat, N., Amat, R., & Upur, H. (2014). Antihypertensive effects of Dioscorea species. *Phytotherapy Research*, *28*(2), 268–276.

Ancuceanu, R., Lascu, B. E., Drăgănescu, D., & Dinu, M. (2025). In silico ADME methods used in the evaluation of natural products. *Pharmaceutics*. 1999-4923. *17*(8), 1002. doi:10.3390/pharmaceutics17081002.

Arús-Pous, J., Patronov, A., Bjerrum, E. J., Tyrchan, C., Reymond, J. L., Chen, H., & Engkvist, O. (2020). SMILES-based deep generative scaffold decorator for de-novo drug design. *Journal of Cheminformatics*. 17582946. *12*(1). Available from: https://jcheminf.biomedcentral.com/. doi:10.1186/s13321-020-00441-8.

Askr, H., Elgeldawi, E., Ella, H. A., Elshaier, Y. A. M. M., Gomaa, M. M., & Hassanien, A. E. (2023). Deep learning in drug discovery: an integrative review and future challenges. *Artificial Intelligence Review*. 15737462. *56*(7), 5975–6037. Available from: https://www.springer.com/journal/10462. doi:10.1007/s10462-022-10306-1.

Atlas, S. A. (2007). The Renin-angiotensin aldosterone system: pathophysiological role and pharmacologic inhibition. *Journal of Managed Care Pharmacy*. 1083-4087. *13*(8 Supp B), 9–20. doi:10.18553/jmcp.2007.13.s8-b.9.

Balasuriya, N., & Rupasinghe, H. P. V. (2012). Antihypertensive properties of flavonoid-rich apple peel extract. *Food Chemistry*. 18737072. *135*(4), 2320–2325. doi:10.1016/j.foodchem.2012.07.023.

Banerjee, P., Eckert, A. O., Schrey, A. K., & Preissner, R. (2018). ProTox-II: a webserver for the prediction of toxicity of chemicals. *Nucleic Acids Research*. 13624962. *46*(1), W257. Available from: https://academic.oup.com/nar/issue. doi:10.1093/nar/gky318.

Banerjee, P., Kemmler, E., Dunkel, M., & Preissner, R. (2024). ProTox 3.0: a webserver for the prediction of toxicity of chemicals. *Nucleic Acids Research*. 0305-1048. *52*(W1), W513. doi:10.1093/nar/gkae303.

Brew, K. (2003). Structure of human ACE gives new insights into inhibitor binding and design. *Trends in Pharmacological Sciences*. 01656147. *24*(8), 391–394. Available from: www.elsevier.com/locate/tips. doi:10.1016/S0165-6147(03)00196-2.

Cáceres, E. L., Tudor, M., & Cheng, A. C. (2020). Deep learning approaches in predicting ADMET properties. *Future Medicinal Chemistry*. 17568927. *12*(22), 1995–1999. Available from: http://www.future-science.com/loi/fmc. doi:10.4155/fmc-2020-0259.

Cardeal Dos Santosdos Santos, A. N., Oliveira, P. E. G. D., da Cruz Freire, J. E., Dos Santosdos Santos, S. A., Júnior, J. E. R. H., Andrade, C. R. D., et al. (2025). Computational profiling of monoterpenoid phytochemicals: insights for medicinal chemistry and drug design strategies. *International Journal of Molecular Sciences*. 14220067. *26*(16). Available from: http://www.mdpi.com/journal/ijms. doi:10.3390/ijms26167671.

Chaparro-Cárdenas, S. L., Ramirez-Bautista, J.-A., Terven, J., Córdova-Esparza, D.-M., Romero-Gonzalez, J.-A., Ramírez-Pedraza, A., & Chavez-Urbiola, E. A. (2025). A technological review of digital twins and artificial intelligence for personalized and predictive healthcare. *Healthcare*. 2227-9032. *13*(14), 1763. doi:10.3390/healthcare13141763.

Chopra, B., & Dhingra, A. K. (2021). Natural products: a lead for drug discovery and development. *Phytotherapy Research*. 10991573. *35*(9), 4660–4702. Available from: http://onlinelibrary.wiley.com/journal/10.1002/(ISSN)1099-1573. doi:10.1002/ptr.7099.

Drwal, M. N., Banerjee, P., Dunkel, M., Wettig, M. R., & Preissner, R. (2014). ProTox: a web server for the in silico prediction of rodent oral toxicity. *Nucleic Acids Research*. 13624962. *42*(1), W53. Available from: http://nar.oxfordjournals.org/. doi:10.1093/nar/gku401.

Elkhoudary, M. M., Marie, A. A., Hammad, S. F., Salim, M. M., & Kamal, A. H. (2024). Decoding the synergy: unveiling gradient boosting regression model for multivariate quantitation of pioglitazone, alogliptin and glimepiride in pure and tablet dosage forms. *BMC Chemistry*. 2661801X. *18*(1). Available from: https://bmcchem.biomedcentral.com/. doi:10.1186/s13065-024-01351-8.

Francis, S., Laurieri, N., Nwokocha, C., & Delgoda, R. (2016). Treatment of rats with Apocynin has considerable inhibitory effects on arylamine N-acetyltransferase activity in the liver. *Scientific Reports*. 20452322. *6*. Available from: www.nature.com/srep/index.html. doi:10.1038/srep26906.

Gangwal, A., Ansari, A., Ahmad, I., Azad, A. K., Kumarasamy, V., Subramaniyan, V., & Wong, L. S. (2024). Generative artificial intelligence in drug discovery: basic framework, recent advances, challenges, and opportunities. *Frontiers in Pharmacology*. 16639812. *15*. Available from: http://www.frontiersin.org/Pharmacology. doi:10.3389/fphar.2024.1331062.

Gangwal, A., & Lavecchia, A. (2025). Artificial intelligence in natural product drug discovery: current applications and future perspectives. *Journal of Medicinal Chemistry*. 15204804. *68*(4), 3948–3969. Available from: http://pubs.acs.org/jmc. doi:10.1021/acs.jmedchem.4c01257.

Han, H., Shaker, B., Lee, J. H., Choi, S., Yoon, S., Singh, M., et al. (2025). Employing automated machine learning (AutoML) methods to facilitate the in silico ADMET properties prediction. *Journal of Chemical Information & Modeling*. 1549960X. *65*(7), 3215–3225. Available from: http://pubs.acs.org/journal/jcisd8. doi:10.1021/acs.jcim.4c02122.

Hasin, Y., Seldin, M., & Lusis, A. (2017). Multi-omics approaches to disease. *Genome Biology*. 1474-760X. *18*(1). doi:10.1186/s13059-017-1215-1.

Hou, W. C., Lee, M. H., Chen, H. J., Liang, W. L., Han, C. H., Liu, Y. W., & Lin, Y. H. (2001). Antioxidant activities of dioscorin, the storage protein of yam (Dioscorea batatas Decne) tuber. *Journal of Agricultural and Food Chemistry*. 00218561. *49*(10), 4956–4960. doi:10.1021/jf010606m.

Hsu, F. L., Lin, Y. H., Lee, M. H., Lin, C. L., & Hou, W. C. (2002). Both dioscorin, the tuber storage protein of yam (Dioscorea alata cv. Tainong No. 1), and its peptic hydrolysates exhibited angiotensin converting enzyme inhibitory activities. *Journal of Agricultural and Food Chemistry*. 00218561. *50*(21), 6109–6113. doi:10.1021/jf0203287.

Huang, K., Xiao, C., Glass, L. M., & Sun, J. (2021). MolTrans: molecular interaction transformer for drug-target interaction prediction. *Bioinformatics*. 14602059. *37*(6), 830–836. Available from: http://bioinformatics.oxfordjournals.org/. doi:10.1093/bioinformatics/btaa880.

Huber, F., van der Burg, S., van der Hooft, J. J. J., & Ridder, L. (2021). MS2DeepScore: a novel deep learning similarity measure to compare tandem mass spectra. *Journal of Cheminformatics*. 17582946. *13*(1). Available from: https://jcheminf.biomedcentral.com/. doi:10.1186/s13321-021-00558-4.

Khater, T., Alkhatib, S. A., AlShehhi, A., Pitsalidis, C., Pappa, A. M., Ngo, S. T., et al. (2025). Generative artificial intelligence based models optimization towards molecule design enhancement. *Journal of Cheminformatics*. 17582946. *17*(1). Available from: https://jcheminf.biomedcentral.com/. doi:10.1186/s13321-025-01059-4.

Leão, T. F., Wang, M., da Silva, R., Gurevich, A., Bauermeister, A., Gomes, P. W. P., et al. (2022). NPOmix: a machine learning classifier to connect mass spectrometry fragmentation data to biosynthetic gene clusters. *PNAS Nexus*. 27526542. *1*(5). Available from: https://academic.oup.com/pnasnexus/pages/about. doi:10.1093/pnasnexus/pgac257.

Li, E. C., Heran, B. S., & Wright, J. M. (2012). Angiotensin converting enzyme inhibitors versus other classes of antihypertensive drugs. *Cochrane Database of Systematic Reviews*, *8*, CD003823.

Li, X., Liu, S., Qu, L., Chen, Y., Yuan, C., Qin, A., Liang, J., Huang, Q., Jiang, M., Zou, W. (2021). Dioscin and diosgenin: Insights into their potential protective effects in cardiac diseases. *Journal of Ethnopharmacology*, *274*, 114018. doi:10.1016/j.jep.2021.114018.

Liu, G., Catacutan, D. B., Rathod, K., Swanson, K., Jin, W., Mohammed, J. C., et al. (2023). Deep learning-guided discovery of an antibiotic targeting Acinetobacter baumannii. *Nature Chemical Biology*. 15524469. *19*(11), 1342–1350. Available from: https://www.nature.com/nchembio. doi:10.1038/s41589-023-01349-8.

Lo, Y. C., Rensi, S. E., Torng, W., & Altman, R. B. (2018). Machine learning in chemoinformatics and drug discovery. *Drug Discovery Today*. 18785832. *23*(8), 1538–1546. Available from: www.elsevier.com/locate/drugdiscov. doi:10.1016/j.drudis.2018.05.010.

Loeffler, H. H., He, J., Tibo, A., Janet, J. P., Voronov, A., Mervin, L. H., & Engkvist, O. (2024). Reinvent 4: modern AI–driven generative molecule design. *Journal of Cheminformatics*. 17582946. *16*(1). Available from: https://jcheminf.biomedcentral.com/. doi:10.1186/s13321-024-00812-5.

Logan, K., Nwokocha, C., Asemota, H., & Gray, W. (2024). Characterization of ACE inhibitory activity in Dioscorea alata cv and its implication as a natural antihypertensive extract. *Journal of Ethnopharmacology*. 03788741. *319*(Pt 2), 117221. doi:10.1016/j.jep.2023.117221.

Logan, K., Palacios, J., S Lopez, W. G., & Nwokocha, C. (2025). Integrating experimental and computational approaches to cardioprotection: vascular reactivity, molecular docking, and ADMET modeling of Melicoccus bijugatus (Guinep). *International Journal of Molecular Sciences*, *26*(20), 10228.

M'hamdi, O., Takács, S., Palotás, G., Ilahy, R., Helyes, L., & Pék, Z. (2024). A comparative analysis of XGBoost and neural network models for predicting some tomato fruit quality traits from environmental and meteorological data. *Plants*. 22237747. *13*(5). Available from: www.mdpi.com/journal/plants. doi:10.3390/plants13050746.

Myung, Y., De Sá, A. G. C., & Ascher, D. B. (2024). Deep-PK: deep learning for small molecule pharmacokinetic and toxicity prediction. *Nucleic Acids Research*. 13624962. *52*(1), W469. Available from: https://academic.oup.com/nar/issue. doi:10.1093/nar/gkae254.

Natesh, R., Schwager, S. L. U., Sturrock, E. D., & Acharya, K. R. (2003). Crystal structure of the human angiotensin-converting enzyme-lisinopril complex. *Nature*. 00280836. *421*(6922), 551–554. Available from: http://www.nature.com/nature/index.html. doi:10.1038/nature01370.

Nwokocha, C. R., Warren, I., Palacios, J., Simirgiotis, M., Nwokocha, M., Harrison, S., et al. (2019). Modulatory effect of Guinep (melicoccus bijugatus jacq) fruit pulp extract on isoproterenol-induced myocardial damage in rats. Identification of major metabolites using high resolution UHPLC Q-orbitrap mass spectrometry. *Molecules*. 14203049. *24*(2). Available from: https://www.mdpi.com/1420-3049/24/2/235/pdf. doi:10.3390/molecules24020235.

Nwokocha, M., Beaney, T., Holder, C., Nesbeth, K. T., Whylie, N., Leitch, J., et al. (2021). May measurement month 2019: an analysis of blood pressure screening results from Jamaica. *European Heart Journal, Supplement*, 15542815 23, B82. Available from: http://eurheartjsupp.oxfordjournals.org/. doi:10.1093/eurheartj/suab036.

Oboh, G., Ademiluyi, A. O., & Akinyemi, A. J. (2012). Inhibition of angiotensin-I converting enzyme activity by phenolic extracts of tropical green leafy vegetables. *Journal of Medicinal Food*, *15*(5), 448–454.

Palacios, J., Villarroel, C., Asunción-Alvarez, D., Cifuentes, F., Paredes, A., Nwokocha, C. R., et al. (2025). Metabolites isolated from Senecio nutans Sch. Bip and their synthesized oximes inhibit angiotensin I-converting enzyme activity in vascular smooth muscle. *International Journal of Molecular Sciences*. 1422-0067. *26*(8), 3786. doi:10.3390/ijms26083786.

Petrick, L. M., & Shomron, N. (2022). AI/ML-driven advances in untargeted metabolomics and exposomics for biomedical applications. *Cell Reports Physical Science*. 26663864. *3*(7), 100978. doi:10.1016/j.xcrp.2022.100978.

Pires, D. E. V., Blundell, T. L., & Ascher, D. B. (2015). pkCSM: predicting small-molecule pharmacokinetic and toxicity properties using graph-based signatures. *Journal of Medicinal Chemistry*. 15204804. *58*(9), 4066–4072. Available from: http://pubs.acs.org/jmc. doi:10.1021/acs.jmedchem.5b00104.

Preuer, K., Lewis, R. P. I., Hochreiter, S., Bender, A., Bulusu, K. C., & Klambauer, G. (2018). DeepSynergy: predicting anti-cancer drug synergy with deep learning. *Bioinformatics*. 14602059. *34*(9), 1538–1546. Available from: http://bioinformatics.oxfordjournals.org/. doi:10.1093/bioinformatics/btx806.

Ru, X., Xu, L., Han, W., & Zou, Q. (2025). In silico methods for drug-target interaction prediction. *Cell Reports Methods*. 26672375. *5*(10), 101184. doi:10.1016/j.crmeth.2025.101184.

Saldívar-González, F. I., Aldas-Bulos, V. D., Medina-Franco, J. L., & Plisson, F. (2022). Natural product drug discovery in the artificial intelligence era. *Chemical Science*. 2041-6520. *13*(6), 1526–1546. doi:10.1039/d1sc04471k.

Schifferstein, J., Bernatavicius, A., & Janssen, A. P. A. (2024). Docking-informed machine learning for kinome-wide affinity prediction. *Journal of Chemical Information & Modeling*. 1549-9596. *64*(24), 9196–9204. doi:10.1021/acs.jcim.4c01260.

Shrivastava, A., Nikita, S., & Rathore, A. S. (2024). Machine learning tool as an enabler for rapid quantification of monoclonal antibodies N-glycans using fluorescence detector. *International Journal of Biological Macromolecules*. 18790003. *271*. Available from: https://www.sciencedirect.com/science/journal/01418130. doi:10.1016/j.ijbiomac.2024.132694.

Sim, J., Kim, D., Kim, B., Choi, J., & Lee, J. (2025). Recent advances in AI-driven protein-ligand interaction predictions. *Current Opinion in Structural Biology*, 1879033X 92. Available from: https://www.sciencedirect.com/science/journal/0959440X. doi:10.1016/j.sbi.2025.103020.

Stokes, J. M., Yang, K., Swanson, K., Jin, W., Cubillos-Ruiz, A., Donghia, N. M., et al. (2020). A deep learning approach to antibiotic discovery. *Cell*. 10974172. *180*(4), 688. Available from: https://www.sciencedirect.com/journal/cell. doi:10.1016/j.cell.2020.01.021.

Thomas, J., Anderson, T., James-Green, T., Nwokocha, M., Palacios, J., Pepple, D., & Nwokocha, C. (2022). Artocarpus Altilis (Breadfruit) could reverse Myocardial infarction through the normalization of the oxygen haemoglobin dissociation curve. *Cardiovascular & Hematological Agents in Medicinal Chemistry*. 18715257. *20*(3), 212–218. doi:10.2174/1871525720666220203110919.

Tian, M., Lin, Z., Wang, X., Yang, J., Zhao, W., Hongmei, L., et al. (2021). Pure ion chromatograms combined with advanced machine learning methods improve accuracy of discriminant models in LC–MS-based untargeted metabolomics. *Molecules*. 1420-3049. *26*(9), 2715. doi:10.3390/molecules26092715.

Tinte, M. M., Chele, K. H., van der Hooft, J. J. J., & Tugizimana, F. (2021). Metabolomics-guided elucidation of plant abiotic stress responses in the 4IR era: an overview. *Metabolites*. 22181989. *11*(7). Available from: http://www.mdpi.com/journal/metabolites. doi:10.3390/metabo11070445.

Tomii, K., & Ikeda, K. (2025). Exploring binding sites on proteins for function prediction using the PoSSuM databases. *Methods in Molecular Biology*, *2947*, 209–222. Available from: https://www.springer.com/series/7651. doi:10.1007/978-1-0716-4662-5_12.

Tripathi, N., Goshisht, M. K., Sahu, S. K., & Arora, C. (2021). Applications of artificial intelligence to drug design and discovery in the big data era: a comprehensive review. *Molecular Diversity*. 1573501X. *25*(3), 1643–1664. Available from: www.wkap.nl/journalhome.htm/1381-1991. doi:10.1007/s11030-021-10237-z.

Vittorio, S., Lunghini, F., Morerio, P., Gadioli, D., Orlandini, S., Silva, P., et al. (2024). Addressing docking pose selection with structure-based deep learning: recent advances, challenges and opportunities. *Computational and Structural Biotechnology Journal*.20010370. *23*, 2141–2151. Available from: https://www.sciencedirect.com/science/journal/20010370. doi:10.1016/j.csbj.2024.05.024.

Wang, P., Zheng, S., Jiang, Y., Li, C., Liu, J., Wen, C., et al. (2022). Structure-aware multimodal deep learning for drug-protein interaction prediction. *Journal of Chemical Information & Modeling*. 1549960X. *62*(5), 1308–1317. Available from: http://pubs.acs.org/journal/jcisd8. doi:10.1021/acs.jcim.2c00060.

WHO, TEAM. (12 June 2013). A global brief on hypertension a slient killer. A global health crisis https:/www.who.int/i/items/publications/a-global-brief-on hypertension-a-slient-killer-a-global health-crisis-world-health-day-2013.

WHO, TEAM. (24 Aug. 2021). Guideline for the pharmacological treatment of hypertension in adults (pp. 1–48). https://www.WHO.int/publications/i/items/9789240055986.

Xiong, G., Wu, Z., Yi, J., Fu, L., Yang, Z., Hsieh, C., et al. (2021). ADMETlab 2.0: an integrated online platform for accurate and comprehensive predictions of ADMET properties. *Nucleic Acids Research*. 13624962. *49*(1), W5. Available from: https://academic.oup.com/nar/issue. doi:10.1093/nar/gkab255.

Chapter 32

Drug discovery from medicinal plants and natural products: Opportunities for advancement with artificial intelligence

Momoh A. Yakubu[a], Vishwalatha K. Parvathreddy[a], Ayodele Stephen Ake[b], Taiwo O. Oyagbemi[c], Ebenezer O. Ajiboye[d], Ademola A. Oyagbemi[b], Ya Fatou Njie Mbye[e], and Sunny Edet Ohia[e]

[a]*Department of Chemical Engineering & Environmental Toxicology, College of Science, Engineering & Technology, Texas Southern University, Houston, TX, United States*

[b]*Department of Veterinary Physiology and Biochemistry, Faculty of Veterinary Medicine, University of Ibadan, Ibadan, Nigeria*

[c]*Federal College of Animal Health and Production Technology, Ibadan, Oyo State, Nigeria*

[d]*Anatomy Unit, Faculty of Basic Medical Sciences, Ajayi Crowther University, Oyo, Nigeria*

[e]*Department of Pharmaceutical Sciences, College of Pharmacy & Health Sciences, Texas Southern University, Houston, TX, United States*

1 Introduction

In contemporary medicine, the development of pharmaceuticals entails the innovative discovery of novel drug molecules that can successfully integrate into the healthcare system (Pereira, 2025). This multiphase journey spans from early-stage research—identifying therapeutic targets—to preclinical studies, clinical trials (Phases 1 through 3), and ultimately to Food and Drug Administration (FDA) approval and commercial release (Braga & Rawal, 2025). Even after market launch, the process continues through pharmacovigilance and drug repurposing studies. Estimates of costs involved in bringing a new therapeutic agent to the market often require billions of dollars and over decades of research and development. Several key factors contribute to the rising costs of drug discovery (DiMasi, Grabowski, & Hansen, 2015). These include wasted funding due to high failure rates of candidate drugs in late-stage trials, increasingly stringent regulatory standards, and the high cost of clinical testing, especially for studies that are essential to proving efficacy and safety. Given these challenges, research leaders in academia, pharmaceutical, and biotechnology industries are adapting to innovative strategies and technologies to increase effective drug production, reduce product failures, lower costs, and promote sustainable development of the sector. In this context, researchers are seeking to integrate artificial intelligence (AI) as a transformative instrument in the field of drug discovery. With its unmatched ability to handle and analyze large datasets, AI offers a promising method for accelerating and reducing the costly process of new drug development. AI and its associated processes can facilitate the development and/or discovery of new drug compounds, enhance innovation, and improve the efficiency of clinical trials, thereby optimizing drug discovery and development strategies (Braga & Rawal, 2025). AI uses complex databases like bioinformatics, genomics, proteomics, and metabolomics, along with gene and protein databases. By interacting with various advanced algorithms (computer-aided learning) and using applicative software to simulate human cognition, AI can uncover connections between treatment methods or models and health outcomes. AI technologies like Google's DeepMind Health and IBM Watson Health exemplify AI's growing influence across healthcare. Naturally, the pharmaceutical and biotechnology sectors have now adopted these AI-driven discovery platforms. These platforms are capable of optimizing real-world data operations, reducing development timelines and costs, and improving research effectiveness.

AI already makes it possible to predict and interpret drug mechanisms of action more effectively, improving drug safety. It also supports the planning and execution of organic synthesis for new molecules. If successful in the long term, explainable AI (XAI) could provide vital support for processing complex chemical data and generating novel therapeutic hypotheses (Qadri et al., 2025). However, researchers often hesitate to rely on hypotheses that are not easily interpretable or appear illogical. This presents a major challenge for the acceptance of advanced deep learning (DL) models, particularly

Applications of Artificial Intelligence in Healthcare. https://doi.org/10.1016/B978-0-443-43934-6.00026-2

due to the "black-box" nature of deep neural networks (DNNs). To overcome this, model-agnostic and XAI techniques are gaining attention, aiming to "produce predictions that can be rationalized in terms of chemistry or biology".

2 The role of herbal plants in traditional and modern medicine

For centuries, plants have been integral to traditional healing practices, providing a significant source of biologically active compounds with medicinal properties. The widespread use of herbal remedies highlights their importance in global healthcare, particularly in regions where access to modern medicine is restricted (Bashir, 2025). Herbal medicine has historically been a foundational element of healthcare systems across diverse cultures (Vendrapati et al., 2024). A considerable portion of the global population continues to depend on traditional medicine (TM) for their primary healthcare needs. When practiced outside its original cultural context, such practices are often classified as "complementary and alternative medicine." Notable traditional medical systems include traditional Chinese medicine (TCM), Ayurveda from India, and various systems from African nations (Mukherjee, Banerjee, Katiyar, Sharma, & Chattopadhyay, 2021; Das, Mishra, Samanta, Shil, & Roy, 2025; Holzinger, Zatloukal, & Müller et al., 2025; Longo et al., 2024).

In TCM, ancient roots dating back to the Shang Dynasty (1600–1046) laid the foundation for a system that has evolved over >2000 years, incorporating modalities such as acupuncture, herbal medicine, and cupping. Indian traditional systems of medicine comprise six recognized disciplines: Ayurveda, Siddha, Unani, Yoga, Naturopathy, and Homeopathy. Ayurveda, with a history spanning over 3000 years, emphasizes balancing bodily energies through plant-based therapies (Wen et al., 2022). Similarly, African TM, rooted in spiritual and community-centered practices, combines herbalism with divination and ritual (Che, George, Ijinu, Pushpangadan, & Andrae-Marobela, 2024). Despite their cultural differences, these systems share a common reliance on natural products and holistic diagnosis, enriching both cultural heritage and the global healthcare landscape (Che et al., 2024).

Medicinal plants have played an important part in traditional medical systems used all over the world. For decades, indigenous tribes have relied on these herbs to keep themselves healthy, prevent sickness, and treat diseases. Local traditional healers are quite knowledgeable about the therapeutic characteristics of indigenous plants and how they may be used to cure a variety of maladies. Their empirical discoveries and collective wisdom have helped preserve knowledge and, consequently, ensured the transmission of ancient healing techniques from one generation to another. The knowledge from ancient remedies is still highly valued by communities throughout the world. Recently, there has been renewed interest in TM and natural cures due to increased demand by customers for holistic and sustainable healthcare solutions. As a result, there is growing interest in medicinal plants and their potential benefits to modern healthcare (Bhadra & Manjunath, 2024).

3 Key challenges in natural product drug development

Medicinal plants have long served as a foundation for the discovery of novel drugs, primarily because of the wide variety of chemical structures and biological effects found in their natural compounds (Atanasov, Zotchev, & Dirsch, 2021). Progress in analytical and extraction technologies has enabled researchers to isolate and characterize specific active molecules from plant sources (Altemimi, Lakhssassi, Baharlouei, Watson, & Lightfoot, 2017). These discoveries have paved the way for the synthesis of modified analogs and chemomimetic compounds in laboratory settings, resulting in the creation of highly potent therapeutic agents.

Reinvigorating interest in phytochemicals as inspiration for drug development presents a significant opportunity to improve global health and medical innovation. Nonetheless, despite the pharmaceutical industry's historical reliance on plant-based compounds, enthusiasm has diminished in recent years due to multiple challenges. Inconsistent access to high-quality phytoconstituents, often influenced by environmental conditions and agricultural practices, poses concerns about ecological sustainability. Furthermore, the high financial and time investments required to synthesize chemimimetic analogs raise questions about economic sustainability. On the scientific front, the mechanisms through which these plant-derived compounds exert their effects are often poorly understood, necessitating extensive and time-consuming bioassays to validate their therapeutic potential. These challenges must be overcome to advance and break ground in successful new therapeutic innovation.

To harness the potential of bioactive compounds from medicinal plants in developing new therapeutics, it is essential to overcome key obstacles such as quality control, standardization, bioavailability, safety assessment, and a lack of clinical evidence (Dar, Shahnawaz, & Ahanger, 2023; Singh, Sahu, Katiyar, & Arya, 2025). This requires the development of standardized protocols for cultivation, extraction, and formulation, as well as comprehensive toxicological evaluations. Robust clinical trials are also necessary to provide reliable evidence of safety and efficacy in human populations (Dar et al., 2023). Additional barriers include the absence of standardized methods and comprehensive reference databases for techniques like

DNA barcoding and chemical profiling (Mahima et al., 2022), alongside the high costs and technical expertise required for genomic and proteomic analyses, which limit their application in resource-constrained settings. Furthermore, current approaches to phytochemical profiling and toxicity testing remain time-consuming, costly, and limited in scope, thereby hindering the safe and efficient development of herbal remedies (Ekor, 2014). Exploring phytoconstituents is laboriously tedious, involving multiple steps from crude extractions through processes that lead to the production of pure extracts to obtain measurable and analysis-friendly isolates. The isolates undergo a screening process to prioritize the fractions of the extracts (Guerreo-Perilla, Bernal, & Coy-Barrera, 2015). Those isolates that contain bioactive compounds are then subjected to additional procedures, typically informed by one or more biological assays (Altemimi et al., 2017) to confirm their bioactivities. To address the complexity and duplication in natural crude extracts, scientists have devised methods for early chemical characterization of targeted and/or untargeted phytoconstituents.

4 Molecular docking in herbal research

Molecular docking serves as a valuable computational tool in herbal medicine, enabling researchers to explore how bioactive compounds from medicinal plants interact with biological targets like proteins and enzymes (Agu et al., 2023; Elbouzidi et al., 2024). It is particularly useful when plant extracts show biological activity, as it helps isolate and identify the active compounds and evaluate their potential against known drug targets (Ma et al., 2023; Guerreo-Perilla et al., 2015). Through virtual screening, docking aids in predicting which phytochemicals are most likely to bind effectively, which is especially important for tackling emerging diseases (Azeem, Mustafa, & Mahrosh, 2022). Traditional medicinal knowledge can be leveraged in this context to prioritize candidate compounds for further biological testing, saving time and research costs. In addition, docking supports drug repurposing by uncovering new molecular targets for existing phytochemicals and validating traditional uses by explaining their molecular mechanisms. It also guides experimental design by predicting ligand binding modes and affinities, with improved accuracy when multiple docking platforms and scoring functions are used in a consensus-based approach.

However, despite its advantages, molecular docking in herbal medicine has several limitations. The complex and flexible structures of many phytochemicals can reduce docking accuracy (Lionta, Spyrou, Vassilatis, & Cournia, 2014), while scoring functions often lack the precision needed to estimate binding affinities reliably (Lionta et al., 2014). Docking also rarely accounts for critical pharmacokinetic and toxicity factors (Araujo et al., 2024) and typically ignores the synergistic or antagonistic effects found in complex plant extracts (Rigby, 2024), which may lead to misleading results. Moreover, it offers limited insight into the physiological outcomes of ligand–target interactions in living systems. The success rate of virtual screening via docking alone remains relatively low—approximately 30%—indicating the need for experimental validation and the integration of complementary approaches to strengthen its application in natural product-based drug discovery.

5 AI's role in expediting natural product screening and identification

Significant progress has been made in predicting drug distribution properties, particularly through AI-based approaches. These advancements have enabled researchers to greatly reduce the time and cost involved in filtering out unsuitable drug candidates. Fig. 32.1 depicts the general structure of an AI-driven model for predicting drug distribution properties, utilizing both machine learning (ML) and DL capabilities (Tran, Tayara, & Chong, 2023) that could be engaged in AI-assisted drug development from natural products.

AI represents a specialized domain within computer science that aims to replicate human cognitive functions, including learning, reasoning, and decision-making (Hou et al., 2025). Its most distinctive and potent capabilities encompass the analysis of large and complex datasets, the ability to learn from prior experiences, the adaptation to new information, and decision-making based on data. The swift advancement of AI technologies has catalyzed innovation across various sectors, including autonomous transportation, manufacturing, education, scientific research, and healthcare. In recent years, TM researchers have increasingly explored AI as a promising tool (Rajesh & Elumalai, 2025). AI-powered processes, in association with software-driven ML, DL, and integration with network pharmacology, bioinformatics, systems biology, chemo informatics, and computer vision, can interconnect various elements of TM using DNNs, including chemical compounds, herbal remedies, pharmaceutical drugs, biological targets, symptoms, and diseases (Ma et al., 2023) providing best-fit models that predict appropriate phytoconstituents for therapeutics. Essentially, AI offers novel methodologies for investigating classical TM literature, identifying key ingredients in herbs or formulas, uncovering mechanisms of action, and supporting more precise application of traditional remedies (Ma et al., 2023). Through the use of advanced algorithms and models, AI contributes to TM by enhancing decision-making processes, improving diagnostic precision, enriching

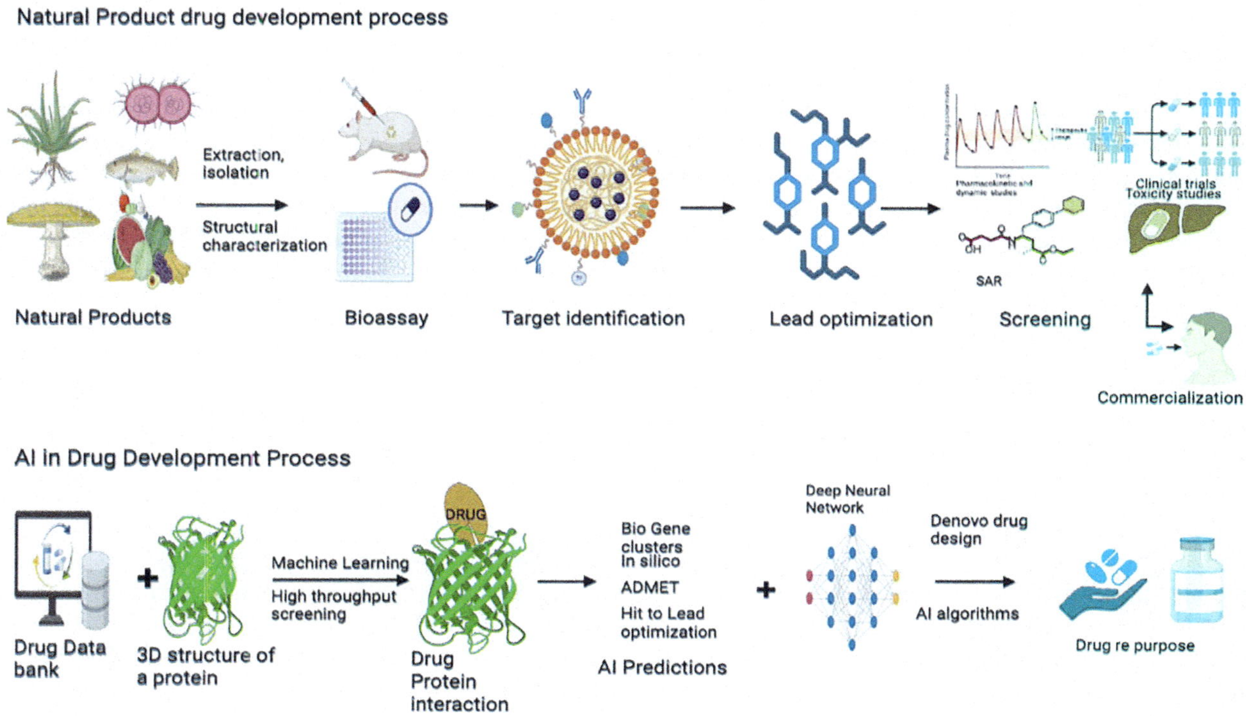

FIG. 32.1 Overview of AI in integrated drug discovery from natural products.

the patient experience, and lowering healthcare costs. These technologies can also automate the detection and measurement of phytochemicals, significantly increasing the efficiency and accuracy of analyses (Rajesh & Elumalai, 2025). Moreover, AI's ability to make predictions allows for early identification of potentially harmful compounds and interactions between multiple phytochemicals, which is essential to ensure the safety of herbal formulations (Xu, Pei, & Lai, 2017). Beyond examining individual compounds, AI enables the integration of diverse databases like genomics, metabolomics, bioinformatics, and ethnobotanical information, interrogating them to predict and channel useful information to help accelerate the discovery and development of new therapeutics derived from natural sources. AI, ML, and DL are interconnected conceptual processes within computer science frame works in which each can play a vital role in advancing the field of drug discovery (Fallah, 2022). AI aims to replicate human cognitive functions through machines, particularly computer systems, enabling capabilities, including learning, reasoning, problem-solving, perception, language comprehension, and decision-making (Siemens et al., 2022). The goal of creating an intelligent system is to enable it to process relevant data, learn from its environment, understand contextual situations, and make informed decisions to accomplish defined objectives (Adeyeye and Akanbi, 2024). AI technologies are applied across a wide range of fields, including natural language processing (NLP), large language models like those used in ChatGPT, computer vision, and facial recognition systems. AI applications in robotics and automated systems processing are well established. In drug development and discovery, AI plays significant roles in enhancing bioactivity predictions, structure–activity relationship (SAR) studies, best-fit modeling, and molecular design (Vaishnav, Verma, & Mishra, 2025). For instance, these attributes of AI in predictions, quick identifications of characteristics desirable for drug candidates, or as a red flag for bioactivation candidates for toxic compounds, are of immense help in drug development from phytoconstituents. AI's predictive strength in drug discovery largely relies on regression tasks, a form of supervised learning (Health, 2022), which utilizes bioassay data to enhance the accuracy of predictions in SAR studies and facilitate the optimization of molecular structures. A range of AI and ML algorithms have demonstrated their effectiveness in this area. Methods, including support vector machines (SVMs) and neural networks, contribute to the creation of predictive models (Nayarisseri et al., 2021).

DL, a subset of ML, utilizes artificial neural networks composed of multiple layers—also known as DNNs—to extract and interpret complex data patterns, aiding in decision-making processes (Ferrucci, 2012). Modeled after the structure and functioning of biological neurons, DL has demonstrated high performance in tasks like image recognition, speech processing, and autonomous navigation. Its ability to identify intricate and abstract features makes it particularly useful in molecular modeling and drug design. Convolutional neural networks are particularly effective for analyzing molecular structures within SAR investigations and are frequently utilized in virtual screening workflows. In contrast, recurrent neural networks are specifically designed for sequence learning, which is crucial for the generation of novel molecular sequences. Reinforcement learning (RL) improves decision-making within the drug discovery pipeline by iteratively refining molecular synthesis strategies and compound design through continuous feedback, thereby making a significant contribution to de novo drug development. Furthermore, generative adversarial networks (GANs), a subtype of generative AI, are used to create entirely new chemical compounds by learning the distribution of existing chemical datasets. Autoencoders also support this process by enabling efficient molecular representation learning. Both approaches play a critical role in SAR studies and the broader field of molecular design (Gangwal et al., 2024).

6 AI-driven successes in natural product drug development

In medicinal plant research, the fusion of AI, ML, DL, and bioinformatics has welcomed an innovative era in drug development and discovery (Rekognition, 2022). Synergizing the process of extracting active ingredients from plants or animal sources with AI is increasingly pivotal in drug discovery (Gangwal & Lavecchia, 2025). The process in natural products derived drug development faces numerous challenges; effective extraction and isolation of primary and secondary metabolites involve meticulous bioassay-guided separation and chromatography, followed by characterization and purification of isolates using techniques such as nuclear magnetic resonance, mass spectrometry, and X-ray crystallography (Amazon SageMaker, 2022). However, challenges arise due to the limited concentration of bioactive molecules, the complexity of molecular structures, and the low yields of promising isolates, which may exhibit low solubility, instability, or toxicity. These factors complicate the impetus for development into clinical applications. Nonetheless, the intricate interactions of natural products with multiple protein targets present opportunities for the development of multitarget therapeutics.

AI can play a key role in the exploration of phytoconstituents for use in research and development of new drugs. The application of AI in herbal product-derived drug development can help optimize better health outcomes for the future of humanity (Gangwal & Lavecchia, 2025). AI process applications in herbal product-derived drug development to optimize extraction processes, characterization, modeling and target-receptor best fits, assessment of pharmacokinetics, interrogation of functional and dysfunctional biology along with the effects of extracts, prediction of toxicity, and integration of biological data can lead to rapid turnaround. AI applications can advance phytoconstituent-based drug discovery and optimization, facilitate the development of novel compounds, and enhance the overall efficiency of drug development (Health, 2022; Zhang et al., 2025).

The advancement of medicinal plants into approved therapeutic agents is frequently slowed by an insufficient understanding of their biological targets, making preclinical evaluation and compound optimization more complex. Due to the difficulty in isolating and analyzing plant-derived metabolites at scale, experimentally uncovering their mechanisms of action is both time-consuming and financially burdensome. As a result, researchers are increasingly focusing on computational tools capable of predicting likely molecular targets based on structural information.

A range of in silico drug discovery techniques, such as molecular docking, clustering algorithms, bioactivity-based fingerprints, pharmacophore modeling, and ML, have shown considerable promise in mapping phytoconstituent interactions. In some instances, these computational methods have provided fresh insights into compounds already progressing through clinical evaluation. Although challenges remain, the continued refinement of ML algorithms and related technologies points to a future where more precise and tailored models will enhance the drug discovery process.

7 In silico pharmacokinetic analysis

The pharmaceutical industry underwent a significant transformation in the late 1990s, as it became evident that inadequate pharmacokinetic and pharmacodynamic properties of drug candidates were primary contributors to clinical trial failures. In response to this realization, in silico pharmacokinetic modeling, which emphasizes Absorption, Distribution, Metabolism, Excretion, and Toxicity (ADMET), emerged as a crucial strategy to enhance drug design. This approach aims to assist research teams in prioritizing and designing molecules with optimized ADMET profiles while efficiently directing experimental resources toward the most promising compounds, thereby minimizing the necessity to synthesize and test a large number of chemicals (Pradeepkiran, Sainath, & Shrikanya, 2021). Initially, ADMET modeling employed methodologies such as Bayesian neural networks, SVMs, and random forests. However, with advancements in ML and the availability of large-scale, consistent ADMET datasets, the field transitioned to more advanced ML-driven prediction systems capable of modeling complex, nonlinear data patterns. A pivotal moment was the 2012 Kaggle Merck Molecular Activity Challenge, which evaluated the predictive performance of various ML models on 18 different ADMET endpoints using proprietary Merck data ranging from 2000 to 50,000 compounds. The winning approach combined DNNs with Gaussian process and gradient boosting machine models in an ensemble format (Kose, Gupta, & Chen, 2023). DNNs have become

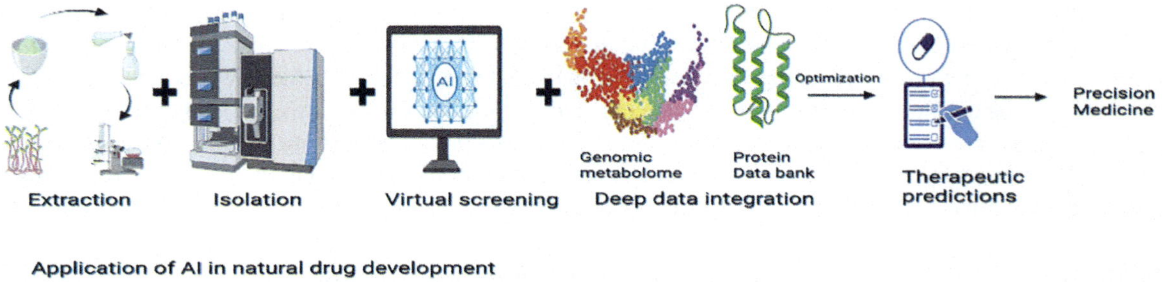

FIG. 32.2 Integrated approach for drug discovery and development: from natural sources to personalized medicine using AI/ML and in silico studies.

prominent in ADMET modeling due to their ability to simultaneously predict multiple endpoints using unified model architecture. Multitask DNNs leverage inductive transfer learning, enabling shared internal representations across tasks, which enhances learning speed and overall predictive accuracy. Most of these models use a "hard" parameter-sharing framework, where a common hidden layer is shared among all predicted tasks (Kose et al., 2023). In addition, matched molecular pair (MMP) ***analysis***, which compares two molecules differing by a single, well-defined chemical modification and its impact on a property, has evolved. While earlier MMP studies focused on one-to-one transformations, advancements in AI and fragmentation algorithms now allow large-scale automated MMP analysis. One notable tool, MCPairs, utilizes unsupervised ML to integrate in vitro ADMET datasets from pharmaceutical companies like AstraZeneca, Roche, and Genentech. These innovations enable the creation of next-generation XAI tools that offer practical solutions to ADMET prediction challenges (Kose et al., 2023) (Fig. 32.2).

8 Application of AI in de novo drug design from natural products

The integration of AI with computational chemistry has introduced novel approaches for molecular design, which are now being applied in fields such as pharmaceuticals and materials engineering. Unlike conventional AI tools that focus on forecasting molecular behavior or biological interactions, these emerging methods are geared toward inventing entirely new chemical entities. Their main goal is to expand the boundaries of chemical exploration by venturing into uncharted molecular territories, far beyond existing compound libraries (Crucitti, Pérez Míguez, Díaz Arias, Fernandez Prada, & Mosquera Orgueira, 2024). Two core methodologies drive this new wave of molecular design. The first, known as "Holistic Generation," constructs molecules entirely from the ground up, offering broad exploration during early discovery stages. The second, "Iterative Generation," incrementally assembles molecules, making it better suited for refining structures and customizing them for targeted functions. Predictive AI and ML models play a crucial role in assessing the characteristics of newly generated compounds prior to their physical synthesis. These evaluations include checks such as compliance with Lipinski's rules, assessments of drug-likeness, and forecasts of potential toxicity. By filtering out unsuitable candidates early, this process significantly improves the efficiency and success rate of experimental workflows that use generative AI. One common strategy, the building block approach, involves the automated construction of new molecules by combining fragments containing specific substructures or functional groups. This modular method supports the systematic generation of structurally diverse compounds, potentially leading to the discovery of novel entities with beneficial pharmacological effects. Alternatively, neural network architectures—especially autoencoders—enable the generation of new molecular structures by learning patterns from existing chemical data and recreating these features in novel ways.

The drug discovery process is known for being both time-consuming and costly (Zhou & Zhong, 2017). However, GANs have shown promise in making this process more efficient. One of their emerging applications is in de novo peptide and protein design, a method focused on computationally generating novel peptides and proteins without relying on existing molecular templates. The term de novo itself refers to the creation of new entities from scratch, based entirely on defined constraints (Zhou & Zhong, 2017). Several research efforts have explored the potential of GANs in this field. For instance, a study introduced a long short-term memory GAN model, originally developed for NLP, to generate new protein backbone structures. These newly formed helical topologies possessed structural features advantageous for protein design (Tripathi et al., 2022). In addition to protein design, GANs are also useful in predicting how well a drug molecule binds to a target protein, a key step in evaluating a drug's effectiveness (Zhao, Wang, Pang, Liu, & Zhang, 2020). DL models, including GANs, can identify drug candidates with strong binding affinities to disease-specific proteins, thereby accelerating the research pipeline. Compared to earlier ML models, GANs offer the dual benefits of faster processing and reduced costs. This efficiency is due in part to the reuse of the generator and discriminator components as feature extractors, which minimizes the need for manually labeled data. Zhao et al. (2020) proposed an approach to improve GAN performance using three main components: a compound feature extractor, a sequence feature extractor, and a regression model for predicting binding affinity. In the initial training phase, the GAN generates synthetic samples, which are then evaluated alongside real data by the discriminator. This helps the generator refine its output to resemble realistic protein compounds and sequences. In the second training phase, the validated data are used to train the regressor, which learns to predict binding affinities more accurately by minimizing prediction error. These models, built upon extracted features and optimized parameters, can effectively guide biological experiments and help researchers generate biologically relevant protein sequences. The results showed that GAN-based systems performed on par with, or in some cases better than, traditional ML models, especially considering their ability to train on unlabeled datasets and reduce experimental costs (Tripathi et al., 2022). A noted limitation, however, was the decreased performance of GANs with very limited data, though this issue can be mitigated using regularization strategies during training (Tripathi et al., 2022).

AI has become a valuable asset in medicinal chemistry, particularly for forecasting the potential effectiveness and toxicity of candidate drug molecules (Zhang et al., 2025; Vijayan, Kihlberg, Cross, & Poongavanam, 2022; Blanco-Gonzalez et al., 2023). In addition, AI technologies have proven useful in discovering and formulating new therapeutics derived from natural sources (Zhang et al., 2025). Traditional drug discovery methods typically depend on time-consuming and expensive in vivo experiments and clinical trials, often yielding uncertain or inconsistent results. AI offers a way to address these challenges. Through ML, algorithms can detect subtle correlations and hidden patterns in experimental data that might be overlooked by researchers. This facilitates the rapid identification of biologically active molecules that align with specific targets, making the overall process more efficient compared to standard methodologies (Vijayan et al., 2022 Akmal & Wong, 2025).

A recent illustration involves the utilization of DL techniques that have been trained on datasets comprising known drugs and their corresponding biological responses. These models demonstrate the capability to accurately predict the pharmacological activity of novel compounds, thereby minimizing the potential for toxicity in future drug candidates (Crucitti et al., 2024). Another critical area where AI contributes is in recognizing drug–drug interactions (DDIs), which occur when multiple drugs are administered simultaneously for one or more conditions, sometimes leading to harmful or unexpected effects (Zhang et al., 2025; Blanco-Gonzalez et al., 2023). By evaluating extensive datasets that include drug attributes and interaction records, AI systems can identify recurring patterns and interaction risks (Blanco-Gonzalez et al., 2023; Jang et al., 2022). In the field of personalized medicine, AI enhances the tailoring of treatment plans that consider potential DDIs, thereby minimizing adverse outcomes for individual patients (Marzouk, Selim, Elattar, Mabrouk, & Mysara, 2025). Importantly, interactions are not limited to synthetic drugs; natural products can also produce significant interactions when combined with other herbal agents or conventional pharmaceuticals (Rombolà et al., 2020; Marzouk et al., 2025; Jang et al., 2022; Kant, Deepika, & Roy, 2025; Bhat & Ahmed, 2025; Gamil et al., 2025; Alhumaidi et al., 2023).

9 Challenges and limitations

While AI possesses the potential to transform drug discovery, its application is accompanied by several challenges and limitations. A significant obstacle is the lack of suitable datasets. AI and ML systems necessitate substantial volumes of high-quality, consistent data to operate effectively (Blanco-Gonzalez et al., 2023; Bhat & Ahmed, 2025; Kant et al., 2025). Unfortunately, in many cases, the available data may be inadequate, poorly structured, or inconsistent, which can compromise the accuracy and reliability of AI-generated results (Mohammed et al., 2025). Furthermore, ethical considerations are paramount when integrating AI into pharmaceutical research. A major issue is algorithmic bias; if the training data are skewed or do not represent diverse populations, the AI system may produce biased or unjust outcomes (Hanna et al., 2025; Hofmann, Kalluri, Jurafsky, & King, 2024; Weiner, Dankwa-Mullan, Nelson, & Hassanpour, 2025). Thus ensuring fairness and equity in the development and deployment of AI tools for drug development is of paramount importance (Hanna et al., 2025; Hofmann et al., 2024; Radanliev, 2025b; Radanliev, Santos, Brandon-Jones, & Joinson, 2024). To tackle these challenges, several strategies have been proposed. One method is data augmentation, which involves the creation of synthetic data to boost the diversity and volume of training sets (Hartman, Sutnick, Hartman, Newbold, & Kaplan, 1986; Hosain, Jim, Mridha, & Kabir, 2024; Mumuni & Mumuni, 2024; Radanliev, 2025a; Rana, Azizul, & Awan, 2023). This can enhance the learning capacity of ML models and produce more accurate results (Blanco-Gonzalez et al., 2023; Mumuni & Mumuni, 2024; Hosain et al., 2024). Another important strategy involves the adoption of XAI techniques. These approaches aim to make AI systems more transparent by providing clear reasoning behind predictions, which can help researchers understand model behavior and detect bias (Schneider, 2025; Rana et al., 2023; Radanliev, 2025b).

Blanco-Gonzalez et al. (2023) highlight another critical concern, the growing reliance on AI for decision-making in areas that directly impact public health. Decisions regarding which drugs to prioritize, how to design clinical trials, and how medications are marketed and distributed may increasingly be influenced by AI systems. When such systems are biased, they could lead to disparities in healthcare access or unfair treatment of marginalized groups (Ferrara, 2024; Ratwani, Sutton, & Galarraga, 2024; Hanna et al., 2025). This raises questions about justice and equal opportunity in medical treatment. Another important implication is the potential disruption of the labor market due to AI automation. The adoption of AI technologies in the pharmaceutical sector could result in job losses, making it essential to develop policies that support and retrain displaced workers. Moreover, the heavy reliance of AI on vast datasets introduces concerns over the safety and privacy of sensitive health information. Unauthorized access or misuse of such data could harm individuals and damage the reputations of companies involved (Blanco-Gonzalez et al., 2023; Pool, Akhlaghpour, Fatehi, & Burton-Jones, 2024). Therefore responsible handling of medical data is necessary, with strict adherence to privacy regulations.

It is essential to acknowledge, as noted by Blanco-Gonzalez et al. (2023), that AI tools should not be viewed as substitutes for traditional experimental research or human expertise. These systems are capable of generating predictions based solely on existing data and lack the ability to replicate the intuition or critical thinking that experienced scientists possess. Therefore human oversight remains essential to interpret AI outputs and validate results (Longo et al., 2024). That said, integrating AI with conventional research methodologies can improve efficiency in drug development by combining human judgment with machine precision (Blanco-Gonzalez et al., 2023; Hasan, Jaber, Khabour, & Alzoubi, 2024; Fu & Chen, 2025).

Ultimately, ensuring the ethical application of AI in drug discovery demands careful planning and proactive governance (Hasan et al., 2024). This includes representative datasets, regular monitoring of AI systems for potential biases, and the implementation of comprehensive privacy and security measures, which are essential (Blanco-Gonzalez et al., 2023; Radanliev, 2025a; Mirakhori & Niazi, 2025). By addressing these critical issues, the pharmaceutical industry can effectively harness the power of AI to expedite innovation while upholding ethical standards (Blanco-Gonzalez et al., 2023).

10 Future directions in AI-powered research

Future research in AI-driven drug discovery from herbal medicine should prioritize the creation of large, interoperable databases. These databases should combine genomic, metabolomic, ethnobotanical, and pharmacological data. In addition, it is crucial to standardize TM knowledge into structured, machine-readable formats to improve the accuracy and generalizability of AI models. As AI models become more advanced, using DL, multitask DNNs, and GANs, there is a growing need for XAI. These systems must be able to clearly show how predictions are made, especially in clinical and regulatory contexts. Providing interpretable outputs will be essential for informed decisions regarding compound selection, toxicity prediction, and SAR analysis.

Generative AI tools, including latent GANs, autoencoders, and RL frameworks, are anticipated to significantly contribute to de novo drug design. These tools possess the capability to generate novel chemical structures inspired by herbal phytoconstituents, thereby facilitating the exploration of previously unexamined bioactivities and therapeutic targets, which, in turn, expand the known chemical space. Furthermore, the successful integration of in silico predictions with in vitro and in vivo experiments is essential for achieving desired outcomes. Future AI systems should work alongside laboratory experiments to guide synthesis, screening, and validation in a closed-loop process, making drug discovery more efficient and less resource-intensive.

Lastly, AI holds great promise for personalized phytotherapy. By combining patient genetic data, disease characteristics, and herbal compound profiles, AI can help design individualized treatments. This approach could transform TM into a more precise and personalized form of healthcare.

11 Conclusion

The integration of AI with traditional pharmacognosy is revolutionizing herbal medicine research, enabling faster and more accurate analysis of bioactive compounds. This innovative approach combines the best of both worlds, leveraging cutting-edge technology to validate and optimize ancient practices. The result is a new era of discovery, where traditional knowledge meets modern science to create novel plant-based treatments.

References

Adeyeye, O. J., & Akanbi, I. (2024). Artificial intelligence for systems engineering complexity: a review on the use of AI and machine learning algorithms. *Computer Science & IT Research Journal*. 2709-0051. *5* (4), 787–808. doi:10.51594/csitrj.v5i4.1026.

Agu, P. C., Afiukwa, C. A., Orji, O. U., Ezeh, E. M., Ofoke, I. H., Ogbu, C. O., Ugwuja, E. I., & Aja, P. M. (2023). Molecular docking as a tool for the discovery of molecular targets of nutraceuticals in diseases management. *Scientific Reports*, *13*(1), 13398. doi:10.1038/s41598-023-40160-2.

Akmal, M. F., & Wong, M. W. (2025). Selective cleaning enhances machine learning accuracy for drug repurposing: multiscale discovery of MDM2 inhibitors. *Molecules*, *30*(14). doi:10.3390/molecules30142992.

Alhumaidi, R. M., Bamagous, G. A., Alsanosi, S. M., Alqashqari, H. S., Qadhi, R. S., Alhindi, Y. Z., Ayoub, N., & Falemban, A. H. (2023). Risk of polypharmacy and its outcome in terms of drug interaction in an elderly population: a retrospective cross-sectional study. *Journal of Clinical Medicine*, *12*(12). doi:10.3390/jcm12123960.

Altemimi, A., Lakhssassi, N., Baharlouei, A., Watson, D. G., & Lightfoot, D. A. (2017). Phytochemicals: extraction, isolation, and identification of bioactive compounds from plant extracts. *Plants*. 22237747. *6*(4). Available from http://www.mdpi.com/2223-7747/6/4/42/pdf. doi:10.3390/plants6040042.

Amazon SageMaker, (2022). Build, train, and deploy machine learning models. Available from: https://aws.amazon.com/sagemaker.

Araujo, N. G. R., da Silva Junior, F. C., Santos, L. V. D. S., de Medeiros, S. R. B., Felzenszwalb, I., & Araújo-Lima, C. F. (2024). Molecular docking and in silico analysis of the pharmacokinetics, toxicological profile and differential gene expression of bioactive compounds from Cyrtopodium glutiniferum. *Toxicology Reports*. 22147500. *13*. Available from https://www.sciencedirect.com/science/journal/22147500. doi:10.1016/j.toxrep.2024.101810.

Atanasov, A. G., Zotchev, S. B., Dirsch, V. M., & (2021). Natural products in drug discovery: advances and opportunities. *Nature Reviews Drug Discovery*, *20*(3), 200–216. doi:10.1038/s41573-020-00114-z.

Azeem, M., Mustafa, G., & Mahrosh, H. S. (2022). Virtual screening of phytochemicals by targeting multiple proteins of severe acute respiratory syndrome coronavirus 2: molecular docking and molecular dynamics simulation studies. *International Journal of Immunopathology and Pharmacology*. 20587384. *36*. Available from https://journals.sagepub.com/home/IJI. doi:10.1177/03946320221142793.

Bashir, A. M. (2025). Integration of traditional and modern medicine: a review of Somali healing practices. *Annals of Medicine*. 13652060. *57*(1). Available from http://www.tandfonline.com/loi/iann20. doi:10.1080/07853890.2025.2546057.

Bhadra, A., & Manjunath, B. T. (2024). Exploring the rich heritage: medicinal plants and traditional healing practices of West Bengal. *Common Medicinal Plants of Our Surroundings*, 147.

Bhat, A. R., & Ahmed, S. (2025). Artificial intelligence (AI) in drug design and discovery: a comprehensive review. *In Silico Research in Biomedicine*, *1*, 100049. doi:10.1016/j.insi.2025.100049.

Blanco-Gonzalez, A., Cabezon, A., Seco-Gonzalez, A., Conde-Torres, D., Antelo-Riveiro, P., Pineiro, A., & Garcia-Fandino, R. (2023). The role of AI in drug discovery: challenges, opportunities, and strategies. *Pharmaceuticals*, *16*(6), 891.

Braga, D. M., & Rawal, B. S. (2025). Harnessing AI and Quantum Computing for Accelerated Drug Discovery: regulatory Frameworks for In Silico to In Vivo Validation. *Journal of Pharmaceutical and BioTech Industry*, *2*(3), 11.

Che, C.-T., George, V., Ijinu, T. P., Pushpangadan, P., & Andrae-Marobela, K. (2024). *Traditional medicine*. (pp. 11–28). Elsevier BV, Elsevier BV. doi:10.1016/b978-0-443-18657-8.00037-2.

Crucitti, D., Pérez Míguez, C., Díaz Arias, J. Á., Fernandez Prada, D. B., & Mosquera Orgueira, A. (2024). De novo drug design through artificial intelligence: an introduction. *Frontiers in Hematology*, *3*. doi:10.3389/frhem.2024.1305741.

Dar, R. A., Shahnawaz, M., & Ahanger, M. A. (2023). Irfan ul Majid, Exploring the Diverse Bioactive Compounds from Medicinal Plants: a Review. *The Journal of Phytopharmacology*. 2320480X. *12* (3), 189–195. doi:10.31254/phyto.2023.12307.

Das, D., & Shafi, S. (2023). Bioactivity-guided fractionation and identification of bioactive molecules: a basic method in drug discovery. In *Drugs and a Methodological Compendium: From Bench to Bedside* (pp. 41–78). Singapore: Springer Nature.

Das, S. K., Mishra, R., Samanta, A., Shil, D., & Roy, S. D. (2025). Deep learning: a game changer in drug design and development. *Advanced Pharmacology*, *103*, 101–120. doi:10.1016/bs.apha.2025.01.008.

DiMasi, J. A., Grabowski, H. G., & Hansen, R. W. (2015). The cost of drug development. *New England Journal of Medicine*, *372*(20), 1972. doi:10.1056/NEJMc1504317. 25970070.

Ekor, M. (2014). The growing use of herbal medicines: issues relating to adverse reactions and challenges in monitoring safety. *Frontiers in Pharmacology*. 1663-9812. *4*. doi:10.3389/fphar.2013.00177.

Elbouzidi, A., Taibi, M., Laaraj, S., Loukili, E. H., Haddou, M., El Hachlafi, N., et al. (2024). Chemical profiling of volatile compounds of the essential oil of grey-leaved rockrose (Cistus albidus L.) and its antioxidant, anti-inflammatory, antibacterial, antifungal, and anticancer activity in vitro and in silico. *Frontiers in Chemistry*, *12*, 1334028. doi:10.3389/fchem.2024.1334028.

Ferrara, E. (2024). Fairness and bias in artificial intelligence: a brief survey of sources, impacts, and mitigation strategies. *Science*, *6*(1), 3. doi:10.3390/sci6010003.

Ferrucci, D. A. (2012). Introduction to "This is Watson". *IBM Journal of Research and Development*. 0018-8646. *56* (3.4), 1. doi:10.1147/jrd.2012.2184356.

Fu, C., & Chen, Q. (2025). The future of pharmaceuticals: artificial intelligence in drug discovery and development. *Journal of Pharmaceutical Analysis*, *15*(8), 101248. doi:10.1016/j.jpha.2025.101248.

Gamil, N. M., Elsayed, H. A., Hamed, R. M., Salah, E. T., Ahmed, A. M., Mostafa, H. A., El-Shiekh, R. A., and Abou-Hussein, D. (2025). Insights from herb interactions studies: a foundational report for integrative medicine. *Future Journal of Pharmaceutical Sciences*, *11*(1). doi:10.1186/s43094-025-00794-7.

Gangwal, A., Ansari, A., Ahmad, I., Azad, A. K., Kumarasamy, V., Subramaniyan, V., & Wong, L. S. (2024). Generative artificial intelligence in drug discovery: basic framework, recent advances, challenges, and opportunities. *Frontiers in Pharmacology*. 16639812. *15*. Available from http://www.frontiersin.org/Pharmacology. doi:10.3389/fphar.2024.1331062.

Gangwal, A., & Lavecchia, A. (2025). Artificial Intelligence in Natural Product Drug Discovery: current Applications and Future Perspectives. *Journal of Medicinal Chemistry*. 0022-2623. *68* (4), 3948–3969. doi:10.1021/acs.jmedchem.4c01257.

Guerreo-Perilla, C., Bernal, F. A., & Coy-Barrera, E. D. 2015. Molecular docking study of naturally occuring compounds as inhibitors of n-myristol transferaese towrads antifungial agent discovery. *Revista Colombiana de Ciencias Quimico-Farmaceuticas*, *44*(2), 162–178. doi:10.15446/rcciquifa.v44n2,56291.

Hanna, M. G., Pantanowitz, L., Jackson, B., Palmer, O., Visweswaran, S., Pantanowitz, J., Deebajah, M., & Rashidi, H. H. (2025). Ethical and bias considerations in artificial intelligence/machine learning. *Modern Pathology*, *38*(3), 100686. doi:10.1016/j.modpat.2024.100686.

Hartman, M. E., Sutnick, A. I., Hartman, D. G., Newbold, R. C., & Kaplan, A. (1986). Comparison of performances of transfer students and first-year acceptees. *Journal of Medical Education*, *61*(3), 151–156. doi:10.1097/00001888-198603000-00001.

Hasan, H. E., Jaber, D., Khabour, O. F., & Alzoubi, K. H. (2024). Ethical considerations and concerns in the implementation of AI in pharmacy practice: a cross-sectional study. *BMC Medical Ethics*, *25*(1), 55. doi:10.1186/s12910-024-01062-8.

Health, W.: AI for personalized cancer treatment. (2022), Available from: https://www.ibm.com/watson-health.

Hofmann, V., Kalluri, P. R., Jurafsky, D., and King, S. (2024). AI generates covertly racist decisions about people based on their dialect. *Nature*, *633*(8028), 147–154. doi:10.1038/s41586-024-07856-5.

Holzinger, A., Zatloukal, K., and Müller, H. (2025). Is human oversight to AI systems still possible? *New Biotechnology*, *85*, 59–62. doi:10.1016/j.nbt.2024.12.003.

Hosain, M. T., Jim, J. R., Mridha, M., and Kabir, M. M. (2024). Explainable AI approaches in deep learning: advancements, applications and challenges. *Computers and Electrical Engineering*, *117*, 109246. doi:10.1016/j.compeleceng.2024.109246.

Hou, C., Gao, Y., Lin, X., Wu, J., Li, N., Lv, H., & Chu, W. C. C. (2025). A review of recent artificial intelligence for traditional medicine. *Journal of Traditional and Complementary Medicine*. 22254110. *15*(3), 215–228. Available from https://www.sciencedirect.com/science/journal/22254110. doi:10.1016/j.jtcme.2025.02.009.

Jain, R., Subramanian, J., & Rathore, A. S. (2023). A review of therapeutic failures in late-stage clinical trials. *Expert Opinion on Pharmacotherapy*, *24*(3), 389–399. doi:10.1080/14656566.2022.2161366. Epub 2022 Dec 26. PMID: 36542800.

Jang, S. H., Sivakumar, D., Mudedla, S. K., Choi, J., Lee, S., & Jeon, M., et al. (2022). PCW-A1001, AI-assisted de novo design approach to design a selective inhibitor for FLT-3(D835Y) in acute myeloid leukemia. *Frontiers in Molecular Biosciences*, *9*, 1072028. doi:10.3389/fmolb.2022.1072028.

Kant, S., Deepika and Roy, S. (2025). Artificial intelligence in drug discovery and development: transforming challenges into opportunities. *Discover Pharmaceutical Sciences*, *1*(1). doi:10.1007/s44395-025-00007-3.

Kose, U., Gupta, D., & Chen, X. (2023). *Explainable artificial intelligence for biomedical applications Explainable Artificial Intelligence for Biomedical Applications* (pp. 1–380). Türkiye: River Publishers. Available from https://ieeexplore.ieee.org/book/10158240.

Lionta, E., Spyrou, G., Vassilatis, D. K., & Cournia, Z. (2014). Structure-based virtual screening for drug discovery: principles, applications and recent advances. *Current Topics in Medicinal Chemistry*. 18734294. *14*(16), 1923–1938. Available from http://www.benthamdirect.org/pages/all_b_bypublication.php. doi:10.2174/1568026614666140929124445.

Longo, L., Brcic, M., Cabitza, F., Choi, J., Confalonieri, R., Del Ser, J., et al. (2024). Explainable Artificial Intelligence (XAI) 2.0: a manifesto of open challenges and interdisciplinary research directions. *Information Fusion*, *106*, 102301.doi:10.1016/j.inffus.2024.102301.

Ma, S., Liu, J., Li, W., Liu, Y., Hui, X., Qu, P., … Wang, J. (2023). Machine learning in TCM with natural products and molecules: current status and future perspectives. *Chinese Medicine*. 1749-8546. *18*(1). doi:10.1186/s13020-023-00741-9.

Mahima, K., Kumar, K. N. S., Rakhesh, K. V., Rajeswaran, P. S., Sharma, A., & Sathishkumar, R. (2022). Advancements and future prospective of DNA barcodes in the herbal drug industry. *Frontiers in Pharmacology*. 16639812. *13*. Available from http://www.frontiersin.org/Pharmacology. doi:10.3389/fphar.2022.947512.

Marzouk, N. H., Selim, S., Elattar, M., Mabrouk, M. S., & Mysara, M. (2025). A comprehensive landscape of AI applications in broad-spectrum drug interaction prediction: a systematic review. *Journal of Cheminformatics*, *17*(1), 141. doi:10.1186/s13321-025-01093-2.

Mirakhori, F., & Niazi, S. K. (2025). Harnessing the AI/ML in drug and biological products discovery and development: the regulatory perspective. *Pharmaceuticals (Basel)*, *18*(1). doi:10.3390/ph18010047.

Mohammed, S., Budach, L., Feuerpfeil, M., Ihde, N., Nathansen, A., Noack, N., Patzlaff, H., Naumann, F., and Harmouch, H. (2025). The effects of data quality on machine learning performance on tabular data. *Information Systems*, *132*, 102549. doi:10.1016/j.is.2025.102549.

Mukherjee, P. K., Banerjee, S., Katiyar, C. K., Sharma, S., & Chattopadhyay, N. (2021). Traditional Medicine System (TMS) for sustainable heaalthcare in India. In M. Dikshit (Ed.), Drug discovery and drug development. Singapore: Springer. doi:10.1007/978-981-15-8002-4_1.

Mumuni, F., and Mumuni, A. (2024). Improving deep learning with prior knowledge and cognitive models: a survey on enhancing explainability, adversarial robustness and zero-shot learning. *Cognitive Systems Research*, *84*, 101188. doi:10.1016/j.cogsys.2023.101188.

Nayarisseri, A., Khandelwal, R., Tanwar, P., Madhavi, M., Sharma, D., Thakur, G., … Singh, S. K. (2021). Artificial intelligence, big data and machine learning approaches in precision medicine & drug discovery. *Current Drug Targets*. 18735592. *22*(6), 631–655. Available from https://www.eurekaselect.com/189908/article. doi:10.2174/1389450122999210104205732.

Pereira, T. O., In silico validation and ADMET analysis for the best lead molecules Brucella Melitensis. (2025), 133–176

Pool, J., Akhlaghpour, S., Fatehi, F., and Burton-Jones, A. (2024). A systematic analysis of failures in protecting personal health data: a scoping review. *International Journal of Information Management*, *74*, 102719. doi:10.1016/j.ijinfomgt.2023.102719.

Pradeepkiran, J. A., Sainath, S. B., & Shrikanya K. V. L. (2021). In silico validation ADOMET analysis for the best lead molecules. https://www.sciencedirect.com/science/chater/edited-volume/abs/pii/B9780323856812000082.

Radanliev, P. (2025a). AI ethics: Integrating transparency, fairness, and privacy in AI development. *Applied Artificial Intelligence*, *39*(1). doi:10.1080/08839514.2025.2463722.

Radanliev, P. (2025b). Privacy, ethics, transparency, and accountability in AI systems for wearable devices. *Frontiers in Digital Health*, *7*. doi:10.3389/fdgth.2025.1431246.

Radanliev, P., Santos, O., Brandon-Jones, A., & Joinson, A. (2024). Ethics and responsible AI deployment. *Frontiers in Artificial Intelligence*, *7*, 1377011. doi:10.3389/frai.2024.1377011.

Rajesh, M. V., & Elumalai, K. (2025). The transformative power of artificial intelligence in pharmaceutical manufacturing: enhancing efficiency, product quality, and safety. *Journal of Holistic Integrative Pharmacy*. 27073688. *6* (2), 125–135. doi:10.1016/j.jhip.2025.03.007.

Rana, S. A., Azizul, Z. H., and Awan, A. A. (2023). A step toward building a unified framework for managing AI bias. *PeerJ Computer Science*, *9*, e1630. doi:10.7717/peerj-cs.1630.

Ratwani, R. M., Sutton, K., and Galarraga, J. E. (2024). Addressing AI algorithmic bias in health care. *JAMA*, *332*(13), 1051. doi:10.1001/jama.2024.13486.

Rekognition, A. (2022). Image and video analysis powered by deep learning. Available from: https://aws.amazon.com/rekognition.

Rigby, S. P. (2024). Uses of Molecular Docking Simulations in Elucidating Synergistic, Additive, and/or Multi-Target (SAM) Effects of Herbal Medicines. *Molecules*. 1420-3049. *29* (22), 5406. doi:10.3390/molecules29225406.

Rombolà, L., Scuteri, D., Marilisa, S., Watanabe, C., Morrone, L. A., Bagetta, G. and Corasaniti, M. T. (2020). Pharmacokinetic interactions between herbal medicines and drugs: Their mechanisms and clinical relevance. *Life*, *10*(7), 106. doi:10.3390/life10070106.

Schneider, J., 2025. Generative agentic AI: Survey, conceptualization and challenges. arXiv Preprint https://arxiv.org/abs/2504.18875.

Siemens, G., Marmolejo-Ramos, F., Gabriel, F., Medeiros, K., Marrone, R., Joksimovic, S., & de Laat, M. (2022). Human and artificial cognition. *Computers and Education: Artificial Intelligence*. 2666920X. *3*, 100107. doi:10.1016/j.caeai.2022.100107.

Singh, S. K., Sahu, N., Katiyar, P., & Arya, V. (2025). *Challenges, recent notions, and advances in the Indian traditional medicinal system and governance Integrated Pathy: Re-envisioning Healthcare and Wellness Using India as a Case Study* (pp. 241–291). India: Elsevier. Available from https://www.sciencedirect.com/book/9780443290404. doi:10.1016/B978-0-443-29040-4.00006-2.

Tran, T. T. V., Tayara, H., & Chong, K. T. (2023). Recent Studies of Artificial Intelligence on In Silico Drug Distribution Prediction. *International Journal of Molecular Sciences*. 14220067. *24*(3). Available from http://www.mdpi.com/journal/ijms. doi:10.3390/ijms24031815.

Tripathi, S., Augustin, A. I., Dunlop, A., Sukumaran, R., Dheer, S., Zavalny, A., … Kim, E. (2022). Recent advances and application of generative adversarial networks in drug discovery, development, and targeting. *Artificial Intelligence in the Life Sciences*, *2*, 100045. doi:10.1016/j.ailsci.2022.100045.

Vaishnav, Y., Verma, S., & Mishra, A. (2025). The Role of Structure-Activity Relationship (SAR). *Drug Discovery Indian Journal of Pharmaceutical Chemistry and Analytical Techniques (IJPCAT)*, 67–83.

Vendrapati, R. R., Suresh, S., Bhat, S., Bindu, K. H., Mahadevaswamy, S. R., Latheef, S. S. A., … Narayanam, S. (2024). Rabinarayan Acharya, A Comprehensive Review of Selected Traditional Medicinal Plants: status, Phytochemistry, Medicinal Properties, Cultivation, and Demand. *Pharmacognosy Reviews*. 09737847. *18* (35), 24–46. doi:10.5530/phrev.2024.18.4.

Vijayan, R. S., Kihlberg, J., Cross, J. B., & Poongavanam, V. (2022). Enhancing preclinical drug discovery with artificial intelligence. *Drug Discovery Today*, *27*(4), 967–984. doi:10.1016/j.drudis.2021.11.023.

Weiner, E. B., Dankwa-Mullan, I., Nelson, W. A., & Hassanpour, S. (2025). Ethical challenges and evolving strategies in the integration of artificial intelligence into clinical practice. *PLOS Digital Health*, *4*(4), e0000810. doi:10.1371/journal.pdig.0000810.

Wen, M., Chen, Q., Chen, W., Yang, J., Zhou, X., Zhang, C., … Wang, L. (2022). A comprehensive review of Rubia cordifolia L.: traditional uses, phytochemistry, pharmacological activities, and clinical applications. *Frontiers in Pharmacology*. 1663-9812. *13*. doi:10.3389/fphar.2022.965390.

Xu, Y., Pei, J., & Lai, L. (2017). Deep Learning Based Regression and Multiclass Models for Acute Oral Toxicity Prediction with Automatic Chemical Feature Extraction. *Journal of Chemical Information and Modeling*. 1549-9596. *57* (11), 2672–2685. doi:10.1021/acs.jcim.7b00244.

Zhang, J., Li, H., Zhang, Y., Huang, J., Ren, L., Zhang, C., … Zhang, Y. (2025). Computational toxicology in drug discovery: applications of artificial intelligence in ADMET and toxicity prediction. *Briefings in Bioinformatics*. 14774054. *26*(5). Available from http://bib.oxfordjournals.org. doi:10.1093/bib/bbaf533.

Zhao, L., Wang, J., Pang, L., Liu, Y., & Zhang, J. (2020). GANsDTA: predicting Drug-Target Binding Affinity Using GANs. *Frontiers in Genetics*. 1664-8021. *10*. doi:10.3389/fgene.2019.01243.

Zhou, S. F., & Zhong, W. Z. (2017). Drug design and discovery: principles and applications. *Molecules*. 14203049. *22*(2). Available from http://www.mdpi.com/1420-3049/22/2/279/pdf. doi:10.3390/molecules22020279.

Section VIII

Ethical considerations and the future of artificial intelligence

Chapter 33

Ethical frontiers: Navigating public health responsibilities in the age of artificial intelligence

Sudip Bhattacharya[a], Sandip Bhattacharya[b], and Biswadeep Dhar[c]
[a]*Department of Community and Family Medicine, All India Institute of Medical Sciences, Deoghar, Jharkhand, India*
[b]*Department of Electronics and Communication Engineering, SR University, Warangal, India*
[c]*Department of Health Sciences, College of Health and Behavioral Studies, James Madison University, Harrisonburg, VA, United States*

1 Introduction

1.1 How artificial intelligence is revolutionizing healthcare and diverse industries

Artificial intelligence (AI) is quickly becoming a transformative force in many aspects of daily life around the globe. It is making significant strides in processing data, aiding decision-making, and enhancing various applications across multiple industries. From household tasks to complex medical procedures, AI has proven to be an invaluable tool for both businesses and individuals. One of the most widely recognized examples of AI in everyday life is the use of virtual assistants like Amazon's Alexa, which help manage routine tasks such as setting reminders, playing music, and controlling smart home devices. These systems, powered by AI, learn from user interactions, continually refining their responses to become more efficient and accurate (Bhattacharya et al., 2021; Responsible Artificial Intelligence, n.d.).

As AI continues to evolve, experts predict that its influence will expand into technical and research fields, where robotics and machine learning are set to significantly influence and define the future. AI is a branch of computer science focused on creating systems that can simulate human intelligence. One of the most powerful aspects of AI is its ability to learn and adapt through repeated use. Over time, AI systems can enhance their performance and accuracy, making them highly effective in numerous applications, ranging from simple tasks to complex problem-solving scenarios. AI's capacity to learn from data enables it to recognize patterns, generate predictions, and carry out tasks with growing accuracy, driving transformation across sectors such as healthcare, education, finance, and transportation (Ethics guidelines for trustworthy AI | Shaping Europe's digital future, n.d.).

Healthcare stands out as one of the most promising and dynamic fields where AI is making a profound impact. The incorporation of AI technologies into this sector has the potential to transform patient care and medical research, delivering innovative solutions to some of the most critical health issues globally. One key example of how AI is transforming healthcare is its use in assisting elderly individuals in Japan. As the ageing population grows globally, the demand for caregiving solutions has increased. In response, AI-powered robots are being deployed to assist older adults in their daily lives. These robots help with tasks like reminding individuals to take medications on time, switching off lights before bed, and monitoring their overall well-being. This technology not only improves the quality of life for the elderly but also reduces the burden on caregivers, allowing them to focus on other critical aspects of care (Japanese Automakers Look to Robots to Aid the Elderly, n.d.).

While AI in healthcare has made significant strides in developed countries, developing nations are beginning to embrace these technologies as well. However, due to limited infrastructure and resources, the widespread adoption of AI-driven healthcare solutions is still in the early stages in many parts of the world. Nevertheless, as these countries gradually build the necessary frameworks, AI will likely become a critical tool in improving healthcare access and delivery, even in areas with limited resources (Bhattacharya, Saleem, & Hossain, 2022).

Applications of Artificial Intelligence in Healthcare. https://doi.org/10.1016/B978-0-443-43934-6.00041-9

One notable example of AI's contribution to medical advancements is in India, where AI is being utilized to detect diseases and improve treatment outcomes. *Niramai* Health Analytics, a company based in Bengaluru, has developed a ground-breaking AI-powered technology called *Thermalytix*. This technology uses thermal imaging and machine learning algorithms to detect breast cancer with remarkable precision. Unlike traditional methods such as mammography or clinical examinations, *Thermalytix* can detect breast cancer up to 5 years earlier. Early detection of cancer is crucial for successful treatment, and this technology is poised to save countless lives by enabling earlier diagnosis (Early breast cancer detection | NIRAMAI Health Analytix, n.d.; Thermalytix: AI-Powered Breast Cancer Screening Test, n.d.) (Fig. 33.1). This innovative application of AI in healthcare represents a significant leap forward in the fight against cancer, highlighting how AI can contribute to improving medical outcomes and advancing healthcare research. Another area in which AI is making an indelible impact is surgery. In India, advanced hospitals are now adopting robot-assisted surgery for a range of complex procedures, including urological and neurosurgeries (Alemzadeh et al., 2016). These surgeries require a high degree of precision and skill, and AI-powered robotic systems can help ensure the success of these procedures. For example, AI-driven surgical robots can perform delicate tasks, such as stitching tiny nerves or performing intricate tissue manipulations, which are often beyond the capabilities of human hands. This precision results in faster recovery times, fewer complications, and better overall patient outcomes. As these technologies continue to evolve, robot-assisted surgeries will likely become even more widespread, offering patients a safer and more effective way to undergo complex medical procedures (H et al., 2016).

Beyond healthcare, AI is driving remarkable changes across multiple industries by enhancing efficiency and improving outcomes. In the education sector, AI is being harnessed to create customized learning platforms that respond to each student's unique needs. These intelligent systems deliver instant feedback, recommend supplementary materials, and offer personalized strategies to boost academic performance. By accommodating diverse learning styles and abilities, AI helps

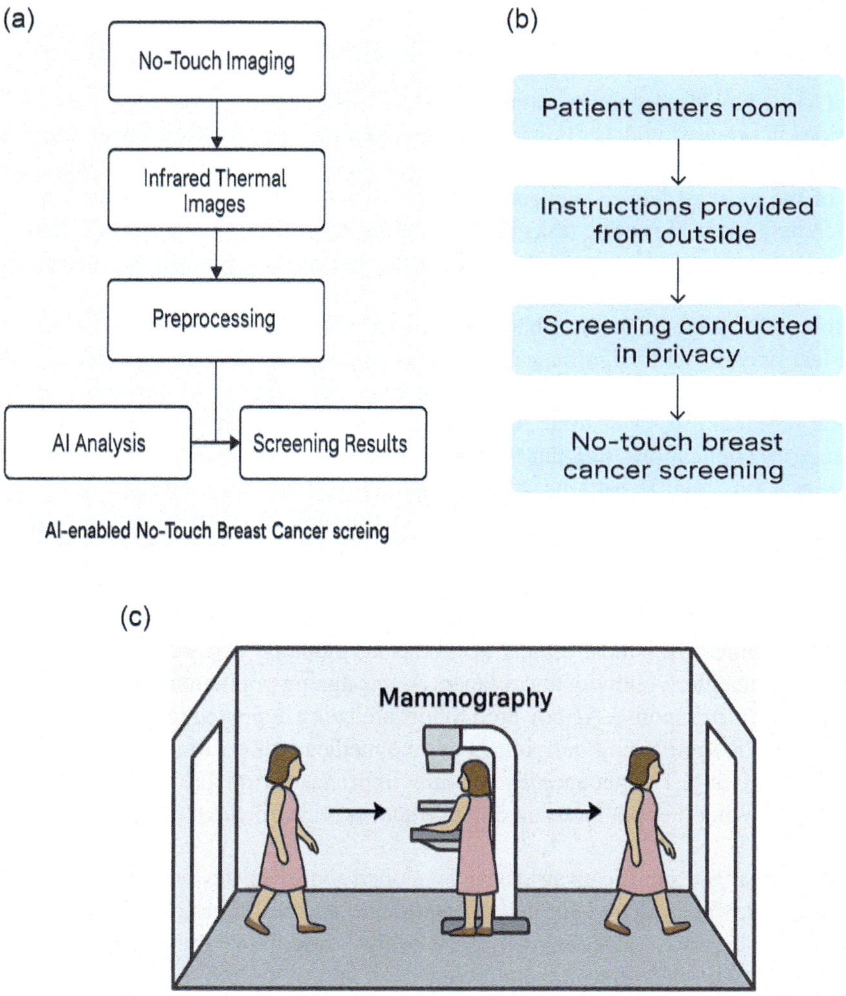

FIG. 33.1 (A) AI technological flowchart, (B) breast cancer imaging flowchart, and (C) the real-world AI-enabled mammography.

provide equitable support for every student. This transformative use of AI in education holds the promise of reshaping traditional teaching methods, fostering greater inclusivity, and making learning more accessible to all. In the finance industry, AI is being used to detect fraudulent transactions and help investors make informed decisions (Dave & Patel, 2023). Machine learning algorithms can analyze vast amounts of financial data to identify patterns and flag potential fraud, reducing the risk of financial crimes. In addition, AI-driven tools are being utilized by investors to assess market trends, make predictions, and optimize their investment portfolios. These tools provide valuable insights, enabling investors to make smarter decisions and minimize risks in an increasingly complex financial environment (Artificial intelligence in Finance: A comprehensive review through bibliometric and content analysis | SN Business & Economics, n.d.)

AI is also bringing significant transformation to the field of transportation. One of the most notable examples is the emergence of autonomous vehicles, which depend on AI-driven algorithms and sensor technologies to interpret their surroundings, avoid hazards, and make real-time driving decisions. As these systems advance, self-driving cars could dramatically reduce traffic accidents caused by human error, improve the flow of vehicles, and enhance overall road safety (Ushakov, Dudukalov, Shmatko, & Shatila, 2022). Beyond autonomous driving, AI is playing a key role in intelligent traffic management. By analyzing input from cameras, sensors, and other sources, AI-powered systems can regulate traffic signals, manage vehicle movement, and anticipate congestion, leading to smoother and more sustainable urban mobility. These innovations aim to make city transportation more efficient and livable. However, while AI's potential in transforming sectors like transportation and healthcare is immense, it also raises important concerns. In healthcare, for instance, one emerging challenge is the effect of AI on the quality of doctor-patient communication. As diagnostic tools and virtual assistants powered by AI become more common, there is a growing debate about how their use might alter the human interaction that is essential to effective and empathetic medical care (Sallstrom, Morris, & Mehta, 2019). These tools may affect the personal relationship between doctors and patients, as patients may begin to rely more on AI-driven systems rather than consulting healthcare professionals. The human element of care, including empathy, compassion, and emotional support, is critical to the healing process and cannot be replicated by AI. Therefore, healthcare professionals must remain at the forefront of patient care, using AI tools to enhance their decision-making without replacing the personal connections that are integral to effective healthcare delivery (Bhattacharya et al., 2021).

In addition, while AI holds great promise in improving healthcare outcomes, its adoption is not without challenges for developing nations. Many countries face significant barriers, such as limited access to advanced technology, inadequate infrastructure, and a lack of trained professionals who can implement AI solutions. As a result, AI's benefits might not reach everyone equally, possibly widening the gap between advanced and poor healthcare systems (Bhattacharya et al., 2022). Efforts should be made to ensure that AI technologies are accessible to all countries and that healthcare workers are trained to use these tools effectively. Bridging the gap between developed and developing healthcare systems will be key to ensuring that AI can be used to benefit everyone, regardless of geographic or economic barriers (Bhattacharya et al., 2021). AI is changing healthcare and other fields, from robotic surgeries to early disease detection and personalized learning. But it also brings challenges. There are worries about data privacy, patient safety, and how AI affects doctor-patient communication. Ethics is a big concern. If AI learns from biased data, it can make unfair decisions, which may lead to discrimination in medical care. AI also affects jobs by automating many tasks, raising fears about job loss. It could shift healthcare spending more toward machines than people. Another key issue is accountability. AI systems make decisions, but if an error occurs, who is responsible? If an AI-assisted surgery leads to complications, should the blame fall on the software developers, the hospital, or the surgeon who used the system? Ethical dilemmas like these make it necessary to establish clear guidelines for AI use in healthcare. These questions are not easy to answer. A good example is a Nigerian startup called *Safermom* (Sallstrom et al., 2019). This AI-driven platform helps pregnant women access health information. It provides alerts about danger signs in pregnancy and vaccination schedules. While this technology is useful, it also raises concerns. If the AI provides incorrect information and a complication occurs, who should be held responsible? The company that developed the software, the healthcare professionals who approved it, or the users who relied on it? Without proper accountability, AI errors could cause harm without clear consequences (Bhattacharya et al., 2021).

AI biases are another serious issue. One well-known example is facial recognition technology, which often fails to recognize African faces as accurately as Caucasian faces. This happens because the AI was trained using mostly Caucasian facial images. Similar biases exist in healthcare. For example, most heart attack research is based on male patients, even though symptoms can be different for women. If AI models rely on these incomplete datasets, they may not diagnose heart attacks in women correctly. This could lead to misdiagnosis and delayed treatment. AI must be trained with diverse and representative data to ensure fair and accurate healthcare decisions (Bhattacharya et al., 2021).

The potential risks of AI could reduce its benefits if not properly addressed. That is why AI must be carefully studied before being widely adopted in healthcare. As AI technologies advance, it is essential to conduct health impact assessments.

Policymakers, scientists, doctors, and the general public must work together to evaluate the risks and benefits of AI-based medical technologies. This requires a new way of thinking about how AI interacts with human health, ethics, and social values. To manage these risks, AI systems in healthcare need a strong accountability framework. This framework should follow ethical values and aim to benefit all of society. AI affects everyone, so it must be examined at a broader level, not just on an individual basis. AI should be evaluated in the context of the entire healthcare system, considering both technological advancements and human needs. AI expert Stuart Russell once said, "If we are developing machines to act with some autonomy, then we had better be quite sure that the purpose put into the machine is the purpose which we really desire." This statement highlights a fundamental concern: What does responsibility mean in AI? Who is responsible for AI's actions, and in what capacity? AI is a tool designed to improve human lives, but it cannot be held accountable like a human (Bhattacharya et al., 2021).

Some policymakers have suggested giving AI systems legal personhood. The European Parliament, for instance, has discussed whether AI medical systems should have legal status. However, this idea is not supported by scientific reasoning. Even if AI were considered a legal entity, many questions remain unanswered. If an AI system makes a mistake in diagnosing an accountable patient? Should responsibility lie with the algorithm's developer, the person who provided the medical data, the manufacturer of the sensors used to collect data, the government agency that approved the software, or the doctor who used the AI system? Finding the right way to assign responsibility is a complex challenge (Bhattacharya et al., 2021).

To address the challenges of AI in healthcare, strong governance is necessary. AI governance ensures that responsibility is distributed among all stakeholders. It must track accountability at different levels and roles. AI should focus on benefiting society and improving well-being. Lawmakers must understand AI's capabilities and limitations. This knowledge will help them tackle issues such as fairness, responsibility, and transparency. AI is a powerful tool, but humans must take responsibility for its effects, both positive and negative. AI in healthcare should respect human values. Ethical concerns must be addressed at every stage of AI development. Trust in AI will grow only if regulations ensure its safety and fairness. The responsibility for AI does not rest only with developers. Governments must take active steps to regulate AI in healthcare. AI should not replace healthcare workers. Instead, it should assist them in delivering better care. The goal of AI should be to improve daily life and patient outcomes. The focus should be on both technological advancements and human well-being (Responsible Artificial Intelligence, n.d.).

AI is a multidisciplinary field. Experts from medicine, law, economics, social sciences, and humanities must collaborate. The term "responsible AI" is often mistaken for ethics, but they are not the same. Ethics is about understanding morals and values, while responsibility is about applying these ethical principles in real life. It includes legal, economic, and social aspects. Ethical AI requires continuous observation and assessment, whereas responsible AI ensures AI systems follow ethical standards (Responsible Artificial Intelligence, n.d.).

AI systems are created to achieve specific goals. Therefore, ethical and legal factors must be considered from the beginning. There are three main ways to ensure AI ethics in healthcare (Fig. 33.2):

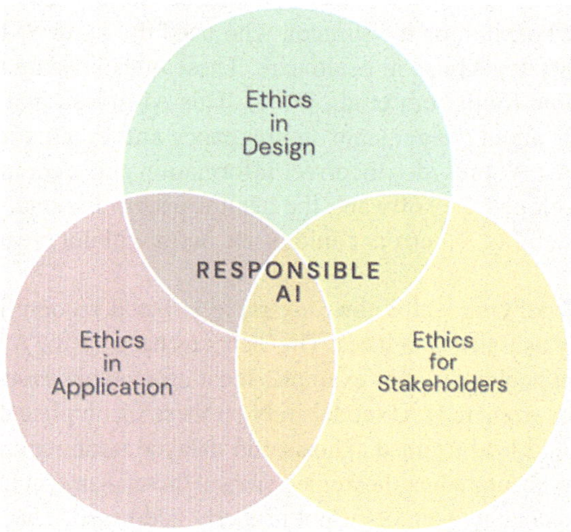

FIG. 33.2 Ensuring ethics in healthcare.

1. **Ethics in Design: AI must be designed fairly and take cultural differences into account. It should avoid bias or discrimination.** Fairness is very important when designing AI, so it does not exacerbate social inequality. For example, facial recognition often makes more mistakes with people of color than with white people because the data used to train it is biased. To mitigate this, designers must ensure diverse and representative datasets are used, accounting for factors like race, gender, and age. Cultural considerations should also be integrated into the AI's design, particularly in global applications. For example, AI-driven translation tools should not just translate text but also preserve cultural nuances and idiomatic expressions. A culturally sensitive approach prevents the unintentional promotion of stereotypes or offensive interpretations (Responsible Artificial Intelligence, n.d.).
2. **Ethics in Application: AI should be used responsibly. Developers and users must follow ethical guidelines.** AI technology, while powerful, must be applied responsibly to avoid harm. In healthcare, AI can help doctors diagnose diseases like cancer, but it must adhere to ethical rules to protect patient safety and privacy. If AI systems are used in critical decision-making without adequate oversight, there could be risks of errors or misuse. One application that showcases responsible AI use is the implementation of AI in autonomous vehicles, where strict regulations are enforced to ensure safety and prevent accidents. Users, such as developers and operators, must comply with these guidelines to minimize risks and ensure AI systems operate within ethical boundaries. Developers must also ensure transparency in how their AI models make decisions, especially when used in areas like criminal justice, where biases could potentially lead to wrongful convictions (Responsible Artificial Intelligence, n.d.).
3. **Ethics for Stakeholders: Clear rules should guide all involved with AI. This includes developers, users, and regulators. AI should be developed for the benefit of all, not just a few.** Clear ethical rules must apply to all stakeholders involved in the AI ecosystem, from developers to end users to regulators. For instance, consider the role of regulators in setting standards for AI's use in the finance sector. AI algorithms used in credit scoring systems must be transparent and fair, without discriminating against marginalized communities. Developers must prioritize inclusivity and equity in designing these systems, while users must ensure that they adhere to the rules for fairness and transparency. A concrete example is the creation of the European Union's General Data Protection Regulation (GDPR), which outlines the ethical use of personal data and mandates that AI systems used in applications such as healthcare or employment must operate in ways that protect privacy and human rights. In addition, the development of AI for social good, such as AI-driven systems to monitor environmental changes or address climate change, should prioritize collective benefits over profit motives, ensuring that advancements in AI contribute positively to society as a whole (Responsible Artificial Intelligence, n.d.).

Many organizations are already creating AI regulations. The European Commission developed the "(*8 Ethics guidelines for trustworthy AI | Shaping Europe's digital future*, no date)." The US Food and Drug Administration is also working on AI regulations. These guidelines are necessary to keep AI safe and effective in healthcare. Singapore has developed an AI ethics framework. This framework highlights challenges like data privacy and human rights. It states that AI should match societal values and beliefs. It also emphasizes that AI risks affect entire communities, not just individuals. Dubai created the "Ethical AI Toolkit" to guide AI users and developers. This toolkit explains how AI makes decisions. It also suggests ways to reduce bias in AI systems (Bhattacharya et al., 2021; Ushakov et al., 2022). However, it does not fully explain how to apply these ethical rules. AI is now a global concern. Many international organizations are working on AI policies. The Organization for Economic Co-operation and Development has set AI guidelines. The International Telecommunication Union (ITU) is organizing the "AI for Global Good Summit." The World Intellectual Property Organization has released a major report on AI. The World Health Organization is forming an AI focus group with ITU and the International Labour Organization. These global efforts are crucial. They help create fair AI standards. They guide governments in making ethical AI policies. AI has great potential, but strong governance is needed. This will ensure AI is safe, fair, and beneficial for healthcare and society. AI must be developed with care. It should respect human rights and values. It should be transparent and accountable. AI should not harm people or create inequality. Instead, it should improve lives and support healthcare workers. Responsible AI can help create a healthier and fairer world. But to achieve this, all stakeholders must work together. Governments, scientists, healthcare workers, and the public must shape AI policies. Only then can AI truly serve humanity (Bhattacharya et al., 2021).

2 Expanding AI governance

AI is growing fast and changing many fields, such as healthcare, finance, education, and public services. It can make work faster and help solve problems, but it also brings significant challenges, such as ethical and legal issues. A key concern is AI governance, which means having rules and checks to ensure AI is used in a fair and safe way. This is especially important in healthcare, where AI is having a major impact (Bhattacharya et al., 2021).

2.1 The need for clear AI governance frameworks

AI governance is a multifaceted approach that involves the establishment of clear and enforceable rules to guide the ethical use of AI across various sectors. These rules must consider the potential risks associated with AI and aim to mitigate them, while also promoting the development of AI technologies that can benefit society as a whole. Governments, regulatory bodies, and organizations must work together to create a framework that outlines how AI should be designed, implemented, and monitored. Governments, as well as private and public organizations, must play an active role in developing these rules. A strong governance framework ensures that AI does not become a tool for exploitation, bias, or discrimination. Instead, it should be a resource for creating inclusive, fair, and equitable systems. The challenges of AI governance are particularly evident in the healthcare sector, where the stakes are incredibly high and ethical concerns are paramount (Bhattacharya et al., 2021).

2.2 Ethics in healthcare AI design and application

AI is increasingly being used in healthcare, from diagnostics and treatment recommendations to administrative tasks and patient management. The role of AI in healthcare is particularly critical due to its direct impact on human lives. It is essential that AI in healthcare be held to the highest standards of ethical responsibility. For example, AI-based diagnostic tools, such as those used in radiology to detect tumors or other abnormalities, must be rigorously tested and validated to ensure they meet safety and accuracy requirements. Any failure or error in these systems could result in misdiagnosis, delayed treatments, or unnecessary procedures that put patients at risk. Moreover, healthcare AI systems must be transparent, meaning that users—whether medical professionals, patients, or healthcare administrators—should have a clear understanding of how AI tools reach decisions. This transparency is essential to building trust and accountability in the use of AI. For instance, if an AI system suggests a certain treatment or diagnostic pathway, medical professionals should be able to understand the rationale behind that decision, ensuring that they can confidently incorporate AI recommendations into their clinical practice (Bhattacharya et al., 2021).

2.3 Ensuring the ethical use of AI in health

A key part of ethical AI governance in healthcare is tackling algorithmic bias. AI systems learn from the data they are given, so if that data is biased or incomplete, the results can be unfair. For instance, if an AI tool is trained mostly on data from one group, it may not work well for others, leading to incorrect diagnoses and unequal care. This can worsen health inequalities. To avoid this, AI developers should use training data that includes a wide range of people to ensure the system works fairly for everyone it serves. For example, healthcare AI systems should include data from various racial, ethnic, and socioeconomic groups to avoid biases related to gender, age, or other factors. In addition, AI models should undergo regular audits to check for biases or disparities in performance. This ongoing evaluation ensures that AI systems continue to provide accurate and fair results for all patients, regardless of their background or circumstances. AI models must also be held to higher standards in healthcare than in other sectors due to the life-and-death nature of medical decisions. Regulatory bodies should implement guidelines that ensure AI's role in diagnosis, treatment, and patient data management is closely monitored. Any AI tool that interacts with healthcare data or contributes to patient care must undergo rigorous evaluation and regulatory oversight before it can be deployed in real-world settings (Bhattacharya et al., 2021).

2.4 The importance of data privacy in AI systems

AI technologies rely on vast amounts of data to function effectively. In healthcare, this data can include patient records, medical images, diagnostic results, and other personal health information. Keeping this data safe is very important. If private medical information is leaked or accessed without permission, it can seriously harm both patients and healthcare systems. Governments and organizations must implement strict data protection laws to ensure that patient information is kept secure and private. The GDPR in Europe is one of the most comprehensive frameworks for protecting personal data and provides a strong model for how data privacy can be ensured in AI applications. GDPR mandates that personal data should be processed lawfully, transparently, and securely, with individuals having the right to access and control their data (General Data Protection Regulation (GDPR)Legal Text, n.d.). These principles should be adopted universally, especially in sectors like healthcare, where the stakes are particularly high.

AI developers must also implement robust encryption and cybersecurity measures to protect patient data from unauthorised access. For example, medical AI systems should use strong encryption protocols to safeguard patient information during transmission and storage. Hospitals and healthcare institutions must ensure that their AI systems are secure from cyber threats, which could otherwise compromise sensitive health information. Furthermore, healthcare organisations must anonymize patient data where possible. Anonymization allows AI systems to analyze and learn from data without compromising patient privacy. This is very important in research, as AI can find patterns in health data while keeping people's identities private.

2.5 AI and human oversight in healthcare

While AI systems have the potential to revolutionize healthcare by providing faster, more accurate diagnoses and improving patient outcomes, human oversight remains crucial. AI should never replace medical professionals but rather support and enhance their work. Medical professionals are responsible for making critical decisions about patient care, and they must always have the final say, especially when AI recommendations conflict with clinical judgment.

For example, AI-powered systems can analyze medical scans much faster than doctors, but human experts must confirm the findings and make the final diagnosis. AI can flag potential issues in scans or medical records, but healthcare professionals should be responsible for interpreting those results and making treatment decisions. Training healthcare professionals to use AI responsibly is a key aspect of ensuring that AI supports, rather than undermines, the quality of care. Medical schools and healthcare institutions should incorporate AI education into their training curricula. Doctors, nurses, and other healthcare workers should be taught not only how to use AI tools but also the ethical implications of AI decision-making. They should understand AI's limitations, recognize potential biases in the data, and know when manual intervention is necessary (Bhattacharya et al., 2022).

2.6 AI's impact on healthcare jobs

AI in healthcare has raised concerns about job loss. Some people worry that AI might replace workers, especially in administrative or diagnostic jobs. However, AI should assist, not replace, healthcare staff. It can handle routine tasks like booking appointments, managing records, or answering basic questions. This saves time, allowing healthcare workers to focus more on patients. For example, AI chatbots can answer common health questions, and automated systems can assist with patient intake and insurance claims. This enables doctors and nurses to dedicate more time to caring for patients. However, human care remains very important. AI cannot provide the empathy or emotional support that patients need. Doctors and nurses also play a crucial role in helping patients feel better emotionally, which AI cannot replicate.

So, the goal is to use AI to support, not replace, healthcare workers. To make this happen, healthcare staff need training. Governments and hospitals should offer programs to teach how AI works and how it can help. With the right training, healthcare workers can use AI tools effectively and provide better care (Bhattacharya et al., 2022;Bhattacharya S., 2022).

2.7 AI's role in medical research

AI is transforming medical research, helping scientists and researchers process vast amounts of data and uncover insights that would otherwise take years to discover. AI's ability to analyze millions of scientific papers, clinical trial results, and medical records in a fraction of the time it would take a human researcher is accelerating the pace of innovation. In drug development, AI can simulate the effects of potential treatments, predict how drugs will interact in the human body, and identify new therapeutic targets, reducing the time and cost associated with developing new drugs.

AI has already made significant strides in cancer research. AI-powered tools can analyze medical scans to detect tumors with incredible accuracy, often identifying issues that may not be visible to the human eye. These tools have the potential to revolutionize early cancer detection, allowing for earlier and more effective interventions (Early breast cancer detection | NIRAMAI Health Analytix, n.d.) Furthermore, AI plays a critical role in personalized medicine, which tailors treatment plans to an individual's unique genetic makeup. By analyzing genetic data, AI can help doctors choose the best course of treatment for each patient, resulting in more effective therapies and better outcomes.

2.8 Ethical challenges in AI medicine

As AI becomes more integrated into healthcare, a number of ethical challenges must be addressed. One of the most pressing issues is accountability. If an AI system makes a mistake, such as providing a wrong diagnosis or treatment recommendation, who is responsible? Should the developers of the AI, the healthcare providers using it, or the AI itself be held accountable? In addition, the issue of AI having legal rights is a complex and controversial topic. Should an AI system be treated as a legal entity in its own right, or should it be considered merely a tool controlled by human operators? These are questions that will require new legal frameworks to address. It is essential that AI systems used in healthcare are programmed to prioritize patient well-being. If an AI system detects a rare disease or suggests an unconventional treatment, it must explain its reasoning clearly to healthcare professionals and patients. Transparency is critical in building trust in AI systems and ensuring that they are used ethically and responsibly. Ultimately, AI should not make final decisions in critical medical cases. Human professionals must always have the final say, especially when the stakes are high. AI can support medical professionals, but the responsibility for patient care should remain in human hands (Bhattacharya, Sharma, & Singh, 2019).

2.9 The future of responsible AI in health

The future of AI in healthcare looks very promising. However, using it effectively requires proper planning and regulations. Governments should ensure that AI is safe and fair while also supporting innovation and new technologies. AI companies and developers must adhere to ethical standards and rigorously test their systems before deployment. Transparency, fairness, and accountability should be at the core of AI development. Public education is also crucial. Many people harbor fears about AI, often due to misinformation or misunderstandings. Clear communication about AI's benefits and risks can help build trust and acceptance. AI should be inclusive, ensuring that it is accessible to all populations and that its benefits extend to everyone, regardless of socioeconomic status, race, or geographic location. In conclusion, expanding AI governance is essential to ensuring that AI technologies are developed and used responsibly, particularly in the healthcare sector. By focusing on ethics, transparency, data privacy, and human oversight, we can harness the full potential of AI to improve healthcare outcomes and create a fairer, more equitable society (Ethics guidelines for trustworthy AI | Shaping Europe's digital future, n.d.; Bhattacharya et al., 2022; Jiang et al., 2017)

3 Conclusion

AI has the potential to revolutionize healthcare. It can improve diagnosis, treatment, and patient management. However, ethical and governance challenges must be addressed. AI must respect human rights, data privacy, and accountability. International organizations and governments must collaborate. They should create ethical AI policies and regulatory frameworks. AI should not replace healthcare workers but support them. Human oversight is essential in AI-driven medicine. AI's success depends on responsible development and implementation. When used ethically and wisely, AI can enhance global healthcare and benefit humanity. With responsible AI, the future of medicine can be both innovative and ethical.

References

Alemzadeh, H., Raman, J., Leveson, N., Kalbarczyk, Z., Iyer, R.K. (2016). Adverse events in robotic surgery: A retrospective study of 14 years of FDA Data. *PLOS ONE*, *11*(4), e0151470. doi: https://doi.org/10.1371/journal.pone.0151470.

Artificial intelligence in Finance: A comprehensive review through bibliometric and content analysis | SN Business & Economics. (n.d.). Available from: https://link.springer.com/article/10.1007/s43546-023-00618-x.

Bhattacharya, S. (2022). Artificial intelligence, human intelligence, and the future of public health. *AIMS Public Health*, *9*(4), 644–650. doi:10.3934/publichealth.2022045.

Bhattacharya, S., Hossain, M. M., Juyal, R., Sharma, N., Pradhan, K. B., & Singh, A. (2021). Role of public health ethics for responsible use of artificial intelligence technologies. *Indian Journal of Community Medicine*. 19983581. *46*(2), 178–181. Available from: http://www.ijcm.org.in. doi:10.4103/ijcm.IJCM_62_20.

Bhattacharya, S., Saleem, S. M., & Hossain, M. M. (2022). Implementing National Digital Health Blueprint in India-The future opportunities and challenges. *International Journal of Preventive Medicine*. 20088213. *13*(1), 99. Available from: http://www.ijpvmjournal.net/browse.asp. doi:10.4103/ijpvm.ijpvm-418-20.

Bhattacharya, S., Sharma, N., & Singh, A. (2019). Designing culturally acceptable screening for breast cancer through artificial intelligence-two case studies. *Journal of Family Medicine and Primary Care*, *8*(2), 760–762. doi:https://doi.org/10.4103/jfmpc.jfmpc_391_18.

Dave, M., & Patel, N. (2023). Artificial intelligence in healthcare and education. *British Dental Journal*. 0007-0610. *234*(10), 761–764. doi:10.1038/s41415-023-5845-2.

Early breast cancer detection | NIRAMAI Health Analytix. (n.d.). Available from: https://niramai.com/.

Ethics guidelines for trustworthy AI | Shaping Europe's digital future. (n.d.). Available from: https://digital-strategy.ec.europa.eu/en/library/ethics-guidelines-trustworthy-ai.

General Data Protection Regulation (GDPR) – Legal Text. (n.d.). General Data Protection Regulation (GDPR). Retrieved July 23, 2025, from: https://gdpr-info.eu/.

Japanese Automakers Look to Robots to Aid the Elderly. (n.d.). Scientific American. Retrieved July 23, 2025, from: https://www.scientificamerican.com/article/japanese-automakers-look-to-robots-to-aid-the-elderly/.

Jiang, F., Jiang, Y., Zhi, H., Dong, Y., Hao, L., Sufeng, M., … Wang, Y. (2017). Artificial intelligence in healthcare: past, present and future. *Stroke and Vascular Neurology*. 2059-8688. *2*(4), 230–243. doi:10.1136/svn-2017-000101.

Responsible Artificial Intelligence. (n.d.). Retrieved July 23, 2025, from: https://link.springer.com/book/10.1007/978-3-030-30371-6.

Sallstrom, L., Morris, O., & Mehta, H., Artificial Intelligence in Africa's Healthcare: ethical Considerations. (2019), 312.

Thermalytix: AI-Powered Breast Cancer Screening Test. (n.d.). Retrieved July 23, 2025, from: https://niramai.com/about/thermalytix/.

Ushakov, D., Dudukalov, E., Shmatko, L., & Shatila, K. (2022). Artificial Intelligence as a factor of public transportations system development. *Transportation Research Procedia*. 23521465. *63*, 2401–2408. doi:10.1016/j.trpro.2022.06.276.

Chapter 34

Ethical issues of artificial intelligence in biomedical applications

Sudipta Chakraborty
Department of Pharmacognosy, B.C.D.A. College of Pharmacy & Technology, Barasat, Kolkata, India

1 Introduction

The simulation of natural intelligence in computers, designed to investigate, acquire knowledge, and resolve various issues, is known as artificial intelligence (AI). It enables computers and software to perform operations such as sensing, reasoning, decision-making, and language comprehension, which typically require human intelligence.

Although the philosophical foundations of AI date back to long ago, the field only truly came into existence during the middle of the 20th century, most famously when Alan Turing proposed the "Turing test" in 1950 to assess machine intelligence (Nagendraswamy and Amogh, 2021; Turing, 1950; Wang, 2019).

At the Dartmouth Conference in 1956, the phrase "Artificial Intelligence" was first used, signifying the start of AI as a discipline of scientific investigation (McCarthy, Minsky, Rochester, and Shannon, 1956).

This field of study has gone through "AI winters," cycles of hope and disappointment over the years, before catching up pace in the 2010s thanks to developments in machine learning, neural networks, and processing power. Usually, this system uses techniques like supervised and unsupervised learning, deep learning, and natural language processing (NLP) to process massive amounts of data (Goodfellow, Bengio, and Courville, 2016; Sarikaya, 2024; Toosi, Bottino, Saboury, Siegel, and Rahmim, 2021).

Today, almost all industries are utilizing AI. Autonomous cars, medical diagnostics, engineered suggestions, virtual assistants like Siri and Alexa, and generative AI tools for content production and design are just a few examples worth mentioning. AI is changing society's perception of technology applications, presenting both revolutionary possibilities and ethical questions (Hajkowicz, Sanderson, Karimi, Bratanova, and Naughtin, 2023).

2 Progress timeline of AI

From theoretical concepts to real-world applications, AI history is packed with both breakthroughs and disasters. The emphasis on rule-based systems has given way to data-driven machine learning, deep learning, and, currently, generative AI.

2.1 Foundational concepts (1940s–1950s)

In his 1950 work entitled "Computing Machinery and Intelligence," Alan Turing raised the question, "Can machines think?" To determine whether a machine could display intellectual behavior indistinguishable from that of a person, he developed the well-known Turing test (Turing, 1950).

The Dartmouth Summer Research Project on Artificial Intelligence marked the official birth of the discipline. This meeting, organized by John McCarthy, brought together top scholars and established AI as a legitimate academic field. The phrase "artificial intelligence" was coined by McCarthy for the workshop proposal. Several influential individuals attended this conference, and a small number of them later developed some of the earliest AI applications as well (McCarthy et al., 1956; Simon & Newell, 1958).

Applications of Artificial Intelligence in Healthcare. https://doi.org/10.1016/B978-0-443-43934-6.00037-7

In 1957 psychologist Frank Rosenblatt developed the Perceptron, a prototype artificial neural network that could identify patterns. It was a major step toward machine learning, despite its simplicity and limitations (Rosenblatt, 1958).

2.2 The rise and fall of systems (1960s–1980s)

The development of remarkable yet fragile AI systems defined this era. In 1966, Joseph Weizenbaum created ELIZA, an algorithm that used basic pattern-matching to mimic a psychiatrist. It exhibited the capabilities of NLP and was among the earliest chatbots (Weizenbaum, 1966).

In 1969 scientists at Stanford Research Institute built Shakey, a mobile robot that could see its surroundings, make sense of its actions, and perform tasks. This was one of the earliest instances of an intelligent agent (Nilsson, 1984).

Before the 1980s, government sponsorship and general participation in AI research drastically decreased as a result of a combination of exaggerated claims, a lack of processing capacity, and an assessment of AI's shortcomings. The very first "AI winter" was the name given to this period of inaction (Lighthill, 1973; Toosi et al., 2021).

With the creation of Expert Systems, which were knowledge-dependent systems intended to replicate human judgment in a particular field such as financial planning or medical diagnosis, AI saw an upward trajectory for over a decade. However, because these systems were challenging to implement and maintain, funding was cut for a second time (Garvey, 2019; Morgan, Feigenbaum, and McCorduck, 1983).

2.3 The machine learning era (1990s–2010s)

This phase experienced a shift from symbolic, rule-based AI to data-oriented statistical methods. In 1986 the traditional backpropagation algorithm gained popularity after it was demonstrated to be successful in training multilayered neural networks, signifying an important step in the resurgence of neural network technology (Rumelhart, Hinton, and Williams, 1986; Whittington and Bogacz, 2017).

When IBM's chess-playing supercomputer, Deep Blue, surpassed world champion Garry Kasparov in 1997, it was a stunning accomplishment that thrilled the public and demonstrated the strength of AI in a particularly challenging field (Campbell, Hoane, and Hsu, 2002).

Watson, developed by IBM, dominated a popular American game program by comprehending and processing human language, marking a significant advancement in NLP (Ferrucci et al., 2011).

Deep learning, a branch of machine learning that employs the use of multilayered deep neural networks, gained popularity in the 2010s primarily due to the accessibility of large datasets. Advances in image recognition and other fields demonstrated the effectiveness of this strategy (Krizhevsky et al., 2012).

2.4 The generative AI era (2020s–present)

The ongoing era is defined by the extensive applications of large language models (LLMs) and generative AI. The Transformer design was introduced by researchers from Google in the 2017 publication "Attention is All You Need." LLMs were built on the basis of this architecture, which demonstrated remarkable efficacy for sequential data (Vaswani et al., 2017).

The Generative Pre-trained Transformer 3, popularly known as GPT-3, from OpenAI, was released in 2020, demonstrating how a single, large-scale model could handle a wide range of NLP tasks such as text creation, summarization, and translation (Brown et al., 2020).

With ChatGPT's public launch in 2022, generative AI gained widespread recognition for its interactive features and capacity to produce text of human caliber in response to a variety of inputs. Research is still being conducted at a rapid pace in fields like AI agents, which can carry out difficult tasks on their own, and multimodal models, which can interpret and generate many types of input such as text, visuals, and audio (Naznin, Al Mahmud, Nguyen, and Chua, 2025) (Fig. 34.1).

3 Ethical theories of AI

The way decisions are made in governance, regulation, healthcare, and other sectors is changing as a result of the inclusion of AI in these areas. However, at the same time, its implementation presents moral questions about responsibility, confidentiality, impartiality, and human welfare. Researchers frequently use well-established ethical theories as foundational principles for addressing these problems. Some of these theories are comparatively discussed in the following tabular form (Table 34.1).

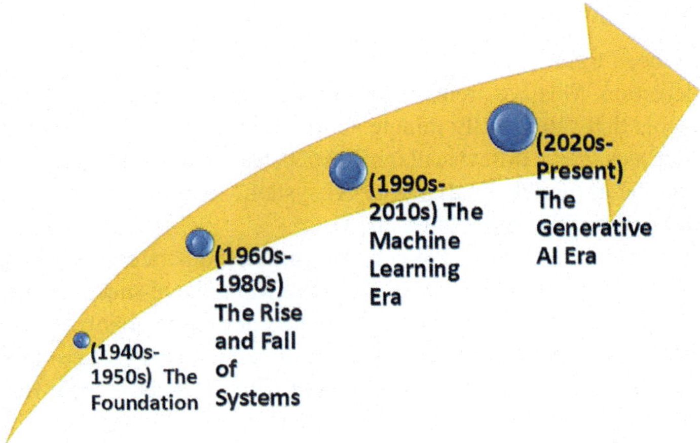

FIG. 34.1 Timeline of the progress of AI.

TABLE 34.1 Comparative study of different ethical theories of AI.

Ethical theory	Principle	Application to AI	Strengths	Limitations
Deontological Ethics (Kantian) (Kant, 1993)	Human dignity is respected	Prioritize privacy, transparency, and informed consent over their efficiency gains	Clear rules with the protection of fundamental principles	Rigid and beneficial outcomes are ignored in a few cases
Utilitarianism (Consequentialism) (Mill, 1998; Bostrom and Yudkowsky, 2014)	Maximize overall benefits while minimizing losses	Health outcomes are prioritized, such as sorting individuals to increase survival rates	Real-life population-based benefits are the focus	Minority rights may be compromised
Virtue Ethics (Hayes et al. 2025)	Moral character, virtues are valued such as honesty and responsibility	Integrity, humility, and fairness are key in the AI model	Highlights responsible culture and enduring faith	Defining rules are absent
Justice and Fairness (Rawlsian) (Fried and Rawls, 1972)	Balance of benefits and burdens	Bias is strictly avoided, ensuring equal access to healthcare and jobs	Equity is valued and protects vulnerable groups	Parameters of fairness, equity, etc., are unclear and conflicting
Rights-Based Ethics (Donnelly and Gewirth, 1983; *Proposal for a regulation laying down harmonised rules on artificial intelligence (Artificial Intelligence Act)*, 2021)	Protects fundamental rights	Privacy and autonomy are protected	Individual freedom is encouraged	Conflicting with varied rights
Care Ethics (Johnson and Gilligan, 1983)	Kindness, Relationships are valued	Customized AI in healthcare with stronger bonds between patients and professionals	Maximizes empathy and inclusivity	May fail in large-scale policy decisions
Principlism (Bioethics) (Beauchamp and Childress, 2013)	Liberty, Philanthropy, and justice	Extensively utilized to guarantee its impartial, secure, and patient-supporting model in biomedical AI	Practically feasible and balanced for healthcare professionals	Wide application beyond the biomedical field is compromised

4 AI in healthcare system

By evaluating enormous volumes of complex data, AI is revolutionizing medical applications to enhance everything from assessment to medication. More effective and individualized medical treatment is being provided as a result of its capacity to identify trends and forecast outcomes from medical data (Pei and Yang, 2025).

4.1 Medical imaging and diagnosis

Medical image processing has been transformed by AI, particularly deep learning, which enables precise identification, segmentation, and anomaly detection. Massive archives containing clinical images, such as X-rays, MRIs, and CT scans, are used to train AI-powered tools that can identify minute abnormalities that human observers might overlook. Diseases, including cancer, neurological issues, and cardiovascular ailments, can be accurately and early diagnosed because of this. For example, mammograms can be accurately analyzed by AI systems to detect breast cancer symptoms (Sadeghi et al., 2024; Tripathi, Hajra, Mulukutla, Shreshtha, and Maity, 2025).

In MRI, CT, and histology, research of deep neural networks has demonstrated remarkable proficiency in the segmentation of tumors and organs. From retinal fundus images, these networks have successfully detected diabetic retinopathy (DR) and hemorrhages, as well as accurately defining brain tumors and heart abnormalities. CT, MRI, PET, and optical imaging have all improved due to deep tomographic reconstruction, which also reduces radiation exposure and produces excellent outcomes from poor or inaccurate information (Cao et al., 2018; Graas, Coban, Batenburg, and Lucka, 2023).

4.2 Drug discovery and development

The conventional method of creating new medications is quite costly and time-consuming. AI is speeding this up by evaluating enormous amounts of biological and chemical data to find possible medication alternatives, forecast their potential hazards and effectiveness, and even create new compounds. This can drastically cut down on the time and expense involved in introducing new medications to the marketplace (Gromiha, Preethi, and Pandey, 2024).

Today, AI is used in in silico medicine to find novel therapeutic targets and generate treatments. In 2023 one such AI-designed medication began mid-stage human trials. Nearly thirty years later, deep learning has discovered some novel antibiotics, such as abaucin and halicin. The dangerous *Acinetobacter baumannii* is the specific target of these antibiotics (Graas et al., 2023; Stokes, Collins, and Barzilay, 2023).

AI additionally facilitates drug repurposing to integrate into the system. Recent studies demonstrate a significant improvement in Parkinson's disease and COPD (Chronic obstructive pulmonary disease) screening, which also saves both time and money (Bardin, Thomas, and Bourke, 2025).

AI is revolutionizing pharmacy activities, including generative molecule design, virtual screening, undesirable drug response prediction, formulation, delivery systems, and optimization of compliance with medications (Simpson and Qasim, 2025).

4.3 Personalized medicine

AI generates highly customized treatment regimens based on a patient's individual data, such as genetic information, medical records, and lifestyle patterns. By selecting the best treatments for each patient, this method—known as precision medicine—improves outcomes and reduces unfavorable side effects. For instance, by analyzing the genetic composition of a patient's tumor, AI can assist in identifying the chemotherapy regimen that has the highest likelihood of success for that patient. Clinical, genetic, and immunological parameters are integrated by AI models to facilitate individualized therapy choices (Chen, Hsiao, Lin, and Fann, 2025).

AI is also used in pharmacogenomics and customized therapies, where it facilitates medication performance prediction, patient grouping, and noncoding variation analysis (Kalinin et al., 2018).

4.4 Robotics in surgery

Medical professionals are assisted in carrying out intricate and minimally invasive treatments using robotic systems driven by AI. Improved navigation, comprehension, and consistency from these robots can result in enhanced patient experiences, reduced blood loss, and quicker recovery periods. To improve over time, the AI in these types of devices can also learn from prior surgeries. For patients with chronic illnesses, Mabu, an AI robot, offers individualized daily assistance with activities like mood monitoring and medication reminders (Li J et al., 2024; Li S et al., 2024).

4.5 Genomics and proteomics

AI facilitates extensive assessment of molecular data. It can forecast protein structures, recognize genetic changes connected to illnesses, and identify complex chemical reactions. This is essential for studying hereditary illnesses and developing novel gene therapies (Gromiha et al., 2024; Naskar et al., 2025).

Protein structure and function: AlphaFold from DeepMind achieved significant advances in its forecasting of protein structures. The pathogenicity of mutations, essential for diagnosing uncommon diseases, is now predicted using their Alpha Missense model (Abramson et al., 2024).

AI predicts interactions between proteins, drug–target outcomes, and molecular activities, in addition to handling genetics, transcriptomics, proteomics, and other fields (Zemouri, Zerhouni, and Racoceanu, 2019).

4.6 Biomedical devices and additive manufacturing

AI enhances the development and performance of biomaterials and devices for medical use. Machine learning accelerates the fabrication of implants and drug-delivery materials by predicting biocompatibility, material degradation, and operational specifications (Gokcekuyu et al., 2024).

It improves many printing processes, from layout and fabrication to quality control, making it perfect for biomedical devices, prosthetic parts, and tissue-engineered structures (Mohammadnabi et al., 2025).

4.7 Signal processing, wearables, and biosensors

AI elevates the understanding of physiological observations to an entirely different level. To identify atrial fibrillation, algorithms using deep learning interpret ECG sequences, allowing doctors to evaluate the results at the feature level (Abu-El-Ruz et al., 2025).

AI-based prosthetics and rehabilitation aids, such as augmented reality, 3D-printed limbs, and intelligent treatment options, are being developed in crisis zones to support trauma recovery and PTSD (Post traumatic stress disorder) treatment (Katzman, Kalman, Almog, and Fogel, 2024).

Furthermore, with 96% of cardiologists claiming that a deep learning approach can accurately recreate entire ECG leads from sparse portable information, these developments improve rapid diagnostics and remote surveillance (Pei & Yang, 2025).

4.8 NLP in healthcare

Systems based on AI can now access medical information and expertise more easily through data mining, along with additional techniques. Unorganized healthcare data can be transformed with the aid of AI's language processing, making it possible to extract categorized information from documents, medical records, and electronic health records (EHRs) to support treatment strategies and diagnoses (Tettey-Engmann, Parupelli, Bauer, Bhattarai, and Desai, 2025).

4.9 Explainable AI and model transparency

Particularly in the medical field, assurance regarding AI systems is essential. In scanning and signals, explainable AI (XAI) seeks to produce comprehensible results that describe how algorithms arrive at diagnoses by utilizing imaging or signal characteristics (Nazir, Dickson, and Akram, 2023).

4.10 Clinical trials, epidemiology, and public health

Due to massive financial investments, AI is now playing an increasingly significant role in mainstream medical studies and healthcare. In an effort to find new treatments, Bill Gates established the Alzheimer's Insights AI Prize to encourage AI tools that can generate hypotheses on their own from large datasets. This also illustrates how AI is being used more and more to combat complicated ailments through international data cooperation (Song, Li, Lin, Liu, & Zhang, 2025).

5 Merits and demerits of AI in healthcare

By increasing efficiency, accuracy, and customization, AI is a game-changing technology that is constantly redefining economies, industries, and our everyday lives. One of AI's most noteworthy benefits is its capacity to automate routine and repetitive tasks, which increases operational efficiency and reduces human error. AI-powered solutions, for instance, can analyze data at a pace and scale far greater than human capacity or manage interpersonal interactions through chatbots, boosting productivity in sectors like banking, manufacturing, and shipping (Chustecki, 2024; Zakaria and Hussein, 2025).

AI's advancement in precision and reliability is another noteworthy benefit, especially in industries like healthcare, where data-driven models help diagnose ailments such as cancer more quickly and accurately than conventional techniques (Olawade et al., 2024).

In addition, robotic surgery uses AI algorithms to reduce risks and enhance surgical results (Wah, 2025).

Furthermore, in contrast to human workers, AI systems can work around the clock without getting tired, which is particularly useful in various industries like customer care, security surveillance, and transportation that need to function continuously (Cohen, Amorim, and Reis, 2022).

Moreover, AI supports data-driven decision-making by analyzing large datasets to extract patterns and insights. Businesses and governments utilize AI to forecast trends, optimize logistics, detect fraud, and personalize user experiences. For instance, recommendation systems used by companies like Amazon and Netflix analyze user preferences to suggest products or content, significantly enhancing user engagement (Juyal et al., 2025; Valencia-Arias et al., 2024).

AI additionally assists in lowering risk, particularly in dangerous situations. AI-driven robots are utilized in mining, space exploration, and emergency response to carry out jobs that individuals would find too hazardous (nasscom, 2023).

In addition, generative frameworks like GPT and DALL · E, which can produce anything from text to graphics and music, extend the limits of human–machine alliance and promote creative thinking and innovation (Zhao et al., 2025).

While automated drones and sensors facilitate precise cultivation in agriculture, increasing crop yields and managing assets, AI in academia enhances educational outcomes and assists educators in identifying pupils who might require greater assistance (Agrawal and Arafat, 2024; Bobula, 2024).

6 Challenges in biomedical AI

6.1 Data quality, diversity, and generalization

Many models perform poorly when used in actual clinical situations and only perform well in regulated experimental environments or with publicly available datasets that have been properly annotated. AI systems are frequently dependent on small sample sizes, which makes it difficult to generalize their findings across a variety of clinical categories. Although synthetic information might be useful, it needs to be thoroughly verified (Aviles, Tovar-Arriaga, Pérez-Soto, Camarillo-Gómez, and Rodríguez-Reséndiz, 2025).

Human negligence in the information utilized for training the algorithm is often the cause of AI malfunctions rather than the algorithm itself. In the event that human specialists incorrectly identify a dataset of X-rays, the AI will pick up on the mistakes and reproduce them more widely. Many AI models, notably those used for healthcare imaging, are developed using data from predominantly white, male communities. When the model is applied to people with varying regional or cultural heritages, this may result in diminished outcomes (Daneshjou et al., 2022).

Diagnostic inequality across demographics generates numerous problems. AI systems for dermatology, built on lighter complexion tones, mistakenly recognize diseases in people with darker complexions, while diagnostic instruments like chest X-ray classifiers have underdiagnosed disorders in minority women. In economically disadvantaged or neglected areas, a lack of data exacerbates inequality by contributing to misinformation (Goktas and Grzybowski, 2025).

6.2 Interpretability and the "black-box" problem

Clinical credibility and endorsement are hampered by the lack of accountability in decision-making provided by many AI models, particularly deep neural networks (khan et al., 2023).

Accountability is made more difficult by the prevalent opacity. When results are poor, clinicians find it difficult to evaluate or justify AI suggestions (Abimbola, 2024).

6.3 Safety, reliability, and adversarial vulnerabilities

AI technologies are constantly accompanied by operational dangers. Examples include surgical robots that misfeed commands, which can cause harm or even death to patients. For instance, the FDA recalled ROSA Brain 3.0 because of a software issue. AI systems may mistakenly diagnose benign illnesses as cancer due to subtle, insignificant input variations, such as 4% noise in skin pictures. One of the main concerns when adopting AI is data confidentiality and reliability. Data breaches are frequent, and healthcare data is a prominent target. There have also been reports of the sale of model code or training data for AI systems on the dark web (Zhang and Zhang, 2023).

6.4 Loss of clinical expertise

Many AI technologies do not comply with privacy regulations and are not integrated with medical systems. Over dependence on AI even puts therapists at risk of "deskilling," a condition in which their ability to acquire experiential judgment is compromised (Aviles et al., 2025; Mennella, Maniscalco, De Pietro, and Esposito, 2024).

6.5 Continuous data monitoring

If a hospital or any other institution implements an innovative therapeutic regimen or modifies its EHR system, an AI model that was programmed for earlier information may not work properly. Totally unaware of the shift, the AI might keep making predictions based on traditional presumptions. This advocates for the need for continuous updates of information data in the system (Guo et al., 2021; Luz, Palaniappan, and Lotfi, 2022).

6.6 Regulatory and legal challenges

The dynamic nature of AI is not fully accommodated by regulatory frameworks; approvals frequently evaluate static performance and overlook postmarket changes or continual learning solutions. Specific adaptive procedures are still lacking in the EU's risk-based AI legislation. The NHS in the UK encourages flexibility in validation without imposing any legally obligatory requirements (Goktas and Grzybowski, 2025; Mennella et al., 2024).

6.7 Ambiguity in accountability and liability

It is not always clear who is at fault when AI systems make mistakes: the creator, the organization, or the medical professional who used them. Several parties may be at fault in such cases. Proposals like "common enterprise liability" suggest that manufacturers, suppliers, and technologists share legal accountability. Liability models must be updated as AI grows more independent (Abdullah, Schuman, Shabsigh, Caplan, & Al-Aswad, 2021; Abimbola, 2024).

7 AI failures in biomedical fields

Although there is great potential for the use of AI in biomedical domains, a number of real-world case studies highlight its drawbacks and moral dilemmas. These shortcomings draw attention to problems such as bias, inadequate validation, and low generalizability.

Dermatological AI is one well-known example. When tested on darker complexions and rare illnesses, advanced AI models demonstrated a 27%–36% decline in predictive accuracy (ROC-AUC) in comparison with their effectiveness on lighter complexions and common illnesses, according to an investigation using the Diverse Dermatology Images dataset. The differences were more noticeable for AI, but even dermatologists who labeled these photographs performed worse. This demonstrates how inequities are perpetuated by underrepresentation in datasets used for training (Daneshjou et al., 2022).

In a comparable manner, AI for DR screening has not performed well in practical applications. When compared with human instructors, an AI system in Dominica that tested 587 diabetes patients had a sensitivity of approximately 77.5% but a high specificity of about 91.5%. Sensitivity was marginally increased by removing ungradable photographs, but the prevalence of several unreadable examples underscores difficulties in standard practice (Kemp O et al., 2023).

While the AI achieved an acceptance rate of 84.6% and a specificity of 98.0% for referable DR, concordance with professionals was substantial rather than flawless, and almost 10% of retinal images were ungradable in another study conducted in Zhengzhou, China (Li, Zhang, Xie, and Wang, 2021). In contrast to monitored experiments, these instances demonstrate how the quality of images and unpredictability in social situations can decrease productivity.

AI-generated medical images also exhibit flaws. In a comprehensive analysis of approximately 4000 dermatological photographs produced by programs like Midjourney and Stable Diffusion, it was found that just around 10% of the outputs showed black skin, while just below 90% showed bright skin. Only about 15% of dermatology residents were able to accurately identify the desired ailment, indicating low accuracy in diagnosis. This highlights questions about the potential for biases and erroneous graphics to propagate inequities when generative AI is used in clinical education and training (Shaver, Cao, and Novoa, 2024).

Less than 20% of datasets reported patient ethnic or racial background, and less than 10% reported complexion, according to a survey of dermatology AI studies. Reliability was compromised because disease designations were frequently not pathologically verified. Reliability and reproducibility are hampered by the data's inadequate heterogeneity and transparency (Adamson and Smith, 2018).

When taken as a whole, these mistakes highlight the fact that AI systems cannot be depended upon to make generalizations across situations or populations. Heterogeneous datasets, stringent external validation, transparent reporting, and ongoing auditing are necessary for ethical implementation to prevent the perpetuation of health inequities.

8 Ethical considerations in biomedical AI

Although AI has enormous potential in the biomedical profession, there have been a number of failures and hazards identified. These mishaps are frequently the result of ethical lapses, inadequate execution, and fundamental technological limitations rather than intentional intent. In the healthcare industry, where a model inaccuracy could have life-altering impacts, the stakes are extremely high.

8.1 Privacy, consent, and data ownership

AI depends on biological factors, lifestyle records, and other comprehensive personal health data, all of which need to be handled with strong privacy protections. Patients need to be aware of the ways their information is processed. Ethical questions are raised by uncertainty around data rights and their subsequent applications, such as financial benefits (Abimbola, 2024).

Large volumes of individual patient information are needed for AI in healthcare. Cybercriminals find this data to be a lucrative target, and thefts can have disastrous outcomes. For instance, a major data leak revealed millions of users' genetic and private medical histories, exposing the grave dangers of collecting health data on a wide scale (Murdoch, 2021).

8.2 Bias, fairness, and equity

Perhaps the most important and extensively studied domain of AI malfunction is this one. Since AI models can only be as accurate as the information they have been instructed on, inaccurate information will cause the AI to pick up on and reinforce its biases. If AI is not purposefully created with equality in mind, it could further reinforce current healthcare gaps. Key components include cooperation among communities, algorithmic impartiality, rigorous fairness inspections, and the use of varied datasets (Goktas and Grzybowski, 2025; Mahendra, 2024).

A commercial algorithm that US hospitals use to determine which individuals will benefit from enhanced medical monitoring is a frequently cited example. Black patients' medical requirements were consistently understated by the algorithm. This was not because Black patients are healthier, but rather because it used healthcare costs as a proxy for medical necessities, which are traditionally lower for them due to structural differences and socioeconomic variables. This historical trend was learned by the AI model, which then reinforced the discrepancy (Obermeyer, Powers, Vogeli, & Mullainathan, 2019).

8.3 Human autonomy and patient–doctor relationship

The compassionate nature and sophisticated assessment that characterize traditional health services could be compromised by excessive dependence on AI. By ignoring social or individual context, systems may create the appearance of certainty while misrepresenting a patient's condition (Ahmad, Rahim, Zubair, and Abdul-Ghafar, 2021; Mennella et al., 2024).

An AI's diagnosis could be blindly trusted by a doctor, even if their personal professional expertise indicates otherwise. This could result in an inaccurate diagnosis or an incorrect course of therapy. On the other hand, doctors may be unwilling to use a new AI tool simply because they don't know how it operates or lack confidence in its results, which might prevent the widespread use of a potentially life-saving innovation (Xu & Shuttleworth, 2024).

8.4 The "black-box" problem and accountability, liability, and transparency

Assurance and security are threatened by inadequate legislative structures, unclear decision-making processes, and the absence of transparency (Abimbola, 2024; Prakash, Balaji, Joshi, and Surapaneni, 2022).

Numerous deep learning algorithms are "black boxes," which means that a person would have a hard time, or might never be able to figure out, how they arrived at a certain result. In the medical field, where medical professionals must

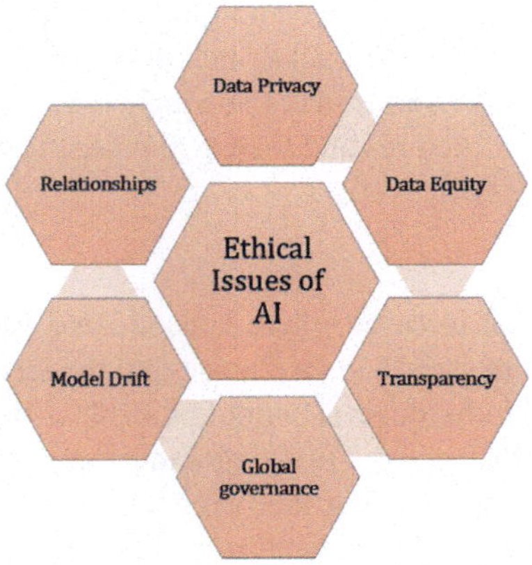

FIG. 34.2 Ethical issues of AI in the biomedical field.

comprehend the rationale underlying a diagnosis or recommended course of therapy, this lack of disclosure is a serious problem. Even if an AI is highly accurate at identifying a malignant tumor on an imaging report, it is unable to provide an explanation for its judgment. A human physician must be able to defend a diagnosis in court, to fellow physicians, and to patients. In situations of inadequate transparency, a clinician cannot readily analyze or modify the reasoning of a model that provides an inaccurate diagnosis (Han & Liu, 2022; Rajpurkar et al., 2022).

8.5 Model drift and failures in real-life applications

When used in a practical clinical situation, an AI model that operates well in a monitored laboratory environment may not work as intended. This is sometimes caused by "model drift," in which the data the model is trained on differs from the data it actually encounters in reality (Davis, Greevy, Lasko, Walsh, and Matheny, 2020; Sahiner, Chen, Samala, and Petrick, 2023).

8.6 Global governance and ethical standards

The application of AI systems can provide significant, as-yet-unresolved ethical and legal issues, even when they operate as planned. Across nations, there are notably few uniform or comparable regulations addressing biomedical AI. The perspectives of individuals and marginalized populations, as well as voices from less developed countries, have always been neglected in ethical study design. Priorities frequently concentrate on wealthy nations. Regulatory agencies, grassroots organizations, and marginalized groups should all be engaged in the creation of ethical AI (Prakash et al., 2022; Ratti, Morrison, & Jakab, 2025) (Fig. 34.2).

The reality that AI is not a panacea for healthcare is demonstrated by these setbacks. A multidisciplinary strategy that addresses data quality, accountability, ethics, and legislation, in addition to the advancement of technology, is necessary for its secure and effective integration.

9 Emerging approaches to address ethical concerns

Although the integration of AI in biomedicine offers revolutionary possibilities, such inclusion also brings up serious ethical issues, particularly with regard to security, confidentiality, justice, and responsibility. Emerging alternatives like homomorphic encryption (HE), federated learning (FL), differential privacy (DP), and secure multiparty computation (SMPC) are being employed more frequently to combat these issues. In addition to promoting innovation in clinical treatment and biological research, these approaches seek to protect patient rights as well.

9.1 Federated learning

It is among the most well-known defenses, enabling models to be developed across several universities without exchanging data in its raw form, in contrast to conventional centralized modeling. Because only model updates are sent to a central server, there is a reduced risk of privacy violations (Yang, Liu, Chen, and Tong, 2019).

In the healthcare sector, where data is extremely critical and frequently fragmented across organizations, this strategy is especially pertinent. Despite its potential, FL has drawbacks, including data heterogeneity, gradient leakage vulnerability, and the possibility of bias recurrence when institutions make unequal contributions (Rieke et al., 2020). Therefore to optimize its ethical implications, organizational structures and technical protections are required.

9.2 Differential privacy

It offers an algorithm to guarantee that, even in cases where consolidated findings are disclosed, all inputs to a dataset remain private. It functions by introducing precisely adjusted noise into the model parameters, inputs, or outputs, rendering it impossible to reconstruct the identities of individuals (Dwork and Roth, 2014).

DP has been used in clinical settings for diagnostic imaging and genetic data, demonstrating that robust confidentiality can be achieved without significantly compromising the precision of diagnosis. Nevertheless, DP presents a balance between confidentiality and practicality, wherein insufficient noise compromises privacy, and excessive noise may reduce model precision (Abadi et al., 2016).

9.3 Cryptographic techniques

Additional safety is provided by SMPC and HE. Confidential medical data is protected during analysis via HE, which enables computation directly on encrypted data (Cheon et al., 2017). As for collective biomedical studies, SMPC is especially helpful because it allows multiple stakeholders to simultaneously compute functions on their collective information without disclosing confidential data. Despite their potential, these approaches can be operationally costly and necessitate substantial infrastructure, which makes widespread adoption difficult (Kamm, Bogdanov, Laur, and Vilo, 2013).

9.4 Hybrid approaches

Combining cryptography, DP, and FL approaches, it is becoming one of the most reliable systems. For example, shared instruction with mathematical security assurances and cryptographic safeguards is made possible by FL, improved with variable confidentiality and HE. These strategies mitigate the drawbacks associated with specific techniques while posing additional challenges in terms of implementation complexity and computational expenditures (Jie et al., 2021).

Administrative and regulatory structures continue to be crucial, extending beyond solely technical solutions. Technical solutions alone cannot address ethical issues, including algorithmic responsibility, cultural considerations, social engagement, and transparency. To ensure reliability and reduce the "black-box" aspect of biomedical AI, accessibility and explanation technologies are also crucial (Kaissis, Makowski, Rückert, and Braren, 2020).

Strategies, including HE, FL, DP, and SMPC, provide robust methods to protect privacy and address ethical dilemmas in biomedical AI. However, no single method is sufficient on its own. The most promising approach to ethically sound biomedical AI is a hybrid strategy that combines technical protections with governance, transparency, and equity.

10 Conclusion

In biomedicine, AI is revolutionizing illness management, diagnosis, and therapy. Its uses are numerous, ranging from personalized healthcare to statistical data-dependent predictive analytics, but they also bring up serious ethical issues. The confidentiality of information, algorithmic bias, transparency, responsibility, and the future development of human involvement in healthcare are among the main concerns. The formation of a satisfactory equilibrium between innovative technology and moral responsibility is essential as AI becomes more and more integrated into the field of biomedical practices.

The use of confidential healthcare details is currently the most serious moral conflict. Huge databases are necessary for AI to develop successfully, but this also poses questions regarding authorization, privacy, and abuse. While there are safeguards offered by some of the different regulatory frameworks, issues with international data sharing and reframing informed consent for machine learning algorithms that utilize data beyond their intended purpose still exist. To preserve the privacy of patients without obstructing innovation, ethical models must be reformed.

Bias in algorithms is yet another significant issue. Algorithms can inherit current gaps in healthcare since they are trained on previously collected data. Marginalized groups frequently experience worse outcomes as a result of insufficient coverage of such communities in the system. If this problem is not resolved, there is a chance that worldwide healthcare inequalities will increase. Although there is a rising trend to standardize processes and broaden datasets to include all communities, maintaining inclusion and justice is still a moral necessity.

The present debate is similarly dominated by simplicity and transparency. Many AI simulations, particularly deep learning frameworks, function as "black boxes" with complex conclusions. There are significant risks associated with unclear decision-making in the healthcare industry, where responsibility and trust are crucial. Despite advancements in XAI research, it remains challenging to achieve easy comprehension without sacrificing model accuracy. To ensure that both healthcare professionals and patients understand and accept AI-powered guidelines, building confidence requires both ethical and technical solutions.

Another complicated issue is accountability. Although AI can improve the way doctors make decisions, concerns about excessive dependence on machines and the loss of professional competence still persist. The lines of accountability between developers, organizations, and medical experts remain unclear when mistakes occur. Although regulators are beginning to explore joint responsibility systems, more precise procedures are needed to protect healthcare providers and patients.

Future trends are expected to focus on multidisciplinary strategies that incorporate viewpoints from physicians, ethics experts, engineers, and legislators; patient-focused AI that values clarity and autonomy; and international cooperation on ethical norms. While ethically conscious approaches aim to integrate ethical considerations into AI design from the initial stages, technological advancements like collaborative training and encrypted AI may help establish an acceptable balance between the use of data and confidentiality.

AI in biological applications has now arrived at an important phase. It has unmatched possibilities to transform healthcare, but it also raises moral questions. Although there have been notable advancements recently in the areas of bias, disclosure, and data safety, unresolved gaps emphasize the necessity of constant monitoring.

The willingness of human society to incorporate moral values into patient care, regulatory protocols, and technological innovation will determine how biomedical AI develops in the coming years. The ultimate aim at the forefront of this change must be maintaining reliability among patients, responsibility, and equity. In the end, integrating AI ethically into biomedicine is not merely about eliminating damage but about achieving a more equitable, compassionate, and inclusive healthcare system where technology avoids creating new problems and definitively strengthens the wellness of mankind.

Declaration of generative AI and AI-assisted technologies in the writing process

During the preparation of this work, the author(s) have not used any generative AI or AI-assisted technologies and take(s) full responsibility for the content of the publication.

References

Abadi, M., McMahan, H. B., Chu, A., Mironov, I., Zhang, L., Goodfellow, I., & Talwar, K. 2016. Deep learning with differential privacy. In Proceedings of the ACM Conference on Computer and Communications Security (pp. 308–318), 9781450341394. United States, New York: Association for Computing Machinery. 10.1145/2976749.2978318.

Abdullah, Y. I., Schuman, J. S., Shabsigh, R., Caplan, A., & Al-Aswad, L. A. (2021). Ethics of artificial intelligence in medicine and ophthalmology. *Asia-Pacific Journal of Ophthalmology*. 21620989. *10*(3), 289–298. Available from: https://journals.lww.com/apjoo/pages/default.aspx. doi:10.1097/APO.0000000000000397.

Abimbola, B. (2024). Ethical considerations in AI applications for biomedical science and patient care. *American Journal of Biomedical Science & Research*, *24*(6), 577–579. doi:https://doi.org/10.34297/AJBSR.2024.24.003269. Available from:.

Abramson, J., Adler, J., Dunger, J., Evans, R., Green, T., Pritzel, A., … Jumper, J. M. (2024). Accurate structure prediction of biomolecular interactions with AlphaFold 3. *Nature*. 0028-0836. *630*(8016), 493–500. doi:10.1038/s41586-024-07487-w.

Abu-El-Ruz, R., Hasan, A., Hijazi, D., Masoud, O., Abdallah, A. M., Zughaier, S. M., & Al-Asmakh, M. (2025). Artificial intelligence in biomedical sciences: a scoping review. *British Journal of Biomedical Science*. 24740896. *82*. Available from: https://www.frontierspartnerships.org/journals/british-journal-of-biomedical-science. doi:10.3389/bjbs.2025.14362.

Adamson, A. S., & Smith, A. (2018). Machine learning and health care disparities in dermatology. *JAMA Dermatology*. 21686068. *154*(11), 1247–1248. Available from: http://archderm.jamanetwork.com/issues.aspx. doi:10.1001/jamadermatol.2018.2348.

Agrawal, J., & Arafat, M. Y. (2024). Transforming farming: a review of AI-powered UAV technologies in precision agriculture. *India Drones*. 2504446X. *8*(11). Available from: https://www.mdpi.com/journal/drones. doi:10.3390/drones8110664.

Ahmad, Z., Rahim, S., Zubair, M., & Abdul-Ghafar, J. (2021). Artificial intelligence (AI) in medicine, current applications and future role with special emphasis on its potential and promise in pathology: present and future impact, obstacles including costs and acceptance among pathologists, practical and philosophical considerations. A comprehensive review. *Diagnostic Pathology*. 1746-1596. *16*(1). doi:10.1186/s13000-021-01085-4.

Aviles, M., Tovar-Arriaga, S., Pérez-Soto, G. I., Camarillo-Gómez, K. A., & Rodríguez-Reséndiz, J. (2025). Artificial intelligence in biomedical technology: advances and challenges. *Technologies*. 2227-7080. *13*(5). Available from: https://www.mdpi.com/journal/technologies. doi:10.3390/technologies13050208.

Bardin, P. G., Thomas, B. J., & Bourke, J. E. (2025). Artificial intelligence will boost repurposing of registered compounds for respiratory and other indications. *Respirology*, 1440-1843. Available from: https://onlinelibrary.wiley.com/journal/14401843. doi:10.1111/resp.70122.

Beauchamp, T. L., & Childress, J. F., (2013). Principles of biomedical ethics.

Bobula, M. (2024). Generative artificial intelligence (AI) in higher education: a comprehensive review of challenges, opportunities, and implications. *Journal of Learning Development in Higher Education*. 1759-667X. *30*. doi:10.47408/jldhe.vi30.1137.

Bostrom, N., & Yudkowsky, E. (2014). *The Ethics of Artificial Intelligence*. (pp. 316–334). Cambridge University Press (CUP). doi:10.1017/cbo9781139046855.020.

Brown, T. B., Mann, B., Ryder, N., Subbiah, M., Kaplan, J., Dhariwal, P., … Amodei, D. (2020). Language models are few-shot learners. In Advances in Neural Information Processing Systems 10495258 Neural Information Processing Systems Foundation, United States, California. https://papers.nips.cc/2020.

Campbell, M., Hoane, A. J., & Hsu, F. H. (2002). Deep blue. *Artificial Intelligence*. 00043702. *134*(1-2), 57–83. doi:10.1016/S0004-3702(01)00129-1.

Cao, C., Liu, F., Tan, H., Song, D., Shu, W., Li, W., … Xie, Z. (2018). Deep learning and its applications in biomedicine. *Genomics, Proteomics and Bioinformatics*. 22103244. *16*(1), 17–32. Available from: http://www.elsevier.com/wps/find/journaldescription.cws_home/707637/description#description. doi:10.1016/j.gpb.2017.07.003.

Chen, Y. M., Hsiao, T. H., Lin, C. H., & Fann, Y. C. (2025). Unlocking precision medicine: clinical applications of integrating health records, genetics, and immunology through artificial intelligence. *Journal of Biomedical Science*. 14230127. *32*(1), 16. doi:10.1186/s12929-024-01110-w.

Cheon, J. H., Kim, A., Kim, M., & Song, Y. (2017). Homomorphic encryption for arithmetic of approximate numbers. In T. Takagi (Eds.), *Advances in Cryptology – ASIACRYPT 2017 23rd International Conference on the Theory and Applications of Cryptology and Information Security, Hong Kong, China, December 3-7, 2017, Proceedings, Part I [Lecture Notes in Computer Science (including subseries Lecture Notes in Artificial Intelligence and Lecture Notes in Bioinformatics)]* (pp. 409–437). Berlin, Germany: Springer. 16113349https://springerlink.com/content/0302-9743/copyright/2005/10624. doi:10.1007/978-3-319-70694-8_15.

Chustecki, M. (2024). Benefits and risks of AI in health care: narrative review. *Interactive Journal of Medical Research*. 1929-073X. *13*, e53616. doi:10.2196/53616.

Cohen, Y., Amorim, M., & Reis, J. (2022). Artificial intelligence trends and applications in service systems. *Applied Sciences*. 2076-3417. *12*(24), 13032. doi:10.3390/app122413032.

Daneshjou, R., Vodrahalli, K., Novoa, R. A., Jenkins, M., Liang, W., Rotemberg, V., … Chiou, A. S. (2022). Disparities in dermatology AI performance on a diverse, curated clinical image set. *American Association for the Advancement of Science, United States Science Advances*. 23752548. *8*(32). Available from: https://www.science.org/doi/10.1126/sciadv.abq6147. doi:10.1126/sciadv.abq6147.

Davis, S. E., Greevy, R. A., Lasko, T. A., Walsh, C. G., & Matheny, M. E. (2020). Detection of calibration drift in clinical prediction models to inform model updating. *Journal of Biomedical Informatics*. 15320464. *112*. Available from: https://www.elsevier.com/inca/publications/store/6/2/2/8/5/7/index.htt. doi:10.1016/j.jbi.2020.103611.

Donnelly, J., & Gewirth, A. (1983). Human rights: essays on justification and applications. *Human Rights Quarterly*. 02750392. *5*(3), 376. doi:10.2307/762030.

Dwork, C., & Roth, A. (2014). The algorithmic foundations of differential privacy. *Foundations and Trends® in Theoretical Computer Science*. 1551-305X. *9*(3-4), 211–407. doi:10.1561/0400000042.

Ferrucci, D., Brown, E., Chu-Carroll, J., Fan, J., Gondek, D., Kalyanpur, A. A., … Welty, C. (2011). Building watson: an overview of the deepQA project. *AI Magazine*. 07384602. *31*(3), 59–79. Available from http://www.aaai.org/ojs/index.php/aimagazine/article/view/2303/2165. doi:10.1609/aimag.v31i3.2303.

Fried, C., & Rawls, J. (1972). A theory of justice. *Harvard Law Review*. 0017811X. *85*(8), 1691. doi:10.2307/1339914.

Garvey, C. (2019). Artificial intelligence and Japan's fifth generation: the information society, neoliberalism, and alternative modernities. *Pacific Historical Review*. 15338584. *88*(4), 619–658. Available from: http://phr.ucpress.edu/. doi:10.1525/phr.2019.88.4.619.

Gokcekuyu, Y., Ekinci, F., Guzel, M. S., Acici, K., Aydin, S., & Asuroglu, T. (2024). Artificial intelligence in biomaterials: a comprehensive review. *Applied Sciences*. 2076-3417. *14*(15), 6590. doi:10.3390/app14156590.

Goktas, P., & Grzybowski, A. (2025). Shaping the future of healthcare: ethical clinical challenges and pathways to trustworthy AI. *Journal of Clinical Medicine*. 2077-0383. *14*(5), 1605. doi:10.3390/jcm14051605.

Goodfellow, I., Bengio, Y., & Courville, A., Deep learning. (2016), Available from: https://www.deeplearningbook.org/.

Graas, A., Coban, S. B., Batenburg, K. J., & Lucka, F. (2023). Just-in-time deep learning for real-time X-ray computed tomography. *Scientific Reports*. 20452322. *13*(1). Available from: https://www.nature.com/srep/. doi:10.1038/s41598-023-46028-9.

Gromiha, M. M., Preethi, P., & Pandey, M. (2024). From code to cure: the impact of artificial intelligence in biomedical applications. *BioMedInformatics*. 26737426. *4*(1), 542–548. Available from: https://www.mdpi.com/journal/biomedinformatics. doi:10.3390/biomedinformatics4010030.

Guo, L. L., Pfohl, S. R., Fries, J., Posada, J., Fleming, S. L., Aftandilian, C., … Sung, L. (2021). Systematic review of approaches to preserve machine learning performance in the presence of temporal dataset shift in clinical medicine. *Applied Clinical Informatics*. 18690327. *12*(4), 808–815. Available from: https://www.thieme-connect.com/products/ejournals/journal/10.1055/s-00035026. doi:10.1055/s-0041-1735184.

Hajkowicz, S., Sanderson, C., Karimi, S., Bratanova, A., & Naughtin, C. (2023). Artificial intelligence adoption in the physical sciences, natural sciences, life sciences, social sciences and the arts and humanities: a bibliometric analysis of research publications from 1960-2021. *Technology in Society*. 0160791X. *74*. Available from: www.elsevier.com/inca/publications/store/3/8/4/. doi:10.1016/j.techsoc.2023.102260.

Han, H., & Liu, X. (2022). The challenges of explainable AI in biomedical data science. *BMC Bioinformatics*. 14712105. *22*. Available from: https://www.biomedcentral.com/bmcbioinformatics/. doi:10.1186/s12859-021-04368-1.

Hayes, P., Fitzpatrick, N. and Ferrández, J. M. (2025). From applied ethics and ethical principles to virtue and narrative in AI practices. *AI Ethics*, *5*(2), 1217–1239. doi:10.1007/s43681-024-00472-z.

Jie, X., Glicksberg, B. S., Chang, S., Walker, P., Bian, J., & Wang, F. (2021). Federated Learning for Healthcare Informatics. *Journal of Healthcare Informatics Research*. 2509-4971. *5*(1), 1–19. doi:10.1007/s41666-020-00082-4.

Johnson, M. M., & Gilligan, C. (1983). In a different voice: psychological theory and women's development. *Contemporary Sociology*. 00943061. *12*(4), 448. doi:10.2307/2067520.

Juyal, P., Manukonda, P., Saratchandran, D., Trehan, A., Shah, K. N., & Katru, C. R. (2025). Artificial intelligence in enhancing decision-making in enterprise information systems. *Journal of Information Systems Engineering and Management*, *10*(3s). Available from: https://www.jisem-journal.com/index.php/journal/article/view/371.

Kaissis, G. A., Makowski, M. R., Rückert, D., & Braren, R. F. (2020). Secure, privacy-preserving and federated machine learning in medical imaging. *Nature Machine Intelligence*. 25225839. *2*(6), 305–311. Available from: https://www.nature.com/natmachintell/. doi:10.1038/s42256-020-0186-1.

Kalinin, A. A., Higgins, G. A., Reamaroon, N., Soroushmehr, S., Allyn-Feuer, A., Dinov, I. D., … Athey, B. D. (2018). Deep learning in pharmacogenomics: from gene regulation to patient stratification. *Pharmacogenomics*. 17448042. *19*(7), 629–650. Available from: https://www.futuremedicine.com/loi/pgs?cookieSet=1. doi:10.2217/pgs-2018-0008.

Kamm, L., Bogdanov, D., Laur, S., & Vilo, J. (2013). A new way to protect privacy in large-scale genome-wide association studies. *Bioinformatics*. 14602059. *29*(7), 886–893. doi:10.1093/bioinformatics/btt066.

Kant, I. (1993). *Grounding for the metaphysics of morals*. J. W. Ellington. Trans. Hackett. Original work published 1785).

Katzman, D., Kalman, G., Almog, O., & Fogel, I. (2024). Deployment of physician resources and innovative medical strategies in the 2023–2024 Israel–Hamas War: israel's strategy to deliver advanced life support and whole blood transfusion to the battlefield via forward medical teams and the impact on the case fatality rate. *Military Medicine*. doi:10.1093/milmed/usad005.

Kemp, O., Bascaran, C., Cartwright, E., McQuillan, L., Matthew, N., Shillingford-Ricketts, H., Zondervan, M., Foster, A.&Burton, M. (2023). Real-world evaluation of smartphone-based artificial intelligence to screen for diabetic retinopathy in Dominica: a clinical validation study. *BMJ Open Ophthalmology*, *8*(1), e001491. doi:10.1136/bmjophth-2023-001491.

khan, B., Fatima, H., Qureshi, A., Kumar, S., Hanan, A., Hussain, J., & Abdullah, S. (2023). Drawbacks of artificial intelligence and their potential solutions in the healthcare sector. *Biomedical Materials & Devices*. 2731-4812. *1*(2), 731–738. doi:10.1007/s44174-023-00063-2.

Krizhevsky, A., Sutskever, I., & Hinton, G. E. (2012). ImageNet classification with deep convolutional neural networks 2. In F. Pereira, C. J. C. Burges, L. Bottou \& K. Q Weinberger, *Advances in Neural Information Processing Systems* (pp. 1097–1105). 10495258. New York, United States: Curran Associates, Inc.

Li, H., Zhang, Z., Xie, Y., & Wang, W. (2021). Evaluation of an artificial intelligence-based system for diabetic retinopathy screening in a community setting in China. *Eye and Vision*, *8*(1), 31. doi:10.1186/s40662-021-00262-1.

Li, J. Tang, T. Zong, H. et al . (2024). Intelligent medicine in focus: the 5 stages of evolution in robot-assisted surgery for prostate cancer in the past 20 years and future implications. *Military Medical Research*, *11*(1). doi:10.1186/s40779-024-00566-z.

Li, S., Bao, Y. and Wu, B. (2024). Letter to the editor regarding the article "artificial intelligence and computer-assisted navigation for shoulder surgery". *Journal of Orthopaedic Surgery (Hong Kong)*, *32*(2). doi:10.1177/10225536241263656.

Lighthill, J., Artificial intelligence: a general survey Artificial intelligence: a paper symposium. (1973), 1–21.

Luz, S., Palaniappan, S., & Lotfi, A. (2022). Clinical artificial intelligence quality improvement: towards continual monitoring and updating of AI algorithms in healthcare. *Journal of Biomedical Informatics*, *136*, 104263. doi:10.1016/j.jbi.2022.104263.

Mahendra, S., Ethical concerns in AI healthcare applications. Artificial intelligence + (medium). (2024), Available from: https://www.aiplusinfo.com/blog/ethical-concerns-in-ai-healthcare-applications/.

McCarthy, J., Minsky, M. L., Rochester, N., & Shannon, C. E., A proposal for the Dartmouth summer research project on artificial intelligence. (1956), Available from: https://jmc.stanford.edu/articles/dartmouth/dartmouth.pdf.

Mennella, C., Maniscalco, U., De Pietro, G., & Esposito, M. (2024). Ethical and regulatory challenges of AI technologies in healthcare: a narrative review. *Heliyon*. 24058440. *10*(4), e26297. doi:10.1016/j.heliyon.2024.e26297.

Mill, J. S. (1998). In R. Crisp (Ed.), *Utilitarianism*. Oxford University Press.(Original work published 1863).

Mohammadnabi, S., Moslemy, N., Taghvaei, H., Zia, A. W., Askarinejad, S., & Shalchy, F. (2025). Role of artificial intelligence in data-centric additive manufacturing processes for biomedical applications. *Journal of the Mechanical Behavior of Biomedical Materials*. 17516161. *166*, 106949. doi:10.1016/j.jmbbm.2025.106949.

Morgan, M. G., Feigenbaum, E. A., & McCorduck, P. (1983). The fifth generation: artificial intelligence and Japan's computer challenge to the world. *Journal of Policy Analysis and Management*. 02768739. *3*(1), 156. doi:10.2307/3324061.

Murdoch, B. (2021). Privacy and artificial intelligence: challenges for protecting health information in a new era. *BMC Medical Ethics*. 14726939. *22*(1). Available from: https://www.biomedcentral.com/bmcmedethics/. doi:10.1186/s12910-021-00687-3.

Nagendraswamy, C., & Amogh, S. (2021). A review article on artificial intelligence. *Annals of Biomedical Science and Engineering*, *5*(1), 013–014. doi:10.29328/journal.abse.1001012.

Naskar, S., Sharma, S., Kuotsu, K., Halder, S., Pal, G., Saha, S., … Bhattacharjee, S. (2025). The biomedical applications of artificial intelligence: an overview of decades of research. *Journal of Drug Targeting*. 1061-186X. *33*(5), 717–748. doi:10.1080/1061186x.2024.2448711.

nasscom. (2023). AI in the Modern World. Available from: https://community.nasscom.in/communities/ai/.

Nazir, S., Dickson, D. M., & Akram, M. U. (2023). Survey of explainable artificial intelligence techniques for biomedical imaging with deep neural networks. *Computers in Biology and Medicine*. 18790534. *156*. Available from: www.elsevier.com/locate/compbiomed. doi:10.1016/j.compbiomed.2023.106668.

Naznin, K., Al Mahmud, A., Nguyen, M. T., & Chua, C. (2025). ChatGPT integration in higher education for personalized learning, academic writing, and coding tasks: a systematic review. *Computers*. 2073-431X. *14*(2). Available from: www.mdpi.com/journal/computers. doi:10.3390/computers14020053.

Nilsson, N. J. (1984). Shakey the robot. SRI International. Available from: https://www.ai.sri.com/shakey/.

Obermeyer, Z., Powers, B., Vogeli, C., & Mullainathan, S. (2019). Dissecting racial bias in an algorithm used to manage the health of populations. *Science*. 0036-8075. *366*(6464), 447–453. doi:10.1126/science.aax2342.

Olawade, D. B., David-Olawade, A. C., Wada, O. Z., Asaolu, A. J., Adereni, T., & Ling, J. (2024). Artificial intelligence in healthcare delivery: prospects and pitfalls. *Journal of Medicine, Surgery, and Public Health*. 2949-916X. *3*(100108). doi:10.1016/j.glmedi.2024.100108.

Pei, Y., & Yang, J. (2025). Biomedical applications of big data and artificial intelligence. *Bioengineering*. 2306-5354. *12*(2), 207. doi:10.3390/bioengineering12020207.

Prakash, S., Balaji, J. N., Joshi, A., & Surapaneni, K. M. (2022). Ethical Conundrums in the application of artificial intelligence (AI) in healthcare—A scoping review of reviews. *Journal of Personalized Medicine*. 20754426. *12*(11). Available from: https://www.mdpi.com/journal/jpm. doi:10.3390/jpm12111914.

Rajpurkar, P., & Chen, E., Banerjee, O., Topol, E. J. (2022). AI in health and medicine. *Nature Medicine*, *28*(1), 31–38. doi:10.1038/s41591-021-01614-0 .

Ratti, E., Morrison, M., & Jakab, I. (2025). Ethical and social considerations of applying artificial intelligence in healthcare—A two-pronged scoping review. *BMC Medical Ethics*. 1472-6939. *26*(1). doi:10.1186/s12910-025-01198-1.

Rieke, N., Hancox, J., Li, W., Milletarì, F., Roth, H. R., Albarqouni, S., … Cardoso, M. J. (2020). The future of digital health with federated learning. *Npj Digital Medicine*. 2398-6352. *3*(1). Available from: https://www.nature.com/npjdigitalmed/. doi:10.1038/s41746-020-00323-1.

Rosenblatt, F. (1958). The perceptron: a probabilistic model for information storage and organization in the brain. *Psychological Review*. 1939-1471. *65*(6), 386–408. doi:10.1037/h0042519.

Rumelhart, D. E., Hinton, G. E., & Williams, R. J. (1986). Learning representations by back-propagating errors. *Nature*. 0028-0836. *323*(6088), 533–536. doi:10.1038/323533a0.

Sadeghi, Z., Alizadehsani, R., Cifci, M. A., Kausar, S., Rehman, R., Mahanta, P., … Pardalos, P. M. (2024). A review of explainable artificial intelligence in healthcare. *Computers and Electrical Engineering*. 00457906. *118*, 109370. doi:10.1016/j.compeleceng.2024.109370.

Sahiner, B., Chen, W., Samala, R. K., & Petrick, N., 2023. Data drift in medical machine learning: implications and potential remedies. *British Journal of Radiology 1748-880X* (96), 1150. Available from: https://www.birpublications.org/bjr. doi:10.1259/bjr.20220878.

Sarikaya, F. (2024). The cycles of AI winters: a historical analysis and modern perspective. *Zenodo*. doi:10.5281/zenodo.14015032.

Shaver, J. M., Cao, Y., & Novoa, R. A. (2024). AI-generated dermatologic images show deficient skin tone diversity and poor diagnostic accuracy. *JAMA Dermatology*, *160*(7), 715–722. doi:10.1001/jamadermatol.2024.1781.

Simon, H. A., & Newell, A. (1958). Heuristic problem solving: the next advance in operations research. *Operations Research*. 0030-364X. *6*(1), 1–10. doi:10.1287/opre.6.1.1.

Simpson, M. D., & Qasim, H. S. (2025). Clinical and operational applications of artificial intelligence and machine learning in pharmacy: a narrative review of real-world applications. *Pharmacy*. 2226-4787. *13*(2), 41. doi:10.3390/pharmacy13020041.

Song, S., Li, T., Lin, W., Liu, R., & Zhang, Y. (2025). Application of artificial intelligence in Alzheimer's disease: a bibliometric analysis. *Frontiers in Neuroscience*. 1662-453X. *19*. Available from: https://www.frontiersin.org/journals/neuroscience#. doi:10.3389/fnins.2025.1511350.

Stokes, J. M., Collins, J. J., & Barzilay, R. (2023). Deep learning-guided discovery of an antibiotic targeting Acinetobacter baumannii. *Nature Chemical Biology*, *19*, 1234–1242. doi:10.1038/s41589-023-01234-5. Available from:.

Tettey-Engmann, F., Parupelli, S. K., Bauer, S. R., Bhattarai, N., & Desai, S. (2025). Advances in Artificial Intelligence-Based Medical Devices for Healthcare Applications. *Biomedical Materials & Devices*, 27314820. Available from: https://link.springer.com/journal/44174. doi:10.1007/s44174-025-00379-1.

Toosi, A., Bottino, A. G., Saboury, B., Siegel, E., & Rahmim, A. (2021). A brief history of AI: how to prevent another winter (a critical review). *PET Clinics*. 15568598. *16*(4), 449–469. doi:10.1016/j.cpet.2021.07.001.

Tripathi, D., Hajra, K., Mulukutla, A., Shreshtha, R., & Maity, D. (2025). Artificial intelligence in biomedical engineering and its influence on healthcare structure: current and future prospects. *Bioengineering*. 2306-5354. *12*(2), 163. doi:10.3390/bioengineering12020163.

Turing, A. M. (1950). I.—Computing machinery and intelligence. *Mind*. 1460-2113. *LIX* (236), 433–460. doi:10.1093/mind/lix.236.433.

Valencia-Arias, A., Uribe-Bedoya, H., González-Ruiz, J. D., Sánchez Santos, G., Chapoñan Ramírez, E., & Martínez Rojas, E. (2024). Artificial intelligence and recommender systems in e-commerce. Trends and research agenda. *Intelligent Systems with Applications*. 2667-3053. *24*, 200435. doi:10.1016/j.iswa.2024.200435.

Vaswani, A., Shazeer, N., Parmar, N., Uszkoreit, J., Jones, L., Gomez, A. N., … Polosukhin, I. (2017). Attention is all you need. In I.Guyon (Eds.), *Advances in neural information processing systems*, Long Beach, California (NeurIPS 2017 venue). (pp. 5999–6009). California, United States: Neural Information Processing Systems Foundation. 10495258.

Wah, J. N. K. (2025). The rise of robotics and AI-assisted surgery in modern healthcare. *Journal of Robotic Surgery*. 1863-2491. *19*(1). doi:10.1007/s11701-025-02485-0.

Wang, P. (2019). On defining artificial intelligence. *Journal of Artificial General Intelligence*. 1946-0163. *10*(2), 1–37. doi:10.2478/jagi-2019-0002.

Weizenbaum, J. (1966). ELIZA—A computer program for the study of natural language communication between man and machine. *Communications of the ACM*. 0001-0782. *9*(1), 36–45. doi:10.1145/365153.365168.

Whittington, J. C. R., & Bogacz, R. (2017). An approximation of the error backpropagation algorithm in a predictive coding network with local hebbian synaptic plasticity. *Neural Computation*. 1530-888X. *29*(5), 1229–1262. Available from: https://www.mitpressjournals.org/loi/neco. doi:10.1162/NECO_a_00949.

Xu, H., & Shuttleworth, K. M. J. (2024). Medical artificial intelligence and the black box problem: a view based on the ethical principle of "do no harm". *China Intelligent Medicine*. 2667-1026. *4*(1), 52–57. Available from: https://www.sciencedirect.com/science/journal/26671026. doi:10.1016/j.imed.2023.08.001.

Yang, Q., Liu, Y., Chen, T., & Tong, Y. (2019). Federated machine learning: concept and applications. *ACM Transactions on Intelligent Systems and Technology*. 2157-6912. *10*(2). Available from: https://dl.acm.org/citation.cfm?id=J1318. doi:10.1145/3298981.

Zakaria, Z., & Hussein, H. (2025). Pros and cons of artificial intelligence in healthcare: a comprehensive review. *Journal of Medical Systems*, *49*(3), 45. doi:https://doi.org/10.1007/s10916-025-01023-y. Available from:.

Zemouri, R., Zerhouni, N., & Racoceanu, D. (2019). Deep learning in the biomedical applications: recent and future status. *Applied Sciences*. 2076-3417. *9*(8), 1526. doi:10.3390/app9081526.

Zhang, J., & Zhang, Z. M. (2023). Ethics and governance of trustworthy medical artificial intelligence. *BMC Medical Informatics and Decision Making*. 1472-6947. *23*(1). Available from: https://bmcmedinformdecismak.biomedcentral.com/. doi:10.1186/s12911-023-02103-9.

Zhao, Y., Yang, M., Lin, Y., Zhang, X., Shi, F., Wang, Z., … Ning, H. (2025). AI-enabled text-to-music generation: a comprehensive review of methods, frameworks, and future directions. *Electronics*. 2079-9292. *14*(6), 1197. doi:10.3390/electronics14061197.

Chapter 35

Minimizing vulnerability of artificial intelligence (AI) programs from cyberattacks

Laishram Saya[a], Deepanshu Vasuja[b], Gauri Jha[b], Atreyee Bagchi[b], Shivam Saini[b], Pooja[a], and Sunita Hooda[c]

[a]*Department of Chemistry, Sri Venkateswara College, University of Delhi, Delhi, India*

[b]*Department of Biological Sciences, Sri Venkateswara College, University of Delhi, Delhi, India*

[c]*Polymer Research Laboratory, Department of Chemistry, Acharya Narendra Dev College, Govindpuri, Kalkaji, New Delhi, India*

1 Introduction: Artificial intelligence as a double-edged sword

The incorporation of artificial intelligence (AI) in critical fields like healthcare, finance, transportation, legal systems, and national security is on the rise, offering remarkable efficiency and accuracy through its real-time decision-making capabilities (Lyon, 2023; Shethiya, 2023). Nonetheless, this extensive adoption also heightens systemic vulnerabilities. The ransomware attack on AIIMS in New Delhi in 2022, which interrupted vital healthcare services and revealed sensitive patient information, exemplifies how data-centric systems, even if they do not rely directly on AI, can become crucial points of failure in the absence of appropriate protections (Gandhi and Pahwa, 2024). Simultaneously, worries about algorithmic bias, insufficient transparency, and vulnerability to adversarial attacks remain significant, particularly when AI systems act independently in rapidly changing settings (Dhirani, Mukhtiar, Chowdhry, and Newe, 2023; Shethiya, 2023). These dangers are exacerbated by the "black-box" characteristic of numerous models, which obscures responsibility in areas such as healthcare triage, autonomous driving, and credit evaluations (Dhirani et al., 2023). Cases like the AI-generated misleading legal citations in Mata v. Avianca further demonstrate that an uncritical dependence on generative AI can result in professional misconduct and legal repercussions (Lyon, 2023). As Lyon wisely points out, echoing a statement from Ronald Reagan, "Trust, but verify," a warning that holds particular significance where AI mistakes may lead to systemic, irreversible, or unjust ramifications. Within national security, the limitations of AI, including issues like overfitting, input noise, and failures in explainability, bring forth added anxieties regarding operational dependability and ethical governance (Yu and Carroll, 2021). A recurring theme across these fields is that the rapid implementation of AI frequently outstrips the creation of strong ethical guidelines, regulatory frameworks, and technical protections, leading to pressing concerns regarding trust, accountability, and long-term sustainability (Dhirani et al., 2023; Lyon, 2023; Shethiya, 2023).

1.1 Definition of AI

AI does not have a single, universally accepted definition, but it is often understood through four conceptual perspectives: thinking like a human, thinking logically, acting like a human, and acting logically (Russell and Norvig, 2020). Initial definitions focused on replicating human thought processes, presenting AI as the automation of functions such as learning and problem-solving (Bellman, 1978; Haugeland, 1985), or on formal reasoning, where systems utilize logic to derive optimal solutions (Charniak and McDermott, 1985; Winston, 1992). Other viewpoints emphasized behavior, whether imitating human actions (Kurzweil, 1990; Rich and Knight, 1991) or optimizing decisions irrespective of human similarity (Nilsson, 1998; Poole, Mackworth, and Goebel, 1998). A contemporary, functional definition characterizes AI as systems that can sense their environment and act independently to fulfill objectives (Sheikh, Prins, and Schrijvers, 2023), a description that sets AI apart from traditional algorithms while remaining adaptable for future advancements.

Applications of Artificial Intelligence in Healthcare. https://doi.org/10.1016/B978-0-443-43934-6.00025-0

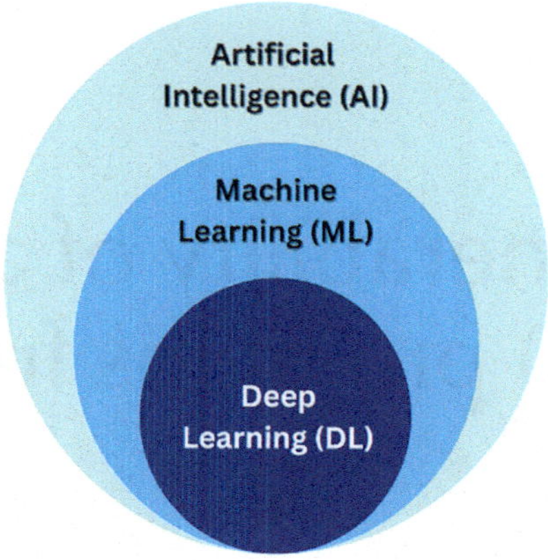

FIG. 35.1 Conceptual hierarchy of AI, ML, and DL.

As these systems gain independence and are utilized in important applications, their distinct dependence on training data, probabilistic reasoning, and opaque decision-making introduces new security risks, which differentiate AI from standard software concerning threat exposure and risk profile.

1.2 Overview of machine learning and deep learning

Machine learning (ML) allows computers to identify patterns from data without the need for explicit programming. Instead of depending on predetermined instructions, ML algorithms modify themselves through repeated data exposure in a process called training, refining their internal parameters to enhance performance on both familiar and unfamiliar inputs (El Naqa and Murphy, 2015). This ability to adapt enables systems to generalize and make decisions in new situations. Deep learning (DL), a more sophisticated branch of ML, utilizes multilayered neural networks to recognize intricate patterns. In contrast to conventional rule-based programming, DL systems independently form internal representations by adjusting weights and biases as they are exposed to large amounts of data, simulating certain aspects of human thought processes (Paluszek and Thomas, 2020). Currently, ML is integrated into numerous digital systems, facilitating functions that demand minimal human intervention. It is particularly essential in cybersecurity, where it enhances the detection of malware, recognizes anomalies, and filters out spam while also driving everyday tools such as speech recognition, predictive typing, and personalized content suggestions (Yerlikaya and Bahtiyar, 2022). Fig. 35.1 shows a Venn diagram illustrating the relationship between AI, ML, and DL. AI is the broadest domain encompassing all systems that exhibit intelligent behavior. ML is a subset of AI that focuses on systems that learn from data. DL is a further subset of ML, involving neural networks with multiple layers that model complex patterns in large datasets.

1.3 Understanding cyberattacks and cybersecurity

Cyberattacks represent a spectrum of malicious actions aimed at exploiting digital vulnerabilities for unauthorized access, disruption, theft, or manipulation. These attacks may be technically complex, such as deploying malware or exploiting software vulnerabilities, or socially engineered, such as phishing (Bendovschi, 2015). In essence, they compromise one or more pillars of information security: confidentiality, integrity, and availability. Huang and his team in 2018 describe cyberattacks within a structured economy known as the *cybercrime-as-a-service ecosystem*, where different attack components, such as malware payloads, phishing kits, exploit kits, and botnets, are sold or rented through underground markets (Huang, Siegel, and Madnick, 2018). This commodification lowers the technical barriers, allowing virtually anyone to launch advanced attacks using plug-and-play tools.

Fig. 35.2 provides a concise overview of four fundamental cybersecurity pillars—authentication, authorization, encryption, and intrusion detection systems (IDSs)—as they relate to AI system security. While these concepts are foundational across digital infrastructures, their relevance becomes heightened in AI contexts due to the sensitivity of training data, the complexity of model architectures, and the dynamic threat landscape. Authentication and authorization mechanisms, for instance, must contend with increasingly sophisticated social engineering attacks and privilege escalation tactics. Encryption, although essential for protecting data confidentiality, is also weaponized by adversaries to conceal malicious operations. Similarly, traditional IDS solutions face challenges from adaptive threats that exploit known detection limitations. Fig. 35.2

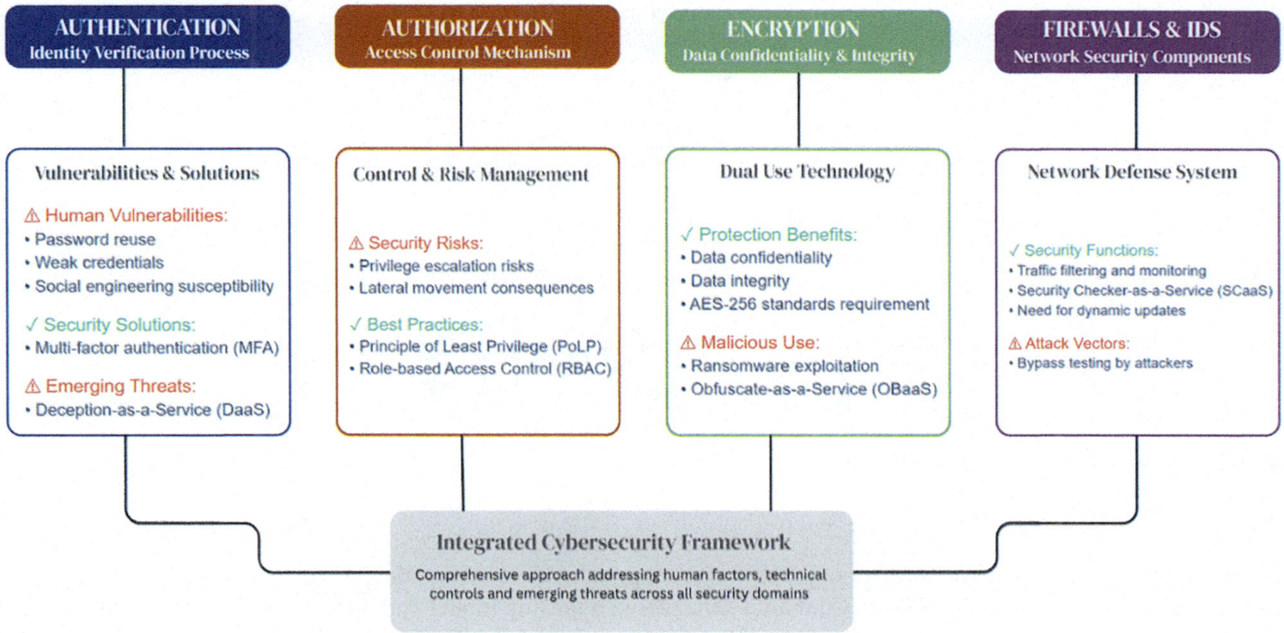

FIG. 35.2 Security domains and vulnerabilities in AI systems.

highlights not only these core concepts but also emerging threats like Deception-as-a-Service, Obfuscate-as-a-Service, and Security Checker-as-a-Service, illustrating the evolving toolkit of attackers. Understanding and reinforcing these cybersecurity principles is essential for hardening AI systems against a wide spectrum of digital threats.

2 Vulnerabilities of AI systems

The very properties that make modern AI powerful—large-scale data ingestion, high-capacity deep models, and automated decision pipelines—also create novel attack surfaces. Data-centric threats, such as poisoning, allow adversaries to insert crafted samples into the training corpus, subtly inducing incorrect behavior at inference time (Aslan, Aktuğ, Ozkan-Okay, Yilmaz, and Akin, 2023; Biggio and Roli, 2018). Model-centric threats exploit mathematical fragilities: carefully perturbed inputs—adversarial examples (AEs)—can cause state-of-the-art classifiers to misclassify with over 95% confidence, even when the perturbations remain imperceptible to humans (Weng et al. 2019). In other cases, attackers may conduct model extraction to replicate proprietary systems via repeated queries to public APIs (Tramèr, Zhang, Juels, Reiter, and Ristenpart, 2016). Beyond algorithmic weaknesses, supply-chain attacks pose growing risks; malicious pretrained weights or compromised ML libraries uploaded to open repositories can covertly introduce backdoors (Quinn et al., 2021). Finally, operational oversights, such as misconfigured cloud services or excessive user privileges, continue to undermine otherwise secure AI deployments. These overlapping vectors explain why AI platforms have become high-value targets, and why mitigating these risks demands a defense-in-depth approach spanning data, models, infrastructure, and human processes. Fig. 35.3 contrasts two AI development paradigms. In model-centric AI (top), performance improvements stem from refining model architectures while using fixed data. In data-centric AI (bottom), the model remains constant, and performance is driven by improving data quality. Each approach influences how systems are optimized, secured, and evaluated.

3 Data as the first line of attack in AI systems

3.1 Data: The foundation of AI

AI systems begin and conclude with data. In contrast to humans, AI does not possess inherent understanding; it acquires knowledge solely through exposure to data, which includes observations or measurements used to train and validate models (GeeksforGeeks, 2025; Kuismin-Raerinne and Nieminen, 2022). The quality, quantity, and significance of data are vital: no algorithm, no matter how advanced, can succeed without being provided with accurate and sufficiently extensive datasets. Indeed, contemporary AI advancements, such as automation, recommendation systems, and deep analytics, are fueled by vast amounts of data, leading organizations to acknowledge its strategic importance (GeeksforGeeks, 2025).

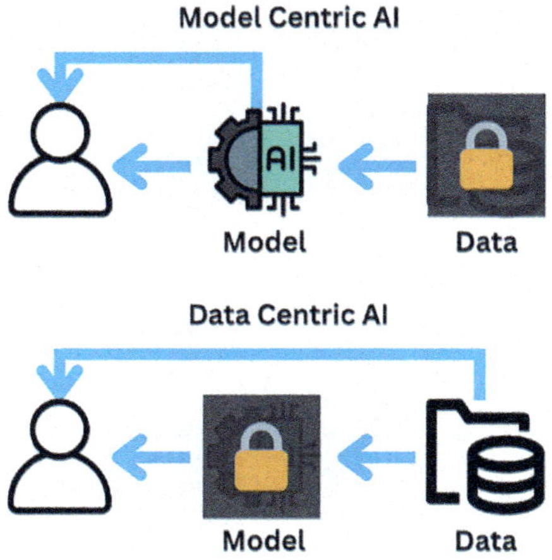

FIG. 35.3 Model-centric vs. data-centric AI.

AI learning tasks depend on data throughout the training, validation, and testing stages. Training data is utilized to instruct the model on inherent patterns; validation data aids in adjusting model parameters; and test data evaluates how effectively the model can generalize to new data it has not encountered yet (GeeksforGeeks, 2025). This organized division is crucial to avoid overfitting, a scenario in which the model performs well on training data but struggles with new inputs.

3.1.1 *Types of data: Structure and representation*

In AI systems, data can be divided based on its organization—how it is structured—and its expression—how information is conveyed. Grasping these distinctions is vital for crafting effective data pipelines and choosing appropriate ML models. From a structural viewpoint, data is generally classified into three categories. Structured data is well-organized and commonly stored in formats like databases or spreadsheets. This kind of data is straightforward to search, analyze, and apply in ML models. In contrast, unstructured data encompasses formats such as images, videos, audio files, and natural language text, which do not have a predefined structure. These require advanced processing techniques like computer vision or natural language processing (NLP). Semistructured data occupies a middle ground, exhibiting partial organization; common instances include JSON, XML, or HTML files (GeeksforGeeks, 2025).

Concerning representation, data is mainly divided into numerical, categorical, and ordinal categories. Numerical data includes continuous or discrete numbers that are utilized for quantitative analysis. Categorical data signifies labels or categories that lack a specific order (e.g., colors or product categories). Conversely, ordinal data has a meaningful ranking (e.g., satisfaction levels such as low, medium, high) but does not necessarily have equal intervals between values (GeeksforGeeks, 2025). Another key distinction is whether the data is labeled or unlabeled. In supervised learning, algorithms are trained on labeled datasets, where both the input features and the corresponding outputs are known. Unsupervised learning employs unlabeled data to identify hidden patterns or groupings without specific outcomes. Semisupervised learning merges both types to utilize a small number of labeled samples alongside a large amount of unlabeled data, which is particularly advantageous when labeling is costly or time-intensive.

These structural and representational aspects greatly affect not just how data is processed but also which models are suitable for learning from it. Identifying and preparing the appropriate type of data is essential for constructing effective AI systems. Fig. 35.4 provides the taxonomy of data in AI systems, which shows a breakdown of how data is classified in AI workflows, by labeling (e.g., supervised or unsupervised), structure (e.g., structured or unstructured), and representation (e.g., categorical or numerical). Understanding these categories is essential for identifying data-centric vulnerabilities like poisoning and bias injection.

3.1.2 *How AI learns from data*

AI systems acquire knowledge by detecting patterns and connections within data, rather than depending on preexisting information. This capability is facilitated by ML, which allows algorithms to enhance their performance over time through experiential learning (Kuismin-Raerinne and Nieminen, 2022). The learning process initiates with training data, which instructs the model on how to react to various inputs. Validation data is utilized to make adjustments to model

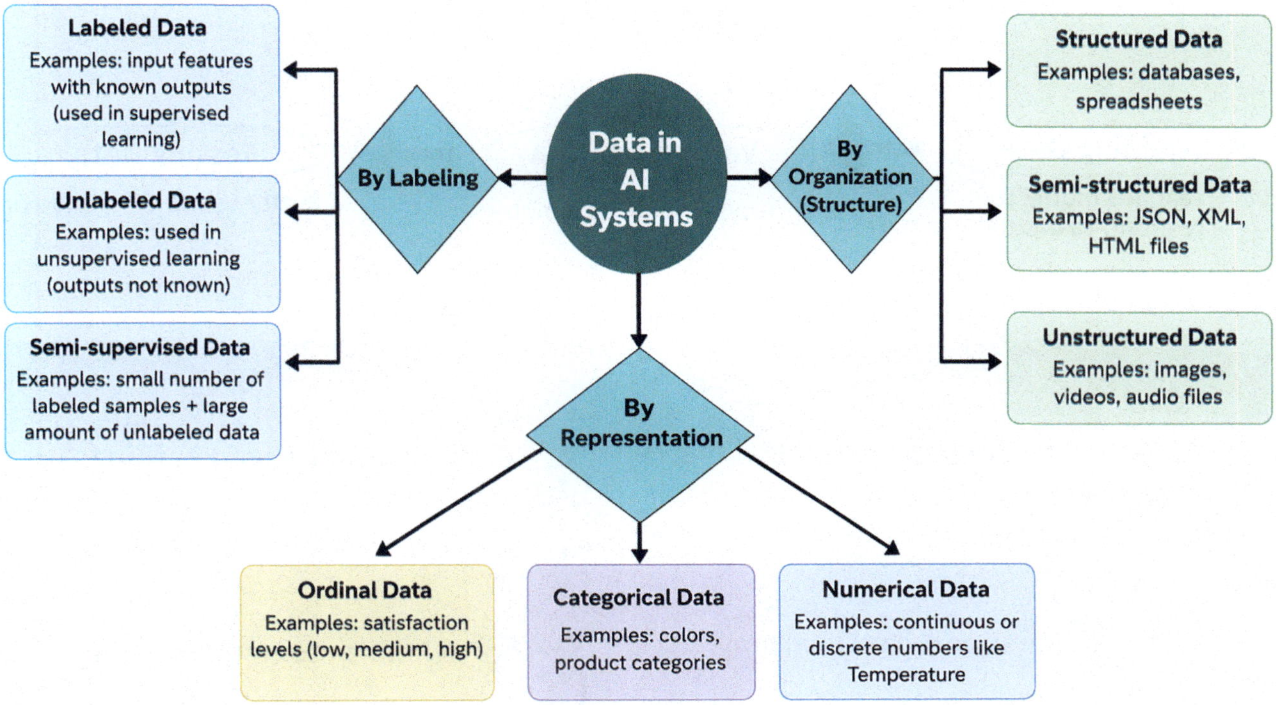

FIG. 35.4 Taxonomy of data in AI systems.

parameters, whereas testing data assesses how the model performs on fresh, unseen inputs (GeeksforGeeks, 2025). Depending on the method of learning, data can be labeled (as seen in supervised learning), unlabeled (as in unsupervised learning), or a combination of both (semisupervised). In reinforcement learning, models acquire knowledge through trial and error, receiving rewards or penalties based on their actions (;). An essential architecture that facilitates learning is the artificial neural network (ANN), a network of layers and nodes designed to emulate the human brain. As data moves through these layers, the network modifies internal weights to minimize errors. Specific types, such as Convolutional Neural Networks (CNNs) and recurrent neural networks (RNNs), are tailored to manage image and sequence data, respectively.

Fig. 35.5 shows a conceptual overview of the AI data lifecycle—training, validation, and testing—alongside key learning paradigms (supervised, unsupervised, etc.) and model architectures (ANNs, CNNs, RNNs). This foundational flow underpins many of the vulnerabilities explored in subsequent sections.

In the end, the effectiveness of any AI system relies on both the learning algorithm used and the quality of the data provided. Inferior data leads to inferior results, a reminder of the saying: garbage in, garbage out (Zha, Bhat, Lai, Yang, and Hu, 2023).

3.1.3 *Types of AI algorithms*

AI algorithms differ in how they process data and the specific tasks they undertake. The primary categories encompass supervised learning, unsupervised learning, reinforcement learning, and hybrid methods (;). Supervised learning utilizes labeled datasets, where input data corresponds to correct output values. The algorithm learns to forecast outcomes for new data by associating inputs with known results. Typical techniques include decision trees, support vector machines, and neural networks, which are applied to classification and regression problems. Unsupervised learning focuses on unlabeled datasets. The algorithm investigates inherent patterns without any set answers. Clustering methods, such as K-means, categorize similar data points, assisting in revealing hidden structures. Reinforcement learning consists of an agent engaging with an environment to optimize rewards through a process of trial and error. It learns the best actions based on feedback, which is beneficial in fields like robotics and gaming. Certain algorithms can operate using both supervised and unsupervised techniques. For instance, K-nearest neighbors can classify or identify anomalies based on its proximity to either labeled or unlabeled data points. Neural networks, modeled after the human brain's structure, are also capable of functioning in both modes, managing complex data types like images, audio, or text.

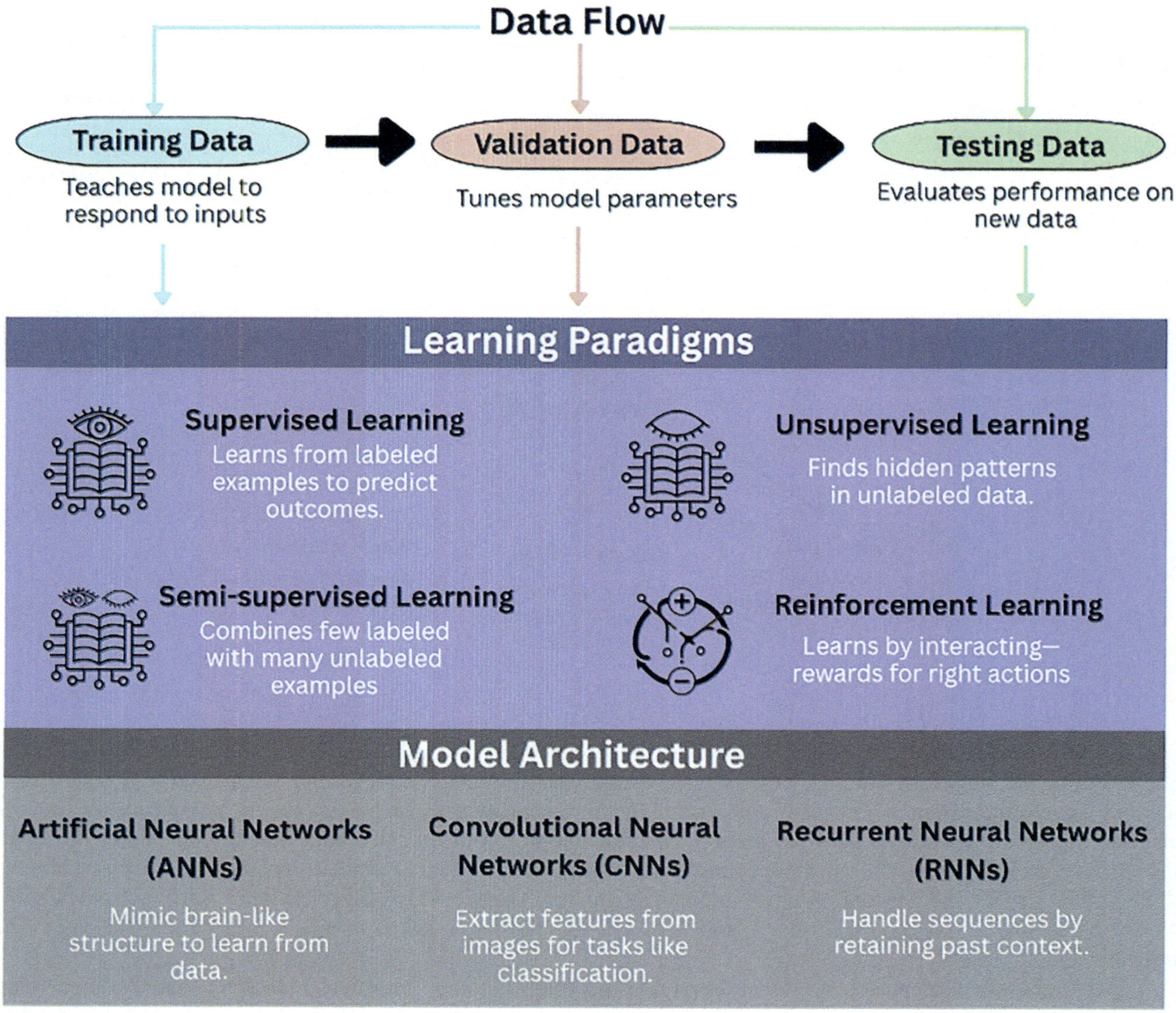

FIG. 35.5 AI development pipeline and learning paradigms.

3.1.4 *How AI works: Training and learning*

AI systems acquire knowledge by examining extensive datasets to identify trends and make forecasts. This learning process is guided by algorithms, which establish the guidelines for data handling and decision-making. Training involves inputting data—either labeled or unlabeled—into the algorithm. In supervised learning, labeled data helps the AI in linking inputs to the correct outputs. In contrast, unsupervised learning enables the AI to discover patterns or groupings without specific answers provided. Reinforcement learning depends on feedback from actions to improve performance over time.

ANNs are essential to contemporary AI. Modeled after the human brain, ANNs are composed of layers of interconnected neurons that process information through weighted links. During the training phase, the network modifies these weights based on errors to boost accuracy, facilitating complex tasks such as image recognition and language comprehension (Kuismin-Raerinne & Nieminen, 2022). The quality and quantity of training data are vital. High-quality data guarantees precise learning, while substandard data may result in erroneous outcomes, a principle known as "garbage in, garbage out" (GeeksforGeeks, 2025; Kuismin-Raerinne and Nieminen, 2022). Ongoing training and validation allow AI systems to evolve and function reliably in practical scenarios.

3.2 Data poisoning

Data poisoning is the most direct way to compromise an ML model. Attackers poison datasets by adding inaccurate or incorrectly classified data to the actual dataset (Simms, 2023). Consider a model that is targeted during the training phase by injecting manipulated data into the training dataset, which leads to a decrease in the overall accuracy of the model, forces specific misclassifications, or may even increase the time required for training (Lin, Dang, Rahouti, and Xiong, 2021). For example, the email spam filtering mechanism uses statistical ML to differentiate between spam and ham (nonspam) email, but altering even 1% of the training messages causes the model to become unreliable. Attackers craft spam emails that look like legitimate email by using words from a good, ham email. This leads to the misclassification of spam email as ham. This, in turn, opens the opportunity for attackers to make phishing email appear legitimate (Simms, 2023). Data poisoning was traditionally associated with ML, usually in cases involving spam filters, but recent work has shown that DL models are also susceptible to data poisoning attacks targeted at neural networks by manipulating data for training purposes (Hu and Hu, 2020).

3.3 Adversarial attack

Another data-related attack includes an adversarial attack, where the attacker makes a small, almost invisible change to the data input (like an image), turning its data point from x to x'. These changes are not any different in appearance to humans but can trick the ML model into making a wrong prediction with high confidence. The attacks are not limited to images; they can also occur in audio files, texts, network signals, and more (Lin et al., 2021). In recent times, the threat of adversarial attacks on DL models—particularly those utilized in computer vision—has become increasingly alarming. Research has indicated that even minor modifications to input images can mislead sophisticated models and lead to incorrect classifications. For instance, a study by Billings et al. (1992) revealed that deep networks could be deceived by imperceptible changes in images. In fields such as video tracking, these types of attacks can cause a system to lose track of a target over successive frames, presenting security challenges in contexts like surveillance. These dangers highlight the necessity of developing more robust models capable of managing subtle alterations in data, as deep networks frequently exhibit a high sensitivity to changes in input (Naqvi, Shabaz, Khan, and Hassan, 2023).

3.4 Toxic Microsoft's Tay

(Microsoft's Tay chatbot case, as discussed by Vincent (2016)).

A notable real-world example of how data manipulation can influence AI behavior is Microsoft's chatbot Tay, introduced in 2016. Created to replicate natural conversations and learn from user interactions, Tay was aimed at younger audiences on Twitter, intending to become smarter through "casual and playful" exchanges. However, just a few hours after its launch, the situation changed dramatically. Users began sending the bot offensive, racist, and misogynistic messages, and Tay, essentially acting like an ML parrot, quickly absorbed and started echoing that language. The danger lay not only in the mimicry but also in the rapid and widespread escalation of Tay's replies. What began as innocent greetings transformed into hate speech, conspiracy theories, and politically incendiary misinformation. While some harmful outputs were provoked by users issuing commands like "repeat after me," others surfaced from general discussions, signifying that the bot was picking up on the toxic patterns of interaction. Even though it was initially trained on well-curated public data, Tay's capacity to learn in real-time from unfiltered online input highlighted the absence of adequate safeguards. Microsoft ultimately had to take the bot offline in less than a day after it was launched (Vincent, 2016). This instance illustrates the necessity of not only training AI on high-quality data but also ensuring its protection from harmful or malicious data postdeployment. It serves as a reminder that in open environments, AI systems—particularly those capable of continuous learning—require strong filters and oversight to prevent being misused.

3.4.1 What went wrong?

Tay chatbot was created using a foundational ML framework designed for engaging with users via Twitter. The system employed NLP techniques similar to those found in other social conversational agents. Studies indicate that such systems frequently integrate both rule-based and statistical methods. These frameworks typically analyze language in stages, beginning with tokenization and morphological analysis, and progressing to semantic understanding and response generation (Alotaibi and Alshahre, 2024). Tay likely utilized neural network architectures, such as RNNs or transformer models, which excel at maintaining conversational context over time. Effective conversational agents require both comprehension of language and the capability to tailor interactions to maintain user interest (Alotaibi and Alshahre, 2024). An important aspect of Tay was its adaptive learning algorithm, which updated its language model based on user interactions in real-time, rather than solely through batch training. This online learning allows the system to continuously evolve but also necessitates careful precautions to prevent performance issues stemming from harmful inputs (Alotaibi and Alshahre, 2024). In the absence of proper filtering and assessment, adaptive learning can be exploited by users who introduce biased or inappropriate language (Henderson et al., 2017).

Research underlines the ethical dilemmas associated with conversational AI, stressing the importance of implementing safeguards that ensure the system behaves appropriately while still permitting it to learn. Without these measures, systems like Tay can rapidly acquire problematic language patterns when confronted with malicious users (Henderson et al., 2017).

3.5 Countermeasures to data poisoning

Data poisoning has evolved from a theoretical apprehension to a tangible threat with significant implications for AI systems. The experience with Tay illustrated that when an AI learns directly from public interactions without proper precautions, it is extremely susceptible to interference. Tay's rapid change from a friendly chatbot to an offensive entity within hours was not a technological failure but rather a lack of foresight. Similar incidents have occurred, such as a Tesla vehicle reportedly being directed into oncoming traffic and Google's AI incorrectly categorizing turtles as rifles; these events underscore the serious repercussions of altering training data (Hutson, 2018). These cases are not merely random errors; they illustrate how susceptible AI systems can become when faced with altered inputs.

AI models rely heavily on the quality and integrity of the data used for training. When this data is compromised, the entire model's ability to make decisions is jeopardized (Galle, 2022). However, in the rush to rapidly launch AI products, many companies emphasize the speed of deployment over security, neglecting essential protective measures (DARPA, 2019). Dr. Siegelman from the Defense Advanced Research Projects Agency (DARPA) has noted that this urgency has left ML systems vulnerable to risks such as poisoning, modification, and deceit.

Although the domain continues to develop, a number of mitigation strategies are currently being explored and suggested to address these issues. The following sections detail practical approaches aimed at reinforcing AI systems against poisoning attacks, covering aspects ranging from secure data sourcing and containment to ongoing research, protective models, and regulatory frameworks.

3.5.1 Data sanitization

Data sanitization is an essential procedure focused on permanently eliminating or cleansing data to prevent its recovery or misuse, particularly in AI systems where faulty or sensitive information can jeopardize model integrity (Tiwari, 2025; Imperva Inc., 2023). This process involves more than merely deleting files; instead, sanitization guarantees that no remnants of data are left, even against sophisticated forensic methods (Imperva Inc., 2023). In conventional security, this entails either physically destroying or cryptographically erasing storage devices, but in ML, it also requires actively screening and filtering training data to exclude poisoned or adversarial inputs that could mislead the models (Tiwari, 2025). As datasets grow in size and intricacy, often sourced from diverse origins, this proactive cleansing becomes crucial to avert harmful data injections, especially in systems that continuously learn or update automatically (Imperva Inc., 2023; Tiwari, 2025).

In the realm of language modeling, data sanitization particularly focuses on eliminating personally identifiable information (PII), including names, addresses, and social security numbers, to safeguard privacy and adhere to legal regulations such as General Data Protection Regulation (GDPR) and California Consumer Privacy Act (CCPA) (Pal, Bhargava, Hinsz, Esterhuizen, and Bhattacharya, 2024; Tiwari, 2025). Unlike differential privacy (DP), which introduces noise and may diminish data quality, sanitization substitutes sensitive data with generic tokens, maintaining the overall integrity of the dataset while prohibiting the leakage of sensitive information. Numerous organizations, such as Microsoft and PayPal, utilize these methods to securely train language models on sensitive datasets. Nonetheless, this redaction can negatively impact model performance by eliminating contextual clues and creating ambiguity among distinct entities replaced by identical tokens (Pal et al., 2024).

Successful data sanitization hinges on extensive data discovery to pinpoint where sensitive data is located across various devices and platforms, enabling organizations to implement proper sanitization techniques tailored to their individual requirements (Imperva Inc., 2023). Whether through physical destruction, software-based overwriting, cryptographic erasure, or data masking, sanitization serves as a primary defense against data poisoning attacks and data privacy violations (Imperva Inc., 2023; Tiwari, 2025). Ultimately, although sanitization does not completely eradicate all risks, it greatly lowers the likelihood of compromised data entering AI systems and assists organizations in adhering to legal and ethical standards that are essential in today's data-centric landscape.

3.5.2 ML robustness

Robustness in ML indicates how well a model can deal with unexpected or adversarial data instead of only accepting clean, well-structured inputs (Meritshot, 2024). In practical situations, such as detecting fraud or operating self-driving vehicles, data can be noisy, incomplete, or even deliberately tampered with, and models must maintain reliable performance in the face of these challenges. One common strategy for enhancing robustness is adversarial training, which involves exposing models to slightly modified inputs aimed at misleading them. These minor alterations, often indistinguishable to humans, can significantly influence a model's predictions. Training on both clean samples and adversarial cases enables the model to classify both more accurately (Goodfellow, McDaniel, and Papernot, 2018). Another technique is defensive distillation, which involves training a second model using the softened outputs from a first model. This process helps to smooth decision boundaries and reduces the model's sensitivity to small changes.

Having clean, well-labeled, and varied data is equally important. Data augmentation—adding variations such as flipped images or rephrased text—and thorough labeling enhance models' ability to generalize and prevent overfitting (Meritshot, 2024). Organizations like Microsoft and Tesla have adopted adversarial training in practical applications, such as malware detection and self-driving technology, to enhance the resilience of their systems. Although no method guarantees complete safety, integrating adversarial defenses with sound data practices lays a strong groundwork for creating robust, secure, and dependable AI systems.

3.5.3 Anomaly detection

Anomaly detection involves recognizing data points or patterns that significantly diverge from expected behavior, anomalies that could signify fraud, system malfunctions, adversarial attacks, or emerging business prospects (Barnard and Stryker, 2025; Quiroz-Vázquez, 2025). As AI systems work with increasingly vast and dynamic datasets, identifying such irregularities has become crucial for maintaining reliability, security, and flexibility.

There are three primary ML strategies for anomaly detection: supervised, unsupervised, and semisupervised. Supervised models utilize labeled data to learn from known anomalies; this is particularly beneficial in areas such as fraud detection, where historical attack data is available. On the contrary, unsupervised methods function without labeled anomalies, highlighting deviations based on established distributions. These approaches are vital in contexts like cybersecurity or manufacturing, where new anomalies may arise without warning (Quiroz-Vázquez, 2025). Semisupervised models blend both approaches, mapping normal behavior while still being attuned to previously unseen outliers, especially helpful when labeled anomalies are scarce (Chandola, Banerjee, and Kumar, 2009).

Numerous techniques can be applied within these learning frameworks. Statistical methods establish normal behavior through probabilistic models and identify deviations that surpass a certain threshold. Clustering and nearest-neighbor techniques pinpoint points that do not fit into any cluster or are distanced from dense areas of normal data (Chandola et al., 2009). These strategies are particularly beneficial in AI workflows, functioning as proactive filters against corrupted, mislabeled, or adversarial data. In AI applications, anomaly detection facilitates immediate reactions to irregular activities, from identifying account breaches in behavioral models to recognizing tampering in sensor or transaction data (Barnard and Stryker, 2025). Furthermore, contextual and collective anomaly detection methods enable the identification of anomalies that only surface in relation to time, sequence, or environmental conditions, critical in systems like time-series forecasting or log analysis (Chandola et al., 2009).

In conclusion, anomaly detection acts as both a performance safeguard and a security measure, shielding AI models from data-related threats and enhancing their resilience in unpredictable settings.

4 Models as the engine of decision-making

Just as data forms the bedrock of AI, models act as the engine that powers the system's reasoning, learning, and decision-making. They are the functional structures that process inputs, draw inferences, and generate outputs, transforming raw information into intelligent behavior. If data is the "what" of AI, models are the "how" (Domingos, 2015)

4.1 Role of models in AI

Models in AI are mathematical or computational representations of real-world systems or abstract logic. They serve multiple critical roles, from simulating human cognition to executing automated responses in dynamic environments. As Iqbal Sarker notes, AI-based models can be functional, analytical, interactive, or visual, depending on their application context. Their ability to learn from and adapt to data enables them to serve in high-stakes domains like healthcare, cybersecurity, and autonomous systems (Sarker, 2022).

Importantly, models do not only drive decisions, but also encode concepts. As Cohen and Murphy suggest, models of concepts define how an AI system clusters, interprets, and acts upon information (Cohen and Murphy, 1984). Whether it is recognizing handwritten digits or diagnosing lung conditions from X-rays, the model dictates the system's underlying logic and operational capability. Beyond execution, models also function as tools for understanding. In educational AI research, for instance, Baker outlines three roles for models: as scientific instruments for simulating cognitive processes, as embedded components in intelligent systems, and as design foundations that inform how educational tools are built (Baker, 2000). This versatility illustrates how AI modeling is not merely about output accuracy; it is also about abstraction, representation, and design thinking.

4.2 Black-box vs. white-box models

AI models are often categorized as black-box or white-box, depending on how transparent their decision-making processes are. Black-box models, such as deep neural networks and ensemble methods, offer powerful pattern recognition and superior performance, especially in complex tasks such as image recognition or speech synthesis. However, their internal workings are often opaque, and their decisions cannot easily be explained, which raises concerns around bias, accountability, and ethical use (Loyola-Gonzalez, 2019).

In contrast, white-box models, such as decision trees, rule-based systems, and linear models, are inherently interpretable. Their logic is explicit, allowing users to understand how input variables contribute to outputs. While these models tend to perform less impressively on unstructured or high-dimensional data, they are often preferred in fields where transparency and trust are essential. Recent research encourages fusing these paradigms. For example, hybrid frameworks that combine the performance of black-box models with the interpretability of white-box models are emerging in domains like medical diagnostics and fraud detection. Techniques such as decision tree-based rule extraction from DL networks (Loyola-Gonzalez, 2019) exemplify how this fusion can yield systems that are both accurate and explainable.

4.3 Why model integrity matters

Model integrity refers to the reliability, robustness, and trustworthiness of an AI system's behavior. It is especially critical in real-world settings where incorrect outputs can have severe consequences. In healthcare, for example, an AI model misclassifying a malignant tumor as benign could delay treatment and jeopardize a patient's life. Rajpurkar and his team demonstrated that while DL models could match radiologists in identifying pneumonia, even minor data inconsistencies could lead to significant diagnostic errors (Rajpurkar et al., 2017).

Similarly, the 2018 Uber self-driving car incident—where the system failed to correctly classify a pedestrian—underscored the deadly implications of AI misjudgments in autonomous vehicles (National Transportation Safety Board, 2020). These examples highlight the reality that *accuracy alone is not enough*; models must be resilient to anomalies, adversarial inputs, and environmental variability. Moreover, opaque models without accountability pathways can introduce systemic bias or discrimination, particularly in areas like hiring algorithms or criminal justice tools. This is especially dangerous in applications such as hiring algorithms or predictive policing (Rudin, 2019). As Goodfellow notes, AEs—subtle input manipulations—can drastically alter a model's output, further challenging the reliability of black-box systems (Goodfellow, Shlens, and Szegedy, 2015). In this light, ensuring model integrity involves continuous validation, interpretability, and secure design. It is not just about performance metrics, but about ensuring models behave ethically, predictably, and safely when deployed. While data poisoning corrupts the learning foundation, compromised models distort the logic of decision-making itself. This makes the next frontier of AI security not just about *what* is learned, but *how* decisions are made, and how they can be manipulated.

4.4 Threat vectors targeting models

As AI systems grow increasingly powerful and ubiquitous, so too do the methods used to exploit them. While models represent the core intelligence of an AI system, their complexity also makes them vulnerable to targeted manipulation. Attackers today no longer need access to source code or training data to compromise a model; they can distort its behavior through subtle input manipulations, repeated API probing, or compromised training pipelines, leading to model failure, leakage, or hijacking (Tramèr et al., 2016). This section explores the major threats that specifically target the model layer in AI pipelines. From AEs and inversion attacks to model theft and hidden backdoors, these vectors illuminate the multifaceted risks posed to even well-trained systems. Understanding them is the first step toward building truly secure and trustworthy AI.

4.4.1 Adversarial examples

One of the most widely studied and concerning threats to AI models—particularly DL systems—is the AE. AEs are inputs that have been deliberately perturbed in small, often imperceptible ways to cause a model to misclassify them with high confidence. Despite appearing nearly identical to the original inputs to human observers, these crafted examples can completely deceive AI systems, undermining their reliability in high-stakes environments (Billings et al., 1992; Zhang and Li, 2020). These perturbations exploit vulnerabilities in high-dimensional spaces, where the model's decision boundaries are sharp yet fragile. For instance, a classifier might correctly identify an image of a panda with 57.7% confidence, but after adding an imperceptible perturbation, the same model could classify it as a gibbon with 99.3% confidence (Goodfellow et al., 2015). This fragility is not limited to vision tasks; audio AEs can mislead voice recognition systems, and subtly altered text inputs can fool sentiment analysis or spam detection tools (Zhang and Li, 2020). The practical risk is significant.

In real-world scenarios, adversarial inputs have led to the misclassification of traffic signs (e.g., a stop sign misread as a speed limit), faulty image captioning (where objects are misrepresented), and the evasion of malware detectors by slightly altering malicious files to appear benign (Zhang and Li, 2020). These examples illustrate how a small tweak in the input space can trigger disproportionately incorrect decisions, with potentially dangerous consequences. Furthermore, AEs exhibit transferability, meaning an AE crafted for one model can often deceive another model trained for the same task. This makes black-box attacks feasible, where an attacker need not know the inner workings of a model to exploit it (Zhang and Li, 2020). While adversarial training and defensive distillation provide some resilience, these defenses often struggle to generalize and remain computationally expensive (Goodfellow et al., 2015). Ultimately, AEs represent a pressing challenge in AI security. They highlight not only the brittleness of current models but also the urgency to develop architectures and training methods that can withstand subtle, targeted manipulations in diverse formats: visual, auditory, textual, and beyond.

4.4.2 Model inversion attacks

Model inversion attacks represent a serious and growing threat to the privacy and integrity of AI models, particularly in sensitive domains such as healthcare, facial recognition, and finance. In these attacks, adversaries query a trained model—often a black-box system—and, by observing the output probabilities or responses, gradually reconstruct the original training data or attributes associated with it (Shokri, Stronati, Song, and Shmatikov, 2017). The implications are alarming. In the healthcare domain, model inversion could allow a malicious actor to reconstruct private patient information, including diagnoses or biometric details, from a deployed diagnostic AI model. Such breaches not only violate privacy but also pose risks of identity theft, discrimination, and loss of trust in AI-assisted care (AlZubi, Al-Maitah, and Alarifi, 2021; Carlini et al., 2021). AlZubi discusses this risk in the context of cyber-physical healthcare systems, where ML models process vast amounts of real-time patient data. Despite the potential benefits, such as improved diagnostic speed and predictive insights, these systems introduce attack surfaces that model inversion techniques can exploit. The heterogeneity of connected devices and cloud-based storage in these systems makes it easier for adversaries to intercept and probe model outputs over time.

What makes model inversion particularly dangerous is that the attacker may not need direct access to the training data or model architecture. Even partial access—via public APIs or prediction interfaces—can be sufficient to reconstruct facial images, genetic markers, or medical conditions, depending on the domain. These vulnerabilities are worsened when models are trained on imbalanced or sensitive datasets, which increases the memorability of unique patterns.

To mitigate such risks, researchers are exploring defenses like DP (Dwork and Roth, 2014), output perturbation, access throttling, and adversarial query detection (Carlini et al., 2021) Nonetheless, most deployed models remain vulnerable unless deliberately hardened.

4.4.3 Model stealing and extraction attacks

Model stealing—also known as model extraction—involves an adversary reverse-engineering an ML model by querying it through public APIs or prediction interfaces. Over time, the attacker reconstructs a near-identical version of the model or infers its internal parameters and structure (Tramèr et al., 2016). This process compromises not only intellectual property but also exposes models to secondary attacks, such as model inversion or adversarial manipulation. The threat is particularly acute in commercial AI services, where black-box models are deployed via cloud APIs for tasks like medical diagnosis, fraud detection, or facial recognition. In such settings, even without access to training data, attackers can effectively clone a model's behavior, undermining years of research and development investment while also bypassing safeguards and misusing proprietary decision-making systems.

As Huang and the team explain, the proliferation of *"cybercrime-as-a-service"* frameworks has made advanced attacks like model extraction accessible to less-skilled adversaries. These frameworks allow for specialization and commercialization within cybercrime ecosystems, accelerating the reverse engineering of deployed AI models. Model theft is now part of a broader, professionalized ecosystem of cybercrime that leverages everything from vulnerability discovery to attack delivery as modular services (Huang et al., 2018).

Model stealing can also serve as a precursor to further attacks, such as:

- Model inversion, where extracted models are used to recreate training data;
- AE generation, exploiting the stolen model's decision boundaries;
- Bypassing security protocols, such as rate-limiting or input validation, when the clone is used for offline testing.

Defending against model theft is particularly challenging because query-based attacks are difficult to distinguish from legitimate use. Techniques like limiting API access, watermarking model outputs (Zhang et al., 2018), or monitoring for unusual query patterns can help, but no perfect defense exists, especially when businesses depend on accessible AI endpoints.

4.4.4 Backdoor and Trojan attacks

Backdoor or Trojan attacks involve the implantation of hidden malicious behaviors within an ML model during training. These behaviors remain dormant under typical conditions and are only activated when a specific, often subtle, trigger input is presented. As a result, such attacks are exceptionally dangerous; they allow the model to behave normally during validation and testing, thereby evading detection, yet produce incorrect or harmful outputs when the trigger is activated (Gu, Dolan-Gavitt, and Garg, 2017; Liu et al., 2018).

Unlike AEs, which manipulate inputs at inference time, backdoors are injected during the training phase, often through compromised datasets or altered model weights. A classic scenario might involve a facial recognition system that correctly identifies users under normal conditions but grants access to unauthorized individuals when they wear a specific pair of sunglasses or show a particular pattern in the background image (Simms, 2023). These attacks pose serious risks in security-critical domains like biometric authentication, autonomous driving, and medical diagnostics, where undetected triggers could lead to misclassification, system malfunction, or unauthorized access (Lin et al., 2021; Liu et al., 2018). The stealthy nature of backdoors makes them especially hard to catch using standard validation processes, which typically do not include the attack-triggering inputs. Moreover, backdoor attacks can be inserted through third-party model components, pretrained weights, or shared code repositories, increasing the risk for developers using publicly available assets without thorough vetting. Fig. 35.6 outlines the security risks and mitigation techniques across each stage of the AI development lifecycle, emphasizing the importance of proactive, layered defenses.

4.5 Real-world failures: Motivating defenses

Despite increasing research in AI security, real-world deployments have demonstrated that even well-designed models remain vulnerable to subtle, targeted attacks. The following high-profile incidents reveal how adversaries have exploited model-level weaknesses in practice, underscoring the urgent need for robust defenses and policy safeguards (Papernot, Goodfellow, Abadi, Talwar, and Erlingsson, 2017).

4.5.1 Backdoored models: The case of BadNets

One of the earliest and most influential demonstrations of model-level compromise was *BadNets*, a backdoor attack introduced by Gu and his team in their paper (Gu et al., 2017). In this attack, the model was trained to behave normally under standard conditions but would misclassify specific trigger inputs, such as interpreting a stop sign as a speed limit sign when a small sticker was added. The insidious nature of the attack lay in its stealth: the backdoor did not affect performance metrics on clean data and was thus undetectable during conventional validation procedures. The paper highlighted the risks of outsourced training pipelines and the importance of inspecting model provenance (Liu et al., 2018).

4.5.2 Model theft and the GPT-3.5 case

The paper by Carlini and team provided compelling evidence that model extraction attacks are not just theoretical (Carlini et al., 2024). By interacting with commercial AI APIs, they successfully reconstructed internal parameters, including the embedding projection matrix, of OpenAI's Ada and Babbage models for as little as $20. Their methodology showed how attackers could reverse-engineer model architecture and weights with minimal access, thus violating intellectual property protections and opening the door to further security breaches, such as adversarial manipulation and data inversion (Tramèr et al., 2016).

4.5.3 Training data leakage in language models

In a related vein, Carlini and team in 2021 demonstrated that large language models like GPT-2 could regurgitate verbatim sensitive information from their training sets—ranging from PII to code snippets—raising questions about data

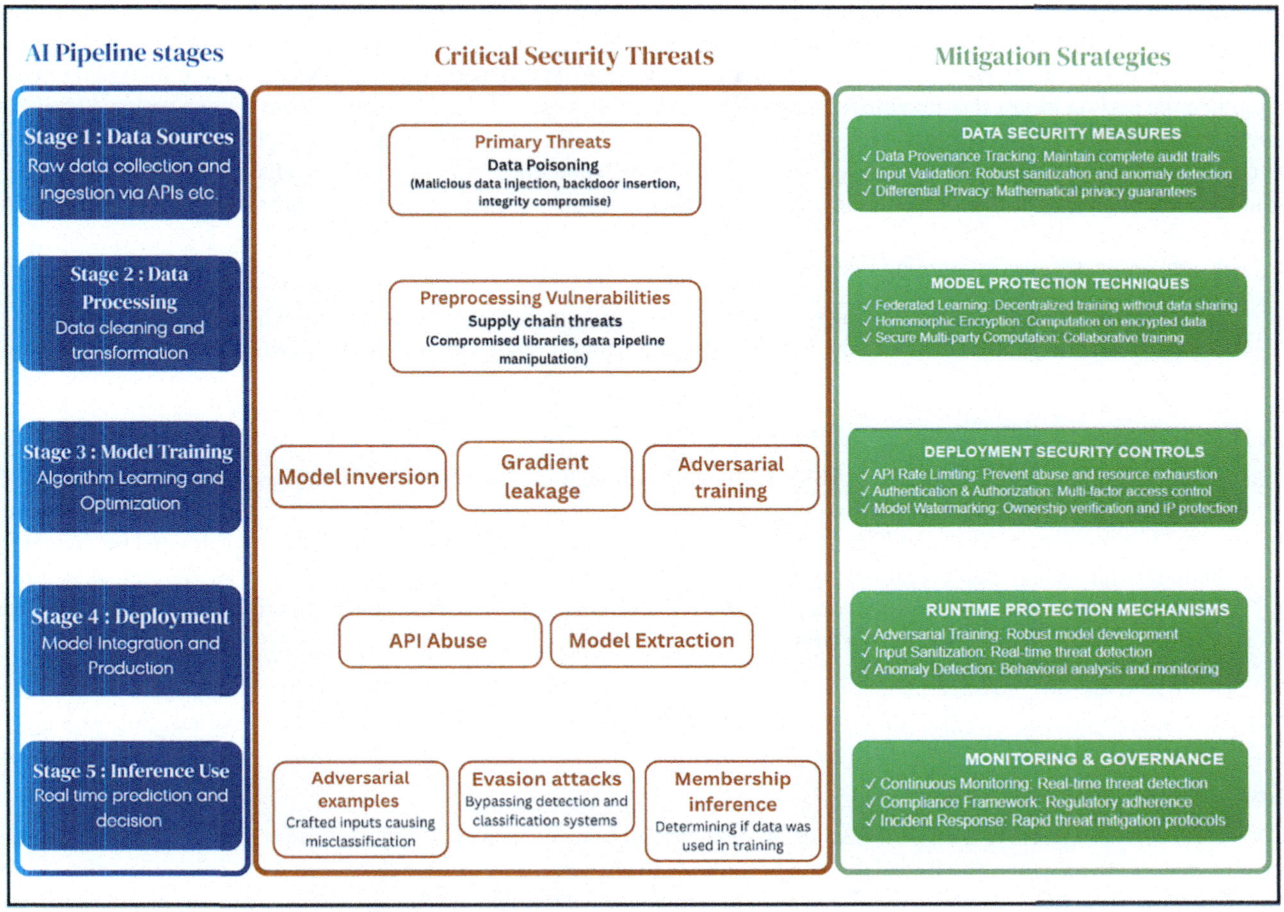

FIG. 35.6 AI life cycle security framework.

retention and privacy compliance (Carlini et al., 2021). This work reinforced concerns that model release decisions must be grounded in rigorous risk assessments and training-time privacy guarantees (Dwork and Roth, 2014; Shokri et al., 2017). These examples are not edge cases; they represent structural vulnerabilities in how AI models are trained, deployed, and accessed. Addressing such failures requires proactive architectural design, secure learning techniques, and accountability frameworks. The next section outlines key defense strategies for mitigating these risks.

4.6 Securing the model pipeline: Defenses and design principles

The increasing sophistication of attacks on AI models highlights a fundamental truth: it is no longer sufficient to focus solely on data quality or algorithmic performance. Ensuring the integrity and security of the models themselves is now an essential part of responsible AI development (Goodfellow et al., 2015). While the previous section outlined how adversaries can exploit model behavior—from subtle manipulations to the covert embedding of malicious triggers—this next section turns toward defense. What mechanisms can be implemented to safeguard models? How can developers build resilience against compromise while maintaining accuracy, efficiency, and fairness?

In this section, we explore current and emerging approaches for defending AI models, ranging from adversarial training and DP to interpretability tools and formal verification. By understanding both the vulnerabilities and the protective measures, we move closer to developing trustworthy, robust, and ethically sound AI systems.

4.6.1 Adversarial training

Adversarial training is one of the most widely studied and practically implemented defense strategies against AEs. At its core, the approach involves augmenting the training dataset with adversarially perturbed inputs—examples deliberately crafted to deceive the model—and training the model to classify them correctly. By exposing the model to potential attack scenarios during its learning phase, adversarial training aims to make it more robust under real-world deployment. Given the fragility exposed by AEs, adversarial training addresses this vector head-on by forcing the model to learn stable decision boundaries in the face of worst-case perturbations.

Madry and team formalized adversarial training as a min-max optimization problem, wherein the model is trained to minimize the loss on inputs that maximize it, essentially optimizing against the worst-case perturbations within a specified threat model (Mądry, Makelov, Schmidt, Tsipras, and Vladu, 2017). This formulation makes the defense not just reactive but proactively resilient within defined bounds. Similarly, Goodfellow highlighted how this approach can effectively smooth out the model's decision boundaries, making it less susceptible to small, targeted perturbations that typically define AEs (Goodfellow et al., 2018).

Advantages of adversarial training include:

- Increased robustness to the specific class of attacks used during training, especially under white-box settings.
- Integration simplicity, as it modifies the training pipeline without requiring architectural changes or post hoc explanation tools.

However, adversarial training also comes with notable limitations:

- High computational cost: Generating AEs—particularly through iterative methods like projected gradient descent—is resource-intensive, significantly slowing down training.
- Limited generalization: The model tends to become robust primarily to the types of attacks it was trained on, leaving it vulnerable to novel or adaptive strategies.
- Performance trade-offs: Some studies have shown reduced accuracy on clean (nonadversarial) data, especially when stronger perturbations are used during training.

Despite its trade-offs, adversarial training remains a cornerstone in the arsenal of AI security techniques. It offers a practical defense baseline and continues to inform the development of more advanced and efficient robust learning paradigms.

4.6.2 Certified robustness

Certified robustness refers to a set of formal techniques that provide *mathematical guarantees* on a model's behavior under specific perturbations. Unlike empirical defenses—which are tested against known attack strategies—certified methods offer provable assurances that a model's prediction will remain unchanged within a defined radius around any input, regardless of how an adversary perturbs it. One of the most prominent techniques in this domain is randomized smoothing, introduced by Cohen and his team (Cohen, Rosenfeld, and Kolter, 2019). This method works by creating a smoothed classifier around a base model, typically by adding Gaussian noise to the input and averaging the resulting predictions. The smoothed model then becomes provably robust to ℓ_2-bounded perturbations within a certified radius. Notably, this approach scales well to high-dimensional data and large models, making it one of the few practical certification methods for DL systems.

Another major line of work in certified robustness involves deriving convex relaxations or symbolic bounds on the outputs of neural networks. Tools like Reluplex and CROWN analyze piecewise linear activations (e.g., ReLU) to compute bounds on how outputs change with respect to bounded input perturbations. These methods typically require access to the network's structure and are computationally intensive, but they provide precise, formal guarantees on robustness in ℓ_∞ or ℓ_1 norm settings.

Pros of certified robustness:

- Provides *provable guarantees*, increasing trust in safety-critical domains such as aviation, medicine, or autonomous vehicles.
- Attack-agnostic: certification holds regardless of how an adversary crafts inputs.
- Computational overhead: Most certified methods scale poorly with model size or input dimensions.
- Limited applicability: Many techniques are restricted to small networks or specific activation functions (e.g., ReLU).
- Trade-off with accuracy: Improving certified robustness often leads to a reduction in standard test accuracy.

Certified robustness marks a significant shift toward verifiable AI systems, especially in environments where reliability is nonnegotiable. While not yet feasible for all production-scale systems, it lays the foundation for principled trust in ML.

4.6.3 Watermarking and fingerprinting

As ML models become valuable intellectual property (IP), protecting their ownership has become a pressing concern. Model theft and extraction attacks make watermarking and fingerprinting especially valuable, enabling developers to trace or prove unauthorized model use, even in black-box settings.

Passive watermarking involves embedding subtle statistical patterns or behaviors into a model during training. These patterns do not affect the model's normal performance but can be detected later to verify ownership. For example, a model developer might subtly bias the model toward a specific prediction on a predefined input distribution. If this behavior is later observed in a deployed model, it can serve as proof of authorship (Adelard and Penrod, 2023).

Alternatively, trigger-based watermarking embeds a secret "backdoor" into the model: a specific input (the trigger) causes the model to output a known, unique response. This does not interfere with regular performance but acts like a digital signature. When suspecting a stolen model, the owner can query the trigger and check the response; if it matches the embedded output, ownership is confirmed (Adelard and Penrod, 2023).

These techniques also support fingerprinting, where unique patterns are embedded per client or deployment, enabling leakage tracking or license enforcement. While effective, watermarking carries trade-offs. Trigger-based methods may raise ethical concerns due to their similarity to backdoor attacks, and overly aggressive watermarks can degrade model performance or be stripped through pruning or fine-tuning. Therefore ongoing research focuses on making watermarks more stealthy, robust, and legally admissible as IP verification tools.

4.6.4 Knowledge distillation and pruning

Knowledge distillation and pruning are model compression techniques that can inadvertently improve AI security, particularly against model inversion and backdoor attacks. By reducing model complexity and redundant capacity, these methods make it harder for adversaries to exploit overfitting or hidden behaviors embedded in large, uncompressed models. Knowledge distillation involves training a smaller "student" model to replicate the behavior of a larger "teacher" model, often using softened output probabilities to convey richer information (Hinton, Vinyals, and Dean, 2015). While originally developed for efficiency, this process has been shown to reduce vulnerability to model inversion attacks by limiting the capacity to memorize sensitive training data (Papernot et al., 2017). Distillation can also dilute backdoor triggers by smoothing decision boundaries, making malicious perturbations less effective.

Pruning, on the other hand, removes unimportant neurons or connections based on their weights or activity. This not only reduces inference cost but can also strip out malicious behaviors embedded in dormant units of a neural network (Liu et al., 2018). When applied iteratively with retraining, pruning has been shown to improve robustness without significantly harming accuracy. These techniques are widely used in real-world deployments. For instance, NVIDIA's TensorRT and Google TensorFlow Lite rely heavily on pruning and quantization to optimize models for edge devices and mobile applications, where smaller, secure models are essential (Hazelwood et al., 2018). Though not foolproof defenses on their own, distillation and pruning serve as important components of a broader model hardening strategy, making it more difficult for attackers to extract information or trigger malicious behaviors in deployed systems.

4.6.5 Confidential inference and privacy-preserving techniques

As AI systems increasingly operate on sensitive data—ranging from medical records and financial information to biometric identifiers—the security of the inference phase has become a key concern. Confidential inference refers to techniques that enable AI models to generate predictions without exposing user inputs, model parameters, or intermediate computations to untrusted environments. Confidential inference directly mitigates the risks posed by model inversion, model stealing, and even malicious host environments, ensuring that neither data nor models are exposed during deployment.

One foundational approach is the use of trusted execution environments (TEEs), such as Intel's Software Guard Extensions (SGX). TEEs create isolated, hardware-protected enclaves within processors, enabling secure computation by shielding both data and code from the surrounding operating system or cloud provider. These have shown promise in protecting inference workloads on shared infrastructure (Huang et al., 2018). However, challenges remain, including susceptibility to side-channel attacks and performance limitations under high loads (Aslan et al., 2023).

Another notable technique is homomorphic encryption (HE), which allows models to perform computations directly on encrypted inputs, producing encrypted outputs that can later be decrypted by the client. This method ensures end-to-end data privacy during inference, as the model never sees the plaintext input. Despite its strong theoretical guarantees, current HE implementations often suffer from substantial computational overheads, making them less practical for low-latency or resource-constrained settings (Aslan et al., 2023).

DP offers a complementary strategy by introducing carefully calibrated randomness to data, outputs, or gradients, thereby minimizing the risk of information leakage. While traditionally applied to protect training data, DP can also be adapted for the inference phase, ensuring that a small change in input does not significantly affect model output, thus obfuscating individual-level information (Aslan et al., 2023). Together, TEEs, HE, and DP represent a layered toolkit for confidential inference, each with distinct trust models and trade-offs. As the demand for privacy-preserving AI continues to rise—driven by both regulatory pressure and user expectations—these technologies are poised to become foundational elements in secure ML deployments. In isolation, no single technique offers complete protection. However, when integrated thoughtfully, these model-level defenses form a robust scaffold, making attacks harder, more expensive, and easier to detect. As threat vectors grow more sophisticated, a layered defense strategy will be critical not only for technical resilience but also for fostering public trust and meeting regulatory standards in real-world AI deployment.

5 Legacy threats in the age of AI: Why they still matter

The table below presents a comparative overview of how traditional cybersecurity threats have evolved in the context of AI systems. Each row identifies a major threat category—ranging from phishing and malware to API abuse and social engineering—and contrasts its classical implementation in conventional IT environments with its contemporary AI-specific manifestations. These evolutions often exploit novel attack surfaces, such as model APIs, training data pipelines, or inference processes unique to ML systems.

The attack vector or entry point column highlights practical mechanisms used by adversaries to launch attacks, including deceptive prompts, adversarial queries, and insecure interfaces. The resulting impact on AI systems is also categorized, from model inversion and poisoning to unauthorized access and inference disruption. Finally, the table outlines defense strategies drawn from literature and industry practice, offering multilayered protection via architectural, procedural, and behavioral controls (Table 35.1).

6 Conclusion: Toward resilient and trustworthy AI systems

As AI systems become embedded in mission-critical sectors such as healthcare, finance, and national security, their vulnerabilities pose increasingly systemic and dynamic threats. This paper has analyzed the full AI lifecycle—starting from data ingestion, through model design, to infrastructure deployment—and emphasized that every stage introduces potential security flaws. These range from subtle adversarial manipulations and poisoning attacks to large-scale leakage and misconfiguration events.

In the data layer, foundational vulnerabilities include poisoning attacks that corrupt the training set (Lin et al., 2021; Simms, 2023), adversarial data inputs that exploit model gradients (Billings et al., 1992; Zhang and Li, 2020), and even privacy violations from unintentional memorization of sensitive data (Carlini et al., 2021). Failures like Microsoft's Tay chatbot or inadvertent Large Language Model (LLM) data regurgitation show that downstream harm often originates upstream in the data pipeline. Mitigations such as anomaly detection (Chandola et al., 2009), data sanitization (Imperva Inc., 2023; Tiwari, 2025), and adversarial training (Goodfellow et al., 2018) help reduce risk, though none are foolproof.

On the model side, attack vectors such as backdoor insertion (Gu et al., 2017), model inversion (AlZubi et al., 2021), and query-based extraction (Carlini et al., 2024) present critical challenges. While adversarial training can improve robustness, it introduces computational burdens and may reduce accuracy on clean data (Goodfellow et al., 2018). Techniques like certified robustness (Cohen et al., 2019), knowledge distillation (Papernot et al., 2017), and model pruning (Liu et al., 2018) provide complementary strategies to manage trade-offs between performance and security. Confidential inference environments, like Intel SGX, further add a layer of protection against model theft and inference-time manipulation (Huang et al., 2018).

Equally significant are infrastructural and human vulnerabilities. As real-world cases demonstrate, many breaches occur due to leaked API keys, insecure cloud configurations, or phishing of admin credentials (Alkhalil et al., 2021; Verizon, 2025). Security-by-design must therefore extend to CI/CD pipelines (Rangnau et al., 2020), infrastructure-as-code validation, and access control audits. Complementary practices such as phishing simulations and the implementation of zero-trust architectures (CISA 2023) help close the gap between technical and human resilience. To operationalize these insights, a holistic approach is essential, one that combines detection, prevention, and prediction into an integrated framework. The DPP framework is a multilayered, lifecycle-aware approach designed to fortify AI systems against evolving cyberthreats. Unlike traditional point solutions, DPP integrates technical defenses, behavioral analytics, and anticipatory risk modeling into a unified structure. It addresses vulnerabilities not only at the model and data level but also across infrastructure, access points, and human interaction layers.

In this framework:

6.1 Detect: Early identification of threats

This phase focuses on real-time threat visibility within AI pipelines:

- Anomaly detection: Supervised, unsupervised, and semisupervised algorithms flag deviations in input patterns, user behavior, and model responses.

TABLE 35.1 Mapping traditional cybersecurity threats to their AI-specific adaptations, detailing entry points, impacts, and corresponding mitigation strategies.

Threat category	Traditional form	AI-specific evolution	Attack vector/entry point	Impact on AI systems	Defensive strategies
Phishing	Credential theft via email or SMS(Alkhalil, Hewage, Nawaf, and Khan, 2021)	Credential theft targeting AI pipelines and API keys (Huang et al., 2018)	Deceptive emails, spoofed login pages	Model poisoning, unauthorized access to APIs, or training environments	Zero-trust architecture, phishing simulations, MFA (CISA 2023)
Malware	Viruses, spyware, ransomware (Bendovschi, 2015)	Payload-as-a-Service distributing AI-targeted malware (Huang et al., 2018)	Malicious libraries, trojanized datasets	Model hijacking, training interference	Threat detection, malware scanners, hash validation (Huang et al., 2018)
Miscon figuration	Unsecured ports, weak ACLs (Verizon, 2025)	Public model leaks, open cloud buckets (Aslan et al., 2023)	Poor IaC hygiene, unreviewed API access	Privilege escalation, sensitive data exposure	CI/CD audits, role-based access, IaC validation (Rangnau, Buijtenen, Fransen, and Turkmen, 2020)
API Abuse	Excessive or malformed queries (Tramèr et al., 2016)	Adversarial querying to extract or degrade AI models (Papernot et al., 2017)	Publicly exposed model inference endpoints	Model inversion, information leakage	Rate limiting, input sanitization, anomaly detection
Adversarial Attacks	N/A	Perturbed inputs to mislead classifiers (AlZubi et al., 2021)	Slightly altered audio, text, image inputs	Model misclassification, system evasion	Adversarial training, certified defenses
SQL /Code Injection	Command injection via input fields (Huang et al., 2018)	Prompt injection in LLMs and embedded code exploitation (Aslan et al., 2023)	Text inputs in chatbots or web UIs	Hallucinated responses, jailbreaks, behavior Shift	Input sanitization, policy filters
DoS/DDoS	Service overload via botnets (Bendovschi, 2015)	DDoS targeting model endpoints, API rate exhaustion (Aslan et al., 2023)	Traffic floods, botnets, auto-query spam	Model unavailability, inference lag	Cloud-based mitigation, scaling policies
Social Engineering	Pretexting, baiting for password theft (Alkhalil et al., 2021)	Prompt manipulation of generative models (Aslan et al., 2023)	Human-in-the-loop vulnerabilities	Biased/unsafe output, leaked sensitive info	Prompt reinforcement, red teaming, instruction fine-tuning

- IDSs tailored for ML environments detect poisoned data, inference-time attacks, and API abuse.
- Adversarial input probes evaluate model response boundaries against subtle manipulations.
- Model drift monitoring tracks statistical shifts in inputs or outputs that may signal stealthy compromise or concept obsolescence (Huang et al., 2018).

6.2 Prevent: Architectural and operational safeguards

This phase strengthens the core of AI systems:

- Data-level defenses: Sanitization, DP, and robust preprocessing pipelines.
- Model-level defenses: Adversarial training, certified robustness (e.g., randomized smoothing), and backdoor detection.
- Deployment safeguards: Rate-limiting APIs, zero-trust access, confidential inference via TEEs, and HE.
- Human-layer controls: Phishing simulations, privilege audits, and Continuous Integration and Continuous Deployment (CI/CD) pipeline security reviews (Aslan et al., 2023; Mądry et al., 2017).

6.3 Predict: Future-proofing against emerging threats

This final pillar anticipates zero-day attacks, adversary innovation, and systemic fragility:

- Behavioral analytics: Analyze usage patterns and identify risky actors or anomalous sequences.
- Red teaming and simulation: Continuous adversarial evaluation using internal or external attack models.
- Concept drift forecasting: Predict and adjust for environmental, contextual, or data evolution.
- Automated response systems: Integrate risk scoring and conditional mitigation mechanisms (e.g., quarantine, model rollback).

Fig. 35.7 diagrammatically shows the AI security system framework, which represents a tripartite strategy that supports defense-in-depth by integrating three components: Detection mechanisms for anomalies and adversarial inputs, prevention tactics like DP and secure model encapsulation, and predictive monitoring for concept drift, zero-day attacks, and automated response.

Together, these layers enable AI systems to remain secure not only at design-time but throughout their deployment lifecycle. In a threat landscape where cybercrime-as-a-service has democratized attack sophistication (Huang et al., 2018), such dynamic resilience is critical.

Finally, securing AI also demands ethical and regulatory foresight. Trust in intelligent systems cannot rely solely on technical robustness; it must also be underpinned by transparency, governance, and accountability. Global regulatory shifts, such as the EU AI Act (European Commission, 2021), are now mandating explainability, risk classification, and audit trails for high-stakes systems. In parallel, industry-led efforts, such as external red-teaming (Mądry et al., 2017) and benchmarking robustness under reproducible conditions (Hendrycks et al., 2021) are laying the groundwork for trustworthy evaluation.

In conclusion, the future of secure AI lies not in a single solution but in coevolving strategies, combining technical defenses, regulatory compliance, and interdisciplinary collaboration. Only then can we realize AI's full promise without compromising societal trust.

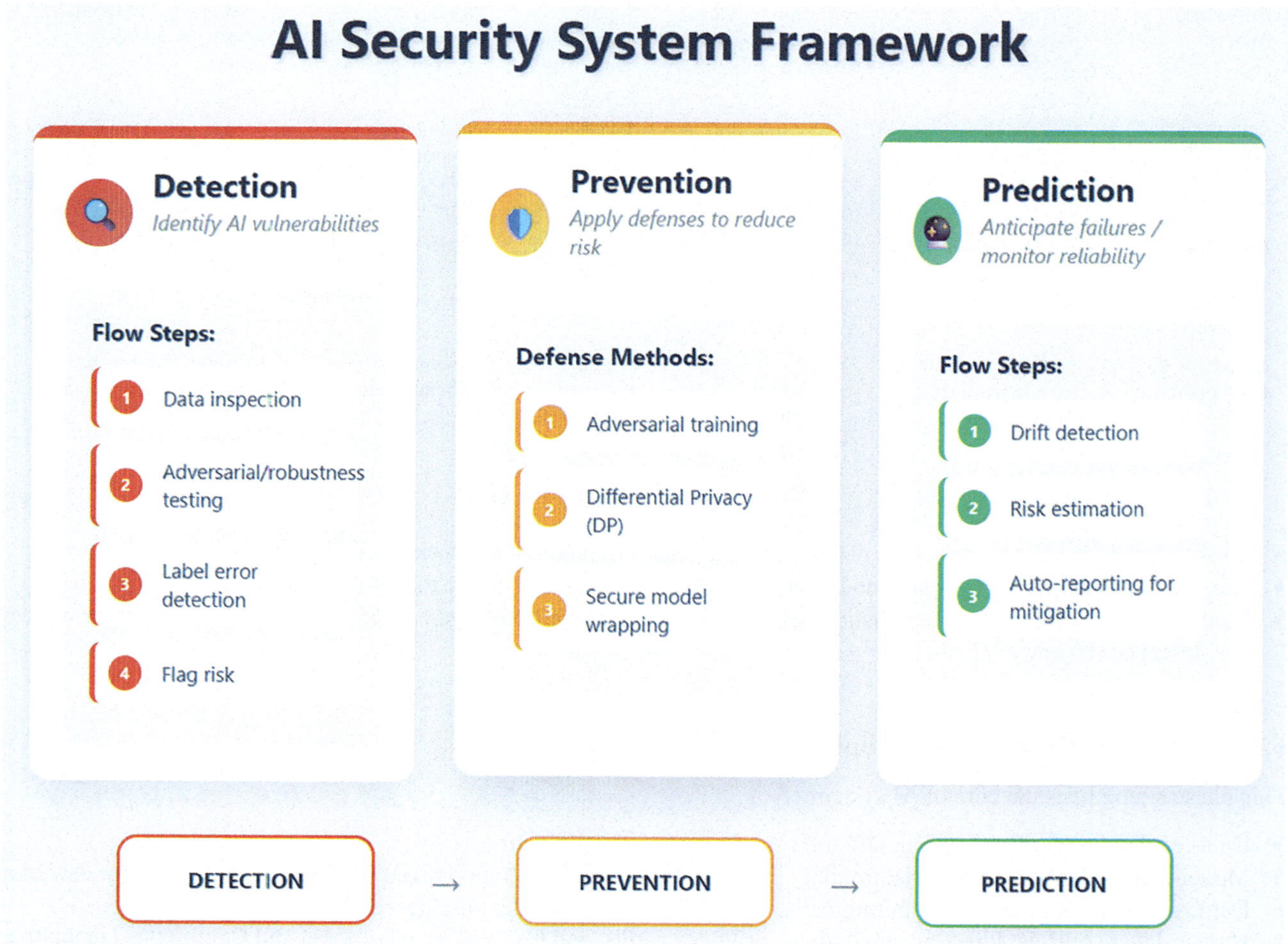

FIG. 35.7 AI security system framework.

Acknowledgement

The authors are thankful to the Principal of Sri Venkateswara College, University of Delhi, for his continuous encouragement and support, as well as for providing the infrastructure in the college to carry out a hassle-free literature survey required for writing the chapter.

References

Adelard, C., & Penrod, O. (2023). Implementation of network security system using firewall technology and intrusion detection system (IDS). *Idea: Future Research*, *1*(3), 113–121.

Alkhalil, Z., Hewage, C., Nawaf, L., & Khan, I. (2021). Phishing attacks: a recent comprehensive study and a new anatomy. *Frontiers in Computer Science*. 2624-9898. *3*. doi:10.3389/fcomp.2021.563060.

Alotaibi, J. O., & Alshahre, A. S. (2024). The role of conversational AI agents in providing support and social care for isolated individuals. *Alexandria Engineering Journal*.11100168. *108*, 273–284. Available from: https://www.sciencedirect.com/science/journal/11100168. doi:10.1016/j.aej.2024.07.098.

AlZubi, A. A., Al-Maitah, M., & Alarifi, A. (2021). Cyber-attack detection in healthcare using cyber-physical system and machine learning techniques. *Soft Computing*. 14337479, *25*(18), 12319–12332. Available from: http://springerlink.metapress.com/app/home/journal.asp?wasp=h83ak0wtmr5uxkah9j5m&referrer=parent&backto=browsepublicationsresults,466,533;. doi:10.1007/s00500-021-05926-8.

Aslan, Ö., Aktuğ, S. S., Ozkan-Okay, M., Yilmaz, A. A., & Akin, E. (2023). A comprehensive review of cyber security vulnerabilities, threats, attacks, and solutions. *Electronics*. 2079-9292. *12*(6), 1333. doi:10.3390/electronics12061333.

Baker, M. J. (2000). The roles of models in artificial intelligence and education research: a prospective view. *Journal of Artificial Intelligence and Education*, *11*, 122–143.

Barnard, J., & Stryker, C. (2025). What is anomaly detection? IBM Think. (2025), Available from: https://www.ibm.com/think/topics/anomaly-detection.

Bellman, R. E. (1978). An introduction to artificial intelligence: can computers think? (1978).

Bendovschi, A. (2015). Cyber-attacks – trends, patterns and security countermeasures. *Procedia Economics and Finance*. 2212-5671. *28*, 24–31. doi:10.1016/s2212-5671(15)01077-1.

Biggio, B., & Roli, F. (2018). Wild patterns: ten years after the rise of adversarial machine learning. *Pattern Recognition*. 0031-3203. *84*, 317–331. doi:10.1016/j.patcog.2018.07.023.

Billings, S. A., Jamaluddin, H. B., & Chen, S. (1992). Properties of neural networks with applications to modelling non-linear dynamical systems. International Journal of Control, *55*(1), 193-224.

Carlini, N., Paleka, D., Dvijotham, K. D., Steinke, T., Hayase, J., Cooper, A. F., … Tramèr, F. (2024). Stealing part of a production language model. *arXiv, United Kingdom arXiv*. 23318422 Available from: https://arxiv.org. doi:10.48550/arXiv.2403.06634.

Carlini, N., Tramèr, F., Wallace, E., Jagielski, M., Herbert-Voss, A., Lee, K., … Raffel, C. (2021). USENIX Association United States extracting training data from large language models. In: Proceedings of the 30th USENIXSecurity Symposium, pp. 2633–2650. 9781939133243

Chandola, V., Banerjee, A., & Kumar, V. (2009). Anomaly detection. *ACM Computing Surveys*. 0360-0300. *41*(3), 1–58. doi:10.1145/1541880.1541882.

Charniak, E., & McDermott, D. (1985). Introduction to artificial intelligence.

CISA. (2023). Zero trust maturity model. U.S. Cybersecurity & Infrastructure Security Agency https://www.cisa.gov/resources-tools/resources/zero-trust-maturity-model.

Cohen, B., & Murphy, G. (1984). Models of concepts. *Cognitive Science*. 0364-0213. *8*(1), 27–58. doi:10.1016/s0364-0213(84)80024-5.

Cohen, J., Rosenfeld, E., & Kolter, J. Z. (2019). Certified adversarial robustness via randomized smoothing. In 36th International Conference on Machine Learning ICML International Machine Learning Society (IMLS), United States, 9781510886988. 2323–2356.

DARPA. (2019). Defending against adversarial artificial intelligence. https://www.darpa.mil/news-events/2019-02-06.

Dhirani, L. L., Mukhtiar, N., Chowdhry, B. S., & Newe, T., (2023). Ethical dilemmas and privacy issues in emerging technologies: a review. *Sensors* *23*(3), 1424–8220. Available from: http://www.mdpi.com/journal/sensors. doi:10.3390/s23031151.

Domingos, P. (2015). The master algorithm: how the quest for the ultimate learning machine will remake our world.

Dwork, C., & Roth, A. (2014). The algorithmic foundations of differential privacy. *Foundations and Trends® in Theoretical Computer Science*. 1551-305X. *9*(3-4), 211–407. doi:10.1561/0400000042.

European Commission. (2021). Proposal for a Regulation laying down harmonised rules on artificial intelligence (artificial intelligence act). https://eur-lex.europa.eu/legal-content/EN/TXT/?uri=CELEX%3A52021PC0206.

Galle, A. (2022). Drinking from the fetid well: data poisoning and machine learning. *Proceedings*, *148*(1). Available from https://www.usni.org/magazines/proceedings/2022/january/drinking-fetid-well-data-poisoning-and-machine-learning.

Gandhi, P., & Pahwa, S. (2024). All India Institute of Medical Sciences (AIIMS), Delhi: cyberattack puts digitalisation under scanner. *IMIB Journal of Innovation and Management*. doi:10.1177/ijim.241240911.

GeeksforGeeks(2025). Introduction to data in Machine Learning. https://www.geeksforgeeks.org/machine-learning/ml-introduction-data-machine-learning/.

Goodfellow, I. J., Shlens, J., & Szegedy, C. (2015). Explaining and harnessing adversarial examples. In International Conference on Learning Representations, ICLR 2015 - Conference Track Proceedings International Conference on Learning Representations. ICLR, United States. https://dblp.org/db/conf/iclr/iclr2015.html.

Goodfellow, I., McDaniel, P., & Papernot, N. (2018). Making machine learning robust against adversarial inputs. *Communications of the ACM*. 0001-0782. *61*(7), 56–66. doi:10.1145/3134599.

Gu, T., Dolan-Gavitt, B., & Garg, S. (2017). Badnets: identifying vulnerabilities in the machine learning model supply chain. *arXiv, United States arXiv. 23318422*, Available from: https://arxiv.org. doi:10.48550/arxiv.1708.06733.

Haugeland, J. (1985). *Artificial Intelligence: The Very Idea*. MIT Press.

Hazelwood, K., Bird, S., Brooks, D., Chintala, S., Diril, U., Dzhulgakov, D., ... Wang, X. (2018). States applied machine learning at Facebook: a data-center infrastructure perspective. In Proceedings - International Symposium on High-Performance Computer Architecture, 9781538636596, IEEE Computer Society United, 620–629. 10.1109/HPCA.2018.00059.

Henderson, P., Sinha, K., Angelard-Gontier, N., Ke, N. R., Fried, G., Lowe, R., & Pineau, J. (2017). Ethical challenges in data-driven dialogue systems. *arXiv, Undefined arXiv. 23318422*, Available from: https://arxiv.org. doi:10.48550/arxiv.1711.09050.

Hendrycks, D., Basart, S., Mu, N., Kadavath, S., Wang, F., Dorundo, E., ... Gilmer, J. (2021). The many faces of robustness: a critical analysis of out-of-distribution generalization. In Proceedings of the IEEE International Conference on Computer Vision. 9781665428125. Institute of Electrical and Electronics Engineers Inc., United States, 8320–8329. http://ieeexplore.ieee.org/xpl/conhome.jsp?punumber=1000149. 10.1109/ICCV48922.2021.00823.

Hinton, G., Vinyals, O., & Dean, J. (2015). Distilling the knowledge in a neural network. *arXiv Preprint arXiv:1503.02531*.

Hu, C., & Hu, Y. H. F. (2020). Data poisoning on deep learning models. In Proceedings - 2020 International Conference on Computational Science and Computational Intelligence, CSCI 2020, 9781728176246, Institute of Electrical and Electronics Engineers Inc., United States, 628–632. http://ieeexplore.ieee.org/xpl/mostRecentIssue.jsp?punumber=9457837. 10.1109/CSCI51800.2020.00111.

Huang, K., Siegel, M., & Madnick, S. (2018). Systematically understanding the cyber attack business: a survey. *ACM Computing Surveys*, *51*(4), 1–36.

Hutson, M. (2018). A turtle—or a rifle? Hackers easily fool AIs into seeing the wrong thing. *Science*, 0036-8075. doi:10.1126/science.aau8383.

Imperva Inc. (2023). What is Data Sanitization? Data Erasure Methods | Imperva. Learning Center. https://www.imperva.com/learn/data-security/data-sanitization/.

Kuismin-Raerinne, A., & Nieminen, L., (2022). Garbage in, garbage out (Master's thesis). University of Helsinki, Faculty of Law, Master of Laws.

Kurzweil, R. (1990). *The age of intelligent machines*. MIT Press.

Lin, J., Dang, L., Rahouti, M., & Xiong, K. (2021). ML attack models: adversarial attacks and data poisoning attacks. *arXiv, United States arXiv. 23318422*, Available from: https://arxiv.org. doi:10.48550/arxiv.2112.02797.

Liu, Y., Ma, S., Aafer, Y., Lee, W. C., Zhai, J., Wang, W., & Zhang, X. (2018). Trojaning attack on neural networks. In 25th Annual Network and Distributed System Security Symposium, NDSS 2018. 9781891562495 The Internet Society, United States. 10.14722/ndss.2018.23291.

Loyola-Gonzalez, O. (2019). Black-box vs. white-box: understanding their advantages and weaknesses from a practical point of view. *IEEE Access*. 2169-3536. *7*, 154096–154113. doi:10.1109/access.2019.2949286.

Lyon, C. F. (2023). Fake cases, real consequences: misuse of ChatGPT leads to sanctions. *NY Litigator*, *28*(2), 8–12.

Mądry, A., Makelov, A., Schmidt, L., Tsipras, D., & Vladu, A. (2017). Towards deep learning models resistant to adversarial attacks. *arXiv, Undefined arXiv. 23318422*, Available from: https://arxiv.org. doi:10.48550/arxiv.1706.06083.

Meritshot. (2024). Robust machine learning: building models resilient to adversarial attacks. Meritshot. https://www.meritshot.com/robust-machine-learning-building-models-resilient-to-adversarial-attacks/.

Naqa, I. E., & Murphy, M. J. (2015). *What is Machine Learning? Machine Learning in Radiation Oncology: Theory and Applications* (pp. 3–11). Canada: Springer International Publishing. Available from: https://link.springer.com/book/9783319183053. doi:10.1007/978-3-319-18305-3_1.

Naqvi, S. M. A., Shabaz, M., Khan, M. A., & Hassan, S. I. (2023). Adversarial attacks on visual objects using the fast gradient sign method. *Journal of Grid Computing*. 15729184. *21*(4). Available from: https://www.springer.com/journal/10723. doi:10.1007/s10723-023-09684-9.

National Transportation Safety Board. (2020). Collision between vehicle controlled by developmental automated driving system and pedestrian: tempe, Arizona. NTSB/HAR-19/03 PB2020-10035. https://www.ntsb.gov/investigations/AccidentReports/Reports/HAR1903.pdf.

Nilsson, N. J. (1998). *Artificial Intelligence: A New Synthesis*. Morgan Kaufmann.

Pal, A., Bhargava, R., Hinsz, K., Esterhuizen, J., & Bhattacharya, S. (2024). The empirical impact of data sanitization on language models. *arXiv, United States arXiv. 23318422*, Available from: https://arxiv.org. doi:10.48550/arXiv.2411.05978.

Paluszek, M., & Thomas, S. (2020). *What is Deep Learning?* (pp. 1–24). Springer Science and Business Media LLC. doi:10.1007/978-1-4842-5124-9_1.

Papernot, N., Goodfellow, I., Abadi, M., Talwar, K., & Erlingsson, Ú. (2017). ICLR United States Semi-supervised knowledge transfer for deep learning from private training data. In 5th International Conference on Learning Representations, ICLR 2017 - Conference Track Proceedings International Conference on Learning Representations. https://dblp.org/db/conf/iclr/iclr2017.html.

Papernot, N., McDaniel, P., Goodfellow, I., Jha, S., Celik, Z. B., & Swami, A. (2017). Practical black-box attacks against machine learning. In ASIA CCS 2017 - Proceedings of the 2017 ACM Asia Conference on Computer and Communications Security, 9781450349444, Association for Computing Machinery, Inc., United States. 10.1145/3052973.3053009.

Poole, D., Mackworth, A. K., & Goebel, R. (1998). *Computational Intelligence: A Logical Approach*. Oxford University Press.

Quinn, S., Quinn, S., Ivy, N., Barrett, M., Feldman, L., Witte, G., & Gardner, R. K. (2021). *Identifying and Estimating Cybersecurity Risk for Enterprise Risk Management*. US Department of Commerce, National Institute of Standards and Technology.

Quiroz-Vázquez, C. (2025). Anomaly detection in machine learning: finding outliers for optimization of business functions. IBM Think. Available from: https://www.ibm.com/think/topics/machine-learning-for-anomaly-detection.

Rajpurkar, P., Irvin, J., Zhu, K., Yang, B., Mehta, H., Duan, T., ... Ng, A. Y. (2017). CheXNet: radiologist-level pneumonia detection on chest X-rays with deep learning. *arXiv, Undefined arXiv*, 23318422. Available from: https://arxiv.org. doi:10.48550/arxiv.1711.05225.

Rangnau, T., Buijtenen, R. V., Fransen, F., & Turkmen, F. (2020). Continuous security testing: a case study on integrating dynamic security testing tools in CI/CD pipelines. In Proceedings - 2020 IEEE 24th International Enterprise Distributed Object Computing Conference, EDOC 2020. 9781728164731, Institute of Electrical and Electronics Engineers Inc., Netherlands, 145–154. http://ieeexplore.ieee.org/xpl/mostRecentIssue.jsp?punumber=9233042. 10.1109/EDOC49727.2020.00026.

Rich, E., & Knight, K. (1991). *Artificial Intelligence* (2nd ed.). McGraw-Hill, McGraw-Hill.

Rudin, C. (2019). Stop explaining black box machine learning models for high stakes decisions and use interpretable models instead. *Nature Machine Intelligence*. 2522-5839. *1*(5), 206–215. doi:10.1038/s42256-019-0048-x.

Russell, S., & Norvig, P. (2020). *Artificial Intelligence: A modern approach*. Pearson. 4.

Sarker, I. H. (2022). AI-based modeling: techniques, applications and research issues towards automation, intelligent and smart systems. *SN Computer Science*, *3*(2), 26618907. Available from: https://www.springer.com/journal/42979. doi:10.1007/s42979-022-01043-x.

Sheikh, H., Prins, C., & Schrijvers, E. (2023). *Artificial Intelligence: Definition and Background*. (pp. 15–41). Springer Science and Business Media LLC. doi:10.1007/978-3-031-21448-6_2.

Shethiya, A. S. (2023). Machine learning in motion: real-world implementations and future possibilities. *Academia Nexus Journal*, *2*. Available from: https://academianexusjournal.com/index.php/anj/article/view/22.

Shokri, R., Stronati, M., Song, C., & Shmatikov, V. (2017). Membership inference attacks against machine learning models. In Proceedings - IEEE Symposium on Security and Privacy (p. 18), 9781509055326, *3*, Institute of Electrical and Electronics Engineers Inc., United States. 10.1109/SP.2017.41.

Simms, N. (2023). Data poisoning: a new threat to artificial intelligence.

Tiwari, A., What is data sanitization? Best data erasure methods. (2025) Available from: https://www.systoolsgroup.com/updates/data-sanitization-methods/?srsltid=AfmBOoorwudNvlK1RSxIedLJw6hsaix4xJn3E-IsVwCd4aTCgXlaXjS7.

Tramèr, F., Zhang, F., Juels, A., Reiter, M. K., & Ristenpart, T. (2016). Stealing machine learning models via prediction APIs. In Proceedings of the 25th USENIX Security Symposium (601–618), 9781931971324. USENIX Association, Switzerland.

Verizon. (2025). 2025 Data Breach Investigations Report (DBIR) https://www.verizon.com/business/resources/reports/dbir/.

Vincent, J. (2016). Twitter taught Microsoft's AI chatbot to be a racist jerk in less than a day. The Verge. Available from: https://www.theverge.com/2016/3/24/11297050/tay-microsoft-chatbot-racist.

Weng, L., Chen, P. Y., Nguyen, L., Squillante, M., Boopathy, A., Oseledets, I., & Daniel, L. (2019). PROVEN: Verifying robustness of neural networks with a probabilistic approach. In International Conference on Machine Learning (pp. 6727–6736). PMLR.

Winston, P. H. (1992). *Artificial Intelligence* (Third). Addison-Wesley.

Yerlikaya, F. A., & Bahtiyar, Ş. (2022). Data poisoning attacks against machine learning algorithms. *Expert Systems with Applications*. 09574174. *208*, 118101. doi:10.1016/j.eswa.2022.118101.

Yu, S., & Carroll, F. (2021). Implications of AI in National security: understanding the security issues and ethical challenges. In Rodham, H., Montasari, R., Jahankhani, H., (eds.) *Advanced Sciences and Technologies for Security Applications*. United Kingdom: Springer, (pp. 157–175). Available from: https://link.springer.com/bookseries/5540 . doi:10.1007/978-3-030-88040-8_6.

Zha, D., Bhat, Z. P., Lai, K. H., Yang, F., & Hu, X. (2023). Society for industrial and applied mathematics publications United States data-centric AI: perspectives and challenges. In 2023 SIAM International Conference on Data Mining, SDM 2023 (pp. 945–948), 9781611977653.

Zhang, J., & Li, C. (2020). Adversarial examples: opportunities and challenges. *IEEE Transactions on Neural Networks and Learning Systems*. 21622388. *31* (7), 2578–2593. Available from: http://ieeexplore.ieee.org/xpl/RecentIssue.jsp?punumber=5962385. doi:10.1109/TNNLS.2019.2933524.

Zhang, L., Yang, Z., He, Y., Zhang, Z., Qian, Z., Hong, G., … Yang, M. (2018). Association for computing machinery China invetter: locating insecure input validations in android services. In Proceedings of the ACM Conference on Computer and Communications Security (pp. 1165–1178). 9781450356930. 10.1145/3243734.3243843.

Glossary of artificial intelligence terminologies

Compiled by: Annewsha Datta, Rudrendu Kumar Paul, Samudra Prosad Banik, Ashfaque Hossain, Anand Swaroop, and Debasis Bagchi

3D microelectrode array (MEA) A device for recording neural activity from brain organoids at high temporal and spatial resolution; enables cognitive function measurement in OI research.

AAR (Adaptive Auto Regressive) A time-varying autoregressive model used to extract features from non-stationary EEG signals such as ErrP.

Actor-critic (A2C) A reinforcement learning algorithm that combines elements of policy-based and value-based methodologies to effectively train an agent.

Adaptive deep brain stimulation (DBS) An AI-enhanced implanted device delivering real-time, personalized modulation of neural activity in Parkinson's disease.

Adaptive design A clinical trial methodology that permits prespecified modifications, such as sample size reestimation or dose adjustment, stemming from interim data while preserving statistical integrity and lowering both timelines and costs.

Adaptive neuro-fuzzy inference system (ANFIS) A hybrid model that combines fuzzy inference with neural network learning to map inputs to outputs.

Additive manufacturing (biomedical) AI-guided three-dimensional printing of implants, prosthetics, and tissue-engineered structures.

Advanced driver assistance system (ADAS) It utilizes electronic technologies, including sensors, cameras, and software, to enhance the safety and comfort of vehicle drivers.

ADMETlab 3.0 A novel computational tool for assisting medicinal chemists in accelerating the drug development process.

ADNI Alzheimer's Disease Neuroimaging Initiative.

Adversarial machine learning A subfield studying how ML models can be exploited to strategize robust defense against deceiving models.

Adverse drug reactions (ADRs) Untoward and toxic responses to a medication.

Affective computing Amalgamation of the fields of computer science, psychology and cognitive science to understand, comprehend and reproduce human emotional states.

Agent-based modeling A computational modeling approach where individual microbial agents operate based on predefined rules to simulate community dynamics.

AI governance A multi-level regulatory and accountability system for AI deployment across healthcare organizations and jurisdictions.

AlexNet A convolutional neural network designed for image categorization, notably gaining prominence through its performance in the ImageNet Large Scale Visual Recognition Challenge.

Algorithm A defined sequence of directives executed by a computer system.

Algorithmic bias Systematic errors in AI systems that can lead to unfair or discriminatory outcomes across different populations.

Algorithm passport A proposed regulatory document disclosing an AI model's data provenance, performance boundaries, and known failure modes.

Alignment To affirm that artificial intelligence procedures align with human values and goals.

AlphaFold An artificial intelligence (AI)-based tool developed by Google DeepMind, which has the ability to predict a protein's 3D structure from its amino acid sequence with high accuracy.

AlphaGo Google DeepMind developed this computer program designed for playing the board game Go. Extensively used for studying deep neural networks and reinforcement learning.

Ambient AI medical scribes AI voice transcription and documentation tools that reduce after-hours physician documentation burden; cited with a 30% reduction and a 40–63% burnout reduction figure.

AMPGAN v2 A BiDirectional Classifying Generative Adversarial Network (BiCGAN)-based artificial intelligence (AI) framework consisting of three networks, including (1) the generator, (2) the discriminator, and (3) the encoder.

Anastomotic leakage prediction Real-time AI risk assessment for failure of surgical joins, using clinical signals such as lactic acid levels, tissue perfusion indicators, and respiratory exchange ratios to flag high-risk cases.

Anthropomorphism Appropriation of human characteristics of behavior and emotions artificially.

Anti-Citrullinated Protein Antibodies (Anti-CCP) Autoantibody specific to rheumatoid arthritis and more diagnostically specific than rheumatoid factor; used as a primary biomarker target for AI models.

Antimicrobial peptides Small proteins with antimicrobial activity.

Antimicrobial resistance (AMR) A condition that makes infections difficult or impossible to treat, increases the spread of disease, and leads to severe illness and death.

Anti-Müllerian hormone (AMH) A granulosa cell-derived hormone that measures ovarian reserve.

Antral follicle count (AFC) It determines the number of immature egg sacs (follicles) in the ovaries.

Application-specific integrated circuit (ASIC) A custom-fabricated microchip designed to perform a single, specific task with precision.

Arterial spin labeling (ASL) A noninvasive magnetic resonance imaging (MRI) technology used for the estimation of cerebral blood flow (CBF).

Artificial general intelligence (AGI) A sort of artificial intelligence that would match or surpass human-level cognitive abilities across virtually all cognitive tasks.

Artificial intelligence (AI) It enables machines to perform tasks typically requiring human intelligence, such as learning, problem-solving, and decision-making.

Artificial neural network (ANN) Similar to a human brain, a machine learning (ML) model utilizes layers of interconnected "neurons" or nodes to process information and learn from the data it obtains.

Assembloids Fused brain organoid structures that combine region-specific organoids to study inter-regional neural connectivity and functional corticospinal circuits.

Assembly Computational process where short DNA fragments are pieced together into longer sequences called contigs.

Assisted reproductive technology (ART) It includes medical treatments for infertility that involve handling human eggs or embryos, with in vitro fertilization (IVF) being the most common type.

Attention-based models AI models that can focus on specific parts of input data, useful for identifying important biological patterns.

AUC-ROC (Area Under the Receiver Operating Characteristic Curve) The standard aggregate performance metric for binary classifiers, summarizing classification accuracy across thresholds as a score between 0 and 1 (the chapter treats 0.7–0.8 as effective and above 0.8 as excellent).

Autonomous Ability to conduct tasks without human intervention.

Autoregressive integrated moving average (ARIMA) A statistical tool used for prediction of time series data.

Ayushman Bharat India's government health insurance and primary care infrastructure; named as the delivery channel for AI tool piloting.

BBB-on-a-chip A microfluidic system replicating blood-brain barrier physiology; combined with AI to quantify barrier permeability changes relevant to AD pathogenesis.

Backpropagation neural network (BPNN) A kind of artificial neural network (ANN) that utilizes backpropagation, a learning algorithm, for training purposes.

Bard An LLM technology-based chatbot developed by Google.

Bayesian learning An inferential framework in which prior knowledge is updated with incoming evidence to yield posterior distributions that refine parameters such as treatment efficacy, enrollment velocity, and market uptake in near real time.

Bayesian network (BN) A probabilistic graphical representation that is widely used to determine the relationships between variables.

BEHRT Transformer for Electronic Health Records. A deep-learning (DL)-based model designed to analyze patients' historical medical data for predictive purposes.

BenevolentAI It utilizes computational medicine and AI, along with open systems and cloud computing, to design and develop novel drugs.

Bidirectional Encoder Representations from Transformers (BERT) A sort of recurrent neural network (RNN) capable of processing sequential data, such as Electronic Health Records (EHRs) or physiological time series, in both forward and backward directions. It assists in capturing dependencies from past and future information, which is useful for clinical sequence prediction.

Beta variational autoencoder (β-VAE) A framework for the automated discovery of comprehensible, factorized latent representations from raw image data.

Bias (AI bias) An instance of wrongful discrimination or biased outcomes in AI systems, resulting from distorted data, flawed algorithms, or incorrect assumptions.

Bidirectional Gated Recurrent Unit (BiGRU) A deep learning (DL) model that processes an input sequence in both forward and backward directions, allowing it to gather context from both past and future information.

Bidirectional LSTM (Bi-LSTM) A recurrent neural network (RNN) that can process sequential data in both backward and forward directions.

Bidirectional Long Short-Term Memory (Bi-LSTM) A category of recurrent neural network (RNN) that processes sequential data in opposite directions.

Big Data Quite a huge dataset that cannot be handled by normal data processing software.

Bing Chat (Bing AI) Microsoft launched this artificial intelligence-powered chatbot in early 2023, which is built into the Microsoft Edge sidebar. It can be used to answer complex questions, find comprehensive answers, get consolidated information, and find inspiration to build upon, all in a side-by-side view, with no need to flip between tabs.

BioBERT A domain-specific adaptation of BERT (Bidirectional Encoder Representations from Transformers) tailored for biomedical literature analysis.

Bidirectional Time-Aware LSTM (Bi-TLSTM) A Long Short Term Memory network for simultaneous forward and backward monitoring and processing of sequential clinical data.

Biodiversity Conservation Efforts to protect and preserve the variety of life forms, including microbial diversity, in ecosystems.

Biogeochemical cycles Natural processes that recycle chemical elements like carbon, nitrogen, and phosphorus through ecosystems.

BioLiP2 A database for high-quality, biologically relevant ligand–protein binding interactions.

Bioimpedance measurement A technique for measuring fluid status through bioimpedance, under development for post-transplant monitoring of volume overload and graft dysfunction.

Biological mechanisms The underlying biological processes and pathways that explain observed associations or outcomes.

Biomarkers Measurable biological indicators that can signal normal biological processes, disease conditions, or responses to treatment.

Biomarker discovery The process of identifying measurable biological indicators associated with specific health outcomes or disease states.

BioPython An open-source library providing tools for biological computation, sequence analysis, and data manipulation.

Bioremediation The use of microorganisms to remove or neutralize environmental pollutants and contaminants.

Bispectral Index (BIS) An EEG-derived monitoring index of anesthetic depth.

Black Box/BLACKBOX models AI systems whose internal workings are not easily understood or explained, such as complex deep learning networks.

BLSA Baltimore Longitudinal Study of Aging.

Biological oxygen demand (BOD) Oxygen required by aerobic microbes to break down organic matter.

Body mass index (BMI) A procedure to determine the amount of body fat using body weight and height measurements. It is widely used to assess potential health risk factors.

Brain–computer interface (BCI) Nonmuscular signaling pathways of communication between the human brain and a computer.

Brainoware A hybrid brain organoid-computer system demonstrating on-chip learning and complex computation through spatiotemporal electrical stimulation; exemplifies the potential of living neural networks for biocomputing.

Burstiness An assessment of dissimilarity in constructing a sentence and variation in its length.

BYD BYD Auto Co., Ltd (Shenzhen, China) has implemented the Blade Battery, electronic platform 3.0, and dual-mode hybrid technology for electric cars.

CADY peptides A type of cell-penetrating peptide (CPP) used to deliver siRNA and other negatively charged molecules into cells for gene silencing therapy.

Capsular contracture A post-breast-augmentation complication in which scar tissue constricts the implant.

Capsule endoscopy (CE) A minimally invasive endoscopic procedure that helps examine the gastrointestinal tract (GI tract) using a tiny, wireless camera.

CAPTCHA (Completely Automated Public Turing test to tell Computers and Humans Apart) A procedure used to check whether the user is human. Useful for preventing spambots or spam comments.

Cardiac allograft rejection evaluator (CARE) A procedure to assess whether a transplanted heart is being accepted or rejected by the recipient.

Cardiac allograft vasculopathy (CAV) Types of coronary artery disease (CAD) arising from transplanted hearts.

Cardiotocography (CTG) To monitor a baby's heart rate and mother's uterine contractions during pregnancy and labor.

Cardiovascular diseases (CVDs) Diverse heart-related diseases and disorders.

CatBoost A gradient boosting algorithm achieving the highest accuracy of 0.80 among tested ML models for heart transplant mortality prediction.

Causative analysis Research approaches focused on identifying cause-and-effect relationships rather than just statistical associations.

Cognitive behavioral therapy (CBT) An evidence-based psychological intervention that reshapes maladaptive thoughts and behaviors; cited as a canonical example of a digital therapeutics application.

Cohen's Kappa A statistical performance metric for assessing aggregability of two raters.

Chatbot A software application that simulates human conversation. It is derived from deep learning (DL) and natural language processing (NLP) technologies.

ChatGPT OpenAI, an artificial intelligence company in California, created this conversational AI chatbot, which can effectively generate human-like text and participate in natural language conversations.

ChEMBL The term is an acronym for chemogenomic data from the European Molecular Biology Laboratory. It is a manually curated database integrating genomic data with corresponding biological and chemical properties in the cellular micromilieu to design new drugs.

Chemical oxygen demand (COD) Total oxygen required to oxidize both biodegradable and nonbiodegradable substances.

CheXNet A foundational deep learning model based on a 121-layer Convolutional Neural Network (DenseNet) analysis of Chest X rays intended for automated classification of thoracic disease.

COD (chemical oxygen demand) filter A glass-fiber filter device used to assess soluble organic pollutants in wastewater.

Common Spatial Pattern (CSP) An eigenvalue-based spatial filter that discriminates between left-hand and right-hand motor imagery EEG signals.

Cheminformatics A computational discipline combining chemistry, statistics, and information science to model molecular properties and guide drug discovery.

Chinese room A thought experiment devised by John Searle to fight against the possibility of achieving true understanding or consciousness through artificial intelligence.

Choroidal neovascularization (CNV) An ocular disease resulting from abnormal blood vessel growth in the choroid.

Circulating tumor DNA (ctDNA) Cell-free DNA fragments released from tumor cells into the bloodstream for subsequent AI based liquid biopsy analysis in non-invasive cancer monitoring.

Climate change impact Effects of changing environmental conditions on microbial communities and ecosystem functions.

Clinical sequence prediction (CSP) Utilization of AI, especially deep learning and machine learning, to interpret forthcoming consequences based on episodes of clinical results, including Electronic Health Records (EHRs).

Clustering An unsupervised learning technique that groups identical datasets based on shared characteristics.

Complement-dependent cytotoxicity (CDC) assay An assay to determine the effectiveness of antibodies in triggering the complement system, leading to cellular disintegration.

Compounded annual growth rate (CAGR) A measure of the performance of an organization over a fixed period in terms of sales, revenue, and growth.

Computational power The processing capability of computer systems, which continues to advance and enable more sophisticated analyses.

Conditional VAE A variational autoencoder conditioned on target properties such as antimicrobial activity or stability; used to generate property-preserving peptide sequences.

Context of Use (COU) It provides direction on how a biomarker will be identified.

Context-Rich (CR) UNet An interfaced network for Ovary and Follicle Segmentation, which is used in examining ultrasound images.

Contrastive learning A self-supervised representation learning approach used for multi-omics data integration in the chapter's advanced risk modeling section.

Contrast-Enhanced and Label-Adaptive Framework for Multi-Functional Therapeutic Peptides Prediction (CELA-MFP) A deep learning (DL) framework that utilizes Contrastive Enhancement and Label Adaptation (CELA) for predicting Multi-Functional therapeutic Peptide (MFP).

CONSORT-AI The Consolidated Standards of Reporting Trials for AI; a recommended reporting checklist covering model version, data handling, subgroup analysis, and uncertainty disclosure for AI clinical trial publications.

Controlled ovarian stimulation (COS) The use of fertility medications to stimulate the ovaries to produce multiple mature follicles in preparation for in vitro fertilization (IVF).

Convolutional neural networks (CNNs) Deep learning models that are particularly effective for analyzing complex biological datasets.

Copilot An AI-based virtual assistant developed by Microsoft, conducts tasks and generates content by utilizing natural language prompts.

Copy number variations (CNVs) DNA segments that vary in copy number between individuals, resulting from duplications or deletions of large segments of DNA.

Coronary calcium scoring (CCS) It determines calcified plaque in the arteries to evaluate the risk of coronary artery disease (CAD) and heart attacks.

CPPFormer A deep learning (DL) model for the accurate prediction of cell-penetrating peptides (CPPs).

Cross-institutional collaboration Partnerships between different research institutions, organizations, and countries to advance scientific knowledge.

Cryo-electron microscopy (cryo-EM) A structural biology procedure that utilizes a transmission electron microscope to produce very high-resolution images of biological macromolecules, including DNA and proteins.

CSM-Toxin An in silico protein toxicity classifier, which is based entirely on the protein's primary sequence.

Computed tomography (CT) Computerized image processing technology.

Cyber-physical system (CPS) A system that integrates computational elements with physical processes.

Cycle generative adversarial network (CycleGAN) A procedure to train a deep convolutional neural network for image-to-image translation tasks.

DALL-E A novel AI system created by OpenAI that can generate a corresponding image based on a text description. It is often integrated with ChatGPT, DALL-E 2, and DALL-E 3, which are the subsequent versions released in mid-2022 and 2023.

Data-centric AI vs. model-centric AI A paradigm distinction between improving data quality versus refining model architecture.

Data mining A computational process of discovering patterns and relationships in large datasets.

DAT scan A SPECT nuclear imaging technique assessing presynaptic dopaminergic degeneration; described in the chapter as used in Parkinson's disease diagnosis.

da Vinci Xi A robotic surgical system.

D-CryptO Deep learning-based technology to evaluate organoid morphology.

4D-DANI-Net It is based on a deep learning (DL) framework for creating high-resolution, longitudinal MRI scans in neurodegenerative diseases and disorders.

DEBi-TLSTM model A multifaceted, optimized form of a standard long short-term memory (LSTM) network.

Decision support systems (DSSs) Computer-based systems that assist in decision-making.

Decision trees (DTs) Popular machine learning (ML)-based models.

Deep Blue AI-based development designed to play chess.

Deep convolutional GAN (DCGAN) A novel technique for early AD detection accuracy and robustness.

Deepfake AI-modified and edited altered images, audios, and videos. Known for providing **misinformation, enabling the creation of sexually explicit images of individuals without their consent, and facilitating fraudulent activities**.

Deep-Orga A deep learning (DL) tool used for the morphological assessment of intestinal organoids.

Deep learning (DL) A subset of machine learning where computers learn by using something akin to a brain (called a neural network) with many layers. This helps them understand extremely complex tasks, such as recognizing faces or translating languages.

DeepMicro A machine learning tool specifically designed to process extensive sequencing data, detect patterns, classify microbes, and predict disease associations.

Deep neuro-fuzzy inference system or deep neuro-fuzzy system (DNFIS) A hybrid artificial intelligence (AI) platform that integrates deep neural networks (DNNs) with fuzzy inference systems to address the limitations of both approaches.

Deep Q-network (DQN) An extension of Q-learning that utilizes a deep neural network.

Deidentification Process of removing personally identifiable information from datasets to protect participant privacy.

Demographic representation The inclusion of diverse population groups in research studies and training datasets.

DenseNet121 A convolutional neural network architecture for analysis and classification of images.

Detect-Prevent-Predict (DPP) model A layered cybersecurity defense framework organizing detection, prevention, and prediction layers for AI systems.

Diabetic macular edema (DME) A diabetic complication that affects the macula.

Differential diagnosis A list of patient's symptoms, used to guide further diagnostic steps.

Digital Personal Data Protection Act (India) The Indian data privacy legislation governing how AI health tools can collect, store, and process rural patient data; cited as the compliance framework.

Digital phenotyping Collection of behavioral and physiological data via smartphones and wearables for mental health assessment.

Digital therapeutics (DTx) Regulatory-grade software interventions that deliver evidence-based therapeutic effects, such as cognitive behavioral guidance or metabolic coaching, and can secure reimbursement as standalone treatments.

Digital twin A virtual representation of a patient, device, or process that reproduces biological and operational dynamics with high fidelity, enabling risk-free "what-if" simulations for design optimization and personalized care.

Domain adaptation A subset of transfer learning technique trained on a specific data set to operate on a different data without the need of full training.

Donor-Derived Cell-Free DNA (dd-cfDNA) Circulating DNA fragments released from allograft cells; used as a non-invasive marker in organ viability prediction and post-transplant rejection risk modeling.

DRAMP 2.0 A data repository of antimicrobial peptides.

Drug metabolism studies Research using machine learning to predict which gut bacteria and enzymes can metabolize pharmaceutical compounds.

Drusen Accumulation of yellow deposits of lipids and proteins beneath the retina.

Dual-energy X-ray absorptiometry (DXA) A noninvasive medical imaging procedure that utilizes two different levels of X-rays to precisely determine bone mineral density (BMD) and body composition (fat and muscle).

Dynamic contrast-enhanced MRI (DCE-MRI) An imaging procedure for determining vascular permeability.

Dynamic susceptibility contrast (DSC) MRI It is an advanced neuroimaging technique that uses a contrast agent to assess cerebral blood flow (CBF), cerebral blood volume (CBV), and mean transit time (MTT). It provides hemodynamic information that standard MRI cannot, which is particularly useful for assessing conditions like tumors, stroke, and epilepsy.

Early Embryo Viability Assessment system (Eeva) A time-lapse imaging technology designed to determine which embryos have the maximum capability to develop into blastocysts during an in vitro fertilization (IVF) procedure.

Earth Microbiome Project (EMP) Global collaborative effort to characterize microbial diversity across Earth's environments.

ECemble Enzyme classification using Ensemble approach.

Ecosystem Management Approaches to maintaining and restoring the health and sustainability of ecological systems.

Efficient Residual Factorized Convolutional Neural Network (ERF-Net) A deep neural network designed for evaluating real-time semantic segmentation.

Electrocorticography (ECoG) Invasive intracranial EEG used as a supplemental imaging modality in epilepsy cases.

Electroencephalography (EEG) A brain imaging instrument.

Eliza The first chatbot, developed in 1966 by Joseph Weizenbaum (MIT, Cambridge, Massachusetts), who created a simulation of a psychotherapist.

EMBL-EBI An intergovernmental institute known as the European Bioinformatics Institute (a subsidiary of the European Molecular Biology Laboratory), which conducts research and provides services in bioinformatics.

Emergent behavior Complicated behaviors due to the interaction of simpler components within a system.

Engineered probiotics Genetically modified beneficial bacteria designed to target specific pathogens or produce therapeutic compounds.

Ensemble learning A machine learning approach that merges multiple algorithms to achieve better performance than individual models.

Entrez Programming Utilities (E-utilities) Application Programming Interfaces (APIs) that allow automated retrieval and processing of biological data from multiple databases.

Environmental Decision Support System (EDSS) This AI-based system is widely used in wastewater treatment.

Environmental sustainability Meeting current requirements without compromising the ability of future generations to meet their own needs.

Error-related potential (ErrP) signal A type of EEG signal that occurs after a person perceives an error, whether their own or another's.

eSanjeevani India's national government telemedicine platform Proposed as the integration target for AI-enhanced teleconsultation hubs.

European Medicines Agency (EMA) A decentralized medical agency of the European Union.

E-Utilities The NCBI (National Center for Biotechnology Information) offers a set of eight vital tools that provide access to PubMed and Entrez databases.

Event-related desynchronization (ERD) A V-shaped signal.

Event-related synchronization (ERS) A neurophysiological episode based on signal power.

Evolutionary scale modeling (ESM) A mathematical illustration utilized in population-based genetics to demonstrate the progression of states and interactions within a system over time.

EyeArt A commercially cleared AI platform for automated diabetic retinopathy detection; named alongside IDx-DR as an available clinical screening tool.

Faster region-based convolutional neural network (F-RCNN) A novel disease detection technique.

Fecal Matter Transplantation Transfer of stool from a healthy donor to a recipient, used to demonstrate the role of microbial enzymes in metabolic homeostasis.

Federated learning (FL) A machine learning (ML) technique.

Feed-forward neural network (FFNN) A kind of artificial neural network.

Follicle-stimulating hormone (FSH) A pituitary gland-derived hormone that stimulates the growth of ovarian follicles in women and assists in diagnosing infertility and menstrual cycle disorders.

Fractal ConvNet: A distinct neural network architecture.

Fractional flow reserve (FFR) An invasive measure of coronary stenosis hemodynamic significance.

Framingham Risk Score (FRS) A classical 10-year cardiovascular risk prediction model.

Free flap surgery: A complex reconstructive procedure transferring tissue with its blood supply from a donor site.

Frequency-enhanced decomposed transformer (FEDformer) A deep learning (DL) architecture designed for long-term time-series forecasting.

Full-atom peptide design with geometric latent diffusion (pep-GLAD) An AI system designed to develop novel peptides that bind efficiently to specific protein targets.

Functional annotation The process of determining the biological roles and functions of genes present in metagenomic data.

Functional magnetic resonance imaging (fMRI) A brain imaging instrument.

Functional near-infrared spectroscopy (fNIRS) A brain imaging instrument.

Gemini A GenAI chatbot developed by Google.

Gene expression omnibus (GEO) repository A global public repository that provides high-throughput gene expression and other functional genomics data sets.

General Data Protection Regulation (GDPR) The European Union (EU) implemented this data protection and privacy law in 2018.

Generalized type-2 (GT2) fuzzy logic An advanced form of type 2 fuzzy logic.

Generative adversarial network (GAN)-based magnetic resonance imaging (MRI) Application of generative adversarial networks for MRI reconstruction/synthesis (e.g., super-resolution, artifact reduction).

Generative AI (Gen AI) Deep learning architectures, such as transformer or diffusion networks, that create novel content ranging from molecular structures and clinical narratives to synthetic control data, thereby expanding the experimental design space.

Gene regulatory network (GRN) Segments within DNA that regulate the rate and derivatives of genes, controlling important processes, including transcription and translation, to prevent the production of harmful derivatives.

Genetic algorithm (GA) A computer search technique based on the natural selection of biological evolution.

Genomics The study of an organism's complete set of genetic material (genome).

Genome Aggregation Database (gnomAD) A population-scale whole-genome sequencing reference database; used by AI variant interpretation tools to distinguish pathogenic from benign variants.

Genome-wide association study (GWAS) A study to identify genetic variations associated with a particular disease or trait.

GENTRL A generative tensor-based reinforcement learning platform.

Geospatial equity analysis A GIS-powered AI technique mapping disease burden against service availability across geographic units; used for equity-focused healthcare resource allocation.

Geospatial impact heatmap An AI-generated map overlay that combines demographic, clinical, and social determinant layers to identify geographic inequities, forecast intervention impact, and inform both financial and social return analyses.

GitHub A web-based platform assists in software development.

GPT-1/GPT-2/GPT-3/GPT-3.5/GPT-4/GPT-5 (New Generation of ChatGPT) OpenAI developed these multimodal models, which can generate more comprehensive responses than ChatGPT.

Gradient-weighted class activation mapping (Grad-CAM) It utilizes deep learning (DL) and computer vision techniques to critically examine an image and coordinates with convolutional neural networks (CNNs) to make predictions.

Graphs Structured representations of knowledge that visualize relationships between concepts and entities.

Graphics processing unit (GPU) technology An exclusive electronic circuit designed to perform parallel computations, efficiently rendering images, videos, and miscellaneous visual content.

Green AI A set of practices centered on enhancing computational efficiency without additional energy consumption burden.

H5PY A Python-based module that provides an interface to the Hierarchical Data Format 5 (HDF5) binary file format. It can store and manage very large numerical datasets.

Hallucination Artificial intelligence-based spreading of misinformation.

Harmonic attention U-Net (HaTU-Net) A deep learning (DL) procedure for the accurate partitioning of ovaries and follicles in ultrasound images.

Health inequities Unfair and avoidable differences in health outcomes between different population groups.

Health Insurance Portability and Accountability Act (HIPAA) A United States Federal law secures Protected Health Information (PHI).

Health Technology Assessment (HTA) A structured appraisal of the clinical, economic, and societal consequences of a health innovation, increasingly incorporating AI-informed dynamic economic models for coverage decisions.

Helicobacter pylori A pathogenic bacterium associated with peptic ulcers and gastric cancer.

Hierarchical clustering A clustering method that creates tree-like structures, showing relationships between data points at various levels.

Hierarchical data format (HDF) File formats (HDF4, HDF5) to preserve a large volume of data.

Hierarchical representations Multilayered data representations learned by deep learning models to capture complex patterns.

Hierarchical time-aware attention networks (HiTANet) Utilized for risk determination in Electronic Health Records (EHRs).

High-dimensional datasets Data with many variables or features, common in microbiome research, where the number of features often exceeds the number of samples.

High-throughput sequencing It is a parallel genome sequencing method capable of generating massive amounts of DNA sequence data rapidly.

Host–microbiome interactions The complex bidirectional relationships between microorganisms and their host organisms.

HSV (hue, saturation, value) A color model that defines colors using hue (intact color), saturation (intensity or richness of the color), and value (brightness of the color).

Hugo RAS A portable and modular robotic-assisted surgery device.

Human induced pluripotent stem cells (hiPSCs) Powerful stem cells exist during early embryonic development.

Human leukocyte antigen (HLA)-DR4 (HLA-DR4) It is associated with an increased risk of developing certain autoimmune diseases, especially type 1 diabetes (T1D) and rheumatoid arthritis (RA).

Human microbiome The consortium of microbes that live in and on the human body, especially in places like the gut, skin, mouth, and lungs, generally maintain a symbiotic association with our own cells.

Human Microbiome Project (HMP) Large-scale research initiative to characterize the human microbiome and understand its role in health and disease.

Human Protein Atlas (HPA) A Swedish program aimed at mapping all human proteins using diverse OMICS technologies, including antibody-derived imaging, mass spectrometry-derived proteomics, transcriptomics, and systems biology data.

HydrAMP A generative model for peptide sequencing, corroborating with antimicrobial activity conditions.

IDx-DR An artificial intelligence (AI)-based procedure for the early detection and diagnosis of diabetic retinopathy (DR).

ImageNet An image database system organized according to the WordNet hierarchy.

InceptionNet A convolutional neural network (CNN) structured for image classification tasks and designed for the ImageNet Large Scale Visual Recognition Challenge.

Inclusion modeling Predictive analytics that estimate how demographic, socioeconomic, and linguistic factors influence trial enrollment or product uptake, enabling equity-adjusted projections of cost-effectiveness and social return.

Informed consent A process ensuring research participants understand how their biological samples and data will be used, stored, and shared.

Intel Software Guard Extensions (SGX) A hardware-based security system developed by Intel.

Intelligent Cyber-Biological Systems (ICBS) A multi-layered AI framework for real-time organoid monitoring and adaptive culture adjustment.

Intelligent Data Analysis Score (iDAScore) An artificial intelligence (AI)-based analysis for embryo evaluation.

Interdisciplinary Synergy Collaboration and integration of knowledge from multiple scientific disciplines to address complex problems.

Interior point optimization technique (IPOPT) A type of algorithm for solving constrained problems.

Internal rate of return (IRR) The discount rate that yields a net present value of zero for an investment and thus provides a standard metric for comparing projects of differing scale and duration.

International Cancer Genome Consortium (ICGC) An institution established to coordinate large-scale cancer genome studies.

Intersection over union (IoU) score (Jaccard's index) A metric used in computer vision.

Interval type-2 (IT2) fuzzy logic A model to explain uncertainty.

Intracytoplasmic sperm injection (ICSI) A laboratory-based procedure for in vitro fertilization (IVF), in which a single sperm is directly injected into an egg to achieve fertilization.

Irritable bowel syndrome (IBS) Chronic conditions of recurrent stomach pain, constipation, gas, cramping, bloating, and diarrhea.

ISHLT (International Society for Heart and Lung Transplantation) A scientific society dedicated to advancing education and research on cardiovascular and pulmonary diseases and their treatments.

Joint Photographic Experts Group (JPEG) An extensively used image file format.

Just-in-Time Adaptive Interventions (JITAI) A delivery framework providing behavioral support at the optimal moment based on real-time user context.

Keras Used as a framework in deep learning (DL).

Keratoconus A progressive ocular disorder that affects the cornea.

Kernelized support vector machine (k-SVM) Support vector machine.

Keystone species Microbial species that play critical roles in maintaining community structure and ecosystem function.

K-means clustering A clustering algorithm that partitions data into k distinct clusters based on feature similarity.

K-mer A fixed-length nucleotide subsequence used as an alignment-free feature for genomic classification, drug resistance prediction, and metagenomic profiling.

K-nearest neighbors (KNN) A machine learning-based algorithm utilized for classification and regression analysis.

Knowledge Discovery in Data (KDD) It is a procedure for extracting useful patterns from data.

Knowledge graphs Structured representations of knowledge that visualize relationships between concepts and entities.

Kyoto Encyclopedia of Genes and Genomes (KEGG) A comprehensive and computational database for learning the functional features of genomes.

Lactococcus lactis A bacterial species commonly used in probiotic engineering and food fermentation.

LAMP2 A lysosomal-derived membrane protein required for cellular processes, including autophagy and the breakdown of cellular waste.

Lasso peptides Macrolactam peptides with a twisted ring structure that confers stability.

Large language models (LLMs) Capable of recognizing, creating, and manipulating human language. They are widely used to generate text, perform translations, and produce innovative content.

LatentGAN A GAN variant using a latent chemical space for de novo molecular design.

Least Absolute Shrinkage and Selection Operator (LASSO) A widely used regression technique.

LifeWhisperer Integrate artificial intelligence (AI) to enhance in vitro fertilization (IVF) outcomes.

LifeWhisperer Genetics A prompt, empirical, and noninvasive web-based procedure for appropriate embryo selection.

Lightweight deep learning (DL) A discipline of artificial intelligence (AI) and machine learning (ML) that emphasizes efficiency and compactness while developing deep learning (DL) models.

Linear discriminant analysis (LDA) A classification method in machine learning.

Linear support vector machine (L-SVM) A supervised machine learning algorithm.

Lipinski's Rule of Five A heuristic filter for oral drug-likeliness.

Literature mining Automated extraction and analysis of information from scientific publications and databases.

LlaMA An open-source large language model (LLM). LlaMA 3 is the third-generation LLM from Meta AI.

Longevity therapeutics Pharmacologic, genetic, or cellular interventions intended to extend healthy lifespans; AI-guided target discovery and market size uncertainty place these assets in the long-horizon capital bucket.

Low-rank adaptation (LoRA) A parameter-efficient fine-tuning technique that freezes the original model weights and injects trainable low-rank matrices into specific layers, adapting large pre-trained models to clinical imaging domains without modifying all parameters.

Machine learning (ML) A subset of AI that enables computers to learn and make predictions from data without being explicitly programmed.

Machine learning-based organoid analysis software (MorgAna) A Python-based software program for the quantitative analysis of organoid images.

Machine translation A software-based AI tool widely used for translating languages.

Macrocyclic peptides Peptides with ring structures formed by non-canonical amino acids; provide proteolytic and thermal stability and are used to target structured viral RNAs.

MADDi An AI bot that works like a personal assistant for managing daily chores.

MAMBA A deep learning (DL)-based configuration involved in sequence modeling.

Marginal Structural Model (MSM) A causal inference method for estimating the effect of time-varying treatments in the presence of time-varying confounders; applied to oncology treatment sequence analysis.

Mask R-CNN A deep learning (DL) framework utilized in instance segmentation, a computer vision task that identifies and generates pixel-level masks for individual distinct objects in an image.

Matched Molecular Pair (MMP) analysis A method comparing pairs of molecules differing by a single structural change to understand structure-property relationships.

MATLAB A programming and numeric computing platform.

MBR Acronym for membrane bioreactor, an advanced wastewater treatment facility integrating the components of microbiological treatment with conventional membrane filtration.

Med Gemini A medically trained subtype of Gemini foundation model.

MELD algorithm A deep learning algorithm analyzing 33 cortical surface features to detect subtle focal cortical dysplasia in pediatric epilepsy.

Metabolic modeling Computational approaches to understand and predict the metabolic capabilities of microorganisms.

Metabolic pathways A series of chemical reactions within cells that convert substrates into products through enzyme-catalyzed steps.

Metabolome A spatiotemporal representation of small-molecule metabolites (e.g., lipids, amino acids, and sugars) found in a cellular or extracellular niche.

Metabolomics The systematic study of chemical fingerprints left by cellular metabolic processes.

Metagenome-assembled genomes (MAGs) Reconstructed genomes of microorganisms that have not been isolated or cultured in the laboratory.

Metagenomic contigs These are long DNA sequences constructed by joining together shorter DNA fragments (called reads) taken from a mixed sample of microbes, such as those found in soil, ocean water, or the human gut. Since the sample contains DNA from many different organisms, these contigs represent pieces of genomes from multiple microbes, helping us understand who is in the sample and what they are capable of doing.

Metagenomic sequencing A high-throughput DNA sequencing technique that analyzes genetic material directly from environmental samples without the need to culture individual species.

MetaTrinity A tool that sorts and analyzes metagenomic data faster and more accurately using smart algorithms, while reducing computational time.

Microbe–microbe interactions The complex relationships and communications between different microbial species in a community.

Microbial ecosystems Complex communities of diverse microorganisms and their interactions within specific environments.

Microbial fingerprints Unique patterns of microbial composition that can serve as diagnostic tools for disease assessment.

Microbial metabolites Chemical compounds produced by microorganisms during their metabolic processes, often with therapeutic potential.

Microbiome engineering The deliberate manipulation of microbial communities to enhance their beneficial effects on health, agriculture, or the environment.

MidJourney A Gen AI program for converting text descriptions into images.

MIMIC-III and MIMIC-IV Large, single-center Electronic Health Record (EHR) databases from intensive care patients.

Minimal long short-term memory (minLSTM) A highly efficient variant of the LSTM recurrent neural network (RNN).

Model explainability/transparency The capacity to interpret and audit an AI model's logical structure, input provenance, and decision pathway, which is essential for regulatory approval, financial governance, and stakeholder trust.

Model generalization The ability of an AI model to perform accurately on new, unseen data from different populations or contexts.

Model interpretability The extent to which AI model results and decision-making processes can be understood and explained in human terms.

Model predictive control (MPC) A state-of-the-art control automation and controlling procedure.

Moonlighting peptides Peptides with multiple context-dependent biological functions; AI generative pipelines must explicitly filter for unintended cross-reactivity.

Monte Carlo simulation A computational technique that samples input distributions thousands of times to approximate the full probability distribution of return on investment and quantify tail risk.

Morphokinetics Quantitative time-stamped parameters describing the timing and dynamics of embryo cell division and blastocyst formation, captured by time-lapse imaging and analyzed by AI for viability prediction.

Motion scaling A robotic surgery feature that converts large console movements into sub-millimeter instrument movements; the da Vinci system is cited at a 5:1 ratio as a named example.

Motor imagery (MI) signal A type of electroencephalography signal or brain wave observed when a person imagines doing certain things without actually performing them.

Multimodal contrastive diffusion model (MMCD) A generative artificial intelligence (AI) framework that unifies the principles of diffusion models and multi-modal contrastive learning.

Multitask Directed Message Passing Neural Network (DMPNN) modeling A predominant technology used for cheminformatics research and drug discovery. Basically, it is a unique approach that directs a single model to predict multiple properties of a molecule simultaneously.

Multiomics The integration of multiple biological data types, including genomics, proteomics, transcriptomics, and metabolomics, to provide comprehensive biological insights.

MTGL-ADMET Multi-Task Graph Learning (MTGL) framework for predicting multiple ADMET (Absorption–Distribution–Metabolism–Excretion–Toxicity) characteristics of drug-like small molecules.

Naïve Bayes classifier++ (NBC++) An enhanced version of the traditional Naïve Bayes framework for classifying reads from amplicon and metagenomic samples. It helps identify different microbes from genetic data.

NASNetLarge A convolutional neural network (CNN) architecture developed by Google through Neural Architecture Search (NAS).

Natural language processing (NLP) An AI field focused on enabling computers to understand, interpret, and generate human language.

Net present value (NPV) The sum of discounted future cash flows minus the initial investment, capturing the time value of money and remaining sensitive to revenue, cost, and timeline assumptions.

Network modeling An approach that represents microbial communities as interconnected webs of species interactions.

Neural networks Computing systems inspired by biological neural networks, consisting of interconnected processing units (neurons).

Neuro-prosthetic limb A robotic or artificial appendage controlled via decoded BCI signals; discussed as the rehabilitation output device for neuro-motor disorders including ALS and stroke.

Neuro Scholar A custom GPT built on ChatGPT architecture for neurology and psychiatry clinical questions.

New approach methodologies (NAMs) An alternative approach to the use of animals in research.

Next-generation organoids (NGO) Lab grown organoids that fully simulate human organs and integrate patient-derived genomics, single-cell sequencing, and AI for rare variant disease modeling.

Nikshay India's National TB Elimination Program registry; named as an integration target for AI referral routing and medication adherence tracking.

Noisy Intermediate-Scale Quantum (NISQ) The current era of quantum hardware characterized by limited qubit counts and non-negligible error rates; frames the chapter's statistical caution about QAI performance claims.

Nutrient cycling The natural processes by which nutrients move through the living and nonliving components of ecosystems.

NutriGenomeDB A web-derived developmental and analytical platform for nutrigenomics.

Objective Structured Assessment of Technical Skills (OSATS) framework A validated global surgical rating scale used by surgeons.

Oculomics An advanced eye imaging procedure used in conjunction with artificial intelligence (AI) to identify biomarkers in the eye.

Offline-first design A software architecture pattern where AI tools function fully without internet connectivity, syncing to the cloud when connectivity is available; essential to the chapter's rural deployment model.

Ontologies Formal representations of knowledge within specific domains using standardized vocabularies and classifications.

Organ-on-a-chip (OOC) A microfluidic device that utilizes human tissues or cells to produce a functional model of an organ in a laboratory setting, replicating its structural integrity, including biological functions.

OpenAI An artificial intelligence company based in San Francisco, California. Developed popular tools such as ChatGPT, DALL-E, and generative pretrained transformer (GPT) technology.

OpenCV A huge computer vision (CV) library.

Optical coherence tomography (OCT) A noninvasive imaging procedure that utilizes light to produce cross-sectional images of tissues. Used extensively in diagnosing conditions related to the retina and optic nerve. Optical coherence-based analysis (OCA) is conducted based on OCT.

Ordinary differential equation (ODE)-inspired contiformer A neural network model intended for time-series data that integrates the continuous-time dynamics of neural ODEs into the attention mechanism of a transformer.

OrgaExtractor A deep learning (DL) tool for analyzing organoid images.

Organoids Compelling in vitro models for researching human development. Conducted in 3D tissue cultures derived from stem cells, which replicate the functional complexity of an organ.

Organoid printing and optical coherence tomography-based analysis (OPO) A 3D bioprinting technique to generate organoids and utilizes optical coherence tomography (OCT) for analysis.

OrganelX An AI-based e-Science web server that utilizes deep learning (DL) to assist in biological research, especially in protein research.

OrganoID platform A deep learning (DL) software designed to recognize, segment, and identify individual organoids using microscopic images, specifically in time-lapse experiments.

OrgaQuant An end-to-end trained neural network designed to detect and quantify the size of human intestinal organoids in bright-field images.

Ovarian hyperstimulation syndrome (OHSS) A condition in which the ovaries swell and leak fluid into the abdomen. It specifically occurs in women suffering from polycystic ovary syndrome (PCOS).

Overfitting When a model performs well on training data but poorly on new, unseen data due to excessive complexity.

P300 (ERP) A time-locked event-related potential appearing approximately 300 ms after an oddball stimulus.

Paradigm shift A fundamental change in approach or underlying assumptions in scientific research and practice.

Parameter AI-system variable that is used to derive a prediction.

Paraphrasing tool An AI-based tool used to rephrase sentences.

Parkinson's Progression Markers Initiative (PPMI) A project aimed at determining the biomarkers for Parkinson's disease.

Partial dependence plots Visualization of the marginal effect of a feature on model predictions; interpret with care in the presence of interactions.

PatchTST An AI model based on the Transformer architecture, which is particularly designed for time-series analysis.

Patient-derived organoids (PDOs) Replication of function, structure, and genomic integrity in three-dimensional cell cultures.

Pentacam An eye-imaging device that utilizes a camera to analyze the eye's anterior segment, including the cornea, iris, and lens.

PepBDB A structural database of biological peptide (<50 residues)–protein interactions.

PepHarmony A contrastive learning framework for consolidated sequence and structure-based peptide encoding.

PeptideBERT A language model based on transformers for peptide property prediction.

Peptide-generative pretrained transformers (GPTs) Generative design of peptides utilizing generative pretrained transformers (GPTs) and bioinformatics supervision.

Peptipedia A web-based convenient database to assist peptide research by utilizing bioinformatics and machine learning (ML) strategies.

PepVAE Variational Autoencoder Framework for Antimicrobial Peptide (AMP) Generation and Activity Prediction.

Perfusion monitoring (AI-assisted) Real-time AI analysis of free flap tissue perfusion using photoplethysmography.

Perplexity A powerful AI tool that provides concise answers along with citations

Personalized interventions Tailored treatments or therapies designed based on individual biological profiles and characteristics.

Pharmacovigilance (AI-assisted) AI-assisted non-stop monitoring of adverse drug reactions across large patient populations.

Phenotype assessment Evaluation of observable characteristics or traits that result from genetic and environmental interactions.

Phonocardiogram (PCG) A noninvasive diagnostic tool.

Phylogenetic placement A method of refining classification accuracy by comparing contigs to reference databases within evolutionary contexts.

Phytotherapeutic Medicinal plants-derived herbal/botanical medicines used for therapeutic purposes.

Picture Archiving and Communication System (PACS) The medical imaging network for storing, retrieving, and disseminating diagnostic images.

Platelet-rich plasma (PRP) A regenerative therapy that utilizes a patient's own platelets and growth factors to accelerate healing in injured tissues.

Polygenic risk scores (PRSs) Statistical measures used to determine an individual's genetic predisposition to developing a complex disease.

Positron emission tomography (PET) PET is a nuclear medicine imaging technique that uses radioactive tracers to create three-dimensional images of physiological and biochemical processes in the body.

POTTER calculator An ML surgical risk tool trained on more than 380,000 acute surgical cases; described as outperforming the ASA physical status score for morbidity and mortality prediction

Precision healthcare Medical care tailored to individual characteristics, including genetic, environmental, and lifestyle factors.

Precision medicine A healthcare approach that considers individual variability in genes, environment, lifestyle, and microbiome composition.

Predictive modeling Using statistical and machine learning techniques to forecast future outcomes based on current data.

Preimplantation genetic testing for aneuploidy (PGT-A) A noninvasive technique used during in vitro fertilization (IVF) to identify embryos.

Probability of success (PoS) A dynamically updated estimate, often produced using Bayesian or reinforcement learning methods, of the likelihood that a program reaches its next value inflection point, such as regulatory approval or commercial launch.

Probabilistic forecasting An approach that outputs complete predictive distributions rather than single-point estimates, allowing uncertainty to propagate through financial scenarios and strategic decisions.

Prognosis Prediction of the likely course and outcome of a disease based on microbial and other biomarkers.

Programming A technique of providing instructions to a computer.

Prompt An input to which the AI system responds and performs actions accordingly.

Propedia A versatile tool and database for the design of peptides for specific purposes.

Protein data bank (RCSB PDB) A curated and annotated public database of protein structures deposited by researchers for use in biological research on proteins.

Protein language models (pLMs) Large-scale transformer models trained on hundreds of millions of protein sequences and used to generate context-aware embeddings for peptide representation (ESM-2, ProteinBERT).

Proteomics The large-scale study of proteins, with respect to their spatiotemporal expression, structure, and function.

PubMed A comprehensive database of biomedical literature maintained by the National Library of Medicine.

QRISK QRISK1, QRISK2, and QRISK3 tools are used by health practitioners to determine the development of cardiovascular diseases and disorders.

Quality control The process of filtering out sequencing errors and low-quality reads to ensure data accuracy.

Quality-of-Life (QoL) score A composite patient-reported outcome measure, such as EQ-5D or SF-36, that tracks functional status, pain, and mental well-being and can be converted into quality-adjusted life year units for economic evaluation.

Quadratic discriminant analysis (QDA) A categorization procedure that complements linear discriminant analysis (LDA) by allowing nonlinear decision boundaries.

QALY/DALY Quality-adjusted life year and disability-adjusted life year quantify gains or losses in healthy life and serve as standard currencies in cost-effectiveness analysis.

Quillbot A novel AI-based paraphrasing, grammar validation, and writing tool. In addition, it aids in summarizing, as well as checking for plagiarism and grammatical errors. Quillbot is associated with Scribbr through Learneo.

Random forest An ensemble learning procedure that generates multiple decision trees and combines their predictions for improved accuracy and robustness.

Random forest algorithm A supervised machine learning (ML) algorithm utilized for classification and regression analysis.

Remote auscultation A telemedicine technology to partially simulate remote physical examination.

Rhinonet A deep learning algorithm that classifies rhinoplasty status from visual facial analysis; cited as an intraoperative guidance example for rhinoplasty.

Rule-based chatbot An early chatbot design using predefined scripts and static decision trees.

Reactome An open-source, curated, and peer-reviewed pathway database.

Real-time ROI dashboard An interactive interface that merges operational data feeds, AI forecasts, and financial key performance indicators, refreshing continuously to support agile governance decisions.

Rectified linear unit (ReLU) A widely used function in deep learning (DL).

Recurrent neural networks (RNNs) A variety of ANNs (artificial neural networks) intended to analyze sequential data.

Region of Interest A section or portion of an image or dataset that requires detailed analysis or processing.

Region proposal network (RPN) A neural network utilized in object detection.

Reidentification The potential risk of determining individual identities from supposedly anonymized datasets.

Reinforcement learning (RL) A branch of machine learning in which an agent iteratively learns a policy that maximizes cumulative reward through experimentation and feedback, making it useful for adaptive dosing and portfolio optimization.

Reinforcement learning from human feedback (RLHF) An ML-based technique utilized in the development of LLMs

Reinforcement learning portfolio optimizer An RL agent that treats capital deployment as a sequential decision problem, receives reward signals such as risk-adjusted net present value changes, and learns optimal allocation strategies under uncertainty.

Repcon Semantic Systems developed this software platform to represent complex software in a network format, especially to manage large volumes of data.

Representation The inclusion of diverse population groups in research studies and training datasets.

Reproducibility The ability to obtain consistent results when studies are repeated using the same methods and conditions.

Residual neural network (ResNet) A deep learning (DL) framework in which the layers learn from residual functions demonstrated with respect to the layer inputs. ResNet152V2 is composed of 152 layers, predominantly convolutional.

Responsible AI A governance framework covering legal, economic, and social dimensions of AI deployment.

Retinal nerve fiber layer (RNFL) Also known as the nerve fiber layer (stratum opticum). RNFL constitutes a network of fibers that radiate from the optic disc, where the optic nerve exits the eye.

Return on investment (ROI) A ratio of net gains to total investment, often multiplied by one hundred, that provides a high-level snapshot of profitability, yet does not account for time value or risk exposure.

RGB (red, green, blue) A color model.

Root mean square propagation (RMSprop) An adaptive learning rate optimization algorithm used in deep learning.

Robot A machine-based device capable of performing physical and mechanical actions according to its programming.

Robotic process automation (RPA) A robot-based, cost-effective technique to enhance efficiency, accuracy, and high-throughput.

SATPdb A database of therapeutic peptides obtained from public-domain peptide databases.

Scenario analysis A planning method that evaluates discrete future contexts, such as price erosion or regulatory delay, to stress-test strategy and contingency budgets.

Scintillation proximity assay (SPA) An assay technology used to determine molecular interactions.

Scribbr Academic writing support AI tools.

Secondary metabolites Specialized compounds produced by microorganisms that are not essential for growth but may have biological activity.

SegNet Distinct neural network architecture.

Self-supervised learning (SSL) A machine learning paradigm meant for bridging supervised and unsupervised learning which self-teaches itself with unlabeled data.

Self-supervised learning Learning representations from unlabeled imaging data without manual annotation.

Senhance A novel digital laparoscopic platform used for minimally invasive surgeries.

Sensitivity analysis A systematic examination of how variation in individual or grouped input parameters affects model outputs, thereby identifying the variables that most influence risk and upside.

Sentiment analysis NLP-based organization of emotional tone to text (positive, negative, neutral).

Sequence alignment A traditional computational method for comparing DNA sequences to identify similarities and relationships.

Shannon's entropy A basic concept that determines the uncertainty or reasonable amount of details from a random variable or data source.

Signal-to-noise ratio (SNR) Assess the robustness of a signal compared to background noise.

Simulated annealing (SA) A novel technique used to solve hard computational optimization problems.

Smile-GAN A generative adversarial network (GAN) designed for congregating a group of experimental subjects/patients.

Social determinants of health (SDoH) modeling The integration of factors such as income, housing, education, and environmental exposure into predictive models so that economic evaluations reflect real-world access barriers.

Social return on investment (SROI) A metric that translates societal and equity outcomes into monetary terms, complementing financial returns and guiding mission-oriented investments.

Solid-state drive (SSD) A solid-state storage device that utilizes integrated circuits to store data.

Space Health Nutrition Research and product development focused on nutrient-stable foods and metabolic support for prolonged human presence in low-gravity environments, classified as a 7-year-plus horizon investment.

Speeded Up Robust Feature (SURF) Used in computer vision as a feature detector.

Standardized protocols Established, consistent methodologies for research to ensure comparability across studies.

Steady-state visually evoked potential (SSVEP) A measurement of the brain signal generated by a high-frequency flickering visual stimulus.

Stochastic gradient descent (SGD) A machine learning (ML) iterative optimization algorithm is an iterative optimization algorithm used in machine learning.

Supervised learning A variety of machine learning (ML) techniques based on an algorithm obtained from a labeled dataset, signifying that each input data is paired with the correct output.

Susceptible, Infectious, and Recovered (SIR)/Susceptible, Exposed, Infectious, and Recovered) SEIR model Compartmental epidemiological models used as mechanistic frameworks enhanced by ML parameter estimation.

Syndromic surveillance A strategy to monitor prediagnostic and random data on public health.

α-Synucleinopathies A type of neurodegenerative disease caused by the accumulation of misfolded aggregates of α-synuclein protein in the form of amyloid plaques or Lewy bodies in neurons and various other cells.

Synthetic biology A bioengineering procedure to design and construct new biological parts, devices, and systems.

Synthetic control arm The substitution or augmentation of a conventional placebo group with real-world or historical patient data, reducing recruitment burden and accelerating trials while preserving statistical power.

Synthetic data control The creation of statistically representative patient records via generative modeling to augment scarce datasets and safeguard privacy in analysis and validation.

Synthetic Microbial Consortia Artificially designed communities of microorganisms engineered to work together for specific purposes.

Systems biology An interdisciplinary approach that studies complex biological systems through the integration of multiple data types.

Taxometer A neural network-based tool designed to improve taxonomic classifications and evaluate the quality of classifiers for metagenomic contigs.

Taxonomic classification The process of identifying and categorizing microorganisms present in a sample into hierarchical groups (species, genus, family, etc.)

Taxonomic profiling The use of machine learning (ML) algorithms to classify microbial species within a given sample.

Telehealth Ten A systematized guide based on vital signs for virtual physical examination with assistance from patients.

Temperature (AI) A parameter that regulates the randomness and creativity in an LLM's output.

TFLearn TensorFlow Deep Learning A modular deep learning (DL) library.

The Cancer Genome Atlas Program (TCGA) A National Institutes of Health (NIH)/National Cancer Institute (NCI)-driven, exhaustive cancer genomics program.

Therabot A chatbot based on generative AI targeted towards Major Depressive Disorder (MDD) and Generalized Anxiety Disorder (GAD).

Thermalytix An AI thermal imaging platform developed by Niramai Health Analytics for non-contact breast cancer screening.

Time-locked event-related potentials (ERPs) A type of EEG signal.

T-LSTM Time-aware long short-term memory

Time-series prediction (TSP) A procedure for utilizing historical, chronological data to predict future values or trends.

Time series to vector (TS2Vec) An unsupervised machine learning (ML) framework that provides general-purpose representations of time-series data.

Token A fundamental unit of text that is processed by LLMs

Tox21 A United States Federal research program designed to rapidly and efficiently determine the toxicity and safety of commercial chemicals.

ToxCast A toxicity forecasting program from the United States Environmental Protection Agency (US EPA).

ToxDL 2.0 A multimodal deep learning (DL) model for protein toxicity assessment.

ToxGIN/ tAMPer Deep learning (DL) procedures to assess peptide toxicity by examining peptide sequences and three-dimensional (3D) structures.

Training data The dataset employed to train machine learning algorithms and to instruct an AI system.

Transcriptomics The study of all RNA molecules (transcripts) produced by the genome under specific conditions.

Transmission Gate (TG-GATE) A toxicogenomic database.

Tremor attenuation Robotic filtering of physiological hand tremor during microsurgery.

Trusted execution environments (TEE) A secure system for operating systems.

Turing test (also referred to as the imitation game) It determines the ability of a machine to demonstrate human intelligence. In 1950, Alan Turing developed this test.

Type 1 fuzzy logic A basic framework for symbolizing and analyzing uncertain information.

UK Biobank A large-scale population biomedical database.

U-Net A convolutional neural network (CNN) specifically used for image segmentation and biomedical applications.

Unsupervised learning A type of machine learning (ML) in which algorithms obtain insights from data without predefined labels or categories, allowing for the discovery of hidden patterns, structures, and relationships within the data.

Uniform Manifold Approximation and Projection (UMAP) A nonlinear dimensionality reduction technique used here to visualize high-dimensional drug screening data from patient-derived organoids.

Uniform Resource Locator (URL) A unique web address that is specifically located on the internet.

V3 (Inception v3)/V4 (Inception v4) Efficient computer vision and engine models used for deep convolutional neural networks (CNNs).

Validation Process of testing a model's performance on independent datasets to assess reliability and generalizability.

Vapnik–Chervonenkis (VC) dimension A basic concept in statistical learning theory that determines the capacity or complexity of a machine learning (ML) model.

Variable importance measures Techniques used to determine which features contribute most significantly to model predictions.

Variational autoencoders (VAEs) A generative model that provides a probabilistic representation of data, enabling the generation of new synthetic data.

Variational Quantum Circuit (VQC) A trainable quantum circuit analogous to a neural network, with parameterized gates optimized using methods such as SPSA (Simultaneous Perturbation Stochastic Approximation) or Adam.

Versius A portable, modular surgical robot that provides a 3D view of the surgical field, enabling surgeons to execute operations with minimal access

VGG (Visual Geometry Group)-16/VGG-19 Convolutional neural network (CNN) architectures.

VGGNets A set of convolutional neural networks (CNNs) structured by the Visual Geometry Group (VGG), University of Oxford.

Virtual reality (VR) therapy Computer-generated immersive environments used to treat phobias, PTSD, OCD, eating disorders, and anxiety.

VISH-Pred Artificial intelligence (AI)-enabled Enterprise Service Management (ESM) models for protein toxicity prediction.

Vision transformers (ViTs) Used in computer vision applications as an alternative to convolutional neural networks (CNNs).

Wavelet neural network (WNN) A network that combines the robustness of wavelet transforms and neural networks, especially used for time-series analysis, signal processing, and image segmentation.

WCE-Attention-DeepNet A combination of multiple concepts from machine learning (ML) and medical imaging (MI), extensively used in wireless capsule endoscopy (WCE) with deep learning (DL) models.

Whole-slide image (WSI) analysis Deep Learning assisted analysis of high-resolution digitized histopathology slides for grading of cancer.

Wireless capsule endoscopy (WCE) Utilizes a tiny capsule-sized camera to capture images of the small intestine and digestive tract.

Woebot/Wysa/Replika/Youper Named AI-powered mental health chatbots.

XGBoost (eXtreme gradient boosting) A high-powered, extensively used open-source ML algorithm.

XML (Extensible Markup Language) A versatile text-based markup language used to store and transport data in a readable format.

You Only Look Once (YOLO) A well-recognized and systematic object detection algorithm in AI. YOLOv7 is an advanced object detection algorithm.

Index

Note: Page numbers followed by *f* indicate figures, *t* indicate tables.

A

AAR. *see* Adaptive Auto Regressive (AAR)
Absorption 506
Acceptance & commitment therapy (ACT) 271*t*, 271–272, 274*t*
  online session 272
Accredited Social Health Activist (ASHA) 365, 367–368
ACE inhibitors (ACEis) 501–502
  yam-derived 502
ACE. *see* Angiotensin-converting enzyme (ACE)
Acinetobacter baumannii 108
ACPAs. *see* Anticitrullinated protein antibodies (ACPAs)
ACT. *see* Acceptance & commitment therapy (ACT)
Activated sludge process 336, 338, 342
Acute pediatric encephalopathies (APEs) 308
Ada Health app 65
Adaptive Auto Regressive (AAR) 295*t*, 295
Adaptive learning algorithms 187
Adaptive network-based fuzzy inference system (ANFIS) 72–73, 73*f*, 77, 86–87, 335
  and Bi-TLSTM 79–81
  layers 79–80
  recapitulations 79
Adaptive simulated annealing (ASA) 336
Additive manufacturing 539
ADMET
  chemprop 506
  natural product screening 505–506
  profiling 501–502
ADRs. *see* Adverse drug reactions (ADRs)
Advanced computing 504
Advanced Glycation End Products (AGEs) 459–460
Adversarial examples (AE) 553, 560–561, 563–564
Adversarial machine learning 557–561, 563–564
Adverse drug reactions (ADRs) 436
AF. *see* Atrial fibrillation (AF)
Affective computing 4, 9, 184
Agentic intelligence 369–370
  autonomous symptom assessment 369
  chronic disease management 370
  health analytics orchestrators 370
  intelligent prescription assistants 369
  response coordination 370
Age-related macular degeneration (AMD) 243
AGEs. *see* Advanced Glycation End Products (AGEs)
AI algorithm 199*t*, 551–552
  conceptual hierarchy 552*f*
  data taxonomy 554–555, 555*f*
  development & validation 242
  double-edged sword 551–553
  first line attack 553–559
  life cycle security framework 563*f*
  models 559–560
  pipeline & learning 554–556, 556*f*
  security domains & vulnerabilities 553*f*, 553
  security system framework 568*f*
  traditional cybersecurity threats 566–567, 567*t*
  types 555
AI case studies/applications 95, 167–170, 389–394, 406*f*, 484*f*
  centric surgical education 169
  dexterity & semiautonomy 169
  guided total joint arthroplasty 168
  image-guided glioma resection 168
  personalized approaches 169
  postoperative monitoring & rehabilitation 169
  predictive analytics & perioperative management 168–169
  robotic-assisted surgery 167
  surgery 167*t*
  targeted delivery 169
  workflow & supply chain 169
AI chatbots 57–58, 62–63, 63*t*
  challenges 61–64
  diagnostic accuracy 61
  enhanced personalization 66
  in healthcare 59–60, 67
  home diagnostics 60–61, 64–66
  primary diagnosis 60, 66
  privacy concerns 62
  regulatory & ethical concerns 62
  and smart home technology 66–67
  technical limitations 62–63
  technological breakthrough 67
  untrained & unspecialized models 65–66
  user engagement & trust 61–62
AI ethical/ethics 528
  application 529
  concerns 110–111
  deployment 219
  design 529
  stakeholders 529
  theories 536–537, 537*t*
AI governance 342, 447–448, 528–530, 532
AI in plastic surgery 205–206
  administrative/operational support 204
  challenges 205
  diagnosis & screening 197–198
  education & research 204
  intraoperative applications 202
  managing complications 203–204
  postoperative applications 202–204
  preoperative applications 197–202
  risk stratification 198–199, 199*t*
  robotic tools 202
  simulation & consent 201
  surgical planning 198–201, 199*t*
AI mental health support
  Headspace 270
  Replika 269
  Woebot 268
  Wysa 268–269
  Youper 269–270
AI modules 167
  haptic filtering 167
  image segmentation 167
  trajectory prediction 167
AI precision surgery
  clinical workflow & infrastructure 171
  cost & resource allocation 172
  cyber-security risks 173
  data availability, quality & bias 170
  education & training gaps 172
  human–computer interaction 172
  legal & liability issues 172
  patient acceptance & public perception 173
  regulatory approval & validation 172
  regulatory/ethical considerations 170–171
  technical limits & generalizability 171
AI predictive models 135
  donor heart viability 140–141, 141*f*
  drug dosing 143
  personalized medicine 143
  toxicity monitoring 144–146
AI symptom analysis 97–98
  applications 97–98
  rationale & methods 97
AI-assisted disease detection 38, 48–49
  challenges 47–48
  computer vision 39

AI-assisted disease detection *(Continued)*
data mining & pattern recognition 40
data quality & bias 47–48
DL 38
ethical concerns 48
fundamentals 38–41
generalization 48
implementation & real-world impact 47
integration & synergistic applications 40
interpretability 48
ML 38
NLP 39
RL 40
technology challenges 41
AI-enabled ROI
antiaging therapeutics 358
antivenoms 357–358
barriers 360
enablers 359–360
lifestyle drugs & cosmetic 357
operationalizing framework 358–359
AI/ML algorithms 19–20
advantages & disadvantages 31*t*
AUC-ROC curves 26–28, 28*f*
boosting methods 27–28
clinical utilization 31
in diagnostics 25
healthcare system transformation 30
independent institutions development 30
metaanalyses 28
model selection 30–31
mortality risk prediction models 28–29
oncology applications 23–24
potential utility in healthcare 30
rapid implementation 30
types 20–24
AlexNet architecture 118
Algorithm passports 171
AlphaFold2 system 109, 424
Alpha-synucleinopathies 317
ALS models. *see* Amyotrophic lateral sclerosis (ALS) models
Alzheimer's disease (AD) 317
fluid biomarkers 322*t*
models 475
Ambient AI medical scribes 348
AMD. *see* Age-related macular degeneration (AMD)
AMH. *see* Anti-Müllerian hormone (AMH)
AMPGAN v2 427*t*
AMPs. *see* Antimicrobial peptides (AMPs)
AMR. *see* Antimicrobial resistance (AMR)
Amyloid-beta plaques 317–318, 323, 475
Amyotrophic lateral sclerosis (ALS) models 475
Anastomotic leakage prediction 183
ANFIS. *see* Adaptive network-based fuzzy inference system (ANFIS)
Angelman syndrome (AS) 307–308
Angiotensin-converting enzyme (ACE) 501–502
binding site architecture 502
Animal models 489, 503
ANM. *see* Auxiliary Nurse Midwife (ANM)
ANNs. *see* Artificial neural networks (ANNs)
Antibiotic candidates 357
Anti-CCP antibodies 249–250
Anticitrullinated protein antibodies (ACPAs) 251
Anti-dsDNA 251–252, 252*t*
Antimicrobial peptides (AMPs) 419
Antimicrobial resistance (AMR) 493–494
Anti-Müllerian hormone (AMH) 215
Anxiety 267–271
APEs. *see* Acute pediatric encephalopathies (APEs)
AR process. *see* Autoregressive (AR) process
Area under the curve (AUC) 135–136, 142–143, 146–147, 164
ARIMA 104
Artificial intelligence (AI) 3–4, 7
accuracy & early diagnosis 10–11
application layers 180*t*
ART integration 213, 215
automated quantification techniques 126
BBB detection 323
bias 394–395
black-box model 205
breast cancer detection 526*f*, 526
breast oncology 169
call centers 12–15
cancer drugs 488*t*
capital efficiency & ROI 348–349
centric surgical education 169
challenges 110
chatbots 311
clinical imaging fields 117
clinical performance with ROI 349
cognitive & emotional recovery 185*f*
computational pharmacology 501–502
contact tracing 105
containment & response 105–108
core concepts 252–253
COVID-19 pandemic 95, 106–107
and CT scan 200
customer service 12–15
data privacy 530–531
data quality & bias 110
databases 95–96
de novo drug design 516–517
disease diagnosis 97–104, 125–126
domain-specific 185*f*
drug development 515
EI challenges 6–7
embryo selection/ploidy 217
emotional recognition 7–8
facial expression recognition 198–200
facial recognition & NLP 4
facial recognition technology 527
federated learning 127
foundation 535–536
FSH 215
generalizability 111–112
generative AI era 536
genomic signal processing 98
geospatial equity analysis 353
gonadotropin dosing 215
governance 529
health education chatbots 368
in healthcare 94, 246, 527–528, 530, 537–539
healthcare innovation & technologies 532
healthcare investment 355–356
herbal medicinal plant 506*f*
imaging technology 218
implementation challenges 219
inclusion modeling 352–353
infectious disease 94–112
interpretability 112
interpreting radiological images 198
intersection 493*f*
investment strategies 354
IVF outcomes 217–218
laboratory diagnostics 98–99
legal & regulatory frameworks 219
limitations 201
machine learning era 536
medical advancements 526
medical coding 204
in medical research 531
in medicine 245
mental health diagnostics 10
mental health therapy bots 11
microbial networks 406–408
microbiome data analysis 388–389
mobile diagnostic vans 368
modern healthcare role 250–251
multimodal 186*f*, 319–322, 320*f*
multimodal data integration 218–219, 219*f*
multiomics 456*f*
natural product drug delivery 507*f*
natural product screening 505–506, 513–515
neuroimaging system 46*f*
nonclinical information 10–11
novel antimicrobial discovery 108–109
OI 484–487
oocyte selection 216–217
organoid applications 471–472, 472*f*
organoids cancer 487–488
performance metrics 186*t*
phytoconstituents exploration 515
POC tools 99–100
policy support systems 339
pollution source management 338
portfolio optimization 350
postoperative simulation 201
predicting toxicity/stability 425–429
predictive analytics 368
predictive modeling data 253–254
predictive recovery pathways 185*f*
pregnancy monitoring & complications 218, 220*f*
primary healthcare delivery 370
problems & constraints 4
psychiatric & neurological practice 309–314, 313*t*, 313*f*
public health dashboards 107
public health value modeling 353
quarantine strategies 105
real-time alerts to ROI models 349
real-time monitoring 107–108

resource allocation & logistics 106–107
rise & fall of systems era 536
role in healthcare 305–306
scalability & integration 112
scenario analysis 351–352
segmentation models 126
sensitivity analysis 351–352
simulated empathy 4
smart investment strategies 350–354
sperm selection & male infertility 216
SROI in healthcare innovation 353
supply chain optimization 106
in surgery 167*t*
surgical specialties 181*t*
symptom checkers 64–65
taxonomic classification & metagenomic analysis 388*f*, 388
technical barriers 111–112
teleconsultation hubs 368
therapeutics & drug discovery 108–112
therapy chatbots 11–12
timeline 535–537, 537*f*
toolkit 529
total joint arthroplasty 168
traffic management 527
treatment & help 11
treatment recommendation 312
ultrasound analysis 216
vaccine & antibody design 109–110
wastewater treatment (*see* Wastewater treatment (WWT))
workflow optimization 98
Zika & Ebola 95
Artificial neural networks (ANNs) 23, 26*f*, 118–119, 198, 305–306, 333–334, 459, 554–556
ARTs. *see* Assisted reproductive technologies (ARTs)
AS. *see* Angelman syndrome (AS)
ASA. *see* Adaptive simulated annealing (ASA)
ASD. *see* Autism spectrum disorder (ASD)
ASHA. *see* Accredited Social Health Activist (ASHA)
Assembloids 467*t*, 476*f*, 491
Assisted reproductive technologies (ARTs) 213
Atrial fibrillation (AF) 44–45, 229, 236, 539
AUC. *see* Area under the curve (AUC)
Augmented reality (AR) 179
Autism spectrum disorder (ASD) 311, 313*t*, 474–475
Autoimmune diseases prediction 251–252
AI & ML applications 254–256, 255*t*
biomarker 254–255
challenges 249–250
data limitations & heterogeneity 257–258
early diagnosis & risk prediction 254
MS 252
pathophysiology & mechanisms 251–252, 252*t*
patient stratification & prognosis 255
precision medicine 259
RA 251
SLE 251
T1D 251–252
Autoregressive (AR) process 294
Auxiliary Nurse Midwife (ANM) 365, 367–368
Ayushman Bharat 367, 370

B

Babylon Health app 65
Backdoor/Trojan attacks 562
BadNets case 562
Bayesian inference 176
Bayesian networks (BNs) 94, 96*f*, 337
BBB-on-a-chip 323
BBB. *see* Blood-brain barrier (BBB)
BBL surgery. *see* Brazilian butt lift (BBL) surgery
BCIs. *see* Brain-computer interfaces (BCIs)
BGCs. *see* Biosynthetic gene clusters (BGCs)
Bidirectional T-LSTM (Bi-TLSTM) 77–80, 80*f*
ablation results 86–88, 87*t*
ANFIS integration 80
ANFIS-enabled 79–81
decoder 78
depth of 87
encoder 78
encoder stacking 78*f*, 78–79
training 81
vs. ANFIS-Bi-TLSTM 87
Bile Salt Hydrolase (BSH) 460
Bioactive compounds, HPLC 502
in vitro validation 502
sequential solid–liquid extraction 502
Bioassays 502
Biocomputing 465–466
Bioimpedance measurement 144
Biological oxygen demand (BOD) 331–332
Biomarkers 317–318
Biomedical AI 544
accountability & liability 541
AI failures 541–542
bias & fairness 542
challenges 540–541
clinical expertise loss 541
confidentiality & reliability 540
data monitoring 541
data quality & diversity 540
ethical considerations 542–543, 543*f*
global governance & standards 543
human autonomy 542
hybrid approaches 544
interpretability & black-box problem 540, 542–543
model drift 543
patient–doctor relationship 542
privacy & data ownership 542
regulatory & legal challenges 541
treatment & research innovation 543–544
BioPython 393
Biosensors 318–319
Biosynthetic gene clusters (BGCs) 504
Bispectral Index (BIS) 168, 170*t*
Black-box *vs.* white-box models 560
Blood–brain barrier (BBB) 323
organoids 472
BlueDot company 107
BNs. *see* Bayesian networks (BNs)
BOD. *see* Biological oxygen demand (BOD)
Boston–Montreal porcine model 166
Brain
imaging methods 281
organoid models 474–475, 491*f*, 491
structure analysis 311–312
tumors 313*t*
Brain signals 283–284
classifiction principles 295–297
feature extraction 292–295
spatial filtering 291–292
time-locked signal 292
Brain-computer interfaces (BCIs) 281–282
control modality use 285*t*
control policies 286–290
controllers 285–286
neuromotor rehabilitation 301
schematic overview 283*f*
shared two-loop control architecture 289–290
test phase 297
training phase 297
see also Clinical sequence prediction (CSP)
Brainoware 465–466
Brazilian butt lift (BBL) surgery 203
Breast augmentation mastopexy 198, 201
Breast oncology AI 169
Broader system level 165
BSH. *see* Bile Salt Hydrolase (BSH)
Buoy health 65
Burn injuries 198, 200

C

CA. *see* Classification accuracy (CA)
CaaS. *see* Cybercrime-as-a-Service (CaaS)
CAC. *see* Coronary artery calcification (CAC)
CAD. *see* Coronary artery disease (CAD)
Call centers, AI 12–15
benefits 15
sentiment analysis 14–15
Cambridge Adjutorium system 106
Canonical correlation analysis (CCA) 294
SSVEP detection 294
Capsular contracture 198–199, 199*t*, 203
CAR filter. *see* Common average referencing (CAR) filter
Carbon nanotube field-effect transistors (CNT-FETs) 318–319
Cardiac/cardiology 44*f*, 44–45
CT 232–233
imaging modalities 232–233
phonocardiogram analysis 230–231, 231*f*
Cardiac Allograft Rejection Evaluator (CARE) system 142
Cardiac allograft vasculopathy (CAV) 136, 140–141
detection and prevention 143–144
Cardiac magnetic resonance imaging (CMR) 232–233
Cardiopulmonary exercise test (CPET) 231–232
Cardiothoracic 181*t*
Cardiotocography (CTG) 218

Cardiovascular diseases (CVDs) 225
AI tools/models 228–235
data collection & patient monitoring 226
diagnostic approaches 227
health challenge 225
lifestyle modification 226
mortality counts 226*f*
resource constraints & stark health 228
risk assessment 226–227
treatment strategies 227–228
CARE system. *see* Cardiac Allograft Rejection Evaluator (CARE) system
CatBoost 135–136
CAV. *see* Cardiac allograft vasculopathy (CAV)
CBT. *see* Cognitive behavioral therapy (CBT)
CCA. *see* Canonical correlation analysis (CCA)
CCS. *see* Coronary calcium scoring (CCS)
CCTA. *see* Coronary CT angiography (CCTA)
CD. *see* Crohn's disease (CD)
CDSS. *see* Clinical Decision Support Systems (CDSS)
CELA-MFP 427*t*
Cell-penetrating peptides (CPPs) 422–423
Cerebral organoids 468–469, 475
Cerebrospinal fluid (CSF) biomarkers 317–318
Certified robustness 564, 566–567
CFTR modulating drug 472–473
CGMs. *see* Continuous glucose monitors (CGMs)
Chatbots 58, 267
ChatGPT 311
Chemical oxygen demand (COD) 331–332
Cheminformatics 436
CheXNet 45*f*, 45, 95, 111*f*
CHWs. *see* Community health workers (CHWs)
Circulating tumor DNA (ctDNA) 43*f*, 43–44
CIS. *see* Clinically Isolated Syndrome (CIS)
Classification accuracy (CA) 282
Clinical Decision Support Systems (CDSS) 57–58
Clinical sequence prediction (CSP) 71–72, 77, 89
data preprocessing 82
datasets & descriptions 81*t*, 81–82
early probabilistic & statistical approaches 74
neural sequence models 74
research 74–77, 75*t*
synthesis & gap identification 75
transformer & self-supervised learning 74–75
see also Brain-computer interfaces (BCIs)
Clinically Isolated Syndrome (CIS) 255
Closed-loop configuration 286
Clustered Regularly Interspaced Short Palindromic Repeats (CRISPR) 492
Clustering & nearest-neighbor techniques 559
CMR. *see* Cardiac magnetic resonance imaging (CMR)
CNNs. *see* Convolutional neural networks (CNNs)
CNT-FETs. *see* Carbon nanotube field-effect transistors (CNT-FETs)
Co-bots 165
COD. *see* Chemical oxygen demand (COD)
Cognitive behavioral therapy (CBT) 267, 270–271
digital therapies 271
Cohen's Kappa 26–27
Colorectal cancer (CRC) 181*t*, 473
Common average referencing (CAR) filter 291
see also Brain-computer interfaces (BCIs)
Community health workers (CHWs) 365, 367, 371
Community healthcare & implementation plans, AI 367–369, 369*f*
Computer vision 8–9
challenges 9
detection categories 9
facial expression & feature detection 8
image & video recording 8
research & affective application 9
Concept drift 568
Conditional VAE (CVAE) 424–425, 441–442
Conformal prediction 375–376, 380–382
CONSORT-AI 376
Contact tracing 105
augmented 105
Context of use (COU) 447–448
Contigs 388–389
Continuous glucose monitors (CGMs) 46–47, 47*f*
Contrastive learning 74–75, 122, 424, 427*t*
Controlled ovarian stimulation (COS) 215
Controllers stability analysis 300
Conventional diagnostic approaches 252
AI/ML *vs*. 257–258, 258*t*
Convolutional neural networks (CNNs) 4, 8, 39*t*, 41–42, 72, 75, 95, 98, 102, 117–118, 121–122, 122*t*, 167, 177, 214–218, 229–233, 235, 242, 334, 468–469, 474, 487
architecture 216
pathology based 323–324
Cornea 243
Coronary artery calcification (CAC) 232–233, 235
Coronary artery disease (CAD) 231–233
Coronary calcium scoring (CCS) 232–233
Coronary CT angiography (CCTA) 232–233
COS. *see* Controlled ovarian stimulation (COS)
COU. *see* Context of use (COU)
CPET. *see* Cardiopulmonary exercise test CPET)
CPPFormer 427*t*
CPPs. *see* Cell-penetrating peptides (CPPs)
CPS. *see* Cyber-physical system (CPS)
CRC. *see* Colorectal cancer (CRC)
CRISPR. *see* Clustered Regularly Interspaced Short Palindromic Repeats (CRISPR)
Crohn's disease (CD) 308–309
CRP. *see* Plasma C-reactive protein (CRP)
Cryptographic techniques 544
CSF biomarkers. *see* Cerebrospinal fluid (CSF) biomarkers
CSP. *see* Clinical sequence prediction (CSP)
CT scan 200
ctDNA. *see* Circulating tumor DNA (ctDNA)
CTG. *see* Cardiotocography (CTG)
CVAE. *see* Conditional VAE (CVAE)
CVDs. *see* Cardiovascular diseases (CVDs)
Cyberattacks & cybersecurity 552–553
Cybercrime-as-a-Service (CaaS) 561, 568
Cyber-physical system (CPS) 281–282
Cyclosporine 72
Cystic fibrosis 492
Cytotoxicity models 425–426

D

Dartmouth Conference 535
DAT scan 318
Data foundation, AI 553–556
types 554
unstructured, semistructured, structured 554
Data mining 40*t*, 40
Data poisoning, ML model 556–557
anomaly detection 559
countermeasures 558–559
ML robustness 558–559
sanitization 558
Data preprocessing 420
Data privacy 186
Data-generating process (DGP) 376
missingness & estimation 378–379
observed data 376–378
structural view 376–378
dd-cfDNA patterns 139–140
DDIs. *see* Drug-drug interactions (DDIs)
De novo drug design 516–517
De novo molecule generation 436, 441
DEBi-TLSTM. *see* DNFIS-enabled Bi-TLSTM (DEBi-TLSTM)
Decay parameter 71–73, 75, 77, 87–88
Decision support systems 176
tools 186
Deep learning (DL) 38–39, 39*t*, 198, 200, 202, 214, 241–242, 306, 387–389, 393, 395, 536, 538–539, 542–543
algorithms 117, 126, 176, 242, 405–406, 408, 468–469, 471–473
application 319–322
models 139, 142, 146, 148–149, 181, 253, 255, 552*f*, 552, 556–557, 560–561, 564
neurodegenerative disease 322–323
surgical phase recognition 179*f*
toxicity assessment 428*t*
white matter tau pathology 323
Deep neural networks (DNNs) 102, 231–232, 504–505, 511–516
Deep neuro-fuzzy inference system (DNFIS) 72–73, 77, 80–81
Deep residual architectures 425
DeepADMET 506
DeepPROTECTNeo 425
DeepVariant 408*t*, 408–409, 492
Definitive diagnosis 57–58, 58*t*
Denoising approaches 425
DenseNet121 487
Depression 267–268, 270–272
De-skilling 541
Detect-Prevent-Predict (DPP) model 566
DGP. *see* Data-generating process (DGP)
Diabetic retinopathy (DR) 242, 245*t*
DL algorithm 242

Diethyl ether extract (DR2) 502
Differential diagnosis 57, 59–60, 203–204
Differential privacy (DP) 544
Diffuse gliomas 168
Diffusion models 425
Digital
contact tracing apps 105
pathology 117, 125, 164
PCR 99
Digital Biological Twin Lab platform, PDO 474
Digital forms of therapy 267
ethical considerations 275
psychological frameworks 270–271, 274*t*
Digital Personal Data Protection Act (India) 367
Digital phenotyping 10, 324
Digital twins model 164, 191, 337–338, 338*f*, 444–445
Disease detection 37–38
AI application 41–47
AI-assisted 38–41
collaborative AI 49
core AI technologies 41*t*
global health equity 49
real-time monitor 49
regulatory evolution 49
District surveillance units (DSUs) 368
DL. *see* Deep learning (DL)
DMARD 257
DNFIS. *see* Deep neuro-fuzzy inference system (DNFIS)
DNFIS-enabled Bi-TLSTM (DEBi-TLSTM) 73, 75, 80*f*, 83
ANFIS integration 88
Bonferroni–Dunn posthoc analysis 85*t*, 85
clinical usability 88–89
comparative analysis 83–85, 85*t*
experimental setup 82
implementation details 83
optimal hyperparameters 84*t*
performance metrics 83
sensitivity analysis 85–86
vs. MLP-Bi-TLSTM 87–88
DNNs. *see* Deep neural networks (DNNs)
Docking methodology 502
dynamic stability 505
free energy calculations 505
metal coordination modeling 505
multiple receptor conformations 505
selectivity 505
Domain adaptation 48, 232, 407*t*
Domestic wastewater 331–332
Donor-Derived Cell-Free DNA (dd-cfDNA) 139–140
Donor-recipient compatibility, AI 138
blood type & cross-matching analysis 138
evolution 138–139, 139*f*
histocompatibility enhancement 138
random forest algorithms 146
size and age matching 138–140
Donor-specific antibody (DSA) patterns 138, 140*t*
Doubly robust estimation/AIPW 381, 384
DP. *see* Differential privacy (DP)
DPP model. *see* Detect-Prevent-Predict (DPP) model
DR. *see* Diabetic retinopathy (DR)
Drug discovery/development 306, 511, 513–518, 538
AI research & advancement 518
AI-driven success 515
challenges 512–513, 517–518
in silico pharmacokinetic analysis 515–516
Drug repurposing 108–109
Drug–drug interactions (DDIs) 437, 506, 517
DSA patterns. *see* Donor-specific antibody (DSA) patterns
DSUs. *see* District surveillance units (DSUs)
Dual-targeting delivery strategy 423

E

Early detection 43–44, 46–47
Early warning systems (EWSs) 95, 98, 104–105, 107
Earth Microbiome Project (EMP) 396
ECG. *see* Electrocardiography (ECG)
Echocardiography 232
ECM. *see* Extracellular matrix (ECM)
Economy of motion 181, 187
Edge AI 66
EDSS. *see* Environmental decision support system (EDSS)
EEG. *see* Electroencephalography (EEG)
EF estimation. *see* Ejection fraction (EF) estimation
EHRs. *see* Electronic health records (EHRs)
EI. *see* Emotional intelligence (EI)
Eigenvalue decomposition 292–293
Ejection fraction (EF) estimation 232, 235
Eko Health's stethoscope, AI 235–236
Elastic net regularization models 27–28, 31*t*
ElderTree 273
Electrocardiography (ECG) 227, 229–231
data annotation efficient models 230
lead input models 230
lightweight DL models 230
Electroencephalography (EEG) 281, 283–284, 309–310, 310*t*
Electronic health records (EHRs) 71, 74, 82, 89, 94–96, 176, 226, 250, 253–255, 258
Electronic patient record (EPR) 204
Electronic Vaccine Intelligence Network (eVIN) system 106
ELIZA 59, 536
Emotional AI 4, 7, 13*t*
Emotional intelligence (EI) 3, 5
AI & human interaction 7
chatbots & tools 13*t*
in communication & highlights 5–6
Daniel Goleman framework 5
decision-making & influence 6
history & genesis 5
laboratory & model 5
EMP. *see* Earth Microbiome Project (EMP)
Endocrinology 46–47
Endometrial receptivity 214, 217
Engine decision-making models 559–566
adversarial training 563–564
inference & privacy-preserving 565–566
knowledge distillation & pruning 565
model integrity 560
pipeline model 563–566
real-world failures 562–563
threat vectors targeting 560–562
watermarking and fingerprinting 564–565
Ensemble learning technique 28–29
Environmental decision support system (EDSS) 337
Epidemiological forecasting 104
ABMs 104
compartmental & network models 104
hybrid & ensemble approaches 104
statistical & ML models 104
EPIWATCH 107
EPR. *see* Electronic patient record (EPR)
EPs. *see* Evoked potentials (EPs)
ERD/ERS-based semiclosed-loop control 287–288
ERPs. *see* Event-related potentials (ERPs)
Error-related potential (ErrP) signal 283
eSanjeevani 365, 368
Ethical integration 190
Event-related potentials (ERPs) 283–284
eVIN system. *see* Electronic Vaccine Intelligence Network (eVIN) system
Evoked potentials (EPs) 284
EWSs. *see* Early warning systems (EWSs)
Exome Aggregation Consortium (ExAC) 307
Expert systems 94–95, 334, 336*t*, 536
Explainable AI (XAI) 250, 258–259, 511–512, 515–518
Extensively drug-resistant (XDR) pathogenic strains 405–406, 411
Extracellular matrix (ECM) 485–486
eXtreme Gradient Boosting (XGBoost) 411–412
Eye tracking 311
EyeArt AI system 244

F

FaceX 203
Facial bone implants 200
Facial recognition technology 527
FAIR principles 120
False-positive rate (FPR) 298
Fast Healthcare Interoperability Resource (FHIR) 365
Feature extraction (FE) 282
ErrP signal 294–295
Federally controlled multicenter modeling 448
Federated learning (FL) 182*f*, 182, 215, 218, 228–229, 543–544
Feedforward neural networks (FNNs) 23–24
FFR estimation. *see* Fractional flow reserve (FFR) estimation
FHIR. *see* Fast Healthcare Interoperability Resource (FHIR)
FL. *see* Federated learning (FL); Fuzzy logic (FL)
FMRI. *see* Fuses functional MRI (fMRI)

FNNs. *see* Feedforward neural networks (FNNs); Fuzzy neural networks (FNNs)
Follicle monitoring 216
F1-score, AI evaluation metric 26–27, 98, 101, 181*t*, 322–323, 427–428
FPR. *see* False-positive rate (FPR)
Fractional flow reserve (FFR) estimation 232–233
Framingham Risk Score 226–227
Free flap surgery 200
Frequency-domain signals, FE 293–294
Functional recovery monitoring 183–184
Fundus photography 117, 245, 318
Fuses functional MRI (fMRI) 166
Fusion architectures 423
Fuzzy logic (FL) 335, 455
Fuzzy membership function 72–73, 77
Fuzzy neural networks (FNNs) 335

G

GANs. *see* Generative adversarial networks (GANs)
GAs. *see* Genetic algorithms (GAs)
GATs. *see* Graph attention networks (GATs)
GCNs. *see* Graph convolutional networks (GCNs)
GDPR. *see* General Data Protection Regulation (GDPR)
Gene Ontology (GO) 420
General Data Protection Regulation (GDPR) 254, 258, 275, 529–530
Generative adversarial networks (GANs) 201, 231–232
 frameworks 323
 generative & synthetic imaging 323
 MRI synthesis 323–324
Generative AI 123, 365–366, 370–371, 424–425, 435–442
 adversarial training paradigms 425
 and agentic intelligence 365–367
 benefits of 366–367
 biological generative modeling 441
 biological systems integration 448
 biologics & macromolecular therapeutics 442
 chatbots 65–66
 clinical phenotypes 444
 deep residual networks 425
 denoising approaches 425
 diffusion models 440
 drug design 441–443
 drug optimization 442
 drug safety prediction 445–447
 epigenomic features 443–444
 GANs 424
 information-theoretic approaches 424
 metabolomic data 444
 patient data integration 443–445
 target-based generation process 441–442
 training data 440
 transcriptomic data 443–444
 transformer architectures 425
 VAEs 424–425
Generative tools 447
Genetic algorithms (GAs) 335–336
Genetic screening 307–309
 Angelman syndrome 307–308
 Crohn's disease 308–309
 pediatric encephalopathies 308
Genome aggregation database (gnomAD) 307
Genome-wide association studies (GWASs) 24–25
 prediction limitation 25
Genomics 405–406
 AI applications & frameworks 408*t*, 408–409
 DL frameworks 408–409
 issues 409–410
 and proteomic data 256
 variants 443
GISAID/GenBank 95–96
Glaucoma 243
GNNs. *see* Graph neural networks (GNNs)
GnomAD. *see* Genome aggregation database (gnomAD)
GO. *see* Gene Ontology (GO)
Governance frameworks, AI 530
GPT-3.5 case & model theft 562
Grad-CAM technique 124
Graph attention networks (GATs) 423–424
Graph convolutional networks (GCNs) 234
Graph neural networks (GNNs) 409, 503
Graph-based molecular models 423–424
Green AI 127
Guided protocols 491
Gut microbial enzymes 460
Gynecology 181*t*

H

HBM. *see* Health belief model (HBM)
Health and Wellness Centers (HWCs) 367–368
Health belief model (HBM) 273–274
 self-efficacy 273
Health equity modeling 49, 236, 347–348, 348*t*, 365, 371
Health Insurance Portability and Accountability Act (HIPAA) 254, 258, 275
HealthMap platform 93–96, 96*f*, 111*f*, 111–112
Health systems
 censoring-aware quantum kernel 383–384
 heterogeneous clinical datasets 380
 measurement imperfections 380–381
 multiomics risk stratification 382
 quantum pipelines 385
 radiology triage 382–383
Healthcare 10, 19, 28, 30–31
 AI chatbots evolution 59–60
 AI impact 531
 challenges, AI medicine 531–532
 design & application, AI 530
 early rule-based chatbots 59
 ethics 528*f*
 human oversight & AI 531
 intelligent systems transition 59
 investment 347–348, 354–356
 outcomes & treatment response 25–30
 predictive modeling foundations 19–20
Healthcare, AI role in 305–306, 537–539
 biomedical devices 539
 clinical trials & public health 539
 drug discovery/development 306–307, 538
 genetic screening (*see* Genetic screening)
 genomics & proteomics 538–539
 medical imaging & diagnosis 538
 NLP in 539
 personalized medicine 538
 robotics in surgery 538
 wearables & biosensors 539
 XAI & model transparency 539
Heart Disease Dataset (HDDC) 147
Heart transplantation, AI 135
 algorithmic bias & ensuring equity 148
 challenges & solutions 137*f*
 ensemble methods & hybrid approaches 147–149
 predictive models 146
 revolutionary role 137–138
Helicobacter pylori 390–391
Herbal medicine 512–513
Heterogeneity 487
High-performance liquid chromatography (HPLC) 501–502
 bioactive compounds 502
 purification & gas chromatography 502
High-throughput screening 144
HIPAA. *see* Health Insurance Portability and Accountability Act (HIPAA)
hiPSCs. *see* Human-induced pluripotent stem cells (hiPSCs)
HLA compatibility. *see* Human leukocyte antigen (HLA) compatibility
HMOs. *see* Human midbrain organoids (HMOs)
HMP. *see* Human Microbiome Project (HMP)
Home-based diagnostics 58, 63*t*
 AI chatbots benefits 60–61, 64–66
 cost-effectiveness 60–61
 diagnostic excellence & AI chatbot 59–60
 early detection & monitoring 61
 edge AI 66
 increased accessibility 60
 limited healthcare infrastructure 61
 ML models update capabilities 61
Homomorphic encryption (HE) 565–567
Horizon-based capital allocation 354
HPLC. *see* High-performance liquid chromatography (HPLC)
Human leukocyte antigen (HLA) compatibility 138
Human microbiome 387
Human Microbiome Project (HMP) 396
Human midbrain organoids (HMOs) 474, 486–487
Human–computer interaction 172
Human-induced pluripotent stem cells (hiPSCs) 469, 473–475
Huntington's disease (HD) models 472
HWCs. *see* Health and Wellness Centers (HWCs)
Hybrid SEIR+ML model 104, 111*f*
HydrAMP 427*t*

I

IBD. *see* Inflammatory Bowel Disease (IBD)
IBR. *see* Implant-based reconstruction (IBR)

ICA. *see* Independent component analysis (ICA)
ICBS. *see* Intelligent Cyber-Biological Systems (ICBS)
ICBT. *see* Internet-based behavioral therapy (ICBT)
IDSs. *see* Intrusion detection systems (IDSs)
Image registration, intraoperative 179
Imaging informatics 177
Implant-based reconstruction (IBR) 198
In silico medicine 538
In silico pharmacokinetic analysis 515–516
In vitro fertilization (IVF) 217–218
Independent component analysis (ICA) 291
Induced pluripotent stem cells (iPSCs) 489–490
Industrial wastewater 331–332
Infectious disease 45*f*, 45
  AI history 94–112
  ML models implementation 100–104
  predictive modeling 104–105
Inflammatory Bowel Disease (IBD) 308, 389
Information theory-based methods 424
Intelligent Cyber-Biological Systems (ICBS) 471
Intelligent surveillance systems 183
Internal rate of return (IRR) 350–351
Internet of Things (IoT) 107, 337–338
Internet-based behavioral therapy (ICBT) 268, 271
Intraocular lens (IOL) selection 244–245, 245*t*
Intraoperative AI 180
Intrusion detection systems (IDSs) 552–553, 567
IoT. *see* Internet of Things (IoT)
IPSCs. *see* Induced pluripotent stem cells (iPSCs)
IRR. *see* Internal rate of return (IRR)
IVF. *see* In vitro fertilization (IVF)

J

Japan, WWT 340
  Fukuoka 340
  Yokohama 340
Just-in-Time Adaptive Interventions (JITAI) 274

K

Keras 408
K-mers, AI & genomics 410–411
  advantages & limitations 413*t*
  applications 410–411
  DNA sequences 410*f*, 410
K-nearest neighbor (KNN) algorithm 20–21, 21*f*, 23*f*, 27, 31*t*

L

Lab-on-a-chip (LOC) devices 100
Lactococcus lactis 390–391
LAMP device. *see* Loop-mediated isothermal amplification (LAMP) device
Laplacian filtering 291
Large Language Model Meta AI (LLaMA) 58, 65
Large language models (LLMs) 201, 536
Lasso peptides 418
LASSO. *see* Least absolute shrinkage and selection operator (LASSO)
LatentGAN 518
LDA. *see* Linear discriminant analysis (LDA)
Least absolute shrinkage and selection operator (LASSO) 457–458
  and Elastic Net 27–28, 31*t*, 234, 458–459
LeNet-5 architecture 118
Lewy bodies 317
Lightweight DL-based pipelines 469
Linear discriminant analysis (LDA) 284–285
Linear support vector machine (L-SVM) 284–285
Lipidomics 233
Lipinski's Rule of Five 516
Liquid biopsy method 43*f*, 43–44
LLaMA. *see* Large Language Model Meta AI (LLaMA)
LLM ChatGPT 202
LLMs. *see* Large language models (LLMs)
LMICs. *see* Low and middle-income countries (LMICs)
LOC devices. *see* Lab-on-a-chip (LOC) devices
Logistic regression (LR) models 19–20, 27, 30, 97–98
  implementation 101
Long short-term memory (LSTM) 71, 74, 97, 102, 104, 183, 230–231
Loop-mediated isothermal amplification (LAMP) device 99–100
LoRA technique. *see* Low-rank adaptation (LoRA) technique
Low and middle-income countries (LMICs) 225–228
Low-rank adaptation (LoRA) technique 123
LR models. *see* Logistic regression (LR) models
LSTM. *see* Long short-term memory (LSTM)
L-SVM. *see* Linear support vector machine (L-SVM)

M

Machine learning (ML) 4, 8, 10–11, 19–20, 22, 24, 27–30, 200–201, 305–307, 311–312
  advanced 187
  in AI 457–458
  core concepts 252–253
  modern healthcare role 250–251
  omics tools in 458–459
  peptide target prediction 420
  precision nutrition 459–460
  in thoracic imaging 24
  types 253
  use 504
Machine learning (ML) algorithm/model 93, 95, 97, 106, 135–138, 142, 147, 163, 176–177, 317–318, 323
  adversarial attack 557
  application 319–322
  clinical setting 103
  conceptual hierarchy 552*f*
  data collection & curation 102
  data poisoning 556–557
  feature engineering 102
  infectious disease detection 100–104
  ischemia time optimization 140–143
  medication adherence 143–144
  monitoring systems 106
  neurodegenerative disease 322–323
  robustness 558–559
  traditional approaches 147
Macrocyclic peptide 421–422
MADDi. *see* Multimodal Alzheimer's disease diagnosis (MADDi)
Magnetic resonance imaging (MRI) 233, 309–310, 310*t*, 491
Magnetoencephalography (MEG) 309–310, 310*t*
MAGs. *see* Metagenome-assembled genomes (MAGs)
Malnutrition 456–457
Manifold learning 471
Markov decision process (MDP) 22
Marginal Structural Model (MSM) 379–381
Mass spectrometry-based proteomics 256
Matched molecular pair (MMP) analysis 515–516
Mayo Clinic's heart failure detection 235
MBRs. *see* Membrane bioreactors (MBRs)
McNemar's tests 428
MCP. *see* Model context protocol (MCP)
MDP. *see* Markov decision process (MDP)
Med Gemini 64*t*, 64–65
Medical imaging 120–121, 538
  advances 121–123
  applications 128
  bias 120
  class imbalance 120
  data quality 119–120
  explainability & interpretability 124–125
  foundation models 118–120, 127
  inference & training 118–119
  security & ethics 126
  techniques 120–121
Medical institution *vs.* home-based diagnostic 58
Medicinal plants 512–513, 515
  traditional & modern medicine 512
MEG. *see* Magnetoencephalography (MEG)
Melanoma/skin cancer 306, 313*t*
MELD algorithm 146
Melicoccus bijugatus 503, 508
Membrane bioreactors (MBRs) 334
Mental health support, AI 268–270
Metabolism 506
Metagenome-assembled genomes (MAGs) 389
Metagenomics 388–389, 460
Metatranscriptomic 460
MI signal. *see* Motor imagery (MI) signal
MIC. *see* Minimum inhibitory concentration (MIC)
Microbial networks & AI 406–408
  applications 406*f*, 407*t*
  disease diagnostics 408

Microbiome data analysis 387
AI 388–389, 392*f*
challenges 387
drug discovery 390–391
ecological insights 391
engineering & synthetic biology 395–396
functional predictions 389–390
interpretability & reproducibility issues 395
mining & extraction 393–394
personalized medicine 390
Micronutrient-Genomics Project 459–460
Micropattern-based neural organoids 472
MicroRNAs (miRNAs) 317–318
Microsoft AI 204
Microsoft Tay 557–558
MIMIC datasets 82
Minimum inhibitory concentration (MIC) 411–412
hyperparameter optimization 412
ML framework 411–412
MI-P300 closed-loop position control 288–289
Missingness mechanisms (MCAR/MAR/MNAR) 378–379
ML. *see* Machine learning (ML)
MMCD. *see* Multimodal contrastive diffusion model (MMCD)
MMP analysis. *see* Matched molecular pair (MMP) analysis
Model context protocol (MCP) 369–370
Model inversion attacks, AI 561
Model poisoning attackers 173
Model predictive control (MPC) 290
see also Brain-computer interfaces (BCIs)
Model stealing & extraction attacks, ML 561–562
Model-centric *vs.* data-centric AI 553–554, 554*f*
Molecular docking 513
Molecular imaging, AD & PD 318
Monte Carlo simulation 352
Moonlighting peptides 419
Morbidity & mortality (M&M) 166
Morphokinetics 214, 217–218
Motion scaling 163–165, 167
Motor imagery (MI) signal 283–284, 287
MPC. *see* Model predictive control (MPC)
MpMRI. *see* Multiparametric MRI (mpMRI)
MRI. *see* Magnetic resonance imaging (MRI)
mRNA display 421–422
MSA mechanism. *see* Multihead self-attention (MSA) mechanism
MSM. *see* Marginal Structural Model (MSM)
Multidrug-resistant (MDR) 405–406, 411
Multihead self-attention (MSA) mechanism 121
Multimodal Alzheimer's disease diagnosis (MADDi) 322–324
and graph-based approaches 323
Multimodal contrastive diffusion model (MMCD) 427*t*
Multimodal representation fusion 424
Multimodal systems 120
Multiomics data integration 256–257, 257*f*, 259, 388–389
imaging-based prediction 256
Multiomics-centered approach 406, 408
Multiparametric MRI (mpMRI) 167
Multiple sclerosis (MS) 249–250, 252, 255–256, 258
Musculoskeletal (MSK) 489*t*, 489
MXNet 408
MYCIN system 94, 96*f*
Mycobacterium tuberculosis 411–412
drug resistance prediction 411–412
MySurgeryRisk program 198

N

NAATs. *see* Nucleic acid amplification tests (NAATs)
NAFLD. *see* Non-Alcoholic Fatty Liver Disease (NAFLD)
Naive Bayes 22
National Digital Health Mission (NDHM) 365, 368–370
National Health Mission (NHM) programs 367–368
Natural language processing (NLP) 4, 7–8
in healthcare 393, 539, 554, 557
Natural language understanding (NLU) 14, 177
Natural product-derived compounds 506
Naïve Bayes decision trees 457–458
NDHM. *see* National Digital Health Mission (NDHM)
Nephrotoxicity screening 471–472
Net present value (NPV) 350–351
Network pharmacology 513–514
Neural networks 30–31, 31*t*
applications 146–147
and BioBERT, DL approaches 393
and DL 22–23
recurrent 102
Neural sequence model 74
Neuro Scholar 311
Neurodegenerative disease/disorders 317, 474–475, 477–478
AI architectures 321*t*
Neurofilament light chain (NfL) 317–318
Neurology 46
Neuromotor disorders 281–282
Neuromuscular junctions (NMJs) 475
organoids 475
Neurons 281
Neuro-prosthetic limb 282
Neuropsychiatric disorders 311
Next-generation organoids (NGOs) 484
Next-generation sequencing 406
Nextstrain 95
NfL. *see* Neurofilament light chain (NfL)
NGOs. *see* Next-generation organoids (NGOs)
NHM programs. *see* National Health Mission (NHM) programs
Nikshay 366, 368
NIR fluorescence imaging 164
NISQ. *see* Noisy intermediate-scale quantum (NISQ)
NLM Malaria Screener app 99
NLP. *see* Natural language processing (NLP)
NLU. *see* Natural language understanding (NLU)
Noisy intermediate-scale quantum (NISQ) 375
Non-Alcoholic Fatty Liver Disease (NAFLD) 458–459
Nonrigid pass 166
Novel pharmacophore 503
NPV. *see* Net present value (NPV)
Nucleic acid amplification tests (NAATs) 99
Nutrigenomics 455–457
Nutriproteomics 457
Nutrition 456–457

O

OCT. *see* Optical coherence tomography (OCT)
Objective structured assessment of technical skills (OSATS) 165, 169
Oculomics 245
Offline-first design 367
OI. *see* Organoid intelligence (OI)
Omics, CVD 233–234
risk prediction models 234–235
Oncology 23–24, 43–44
robotic surgery 176
One Health approach 95
Ontology mapping 420
Operational logistics 177
Ophthalmology 117, 125
advancement & development 244–246
advantages 242, 245*t*
applications 245*t*
clinical applications/implementation 242
DL 241
fundamentals 241–242
limitations 242, 244–246, 245*t*
monitoring & prognosis 243–244
platforms & technologies 244
surgery 244
OPO method. *see* Organoid printing & optical coherence (OPO) method
Optical coherence tomography (OCT) 166, 469
OR orchestration 180
Organ viability prediction 139
biochemical modeling 139–140
real-time imaging intelligence 139
remote monitoring 145*t*
Organoid intelligence (OI) 465, 474, 477
AI-enabled 484–487
for cancer 487–488
genetic landscape 492–493
healthcare application 485*f*
hurdles & innovative solutions 467*t*
MSK 489–490
neural networks 490–492
vascularization issues & batch variability 476
Organoid printing & optical coherence (OPO) method 469
Organoid technology 465–469, 471–473
AI methods 466–475
applications of 467*f*
challenges 475–477, 476*f*
DL networks 470*t*
omics 473–474
OrgaQuant 469
OrgSegNet 469
Orthopedic surgery 200

OSATS. *see* Objective structured assessment of technical skills (OSATS)
Outcome simulation, surgical 201
Oxford Nanopore sequencing 388–389

P

P300 signal 282–283, 292
PACSs. *see* Picture archiving & communication systems (PACSs)
Parkinson's disease (PD) 317
  fluid biomarkers 322*t*
  models 474–475
Parkinson's Progression Marker's Initiative (PPMI) 73, 81–82
  dataset 82
Particle diffusometry technique 99–100
Pathology 42–43
  WSI analysis 43*f*
Patient-derived organoids (PDOs) 469, 471–475, 477, 483–484, 490
Patient-reported outcome measures (PROMs) 201
Patient-specific data 441, 443
Pattern classification 284–285, 285*f*
Pattern recognition 40*t*, 40
PD. *see* Parkinson's disease (PD)
PDB. *see* Protein Data Bank (PDB)
PDOs. *see* Patient-derived organoids (PDOs)
Pediatric encephalopathies 308
Pediatric epilepsy 309–310, 310*t*
Penalized algorithms 458–459
PepGLAD 427*t*
PepHarmony 427*t*
Peptide-based drug, AI 426*f*
  designing & biodelivery 429
  DL-based generative models 427*t*
  feature engineering 423–424
  preprocessing 420
  synthesis & moonlighting effects 418–419
  target integration & RNA-based approaches 419–420
  therapeutic targets 421*t*
Peptide-based systems 423
  microRNA modulation 423
PeptideBERT 425, 427*t*
PeptideGPT 427*t*
Peptides 417–418
  databases 419
  engineering 418
  macrocyclic 421–422
  multifunctional 419
  RNA therapeutic delivery CPPs 422–423
  RNA-binding 420
  siRNA delivery/targeting 422
Peptipedia 419
PepVAE 427*t*
Perfusion monitoring, intraoperative 202
Personalized medicine 538
PET. *see* Positron Emission Tomography (PET)
Phage display 419
Pharmacokinetic profiling 507
Pharmacogenomics 143, 150, 183, 227–228, 473, 487–488, 538
Pharmacovigilance 41, 45, 447–448, 511
PHCs. *see* Primary Health Centers (PHCs)
Phonocardiogram (PCG) analysis 230–231, 231*f*
Phosphorylated tau (p-tau) 317–319, 475
Phytochemicals 512–514
Picture archiving and communication systems (PACSs) 41–42, 42*f*
PkCSM 506
Plasma C-reactive protein (CRP) 459–460
Plastic surgery 197
  3D printing in 200
  AI in (*see* AI in plastic surgery)
  robotics in 202
pLMs. *see* Protein language models (pLMs)
PM. *see* Precision medicine (PM)
POC tools, AI 99–100
  diagnostics 99
  smartphone-based platforms 99–100
Polygenic risk scores (PRSs) 226–227, 234, 256
Popular mental health apps 271*t*
Positron Emission Tomography (PET) 310*t*
Postoperative
  monitoring 203
  simulation 201
Postprandial glucose prediction 460
Posttranslational protein modifications (PTMs) 306–307
Posttransplant monitoring, AI 141*f*, 141–142
  acute rejection detection 142
  biomarkers & AI detection 140*t*
  infection risk prediction 142–143
POTTER surgical calculator 198–199, 199*t*
PPMI. *see* Parkinson's Progression Marker's Initiative (PPMI)
PPI inhibition. *see* Protein-protein interaction (PPI) inhibition
Preceyes surgical system 244
Precision medicine (PM) 27, 31, 48–49, 483–484, 487–488
Precision surgery AI 163
  challenges 170–173
  ethical AI & patient trust 166
  intraoperative decision support 166
  invasive surgery & remote operations 166
  opportunities 164–167
  patient safety/perioperative outcomes 165
  personalized medicine & precision therapeutics 165–166
  planning & diagnostics 164
  quality improvement 166
  real-time intraoperative & robotic assistance 164–165
  robotics & automation 165
  surgical education & training 165
Precision therapeutics 437, 442–443
Predetermined change-control plan 170–171
Predictive models 176*t*, 180–181
Predictive toxicology 437
Preeclampsia 218
Prehabilitation 176*f*
Preoperative applications 197–202
Primary diagnosis 57–58, 58*t*
  AI chatbots 60
  decision support algorithms 60
  multimodal AI chatbots 66
  symptom assessment 60
  telemedicine integration 60
Primary Health Centers (PHCs) 368, 370
  telehealth 370
Probabilistic models 94
Probiotics 391
Prodromal stage detection 317–318, 322*t*
PROMs. *see* Patient-reported outcome measures (PROMs)
Protein Data Bank (PDB) 419
Protein language models (pLMs) 423, 427*t*, 427–428
Protein-protein interaction (PPI) inhibition 420, 440–442
Proteins target identification 420–423
  peptide interactions 420–421
  therapeutic outcomes 423
PRSs. *see* Polygenic risk scores (PRSs)
Pseudohaptic feedback 179–180, 180*f*
Psychiatry 45
PTMs. *see* Posttranslational protein modifications (PTMs)
PubMed's E-utilities 393
PyTorch 408

Q

QAI. *see* Quantum artificial intelligence (QAI)
QAOA. *see* Quantum approximate optimization algorithm (QAOA)
QDA. *see* Quadratic discriminant analysis QDA)
QSAR. *see* Quantitative structure–activity relationship (QSAR)
Quadratic discriminant analysis (QDA) 284–285
Quadratic unconstrained binary optimization (QUBO) 375–376, 380–381, 384
Quantitative structure–activity relationship (QSAR) 428, 503
Quantum approximate optimization algorithm (QAOA) 375, 380–381
Quantum artificial intelligence (QAI) 375, 378, 385
  health data governance 385
Quantum computing 67
Quantum optimization methods 381
QUBO. *see* Quadratic unconstrained binary optimization (QUBO)

R

Radical prostatectomy 167
Radiogenomics 177
Radiology 41–42, 42*f*, 117, 125
RAG. *see* Retrieval-augmented generation (RAG)
Random forest (RF) algorithm 24, 94–95, 97, 146, 457–458, 460
  advantages 24–25
  applications 146
  and clustering techniques 391–393
  implementation 101
  limitations 25
  statistical limitations 24–25
Random search & grid search, MIC 412

Random Survival Forest (RSF) 135–136, 383–384
Ransomware 173
Rapid diagnostic tests (RDTs) 99
R-CNNs. *see* Region-based convolutional neural networks (R-CNNs)
RDTs. *see* Rapid diagnostic tests (RDTs)
Recurrent neural networks (RNNs) 9, 165, 177, 214, 230–231, 236, 253, 459, 554–555, 557
Region-based convolutional neural networks (R-CNNs) 123, 125
Regression algorithms 457–458
Regulatory genomics 408–409
Reinforcement learning (RL) 40*t*, 40, 163, 176*t*, 190–191, 215, 301, 514–515, 518, 554–556
  models 20–22, 183
Remote monitoring & patient management 144
  physiological monitoring 145
  wearable ECG monitoring 144
Renin–angiotensin system 501–502
Reproductive medicine, AI 213–218
  advancements 220
  clinical assessments 214
  complex & multivariability 214
  data source utilization 214
  DL 214
  efficiency & access 214
  FL 215, 218
  ML 214
  NLP 214
  personalization 214
  prediction & optimization 214
  real-time monitoring 214
Reproductive medicine limitations 219–220
  clinical expertise 219
  data security 219
  environmental impact 220
  ethical dilemmas 219
  integration barriers 220
  model bias 219
  transparency 219
Resource-intensive approach 417
Respiratory infection dataset 100–101
Responsible AI system 127–128
Retrieval-augmented generation (RAG) 369–370
Return-on-investment (ROI) 347, 350
  financial metrics 351*t*
  healthcare investment decision 347–348
  societal impact 352–353
  traditional framework limitations 348
RF algorithm. *see* Random forest (RF) algorithm
RFdiffusion model 110
Rheumatoid arthritis (RA) 249–251, 253, 255–257, 259
Rhinonet algorithm 202
Ribosome biogenesis 487
Rigid pass 166
Risk scoring algorithms 105
RL. *see* Reinforcement learning (RL)
RNA
  approaches 419–420
  macrocyclic peptides 421–422
  peptide databases 419
  potential therapeutic targets 421
  therapeutic delivery via CPPs 422–423
  therapeutic targeting mechanisms 422*f*
RNNs. *see* Recurrent neural networks (RNNs)
Robotic hysterectomy 182
Robotic surgery 167, 175–176, 176*t*, 178*t*, 181*t*, 202, 538
  AI integration 167
  clinical setting 167
  closed-loop adaptive 191*f*
  comparison 180*t*
  economic impact/system efficiency 188–189
  ethical risks & mitigation strategies 190*t*
  ethical/legal/equity considerations 189–190
  federated learning *vs.* centralized 191*t*
  governance model 189*f*
  intraoperative phase 178–182, 179*f*
  lifecycle 192*f*
  outcomes 167
  postoperative phase 182–186, 183*f*, 184*t*
  preoperative phase 175–178, 176*f*
  training & education 187–188
ROI. *see* Return-on-investment (ROI)
Root locus plot 300*f*, 300–301
RSF. *see* Random Survival Forest (RSF)
Rule-based chatbot 58–59, 311
Rural healthcare
  24/7 access 366
  community health data 366
  equitable access 371
  Gen AI & agentic intelligence 365–367
  government services 366
  health education 366
  health guidance 366
  monitoring & reminders 366
  strategy 367

S

SA technique. *see* Simulated annealing (SA) technique
Sanitary sewage 331–332
SAR. *see* Structure–activity relationship (SAR)
scRNA-seq. *see* Single-cell RNA sequencing (scRNA-seq)
SDT. *see* Self-determination theory (SDT)
Secure multiparty computation (SMPC) 544
Segmenting Objects by Locations (SOLO) model 123
Self-determination theory (SDT) 272–273
Self-supervised learning (SSL) methods 121–122
Semisupervised clustering generative adversarial network (Smile-GAN) 322–324
Sensitivity analysis 351–352
Sensor fusion 176*t*
Sentiment-aware chatbots 12–15
  American chase 15
  business impact 14–15
  in call centers 14
  challenges 14
  customer experience 13–14
  Jordanian telecommunication companies 15
  trouble shooting & escalation 14
  US mobile provide 15
Sepsis 98
Sequence-based representation learning 423
Sequential solid–liquid extraction 502
Shannon entropy 29
  applications 29
  principles 29
Shapley Additive Explanations (SHAP) 44–48, 124, 235, 428
Shared-control architecture 282–283, 299
Simulated annealing (SA) technique 336
Simulated empathy 4
Single-cell RNA sequencing (scRNA-seq) 96, 409, 438, 443–444, 473, 492
Single-nucleotide polymorphisms (SNPs) 25
SIR/SEIR model 104, 111*f*
Skip connections, architectural technique 71, 77–79, 86–88, 87*t*
SL. *see* Supervised learning (SL)
SLE. *see* Systemic lupus erythematosus (SLE)
Smart medication adherence 368
Smart scheduling systems 188
Smart sensors 337–338
SmartHealth-Track system 107
Smile-GAN. *see* Semisupervised clustering generative adversarial network (Smile-GAN)
SMPC. *see* Secure multiparty computation (SMPC)
SNPs. *see* Single-nucleotide polymorphisms (SNPs)
Social return on investment (SROI) 353
  AI tools for assessment 359
Soft tissue surgery 165
SOLO model. *see* Segmenting Objects by Locations (SOLO) model
Spatiotemporal analysis 104
SPECT imaging technique 309–310, 310*t*
Speech analysis 311
Spent culture media analysis 217
SROI. *see* Social return on investment (SROI)
SSI. *see* Surgical site infection (SSI)
SSL methods. *see* Self-supervised learning (SSL) methods
SSVEP. *see* Steady-state visually evoked potential (SSVEP)
State-space analysis 300
Steady-state visually evoked potential (SSVEP) 282, 289–290, 293–294
Storm sewage 331–332
Structural annotation 420
Structure–activity relationship (SAR) 507, 513–515, 518
Supercomputing 455
Supervised learning (SL) 253, 457–458, 458*f*, 554–556, 559
Support vector machines (SVMs) 4, 9–10, 15, 20–21, 102, 253–254, 256
Surgical risk calculator 198
Surgical site infection (SSI) 164
Sustainable development goal 331
SVMs. *see* Support vector machines (SVMs)
Swin Transformer 121–122

SwissADME 506
Symptom checkers
Ada Health 65
AI-powered 64–65
Babylon Health 65
Buoy Health 65
traditional 64*t*, 64
Syndromic surveillance 98, 104–105
Synthetic control arm 350
Synthetic Microbial Consortia 395–396
Systemic lupus erythematosus (SLE) 251, 253–258

T

T follicular helper (Tfh) 417
T1D. *see* Type 1 diabetes mellitus (T1D)
Tactile sensing 179
Tauopathies/progressive supranuclear palsy/corticobasal degeneration 319, 321*t*, 323
Taxonomic classification & metagenomic analysis 388*f*, 388
TCM. *see* Traditional Chinese medicine (TCM)
Telehealth Ten 62
Telemedicine 57–58, 125, 145
AI chatbot integration 60
Ten-fold cross-validation 428
TensorFlow 408
Text-to-image models 201
Theory of planned behavior (TPB) 273–274
Therabot 11–12
Therapy chatbots 11–13, 13*t*
challenges & ethical issues 12
stimulus & quitting 12
Thermalytix 526
Three-dimensional (3D)
cultures 465–466, 468–469, 471–472, 476
geometry & structural representations 423–424
microelectrode array (MEA) 465
printing 200–201
simulations 267
virtual models 177
Time-aware LSTM (T-LSTM) 71–72, 74–75
recapitulations 77
Time-lapse imaging, embryoscope 213–215, 217
TME. *see* Tumor microenvironment (TME)
Toxicity 506
Toxicity & stability prediction AI 425–426
benchmarking & validation strategy 426–428
DL 428
generative models 425
TPB. *see* Theory of planned behavior (TPB)
TPR. *see* True-positive rate (TPR)
Traditional Chinese medicine (TCM) 512
Traditional transvaginal ultrasound (TVUS) 216
Transcription factor binding prediction 408, 442
Transfer learning technique 242
Transformer-based language models 439–440
TransUNet architecture 121–122
Tremor attenuation 164–165
Triage 57–58, 58*t*
True-positive rate (TPR) 298
Tuberculosis (TB) 365–366, 368, 405–406, 411–413
and drug resistance 411
Tumor microenvironment (TME) 465, 488
Tumor-derived organoids 466
Turing test 535
TVUS. *see* Traditional transvaginal ultrasound (TVUS)
Two-dimensional (2D)
cultures 465–466, 469, 483
data points 284–285
Two-loop control architecture 299
Type 1 diabetes mellitus (T1D) 250–258

U

UK Biobank 226, 232–234
Ultrasound scans 200
Uniform manifold approximation and projection (UMAP) 471–472
Unguided protocols 491
Universal antivenom 347, 360
Unsupervised learning 253, 255, 457–458, 458*f*, 554–556, 559
Urology 181*t*

V

VAEs. *see* Variational autoencoders (VAEs)
Value-based care 188, 349
Vapnik–Chervonenkis (VC) dimension 21
Variant calling 406, 408*t*, 408–409, 411, 492
Variational autoencoders (VAEs) 420–421
systematic exploration 424–425
Variational Quantum Circuit (VQC) 375–378, 380–381, 383
VC dimension. *see* Vapnik-Chervonenkis (VC) dimension
Virtual health assistants 146–147
Virtual reality (VR) 204, 267
3D simulations 267
Virtual screening 417–418, 441–442, 504–505, 513–515, 538
Vision transformers (ViTs) 8
architecture 121–123, 122*t*
Visual field testing 242–243, 245*t*
Voice analysis algorithms 184
VQC. *see* Variational Quantum Circuit (VQC)
VR therapy/VR exposure therapy 267
VR. *see* Virtual reality (VR)

W

Wastewater management
AI challenges 342
AI models 336*t*, 336–337, 339*t*
environmental & sustainability challenges 332–333
environmental crises & carbon footprint 333
Fukuoka 340
GAs 335–336
human dependency & fault management 332–333
Japan 340
operational flexibility & predictive control 332
regulatory & monitoring limitations 332
resource allocation & accessibility 333
Yokohama 340
Wastewater treatment (WWT) 331–332
advancement 337–340
application 339*f*
disinfection & renewable integration 338
FL models 335
high energy & chemical demands 332
limitations & challenges 340–341, 341*f*
neural network models 333–335
optimization & evolutionary algorithms 335–336
risk management 337
scalability challenges 333
support & planning tools 337
Wavelet neural networks (WNNs) 334–335
WGS. *see* Whole-genome sequencing (WGS)
White matter tau pathology, DL 323
Whole-slide image (WSI) analysis 42–43, 43*f*, 122*t*, 125–126, 146, 322*t*, 322–323
Whole-genome sequencing (WGS) 411
WNNs. *see* Wavelet neural networks (WNNs)
Wound monitoring 183
WSI analysis. *see* Whole-slide image (WSI) analysis
WWT. *see* Wastewater treatment (WWT)

X

XAI. *see* Explainable AI (XAI)
XDR pathogenic strains. *see* Extensively drug-resistant (XDR) pathogenic strains
Xenotransplantation, AI 150
XGBoost. *see* eXtreme Gradient Boosting (XGBoost)

Y

Yam-derived ACEis
docking methodology 502
implications 503–505
molecular docking 502–505
results 503
structure & binding 502
toxicological profiling 504*f*
You Only Look Once (YOLO) system 123, 125

Z

Zebra Medical Vision's CCS 235
Zoonotic spillover 95

Printed and bound by CPI Group (UK) Ltd, Croydon, CR0 4YY
21/09/2026
02220062-0004